850

McGRAW-HILL PUBLICATIONS IN THE
AGRICULTURAL SCIENCES
LEON J. COLE, Consulting Editor

DESTRUCTIVE AND USEFUL INSECTS

Selected Titles From

McGRAW-HILL PUBLICATIONS IN THE AGRICULTURAL SCIENCES

Leon J. Cole, *Consulting Editor*

Adriance and Brison · Propagation of Horticultural Plants
Boyle · Marketing of Agricultural Products
Brown · Cotton
Cruess · Commercial Fruit and Vegetable Products
Eckles, Combs, and Macy · Milk and Milk Products
Fawcett · Citrus Diseases
Fernald and Shepard · Applied Entomology
Gardner, Bradford, and Hooker · Fruit Production
Gustafson · Conservation of the Soil
Gustafson · Soils and Soil Management
Hayes and Garber · Breeding Crop Plants
Hayes and Immer · Methods of Plant Breeding
Heald · Manual of Plant Diseases
Heald · Introduction to Plant Pathology
Hutcheson, Wolfe, and Kipps · Field Crops
Jenny · Factors of Soil Formation
Jull · Poultry Husbandry
Laurie and Ries · Floriculture
Leach · Insect Transmission of Plant Diseases
Maynard · Animal Nutrition
Metcalf and Flint · Destructive and Useful Insects
Paterson · Statistical Technique in Agricultural Research
Peters · Livestock Production
Rather · Field Crops
Rice · Breeding and Improvement of Farm Animals
Roadhouse and Henderson · The Market-milk Industry
Robbins, Crafts, and Raynor · Weed Control
Schilletter and Richey · Textbook of General Horticulture
Thompson · Vegetable Crops
Waite · Poultry Science and Practice

There are also the related series of McGraw-Hill Publications in the Botanical Sciences, of which Edmund W. Sinnott is Consulting Editor, and in the Zoological Sciences, of which A. Franklin Shull is Consulting Editor. Titles in the Agricultural Sciences were published in these series in the period 1917 to 1937.

DESTRUCTIVE

AND

USEFUL INSECTS

THEIR HABITS AND CONTROL

BY

C. L. METCALF, M.A., D.Sc.

*Professor of Entomology and Head of the Department,
University of Illinois*

AND

W. P. FLINT

*Chief Entomologist, Illinois Agricultural Experiment
Station and Illinois State Natural
History Survey*

SECOND EDITION
SIXTH IMPRESSION

McGRAW-HILL BOOK COMPANY, INC.
NEW YORK AND LONDON
1939

TO

STEPHEN ALFRED FORBES
Dean of American Economic Entomologists

AND

HERBERT OSBORN
Master Teacher of Entomologists

THIS BOOK IS AFFECTIONATELY DEDICATED

PREFACE TO THE SECOND EDITION

None can have an adequate appreciation of the tremendous progress that is being made in the science of entomology, who does not in some way attempt to digest the vast literature that is continually appearing in state and federal bulletins and circulars, in the entomological journals, and in many other serial publications. Such sweeping changes have taken place in entomological practices in the more than a decade since the first edition of DESTRUCTIVE AND USEFUL INSECTS appeared, that a complete revision, essentially a rewriting of the book, has been demanded. The greatest change has taken place in the field of insecticide control, due in part to the poison-residue regulations which have made it necessary to discover new insecticides to replace the extensively used lead, arsenic, and fluorine compounds that are under the ban. Astonishing progress has also been made in our knowledge of the life histories, bionomics, morphology, and physiology of our most important insect species and the better characterization especially of the immature stages of these pests. Of the 918 pages of the original edition, only a few remain unchanged in the present edition. Much new material has been added, and this edition contains about twenty per cent more material than the first edition.

One of the essentials in all work with insects is to determine the identity of the species under consideration. As an aid to this end, a Key to the Orders of Insects has been incorporated in this edition. A new departure in general textbooks of entomology is the presentation of a key to the orders of insects *in their immature stages*. This key is based, in part, upon one developed in the Department of Entomology at the University of Illinois by the late Dr. A. D. MacGillivray. The young of insects are of greater economic importance than the parent stage. They are generally so different in appearance, habitat, and behavior that a knowledge of the adults is of little or no value in recognizing the nymphs or larvae. It is often of great practical importance to recognize an insect outbreak in its incipiency, or to distinguish insect pests from those of little importance, when none of the better known adult stage is present. The authors are aware of the extreme difficulty of constructing a satisfactory key to the immature stages, which have so few and such obscure recognition marks, as compared with the more clearly differentiated adult insects. It is hoped at least that the key will be useful, though it is not perfect, and that it will stimulate greater interest in the almost virgin field of the taxonomy of immature insects. With the keys to orders, the beginner should be able to determine the great group of insects to which a given species belongs. When the plant or animal host, or the nature of damage is known, the Field Keys to economic insects that head each chapter from XI to XXIII should facilitate the determination of the exact species involved. Recognition of the insect responsible for

vii

any given damage is an essential index to the possible control measures. The latter are so specific that, without precise knowledge of what the pest is, attempts at control are likely to be futile.

A considerable rearrangement of the material within the chapters in the latter half of the book has been made in order to provide a more logical plan throughout. Under each economic group of plant pests the following sequence of presentation is followed:

a. Insects that chew the leaves, buds, or stems, above ground.

b. Insects that suck the sap of parts of the plant above ground.

c. Insects that bore in the twigs or stems or beneath bark or in the heartwood.

d. Worms, weevils, or maggots living inside fruits, nuts, or seeds.

e. Leaf miners and gall insects.

f. Subterranean insects that attack roots or other parts of the plant below ground.

g. Insects that injure plants by laying eggs in them.

In the chapters on animal pests are considered:

a. Free-living bloodsuckers or predators.

b. External parasites.

c. Internal parasites.

d. Disease carriers that are not also bloodsuckers or parasites.

Further, under each of the sections just outlined, the sequence followed is that of the arrangement of the orders of insects, as tabulated on page 172. That is, in discussing the chewing insects of any group of crops the lower orders are first described, followed by any representatives of the higher orders in regular sequence; and in each of the other methods of attack, the same sequence of pests is followed, according to the orders to which they belong. This need not disturb anyone not interested in the classification of the pests, but, when it is understood, it will be a great convenience to students and teachers in locating pests of any taxonomic or economic category.

To facilitate reference and cross reference in the latter half of the book, each species or group of species discussed is serially numbered in the heading of that section. After the number of each section, there appear in parentheses the numbers under which the same species or group is discussed in other chapters. For example, the corn earworm is discussed on page 368 as a pest of corn. The combinations of numbers 12 (63, 71, 111, 260), which appears before the name, calls attention to the fact that, in order to get a complete picture of this species as an economic problem, the reader should refer also to number 63, where it is discussed as a pest of cotton; to number 71, where it is discussed as a pest of tobacco; to number 111, where it is discussed as the tomato fruitworm; and to number 260, where there is reference to the same insect as a greenhouse pest.

We regret that the limitations of space have made it impossible to acknowledge adequately our indebtedness to research workers and writers for the enormous storehouse of information from which we have drawn much of the material presented. The references given in connection with the various chapters and sections may serve in an impersonal way to acknowledge our obligations. They also provide an index to the most significant of the literature on American economic entomology of the past quarter century.

Although the authors have made every effort to incorporate in this volume only tested and proved recommendations for the control of the insects discussed, the control of insect pests is a complicated problem, and results may be affected by unusual climatic and other local conditions that cannot be anticipated or detailed in this book. Any use made of the remedies suggested or the recommendations made will be at the sole risk of the reader.

We shall be happy if the present volume shall merit as cordial and continued reception as has been accorded the original edition for over ten years.

C. L. METCALF.
W. P. FLINT.

URBANA, ILLINOIS,
   August, 1939.

PREFACE TO THE FIRST EDITION

This book is intended as a text for the beginning student in entomology and also as a guide or reference book for practical farmers, gardeners, fruit growers, farm advisers, physicians, and general readers who desire up-to-date and reliable information about the many kinds of insect pests. It is the outgrowth of twenty-five years of practical entomological work in combating destructive insects, on the part of one of the authors; and fifteen years' experience in teaching large classes of university students of entomology, by the other author. It aims to present what the authors believe to be the essentials of economic entomology, in language which any reader can understand.

As a textbook, it is adapted to the two types of introductory courses commonly given in American universities and colleges. In the first ten chapters, enough of the fundamentals of technical entomology is given to serve as a basis for further special study in the subject or for students of biology who desire an introduction to entomology. The later chapters of the book are devoted to an analysis of the more important insect pests of the major crops in the continental United States and southern Canada. It is hoped that these discussions will be a ready reference for the practical worker who wishes to determine particular pests and learn their control; and serve classes seeking a broad course in applied entomology, as a part of an agricultural education. Material enough is available so that, by selection on the part of the teacher, practical courses adapted to the needs of a great variety of special classes, or different sections of the country, may be given.

A table or key to help the student or practical worker to identify a particular pest, when the plant or animal attacked is known, introduces each chapter in the latter part of the book. In the discussion of the several pests, a uniform system has been followed: first, the recognition marks and type of injury; second, a brief statement of the life history, with descriptions of the stages and significant habits; third, the control measures known or believed to be practical; and fourth, a few references, which may serve in part to acknowledge our indebtedness to our many colleagues in entomological work and also to point the reader to more extensive sources of information. In giving control measures we have rigorously selected only those which appear to have been tried and proved effective. On account of the limitations of space, the natural enemies of the various species are discussed only where they have been of especial significance in control. In discussing the life cycle, the overwintering stage forms the starting point in all cases. Unless otherwise stated the dates given are based on observations in central Illinois.

In an endeavor to avoid as many errors as possible, the authors have submitted the several chapters of manuscript to authorities in their special fields. We desire to thank the following persons for their invaluable aid in criticising parts of the book and in furnishing illustrations. A. J. Ackerman, George Ainslie, E. A. Back, W. V. Balduf, J. H. Bigger,

F. C. Bishopp, Fred E. Brooks, A. F. Burgess, D. J. Caffrey, B. R. Coad, C. C. Compton, R. A. Cooley, J. J. Davis, George A. Dean, J. E. Dudley, E. O. Essig, F. A. Fenton, C. L. Fluke, T. H. Frison, B. B. Fulton, Hugh Glasgow, R. D. Glasgow, P. A. Glenn, J. E. Graf, Fay Guyton, Leonard Haseman, W. P. Hayes, W. B. Herms, J. S. Houser, Neale F. Howard, J. A. Hyslop, M. C. Lane, W. H. Larrimer, Philip Luginbill, A. E. Mac-Gregor, S. Marcovitch, J. W. McCulloch, Z. P. Metcalf, Herbert Osborn, T. H. Parks, P. J. Parrott, Edith M. Patch, Antonio M. Paterno, Alva Peterson, R. H. Pettit, W. J. Phillips, B. A. Porter, H. J. Quayle, George I. Reeves, G. A. Runner, H. H. Severin, Franklin Sherman, Loren B. Smith. J. R. Watson, R. L. Webster, C. A. Weigel, V. L. Wildermuth, and H. N. Worthley.

Grateful acknowledgment for the illustrations is made in connection with each figure.

The authors will be grateful to any readers who will advise them of errors or criticisms that come to their attention.

C. L. METCALF.
W. P. FLINT.

URBANA, ILLINOIS,
  August, 1928.

CONTENTS

TABLES, SYNOPSES, AND OUTLINES

DESTRUCTIVE AND USEFUL INSECTS

CHAPTER I

INSECTS AS ENEMIES OF MAN

"The struggle between man and insects began long before the dawn of civilization, has continued without cessation to the present time, and will continue, no doubt, as long as the human race endures.   It is due to the fact that both men and certain insect species constantly want the same things at the same time.   Its intensity is owing to the vital importance to both, of the things they struggle for, and its long continuance is due to the fact that the contestants are so equally matched.   We commonly think of ourselves as the lords and conquerors of nature, but insects had thoroughly mastered the world and taken full possession of it long before man began the attempt.   They had, consequently, all the advantage of a possession of the field when the contest began, and they have disputed every step of our invasion of their original domain so persistently and so successfully that we can even yet scarcely flatter ourselves that we have gained any very important advantage over them.   Here and there a truce has been declared, a treaty made, and even a partnership established, advantageous to both parties of the contract—as with the bees and silkworms, for example; but wherever their interests and ours are diametrically opposed, the war still goes on and neither side can claim a final victory.   If they want our crops, they still help themselves to them.   If they wish the blood of our domestic animals, they pump it out of the veins of our cattle and our horses at their leisure and under our very eyes.   If they choose to take up their abode with us, we cannot wholly keep them out of the houses we live in.   We cannot even protect our very persons from their annoying and pestiferous attacks, and since the world began, we have never yet exterminated—we probably never shall exterminate—so much as a single insect species.   They have in fact, inflicted upon us for ages the most serious evils without our even knowing it."[1]

"It is difficult to understand the long-time comparative indifference of the human species to the insect danger . . . Men and nations have always struggled among themselves.   But . . . there is a war, not among human beings, but between all humanity and certain forces that are arrayed against it.   Man . . . has subdued or turned to his own use nearly all kinds of living creatures.   There are still remaining, however, the bacteria and protozoa that cause disease and the enormous forces of injurious insects which attack him from every point and which constitute today his greatest rivals in the control of nature. . . .   If human beings are to continue to exist, they must first gain mastery over insects . . . Insects in this country continually nullify the labor of one million men.   Insects are better equipped to occupy the earth than are humans, having been on the earth for fifty million years, while the human race is but five hundred thousand years old."[2]

If the reader has never experienced or witnessed any great injury by insects, these statements may sound extreme.   Almost everyone, how-

[1] FORBES, "The Insect, the Farmer, the Teacher, the Citizen, and the State."
[2] HOWARD, "The War against Insects," and other writings.

1

ever, has learned to appreciate the destructive capacity of at least a few kinds of insects. Perhaps he has seen a field of corn devoured by army-worms or grasshoppers, an orchard killed by scale insects, a building undermined by termites, a bin of grain consumed or contaminated by weevils, or some valuable garment ruined by clothes moths. Few, who have not studied the matter carefully, will have any idea how many and how varied are the ways in which these minute creatures injuriously affect us.

METHODS OF INJURY BY INSECTS

A. *Insects destroy or damage all kinds of growing crops and other valuable plants:*
1. By chewing leaves, buds, stem, bark, or fruits of the plant.
2. By sucking the sap from leaves, buds, stem, or fruits.
3. By boring or tunneling in the bark, stem, or twigs ("borers"); in fruits, nuts, or seeds ("worms" or "weevils"); or between the surfaces of the leaves ("leaf miners").
4. By causing cancerous growths on plants, within which they live and feed ("gall insects").
5. By attacking roots and underground stems in any of the above ways ("sub-terranean" or "soil insects").
6. By laying their eggs in some part of the plant.
7. By taking parts of the plant for the construction of nests or shelters.
8. By carrying other insects to the plant and establishing them there.
9. By disseminating organisms of plant diseases (fungi, bacteria, protozoa, and viruses), injecting them into the tissues of the plant as they feed, carrying them into their tunnels, or making wounds through which such disease organisms may gain entrance.
10. By bringing about cross-fertilization of certain rusts, which cause diseases of plants, without which help their aecia will not develop.

B. *Insects annoy and injure man and all other living animals, both domesticated and wild:*
1. *Causing annoyance:*
   a. By their presence in places where we object to them.
   b. By the sound of their flying about or "buzzing."
   c. By the foul odor of their secretions or decomposing bodies.
   d. By the offensive taste of their secretions and excretions left upon fruits, foods, dishes, tableware.
   e. By irritations as they crawl over the skin.
   f. By chewing, pinching, or nibbling the skin.
   g. By accidentally entering eyes, ears, nostrils, or alimentary canal, causing myiasis.
   h. By laying their eggs on the skin, hairs, or feathers.
2. *Applying venoms:*
   a. By means of a stinger.
   b. By means of piercing mouth parts.
   c. By the penetration of nettling hairs.
   d. By leaving caustic or corrosive body fluids on the skin, when they are crushed or handled.
   e. By poisoning animals when they are swallowed.
3. *Making their homes on or in the body, as external or internal parasites, injuring the host animal:*
   a. By causing nervous irritation in crawling about.
   b. By causing inflammation in chewing or piercing the skin.
   c. By contaminating fur or feathers with their eggs and excreta.
   d. By sucking the blood.
   e. By tunneling into muscles, nasal, ocular, auricular, or urogenital passages, causing mechanical injury and promoting infections.

    f. By anchoring to stomach or intestinal lining, mechanically blocking food passages, disturbing nutrition, causing an ulcerous condition, or secreting toxins.

4. *Disseminating diseases (bacteria, protozoa, parasitic worms, fungi, or viruses) from sick to healthy animals; from some wild animal (the "reservoir") to man or domestic animals; from a diseased parent or antecedent life stage in one of the following ways (the pathogen may merely cling to the body of the insect, it may increase its numbers in the internal organs of the insect, or it may undergo in the insect an essential part of its life cycle that cannot take place anywhere else.):*

    a. By accidentally conveying pathogens from filth to food.

    b. By transporting pathogens from filth or diseased animals to the lips, eyes, or wounds of healthy animals.

    c. By the insect host of a pathogen being swallowed by the larger animal, in which the pathogen causes disease.

    d. By inoculating the pathogen hypodermically as the insects bite animals.

    e. By depositing the pathogen upon the skin, in feces, or through its proboscis, or in its crushed body; and the pathogen enters through the bite of the insect, or a scratch, or the unbroken skin.

C. *They destroy or depreciate the value of stored products and possessions including food, clothing, drugs, animal and plant collections, paper, books, furniture, bridges, buildings, mine timbers, telephone poles, telegraph lines, railroad ties, trestles, and the like:*

1. By devouring these things as their food.

2. By contaminating them with their secretions, their excretions, their eggs or their own bodies, even though the product may not be eaten.

3. By seeking protection or building tunnels or nests within or on these substances.

4. By increasing the labor and expense of sorting, packing, and preserving foods.

A. INSECT INJURY TO GROWING PLANTS

Nearly all the injury done by insects results directly or indirectly from their attempts to secure food. They are undoubtedly man's chief rivals for the available food supply of the world. When an insect desires as its food something that man also desires, it becomes his enemy, and we say it is an injurious insect. Because of the great numbers of insects and their unequaled variety we find that there are one or more species adapted to take as food, apparently every kind of organic material in the world—plant or animal, living or dead, dry or decomposing, raw or manufactured, sweet or sour, hard or soft.

Injury by Chewing Insects

Insects take their food in a variety of ways. A primitive and very important method is by chewing off the external parts of a plant, grinding them up, and swallowing them, solids and liquid parts together, very much as a cow or a horse grazes, though of course, taking infinitely smaller bites. Such insects we call *chewing insects* (see Fig. 62, page 113). No one can fail to see examples of this injury (Fig. 1). Perhaps the best way to gain an idea of its prevalence is to seek to find leaves of plants absolutely perfect in their freedom from such attack. Cabbageworms, armyworms (Fig. 62,*K*), grasshoppers (Fig. 62,*C*), the Colorado potato beetle (Fig. 62,*J*), the pear slug (Fig. 62,*H*), and the cankerworm are common examples causing injury by chewing. The common, familiar Colorado potato beetles find nearly every potato patch east of the Rocky Mountains every year and, unless checked by poison, may soon strip the leaves from the plants and make the carefully planted and cultivated crop a total failure.

Grasshoppers have periodically overwhelmed American farmers from the earliest pioneer days down to the present. In 1923, these insects completely destroyed the crops in one area in Montana larger than an average state in the East.

"Several counties near the Canadian boundary were completely denuded. Many train loads of livestock were shipped out of this territory because of actual lack of forage to keep them alive. Numerous farmers lost everything and moved out. In 1922, the farmers of this state used more than 5,500 tons of poisoned bran mash to destroy the grasshoppers."

Almost the same situation has prevailed in many states and in other years. In 1920 the Canadian entomologists directed the treatment of

Fig. 1.—Two heads of cabbage from adjoining plats. *A*, sprayed to protect it from insects; *B*, not sprayed and badly injured by chewing insects. (*From Wilson and Gentner, Jour. Econ. Ento.*)

more than 1,400,000 acres of wheat in Saskatchewan with the saving of $20,000,000 worth of grain, otherwise certain to have been destroyed. Paying a bounty for dead grasshoppers at the rate of 60 cents a bushel, one county in Utah paid out more than $5,000 in a single year, accounting in this way for 274 tons of grasshoppers averaging about 8,000,000 hoppers to the ton. More than 150,000,000 pounds of poisoned bran bait were used against grasshoppers in the United States in 1937, at an estimated cost of $2,000,000. It is believed that over $100,000,000 worth of crops were saved in this way, giving an average return of about $50 for each dollar expended.

Armyworms, like grasshoppers, appear in countless numbers in certain years and practically devastate large areas of the country. Notable outbreaks of this kind have occurred in 1743, 1861, 1896, 1914,

1924, and 1936. The numbers of caterpillars that occur in such outbreaks can hardly be overestimated. Whole fields in which a man could scarcely put his foot to the ground without covering 10 or 12 worms are commonly observed. Since these insects feed chiefly at night, they may, unless noticed in the early stages of the outbreak, destroy a farmer's entire crop before he has time to apply control measures.

In 1868, the gypsy moth, a very destructive leaf-eating caterpillar of shade and forest trees, accidentally escaped into the woodlands of Massachusetts and was soon stripping the leaves and killing fruit and forest trees over thousands of square miles. A thousand men at a time have been employed to fight the pest with spray guns, fire, axes, and parasites. By 1927, $25,000,000 had been spent in keeping this pest in control, which at one time could almost certainly have been exterminated at a cost of a few hundred thousands. In 1923 a "dead line" 250 miles long was established along the Hudson River and Lake Champlain valley, beyond which the entomologists are determined that this pest shall not pass. Here the pest has been held for 15 years. If the gypsy moth should become established in the great forests of New York state, its spread over the entire North American continent would seem to be inevitable. The state of Massachusetts alone spends $3,000,000 a year fighting this pest. If it should become generally established over this country and Canada, the millions required to fight it and save our trees would be a terrible tax upon public and private funds.

Injury by Piercing-sucking Insects

A second very important way in which insects feed on growing plants is by piercing the epidermis ("skin") and sucking out the sap from the cells within. In this case, only internal and liquid portions of the plant are swallowed, although the insect itself remains externally on the plant. Such insects we call *piercing-sucking insects*. Their work is accomplished by means of an extremely slender and sharp-pointed portion of the beak (see Fig. 65) which is thrust into the plant and through which the sap is sucked. This results in a very different looking, but none the less severe, injury. The hole made by the beak is so small that it is never seen, but the withdrawal of the sap results in either minute spotting of white, brown, or red on leaves, fruit, or twigs; curling of the leaves; deforming of the fruit; or a general wilting, browning, and dying of the whole plant (Fig. 2). Aphids, scale insects, the chinch bug, the harlequin cabbage bug, leafhoppers, and plant bugs are well-known examples of piercing-sucking insects (see Fig. 64, page 116).

Aphids (plant lice) (Fig. 64,*A* and *B*) are probably the most universal group of plant-feeding insects. There is scarcely a kind of plant, cultivated or wild, but what supports from one to several species of aphids, and a large percentage of the individual plants will be found infested each summer. The innumerable beaks of these little pests continuously pumping sap from the plants constitute a very severe drain on their vitality. It curtails growth, and interferes with the size and flavor of the fruit if, indeed, the plant is not killed outright. The pea aphid, for example (Figs. 304 and 305), caused in 1937, in Wisconsin, a 50 per cent loss of the pea crop over 50,000 acres. Even when the quantity of the yield is not appreciably reduced, the quality and flavor of the peas are depreciated, making it necessary for the commercial canners

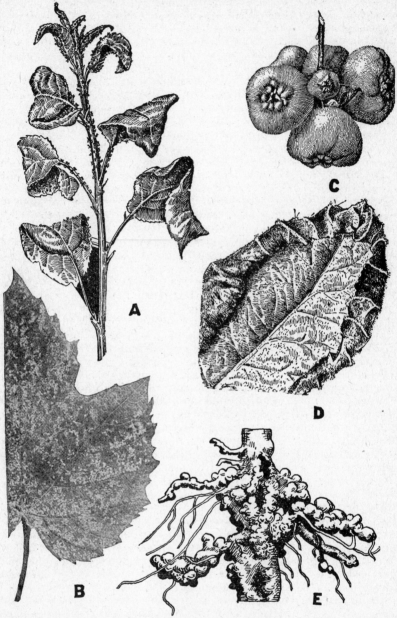

FIG. 2.—Examples of injury by piercing-sucking insects. *A*, curling of leaves and stunting of terminal growth by the green apple aphid (*from Quaintance and Baker, U.S. D.A.*); *B*, minute, white spotting, caused by the feeding of grape leafhoppers (*from Slingerland*); *C*, aphid apples, the result of the feeding of rosy apple aphids (*from Fulton*); *D*, hopperburn or tipburn caused by the feeding of apple leafhoppers on potato (*from Dudley, U.S.D.A.*); *E*, galls on roots of apple caused by the feeding of woolly apple aphids (*original*).

to add more sugar in an attempt to make up the deficiency. In one experiment nearly two million aphids were found to the acre of peas. The pea aphid is only one of many dozens of destructive aphids. Any one of the following kinds certainly occasions as great or greater annual losses: the corn root aphid, the rosy apple aphid, the woolly apple aphid, the green bug, or the melon aphid.

Another sap-sucking insect, the San Jose scale (Fig. 64,*E* and *F*), has killed tens of thousands of acres of fruit trees since its introduction to the eastern states in 1886 and 1887. Even in recent years, when the lime-sulphur spray has held it in check very generally, it has proved to be seriously destructive in some sections. For example, more than one thousand acres of commercial apple orchard were killed by this pest in Illinois alone during the years 1921 and 1922.

The chinch bug (Fig. 64,*H*), since its first recorded outbreak in the United States in 1783, has destroyed more than one billion dollars worth of grain crops. In 1914, this insect caused the loss of more than six million dollars worth of corn, wheat, and oats in 13 Illinois counties. There is scarcely a year when the farmers of the Mississippi Valley do not have to reckon with this insect in the production of their crops. In 1934 this pest was so prevalent that a federal appropriation of one million dollars was made to aid farmers in fighting it.

These two groups of insects, the chewing and the piercing-sucking, are the ones for which most spraying is done. It would be difficult to say which group is the more injurious on the whole; but it may be said that the piercing-sucking kinds are generally more difficult to control.

Injury by Internal Feeders

So long as an insect feeds externally upon crops, it can usually be destroyed by the application of the proper insecticide. But many of our worst pests feed *within* the plant tissues during a part or all of their destructive stages. They gain entrance to the plant either by having the egg thrust into the tissues by the sharp ovipositor of the parent insect or by eating their way in after they hatch from the eggs. In either case, the hole by which they enter is almost always very minute, often invisible. A large hole in a fruit, seed, nut, twig, or trunk generally indicates where the insect has come out and not the point where it entered.

The chief groups of internal feeders are indicated by their common group names: (*a*) "borers," in wood or pith, (*b*) "worms" or "weevils," in fruits, nuts, or seeds, (*c*) "leaf miners," and (*d*) "gall insects." Each group except the third contains some of the foremost insect pests of the world. In nearly all of them the insect is *internal* in only a part of its life stages, sooner or later emerging for a period of free living, usually as adults. This often affords an opportunity to control internal insects by dusting or spraying before their progeny gains entrance to the plant again.

Borers (Fig. 3) may attack any plant or part of a plant large enough to contain their bodies. Fruit and shade trees and many herbaceous plants suffer severely in this way. Various bud moths eat out the succulent tissues of swelling buds of trees. The bark beetles, the flatheaded borers, and the peach tree borer work chiefly in the vital cambium layer ("inner bark") of twigs or trunk. The roundheaded borers tunnel through the heartwood, as well as the cambium, greatly weakening the tree and damaging it for lumber.

The European corn borer (Fig. 215) and stalk borers of several kinds tunnel throughout the stems of corn plants from tassel to roots.

In 1927 the Congress of the United States appropriated $10,000,000 for the study and control of the European corn borer—the most extensive campaign against an insect pest ever attempted anywhere in the world. The corn earworm (Fig. 221) feeds on growing corn kernels underneath the husks at the tip of the ear. It has been estimated[1] that in its worst years, this insect attacks more than 70 per cent of the ears of field corn the country over, with the actual consumption of from 1 to 17 per cent of the grain in the infested ears.

Borers in fruits, including nuts and seeds, are generally called *worms* or *weevils*. Notorious examples are the codling moth (Fig. 62,*E*), bean weevils, the cotton boll weevil, the plum curculio, the melonworm, the apple maggot, and the chestnut weevil. Sometimes only one life stage is spent in the fruit, as with the codling moth and apple maggot; in other cases egg, larva, and pupa are all thus concealed from external attack; while in the bean weevil and granary weevil almost the entire life history is spent inside the seeds.

FIG. 3.—Seven-inch trunk of black oak tree tunneled by larva of one of the long-horned beetles. The broad tunnel in the inner bark, below, represents the work of the first year of the insect's life. (*From Ill. State Natural History Surv.*)

The cotton boll weevil (Fig. 279) inserts its eggs into holes made by its long snout in the tissues of the developing bolls from which the cotton lint should later unfold. The grubs that hatch from these eggs devour the immature lint so that no cotton is secured from the infested bolls. Entering this country into southeastern Texas from Mexico in 1890, by 1900 this weevil had increased and spread to such an extent that whole counties were destitute. Their one crop, cotton, having failed, their credit was gone; families became needy; farms were deserted; merchants went bankrupt; and banks failed. The loss increased rapidly in succeeding years until it reached the stupendous sum of $1,000,000,000 in a single year. The beetle spread northward and eastward at an average rate of about 60 miles a year, until all the important cotton-growing states were invaded (see Fig. 280). Recent discoveries of improved methods of control have greatly checked these losses, but still in certain years the cotton boll weevil harvests more than half of all the cotton

[1] See *U.S.D.A., Farmers' Bul.* 872, 1922.

planted in the United States. This one insect has collected a toll of nearly $3 a year from every acre of cotton land in the United States.

In April, 1929, the Mediterranean fruit fly, a serious maggot pest of both citrus and deciduous fruits, whose introduction to the United States entomologists had feared for 20 years, was discovered in Florida. The federal Congress appropriated $4,250,000 for its eradication. A strict

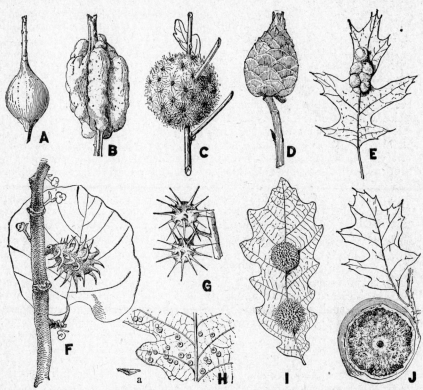

Fig. 4.—A group of insect galls. *A*, goldenrod ball gall, caused by a fly, *Eurosta solidaginis* Fitch; *B*, blackberry knot gall, caused by a gall wasp, *Diastrophus nebulosus* O. S.; *C*, wool sower gall on oak twig, caused by a gall wasp, *Andricus seminator* Harr.; *D*, pine cone gall, a common growth on willow, caused by a gall fly, *Rhabdophaga strobiloides* Walsh; *E*, dryophanta galls on oak leaf, caused by a gall wasp, *Dryophanta lanata* Gill; *F*, spiny witch hazel gall, caused by an aphid, *Hamamelistes spinosus* Shim.; *G*, spiny rose gall, caused by a gall wasp, *Rhodites bicolor* Harr. *H*, oak spangles caused by a gall fly, *Cecidomyia poculum* O. S., one gall shown in section at *a*; *I*, spiny oak gall, caused by a gall wasp, *Philonix prinoides* Beutm.; *J*, large oak apple, caused by a gall wasp, *Amphibolips confluens* Harr. (*From Felt, "Key to American Insect Galls," N.Y. State Museum Bul.* 200.)

quarantine preventing the shipment of citrus fruits out of the state was enforced. By the wholesale destruction of every orange, lemon, and grapefruit, both fallen and on the trees, within the infested area, and the spraying of trees of all kinds, the miracle of complete eradication was accomplished. Not a single Mediterranean fruit fly has been found in the United States since July, 1930.

A number of internal feeders are small enough to find comfortable quarters and an abundance of food between the upper and lower epidermis of a leaf. These are known as *leaf miners*. Surely these are the things Lowell had in mind when he said "There's never a leaf nor a blade too mean to be some happy creature's palace." Among the injurious forms are the apple leaf miners, beet leaf miner, spinach leaf miner, and many others.

The *gall insects* "sting" the plant and make it grow a home for them, within which they find not only shelter but also suitable and abundant food. This is probably the most marvelous instance in biology of the profound influence exerted over one organism by another. We do not know as yet exactly what it is that makes the plants, when attacked by the insect, grow these curious, often elaborate structures (Fig. 4) which are absolutely foreign to them in the absence of the gall insect. A strange feature of the work of gall insects is that the same species of insect on different species of plants causes galls that are similar; while several species of insects attacking the same plant cause galls that are greatly different in appearance. Although the gall is entirely plant tissue, the insect in some unknown manner controls and directs the form and shape it shall take as it grows. There is a marvelous variety of such homes for insects built by the "unwilling" but helpless plants.[1] Many of these galls seem to be practically harmless to the plant that grows them. The wheat jointworm, however, one of our worst pests of wheat, is a gall insect, and the grape phylloxera has destroyed thousands of acres of the most valuable vineyards in Europe and America.

INJURY BY SUBTERRANEAN INSECTS

Almost as secure from man's attack as the internal feeders are those insects that attack plants below the surface of the ground. These include chewers, sap suckers, root borers, and gall insects, the attacks of which differ from the aboveground forms just described, only in their position with reference to the soil surface. The subterranean insects may spend their entire life cycle below ground, *e.g.*, the woolly apple aphid. This insect, as both nymph and adult, sucks the sap from the roots of apple causing the development of ugly tumors (Fig. 2,*E*) and the subsequent decay of the roots at the point of attack. More often there is at least one life stage of the insect that has not taken up the subterranean habit, as in the case of the white grubs, wireworms, Japanese beetle, root maggots, and grape and corn rootworms, in all of which the larvae are root feeders while the adults have largely retained the more primitive life aboveground. Interesting gradations and adaptations to the subterranean life are seen in the way in which the eggs of these insects are laid and in the place of pupation. In general it may be said that the more of its life stages the insect spends underground, the more difficult it is to control.

INJURY BY LAYING EGGS

Probably 95 per cent or more of the direct injury to plants is caused by insects feeding in the various ways just described. Another instinct, almost as powerful, is the urge to provide for the welfare of the offspring. While, in general, the maternal

[1] See FELT, E. P., "Key to American Insect Galls," *N. Y. State Museum Bul.* 200, 1917.

instinct among insects is not developed to the point where they care for their young after birth, most insects have a marvelously effective instinct to lay their eggs in exactly the right place so that their young will have the best chance to survive; and

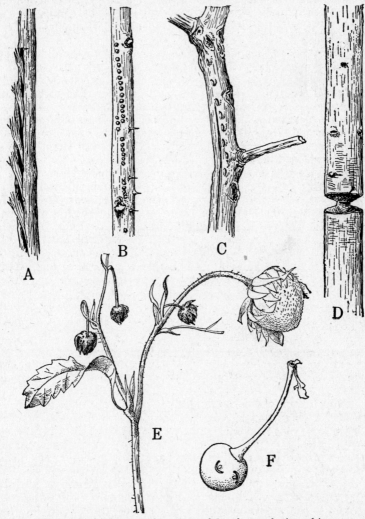

Fig. 5.—Examples of injury to plants caused by the egg-laying of insects. *A*, twig split by egg-laying of the periodical cicada; *B*, holes in stem of raspberry made by egg-laying of a tree cricket; *C*, slits in bark of apple twig beneath which a treehopper has thrust her eggs; *D*, twig of pecan cut nearly in two by egg-laying of female twig girdler; *E*, fruit buds of a strawberry, partially severed by strawberry weevil after laying an egg in the buds; *F*, cherry showing two egg punctures of the plum curculio. (*A*, *B*, *C*, *D*, *and F*, *original; E, from N.J. Agr. Exp. Sta., after U.S.D.A.*)

there are some very striking cases of great effort and care in the preparation of a nest or the deposition of the eggs. Sometimes this provision for the young leads to serious injury to man's possessions. The periodical cicada[1] deposits her eggs in the 1-year-old

---

[1] Also called "seventeen-year locust."

growth of fruit and forest trees, splitting the wood so severely that the entire twig beyond this point often dies (Fig. 5,*A*). The treehoppers and tree crickets split and ruin the bark or twigs of raspberry, currant, and apple in pushing their eggs into the plant tissues (Fig. 5,*B* and *C*). It is interesting to note that these are purely nesting sites. As soon as the young hatch, they desert the twigs and injure the plant no further. In other cases, the young, at least, subsequently feed upon the plant attacked by the egg-laying female; but we wish to emphasize at this point the injury *by the egg-laying act*, quite independent of any subsequent feeding of the young. Thus the plum curculio ruins the fruits of apple, plum, peach, and cherry by her characteristic egg-laying punctures (Fig. 5,*F*). The strawberry weevil, after laying an egg in the unopened bud, cuts the blossom stem partly off so that the flower never opens (Fig. 5,*E*). One of the most extreme cases of devotion to the welfare of the young is that of the twig girdler. In order that the larvae of this insect may have wood in a suitable condition of moisture and decay, the female laboriously chews off twig after twig of oak, hickory, pecan, elm, persimmon, or other tree in which to lay her eggs (Fig. 5,*D*). The severing of a single twig requires several days of work by the female.

The Use of Plants for Making Nests

Besides laying eggs in plants, insects sometimes remove parts of the plant for the construction of nests or for provisioning nests elsewhere, though they do not feed on these materials. This injury is more interesting than it is serious. Leaf cutter bees thus nip out rather neat circular pieces of rose and other foliage which are carried away and fashioned and cemented together to form thimble-shaped cells one above the other in a tunnel previously made in the stem of a plant. Each cell when completed contains a mass of nectar and pollen and an egg completely surrounded by bits of leaf; in this nest the young bee develops. The tropical leaf-cutting ants strip millions of leaves from trees or herbaceous plants and carry them into their nests where they are cut into fine pieces, sometimes are mixed with bits of their own or other insects' excreta, and form the medium upon which fungi are grown, as the only food of both larvae and adults. Other kinds of ants hollow out the stems or thorns of plants in which they dwell; but this phase of injury is not serious to man.

Insects That Care for Other Insects

Ants and some other kinds of insects, which are not in themselves serious pests, become injurious because they bring to our cultivated crops (corn, asters, citrus fruits) such noxious forms as aphids and mealy bugs, which they care for and protect because they like to eat the honeydew secreted by these pests. In some cases the most intricate and intimate interrelations have grown up between the ants, on the one hand, and the aphids, on the other. In general, ants furnish protection and a feeding place for the aphids, and the aphids furnish food ("honeydew") for the ants. Such cases of mutual dependency of two organisms upon each other are known as *mutualism*. One of the best examples is furnished by the corn field ant and the corn root aphid. This destructive aphid (Fig. 223) has become totally dependent upon the ants, which care for the aphid eggs over winter, and in the spring and throughout the summer carry the young aphids in their mouths through underground tunnels and actually place them on the roots of corn and weeds on which the aphids can feed. The ants are paid for this solicitous care with the sweet honeydew, which the aphids continually excrete and which serves as food for the ants. The cornfield ant is thus a menace to the corn crop although the ants themselves probably never injure the corn plant in any way.

INSECTS AS DISSEMINATORS OF PLANT DISEASES[1][2]

A serious phase of insect injury and one which may rival in importance the destruction caused by their direct feeding is the connection of insects with the ravages of plant diseases. Everyone has known something about insects that transmit human and animal diseases, but few have realized that other insects are engaged in spreading very disastrous diseases of plants. Since 1892, when it was first proved that a plant disease (fire blight of fruit trees) may be spread by an insect (the honeybee), the knowledge of this subject has grown rapidly, and at present there is apparently good evidence that more than 100 such diseases are disseminated by insects. The majority of them—60 to 80—belong to the group known as filterable viruses, 20 or more are due to parasitic fungi, 10 or more are bacterial diseases, and a few are caused by protozoa. The essential facts regarding a few of the most important of them are given in Table I.

Insects are responsible for favoring plant diseases in a number of different ways: (a) By feeding or laying eggs or boring into plants, they may make an entrance point for a disease that is not actually transported by them. (b) They have been found actually to disseminate the pathogens on or in their bodies, from one plant to a susceptible surface of another plant, such as a blossom or to a wound made by some other agent. (c) They carry pathogens on the outside or inside of their bodies and inject them hypodermically into the plant as they feed. (d) They harbor the pathogens inside their bodies during adverse periods, as overwinter or through a period of drought or host-plant scarcity, protecting them from the adverse climatic condition and from natural enemies. (e) They may be essential hosts for an incubation period, for increase in numbers of the pathogen, or for some part of its life cycle that cannot be completed elsewhere; or they may be essential for dissemination that is not normally accomplished otherwise.

Agents of Invasion and Passive Transmission.—The epidermis and bark of plants, like the skin of animals, have a highly protective function. When either is broken, an opportunity is afforded for the ingress of various injurious organisms. The attacks of the corn earworm (Fig. 221) are almost always followed by destructive molds and rots which spoil the corn and some of which are dangerous to animals that eat them. Many of these would not normally gain entrance to the ear were it not for the pathways formed by the tunnels of the worms. The fungus causing early blight of potatoes is similarly favored by the numerous holes made in the leaves by flea beetles (Fig. 62,I), and chestnut blight by the various bark beetles and borers attacking the bark of that tree.

Active Mechanical Transmission and Inoculation.—In addition to thus passively favoring plant diseases, certain kinds of insects actually carry the pathogens on or in their bodies from plant to plant. Bacteria are especially likely to be disseminated mechanically, because they adhere readily to the insects. Some fungous spores are sticky or moist, and some

[1] See *Phytopathology*, **10**: 189–231, 1920, and **12**: 225–240, 1922; *Bot. Rev.*, **1**: 448–466, 1935; SMITH, K. M., "A Textbook of Plant Virus Diseases," Blakiston, 1937; *Jour. Econ. Ento.*, **31**: 11–44, 1938.

[2] The authors are indebted to Dr. Neil E. Stevens and Miss Jessie I. Wood for a critical reading of this section of the manuscript.

TABLE I.—A SUMMARY OF CERTAIN INSECT-BORNE DISEASES OF PLANTS

Name of the disease	Plants affected	Insect carriers	Method of transmission	Causal organism or pathogen
A. Fungous Diseases of Plants				
Dutch elm disease	Various kinds of elm	The bark beetles, *Scolytus scolytus*,[1] *Scolytus multistriatus*, and *Hylurgopinus rufipes*	Spores introduced to cambium as beetles feed or make egg tunnels	*Ceratostomella* (=*Graphium*) *ulmi*
Chestnut blight...	Chestnut	Various beetles	Spores carried by insects or by birds. Rain washes spores into insect holes	*Endothia parasitica*
Blue stain........	Norway pine	The bark beetles, *Ips pini*, and *Ips grandicollis*	Spores introduced to cambium as beetles make egg tunnels	*Ceratostomella ips*
Brown rot........	Peach, plum, cherry	Plum curculio	Punctures made while feeding or egg laying inoculate or permit entrance of pathogen	*Sclerotinia fructicola*
Black rot.........	Apple	Codling moth, plum curculio, apple curculio	Spores inoculated beneath skin of fruits by mouth parts of insects	*Physalospora obtusa*
Perennial canker..	Apple	Woolly apple aphid	Feeding of insects causes cracking of callus and permits entrance of pathogen	*Gleosporium perennans*
Early blight......	Potato and tomato	*Epitrix cucumeris* and other flea beetles	Introduced to leaves as beetles feed	*Alternaria solani*
Downy mildew...	Lima beans	Bees	Transferred to blossoms as bees touch the floral organs	*Phytophthora phaseoli*
Black leg........	Cabbage	Cabbage maggot, wireworms	Carry spores on bodies and make wounds for ingress	*Phoma lingam*
Root rot.........	Sugarcane	Carrion-loving Muscid and Sarcophagid flies	Flies eat spores and deposit them in feces; also carry externally	*Ithyphallus coralloides*
Bud rot..........	Carnations	The mite, *Pediculoides dianthophilus*	Mites bearing spores crawl down into buds to feed	*Sporotrichum anthophilum*
Ergot	Rye, barley, wheat, many grasses	Flies, bees	Carried in alimentary canal and externally by flower-visiting insects	*Claviceps purpurea*
B. Bacterial Diseases of Plants				
Cucurbit wilt.....	Cucumbers, melons, and related plants	Striped and spotted cucumber beetles	Winters in alimentary canal of insect; deposited as they feed, or in feces over wounds	*Bacillus* (=*Erwinia*) *tracheiphilus*
Bacterial wilt or Stewart's disease	Corn, Job's-tears, teosinte	Corn flea beetle, northern and southern corn rootworms, and seed corn maggot	Winters in alimentary canal of flea beetles; deposited in wounds made by feeding	*Aplanobacter stewarti*
Fire blight.......	Apple, pear, quince	Bees, wasps, flies, aphids, leafhoppers, tarnished plant bug, bark beetles	To blossoms from mouth parts of nectar feeders. From cankers to new growth by direct inoculation	*Bacillus* (=*Erwinia*) *amylovorus*
Bacterial soft rot..	Cabbage, potato, and other vegetables	Cabbage maggot, seed corn maggot	Maggots inoculate seeds and roots. Persists through metamorphosis and adult flies spread while ovipositing	*Bacillus* (=*Erwinia*) *carotovorus*
Olive knot........	Olives	Olive fly, *Dacus oleae*	Persists through metamorphosis and adult flies spread while ovipositing. Infection also passes through the egg stage	*Bacillus savastonoi*
Black rot........	Cabbage	Aphids, cabbage looper	By direct inoculation and contaminated mouth parts	*Bacterium campestre*

[1] Not known to occur in America.

TABLE I.—(*Continued*)

Name of the disease	Plants affected	Insect carriers	Method of transmission	Causal organism or pathogen
		C. Virus Diseases of Plants		
Curly top........	Sugarbeet	Beet leafhopper and *Agallia stictocollis*	Virus is inoculated by feeding of the insect	A filterable virus
Peach yellows and little peach.....	Peach	The leafhopper, *Macropsis trimaculata*	The same	The same
False blossom....	Cranberries	The leafhopper, *Euscelis striatulus*	The same	The same
Aster yellows.....	Asters and many other hosts	The leafhopper, *Macrosteles divisus*	The same	The same
Streak disease of maize..........	Corn	The leafhoppers, *Cicadulina mbila* and *C. zeae*	Virus overwinters in insects and is inherited. Inoculated through stylets to phloem. Incubation period required	The same
Stunt or dwarf disease.........	Rice	The leafhopper, *Nephotettix apicalis*	Virus inoculated by feeding of the insect. Passes to young through the egg	The same
Spindle tuber.....	Potato	Potato and pale-striped flea beetles, Colorado potato beetle, grasshoppers, tarnished plant bug	Virus is inoculated by feeding of the insect	The same
Potato leaf roll and crinkle disease..	Potato	Green peach aphid and *Aphis rhamni*	The same	The same
Cowpea mosaic...	Cowpeas	Bean leaf beetle, aphids, and mealy bugs	The same	The same
Sugarcane mosaic	Corn, sorghum, sugarcane	The aphids; *Aphis maidis, Hysteroneura setariae,* and *Toxoptera graminum*	The same	The same
Mosaic of red clover.........	Alsike, red, white, and crimson clover, peas	The pea aphid	The same	The same
Bean mosaic.....	Beans	*Myzus persicae, Macrosiphum gei,* and a dozen other aphids	The same	The same
Mosaic of Cruciferae...........	Cauliflower, cabbage, turnip, mustard	Cabbage and turnip aphids	The same	The same
Cucumber mosaic.	Cucumber, tobacco	Melon aphid, green peach aphid, striped and spotted cucumber beetles	The same	The same
Spotted wilt or yellow spot.....	Tomato, pineapple	Tobacco thrips, onion thrips, and *Frankliniella insularis*	Must feed on diseased plants in the nymphal stage	The same
Spinach blight....	Spinach	Potato and green peach aphids and tarnished plant bug	Virus is inoculated by the feeding of the insect	The same
Tobacco mosaic...	Tobacco	Green peach aphid	The same	The same

are very fine, dusty, or spiny, or they have an electric charge opposite to that of the insect body, so that they cling well to the hairy bodies of insects. Bacteria and fungous spores may, therefore, be carried on the outside of the body of almost any insect and, if deposited upon a susceptible plant tissue, may start disease. Insects that pierce plant tissues with their mouth parts and those which tunnel beneath the surface, when they attack a diseased plant, are very likely to pick up externally, or take internally with their food, the germs causing that disease. As they pass to fresh plants to feed, the pathogens may be deposited on these plants with their excrement, whence they may enter any wound or the germinating spores may penetrate the plant unaided. In other cases the disease organisms may actually be injected into the tissues by contaminated mouth parts or ovipositors; or, in the case of borers, by their tunneling into the plant. The active flying and feeding insects serve admirably to disseminate, widely and rapidly, and to act as agents of ingress for, the inactive disease organisms which by their own efforts could seldom get from one plant to another. The habit of many insects of visiting only certain kinds of plants and even specific organs of those plants aids in their efficiency as inoculators. Fire blight of apple and pear is carried by aphids, bees, and other insects, which, as they feed, spread the destructive bacilli. The Dutch elm disease, which has recently been imported into the United States with results alarming to lovers of shade trees, is disseminated from tree to tree by the European bark beetle (see page 720). Since bark beetles prefer to make their breeding tunnels in trees that are in a weak or dying condition, they flock to trees dying of this fungus and later spread the malady far and wide as they seek new, vigorous trees upon which to feed. The egg punctures and feeding punctures of the plum curculio (Fig. 5,*F*) are commonly starting points for the brown rot of the peach, and the amount of brown rot in any season is highly correlated with the abundance of curculios. Black rot disease of apples often starts around holes made by codling moth larvae or curculios. The areas around the punctures of aphids, bugs, and leafhoppers frequently thicken, and the leaves become diseased, curl, and drop off, troubles that are often inseparable from and specific for the particular insect.

In feeding upon a plant, the amount of damage that the insect can do is more or less limited by the amount of tissue that it can devour; but if the insect's mouth parts are contaminated with disease organisms, the organisms may be established on the plant and not cease their attack until the entire plant is killed. In this connection it is noteworthy that much more effective control is needed for these disease-carrying insects than for insects which harm the plant only by their feeding. Nothing short of absolute control is satisfactory for some of the disease carriers, because a single insect may deal the plant a death blow by inoculating it with a disease organism, whereas the *feeding* of one insect would be ordinarily insignificant. It is also extremely important to prevent disease-carrying insects from starting to feed upon a crop.

Biological Transmission.—In most of the above cases, the disease may and probably does survive and spread to some extent without the help of the insects. In the following cases, however, it seems that the normal means of spread of the disease from one plant to another is always by the intervention of some particular insect. In at least a part of these cases it appears that the insect is necessary to the continued development and life of the pathogen, some essential part of its life cycle taking place in the insect's body, usually with a concomitant increase in numbers. The best known case in this category is the cucurbit wilt disease carried by the striped cucumber beetle (Fig. 308) and the spotted cucumber beetle (Fig. 232). The causal organism of the disease, a bacterium, spends the winter in the digestive tract of the hibernating beetle. The infected beetle, when it begins feeding upon a young cucumber plant in the spring, deposits in its feces some of the wilt bacteria. These are later washed over the surface of the leaf by dew or rain, and wherever there is a fresh wound opening into the vascular system of the leaf, the disease may become established. Probably the wounds, which are necessary to infection, are chiefly made by the insects in feeding. After the disease is started in this way, any cucumber beetle feeding on the plant may contaminate its mouth parts and then infect the next plant on which it feeds. No other means of spread for this disease are known. A similar relation exists between Stewart's disease or bacterial wilt of corn and the corn flea beetle (Fig. 238). The bacterium of this disease[1] may be found in winter in the viscera of this beetle, becoming an almost pure culture by spring. It has been shown that this disease usually becomes more destructive after mild winters but is not important after cold winters when, nor in regions north of an isotherm where, the sum of the mean monthly temperatures for December, January, and February is below 80°F. It is believed that this results from the effects of cold, not upon the bacterium that causes the disease, but upon the insect carrier. In the spring the diseased beetles fly to fields of corn and transfer the wilt organism directly to the leaves by their feeding.

The viruses which cause the mosaics and related diseases of plants, are nearly or quite ultramicroscopic and capable of passing through filters fine enough to sieve out the smallest known bacteria. They are either a precellular form of life, or nonliving gigantic protein molecules close to the border line between living and nonliving. They are not able to multiply in the absence of living cells and so cannot be grown in artificial media, but they have the ability to cause serious diseases when in contact with living cells and are regenerated and reproduced in the process. They are not destroyed by being crystallized by chemical treatment. They have great tenacity and longevity. It has been claimed that tobacco mosaic has been kept viable in dried tobacco leaves for 24 years and bean mosaic in stored seeds for 30 years.

Insects are known to be the carriers of more of the virus diseases than of bacteria, fungi, and protozoa combined. Some of these virus diseases appear capable of being transmitted mechanically by almost any insect that will feed upon the affected plant. It is difficult to distinguish some of them from disease-like injuries by insects in which no virus or other pathogen has been demonstrated. Thus the psyllid yellows disease of potato and tomato appears upon the plant only when it has been attacked by the psyllid, *Paratrioza cockerelli*. The tipburn of potatoes (one of the most serious of potato diseases, Fig. 2,*D*) and similar injury to alfalfa, clovers, soybeans, and peanuts appear only upon leaves that have been pierced by the potato leafhopper (page 510). The disease has been attributed to interference with the translocation of food materials within the plant by the mechanical injuries caused as the leafhoppers feed or to a toxin or enzyme injected into the leaf by the mouth parts of the insect. In some cases the virus diseases apparently can be transmitted by only one kind of insect, probably because some stage in its development requires incubation or nourishment in the body of that particular host. Thus the curly top disease of sugar beets is contracted by the plant only when it is punctured by the mouth parts of the beet leafhopper (page 545), which also carries the disease over the winter. False blossom of cranberries, which practically exterminated certain, very susceptible varieties of cranberries in the United States about 1895 to

---

[1] *Aplanobacter stewarti* (Smith) McCulloch. See *U. S. Bur. Plant Ind., Disease Reporter*, **18**: 141–149, 1934, **19**: 286–288, 1935, **20**: 109–113, 1936, and **21**: 102–107, 1937.

1905, is transmitted by the leafhopper, *Euscelis striatulus* (Fall).   Control measures for the leafhopper, such as flooding, dusting, and the use of resistant varieties, have greatly reduced the rate of spread of the disease. In some cases, such as the Japanese stunt disease of rice, a virus disseminated only by the leafhopper, *Nephotettix apicalis* Motsch, if the insect feeds upon a diseased plant, the eggs become infected, and the young subsequently born may be infective, sometimes for three or four generations.

With reference to the kinds of insects which disseminate plant diseases, it is noteworthy that the piercing-sucking insects are much more important than chewing insects, probably because the superficial wounds made by chewing insects dry out so quickly that the organisms fail to gain entrance to the plant, whereas the slender deep-seated punctures made by the piercing insects protect the pathogen from desiccation until it has established itself in the new host.   Only a few Orthoptera and Coleoptera have been incriminated in disease transmission.   There are several species of thrips, lace bugs, and plant bugs connected with definite diseases; but the Hemiptera are, so far as we know, much less extensive carriers of plant infections than the closely related order Homoptera (see page 198).   The explanation suggested for this is that the Hemiptera often kill the cells adjacent to their punctures by toxic saliva, so that the pathogen cannot establish itself.   The leafhoppers (page 200) and especially the aphids (page 5) are the known disseminators of more plant diseases than all other kinds of insects combined.   More than two dozen plant viruses have been described as disseminated by an equal number of aphid species.   Some of the diseases, such as cranberry false blossom and peach yellows, seem to have a very restricted host range, while others, *e.g.*, aster yellows and curly top of sugar beets, affect a great number of host plants.

The importance of the subject opened up by the recent discoveries just cited, time alone can tell.   However, the facts that these mosaic diseases attack so many of our most important crops, that they are very destructive, that they appear to be increasing in importance year after year, and that insect carriers have been discovered for these troubles in more than 60 different kinds of plants point to the probability that this is one of the most serious methods of injury by insects, constituting one of the most promising and at the same time most difficult fields for future investigation.

A new light on the intricate ramifications of insect injuries, is the recent revelation that for black stem rust of wheat (*Puccinia graminis*), growing on barberry, and the related *P. helianthi* on sunflower, if the sori are protected from the visits of insects, the aecia never produce spores, while the visits of insects usually bring about spore production within a few days.   Probably many rusts have two sexual phases, incapable of self-fertilization.   It seems probable that insects favor the ravages of these destructive fungi, much as they do the growth of valuable fruits, vegetables, and flowers (see pages 50 to 53).   Thus is added new confirmation of Dr. Asa Fitch's surmise: "There is no kind of mischief going on in the world of nature around us, but what some insect is at the bottom of it."

B. INSECT INJURY TO MAN AND OTHER LIVING ANIMALS

The second great group of things that fall prey to insects is all manner of animal life, from protozoa to man.   We find no records of insects that feed upon the marine animals known as echinoderms (starfish, sea cucumbers, etc.).   With this exception all the principal branches of the animal kingdom are attacked.   Insects in their relation to us make no distinction between man and other animals, and we shall make none in this discussion.

ANNOYING INSECTS

There are, first of all, a number of minor ways in which insects conflict with man's comfort and pleasure.   They are annoying by their presence, by their sounds, by the bad odors and tastes of their secretions, by crawling over one's body, by getting into the eyes or ears, or by laying their eggs upon animals.   All of us have experienced great annoyance from flying, buzzing, or crawling creatures at times, particularly when we desired to rest or apply ourselves to some exacting task.   The unpleasant taste left by certain stink bugs on berries and other fruits, the disgusting odor of cockroaches about the table service of some restaurants, and the sharp pain caused by getting certain minute insects[1] into the eye when driving at night are familiar examples of annoyance by insects.   The accidental invasion by living insects of the ears, nostrils, or stomach is usually serious but fortunately a rare experience.   Animals suffer from the attempts of certain flies (the botflies) to lay eggs on their bodies.   This is sufficient to cause the wildest stampeding of cattle, horses, and deer (Fig. 6).   While these annoyances constitute the least important of all the phases of insect injury to animals, still they are sufficient to account for a great deal of monetary loss, discomfort, and inefficiency.

VENOMOUS INSECTS

Insects are not popular.   The innate abhorrence of "crawling things" possessed by some persons is unfortunate, however, because it prevents them from learning anything a b o u t insects and interferes unnecessarily with their enjoyment of out-of-door life.   An extreme fear of bugs, caterpillars, spiders, and bees is not warranted by the facts.   In temperate climates there are very few kinds of insects that can harm the body s e r i o u s l y.   Nevertheless, there are a number of kinds that can bite and sting

FIG. 6.—A cow being chased by an ox-warble fly, intent on laying its eggs.   Note the terrified look of the eyes.   (*From Hadwen, Can. Dept. Agr., Health of Animals Branch, Sci. Ser. Bul. 27.*)

painfully, and these wounds may become infected with serious results.   It is so often true as to be almost a rule that the worst-looking forms are generally harmless; while some of the most painful experiences result from contact with very innocent-looking specimens.

Bodily pain and illness may be caused by the venoms of insects applied to the body in the following ways: (*a*) by stinging, *i.e.*, penetrating the skin with a defensive and offensive organ located near the tip of the abdomen; (*b*) by biting, *i.e.*, with the mouth parts—generally inserted to secure food, but sometimes used in a defensive way when certain insects are handled; (*c*) by nettling with hollow poison hairs located on the bodies of certain caterpillars, which inject venom after the manner of the common nettle plant; (*d*) by the application of caustic or corrosive fluids to the unbroken skin;

[1] For example, staphylinid beetles and certain small Hemiptera.

(e) by poisoning animals when they are swallowed, accidentally, or with food.   Chewing insects are rarely able to cause much pain.   The "pinching bugs," which are about the worst of this kind, can scarcely break the skin.   The really painful bites are made by insects with piercing mouth parts, are accompanied by the introduction of a venom, and are, therefore, a chemical injury.   The nature of the venom appears to vary but has the common characteristic that it is in some way toxic to animal tissues and so causes pain.   It is interesting to note, for example, that Schaudinn[1] seems to have proved that the irritation from a mosquito bite is due to an enzyme from a kind of yeast plant that lives as a commensal in the lobes of its stomach and that is expelled into our flesh as the insect bites.

Besides the spiders, ticks, and centipedes (pages 165–168), the following kinds of true insects are notorious for the injury inflicted by their bites: the Diptera or two-winged flies, including mosquitoes, black flies, horse flies, the stable fly, tsetse flies, the sheep tick, etc.; the Hemiptera or true bugs, including the bedbug, assassin bugs, back swimmers, water scorpions, etc.; the Anoplura or bloodsucking lice; the Siphonaptera or fleas.

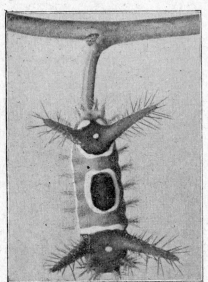

FIG. 7.—A caterpillar with poisonous hairs, the saddle-back, *Sabine stimulea* Clemens, about natural size. (*From* U.S.D.A., *Farmers' Bul.* 1495.)

Many insects regularly feed on animal blood, including that of man, as their only food; and, while so feeding, usually introduce a poison that causes a painful irritation.   Many sections of the mountains and woods are rendered temporarily uninhabitable by swarms of black flies in early summer.   Forbes[2] quotes from Agassiz's "Lake Superior" as follows: "Nothing could tempt us into the woods so terrible were the black flies.   One, whom scientific ardor tempted a little way up the river in a canoe, after water plants, came back a frightful spectacle, with blood-red rings around his eyes, his face bloody, and covered with punctures.   The next morning his head and neck were swollen as if from an attack of erysipelas."

The stable fly (Fig. 541), that dreaded but constant companion of horses, mules, cattle, and hogs, all summer long, inflicts such painful bites and withdraws so much blood that animals are sometimes killed outright.   Unable to don protective clothing, to retreat into screened houses, or even to "swat" efficiently, domestic animals must suffer beyond our comprehension from these many, bloodthirsty pests.   This suffering is translated into losses to the livestock farmer in decreased milk yield, loss of flesh, unsatisfactory growth, inefficiency and unmanageableness of work animals, and in greater susceptibility of the weakened animals to diseases.

The other methods of applying a venom are much less important than that by biting.   The stinging insects are, so far as man and the larger animals are concerned, largely a peaceable and defensive lot, inflicting their punishment almost exclusively on creatures that have injured them or disturbed their nests.   Herms,[3] however, records that a California farmer claims to have lost 400 small pigs during one year,

[1] See RILEY and JOHANNSEN: "Medical Entomology," p. 223, 1932.

[2] FORBES, S. A.: "The Insect, the Farmer, the Teacher, the Citizen and the State," *Ill. State Lab. Natural History*, 1915.

[3] HERMS, W. B., "Medical and Veterinary Entomology" 2d ed., Macmillan, 1926.

due to the stings of myriads of ants. Some of the ants possess a venom but have lost the stinger with which to inject it. According to Wheeler, these spray the poison from the tip of the abdomen into a wound made by the mouth parts. Certain beetles have a similar method of defense. The bombardier beetles[1] eject an acrid fluid which is discharged with a distinct popping sound and a small cloud of vapor that looks like the smoke from a miniature cannon.

Among the most interesting protective structures that insects possess are the nettling hairs of many caterpillars. These structures are similar to the poison hairs of the nettle plant. Not all the hairs of the body are of this type but only certain ones are hollow and connect at their base beneath the cuticula with poison gland cells. When these hairs penetrate the human skin the poison is released at a broken point and may create a serious skin cruption accompanied by intense itching and intestinal disturbance. The following are the best known of the nettling caterpillars: the

Fig. 8.—An entirely harmless, though evil-looking caterpillar, the hickory horned devil (*Citheronia regalis* Fab.), about ⅕ smaller than natural size. (*From Houser, after Packard.*)

brown-tail moth (Fig. 467), the io moth, the saddle-back caterpillar (Fig. 7), the flannel moth, the hag moth, and the buck moth. There is nothing distinctive about the appearance of these stinging caterpillars as a group. One has simply to learn to recognize each of them. As pointed out already, there are many more formidable-looking kinds that are totally harmless. For example, the hickory horned devil (Fig. 8) with its many thorny spines, some of them ¾ inch long; the common tomato worm (Fig. 326) and other sphingid larvae with a pointed horn near the tail end of the body; the celery caterpillar (Fig. 148,*b*) with a pair of soft yellow horns near the head that are erected and thrust out when it is disturbed and give off a peculiar odor; these and many other dangerous-looking forms are absolutely incapable of harming a person.

There are certain insects that carry a venomous substance diffusely throughout the body, especially in the blood, rather than confined to particular glands. In some cases, notably the blister beetles[2] (Fig. 264), this poison possesses caustic or

[1] *Brachinus* spp., Order Coleoptera, Family Carabidae.
[2] Order Coleoptera, Family Meloidae.

blistering properties when the insects are accidentally crushed on the body. They are also poisonous if taken internally, as when cattle eat them while grazing. Chickens are often killed by feeding upon the rose chafer (Fig. 423) in localities where it is abundant, and Lamson has shown that death results from an unknown poison contained in the bodies of these beetles.

External and Internal Parasites

The most loathsome attack we suffer at the hands of insects is their very common habit of taking up their residence on or within our bodies or those of domesticated animals. Such insects are called *zoophagous parasites*. Many insects, such as lice, lay their eggs on animals and live continuously on their hosts, generation after generation, never leaving them except as they instinctively transfer from older to younger animals of the same species or as they are forced to migrate when their host dies. Others spend certain life stages or certain parts of the day on the host and are free living the rest of the time. Some of these lay their eggs on the host and live there for a time but desert it in their later stages (*e.g.*, botflies); others lay eggs and develop away from the host and then become parasites only as adults (*e.g.*, fleas).

Three entire orders of true insects, *viz.*, the chewing lice (Mallophaga), the bloodsucking lice (Anoplura), and the fleas (Siphonaptera), a total of more than 3,000 described species, are entirely parasitic; besides hundreds of species from among the flies (Diptera) and true bugs (Hemiptera) and hundreds more of the ticks and mites of the class Arachnida.

The greater number of these species live externally on the surface of the skin. Their constant crawling about on the skin causes nervousness, restlessness, loss of sleep, failure to feed, and thus a general "run-down" condition and increased susceptibility to diseases. Their excreta and eggs mat the coat and create a foul condition that interferes with the excretory function of the skin. All these external parasites, except the order Mallophaga, feed by inserting their mouth parts and pumping out the blood. When they are abundant, this irritation, intensified by the animal's rubbing or scratching, may lead to great sores as in the case of the scab mite of sheep (Fig. 564) or the body louse ("cootie") of man. In other cases serious constitutional disturbances result. The mites known as chiggers or harvest mites often give rise to chills, nausea, and vomiting. The insertion of the mouth parts of the Rocky Mountain spotted-fever tick at the base of the head of man or sheep causes an ascending motor paralysis involving complete loss of the use of the limbs and finally death, unless the tick is removed.

External parasites that suck the blood of the higher animals include such common pests as the hog louse, the sheep "tick," fleas, bedbugs, and the cattle tick. There is one kind of lice that does not suck blood. These are the so-called bird lice or chewing lice,[1] including all common poultry lice (Fig. 567) and certain species of lice found on horses, cattle, and other mammals. Their mouth parts are formed to cut off and ingest solid particles rather than to draw blood. They feed upon the dry skin, parts of feathers or hairs, clots of blood, and the like, and their injury is probably chiefly due to nervous irritation from nibbling at the skin and running about over it.

[1] Order Mallophaga, p. 195.

Internal insect parasites of the higher animals are of few kinds. But they are so troublesome to our livestock that they constitute a group probably more destructive than the external parasites. These internal parasites are either mites (Acarina) or true flies (Diptera). With a few possible exceptions all internal insect parasites are *transitory, i.e.,* they pass only a part of their life cycle inside the body. For example, in flies it is always the maggot stage that lives within the animal body;

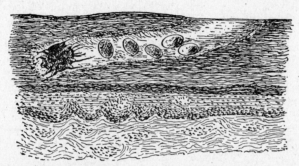

Fig. 9.—The itch mite tunneling into the skin of man to lay its eggs, several of which are shown. Diagrammatic and greatly enlarged. (*From Riley and Johannsen, "Handbook of Medical Entomology," Comstock Publishing Co.*)

Fig. 10.—A piece of grubby hide, after tanning, showing holes made by ox-warble larvae in the most valuable part of the hide, which render the leather practically useless. (*From Hadwen, Can. Dept. Agr., Health of Animals Branch, Sci. Ser. Bul. 27.*)

the adult, at least, living away from the host. In the mites the young may be internal, but they live as external parasites on the surface of the skin for at least a part of the adult life.

Just as there are borers in plants, so there are insects that live as borers in the animal body. The itch mite, scaly-leg mite and their kind dig tunnels into the flesh, in which their eggs are laid (Fig. 9). This explains the intolerable itching that is the most prominent symptom of their attack. Before the cause of the disease was known, the "seven-year itch," as it was often called, was a most loathsome and persistent affliction.

We have spoken of the injury insects cause when they try to lay their eggs on the bodies of animals. The larvae from such eggs are very serious parasites. For example, ox warbles cause serious damage to the hides of cattle which are rendered more or less useless for making leather (Fig. 10). The loss in value of the hides (augmented by the pain suffered by the animals, the depreciation in value of the carcass for beef, the loss of milk flow, and decreased growth) amounts to about $65,000,000 a year in the United States. In Africa the tumbu fly

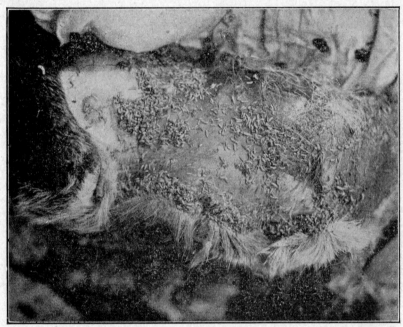

Fig. 11.—Larvae of a fly (*Phormia terrae-novae* Desv.) found under the skin of a reindeer. When heavily infested, the animal usually dies. (*From Hadwen, U.S.D.A., Dept. Bul.* 1089.)

attacks man in a manner similar to that of the ox warble in cattle. Another species of botfly causes lumps in the necks of rabbits, and there is a species that emasculates squirrels by living in the scrotum.

The screwworm and other flies live in the flesh of cattle, horses, hogs, man, and other animals during the maggot stage (Fig. 11). These flies are attracted by any wound such as a barbed-wire cut, a dog's bite, dehorning or branding wounds, even the spot where a tick has bitten an animal, or by foul secretions, nasal catarrh, bad breath and the like. In such situations the eggs are laid, and the larvae tunnel about and feed, greatly aggravating the inflammation and suppuration of the wound and preventing its healing. It is necessary in the Southern half of the United States to treat all wounds of animals with a repellent, antiseptic dressing to prevent contamination by this pest. Fortunately such attacks upon human beings are rare.

In addition to the mechanical injury from these internal larvae, Hadwen[1] appears to have shown that the larvae of ox warbles secrete a toxin into the lymph of the host that may be very injurious.

A few kinds of insects, habitually, and some others, accidentally, live in the alimentary canal of animals. These should not be confused with the intestinal worms[2] which are not insects. The best known of intestinal insects are the several species of horse bots (Fig. 546). A horse that is heavily infested with bots generally presents a badly run-down condition. There are no insects that habitually live in the alimentary canal of man, but the eggs or small larvae of bluebottle flies, flesh flies, the house fly, rattailed maggots, and others may be swallowed with impure drinking water, milk, or infested food. Their presence in the stomach generally causes symptoms of nausea, vomiting, and fever.

INSECTS AS CARRIERS OF ANIMAL DISEASES

The most complicated way in which insects injure man and his animals and the most sinister of all their attacks are as carriers of disease organisms. Any disease in order to persist must continually find new hosts to supplant those lost by death, for the death of a host is calamitous to the parasites that caused it and must often be followed by the death of all the parasites on or in its body. The transfer of disease organisms to new hosts may take place (a) during close association or bodily contact of a healthy animal with a diseased one, as in leprosy or smallpox; (b) by way of the air, as in tuberculosis; (c) through contaminated food or drinking water, as in typhoid fever; (d) by contact with infected soil, as in hookworm; (e) by fomites, *i.e.*, contaminated clothing, bedding, or towels, as in scarlet fever; or (f) by being carried by insects or other small anthropods.

Of all these methods of transmission, none are so effective as the insects. Most of the pathogens are helpless and inactive, often so delicate that they cannot even withstand exposure to dry air. The insects which carry them are active, hardy, and ubiquitous; and they instinctively seek out for their own food the very animals upon which, alone, the disease organisms can survive. The insects thus provide rapid, selective transportation; they also shelter the organisms in their bodies, often over winter; and in many cases are hosts for a part of the life cycle of the germ that cannot take place anywhere else. When these insect carriers bite man or other animals, the disease germs pass down the mouth parts into the blood or upon the skin; or the insects' excrement, laden with germs, may be deposited over the wound.

The bite of an insect, of itself, rarely kills man, but when their bodies and mouth parts are contaminated with disease germs, the slightest bite may mark the victim for death, as surely as the bite of a mad dog may cause hydrophobia or the slightest wound from a poisoned arrow prove fatal. For the pathogens of disease are *living poisons*, which, in the body of a suitable host, may multiply until the host's body is completely destroyed by them. Thus the damage the insect can do is multiplied a thousand-fold.

---

[1] HADWEN, S., *Can. Dept. Agr., Health of Animals Branch, Sci. Ser. Bul.* 27, July, 1919.

[2] Phylum Nemathelminthes.

TABLE II.—SOME OF THE MORE-IMPORTANT INSECT-

Name of the disease	Animal affected	Carrier of the disease	Classification of the carrier	Pathogenic organism
Yellow fever....	Man	The yellow fever mosquito, *Aedes aegypti*[1]	Class Hexapoda, Order Diptera, Family Culicidae	Unknown; a filterable virus
Malaria.........	Man	The malarial mosquitoes[2]	The same	*Plasmodium malariae, P. vivax,* and *P. falciparum*
Filariasis or elephantiasis.	Man	About a dozen species of mosquitoes, including some Culex, Aedes, and Mansonia	The same	*Wücheria* (=*Filaria*) *bancrofti* and other species
Dengue.........	Man	The mosquitoes, *Aedes aegypti, A. albopictus,* and *Armigeres obturbans*	The same	Unknown; a filterable virus
African sleeping sickness: two kinds, Gambian and Rhodesian.	Man	The tsetse flies, *Glossina palpalis* and *G. morsitans* and other Glossina	Class Hexapoda, Order Diptera, Family Muscidae	*Trypanosoma gambiense* and *T. rhodesiense*
Nagana.........	Domestic animals, wild game	The tsetse flies, *Glossina morsitans, G. longipalpis* and other Glossina	The same	*Trypanosoma brucei*
Typhus fever....	Man	The human louse, *Pediculus humanus*[3]	Class Hexapoda, Order Anoplura, Family Pediculidae	*Rickettsia prowazeki*
Trench fever or Wolhynian fever.	Man	Human louse, *Pediculus humanus*	The same	*Rickettsia quintana*
Chagas disease..	Man	The assassin bugs, *Triatoma megista, Rhodnius prolixus,* and *R. pictipes*	Class Hexapoda, Order Hemiptera, Family Reduviidae	*Schizotrypanum cruzi*
Bubonic plague..	Man, rat, ground squirrel	Eleven species of fleas, especially *Xenopsylla cheopis*	Class Hexapoda, Order Siphonaptera, Family Pulicidae	*Pasteurella pestis*
Typhoid fever...	Man	The house fly, *Musca domestica*	Class Hexapoda, Order Diptera, Family Muscidae	*Bacillus typhosus*
Summer diarrhea	Man	The house fly, *Musca domestica*	The same	Bacillus of Morgan
Texas fever or splenic fever.	Cattle	The cattle tick, *Margaropus annulatus*	Class Archnida, Order Acarina, Family Ixodidae	*Piroplasma bigeminum*
Rocky Mountain spotted fever.	Man	The spotted-fever tick, *Dermacentor andersoni* and other ticks[4]	The same	*Dermacentroxinus rickettsi*
American relapsing fever.[5]	Man	The ticks, *Ornithodorus turicata, O. hermsi, O. talaje,* and others	The same	*Spirochaeta turicatae*
Fowl spirochaetosis.	Chicken, turkey, goose	The fowl tick, *Argas persicus*	Class Arachnida, Order Acarina, Family Argasidae	*Spiroschaudinnia marchouxi,* or *Spirochaeta gallinarum*
Tularaemia......	Man, rabbits, ground squirrels	Deer flies, fleas, rabbit louse, spotted-fever tick, rabbit tick	Tabanidae, Pulicidae, Haematopinidae, Ixodidae	*Bacterium tularense*

[1] A score of other species, mostly *Aedes* spp., but also representatives of three other genera, may transmit yellow fever. *Haemagogus aquinus* is thought especially important in maintaining the "jungle fever" among monkeys, in areas where human beings and *Aedes aegypti* are both absent.

[2] Especially *Anopheles 4-maculatus* and *A. maculipennis;* but at least 30 other species.

[3] A bloodsucking louse, *Polyplax spinulosus,* and several ticks keep the disease circulating among rats, squirrels, and rabbits; fleas and the tropical mite, *Liponyssus bacoti,* may spread it from these animal reservoirs to man; but cooties and head lice are the chief disseminators from person to person.

In 1893 Smith and Kilbourne published their proof that the serious disease of cattle known as Texas fever (see page 835) is caused by a protozoan parasite that lives within, and destroys, the blood corpuscles and is spread exclusively by the bites of the cattle tick (Fig. 550). This, considered one of the greatest scientific discoveries of all time, was a

BORNE DISEASES OF MAN AND DOMESTIC ANIMALS

Classification of the pathogen	Distribution of the disease	Methods of transmission	Other ordinary ways of getting the disease
Unknown; probably a protozoon	The tropics and subtropics of Africa and America	Directly inoculated into blood by mouth parts of mosquito	None. Exclusively insect borne
Phylum Protozoa, Subphylum Sporozoa, Class Telosporidia, Order Xenosporidia	In a broad belt around the globe between 4th i.l.N. and 16th i.l.S.	The same	None. Exclusively insect borne
Phylum Nemathelminthes Class Nematoda Family Filariidae	Southern United States, southern China, Africa, West Indies, Samoa, Tahiti	The same	None. Exclusively insect borne
Unknown	Around the globe in the tropics and subtropics	The same	None. Exclusively insect borne
Phylum Protozoa, Subphylum Mastigophora Order Trypanosomatida	Equatorial Africa	Directly inoculated into blood by mouth parts of the fly	Possibly other species of tsetse flies
The same	Equatorial Africa	The same	Possibly other species of tsetse flies
Bacteria Family Bacteriaceae	Europe, Mexico	Deposited in feces by louse and scratched into skin, or by bites	None. Exclusively insect borne
The same	Europe	Feces or body fluids of lice scratched into skin	None. Exclusively insect borne
Phylum Protozoa, Subphylum Mastigophora Order Trypanosomatida	South and Central America	The same or inoculated into blood by mouth parts or through eyes or respiratory passages	Possibly by other bloodsucking Hemiptera
Bacteria Family Bacteriaceae	Nearly cosmopolitan	From feces or vomit spots, by bites, or through abrasions	By breath of a plague victim or from fomites
Bacteria Family Bacteriaceae	Cosmopolitan	Carried externally or deposited in fly feces or vomit spots	From contaminated milk, water, or food
Bacteria Family Bacteriaceae	Cosmopolitan	The same	The same
Phylum Protozoa, Subphylum Sporozoa, Class Telosporidia, Order Xenosporidia	Southern United States, Central and South America, South Africa, Philippines, Europe	Directly inoculated into blood by mouth parts of the tick	None except by other species of ticks
Unknown	Entire United States and Alaska; Bitter Root Valley of Montana	The same	None. Exclusively tick borne
A spirochete	Texas, Arizona, California, Kansas, northwestern United States	By bites or from feces or coxal fluid	None. Exclusively tick borne
A spirochete	India, Australia, Brazil, North America, Persia, Egypt, etc.	The same	None. Exclusively tick borne
Bacteria Family Bacteriaceae	General in United States, Russia, Japan	Directly inoculated into blood by mouth parts of the insects	Butchering rabbits. Eating undercooked flesh of rabbits. Bites of rodents.

[4] The spotted-fever tick and the dog tick (*D. variabilis*) are the most important. *Dermacentor occidentalis, Amblyomma americanum,* and *A. cajannense* transmit the disease under laboratory conditions and probably also in nature.

[5] A related disease, African relapsing fever, caused by *Spirochaeta duttoni*, is disseminated by *Ornithodorus moubata*.

startling revelation of most far-reaching consequences. Following this revelation, other similar discoveries came rapidly, so that, within thirty-five years after the first proof of the insect transmission of any animal disease, scientists have exposed so long and alarming a list of such cases that one is led to look upon these small arthropods as particularly menacing (Fig. 12).

A study of the accompanying table (Table II) will show the essential facts regarding some of the most important diseases of man and domestic animals which are known to be carried by insects.

Many and varied are the ways in which these microscopic germs "ride" on, or in, the bodies of insects to a position where they can again increase in numbers and complete their life cycle at the expense of the health of some animal. They may "ride" on the hairy legs, wings, or mouth parts of the insect, or they may make the journey within the body of the insect—in the alimentary canal, body cavity, salivary glands, muscles, or Malpighian tubules.

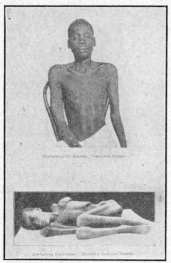

FIG. 12.—A victim of African sleeping sickness, a highly fatal disease carried to man exclusively by the bites of certain flies known as tsetse flies. (*From Osler's "Modern Medicine," Lea.*)

These disease organisms may "alight" on either the food or drink of the next victim and attack its body after having been swallowed. Or they may alight from the insect directly on the body of the new host in some receptive place, such as the lips, the surface of the eye, a wound, or a sore. They cannot, as a rule, attack through the healthy unbroken skin. They are often injected hypodermically into the blood or lymph of an animal when the insect that carries them on its mouth parts pierces the skin to suck blood. They may drop or be rubbed off the insect body when the journey is completed, pass out with the feces, be regurgitated in vomit spots, be pumped out with the saliva, or be swallowed by the new host when, as sometimes happens, the insect carrier is swallowed.

They may, during the journey to a new host, remain unchanged in the insect body. They may increase their numbers without a change of form. They may increase in numbers by a metamorphosis, passing through definite stages of their life cycle in the body of the insect which they cannot undergo in the body of any other animal or in any other situation. In the latter case we say the insect is *an essential host* (besides being a carrier) of the disease-producing organism. The complication of these problems is increased in some cases by the fact that, besides the insect carrier, the disease organism may have at least two other hosts, one in which it causes a definite disease and another in which it appears to be more or less harmless. The latter animals are known as *reservoirs* of the disease. Thus, in the case of Rocky Mountain spotted fever, the spotted-fever tick (Fig. 579) and other ticks are the carriers, and man is the victim of the disease; but a number of other animals, such as snowshoe rabbits, ground squirrels, gophers, woodchucks, and mice are reservoirs of the disease. It has recently been discovered that certain kinds of monkeys, in the vast hinterlands of South America and Africa into which man rarely penetrates, harbor a type of yellow fever, known as *jungle fever*, which may be carried to villages and cities by an occasional human traveler. This inaccessible, permanent source of yellow fever renders the hope of eradicating the disease from the earth (an accomplishment felt, until 1929, to be within the grasp of medical science), a dream of the remote future, if not, indeed, hopelessly impossible. Again the problem is complicated by the fact that the disease-causing organism may pass from an infected insect through its eggs to the succeeding generation or generations of the insect to reappear in a virulent form again after a long period of latency. Thus the infection of one insect carrier may mean that its offspring, in some cases to at least the fourth generation, will be infective without further exposure to the disease.

It will be obvious that in the relation of insects to animal diseases, just as in their connection with the spread of plant diseases, we have a subject fraught with the greatest possibilities. The fact that man did not discover, until within the past 50 years, the very obvious connection of mosquitoes with malaria and yellow fever, of fleas with bubonic plague, of lice with typhus, and tsetse flies with sleeping sickness—all of which have doubtless been in operation for thousands of years—and the fact that scarcely a year passes without the discovery of some new association between insects and human and animal diseases, forces one to the conviction that the arthropods are the most important phylum of the animal kingdom, next to the vertebrates, and that these three-party relationships, involving man and the two great unconquered biotic groups, insects and microorganisms, constitute the greatest medical problem of the future. The innumerable unsolved problems of insect habits, host relationships between the insects and the pathogens, immunity, virulent, negative and latent strains of the viruses, animal reservoirs of the diseases, and the difficulties of control call for prolonged and intensive study, with the closest cooperation of expert entomologists, parasitologists, and physicians, if man and his domesticated animals are speedily to be freed from many of their worst ailments.

C. INSECTS AS DESPOILERS OF STORED PRODUCTS AND OTHER MATERIALS

We have reviewed the ways in which insects are injurious to living plants and to living animals, including man. The third great phase of insect injury arises from the fact that they compete with us for the possession and use of practically all our stored products—both stored foods and the many other articles with which we habitually surround ourselves. We find here, exactly as in the case with living plants and living animals, that the attacks of insects are motivated mostly by hunger. To a lesser degree, damage results from their efforts to make provision for their eggs or young or in seeking shelter or building nests for themselves.

Since hunger is the principal motive involved and since insects eat organic matter of every kind, and practically only organic matter, it follows that the chief stored products to be protected are things of plant or animal origin, including grains, seeds, flour, meal, candies, nuts, fruits, vegetables, meats, fats, milk, cheese, honey, wax, tobacco, spices, drugs, feathers, furs, leather goods, woolens, paper, books, labels, photographs, boxes, furniture, wooden buildings, bridges, piling, mine props, telephone poles, railroad ties and trestles, and collections of insects, plants, and animals. These are some of the things that must be guarded from insect depredations.

Many other things not of an organic origin are generally immune from attack, *e.g.*, jewelry, metals of all kinds, pottery, statuary, brick, stone, and cement work. Even these inorganic objects are not entirely inviolate by insects. A species of powder post beetle has been given the name of lead cable borer[1] because of its troublesome habit of eating holes through the lead sheathing of aerial telephone cables (Fig. 13). These holes admit moisture and cause short circuits; often the insulation becomes water-soaked and ruined for an appreciable length, necessitating splicing and resheathing. In southern California this type of insect

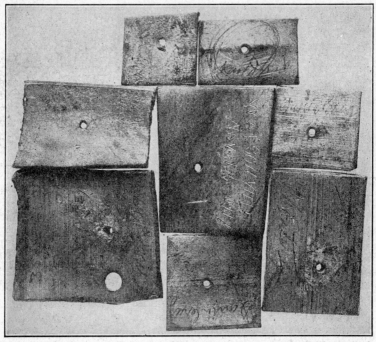

Fig. 13.—Holes bored in the lead sheathing of aerial cables by the California lead cable borer (*Scobicia declivis* Lec.), from various localities in California. (*From U.S.D.A., Dept. Bul.* 1107.)

injury is reported as causing about one-fifth of all aerial cable troubles. As many as 125 holes to a span of 100 feet have been found. A single hole may put from 50 to 600 telephones out of use for from 1 to 10 days. Termites similarly have bored through the lead pipe and the cotton insulation enclosing underground cables, thus ruining them within a year after they were laid down. Beetles of at least a dozen different families, besides the caterpillars of several moths and adult wasps of several kinds, have been recorded as boring through metal. Some years ago, one of the large railroads discovered that a mud-dauber wasp was causing great trouble and expense by building its mud nests in the exhaust port of the pressure-retaining valves of the Westinghouse air brake. It became necessary to change the shape of the valve and the form of its opening to overcome the trouble.

[1] *U.S.D.A., Dept. Bul.* 1107, 1922.

FOOD-INFESTING INSECTS

While the above cases are spectacular, it must be borne in mind that the total injury by all the insects that attack inorganic articles in all time past is probably exceeded in a single year by the injury of any one of a dozen or more pests of stored foods. The pests of stored products are the most expensive of all insects to feed, because they feed upon products that have been grown, harvested, stored, and, in many cases, have incurred further expense through manufacturing, advertising, selling, and distributing processes.

In times of stress, attacks by insects upon man's food supply may mean death to thousands. In earlier times one of the critical duties of sailors upon long ocean voyages was to guard the ship's biscuits from the ever-present and ravenous cockroaches. During the World War, large quantities of wheat, badly needed by the European nations, were destroyed by weevils in Australia, and entomologists were hurriedly dispatched to check the destruction. Seeds are among the most concentrated foods known, and a large part of the injury to stored products is to seeds of our cereal and leguminous crops. Of the many animals that compete for this valuable food material in any community, such as birds, rats, mice, insects and man, insects probably get the largest share, next to man.

The former State Entomologist of Connecticut estimated in 1917 that it costs the American people $200,000,000 a year to feed the insect pests of stored foods. Every individual can recall some case where an article of food had to be discarded because insects "beat man to it." These small losses make in the aggregate a heavy total. To the total of small losses must be added all-too-frequent cases where an entire crop in storage, the contents of a large elevator

Fig. 14.—Ear of corn showing the silk webs formed by larvae of the Indian meal moth, about 1/3 natural size. (*Reduced from U.S.D.A., Farmers' Bul.* 1260.)

or a shipment of foodstuffs has been rendered unfit for human consumption.

The method of attack is varied. Some species hide their eggs in the developing seeds as they grow in the field, and the injury becomes apparent only after the immature insects have been carried into the storehouse. Such is the case with the bean and pea weevils (Figs. 535 and 536) and the Angoumois grain moth (Fig. 531). Other kinds enter by stealth into kitchens, granaries, or factories and deposit their eggs on cured meats, harvested seeds, or any of the products manufactured from the raw-food materials. Some of the grain insects make their homes inside of single whole grains during all their growth; others attack only the broken or ground seeds, roaming about in flour, meal, and other foods and contaminating much more than they eat, with their excreta or the silk that they spin (Fig. 14). Many other kinds do not breed in the stored foods, either having nests outside and entering our foods only on foraging expeditions, *e.g.*, the ants; or leading a gypsy life, *e.g.*, the cockroaches, which are objectionable more on account of the filth and disease germs that they probably carry than because of the amount of food consumed.

Two orders of insects are of prime importance as pests of stored foods: (*a*) the Coleoptera, including such notorious pests as the granary weevil (Fig. 526), the confused flour beetle, the saw-toothed grain beetle, the pea weevil, the bean weevils, the larder beetle, and many others; and (*b*) the Lepidoptera, including the Angoumois grain moth, the Mediterranean flour moth (Fig. 533), the Indian meal moth, and others. In the former of these groups both the grubs or larvae and the adult beetles feed on the stored materials, while among the moths only the caterpillars are directly injurious. In addition to these two groups of most importance, such pests as the book-louse, the cheese skipper, the cheese and ham mites, sporadically destroy large quantities of food (see pages 792, 796).

No other economic group is more widely and equitably distributed than these insects of stored products. Many of the worst kinds are quite cosmopolitan. Some of them have been said to have spread over an entire continent in two or three years. They crawl and fly about seeking the concentrations of attractive foods that we bring together; they enter our storehouses on the crops we harvest; they are distributed in the seeds we purchase for planting; they go to market with the grains we sell; and they come back to us in the flour, breakfast cereals, cakes, and crackers from our grocers.

Pests of Wood and Wooden Articles, Clothing, and Drugs

A particularly insidious pest is the termite, or white ant (Fig. 512), whose fondness for a diet of woody tissues leads to most surprising invasions of dwellings (Fig. 15), libraries, trestles, fence posts, and indeed any article of wooden origin, such as stores of paper stock, cardboard boxes, library books (Fig. 16), and the like. Since they avoid exposure to the air, they are seldom seen until great damage has been done. Living in the ground, they make an opening into any timber that touches the soil and, from that tiny entrance, excavate a honeycomb of connecting passageways, working always in the interior of the structure invaded, without breaking the surface, until the timbers are so weakened that they break through before the occupants are aware that an enemy has been at work beneath their feet. Many cases are on record of extensive damage to private and public buildings by this insect, and its numbers seem to be on the increase. The powder post beetles (Fig. 514) do similar injury, but their dust-filled tunnels are easily distinguished from the frass-free runways of the termites.

In a somewhat different way the silverfish (Fig. 83) usually establishes itself in new quarters before it is suspected, because it is nocturnal and hides in cracks during the daytime. It has a fondness for starchy material and glue, which leads it to eat at book bindings, photographs, wall paper and all kinds of labels. In a large engineering laboratory it was discovered that the silverfish, in order to get the sizing in the paper, had effaced the numbers on inventory cards that gave the only clue to valuable apparatus out on loan.

Everyone has had the discomfiting experience of finding that furs, rugs, upholstered furniture, and winter clothing, stored during the summer, have been so eaten by clothes moths as to render them useless

FIG. 15.—A public building, the foundation timbers of which have been badly damaged by termites, necessitating extensive repair work. (*From photo by Snyder, U.S.D.A.*)

FIG. 16.—A book from an Arkansas library, ruined by the feeding of termites. (*From U.S.D.A., Farmers' Bul. 759.*)

TABLE III.—THE INSECTS OF THE UNITED STATES IN ACCOUNT WITH
THE AMERICAN PEOPLE
Debit
Most Important Items for the Year 1936[1]

Damage to—		Per cent	Estimated loss due to insects
Quantity	Kind of crop		
860,000 bu.	Alfalfa seed crop..................	10	$ 933,700
219,839,000 bu.	Barley, rye, rice crops.............	8	14,344,720
1,112,200,000 lb.	Bean crop (dry edible)............	12	6,106,080
39,000 tons	Broomcorn crop..................	10	465,500
6,218,000 bu.	Buckwheat crop..................	5	246,950
34,392,000 gal.	Cane and sorgo sirup crop.........	7	1,161,230
1,975,000 bu.	Cloverseed crop (red, alsike, and sweet clover)....................	28	5,475,680
1,524,317,000 bu.	Corn crop.......................	9	136,278,270
12,407,000 bales	Cotton lint crop..................	15	114,057,900
5,513,000 tons	Cottonseed crop..................	15	29,279,250
37,242,000 bu.	Cowpea and soybean crops........	5	2,200,300
5,908,000 bu.	Flaxseed crop....................	10	1,151,000
55,701,000 bu.	Grain sorghums crop..............	7	3,318,490
70,224,000 tons	Hay crop.........................	11	85,207,100
23,310,000 lb.	Hops crop........................	12	765,600
789,100,000 bu.	Oats crop........................	5	17,430,500
1,300,540,000 lb.	Peanut crop.....................	3	1,324,500
4,432,000 bu.	Pea crop (dry field).............	8	544,320
329,997,000 bu.	Potato crop......................	15	55,110,900
9,177,000 tons	Sugarbeet crop...................	15	8,284,350
5,949,000 tons	Sugarcane for sugar...............	20	3,981,400
64,144,000 bu.	Sweetpotato crop.................	5	3,001,750
2,915,000 tons	Sweet sorghums crop.............	8	1,920,800
1,038,000 bu.	Timothy seed crop...............	10	269,300
1,167,068,000 lb.	Tobacco crop.....................	10	25,036,400
895,000 tons	Velvet bean crop.................	12	1,434,960
626,461,000 bu.	Wheat crop......................	9	56,190,420
Total estimated damage to staple crops by insects			$ 575,521,370
864,000 boxes	Artichoke crop...................	10	172,800
6,042,000 crates[2] +59,100 tons[3] 745,000 bu.[2]	Asparagus crop..................	5	669,750
+19,900 tons[3] 12,019,000 bu.[2]	Lima bean crop..................	10	208,800
+70,600 tons[3] 1,937,000 bu.[2]	Snap bean crop..................	8	1,427,040
+42,400 tons[3] 984,000 tons[2]	Beet crop........................	5	65,200
	Cabbage crop		
+105,100 tons	Kraut...........................	20	4,221,000
13,148,000 crates	Cantaloupe crop..................	20	2,614,200
13,535,000 bu.	Carrot crop......................	10	761,000
7,198,000 crates	Cauliflower crop..................	10	594,300
9,376,000 crates	Celery crop......................	10	1,664,600
122,400,000 ears[2] +605,100 tons[3]	(N. J. only) Sweet corn crop..................	10	741,900
3,724,000 bu.[2] +6,314,000 bu.	Cucumber crop Pickles...........................	20	1,459,800
820,000 bu.	Eggplant crop...................	15	74,100
358,000 bu.	Kale crop (Virginia only)..........	15	16,050
21,820,000 crates	Lettuce crop.....................	5	1,590,900
17,322,000 sacks	Onion crop.......................	20	2,487,400

TABLE III.—(*Continued*)

Damage to—		Per cent	Estimated loss due to insects
Quantity	Kind of crop		
9,168,000 bu.[2] +187,400 tons[3]	Pea crop.........................	12	$ 2,472,480
4,033,000 bu.	Peppers crop.....................	10	270,200
13,900 tons[3]	Pimento crop....................	5	20,800
35,960,000 bu.	Potato crop (early)..............	10	4,758,800
13,100,000 bu.[2] +63,500 tons[3]	Spinach crop....................	10	601,000
20,346,000 bu.[2] +1,975,900 tons[3]	Tomato crop.....................	7	3,667,370
63,339,000 melons	Watermelons.....................	15	1,208,850
Total estimated damage to truck crops by insects			$ 31,768,340
108,031,000 bu.	Apple crop......................	20	21,815,600
106,000 tons	Cherry crop (12 states)...........	10	813,700
515,000 bbl.	Cranberry crop..................	6	413,640
1,879,000 tons	Grape crop......................	15	6,013,650
27,383,000 boxes	Grapefruit crop (4 states).........	10	2,190,000
8,316,000 boxes	Lemon crop (California)...........	10	2,494,800
60,891,000 boxes	Orange crop (7 states)............	10	8,118,600
46,118,000 bu.	Peach crop......................	20	8,687,000
24,128,000 bu.	Pear crop.......................	8	1,301,760
34,760,000 lb.	Pecan crop......................	10	419,100
134,000 tons	Plums and prunes (fresh, 5 states)..	8	305,600
177,000 tons	Prunes (dried, 3 states)...........	8	1,10.,520
10,010,000 crates	Strawberry crop.................	10	2,858,000
Total estimated damage to fruit crops by insects			$ 56,536,970
$92,127,484 worth	Flowers grown under glass........	20	18,425,496*
$ 7,037,038 worth	Flowers grown in open............	15	1,055,555*
$12,384,157 worth	Vegetables and vegetable plants grown under glass................	15	1,857,623*
$66,116,550 worth	Nursery stock...................	12	7,933,986*
Total estimated damage to nursery and greenhouse products			$ 29,272,660
67,968,000 head	Cattle and calves, including milk cows and heifers of all ages........	4	92,694,800
11,635,000 head	Horses and colts.................	½	5,632,285
4,684,000 head	Mules and mule colts.............	½	2,818,905
52,022,000 head	Sheep and lambs.................	3	9,957,660
42,837,000 head	Swine, including pigs.............	¼	1,362,277
426,145,000 head	Chickens.......................	5	15,777,300
Total estimated loss in livestock production by insects..........			$ 128,243,227
Total estimated damage to all products in storage..............			300,000,000*
Total estimated damage to forest trees and forest products.......			130,000,000*
Injury by transmission of malaria by mosquitoes...........		50,000,000*	
Injury by transmission of typhoid fever, tuberculosis, enteritis, diarrhea, etc., by house flies...........................		25,000,000*	
Injury by transmission of spotted fever by ticks...............		100,000*	
Total economic loss by insects that carry human diseases........			$ 75,100,000*
Grand Total..			$1,326,442,567

[1] In the case of the items starred (*) estimates are for the latest year available. In the case of nursery and greenhouse products' losses, this estimate is for the year 1929.

[2] For market.

[3] For manufacture.

(Fig. 17). A few insects have the surprising habit of feeding upon such things as drugs and tobaccos. The tobacco beetle riddles cigars and cigarettes with fine holes (Fig. 18). It also does great damage to upholstered furniture and in wholesale houses sometimes causes losses by tunneling through the leather soles of boots and shoes. The drug store beetle is most catholic in its tastes, having been found feeding in at least 45 different kinds of drugs, some of which are poisonous to man, and also on such widely different articles as books, sheet cork, chocolate, red and black pepper, ginger, and yeast cakes.

Conclusion

It has been seen how widespread and diverse is the conflict between insects and man. Our growing crops must struggle against insect attacks from the time the seed is planted until the crop is safely harvested—attacks upon leaves, branches, stems, roots, buds, blossoms, and fruits. Our domesticated animals, and man himself, are harassed and bitten and worried and their bodies infested with maggots and inoculated with disease. Our foods are con-

Fig. 17.—Overstuffed winged chair damaged by clothes moths. The bare spots were caused by the moths eating beneath the cover, resulting in the falling out of the mohair pile. (*From Back and Cotton, U.S.D.A.*)

Fig. 18.—Cigar ruined by the feeding of the tobacco beetle. (*Original.*)

taminated, our clothing ruined, our books and papers consumed, the wires that carry our messages rendered ineffective, and the timbers of our houses eaten piecemeal.

No one knows how much better off man might be were all his insect enemies destroyed. We can only conjecture and, by piecing together scattered records from many individuals of authentic losses, arrive at estimates of the total. Such an attempt to indicate what our insect enemies cost us is given in Table III. Estimates of this kind have usually been based on the commonly accepted belief that insects destroy, on the average and the country over, at least 10 per cent of every crop every year. Realizing that damage varies greatly from crop to crop and among the different species of animals, the authors have attempted, as a substitute for the flat rate of 10 per cent, to estimate in a more specific manner the damage to the various crops and animals. The percentage estimates given in Table III are based upon field observations and records kept in the Illinois State Natural History Survey for more than seventy years.[1]

The extent and value of the various crops and products are taken chiefly from statistics of the U. S. Department of Agriculture and are for the year 1936 unless otherwise credited. In the case of items starred (*), estimates are for the latest year for which we have statistics.

[1] See also "An Estimate of the Damage by Some of the More Important Insect Pests in the United States," compiled by J. A. Hyslop, in charge of Insect Pest Survey in the U.S.D.A., and *U.S.D.A. Bur. Ento., Cir.* E-444, 1938.

CHAPTER II

THE VALUE OF INSECTS TO MAN

So much of the writings about insects must necessarily deal with their destructiveness that we may be in danger of forgetting that many insects have beneficial attributes and habits the value of which we can hardly overestimate. It is a little startling to discover that this humble class of animals contributes to the world's commerce products that sell for more than $125,000,000 each year, in the United States alone;[1] or to read from the pen of an American entomologist[2] that:

"Except for the check put upon insect multiplication through warfare within the insect household, by which one species of insect destroys its relatives, no informed naturalist would expect the survival of the human race for a longer period than 5 to 6 years. Not only would man's food supply be appropriated by his insect enemies, but it would be impossible for him to withstand the withering march of malaria, yellow fever, typhoid, bubonic plague, sleeping sickness, and other maladies transmitted by insect carriers."

We need to study these creatures very carefully, in order that we may be able to distinguish insect friends from insect enemies. Almost any entomologist can tell from his own experience of incidents where people have gone to great trouble and expense to destroy quantities of insects, only to learn later that the insect destroyed was not only harmless but was actually engaged in saving their crops by eating the destructive form. Certainly most entomologists have had correspondents send in the larvae of Syrphidae or lady beetles with the complaint that they were injuring plants; at the same time overlooking the smaller aphids which were causing the injury and which these larvae were continually devouring.

Each citizen owes it to himself to know, as well as possible, the sundry ways in which beneficial insects affect the complex currents of plant and animal life to his advantage. The curious facts of honey production, silk production, and shellac production; the wonderfully intricate mechanisms of pollinization; the nature of food and the means of getting it, of insectivorous game, fur-bearing animals, fish, fowl, and songbirds, which subsist so largely on insect food; the possibilities of greatly increasing the quantity of certain fish and game, at present only a delicacy on the tables of the wealthy (or the lucky!); the overwhelming possibilities for human ill or welfare leashed in the prodigious hordes of bugs that eat each other —all these things are entomological topics that deserve our earnest consideration.

THE WAYS IN WHICH INSECTS ARE BENEFICIAL OR USEFUL TO MAN

A. *Insects produce and collect useful products or articles of commerce:*
    1. The secretions of insects are valuable:
        a. The saliva of the silkworm is the true silk of commerce.

[1] See *U.S.D.A.*, *Yearbooks*, statistics on silk, honey, beeswax, shellac, etc.
[2] GOSSARD, H. A., "Relation of Insects to Human Welfare," *Jour. Econ. Ento.*, **2:** 313–332, 1909.

    b. Beeswax is a secretion from hypodermal glands on the underside of the honeybee's abdomen.

    c. Shellac is the secretion from hypodermal glands on the back of a scale insect of India.

    d. The light-producing secretion of the giant firefly of the tropics is used in minor ways for illumination and may point the way to the synthesis of a substance giving brilliant light with almost no accompanying heat.

  2. The bodies of insects are useful or contain certain useful substances:

    a. Cochineal and crimson lake are pigments made by drying the bodies of a cactus scale insect of the tropics.

    b. Cantharidin is secured from the dried bodies of a European blister beetle known as the "Spanish fly."

    c. Insects, such as the hellgrammite or dobson, are widely used as fish bait, and the best artificial flies are modeled after insects.

  3. Insects collect, elaborate, and store plant products of value:

    a. Honey is nectar assembled from blossoms—concentrated, modified chemically, and sealed in waxen "bottles" by the honeybee.

  4. Insects cause plants to produce galls, some of which are valuable:

    a. Tannic acid from insect galls has been used for centuries to tan the skins of animals for leather or furs.

    b. Many insect galls contain materials that make the finest and most permanent inks and dyes.

B. Insects aid in the production of fruits, seeds, vegetables, and flowers, by pollinizing the blossoms:

  1. Most of our common fruits are pollinized by insects. The growing of Smyrna figs is dependent upon a small wasp that crawls into the flower cluster.

  2. Clover seed does not form without the visit of an insect, usually some kind of a bee, to each blossom.

  3. Peas, beans, tomatoes, melons, squash, and many other vegetables require insect visits before the fruits "set."

  4. Many ornamental plants, both in the greenhouse and out-of-doors, are pollinized by insects, *e.g.*, chrysanthemums, iris, orchids, and yucca.

C. The bodies of insects serve as food for many animals that are valuable to us:

  1. Many of our food fish subsist largely upon aquatic insects.

  2. Many highly prized song and game birds depend upon insects for a large percentage of their food.

  3. Chickens and turkeys naturally feed upon insects and, under proper conditions, can be raised almost exclusively on such a diet.

  4. Hogs may feed and fatten upon white grubs rooted from the soil.

  5. A few of the wild, fur and game animals eat insects, *e.g.*, skunk and raccoon.

  6. In many parts of the world, from ancient times to the present day, insects have been eaten extensively by human beings. Grasshoppers, crickets, walkingsticks, beetles, caterpillars and pupae of moths and butterflies, termites, large ants, aquatic bugs, cicadas, and bee larvae and pupae are prized as food by most of the more primitive races of men.

D. Many insects destroy other injurious insects:

  1. As parasites, living on or in their bodies and their eggs.

  2. As predators, capturing and devouring other insects.

E. Insects destroy various weeds in the same ways that they injure crop plants.

F. Insects improve the physical condition of the soil and promote its fertility:

  1. By burrowing throughout the surface layer.

  2. Their dead bodies and droppings serve as fertilizer.

G. Insects perform a valuable service as scavengers:

  1. By devouring the bodies of dead animals and plants.

  2. By burying carcasses and dung.

H. Certain insects are indispensable in scientific investigations:

  1. The ease of handling, rapidity of multiplication, great variability, and low cost of keeping and rearing have made the pomace fly invaluable in the study of genetics.

2. Studies of variation, geographical distribution, and the relation of color and pattern to surroundings have been greatly advanced through the study of insects.

3. Principles of regeneration and parthenogenesis have been discovered by the study of insect physiology.

4. The behavior and psychology of higher animals have been illuminated by a study of the reactions of insects, whose behavior can be analyzed into very simple tropisms. Valuable lessons in sociology have been deduced from a consideration of the economy of social insects.

I. *Insects have aesthetic and entertaining value:*

1. Their shapes, colors, and patterns serve as models for artists, florists, milliners, and decorators.

2. The more highly colored and striking forms are much used as ornaments in trays, pins, rings, necklaces, and other jewelry.

3. Moths and butterflies are universally admired, while those who use the microscope find much to admire in the colors and patterns of many of the smaller insects.

4. The songs of insects have been found highly interesting.

5. Insects have served as subject matter for hundreds of poems.

6. The inimitable variety found among insects, and their curious habits, afford entertainment and diversion for thousands who collect and study them.

7. The Orientals gamble on crickets trained for fighting, and fleas have often been used for circus stunts.

J. *Insects are useful in the practice of medicine and surgery:*

1. The maggots of certain flies, reared aseptically, have unique value in the treatment of wounds, especially deep-seated bone injuries contaminated with osteomyelitis.

2. The stings of honeybees have remedial value for diseases such as rheumatism and arthritis.

3. Extracts from the bodies of cockroaches, honeybees, and some other insects are used to some extent as medicines.

Useful Insect Products

The most obvious and tangible of the benefits that arise from insect activities is the utilization of the things that insects make, collect, or produce, such as silk, honey, beeswax, shellac, paints, dyes, and medicines.

Silk.[1]—Preeminent among insects valuable in this way is the silkworm. Very few persons know the silkworm by sight, but everyone knows the product it manufactures. Those who use this product most, seldom, if ever, think of its lowly origin; many doubtless do not know that silk is caterpillar's spittle.

The parent of the silkworm is a creamy white moth (Fig. 19) about 2 inches across the open wings. It is fat-bodied and feeble winged; it scarcely ever flies; it takes no food and lives only 2 or 3 days, but long enough to mature and lay 300 or 400 eggs. This insect has been a creature of domestication from time beyond memory and history, a captive and slave to man. For more than 35 centuries it has toiled ceaselessly for man; countless generations laying their eggs, eating the mulberry leaves provided for them in the larval stage, spinning their cocoons, and then dying—a perpetual sacrifice to the demand of men and women for adornment. For more than 2,000 years, only the Chinese people knew what silk is and where it comes from.[2] They punished with death

---

[1] "Silk, Its Origin, Culture, and Manufacture," Corticelli Silk Company.

[2] The discovery of its usefulness has been credited to Lotzn, Empress of Kwang-ti, about 2697 B.C.

anyone who tried to take the eggs or silkworms out of their country, and it is said that silk was, for a time, valued at its weight in gold. But in the year A.D. 555 two monks, sent as spies to China, discovered the nature and source of silk, and brought back to Constantinople some eggs of the silkworm moth concealed in a staff that they carried. In this humble way silk culture was introduced into Europe.

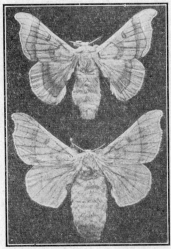

Silkworms are the larvae, or "worm stage" of the moth, *Bombyx mori* Linné. They hatch from the eggs laid by the moth and in 3 or 4 weeks have grown to fat caterpillars about 3 inches long (Fig. 20). Then comes the marvelous metamorphosis that converts the caterpillar into a moth like its parents. That this transformation may take place most safely, the last act of the larva is to enclose itself in a silken house. This house is constructed out of a curious saliva that hardens at once upon exposure to the air to form soft delicate threads of remarkable strength and pliability. These silken threads are formed by a silkworm at the average rate of about 6 inches a minute. The house building or cocoon making occupies 3 days. When its cocoon is completed, the larva changes to a pupa or chrysalid, which, if allowed to live,

FIG. 19.—The parent or moth stage of the silkworm, about natural size. Male above, female below. (*From Slingerland.*)

transforms to a moth 2 or 3 weeks later. When the moth is fully developed and ready to emerge, it secretes an alkaline fluid that softens one end of the cocoon, and, breaking the strands of silk, it squeezes out—a soft, feeble, crumpled adult.

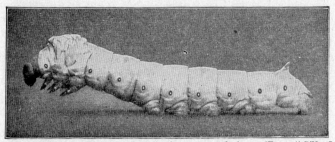

FIG. 20.—The full-grown silkworm (larva) about natural size. (*From "Silk, Its Origin, Culture and Manufacture," Corticelli Silk Company.*)

The substance that is so skillfully wound together to make the cocoon is produced from true salivary (labial) glands (Figs. 55,*sg* and 86,*sg*) that open through the mouth. All during the growth of the larva, these large glands are being filled with the relatively enormous quantity of fluid necessary to make the cocoon. This secretion has a commercial value of from $200,000,000 to $500,000,000 a year. From 50,000,000 to 70,000,-000 pounds of it are marketed each year as raw silk. It requires more than 25,000 cocoons (Fig. 21) and the consumption of about a ton of mulberry leaves to make a pound of silk. The production of this vast quantity of silk floss, therefore, is accompanied by the sacrifice of the lives of at least 1,000 billion caterpillars each year.

Each cocoon is composed of a single, continuous thread commonly averaging about 1,000 feet in length. If the moth were allowed to emerge, this single, unbroken thread would be dissolved and broken into

Fig. 21.—Cocoons of the silkworm from which the true silk of commerce is unwound, about ½ natural size. (*From Garman, Ky. Agr. Exp. Sta.*)

hundreds of tiny, useless pieces. The silk grower needs comparatively few moths to provide a new stock of eggs. He does need millions of cocoons. The other cocoons must be saved from injury by the moths. Accordingly, about 10 days after they have made their cocoons, the caterpillars or their pupae are killed by dropping the whole lot into hot water, by steaming them, by dry heat, or by fumigation. The cocoons are then assorted according to color and texture, the loose outer threads removed, the cocoons soaked in warm water to soften the gum that binds the silk threads together, and skillfully unwound by expert operators. The threads from several cocoons are wound together on wheels to form the reels of raw silk. Subsequently, the raw silk is boiled, scoured, steamed, stretched, purified by acids or by fermentation, washed and rewashed, to remove the gum and bring out the much-prized luster; and finally, combed and untangled, it is ready for spinning into the beautiful fabrics that eventually appear upon the market to beautify our homes and adorn our bodies.

Many kinds of moths (and insects of some other orders) thus spin cocoons for the protection of the pupal stage (see Fig. 90). But one species chiefly is cultivated for this purpose, largely because of the ease with which the larvae and adults are handled and kept captive and the readiness with which this particular cocoon can be unwound. Sericulture has been given serious study and is an important industry in Japan, China, Italy, France, and Spain, where the silk-growing peasants rear, feed, and tend these worms from eggs to cocoons as carefully as farmers in America rear sheep, cattle, hogs, or poultry. The chief reason that silk growing is not followed in America is that the higher cost of American labor does not enable us to compete with the Orient.

Honey.—Man has domesticated many kinds of animals and directed their activities so that the products resulting from their life processes might be available for his use. Of the many species of insects only two, the silkworm and the honeybee (*Apis mellifica* Linné), have been domesticated. These are so remarkably successful that one wonders if there are not many other kinds that could be domesticated with profit. Many kinds of bees and wasps store honey as food for themselves and their young. But here again, as with the silkworm, only one species of outstanding merit has found a place in the husbandry of man.

Plants secrete nectar in profusion, but in such numerous and infinitesimally small portions that man unaided could never afford to collect it. It requires from 40,000 to 80,000 trips of a honeybee, and visits to many times this number of flowers, to find and assemble nectar enough to

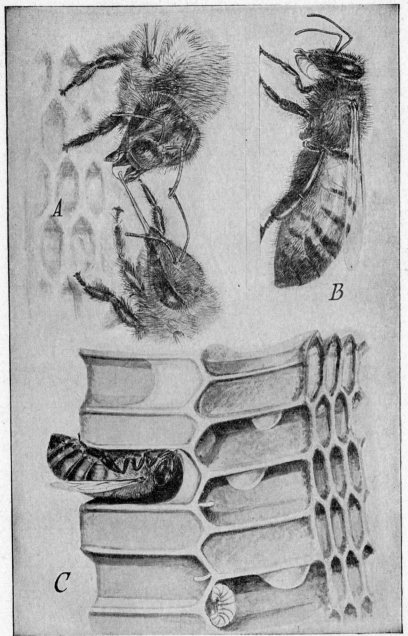

Fig. 22.—A bit of honeybees' comb in section, showing the waxen "bottles" in which honey is being stored and which serve also as cells for the protection of the egg, larval, and pupal stages of the bees. Note the two eggs and the partly grown larva. Above at left a field bee is transferring her load of nectar to a hive bee. At the right a hive bee is ripening the nectar by forcing it in and out of her mouth. (*After Park and Joutel, in Jour. Econ. Ento., vol.* 18.)

make a pound of honey. The average trip is thought to be about 1 or 1½ miles. Hence, for a single bee to collect nectar enough for 1 pound of honey would mean traveling at least twice the distance around the world.

Driven by an instinct to provide for their young and to fortify themselves against times of want, especially to lay up a winter supply of food, and operating in almost countless numbers, day after day, trip after trip, bit by bit, the bees gather these sugar-bearing secretions of plants and store their treasures in minute waxen "bottles" of their own making. In this way 150 to 200 million pounds of this product are collected and

FIG. 23.—Scales of wax on the ventral side of the abdomen of the honeybee. Beeswax is a secretion from hypodermal glands that open through the cuticula of the bee. (*From H. F. Wilson.*)

stored annually in the United States. Billions of pounds of nectar go to waste each year for want of enough bees to gather it.

The honeybee does not make honey in the same sense that the silkworm makes silk. Nevertheless the bees are more than harvesters, being an essential intermediary between nectar-bearing plants and man. Honey is truly an insect product. The nectar that is swallowed undergoes chemical changes due to intermixture with the saliva.[1] Some ingredients of the bees' own elaboration are probably added to the secretions of the plant, and the large and variable water content of the different nectars is greatly reduced.

The nectar obtained from flowers, after being mixed with saliva and swallowed, is carried in the honey sac (crop) until the bee reaches the hive. The honey sac is a kind of first stomach surrounded by muscles

[1] Sucrose (cane sugar) is hydrolyzed to glucose (grape sugar) and fructose (fruit sugar). These "invert" sugars are more readily assimilated by man.

and provided with valves so arranged that its contents may be passed on into the stomach or, by compression, emptied back through the mouth parts into the cells of the hive (Fig. 22). The nectar is concentrated by evaporation of a large part of its water content in a strong current of air produced over the cells by the rapid beating of the wings of the worker bees. When a cell is filled with properly "ripened" honey, it is capped over with wax and appears in the form familiar to everyone as

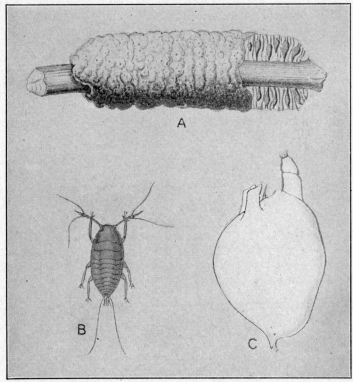

Fig. 24.—The lac insect. *A*, a piece of twig thickly encrusted with lac. The part on the right is broken open to show the worm-like lac insects in their cells. Each living insect communicates with the exterior through three small apertures, shown on the surface of the incrustation (about natural size). *B*, the young lac insect (first instar nymph or "crawler") greatly enlarged. *C*, the body of an adult female lac insect, freed from its resinous secretions, as seen from the side. At the upper end is the segmented abdominal extremity of the insect, terminating in the fringed anus. At the left is the dorsal spine and overhanging it the two respiratory processes which guard the spiracles. The mouth parts are at the lower extremity and, during life, inserted in the bark. About 12 times natural size. (*Modified from Green, "Coccidae of Ceylon."*)

comb honey. Nectar or honey forms the principal food of adult and larval bees. Honey is extensively used as a natural sweet, as a spread, and in making candies, cakes, and bread. Whether or not it has value in the dietetics of persons having diabetic tendencies, it is a rich and wholesome food. For example, deep-sea divers find large amounts of honey one of the few foods suitable to their strenuous physical occupation.

Beeswax.—The life processes of the honeybee also give us the useful material known as *beeswax*. Besides forming the cells in which the honey is stored and serving as "cradles" for the bees during their development (Fig. 22), beeswax is extensively used in many arts and trades. From 5 to 15 million pounds of it, worth several million dollars, are used in the United States each year.   It is a common mistake to suppose that honeybees convert the pollen they collect from flowers into beeswax or honey.   This material is a natural secretion of the worker bees that is poured out in thin delicate scales or flakes, from glands that open on the underside of the abdomen (Fig. 23).   Its production directly follows the digestion of a quantity of honey, a pound of wax resulting from the consumption by the worker bees of from 3 to 20 pounds of honey in about 24 hours' time.   About one million pounds of the annual crop is pressed into "foundation" and returned to the bee hives as a basis for the comb honey.   Several million pounds are used in manufacturing candles, which are nearly smokeless and do not bend over with the heat.   Much beeswax is used in shaving creams (to keep the lather from drying quickly), in cold creams, cosmetics, polishes, floor waxes, patterns for castings, models, carbon paper, crayons, and electrical and lithographing products.[1]

Shellac.—A tiny species of scale insect (related to our fruit pest, the San Jose scale) yields the substance from which shellac is made.   This substance is extensively used in making varnishes and polishes, for finishing woods and metals; for stiffening hat materials; as an ingredient of lithographic ink; as sealing wax; as an insulating material in electrical work; and in making phonograph records, airplanes, linoleum, buttons, shoe polishes, pottery, toys, and imitation fruits and flowers.   That this substance is derived from an insect is realized by very few persons.

The lac insect, *Laccifer* (= *Tachardia*) *lacca* Kerr[2] (Fig. 24), lives on native forest trees in India and Burma.   The natural function of the lac is to protect the motionless insect from adverse weather and natural enemies.   On contact with the air the resinous secretion hardens, and, where the insects are closely crowded together, it forms a continuous layer over the branches (Fig. 24,*A*).   In this condition the substance is known as *stick lac*.

About 40 to 90 million pounds of stick lac are collected annually.   It is ground in crude, hand-operated mortars and the resulting material separated into (*a*) granules of lac known as *seed lac*, (*b*) dust, which is used to make toys, bracelets, and bangles, and (*c*) wood, which is used as fuel.   The seed lac is next soaked in water and trodden by foot, which crushes and washes out the wine-colored pigments that were formerly sold as dyes but no longer have any commercial value.   The granular lac, after drying and bleaching in the sun, is placed in slender cloth bags, 10 or 12 feet long, is heated by open charcoal fires, and, by the twisting of the ends of the bag, is forced out as it melts, dropping upon the floor (Fig. 25).   Before these pads of melted lac have had time to congeal, they are grasped by natives who stretch them with their hands, teeth, and feet to extremely thin sheets.[3]   After drying, these sheets are broken into thin, small flakes; in this condition they are shipped.   Dissolving the

[1] GROUT, ROY, "Current Uses of Beeswax," *Amer. Bee Jour.*, September, 1935.

[2] Order Homoptera, Family Coccidae.

[3] See "Shellac: A Story of Yesterday, Today and Tomorrow," James B. Day and Company; GLOVER, P. M., "Lac Cultivation in India," *Indian Lac Res. Inst.*, 1937; PARRY, E. J., "Shellac," Pitman, 1935.

flake lac produces the familiar white or orange, liquid shellac. It requires about 150,000 lac insects to make a pound of lac.

The lac industry is an ancient one, certainly several thousand years old. The material has been used as a varnish at least since 1590. In 1709 Father Tachard described the insect that produces lac. The Mohave Indians used the secretion of a North American lac insect to make baskets watertight for cooking. The insect is not domesticated or even cultivated but lives in such abundance on the forest trees that millions of the poorer classes of India find in this industry their sole means of livelihood. No superior modern substitute has been found. From 10 to 20 million dollars worth of this product is used in the United States each year. Of late, extensive fluctuations in the abundance of the insect have led to the suggestion of simple routine methods of manipulation intended to reduce the extent of parasitism and increase the "set" of young insects on the host trees; and machine methods are now being used

Fig. 25.—Indian natives melting lac. One step in the preparation of shellac for the market. (*From "Shellac: A Story of Yesterday, Today, and Tomorrow," James B. Day & Co.*)

to some extent in producing the lac. Doubtless many other improvements will be initiated from time to time as the demand for the product makes necessary a more economical method of production. Recently synthetic lacquers have been produced which will doubtless supplant the true shellac for many purposes.

Living Light or Light without Heat.—One of the most remarkable phenomena in nature is the property that many animals possess of emitting light, "phosphorescence" or *bioluminescence*. This phenomenon is known among the Protozoa, Coelenterata, Mollusca, Annelida, fishes, birds, Crustacea, bacteria, fungi, and others, but doubtless the best-known examples are among the beetles, commonly known as fireflies or lightning-bugs and glowworms[1] (see Fig. 133).

The substance that gives rise to the light rays is not phosphorus but has been called *luciferin*. It is formed in certain specialized cells of the insect body. These cells are abundantly supplied with tracheae or breathing tubes. When air is admitted under the control of the insect, the combustion of the luciferin takes place under the

[1] The phosphorescent insects belong in the genera Photinus and Lampyris of the family Lampyridae and in the genus Pyrophorus of the family Elateridae.

influence of an enzyme, called *luciferase*, to form *oxyluciferin*, in the cells that produce it, and the production of light is instantaneous.   The insect body is generally provided with a reflector for the light, formed by a white material (probably ammonium urate), secreted by the cells directly behind the photogenic tissues.   In some organisms this reaction is reversible and the oxyluciferin can be reduced again, with loss of oxygen, and the luciferin used over and over.

The light emitted by our fireflies is most remarkable in having the maximum of visibility (from 92 to nearly 100 per cent light rays) and practically no heat rays or ultraviolet rays.   In the ordinary gas flame, by contrast, only 2 per cent of the energy is converted into light rays, the rest being lost as low heat rays; in the electric arc only 10 per cent of the energy produces light; while sunshine is only 35 per cent light.

Some slight use has been made of these light-giving insects as ornaments, as an artificial illuminant, and in photography.   But the significant benefit we may hope to derive from them is guidance to the synthesis of a luciferin in the chemical laboratory which may give to the world an artificial illuminant many times as efficient as the best artificial lights of the present day.

Cochineal.—Cochineal, a beautiful carmine-red pigment or paint, is the dried, pulverized bodies of a kind of scale insect, *Coccus cacti* Linné, that lives on the prickly pear, *Opuntia coccinellifera*.   Cochineal is now used principally as a cosmetic or rouge; for decorating fancy cakes; for coloring beverages and medicines; for dyeing where unusual permanence is desired; and, because of its property of allaying pain, for treating whooping cough and neuralgia.

It was used by the Aztecs in Mexico before the Europeans discovered America and, until the aniline dyes came on the market, was a very important product.   The cochineal insect is now cultivated principally in Honduras and the Canary Islands, though still a product from Mexico, Peru, Algiers, and Spain.   The insects are carefully cultivated.   The Mexicans keep them indoors over winter on branches cut in the fall from the prickly pear.   In spring they put the females out in little straw nests fastened to the cacti.   The young bugs settle on the cactus, and in 3 months' time the cochineal insects are fully developed and ready to harvest.   The branches are broken off, the insects brushed from them into bags and killed by hot water, steam, dry heat, or drying in the sun.   The impurities are then removed and the product is ready for the market.   It requires about 70,000 insects to make a pound of cochineal.

Insects as Medicine.—Many kinds of insects have been reputed to possess medicinal properties.   In fact, during the seventeenth century some curative power was attributed to almost every known insect.   At that time the belief prevailed that every creature possessed some special usefulness to man.   No other virtue being apparent for many of the insects, special therapeutic qualities were ascribed to them. Of course, the vast majority of such recommendations are pure quackery and are founded on the rankest of superstition.   Examples are the reputed belief that the bite of a katydid or cricket will remove warts; that cockroaches, crickets, or earwigs, variously bruised, burned, or boiled and properly compounded and applied, will cure earache, weak sight, ulcers, and dropsy.

Certain insects, however, do have a real medicinal value, notably the maggots of certain flies, the honeybee (*Apis*), and blister beetles.

During the World War, Dr. W. S. Baer noticed that the wounds of soldiers who had been lying on the battlefield for hours did not develop infections, such as osteomyelitis, as did those whose wounds had been treated and dressed promptly after they were made.   The difference was found to be due to the fact that the older wounds were always infested with maggots developed from eggs laid about the wounds by certain flies.   The remarkable discovery that these maggots could clean up the infection in deep-seated wounds much better than any known surgical or medicinal treatment has led to the practice of rearing maggots of the house fly and certain bluebottle flies under sterile conditions and introducing these surgically clean maggots into wounds to eat out every microscopical particle of putrid flesh and bone.   In one survey 92 per cent of 600 physicians who had used this treatment reported favorably upon it.   Dr. William Robinson[1] has isolated a substance from the secretions of the maggots.

[1] *Jour. Parasitol.*, **21**: 354, 1935.

which has the same property of healing in infected wounds as the maggots themselves. This material, known as *allantoin*, is commercially available and is described as a harmless, odorless, tasteless, stainless, painless, and inexpensive lotion which, when applied to chronic ulcers, burns, and similar pus-forming wounds, stimulates local, rather than general granulation and so is of especial value in treating deep wounds such as bone-marrow infections, where the internal parts of the wound must be healed first. Robinson does not believe, however, that the allantoin solutions will replace the living maggots in the treatment of bone infections, because the maggots actually eat out the necrotic tissue, kill the pus-forming bacteria by digesting them, and continually apply minute quantities of the allantoin in their excreta to the very depths of the wound, more effectively than can possibly be done by instrumentation.

The best known of the blister beetles is the so-called Spanish fly, *Lytta vesicatoria* Linné, that occurs in great abundance in France and Spain. This is a relative of our American blister beetles or "old-fashioned potato beetles." These insects have in their blood and internal organs a substance known as *cantharidin*, $C_{10}H_{12}O_4$. It was formerly greatly used as an external local irritant or blister. Lloyd[1] says: "The barbarisms practiced upon the American people during the nineteenth century by the application of cantharis blisters for all sorts of ailments, overtopped the misery endured by those who suffered in the war of the Revolution." This material was also much used as an aphrodisiac before its dangerous nature was appreciated. At present it has a place as an internal treatment in certain diseases of the urinogenital system and in animal breeding.

While the use of cantharidin appears to be dying out, the use of the honeybee in medicine has increased in the past century. The preparation known as "specific medicine Apis" is extracted from the bodies of honeybees by killing them in alcohol while they are intensely excited and by digesting their bodies in this medium for a month at a warm temperature. It is finally brought to a strength representing 2 ounces of bees to 1 pint of the medicine. In 1858 this preparation was said to be the most universally useful remedy next to aconite. It has been used by many physicians for the treatment of "hives," diphtheria, scarlet fever, erysipelas, dropsy, urinary irritation, and all kinds of edema accompanied by swelling and burning.[1] In an article entitled "The Remedial Value of Stings," Root[2] summarizes a vast amount of testimony of medical men and others as to the curative properties of bee venom. It appears that this remedy is extensively used especially in the Old World. The bee venom has also been placed upon the market in ampoules to be injected hypodermically, thus providing the same effects as the natural stings without the pain.

Use of Insect Galls.—The injury to plants by insect galls was discussed in Chapter I. These galls contain certain valuable products that have been used in a variety of ways. Many superstitions have attached to these remarkable growths on plants.[3] However, certain galls are reputed to have genuine medicinal or curative properties. The Aleppo gall, or gall nut, of western Asia and eastern Europe has been used in medicine since the fifth century B.C. It is a powerful vegetable astringent, tonic, and antidote for certain poisons.

Other galls have been used as dyes. The African Somali women use them as a tattooing dye. The Turks secure a fine scarlet color from a reddish gall on oak. Turkey red is dyed from the "mad apple" in Asia Minor. The ancient Greeks used the Aleppo gall for dyeing wool, hair, and skins; more recently great quantities of it[4] have been used in dyeing leather and seal skins. Tannic acid occurs in high percentages (30 to 70 per cent) in many of these galls, which are the richest of all sources of this material.

---

[1] Lloyd Brothers, *Drug Treatise* XXI, Cincinnati, Ohio.
[2] Root, E. R., *Gleanings in Bee Culture*, **75**: 16–20, 84–87, 1925; Beck, B. F., "Bee Venom, Its Nature and Effect on Arthritic and Rheumatoid Conditions," Appleton-Century, 1935.
[3] See Fagan, Margaret M., "The Uses of Insect Galls," *Amer. Naturalist*, **52**: 155–176, 1918.
[4] In 1914, $17,000 worth was imported to the United States from Bagdad.

Galls are used in the preparation of very durable and permanent inks. In some countries the laws require that certain records be made with ink compounded of gall nuts.

INSECTS AS POLLINIZERS

The maintenance of plant life generation after generation may be accomplished by *asexual reproduction* (the formation of buds, bulbs, and tubers) or by *sexual reproduction*. In the latter case one specialized reproductive cell, the male gamete (sex cell or sperm), unites with another, the female gamete (sex cell or egg), and from this union a new individual arises. In the higher plants sexual reproduction is made possible by the process known as *pollination*. The essential carrier of the pollen (male sex cells) from the anthers of one flower to the stigma of another is in most cases either the wind or an insect. Well-known examples of wind-pollinated flowers are corn, wheat, and other cereals, nut trees, willows, oaks, pines, etc. Wind-pollinated plants generally have flowers that are small and inconspicuous, with poorly developed petals, unisexual flowers, no nectar, dry and light pollen, and brush-like stigmas.

Most of our fruits and ornamental flowers; many of our vegetables such as beans, peas, tomatoes, melons, squash; and such field crops as clover, buckwheat, cotton, and tobacco depend mainly upon the visits of insects to carry the pollen to the stigma and so make possible a fertilization without which no seed or fruit would form. Flowers that depend upon insects for pollination can be recognized generally by their well-developed corollas of conspicuous size, by showy colors, or by a marked odor. They have sticky pollen grains, sticky stigmas, and nectaries that secrete a sweet liquid attractive as food for the insects.

Plants do not develop beautiful blossoms and sweet odors to delight the senses of man. They serve to attract insects. Plants have many remarkable modifications of their structures which compel the insects that come for nectar to carry away with them to the next flowers visited a load of pollen. As they crowd their way in and out of flowers, their bodies become covered with the fine pollen dust. In the honeybee this is removed from the general body hairs by a highly specialized brush on one of the segments of the hind leg (see Fig. 47,*C, Tarsus*). When the brush becomes filled, the hind legs are crossed and the pollen grains from one leg are scraped into the pollen basket on another segment (Fig. 47,*C, Tibia*) of the opposite leg. In the pollen baskets it is carried to the hive, where a spine on the end of the middle leg is used to pry it off and it is stored in the cells. While much of the pollen is thus collected and used as food by the bees, some of it brushes off when the bee or other insect crowds into the next flower visited. Many flowers are so constructed that an insect can hardly get the nectar from them without dusting some pollen from previously visited flowers upon their stigmas. In this way much of the work of fertilizing flowers, without which we should not have much fruit, tomatoes, melons, clover seed, tobacco, coffee, tea, chocolate, linen, or cotton, is made possible.

Bumblebees (Fig. 26) are worth tens of millions of dollars a year to the American farmer. Farmers generally rely upon the second crop of red clover for the production of seed. Bumblebees are the most important pollinizers of second-crop clover. The honeybee is not important as a pollinizer of red clover because the length of its tongue (see Fig. 76,*D*) is such that it can generally not reach the deep-seated nectary of this flower. The bumblebees winter only as young queens, and it takes several months in spring and summer to build up colonies to a point where they are abundant enough to visit every flower head over acres of clover and so make possible the setting of seed. In New Zealand it was found impossible to obtain seed from red clover until the bumblebees were imported into that country. Certain wild solitary bees[1] have been found also to be efficient pollinizers of red clover. In localities and seasons when these bees are abundant, the first crop of red clover will usually set a large amount of seed, often producing a greater yield than that from the second

Fig. 26.—Queens of bumblebees visiting and pollinizing flowers in spring. Six individuals of three different species are shown: *Bombus americanorum* Fab., lower center; *Bombus separatus* Cress., right center; and *Bombus auricomus* Robt., at right. (*From T. H. Frison.*)

crop. It has been suggested that races of honeybees with longer tongues, or strains of red clover with shorter corollas, should be bred, so that the pollination of this crop could be accomplished by the honeybee. Since entomologists have recently learned how to control matings in the honeybee, this and other improvements, such as the capacity to carry greater loads of pollen and nectar and the inability or reluctance to sting, are distinct possibilities.

Too often the impression prevails that bees are the only insects of importance in cross-pollination. Many kinds of flies, butterflies, moths, and beetles also share in this work. Many of these insects, however, produce injurious caterpillars, grubs, or maggots which may largely offset the benefit derived from the parents. Bees and wasps develop no objectionable progeny. The honeybees can be introduced into any neighborhood, orchard, or garden, and their numbers increased as we wish; or should they no longer be desired, the whole population could

[1] *Tetralonia* and *Melissoides* spp., Order Hymenoptera, Family Anthophoridae.

be exterminated at will.   They are under man's control as no other insect is, except the silkworm.

Most people think of honeybees only as a source of honey. It seems that their most important contribution to man is in the production of fruits and seeds.   It has been said that every time a colony of bees makes $5 worth of honey they have made $100 worth of seeds or fruits.   In orchards where colonies of honeybees have been placed at blossoming time, the yields of fruits have been increased almost beyond belief.   In

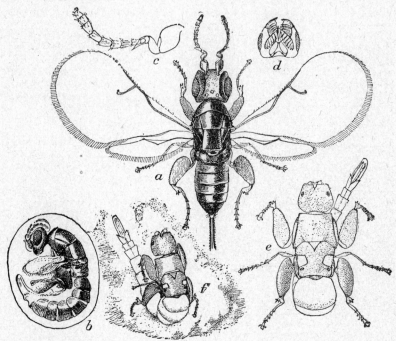

Fig. 27.—The fig wasp.   *A*, adult female with wings spread; *b*, female not entirely issued from pupal skin and still contained in gall of the fig flower; *c*, antenna of female; *d*, head of female from below; *e*, and *f*, adult males.   All much enlarged; *a*, about 16 times, *f*, about 27 times.   (*From U.S.D.A., Dept. Bul. 732.*)

a Michigan apple orchard where the largest crop in eight years had been 1,500 bushels, 40 colonies of bees were introduced in 1927, and 5,200 bushels of apples were harvested.   In a cherry orchard, the beekeeper made $100 from the honey and rent for his bees, and it was estimated that the owner of the orchard made $10,000 more from his cherry crop than he would have without the bees.   In another case, an alfalfa grower harvested alfalfa seed at the rate of 1,200 pounds per acre where bees were abundant, and only 300 to 400 pounds without the bees.

The Fig Insect.—One of the most striking illustrations of the dependence of plants upon insects is found in the history of the introduction of Smyrna fig culture into the United States.   Previous to 1900 figs grown in this country were of quality and flavor very inferior to those of figs imported from Asia Minor.   A careful study of the situation revealed the remarkable fact that the palatability of the Smyrna fig, in Asia Minor, was dependent upon the pollination of the flowers by a small wasp.[1]

A fig is a hollow pear-shaped receptacle that bears a very large number of minute flowers lining its inner surface. The only entrance to the flowers is a tiny opening at the free end of the fig. If the flowers are not fertilized, the seeds do not form, and the fleshy, nearly closed receptacle that bears them does not develop the sweet, nutty flavor that characterizes the perfect fruit. The Smyrna fig is exclusively female and produces no pollen. A pollen-producing variety, known as the *caprifig*, produces inedible fruits but an abundance of pollen. Pollination is performed only by the female of this tiny fig wasp (Fig. 27,*a*). These insects lay their eggs in the flowers of wild figs, or *caprifigs*, and their larvae develop in small galls at the base of the caprifig flowers. The males that are formed are wingless (Fig. 27,*e* and *f*) and never leave the wild fig in which they develop. They crawl about, gnaw into the galls in which females are developing, and fertilize the females through the puncture. After mating, the female escapes from the gall, becomes covered with pollen from the stamens of the caprifig, then squeezes her way out of the small opening at the free end of the caprifig, and flies about among the trees. In seeking places to lay their eggs, the females enter Smyrna figs as well as the caprifigs, and, although they are said not to lay their eggs in the former, because the ovaries are so deep-seated that their ovipositor cannot reach them, they crawl over the minute flowers and scatter pollen over them from the flowers of the caprifigs in which they developed.

When the part that this minute insect plays in the culture of figs was known, efforts were made to bring some of the wasps from Algeria into California. There were many failures, but a decade of effort finally resulted in the establishment of the insect and the subsequent production in America of figs equal to those grown in Asia Minor. It is necessary to grow caprifigs as well as Smyrna figs to keep up the supply of wasps, since the insect does not reproduce in the edible fig. Figs containing mature fig wasps are removed from the caprifig trees, strung on fibers, and suspended among the branches on the Smyrna-fig trees when the latter are ready for fertilization.

Recently the fig wasp threatened ruin to the fig growers in California by spreading an endosepsis or internal rot disease to the Smyrna figs from infected caprifigs in which they develop. The little wasps are being reared by millions in sterile incubators and released in the orchards free from brown rot germs in an attempt to prevent this contamination.

INSECTS AS FOOD

Although of small size, insects, because of their prodigious numbers, probably exceed in weight all other animal matter on the land areas of the earth. This great mass of material possesses genuine food value. Chemical analyses of white grubs and May beetles, for example, have shown[2] that these insects compare favorably with tankage in food value. Turkeys, hogs, and other domestic animals will often fatten on insects.

It has been said that insects make up, on the average, about two-thirds of the food of our common land birds.[3] The extensive investigations carried on by Professor Forbes and his associates in Illinois led him to the conclusion, from the examination of over 1,200 fishes from all kinds of Illinois waters, that fully two-fifths of the food of adult fresh-water fishes is insects.[4] The insects of most importance as fish food are (*a*) small,

[1] *Blastophaga psenes* (Linné), Order Hymenoptera, Family Chalcididae (see *Calif. Agr. Exp. Sta. Bul.* 319, 1922; *U.S.D.A., Dept. Bul.* 732, 1918).

[2] *U.S.D.A., Farmers' Bul.* 940, 1918.

[3] *U.S.D.A., Farmers' Bul.* 630, 1915.

[4] *Bul. Ill. State Lab. Natural History,* **1**: 75, and **2**: 475–538, 1888.

slender midge larvae known as bloodworms,[1] (*b*) May-fly nymphs,[2] and (*c*) caddice-worms.[3]

In many parts of the world considerable quantities of insects are regularly eaten by human beings. These are generally looked upon as great luxuries by the less civilized races. In Mexico the eggs of certain large aquatic bugs are regularly sold in the city markets. The eggs are about the size of bird shot. The Mexicans sink sheets of matting under water upon which the eggs are laid by millions. These are then dried and placed in sacks, sold by the pound and used for making cakes. The people of Jamaica consider a plate of crickets a compliment to the most distinguished guest. Ox warbles are eaten raw by the Dog Rib Indians. Natives of Australia collect quantities of the bugong moth, *Agrotis infusa*, in bags, roast them in hot coals, and claim that they taste like nuts and abound in oil. The Indians and semicivilized natives of many countries catch quantities of ants, grasshoppers, and the larvae and pupae of bees, moths, crane flies, and wood-boring beetles and eat them raw, dried, or roasted.

From the actions of wild animals and the testimony of those persons who have tried insects as food, it seems that much of this material is palatable. It would, in fact, be difficult to give any sound reasons why we should consume quantities of oysters, crabs, and lobsters, and disdain to eat equally clean, palatable, and nutritious insects. Perhaps the economists of the future, if hard pressed to maintain an ever-increasing population, may well turn their attention to the utilization of certain kinds of insects as human food.

PREDACEOUS AND PARASITIC INSECTS

Many of the benefits from insects enumerated above, although genuine, are insignificant compared with the good that insects do by fighting among themselves. There is no doubt that the greatest single factor in keeping plant-feeding insects from over-

FIG. 28.—An insect predator: a robber fly with its prey. (*From Howes' "Insect Behavior," Badger*.)

FIG. 29.—An insect predator: a rove beetle devouring a fly. (*From Howes' "Insect Behavior," Badger*.)

whelming the rest of the world is that they are fed upon by other insects. It is easy to see how the industry of insects and their devotion to purpose, when coupled with almost unlimited numbers of individuals, can work a miracle for us when their instincts lead them to seek and devour myriads of pests scattered over a farm or a forest. Man will probably never be

[1] Order Diptera, Family Chironomidae.
[2] Order Ephemeroptera, Family Ephemeridae.
[3] Order Trichoptera.

able to do so much in controlling his insect enemies as his insect friends do for him.

These insect eaters, or *entomophagous insects*, as they are called, are advantageously considered in two groups known as (*a*) *predators* and (*b*) *parasites*. Predators (Figs. 28 and 29) are insects (or other animals) that catch and devour smaller or more helpless creatures (called the *prey*), usually killing them in getting a single meal. The prey is generally either smaller, weaker, or less intelligent than the predator. *Parasites* are forms of living organisms that make their homes on or in the bodies of other living organisms (called the *hosts*) from which they get their food, during at least one stage of their existence (Fig. 30). The hosts are usually larger, stronger, or more intelligent than the parasites and are not killed promptly but continue to live during a longer or shorter period of close association with the parasite. An important difference between these two groups is that, in parasitism, the host makes or determines the *habitat* for the parasite; whereas the prey does not necessarily fix the habitat for the predator, which lives quite independently of its victims during the intervals between meals. Another useful distinction is that generally a parasitic larva requires only a single individual of the host species to nourish it to maturity, while a predaceous insect takes many individual victims in completing its development.[1] Predators are typically very active and have long life cycles; parasites are typically sluggish, often sessile, and tend to have very short life cycles.

FIG. 30.—Insect parasites: larvae of a parasitic wasp, *Apanteles fulvipes* Haliday, leaving the body of the still living, but doomed, host (a caterpillar of the gypsy moth) after having fed within its body for several weeks. (*From U.S.D.A., Bur. Ento. Bul.* 91.)

Phytophagous parasites are those which live upon plant hosts. Insects that parasitize the larger animals (especially domestic animals) may be called *zoophagous parasites*. Insects that parasitize other insects are called *entomophagous parasites*. Zoophagous parasites as a group are highly injurious to man (see page 813). They are often permanent parasites and rarely kill the host. The entomophagous parasites are largely beneficial to us. They are almost always transitory and generally kill their hosts.

There are numerous kinds and gradations of parasitism. The term *permanent parasite* is used for those parasites (such as the bloodsucking lice) which spend all their time and all life stages on or in the body of the host. *Transitory parasites* are those that pass certain life stages with one host and during other life stages are either free-living, like the horse bot, or parasitic in the body of an alternate host of a different species, such as the protozoon that causes human malaria. The term *intermittent parasitism* is used by some authors for such attacks as those of mosquitoes or bedbugs, which approach the host only at the time of feeding and, after the meal, leave the host for a period of free living. It is probably better to call such an attack *predatism*. *Obligatory parasites* are those which can live only as parasites and usually on only one species of host, *e.g.*, many of the chewing lice. *Facultative parasites* are those which, like the common flea of cats and dogs, can live free from the hosts part of the time and shift successfully from one individual or species of host to others. Parasites that live on the outside of the body are known as *ectoparasites*, while those that enter the body

[1] See SMITH, HARRY S., "An Attempt to Redefine the Host Relationships Exhibited by Entomophagous Insects," *Jour. Econ. Ento.*, **9**: 477–486, 1916; SWEETMAN, H. L., "The Biological Control of Insects," Comstock Publishing Co., 1936.

or eggs of their hosts are known as *endoparasites*. *Monophagous parasites* are those which are restricted to one species of host; *oligophagous parasites* those which are capable of developing upon a few closely related host species; and *polyphagous parasites* those which are capable of parasitizing a considerable number of host species. *Simple parasitism* refers to the condition resulting from a single attack of the parasite, whether one or many eggs are laid; *superparasitism* refers to the condition resulting from several different attacks by individuals of the same parasite species; while *multiparasitism* describes the condition of a host which is suffering simultaneously from the attacks of two or more species of primary parasites. A *monoxenous parasite* requires only one host for its complete development; a *heteroxenous parasite* requires several or different hosts for its complete development.

A classification of parasitism that should be kept in mind is that, with respect to a given host, a parasite or predator may be primary, secondary, tertiary, or quaternary. Not all parasites that kill insects are beneficial to man, and indeed it is sometimes almost impossible to tell whether the presence of a parasitic species in a given territory would be beneficial to man or injurious. If we have an injurious insect such as the cotton boll weevil, any parasite attacking it is *primary* for the cotton boll weevil and helpful to man. That parasite may in turn be attacked by a parasite, which is then *secondary* to the cotton boll weevil and inimical to man. A parasite attacking such a secondary parasite would be known as a *tertiary* parasite of the cotton boll weevil and would in this capacity be beneficial to man. All parasites whose hosts are also parasites are collectively known as *hyperparasites*. When abundant, the hyperparasites may completely offset the beneficial work of primary parasites. How complicated the interrelations of good and bad parasites may be is illustrated by the accompanying diagram (Fig. 31) of the various parasites and predators associated with the cotton boll weevil.

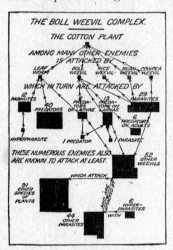

FIG. 31.—Diagram illustrating the interrelations of some helpful and harmful parasites and predators associated with the cotton boll weevil. (*From U.S.D.A., Bur. Ento. Bul.* 100.)

The practical man should make an effort to become acquainted with the typical appearance and manner of life of the more important groups of predators and parasites; and those scientifically inclined can hardly find more fascinating groups for study.

Insect Predators.—Among the best-known entomophagous predators are the following:

Dragonflies, Order Odonata (see page 190).
Aphid-lions, Order Neuroptera, Family Chrysopidae (see page 57).
Ground beetles, Order Coleoptera, Family Carabidae (see page 57).
Lady beetles,[1] Order Coleoptera, Family Coccinellidae (see page 58).
Flower flies, Order Diptera, Family Syrphidae (see page 59).

Tillyard said of the dragonflies: "They are the most powerful determining factor in preserving the balance of insect life in ponds, rivers, lakes, and their surroundings." Although popularly called "snake-feeders," "snake-doctors," "devil's darning

[1] Also called "ladybirds" and "ladybugs."

needles," and other unsavory names, neither the adults (Fig. 111,*5*) nor the nymphs (Fig. 111,*1* and *3*)—which develop for a year of two in the water before the winged stage appears—bite, sting, feed snakes, perform sorcery, or harm man in any way whatever. They scull, soar, and dart about, near and over ponds and streams in a manner to arouse the envy of the most daredevil aviator. The swiftest of them attain a speed of nearly 60 miles an hour.[1] They both catch and eat their prey while on the wing. Tillyard found over a hundred mosquitoes in the mouth of a dragonfly at one time. The same author fed mosquito larvae to a hungry dragonfly nymph which swallowed 60 of them in 10 minutes. It seems likely that these insects are a help in keeping down mosquitoes. The adults also are known to catch flies, beetles, moths, and wasps, many of which are doubtless injurious to man.

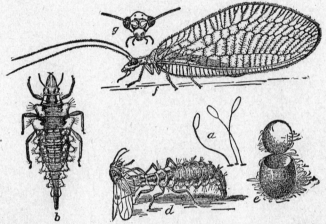

Fig. 32.—An aphid-lion or lace-winged fly: *a*, several eggs, showing the long pedicels that elevate them; *b*, larva dorsal view; *d*, larva feeding on a plant louse; *e*, empty cocoon, showing lid through which the adult has escaped; *f*, adult side view; *g*, head of an adult front view. (*From Marlatt, U.S.D.A.*)

Aphid-lions.—These voracious creatures are the young of delicate, gauzy-winged, weak-bodied insects called lace-winged flies or golden-eyed flies (Fig. 32,*f*). In the adult stage (*f*) these insects probably have little importance to man, but their larvae (*b*, *d*) are very beneficial. Their white eggs (*a*) are curious objects placed on long slender stems attached to the leaves or stems of trees, vegetables, or field crops (Fig. 78,*R*). The spindle-shaped larvae that hatch from them may be found scurrying about on the plants in search of aphids. They have very long, sharp-pointed jaws or mandibles, with which they grasp and puncture the bodies of aphids or other small, soft insects, or their eggs. These mandibles have grooves along their ventral surface against which the maxillae fit to make two closed tubes through which the juices of the victim are sucked into the mouth. The aphid-lions spend the winter in small, white, silken, spherical or oval cocoons (Fig. 32,*e*) about the size of a common elderberry. In the spring the adult cuts off a circular lid through which it makes its escape.

Ground Beetles.—The exact food habits of the *ground beetles* (Fig. 33) are less well known. One finds the general statement "they are beneficial because of their predacious habits"; or, since "both larvae and adults feed on many of our most noxious insects, ground beetles must rank among the farmer's best friends." At least a few of the species are injurious by feeding on seeds and berries, and the food habits of most of the 1,200 or more American species have never been recorded. But certain species are known to be very valuable;[2] and, in general, these flattened, black or brown, long-

[1] TILLYARD, "The Biology of Dragon Flies," Cambridge University Press, 1917.
[2] A European species, *Calosoma sycophanta* Linné, has been introduced into New England to prey on the gypsy and brown-tail moths (see *U.S.D.A., Bur. Ento. Bul.* 101, 1911).

legged, swift-running, strongly built "caterpillar hunters" are probably helpful to man. They hide during the day under stones, boards, logs, in the grass, or below the surface of the soil, and hunt chiefly at night.   The larvae are slender, a little flattened, slightly tapering to the tail, which terminates in two bristly, hair-like or spine-like processes.

Lady Beetles.—The lady beetles (Fig. 34) will need introduction to very few persons. Their bright bodies, their active habits, their great abundance and equitable distribution, and popular songs and stories, all combine to insure that nearly all of us make their acquaintance at an early age.   They are nearly hemispherical in shape.   The commonest species are red, brown, or tan, usually with black spots; a few are black, sometimes spotted with red.   In size they are commonly from $\frac{1}{16}$ to $\frac{1}{4}$ inch long

Fig. 33.—European ground beetle, *Calosoma sycophanta;* adult, at right, feeding on pupa of gypsy moth and larva, at left, feeding on caterpillar of gypsy moth.   (*From Mass. State Forester.*)

and about two-thirds as broad.   Some of the destructive leaf beetles look much like lady beetles.   The student need never confuse these two groups if he will remember that the lady beetles have three-segmented tarsi, the leaf beetles four segments in each tarsus.

Both the adults and the larvae of the lady beetles feed on scale insects, aphids, or other small, soft-bodied creatures or their eggs.   The larvae of lady beetles (Fig. 35, *center*) are carrot-shaped and resemble somewhat the aphid-lions in their flattened, gradually tapering bodies, distinct body regions, long legs, and warty or spiny backs. They do not have the inordinately long mandibles or extra-wide thoracic segments of the aphid-lions, and are generally more conspicuously colored with patches of blue, black, and orange.   The pupae (Fig. 35, *right*) are not enclosed in cocoons but are exposed on the leaf to which the tips of their abdomens are cemented.   When disturbed, they have the curious habit (possibly protective) of lifting the body into a vertical position and soon dropping back again.   The orange eggs of many lady beetles are placed in small masses of a dozen or two, the individual eggs standing on end in contact with each other.   They should not be destroyed.

Syrphid Flies.—Another important group of predators that rival in importance the lady beetles and the aphid-lions are the creatures known as syrphid flies, flower flies, or sweat flies (Fig. 36). They are predaceous only in the larval stage; the adult flies never attack other insects. It is a rare aphid colony that does not have from one to many of the elongate, footless, slug-like, tan, or greenish maggots of these flies

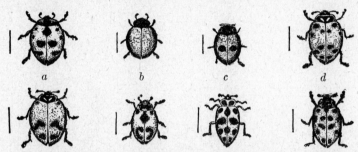

FIG. 34.—Some common lady beetles: *a*, the nine-spotted lady beetle, *Coccinella novemnotata* Hbst.; *b*, the red lady bettle, *Cycloneda munda* Say; *c*, the two-spotted lady beetle, *Adalia bipunctata* Linné; *d*, the convergent lady beetle, *Hippodamia convergens* Guer.; *e*, the glacial lady beetle, *Hippodamia glacialis* Fab.; *f*, the parenthesis lady beetle, *Hippodamia parenthesis* Say; *g*, the spotted lady beetle, *Megilla fuscilabris* Muls.; *h*, the thirteen-spotted lady beetle, *Hippodamia tredecimpunctata* Linné. The ground color of the spotted lady beetle is pink; of the others yellowish or tan. (*From Conn. Agr. Exp. Sta. Bul.* 181.)

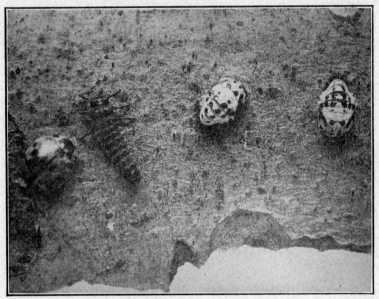

FIG. 35.—The fifteen-spotted lady beetle; adult at left, larva next, two pupae on the right. (*From Conn. Agr. Exp. Sta. Bul.* 181.)

(Fig. 36,*C*) preying upon it. Hidden among the aphids or quietly looping about over the surface of the plant, these larvae grasp aphid after aphid by their pointed jaws, raise it in the air, and slowly pick out and suck out all the body contents, finally discarding the empty skin. A syrphid fly larva often destroys aphids at the rate of one a minute over considerable periods of time. It is difficult to measure the great service thus performed. The adults lay their glistening-white elongate eggs (Fig. 36,*B*) usually one in a place, among groups of aphids. They themselves feed on nectar and pollen and have considerable value as pollinizers. Adult syrphid flies are often confused with wasps and bees. They are generally banded or spotted with bright yellow or covered with long black and yellow hairs that give them a very striking general resemblance to stinging insects. They can be distinguished by having only one pair of wings, by their hovering or poising flight, and by the presence of the false vein (*x*, Fig. 36,*A*) in their wings.

The number of predators working in a field is often astonishingly large. W. G. Johnson records that in packing peas in southern Maryland in 1899, the separators

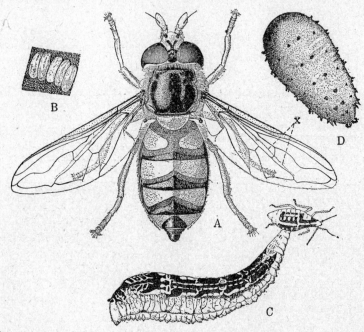

Fig. 36.—A syrphid fly: *A*, adult of *Didea fasciata* Macq.; *x*, the false vein in the wing that characterizes flies of this family (*after Metcalf, from the Ohio Naturalist); B*, eggs of *Melanostoma mellinum* Linné (*after Metcalf from Maine Agr. Exp. Sta. Bul.* 253); *C*, larva feeding on an aphid (*redrawn after Jones, Colo. Agr. Exp. Sta.*); *D*, puparium of *Didea fasciata* Macq. (*after Metcalf, from the Ohio Naturalist*).

sieved out in a few days about 25 bushels of larvae of Syrphidae, chiefly one species. They were so abundant that they almost completely destroyed the pea aphids in the fields. J. E. Dudley reports that in experimenting with an aphidozer or trap to collect pea aphids from pea fields in Wisconsin, he collected with the aphids from 2½ acres, 1,523 syrphid fly larvae, 173 adults and larvae of lady beetles, and 42 other predators.

Insect Parasites.—The most valuable *entomophagous parasites* are probably contained in the following families:

Tachinid flies, Order Diptera, Family Tachinidae (see page 61).

Ichneumon wasps, Order Hymenoptera, Family Ichneumonidae (see page 63).

Braconid wasps, Order Hymenoptera, Family Braconidae (see page 64).

Chalcid wasps, Order Hymenoptera, Family Chalcididae (see page 62).

Egg parasites, Order Hymenoptera, Family Scelionidae (see page 62).

These families of entomophagous insects differ from the predators just discussed in that they enter the body of the victim and live inside of it, feeding for a period of time on blood or tissues, instinctively avoiding vital parts, usually until the parasite is full grown. By that time or shortly afterward the victim dies and the parasite completes its transformation to an adult either within or without the dead body of the host. Representatives of all insect orders and all life stages from egg to adult may be thus attacked by the various parasites.

Tachinid Flies.—One of the most important families of entomophagous parasites is the group of flies known as Tachinidae. Many of the species resemble, superficially, a much overgrown house fly, being very bristly, usually grayish, brownish, or black-mottled flies, without bright colors (Fig. 37). They may be distinguished at once from the house fly family, however, by the entirely bare bristle on the antenna. The adults are found chiefly resting on foliage or about flowers upon which they feed, but may often be seen attacking the caterpillars of moths and butterflies which are the commonest hosts of their larvae.

The eggs of the tachinid flies are glued to the skin of the host (Fig. 199) or laid on foliage where the host insect may ingest them; or already hatched larvae are deposited on the body of the victim or beneath its skin. They feed at the expense of muscles and fatty tissues, not absolutely necessary to life, and the host caterpillar may complete its growth and form a chysalid or cocoon before it dies. From the cocoon, however, emerges the parasitic fly instead of the moth or butterfly. Whenever armyworms become abundant, one of the noticeable features of the outbreak is the great number of flies that are buzzing about.

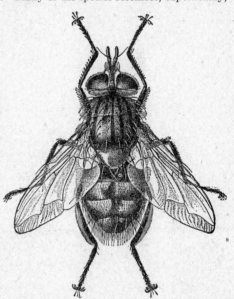

FIG. 37.—A tachinid fly, *Winthemia quadripustulata* (Fab.); a fly that lays eggs on armyworms and whose larvae destroy the worms. About 5 times natural size. (*From U.S.D.A., Farmers' Bul.* 752.)

Farmers sometimes think that flies "make" the armyworms. But the parents of the armyworms are tan-colored moths (Fig. 200); and these flies[1] are one of their principal natural enemies. If you examine the back of an armyworm, you will often find the small white eggs of this fly glued to the skin, in numbers ranging from 1 to 50 (Fig. 199, *left*). Few of the eggs are found farther back on the body than the thorax—seemingly a nice protective instinct on the part of the parent fly, since the caterpillar could bite off the eggs deposited toward the rear end of its body. Upon hatching, the maggots tunnel directly through the skin and within a short time may have killed the caterpillars by thousands.

The remaining parasites listed above are all Hymenoptera. Although they are commonly spoken of as "flies," the authors prefer the terms ichneumon wasps, brac-

---

[1] *Winthemia quadripustulata* (Fab.), Order Diptera, Family Tachinidae.

onid wasps, and chalcid wasps, as these insects are really small wasps. They are, therefore, four-winged insects; the females are provided with a sharp ovipositor (homologue of the "stinger" of bees), with which the eggs are generally thrust into the flesh of the host beneath the skin. These families of parasitic Hymenoptera are both extremely large in numbers of species and extremely difficult to classify. For the practical man, it is probably sufficient to know that most of these parasites can be recognized by having a divided or two-segmented trochanter (Fig. 47,*D*) and a slender petiole or fore part of the abdomen, next to the thorax.

Egg Parasites.—The family Scelionidae includes some of the smallest of known insects, many of them egg parasites and of such a size that they derive all their nour-

Fig. 38.—The chinch bug egg parasite; adult female greatly enlarged; antenna of male above. (*From Proc. U. S. Nat. Museum, vol.* 46.)

ishment and grow to maturity inside of a single egg of another insect. They are mostly slender, nonmetallic, minute wasps with usually straight antennae, few wing veins, and the ovipositor attached at the tip of the abdomen.

The chinch bug egg parasite[1] (Fig. 38) greatly reduced the hatch of chinch bugs in Kansas during 1913. The extent of parasitism for the state of Kansas was calculated at about 16 per cent, *i.e.*, 16 out of 100 chinch bug eggs were killed by this

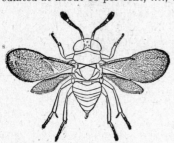

Fig. 39.—A chalcid wasp parasite, *Aphycus eruptor* Howard. (*From Comstock's "Introduction to Entomology."*)

enemy. The next season was generally favorable for chinch bugs, but, largely because of the good work of this parasite, only 1 bug was found in winter quarters in 1913–1914, where from 25 to 100 had been found the winter before. The life cycle of this parasite occupies only 2 or 3 weeks, so that four or five generations may operate on a single generation of the chinch bug. In 1914 Flint brought from Kansas to Illinois over 100,000 chinch bug eggs, many of which were parasitized, and colonized the parasites in about thirty localities in the latter state. By concentrating the parasites, the percentage of infestation of the chinch bug eggs was raised on one farm from less than 10 per cent in 1914 to 53 per cent in 1915. Other species of Scelionidae live in the eggs of many moths, true bugs, grasshoppers, spiders, and flies, such as the Hessian fly.

Chalcid Wasps.—The chalcid wasps are mostly parasitic, though some of the species feed on plants, *e.g.*, the jointworm and the fig wasp. Many species live within the minute bodies or the eggs of scale insects, aphids, caterpillars, and flies (Fig. 39). A very beneficial species is *Pteromalus puparum* (Linné), which attacks our common imported cabbageworm. Although only $\frac{1}{16}$ inch long, it occurs in such numbers in many sections as greatly to reduce the numbers of the cabbageworm. More than 3,000 individuals have been reared from a single specimen of the cabbageworm. Not

[1] *Eumicrosoma benefica* Gahan (see *Jour. Econ. Ento.,* **8:** 248, 1913).

all the parasitic chalcids are beneficial to us, because there are many hyperparasites or secondary parasites.

These wasps are generally of a metallic luster, with elbowed antennae and with the ovipositor hidden but attached to the ventral surface of the abdomen some distance before the tip. The wings generally show a single vein parallel with the costa and slightly forked at its tip (Fig. 39).

Ichneumon Wasps.—This family includes the largest of our parasitic species, as well as some very minute ones. They are often

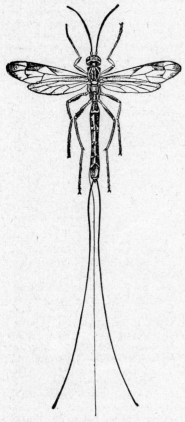

Fig. 40.—The long-sting, *Megarhyssa lunator* (Fab.), an ichneumonid parasite of the pigeon tremex. The long appendage is not a stinger but an egg-laying organ for inserting the eggs into the burrows of the pigeon tremex (Fig. 153), in a tree trunk. (*From Kellogg's "American Insects."*)

Fig. 41.—A sphinx caterpillar covered with the cocoons of a braconid parasite. The larvae of the parasite develop inside the body of the caterpillar, which eventually dies as a result of the attack. (*From Sanderson and Jackson, "Elementary Entomology."*)

brilliantly marked. When active they "can generally be recognized by their short, jerky flight and constantly vibrating antennae" (Lefroy). The very large long-sting or Megarhyssa (Fig. 40), 1½ inches long, with tail-like ovipositor nearly 3 inches long, may often be found fastened by its ovipositor in the side of a tree trunk. They thrust this marvelously thin and strong ovipositor into the wood until it strikes the burrow of a wood-boring larva (the pigeon tremex, see Fig. 153) when an egg is inserted. The parasitic larva upon hatching from the egg crawls along the tunnel until it encounters the wood-boring larva which it then attacks. Possibly the adult long-sting locates the position of the wood-boring larva by the sounds it makes in tunneling through the wood.   Most ichneumons attack caterpillars, with resultant great benefit to man.

Braconid Wasps.—Commonly one sees specimens of tomato hornworm or catalpa sphinx whose backs are covered with small, elongate, white objects glued by one end to the skin of the caterpillar (Fig. 41).   They are commonly called "eggs."   In reality they are silken cocoons enclosing the pupal stage of a species of braconid wasp of the genus Microgaster.   The parent wasp had previously thrust many eggs through the skin of the larva; her larvae had fed within the tissues of the caterpillar, and, when full grown, they had eaten out through the body wall again, to spin their cocoons fast to the host.   Some other species leave the host and form their cocoons in masses on a leaf near their dead victim.

If one examines carefully almost any leaf that is infested with aphids, one will see some specimens of the aphids whose bodies are distended, shiny, and brown and

Fig. 42.—Aphids parasitized by a braconid wasp, *Lysiphlebus testaceipes* (Cresson). Adult wasps have emerged through the circular holes after having killed the corn leaf aphids by developing in their bodies.   Nearly 100 per cent of the aphids were parasitized. (*From Essig, "Insects of Western North America," copyright, 1926, by Macmillan; reprinted by permission.*)

many of them with neat circular holes cut in their backs (Fig. 42).   This illustrates the work of some of the smallest of the braconid wasps, *e.g.,* the genera Aphidius and Lysiphlebus.   Still others of the braconid wasps attack dipterous larvae, ants, cockroaches, beetles, bees, and wasps, and some even swim with their wings beneath the water to parasitize such aquatic insects as caddice-worms.

INSECTS AND WEEDS

Whether an insect is injurious or beneficial depends more often than anything else upon the economic status of the thing or material it eats.   That great horde of insects that feeds on weeds or attacks noxious animals must perform a very valuable service.   Anyone who stoops to notice the abundance and variety of insects on almost any bit of wild vegetation will need no arguments to convince him that these plants suffer grievously from insect attack. To a corresponding degree they help the farmer, who has to contend with these undesirable plants.

This beneficent work is unfortunately not without its drawbacks. Too often insects that have increased at the expense of weeds subsequently shift their attack to cultivated crops. This is especially likely to occur when the weeds are botanically close relatives of the cultivated crop. For example, the writers have often seen numbers of flea beetles and tomato hornworms developing in hog lots, barnyards, fence rows, and fields, where Jimson weed, horse nettle, and morning-glories were allowed to grow in profusion. A little later these pests shifted to adjoining potato, tomato, or tobacco plants. Forbes has shown that the corn root aphid is dependent in the early spring upon such weeds as smartweed, foxtail, ragweed, purslane, and crab grass, before corn is planted. The common stalk borer, *Papaipema nebris nitela* Guen., is seldom destructive to corn or other cultivated crops except along the margins of fields where the larvae have begun development on grasses or weeds. The bean aphid, *Aphis rumicis* Linné, is often very abundant on pigweed and dock, from which it easily migrates to beans, dahlias, euonymus, etc. The spinach flea beetle increases in the early part of the season at the expense of such weeds as chickweed and Chenopodium. The tarnished plant bug occurs in such numbers on ragweed, pigweed, common mallow, goldenrod, evening primrose, and wild asters as to constitute a considerable check to their development. But while they are checking these weeds, they are also increasing their own numbers for a probable later invasion of our nurseries, orchards, and flower gardens. The thistle butterfly or painted lady, in periods of abundance, is sometimes hailed as a destroyer of Canada thistle. But it is also likely to feed injuriously upon some cultivated crops. Even insects that are not at present known to attack any cultivated crop may change their food habits, desert their weed hosts, and adapt themselves almost exclusively to a new and valuable food plant. Such a shift is known to have occurred in the case of the Colorado potato beetle (see page 507).

We may fairly sum up this phase of the subject with the statement that, while we are not unmindful of the vast service performed by insects in checking weed growth, we cannot encourage their development or even look with favor upon the great natural increase of any species, especially in proximity to cultivated crops, because of their potential injury to valuable plantings (see also pages 304, 305).

INSECTS AS SOIL BUILDERS[1]

In the production or maintenance of productive soils insects play an important part. They help to break up the rock particles and, by bringing them to the surface, expose them to the action of water and other weathering influences. The numerous tunnels made by the insects facilitate the circulation of air into the soil, so essential to the health of plants. Insects burrow to depths ranging up to at least 5 feet for white grubs and 10 feet for the nymphs of cicadas. These burrows doubtless have considerable importance in the movements of capillary water. Finally, insects are of inestimable value in adding humus or organic matter to the soil. This is accomplished in several ways. The dead bodies of the insects themselves accumulating on the surface are a fertilizing element. The excreta of insects is a rich manure, comparing favorably in chemical content with that of the larger animals and in total amount undoubtedly exceeding the latter. In burrowing through the ground, insects bring up the subsoil particles, and cover plant and animal materials lying on the surface. Others carry plant and animal particles into the soil in connection with their feeding or

[1] McCOLLOCH and HAYES, "The Reciprocal Relation of Soil and Insects," *Ecology*, **3**: 288–301, 1922.

nesting activities: as termites, ants, cutworms, burying beetles, dung beetles, and predaceous wasps. This is somewhat similar to plowing under a cover crop.

Here, as in most of the other relations of insects, their small size is abundantly offset by their unparalleled numbers. Wheeler says that ants outnumber in individuals all other terrestrial animals. The earth worm as a soil builder has been brought prominently to our attention by the writings of Darwin; it seems to the writers that insects as a whole must equal or excel earth worms in the formation, fertilization, and renovation of soils. A great variety of insects is found in the soil, representatives of practically all the natural orders. The most abundant are ants, bees, wasps, beetles, the larvae of flies, cutworms, the pupae of moths, cicadas, crickets, and springtails. These creatures have been found in the soil in abundance beyond our comprehension.

INSECTS AS SCAVENGERS

One of the most interesting ecological groups of insects is that group that feeds on the decaying substances of plants and animals. Their service is twofold: in the first place, they help to remove from the earth's surface the dead and decomposing bodies of plants and animals, converting them into simpler and less obnoxious compounds, and removing what would otherwise be a menace to health; and in the second place, they play a very important part in converting dead plants and animals into simpler substances which can be used as food for growing plants. Repulsive as they may be, we can not scorn the beneficent work they do. In this work the larvae of many flies and the larvae and adults of beetles are especially important. A fascinating story of the lives of some of the beetles is told by Fabre in "The Life and Love of the Insect."

AESTHETIC VALUE OF INSECTS

The aesthetic value of insects is the least tangible of all, and perhaps most readers will think it the least important. Insects rival birds and flowers in beauty; and there are many more of them. Moths and butterflies are universally admired. But there are thousands of kinds of beetles, bees, flies, leafhoppers, planthoppers, dragonflies, mantids, and others that will be found every bit as handsome as moths and butterflies by anyone who will examine them through a microscope or magnifying glass.

Much practical use has been made of the beauty of insects. Artists, florists, milliners, and designers have drawn extensively from the Lepidoptera and might profitably draw much more extensively upon the inexhaustible grace of line and beauty of color, so lavishly displayed in the less well-known kinds of insects. Insects are widely used as ornaments. In larger cities one finds stores dealing in trays, pins, necklaces, etc., of which the sole claim to beauty is the actual bodies of preserved insects. The reader should not gain the idea that many insects have a very great commercial value. One of the drawbacks of being an entomologist is the frequent necessity of disillusioning some trusting youngster who has reared his first cecropia moth and has hastened to the nearest "bugman" to sell his valuable (?) find.

In the Bahamas, South Africa, and Australia the natives often wear strings of what are called "ground pearls," which have been found to be the case or shell secreted by the nymphs of a kind of scale insect. They are irregular in shape, of a variety of colors, and often have a beautiful luster or iridescence. In the Caribbean Islands living fireflies are enclosed in gauze and worn as hair ornaments.

The songs as well as the form and colors of insects have been of much interest from earliest times. The natives of South America, Africa, Italy, and Portugal cage katydids and crickets for the sake of their songs; and highly ornamented cages containing these little songsters are sold in the streets.[1]   One of our entomologists[2] has taken much pains to assemble the poetry based on insects, in the English language alone, from A.D. 1400 to 1900.   He records having found over 1,200 separate excerpts, including about 75 complete poems.

The curious habits, structures, sounds, and interactions of insects continually afford profitable diversion or hobbies to a small army of amateur entomologists, ranging from carefree children to staid scientists and millionaires.   The Chinese have elaborate cricket fights, contested by champions that are as carefully fed and trained as American race horses and that sell for $5, $10, or even $50 each.   Many persons have been amused and deceived by the seemingly intelligent performances in trained-flea circuses.

THE SCIENTIFIC VALUE OF INSECTS

Insects have taught man a great many things.   They have helped to solve some of the most puzzling problems in natural phenomena. They have led to some remarkable inventions.   They have contributed to the sum of human knowledge in physiology, psychology, and sociology. Instead of "Try it on the dog," the modern slogan is "Try it on the insects."

The use of insects as an index to stream pollution has been a valuable aid in the science of conservation of natural resources.   When toxic factory wastes and large amounts of raw sewage are dumped into streams, aquatic life, such as fish, is destroyed and the pollution also becomes a menace to the health of man and other animals.   The degree of pollution can be measured by the amount and type of insect life that is able to survive in the polluted water.   Streams and lakes having few insects are highly polluted.   It may eventually be possible to measure the amount of pollution with mathematical accuracy by observing the insect life that is present in any stream.

The great abundance and variety of insects have resulted in their selection for many scientific studies and fundamental biological investigations.   Among the many benefits that have accrued from observations of, and experimentation with, these small animals, perhaps the greatest is their contribution to cytology, genetics, and eugenics.   The cells of insects are extraordinarily large, so that they are exceptionally good for cytological work.   Plant and animal breeders and those interested in the future of the human race, have demanded precise and explicit knowledge of how physical and mental traits are inherited when selected mates of various kinds are crossed.   Previous to the discovery of the suitability of the tiny pomace fly, Drosophila, for this work, genetic experimentation with guinea pigs, rats, pigeons, corn, peas, and the evening primrose involved heavy expenses for feed, cages, and caretakers.   Even more serious was the limited number of generations or crosses that one scientist could rear for study during his lifetime.   Finally it was realized that the fundamental principles of inheritance, variation, and race improvement could be revealed as clearly by insects as by the more expensive, more cumbersome, slow-breeding plants and animals.   Drosophila can be handled with the greatest of ease.   Thousands of them can be reared on a bit of fermenting banana in small vials.   Little space and few caretakers are required. This little fly is subject to tremendous variations in visible external

---

[1] CAUDELL, A. N., "An Economic Consideration of Orthoptera Directly Affecting Man," *Smith. Inst. Ann. Rept.*, 1917, pp. 507–514.

[2] WALTON, *Proc. Ento. Soc.*, *Washington*, October and November, 1922

characters. Its cells have only four pairs of chromosomes (see page 136), and its salivary glands have chromosomes large enough that the actual genes (page 136) can apparently be seen under the microscope. Most

TABLE IV.—THE INSECTS OF THE UNITED STATES IN ACCOUNT WITH
THE AMERICAN PEOPLE
Most Important Items for the Year 1929
Credit

To products made or collected by or from insects:

87,170,000 lb. Silk imported.................. @$4.86	$423,646,000
200,000,000 lb. Honey produced.............. @ 0.08	16,000,000
6,000,000 lb. Beeswax....................... @ 0.35	2,100,000
31,548,000 lb. Shellac imported.............. @ 0.40	12,609,000
185,000 lb. Cochineal imported........... @ 0.55	100,000
5,000 lb. Cantharis and other blistering beetles imported.............. @ 0.30	1,500
3,068,000 lb. Nutgalls and gallnuts imported. @ 0.10	306,000

Total value insect products in United States...................... $454,762,500

To pollination of fruits, vegetables, and flowers:[1]

The following crops depend almost exclusively upon insects for the production of fruits and seeds:

Apples	Oranges	Eggplant
Pears	Lemons	Tomatoes
Peaches	Grapegruit	Peppers
Plums	Figs	Clovers
Prunes	Melons	Alfalfa
Cherries	Cucumbers	Soybeans
Strawberries	Pumpkins	Cowpeas
Raspberries	Squash	Sweet clover
Blackberries	Beans	Cotton
Cranberries	Peas	Certain flowers

Total value insect-pollinated crops................ $2,087,833,000

To serving as food for other animals:
    Two-fifths of the food of fresh-water fishes is insects[1]
    One-third of the food of wild song and game birds is insects[1]
    Hogs, turkeys, certain fur-bearing animals eat insects[1]
To services in controlling destructive insects:
    Unless checked by parasites and predators injurious insects would soon make it impossible for man to exist.[1]
To services in aerating and fertilizing the soil with their dead bodies and excreta[1]
To services as scavengers:
    Transforming plant and animal refuse into forms available for utilization by growing plants[1]
To services in scientific investigations[1]
To services as models, ornaments, and for entertainment and diversion[1]
To services in medicine and surgery[1]

[1] It is impossible to estimate in dollars the enormous benefits to mankind from these services.

important of all, this insect will complete a generation in as short a time as 10 days, so that the geneticist can have as many different generations and crosses as he can possibly study during his lifetime. It is safe to say that a large part of our present knowledge of genetics and eugenics would never have been possible without the use of this tiny fly.

SUMMARY

We see, therefore, that many comforts and luxuries, much food, certain dyes and medicines, and our choicest clothing come from insects; that insects are useful in scientific investigations, in surgery, in improving the soil, in pollinizing fruits, flowers, and vegetables, in checking weed growth, as scavengers, and as food for birds and fishes and even for men; that the future of our race may be greatly benefited by knowledge gained from the study of inheritance in insects; and that our very tenure on the earth is probably dependent upon the friendly ones (parasites and predators) among our most numerous animal relatives, the insects.

CHAPTER III

THE EXTERNAL MORPHOLOGY OF INSECTS[1]

Insects are among the humblest and lowliest of animal creatures; man belongs to a species that has been called the dominant species. Yet it has been shown in the preceding chapters that insects dispute with man for the possession of most of the things he values and often succeed in getting the greater part of the product of his labors. The present chapter is an attempt to analyze the success of insects as a group of animals living in keen competition with thousands of others.

Insects are the most abundant of terrestrial animals in both numbers of species and numbers of individuals. According to the theory of *organic evolution* or *biogenesis*, this has come about because they are better adapted to their surroundings or environment than other groups, many of which have become extinct while the insects have continuously diversified and multiplied. In general, every kind of plant and animal produces more offspring every year than could possibly find food and room to survive (*overproduction*). The competition for every necessity of life is so intense that the great majority of all creatures born into the world die before reaching maturity (*the struggle for existence*). Among the individuals of a species, even those from the same parents, there are generally noticeable variations, sometimes slight (*continuous variations*) and sometimes striking ones (*mutations*). There is good evidence for believing that death due to the struggle for existence is not indiscriminate, but that those individuals possessing variations that are advantageous will in the long run survive to perpetuate their kind, while those less favorably equipped for the struggle of life will be eliminated (*elimination of the unfit* or *survival of the fittest*). It follows that any group that has attained the numerical superiority of the insects, both in species and in individuals, must be unusually well fitted for life. We shall be in a better position to fight those insects that are pests, when we understand their structure and the most important characteristics that fit them for life.

Characteristics That Enable Insects to Compete with Man

The Size of Insects.—One of the greatest factors in the success of insects is undoubtedly their small size. While the largest insects reach a length of 6 to 10 inches, these are very exceptional. The smallest known kinds are less than $\frac{1}{100}$ of an inch long. W. T. M. Forbes is quoted as saying that the moth having the greatest wing expanse is probably *Thysania agrippina*, which measures 11 inches from tip to tip of wings, and the smallest probably *Nepticula gossypiella*, whose wing tips reach slightly over $\frac{1}{10}$ inch. The average weight of insects is probably not over $\frac{1}{5000}$ ounce (0.00572 gram) to the individual. House flies weigh from 15 to 30 milligrams, or about 20,000 flies to the pound.

[1] General reference: Snodgrass, R. E., "Principles of Insect Morphology," McGraw, 1935.

Yellow fever mosquitoes are said to range from 0.0015 to 0.00175 gram, unfed, and from 0.0036 to 0.0039 gram when engorged with blood. This is equivalent to from 120,000 to 280,000 mosquitoes per pound. Full-grown silkworm larvae average 3.5 grams. Richardson reported the average weight of the grasshopper, *Melanoplus differentialis*, to be 1.25 grams for females and 0.8 gram for males. At first thought, small size might appear to be a disadvantage. However, it enables insects to live in cracks and crannies of the plant and animal communities, where competition with the larger animals is not so great. Many insects can subsist on the portions left from the feeding of larger animals. They can retreat into protected places where larger animals cannot follow. They often escape death because they are completely overlooked: very often the first intimation a grower has of the presence of the San Jose scale in his orchard, for example, is the red spotting of the fruit at harvest time.

A small body makes possible feats of strength that, in comparison to size, seem marvelous. Weight-pulling contests for horses are frequently held in the Middle West. A team that can pull the equivalent of its own weight a few dozen feet is generally good enough to win. A female stag beetle, *Lucanus dama*, weighing 1.2 grams dragged, by a silk thread about her waist, two fountain pens, a pencil and a watch, weighing 106 grams or ninety times her own weight, 30 times her own length in 25 minutes. Another dragged a loaded paper box, not on wheels, weighing 120 times her own weight a short distance; and, while suspended by her claws, held a weight of 200 grams attached to her waist. In proportion to size, a man weighing 160 pounds should be able to hold on by hands and feet with $13\frac{1}{2}$ tons tied around his waist. Many insects can carry from ten to twenty times their own weight. A flea whose legs are about $\frac{1}{20}$ inch long can jump as far as 13 inches horizontally and $7\frac{3}{4}$ inches high. If length of legs were the only factor involved, we should expect an athlete with legs 3 feet long to make a broad jump of 700 feet, and a high jump of at least 450 feet. The smaller the muscle (other factors being equal), the greater the proportionate work it can do.

The Abundance of Insects and Their Rapidity of Reproduction.—The great reproductive capacity of insects, in general, is unquestionably responsible, in a large measure for their success. Different counts of the number of insects found in and on the soil indicate a frequency of 1,000,000, 3,500,000, and 10,000,000 individuals per acre, under various conditions. Margaret Windsor found insects occurring on and in the forest soil in Illinois, during the winter, to a depth of 18 inches, at an average frequency of 65,000,000 to the acre. H. Elliott McClure, from extensive trapping experiments, concluded that there is frequently an average population of about 3,000 insects in flight over each acre of surface, or about 1,850,000 per square mile in the morning, and about 11,000 per acre, or nearly 7,000,000 per square mile, flying in the evening. By way of comparison, there is an average of one human being to each 16 acres of dry land on the earth's surface; an average of about two head of horses, mules, cattle, sheep, hogs, and chickens, combined, per acre of land in the United States; and an average of three birds per acre in Illinois fields, woods and orchards, during the summer months. The reproductive capacity of insects is further discussed on pages 139 to 141.

The Adaptability of Insects.—In the adaptability of insects we see another superior quality. New structures and new habits are continuously developing. New forms are evolving and old forms changing in accommodation to the constantly changing face of the continent. Within the memory of those still living, certain insects have adapted themselves to new host plants, thus changing from insignificant bugs to major pests. Insects have not restricted themselves to one medium, like the fish or the birds, or to one kind of host, like the parasitic worms; they occur on and in the water, in the air and soil, on animals and plants, and within the bodies of both; in houses, ships, and mills, and in almost all sorts of organic and inorganic substances. The problems in insect control are more serious in areas where the biological and physical environment is undergoing rapid changes than they are in areas where conditions have become more stable by a long and gradual development. It is one of the greatest tributes to the remarkable variety and plasticity of the insect class that, in the face of rapid and radical changes, *some* species of insect is certain to adapt itself with surprising promptness to the new opportunity and to accomplish this so successfully that it very soon becomes a serious pest.

The Persistence of Insects.—In the instinctive behavior of insects[1] we see an explanation of their apparent fixity of purpose. They lack both reason and judgment. This would seem to put them at a great disadvantage compared to man; and so it does, individual to individual. But considered in the mass, it means on the part of the insect, unfaltering pursuit of the work for which it is adapted. The everyday behavior of the flies about an animal, or in a room when one is trying to sleep, or of chinch bugs trying to cross a barrier from a wheat field to a cornfield, illustrates this point. They cannot be frightened away, or discouraged by repeated assault; they recognize no defeat. So long as life persists within them they continue unflinchingly to procure a living for themselves and to prepare for the next generation.

The Exoskeleton of Insects.—A characteristic of insects that we may be sure has been important in their evolution is the nature of the body wall. Insects have no bones, but are covered all over the outside with a very remarkable hard shell. This shell is not heavy. It is much lighter and also stronger than bone. It is also remarkably resistant to solution or corrosion, not being visibly affected by any of the ordinary chemicals. Water, ether, alcohol, strong acids and alkalies, the digestive fluids of animals, and other solvents have no noticeable effect on the outer body wall of insects. Even boiling potassium hydroxide, which quickly dissolves flesh and horn, does not destroy, or change the appearance of, the skin of the insect, unless the treatment is continued for a long time. Because of the unusually stable character of the body wall, insects can be kept, like mummies, for hundreds of years without any preservative; retaining a life-like appearance after death. The practical importance of this we recognize when we attempt to control insects by contact sprays. The great difficulty has always been to find a substance that would attack the body of an insect and kill it without at the same time killing or injuring the plant on which it was feeding.

It is interesting to know how an insect grows the hard shell which covers every part of the outside of its body, even the eyes, feelers, claws, and mouth parts. A very thin layer of it also extends, as an inner lining, down the throat, over the posterior part of the alimentary canal, and

[1] See footnote, page 160.

throughout the length of the breathing tubes (the *tracheae*, but not the *tracheoles*, see page 96). It is formed first as a secretion which is semi-fluid and covers the entire body like a soft leather glove. The secretion which is known as the *cuticula* comes from a layer of cells just beneath it, called the *hypodermis* (Fig. 43). The hypodermis is composed of a single continuous layer of living cells, terminated at their inner ends (entad) by a continuous sheet of tissue, the *basement membrane*, and at their outer ends (ectad) by the cuticula, which is lifeless, somewhat striated, but noncellular material like our nails or hair. Both cover the outside of the body completely. The cuticula may remain soft and flexible, as in caterpillars, maggots, and similar larvae, but in most insects the outer portion of it (the *exocuticula*) begins to harden or "set," much like cement, shortly after it is secreted. In the course of an hour or so, hard plates

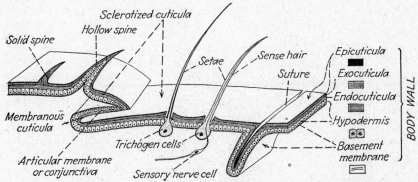

Fig. 43.—Diagrammatic section of the body wall of an insect. The living cells called the *hypodermis* secrete the three layers of cuticula over their outer (ectal) surface, while internally (entad) the hypodermis is limited by a basement membrane. The cuticula is usually greatly hardened or sclerotized over most of the body, but at infoldings called *conjunctivae* it remains soft and flexible, as membranous cuticula. A solid and a hollow spine, an ordinary seta, and a sense hair are also shown; and, at the lower right, subtending a suture, is shown an apodeme to which muscles may be attached. (*In part after Comstock and Snodgrass.*)

have formed over most of the body wall, like pieces of metal armor. The formation of these hardened plates or *sclerites* is due to a chemical process (*sclerotization*), but the nature of the chemical which makes the body wall hard is not known. Between these plates, the cuticula or "skin" is continuous and uninterrupted from one sclerite to another, but remains soft and flexible and is often infolded. Definite impressed lines or internal ridges between sclerites are known as *sutures*. Sutures surround sclerites as seams surround patches. Flexible infoldings of the body wall are called *conjunctivae*, and definite joints, as between segments of the leg, which permit rotary or hinge-like movements, are called *articulations*. The sclerites may often be moved very freely with reference to each other, by the action of muscles attached to their internal faces. The cuticula or skin is penetrated by fine submicroscopic pores or canals, so that small amounts of gases and liquids may penetrate it, permitting the detection of odors, the discharge of secretions from hypodermal glands, and the absorption of substances through the walls of the fore- and hind-intestines and the breathing tubes. Certain insecticides (nicotine, derris, sodium fluoride) are known to penetrate through the body wall, but contact poisons and gases are generally absorbed through the tracheal walls after having penetrated through the spiracles into the breathing tubes.

The most characteristic chemical in the cuticula is called *chitin* (pronounced ky'-tin). Chitin is composed of carbon, hydrogen, nitrogen, and oxygen. The formula, according to Brach, is $(C_{32}H_{54}N_4O_{21})_x$.

On account of this covering of the body and the absence of bones, within, insects are said to have a *chitinous exoskeleton* instead of a bony endoskeleton (Fig. 44). The exoskeleton serves two very important functions. It protects the delicate muscles, nerves, and other organs from mechanical injury. It serves as a framework for the attachment of muscles. In this capacity an exoskeleton seems to offer certain advantages over an endoskeleton. It gives vastly greater area for the attachment of muscles and certain opportunities for very effective leverage. As a protective armor it could hardly be improved upon. It protects the body from becoming soaked with water, from excessive drying, and from the attacks of many disease organisms and is probably the chief thing that enables insects to live in the greatest variety of conditions. The typical insect is built with its shell somewhat in the form of a hollow cylinder, which is the strongest type of construction possible with a given amount of material. The ends of the cylinder are closed with more or less convex caps, so that from almost any point of attack the insect body presents an arched construction. This protects insects from injury by ordinary blows or falling; a point of great significance when we consider the very active, apparently reckless, sort of lives most of them lead. It should be understood that all insects in all stages, however soft the body wall may be, are completely covered by a chitinized exoskeleton.

The body wall has many outgrowths. The commonest are unicellular hair-like outgrowths, known as *setae*. Flat, scale-like setae are generally called *scales*, while others take the form of hooks or pegs. Multicellular outgrowths are known as *spines*, if they are immovable, and *spurs*, if they are articulated and movable (see Fig. 43).

The Segmentation of the Body.—Many animals that are covered with a hard outer shell are sluggish, inactive, and sedentary in habits, like the clams, snails, and barnacles. This is not true of most insects. The latter have achieved the happy combination of an armor plate and great freedom of movement. This is made possible by the characteristic known as *segmentation*. As one examines the bodies of insects (especially caterpillars), one sees that the external wall does not present a smooth, unbroken surface (see Figs. 55,*1* and 46). On the contrary, it is divided by constrictions into a series of ring-like pieces, all connected, yet moving rather freely on each other. The word insect is from the Latin *insectum*, "cut into," and refers to the manner in which the parts of the body are separated by constrictions.

Segmentation of the body is an important advantage. If we contrast it with the condition shown in other shelled animals such as snails and clams, we see that it permits great freedom of movement and activity. In the second place, it facilitates specialization, which makes for efficiency. The body so divided into segments may devote one part to securing food, another to locomotion, another to reproduction, another to defense, and so on. It permits *division of labor*, which has always made for success

Fig. 44.—Skeleton of a mammal and an insect compared: *A*, an animal (cat) with a bony endoskeleton; *B*, an animal (honeybee) with a chitinous exoskeleton. For emphasis certain ingrowths of the cuticula (the endoskeleton of the bee) have been omitted. (*A, modified from Davison's "Mammalian Anatomy," Blakiston; B, redrawn after Snodgrass.*)

and progress the world over, whether it be in a force of factory workers, in a football team, or in the body of a bumblebee.

The segments of the body are called *somites* or body segments, to distinguish them from the segments of legs, antennae, and other appendages. The term "joint" properly refers to the constriction between two segments, and should not be used for the segment. The flexible portion of the cuticula connecting the hard ring-like portions of any two segments is called a *conjunctiva* or *articular membrane* (Fig. 43). Each body segment is made up of (at most) four exposed faces. The dorsal or upper face, above the bases of the legs, is called the *tergum* or *notum;*

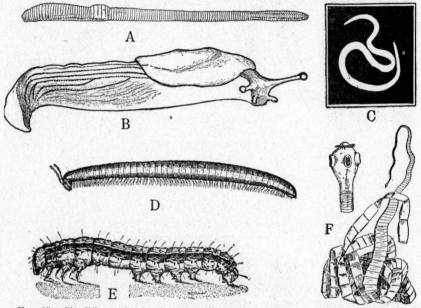

Fig. 45.—Six different kinds of animals often called "worms," that should be sharply distinguished. *A*, an earth worm, phylum Annelida (*from Sanderson and Jackson*); *B*, a slug, phylum Mollusca (*redrawn after Lovett*); *C*, a roundworm, phylum Nemathelminthes (*from Herms*); *D*, a millipede, phylum Arthropoda, class Diplopoda (*from U.S.D.A.*); *E*, a caterpillar, phylum Arthropoda, class Hexapoda (*from U.S.D.A.*); *F*, a tapeworm, phylum Platyhelminthes, the head greatly enlarged at the left (*from Jordan and Kellogg*).

the ventral face (the part between the legs and next to the ground when the insect is in its normal position) is called the *sternum;* and the lateral faces or side pieces, containing the bases of the legs, are called the *pleura.* Each of these faces may be made up of several sclerites, which are then collectively called *tergites, sternites,* and *pleurites,* respectively, and given individual names. The pleura, however, are generally nearly or entirely membranous. The typical number of segments in the insect body is about 20 or 21. This number is greatly obscured and reduced in most insects by the fusion of some of the segments and the degeneration of others. So that ordinarily one will recognize only from 8 to 12 obvious body rings in most insects (Figs. 36,*A* and 46).

The Body Regions of Insects.—Other animals besides insects have the body externally segmented, notably the true worms, thousand-legged worms, crayfish, and their relatives.   Members of the phylum Arthropoda, to which insects, spiders, crayfish, and similar creatures belong, differ from the true worms (phylum Annelida) by having a pair of jointed appendages on at least a part of the body segments.   In this way we distinguish such things as caterpillars, maggots, and grubs (often called "worms," but really young insects) from the true worms such as the earth worm, and from parasitic worms like the tapeworm.   The latter have no legs (Fig. 45).

In the true insects, there are never more than one pair of jointed appendages on any body segment.   Hence, even if the conjunctivae are lost and the segments fused together, as in the head, we can usually tell how many segments are represented by noting the number of paired, jointed appendages present.

Six of the 20 or 21 segments of the insect body are fused into what is called the *head*.   Three of these lie in front of the primitive mouth and three behind it.   The next 3 segments comprise what is called the *thorax*, which is distinguished as *that part of the insect body which always bears the jointed legs, and the wings*, if they are present.   The remaining part of the body (typically eleven or twelve segments) bears no jointed legs and is called the *abdomen*.   The eleventh true abdominal segment bears the appendages called the *cerci* (see Figs. 46 and 54).   Behind this the abdomen is terminated by a twelfth segment that never bears appendages, known as the *telson* or *periproct*.   Primitively the anus is supposed to have opened on the telson; the female reproductive organs between the seventh and eight segments, on the ventral side; and the male reproductive organs on the posterior border of the ninth abdominal segment, ventrally.   These body openings, however, often become greatly shifted as the body regions become specialized.   We say, therefore, that the segments of the body of an insect are grouped into three *body regions*, known as head, thorax, and abdomen (Fig. 46).   This is a further development of the specialization and division of labor spoken of in connection with the segmentation of the body.   The head of the insect takes over the function of locating and taking in food, as well as most of the work of sensing danger and recognizing friends and enemies. The thorax practically always performs locomotion, while the organs of reproduction are borne by the abdomen.   The one thing that marks off the thorax from the head and abdomen is that it bears the legs.   Hence we find the thorax and determine its limits by finding where the six legs are attached, even in such insects as beetles, where the thorax *appears* to end just back of the front legs.   The three segments of the thorax are given distinctive names: the one bearing the first pair of legs is known as the *prothorax*, the segment bearing the middle pair of legs is the *mesothorax*, and the segment to which the hind legs attach is the *metathorax*.

The Legs of Insects.—The most characteristic single thing we can name for insects is the presence of three pairs of jointed legs.   These are practically always present in adult or mature insects, and generally present in the other stages.   However, a good many maggots, grubs, and other larvae are entirely legless (see Figs. 477, 536,*b* and 556).   Some insect larvae, notably the caterpillars, have in addition to the three pairs of jointed legs on the thorax, anywhere from two to eight additional pairs of fleshy, *unjointed* projections on the abdomen, which are used as legs and are known as *prolegs* (see Figs. 87 and 88,*A*,*F*).

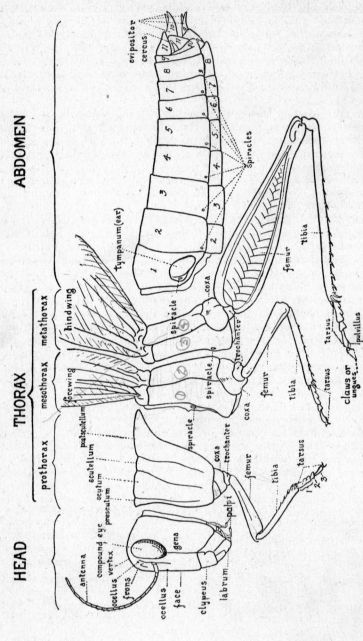

FIG. 46.—Outline of body of a grasshopper as seen from the side, dissected to show the three body regions and the parts of the body commonly referred to in books and bulletins. (*Slightly modified from Herms, "Medical and Veterinary Entomology."*)

Insects are the six-legged Arthropoda. It is this character that has given them their class name Hexapoda (meaning six legs). The spiders, mites, and ticks (Figs. 93 and 94) have four pairs of legs. The crayfish (Fig. 97), lobsters, crabs, and their relatives have five pairs of walking legs. The hundred-legged-worms (Fig. 92) have a pair to each body segment, anywhere from 12 to 60 pairs in all. While the thousand-legged worms (Fig. 45,*D*) have two pairs to each apparent segment, sometimes as many as 213 pairs.[1]

No one can study active insects long without being impressed by the extensive use they make of their legs. In insects the legs perform many of the functions for which we would use our hands, though sometimes the mouth parts are used for digging, carrying, fighting, and the like. In addition to walking and jumping, insects often use their legs for digging, grasping, feeling, swimming, carrying loads, building nests, and cleaning parts of the body. The cricket and katydid have "ears" on their front legs (Fig. 47,*A*).

Perhaps there is something significant in the number of legs that we find in this most abundant group of animals. If we study the movements of the legs, we find that the insect does not move the three on either side together and alternately with those on the other side; nor does it move the two of any pair in unison. Instead they go in tripods, the middle one of either side being raised and advanced about the same time as the front and hind ones of the other side. While these three are being advanced, the other three are supporting the body. Since three supports is the smallest number that will give a stable equilibrium, we see that insects, unlike two-legged and four-legged animals, are in a state of stable equilibrium, whether standing or moving. This requires less muscular effort and practically eliminates that hazardous period that many animals undergo while learning to walk. As to a larger number of legs we need only quote the following ditty:

> A centipede was happy, quite,
> Until a toad in fun
> Said, "Pray, which leg moves after which?"
> Which raised her doubts to such a pitch,
> She fell exhausted in the ditch,
> Not knowing how to run.

The legs of insects are hollow, more or less cylindrical outgrowths or continuations of the body wall at the point where pleura and sterna meet. Nerves, tracheae, blood spaces and muscles occupy their internal cavities. The coxa of the leg is articulated to pleurites, some of which are considered to have been primitive segments of the leg itself, or to both a pleurite and a sternite; and the segments of the leg are articulated to each other by hinge-like joints or in such a way that some axial revolution is possible. The legs are moved by muscles that extend from one segment of the leg to more distal ones or that enter the base of the leg from adjacent parts of the body wall.

All insect legs are made up of six true, independently movable segments and these parts always occur in the same order. Beginning next to the thorax, the names of the parts are (Figs. 46, 47, and 54):

---

[1] *Parajulus pennsylvanicus.*

> Coxa (plural coxae).
> Trochanter (plural trochanters).
> Femur (plural femora).
> Tibia (plural tibiae).
> Tarsus (plural tarsi).
> Pretarsus (plural pretarsi).

These names are constantly used in describing and determining insects, and they should be learned once for all. All these parts are single segments except (a) the tarsus, or "foot," which is commonly divided into from two to five subsegments, besides the claws and pads at the end of the leg which are not counted as segments of the tarsus, but constitute the *pretarsus* (Fig. 47,*H*); and (b) the trochanter, which in a few cases, notably the parasitic wasps and the Odonata has two segments (see Fig. 47,*D*). The femur and the tibia, which form, between them, the knee joint of the leg, are usually much longer than any other segments in the leg, and of these the one nearer the body, and usually the thicker one, is the femur; the slenderer, outer one is the tibia. Between the femur and the body there are always two (rarely three) small pieces, the one nearest the femur being the trochanter, and the one next the body the coxa. All that part of the leg beyond the end of the tibia (the tarsus and the pretarsus) is generally placed flat upon the ground when the insect is walking, and the end of the tibia generally has prominent spines or spurs that help to maintain a footing. The pretarsus (Fig. 47,*H*) usually bears two sharp curved hooks or claws (the *ungues*), though there may be only one; and some complicated pads (known as *pulvilli, arolia, empodia*, etc.), which are very important in locomotion. For example, in the house fly there are many microscopic hollow hairs on these pads through each of which a sticky substance exudes that enables the fly to walk upside down and up very smooth surfaces.

The Wings of Insects.—A characteristic of insects that we may be sure has been of very great advantage in their struggle for existence is the possession of wings. Wings enable insects (a) to forage far and wide to find suitable food, (b) to flee quickly from enemies and other dangers, (c) to disperse widely and intimately to find mates and lay their eggs, (d) often to select nesting sites not accessible to many of their animal enemies.

Insects are the only winged invertebrates. That is to say, if one finds an animal that has wings and does not have a backbone one may be sure it is an insect. Wings have been developed also in two groups of vertebrate animals, the birds (Aves) and the bats (Mammalia). But insects were almost certainly the first "flying machines," because we know from fossil records that winged insects were present on the earth in the Carboniferous period and almost certainly in the Devonian and Silurian, long aeons before either birds, flying reptiles, or bats made their appearance upon our globe, in the Jurassic, Cretaceous, and Tertiary periods.

While adult insects regularly have six legs, the number of wings varies among the different kinds. Insects never have functional wings until they are full-grown or adult, and many adult insects do not have wings. Silverfish and springtails (pages 184 and 185) represent wingless-insect groups, whose ancestors apparently never had wings. Others, such as fleas and lice and certain ants and aphids, are considered to be degenerate forms whose distant ancestors possessed wings which have been

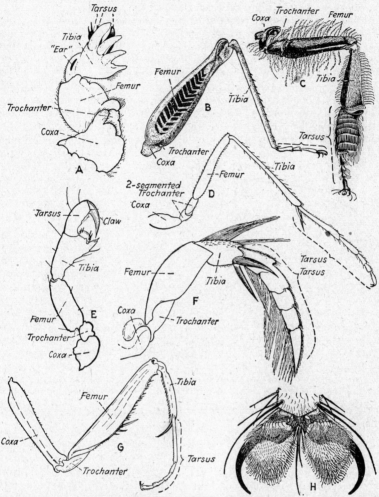

FIG. 47.—Legs of insects showing modifications for different functions. *A*, digging leg of a mole cricket. Note the rake-like tibia with the three-segmented tarsus beneath it; also the slit-like "ear" or tympanum toward the base of the tibia. *B*, jumping leg of a grasshopper (*from Univ. Kan.*). *C*, hind leg of worker honeybee adapted for assembling and carrying food substances; the rows of regular hairs on the basal segment of the tarsus are used for gathering pollen; the large marginal bristles of the tibia form, on the side opposite that shown, the pollen basket for carrying pollen to the hive (*from Cheshire*). *D*, walking leg of an ichneumon wasp. Note the two-segmented trochanter. *E*, clinging leg of the hog louse. Note the one-segmented tarsus with a single claw adapted for clinging around a hair. *F*, swimming leg of a predaceous diving beetle. Note the numerous long hairs used in rowing and the coxa which flattens out on the body wall. *G*, grasping leg of a praying mantis showing the very long coxa to extend the reach and the spiny femur and tibia between which insects are caught to be devoured. *H*, foot of the common house fly, showing claws, pulvilli, and the tenent hairs that make it possible for flies to walk upside down (much magnified). (*H from Kellogg's "American Insects;" A, D, E, F, G, original.*)

lost in adaptation to a more quiescent life in the ground or on the bodies of animals. Some of the beetles that possess wings have not used them, and they have become atrophied and useless. Other insects, such as termites and certain ants, break off or tear off their wings after a single nuptial flight and before beginning their life in the soil. No insect has more than four wings, or two pairs. This is the typical number. Some insects have only one pair. A good many are wingless throughout life. Frequently one sex has wings and the other sex is wingless.

The wings, like the legs, are always attached to the thorax, the front pair to the mesothorax and the hind pair to the metathorax (Figs. 44,*B* 46, and 54). When one pair only is present (Fig. 36), it is the first

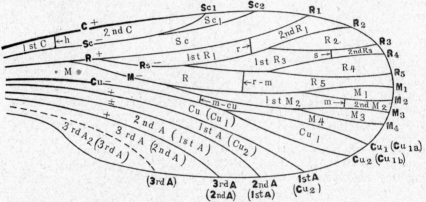

Fig. 48.—The hypothetical, primitive wing of insects, showing the type of venation from which all modern insects' wings are supposed to have evolved. A slightly different interpretation, applicable especially to the Mecoptera, Trichoptera, Neuroptera, Lepidoptera, and Diptera, is indicated by the abbreviations in parentheses. *The names of veins* are indicated on the veins and at the ends of the branches, in **bold-faced type.** *The names of cells* are indicated in the spaces: cells take their names from the veins along their anterior margin. **C** = costa, **Sc** = subcosta, **R** = radius, **Rs** = radial sector, **M** = media, **Cu** = cubitus, **A** = anal. Sc_1 is read first subcostal vein; $1st$ R_1 is read first radial one cell; $2nd$ M_2 is read second medial two cell, etc. The cross-veins are named *r*, radial; *s*, sectoral; *r-m*, radio-medial; *m*, medial; and *m-cu*, medio-cubital. (*Drawn by Robert L. Metcalf.*)

pair, and it is borne by the mesothorax. We see thus that the thorax has all the organs of locomotion. This is ideal because it is near the center of mass of the body.

The wings of an insect are remarkably different structurally from those of a bird or bat. They contain no bones, muscles, joints, or feathers; and nerves and blood vessels are wanting or greatly reduced in activity. They are simply thin sheets of parchment-like cuticula that are moved by the action of muscles attached to the base of the wing, inside the body wall.

The wings develop, as anyone can easily determine by examining a young grasshopper or bug, a moth newly emerged from its cocoon, or a teneral fly, each as a hollow sac folded out from the body wall (Fig. 86). Promptly after the insect breaks out of its pupal or last nymphal skin, these wing sacs enlarge, and their walls flatten together and unite, so that the upper and under walls of the sac become fused and indistinguishable from a single membrane. They fuse closely in this way, except along certain lines where the two walls remain separated slightly and become thickened to make a kind of framework of hollow ribs, between which the wing membrane stretches. The hollow linear ribs (Fig. 49)

which make up the framework of the wing are known as *veins*, though they have nothing to do with the circulation of the blood. The areas, of various shapes, enclosed between the veins are called *cells; closed cells* if the area is entirely surrounded by veins, *open cells* if the area extends to the wing margin without an intervening vein (Fig. 49). The exact number, branching, and arrangement of the veins of the wings are extensively used in classifying the various suborders, families, genera, and species of insects. Attempts have been made to homologize the veins of all the orders.[1] What is believed by some authors to be near the

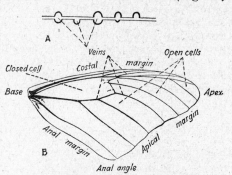

Fig. 49.—Front wing of the monarch butterfly, showing names applied to various parts of an insect wing. (*Original.*) *A*, diagrammatic cross-section of an insect wing to show how it is formed of two membranes and how the veins develop as hollow rods by the thickening of one or both membranes along certain lines. (*Redrawn after Woodworth.*)

condition of the veins in an ancestral form of the winged insects is reproduced in Fig. 48. The student should note that each wing cell is given the same name as the wing vein along its anterior border.

In outline (Fig. 49) the wings are often somewhat triangular. The front side of the triangle is known as the *costal margin,* and the outer side the *apical margin,* while the third side is called the *inner* or *anal margin.* Generally the veins are heavier or more closely placed toward the costal margin, since the greatest stress during flight is on this area of the wing.

Because they vary so widely in form and appearance in different groups, wings are very important structures in the classification of insects. As we shall learn later, most of the order names of insects end in *ptera,* meaning wing. Thus the Diptera (flies) are the "two-winged" insects, the Coleoptera (beetles) are the "sheath-winged" insects, the Lepidoptera (moths and butterflies) are the "scale-winged" insects, the Hemiptera (true bugs) are the "half-winged" insects, the Hymenoptera (wasps, bees) are the "membrane-winged" insects, and the Orthoptera (grasshoppers, etc.) are the "straight-winged" insects (see Table VIII).

There is a great range in size of wings from those of the largest known grasshoppers and moths, whose two wings spread 10 or 11 inches, to those of minute egg parasites which barely expand $\frac{1}{100}$ inch from tip to tip. A large wing does not necessarily mean that its possessor is a rapid flier. Indeed, in general, the swiftest flying insects are those with small- to moderate-sized wings. The rapidity with which the wings beat is a much more important factor in the speed of flight. Butterflies in flying commonly beat the wings at about 9 strokes a second, dragon-

---

[1] COMSTOCK, J. H., "The Wings of Insects," Comstock Publishing Co., 1918.

flies about 28, the house fly from 180 to 350, and the honeybee from 200 to 400 strokes per second—so rapidly that the wings become completely invisible.

The highest speed of insect flight that has been carefully measured is about 35 miles per hour for certain hawk moths and horse flies, although Tillyard claimed that certain dragonflies may speed nearly 60 miles per hour. The honeybee does only 5 to 8 miles per hour and the house fly scarcely 5. In proportion to size, a giant bombing plane of the latest type would have to be capable of flying at a speed of 60,000 miles per hour to equal the aeronautical achievement of the horse fly.

The wings of insects provide lifting force, forward driving power, and efficient steering apparatus, enabling them to take off with surprising quickness, to change direction with lightning-like rapidity, to dart side-wise, to dip, to bank, and, most remarkable of all, to hover in one spot and even fly backward without changing the position of the body.

The Antennae of Insects.—The way in which segmentation of the body facilitates division of labor is well shown in the paired jointed appendages on the heads of insects. These appendages of the head arise during embryonic development in exactly the same way as the legs and for a time are indistinguishable from them. But in the active insect these appendages have become greatly differentiated for several distinct functions. One group of them comprises the mouth parts which are discussed more fully in Chapter V. Another pair of append-ages, homologous with the legs, has become modified to form the "feelers" or "horns," or, as they are properly called, the *antennae* (Figs. 46 and 50).

None of the higher animals has anything to compare with the antennae of insects and their relatives. With them various insects feel their way, detect danger, locate their food, find their mates, and, at least in some cases, use them to communicate with others of their own kind, *e.g.*, the ants; or bear end organs of smell (as in the flies); or use them in hearing, *e.g.*, as the male mosquito; or rarely for grasping a mate or the prey. A pair of long, flexible, highly sensitive "feelers," with which the insect can sound out the environment ahead, must be of very great advan-tage to these active animals. Most insects show great distress and some-times helplessness when the antennae are removed or injured.

All true insects have one pair of antennae. They are the appendages of the second head segment (see pages 109, 111). This is a useful dis-tinguishing mark; for the spiders, mites, ticks, and scorpions have no antennae, the crayfish, lobsters, and crabs have two pairs, while the centipedes and millipedes agree with the insects in having one pair.

Types of Antennae.—The antennae of insects vary greatly in size and form, and are much used in classification. The following special names have been applied to certain of the common types (Fig. 50). A *filiform* or thread-like antenna is one in which all the segments are of about the same thickness and have no prominent con-strictions at the joints. A *moniliform* antenna is one made up of somewhat globular segments, with prominent constrictions between them, the whole suggesting the shape of a string of beads. A *setaceous*, or bristle-like, antenna is characterized by a notice-able decrease in the size of segments from the base to the apex, so that the antenna tapers from a rather thick base to a very slender tip. A *clavate*, or club-shaped, antenna enlarges gradually toward the tip, the segments near the end being larger than those near mid-length. An antenna in which the enlargement toward the tip is more abrupt and greater than in the clavate type is called *capitate* or knobbed. If the enlargement at the end is almost entirely toward one side from the axis of the antenna and forms broad, somewhat flattened plates, we call the antenna *lamellate*. Sometimes all the segments show projections to one or more sides.　If the segments have short triangular projections to one side, we call the antenna *serrate* or saw-toothed.

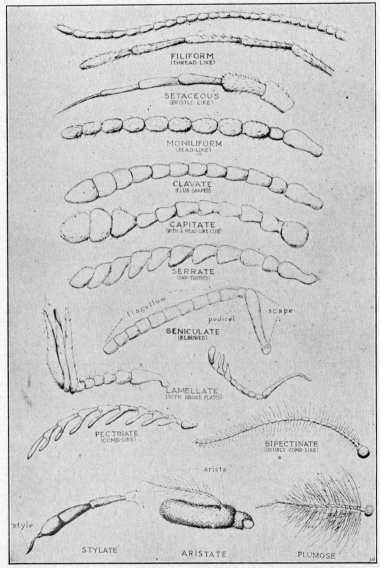

FIG: 50.—The commoner types of insect antennae with the names applied to each type. (*Redrawn from various sources.*)

If these side projections are long, the antenna is said to be *pectinate* or comb-like. Sometimes each segment has two or even three such projections, which form two or three rows of teeth along the sides of the antenna, when it is called *bipectinate* or *tripectinate*.　An antenna which has whorls of hairs coming off at or near the joints is called *plumose* or plume-like. In many of the flies, the antennae bear on the upper side of the third and last segment a heavy bristle known as the arista; this kind of antenna is called *aristate*. A similar bristle or appendage at the end of the antenna is called a *style*, and such an antenna is said to be *stylate*. A *geniculate* antenna is one that has a sharp bend like a flexed arm.

The Eyes of Insects.—Insects have eyes of very complicated structure and of at least two very distinct kinds. Yet we believe that vision in most insects is poor, and probably subordinate to smell and touch as a guide to them in their reactions to the environment.

The two kinds of eyes are called *compound eyes* and simple eyes or *ocelli* (Fig. 51). The latter, though they look much alike, are again of two kinds, being innervated from different parts of the brain: (*a*) *dorsal ocelli* of nymphs and adults and (*b*) *lateral ocelli* of larvae. Typically in adults and nymphs there are two compound eyes and three simple eyes. The simple eyes are so very small as to require careful examina-

 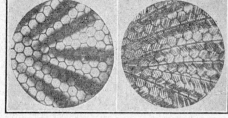

Fig. 51.   Fig. 52.

Fig. 51.—Head of a fly, *Didea fasciata* Macq., dorsal view, showing the large compound eyes which occupy the entire sides of the head; between them the three minute simple eyes or ocelli arranged in a triangle; and the aristate antennae. This is a female fly; many male flies have the compound eyes touching each other on top of the head. Much enlarged. (*From Metcalf, in the Ohio Naturalist.*)

Fig. 52.—Photomicrographs taken through the cornea of a fly. Note the honeycomb-like margins ("sash") which divide the cornea into a number of hexagonal facets ("panes"). In one case the object photographed is extremely close to the eye, in the other case moved a slight distance farther from the eye. (*From Howes' "Insect Behavior," Badger.*)

tion to find them. They are often arranged in a triangle on the head, somewhere between the compound eyes. Butterflies have no ocelli; moths have two, one near the base of each antenna. In many other insects there are no ocelli. Larvae, such as caterpillars, maggots, and grubs, never have compound eyes, though they generally have simple eyes (lateral ocelli), from one to six or more on each side of the head (Fig. 55,*2,3,o*). Nymphs, the young of true bugs, grasshoppers, and the like, (Fig. 84) have compound eyes and dorsal ocelli like the adults.

The compound eyes are usually the most conspicuous objects on the wall of the head. They are convex, round, oval, or kidney-shaped areas, one on each side. They usually appear shiny and suggest their function even at a glance. But when examined through a microscope, they are found to be hardly eye-like at all. There are no eyelids or eyelashes protecting a delicate moist surface, as in the vertebrate eye; but the hard resistant cuticula of the general body wall continues without a break across the surface of the eye. It is, however, transparent over the eye surface and admits the light, in this respect being something like a window fixed in the wall of the otherwise dark skull case, a window composed of very many minute hexagonal panes fitted closely together (Fig. 52). Each of these hexagonal areas is called a *facet* and is the exposed face of an independent lens (*cornea*). Ants have from 50 to 400 facets or corneas in each eye; the house fly has about 4,000 to each eye; a swallowtail butterfly, 17,000; while in certain of the sphinx moths and dragonflies more than 50,000 facets occur in the two eyes (see also page 102).

While most insects have very complicated organs of vision, there are some kinds that appear to be entirely blind.

SUMMARY

In this discussion of the structures that adapt insects for so successful a competition with other animals, most of their *distinguishing characteristics* have been given, *i.e.*, the structural peculiarities that serve to separate them from all other animal groups. These may be summarized as follows:

A. *Characteristics which the insects share with their near relatives, the other classes of the phylum Arthropoda* (see also pages 108 and 163):
   1. Insects have a segmented body.
   2. Many of the segments bear a pair of jointed appendages.
   3. The body of insects has a chitinous exoskeleton, instead of a bony endoskeleton.
   4. The body is bilaterally symmetrical.
B. *Characteristics which distinguish the insects from their nearest relatives, the other Arthropoda:*
   1. Insects have three pairs of legs in the adult stage.
   2. Insects usually have two pairs of wings in the adult stage, sometimes only one pair, although many species are wingless.
   3. The segments are grouped into three body regions, known as head thorax, and abdomen.
   4. The head is provided with a single pair of antennae.
   5. Insects usually have a pair of compound eyes, in nymphal and adult stages, and sometimes simple eyes in addition.

CHAPTER IV

THE INTERNAL ANATOMY AND PHYSIOLOGY OF INSECTS[1]

A living insect performs all of the functions that are common to animals and, so far as we can determine, none that are peculiar to insects, *i.e.*, none that are not performed also by the body of man or any other animal. The smallest insects are as perfectly formed in the structure of tissues and organs as man or an elephant. Few, if any, other animals have so complex an organization in so small a body.

The more important functions, together with the principal organ systems that perform them, may be listed as follows:

Functions	*Organs*
1. Ingestion (feeding)	Mouth parts and pharynx
2. Egestion (evacuation)	Rectum and anus
3. Nutrition (digestion, absorption, and assimilation)	Alimentary system
4. Excretion	Excretory system
5. Secretion	Glands
6. Respiration (oxidation)	Respiratory system
7. Circulation	Circulatory system
8. Sensation, conduction, and coordination	Nervous system
9. Motion and locomotion	Muscles and skeleton
10. Reproduction	Reproductive system
11. Growth and development (discussed in Chapter VI)	
12. Protection	Various organs and habits

METABOLISM

Living animals of most kinds, and insects in particular, are noteworthy for their great activity, the continual changes of their physical bodies, and their ability to do "work." This work involves especially moving about to secure food and congenial surroundings, to escape enemies, to find mates, to construct nests, and to lay eggs for another generation. The ability to do work depends upon the availability of energy. This is secured from food substances in which energy from the sun has been stored by green plants by the process known as *photosynthesis*.

Metabolism is the term applied to all the chemical changes taking place in the living body. It has two phases: (*a*) *anabolism*, a building-up process, which begins with digestion and reaches its climax after the assimilation of the digested foods and their synthesis into various bodily *secretions* and the complex proteid compounds which make up *protoplasm*, the living substance. The complex and unstable protoplasm is continuously being attacked by breaking-down reactions (especially oxidation), which are known collectively as (*b*) *katabolism*.

Katabolism is a very important process. The trite comparison of katabolism to a furnace fire should help us to understand its significance.

[1] General reference: SNODGRASS, R. E., "Principles of Insect Morphology," McGraw, 1935.

88

Coal in the furnace and food in the animal body are alike useless unless they are oxidized (burned). The fire results in the release of energy in the form of heat. Katabolism is the only way in which the energy essential for all life activities can be released from the food, in a usable form, inside the body. The fire inevitably produces both gaseous and solid wastes. Katabolism, likewise results in the formation of wastes or by-products. Some of these (*e.g.*, water) may be used again by the body. Gaseous wastes, such as carbon dioxide (the smoke), are removed by the tracheae in the process known as respiration (see below). Other wastes (the ashes) are taken by the blood and carried in solution until they are eliminated by the process known as excretion (see below).

Ingestion and Egestion.—The *ingestion* or taking in of foods is treated in the chapter on mouth parts of insects (Chapter V). Most food materials when taken into the alimentary canal cannot pass through its walls into the blood and be carried to the various tissues requiring them until they have undergone certain modifications known as *digestion*, which render them soluble or absorbable. Much of the material swallowed is indigestible and is *egested* from the alimentary canal, never having been a part of the living body of the insect. This act of voiding material that has not been digested (*egestion or evacuation*) must not be confused with *excretion*, discussed later. Often poisons designed to kill insects fail because, even though eaten, they pass unchanged through the body. It is an interesting fact that certain insects, such as the honey-bee and parasitic wasps, while young, have no connection between the mid-intestine and the hind-intestine and cannot void any excreta until they are full-grown larvae, before spinning cocoons, or as adults they leave the body of the host.

Digestion, Absorption, and Assimilation.—The digestive processes of insects have never been adequately studied. Apparently insects have no organs like the liver and pancreas of higher animals. Yet some insects digest substances that the human system cannot assimilate, such as dry wood (cellulose) and wool. And others may develop on a pure diet of substances such as black and red pepper, ginger, ergot, or tobacco that to us would be inacceptable or poisonous. In some cases (see termites, page 192) it appears that the digestion of substances like cellulose is really brought about by minute animals, protozoa, that make their home in the alimentary canal of the insect. Without these protozoa the insects cannot live. In such cases the insect may live on the excreta of the protozoa, or the latter may function to utilize atmospheric nitrogen, an essential element that is lacking in the cellulose.

The digestive enzymes produced in the several parts of the alimentary canal and in the various insect species may be expected to vary greatly in correlation with the nature of the food. In some species the secretion of the salivary glands (Fig. 54) contains *amylase* which converts starch into maltose or glucose, as does the saliva of man. The saliva of some bloodsucking insects contains an anticoagulin. In predaceous forms it may have a paralyzing effect upon the prey, and in a number of insects it is injected into the food to partially predigest it before it is swallowed. Other enzymes produced by the epithelial cells of the canal walls or of the gastric caeca (Fig. 54) act upon the food substances as the peristaltic movements of the canal force the pabulum slowly along. In the stomach the digestion of carbohydrates into maltose, invertose, and lactose is completed, while *trypsin* and *erepsin* attack the proteins and *lipase* digests the fats in the food. Pepsin and hydrochloric acid, produced in the stomach of man, have not been detected in insects. Gradually a portion of the material consumed is rendered soluble and absorbed through the canal walls into the blood stream, to be carried to all parts of the body.

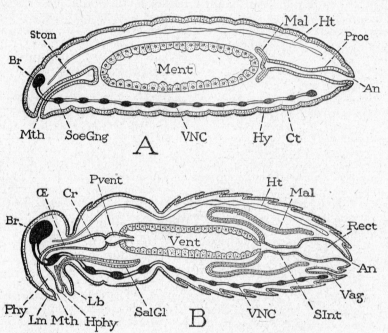

Fig. 53.—Diagrammatic sagittal sections of the body of an insect to show the arrangement of the principal organs and the formation of the alimentary canal. *A*, section of an insect in the embryonic stage, showing the anterior invagination of the ectoderm which forms the mouth (*Mth*) and fore-intestine or stomodeum (*Stom*); the posterior invagination which forms the anus (*An*) and hind-intestine or proctodeum (*Proc*); and the endodermal development of the mid-intestine called the mesenteron (*Ment*). *B*, section of a mature insect, after fore- and hind-intestines have united with the mid-intestine to form a complete canal. The fore-intestine has differentiated to form the pharynx (*Phy*), esophagus (*Œ*), crop (*Cr*), and gizzard or proventriculus (*Pvent*). The hind-intestine has differentiated into small intestine (*SInt*) with the Malpighian tubes (*Mal*) arising from its anterior extremity, and rectum (*Rect*). The mesenteron has become the stomach or ventriculus (*Vent*) of the adult. *An*, anus; *Br*, brain; *Cr*, crop; *Ct*, cuticula; *Hphy*, hypopharynx; *Ht*, heart; *Hy*, hypodermis; *Lb*, labium; *Lm*, labrum; *Mal*, Malpighian tubules; *Ment*, mesenteron; *Mth*, mouth; *Œ*, esophagus; *Proc*, proctodeum; *Pvent*, proventriculus; *Rect*, rectum; *SalGl*, salivary gland; *SInt*, small intestine; *SoeGng*, subesophageal ganglion; *Stom*, stomodeum; *Vag*, vagina; *Vent*, ventriculus; *VNC*, ventral nerve cord. (*From Snodgrass, "Anatomy of the Honeybee."*)

The alimentary canal of insects is a tube leading from the mouth to the anus at the tip of the abdomen. The mouth primitively is located between the third and fourth head segments, and the anus opens upon the twelfth abdominal segment (the *telson*). The structure of the insect has aptly been likened to a tube within a tube, the outer tube being the body wall, the inner tube the alimentary canal. The space between these two tubes is called the body cavity, blood cavity, or *hemocoele*. In insects it is largely filled with blood. The length of the alimentary canal and the complexity of its structure depend on the food habits of the insect, and they vary greatly.

During embryonic development, the alimentary canal forms in three sections (Fig. 53): the *fore-intestine* and the *hind-intestine* grow as invaginations from the outside, and consequently are lined with cuticula similar to that on the outside of the body. This cuticular lining performs an important function in some insects, such as the grasshoppers, in which it is developed, in the region of the proventriculus or gizzard, into strong teeth. The cuticula of the fore- and hind-intestines is molted each time the external skin is shed. The *mid-intestine* develops internally (Fig. 53,*Ment*) and consequently lacks the cuticular lining.

In some insects the fore-intestine consists only of the mouth or *buccal cavity*, the *pharynx* (Fig. 53,*Phy*), and a straight thin-walled tube, the *esophagus* (*Œ*), leading back to the *stomach* (*Vent*). In other kinds the fore-intestine becomes greatly specialized. The pharynx, especially of insects having piercing, siphoning, or sponging mouth parts, is generally developed into a pump or sucking device, closed by the elasticity of its cuticular lining and by muscles in its walls, and opened by muscles extending from these walls to the inside of the skull case (Figs. 65 and 69). The middle region of the esophagus is often enlarged into a *crop* (Fig. 53, *Cr*, and Fig. 54) where food is held temporarily; and the far or caudal end, especially in chewing insects, is specialized into a *gizzard* (*Pvent*) for grinding and straining the food and, by a sphincter-like action, for controlling its passage into the stomach.

The hind-intestine, which begins near the point where the Malpighian tubules attach, may be a simple tube leading from stomach or mid-intestine to the anus. In other species it may be variously subdivided into a *pylorus*, containing the bases of the Malpighian tubules; an *ileum;* a *colon;* a *rectal sac;* and a *rectum*.

Between fore-intestine and hind-intestine, the *mid-intestine* (Figs. 53, *Ment, Vent* and 54) is developed internally from the embryonic endodermal tissues. This tube eventually meets and connects with the fore- and the hind-intestine to make a continuous canal. Near the anterior and posterior ends of the mid-intestine are valves (called *cardiac* and *pyloric*, respectively) that regulate the passage of food materials into and from the stomach. The boundaries of the mid-intestine can also be recognized by the thinness or absence of a cuticular lining. The muscles surrounding this portion of the canal are less well developed and, instead, the cells lining it are large active cells that secrete digestive juices. Their number, and accordingly the amount of the secretion, is often increased by outpocketings of the wall into short or long, blind tubes known as *gastric caeca* (Fig. 54), the number of which varies greatly. It seems likely that it is in the region of the mid-intestine that the digestion and absorption of food and its transmission to the blood chiefly occur, although some digestion probably begins in the crop and much absorption doubtless takes place also in the hind-intestine. It seems certain that it is chiefly through the thin-walled cells of the mid-intestine that stomach poisons, swallowed by insects, attack them. Food products are kept moving along in the alimentary canal by circular and longitudinal muscles that surround the tube like a sheath. There are also many muscles that extend from its walls to the inner face of the body wall.

Excretion.—The beginning of the hind-intestine is marked by the attachment of a variable number (1 to 150) of tubes, often very long and slender, sometimes branched, that are known as *Malpighian tubes* (Figs. 53,*Mal* and 54). These are believed to have a function similar to the kidneys of higher animals and the nephridia of Annelida, in sorting out from the blood stream the waste organic compounds ("ashes") that result from katabolism. The tubes are always closed at the free (distal) end and have a *lumen* that is surrounded by the secreting cells. Their cavities or lumina open into the cavity of the hind-intestine. They differ from the kidneys of mammals, not only in form and in having no separate opening to the exterior, but also in the nature of the urinary products formed. In the higher animals these products (*e.g.*, urea) are soluble and are eliminated in solution as urine. In the insects the Malpighian tubes pour their semisolid excretory products (principally uric acid) into the hind-gut, through which they are carried from the body with the feces. Besides the excretory work of the Malpighian tubules, the body wall of the insect is sometimes responsible for the elimination of nitrogenous excreta, calcium salts, carbon dioxide, and other useless or harmful substances. So far as we know the equivalent of perspiring or sweating is not performed by insects, but the waste materials may be incorporated in the body wall where they are not only harmless but may even serve the useful purpose of strengthening or coloring the cuticula, until they are finally disposed of in the process of molting (see page 147). The walls of the mid-intestine and certain special cells in the body known as *nephrocytes, pericardial cells,* and the *urate cells* in the fat-body also aid in excretion. It will be remembered that the gaseous wastes, such as carbon dioxide and water, are eliminated largely through the tracheal system. The Malpighian tubules are wanting in a few kinds of insects such as aphids and certain springtails. In some insects (Neuroptera and ground beetles) they secrete a kind of silk through the anus, which is used in making cocoons.

Secretion.—The metabolism of insects results in the formation, not only of the living protoplasm, but also of various other chemical substances that serve a useful purpose in the body. Such substances are called *secretions.* Cells of the body that form secretions, or groups of such cells, are known as *glands.* Many of these glands are formed as infoldings or other modifications of the hypodermis; others from the epithelium that lines the internal surfaces of the body. Examples among the insects are the salivary glands which secrete saliva (including silk, page 40). Scent glands, which give the odor to so many bugs, may be repellent or alluring. The repellent glands ward off enemies. The alluring glands doubtless serve an important function in recognition and communication among members of the same species or colony. For example, among bees and ants, many of their seemingly intelligent acts may be accounted for by the recognition of telltale odors. Alluring glands, serving by the odor of their secretions to attract the opposite sex, occur on patches of peculiar scales (*androconia*) on the wings of male butterflies, or as tufts on the legs of the males, or have their openings near the tip of the abdomen of females. Hypodermal gland cells secrete the cuticular covering that functions as the skeleton of the insect; also the waxy covering of scale insects (including lac, page 46); also a molting fluid used in casting old skins. The wax glands of the bees form a

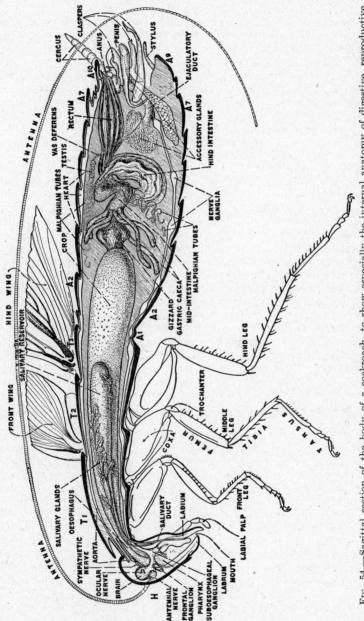

Fig. 54.—Sagittal section of the body of a cockroach, to show especially the internal anatomy of digestive, reproductive, nervous, circulatory, and excretory systems. None of the tracheae are shown. *H*, head; *T₁*, prothorax; *T₂*, mesothorax; *T₃*, metathorax; *A₁* to *A₁₀* first to tenth abdominal segments. (*Original; drawing by Antonio M. Paterno; in part after Miall and Denny.*)

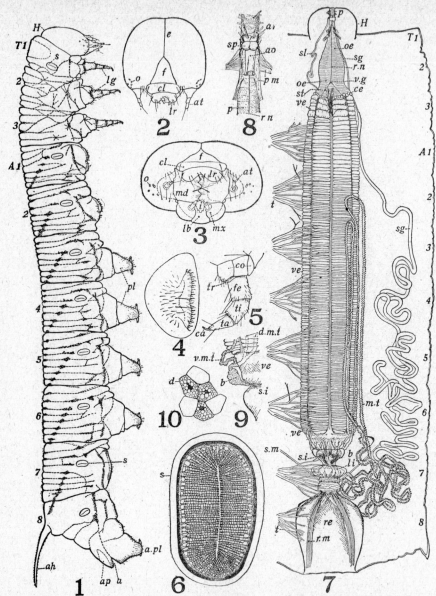

Fig. 55.—Anatomy of the tomatoworm or hornworm. 1, Side view of entire larva; 2, front view
of the head; 3, ventral view of the head to show chewing mouth parts; 4, a proleg, ventral view, showing
crochets; 5, a true leg, or thoracic leg, front view; 6, a single spiracle, greatly magnified; 7, body of the
larva opened from the dorsal side to show the alimentary canal. On the left side are shown the salivary
gland and the terminal branches of tracheae which enter the canal; on the right the silk gland and
Malpighian tubes of that side of the body are represented; 8, the pharynx from above in the region of
the brain; 9, enlarged view of the alimentary canal at the point where the bladder of the Malpighian
tubule is attached; 10, a bit of the adipose tissue or fat-body. The following abbreviations are used:
A1 to A8 abdominal segments one to eight; a, anus; ah, anal horn; ao, aorta; ap, anal plate; a. pl, anal
proleg; ar, arched nerve; at, antenna; b, bladder of Malpighian tubule; ca, claw; ce, caeca; cl, clypeus;
co, coxa; d, adipose tissue; d.m.t., dorsal Malpighian tubule; e, epicranial suture; f, front; fe, femur; H,
head; i, spinneret; lb, labium; lg, leg; li, large intestine; lr, labrum; md, mandible; mt, Malpighian tubule;
mx, maxilla; o, lateral ocelli; oe, oesophagus; p, pharynx; pl, proleg; pm, pharyngeal muscle; re, rectum;
rn, recurrent nerve; rm, rectal muscle; s, spiracle; sg, silk gland; si, small intestine; sl, salivary gland;
sp, brain; st, part of sympathetic nerve; T1, T2 and T3, thoracic segments, prothorax, mesothorax
and metathorax, respectively; t, tracheae; ta, tarsus; ti, tibia; tr, trochanter; ve, ventriculus; v.m.t.,
ventral Malpighian tubule. (From Peterson in Ann. Ento. Soc. Amer., vol. 5.)

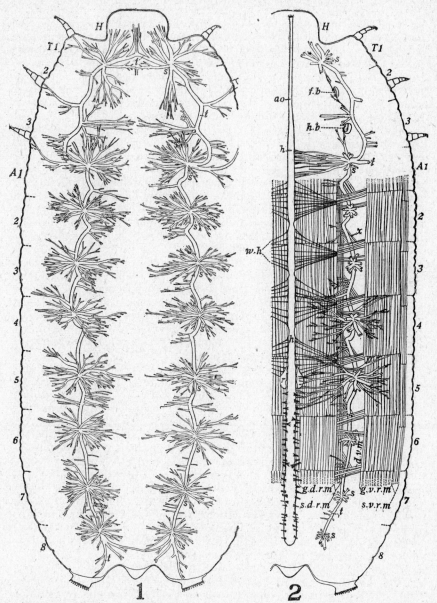

Fig. 56.—Anatomy of the tomatoworm or hornworm. The body wall of the larva has been opened along the ventral side and spread to show, on the left, the principal branches of the tracheae, all finer branches omitted; *s*, indicates the points at which spiracles open to the outside through the body wall. On the right are shown the median heart (*h*) and aorta (*ao*) with the wing-shaped muscles (*w.h.*) that hold the heart in place; the longitudinal tracheal trunk of that side (*t.t.*); the buds from which the wings of the adult are subsequently developed—(*f.b.*), the front wing bud, and (*h.b.*) the hind wing bud; the reproductive organs (*r*) and some of the many muscles of the larva (*d.v.m., g.d.r.m., g.v.r.m. s.d.r.m., s.v.r.m., x*). (For the explanation of other letters see under Fig. 55.) (*From Peterson in Ann. Ento. Soc. Amer., vol. 5.*)

material used in making nests (page 45). The tenent hairs of the fly's foot have already been described. The stings of Hymenoptera and the nettling hairs of caterpillars (page 21) have poison glands associated with them. Doubtless there are many other glands in the body of the insect that secrete important enzymes and hormones necessary to regulate and control metabolism and activity, but little study of this subject in insects has been made. The digestive, excretory, and reproductive glands are briefly discussed under those headings.

The fat-body of insects (Fig. 55,10,d) is an important group of cells that occupies much of the space between the larger organs in the body cavity. In these cells are stored fat, glycogen, and other nutrient substances not immediately required to maintain life. This store is drawn upon for the histogenesis necessary when an insect changes from larva to pupa and adult; for the development of the eggs of the adult female; and as a source of energy to maintain life during periods of hibernation or other stressful times. Since no food is taken during the pupal stage, the most important function of the fat-body seems to be to accumulate, during the feeding period of the larva, a sufficient reserve to furnish energy for the metamorphosis to the adult. Rogers[1] states that the fat gives rise to glycogen during the pupal stage of the silkworm, which in turn becomes transformed into sugar by the time the adult stage is reached. In some insects, especially those that do not have Malpighian tubes, it is believed that the fat-body serves as an important organ of excretion. The excretory products may be held as more or less permanent concretions in the fat-body or discharged through the alimentary canal as a *meconium* when the adult emerges from the pupa (Imms).

Respiration.—It is important to distinguish between the *ventilation* of the air passages of an animal (*tracheal tubes* or *lungs*) by *breathing processes* and *true respiration* which is *the oxidation of the tissues*, already referred to under metabolism (page 88). The fundamental respiratory process is probably the same in all animals and even in plants, but the methods of ventilating the respiratory passages vary enormously in different animals. Many of the simplest animals and some insects, such as springtails and certain aquatic and parasitic larvae, ventilate the body directly through the thin body wall or integument. Most insects, however, have special ventilating organs. Unlike the higher animals, they have neither nostrils nor lungs. They breathe, not through the mouth or other openings on the head, but generally through a series of paired holes along the sides of the body which are called *spiracles* (Figs. 46, 55,1,s, and 6, 20, 410). These openings occur on thorax and abdomen, but not on the head; usually there are two pairs on the thorax and from six to eight pairs on the abdomen.

The spiracles lead into tubes called *tracheae* (Fig. 57). The tracheae from different spiracles connect with each other, forming longitudinal *trunks*, and divide and subdivide and continue into the minutest branches, called *tracheoles* (Fig. 57,*tra*), which reach ultimately to every organ, tissue, and cell of the body. These branching tubes form an intricate network throughout the entire body, and carry the air directly to every part of the insect (Fig. 56,*t*) instead of utilizing the blood for the transportation of oxygen and carbon dioxide, as most animals do. Through the thin walls of the tracheoles the living cells withdraw the oxygen

[1] ROGERS, "Textbook of Comparative Physiology," McGraw, 1927.

necessary to respiration, by diffusion, and in the same manner the waste gases that result from metabolism are returned to the tracheae and so out through the spiracles.

The spiracles (Fig. 55,6) are often guarded by hairs, lips, or plugs that serve to exclude foreign matter, and the trachea just within the spiracle has in some insects a valve-like device or pinchcock for closing it. The tracheae, but not the tracheoles, are lined with cuticula continuous with that on the outside of the body. This takes the form of fine, spirally arranged threads (Fig. 57,*A* and *B,Tae*) that serve to keep the tracheal tubes from collapsing and that give to tracheae a characteristically cross-striated appearance. This is sometimes useful in distinguishing even minute bits of insect tissue, which will contain tracheae, from the tissues of many other animals which will lack it.

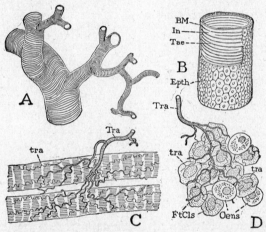

Fig. 57.—Structure and terminal branches of tracheae. *A*, a piece of trachea showing characteristic cross-striated appearance due to spiral taenidia, and method of branching. *B*, structure of tracheal tube, showing epithelium (*Epth*) of flat cells, outer covering of basement membrane (*BM*), and inner cuticular lining or intima (*In*) with spiral thickenings of taenidia (*Tae*). *C*, tracheal branches (*Tra*) ending in tracheoles (*tra*) on muscle fibers. *D*, tracheation of piece of fat-body, showing tracheoles on fat cells (*Ft Cls*), but not on oenocytes (*Oens*). (*From Snodgrass, "Anatomy of the Honeybee."*)

The sprays known as contact poisons (page 256) and fumigants (page 274) generally attack the insect by way of the respiratory system. To smother an insect, it would probably be necessary to close practically all the spiracles. Contact insecticides seldom kill by smothering except in the case of very small insects. The contact poisons sometimes penetrate directly through minute pores in the body wall, but usually penetrate through the tracheae as gases or liquids and produce a chemical effect on the tissues which may result in the death of the insect. The highly resistant cuticular lining of these breathing tubes, though extremely thin, must offer considerable resistance to injury by insecticides, unless the latter penetrate to the minute tracheoles, which are not lined with cuticula.

Much is yet to be learned about the way in which insects respire; particularly how the air is forced the full length of the extremely delicate tubes. Some insects have been said to use certain of their spiracles for

inhalation and others for exhalation.   The movement of the gases in the
tracheae may be largely by diffusion, but many insects perform breathing
movements by expansion and contraction of the dorsal and ventral walls
of the abdomen, which serve to send tidal waves of the gases along
the larger tracheal trunks.   In insects inhalation is a passive act, while
exhalation requires muscular effort—the opposite of the condition in
man.   It has been shown that the distal ends of the tracheoles are
partially filled with liquid, which may be quickly sucked through their
walls by the surrounding tissues when they are undergoing great metabolic
activity.   This allows air from the tracheae to penetrate farther toward
the tips of the tracheoles.   Whether the liquid, as it is withdrawn from
the tracheoles, actually carries the oxygen in solution into the tissues or
whether the gases diffuse directly to the adjacent cells when the liquid
is removed seems to be unknown.   In the latter case the movement of the
liquid would be merely a regulatory device, increasing or decreasing the
amount of surface in the tracheoles from which the oxygen can diffuse,
so as to supply the right amount of oxygen for the release of the necessary
energy.

Insects that live in the water come to the surface to breathe at
intervals, or carry air under water clinging to special parts of their
bodies, or respire by means of gills.   In the latter case, the spiracles have
grown shut, and thin-walled outpocketings of the body wall, containing
tracheae and tracheoles, have specialized to extract, by osmosis, *the
oxygen dissolved in the water* and return carbon dioxide to it.   These
tracheal gills of insects (Fig. 110,*B*) differ from the blood gills of fish and
other animals in that the oxygen taken through their walls enters tracheoles and tracheae and is transported in gaseous form, instead of being
received by blood capillaries to be transported in liquid.   Blood gills
are rarely found in insects.

Circulation.—Tracheal respiration is much more direct and rapid
than the method employed by the higher animals.   In the higher animals
the lungs serve as a place where the blood is purified by exchanging the
waste gases of the body for oxygen.   The red blood corpuscles carry these
gases between the lungs and all other parts of the body.   The blood of
insects has no red corpuscles and, ordinarily, has little or nothing to do
with respiration.[1]   Phagocytic cells ("white corpuscles") are, however,
common in insect blood.   The blood is yellowish or greenish in color,
slightly alkaline, and will often clot, like mammalian blood.   Its chief
function is nutritive, carrying food materials from the digestive system to
all parts of the body.

Structurally the circulatory system is very simple in insects, consisting
of a single tube, lying close under the body wall down the middle of the
back.   This tube, the abdominal portion of which is often called "the
heart" (Figs. 53,*Ht*, 54, and 56,*h*), is closed at the posterior end but has
small openings (*ostia*) at regular intervals along the sides.   The blood
may enter these openings, but an arrangement of valves prevents its
flowing back out of them.   An anterior, nonpulsating, thoracic part of
the blood vessel may be called the *aorta* (Fig. 56,*ao*).   The heart is held
in place by triangular sheets of muscles extending to the body wall.

[1] In the bloodworms (Diptera, Chironomidae) the blood is red from the presence
of hemoglobin, the tracheal system is rudimentary, and apparently in these insects
the blood has a respiratory function.

These often separate a definite *pericardial cavity* or *dorsal sinus* from the general *perivisceral cavity* below them.

The blood may flow either way through the heart, but is generally forced forward in systolic waves by the contraction of the tube, to the region of the head, where it leaves the aorta and enters the general body cavity, filling the spaces between brain, nerves, tracheal tubes, and other viscera and the body wall. The dorsal diaphragm also pulsates in an anterior direction. As the heart tube expands, blood is sucked into it through the ostia. A low pressure area is thus formed in the dorsal sinus, and the blood filters backward and upward through the perivisceral cavity to equalize it. Such an open and imperfect circulatory system probably could not circulate the blood out into appendages, such as the wings and long slender antennae and legs, were it not for supplementary pulsating membranes located near their bases. As the blood returns from the appendages, it tends to collect in the lower part of the thoracic and abdominal cavity, where a ventral diaphragm, pulsating in a posterior direction, pushes it backward and upward through the perivisceral spaces and into the dorsal sinus again. As the blood bathes the alimentary canal, it receives by osmosis through the intestinal walls the digested food; as it passes muscles, glands, and other tissues, these nutrient substances from the food are given up, to be transformed into protoplasm or secretions; and as it flows over the Malpighian tubes it gives up to them the waste products of metabolism ("ashes") which have not been passed out in gaseous form through the tracheae.

SENSATION, CONDUCTION, AND COORDINATION

It is one of the inherent characteristics of living substance to be sensitive or irritable to the various stimuli that act upon it and to respond by altering its behavior in some way. In all but the simplest of animals this is accomplished by a nervous system. Nerves are composed of cells, very highly specialized for sensation, conduction, and coordination. Each nerve cell is known as a *neuron*, and may be a *sensory* (afferent) *neuron*, which has its cell body lying in the hypodermis and which conducts impulses *inward* from a sense organ; a *motor* (efferent) *neuron*, which has its cell body lying in the central nervous system and which conducts impulses from the ganglia *outward* to muscles or glands; or an *association neuron*, which stands between a sensory and a motor neuron, and may modify, direct, or harmonize impulses received from one or several sensory cells, so as to coordinate the response of the organism as a whole. The association neurons are, therefore, believed to be the seat of consciousness and intelligent behavior. Each neuron has a cell body, including the nucleus, from which extremely elongate, branched processes extend. The nervous system in insects is composed of groups of nerve cells arranged in small, white, compact masses known as *ganglia* (Fig. 54); of *connectives* that extend from one ganglion to another; and of *nerve fibers* which extend from ganglia to all other organs of the body. These fibers connect eyes, antennae, palps, and other sensitive parts of the body through sensory neurons to the ganglia, whence motor neurons carry the impulses to the muscles or glands. In this way, stimuli from the receptive end organs are transmitted (as controlling and directing impulses) to the effective muscles or glands which respond by altering their behavior in some way. Stimuli are changes of any kind, either within the body of the insect or in the environment surrounding it, which produce changes in the body of the insect. Their general effect appears to be to destroy certain chemical substances in the nerves, resulting in an increased metabolism that may pass far along the nerves (*conductivity*) to effect changes in tissues or organs remote from the source of the impulse.

The part of the nervous system that is composed of the association neurons together with the cell bodies of the motor neurons is called the *central nervous system*. In insects it is composed typically of a series of ganglia which are connected by a double nerve cord running from end to end of the body (Figs. 53, *VNC* and 54) and is readily visible upon dissection. There are typically two ganglia, a right and a left, in each segment, but they are usually fused together, and the double cords also fuse so that the typically ladder-like structure looks more like a thread with a knot in each body segment. Except in the head, the whole system lies in the bottom of the body cavity along the ventral side of the body, instead of down the back as in vertebrates. In the head a small mass of ganglia (those belonging to the first three head segments) is known as the *brain* (Figs. 53, *Br* and 54). It lies above the esophagus. Another ganglionic mass (made up of the ganglia from head segments 4 to 6) is called the *subesophageal ganglia*. These lie below the esophagus. The alimentary canal (esophagus) passes between the two connectives that extend from brain to subesophageal ganglia.

The brain of an insect is a relatively much less important organ than the brain of a vertebrate. It is believed that in insects there is much less of associative processes, adjusting, coordinating, deliberating, and choosing, than in higher animals; much of the nervous activity being on a simple receptor-effector basis. As a result of this, each body segment exhibits considerable independence of other segments, its ganglion largely controlling the movements of the appendages to which its nerve fibers run. Insects that have had the brain entirely removed may live, walk, and fly for some time, may even mate and lay eggs, though their movements are usually erratic and not coordinated. In many adult insects, however, the ganglia from a number of segments fuse into one mass in the thorax. When such a thoracic mass occurs, it is usually highly sensitive to injury, and any mutilation of it results in death more promptly than similar injuries to the brain.

The *peripheral nervous system* consists of the sensory neurons and their end organs and the motor neurons, except the cell bodies of the latter which lie in the central nervous system. Their far-reaching, delicate fibers are largely invisible upon gross dissection. The brain, for example, sends nerves to the eyes, antennae, and upper lip; the subesophageal ganglion to the other mouth parts (see Fig. 54); and each other ganglion typically to the appendages of the segment in which it lies. In addition to the central nervous system and the peripheral nervous system, just described, there is a small but rather complex *sympathetic system* (Fig. 54), which is thought to control the movements of the heart, the digestive system, and the muscles controlling breathing.

The Sense Organs of Insects.—The function of a nervous system is to acquaint the insect with changes in its environment. To do this it must have certain *receptors* (sensory end organs) which detect and interpret these changes (*stimuli*) and translate them into impulses that traverse the nerves and eventually produce a response in the muscle or gland to which they run. Some of the sense organs of insects are affected by light, others by pressure or contact with other objects; some by sound waves, and others by chemicals in solution. Each kind of sense organ is adapted to receive only one kind of stimulus, ignoring or excluding all others. Those for touch, taste, and smell are typically single cells and are considered to be modified setae. Those for hearing and sight are much more complex and are composed of many sense cells, together with greatly modified areas of the body wall. The sense organs of touch are chiefly certain *sense hairs* on the surface of the body which differ from ordinary hairs, or *setae*, by having a minute nerve fiber from a sensory cell running into them (see Fig. 43). Since these tactile sense hairs occur on nearly all parts of the body and its appendages and are surrounded at the base by a very thin ring of cuticula, they move very freely and make the entire body of the insect extremely sensitive to tactile stimuli. The sense organs for the perception of chemicals are generally described as very thin-walled cones or plates covering a minute pore in the cuticula and kept moist by a secreting cell, the theory being that the contact of such a sense organ with *solid* parts of food or other chemical results in some of the latter passing in solution through the thin wall of the cone or plate by osmosis and so stimulating the nerve ending in a manner which we call *taste*. If *gases* or minute *drifting particles* of food or other chemicals are dissolved on the moist end organ, the result is similar, but we speak of it as *smell*. The sense organs of taste and smell are not so restricted in distribution as in higher animals. Organs of taste are generally thought to occur on palps, epipharynx, and hypopharynx; but at least certain butterflies and flies are able to taste very sensitively through their feet (tarsi). Organs of smell have been described as occurring on the antennae, the palps, and, in some cases, on the cerci (Fig. 54), near the posterior end of the body.

Smell is probably a very important function among insects, and there seems every reason to believe that they detect odors that are imperceptible to us. This sense is important in locating foods, in locating suitable places to deposit eggs, and in finding mates. If the females of our larger moths, newly emerged from the cocoon, are exposed out-of-doors their odors usually attract males of their species from considerable distances. In one instance 73 males of the cecropia moth were thus attracted by a single female exposed in a large screened cage during one night.

Organs of Hearing.—Insects have no ears on the sides of the head. However, they do have various organs that are believed to serve for the perception of sound waves and other vibrations. These always consist of a more or less complex group of characteristic sense cells, known as *scolophores*, or *chordotonal organs*, which may have greatly modified parts of the integument associated with them. The best known of these are the so-called "ears" of grasshoppers, which are conspicuous oval plates, *tympana*, one on each side of the first segment of the abdomen (Fig. 46). Over these areas the body wall is much thinned and apparently fitted to be set in vibration by sound waves. Internal to the tympana are certain complicated structures that doubtless serve to translate vibrations into impulses which traverse the auditory nerve to the thoracic ganglia. In crickets and katydids, a smaller tympanum is found on each front leg near the base of the tibia (Fig. 47,*A*).

Another very interesting organ of hearing has been described from the antennae of the male mosquito. In the male mosquito, the whorls of hairs that come off from the antennal segments are much longer and more numerous than in the female. The second segment from the head is very large in the male and the rest of the antenna arises from a cup-shaped cavity on its outer, or *distal*, face (Fig. 50,*Plumose*). The whorls of hairs are of different lengths, and vibrate to sound waves of different tones. It is believed that these vibrations are transmitted through the large basal segment to the auditory nerve. Since the hairs vibrate extensively to tones of the same pitch as that of the hum made by the female mosquito in flight, the theory has been advanced that the male turns his body until the two antennae are stimulated equally by such a sound, when, by flying straight ahead, he is able to find his mate.

Vision.—In the preceding chapter, page 86, the external appearance of compound eyes with their many hexagonal facets was described. Each facet is the *cornea* or end lens (Fig. 58,*c*) of a functionally independent eye unit, called an *ommatidium*, which runs inward like a tube or rod toward the center of the head (Fig. 58). From the inner end of each ommatidium a nerve fiber (*n*) extends to the brain. The light admitted through the six-sided area of transparent cuticula at the outer end (the *cornea*) may next be focused or concentrated by cone-shaped bodies usually crystalline in nature (the *crystalline cones*). It then traverses the length of the "tube" and falls upon retinal or visual end organs (the *retinulae*), usually six to eight to each ommatidium, which generally secrete an axial

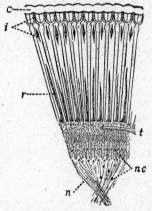

FIG. 58.—Portion of the compound eye of a blow fly, *Calliphora vomitoria*, in radial section. *c*, cornea which is modified cuticula; *i*, iris pigment; *n*, nerve fibers leading to the brain; *nc*, nerve or retinal cells; *r*, retinal pigment; *t*, trachea. (*From Folsom's "Entomology" after Hickson, Blakiston.*)

rhabdom, or optic rod, which is apparently the part of the eye that translates the wave lengths into stimuli that pass through the optic nerve to the brain. It is believed that each ommatidium does not form an image of the whole object, but only preserves the intensity, pattern, and color of the light coming from the particular small part of the object that is in line with its long axis. Indeed, the several ommatidia, or "tubes," are usually so isolated from each other by pigment (Fig. 58,*i* and *r*) that no light can pass from one to the other. The combined effect of thousands of such minute areas of light and shade makes the stimulus that the insect recognizes as the object before it. This is called "mosaic vision" and may be thought of as something like a cutout picture or "jigsaw" puzzle; each unit of the puzzle, with its particular areas of light and shade, is meaningless in itself but, when properly fitted to the others, forms an image (Fig. 52).

We really do not know how well an insect can see with such eyes. They are sometimes so big and bulging that the insect must be able to see in front of it, to each side, above and below, and even to some extent behind. On the other hand, it cannot move its eyes, it cannot focus them upon objects at various distances, and, of course, it cannot close them. In all probability, the closer an object, the better it is seen, because more of the ommatidia will cover an object close at hand than one far away. It is believed that the compound eye is especially good at detecting objects in motion, because different independent ommatidia are stimulated in succession. It has been claimed that insects such as dragonflies may respond to objects in motion as far as 60 feet away. The perception of the shape of objects is probably limited to a distance of a few feet. The compound eyes of insects doubtless enable their possessors also to measure distances, and it has been proved that certain insects can see colors, often including great bands of ultraviolet rays that are completely invisible to man.

The *ocelli* or simple eyes are of several kinds, but all agree in differing from the compound eyes in having a single facet in their cornea, the shining convex lens through which light is thrown upon retinal or visual cells. Such an eye, in which the shape of the lens and the distance from lens to retina are fixed, must see things clearly only at one definite distance away. Most objects must be out-of-focus. The lenses are very convex; hence the focal distance is short and these simple eyes must be nearsighted ones. They may serve simply to distinguish the lighter from the darker parts of the environment.

MOTION AND LOCOMOTION

Muscles are composed of cells highly specialized for contractility. They are grouped together into *fibers*, and the fibers in turn into *muscles* which act as units in moving the body and its appendages (Fig. 56). The number of muscles in an insect body is very large, sometimes at least 4,000, in contrast with 400 or 500 in the body of man. The ends of the muscles which move the jaws, legs, and wings of insects are secured to the inner surface of some portion of the body wall, since there are no bones to serve for this attachment (Figs. 65 and 69).

Insect muscles, of both the voluntary and involuntary kinds, are of the striated type, in which alternate light and dark bands cross the fibers (see Fig. 57,*C*). They differ from the muscles of the larger animals in being yellowish or colorless; and, the fibers not being enclosed in tendinous sheaths, they seem softer than those of vertebrates. We have already indicated, however, that they are very efficient in operating the small bodies of insects. They are also capable of remarkable endurance, as indicated by the long-sustained flight of many insects, during which the wings may vibrate several thousand times a minute. This, like the rapid reproduction of insects, is dependent upon a high rate of metabolism, *i.e.*, the rapid conversion of food materials into living protoplasm and the equally rapid breaking down of the living tissues in order to release the energy necessary for all bodily activities. Most insects literally lead a fast life.

THE REPRODUCTIVE SYSTEM OF INSECTS

In some animals each individual may be both male and female (as in the common earth worm); but among the insects, male and female reproductive organs are always borne by different individuals. Male and female insects are often equally injurious. In some species the females are more injurious, because the adult males are short-lived and do not feed (as in scale insects), because the female inflicts injury in the egg-laying act (*e.g.*, the tree crickets), or because females must consume more food to mature large numbers of eggs.   Male insects are generally smaller than the females.   The sexes are sometimes indistinguishable on external appearance but generally show minor differences by which they can be identified.   In many species of flies the sexes can be told apart by the difference in the eyes, as shown in Fig. 51.   Frequently an examination of the tip of the abdomen will show differences sufficient to distinguish the sexes.   In some cases the two sexes are so different looking as to appear like widely separated kinds (*cf.* Figs. 135 and 136).

The reproductive organs are usually found toward the tip of the abdomen (Fig. 54), although, when the eggs are developing, they often pack the entire body cavity of the female.   The opening from the reproductive organs is near the posterior end of the body (on the eighth or ninth sternite) in all insects and is commonly surrounded by external genitalia.   In the female, there may be an *ovipositor* (Figs. 46 and 105), an organ for thrusting eggs into the ground, the tissues of plants, or the bodies of other animals.   In the male, there are *claspers* (Fig. 54), used to hold the female during mating, and the *penis*.   The external genitalia are coming to be used very extensively in the determination of species; the small, and often complex, parts showing perhaps more distinct differences between species than any other group of structures.

In either sex the *gametes* or germ cells (eggs and sperms, respectively) are developed from cells of ordinary appearance, and prepared for union in the sex glands.   These glands (Fig. 59) are commonly more or less compact bodies lying among the other organs of the abdomen or suspended by filaments or ligaments from the inside of the dorsal wall of the body.   The *testes* and *ovaries* differ so much in form that no brief general description can be given.   Each consists internally of a number of minute tubes or bead-like strings of cells, the *ovarian tubes* in the female (ranging in number from 1 to as many as 2,400 in each ovary, but commonly 4 to 8) and the *testicular follicles* in the male.   In these tubes the sex cells (*gametes*) become more and more highly specialized, changing from primary germ cells to *oögonia* or *spermatogonia* and then to *oöcytes* or *spermatocytes* and, finally, after the reduction division of the chromosomes (see page 137), emerging as matured eggs or sperms.   The growth of the eggs is made possible by the secretions (yolk) of adjacent cells known as *nurse cells*.   The entire ovary or testis may or may not be enclosed in a sheath of connective tissue.   As the eggs descend the ovarian tubes, the shell is secreted about them by the small *follicular cells* that form the walls of the tube.   From the sex glands the gametes are conveyed toward the exterior of the body through tubes known as *vasa deferentia* or *seminal ducts* in the male and *oviducts* in the female.   Pending mating, the accumulated sperms are held in the body of the male in dilatations of the vasa deferentia known as *seminal vesicles*.   At the time of mating, the sperms are usually received into a special pouch of the female system known as the *seminal receptacle* or *spermatheca*, which is attached to the oviducts or vagina.   Later, as the eggs are being laid, some of the stored sperms are forced out upon them when they pass the opening of the seminal receptacle, thus bringing about fertilization.   In the case of the honeybee queen, the sperms retain their viability in the seminal receptacle throughout her lifetime of several years.   Eggs may be fertilized by these sperms years after the mating took place. The two oviducts usually unite to form a common duct, the *vagina*, opening ventrally on the seventh, eighth, or ninth abdominal segment. The two seminal ducts usually fuse into a single *ejaculatory duct* opening ventrally on the ninth abdominal segment. *Accessory glands* of the male (Fig. 54) secrete the fluid with which the sperms are mixed, and sometimes sac-like coverings for packets of sperms known as spermatophores. In the female the accessory glands secrete a substance for cementing eggs together, or fastening them to leaves and other objects; or capsules to enclose a number of eggs; and, in the honeybee, one of the poisons used in the stinger. The correspondence of the organs of male and female is shown in the following table (after Folsom, compare with Fig. 59):

Male	*Female*
Paired testes[1]	Paired ovaries
Sperm tubes (testicular follicles)	Egg tubes (ovarian follicles)
Paired vasa deferentia (seminal ducts)	Paired oviducts
Median ejaculatory duct	Vagina (median oviduct)
Penis (phallus or aedeagus) and claspers	Genital chamber (bursa copulatrix) and ovipositor or stinger
Seminal vesicle	Seminal receptacle (spermatheca)
Accessory glands	Accessory glands

[1] Sometimes united into a single testis, as in moths and butterflies.

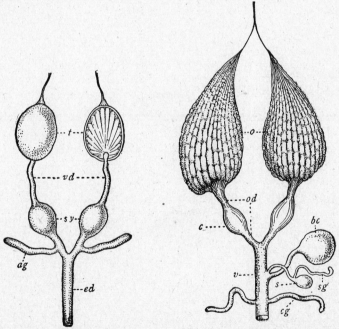

Fig. 59.—Male reproductive organs, on the left, diagrammatic; right testis is shown in section to expose the testicular follicles. *ag*, accessory glands; *ed*, ejaculatory duct; *sv*, seminal vesicles; *t*. testes; *vd*, vasa deferentia.

Female reproductive organs, on the right, diagrammatic. *o*, ovary; *od*, oviduct; *c*, egg calyx; *v*, vagina; *s*, spermatheca; *bc*, bursa copulatrix; *sg*, spermathecal gland; *cg*, colleterial gland. (*From Comstock's "Introduction to Entomology."*)

The growth and development of the insect following the formation and fusion of the egg and sperm are discussed in Chapter VI.

PROTECTION

Protection from natural enemies and adverse climatic conditions is as necessary to safeguard the individual as reproduction is to perpetuate the species. The keen struggle for existence among animals has resulted in the perfection and adoption of a great variety of protective structures and devices, which are nowhere better illustrated than among insects.

The methods of protection among insects may be classified into:

    a. Protective structures.
    b. Protective constructions.
    c. Protective size, form, and color.
    d. Protective positions.
    e. Protective behavior or reactions.

The importance of a sclerotized exoskeleton as a protection to insects has already been emphasized in Chapter III. Such a hard body wall is characteristic of the great majority of insects. The cuticula is often prolonged into bristles, spines, hairs, and scales which further protect the insect from mechanical injury, from excessive heat or evaporation, or from natural enemies, which find the hairy or spiny creatures unpalatable.

In certain caterpillars the protective value of the hairs is further increased by venom which fills them, and which nettles or poisons other animals that touch the hairs (see page 21). The body fluids of blister beetles and some other insects are corrosive or poisonous. Many insects, especially in the bug order (Hemiptera), have odors that are repulsive. Some, like the celery caterpillar, give off odors or bitter secretions from eversible glands (Fig. 148,*b*), the sudden erection of which may have value in frightening certain enemies.

Insects build many curious constructions to protect themselves or their young. The best known of these are the cocoons of moths, which are formed of silk secreted from the mouth. A great variety of cases or nests is used to protect especially the motionless pupal stage (see Fig. 90). Sometimes the larval stage is protected by the cocoon throughout growth, as in the instance of the casemaking clothes moth (Fig. 522,*b*), the bagworm (Fig. 457), and the caddice-worms. Soil, leaves, small pebbles, shavings of wood, and many other substances are used to cover the body. Or a special secretion or excretion may be poured out through the body wall; as the waxy, woolly covering of mealy bugs and many aphids. The social insects build elaborate nests, *e.g.*, the paper globe of the bald-faced hornet, the earthen mounds of ants, the soil pyramids of termites, or the overwintering nests of the brown-tail moth larvae.

A curious method of protection is illustrated by the larvae of the lace-winged flies and the larvae of sweetpotato beetles; these disgusting insects pile their own excrement and shed skins on their backs for concealment. In many flies the next to the last exuviae of the larvae are not shed but are retained about the body during the last larval and pupal stages, forming an excellent protective case (Fig. 90,*C*). The elevating threads that bear the eggs of the lace-winged fly (Fig. 78,*R*) illustrate another device which is probably of protective value.

It is probable that size may have protective value from certain enemies, extremely small insects being overlooked by larger enemies, and unusually large insects appearing too formidable for certain of their smaller enemies. Some insects have a very grotesque appearance, which may well be frightening to some of their enemies. Concealing form and coloration (camouflaging) is pronounced among the insects. It takes two special forms. One condition, known as *protective resemblance*, is well illustrated by walking-sticks (Fig. 107), which look so much like the twigs among which they live as to be difficult to detect except when they move. The other condition is known as *mimicry*. Many butterflies, flies, and other insects which are edible to birds and toads resemble in shape or color or both other butterflies which are poisonous or bitter to taste, or wasps which have a sting. It is believed by many naturalists that the palatable kinds gain a valuable protection from their natural enemies by this deceptive appearance. Such resemblance to another animal is known as *protective mimicry*. If the camouflaging or mimicry enables its possessor to stalk its prey or lie in ambush more successfully, it is called *aggressive resemblance* or *aggressive mimicry*. Many insects that have stings or bad tastes, so as to be distasteful to predators, are gaudily marked, *e.g.*, the bright-banded wasps. This is called *warning coloration*. .

The positions in which an insect normally lives and feeds may make other protection less necessary. Insects which burrow in the soil, or tunnel in trunks of trees, or live in fermenting organic material, or swim through the water, or live on the skins or in the bodies of animals, gain a greater or less degree of security from extremes of temperature, excessive evaporation, and storms. They also incidentally gain security from a great many of the parasites and predators that would molest them if they fed in exposed situations.

Some insects which normally feed in exposed places have learned to take shelter or hide at the approach of certain enemies. Others depend upon their legs or wings for escape. Running away, flying away, jumping, swimming, or diving is perhaps the commonest of all protective measures. A few insects carry away their young or eggs when forced to retreat. An interesting method of escape, rather widely exemplified, is by insects feigning death or "playing 'possum" when danger threatens. Leaf beetles, click beetles, measuring-worms, sphinx larvae, cuckoo wasps, and many curculios are examples of insects that behave in this manner.

In contrast to those just mentioned are the pugnacious insects that stand·their ground or take the offensive when danger threatens them or their nests. The stinging Hymenoptera illustrate this best, although many kinds pinch or pierce with the mouth parts when handled. Others, which have no weapons, threaten or show fight and doubtless succeed with their bluffing especially if they resemble well-protected kinds.

Just as some kinds enjoy a measure of immunity by living in situations removed from the beaten paths of insect life, so others, by using the less-crowded periods of the day, escape attack from some enemies though they expose themselves to others. Insects which remain quiet during the day and become active at dusk are called *crepuscular* insects; those which confine their activity to the darkness of night are called *nocturnal*.

Summary

Insects, although very different structurally from man, perform the same functions or life processes that are familiar to students of human physiology; and, although very small, their internal organization is fully as complex as that of the larger animals.   The most important differences in internal anatomy and physiology between insects and men are the following:

1. Insects have no lungs, but breathe through intricately branching tracheae that carry air to every part of the body.   They inhale and exhale, not through mouth or nostrils, but through small holes, known as spiracles,[1] along the sides of the body.

2. The blood has no red corpuscles and ordinarily does not transmit the gases of respiration.   The circulatory system is an open one; the principal vessel or "heart" lies dorsal to the alimentary canal, and the blood elsewhere flows in the body cavity, without veins, arteries, or capillaries.[1]

3. The excretory organs ("kidneys") of insects, known as Malpighian tubes, open into the front part of the hind-intestine, instead of having separate openings to the exterior.[1]   They form semisolid excretory products, rather than urine.

4. Insects have no digestive organs corresponding to the liver and pancreas of higher animals, yet they are capable of digesting a remarkable variety of substances.

5. A great variety of glands, which open to the outside of the body or into the alimentary canal, form secretions of great importance in the metabolism, economy, and ecology of the insect.

6. The central nervous system is a ladder-like or string-like chain of ganglia along the ventral side of the body, except one complex nerve center in the head (the "brain") which is dorsal to the alimentary canal.[1]

7. The organs of vision, although diverse and conspicuous, are in most cases probably not highly effective.   They are totally different structurally from the eyes of other animals.   Insects have widely scattered and intricate organs for hearing, sometimes on the abdomen, sometimes on the legs, sometimes associated with the antennae, but never as evident ears on the sides of the head.   Organs of smell and taste in insects are not so localized as in man but may occur on antennae, mouth parts, tarsi, and cerci.   Such diffuse chemical detectors are probably very important to the lives of insects.

8. The small and relatively unintelligent insects are aided in their persistence by a great variety of protective structures, constructions, forms and colors, positions and habits.

[1] These characteristics should be added to those of external anatomy cited on p. 87 to distinguish insects from other animals.

CHAPTER V

THE MOUTH PARTS OF INSECTS

The most important way in which insects inflict losses and injury upon man and his possessions is by eating or feeding. Since insects feed in various ways, it is evident that a knowledge of insect mouth parts is of prime importance in the study of entomology. By those interested in controlling insects, this part of insect anatomy receives very careful consideration; and the homologies of the regions and appendages of the heads of various insects have presented some of the most interesting and puzzling problems in insect anatomy.

THE SEGMENTATION OF THE HEAD

The head of an insect is believed to be composed of six segments, or twice the number in the thorax, although it looks like a single segment. The reasons for con-

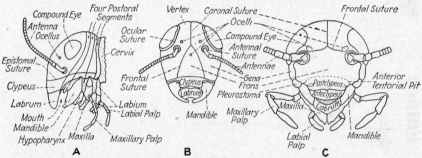

FIG. 60.—Diagrams illustrating the fundamental structure of the head of insects and the principal named parts. *A*, suggests the origin of the head from six primitive segments. (*Redrawn from Snodgrass, "Principles of Insect Morphology," McGraw.*)

cluding this are that in the embryos of certain insects, six nerve ganglia, six pairs of rudimentary appendages, and six pairs of coelomic sacs or primary divisions of the body cavity can be recognized in the part that subsequently forms the head. Since the typical arthropod segment bears one ganglion, one pair of jointed appendages, and a pair of coelomic sacs, this can only mean that six such segments have fused to form the head of the insect. At this stage of development, the mouth opening is found about mid-length of the head on the ventral side between the third and fourth segments. Probably in the ancestors of present-day insects these six segments were as distinct as those of the abdomen in the insects living now; and each segment bore a pair of appendages, all very much alike and similar to the legs.

For the function of walking, we concluded (page 79) that it is an advantage to have the segments distinct and the appendages thereof widespread to form two tripods. But for chewing and ingestion of food the appendages must be close together where they can work against each other, to cut off, to hold, and to masticate the food. So the head of the insect (Fig. 61) has come to be more and more compact, its six segments have fused together, and almost all trace of their conjunctivae and segmentation has been lost. Its walls have become thick and firm to serve as an adequate base for the attachment of the powerful muscles needed to operate the mouth parts. In this way the skull case (called the *cranium* or *head capsule*, in insects) has been developed. This strong cranium supports the eyes, the antennae, and the mouth parts. It encloses the brain, the mouth or buccal cavity, the pharynx, and the muscles that operate the mouth parts. In insects the size of the head is not an indication of the development of the brain but is correlated largely with the size and strength of the jaws.

Differences in the various appendages, sutures, and sclerites of the head are much used in determining the names of insects. To one who understands these parts, they will tell almost the whole story of the insect's food habits. So it is important to learn the names of the appendages and sclerites that follow.

THE SCLERITES OF THE HEAD

The most evident landmarks on the wall of the head of adult insects are the large compound eyes (Figs. 51, 66, and 74), described in Chapter III. These represent the first of the six head segments. In some insects there is an impressed line or suture running over the top of the head, between the compound eyes on the median line, and forking into two branches toward the front, the whole thing being like the letter Y, with its stem toward the thorax and its arms toward the front. This suture is called the *epicranial* or *coronal suture* (Fig. 60). The area of the wall of the head between the branches of this suture, or the equivalent area in insects where the suture cannot be found, is known as the *front* or *frons*. It usually bears the median ocellus. Below the front are two other areas or sclerites typically separated from each other by transverse sutures. The upper one of these is called the *clypeus*, sometimes divided into two parts (Fig. 60), and the lower one the upper lip or *labrum* (Fig. 63). One *condyle* or root of the mandible articulates to the clypeus, while the labrum is hinged along its entire base to the clypeus but is otherwise free and movable as an upper lip. The area between the compound eyes on the upper part of the head is called the *vertex*. Here the other two ocelli (Figs. 60 and 61) are usually located, and from this area and slightly below the ocelli the antennae (Figs. 50, 60, and 66) arise. The cheeks or sides of the head below and behind the compound eyes and between them and the mandibles are called the *genae*. Sometimes a narrow strip along their lower margins is marked off by a suture and is known as the *subgena*. The part of the head that abuts against the thorax is called the *occiput* (Fig. 61). It generally consists of a somewhat horseshoe-shaped arch the lower or lateral parts of which (the heels of the horseshoe) are called the *postgenae*. The occiput surrounds the small posterior opening or passageway in the skull case through which all the food, as it is swallowed, and also the nerve cords, dorsal blood vessel, and tracheae, pass from head to thorax. This passageway is called the *occipital foramen* (see Fig. 68,*B*). The *tentorium* is an internal structure in the head of insects formed by two pairs of ingrowths of the body wall (*anterior and posterior arms of the tentorium*) that unite to form a cuticular framework or plate supporting the pharynx and esophagus and arching over the subesophageal ganglion. The tentorium may be further braced by a pair of *dorsal arms*, extending from the cranium to the anterior arms, above the bases of the antennae. These arms and their intersections brace the cranial walls and serve as a place of origin for muscles that operate the mouth parts and the antennae. The pits or slits where the *anterior arms* of the tentorium invaginated (Fig. 60,*C*) lie near or in the lateral ends of the frontoclypeal (epistomal) suture, just above the anterior articulation of the mandibles. The pits or slits where the *posterior arms* of the tentorium invaginated lie at the caudoventral corners of the postgenae, in the lower ends of the suture separating the fifth and sixth head segments. When the dorsal arms of the tentorium are present, their points of origin are marked by spots or depressions somewhat above the base of the antennae and near the arms of the epicranial suture.

The compacting and fusion of the originally separated segments of the head have not always been strictly in the direction of the long axis of the body, but a distortion has sometimes carried the foremost segments, with their appendages, toward the top of the head, leaving the mouth at the very front of the body (*prognathous*); or sometimes, by reason of an opposite tendency, the mouth has been shifted backward and comes to occupy a place at the lowermost part of the head (*hypognathous*).

The Appendages of the Head

The appendages of the first one of the head segments have become replaced by, or specialized into, the compound eyes (Figs. 51, 66, and 74). Those of the second segment are the antennae or feelers, (Fig. 50, see page 84). The third pair of the original appendages (second pair of antennae) has been lost in the insects, but we have in this region an unpaired sclerite of the head, developed somewhat like an appendage, the *labrum* or upper lip (Figs. 61 and 63). The appendages of the fourth head segment are the *mandibles* or first pair of jaws (Figs. 61 and 63); those of the fifth segment, the *maxillae* or second pair of jaws (Figs. 61 and 63); and those of the sixth segment, the *labium* or lower lip (Figs. 61 and 63). In addition to the above parts, portions of the wall of the mouth or preoral cavity are sometimes projected outward, or otherwise so specialized that they form an essential part of the mouth parts; the one from the dorsal wall (roof of the mouth) is called the *epipharynx* (Figs. 63 and 66), the one from the ventral wall (floor of the mouth) is called the *hypopharynx* (Figs. 63 and 66). The typical arrangement of these parts with reference to the mouth opening may be indicated by the following diagram. It is plotted to represent the arrangement of parts as the

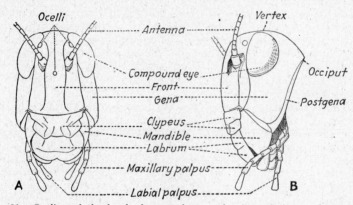

Fig. 61.—Outline of the head of a grasshopper showing sclerites and appendages. *A*, as seen from in front; *B* from the side. (*Rearranged from Folsom's "Entomology," Blakiston.*)

Right compound eye	Vertex	Left compound eye
Right ocellus		Left ocellus
Right antenna	Median ocellus	*Left antenna*
	Frons	
Right gena and subgena	Clypeus	Left gena and subgena
	(Labrum)	
	(Epipharynx)	
(*Right mandible*)		(*Left mandible*)
	MOUTH	
	(PREORAL CAVITY)	
(*Right maxilla*)		(*Left maxilla*)
	(Hypopharynx)	
	(*Labium*)	
Right postgena	Occiput	Left postgena

insect would face the reader. The parts that are printed in italics are believed to be homologous to the legs of the thoracic segments. The parts written in parentheses constitute the mouth parts.

The eight parts which are collectively called the mouth parts vary extremely in different insects for different kinds of feeding. They are first described as they are found in an insect that chews solid foods, such as a grasshopper, cricket, beetle, or caterpillar. Insects of this type of mouth parts (Fig. 62) inflict losses upon American agriculture in excess of half a billion dollars a year.

THE CHEWING TYPE OF MOUTH PARTS

1. The Labrum (Figs. 61 and 63).—This so-called upper lip covers the mandibles and closes the mouth cavity from in front, much as our upper lip covers our teeth. It helps to pull food into the mouth. There is often a notch at the middle of the labrum which is of service in holding the edge of a leaf in position so that the mandibles can bite across it effectively. Very often the insect places itself in such a position at the edge of the leaf that the sagittal plane of its body (the plane that divides the body into two equal halves) is parallel to the plane of the leaf. Some insects, however, can eat directly into the face of a leaf away from its margin.

2. The Epipharynx (Fig. 63).—If we had only the chewing insects to consider, it would hardly be worth while to give a separate name to this part. But in some other types of mouth parts it becomes differentiated into an important structure. In the grasshoppers and crickets, the epipharynx is inseparably attached to the labrum forming the inner, under, or posterior face of the labrum and is continuous with the roof of the mouth and thence into the esophagus. It is a sensory area believed to contain end organs of taste.

3. The Mandibles (Figs. 61 and 63).—The mandibles, teeth, or first pair of jaws are in chewing insects the most important part of the mouth structures. Besides masticating the food, they are the structures that cut it off or tear it off the leaf or other object on which the insect is feeding. In different insects they function to carry things, to fight with, or to mold wax. In the chewing insects there are generally several projections or small teeth on each mandible, which work against those of the opposite side and so make very efficient grinders. It should be noted that the action of the mandibles and maxillae in insects is transversely, or from side to side, instead of longitudinally, or up and down, as in man.

Each mandible is typically a single, nearly solid piece of chitin, roughly shaped like a pyramid with three faces, one of which is continuous with the gena or cheek (see Fig. 61,*B*). These heavy teeth articulate, in front, by a socket to convex processes from the lateral corners of the clypeus where it joins the genae and, behind, by a rounded process that fits into a socket of the genae, postgenae, or subgenae. Two sets of muscles operate them, one closing them against each other, the other pulling them apart.

4. The Maxillae (Figs. 61 and 63).—These are the second pair of jaws, much more complicated in structure than the first pair, but working from side to side in a manner much like the mandibles. The exact shape differs with the kind of food and the manner of feeding, but the following parts are typically represented. There is a central body of three or four

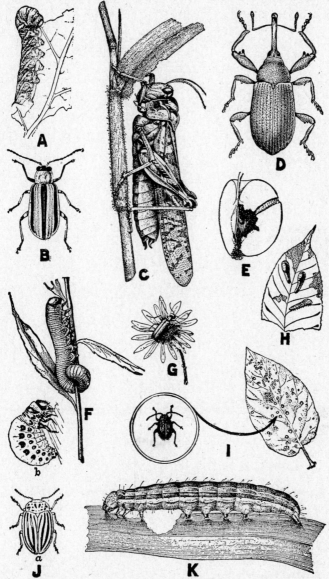

Fig. 62.—A group of common insects that have chewing mouth parts. *A*, tomato hornworm (*from Conn. Agr. Exp. Sta.*); *B*, striped cucumber beetle (*from U.S.D.A.*); *C*, grasshopper (*redrawn after Walton, U.S.D.A.*); *D*, cotton boll weevil (*from U.S.D.A.*); *E*, codling moth larva in injured apple (*redrawn after Conn. Agr. Exp. Sta.*); *F*, elm sawfly (*from Riley*); *G*, black blister beetle (*from Conn. Agr. Exp. Sta.*); *H*, pear slugs, skeletonizing a leaf (*from Conn. Agr. Exp. Sta.*); *I*, potato flea beetle and characteristic work on potato leaf (*original*); *J*, Colorado potato beetle, *a*, adult; *b*, larva (*from U.S.D.A.*); *K*, armyworm feeding on leaf (*from U.S.D.A.*)

Such chewing insects are regularly controlled by spraying or dusting with a stomach poison such as lead arsenate.

sclerites (*cardo, stipes, palpifer*), from which three appendages arise. One of the appendages is antenna-like in shape, of from one to five or six segments, and is a kind of sense organ bearing tactile hairs and probably

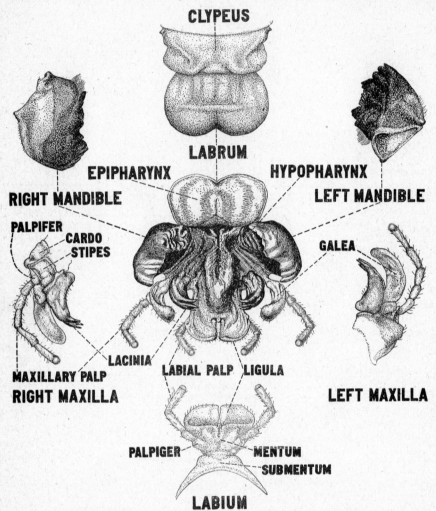

CLYPEUS

LABRUM

EPIPHARYNX   **HYPOPHARYNX**

RIGHT MANDIBLE   **LEFT MANDIBLE**

PALPIFER

CARDO
STIPES

GALEA

LACINIA   **LABIAL PALP**   **LIGULA**

MAXILLARY PALP

RIGHT MAXILLA   **LEFT MAXILLA**

PALPIGER   **MENTUM**

SUBMENTUM

LABIUM

Fig. 63.—The chewing type of mouth parts as found in a grasshopper. The central figure is looking into the mouth with all the appendages widespread. The upper left shows ectal view of right mandible; upper right ental view of left mandible; upper center the labrum with the clypeus to which it attaches; lower left is ectal view of the right maxilla; lower right, ental view of left maxilla; lower center, ectal view of labium. The hypopharynx is shown near the center of the central figure. (*Original; drawings by Antonio M. Paterno.*)

also organs of smell or taste. It is known as the *maxillary palp* or *palpus* (Fig. 63) and is regularly the longer of the two pairs of palps found in chewing insects. The second appendage, called the *galea*, is very variable in form (helmet-like in the grasshopper); and the third, known as the *lacinia,* is the tooth part of the maxilla, modified often for cutting, grasping or grinding the food. The central body of the maxilla articulates to the lower part of the posterior wall of the head (the *subgena*) (see Fig. 61,*B*), and its several parts are freely movable by muscles that run out from inside the wall of the head and the tentorium.

5. The Hypopharynx (Fig. 63).—This is a tongue-like prolongation of the floor of the mouth or preoral cavity, usually attaching to the inside (anterior) wall of the labium. It is of interest as the part through which the salivary glands of insects open, these glands being especially significant in the silkworm (see page 41) as also in some disease-carrying insects.

6. The Labium[1] (Figs. 61 and 63).—A good idea of the function of this part is conveyed by its common name, lower lip. It stands apposed to the upper lip, closing the mouth below or behind. It is the most complicated of all the parts, but we easily analyze its structure when we understand its origin. It has developed from two maxilla-like pieces by the two growing together along the middle line. It therefore consists of a large central body, more or less produced into lobes or unsegmented appendages (typically four) at its free end, and gives off at each side a short antenna-like appendage known as the *labial palp* (Fig. 63). The four unsegmented appendages are a median pair of *glossae* and a lateral pair of *paraglossae.* These four parts, when more or less fused, are called the *ligula.* The labial palps are regularly shorter than the maxillary palps, the number of segments varying from one to three or four. Their function is similar to that of the maxillary palps. In some insects care will be needed not to confuse the palps with antennae, especially if the antennae are concealed. The labium attaches to the neck of the insect, articulating with the lower ends of the *occiput,* between the maxillae at the back part of the head.

How Insects Feed

Most of our serious insect pests feed in one of the following ways: (*a*) They tear or pinch off, chew up, and swallow bits of plant and animal tissue, much as a cow eats the leaves off a stalk of corn. In other words they are "chewers," like grasshoppers and caterpillars, and are said to have chewing mouth parts[2] (see Fig. 62). (*b*) Or they extract from beneath the surface of a plant or animal body the body liquids (without swallowing the tissues), just as a person might insert a straw into a piece

---

[1] The student should notice the very similar spelling of the names for upper lip, *labrum,* and lower lip, *labium;* the latter name having an *i* where the former has an *r*—otherwise the same.

[2] In most textbooks these two general feeding habits have been called "biting" and "sucking." A little reflection will show that "biting" is a particularly ambiguous term, which in common usage conveys exactly the opposite meaning to that intended in the classification of insect mouth parts. That is, when one speaks of a *biting insect* the average person thinks—not of a beetle or a caterpillar chewing up leaves—but of a "biting" mosquito or flea or horse fly, whose mouth parts, according to the usual classification, are not *biting* at all but *sucking.* Thus the "biting house fly" would have, according to such a classification "sucking" and not "biting" mouth parts. The term "sucking" is also objectionable because it is too general; as shown in the table at the end of this chapter, there are several, radically different, kinds of mouth parts which are all "sucking" or suctorial in function. We therefore urge the adoption of the term *chewing* instead of "biting" and *piercing-sucking* instead of "sucking," for the two commonest types of mouth parts.

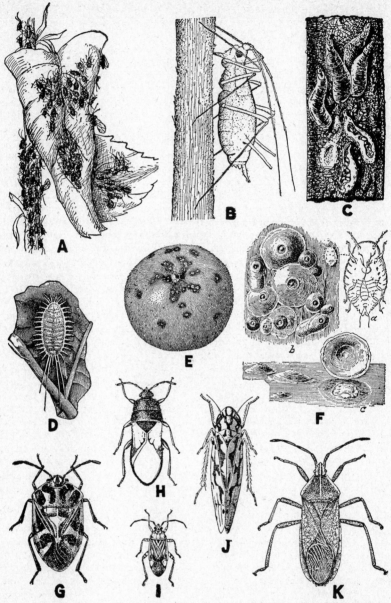

Fig. 64.—A group of common plant-feeding insects that have piercing-sucking mouth parts. *A*, aphids or plant lice clustered on stem and leaf, about natural size (*original*); *B*, a single potato aphid in feeding position, greatly enlarged, note beak appressed to stem (*from Ohio Agr. Exp. Sta.*); *C*, oyster-shell scales on bark, the lower two turned over to expose eggs from beneath the scale, enlarged (*original*); *D*, long-tailed mealy bug on leaf, much enlarged (*from Comstock*); *E*, San Jose scale on fruit of apple showing spotting due to its feeding (*from Fulton*); *F*, San Jose scale, *a*, first instar nymph, or "crawler," greatly enlarged; *b*, a group of scales as seen on the bark, much enlarged; the large round ones cover

of wet sponge and suck out the liquid or into a cocoanut and suck out the "milk." In other words, they are "drinkers," like mosquitoes or aphids, and are said to have piercing-sucking mouth parts[1] (see Figs. 2 and 64).

The condition of the mouth parts in a typical chewing insect has been described in some detail because it is the most primitive type and the kind from which all other mouth parts have been derived. Insects with chewing mouth parts may generally be controlled by spraying or dusting with a stomach poison.

The Piercing-sucking Type of Mouth Parts

The piercing-sucking mouth parts take the most valuable liquids in the world—the sap of growing plants and the blood of living animals. In order to get this food, they puncture or pierce the skin of the animal or the epidermis of the plant, making a very tiny, invisible hole with their mouth stylets through which they suck the sap or blood from beneath the surface. There are two very distinct operations involved in this act— (a) piercing and (b) sucking—hence we call these mouth parts piercing-sucking. The insects that have piercing-sucking mouth parts include both serious animal parasites, such as many flies and mosquitoes, the fleas, the bloodsucking lice, and destructive crop pests, especially the true bugs of the orders Homoptera and Hemiptera (Fig. 64). Insects of this type of mouth parts are therefore pests of importance to both the livestock raiser and the grain farmer or gardener.

The appearance of the mouth parts of this type is totally different from those described above. We do not find a complex group of appendages surrounding an evident mouth. In fact it is sometimes hard to tell just where the mouth opening is, it is so small or well hidden. We do not find hard, tooth-like mandibles for grinding. What we do find (see Figs. 65,*C* and 66) is a long, needle-like beak, slender, cylindrical, usually jointed, which may point forward, downward, or backward, but, when not in use, is generally found laid back on the breast between the front legs. In many cases there are no palps at all, in other subtypes one pair, and rarely both pairs are present.

In the true bugs, such as chinch bugs, cicadas, bedbugs, aphids, and squash bugs, we find a jointed slender beak of three or four segments (Fig. 65,*C*), inside of which lie four extremely slender, pointed stylets (Figs. 65,*B*,*C*), that normally cling together to form what appears like a single, slender, brown bristle. The jointed beak is not a closed cylinder but is open down the entire front side and at the end, like a trough, or the handle of a pocketknife. This largest outside piece is the *labium* (Fig. 65,*C*). *It has nothing to do with puncturing the plant or sucking up the sap.* Lying inside the groove of the labium are four very sharp, chitinous "stabbers" or "needles," the *stylets*, which do the work of piercing the plant and drawing out the sap. These four pieces are the two *mandibles*

mature females, the elongate ones cover males, and the small ones cover the second instar nymphs; *c*, a group of scales from the side, one of them lifted to show the body of the female insect (*from Quaintance, U.S.D.A.*); *G*, harlequin cabbage bug, enlarged (*from U.S.D.A.*); *H*, chinch bug, enlarged (*from S. D. Agr. Exp. Sta.*); *I*, tarnished plant bug, enlarged (*from U.S.D.A.*); *J*, grape leafhopper, enlarged (*from U.S.D.A.*); *K*, squash bug, enlarged (*from Iowa Agr. Exp. Sta.*).

Such piercing-sucking insects cannot be controlled by spraying or dusting with stomach poisons; but many of them can be destroyed by applying contact insecticides to their bodies.

[1] See footnote on p. 115.

and the two *maxillae* (Fig. 65), all extremely modified from their condition in chewing insects.   The *labrum* in this type (Fig. 65,*B,C*) is a short flap that covers the groove in the labium toward the base of the latter.

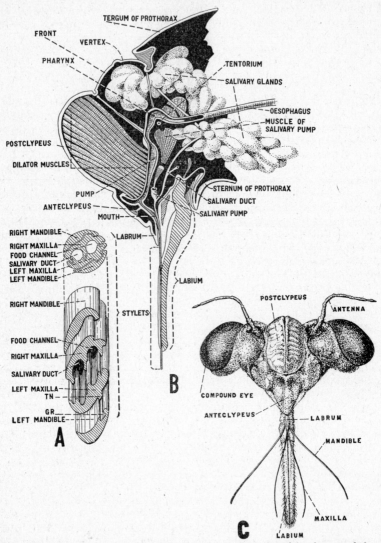

Fig. 65.—The piercing-sucking type of mouth parts as found in the squash bug and cicada.   *A,* cross-section and isometric projection of the stylets as described by Tower, greatly magnified (*original*); *B,* sagittal section of the head of the periodical cicada showing the relation of the stylets to the mouth opening, the pump, the pharynx, and the salivary glands, duct and pump, much enlarged (*redrawn after Snodgrass, Proc. Ento. Soc., Wash.*); *C,* front or dorsal view of head and mouth parts of a dog-day cicada, much enlarged (*original*).

Sometimes the mandibles and maxillae have little sharp barbs near the apex.   As the stylets are alternately thrust out from the head at a rapid rate, the barbs catch in the leaf tissue or flesh, and, anchored in this way against a backward pull, they help to sink the stylets deeper into the wound at each thrust, until the level of sap or blood is reached. In other species, such as the scale insects and whiteflies, the barbs are wanting, and the stylets are kept from being pulled out of the wound at each counterthrust by a muscular clamp in the groove of the labium, which alternately grips the stylets and allows them to slip deeper into the wound. Each maxilla is doubly grooved from end to end along its inner face (Fig. 65,*A*), and these concave faces of the two maxillae fit tightly together to make two closed tubes known as the *food channel* and the *salivary duct*. A mandible fits closely against each side of the apposed maxillae, and in some species, such as the squash bug, a tongue (*Tn*) on the outer face of each maxilla locks into a groove (*Gr*) on the inner face of each mandible so that they can slide in and out on each other. In this way the delicate hollow apparatus is greatly strengthened for the thrust into the tissues. The stylets are sometimes many times as long as the labium and even longer than the entire body. The extra length needed to reach the deeper tissues of plants or animals is accommodated when the insect is not feeding by being thrown into loops or coils inside the head, the thorax, or the base of the labium. The slender hair-like stylets, just described, have their bases sunken in the head and enclosed in pockets or pouches, whose lining is continuous with the integument of the cranium. In addition to the highly functional stylet parts, the mandible and the maxilla each has a plate-like basal part on the side of the head capsule, below the compound eye. A cuticular lever reaches from each mandibular and maxillary plate to the expanded base of the stylet inside the head, and muscles originating on the inner faces of the mandibular and maxillary plates are inserted upon these levers. As these muscles contract, they force the stylets outward or downward into the food tissue. Other muscles, originating near the top of the head and inserted upon the bases of the stylets, themselves, retract or pull the stylets upward and out of the wound. By rapid alternate protraction of right and left mandibles, followed by right and left maxillae, the stylets are forced into the food tissue rapidly, although by very short thrusts.

When the stylets have reached the sap beneath the epidermis of leaf or bark, or the blood in bloodsucking forms, the liquid food is sucked up through the microscopic tube called the food channel (Fig. 65,*A*) by expansion of a pump, formed by the walls of the buccal or mouth cavity and operated by powerful dilator muscles (Fig. 65,*B*). The contraction of these muscles pulls the dorsal wall of the pump away from the ventral wall, thus producing the necessary suction clear to the tip of the food channel in the stylets. As the muscles relax, the elastic dorsal wall springs back into place, its lower or anterior end closing first, and so forcing the sucked-up liquid backward and upward into the *esophagus* (Fig. 65,*B*). The labium shortens up or bends back out of the way and does not enter the plant tissues or flesh. There may be sensory hairs at the end of the labium that serve to sample the food and select the spot for feeding, and this part may also be used as a kind of fulcrum to steady the head and the stylets while they are piercing. However, its chief function appears to be to act as a kind of scabbard for the four-parted dagger that we call the stylets. The saliva that is pumped through the salivary duct into the plant may soften the cell walls and even predigest the liquid food. *Both pairs of palps are wanting in the Homoptera and Hemiptera.*

Any insect having mouth parts of this nature is open to suspicion as a serious enemy of crops or animals. If leaves show minute pale or brown spots, or are curled or wilted (see Fig. 2, page 6), this kind of an insect should be sought as the probable cause. Many piercing insects feed on other insects in a manner helpful to us, but as a group they are predominantly destructive.

One important principle should be fixed in mind at this point. *No insect with piercing-sucking mouth parts can be killed by applying a stomach poison, like lead arsenate or paris green, to the plant on which it is feeding.* These insects cannot take up solid particles as food, and the minute stylets may penetrate through a coating of lead arsenate on leaf or fruit and get none of it, since they do not suck up anything off the outer surface. The only way we could kill piercing-sucking insects with a stomach poison, as they are feeding on plants, would be to poison the sap of the plant. Experiments have been tried along that line, but so far there is no material known that can be inserted into the trunk or sap stream of the tree, that will kill the insects feeding on leaves and twigs, without injuring the tree.

VARIETIES OF PIERCING-SUCKING MOUTH PARTS

Besides the piercing-sucking mouth parts of the true bugs, there are several variations of the same functional type, especially among the bloodsucking insects. These are spoken of as subtypes and their chief structural features are indicated in the table of mouth parts below. For example, in the piercing apparatus of the mosquito (Fig. 66), we note that there are six stylets instead of four, the *labrum-epipharynx* and the *hypopharynx* being long and slender like the *mandibles* and the *maxillae*. These stylets are not protractile and retractile, like those of the Homoptera and Hemiptera, but are imbedded, when the insect bites, by a strong downward or forward thrust of the body. We find also a pair of palps (entirely wanting in the bugs), which are the *maxillary palps*. The food channel (Fig. 66,*B*) is formed chiefly by the labrum-epipharynx, which is the heaviest of the six stylets. The groove along its lower or posterior face is closed by the apposition of the hypopharynx or the flattened mandibles. The hypopharynx carries throughout its entire length, in the *salivary duct*, the saliva that causes the irritation when a mosquito bites and the malarial organism from an infected mosquito to the blood stream of man. The labium is not jointed except for the differentiation of a pair of oral lobes or labella (Fig. 66) at its tip.

Among the flies of the house fly family are some species, such as the stable fly, that bite, although the house fly never does. These have mouth parts of the type illustrated (special biting-fly subtype) in Fig. 67. The *labium*, the *labella*, the *labrum-epipharynx*, the *hypopharynx*, and the *maxillary palps* are much like those of the mosquito. But the mandibles and maxillae are entirely wanting, there being therefore only two stylets. Patton and Cragg[1] state that piercing by these insects is accomplished by the rapid protraction and retraction of the labella, which bear sharp teeth on their inner faces (Fig. 67,*C*), and that the labium as well as the stylets follows these teeth into the flesh. The food channel is formed by the labrum-epipharynx and hypopharynx and the salivary duct traverses the hypopharynx (Fig. 67,*A*).

The mouth parts of the fleas show another variation in the combination of stylets that are used for piercing and sucking (Fig. 68). There are three stylets: the *labrum-epipharynx* and the two *mandibles*. The mandibles have toothed edges and are the chief cutting organs. The labrum-epipharynx is grooved on its ventral or posterior face, much as in the stable fly (Fig. 68,*C*; compare Fig. 67,*A*). In order to make this

---

[1] PATTON and CRAGG, "*A Textbook of Medical Entomology*," Christian Literature Society for India. London, 1913.

groove or food channel tight enough to suck up blood, the slit is probably closed by the apposition of the edges of the mandibles. Each mandible has in its posterior (mesal) edge a tiny groove (Fig. 68,*C*); these two grooves when pressed together, form a second tube, the *salivary groove*, to carry the saliva to the tips of the stylets. The maxillae do

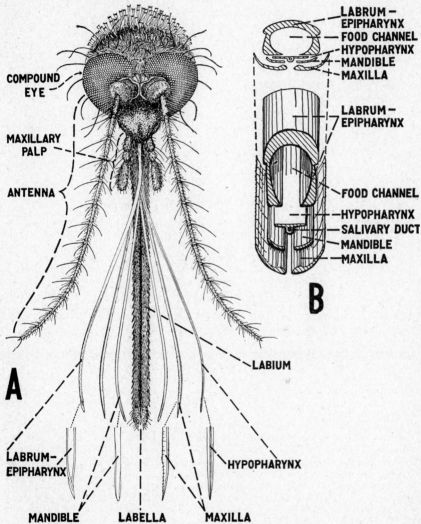

Fig. 66.—Piercing-sucking mouth parts as found in a female mosquito; common biting-fly subtype. *A*, front or dorsal view of head and mouth parts with the stylets spread out of the labium and their tips more enlarged below. *B*, cross-section and isometric projection of the stylets as described by Howard, Dyar, and Knab. Much enlarged. (*Original; drawing by Antonio M. Paterno.*)

not enter the wound but are broad flaps, said to serve as levers or fulcrums to steady the head during the piercing and sucking operations. They have a pair of prominent *maxillary palps*. The labium suggests somewhat the condition in chewing insects, consisting of a basal piece called the *mentum* and two, slender, jointed pieces called the *labial palps*.   These palps are concave on their sides which are next to the stylets, and the two together form a protective sheath for the mandibles and labrum-epipharynx.

The piercing-sucking organs of the bloodsucking lice are anomalous, and the homologies of the parts are not clear.   There is no external evidence of the mouth parts except a few teeth (mouth hooks) around the mouth opening (Fig. 69), which serve to anchor the head of the louse against the skin as it prepares to bite.   Between

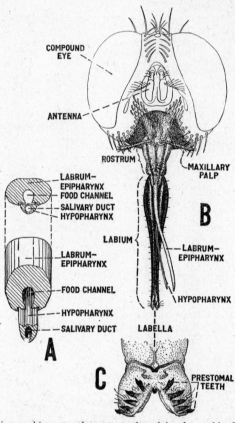

Fig. 67.—Piercing-sucking mouth parts as found in the stable fly; special biting-fly subtype.   *A*, cross-section and isometric projection of the stylets to show the food channel and salivary duct.   *B*, front or dorsal view of the head and mouth parts with the stylets spread out from the labium.   *C*, the labella more magnified to show the prestomal teeth, which are cutting organs, according to Patton and Cragg.   Much enlarged.   (*Original; drawings by Antonio M. Paterno.*)

these teeth an opening leads into the mouth cavity, called the *buccal funnel* or *preoral cavity*.   When not in use, the piercing structures are entirely withdrawn into a sac, the *stylet sac*, that branches off ventrally from the floor of the buccal funnel near the front of the head and ends blindly beneath the esophagus near the back of the head. At the caudal end of this sac are attached three slender stylets.   They are forked at the caudal (proximal) end and imbedded in the walls of the stylet sac.   Their cephalic or distal ends extend nearly to the mouth opening and in the region of the buccal funnel are ensheathed in a trough called the *sac tube*.   Although morphologists are not agreed as to their homologies, it is possible that the largest and most *ventral stylet*

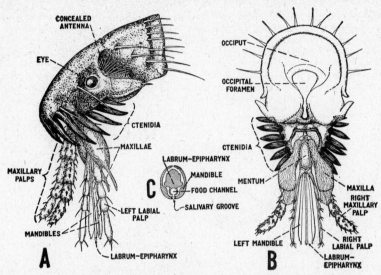

FIG. 68.—Piercing-sucking mouth parts as found in a flea.   *A*, head and mouth parts in side view.   Note the antenna nearly concealed in a groove in the side of the head, the small eye, the ctenidia or comb of heavy spines, and the way the labial palps shield the stylets.   *B*, caudal view of head and mouth parts of a flea, the head having been removed from the thorax.   *C*, cross-section of mandibles and labrum-epipharynx, as described by Patton and Cragg, to show how the food channel and salivary groove are formed.   All much enlarged.   (*Original; drawings by Antonio M. Paterno.*)

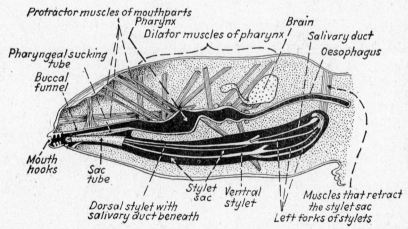

FIG. 69.—Piercing-sucking mouth parts as found in the human body louse.   Note the stylet sac lying beneath the pharynx in which are two prominent stylets.   In order to feed, the stylets are thrust through the mouth opening and forced into the flesh, and the blood as drawn is sucked up through the pharyngeal sucking tube.   (*Redrawn after Imms and Peacock.*)

is the labium and that a delicate hypopharynx surrounds the salivary duct as a *median stylet*, while the so-called *dorsal stylet* may represent the interlocked maxillae. The mandibles and both pairs of palps are entirely wanting. When the louse is ready to feed, protractor muscles (some having their origin on the inner wall of the cranium and insertion on the walls of the stylet sac, and others arising on the walls of the stylet sac and being inserted on the basal forks of the stylets) force the stylets out of the mouth and into the flesh. As the blood exudes, it is sucked up through the short, trough-like or funnel-shaped passageway and forced into the esophagus by the action of a powerful pharyngeal pump, operated by the action of dilator muscles. The saliva possibly serves to prevent coagulation of the blood.[1]

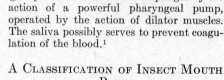

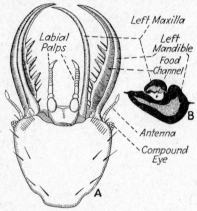

FIG. 70.—Mandibulo-suctorial subtype of mouth parts, as found in certain Neuroptera, ventral view, showing the greatly elongated, sickle-shaped mandibles and maxillae, which fit together to form the food channels. These extend from the tip of each jaw to its base, where an adventitious mouth leads into the pharynx. In *A* the left maxilla has been displaced to show the groove in the mandible. *B*, diagrammatic cross-section of the mandible and maxilla to show how the food channel is formed by their close apposition. (*Drawn by Kathryn Sommerman*.)

A Classification of Insect Mouth Parts

While the chewing and piercing types of mouth parts are the commonest and most important, there are a number of other types found in insects, as indicated in the following table, which is a summary of this chapter.

A. The Chewing Type (Fig. 63). Generalized mouth parts (consisting of eight named parts surrounding an evident mouth opening), the essential features of which are two pairs of tooth-like jaws; the mandibles and maxillae, fitted to work transversely (and used for tearing off and masticating food, or for carrying things, for fighting, etc.); and an upper and a lower lip. A further characteristic is the presence of *two pairs of jointed palps.*

This is the commonest type of mouth parts and is found in the silverfish, grasshoppers, crickets, earwigs, termites, book-lice, chewing lice, beetles, weevils, some Hymenoptera, and many insect larvae, especially grubs and caterpillars, besides many others of little importance to man.

I. *The Grinding or Masticatory Subtype* (Fig. 63).—Mandibles provided with a basal, molar, or grinding area suited for masticating solid plant or animal tissues, in addition to the apical cutting edges. EXAMPLES: the majority of chewing insects including nearly all caterpillars, most beetles, and most of the Orthoptera.

II. *The Grasping or Predaceous Subtype* (Fig. 134).—Mandibles elongate, curved, with one or more sharp distal points for catching and holding the prey, and without a well-developed molar area. EXAMPLES: most predaceous beetles, soldier ants, and such beetles as the stag beetles, in which the mandibles serve for holding the female during mating.

III. *The Grasping-sucking or Mandibulo-suctorial Subtype* (Fig. 70).—Mandibles long, slender, and sickle-shaped for grasping and piercing the prey and with a

[1] See PEACOCK, *Parasitology*, **2**: 98–117, 1918.

groove or closed canal extending from near their tips to near their bases through which the blood of the prey is sucked. The groove may be closed by overgrowth of the mandible itself, by the application of accessory lobes of the mandible, or by the close application of the elongate, flattened, maxillary blades. The mouth opening (preoral cavity) is usually partly or completely closed, the blood of the victim being sucked into the mouth by a pharyngeal pump through its lateral corners, or by secondary "mouths" applied to the opening of the groove near the base of each mandible. Frequently a paralyzing and digestive fluid is pumped into the body of the prey

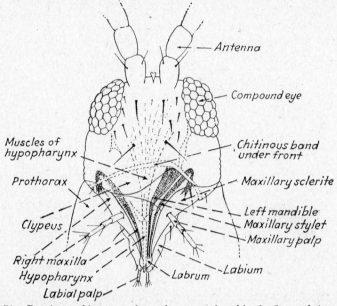

Fig. 71.—Rasping-sucking type of mouth parts as found in the flower thrips. View of the head from in front with only the bases of the antennae shown. The chitinous band serves to connect the mouth parts to the front. It also sends a branch to the left eye while the one to the right eye is a triangular rudiment. Note the two pairs of palps, the three or four stylets, and the cone formed by labrum, maxillae, and labium for sucking up the sap. (*Redrawn after Borden, Jour. Econ. Ento.,* 1915.)

before sucking begins. EXAMPLES: larvae of ant-lions (Fig. 138), aphid-lions (page 57), and some adult beetles such as the predaceous diving beetle and fireflies.

IV. *The Brushing, Spatulate, or Scraping Subtype.*—Mandibles without incisor or molar teeth, densely covered with stiff hairs; or flat, thin, and spatula-like for molding wax, mud, or dung. EXAMPLES: pollen-feeding beetles and dung beetles.

B. The Rasping-sucking Type (Fig. 71).—Mouth parts which are somewhat intermediate in structure between the piercing-sucking type and the chewing type, but are rasping and sucking in their action, serving to lacerate the epidermis of plants and to suck up the exuding sap. The right mandible is reduced, making the head and the mouth parts somewhat asymmetrical. The left mandible, the maxillae, and, according to Borden,[1] the hypopharynx are elongate, suggesting the stylets of the

[1] *Jour. Econ. Ento.,* **8**: 354, 1915.

piercing type and adapted to move in and out through a circular opening at the apex of the cone-shaped head. The stylets are contained, each in a separate pouch, that has invaginated from the surface to give them their internal position. The stylets apparently do not form a food channel, external to the wall of the head, nor do they enter deeply into the wound. The sap as it exudes on the surface is sucked up by the cone-shaped mouth rather than by the stylets. According to Snodgrass, the food channel, within the head capsule, lies between the labrum and the hypopharynx, and the saliva passes to the tip of the stylets between the hypopharynx and the labium. Both pairs of palps are present. The rasping-sucking mouth parts are characteristic of the thrips (page 196).

C. The Piercing-sucking Type.—Specialized mouth parts characterized by a tubular, usually jointed beak, enclosing several needle-like stylets. The outer tube is formed by the *labium*, which is simply a protective structure for the other parts and has nothing to do with piercing the tissues or drawing up the liquid food. The mandibles and maxillae, sometimes supplemented by or replaced by the labrum-epipharynx and hypopharynx, are greatly elongated and slender structures, which serve for piercing the skin of an animal or the epidermis of a plant, and also as the food channel, *i.e.*, an inner tube up which the liquid blood or sap is drawn.

There are several structural variations of this type, of which the following must be noted:

I. *The Bug or Hemipterous Subtype* (Fig. 65).—No palps. Four stylets, two *mandibles* and two *maxillae*, the latter partly fused. Food channel and salivary duct formed by the maxillae. EXAMPLES: chinch bug, aphids, scale insects, and bedbug.

II. *The Louse or Anoplurous Subtype* (Fig. 69).—No palps. Two prominent "stabbers" or stylets, and some associated structures. The proper names and homologies are not well understood, but the stylets are believed to represent the maxillae (interlocked), the hypopharynx, and the labium. This type differs from all other insect mouth parts in that, when not in use, all the parts are entirely withdrawn into a long, slender pocket in the head beneath the pharynx. EXAMPLES: human lice or "cooties," hog lice, and other bloodsucking lice.

III. *The Common Biting-fly or Dipterous Subtype* (Fig. 66).—Maxillary palps present. Six stylets; two *mandibles*, two *maxillae*, the *labrum-epipharynx*, and the *hypopharynx*, the last two parts forming the food channel and the hypopharynx surrounding the salivary duct. The stylets are not protractile and retractile as in the Hemiptera and Homoptera but are imbedded, when the insect bites, by a strong downward or forward thrust of the body. EXAMPLES: mosquitoes, horse flies, black flies, and "no-see-ums."

IV. *The Special Biting-fly or Muscid Subtype* (Fig. 67).—So far as structure is concerned, derived, according to Patton and Cragg,[1] from the sponging type (see below) by the reduction in size of the labella and the attenuation and sclerotization of the labium, which in this subtype is rigid and not retractile. The labella are provided with cutting teeth, and this type differs functionally from all other piercing insects in that the *labium* itself enters the puncture.[1] The *labrum-epipharynx* and *hypopharynx* are similar to those of the sponging type, together forming the food channel; the mandibles are wanting and the maxillae represented only by a pair of palps. This type is exemplified by the bloodsucking species of the order Diptera, family Muscidae, such as the stable fly, horn fly, and tsetse flies.

[1] PATTON and CRAGG, "A Textbook of Medical Entomology," Christian Literature Society for India, London, 1913.

V. *The Flea or Siphonapterous Subtype* (Fig. 68).—Maxillary palps present. Only three stylets (two *mandibles* and the *labrum-epipharynx*) enter the wound. The maxillae are triangular plates that serve as levers while biting. The labium bears two segmented parts, which are probably the *labial palps*. The food channel is formed by the labrum, epipharynx and the apposition of the caudal edges of the mandibles. The latter also form a salivary duct (Fig. 68,*C*). This type of mouth parts is found in the fleas.

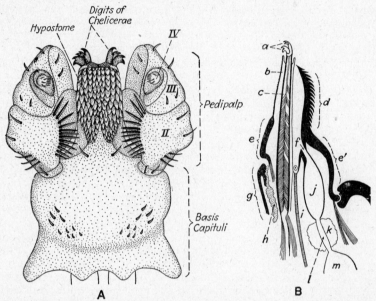

Fig. 72.—Anchoring subtype of mouth parts, as found in ticks. *A*, ventral view, showing the median *hypostome*, with its many retrorse teeth; projecting slightly beyond the hypostome the two *chelicerae*, each with two articles bearing several *digits* or teeth; and at either side the *pedipalp*, with its four segments marked *I, II, III*, and *IV*. All the above structures arise from the sclerotized ring-like false head or *basis capituli*. *B*, diagrammatic sagittal section of the same showing at *d*, the *hypostome*; at *a*, the *digits of the chelicerae*; *b*, the *sheath of the chelicera*, which encloses the *rod of the chelicera*, *c*. Within the rod of the chelicera are muscles that operate the tendons extending to the digits, and to the base of which other muscles attach that manipulate these cutting teeth; *e* and *e′*, the *basis capituli*, sectioned above and below the *mouth opening*, *f*, which leads to the *pharynx*, *j*; *l*, *esophagus*; *m*, *stomach*. The *brain*, *k*, surrounds the esophagus. The *salivary duct*, *i*, opens just above the mouth. Between the *scutum*, *g*, and the basis capituli on the upper side of the body is the opening from the gland called *Gene's organ*, *h*, the secretion from which is used in laying the eggs. (*Redrawn in part from Nuttall*, "*Monograph of the Ixodoidea*.")

VI. *The Tick or Anchoring Subtype* (Fig. 72).—Attached to a false head or *basis capituli* is a pair of four-segmented pedipalps which may function both as sensory organs and for the protection of the more vital parts between them. The latter consist of a labium-like *hypostome* which arises from the basis capituli below the mouth opening. Its ventral surface is usually provided with a number of hooks or teeth projecting backward, and it is the structure with which the tick anchors to the skin of its host. The cutting organs consist of a pair of mandibles or *chelicerae* which arise from the basis capituli above the mouth opening. Each consists of a firm sclerotized rod or shaft enclosed by a sheath and extending back inside the false head where muscles attach that control its movements. The distal end of the shaft protrudes from the sheath and bears two movable digits, each bearing a few very sharp teeth. By movement of these digits the hole is cut in the skin into which the hypostome is thrust and through which blood is sucked. This method of making the wound suggests somewhat that employed by the teeth on the labella of the Special Biting-fly Subtype. This type is found in the cattle tick, the spotted-fever tick, the American dog tick, and their relatives.

VII. *The Chelate or Mite Subtype* (Fig. 73).—Similar to Subtype VI, but the hypostome is not developed. The chelicerae often much more elongated to form needle-like piercing structures. The third or terminal segment of the chelicera sometimes articulated against the side of the second segment to form a pair of minute pinchers; at other times wanting, leaving the chelicerae or mandibles like those of a mosquito, simple slender needles. This type is found in many mites such as the poultry mite, the tropical rat mite, mange mites, scab mites, chiggers, and others.

D. The Sponging Type (Fig. 74).—This type of mouth parts is well illustrated by the condition in the common house fly. We find on the

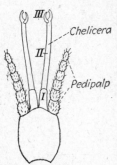

FIG. 73.—The chelate subtype of mouth parts as found in many mites. *I, II, III,* the three segments of the chelicerae. (*Drawn by Kathryn Sommerman.*)

lower side of the head a fleshy, elbowed, and retractile proboscis, which is the *labium.* The basal segment of this elbowed proboscis is the *rostrum* (which contains a part of the clypeus and the basal maxillary plates), and the distal segment is called the *haustellum.* The end of the labium is specialized into a large sponge-like organ, the *labella* (Fig. 74,*A*). The labella are traversed by a series of furrows or channels, narrowly open all along the exposed edge, the *pseudotracheae* (Fig. 74,*C*). These insects feed on exposed liquids, such as nectar or sap; or by dissolving solid substances, such as sugar, in their saliva. When the labella are appressed to such liquids, the pseudotracheae fill with the liquid by capillary attraction. These little channels all converge at one point on the labella, and from this point the liquid food is drawn up through the food channel into the esophagus. The food channel is formed by the labrum-epipharynx and the hypopharynx as in other flies (Fig. 74,*B*). These are the only two stylets present, and in this type they are entirely incapable of piercing the skin. The mandibles are wanting and the maxillae are represented by only a pair of maxillary palps.

This type is found in the house fly and other nonbloodsucking Muscidae, in the Syrphidae, and in many other of the Diptera.

E. The Siphoning Type (Fig. 75).—This is a very much specialized type, in which the labrum is greatly reduced, the maxillary palps rudimentary, and the mandibles usually entirely wanting. The labium is represented only by the large, hairy or scaly, three-segmented labial palps and a very small basal plate. The essential working parts are formed by the *maxillae,* parts of which, the *galeae,* are greatly elongated and joined to form a slender hollow tube which is coiled up under the head like a watch spring when not in use. Its structure suggests that of a piece of flexible metal tubing. Innumerable tiny short muscles extend from one ring of the tube to another, inside each half of the proboscis, which Snodgrass believes serve to coil the tube; while it is extended or straightened out to feed, possibly by the pressure of blood forced into it from the body cavity. This proboscis is not capable of piercing the skin of an animal or the epidermis of a leaf or fruit, except in rare instances. Feeding is accomplished by uncoiling this tube and projecting the tip of it into some exposed liquid (commonly the nectar in the nectary of a flower) and then sucking the liquid up through the *food channel* (Fig. 75,*B*), which runs full length through the proboscis.

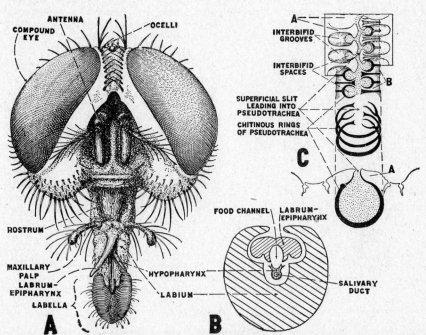

Fig. 74.—Sponging mouth parts as found in the house fly. *A*, dorsal or front view of head and mouth parts with the proboscis extended and the stylets spread out from the labial gutter. Note the pseudotracheae and sensory hairs on the labella. Much enlarged (*original drawing by Antonio M. Paterno*). *B*, diagrammatic cross-section of the proboscis to show composition of food channel, salivary duct, and labial gutter. *C*, details of a single pseudotrachea, greatly magnified. In the upper part of the figure a surface view of the pseudotrachea shows the superficial slit and interbifid spaces through which liquid foods enter the pseudotrachea, and on the left two interbifid grooves leading to the interbifid spaces. On the right the chitinous rings are shown as seen by transmitted light. At *B* the integument of the oral surface of the labellum has been removed to expose the membrane that lines the interior of a pseudotrachea and stretches between the chitinous rings with their alternate bifid and flattened extremities. *A* is the integument of the oral surface of the labellum; *B* is the membrane lining the interior of the tube. At the center are represented three consecutive chitinous rings to show how their bifid and flattened extremities alternate on each side of the superficial slit. In the lowest part of figure *C* is represented a transverse section of a part of the oral surface of the labellum, cutting across a single pseudotrachea. Greatly magnified. (*B* and *C* redrawn after Graham-Smith and Hewitt.*)

This type of mouth parts is found in practically all adult moths and butterflies, the order *Lepidoptera*.

F. The Chewing-lapping Type (Fig. 76).—This type, which is so well illustrated by the honeybee or bumblebees, is a kind of combination type in which the labrum and mandibles are of the same structure as in the chewing type, but the maxillae and the labium are elongated, and closely united, to form a sort of lapping tongue (Fig. 76,A,D). Both the maxillae and the labium are suspended from the cranium, and articulated to it, chiefly through the base of the maxillae. Both pairs of palps are present, the *labial palps* long and conspicuous, but the *maxillary palps* (Fig. 76,D) are very small. The glossae of the labium are greatly elongated to form a hairy, flexible tongue that can be rapidly protracted and retracted to reach deep into the nectaries of tubular flowers. According to Snodgrass, a temporary *food channel* is formed by the concave inner surfaces of the galeae, roofing over the *glossa* and fitting snugly lengthwise against the labial palps, which in turn lie tightly against the sides of the glossa. Through such a complexly formed tube ("held, like a straw in one's mouth, by the mandibles grasping the bases of the galeae while the epipharynx plugs the gap where the ends of the galeae diverge toward the head") a drop of honey may be sucked up.

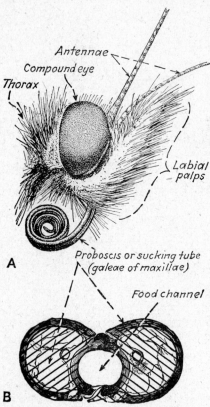

A

B

FIG. 75.—Siphoning type of mouth parts as found in a moth or butterfly. *A*, side view of the head with the proboscis partly coiled. Note the labial palps which are so covered with hairs that the segmentation cannot be distinguished. *B*, cross-section of the proboscis to show how the right and left galeae lock together to form the food channel. Highly magnified. (*Redrawn after Comstock.*)

According to George E. King, in securing nectar from the open nectaries of flowers, the bee thrusts out the glossa or tongue and licks the nectar with the tip of it. The glossa, thus smeared with nectar, is rapidly retracted between the labial palps and galeae, and the nectar is squeezed off the tongue by the galeae and deposited so as to accumulate in the small cavity formed by the paraglossae at the base of the glossa. Then by the bending of the labium upward near midlength, the base of the glossa is brought into close apposition to the mouth cavity, and the accumulated nectar is sucked into the esophagus by the action of a pharyngeal pump. Imms states that the liquid food ascends by capillary action through the ventral canal of the glossa (Fig. 76,B,c); and that the glossa is shortened by a muscular pull on the inner rod, squeezing the nectar upward to enter the space between the glossa and base of the paraglossae (Fig. 76,A); but Snodgrass finds that the liquid food is sucked up more rapidly than could be accomplished in this way. The nectar thus gathered serves as food for the bees, and the surplus is stored as honey.

The inner channel of the rod or the ventral canal of the glossa, or both, may serve as a salivary groove to conduct saliva from the salivary duct, which ends near the base of the glossa, to the tip of the tongue, where it may be used to dissolve solids, such as sugar, preparatory to swallowing. The mandibles are used for carrying things and, in the honeybee, for molding wax into cells.

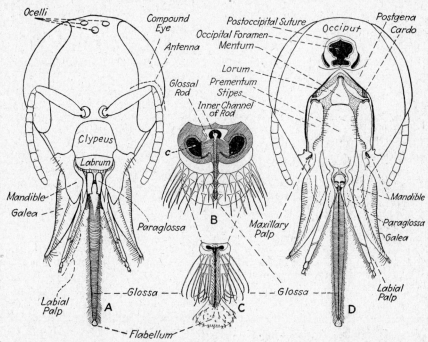

FIG. 76.—Mouth parts of the chewing-lapping type as found in the honeybee. *A*, cephalic or front view of head and mouth parts. Note that labrum and mandibles are of the chewing type, the latter specially shaped for molding the waxen combs, while the galeae, labial palps, and glossa make a five-part tongue for lapping up nectar. *B*, cross-section and isometric view of a portion of the glossa, greatly magnified, showing the ventral groove which communicates with the canal, *c*, the rings of long hairs, the glossal rod embedded in its anterior wall, and the hair-guarded inner channel of the latter. The ventral canal may be used to conduct saliva to the tip of the tongue but is not believed to be important in sucking up nectar. *C*, tip of the tongue or glossa, greatly magnified, ventral or posterior view, showing the close-set hairs that guard the ventral canal and the spoon-shaped flabellum at its tip. *D*, caudal view of head and mouth parts, showing the cardo, stipes, maxillary palp, and galea of the maxilla; and the lorum, mentum, prementum, paraglossae, and glossa of the labium. (*Redrawn, in part, from Snodgrass, "Anatomy and Physiology of the Honeybee."*)

This type of mouth parts is found in some species, only, of the order Hymenoptera, which includes the bees, wasps, and ants.

G. Degenerate Types.—The mouth parts of nymphs are generally very similar to those of the adults. The mouth parts of larvae are fundamentally of the chewing type regardless of the nature of the mouth parts of their parents. These become somewhat reduced in the larvae of certain orders.

I. *Hymenopterous, Trichopterous,* and *Lepidopterous Larvae.*—Maxillae, hypopharynx, and labium associated closely or united to form a combined lower lip. Ligula and hypopharynx form a median lobe through which silk for spinning cocoons issues through a *spinneret.* The typical parts are often reduced in size and simplified, the labial palps, especially, often wanting.

II. *Dipterous Larvae.*—In the lower (Orthorrhaphous, see page 234) Diptera the mouth parts are essentially of the chewing type, but in the higher (Cyclorrhaphous, see page 235) Diptera, such as the house fly and flower fly maggots

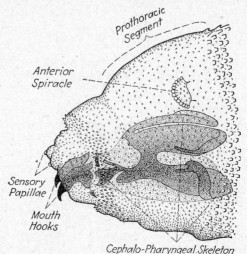

F IG. 77.—Degenerate head and mouth parts of a dipterous larva showing the mouth hooks, the anterior or prothoracic spiracles, and, inside the spiny body wall, the cephalo-pharyngeal skeleton or hypopharyngeal sclerite. (*Drawn by Kathryn Sommerman.*)

and many others, the true mouth parts are entirely wanting, the true head segments have been invaginated ("sucked down") into the throat, and special mouth parts, which work vertically instead of laterally, special sense organs, mouth lips, and a mouth cavity or atrium have been developed to serve, instead of the true head parts, during larval life. Parts of the true hypopharynx, the clypeus, and associated sclerites of the head form a conspicuous sclerotized organ surrounding the beginning of the alimentary canal which is known as the *cephalo-pharyngeal skeleton* or *hypopharyngeal sclerite* (Fig. 77).

CHAPTER VI

DEVELOPMENT AND METAMORPHOSIS

That "truth is stranger than fiction" is well illustrated by the life cycles of insects. The life cycle or life history of an insect means the record of all that the insect habitually does and all the changes in form and habits that it undergoes from the beginning of its life until its death, including the situations where each life stage and every season is spent and the length of time occupied by each stage. This is an important and fascinating study. No one could possibly predict the life cycle of even the most humdrum bug with half of its manifold interrelations and complications. Each species must be studied *by observation*, to determine what normally happens in every region where it lives, and *by experiment*, to determine the effect upon it of unusual or varying conditions.

WHERE INSECTS COME FROM, HOW THEY GROW AND DEVELOP AND PROVIDE FOR SUCCEEDING GENERATIONS

Since the death and disintegration of every individual are inevitable, in order to preserve any line of descent or kind of animal or plant from extinction, a certain part of its living stuff must be freed from the individual before death overtakes it; and under such conditions that this bit of living material will not only survive but grow into a whole new individual, capable of repeating the reproductive process when it in turn has reached maturity. Among insects reproduction is accomplished *sexually, i.e.*, by the release of single cells, known as *gametes*. Gametes are of two kinds: *eggs* from females and *sperms* from males. These are the perfect or complete cells from which all other kinds of cells of the entire body may be produced during development.

All insects, like all other animals, begin life from such a single cell known as the egg (Fig. 78). They do not appear spontaneously or spring up out of nothing, as people sometimes suppose, but come from eggs previously hidden about us by insects of the same kind. Only occasionally do swarms of insects invade a locality from some distant point. It is probably safe to say that nine-tenths of the insect troubles of a given farm come from eggs laid on that farm. Each farmer, or certainly each community, raises its own insect pests, with some exceptions. It is part of the function of entomology to teach us to recognize insect outbreaks in their incipiency and the stages of pests which are harmless, as well as those that do damage.

Before development can begin, it is usually necessary that the insect egg be fertilized by union with a sperm from the male insect (see page 137). But many cases have been found among insects in which fertilization is not necessary, the female insect producing living, normal young without the necessity of mating. This is known as *parthenogenesis*. In the honeybee the fertilized eggs produce "workers" or "queens" (*i.e.*, females); the unfertilized eggs invariably produce males or "drones."

133

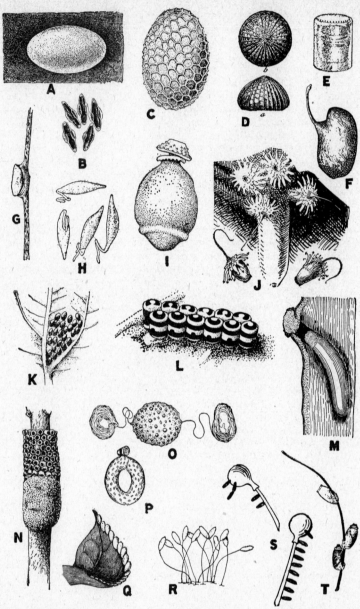

FIG. 78.—Eggs of various insects to show something of the variety in shape, pattern, sculpturing and arrangement. All much enlarged. *A*, egg of Japanese beetle; *B*, a group of eggs of the malarial mosquito; *C*, egg of honeysuckle miner, *Lithocolletes fragilella* (*from Crosby and Leonard*); *D*, egg of the fall armyworm; *a*, side view, *b*, from above; *E*, egg of southern green plant bug, side view; *F*, egg case of great water scavenger beetle which encloses 50 to 100 eggs (*redrawn after Kellogg*); *G*, egg of a ground beetle, *Chlaenius tricolor*, in its mud cell on the stem of a sedge (*redrawn after King in Ann. Ento. Soc. Amer.*); *H*, eggs of the apple seed chalcid; *I*, egg of a stone-fly, *Perla immarginata* (*from Smith in Ann. Ento. Soc. Amer.*); *J*, eggs of poultry lice (*from Ohio Agr. Exp. Sta.*); *K*, egg mass of the squash

In aphids all summer generations are exclusively females developed from unfertilized eggs, males appearing only in the fall and fertilizing only the overwintering eggs. In a considerable number of species, no males have been found, or in certain generations no males are produced. For example, 98 successive generations of aphids have been produced under observation without a single fertilization. Other things being equal, a parthenogenetic species is likely to be a worse pest than one in which mating is necessary; because the hazard of not finding a mate is removed and successful reproduction is that much more sure. In spite of these exceptions, the normal thing among insects is for fertilization to occur.

Eggs and sperms are living cells highly specialized for the particular function of generating complete new individuals, just as nerve cells are highly specialized for sensation and conduction, or muscle cells for contractility. In order that we may understand the specialization that these particular cells undergo, and their subsequent history, we should examine the essential parts of the insect egg and sperm.

The Egg.—The eggs of insects (Fig. 78) are not so varied in size, shape, and appearance as the insects that lay them. They are very small; the characters by which they may be distinguished are often most obscure and elusive; and very little time has yet been given by entomologists to the study of insect eggs. Nevertheless, it is often possible to tell from an examination of the egg the exact kind of insect that will develop from it. This may at times be of the greatest importance in forecasting the appearance of the destructive stages of a particular insect pest. For example, grasshopper or fruit tree leaf roller epidemics may be predicted 6 months before any damage will begin, and materials and an organization to combat them may be perfected long before the voracious insects hatch.

In attempting to recognize the kind of insect from its eggs, one should note the size, shape, and color of the egg; the place in which it is found; the way in which it is laid or attached—whether on its end (Fig. 78,S), its side (K), or on an elevating stem (R); or, if inserted into the tissues of plants or animals, the kind of scar made by the laying (Fig. 5); whether laid singly, in indefinite masses, or in accurately spaced or definitely ranked groups (Fig. 78,LNQ); and the arrangement of the eggs with respect to each other, whether free, cemented together, or covered over with hairs or a cement-like secretion. Of special usefulness is the sculpturing of the eggshell as seen under a microscope. A great variety of impressions, elevations, or depressions are found, the exact shape and arrangement of which will often serve to distinguish one species from another (Fig. 78,C,D,E,H,I,J,O).

Parts of the Insect Egg.—Regardless of shape or size, the following parts of the egg will usually all be represented (Fig. 79):

---

bug (*original*); L, egg mass of the harlequin bug; M, egg of snowy tree cricket (*from Parrott and Fulton*); N, eggs of tent caterpillar forming a collar about a twig. In the upper portion of the mass the eggs have not been covered with the glue-like secretion (*original*); O, egg of a May-fly, *Heptagenia interpunctata*, showing skein of thread at each end which anchors egg on surface of water by winding about sticks or plants (*from Morgan in Ann. Ento. Soc. Amer.*); P, egg mass of a caddice-fly, *Phryganea interrupta* (*from Lloyd*); Q, egg mass of angular-winged katydid on edge of leaf (*from Comstock*); R, eggs of a lace-winged fly (*original*); S, eggs of asparagus beetle; T, eggs of little red louse on hair of cow (A, B, D, E, H, L, S, and T from *U.S.D.A.*).

a. The *chorion* or eggshell, a tough, nonchitinous, protective covering, secreted by the cells lining the ovarian follicles (see page 104).

b. The *micropyle,* a small opening through the chorion usually at one end of the egg, through which the sperm may enter the egg to fertilize it.

c. The *vitelline membrane,* or cell wall of the egg, a delicate membrane completely lining the shell within, and enclosing the following parts:

d. The *cytoplasm,* or general living substance of the egg, the clear, watery, "cell sap."

e. The *yolk,* deutoplasm, or lifeless food material, which is not usually assembled into one mass as in bird eggs but is scattered throughout the cytoplasm.

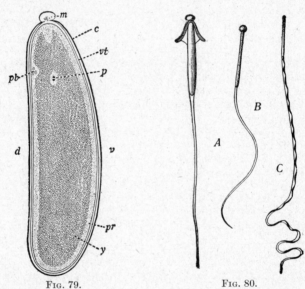

FIG. 79.   FIG. 80.

FIG. 79.—Sagittal section of an egg of the house fly in process of being fertilized. *c,* chorion; *d,* dorsal side; *m,* micropyle, with exudation; *p,* nuclei from sperm and egg about to unite; *pb,* polar bodies; *pr,* peripheral protoplasm; *v,* ventral side; *vt,* vitelline membrane; *y,* yolk. Greatly magnified. (*From Folsom's "Entomology," after Henking and Blochmann, Blakiston.*)

FIG. 80.—Sperms of insects. *A,* of grasshopper; *B,* of cockroach; *C,* of a scarabaeid beetle, greatly magnified. (*From Folsom's "Entomology," after Bütschli and Ballowitz, Blakiston.*)

f. Finally, and most significant of all, the *nucleus,* a highly organized dynamic part of the cell, containing, besides other significant parts, the *chromatin,* which at certain regular times forms into the minute bodies known as *chromosomes.* The chromosomes are composed of a great number of minute granules called *genes,* which (perhaps in conjunction with other portions of the germ cells in which they are borne) are the bearers of hereditary characters and in some unknown manner determine that the insect developing from an egg laid by a bumblebee, for example, shall be a bumblebee and not a grasshopper or a house fly; in general, a creature like its parents, with the peculiar, fundamental, structural characters of the species to which it belongs.

The Sperm.—The sperms of insects are in a general way similar to those of other animals. They are elongate, extremely slender cells, with a whip-like, vibratile tail, by which they may swim actively to find an egg. When examined at high magnifications, three different parts may be distinguished (Fig. 80): (*a*) a slender, rod-like *head* which contains the nucleus and carries the chromosomes and is believed to be the part that bears the hereditary characters of the male parent to the egg; (*b*) a *middle piece* which is thought to contain an "attraction sphere," of significance in the division and development of the egg after fertilization; and (*c*) the *tail*. Such a curious cell is developed from an ordinary-appearing cell in the testis, during the process known as maturation.

Maturation and Fertilization.—The nucleus of the egg and of the sperm is the portion that contains "the germ of life." Every other part of the egg is subservient to the nucleus, serving to protect or nourish this vital part. Before fertilization the nucleus of the egg and that of the sperm each undergoes certain complicated changes known as *maturation*, during which a part of their substance, especially chromatin, is cast aside. In the maturation of the sperm cell, most of the cytoplasm is discarded and the cell specialized for locomotion; in the development of the egg cell most of the cytoplasm is retained and the cell specialized by the inclusion of yolk to nourish the embryo. In both cases only half of the chromatin substance of a normal body cell is retained.

At fertilization the sperm burrows into the egg cell (through the *micropyle*), at least the head and middle piece entering its substance to perform essential functions in the construction of the *fusion nucleus*, from which the entire new insect develops. It is the fusion of these two "half cells," the maturated sperm from the male and the maturated egg from the female, that constitutes *fertilization*. All cells of the new body are direct descendants of the "perfect cell" so formed.

Development.—The life cycle *begins* with fertilization, the fusion of the sperm and egg into a single cell. The life cycle *ends* with a body composed of millions of cells, highly organized into a complex, living machine. All that takes place between the fertilization of the egg and the perfection of the full-grown insect, we call *development* and *growth*. It is sharply divided into two phases by the act of hatching or escape from the eggshell. That part of the development that occurs before hatching, or birth, is called the *embryology* (embryonic development), and all that takes place after hatching or birth is *postembryonic development*.

The embryology of insects is a subject too technical to attempt to cover here, though one of fascinating interest. We would simply emphasize that life begins as a single cell, the fertilized egg, in which cannot be recognized any of the features of the creature into which it later develops. This cell, under the influence of a force we do not understand, divides, and the succeeding cells divide and further multiply and differentiate. The cells formed by division of the fertilized egg migrate to the periphery of the egg, just beneath the vitelline membrane, and form a complete layer of cells known as the *blastoderm*. Along the ventral side of the egg some of the blastoderm cells become thicker to form the *germ band* or *ventral plate*. Lengthwise of the germ band an invagination or groove forms that carries some of the blastoderm cells to an internal position. These form the *mesoderm*, while those remaining in a superficial position are known as the *ectoderm*. Folds of the blastoderm now grow over the germ band from each side to form a double protective covering, the *embryonic membranes*. When these folds meet and fuse, the outer layer of the folds forms the *serosa* and the inner layer forms the *amnion*. Thus the developing insect becomes covered or separated from the vitelline membrane by two other cell layers.

In this sheltered position, and drawing constantly upon the yolk for substance and energy, the following developments take place rapidly and synchronously. Transverse furrows of the germ band divide it into a series of segments. Invaginations or ingrowths from the ectoderm form the fore- and hind-intestine, while internally the mesoderm develops the tube known as the *mid-intestine* to connect the other two. Outgrowths from the ectoderm form the paired appendages, such as antennae, legs, and mouth parts. The germ band grows up around the sides, gradually eliminating the blastoderm until the edges of the germ band meet and unite, thus completely enclosing the yolk and giving cylindrical form to the body. From the mesoderm develop the muscles, the heart, and the mid-intestine. From the ectoderm the body wall is formed, and all the appendages are evaginated; and the central nervous system, the tracheal tubes, and the salivary glands are formed by invaginations of the ectodermal layer. As the cells increase in numbers, all take their place in an orderly and definite manner to form the various tissues and organs of the future insect; until, when it hatches, we have within the eggshell (Fig. 82,*A*,*F*,*K*) an organism capable of discharging all the necessary functions of life. During this time the egg is quiescent. The insect does not feed externally or move during the egg stage. In the egg stage, therefore, insects are never injurious and never beneficial.

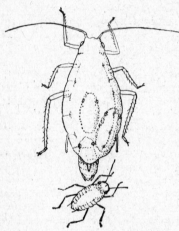

Fig. 81.—A wingless parthenogenetic green bug, giving birth to living young, ovoviviparously. (*Drawn from photographs by Hunter and Glenn, Univ. Kan.*)

Types of Reproduction.—The development within the egg requires food. This may be derived from food material, *yolk, stored inside the shell* by the parent insect before the egg is laid. The parent then nourishes it no further. This method of reproduction is known as *oviparous, i.e.*, bringing forth eggs. It is paralleled by the condition in birds, although it should be noted that insect eggs do not require incubation or other attention from the parents after they are laid.

In contrast with this is the condition in mammals where the young, during embryonic development, establish a definite connection (the *placenta*) with the blood system of the mother and receive their nourishment, moment by moment, from the circulatory system of the parent. Food as well as oxygen for an embryonic mammal passes from mother to young by diffusion or osmosis through the so-called fetal membranes. This condition is known as *viviparous* reproduction, *i.e.*, bringing forth active young.

A somewhat intermediate condition is found among insects, some of which do not lay the eggs but retain them until after they have hatched and then bring forth active young (Fig. 81). This is not at all equivalent to viviparous reproduction, however, because the young insect receives its nourishment from the yolk of the egg and not from the parent's circulation. No organic connection is established. It is simply premature hatching or delayed oviposition. It is distinguished as *ovoviviparous reproduction*. The two kinds of reproduction common among insects are oviparous and ovoviviparous. A condition somewhat analogous to that found in man and other mammals (viviparous reproduction) is known only in the case of a few flies (the sheep tick, see page 853, Fig. 560, and the tsetse flies) in which the young are actually nourished by special nutritive glands in the uterus of the parent fly.

Number of Eggs and Methods of Deposition.—Insect eggs (Fig. 78) are generally small and consequently are seldom noticed except when laid in masses or groups that are conspicuous. Some insect eggs are so small that several dozen could be placed side by side on the head of a common pin; the largest eggs of our common insects are not over $\frac{1}{8}$ inch in diameter.

The number of eggs laid by one insect is as varied as their shape and size. A single female may lay as few as 1 egg (in exceptional cases, like the true females of certain aphids), or at the other extreme 1,000,000 or more. The honeybee queen lays 2,000 or 3,000 eggs a day, actually producing several times her own weight of eggs each day for weeks at a time. The termite queen lays as many as 60 eggs a minute until millions have been produced. The average number for all insects is probably over 100.

The eggs may all be laid at one time, as in the tussock moth; they may be laid a few a day for many days, as in the bloodsucking lice; or there may be a number of successive "batches" of eggs produced at intervals, as in the case of the common house fly, which lays from two to seven lots of eggs at intervals of 2 to 5 days, each lot consisting of about 125 eggs.

Insect eggs are generally laid in such a situation that the young, upon hatching, may find suitable food with the minimum of effort or discriminative action. More often than not, they are simply extruded in a suitable place, which is generally selected with remarkable care. This done, the mother pays no further attention to them, usually dying shortly afterward. There is ordinarily very little of parental care or family life among insects. The eggs hatch (Fig. 82) without attention or control by their parents, and from the moment of hatching the young insect must ordinarily lead an independent self-supporting existence.

There are some very interesting adaptations of egg laying to the subsequent life of the young insect. The chestnut weevil has a beak longer than her own body, with which she reaches through the chestnut bur to chew a hole into the developing nut. In this hole is laid the egg from which the chestnut "worm" develops. The plum curculio and the strawberry weevil make remarkable provision to assure the success of the eggs and young (Fig. 5,*E,F*; see the discussion of egg laying under these insects). The horse botfly, whose young must reach the stomach of the horse in order to develop, lays its eggs, not in the mouth of the horse, but on the hairs of the legs of the horse. Their success in reaching their feeding grounds is thought to depend upon the activities of another insect, the stable fly. The bites of the stable fly cause the horse to nibble at its legs. The eggs of the horse bot hatch almost instantly when stroked by the moist lips or tongue, to which the small larvae cling, and are subsequently swallowed.

One of the most remarkable cases of provision for the young is exhibited by the twig girdler. The twig girdler lays her eggs one in a place in holes which she chews into the soft bark of the terminal twigs of such trees as hickory, oak, pecan, persimmon,

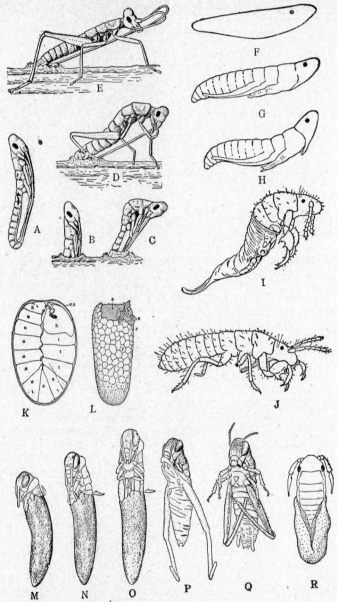

Fig. 82.—The hatching of insects. *A, B, C, D,* and *E* are of a tree cricket: *A* shows the position of the embryo in the egg; *B, C, D,* and *E* successive stages in the hatching of the nymph from an egg sunken in the wood (*from Parrott and Fulton*). *F, G, H, I,* and *J* are the periodical cicada: *F,* the egg with the eye of the embryo showing through the chorion; *G,* the newly hatched nymph; *H,* the same in motion; *I,* the same, shedding embryonic membrane; *J,* the same free from embryonic membrane (*from Snodgrass*). *K* shows the embryo peach borer larva in its U-shaped position in the egg before hatching, and *L* the empty eggshell from which the larva has hatched (*from Peterson*). *M, N, O, P,* and *Q* show successive stages in the hatching of a grasshopper (*from S. D. Agr. Exp. Sta.*). *R* is of an apple aphid showing the nymph partly hatched from the egg. (*From Peterson.*)

and many others. Before the eggs are laid, however, she completely girdles the twig, usually to a depth of ⅛ inch or more, by chewing out the wood, bit by bit, in a band around the twig (Fig. 5,*D*). This girdling requires many hours of work on the part of the female (commonly 40 or 50 hours), and, when it is completed, the insect lays in the partially severed twig perhaps 12 or 20 eggs. Other twigs are then attacked in a similar manner, and the female busies herself in this way all during the long autumn months. The twigs subsequently break off in the wind and the larvae develop in the dead, decaying wood at the surface of the soil. This represents probably the extreme of parental care on the part of an insect, in the mere act of egg laying. A closely related species in the subtropics often cuts off branches 1 or 1½ inches in diameter, several females working together to sever so large a branch.

Many other insects, especially the Hymenoptera, build elaborate nests and provision them with paralyzed insects or honey to serve as food for the young (see Figs. 22 and 159). The European earwig actually broods over her eggs and young after the manner of birds (Fig. 109). But the great majority of insects lay their eggs and die without ever seeing them again. There is in almost every case, a new "crop" of insects each year. It is the exceptional species, like the white grubs, wireworms, periodical cicada, ants, and honeybee, in which the same individuals live longer than one year; even in these cases the adult is generally short-lived, and only the young or the queens persist into the following seasons.

The Rapidity of Insect Increase.—Insects multiply very rapidly. This great increase may be due to either one or both of the following factors: (*a*) a great number of eggs or young in a family or generation and (*b*) a short life cycle and the rapidity with which generations succeed each other. Compared with the half dozen or dozen children that characterize the families of man and our domestic animals and fowls, the hundreds of the average insect family are impressive. Again, the life cycle or generation in the larger animals is from a few months in the case of smaller rodents to as much as 30 years in the case of man, while the shortest known life cycle among insects is about 10 days!

Either one of these factors operating independently may result in a tremendous population of insects from one or a few individuals that may begin the season. For example, the corn root aphid has a family averaging from 12 to 16 young. But each of these begins reproducing at the age of 8 days, and a generation may be completed in 16 days. In this way it would be possible, theoretically, for a single female to produce in 1 year, if all her descendents survived, a chain of these aphids long enough to encircle the earth. The San Jose scale has fewer generations—from two to four in the northern states—yet, because of the large number of young produced by one female (400 to 500), a single pair might be the progenitors, if all their descendants survived, of more than 1,000,000 in a single season. In the case of the house fly both of the above factors operate to produce the alarming increase in numbers of flies as summer comes on. According to Hodge: "A pair of flies beginning operations in April, might be progenitors, if all were to live, of 191,010,000,000,-000,000,000 flies by August. Allowing ⅛ cubic inch to a fly, this number would cover the earth 47 feet deep." Needless to say, owing to the factors of natural control, no such rate of increase ever actually occurs.

The first life stage of all insects is the egg (sometimes concealed within the mother's body). The time spent within the egg may be as short as 8 hours, as in the case of the house fly; is commonly a week or two; and very often insects go through the winter in this stage, all other stages then usually dying off before winter is passed.

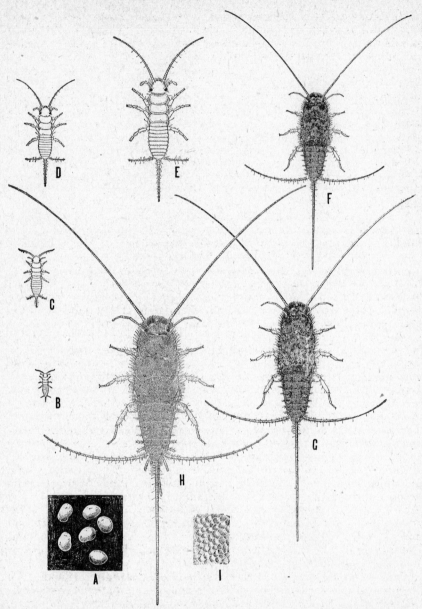

Fig. 83.—Development of the silverfish, *Thermobia domestica* Packard, illustrating development without metamorphosis. There are more instars than illustrated, the intervening ones being of intermediate sizes but otherwise very similar to those shown. The covering of scales first appears at the third or fourth instar. *A*, a group of eggs; *B, C, D, E, F, G*, nymphs or young of various sizes; *H*, adult female, as recognized by the styli and ovipositor at tip of abdomen; *I*, sculpturing of the egg-shell, as faintly visible under the microscope. (*Original, drawings by Ruth Slabaugh and Kathryn Sommerman.*)

Insects without a Metamorphosis.—The egg may lie dormant for a long period of time, but during at least a part of the egg stage there is great activity within the eggshell, as a result of which a perfectly formed young insect is finally disclosed by hatching[1] (Fig. 82). When a chicken or a duckling hatches from its egg, it resembles in most respects (except size) the full-grown chicken or duck. A few insects, when they hatch, also are so much like their parents that anyone would know they belonged to the same kind of insect (Fig. 83). This is especially true of a small number of wingless insects belonging to the orders Thysanura and Collembola, the fishmoths and springtails. Such insects are said to undergo *no metamorphosis*, and the two orders just mentioned are collectively called the *Ametabola* (which means *without change*). Their growth from smallest to largest is hardly accompanied by greater changes in appearance than those that take place from infancy to manhood (see Table V, page 155).

Insects with a Gradual or Simple Metamorphosis.—If the full-grown insect has wings, the young never resemble it completely at hatching, for (unlike birds) *no insect has visible wings when it emerges from the eggshell* (see Fig. 82). All winged insects, therefore, undergo a metamorphosis during their development. *Metamorphosis* may be defined as *a conspicuous change in the form and appearance of an animal between birth (or hatching) and maturity*. Frogs or toads, with their curious young tadpole stages, have a metamorphosis, while birds, rabbits, men, and the like, have no metamorphosis during their postembryonic development.

In many insect species the young are very similar to the adult except for the complete absence of wings and genitalia; in other cases there are striking differences in the color or shape, or in the structure of some of the appendages. In either case, after a period of growth the wings may appear attached to the outside of the body as small *wing pads* which become larger and larger. The more developed the young insect becomes, the more it resembles its parents (Fig. 85). Such a development is called a *gradual* or *simple metamorphosis*. The young of such insects are called *nymphs* (Fig. 84). They commonly have the same habits as their parents, and the old ones and young ones may frequently be seen feeding together, not unlike a hen and her chicks. Grasshopper nymphs and adults both eat grasses and clovers and may be found hopping about together in the pastures. Squash bug nymphs and adults both suck the sap of the squash plant. Bedbug nymphs and adults all suck human blood. This group as a whole is known as the *Heterometabola* (meaning *different change*). It includes many important insects of the orders Orthoptera, Isoptera, Mallophaga, Thysanoptera, Homoptera, Hemiptera, Anoplura, and others (see Table V, page 172).

Insects with a Complete or Complex Metamorphosis.—Finally, we have a large group of insects, most of which have different habits when they are young than when they are full grown. For example, the young may swim in the water and the adults live in the air, like the mosquitoes. Or the young may tunnel through the soil and eat grass roots like the white grubs, while their parents, the May beetles, fly about and feed on the leaves of trees. Or the young may live in the stomach of a horse like

---

[1] The term *hatching* or *eclosion* properly refers only to breaking out of the eggshell or embryonic membranes. The escape of a winged insect from its cocoon or pupal case should be called *emergence*, not hatching.

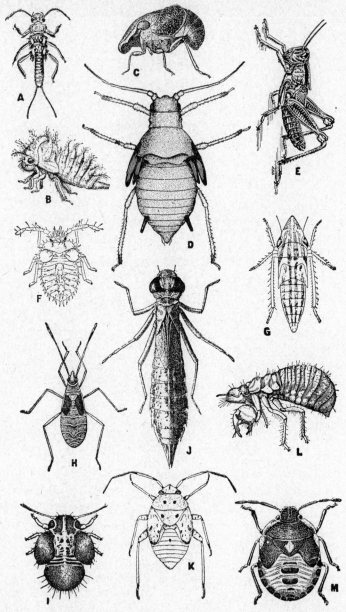

Fig. 84.—*Nymphs:* examples of insects that develop the wings externally during the growing period and transform to the adult usually without a pupal stage. *A,* nymph of a stone-fly, order Plecoptera (*from Kellogg's " American Insects,"*); *B,* nymph of a treehopper, *Ceresa basalis* Walk., order Homoptera (*from N.Y. State College Forestry*); *C,* nymph of a fulgorid, *Bruchomorpha oculata* Newm., order Homoptera (*from N.Y. State College Forestry*); *D,* nymph of an aphid, *Aphis cucumeris,* order Homoptera (*from U.S.D.A.*); *E,* nymph of a grasshopper, order Orthoptera (*from U.S.D.A.*); *F,* nymph of a lace bug, *Corythuca pergandei* Heid., order Hemiptera (*from Ohio Biol. Sur. Bul.* 8); *G.,* nymph of a leafhopper,

the bots and the parents fly freely in the air. Obviously these insects could not exist in such different environments unless they were very different in structure when young and when full grown. Indeed, they are generally so very different in appearance in their several life stages that no one, without previous information, would ever suspect them of being even closely related animals, much less the successive stages of *the same individual*. The young of this group show no trace of wings externally during any period of growth, although the wing buds may be found, by dissection, inside the body wall (see Figs. 56, *f.b* and *h.b*, and 86, *II* and *III*). Furthermore, the oldest larva shows no greater resemblance to the adult than the smallest one, except in size (Fig. 87). Such young insects are known as *larvae* (singular *larva*), in contrast with nymphs (Fig. 88).[1]

When a larva is full grown, a striking change takes place, and the insect, after shedding its skin, appears with the wings exserted as large wing pads and with the usually long legs and feelers of the adult now recognizable. But legs, wings, and antennae always remain functionless for a definite period of time while the internal organs are being transformed to the adult condition. This transformation stage is known as the *pupa* (plural, *pupae*) (Fig. 89). At its completion, the pupal skin is shed and the adult formed by rapid expansion of the wings to full size, the general hardening of the body wall, the development of the color pattern, and numerous other changes. Such development is called a *complete* or *complex metamorphosis*. The largest orders of insects have such a complete metamorphosis, *e.g.*, the Coleoptera, Lepidoptera, Hymenoptera, and Diptera, besides some smaller orders like the Siphonaptera, Neuroptera, and others (see Table VIII, page 172). All these insects that have a complete metamorphosis are referred to collectively as *Holometabola* (meaning *complete change*).

Life Stages and Instars.—For a further understanding of the growth and metamorphosis of insects, it is essential to have clearly in mind what is meant by *life stages* and by *instars*. The life stages are those several periods of an insect's life which are radically different from each other in appearance and usually also in behavior or activity. Thus the insects with a complete metamorphosis have *four life stages;* the *egg*, the *larva*, the *pupa*, and the *adult*. While insects with a gradual metamorphosis and those without metamorphosis have *three life stages*, known as the *egg*, the *nymph*, and the *adult*. *Among insects all increase in size takes*

Draeculacephala mollipes Say, order Homoptera (*from U.S.D.A.*); *H*, nymph of the cotton stainer, *Dysdercus suturellus* H. Schf., order Hemiptera (*from Insect Life*); *I*, nymph of the pear psylla, *Psyllia pyricola* Linné, order Homoptera (*from Slingerland*); *J*, nymph of a dragonfly, *Anax junius*, order Odonata (*from Ill. State Natural History Surv.*); *K*, nymph of the tarnished plant bug, *Lygus pratensis* Linné, order Hemiptera (*from Ill. State Natural History Surv.*); *L*, nymph of the periodical cicada, order Homoptera (*from U.S.D.A.*); *M*, nymph of the green plant bug, *Nezara hilaris* Say, order Hemiptera (*from U.S.D.A.*).

[1] The authors realize that the term larva is often used for the first stage of any insect or other animal having a metamorphosis and wish that a special term could have been employed for the young of insects with a complete metamorphosis. However, the term larva is so firmly established as a general term for caterpillars, grubs, maggots, and the like that there would be no hope of any special term for this purpose becoming adopted. *Larva* as used in this book, therefore, means the young of insects that have a complete or complex metamorphosis.

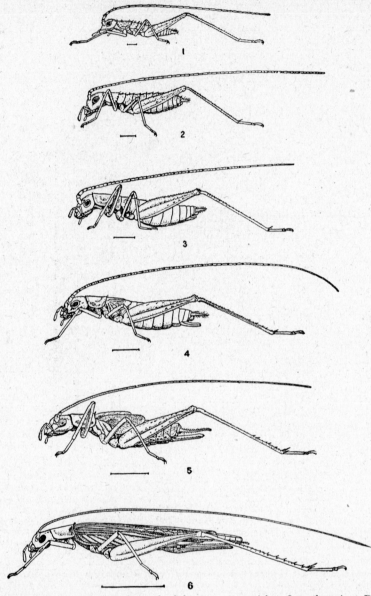

Fig. 85.—Instars or stages of growth of the snowy tree cricket, *Oecanthus niveus* De G., illustrating a gradual metamorphosis.   *1* to *5* the first to fifth nymphal instars, respectively; *6*, the adult.   Note the appearance of the wing pads at *4* and their expansion to full size at *6*; also the very gradual assumption of the adult condition.   (*From N. Y. (Geneva) Agr. Exp. Sta. Bul.* 388.)

place in the life stage that immediately follows hatching, i.e., either as a nymph (if the insect has a gradual metamorphosis) or as a larva (if the insect has a complete metamorphosis). *No growth occurs in the adult stage after the insect once acquires functional wings and none in the pupal stage.* Little flies do not ever grow into large flies or little moths into larger ones. With the appearance of the full-spread wings the size of the insect is fixed for the rest of its life except as the body expands to accommodate a large meal or developing eggs. The enormous increase in size of the insect from birth to maturity is astonishing. In man, the full growth of the individual rarely increases the weight at birth more than twenty or twenty-five times. Among insects, the full-grown individual may weigh from 1,000 to 70,000 times its weight at hatching.

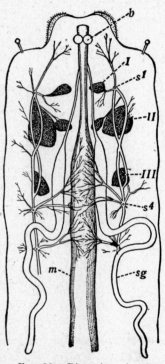

FIG. 86.—Dissection of a full-grown caterpillar, *Pieris* sp., from above, to show the wing buds which are developing inside the body wall. *b*, brain; *m*, alimentary canal; *s1*, prothoracic spiracle; *s4*, first abdominal spiracle; *sg*, silk gland; *I*, bud of the prothoracic segment; *II*, bud of the front wing; *III*, bud of the hind wing. (*From Folsom's "Entomology,"* Blakiston, *after Gonin.*)

Since growth occurs exclusively during the nymphal or larval period, it follows that there must be various sizes of nymphs or larvae (Figs. 85 and 87) in the case of every species and in the development of every individual. An insect does not grow by regular, gradual, imperceptible degrees like a child. Its body wall will not expand like mammal skin to permit this. It has been pointed out (Fig. 43) that the body wall is composed of a double outer layer, the cuticula (itself consisting of an epicuticula, an exocuticula, and an endocuticula); and beneath the cuticula a layer of living sensitive cells known as the hypodermis and that the cuticula becomes sclerotized and inelastic over most of the outside of the body to form a stiff, hard, external armor for the insect.

It follows, therefore, that growth inside this inexpansible shell cannot be regular and continuous. In order to make any considerable increase in size, the shell must be split off. This process is known as *molting* and the old cuticula so cast off is known as the *exuviae* (meaning *clothes*).

Before the old skin or cuticula is split off, a new epicuticula and exocuticula are secreted inside of it by the hypodermal cells. Then a fluid, known as the *molting fluid*, is poured out by the cells in the hypodermis or by special exuvial glands. The molting fluid loosens the old cuticula from the new cuticula by dissolving the inner part of the old endocuticula. The molting fluid also softens the remainder of the old skin and separates it from the new one by a thin film of fluid. The old skin is then split open by pressure from within and the insect squirms and crawls out of it, pulling its appendages free. At this point there is a considerable expansion in the size of the insect, before its new exocuticula becomes sclerotized or "set" to the definite size of the next instar. The molting process is completed by the formation of a new endocuticula beneath the hardened exocuticula. Subsequently there is a relatively long period during which the insect is feeding and accumulating reserve materials within its body, but without any noticeable increase in size. This is followed by another molt and period of constancy in size, and so on.

The molts occurring during the growing period divide this life stage (nymph or larva as the case may be) into a number of sharply separated sizes or steps that are called *instars* (Figs. 85 and 87). Upon hatching from the egg, the insect is said to be in the first instar. This instar is

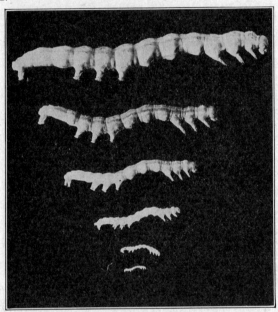

Fig. 87.—Instars or stages of growth of the green clover worm, *Plathypena scabra* Fab., illustrating a complete metamorphosis. Note the entire absence of wing pads even in the largest instar, the fact that the largest larva is no more like the adult moth in form than the smallest, and that a pupal stage intervenes between the last larva and the adult. (*From U.S.D.A., Farmers' Bul. 982.*)

terminated by the first molt, which ushers in the second instar, distinctly marked off from the first at least by its larger size, and often also by differences in structure or color. The second molt introduces the third instar and so on, until commonly 3, 4, 5, or 6, and sometimes as many as 20 molts, have occurred (Table V).

When growth (increase in size) is completed, if the insect is a nymph, a final molt discloses the adult; if it is a larva, the corresponding molt gives rise to the pupal stage; and, when the transformations of this stage are perfected, a final molting of the pupal epidermis discloses the adult.

Any insect during its development passes through either three or four *life stages*, one, and only one, of which is always made up of a number of *instars*. Any considerable difference in size without much change in appearance indicates a different *instar;* while a radical change in structure

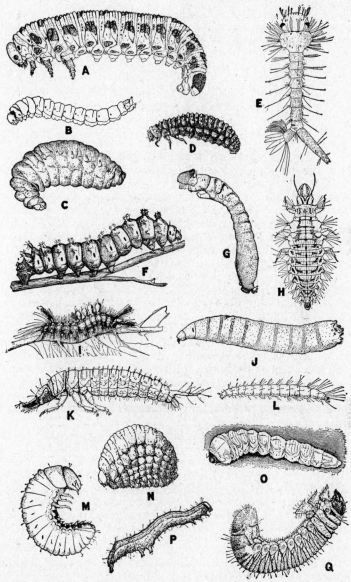

Fig. 88.—*Larvae:* insects that develop the wings internally during the growing period, that are very different in appearance from the adults, and that have a pupal stage intervening between larva and adult. *A,* larva of a sawfly, *Neodiprion lecontei* (Fitch), order Hymenoptera (*from Middleton in Jour. Agr. Res.*); *B,* larva of the wheat stem sawfly, *Cephus pygmaeus* (Linné), order Hymenoptera (*from Ries in Jour. Agr. Res.*); *C,* larva of a black digger wasp, *Tiphia* sp., order Hymenoptera (*from Davis, in Bul. Ill. Natural History Surv.*); *D,* larva of the beet leaf beetle, *Monoxia puncticollis* Say, order Coleoptera (*from U.S.D.A.*); *E,* larva of a mosquito, *Culex territans* Walker, order Diptera (*from Bul. Ill. State Lab. Natural History*); *F,* larva of the cecropia moth, *Samia cecropia* (Linné), order Lepidoptera (*from Saunders*); *G,* larva of a black fly, *Simulium venustum* Say, order

and appearance, without much change in weight, indicates a different *life stage*.

Sometimes it is difficult to decide whether the metamorphosis of a given insect should be called complete or gradual. We must bear in mind that nature does not make sharp division lines; that there is likely to be every conceivable gradation from one condition to another which is remarkably different. It is so with the metamorphosis of insects. The groups described above, however, are very important for convenience of study and reference, and most insects can easily be fitted into one or another of them.

Nymphs vs. Larvae.—The best criterion to divide the winged insects into two groups, in respect to metamorphosis, is (*a*) whether the wing pads are borne externally during the growing stage or concealed beneath the body wall. If the wing pads are developed on the outside of the body wall (see Figs. 84 and 85), we call the growing stage a *nymph*, and we say that insect has a *simple* or *gradual metamorphosis*. If the wing pads are developed internally during the growing stage (Figs. 56,*f.b.* and *h.b*, and 86,*II* and *III*), we call that stage a *larva* (see Figs. 87 and 88) and say that insect has a *complete* or *complex metamorphosis*. A nymph, then, is the growing stage of such insects as have a gradual or simple metamorphosis, and it develops its wings (if it has any) on the outside of the body wall as visible pads. Other general differences between nymphs and larvae are that (*b*) the nymph generally has a shape and body construction similar to that of the adult; (*c*) each successive instar usually looks more like the adult than the one that preceded it; (*d*) nymphs have very few organs that are not also possessed by the adult; (*e*) a nymph has compound eyes unless its parents are without compound eyes; (*f*) it always has the same type of mouth parts as the adult; (*g*) it generally occupies the same kind of habitat, takes the same kind of food, and leads the same manner of life as the adult; and, finally, (*h*) the nymphal period generally passes over into the adult period without any prolonged inactive or pupal stage intervening.

In contrast with the nymph, the larva, or growing stage of insects with complete or complex metamorphosis, (*a*) develops its rudimentary wings during this stage inside the body wall of the thorax; (*b*) it generally has a more or less worm-like form of body, often strikingly different from that of the adult; (*c*) the later instars are no more like the adult, as a rule, than the earlier ones; (*d*) the larva often has provisional structures or organs, of use only in this stage and which are lost or supplanted before the adult stage is reached; (*e*) the larva never has functional compound eyes, though

---

Diptera (*from H. Garman*); *H*, larva of a lace-winged fly, *Chrysopa quadripunctata* Burm., order Neuroptera (*from R. C. Smith*); *I*, larva of the California tussock moth, *Hemerocampa vetusta* Boisd., order Lepidoptera (*from Volck, Calif. Agr. Exp. Sta.*); *J*. larva of the apple maggot, *Rhagoletis pomonella* Walsh, order Diptera (*from Pa. State Dept. Agr.*); *K*, larva of a ground beetle, *Harpalus pennsylvanicus* Dej., order Coleoptera (*from Davis, Ill. State Natural History Surv.*); *L*, larva of a flea, order Siphonaptera (*from Bishopp, U.S.D.A.*); *M*, larva of *Colaspis brunnea* (Fab.), order Coleoptera (*from Ill. State Natural History Surv.*); *N*, larva of the granary weevil, *Sitophilus granarius* Linné, order Coleoptera (*from U.S.D.A.*); *O*, larva of the giant root borer, *Prionus laticollis*, order Coleoptera (*from N. J. Agr. Exp. Sta.*); *P*, larva of the alfalfa looper, *Autographa gamma californica* (Speyer), order Lepidoptera: note reduced number of prolegs (*from Hyslop, U.S.D.A.*); *Q*, larva of a scarab beetle, *Adoretus caliginosus* (O.S.), order Coleoptera (*from T. B. Fletcher*).

it may have simple eyes or ocelli; (*f*) it may occupy the same habitat as the adult but very often lives in a totally different sort of situation; (*g*) the larva commonly has a different type of mouth parts than the adult and often takes a wholly different kind of food; and, finally, (*h*) the larva

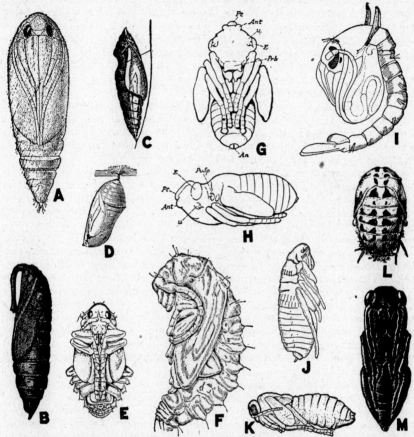

Fig. 89.—Pupae of insects: *A*, *B*, *C*, and *D* of Lepidoptera; *E*, *F*, and *L* of Coleoptera *G*, *H*, and *I* of Diptera; *J* of Siphonaptera; *K* and *M* of Hymenoptera. *A*, pupa of the pink bollworm, ventral view (*from Heinrich in Jour. Agr. Res.*); *B*, pupa of southern tobacco worm, side view (*from U.S.D.A.*); *C*, naked pupa or chrysalis of the alfalfa caterpillar: note how it is suspended by a thread or girdle of silk (*from U.S.D.A.*); *D*, pupa or chrysalis of the monarch butterfly as it hangs suspended by the posterior end (*from French*); *E*, pupa of the beet leaf beetle, ventral view (*from U.S.D.A.*); *F*, pupa of the cherry leaf beetle, side view (*from U.S.D.A.*); *G*, pupa of the apple maggot, ventral view; *H*, the same in side view (*from Snodgrass, in Jour. Agr. Res.*); *I*, pupa of the house mosquito, a pupa that swims in water (*from U.S.D.A.*); *J*, pupa of the dog flea (*from U.S.D.A.*); *K*, pupa of the pear slug (*from Iowa Agr. Exp. Sta.*); *L*, pupa of the convergent lady beetle; it is fastened by silk to the leaf and can rise up on the rear end when disturbed (*from U.S.D.A.*); *M*, pupa of a hymenopterous parasite, *Pardianlomella ibseni* Gir. (*from U.S.D.A.*).

is always separated from the adult by a pupal stage during which the insect takes no food and is usually quiescent.

The Meaning of a Complete Metamorphosis.—The explanation of a complete metamorphosis is probably that growth is all confined to one life stage (the larva) and reproduction to another (the adult). For the functions of eating, growing, and storing up energy a simple cylindrical body with few appendages is well adapted. Since the parent insect generally places the young in the midst of an abundance of food, highly specialized sense organs are not usually required. But to secure a mate, to locate a suitable place to deposit the eggs, and to care for the necessary dispersal of the species, a highly sensitive, active, complex body is required. Hence we have on the one hand the sluggish, stupid, gluttonous caterpillar of simple structure (Fig. 88) and, on the other, the alert, highly specialized fly, bee, moth, or beetle. So very different have these stages become in many cases that the change from one form to the other is profound. In some insects nearly all the larval tissues disintegrate, and the corresponding adult tissues and organs are built up anew from small groups of cells (histoblasts) that have remained dormant and rudimentary during larval life but are now able to multiply rapidly by utilizing the nutritive products resulting from the histolysis of the larval cells and from the fat-body.

What takes place may be likened to a proposal to convert an automobile into an airplane. It is conceivable that there might be about the same amount of building materials in the two vehicles, but to change one into the form of the other would involve a complete reconstruction of all parts. Certainly, while such reconstruction was taking place, the vehicle could not be used. So it is with the insect. So profound is the change from larva to adult in many cases that the organism can accomplish no other functions while it is going on. Locomotion ceases, feeding is suspended, respiration is reduced, and the insect undergoes a transformation period, externally quiescent but internally probably as active as any period subsequent to embryonic development. All available energy is devoted to the development of the wings, legs, eyes, antennae, mouth parts, and other appendages of the adult and to the maturing of the reproductive system and changes in other internal organs. The insect during this period is known as a *pupa* (Fig. 89).

Because it neither feeds nor moves about, the pupal stage is neither injurious nor beneficial to man.

It appears that larvae represent an earlier ontogenetic period of development than do nymphs. The period in gradual metamorphosis corresponding to the larvae of complete metamorphosis is passed in the embryo before the nymph hatches. The pupal period corresponds more nearly with the nymphal stages, and pupae may, in a way, be thought of as nymphs which have lost their activity. Nymphs accomplish both growth and transformation to the adult. In complete metamorphosis growth has all been relegated to the larval stage and transformation has been greatly telescoped or abbreviated into a single period, the pupa.

Methods of Protection for the Pupa.—The life of the insect during the helplessness of the pupal stage is generally safeguarded by the larva. Sometimes the pupa is found naked and exposed, as with many butterflies, lady beetles, and the like. These (Fig. 89,*A,C,D,L*) usually have the tips of their bodies fastened to a leaf but are not covered in any way. Commonly the larva retreats into a protected situation under overhanging bark; into logs, stones, grass or leaf mold; or into the soil before entering upon this defenseless period of its life (Fig. 90,*H,K*). Often a case is

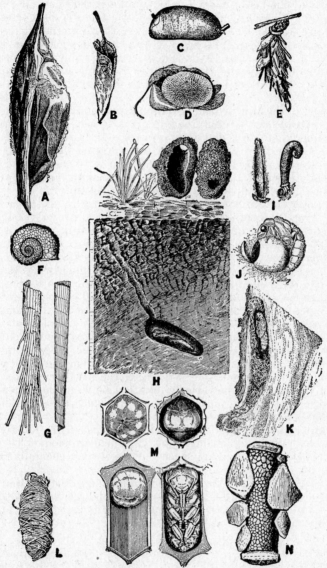

FIG. 90.—Some methods of protection for insect pupae. *A*, cocoon of the cecropia moth (*from Saunders, "Insects Injurious to Fruits,"); B*, folded leaf in which the apple leaf sewer, *Ancylus nebeculana* Clemens, feeds as a larva and which also protects the pupal stage: empty pupal shell from which adult emerged projects at upper left (*from U.S.D.A,*); *C*, puparium of *Tropidia quadrata* Say; a protective case formed from the larval skin (*from Metcalf, Maine Agr. Exp. Sta. Bul. 253*); *D*, cocoon of the clover leaf weevil, partly surrounded by clover leaves (*from Tower and Fenton, U.S.D.A.*); *E*, the bagworm, a case of silk covered with spruce needles carried about by larva during its life and later closed to protect pupa (*redrawn after Riley*); *F*, larval case of a caddice-fly, *Helicopsyche borealis*, formed in the shape of a snail shell and covered with grains of sand (*from Lloyd, "North American Caddice Fly Larvae"*); *G*, larval and pupal cases of a caddice-fly, *Phryganea vestila*, made of slender sticks and bits of leaves arranged in spiral form (*from Lloyd*); *H*, pupa of

formed about the larval body before pupation. The case may consist of a folded leaf (Fig. 90,*B*), fine pebbles (*F*, *N*), fine shavings of wood, bits of soil, hairs from the body of the larva, or other materials that surround the larva as pupation approaches (*E*, *G*, *K*, *L*). These are generally cemented or tied together by a silken secretion from the mouth of the larva; in many moths and Hymenoptera the silk is abundant and forms a complete, sometimes dense, case about the pupa, known as a *cocoon* (Fig. 90,*A*,*D*,*J*).

In many of the flies, instead of spinning a silken cocoon or constructing a case of extraneous material, the larva practices an interesting economy by retaining about itself one of its own cast, dry skins, which is slightly modified by inflation and hardening, to form a waterproof and airtight case known as a *puparium* (Fig. 90,*C*). This next-to-the-last larval skin is not discarded at the time of pupation as in most insects but is retained until the adult breaks out of the pupal skin.

It is important to note that the actual change to the pupa does not coincide with the completion of the active larval life or the formation of the pupal case. The plum curculio buries itself in the soil for about 4 weeks; but 2 weeks of the time it remains as a larva before it pupates. The Hessian fly larva may remain many months after the puparium is formed before the pupa is formed. Many moths and sawflies that winter in a cocoon do so in the larval stage, not pupating until spring. The pupal stage may be said to begin when the larval skin has been molted off, but a great portion of the transformation toward the adult has already taken place before this molt occurs, notably the eversion of the wing pads from their position inside the body wall to a position outside the pupal integument but still enclosed by the unshed larval skin. In the Diptera the formation of the motionless puparium may be called *pupariating*, to distinguish it from the shedding of the final larval skin, or true *pupating*. This period of variable duration between the retreat to the pupal position or formation of the pupal case and the actual change to the pupa is known as the *prepupal period*.

The order to which a given pupa belongs can usually be told by characteristics which are given in connection with the discussion of the orders in Chapter VIII, and the actual species of insect may generally be recognized, by an expert, from the pupa.

The relation of the life stages to each other and to the different instars, their significance to the insect and to man, and the correspondence of life stages in the three types of metamorphosis discussed above are shown graphically in Table V.

---

southern tobaccoworm in its earthen cell formed by the caterpillar. The depth is indicated by the inch marks at the left. Above are shown two earthen cells removed from the ground; note hole by which the larva entered (*from Morgan, U.S.D.A.*); *I*, casebearers; at left the cigar casebearer, at right the pistol casebearer with head and thorax of larva projecting below (*from N.Y. (Geneva) Agr. Exp. Sta.*); *J*, cocoon of the California green lacewing-fly; the pupa has just emerged through the circular hole in the cocoon (*from Wildermuth, Jour. Agr. Res.*); *K*, pupa of the roundheaded apple borer; note the sawdust and shavings with which the larva had closed its tunnel behind the pupa, and also plugged the exit hole for the adult, near the head of the pupa (*drawn from photo by Slingerland*); *L*, cocoon of the Hawaiian sugarcane borer (*from Van Dine, U.S.D.A.*); *M*, larvae and pupae of the honeybee in the hexagonal cells made of wax by the worker bees; at left, end and side views of larvae; at right, end and ventral views of pupae (*from White, U.S.D.A.*); *N*, pupal case of a caddice-fly, *Neophylax concinnus*, made of minute grains of sand cemented together with a few heavier pebbles for ballast (*from Lloyd*).

TABLE V.—THE LIFE STAGES OF INSECTS

Group name and examples	The period of inception. Quiescent. Does not feed. Neither injurious nor beneficial	The period of growth. The insect is always active during this period. The insect always feeds during this period. Most insects are either injurious or beneficial during this period. This is the only period in which different instars occur							The period of transformation. Usually quiescent. Never feeds. Neither injurious nor beneficial			The period of reproduction. Active. Usually feeds. Often injurious or beneficial
Ametabola (without a metamorphosis): Silverfish, spring-tails	EGG — Hatching	YOUNG	Molt	YOUNG	Molt	YOUNG	Molt	YOUNG	Molt			ADULT
Heterometabola (with a gradual metamorphosis): Grasshoppers, termites, thrips, chewing lice, blood-sucking lice, aphids, cicadas, chinch bug, squash bug	EGG — Hatching	NYMPH	Molt	NYMPH	Molt	NYMPH	Molt	NYMPH	Molt			ADULT
Holometabola (with a complete metamorphosis): Beetles, moths, butterflies, ants, wasps, bees, fleas, flies, mosquitoes	EGG — Hatching	LARVA	Molt	LARVA	Molt	LARVA	Molt	LARVA	Molt	PUPA	Molt	ADULT

It should be emphasized that the number of molts during the growth period varies greatly in different species, though usually the same in any given species. Four instars are represented here simply to illustrate the successive moltings and increase in size. The *exact* number shown has no particular significance and may vary from 1 to as many as 23.

CHAPTER VII

THE PLACE OF INSECTS IN THE ANIMAL KINGDOM

The tendency to classify is inherent in the human mind. We are constantly classifying the inexhaustible assemblage of objects about us, whether we are aware of it or not. We say that certain things are useful and others of no use, some ugly and some beautiful, some hard and some soft, and so on. This ability to associate in our thinking things and ideas that are alike, and to differentiate the unlike from them, is a very valuable attribute upon which much of human progress has depended. In science, law, business, and every other human activity where a large number of objects or ideas of varied kind must be handled, a *classification*, *i.e.*, a logical sorting and arrangement of the objects or concepts, is essential to efficient progress.

Too often classifications are made hastily, with little thought, and based upon the most obvious and superficial characters. For example, to group together in a chemical storeroom all bottles of a certain height and all boxes of a certain size, regardless of whether they contain acids, bases, salts, metals, or organic compounds; or to group together a flea, a toad, and a kangaroo because they all jump; and a bird, a bat, and an airplane because they all fly are obviously not the most fundamental, important, or useful groupings of these objects that can be made. Such groupings are known as *artificial classifications*.

No class of objects on earth presents so great and varied an assortment as the living things that inhabit its surface. The biggest job of classifying ever undertaken is the systematic study of plants and animals, including insects; probably no other class of things or ideas has received such painstaking and exhaustive study as the classification of these living things. From the time of Aristotle (384 to 322 B.C.) and Pliny (A.D. 23 to 79), through all the generations since, scores of men the world over have given their entire lifetimes to the great problems of systematic biology. In spite of the labors of these thousands of students, the task is still very far from completion.

The consensus of opinion of students of the present day is that life probably originated only once upon the earth and therefore that all living things have had a common origin and have grown to their present complexity of forms by an orderly and extremely slow process of ascent and differentiation; or, as we say, by *evolution*, in which the principal potent factors have probably been variation, natural selection, and heredity. If life originated but once on the earth and all plants and animals have been derived from a common ancestor by a gradual process of specialization and differentiation, it follows that they must present but one true arrangement with respect to their blood relationships to each other. To discover and record this arrangement and reveal the pathways along which each creature has developed is to make a *natural classification*— the ideal of systematic botanists, zoologists, and entomologists.

The most fundamental grouping of all objects is into (a) *the living or organic world*, composed of all those things which possess life, as characterized by the performance of the functions listed on page 88; and (b) *the lifeless or inorganic world*, made up of the innumerable things which do not possess life and perform none of the functions that characterize the living. All persons are in agreement also about the second step in such a natural classification, *i.e.*, to make two great categories of living things, which are called the *animal kingdom* and the *plant kingdom*. If plants and animals all arose from the same original spark of life, the divergence of these two main branches of living things must have come very early, because they have so many points of difference. Yet, in such fundamental matters as respiration, cell structure, mitosis, and reproduction, plants and animals are so peculiarly alike that no one who understands these processes can doubt that they are the result of a common ancestry. It is generally easy to distinguish the common plants from the common animals, though less easy to point to distinct differences by which we can separate them all. Indeed, there are some minute, intermediate forms of life that botanists and zoologists cannot definitely place as either plant or animal. The truth probably is that they are neither plant nor animal but intermediate forms partaking to some extent of the characteristics of both groups. This fact adds support to the great truth of evolution and the theory of the common origin of all life.

Differences between Plants and Animals

Animal Kingdom.—Characteristically free-moving organisms

Plant Kingdom.—Characteristically sessile

Generally assimilate organic foods

Generally take inorganic foods

Very rarely possess chlorophyll

Generally possess chlorophyll

Have protoplasmic or protein cell walls

Have cellulose or hydrocarbon cell walls

The last of the characteristics mentioned above appears to be the most universal difference between all animals and all plants. Since insects are animals, the plant kingdom is dismissed from further consideration at this point.

The Phyla of Animals

Taking the animal kingdom as a whole, we find that over three-fourths of a million species or kinds have been discovered and named. Practically all the known animals fall naturally into about a dozen important groups or branches, which are called by the Latin name that means branch, namely *phylum* (plural phyla). Thus one large phylum (the Chordata) includes all the animals that have a backbone; such as birds, fishes, and man. Another, the phylum Mollusca, includes such aquatic shelled animals as oysters and clams, and also the slugs and snails. The phylum Echinodermata embraces those radially arranged marine animals known as starfishes, sand dollars, and many others. The phylum Annelida includes true worms such as earth worms and leeches. Other phyla, somewhat less well known, are outlined in Table VI, which also gives the well-known examples and a general idea of the size of each phylum. The species placed in each branch or phylum all have what are thought to be fundamental characteristics in common, *i.e.*, they are kin, and resemble each other in many respects, as a horse does a cow. It may seem in many cases that the resemblance is rather slight, such as that between a bird and a cat. But in all cases it is much

TABLE VI.—THE ANIMAL KINGDOM

Phylum	Class	Examples	Estimated Number of Living Species Described
		VERTEBRATES	
Chordata			
	Mammalia	Man, cat, horse, bat, whale	10,000
	Aves	Birds, fowls...........................	15,000
	Reptilia	Turtles, snakes, lizards, alligators.........	5,500
	Amphibia	Frogs, toads, salamanders...............	2,000
	Pisces	Fishes................................	25,000
		INVERTEBRATES	
	Minor Classes	Tunicates, Balanoglossus, etc.............	2,500
		Total Chordata...........................	60,000
Arthropoda		**(See Table VII for Classes and Examples)**	**713,500**
Mollusca		Snails, slugs, clams, oysters.....................	80,000
Echinodermata		Starfish, sand dollar, sea urchin.................	5,500
Annelida (Annulata)		Earth worm, leeches...........................	8,000
Bryozoa (Polyzoa)		Moss animals, sea mats........................	3,100
Brachiopoda		Lamp shells..................................	500
Nemertinea		Nemertines...................................	600
Nemathelminthes		Roundworms, Trichina, Filaria..................	5,500
Platyhelminthes		Flatworms, flukes, tapeworms..................	7,000
Trochelminthes		Rotifers, wheel animalcules.....................	1,750
Ctenophora		Sea walnuts, comb jellies......................	100
Coelenterata		Jellyfishes, coral animals, Hydra...............	10,000
Porifera		Sponges.....................................	3,250
Protozoa		Amoeba, Paramoecium, Euglena, malarial organisms, trypanosomes..............	17,000
Minor phyla..			200
		Grand total.............................	916,000

greater between the animals of the same phylum than between those of different phyla. Thus the bird and the cat have fundamental points of likeness in the eyes, the heart, the backbone, and the limbs, which are obviously greater than any resemblance between either of them and a starfish, a worm, a sponge, or an insect. Each of the latter is accordingly placed in a different phylum from that to which the bird, cat, and other vertebrates belong.

The phylum that is most familiar to us is the one to which we belong, the phylum Chordata, the better-known part of which is the vertebrates. The most evident characteristic of this phylum is that the animals in it have a backbone or dorsal chain of bony vertebrae surrounding the spinal cord. The vertebrates include, besides man and all the other mammals: birds; snakes, lizards, alligators, and their kind; frogs and toads; and fishes. Certain of these obviously resemble each other much more closely than they do the others. Thus a cow, a horse, and a cat are clearly more alike than a cow, a chicken, a snake, and a fish. To express these differences between animals placed in the same phylum, secondary groupings known as *classes* are used. Thus in the phylum Chordata (the vertebrates) we have the following important *classes:*

CLASSES OF VERTEBRATE ANIMALS

Class Mammalia, the hairy, four-footed, milk-secreting animals, such as man, monkeys, horses, whales, elephants, bats, squirrels, dogs, and anteaters.

Class Aves, the winged and feathered animals, such as the chicken, sparrow, robin, eagle, parrot, ostrich, and all other birds.

Class Reptilia, the cold-blooded, scaly, air-breathing animals, such as snakes, lizards, crocodiles, and turtles.

Class Amphibia, the cold-blooded, soft-skinned vertebrates, whose young breathe by gills and adults by lungs, such as frogs, toads, and salamanders.

Class Pisces, the cold-blooded, aquatic animals, typically covered with scales and having fins, such as carp, salmon, eel, bass, trout, and all other fishes.

ORDERS, FAMILIES, AND GENERA

The examples given under the class Mammalia show clearly that, while these animals are fundamentally alike in the possession of hair and mammary glands, there are still wide differences between them. For example, some of the Mammalia lay eggs, like the anteaters, while most bring forth the young alive, not enclosed in an eggshell. Some are adapted for life in the water, like the whales and porpoises; the appendages of some are clawed, others are hoofed, and others bear nails. These differences are used as the basis of third-rate categories known as *orders*. Man belongs to one of the 15 or 20 *orders* of the class Mammalia, called the order Primates.

The order Primates is divided into a number of smaller groups, such as the lemurs, the apes, several kinds of monkeys, and man. The name for the divisions of an order is *family*. Unfortunately the word family has two entirely different meanings. It is used in a social sense to mean a pair of individuals and their immediate offspring, *e.g.*, the John Smith family, or "a bluebird family," and it is used among biologists to designate a category of lesser scope than an order, often embracing thousands of species, and millions of individuals. Thus, while the average social family comprises perhaps 4 or 5 individuals, the taxonomic *family* Hominidae embraces all men on the earth of whatsoever race or color, and includes, therefore, probably 1,750,000,000 individuals.

Two other categories of successively smaller scope than the family must also be understood, *viz.*, the *genus* and the *species*. The family to which man belongs has only one living genus, the genus Homo; but several fossil forms of primitive and ape-like men have been placed in other genera of this family Hominidae. Thus the Pliocene ape man, found in Java, is placed in a genus Pithecanthropus; the early Pleistocene dawn man from Sussex, in a genus Eoanthropus; while the middle Pleistocene, Neanderthal man has been considered a distinct species of the genus Homo, called *Homo neanderthalensis*, to distinguish it from all living persons of every race which are called *Homo sapiens*. Most families of animals have a number of genera, such as the deer family, which is known as the family Cervidae. The American elk is placed in the genus Cervus. The Virginia or white-tailed deer, the mule deer, and the black-tailed deer belong in the genus Odocoileus, the moose in the genus Alces, and the several kinds of caribou in the genus Rangifer. *The family name is always formed by adding the letters idae to the stem of the name of the typical genus.* Thus: Cervus, Cerv*idae;* Homo, Homin*idae;* Musca, Musc*idae.*

The Species

The innumerable insects and other living things about us naturally group themselves into kinds or species, as everyone knows who has said "Here is a new kind of flower," or "What kind of weed," or "What kind of bird is this?" Not only do the progeny of a single pair associate together closely, but many other individuals of the same kind that look exactly like them, mingle with them, behaving in the same manner, eating the same kind of food, building nests that are very much alike, mating with each other, and continually bringing forth new individuals like themselves.

These natural kinds which normally associate in nature are known as *species*. We should recognize that a species is a *real group* just as truly as the individual animal or plant is real. To define species, however, is difficult. One important criterion is that the members of a species generally interbreed, the matings producing fertile offspring; while members of different species seldom interbreed and, if they do, the offspring are generally not fertile. One species differs from another by characteristics greater than the differences between the children of the same parents. These specific characteristics are constant from one generation to another and throughout the natural range occupied by that species.

The species is the natural reproductive unit among animals. Species do not originate in the cabinets of the museum worker or under the microscope, though the *names* of species may be originated there. Species are formed by the hand of nature, and many of them were present on the earth in remote times before systematic zoologists or botanists or entomologists came into being, just as truly as they are today, when systematists have discovered, named, and labeled over three-fourths of a million of them.

Because we often apply inadequate standards in recognizing species, and because resemblances which seem important to one naturalist may seem trivial to another, there are great differences of opinion about how many species exist. Too often insects have been classified according to their structure alone, and indeed very often according to the structure of only one part or one set of organs. If we are really to know the groups as nature has developed them, we should study, not only the structure of every part, but also their habits, mating, development, distribution, fossil history, physiology, and perhaps even their psychology.

Because of the great number of insects, the species is a fundamental conception without which little progress in economic entomology would be possible. If we had to cope with each *individual* insect, as a separate problem, first determining its status as friend or enemy, and then working out a scheme of control for it, as determined by its structure and habits, our task would be utterly hopeless. But all the individuals of a species *look alike, act alike, eat the same kind of food in the same manner, and are controllable in exactly the same way.* This is because, for the most part, insects lack initiative or intelligence. That is, they do not have the ability to profit by experience, but depend upon instinct[1], and do every-

---

[1] Instincts, which are so characteristic of insect activities, are described as coordinated reflex behavior. They may be distinguished from intelligence and habits by the following characteristics: (*a*) They are performed in the same essential way by all the individuals of a species; (*b*) they are done in a nearly perfect manner without learn-

thing in the same way their ancestors have done for thousands of generations. All chinch bugs feed on grass plants, fly to the same type of winter quarters, and crawl on foot from small grains to corn in early summer, and, if a barrier or trap is provided, they fall into it just as their ancestors have done ever since man has been trapping them. In the same way all boll weevils, all codling moths, all Colorado potato beetles present one problem, and not a million different ones, although each species may be represented by millions of *individuals*.

In human relationships we deal with individuals, each of whom has individuality of behavior and therefore must have a distinctive name; but among insects these separate instinctive individuals do not require separate names. In dealing with insects, *the unit is the species* instead of the individual, and consequently every *species* of insect about which we wish to write or speak must have a name. This may be a common name or a scientific name. In the past, common names have been applied by anyone who chose to apply them and without rules or regulations. Accordingly the same kind of insect may have many different common names (*synonyms*) in different communities, and especially in different countries. Thus the corn earworm is known in different sections of this country as the tobacco budworm, cotton bollworm, tomato fruitworm and vetchworm. The squash bug is also commonly called pumpkin bug and stink bug. Furthermore the same common name is often applied to several very different kinds of insects (*homonym*). For example, the name *locust* in Biblical times, and at present in Europe, means a grasshopper, while to many of us it means a kind of cicada. Recently certain rules have been formulated for the application of common names of insects, and a list of about 1,000 names, approved by the American Association of Economic Entomologists, has been published.[1]

Dr. F. E. Lutz, in his delightful "Fieldbook of Insects," has termed the common names of insects "nicknames." He says:

"There are thousands of kinds of native-born, United States insects which have been really named but not nicknamed . . . Often real names are no longer or harder than the "common" names. An insect is considered to be christened when some student, who has found a kind which he thinks has never been named, publishes a description of it, and gives it a properly formed name. If somebody had previously named the same kind, *the prior name usually holds*."

It is evident that no sound progress could be made by using names as various and ambiguous as the nicknames of insects, because we could never be sure just what kind of insect was being considered by a speaker or writer. There is at least as much reason for insisting that each impor-

ing, *i.e.*, essentially as well the first time they are tried as the second or the hundredth; (c) they are done without consciousness or awareness of what is being done or knowledge of its purpose; (d) the separate acts of a complicated performance must take place in a definite, fixed sequence, each step, except the first, requiring the stimulus of the preceding step to set it going and, once started, tending to go ahead to completion, even though something may have happened in the meantime to make the rest of the performance useless and even ridiculous; (e) they are always acts necessary to the life or welfare of the individuals and are done with remarkable perfection so long as everything goes successfully, but the procedure cannot be changed, even slightly, to meet emergencies and save the act from failure or disaster.

[1] *Jour. Econ. Ento.*, **30**: 527–560, 1937.

tant species of insect pass under a single name as for objecting to the use of aliases and pseudonyms by our fellow men. This points to the necessity for *scientific names*. Scientific names are simply names applied under a set of rules and regulations to which the majority of systematic workers have agreed. Scientific names are written in Latin form since Latin is the most nearly universal language. Also, being a dead language, its form is fixed, and the scientific names are therefore the same in all countries and all languages. If these names seem strange, unnecessarily long, and difficult, we should remember that one cannot have 640,000 different names (see page 172) using only four letters and one syllable! We should be willing to concede something to the necessity for exactness and uniformity and to the importance of making these names available to all nationalities. In the latter part of this book the scientific name of each economic species discussed is given in a footnote.

By general agreement the system called *binomial nomenclature* has become universal. By this system every species is given a scientific name consisting of two parts: the first part is the *genus* name, which is common to from one to many similar species in much the same way that our family names or surnames, such as Johnson or Andrews, are common to all members of the same social family; and the second part is the *species* name, which must never be given to more than one kind of insect in the same genus,[1] in the same way that one family would not have more than one child with the same Christian name or given name, such as William or Richard. By using a Christian name with a surname, we designate each of our friends; by using the correct species name with the correct genus name, each insect kind may be clearly designated. A third part of the scientific name is usually added for convenience in finding the original description of the insect, *i.e.*, the name of the person, called the *author*, who first applied that particular name to the insect along with a published description of it.

There are a few simple rules about scientific names that all should observe. The genus and species parts should always be printed in *italics* or, when written, should be underscored once. The author's name is not printed in italic letters or underlined. The genus part of the name comes first and is always written with an initial capital letter, the species part of the name is given second and should always begin with a small letter. The three parts of the name are written without any punctuation. Thus *Anasa tristis* DeGeer is the scientific name of the squash bug. This name tells us at once that the squash bug belongs in the genus *Anasa* and the species *tristis;* and, further, that it was first named and described by DeGeer, who was a Swedish naturalist of the eighteenth century.

The various categories of successively smaller and smaller size, from the largest (the animal kingdom) to the smallest living unit (known as the *individual*), are given in the first column below. These are used somewhat like the divisions of the earth's surface, given in the last column below, in which each larger unit includes one to many smaller units. By way of illustration the complete classification of two animals, man and the house fly, are also given in parallel columns:

---

[1] Two names applied to the same species of insect are called *synonyms*, and only the first used is valid. The same name applied to two different species is called a *homonym*, and its second application becomes invalid.

Kingdom	Animal	Animal	The Earth
Phylum	Chordata	Arthropoda	Hemispheres
Class	Mammalia	Hexapoda	Continents
Order	Primates	Diptera	Nations
Family	Hominidae	Muscidae	States
Genus	Homo	Musca	Counties
Species	sapiens	domestica	Townships
Variety	Caucasian		Sections
Individual	John Smith	Any house fly	Farms

The relation of the different categories to each other, and some conception of their relative size, may be gained by a study of Fig. 91. In this figure the heavy black lines mark off the *phyla* of animals. A definite area on the paper has been allotted to each group in proportion to the number of species known to scientists. The lighter, solid lines mark off the proportionate size of the *classes* of Chordata and Arthropoda, while the broken lines further subdivide the class Arachnida and the class Hexapoda into their named *orders*. It may be noted, for example, that the *order* Coleoptera is not only the largest order of animals, but is actually several times as large as the *entire phylum* Chordata or Mollusca.

THE PHYLUM ARTHROPODA

In Chapters III and IV the important structural and functional characteristics of insects were described. No single one of these characteristics will define an insect and distinguish all insects from all other kinds of animals. For example, *the segmented body, bilateral symmetry, paired jointed appendages* usually terminating in claws, *chitinous exoskeleton, ventral nervous system,. and dorsal heart* are the characteristics of the entire phylum Arthropoda, which includes besides the true insects many other creatures such as crayfish, crabs, lobsters, sowbugs, centipedes, millipedes, spiders, mites, ticks, scorpions, harvestmen, and many others.

TABLE VII.—THE PHYLUM ARTHROPODA

Classes	Examples	Estimated Number of Living Species Described
Hexapoda (Insecta)	**All true insects**......................	**640,000**
Chilopoda	Centipedes or hundred-legged-worms...	1,200
Diplopoda	Millipedes or thousand-legged-worms...	1,300
Arachnida		
Orders of the Class Arachnida		
Scorpionida	Scorpions............................	600
Phalangida	Harvestmen or daddy-long-legs.........	1,900
Araneida	Spiders.............................	22,000
Acarina	Mites and ticks......................	18,000
Minor Orders	Pseudoscorpions, whip-scorpions........	2,500
	Total Arachnida....................	45,000
Crustacea	Crayfish, lobster, crab, sowbug, barnacles, water fleas, cyclops............	24,750
Minor classes	Peripatus, king crab, Symphyla, sea spiders, bear animalcules, Pauropoda..	1,250
	Total..............................	713,500

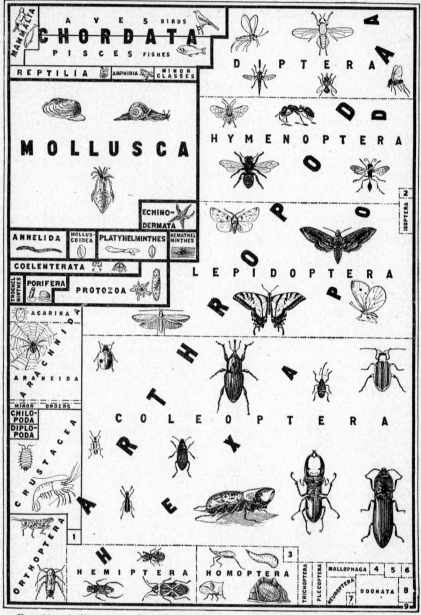

FIG. 91.—A diagram to show the relative size, in numbers of known species, of the various animal phyla, of the classes of two of these phyla (the Chordata and the Arthropoda), and of the orders of two of the classes (the Arachnida and the Hexapoda). Each square inch of this diagram represents about 32,000 known kinds of animals. The estimated number of species in each group is given in Tables VI, VII, and VIII, which should be studied in connection with this diagram.

The figure *1* represents minor classes of the Arthropoda; *2*, the order Siphonaptera; *3*, the order Corrodentia; *4*, the order Ephemeroptera; *5*, the order Dermaptera; *6*, the order Thysanoptera; *7*, the orders Anoplura, Strepsiptera and Mecoptera; *8*, the order Collembola; and *9*, the order Thysanura. (*Original*.)

The *phylum Arthropoda* is the largest phylum in the animal kingdom and, aside from the vertebrates, the phylum of most importance to man. This phylum embraces five important and well-known classes, of which insects (the class Hexapoda) is one (see Table VII).   More than 75 per cent of all the animal kinds hitherto found and named belong in the phylum Arthropoda, and about 90 per cent of these are true insects. The class Hexapoda is further analyzed in the following chapter, and, except for a number of references to mites and ticks, the remainder of this book is devoted to a discussion of important examples of insects.

The possession of three pairs of legs, three body regions, and wings are characteristic things that mark off the insects from the other arthropods.   Each of the four other classes is distinguished by certain structural features not possessed (at least in their entirety) by the rest of the arthropods.   In common usage, many of the representatives of these other classes are considered to be "bugs," and the species of economic importance are dealt with chiefly by entomologists.   Hence a brief discussion is given here of the classes Chilopoda, Diplopoda, Arachnida, and Crustacea (see also Table VII).

Class Chilopoda: The Centipedes or Hundred-legged-worms (Fig. 92).—The closest relatives of the true insects are the centipedes.   Like the insects, they have a single pair of antennae, they breathe by tracheae, and the reproductive organs open at the posterior end of the body.   They differ from insects in having neither thorax nor wings and in the large number of legs, typically one pair to each body segment.   Compound eyes are rarely developed.   They are worm-like in form but differ from the true worms (*cf.* Fig. 45,*A*) in having a distinct head and definite jointed legs.   They are usually somewhat flattened.   There are a pair of poison claws or legs, on the first segment behind the head, that are used to paralyze insects and other prey that they devour.   The centipedes may as a group probably be considered beneficial, although some of the species, especially the larger ones, which in the tropics may reach a length of 18 inches, sometimes inflict very painful bites upon man.

Class Diplopoda: The Millipedes or Thousand-legged-worms (Fig. 45,*D*).—Millipedes are superficially much like centipedes but differ in the following important respects.   The legs are still more numerous than in the centipedes, each apparent body segment having two pairs of legs.   The body is typically round in cross-section, not flattened; there are no poison legs; the antennae are short; and the reproductive organs open far forward close to the head.   The millipedes generally feed on decaying vegetable matter, but some species attack growing crops in damp soil, eating either the roots or the leaves that lie close to the ground.   They are sometimes mistaken for wireworms and may be serious pests in fields and greenhouses (see page 758).   Many of the species have an offensive odor.

Class Arachnida: The Spiders, Ticks, and Their Relatives.—Next to the insects, the largest class of Arthropoda is the class Arachnida,[1] to which the spiders, scorpions, mites, ticks, and harvestmen belong.   The Arachnida resemble insects in their small size, in their predominantly terrestrial habits, and in the possession of tracheae and Malpighian

---

[1] Care should be taken not to confuse the name of this class, Arachnida, with the name of the phylum, *Arthropoda*, since the names somewhat suggest each other to the beginner.

tubes. They differ radically, however, in having four pairs of legs; in having no antennae, true jaws, or compound eyes; in having only two body regions, the head and thorax being grown together into one region; in the curious "book-lungs" used for respiration; in not having a conspicuous metamorphosis; and in the position of the openings from the reproductive organs which are near the front of the abdomen. There are a number of orders in this class of which the following need mention here.

Class Arachnida: Order Araneida.—This large order includes all the spiders—a group of animals rivaling the snakes in their ability to frighten people. Comstock, the author of "The Spider Book," has well

Fig. 92.—A giant centipede from the southeastern United States, feeding on a white grub. Note the size as compared with chair. (*Photograph from life by A. R. Cahn.*)

said: "Few groups of animals are more feared, and few deserve it less." All spiders have a pair of venomous jaws and live on insects which they poison with their bites. They can bite; and occasionally such bites may become infected and cause serious results. But probably in all the world there are not more than a few species, if any, that are capable of killing man by their bites. Comstock assures us that there are no species in the northern United States that we need to fear. The large "tarantula" (Fig. 93) which comes into our midst in bunches of bananas, is capable of killing birds and small mammals by its bite. It apparently cannot kill a man, and besides, it seems hard to persuade it to bite a person. This has been shown by the experiments of Baerg in Arkansas.[1]

[1] *Ann. Ento. Soc. Amer.*, **18**: 471–478, 1925.

Among our native spiders, the one having the worst reputation is the "hourglass spider" or "black widow," *Latrodectus mactans* (see page 884).

Not only are most spiders harmless to man, but, because of the insects they devour, they are exceedingly beneficial. There are no species injurious to crops. But everywhere in the garden, the meadow, the fence row, the orchard, or in buildings one finds them quietly sitting in wait. Who shall estimate the number of pests they devour before the latter have had opportunity to damage our plants or our animals? Anyone who can overcome his prejudices against these delightful little creatures will find them most interesting entertainers. There are few better "shows" than to watch a spider spin her web, or to watch the elaborate courtship of the males. One is much impressed by the disadvantage in size which the male shows, for he is always much smaller. Someone has figured out that the disproportion would sometimes be equivalent to that between a man 6 feet tall and weighing 150 pounds,

Fig. 93.—Tarantula, about ½ natural size. (*From Herrick's "Insects Injurious to the Household." Copyright 1914, by Macmillan, reprinted by permission.*)

and a woman 90 feet tall and weighing 100 tons. And when one considers that her disposition often is to kill the male immediately after mating with him, one appreciates the poor chance male spiders have for a long life.

One of the most characteristic things about spiders is their habit of spinning silk. This is used in a variety of ways. (*a*) Chiefly it serves as a snare to capture food. It is truly a wonderful thing for a dumb animal to manufacture and set a trap. We know of none of the higher animals except man that do this, although it is done by some of the insects. (*b*) It forms tubes or tents for protection. (*c*) It forms sacs for protection of the eggs and newly hatched young. (*d*) It is used for locomotion. Spiders descend from higher to lower levels by spinning out a thread as they let themselves slowly down. Some spiders climb to a high point, and resting on their front legs begin to spin silk, supporting it by the hind legs until the loose end is caught by the breeze. More and more is thrown out until finally this simple kite exerts pull enough to carry the spider away. This can usually be observed in the open country on any bright autumn day, and the threads of "ballooning" silk, revealed by the descending sun, often seem to carpet the grass.

Class Arachnida: Order Acarina (Fig. 94).—Mites and ticks can usually be told at a glance from spiders or insects, because the body is all one region, there being little indication of either body regions or segments. They are like the spiders in the matter of appendages, and some of them spin silk. A curious feature is that the newly hatched young have only three pairs of legs. They breathe either by tracheae or directly through the skin. The only difference between mites and ticks is a difference in size *i.e.*, the larger members of this order (Fig. 94,*A*) are called *ticks*, while the smaller ones (*B*) are called *mites*.

The economic importance of the mites and ticks is at least fourfold: (*a*) Some of them injure plants, *e.g.*, the red spider mites (page 748) and the gall mites. (*b*) A number of species are found on or in the bodies of insects. Some of them are said simply to be riding upon the insects to a new feeding ground, but at least some species are parasitic upon the insects. Thus the Isle of Wight disease among honeybees is caused by a kind of mite that lives in the tracheae of the bee. (*c*) Many species are parasitic upon other animals, including man. Here the most notorious examples are the cattle tick (page 835), the Rocky Mountain spotted-fever tick (page 892), the poultry mite (page 868), itch mite (page 900),

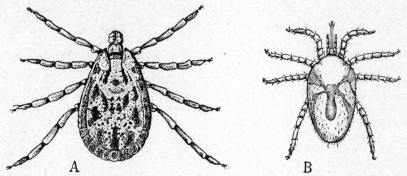

A   B

Fig. 94.—A tick, the spotted-fever tick (*from Cooley*) and one of the smaller members of the same order, usually called a mite, the tropical fowl mite. Much enlarged. (*From Cleveland.*)

scab mite (page 855), and scaly-leg mite (page 870). (*d*) Many of the parasitic species are to be feared because they are the known and only carriers of some animal diseases. Thus Texas fever is transmitted chiefly by the bite of the cattle tick, Rocky Mountain fever of man by the spotted-fever tick, and fowl spirochetosis by the fowl tick (see also Table II, pages 26, 27). The important species of Acarina are further discussed in the following chapters, in connection with the crops and animals they injure.

Class Arachnida: Order Scorpionida (Fig. 95).—The scorpions are common in the southwestern part of the United States and other subtropical and tropical regions. They are well known at least by name to nearly everyone, because of their reputation as stingers. The sting is borne at the tip of the abdomen. The latter is unusually long and the terminal half of it much more slender than the basal half. In addition to the four pairs of walking legs, scorpions have the pedipalps developed to very large size and provided with a pair of pinchers so that they appear to have five pairs of legs. The pedipalps are used to grasp prey, and the abdomen is then curled forward over the back and the stinger plunged into the victim to paralyze it.

The young are born after hatching from the eggs, and they are said to be carried about by the mother for a time after birth, clinging with their

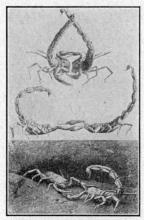

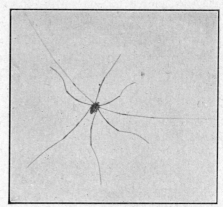

FIG. 95.—A pair of scorpions in the attitudes assumed during courtship. About natural size. (*From Fabre.*)

FIG. 96.—A harvestman or daddy-long-legs. Natural size. (*From Slingerland.*)

pinchers to her body. They are nocturnal creatures that forage about at night, catching and stinging spiders and insects. They may probably be considered a beneficial group. Their sting, although capable of causing a painful wound to man, is probably never fatal.

Class Arachnida: Order Phalangida (Fig. 96).—The harvestmen or daddy-long-legs are familiar to all out-of-door persons. They look much like very long-legged spiders, but close examination will show that the body is not divided by a slender waist. The legs are carried with the "knees" high, and the body swung low between them. The creatures have a noticeable odor that probably discourages many enemies, and Comstock suggests that the ease with which the legs separate from the body is a protective adaptation, enabling them to get away from predators that grasp them by a leg—minus that leg!

The food of the harvestmen is not well known. Some authors state that they feed largely upon insects, others that

FIG. 97.—A crayfish about natural size. (*From Fernald's "Applied Entomology."*)

they take only dead insects, soft fruits, and other plant tissues. At any rate they are not known to have any injurious or objectionable habits.

Class Crustacea: The Crayfish, Crabs, Sowbugs, Barnacles, and Their Relatives (Figs. 97 and 98).—The crayfish, lobsters, and crabs are the

largest and best known representatives of this class.   They are primarily
aquatic in habit and farthest removed from the insects of any of the
classes of arthropods discussed in this book.   They have five pairs of
walking legs, paired jointed appendages on the abdomen, two pairs of
antennae, a pair of compound eyes, and only two body regions.   The
legs are forked or branched into an outer branch, the *exopodite*, and an
inner branch, the *endopodite*.   The coxopodite, from which these branches
arise, usually bears a gill.   Unlike insects they have no tracheae and
breathe by blood gills or through the skin.   The excretory organs lie

Fig. 98.—The greenhouse pillbug.   Left, extended; right, rolled into a ball.   Enlarged.
(*From U.S.D.A., Farmers' Bul.* 1362.)

in the head, opening at the base of the antennae.   The reproductive
organs open at the base of the walking legs.   The forms most likely to
be confused with insects are the small terrestrial sowbugs and pillbugs
which abound under boards, logs, in greenhouses, and other damp places.
The pillbugs have the habit of rolling themselves into a nearly perfect
sphere when disturbed.   They are sometimes injurious in greenhouses.

In this chapter we have seen that all the insects constitute one class
(the class Hexapoda) of one of the phyla (the phylum Arthropoda) of the
animal kingdom.   We have also seen what are their general relationships
to the other important classes of this phylum.   In the following chapter a
third step in classification, the division of the insect class into its orders, is
explained.

CHAPTER VIII

THE ORDERS OF INSECTS

In the preceding chapter the place of insects in the animal kingdom was discussed, and it was pointed out that the insects constitute the largest of all animal groups. Indeed, about 75 per cent of all the known kinds of living animals are insects. In spite of their great numbers, insects have so many important characteristics in common that they are all included in one class, the class Hexapoda. Between 600,000 and 700,000 different kinds have been discovered, properly named, and described. Thousands of new kinds are being found every year, indicating that we are far from having reached a full knowledge of this class. The number of kinds that are probably living in all parts of the earth according to Gossard.[1]

". . . is variously estimated at from 2,500,000 to 10,000,000, with the probabilities favoring the latter figure as the more nearly correct. Assuming the maximum figure to be correct, in what a field does the entomologist find himself! Suppose that he attempts to familiarize himself with each species so that he will recognize it the next time he sees it. Since his task is obviously great, we will start him at it at the age of 5 years and allot him 5 minutes in which to study each species giving him one-half of the time to a male specimen and one-half to a female. Lest he should become lazy, we will provide him with electric lights and keep him working day and night and lest he should become fat, we will forbid him to eat except as he is able to snatch mouthfuls from the 5-minute intervals during which he is expected to fix in his memory the anatomical characters, color patterns, etc., which differentiate each species from every kindred one. Working in this manner and at this rate, the rains of nearly 100 summers will have fallen on his roof before the last representative of the long procession of insects has passed before him."

Obviously none of us shall ever learn to know all of the kinds of insects! Obviously, too, no one need be surprised if an entomologist cannot tell him, offhand, the name of every insect encountered. The field is so vast that it is, in its finer aspects, beyond the comprehension of any one man.

A good working knowledge of the groups of insects, however, is within the grasp of any earnest student. The largest groups of insects are known as *orders*. About 25 orders are commonly recognized by entomologists. Some of these contain only a few species which are so rare as to be almost never encountered. A number of others contain many species common and interesting but of no great importance to man. In the following table (Table VIII) are listed 23 orders, and from these are selected for special study 13 orders which contain species of importance to man; these are printed in **black face type.**

[1] Gossard, H. A., *Jour. Econ. Ento.*, **2**: 314, 1909.

TABLE VIII.—THE ORDERS OF THE CLASS HEXAPODA

Types of Mouth Parts	Orders	Examples	Estimated Number of Species Described	Number of Wings in Adult
		Insects without Metamorphosis *Primitively Wingless Insects*		
	Thysanura	Bristletails, silverfish	325	No wings
	Collembola	Springtails, snow-fleas	1,250	No wings
		Insects with a Gradual or Simple Metamorphosis. Nymphs Have Compound Eyes		
		Wings Develop Externally on Nymphs. Nymphs Have Compound Eyes		
Chewing in nymphs and in adults	Orthoptera	Roaches, crickets, grasshoppers, katydid, walking-sticks	20,000	Four wings (rarely none)
	Dermaptera	Earwigs	900	
	Ephemeroptera	May-flies, shad-flies	800	
	Odonata	Dragonflies, damselflies	4,900	
	Plecoptera	Stone-flies, salmon-flies	2,000	
	Isoptera	Termites, white ants	2,000	Four wings or none
	Corrodentia	Book-lice, bark-lice	850	
	Mallophaga	Chewing lice, bird lice	2,300	No wings
Rasping-sucking (transitional?) in nymphs and in adults	Thysanoptera	Thrips	1,500	Four wings or none
Piercing-sucking in nymphs and in adults	Homoptera	Aphids, scale insects, cicadas, leafhoppers	26,500	Four wings or none
	Hemiptera	Chinch bug, squash bug, bedbug, stink bugs, leaf bugs, aquatic bugs	31,000	
	Anoplura	Bloodsucking lice, "cootie," hog louse, cattle lice	400	No wings
		Insects with a Complete or Complex Metamorphosis *Wings Develop Internally in Larvae. Larvae Do Not Have Compound Eyes*		
Chewing in larvae and in adults	Coleoptera	Beetles, weevils	250,000	Four wings (rarely none)
	Strepsiptera	Twisted-wing parasites	175	
	Neuroptera	Aphid-lions, ant-lions, dobson-flies	4,000	
	Mecoptera	Scorpion-flies	300	
	Trichoptera	Caddice-flies	2,850	
Chewing in larvae; siphoning in adults	Lepidoptera	Butterflies, moths, skippers	120,000	Four wings or none
Chewing or reduced in larvae; chewing or chewing-lapping in adults	Hymenoptera	Bees, wasps, ants, sawflies	89,000	Four wings or none
Chewing or reduced in larvae; piercing-sucking or sponging in adults	Diptera	Flies, mosquitoes, gnats, sheep "tick"	78,000	Two wings or none
Chewing in larvae; piercing-sucking in adults	Siphonaptera	Fleas	850	No wings
	Minor Orders[1]	Zoraptera, Embioptera, and other lowly forms[1]	100	No wings
	Total Insects		640,000	

[1] These forms are so rare that they are not likely ever to be seen by the readers of this book.

KEYS TO THE ORDERS OF INSECTS

A. Insects usually provided with 1 or 2 pairs of wings and capable of flying. Head, thorax, abdomen, and jointed legs nearly always distinct. Tarsi generally of more than 1 segment: if 1 tarsal segment and 1 claw, see page 203. Body wall generally firm, hard, often highly colored. Individuals nearly always active, with little variation in size; often engaged in mating or egg laying. Mouth parts often modified for siphoning (Fig. 75), lapping (Fig. 76), or sponging (Fig. 74), sometimes entirely wanting or functionless..............**Adult Insects,** see **A,** couplet 1*a*, 1*b*.
B. Insects without functional wings, either entirely wingless or with short wing pads incapable of transporting the body. Legs sometimes normally developed, but often short or entirely wanting. Rarely entirely sessile or incapable of locomotion (see page 198). Tarsi frequently consisting of a single segment and bearing a single claw. If legs of type shown in Fig. 47,*E*, and no exposed mouth parts, see couplet 7. Sometimes without a distinct head, often the thorax not distinct from abdomen. Often enclosed in a case of some kind..
.............................**Immature Insects,** see **B,** couplet 41*a*, 41*b*, 41*c*.

A. Key to Orders of Insects in the Adult Stage

1*a*, Insects with wings. (If wings very short, try also couplet 41*a* and 41*c*)........20.
1*b*—Wingless insects...2.

WINGLESS ADULTS

2*a*, Tip of abdomen with 2 or 3 prominent appendages which are at least ⅕ as long as the body..3.
2*b*—Tip of abdomen without any appendages, or with a single prominent appendage, or with a short inconspicuous pair of processes..................................4.
3*a*, With 2 stiff, terminal, abdominal appendages, forceps- or pincher-like in shape (Fig. 99). Tough-skinned insects, not covered with scales. Tarsi 3-segmented...
....*Order Dermaptera,* page 189.
3*b*—Nearly always with 2 or 3 slender, flexible, antenna-like appendages at tip of abdomen. (If abdominal appendages forceps-like, the tarsi are 1-segmented, Order Thysanura, Family Japygidae.) Thin-skinned, delicate, carrot-shaped insects, covered with scales...............................*Order Thysanura,* page 184.
4*a*, Very small, delicate insects, rarely over ⅕ inch long, not much flattened, with not more than 6 abdominal segments, the first usually bearing a short, forked "sucker," and the fourth a long forked "spring" used to flip the body in jumping. Mouth parts sunken in the head. Antennae never more than 6-segmented.............
....*Order Collembola,* page 185.
4*b*—Abdomen of more than 6 segments, though some of them may be obscure; not provided with a terminal, ventral appendage for use in leaping..................5.
5*a*, Mouth parts of the chewing type, with a pair of teeth or mandibles adapted for chewing or pinching transversely, and usually 2 pairs of mouth palps (see also Thysanoptera, 14*a*)...6.
5*b*—Mouth parts not adapted for chewing, but generally with an elongate tongue, beak, or proboscis for piercing, lapping, or sucking. At most 1 pair of palps present (except in Thysanoptera, see 14*a*). (If there are no evident mouth parts see Anoplura, 18*a*)...12.
6*a*, Body louse-like, *i.e.*, rarely over ¼ inch long, flattened, usually somewhat oval in outline, and generally thin-skinned. Legs short. Antennae short, never more than 5-segmented. Live as parasites among clothing, hairs, or feathers of animals....7.
6*b*—Body not louse-like; if flattened, larger. Antennae always of more than 5 segments. Not parasites on the skin of animals..................................8
7*a*, Head as wide, or nearly as wide, as the body and broadly rounded or blunt in front. Tarsi with 1 or 2 segments and 1 or 2 claws. Spiracles of thorax on ventral side. From birds or mammals..................*Order Mallophaga,* page 195.
7*b*—Head narrow, especially toward the front, where it is somewhat pointed. Legs thick, tarsus with a single segment and only 1 claw. Mouth parts piercing-sucking

but not visible externally. Thoracic spiracles on dorsal side. From mammals only...*Order Anoplura*, page 203.

8*a*, Body somewhat ant-like in form, *i.e.*, small, but not flattened; nearly as deep as broad, with moderately long legs and antennae (Fig. 100).....................9.

8*b*—Body not ant-like in shape, generally much larger insects, with firm, dark-colored body wall...11.

9*a*, Waist slender, *i.e.*, a very slender portion of the body between thorax and abdomen. Mouth parts chewing or chewing-lapping. Tarsi 5-segmented. Generally hard-skinned insects..................................*Order Hymenoptera*, page 220.

9*b*—Waist not conspicuously slender, the body at mid-length nearly or quite as thick as elsewhere. Tarsi 2-, 3-, or 4-segmented. Pale, thin-skinned, delicate insects with large broad heads (Fig. 101)...10.

10*a*, Tarsi 2- or 3-segmented. Very small insects rarely over $\frac{1}{10}$ inch long, with compound eyes, very small prothorax, and no cerci. Usually solitary in habit.......
....*Order Corrodentia*, page 194.

10*b*—Tarsi 4-segmented. Small blind insects up to $\frac{1}{4}$ inch in length, with a pair of very small cerci. Live in soil or in wood or paper usually in great colonies.......
....*Order Isoptera*, page 192.

11*a*, Cerci and a conspicuous ovipositor usually present. Prothorax saddle-like. Ocelli present. Antennae usually of many segments, at least 15...............
....*Order Orthoptera*, page 185.

11*b*—Never with cerci or a firm ovipositor. Ocelli wanting. Antennae of not more than 11 segments....................................*Order Coleoptera*, page 205.

12*a*, Motionless or sessile insects without distinct head, thorax, and abdomen, and often without legs, eyes, or antennae (see also *B*, Key to Insects in Their Immature Stages, couplet 41*a*, 41*b*, 41*c*)............................13*a*, 13*b*, 13*c*.

12*b*—Insects capable of locomotion: jointed legs, antennae, and body regions more or less evident...14.

13*a*, Body of the insect covered with a mealy powder, or cottony tufts, or a firm, thin, separable shell or scale. Mouth parts piercing-sucking, with long stylets but no encasing, jointed labium or palps. Living on plants. Female scale insects.......
....*Order Homoptera*, page 198.

13*b*—Living on bodies of bees and wasps, partly enclosed by body wall of the host from which they project like a wart or tumor........*Order Strepsiptera*, page 210.

13*c* Worm-like or caterpillar-like, sometimes without legs, antennae, or mouth parts. Enclosed in tough silken cases, covered with bits of leaves or evergreen "needles" and suspended from twigs of trees by a silken loop..........................
...........................Female bagworms, *Order Lepidoptera*, page 213.

14*a*, Tarsi usually without claws, ending in an inflatable hoof or bladder-like segment (Fig. 102,*F*). Head cone-shaped, with 2 pairs of palps, but no segmented beak; the mandibles and maxillae stylet-like and retractile in the head capsule. Very small slender insects, rarely over $\frac{1}{8}$ inch long, especially prevalent in flowers..........
....*Order Thysanoptera*, page 196.

14*b*—Tarsi always with terminal claws, though they may be obscured by surrounding hairs or scales; no protrusible terminal membrane (Fig. 102,*A–D*).............15.

15*a*, Bodies thickly covered with hairs and flattened scales. Mouth parts, if present, consisting of a proboscis, coiled up under the head like a watch spring, with a pair of palps margining it. Prothorax fused with mesothorax......................
....*Order Lepidoptera*, page 213.

15*b*—Body not thickly covered with scales and not more than moderately hairy. Mouth parts never a coiled proboscis....................................16.

16*a*, Tarsi 5-segmented (Fig. 102,*A*). Antennae very short, usually completely concealed. Palps present..17.

16*b*—Tarsal segments fewer than 5. No palps..............................18.

17*a*, Body strongly compressed or flattened from side to side (thin horizontally). Small, spiny, jumping parasites of animals. Prothorax distinct. Coxae enormously large, those of same pair contiguous. Two pairs of palps...............
....*Order Siphonaptera*, page 236.

17*b*—Body depressed or flattened from above downward (thin vertically). Body leathery and hairy, but not spiny. Coxae of same pair well separated. Not jumping insects. Abdomen not distinctly segmented. Palps present............ ....*Order Diptera* (in part) *Pupipara*, page 236.

18*a*, All 3 thoracic segments fused. Legs short, stout, with 1-segmented tarsi bearing a single claw (Fig. 102,*E*). Head pointed, narrow, no trace of mouth parts externally; stylets within the head apparently 3 in number. Abdomen broad. Parasitic on mammals (if head broad and mandibles minute, see Mallophaga)..... ....*Order Anoplura*, page 203.

18*b*—Beak evident, segmented. No palps. Four stylets, external to head, usually concealed in a segmented beak.......................................19.

19*a*, Beak arising from posterior (caudal) part of head. Frequently very small and delicate insects. Prothorax not unusually large except in treehoppers.......... ....*Order Homoptera*, page 198.

19*b*—Beak arising from front (cephalic) part of head. Prothorax large and distinct. ....*Order Hemiptera*, page 200.

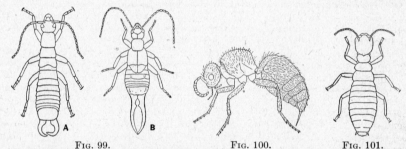

FIG. 99.   FIG. 100.   FIG. 101.

FIG. 99.—*A*, a wingless and *B*, a winged earwig, order Dermaptera, showing the forceps at tip of abdomen. (*Drawn by Kathryn Sommerman.*)

FIG. 100.—A velvet-ant, order Hymenoptera, in side view. (*Drawn by Kathryn Sommerman.*)

FIG. 101.—A termite worker, order Isoptera. (*Drawn by Kathryn Sommerman.*)

WINGED INSECTS

20*a*, With only 1 pair of wings, the hind wings entirely wanting or represented only by a pair of minute knobbed hairs or balancers or halteres...................21.

20*b*—With 2 pairs of wings, *i.e.*, 2 pairs of readily visible projections from the thorax in addition to the legs. (The sheath- or shield-like wing covers of beetles and earwigs and the club-like, mesothoracic projections of Strepsiptera are wings)...23.

21*a*, Tip of abdomen with 1 or more long or prominent processes. Mouth parts wanting...22.

21*b*—Tip of abdomen without prominent projections. Mouth parts generally with a conspicuous proboscis. Hind wings represented by a pair of halteres........... ....*Order Diptera*, page 230.

22*a*, Wings netted-veined, with numerous cross-veins. Antennae very short. No halteres.....................................*Order Ephemeroptera*, page 190.

22*b*—Wings with few veins and no cross-veins. Antennae prominent, long. A pair of hook-like halteres present. Males of scale insects...*Order Homoptera*, page 198.

23*a*, Front wings distinctly thicker, stiffer, and/or less transparent than the hind pair; often horny or leathery. If the front pair is only slightly thicker, the hind wings are broad and folded fan-like. Prothorax distinct from rest of thorax, usually large...24.

23*b*—Front wings of the same texture, stiffness, and color as the hind pair, generally thin and transparent like cellophane...................................30.

24*a*, First pair of wings very small, mere short, blunt clubs; the hind pair large, triangular, folding fan-like and without cross-veins. Minute, rare insects with

protruding eyes and short, flabellate antennae. No cerci.......................
....*Order Strepsiptera* (males), page 210
24*b*—First pair of wings wide enough to cover the abdomen....................25.
25*a*, Front wings very stiff and horny, the veins completely obscured (though there may be parallel ridges on them). The hind wings longer than the front pair and *folded crosswise*, to be completely covered by the front ones when at rest.....26.
25*b*—Front wings leathery or parchment-like (at least at the tip), rather than horny. Hind wings not longer and never folded crosswise at rest....................27.
26*a*, Tip of abdomen without conspicuous appendages. Hind wings folding transversely. Front wings generally covering all or most of the abdomen like a sheath.
....*Order Coleoptera*, page 205.
26*b*—Tip of abdomen with a pair of prominent pincher-like or forceps-like appendages (Fig. 99,*B*). Front wings always much shorter than the abdomen. Hind wings folding radially.....................................*Order Dermaptera*, page 189.
27*a*, Front wings stiff and horny or leathery at the base, the tips (distal third or half) abruptly thinner and membranous, and lying flat, one over the other, above the tip of the abdomen, when at rest. Head usually horizontal in position, with a piercing-sucking beak arising near its front end.........*Order Hemiptera*, page 200.
27*b*—Front wings almost never stiff and horny, though they may be somewhat leathery and colored; always of about the same texture from base to tip. Head usually vertical in position..28.
28*a*, Mouth parts of the chewing type. Hind wings larger than front wings and folding fan-like. Often with modifications of legs and wings in the males for singing or chirping. Generally large insects with long antennae, distinct large prothorax, and a pair of cerci......................*Order Orthoptera*, page 185.
28*b*—Mouth parts of the piercing-sucking type, the labium attached near posterior end of head, close to front legs. No cerci.................................29.
29*a*, Head horizontal. Pronotum large and distinct. Beak attached farther forward on the head..*Order Hemiptera*, page 200.
29*b*—Head vertical. Pronotum often short, but enormous in treehoppers (Membracidae)..*Order Homoptera*, page 198.
30*a*, Both pairs of wings largely or entirely covered with minute scales (short flat hairs) on both upper and undersurface; wings very large and usually varicolored. Mouth parts consisting of an elongate, slender, sucking tube, coiled like a watch spring when not feeding, and a pair of palps margining it; or mouth parts wanting. Prothorax small. Wing vein M_4 distally fused with vein Cu_1....................
....*Order Lepidoptera*, page 213.
30*b*—Wings not shingled with scales, though they may be hairy. Mouth parts not of the siphoning type..31.
31*a*, Membranous part of both wings very narrow, stick-like, but margined with very long stiff hairs or setae. The tarsi 1- or 2-segmented, without claws, ending bladder-like or hoof-like (Fig. 102,*F*). Mouth parts of rasping type. No cerci..........
....*Order Thysanoptera*, page 196.
31*b*—Wings always with broad membranous expansions and bearing veins. Tarsi always with claws (which may be concealed by hairs) (Fig. 102, *A–D*)........32.
32*a*, Mouth parts of the piercing-sucking type. No palps. The beak or stylets arising far back on the underside of the head. No cerci......................
....*Order Homoptera*, page 198.
32*b*—Mouth parts not of the piercing-sucking type. At least the mandibles developed to work transversely for chewing, pinching, or grinding; or mandibles absent. Palps always present...33.
33*a*, Antennae very inconspicuous, scarcely as long as the head. Wings netted-veined. Insects found chiefly near water in which their young develop........34.
33*b*—Antennae usually very well developed. Always longer than the head, or, if rarely very small, the wings generally with few cross-veins (the finely reticulated wings of Isoptera are an exception, see 36*a*)...............................35.
34*a*, Front wings much larger than hind pair; at rest, folded vertically over the back as in butterflies. Two or three long antennae-like processes at tip of abdomen.

Delicate, short-lived insects, found near water in which their young develop......
....*Order Ephemeroptera*, page 190.

34b—Front and hind wings of about equal size, with a nodus and pterostigma; held outspread at sides of thorax when insects come to rest. Abdomen generally very long and slender, with only short or inconspicuous appendages at its tip.........
....*Order Odonata*, page 190.

35a, Head prolonged downward into a short, thick trunk or beak, 2 or 3 times as long, and about ½ as thick, as the head; bearing chewing mouth parts at the end. Two pairs of wings of about the same size, netted-veined, often spotted. Males with a pair of swollen, pincher-like appendages at tip of abdomen (besides the short cerci), suggesting the sting of a scorpion....................*Order Mecoptera*, page 212.

35b—Head not prolonged into a beak. Males without swollen terminal appendages suggesting the sting of a scorpion..36.

36a, The two pairs of wings of the same size and shape, with a suture near the base, along which they easily break off; finely but faintly netted-veined. Tarsi 4-segmented. Often appear in swarms from buildings or from wood.................
....*Order Isoptera*, page 192.

36b—Front and hind wings rarely of the same shape and size; never with a suture for dehiscence. Tarsi not 4-segmented..37.

37a, Tarsi 5-segmented. No cerci or caudal appendages.............38a, 38b, 38c.

37b—Tarsi of fewer than 5 segments, 2- or 3-segmented (Fig. 102,C,D).........40.

38a, Wings generally covered with rather fine, long, silky hairs (*never broad scales*), giving the insects a moth-like appearance. Wing spread not over 2 inches. A bare, semitransparent, whitish spot near the center of each wing. Not many cross-veins along the costal margin. Wings held roof-like over the back when at rest; vein M_4 of the fore wing never fused with vein Cu_1. Prothorax small, weak. Mandibles reduced or wanting; no long proboscis. Rather delicate shy insects found about water in which the young develop.......*Order Trichoptera*, page 212.

38b—Wings appearing transparent, like cellophane. Mandibles generally well developed..39.

38c Wings neither hairy nor transparent, generally colored like the body; hind pair folding fan-like. Prothorax distinct. Cerci well developed. The female with long stiff ovipositor. Crickets.....................*Order Orthoptera*, page 185.

39a, Hind wings of much smaller area than the fore wings and never folded fan-like in the anal area; frequently fitted against, and hooked to, the front pair. Wings with few veins and few cross-veins, which are never numerous in the costal area. Prothorax not distinct from mesothorax. Often slender-waisted insects. Females often with a stinger or a prominent ovipositor. Mouth parts often of chewing-lapping type.....................................*Order Hymenoptera*, page 220.

39b—Hind wings often nearly or quite as large as the front pair; generally with many veins and cross-veins especially in the costal area. Radial sector of wings generally pectinately branched. Prothorax well developed. Many large species. Mouth parts of chewing type.............................*Order Neuroptera*, page 210.

40a, Hind wings always distinctly smaller than front pair; no folded anal area; veins few; wings usually held roof-like over abdomen, at rest. Prothorax small. Cerci wanting or very short. Very small insects..........*Order Corrodentia*, page 194.

40b—Hind wings generally as large as or larger than the front pair, the anal area folding fan-like. Wings flat over abdomen at rest. Prothorax well developed. Abdomen generally with two short appendages at tip. Body somewhat flattened. Found chiefly near water in which the young develop..*Order Plecoptera*, page 191.

B. Key to the Orders of Insects in Their Immature Stages

41a, **Nymphs:** Wing pads often present in the later instars or larger individuals, but never capable of flight. Compound eyes present, unless the adults are eyeless. Entire body usually fairly well sclerotized, tough-skinned, sometimes brightly colored. Abdomen seldom twice as long as head and thorax together. Shape and appendages much like the adults, with which they are often associated. Legs usually long, rarely wanting, nearly always with two claws on the tarsus. A feeding stage: food and mouth parts of same type as the parents, either chewing or piercing-sucking. Individuals show great variation in size, the largest molting directly to the adult...**Nymphs,** couplet 42.

41b—**Larvae:** Never any trace of wings or wing pads. Compound eyes never present. Abdomen and upper part of thorax, or entire body, often soft, thin-skinned, or weakly sclerotized; often whitish or yellowish, but sometimes brightly colored. Abdomen usually 3 or 4 times as long as head and thorax together. Shape often very different from adults, often cylindrical or spindle-shaped. Legs often very short or wanting. A feeding stage—mouth parts always chewing or reduced, sometimes secondarily modified for sucking. Food and habitat often extremely

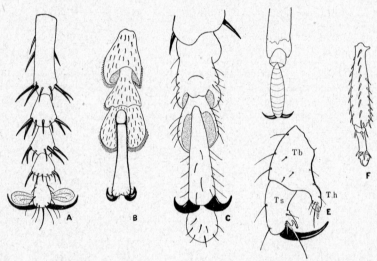

Fig. 102.—Tarsi of various insects. *A,* five-segmented tarsus of a robber fly, showing terminal claws, pulvilli, and the median bristle-like empodium; *B,* tarsus of a leaf beetle, showing the condition described as apparently four-segmented, the fourth segment being very small and hidden in the partially divided third segment; *C,* three-segmented tarsus of a grasshopper, with arolium and claws at end; apex of tibia is also shown; *D,* two-segmented tarsus of an aphid, with apex of tibia shown; *E,* one-segmented tarsus *(Ts)* of the human louse, bearing a single claw; the tibia *(Tb)* with its thumb *(Th)* is also shown; *F,* tibia and two-segmented tarsus of a thrips, terminating in a "hoof" or "bladder," without claws. *(Original.)*

different from that of adult, very often in the ground, or inside living or dead plant or animal tissues. Individuals show great variation in size, the largest followed by a pupal stage (see 41c) before the adult....................**Larvae,** couplet 58.

41c—**Pupae:** Functionless legs and wing pads encased in an extra membrane; never used for locomotion; usually incapable of being moved. Sometimes wriggle abdomen and rarely swim actively. Compound eyes visible unless adults are eyeless. Often entirely surrounded or encased by a silken cocoon, a dried larval skin (puparium), or cell of wax, paper, soil, or other material from the environment: then the entire body usually very soft and pale colored. If without a case, the body wall often very hard and dark-, usually somber-colored. Abdomen not unusually elongate; shape and appendages something like adult. A nonfeeding "resting" stage. Mouth parts usually foreshadow the type of the adult. Habitat often very different from both larva and adult. Individuals show little variation in size. Followed directly, upon molting, by the adult stage....................**Pupae,** couplet 70.

NYMPHS

42a, Mouth parts chewing in type, frequently sunken into the head, so that the mandibles and maxillae are concealed by the walls of the head as seen from the side. Legs always present, the tarsi usually consisting of a single segment, rarely 3- or 4-segmented, frequently with two claws (Fig. 102). Wing pads never present. Thoracic segments larva-like, *i.e.*, pleura and sternum not divided into distinct smaller sclerites (Subclass Apterygota)..............................43.

42b—Mouth parts chewing in type (Fig. 63), greatly elongated (Fig. 65), or sometimes wanting. If of chewing type, always exposed, never concealed by the extensions of lateral aspects of the head. Legs usually present, the tarsi variable, consisting of 1 to 5 segments; if only 1 segment, usually with a single claw (Fig. 102,*E*). Wing pads frequently present (Fig. 84). Thoracic segments usually with pleura and sternum divided into smaller sclerites by distinct sutures (Fig. 46) (Subclass Pterygota)......................................44.

43a, Antennae long, consisting of 10 segments or more (Fig. 83). Tip of abdomen with 2 or 3 long, jointed, antennae-like appendages (cerci) (Fig. 83), or modified into a pair of forceps-like appendages (Fig. 99). (Do not confuse with Dermaptera, page 189.) Abdomen never with a furcula (Fig. 104) or a collophore (Fig. 104). Prothorax never concealed by the overlapping mesothorax. Slender rapid-running insects. Body covered with scales..................*Order Thysanura*, page 184.

43b—Antennae short, never with more than six segments (Fig. 104). Cerci always wanting, the tip of the abdomen never with a pair of forceps-like appendages. Abdomen usually with a jointed furcula, used for jumping, on underside of the fourth segment (Fig. 104) and a collophore or adhesive organ on the underside of second segment (Fig. 104). Prothorax usually small and usually concealed by the mesothorax. Minute, chunky, jumping insects found in damp places............
....*Order Collembola*, page 185.

44a, Tarsi usually consisting of 2, 3, or 4 segments, rarely of 5, and very rarely of a single segment. Legs very rarely wanting. Thorax with all three segments exposed and generally different in form (Fig. 84); pleural and sternal sclerites usually distinct and never concealed (Fig. 121); wing pads usually present (Fig. 84) on dorsal and lateral aspects of body. Epicranial suture does not extend to the clypeus. External genitalia may be evident in later instars. (*Nymphs*)................45.

44b—Tarsi usually consisting of a single segment (Fig. 55,*5*), or legs wanting, or segmentation of tarsi difficult to determine; more rarely tarsi of 2, 3, or 4 segments. If legs of type shown in Fig. 47,*E* and no exposed mouth parts, see couplet 55. Thorax with all three segments similar in form (Fig. 88) and wing pads wanting; or, wing pads present, laterally and ventrally, the thoracic segments not exposed; the pleural and sternal sclerites never distinct, either not differentiated from notum or concealed by legs and wing pads (Fig. 89). Epicranial suture usually extends to clypeus. External genitalia not evident. (*Larvae and pupae*)................57.

45a, Tarsi almost always without claws, the last segment bladder-like or hoof-like in form (Fig. 102,*F*). Body cylindrical, with the tip of the abdomen pointed (Fig. 116,*4*). Mouth parts with 2 pairs of palps; no segmented beak....*Order Thysanoptera*, p. 196.

45b—Tarsi always with 1 or 2 claws, the last segment not bladder-like or hoof-like (Fig. 102)..46.

46a, Mouth parts fitted for chewing (Fig. 63), labium never modified into a tube-like beak, always exposed; maxillary and labial palps rarely wanting................47.

46b—Mouth parts fitted for piercing and sucking (Fig. 64); maxillary and labial palpi always wanting...55.

47a, Labium or lower lip 5 or 6 times as long as broad, elbowed at middle, folded beneath the head and over the mouth like a mask when at rest, but capable of being extended a considerable distance beyond the head. Antennae very small. First pair of legs shortest or about equal to the others. Abdomen broad, entirely without appendages or with 3 leaf-like or finger-like gills at the end, never with long antennalike or bristle-like tails or cerci. Wing pads subequal. Always aquatic.......
....*Order Odonata*, page 190.

47*b*—Labium normal in form, not greatly elongated, extensible, or folded elbow-like beneath the head...48.

48*a*, Abdomen with a series of plate-like or finger-like tracheal gills along each side of the body (Fig. 110). Two or three long cerci or antenna-like filaments at tip of abdomen, often "feathered" or fringed with long setae. Tarsi with a single claw. Antennae small, 1 to 3 times length of head. Mesothoracic wing pads larger than the metathoracic pair. Always aquatic............*Order Ephemeroptera*, page 190.

48*b*—Abdomen never with a series of plate-like, tracheal gills along each lateral margin. Cerci wanting, or short, or if long not fringed with prominent hairs. The tarsi almost always with 2 claws, except in Mallophaga from mammals.....49.

49*a*, Antennae never consisting of more than 5 segments, short. Body always strongly depressed or flattened dorsoventrally. Head flat, prognathous, broad and rounded in front, eyes simple. Prothorax small, but distinct; meso- and metathorax somewhat fused. Ectoparasites on birds or mammals.....*Order Mallophaga*, page 195.

49*b*—Antennae always with more than 5 segments..............*...........50.

50*a*, Prothorax always much shorter and smaller than the other thoracic segments. Head hypognathous, eyes compound. Body not flattened. Not over ⅛ inch long. Antennae long. Clypeus swollen. (Fig. 114)....................
....*Order Corrodentia*, page 194.

50*b*—Thorax with the 3 segments about equal, or the prothorax or mesothorax the largest, the pronotum quadrangular or subquadrangular.....................51.

51*a*, Head vertical with the mouth ventral in position (hypognathous). Ligula 4-lobed. Antennae attached on cephalic or front aspect of head. Pronotum quadrangular, pleura on ventral aspect, with the sterna. Segmented cerci usually present. Hind legs often developed for jumping. Wing pads, when present, with the hind pair overlapping the front pair..............*Order Orthoptera*, page 185.

51*b*—Head horizontal, with the mouth cephalic in position (attached to front end of head—prognathous). Antennae usually attached to dorsal aspect of head. Wing pads not as in 51*a*..52.

52*a*, Tarsi with four segments. Head prognathous, usually distinctly longer than broad. Labium very long. Thin-skinned, pale insects. Wing pads, if present, subequal. Cerci of 3 short segments, very small.......*Order Isoptera*, page 192.

52*b*—Tarsi with two or three segments. Head distinctly broader than long.......53.

53*a*, Basitarsus of first pair of legs about as long as tibia or longer, strongly dilated and provided with openings of silk glands on the ventral surface. Basitarsi of other legs normal in form. Head prognathous........................*Order Embioptera*.

53*b*—Basitarsus of first pair of legs never so long as the tibia, not different in form from the basitarsi of the other legs.....................................54.

54*a*, Thorax and abdomen never with tracheal gills. Terrestrial insects, usually with cerci consisting of a pair of unsegmented forceps. Head hypognathous. Ligula 2-lobed.....................................*Order Dermaptera*, page 189.

54*b*—Thorax and cephalic (front) segments of abdomen with tufts of slender finger-like tracheal gills near bases of legs. Usually no other appendages on sides of abdomen, but generally a pair of long antennae-like tails at its tip. Legs fringed with strong hairs. Aquatic insects, never with a pair of abdominal pinchers or forceps........
....*Order Plecoptera*, page 191.

55*a*, Wing pads always wanting. Labium never an exposed beak, all the mouth parts withdrawn into head at rest. Head pointed in front. Tarsi adapted for clinging to hairs, 1-segmented, and with a single claw (Fig. 102,*E*). Pleura and thoracic spiracles on the dorsal aspect............................*Order Anoplura*, page 203.

55*b*—Wing pads usually present, labium almost always exposed, and segmented. Tarsi never adapted for clinging to hairs. Pleura on ventral or lateral aspect. Thoracic spiracles concealed.......................................56.

56*a*, Labium attached to front end of the head. Pronotum large, distinct..........
....*Order Hemiptera*, page 200.

56*b*—Labium attached to caudal end of the head, often very close to front coxae. Pronotum often short, but enormous in treehoppers. Body sometimes without trace of legs or wings, flattened, scale-like, covered with a powdery or plate-like secretion, beneath which the insect lives, motionless..*Order Homoptera*, page 198.

57*a*, Thorax never with exposed wing pads present. Tarsi consisting of a single segment. Legs usually short and frequently wanting entirely. *Larvae*..........58.

57*b*—Thorax with 3 segments distinct, usually with wing pads. Legs always present, long, and well developed, folded against sternal and pleural surfaces, sometimes fused with the sternal and pleural surfaces of the body and with each other. Tarsi always consisting of more than a single segment, varying from 2 to 5, but the number of tarsal segments sometimes difficult to determine. Generally quiescent stages, often immobile, and frequently enclosed in silken cocoons or in cases made of soil, plant particles, and the like. Usually no epicranial suture. External genitalia often indicated. *Pupae*..70.

LARVAE

58*a*, Three pairs of jointed legs on the thoracic segments, the legs often small and inconspicuous, rarely indistinctly segmented..............................59.

58*b*—Thoracic segments never with legs......................................66.

59*a*, Abdominal segments 1 to 8 without true prolegs or larvapods, except rarely 1 pair at end of abdomen, the other segments at most with ventral folds or wrinkles. First pair of spiracles usually located on the mesothorax. Head often flat and depressed, the mouth directed cephalad (prognathous) (Fig. 137)........................60.

59*b*—Abdomen always with true prolegs or larvapods on several or all of the segments 1 to 8. First pair of spiracles located on the prothorax or apparently wanting. Tarsal segments with a single claw. Head globular..........................64.

60*a*, Mesothoracic and metathoracic legs (second and third pairs) noticeably larger than prothoracic (first) pair..61.

60*b*—Legs all about equal in length, rarely, if ever, all 3 pairs directed cephalad..62.

61*a*, Prothoracic legs directed ventrally, meso- and metathoracic legs much larger, directed laterally; never consisting of more than 4 segments and a single claw. No gills. A cluster of several ocelli on side of head. Pronotal plate wanting. Body strongly curved............*Order Mecoptera* (*in part, Family Boreidae*), page 212.

61*b*—Legs all directed cephalad; consisting of 5 or 6 segments and a single claw. Aquatic insects; abdominal segments usually with hair-like tracheal gills, but without firm, hairy, or feathered filaments. Usually a single pair of abdominal prolegs at tip of abdomen, bearing strong hooked claws. First abdominal segment often with a dorsal, and 2 lateral, fleshy tubercles. Pronotal plate present. Antennae very short. Maxillary palps 4- or 5-segmented. Usually found on bottom of streams or ponds, surrounded by a silken case of varied shapes, covered with pebbles or sticks and open at the head end.....*Order Trichoptera*, page 212.

62*a*, Tarsi of thoracic legs with 2 claws (except Sisyridae). Antennae and mandibles usually as long as, or longer than the head, the latter usually sharp, often sickle-shaped. Abdomen terminating in hooked claws or anal prolegs or an unpaired, median filament, never a pair of cerci. Head usually with a gula. No spinneret. Body usually tapering toward both ends. Often aquatic; then with gills or firm bristle-like or feathered filaments (often segmented) along sides of abdomen, but no prominent terminal appendages. Terrestrial forms without maxillary palps, the maxillae closely fitted to mandibles to form a pair of sucking tubes (Fig. 70)......
....*Order Neuroptera*, page 210.

62*b*—Tarsi of thoracic legs usually with a single claw; if with 2 claws, the abdomen has a pair of anal cerci, sometimes retractile. Antennae and mandibles rarely, if ever, longer than the head. Abdomen never with long anal prolegs with hooked claws.63.

63*a*, Head flat, depressed, the mouth directed cephalad (prognathous). Head never with adfrontal sclerites.............................*Order Coleoptera*, page 205.

63*b*—Head usually globular, the mouth directed ventrad (hypognathous). Whether flattened or globular, always with adfrontal sclerites (Fig. 145), at least in the older larvae. Spinneret present. Thoracic legs with not more than 5 segments, 1 claw. Antennae contiguous to base of mandibles.....*Order Lepidoptera*, page 213.

64a, Head on each side with a group of 12 to 20 or more ocelli, resembling a compound eye. The thoracic legs 3-segmented with 1 tarsal claw. Six to eight pairs of prolegs, without crochets..
.......*Order Mecoptera* (in part, Families Panorpidae and Bittacidae), page 212.
64b—Ocelli on each side of head never more than 10, usually fewer, and sometimes wanting..65.
65a, Prolegs usually 5 pairs, on segments 3 to 6 and 10; sometimes 2 or 3 pairs only. The prolegs always provided with crochets. Ocelli on each side of head either more than one or wanting. Antennae attached to articulating membrane at base of mandibles. Labium usually with a protruding, median spinneret................
....*Order Lepidoptera*, page 213.
65b—Prolegs usually 6 to 8 pairs on segments 2 to 8 and 10, or 2 to 7 and 10, or 2 to 6 and 10. The prolegs never provided with crochets. Only 1 ocellus on each side of head, or none. Antennae attached to frons.....................................
..................*Order Hymenoptera* (in part, Family Tenthredinidae), page 220.
66a, Head always with distinct adfrontal sclerites, at least in the last instar. Labium with a projecting median spinneret..........*Order Lepidoptera* (in part), page 213.
66b—Head never with adfrontal sclerites. No projecting spinneret.............67.
67a, Head always present, usually much darker in color and easily recognized as a distinct region from the rest of the body; never retracted within the prothorax. Head always with recognizable antennae. Maxillary palps of 2 or more segments.
....68.
67b—Head frequently somewhat or completely retracted within the prothorax, often apparently wanting, the body then distinctly tapering and pointed at one end. If the head is exposed, it is usually globular and of the same color as the thorax and abdomen, its transverse width usually much less than that of the prothorax. Antennae and ocelli usually wanting. If antennae distinct, the spiracles of the eighth abdominal segment usually larger than those of other segments and sometimes located at ends of breathing tubes of varying length. Spiracles of all other abdominal segments may be wanting.....................................69.
68a, Body generally short and frequently U-shaped. Ocelli usually present. Antennae and mandibles generally shorter than the head. Each of the principal abdominal segments with a pair of easily recognized spiracles. Abdomen often with fewer than 10 distinct segments and without prominent subanal processes....
....*Order Coleoptera* (in part), page 205.
68b—Body long and unusually slender, less than ⅓ inch long, never U-shaped. Eyes always wanting. Head usually light in color. Antennae distinct, though short, usually of 3 segments. Maxillae brush-like. Spiracles of abdomen minute and inconspicuous or wanting. Abdomen with 10 distinct segments, terminating in a pair of subanal processes; each segment with about 12 stiff, erect setae, longer on the posterior segments........................*Order Siphonaptera*, page 236.
69a, Abdominal segments usually with several pairs of spiracles, at least in the later instars; the last pair usually the same size as those on other segments; if larger, never situated close together on the dorsomeson. If any abdominal spiracles are wanting, all are usually wanting. Antennae always wanting. Maxillary palps never of more than 1 segment. Mandibles opposable, of chewing type. Generally sluggish, soft, white or yellow worms or grubs, usually tapering somewhat toward both ends, very commonly found in individual cells of wax, or paper, or in bodies of other insects or in galls..
..*Order Hymenoptera* (in part, bees, wasps, ants, parasites, and gall wasps), page 220.
69b—Frequently only 1 pair of spiracles on the abdomen; usually large, complex, and located adjacent to each other on the dorsomeson of the eighth segment or sometimes at the end of short subconical to long cylindrical tubes. If a number of abdominal spiracles of about the same size are present, the antennae distinct, or a spatula-shaped "breastbone" on the thorax. Mouth parts often consisting of a pair or group of hooks, not opposable for chewing and articulated to a pharyngeal skeleton (Fig. 77). Head often greatly reduced or apparently wanting, that end of body pointed. Very often found in dead plant or animal refuse or in bodies of living insects or other animals........................*Order Diptera*, page 230.

PUPAE

70*a*, Mandibles, maxillae, and labium recognizable on the head and of the form usually found in the chewing type of mouth parts. (If at end of long prolongation of head, bearing the antennae on its sides, *Order Coleoptera*, in part, *Rhynchophora*).....71.

70*b*—Mandibles, maxillae, and labium wanting, or, if present, with some or all of them modified into tubular, piercing or sucking organs. (If a tubular immovable beak-like prolongation of the head with antennae attached on its sides, *Order Coleoptera*, in part, *Rhynchophora*)...74.

71*a*, Antennae elongate, always with more than 12 segments; wing pads never elytra-like, *i.e.*, the front pair not unusually thick, generally with a number of veins distinct. ....72.

71*b*—Antennae either much shorter than the body, with fewer than 12 segments, or much longer than the body with numerous stout segments. Wing and leg cases rarely fused to the body. The antennae usually lie against the sides of the body curved around above the knees. Wing pads always elytra-like, with few or no veins. Prothorax large, and distinct from mesothorax..*Order Coleoptera*, page 205.

72*a*, Head normal in form, provided with gula. Clypeus or labium and mouth parts not elongated to form a beak or trunk. Prothorax distinct from mesothorax....73.

72*b*—Head abnormal in form, clypeus or labium and mouth parts greatly elongated to form a trunk-like immovable proboscis.............*Order Mecoptera*, page 212.

73*a*, Mandibles stout, curved, subcylindrical, and overlapping or crossing each other. Thorax and abdomen frequently with finger-like or filamentous tracheal gills. Pronotum small and inconspicuous. Nearly always aquatic; often in silken cases covered with sticks, pebbles, and the like.............*Order Trichoptera*, page 212.

73*b*—Mandibles large and stout, but never overlapping or crossing each other. Pronotum large and quadrangular. Wing and leg cases not tightly folded against body. The antennae not lying against body above the knees.................. ....*Order Neuroptera*, page 210.

74*a*, Mouth parts wanting. Front legs directed forward beneath the head. Body in a cocoon or enclosed in old, molted skins, or covered with a waxy separable shell ....*Order Homoptera* (Male Coccidae), page 198.

74*b*—Mouth parts usually present. Front legs not extended forward under head..75.

75*a*, Antennae, mouth parts, legs, and wings usually immovable, firmly grown fast or fused to the pleural and sternal surfaces and to each other. The maxillae extend as a pair of long, slender, adjacent plates along the ventromeson, forming a long proboscis. Pronotum small. Wing pads very large. Antennae usually lie parallel with ventral margin of wing pads. Either in a dense cocoon, in a cave in soil, or attached to some object by tail and girdle of silk around center of body... ....*Order Lepidoptera*, page 213.

75*b*—Appendages of the head and thorax freely movable without tearing, never grown fast to each other or to pleural and sternal surfaces. The maxillae never forming 2 long, slender plates extending along the ventromeson. Proboscis, if present, shorter than in 75*a*...76.

76*a*, Pupae rarely in silken cocoons, but often enclosed in a firm seed-like case or puparium, completely concealing all appendages. The puparium composed of a dried larval skin and bearing the large adjacent pair of larval spiracles or stigmal plates near one end, and often pupal respiratory horns on thorax. If not in a puparium, or when removed therefrom, wing pads very rarely wanting, consisting of a single pair which, with legs and antennae, are not usually grown fast to the body. Pronotum small, not distinct from mesothorax....*Order Diptera*, page 230.

76*b*—Pupae never enclosed in a puparium formed of the larval skin and bearing the larval spiracles at the end. If enclosed in a cocoon or case of any kind, the latter of silk without a pair of large adjacent spiracles or stigmal plates at one end. Appendages of head and thorax always exposed and movable except as they may be concealed by case or cocoon. Wing pads, if present, four in number.......77.

77*a*, Body subcylindrical, often a slender waist between thorax and abdomen. Wing pads usually present, the mesothoracic (first) pair veined and larger than meta-thoracic pair. Antennae always longer than the head. Compound eyes distinct.

Mandibles of chewing type, the maxillae and labium often elongate. Prothorax small, fused with mesothorax. Sometimes in cocoons which are usually parchment-like. Often in nests of many individuals......*Order Hymenoptera*, page 220.
77*b*—Body strongly compressed or flattened from side to side, not over ¼ inch long. Wing pads always wanting. Antennae always minute and shorter than the head. Compound eyes never present; sometimes with simple eyes. Mandibles long and slender, fitted for piercing. Pronotum large and conspicuous. Nearly always in dust- and trash-covered cocoons...................*Order Siphonaptera*, page 236.

ORDER THYSANURA

The Bristletails, or Silverfish

This is one of the smallest orders in number of known species, but it is included here for two reasons. It contains a few species known as silverfish, fishmoths, slickers, or firebrats that are great household pests. These are further discussed in the chapter on Household Insects (see page 763), and it need be said here only that these carrot-shaped swift-running somber-colored nocturnal pests (Fig. 103) are often injurious to stores of paper stock, book bindings or lettering, card labels or indices, rayons, wall paper, and similar starched or sized articles which they eat. Another reason for placing the Thysanura in the list for special study is that they may represent a very lowly offshoot from the insect family tree. Together with the Collembola, they make a group, called the *primitively wingless insects*, which are very different in structure and metamorphosis from the higher insects. Many insects such as fleas, lice, some ants, and aphids are wingless throughout all the stages of their life. Study has shown that the kinds just named are wingless by specialization or degeneration. The members of the order Thysanura, however, are believed to be insects that never had wings in their ancestry, having branched off from the insect stock before the latter evolved wings.

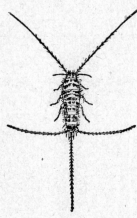

Fig. 103.—A common household silverfish (*Thermobia domestica* Packard), a little larger than natural size. (*From Kellogg's "American Insects," after Howard and Marlatt.*)

Because of the absence of wings in the Thysanura and their direct development, they have no metamorphosis. The young, also called nymphs, grow gradually toward the adult condition without any appreciable change in form or appearance except the change in size (Fig. 83). The mouth parts are of the chewing type, sometimes curiously set into the head cavity so that only the tips of the parts project from the surface. Compound eyes are present in some species, degenerate in others, and wanting in some; ocelli are usually wanting. The antennae are long and many-jointed. Most species have at the tail end of the body two or three bristle-like many-jointed appendages something like antennae, from which the common name bristletails is given. Sometimes these appendages are unsegmented and forceps-like. Some of the species have leg-like structures on the segments of the abdomen, a condition unique among insects. The body is very soft but is covered with scales or hairs that give it a shiny appearance and also account for the name fishmoths. Thysanura live a hidden life, being found in cracks and crevices about buildings, under stones, and in the soil among leaf mold. They are active chiefly at night or in darkness. When disturbed they scuttle about with great rapidity. Most of the species are thought to be scavengers.

Primitively wingless insects. Mouth parts chewing. Abdomen of 10 or 11 segments. Antennae very long, many-segmented. Usually a pair of long cerci at the posterior end, and sometimes three such antenna-like tails. Sometimes rudimentary legs on the abdominal segments. Tarsi one- to four-segmented. Malpighian tubes sometimes wanting. No metamorphosis.

ORDER COLLEMBOLA

THE SPRINGTAILS AND SNOW-FLEAS

The springtails are minute insects, rarely ⅕ inch long, often occurring in enormous numbers on the surface and in the soil of woodlands, in decaying vegetable matter, on the surface of stagnant water, on snow, in mushroom houses, and other damp places. They are seldom noticed except by those who seek them. The points of principal interest about them are that they are entirely and primitively wingless, that they generally have a forked muscular appendage at the tip of the abdomen which is used in springing into the air (Fig. 104), and that they occasionally become pests about maple-sap buckets, in mushroom beds, or on seedlings in greenhouses. They are often deeply colored.

Fig. 104.—The spotted springtail, *Papirius maculosus* Schött, about 6 times natural size. Note the adhesive tube between first and second pairs of legs, and the extended spring behind. (*From Kellogg's "American Insects."*)

Primitively wingless insects. Mouth parts chewing; sunken into the head. Compound eyes degenerate. Malpighian tubes wanting and the tracheal system very slightly developed. Never more than six abdominal segments, the first with forked adhesive organ or ventral tube, shown between the first and second pairs of legs in the figure, and the fourth with a forked spring. Antennae of few segments. Tarsi one-segmented. Development without a metamorphosis.

ORDER ORTHOPTERA

GRASSHOPPERS, CRICKETS, KATYDIDS, AND OTHERS

This large order includes some very primitive insects, such as the roaches; some that are well known to everyone, such as the katydids and crickets; some of the most curious of all insects, such as the walking-sticks and mole crickets, and some that are very destructive to crops, such as the grasshoppers. They are mostly large insects, and many of the species make sounds, so that they attract a great deal of attention. The groups that make noises have the hind legs unusually long and powerful, and progress by jumping. The sounds are not voices. They do not come from the mouth, but are produced by rubbing rough surfaces of the body together. In the crickets and katydids, specialized parts of the front wings are rubbed together to make the sounds. In some grasshoppers the front and hind wings are rubbed together; others rub the inner surface of the hind leg (the femur) over the outer edge of the front wing. It is the males that produce these sounds, the females almost never having special organs for this purpose. Both sexes, however, commonly have sound-perceiving organs or "ears" (see page 101).

All the Orthoptera have well-developed chewing mouth parts.   Many kinds feed on plants, others on small animals, and still others are scavengers.   The metamorphosis is a gradual one, the nymphs generally passing through five instars.   Most species live exposed on plants or hidden on the surface of the ground, but a few burrow into the soil, a few live in houses, and a very few take to the water.   The wings when present are four in number, the front pair narrow and thickened, but with the veins showing, and capable of bending without breaking, somewhat like a piece of leather.   The hind wings are thin; often brightly colored and with many veins; broadly triangular, and, when brought to rest, they fold along radiating straight lines from the base, like a fan and are laid back over the abdomen, so that they are covered by the front wings (see Fig. 197).   The wings are often incompletely developed. The antennae are generally long and prominent, the legs are long, the prothorax is large, and the tip of the abdomen is provided with a pair of cerci and often with a prominent ovipositor in the female (Fig. 105).

Fig. 105.—A common short-winged cricket, *Gryllus assimilis* (Fab.), female. Note the long, spear-shaped ovipositor and, just above its base, the two cerci. (*From Kellogg's "American Insects."*)

Here belong some of the largest of all insects, a Venezuelan grasshopper that measures $6\frac{1}{2}$ inches in length, and African walking-sticks 10 inches long.   In temperate latitudes most species spend the winter in the egg stage.

Mouth parts chewing.   Ligula four-lobed.   Prothorax large and distinct. Wings four, sometimes greatly reduced or wanting.   The front pair narrow, somewhat thickened and usually colored like the rest of the body but distinctly veined.   The hind pair membranous, broad, and folded fan-like when at rest.   Cerci and an ovipositor generally present.   Metamorphosis gradual.

Important economic species of this order are grasshoppers (pages 335, and 468), tree crickets (page 662), and roaches (page 781).

THE FAMILIES OF ORTHOPTERA

The families of Orthoptera are so distinct that certain writers have proposed separate orders for some of them.   The first three families discussed have been called the singing, jumping Orthoptera (Saltatoria), since most of the males have stridulating organs, and all have the hind legs noticeably longer and stouter than the others. In the females the ovipositor is usually well developed.   The tarsi have fewer than five segments in these three families.

Family Locustidae (formerly called Acrididae).   The Grasshoppers or Locusts (Figs. 194 to 197).—These are moderately long insects, a little *deeper than wide, of dark colors, variously mottled*, with prominent heads and large eyes, all characteristically active in the daytime.   The antennae are always much shorter than the body, the tarsi have three segments, and the "ears" (see Fig. 46) are found on the sides of the first abdominal segment.   The ovipositor consists of four short, finger-like pieces, well separated (Fig. 46).   These are used to thrust the eggs into the soil or into soft wood, $\frac{1}{2}$ inch or 1 inch below the surface (Fig. 194).   The eggs are formed into definite masses of 20 to 100 or more each, and are surrounded by a frothy, gummy substance which hardens to form a protective case for them.   There is usually only one generation a year.

Family Tettigoniidae (formerly called Locustidae). The Long-horned Grasshoppers, Green Meadow Grasshoppers, Katydids (Fig. 106), Cave and Camel Crickets.—These are more delicate and less hardy insects than the true grasshoppers, most commonly green in color, but of somewhat similar form. They are often nocturnal and can be distinguished from the grasshoppers by their very long antennae, often longer than the body, and by having four segments in each tarsus. The "ears" are on the base of the front tibiae (see Fig. 47,*A*), and the ovipositor is *sword-shaped*, the four pieces being flattened and closely appressed. The eggs are laid singly or in rows often on or in leaves, stems, twigs, or sometimes in the soil. There is only one generation a year. As a rule they eat either plants or small animals. Only a few of them are serious pests.

Family Gryllidae. Crickets (Fig. 105), Tree Crickets (Fig. 440), Mole Crickets.—These are usually somewhat short dark-colored Orthoptera with the tarsi three-segmented like the grasshoppers, but the antennae very long. In those kinds

FIG. 106.—The angular-winged katydid, *Microcentrum laurifolium* (Linné), and its eggs. Natural size. (*From Sanderson and Jackson, "Elementary Entomology," after Riley.*)

that have wings and produce sounds, the "ears" are found on the front tibiae (Fig. 47,*A*). The ovipositor is a long *spear-shaped* tube. The front margin of the wings is bent sharply down over the sides of the abdomen, like the edge of a box lid. The eggs are laid in groups in the soil or inserted into the stems of plants.

Crickets are nocturnal and negative to light. They feed upon a great variety of substances. The tree crickets are slender greenish insects that live among tall weeds, trees, or bushes and sometimes cause damage by slitting twigs and depositing their eggs in them (see Fig. 5,*B*). A curious fact is that the male has glands opening on the upper side of the thorax from which the female feeds at the time of mating. The mole crickets often burrow and make nests in the soil in the vicinity of water. They eat plant roots, other insects, and earth worms. The body is covered with fine, brown, velvety hairs. The front legs (Fig. 47,*A*) are remarkably developed, both for digging and to act like a pair of scissors, for cutting off small roots that are in their way. The ordinary black field crickets and the "cricket of the hearth" are other well-known representatives of this family. Crickets often do damage in grain fields by cutting the twine used to bind sheaves.

The last three families of Orthoptera have been called the mute nonjumping Orthoptera, since they make no particular sounds and do not have the hind legs enlarged for leaping. The tarsi are always five-segmented in these families, and the ovipositor is concealed or wanting.

Family Phasmidae (Sometimes Classified as a Distinct Order, Phasmoidea). Walking-sticks, Walking-leaves, Devil's Darning Needles (Fig. 107).—Our common representatives of this family are extremely elongate (the mesothorax especially long), cylindrical, wingless, with long stiff legs and very long, slender antennae. They are found feeding upon the foliage of trees, but also often resting about buildings. They move but little and in a very stealthy manner; they are doubtless often overlooked by their enemies because they look so much like slender sticks, being good examples of the phenomenon known as *protective resemblance*. The eggs are simply dropped by the females, one at a time, as they rest among trees, and have often been said to make a noise like the patter of rain drops, when the insects are abundant. They are harmless to man except that they may injure trees by eating the leaves. Some tropical species have well-developed wings, and some of them are broad, flat, green insects that look astonishingly like the leaves among which they live.

Family Blattidae (Sometimes Classified as a Distinct Order, Blattoidea). The Roaches or Cockroaches (Fig. 516).—This family has been called the running Orthoptera. The body is flattened, the head is bent downward and backward and is not prominent. The prothorax is very large, the legs are long and bristly, the hind pair only moderately larger than the others, but the coxae are all very large. The wings may be well developed, short, or wanting; they lie flat over the back, crossing over somewhat toward the tip.

Cockroaches usually have a bad odor, and they frequent all sorts of filthy places and dusty crevices, so that they are inexpressibly dirty. They are common in kitchens, bakeries, and restaurants and may be carriers of disease germs. The eggs are formed into packets of 16 to 40, enclosed in seed-like cases, and these so-called

Fig. 107.—A walking-stick in its usual environment among leaves and twigs. The head is just above the center of the picture, with the long front legs projecting straight forward toward the upper right. (*From Slingerland.*)

oöthecae are sometimes carried about by the female partly extruded from the abdomen until the nymphs hatch from them (see Fig. 516).

Cockroaches are very sensitive to cold. Many species live out-of-doors and in moist tropical countries are very abundant. A half dozen species have the habit of living in dwellings and other buildings. These will be further considered under the discussion of household insects.

Family Mantidae (Sometimes Classified as a Distinct Order, Mantoidea). Praying Mantes, "Mule Killers" (Fig. 108).—These remarkable creatures have curious habits and odd structures. The common name comes from the manner in which they hold up the fore part of the body, with its enormous front legs, as though in an attitude of prayer. They might also be called *preying* mantes, for they are the only family of Orthoptera that seems to be exclusively carnivorous, eating other insects.

The body is elongate, the prothorax and fore coxae especially long, and the front legs so modified that they can grasp small insects between the spiny tibiae and femora, the former closing against the latter like a knife blade against its handle (see Fig. 47, *G*). The wings are usually well developed, but the mantes commonly remain quiet in one place until some insect comes within reach. They sometimes cautiously stalk their prey. The eggs are laid in large masses an inch or so long, in a frothy, gummy substance, on the twigs of trees. These ferocious but beneficial insects are not commonly found north of the fortieth parallel of latitude. They never injure man or the large

FIG. 108.—A praying mantis, *Stagmomantis carolina* (Linné). About natural size. (*From Comstock's "Introduction to Entomology."*)

animals. Some tropical species of this family are very broad and have the fore wings so modified as to resemble leaves or flowers in shape, color, and venation.

ORDER DERMAPTERA (EUPLEXOPTERA)

THE EARWIGS

The earwigs are beetle-like insects easily distinguished from the Coleoptera by the prominent forceps at the rear end of the body and by their gradual metamorphosis.

FIG. 109.—A female earwig brooding over her nest of eggs. (*From Fulton, Ore. Agr. Exp. Sta. Bul.* 207.)

The mother broods over her nest of eggs in the soil and guards the young nymphs (Fig. 109). The food is variable, some species attacking plants in an injurious manner, others catching insects, and others feeding on decaying matter. Sometimes they become serious pests in and about houses, although the superstition that they attack people's ears is absurd.

Front wings horny, veinless, beetle-like, meeting in the middorsal line, but very much shorter than the abdomen. Hind wings membranous, ear-shaped, the veins radiating from the middle of the costal margin, folding both radially and transversely. Often wingless. A conspicuous pair of hooks or forceps at the end of the abdomen. Mouth parts of the chewing type; ligula two-lobed. Tarsi three-segmented. Compound eyes present; ocelli absent. Metamorphosis gradual.

ORDER EPHEMEROPTERA (EPHEMERIDA, PLECTOPTERA)

The May-flies, Lake-flies, or Shad-flies

These delicate, defenseless creatures (Fig. 110) often appear in surprising myriads in cities near lakes or streams, being strongly attracted by lights. They live but a few hours or a few days as adults, but this is often preceded by 1, 2, or 3 years of life beneath the water as nymphs. This ephemeral adult life is responsible for the order name.

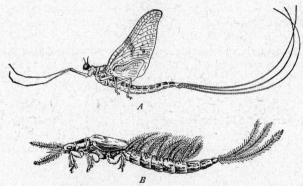

Fig. 110.—A May-fly (*Ephemera varia* Eaton). *A*, the adult; *B*, the nymph, about natural size. Note the tracheal gills on abdomen of the nymph. (*From Comstock, after Needham, Comstock Publishing Co.*)

The nymphs (Fig. 110,*B*) feed mostly on living or dead aquatic vegetation and breathe through tracheal gills. They are important as food for fishes. The adults have rudimentary mouth parts and take no food but are frequently a nuisance because their dead bodies accumulate in windrows on streets and about watering places and have an offensive smell.

Slightly chitinized adults, living but a short time, and molting once after reaching the winged, adult stage. Four, triangular, net-veined, gauzy wings, folding vertically over the back as in butterflies, when at rest; the hind pair much smaller, rarely wanting. Mouth parts of chewing type, but degenerate or wanting in adults. Antennae very short. Compound eyes and ocelli present. Mesothorax large. Two or three, very long, slender, many-jointed "tails." Genital openings double. Metamorphosis gradual, sometimes as many as 20 molts, the nymphs aquatic, elongate; the abdomen with two or three slender tails and seven (or fewer) pairs of tracheal gills along the sides and on the back of the abdomen. Tarsi one- to five-segmented. A single claw on each nymphal tarsus.

ORDER ODONATA

The Dragonflies (Fig. 111) and Damselflies

The adults are aerial; expert flyers; abounding about ponds and streams; catching and eating other insects on the wing (see page 56). The nymphs (Fig. 111,*1,2,3*) are aquatic, walking or hiding on the bottom of ponds and streams and catching other small animals for food. The development of a generation generally requires about a year, but some require several years. Odonata are of value to man by their feeding on horse flies and mosquitoes and as food for fish.

Large, often beautifully colored insects. Head vertical. Chewing mouth parts. Four membranous, slender, finely net-veined wings of about equal size, often not laid over the back when at rest. Near the middle of the front margin is a short, heavy cross-vein and a slight notch, like a joint; and near the tip of the front wings is a dark stigma. Antennae very small. Compound eyes large, ocelli present. Tarsi three-segmented. Abdomen very long and slender. Copulatory organs of male on the second abdominal segment, separate from the openings of the vasa deferentia. Metamorphosis gradual, the nymphs developing in the water and being provided with a very long extensible labium, used in capturing prey and folding like a mask over the face when not in use (Fig. 111,1, cf. 4). Molts numerous, 10 to 15.

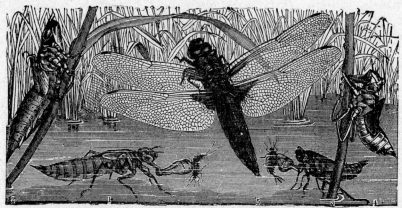

Fig. 111.—A dragonfly, adult and nymphs. At 1 and 3 two nymphs are shown catching prey with the extended labium. 2 is a mature nymph ready to change to the adult. 4 is the shed skin (exuviae) from which the adult (5) has emerged. (*From Sanderson and Jackson, after Brehm.*)

These are two suborders differing as follows:

The Anisoptera or Dragonflies	The Zygoptera or Damselflies
Hind wings broader at base, not folded but held in a horizontal position at sides of body when at rest. Strong flyers	Two pairs of wings of same size and shape, narrow at base; folded back over the abdomen or up over the back, like those of a butterfly, when at rest. Feeble flyers
Eyes do not project from the side of the head	Eyes projecting, constricted at the base
Eggs laid on the water or on aquatic plants or rarely in their stems	Eggs thrust into the stems of aquatic plants, often beneath water
Nymphs respire through tracheal gills inside of the rectum, and the forcible ejection of water from the anus propels the nymphs forward	Nymphs respire by three leaf-like tracheal gills, projecting from the end of the abdomen

ORDER PLECOPTERA

STONE-FLIES

These retiring insects are seldom seen except by those who seek them. They live near streams, resting on stones, trees, and bushes, or flying over the water. Some species are not attracted to lights. The adults (Fig. 112) have rather long antennae, but the tail-like cerci are much shorter than in the May-flies; and the hind wings are much broader than the front pair. The wings fold flat over the back, giving the insects a straight-sided, square-shouldered appearance when resting. The nymphs develop in the water, being common on the surface of stones in swift streams. They feed mostly on diatoms, algae, and other small plant and animal forms. In contrast with May-fly nymphs, they usually bear their gills on the thorax and have two claws on each leg (Fig. 84,A). Their only importance to man is as food for fishes.

Wings four, netted-veined; front pair narrow, hind pair very broad, folding like a fan; abdomen thinly chitinized. Compound eyes and ocelli present. Antennae long, filiform. Tarsi three-segmented. Mouth parts of the chewing type but often reduced and weak. Metamorphosis gradual. Nymphs with long antennae, and usually long cerci. Tracheal gills on the thorax. Tarsi with two claws.

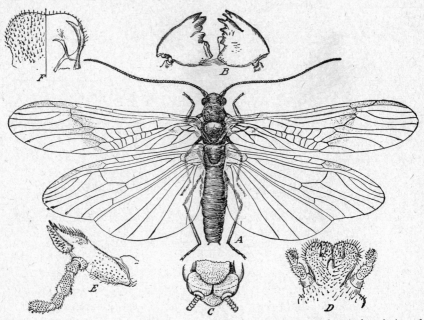

Fig. 112.—A stone-fly, *Taeniopteryx pacifica* Banks, much enlarged. *A*, dorsal view of adult about 4 times natural size; *B*, mandibles; *C*, last segment of female from below; *D*, labium; *E*, maxilla; *F*, labrum, dorsal half at left, ventral half (epipharynx) at right. (*From Newcomer, in Jour. Agr. Res., vol. XIII.*)

ORDER ISOPTERA

Termites or "White-ants" (Fig. 113)

These are the yellowish-white, soft-bodied "wood-ants" that are seen so often in countless numbers in logs, stumps, timbers of buildings, or wood lying in contact with the soil. They are not ants at all, being very different from ants in structure and in metamorphosis. One easy way to distinguish them from ants (see order Hymenoptera) is to note that the base of the abdomen is broadly joined to the thorax and not by a slender petiole.

Their chief resemblance to ants is in their colonial or social life. This is a curious condition, found in this order and in some of the Hymenoptera, in which there are individuals of several *castes*, that differ in structure and in duties in the same species and only a few of which become parents, all the others devoting themselves to the care of the thousands of offspring from the few "kings" and "queens."

In the termites the castes consist of: (*a*) dark-bodied males and females with four long wings, known as kings and queens; (*b*) short-winged males and females; (*c*) wingless males and females; (*d*) wingless workers; and (*e*) wingless soldiers. The kings and queens can reproduce all the castes, including others like themselves, which at certain seasons swarm out of the nest in enormous numbers, often appearing in or about infested buildings and giving the first warning or indication that termites are present. Their wings are long, narrow, whitish, or semitransparent, with many indistinct veins; and the two pairs are almost exactly alike in size and appearance. The wings are used only for a single, wedding flight, after which they are broken off along a suture of weakness near

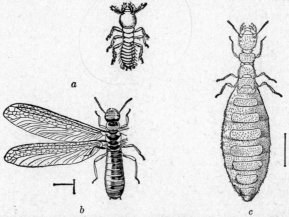

Fig. 113.—A common North American termite. Above at left a worker; below a male with right wings removed; at right a queen after she has lost her wings and her eggs have developed, distending the abdomen. Note the thick-waisted condition of all. (*From Kellogg's "American Insects."*)

the base. Then a nest may be started, copulation takes place, and the queen subsequently becomes enormously enlarged with developing eggs.

These swarming, winged kings and queens are the only termites that normally appear in the open air. The young and all the other castes stay in the same nest throughout their lives and are very delicate, thin-skinned, pale-colored, and blind. The kings and queens are not rulers in any sense, but simply reproductive individuals that do not even feed themselves. Among the termites, in contrast with the ants, bees, and wasps, the males live a long time and mate repeatedly with the queens, and all termite eggs are fertilized. The nymphs are all similar in appearance to workers but they early begin to show minor differences and develop into the different castes at the final molt.

Unlike the ants, bees, and wasps in which workers and soldiers are exclusively females, the termite soldiers and workers are of both the male and female sex but have the reproductive organs underdeveloped, and they never mate or lay eggs. One might almost say that they never pass beyond the nymphal stage. The short-winged and wingless males and females, which are sometimes present, may reproduce their own kind and workers and soldiers, but never the long-winged kings and queens, and so they cannot start new colonies.   All the wingless castes have a strong aversion to being exposed to the open air, and rarely show themselves outside of the nest, the soil, or the wood in which they are working.   This has usually been attributed to a negative reaction to light, but they appear to have little aversion to light, if they are completely protected from the open air.   If they must cross an exposed place, a covered runway of soil and body secretions is built.   The workers build nests, supply food, care for the eggs, feed the extremely young nymphs and the queens and kings, and perform all other duties for the colony, except reproduction and defense.   The mouth parts are of the chewing type; the mandibles of the soldiers become enormously enlarged, and are borne on very large heads.   Compound eyes and ocelli are wanting or greatly reduced, except in the kings and queens, and in the workers of certain tropical species, which forage in the open air for food.

The food of our common termites is primarily wood, and often dead, hard wood, in which there is apparently little except cellulose, a material not usually digested by the larger animals.   The digestive tracts are packed with protozoa, and it is believed that the termites live on the products of the digestion by the protozoa, and not on the cellulose directly.   Termites are great pests of all kinds of wood and products of wooden origin (see page 32 and Figs. 15 and 16).   They also attack living plants, hollowing out the stems.

Termites are really a tropical group.   In Africa and other tropical lands they build enormous nests, 12 feet tall, containing incalculable numbers of workers.   The queens of tropical species may reach a length of 4 or more inches, and they are said to produce eggs at the rate of 60 or more a minute.   Many tropical species cultivate mushroom beds or fungous gardens in their nests, which furnish their food.

Moderate-sized, thin-skinned, slender, social insects, consisting of several castes, living together in great nests or colonies, like ants.   Tarsi nearly always four-segmented.   Abdomen wide where it joins the thorax and has a pair of small cerci at the end.   Mouth parts chewing.   Metamorphosis simple.   Only the kings and queens normally reproduce; they are four-winged, often dark-colored.   The wings are equal in size, long, narrow, with membrane somewhat opaque and the veins indistinct, except along the costal margin and in the anal region; laid flat over their backs when not in flight, and usually broken off after the pairing flight, along a joint of weakness near the base.   The workers and soldiers are of both sexes, but do not regularly reproduce.   They are wingless, usually pale-colored, soft-bodied and eyeless.

ORDER CORRODENTIA (PSOCOPTERA, COPEOGNATHA)

Book-lice, Dust-lice, Bark-lice, "Deathwatches"

Often, as one opens an unused book or disturbs some old papers, a very small yellow insect runs across the page.   If one examines such an animated speck under a lens, one finds a wingless, soft-bodied insect with well-developed head, chewing mouth parts, small compound eyes, antennae nearly as long as the body, and six large legs (Fig. 114).   Relatives of this book-louse live on the bark of trees, vegetation such as corn stalks and greenhouse plants, and some of them have four membranous wings (rarely covered with scales like moths).   They look like aphids but have the veins of the wing peculiarly kinked as shown in Fig. 115. At rest the wings are held roof-like over the back. A structural peculiarity of this order is the "pick" or rod attached to each maxilla and working in and out of the softer part of the maxilla like "a piston sliding to and fro in its cylinder."[1]

The known importance of this order is not very great. Those that live indoors may occasionally become pests by feeding on paper, starch, grain, and other substances in damp places, and one species has been accused of spreading plant diseases. Some species are active at very low temperatures.

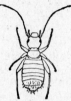

FIG. 114.—A book-louse, greatly enlarged. (*From Comstock's "Introduction to Entomology."*)

FIG. 115.—A winged bark-louse, 13 times natural size. (*From Kellogg's "American Insects."*)

Minute insects; wingless or with four membranous wings with few veins; the first pair larger and at rest held roof-like over the abdomen. Mouth parts chewing, with a curious rod in the maxilla. Compound eyes present, but ocelli wanting in the wingless forms. Prothorax small, tarsi two- or three-segmented and cerci wanting. The metamorphosis is very simple.

ORDER MALLOPHAGA

CHEWING LICE OR BIRD LICE[2]

There are two groups of lice that make their homes continuously on the bodies of warm-blooded animals. These two orders (Mallophaga and Anoplura) agree in never showing any trace of wings; in being flattened, oval, tough-skinned, external parasites; in gluing their eggs to the hairs or the feathers of the host; and in spending all their lives generation after generation on the same host animal. The members of the other order, the Anoplura, suck blood; but the Mallophaga have chewing mouth parts and subsist on bits of hair or feathers, skin scales, or the dried blood from scabs. Another noteworthy difference is that, whereas the Anoplura are confined to the mammals or hair-bearing animals, the Mallophaga occur on both birds and mammals. By far the greatest number, however, are found on birds, nearly every kind of wild or domesticated bird being attacked by one or more kinds of chewing lice. As a general rule, one species of louse will seldom live on more than one species of bird; the chicken, however, has seven common species. Cattle, horses, sheep, dogs, and cats, also are attacked, but none of the chewing lice live on hogs or man. The species that live on birds normally run very rapidly when they are exposed. But the ones that live on mammals, have the tarsus highly modified into a clamp for clinging to the hairs, and move about but awkwardly. The nature of this clamp is explained in the

[1] MAXWELL LEFROY, "Manual of Entomology," Longmans, 1923.
[2] These lice have often been called *biting lice* in an attempt to distinguish them from the Anoplura. This seems absurd, for, if the victims of the attacks were asked which kind of lice "bite," they would undoubtedly say the Anoplura, which insert sharp stylets into the skin to suck the blood, and not these forms, which, at most, only nibble at the skin.

discussion of the Anoplura, which have a similar structure (see Fig. 47,*E*). The metamorphosis is a very simple one; almost the only change from hatching to maturity being an increase in size, in thickness of the body wall, and in darkness of coloring.

The chewing lice are small, wingless, oval or elongate, flattened insects (see Figs. 566 and 567), *mostly ⅕ to ¹⁄₂₅ inch long, with large, broad heads, rounded in front, and bearing short, three- to five-segmented antennae often hidden in grooves of the head. The eyes are degenerate. The legs are not very large, tarsi one- or two-segmented, with one or two claws. The prothorax is distinct, the meso- and metathorax often more or less united. Thoracic and abdominal spiracles located on the ventral side. The skin is tough, often with heavily sclerotized plates of dark color and with scattered hairs. Mostly parasitic on the bodies of birds, but some species on mammals. Mouth parts chewing; not bloodsuckers; the labial palps, sometimes all four palps, wanting. No cerci. Metamorphosis gradual or wanting.*

Important species of Mallophaga are chicken lice and other lice of poultry (see page 862), the little red louse or cattle chewing louse (see page 831), the horse chewing louse (page 819), and the red-headed sheep louse (page 851).

ORDER THYSANOPTERA (PHYSOPODA)

The Thrips (Fig. 116)

The thrips are minute, slender, agile bugs, rarely as long as ⅛ inch. They live in flowers or on other parts of plants, feeding on the sap. Many species are serious pests of fruits, vegetables, flowers, and field crops.

With the order Thysanoptera we take up the first insects with other than the chewing type of mouth parts. The mouth parts of thrips (see Fig. 71) are unique, in some respects being intermediate between the chewing and the piercing-sucking types. They have two pairs of palps, like the former, but the mandibles and maxillae are suggestive of the form in the Hemiptera and Homoptera. The head bears well-developed compound eyes and ocelli, and well-developed but not extremely long antennae of six to nine segments. The head capsule tapers downward in the shape of a cone to a small mouth opening at its lowermost part (Fig. 116,7). Around this opening are the two maxillary and the smaller two labial palps. The labium is not elongated into a beak, but fused into the head cone as are the bases of the maxillae and the labrum. In and out of this funnel-like opening three slender jabbers or stylets operate by end thrusts to lacerate the epidermis of the plant. The sap which exudes is then sucked up through the mouth of the cone, there being no long food channel. The three stylets consist of the left mandible (the right being degenerated) and the two maxillae.

Two other structural peculiarities of thrips, the wings and the tarsi, need explanation. In some species one of the sexes and in other species both sexes, are wingless. Many of the species have four wings which are extremely narrow, almost without veins and laid back over the abdomen at rest. The wing membrane would scarcely be sufficient to sustain the insects in flight. The wing, however, is fringed with close-set long hairs to furnish resistance to the air in flight, somewhat like the long feathers on the wing of a bird. This gave the order the name Thysanoptera, which means *bristle wings*. The other peculiarity is in the foot. The tarsus has one or two segments but usually *no claws*. It ends in a hooflike or cuplike depression surrounding a small bladder that can be protruded or withdrawn by the insect. This characteristic gave the insects the name Physopoda (*bladder-footed*), which is sometimes used instead of Thysanoptera.

Thrips are very active insects, at least when disturbed. They spring or fly readily, they turn up their tails at one, as if to sting, and, according to one writer, they spend most of their time combing the hairs of the body. The compound eyes are small; ocelli usually present.

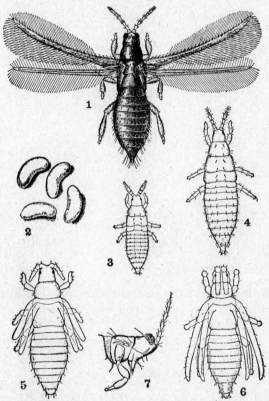

Fig. 116.—The pear thrips (*Taeniothrips inconsequens* Uzel). 1, adult; 2, eggs; 3, first instar nymph; 4, second instar nymph; 5, third instar nymph; 6, last instar nymph; 7, head of adult from the side. All greatly enlarged. (*Reduced from Moulton, U.S.D.A., Dept. Bul.* 173.)

The eggs are laid on the tissues of plants or, in some species, inserted into slits made by a sharp ovipositor. Parthenogenesis is common. There are four or more nymphal instars. The last two do not feed and may be quite inactive—a foreshadowing of the complete metamorphosis of the higher orders. There are a number of destructive species, such as the onion thrips (page 526), camphor thrips, greenhouse thrips (page 736), pear thrips, and oat thrips.

Small to minute, mostly phytophagous, slender-bodied insects, wingless or with two pairs of very slender, nearly veinless, equal wings fringed with long

*hairs and laid longitudinally over the back when not in use.   Mouth parts
rasping, asymmetrical, with two pairs of palps but only one mandible.   The
tarsus consists of one or two segments and terminates in a protrusile bladder.
Cerci wanting.   Metamorphosis gradual, but the larger nymphal stages
quiescent.*

ORDER HOMOPTERA (HEMIPTERA OR RHYNCHOTA, IN PART)

THE CICADAS, APHIDS, SCALE INSECTS, LEAFHOPPERS, TREEHOPPERS, AND OTHERS

For many years authors included the above kinds, together with the
leaf bugs, stink bugs, bedbugs, squash bugs, aquatic bugs, and their

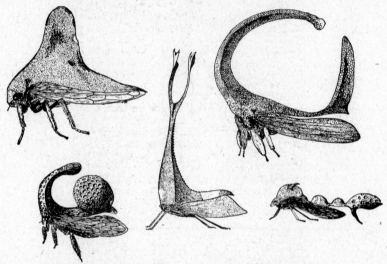

FIG. 117.—A group of treehoppers or membracids showing the remarkable development of
the pronotum.   *(From Funkhouser, N.Y. (Cornell) Agr. Exp. Sta.)*

relatives, all in the order Hemiptera.   They have important points in
common.   *They have a gradual metamorphosis.   The mouth parts of all of
them are of the piercing-sucking type,
with four stylets, and without palps.
There are four wings or none.   The
antennae are of few segments.   Compound eyes and ocelli are usually present and cerci wanting.*

FIG. 118.—A fulgorid, *Stobaera tri-
carinata* Say.   *(From Z. P. Metcalf, Jour.
Elisha Mitchell Sci. Soc., vol. 38.)*

They have such important differences, however, that most recent
authors separate the above groups
into two orders, Homoptera and
Hemiptera.   Except for the cicadas,
the Homoptera are mostly small, inconspicuous insects.   Some, however,
are brilliantly colored and many are of grotesque shapes.   Many of them
are wingless, at least in the female sex or under certain conditions.   *When
wings are present, they are four in number, of a nearly uniform, membranous,
or sometimes somewhat leathery texture; the front pair is longer, the hind pair*

often wider; they usually stand sloping roof-shaped over the abdomen when at rest: their bases are never abruptly thicker than their tips; and they do not overlap much at the tip. An ovipositor sometimes well developed. All terrestrial in habit. Male scale insects are an exception, having only one pair of wings (Fig. 364) and a complete metamorphosis. Some are very

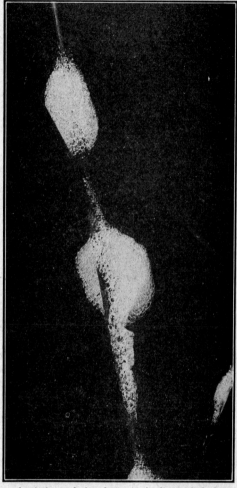

Fig. 119.—Masses of spittle made by the grass-feeding spittle bug, as a protection for the nymphs which live within it. About 3 times natural size. (*From Osborn, Maine Agr. Exp. Sta. Bul.* 254.)

degenerate in form; the female scale insects having neither body regions, eyes, wings, nor legs (Fig. 64,*Fc*).

The labium or beak of the Homoptera attaches to the head near its hinder part, often seeming to arise from between the front legs, which touch the head; sometimes, in sessile forms, the labium is very short, apparently wanting. The food is exclusively the sap of plants. There are many extremely destructive species, some carrying diseases from plant to plant. The following are among the more destructive species of this order: the San Jose, scurfy, and oyster-shell scales (see pages 575–580), the terrapin scale (page 625), greenhouse scale insects (page 739), apple aphids (page 580), corn root aphid (page 371), aphids of many vegetable crops (pages 477, 490), greenhouse aphids (page 745), greenhouse and citrus whiteflies (page 747, 681), apple and beet leafhoppers (pages 584, 545), the buffalo treehopper (page 610), and the periodical cicada (page 607). The figures and descriptions of these typical species of Homoptera should be studied to gain an idea of the nature and diversity of this order.

Fig. 120.—Adult spittle bug, parent of the nymphs which make the masses of spittle. (*From Garman, Conn. Agr. Exp. Sta. Bul.* 230.)

The important families are listed in the following:

SYNOPSIS OF THE ORDER HOMOPTERA

A. **Suborder** *Auchenorhynchi.* The Free Beaks. Labium plainly attached to head. Tarsi three-segmented. Active, free-moving insects. Antennae very small, ending in a bristle. Medium-sized to very large insects. Females generally with a stiff ovipositor.

Family 1. The cicadas ("locusts"), Family Cicadidae.

Family 2. The spittle bugs or froghoppers, Family Cercopidae (Figs. 119, 120)

Family 3. The treehoppers, Family Membracidae (Fig. 117).

Family 4. The leafhoppers, Family Cicadellidae (Jassidae).

Family 5. The planthoppers, Family Fulgoridae (Fig. 118).

B. **Suborder** *Sternorhynchi.* The Fused Beaks. Labium appears to attach to the thorax between the front legs. Tarsi one- or two-segmented. Females sluggish or sedentary, without a stiff ovipositor. Antennae larger, not ending in a bristle, sometimes wanting. Nearly all very small insects.

Family 6. The jumping plant lice, Family Chermidae (Psyllidae).

Family 7. The plant lice or aphids, Family Aphididae.

Family 8. The whiteflies, Family Aleyrodidae.

Family 9. The scale insects, Family Coccidae.

ORDER HEMIPTERA (HETEROPTERA, OR RHYNCHOTA, IN PART)

The True Bugs

These bugs (Fig. 121) *are like the Homoptera in having a gradual metamorphosis; two pairs of wings; antennae generally of five or fewer segments, though they may be long, often concealed in grooves; piercing-sucking mouth parts without palps; and the hind wings usually shorter and wider than the front pair. They are distinguished from the Homoptera by having the front pair of wings thickened and quite stiff about the basal half, the distal half abruptly thinner, usually membranous. When folded back they lie horizontally, or flat, over the back, and the membranous tips of the front pair overlap.* There are many wingless forms and some adults with short wings often in the same species with long-winged ones.

The labium or beak of the Hemiptera attaches well forward near the front end of the head, and the head is free from the coxae and longer than it is in the Homoptera. The prothorax is large and distinct, and the triangular scutellum (Fig. 121) *of the mesothorax separates the bases of the wings and is sometimes very large. The tarsi are three-segmented.* All this order are

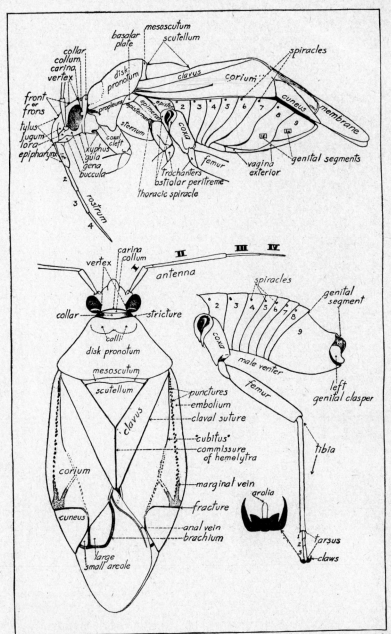

Fig. 121.—A bug of the order Hemiptera (*Deraeocoris fasciolus* Knight), with the principal parts of the body named. The part labeled epipharynx is called the *labrum*, and the rostrum, the *beak*, by some authorities. The parts of the wing labeled *clavus*, *corium*, and *cuneus* are thick and horny; the part labeled *membrane* is thin and transparent. (*From Knight, Minn. Agr. Exp. Sta. Tech. Bul.* 1.)

insect-like in form, typically flattened, there being no extremely degenerate species.   *They commonly possess scent glands that give them a distinct*

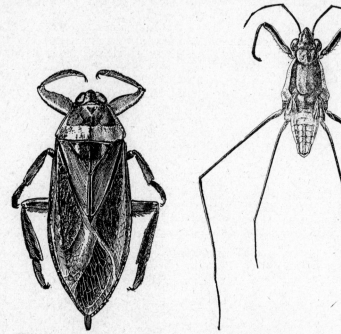

FIG. 122.—A giant water bug or electric-light bug, *Lethocerus americanus* (Leidy). Natural size. (*From Kellogg's "American Insects."*)

FIG. 123.—A water strider. The line shows actual length of body. (*From Osborn, "Agricultural Entomology."*)

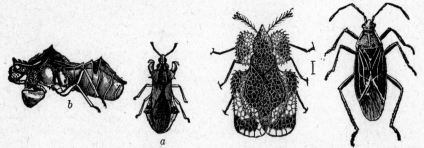

FIG. 124.   FIG. 125.   FIG. 126.

FIG. 124.—An ambush bug, *Phymata erosa* (Linné). *a*, adult from above; about twice natural size; *b*, the same from the side, more enlarged; note the enormous femur of the grasping front leg. (*From Fernald's "Applied Entomology," after Riley.*)

FIG. 125.—A lace bug (*Corythuca arcuata* Say). The line shows actual length. (*From Comstock's "Introduction to Entomology."*)

FIG. 126.—The box-elder bug, *Leptocoris trivitattus* (Say). Twice natural size. (*From Kellogg's "American Insects."*)

odor, usually offensive to man and probably defensive against their natural enemies.   The various species attack a wide variety of both plants and animals, always feeding on liquid parts only. Those that attack plants generally have the labium long, cylindrical, and straight; those that prey on other insects or on larger animals, have a short, curved, pointed labium. Many are adapted to live in the water.

A few of the most important species of Hemiptera are: the chinch bug (page 351), squash bug (page 499), harlequin cabbage bug (page 537), tarnished plant bug (page 479), apple red bug (page 585), bedbug (page 886), and assassin bugs (page 888). The figures of these typical Hemiptera should be examined to illustrate the above characteristics.

The important families are listed in the following:

SYNOPSIS OF THE ORDER HEMIPTERA

A. Suborder *Cryptocerata.*—The Short-horned Bugs. Antennae shorter than the head, nearly concealed on the under side of the head. Stink glands wanting.

1. *Aquatic Predaceous Bugs.*
   Family 1. Water boatmen, Family Corixidae.
   Family 2. Back swimmers, Family Notonectidae.
   Family 3. Water-scorpions, Family Nepidae.
   Family 4. Giant water bugs, Family Belostomidae (Fig. 122).

B. Suborder *Gymnocerata.*—The long-horned Bugs. Antennae at least as long as head and plainly visible at its sides.

2. *Semiaquatic Predaceous Bugs.*
   Family 5. Water striders, Family Gerridae (Fig. 123).

3. *Terrestrial Predaceous Bugs.*
   Family 6. Assassin bugs, Family Reduviidae.
   Family 7. Bedbugs, Family Cimicidae.
   Family 8. Damsel bugs, Family Nabidae.
   Family 9. Ambush bugs, Family Phymatidae (Fig. 124).
   Family 10*a*. Stink bugs, Family Pentatomidae (in part).
   Family 11*a*. Leaf bugs, Family Miridae (Capsidae) (in part).

4. *Terrestrial Plant-eating Bugs.*
   Family 10*b*. Stink bugs, Family Pentatomidae (in part).
   Family 11*b*. Leaf bugs, Family Miridae (Capsidae) (in part).
   Family 12. Lace bugs, Family Tingidae (Fig. 125).
   Family 13. Chinch bugs and others, Family Lygaeidae.
   Family 14. Squash bug, leaf-footed bugs, and others, Family Coreidae (Fig. 311).

ORDER ANOPLURA (SIPHUNCULATA OR PARASITA)

THE BLOODSUCKING LICE

These lice should be distinguished from the chewing lice. These are much the more serious kind. They live exclusively by sucking the blood of the host, and in so doing are very likely to be carriers of diseases. The members of this order are small, wingless, tough-skinned, flattened, usually dark-colored, external parasites. They attack all kinds of wild and domesticated mammals, but none are known to live on any other class of host. Two species attack man—the cootie, body louse or head louse (page 901), and the crab louse (page 904). The first of these is the carrier of typhus fever, trench fever and European relapsing fever. The largest species are about ¼ inch long. They can be distinguished from Mallophaga by the difference in mouth parts and by the relatively narrower, more pointed head. The antennae are short, but usually conspicuous, three-, four-, or five-segmented. The eyes are very degenerate or wanting and ocelli absent. The three thoracic segments are fused together and not well separated from the abdomen. The pleura are usually more strongly sclerotized than the rest of the segment. The legs are relatively very heavy and are highly specialized to enable the insects to hold on with a death-like grip to the hairs. The tarsus is a single segment with one enormous claw (Fig. 47,*E*). This is drawn around by muscles to clamp against a thumb-like projection from the end of the tibia, gripping the hair between the claw, the tarsal segment, the end of the tibia, and its thumb. In some species a spiny pad is forced out from the end of the tibia, further to strengthen the grip.

The mouth parts (Fig. 69) are unique. They are of the piercing type, but when not in use are completely withdrawn into the head of the louse. All one can usually see externally is a fringe of minute teeth at the foremost part of the head. At a short distance inside this fringe the mouth cavity or *buccal funnel* gives rise to two tubes. The upper tube is the pharyngeal duct, which leads to the stomach; the lower tube is only about as long as the head and ends blindly like an introverted glove finger. In this blind tube or *stylet sac* are three slender stylets believed to represent a pair of maxillae fused together, the hypopharynx, and the labium. They are all attached near the blind posterior end of the stylet sac, and their free ends reach to the mouth opening. Muscles are arranged to pull the stylet sac forward and drive the stylets into the flesh. Saliva is carried into the wound through the stylets, and blood is sucked up through the stylets, aided by a short trough-like passageway inside the mouth, and so to the pharynx and esophagus.

The eggs of Anoplura, often called "nits," are glued fast to the hairs or, in one species, laid in seams of clothing while it is being worn. The metamorphosis is very simple. The entire life is spent on the body of the host. For the most part, each species attacks one species of host and can live on no other.

Small, wingless, flattened, external parasites of mammals. Mouth parts retractile, piercing-sucking. Head narrow, pointed in front; eyes wanting or degenerate. All three thoracic segments fused; tarsi one-segmented, with a single, grasping claw. Thoracic and abdominal spiracles located on the dorsal side. Cerci wanting; ovipositor not elongate. Metamorphosis simple.

Only about 400 species are known, but their habits give to this order an importance out of all proportion to its size. The species attacking man belong in the family Pediculidae (see page 901). The family Haematopinidae includes the species attacking domestic animals. Important species are the hog louse (page 847), the bloodsucking horse louse (page 820), the short-nosed and long-nosed ox lice (page 832), and the bloodsucking sheep lice (page 851).

INSECTS WITH A COMPLETE METAMORPHOSIS

With the Anoplura we completed our discussion of the orders that have a gradual or simple metamorphosis. All the orders that follow have a complete or complex metamorphosis. A glance at Table VIII, on page 172, will show that the four largest orders are in this group, and a little checking of the species of economic importance, discussed in the following chapters, will show that most of our destructive insects are representatives of one of the orders with a complete metamorphosis. Apparently the complete metamorphosis is a very successful thing, in spite of the fact that insects undergoing such transformations must usually become adapted to three distinctly different environments during the course of the life cycle and have a helpless pupal stage. In the case of gradual metamorphosis, the same environment usually serves for both nymphs and adults; but in complete metamorphosis the larva must find suitable food, the pupa must be protected from enemies and adverse physical conditions, and the adult must again find food (generally different from that of the larva) or, if it does not feed, at least a suitable place to deposit eggs.[1]

ORDER COLEOPTERA

The Beetles and Weevils

This is the largest of all orders of insects. Two out of every five kinds of insects that have been discovered and named are beetles. This may be due in part to the fact that most of the orders have not been studied so extensively as the beetles have. Nevertheless ordinary observation will show that these insects are so numerous and ubiquitous, and of such diverse form, habit, and appearance, that not to know something about the beetles is to remain in ignorance of a large and very interesting part of our environment. It is well to study beetles also in self-defense, for they attack us at many points, feeding on growing crops of all kinds, from forest trees to greenhouse plants, as well as on stored foods and other possessions. It is noteworthy that there are practically no beetles that attack the larger animals, and very few *parasites* on any group, although many of them are predaceous on insects and other small animals.

Fig. 127.—A beetle, *Ligyrodes relictus* (Say), with wings spread; the left one showing how the large hind wings fold beneath the elytra or front wings as they are laid back over the abdomen. (*Original.*)

The most characteristic thing about beetles is their wings (Fig. 127). The mature insects have the front wings specialized into what are called *elytra* (pronounced ell'-it-ra). These are generally thickened by chitin so that they usually show no veins, and they are not flapped in flight, but serve as a pair of convex shields to cover the hind wings and the rather delicate-walled abdomen from above. When the insect is not flying they lie close over the abdomen, the inner edges of the two coming together to form a straight line from the prothorax back to the tip of the wings, neatly covering the mesothorax, the metathorax, and at least part of the abdomen (note Figs. 130 to 133). A small part of the mesothorax, known as the *scutellum*, remains exposed as a little triangle between the bases of the elytra but is never so large as it is in many of the Hemiptera. On account of the nature of the front wings, the prothorax is unusually distinct from the rest of the thorax and is often wrongly called "the thorax."

The hind wings, or second pair, are the real organs of flight. When a beetle flies, the front wings are held stiffly out at the sides of the body, while the hind wings, only, beat so rapidly that one can scarcely see

[1] See "The Life Cycle in Insects," *Ann. Ento. Soc. Amer.*, **13**: 133–201, 1920.

them.   These wings are about as wide as the elytra but commonly
one-fourth or one-third longer.   They do not project beyond the elytra
when the insect comes to rest, but by a remarkable automatic "joint"
the distal third or fourth is bent under, folding transversely but not
longitudinally, as the wings are laid back to the resting position (Fig. 127).
The second pair of wings are thin and membranous, with a few veins.
Sometimes these hind wings, and rarely both pairs, are wanting, or the
wings may be grown together so they cannot be moved.

The mouth parts of beetles are of the chewing type in both larvae and
adults, the same parts being recognizable as in grasshoppers or crickets

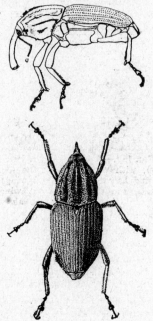

Fig. 128.—A snout beetle, *Calendra
aequalis* Gyllenhal, to show prolongation of
the head into a snout; above in side view,
below, dorsal view. (*From Ill. State
Natural History Surv.*)

Fig. 129.—A tiger beetle and, in the
tunnel, a larva of the same. About natural
size. (*From Sanderson and Jackson after
Linville and Kelly.*)

(see page 112).   One can usually tell something of the habits of a beetle by
examining its teeth.   If the mandibles are short, chunky, and with a
small number of blunt denticles on the mesal face, the grinding subtype
(see page 124), it indicates a species that takes plant food.   If the mandibles are elongate, come out to one or two sharp points toward the end, or
have the inner edges sharp for cutting, the grasping subtype (see page 124),
the insect is carnivorous and probably benefits us by eating other insects.
If the mandibles lack distinct teeth and are covered with stiff hairs, the
brushing subtype (see page 125), the insect is a harmless pollen feeder.

In one group of this order (the snout beetles) the head is prolonged
forward and downward into a cylindrical snout (Fig. 128) that varies
in length from shorter than the rest of the head to several times the
length of the whole body.   The student must be careful not to confuse

such a snout with the beak of the Homoptera or Hemiptera. This snout is not the mouth parts, but a part of the cranium, as indicated by the attachment of the antennae to it. It is not jointed, it is not furrowed down the front, and it, of course, contains no stylets. The mandibles, maxillae, and other mouth parts are found on the end of the snout, being very small but functioning as chewing mouth parts. This long snout is used for making a hole deep into the tissues of plants, in which the eggs are laid. It also enables its possessor to eat the tissues beneath the surface.

Beetles generally have no ocelli in the adult stage, though the compound eyes are well developed. The reverse is true of the larvae, which have a small group of ocelli at each side of the head but never compound eyes. The coxae of the hind legs (Fig. 47,*F*) are flattened out like a part of the body wall, instead of articulating into a socket of the latter, so that these legs at first sight appear to lack the coxae. There are no cerci.

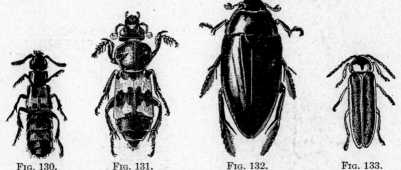

FIG. 130.   FIG. 131.   FIG. 132.   FIG. 133.

FIG. 130.—A rove beetle, *Creophilus maxillosus* (Linné). Enlarged ½. (*From Kellogg's "American Insects."*)

FIG. 131.—A burying beetle, *Necrophorus marginatus* (Fabricius). Enlarged ½. (*From Kellogg's "American Insects."*)

FIG. 132.—The giant water scavenger beetle, *Hydrous triangularis* (Say). Natural size. (*From Kellogg's "American Insects."*)

FIG. 133.—A firefly or lightning beetle (*Photinus scintillans* Say). Three times natural size. The insect gives off flashes of light, while flying at night, from the end of the abdomen on the under side. (*From Kellogg's "American Insects."*)

The Coleoptera embrace many very small insects and, from these, range upward in size to some tropical kinds that are several inches in length. There are very few insects likely to be confused with the Coleoptera. A few kinds of Hemiptera (Fig. 339) have a superficial resemblance to beetles, but the shield over the back is not formed of two wings but is a single piece, an overgrown thorax (*scutellum*). The earwigs look much like the rove beetles (*cf.* Figs. 109 and 130), but always show at the tip of the abdomen a pair of heavy forceps or pinchers.

Beetles all have a complete, sometimes a very complex, metamorphosis, known as a hypermetamorphosis (see page 570). The larvae of beetles (Figs. 224, 229, 315,*b* and 536) are generally called *grubs*, sometimes *borers*. They are very diverse in shape but can be recognized by the following characteristics: They commonly have six thoracic legs each ending in one or two claws, but some are entirely legless. There is never a series of prolegs, at most one pair at the tip of the body, and these have no crochets on the end of them. The head never has the *adfrontal area* that characterizes Lepidopterous caterpillars (Fig. 145). The spinneret at the middle of the labium also is wanting. The head is always distinct, usually dark-colored, bearing definite though often minute antennae. There are always a number of pairs of spiracles on the abdominal segments.

The pupae of the beetles (Fig. 224 and 279,h) can be distinguished from other pupae by noting that all the mouth parts are of the chewing type and that the antennae have fewer than 12 segments. The membranous sacs that enclose the antennae, legs, and wings are not grown fast ("cemented down") to the sides of the body but are free and can be moved about. Beetle pupae are not protected by the dense silken cocoons characteristic of moth pupae. They may have thin cocoons but are often openly exposed on leaves; at other times hidden in the soil, in burrows in wood, or covered over with foreign material accessible to the larva as it prepares for pupation.

In this order both larvae and adults are commonly injurious and in the same manner. Sometimes, however, the habits of larvae and adults are different, so that they may be injurious in two totally different ways; or one or the other, only, may be destructive. In a few cases one life stage is harmful to us and the other beneficial; thus the blister beetles (see page 509) as adults feed on the foliage of potatoes, chard, asters, and

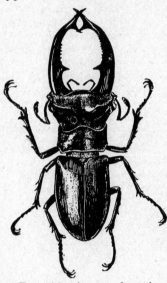

FIG. 134.—A stag beetle, *Lucanus elaphus* Fabricius, male, natural size. (*From Kellogg's "American Insects."*)

other plants, while their larvae are helpful to man by eating the eggs of grasshoppers in the soil. The lady beetles, except for a few species, and many species of ground beetles are highly beneficial, both as larvae and as adults, by devouring injurious insects (see page 58). There are numerous aquatic species.

The Coleoptera are minute to very large, usually heavily sclerotized, robust insects. The front wings are much thickened, veinless, and meeting in a middorsal straight line; the hind wings membranous, with few veins, and the apex folded under transversely when at rest; sometimes wanting. Mouth parts of typical chewing type, although in the snout beetles they are reduced and placed at the end of a slender trunk-like snout easily mistaken by the tyro for piercing mouth parts. Ocelli generally wanting, antennae mostly of 10 or 11 segments. Prothorax very distinct from meso- and metathorax, and freely movable against the mesothorax; meso- and metathorax somewhat fused and united with the abdomen. Tarsi mostly of four or five segments rarely three. Hind coxae plate-like, immovable. Metamorphosis complete. Larvae worm-like or shaped like Thysanura, sometimes with prominent cerci; usually with six thoracic legs and not more than one pair of prolegs; rarely apodous; the legs in most species with a single tarsal claw. Spiracles on principal segments. No adfrontal area. Pupae with appendages nearly always free (obtect in Staphylinidae and Coccinellidae); the body wall generally thin, soft, and pale-colored; rarely in cocoons. Cerci wanting in adults and often in larvae also. No firm ovipositor.

Among the many, very destructive species are the following, to which the reader should refer for figures and descriptions of typical species: white grubs (page 374), wireworms (page 377), billbugs (page 357), corn rootworms (page 380), clover bud weevil (page 422), alfalfa weevil (page 420), cucumber beetles (page 495), Colorado potato beetle (page 507), flea beetles (page 468), Mexican bean beetle (page 487), bean and pea weevils (page 810), sweetpotato beetles (page 518), asparagus beetle (page 550), bark beetles (page 718), flatheaded and roundheaded apple borers (page 590), Japanese beetle (page 621), strawberry weevil (page 670), plum curculio (page 594), tobacco beetle (page 785), buffalo beetles or "moths" (page 787), grain beetles (page 798), granary and rice weevils (page 796), mealworms (page 797), and larder beetle (page 792).

Only the most important families are listed in the following:

SYNOPSIS OF THE ORDER COLEOPTERA

A. **Suborder** *Adephaga.*—The Predaceous Beetles. Mostly feed on other insects. Largely beneficial to us. Tarsi of five segments. Antennae generally filiform. Hind wings with most of the typical veins preserved, and with some cross-veins. Ventral part of first segment of abdomen divided into three areas by the hind coxae—a very small median piece between the coxae and two large side pieces. Larvae shaped like Thysanura, active, carnivorous, with six segments in each leg and two claws at the end of the leg (Fig. 88, *K*).

   1. *Terrestrial Predaceous Beetles.*
      Family 1. Tiger beetles, Family Cicindelidae (Fig. 129).
      Family 2. Ground beetles, Family Carabidae.
   2. *Aquatic Predaceous Beetles.*
      Family 3. Predaceous diving beetles, Family Dystiscidae.
      Family 4. Whirligig beetles, Family Gyrinidae.

B. **Suborder** *Polyphaga.*—Beetles of a great variety of form, habit, and economic importance. All have the first ventral segment of the abdomen in a single piece, not divided by the hind coxae. Hind wings with the venation much reduced; cross-veins often wanting. The legs of the larvae always end in a single claw and have five segments or fewer. Larvae of extremely varied habits, often worm-like in shape.

   3. *The Short-winged Beetles or Brachelytra.*—Elytra short, exposing much of the abdomen. Tarsi five-segmented.
      Family 5. Rove beetles, Family Staphylinidae (Fig. 130).
      Family 6. Carrion or burying beetles, Family Silphidae (Fig. 131).
   4. *The Club-horned Beetles or Clavicornia.*—Antennae clavate. Tarsi five- or three-segmented.
      Family 7. Water scavenger beetles, Family Hydrophilidae (Fig. 132).
      Family 8. Flat bark beetles, Family Cucujidae.
      Family 9. Lady beetles, Family Coccinellidae (Tarsi three-segmented).
      Family 10. The skin beetles, Family Dermestidae.
   5. *The Saw-horned Beetles or Serricornia.*—Antennae serrate. Tarsi five-segmented.
      Family 11. Checkered beetles, Family Cleridae.
      Family 12. Firefly or lightning beetles, Family Lampyridae (Fig. 133).
      Family 13. Soldier beetles, Family Cantharidae.
      Family 14. Metallic wood borers or flatheaded borers, Family Buprestidae.
      Family 15. Click beetles, Family Elateridae.
   6. *The Beetles with Different-jointed Tarsi or Heteromera.*—Five segments in tarsi of front and middle legs; four segments in hind tarsi.
      Family 16. Darkling beetles, Family Tenebrionidae.
      Family 17. Blister beetles, Family Meloidae.

7. *The Leaf-horned Beetles or Lamellicornia.*—Tarsi five-segmented throughout. Antennae lamellate, a cylindrical basal part and a number of flattened leaf-like segments at the tip.

Family 18. The stag beetles, Family Lucanidae (Fig. 134).

Family 19. The lamellicorn beetles, Family Scarabaeidae.

8. *The Plant-eating Beetles or Phytophaga.*—Tarsi apparently four-segmented.

Family 20. The long-horned beetles, Family Cerambycidae.

Family 21. The leaf beetles, Family Chrysomelidae.

Family 22. The pea and bean weevils, Family Mylabridae (Bruchidae).

9. *The Snout Beetles or Rhynchophora.*—Head often produced into a long snout. Tarsi apparently four-segmented. Antennae clubbed and elbowed.

Family 23. Typical weevils, Family Curculionidae (Fig. 128).

Family 24. Engraver or bark beetles, Family Scolytidae (Ipidae).

ORDER STREPSIPTERA

STYLOPS OR TWISTED-WING PARASITES

This is a small order of minute, internal parasites of insects, of little or no importance to man because they destroy mostly wild bees, wasps, and a few leafhoppers and planthoppers. They are so abnormal in structure as to constitute an order of very distinct characteristics.

Males (Fig. 136) *with stalked eyes, flabellate (branched) antennae, and degenerate chewing mouth parts. No ocelli. Wings four, but the first pair reduced to mere short clubs, the hind pair large, triangular, folding fan-like and without cross-veins. Tarsi often without claws. Metathorax extraordinarily large. No cerci. Female* (Fig. 135) *without legs, wings, eyes, or antennae, and the mouth parts mere vestiges; worm-like and living in the interior of insects throughout life except for the fused head and thorax, which project like a lump, or tumor, between two segments of the host. Her body is enclosed in the last larval skin, which is open just behind the head into the brood chamber. This communicates with the unpaired genital pores on several abdominal segments, and serves for fertilization and also for the escape of the larvae. Metamorphosis is complete, with a hypermetamorphosis.*

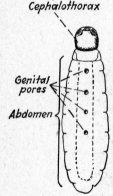

Cephalothorax

Genital pores

Abdomen

FIG. 135.—A full-grown or adult female strepsipteron (*Xenos vesparum*). (*Redrawn from Imms, "General Textbook of Entomology," after von Siebold.*)

ORDER NEUROPTERA (MEGALOPTERA AND PLANIPENNIA)

APHID-LIONS, DOBSON-FLIES (FIG. 137), ANT-LIONS (FIG. 138), AND OTHERS

This order includes species which are mostly predaceous and beneficial to man. The aquatic forms serve as food for fish, and the terrestrial ones are very beneficial in their larval stages by preying upon aphids, ants, and other small insects (see also page 57).

Insects of variable size, large or small, soft-bodied, with four large leaf-like wings of nearly equal size and texture; often colored; generally finely net-veined and held roof-like over the abdomen at rest. Often with many, extra, forked veins around the margin of the wings, especially in the costal region, and with the radial sector pectinately branched. Tarsi five-segmented. Mouth parts of the chewing type, though secondarily adapted in many of the larvae (Fig. 32,b,d) for sucking the blood of other small animals (grasping-sucking subtype) (see Fig. 70), each mandible having a ventral groove closed by the maxilla to form a sucking tube. Antennae generally long, cerci wanting, tarsi of five segments. Larvae spindle-shaped, much like Thysanura, carnivorous, many of them aquatic, with tracheal gills and sometimes with paired, lateral, often jointed, filaments on many of the abdominal segments (Fig. 137). No cerci. Aquatic or terrestrial in habit. The pupae with appendages free; sometimes in cocoons of silk spun from the anus of the larvae.

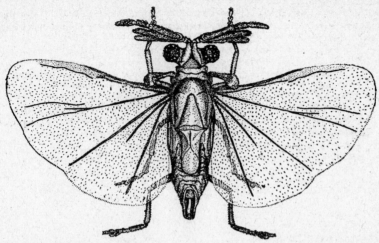

Fig. 136.—An adult male strepsipteron (*Muirixenos dicranotropidis*). Note the flabellate antennae, the club-like front wings, and the veinless hind wings. (*From Pierce, Proc. U. S. Nat. Museum, vol. 54.*)

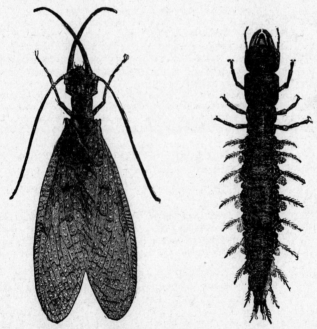

Fig. 137.—The dobson-fly, *Corydalis cornuta* Linné; adult male on the left and its larva, the hellgrammite, on the right. Natural size. Note enormous mandibles of the male and the short hair-like tufts of tracheal gills on the abdomen of the larva. (*From Sanderson and Jackson, "Elementary Entomology," after Comstock.*)

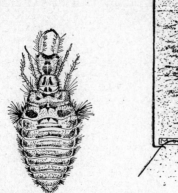
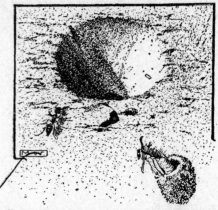

Fig. 138.—The ant-lion or "doodle bug;" *Myrmeleon* sp.; larva on the left, 3 times natural size. Note the elongate mandibles of the mandibulo-suctorial type (*cf.* Fig. 70). On the right is shown the sand pit or trap made by the ant-lion to catch ants, at the bottom of which the larva lives; about ½ natural size. Below is the empty pupal skin and the sand-covered cocoon from which the adult has escaped. (*From Kellogg's "American Insects."*)

ORDER MECOPTERA (PANORPATAE)

The Scorpion-flies

This order includes some seldom-noticed, terrestrial, and predaceous insects of little or no economic importance. The common name is given because the males of some of the species (Panorpa) have the tip of the abdomen swollen and carry it curved upward in a manner suggestive of the sting of a scorpion (Fig. 139, *cf.* Fig. 95).

Moderate to small insects with elongate bodies and long, many jointed antennae. The most distinctive characteristic is the prolongation of the head, maxillae, and labium into a broad snout two or three times as long as the width across the eyes. At the end of this snout are the chewing mouth parts. The four, long, rather narrow wings (rarely wanting) have rather numerous cross-veins, are often spotted, and are laid over the abdomen at rest. Large compound eyes and three ocelli usually present. Tarsi five-segmented. Cerci small. The larvae are caterpillar-like but are distinguishable from Lepidoptera because some of them have eight pairs of prolegs without crochets, while others have no prolegs. They live in soil or in moss. The pupae have the wing cases and leg cases free from the sides of the body.

Fig. 139.—A scor-pion-fly, *Panorpa rufescens* Ramb., twice natural size. Note the "sting-like" or scorpion-like tail. (*From Kellogg's "American Insects."*)

ORDER TRICHOPTERA

Caddice-flies (Fig. 141) and Caddice-worms (Fig. 140)

This order includes species of soft-bodied, weak-flying, moth-like insects that are seldom seen except along streams or lakes, in the water of which the larvae develop. The adults are commonly less than 1 inch in length and have four brownish wings covered with hairs and held sloping roof-like at the sides of the body. The mouth parts are greatly reduced. There are no mandibles and probably most adults take no food.

The eggs (Fig. 78,*P*) are laid in ropes or masses of gelatin-like or cement-like substance in or near the water, often under stones in streams. The larvae are better known than the adults. They make cases that cover their bodies, except the head and legs (Fig. 90,*F*,*G*,*N*) and often drag these protective cases about with them on the bottom of ponds or streams. They feed on small animal life or bits of vegetation and are among the commonest of insects in the water. The cases are very curious, variable, and characteristic for the species. They are lined with silk and more or less open at each end; in shape cylindrical, ovoid, or spiral, and covered with pebbles, sticks, or pieces of leaves, or tiny snail shells. A few kinds make miniature, "fish nets," which are constructed in swift-flowing water and which strain out small particles of food from the current.

Medium-sized insects with four, similar, membranous wings clothed with rather long hairs, the hind pair usually shorter and broader. The wings stand roof-like over the abdomen in repose; longitudinal veins numerous, but cross-veins few. A semitransparent, whitish spot, devoid of hairs, near the center of each wing. Compound eyes generally small; ocelli three or none. Mouth parts modified from chewing type by reduction, the

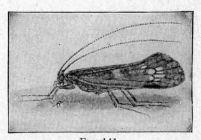

Fig. 140.   Fig. 141.

Fig. 140.—Caddice-fly larvae; above, larva in its case with head and legs projecting in normal position; below, larva removed from its case to expose the tracheal gills on the abdomen. (*From Fernald's "Applied Entomology," after Leuckhart.*)

Fig. 141.—An adult caddice-fly. (*From Kellogg's "American Insects," after Needham.*)

mandibles often being absent, but the palps well developed. Antennae long, filiform; legs long, coxae large, tarsi five-segmented, tibiae with spurs. Metamorphosis complete; larvae and pupae aquatic, living in cases and breathing by abdominal gills. Larvae worm-like, with three pairs of long thoracic legs and one pair of hook-like prolegs on the last segment of the abdomen. Larval tarsi one-segmented with a single claw. Head and thorax more sclerotized than the abdominal segments. Legs rather long. Antennae and labial palps very small. Most larvae with thread-like tracheal gills on abdominal segments. Pupae with antennae, legs, and wings free, and a pair of large mandibles.

The principal importance of the Trichoptera is as food for fish. Both larvae and adults are of great value as food for trout. One species is said to be a pest in water cress beds in England.

ORDER LEPIDOPTERA

Moths, Millers, Butterflies, and Skippers

This is the second largest order of insects, and one of the most destructive. Its members are well marked, both as adults and as larvae, and scarcely likely to be confused with any other order. Nor are the insects of another order likely to be classed as Lepidoptera by mistake. The name means *scaly-winged*, and the most characteristic thing about these insects is the layer of short, flattened, hairs, or scales that typically covers both surfaces of the wings and practically all other parts of the body. The scales (Fig. 143) are of many shapes, and one can find almost every gradation from broad, flat, plate-like scales to slender, cylindrical hairs.

All of them have a projection or pedicel at one end which fits into a cup-like cavity in the cuticula of the wing membrane. In the lower moths they are irregularly scattered over the wings, but in the most specialized ones they present a very perfect arrangement, overlapping on both the sides and the ends, like slates or shingles on a roof (Fig. 142).

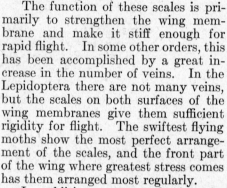

The function of these scales is primarily to strengthen the wing membrane and make it stiff enough for rapid flight. In some other orders, this has been accomplished by a great increase in the number of veins. In the Lepidoptera there are not many veins, but the scales on both surfaces of the wing membranes give them sufficient rigidity for flight. The swiftest flying moths show the most perfect arrangement of the scales, and the front part of the wing where greatest stress comes has them arranged most regularly.

FIG. 142.—A piece of the wing of a butterfly, *Danaus archippus* Fab., showing how the scales overlap at sides and ends. On the right some of the scales have been removed to show the cup-like pits out of which the pedicels of the scales grow. Greatly magnified. (*From Kellogg's "American Insects."*)

In addition to strengthening the wing membrane, the scales give protection to most parts of the body, and in them are resident the colors for which moths and butterflies are justly celebrated. If the "dust" (scales) is rubbed from the wings of a moth or butterfly, its characteristic color is lost. The scales are often ornamented with longitudinal ridges or striae, which may occur as closely as 35,000 ridges to the inch. They are very regular, and many of the brilliant colors of these insects are produced by the diffraction of light rays by the striae or by lamellae in the scales, rather than by pigments.

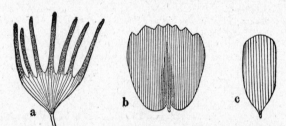

FIG. 143.—Scales from three different species of moths and butterflies. Note the pedicels that attach to cup-like sockets of the wing cuticula and the more or less parallel striae or ridges, which often produce brilliant colors. (*From Kellogg's "American Insects."*)

Aside from the possession of these scales, the wings of Lepidoptera are not highly specialized (Fig. 49). They are usually very broad, and subtriangular in outline, the front pair somewhat larger. They are often too large to be very effective in flight, since they cannot be flapped up and down rapidly enough. In general we note that the swiftest flying insects are those with small wings, and, in this order, the swift-flying ones are such as have rather narrow wings, *e.g.*, the hawk moths (Fig. 144) and clear-winged moths (Fig. 411).

The most highly specialized thing about the Lepidoptera is the mouth parts of the adults. These are probably the most highly specialized of all insect mouth parts. One finds as he examines the head (Fig. 75) a pair of very hairy or scaly palps (the labial palps) projecting forward

Fig. 144.—A swift-flying hawk moth, *Theretra tersa* Linné, natural size. (*From Ill. State Natural History Surv.*)

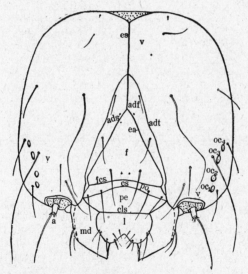

Fig. 145.—The head of the armyworm, larva. *a*, antenna; *adf*, adfrontal sclerite; *adt*, adfrontal suture; *cls*, clypeolabral suture; *cs*, clypeal suture; *ea*, epicranial arm or frontal suture; *es*, epicranial stem or coronal suture; *f*, front; *fcs*, frontoclypeal suture; *l*, labrum; *md*, mandible; *oc₁* to *oc₄*, ocelli; *pe* and *po*, the clypeus; *v*, vertex. (*Adapted from Ripley, Ill. Biol. Monograph, vol.* 8.)

at the sides of the mouth. Between them arises a long slender tube for sucking up liquid foods. Its real make-up is evident from a cross-section (Fig. 75,*B*). From the figure it can be seen that it is double in nature, each half having a groove along its inner (mesal) face, and the two halves are so closely locked together that they form an airtight tube up which exposed liquids can be drawn by suction. This is called the *siphoning type* of mouth parts (page 128).   When not in use, this tube is carried coiled up like a watch spring, so closely beneath the head that it is inconspicuous. Some moths take no food during the adult stage, and in these cases the tube is wanting.

It must be emphasized that this proboscis is not a *piercing* structure. It is too flexible to be thrust into plant or animal tissues.   This means that these insects must satisfy themselves with liquids freely exposed. They have become adapted to taking a very special type of food—the nectar concealed in open cups in the corollas of flowers, particularly those too deep to be reached by other kinds of insects.   The exact length of it is correlated with the kind of flowers the particular moth or butterfly visits. It is often longer than the body, and in a number of cases measures 5 or 6 inches.   It is said that an African moth was once taken with a proboscis

Fig. 146.—The cecropia moth (*Samia cecropia* Linné), a common and beautiful, giant silk-worm moth.   (*From Fernald's "Applied Entomology."*)

measuring 10 inches in length.   From this a biologist predicted that a flower would be found with a corolla 10 inches deep; this was subsequently found.   The removal of this nectar does not injure the plant; in fact the plant usually benefits by the visits of these insects to its flowers, since in this way cross-pollination is generally brought about.

In rare cases, the tip of the proboscis is provided with stiff spines, sharp enough to lacerate the skin of a ripe fruit.   Thus cotton leafworms (Fig. 278) develop only on cotton, but in late summer the adults sometimes fly northward in great swarms, at least as far as Canada.   If they alight in orchards or vineyards, they may do much damage by puncturing the ripe fruits with their "tongues."   This is very exceptional, however, and usually moths and butterflies can do no injury in the adult stage.   When we speak of an injurious moth, such as the clothes moth or the gipsy moth, we refer to the injury caused by the larva or caterpillar of that species.   The economic importance of the Lepidoptera arises almost entirely from the activities of the larvae.   This simplifies the problem of control somewhat, as compared with beetles, for example, since only one injurious stage need be considered.

The larvae of Lepidoptera are called caterpillars, and very often "worms." They have chewing mouth parts and are among the world's greatest pests. All the thousands of kinds known are remarkably similar in structure. The shape is generally nearly cylindrical (see Figs. 55 and 268). The body is composed of 13 segments besides the head. Of these the first three, or thoracic, segments have each a pair of jointed legs, terminating in a single claw. The abdominal segments bear unjointed, soft, fleshy projections of the body called *prolegs*, typically one pair each on the third, fourth, fifth, sixth, and tenth segments of the abdomen. Frequently some of these pairs and, in some cases, all the prolegs are wanting (Fig. 344). Insects of several other orders have larvae with prolegs, but these are the only insects that have the prolegs

Fig. 147.—A skipper butterfly (*Epargyreus tityrus* Fabricius) natural size. Note the recurved or hooked antennae that characterize skipper butterflies. (*From Fernald's "Applied Entomology."*)

armed with a number of fine hooks known as *crochets* (Fig. 55,*4*). These crochets, which enable the insect to hold on so tenaciously to a leaf or twig, are arranged in circles or rows across the apex of the proleg.

The head of a caterpillar is usually well developed (Figs. 55 and 145). There is a group of simple eyes at each side of the head (Fig. 145,*oc₁* to *oc₄*), the number varying from two to six pairs. The antennae are very small. A very characteristic thing about the larvae of this order is the presence, in at least the last instar, of the *adfrontal areas* (*adf*)—slender sclerites bordering the epicranial suture (*es, ea*) that do not occur in insects of other orders. Another characteristic of Lepidopterous larvae is the presence, near the end of the labium, of the spinneret from which the silk exudes.

The pupae of moths are typically enwrapped in a silken case, called a *cocoon*, which is made from saliva secreted by the full-grown caterpillar (see Fig. 90,*A,E,I*). Some of them lie buried in the soil, and a few are formed in tunnels in wood or in other larval habitats. The pupae of butterflies are generally naked "chrysalids" fastened to plants or other supports by a small pad of silk or a girdle (Figs. 89,*C*, 265,*C*). Regardless of the nature of protection, the pupae of Lepidoptera (Figs. 89,*A,B,C,D*) can be recognized from other orders by these two features: (*a*) the leg cases, antennal cases, and wing pads are fastened down to the sides of the body and immovable; (*b*) the long maxillae in the pupal stage appear as two long slender sclerites along the mid-ventral line.

Adult Lepidoptera are minute to very large, soft-skinned, fragile insects, well characterized by the coiled, siphoning mouth parts (sometimes absent) and the scaly wings (sometimes wanting in females). Wings enormous in proportion to size of body; membranous but not transparent, because characteristically covered on both upper and under sides by minute overlapping scales which often form beautiful color patterns and which easily rub off. Similar scales on most other parts of the body. Wings broad, triangular, the front pair larger, few cross-veins; not folding much at rest. Compound eyes large. Ocelli two or none. Legs relatively small, tarsi five-segmented. Prothorax very small. No cerci. A firm ovipositor very rarely present. Metamorphosis complete. The larvae typically worm-like, with chewing mouth parts; paired spiracles on the prothorax and first eight abdominal segments; two to five pairs of prolegs, with crochets, on the abdomen; and adfrontal areas on the head (Fig. 145,*adf*). *Pupae with appendages usually fast to the body wall, often in cocoons. Abdominal segments often capable of much wriggling.*

Among the many pests of crops and stored products, the following should be noted for figures and descriptions of typical species; armyworm (page 341), cutworms (page 346), European corn borer (page 360), tomato hornworm (page 523), swallowtail butterfly (page 545), cabbage butterfly

Fig. 148.—The celery caterpillar or black swallowtail butterfly (*Papilio polyxenes* Fabricius) and its life stages. *a*, full grown caterpillar from the side; *b*, the same from in front showing osmeteria protruded; *c*, the male butterfly; note clubbed antennae; *d*, outline of the egg, greatly enlarged; *e*, a young larva; *f*, the pupa or chrysalis. The artist has shown the pupa with the wrong side toward its support. All about natural size except *d*. (*From U.S.D.A., Farmers' Bul.* 856.)

(page 531), fall webworm (page 562), gypsy moth (page 701), peach tree borer (page 629), greenhouse leaf tyer (page 733), clothes moths (page 790), and flour and meal moths (pages 802 to 806).

For our purposes, probably the most useful division of the order is the popular one into moths and butterflies. The principal families of each suborder are outlined in the following:

SYNOPSIS OF THE LEPIDOPTERA

A. Suborder *Jugatae or Homoneura.*—*Moths or millers* in which the two wings of each side are held together and made to work as a single wing, in flight, by an angular or finger-like process which projects from the base of the *front* wing over the front margin of the *hind* wing. The venation of the hind wing is very similar to that of the front wing. Mouth parts never of the siphoning type with a coiled proboscis.

Family 1. Mandibulate moths, Family Micropterygidae.

Family 2. Swifts, Family Hepialidae.

Suborders B and C are alike in the following characters, by which they differ from suborder **A**. They differ from each other in the characteristics given under **B** and **C**. Lepidoptera in which the action of the two wings on each side is synchronized by one or several strong bristles that project from the base of the *hind* wing over the anal margin of the *front* wing. Such bristles are called a *frenulum*. Sometimes the frenulum is supplanted by an expansion of the base of the hind wing membrane which has taken the place of the bristles. The venation of the hind wing is different from that of the fore wing. Mouth parts siphoning except where they have been reduced by degeneration.

B. Suborder *Heterocera (Frenatae or Heteroneura, in Part).—Moths and Millers.* Mostly night fliers. Antennae of varied form, filiform, pectinate, or otherwise, but never enlarged at the tip to form a club. Abdomen heavy. Wings usually lie horizontally or roof-like at sides of the abdomen or wrapped around the abdomen when at rest. Often have two ocelli. Pupae very often protected by cocoons.

    Family 3. Yucca moths and others, Family Incurvariidae.
    Family 4. Nepticulid moths, Family Nepticulidae.
    Family 5. Carpenter moths, Family Cossidae.
    Family 6. Flannel moths, Family Megalopygidae.
    Family 7. Slug-caterpillar moths, Family Eucleidae.
    Family 8. Clothes moths and others, Family Tineidae.
    Family 9. Bagworm moths, Family Psychidae.
    Family 10. Ribbed cocoon makers and others, Family Lyonetiidae.
    Family 11. Leaf blotch miners and others, Family Gracilariidae.
    Family 12. Casebearers, Family Coleophoridae.
    Family 13. Parsnip webworm and others, Family Oecophoridae.
    Family 14. Gelechiid moths, Family Gelechiidae.
    Family 15. Yponomeutid moths, Family Yponomeutidae.
    Family 16. Clear-winged moths, Family Aegeriidae (Sesiidae).
    Family 17. Leaf roller moths, Family Tortricidae.
    Family 18. Olethreutid moths, Family Olethreutidae.
    Family 19. Pyralid or snout moths, Family Pyralididae.
    Family 20. Hawk moths or sphinx moths, Family Sphingidae.
    Family 21. Measuring-worm moths, Family Geometridae.
    Family 22. The prominents, Family Notodontidae.
    Family 23. Tussock moths, Family Lymantriidae.
    Family 24. Owlet or cutworm moths, Family Noctuidae.
    Family 25. Tiger moths, Family Arctiidae.
    Family 26. Lappet moths, tent caterpillars, and others, Family Lasiocampidae.
    Family 27. Royal moths, Family Citheroniidae.
    Family 28. Giant silkworm moths, Family Saturniidae.

C. Suborder *Rhopalocera (Frenatae or Heteroneura, in Part).—Butterflies and Skippers.* Day fliers. Antennae clubbed or enlarged near the tip. Wings usually held vertically above the body when at rest with the upper surfaces of the two pairs in contact. When in flight, the wings of the same side are held together usually by an expansion of the membrane of the hind wing near its base. No ocelli.

The Skippers.—Abdomen heavy. Antennae with a recurved hook at the tip, their bases wide apart. Larvae with a distinct neck-like constriction just behind the head.

    Family 29. The common skippers, Family Hesperiidae.

The Butterflies.—Abdomen slender. Antennae without a recurved hook on the terminal club, their bases inserted close together. Pupae never protected by cocoons.

    Family 30. Swallowtails, Family Papilionidae.
    Family 31. White and sulphur butterflies, Family Pieridae.
    Family 32. Four-footed butterflies, Family Nymphalidae.
    Family 33. Hair-streaks or gossamer wings, Family Lycaenidae.

ORDER HYMENOPTERA

Bees, Wasps, Ants, Sawflies, Parasitic Wasps, and Others

Everyone knows at least three kinds of Hymenoptera—bees, wasps, and ants. But not only are there numerous families of bees, wasps, and

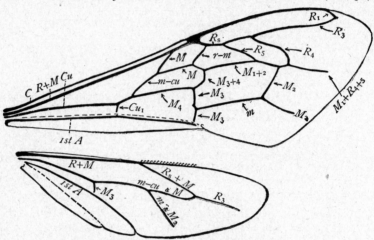

Fig. 149.—Wings of the honeybee showing the minute hooks or hamuli that serve to lock the two wings together. (*From Comstock's "Introduction to Entomology."*)

ants; in addition to them the order includes a much greater number of species of other habits, such as the parasitic wasps, the gall wasps, and the sawflies, which are fully as important to us and altogether as interesting as the better-known kinds.

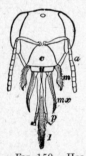

This order is listed as third in point of size (see Table VIII, page 172), but it has been so little studied in comparison with the beetles, moths, and butterflies that it would not be surprising to see it surpass these orders when its unnumbered species are once thoroughly studied.

Most authors place the Hymenoptera at the top of the list of orders, in much the same way that man is placed at the pinnacle of the vertebrate animals; and for much the same reason. This order appears to exhibit instinctive behavior in its highest state of perfection. In some of its representatives is found at least a low grade of intelligence, *i.e.*, the ability to learn or profit by experience, to choose and to form concepts.

Fig. 150.—Head of a honeybee showing chewing-lapping mouth parts from in front. *a*, antenna; *c*, clypeus; *u*, labrum; *m*, mandible; *mx*, maxilla; *p*, labial palp; glossa of labium. (*From Comstock's "Introduction to Entomology."*)

As a basis for the social organization which is so elaborately perfected in the better-known Hymenoptera, *care of the young* is widespread and often solicitous, in contrast with most insects which ordinarily pay no attention to their young after the eggs are laid. The larvae of many groups of this order are completely dependent upon their parents (or other adults) for food. Among the parasitic and gall-making species this obligation is discharged by the females when they lay their eggs in the midst of an abundance of food. The solitary wasps and bees generally gather and store a quantity of food of a suitable kind, available to the larvae in the nest when they hatch from the eggs. But some of the social wasps and bees, and the ants, bring food to the larvae day by day, during their entire lives, and often feed, clean, guard, and care

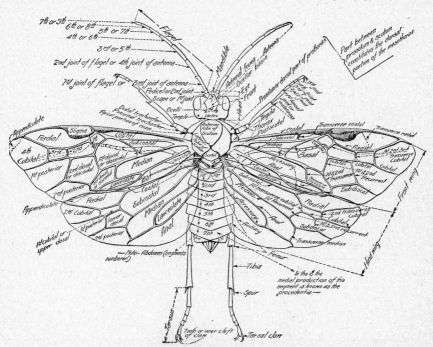

FIG. 151.—Dorsal view of a sawfly to show names of the principal parts of the body which are used in classification. (*From Conn. State Geol. and Natural History Surv. Bul. 22.*)

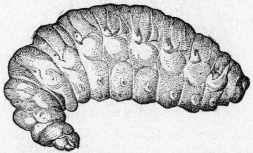

FIG. 152.—Larva of a digger wasp, *Tiphia* sp. Note reduced head, spiracles on the sides of the body, and absence of legs. (*From Ill. State Natural History Surv.*)

for the young in a manner highly suggestive of the maternal care that is general among the higher vertebrate animals. All the ants and many of the wasps and bees live together in great colonies, leading a complex social or cooperative life, the wonders of which increase, the more intimately man exposes their details to common knowledge.

The social life of the Hymenoptera is rivaled by that of the termites already described. As in that case, the reproductive function is limited to a few specialized individuals (kings and queens), while the vast majority of the adults remain infertile. These barren individuals work, not for themselves, but for the common good, building and cleaning the nests, foraging for food, fighting off invaders, and taking complete care of the prodigious number of young that hatch from eggs laid by the queens. In contrast with the termites, the workers and soldiers of the Hymenoptera, when these castes are developed, are *exclusively females*. The males accordingly have earned the name *drones*, their sole usefulness to the species apparently being to insure the fertility of the queen's eggs.

Fig. 153.—One of the horntails, the pigeon tremex, *Tremex columba* (Linné), natural size. (*From Kellogg's "American Insects," after Jordan and Kellogg.*)

The unusual mental development of the higher Hymenoptera is a strong reason for listing this order as the most specialized. But it should be considered as only one criterion and not be allowed to overshadow other considerations. In the matter of wing specialization the Hymenoptera are surpassed by the Diptera, the Coleoptera, and the Hemiptera. In the specialization of the mouth parts, the Diptera, Siphonaptera, Anoplura, Hemiptera, Homoptera, and especially

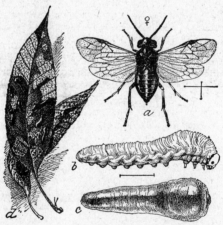

Fig. 154.—The pear slug, *Eriocampoides limacina* (Retz.) *a*, adult female sawfly; *b*, larva with slime removed, side view; *c*, slime-covered larva, dorsal view; *d*, leaves being skeletonized by the larvae, natural size. Natural size of *a*, *b*, and *c* is indicated by the lines. (*From Sanderson and Jackson, "Elementary Entomology," after U.S.D.A.*)

the Lepidoptera, have gone much farther. These are some of the reasons we have for giving the Hymenoptera a middle position in the series of insects with a complete metamorphosis.

The Hymenoptera have a complete or complex metamorphosis. The larvae of this order vary much more in form than those of the beetles or moths, ranging from caterpillar-like sawflies, with distinct head, well-developed legs and prolegs, and independent active habits, to the legless and practically helpless progeny of bees, wasps, and ants. The sawfly larvae (Fig. 88,*A*) can be distinguished from caterpillars (Lepidopterous larvae), which they most resemble, by the number of prolegs, which is from six to eight pairs, whereas Lepidopterous larvae (Fig. 55) never have more than five pairs; also the prolegs are not provided with crochets like those of Lepidoptera; the adfrontal sclerites (Fig. 145) are never present in Hymenoptera; and the ocelli are at most one pair in Hymenoptera, always more than one pair, or none, in Lepidopterous larvae. The more

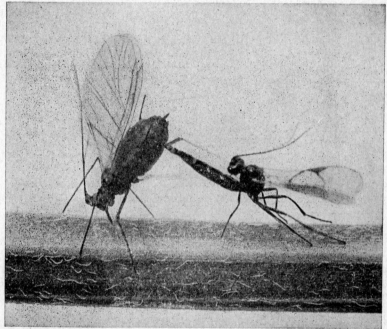

Fig. 155.—A wasp, *Lysiphlebus testaceipes* (Cresson), depositing its eggs in the body of the alfalfa aphid, much enlarged. (*From Essig's "Insects of Western North America," copyright, 1926, Macmillan. Reprinted by permission.*)

specialized larvae (Fig. 152) differ from Dipterous larvae, with which they are most likely to be confused, in having a recognizable head (although it may be reduced in size), with distinct mouth parts. Also, in contrast with most Diptera, the larvae usually have a pair of small spiracles on each of the principal abdominal segments, rather than a large complex pair close together on the last segment. The antennae are usually wanting, and the ocelli wanting or a single pair. In the higher families the head of the larva is opaque white like the rest of the body and often very small, but the true mandibles are retained.

The pupae (Figs. 89,*K* and *M* and 90,*M*) resemble beetle pupae in having the appendages not immovably fastened to the general body wall. The antennae of the pupa are always longer than the head, and the mandibles can be recognized as of the chewing type. The labium and maxillae are often elongate, the compound eyes show distinctly, and the

pupa is generally surrounded by a silken cocoon.   In some of the Hymenoptera, the metamorphosis is complicated by hypermetamorphosis
(see page 510) and by *polyembryony*.   Polyembryony is a remarkable
condition in which anywhere from 2 to over 150 individuals develop
from a single egg by the splitting up of the egg at an early stage of development to form a number of embryos.   The life cycle may be further
complicated by *parthenogenesis* and an *alternation of generations*.   In some
species a given generation may be all females; these lay unfertilized eggs
on a kind of plant, say an oak, and a characteristic gall grows on the plant
to shelter the insect.   When this generation of insects is mature, they
may be of both sexes, and the females lay fertilized eggs on another kind
of plant, for example a rose.   As a result, the rose develops a gall altogether different in appearance from that of its parents on oak.   When
these young are mature, all prove to be parthenogentic females that seek
the oak again and produce galls like those in which their grandparents
developed.   Until such species are carefully watched through several
generations, they are sure to be described as two distinct species, so
different are they, both in habitat and in appearance.

Fig. 156.—A sphecid wasp, *Ammophila* sp., putting a paralyzed measuring-worm into
her nest to serve as food for her young.   When the nest is completed and the eggs laid,
the wasp takes a small pebble in her mandibles and packs down the earth with which she
closes her burrow.   (*From Kellogg's "American Insects."*)

The wings in this order (Fig. 149) are four in number.   They are
generally small, transparent, and with comparatively few veins.   The
front pair is distinctly larger.   The hind wing usually fits exactly against
the hind margin of the front wing, to which it is fastened by a row of very
minute hooks.   The student should note this carefully, otherwise he may
mistake certain ones of this order for Diptera, since the two pairs of wings
are so closely fitted together as easily to be mistaken for a single pair.
     The mouth parts vary from the chewing type to a combination of
chewing and lapping structures (Fig. 150).   In all cases the labrum and
the mandibles are essentially like those of Orthoptera and Coleoptera.
The maxillae and labium are also essentially of the chewing type in the
less specialized families, but in the bees, wasps, and ants, these two paired
structures become progressively longer and longer, to form a hairy,
lapping tongue by means of which liquids are lapped up, as the insect
feeds (see page 129 and Fig. 76).   The maxillae are more or less united
with the labium in both adults and larvae.
     One specialization peculiar to the Hymenoptera is the modification of
the ovipositor into a defensive and offensive weapon known as a "stinger."
This organ of defense and offense is found only among the insects of the
order Hymenoptera—the bees, many of the wasps, and certain kinds of
ants, besides the distantly related scorpions.
     The stinger of a bee or wasp is a very complex and beautifully adapted
organ.   It is known to be the equivalent (homologue) of the egg-laying

organ in other insects, and it follows therefore that *only the females of insects can sting.* In many of the Hymenoptera (Fig. 40), a greatly elongated, sting-like organ serves the function of thrusting eggs into plant parts but is never used in defense. The stinger consists of a similar mechanism for penetrating the skin to a depth of perhaps $\frac{1}{10}$ inch and of a system of glands to secrete the venom that is injected into the wound. In the honeybee the sting proper consists of two extremely sharp, highly polished, brown spears or darts which appear as one. Their concave inner surfaces make, between them, a fine tube down which the venom is forced, to emerge at their tips. As the insect stings, these two darts are alternately and very rapidly thrust outward or downward on guide rails of a surrounding sheath. Each dart has near its tip 9 or 10 recurved hooks which hold it firmly until the next thrust carries it still deeper.

FIG. 157.—A mud-dauber, *Sceliphron cementarius* (Drury). (*From Sanderson and Jackson, "Elementary Entomology," after S. J. Hunter.*)

FIG. 158.—The bald-faced hornet, *Vespa maculata* Kirby, about natural size. (*From Sanderson and Jackson, "Elementary Entomology."*)

Because of these hooks, the honeybee can seldom remove her stinger, and it, with more or less of the viscera, is torn away as she escapes. Other bees, wasps, hornets, and ants may sting repeatedly.

The pain of the sting is due to the venom. In some of the Hymenoptera this venom is deadly to other insects and small animals receiving it. In others it has only a paralyzing effect and is used to stupefy flies, spiders, crickets, caterpillars, or beetles upon which eggs are laid and the helpless plunder is then sealed up in their nests as food for the forthcoming young (Fig. 156).

Another peculiarity of the adults of this order is that one abdominal segment is fused with the thoracic mass, so that what appears to be the first abdominal segment is in reality the second.

Wings typically four, small, membranous, with few veins; hind wings smaller, often hooked to front pair; females sometimes wingless, the males rarely so (Fig. 27). Venation highly specialized; often much reduced. Mouth parts chewing or chewing and lapping, the mandibles always of the chewing type; but the maxillae and labium often elongate (see page 130). Compound eyes usually well developed; three ocelli usually present. Pronotum fused to the mesothorax. Tarsi usually five-segmented. Abdomen often with a slender waist and its first segment united with the thorax. Female provided with an ovipositor or stinger. No cerci. Metamor-

phosis complete. Larva either caterpillar-like or legless; with distinct head, and spiracles on the principal segments. Prolegs, when present, usually more than five pairs and without crochets. No adfrontal areas. Pupae with appendages free, commonly encased in cocoons. Many species live in societies.

The destructive species are comparatively few, the following being the best known: wheat jointworm (page 406), pear slug (page 641), sawflies (page 405), currantworm (page 656), cornfield ant (page 371), house ants (page 767).

The classification of the Hymenoptera appears to be in a very unsettled condition. The following outline will indicate the principal

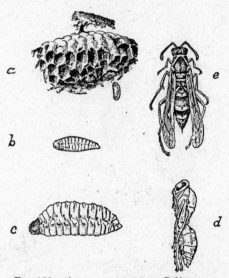

Fig. 159.—Paper nest of the bald-faced hornet, *Vespa maculata* Kirby. Actual size 8 to 15 inches in diameter. (*From Kellogg's "American Insects."*)

Fig. 160.—A paper-nest wasp, *Polistes* sp. *a*, nest and egg; *b*, young larva; *c*, older larva; *d*, pupa; *e*, adult. All enlarged ½, except nest, which is reduced. (*From Kellogg's "American Insects."*)

families and something of their natural grouping. We have followed Comstock's[1] account in many respects but have varied from the opinions of specialists, where by so doing it seemed possible to give the nontechnical student a more useful working conception of these insects.

A SYNOPSIS OF THE HYMENOPTERA

A. Suborder *Chalastogastra, Symphyla, or Sessiliventres.*—Foremost segments of abdomen as broad as the following segments and joined to the thorax by their full width (*i.e.*, without a slender waist). Trochanters two-segmented. Ovipositor adapted for sawing or boring, never a sting. The larvae feed on or in plants and are caterpillar-like in appearance. They have a single pair of ocelli and usually well-developed legs and prolegs, the latter without crochets.

[1] COMSTOCK, "An Introduction to Entomology," The Comstock Publishing Company, 1924.

Family 1. The horntails, Family Siricidae.
Family 2. The stem sawflies, Family Cephidae.
Family 3. The sawflies, Family Tenthredinidae.
B. Suborder *Clistogastra, Apocrita, or Petiolata.*—Foremost segments of abdomen narrower than those which follow, making the connection to the thorax a slender petiole or "waist." The thorax appears to bear 3 pairs of spiracles, but the last pair is really on the first abdominal segment which is completely fused with the thorax. The larvae are of varied habits, always legless and grub-like, with head and mouth parts reduced, antennae and palps at most of one segment, and ocelli usually wanting.

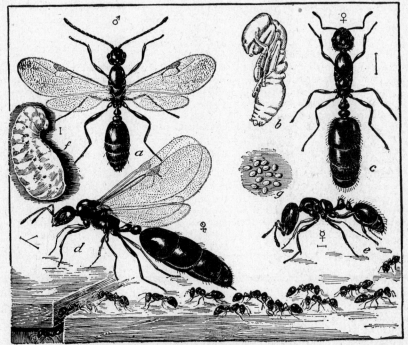

FIG. 161.—The little black ant, *Monomorium minimum* Buckley. *a*, male; *b*, pupa; *c*, female after losing wings; *d*, winged female; *e*, a sterile female or worker; *f*, larva; *g*, eggs; below, a group of workers in line of march. The lines indicate natural size. (*From Marlatt, U.S.D.A., Farmers' Bul.* 740.)

WASPS OR WASP-LIKE FORMS

Hairs of body not branched. First segment of hind tarsus usually slender, cylindrical. No tubercles on petiole of abdomen.
a. Solitary species consisting of only males and females; each female provides for her own young.
 1. Larvae mostly parasitic on other insects. Eggs generally laid on or through the body wall of the active host. Trochanters two-segmented.
  Family 4. Ichneumon wasps, Family Ichneumonidae.
  Family 5. Braconid wasps, Family Braconidae.
  Family 6. Ensign wasps, Family Evaniidae.
  Family 7. Chalcid wasps, Family Chalcididae.
  Family 8. Egg-parasite wasps and others, Family Proctotrupidae (also called Proctotrupoidea and Serphoidea).

2. Larvae mostly live in galls which they cause to grow on plants (Fig. 4).   Trochanters apparently two-segmented.

Family 9.   Gall wasps, Family Cynipidae.

3. Larvae mostly parasitic on other insects or spiders.   Eggs laid on paralyzed ("stung") caterpillars, grubs, or spiders, which are buried in the ground or stored in mud cells, burrows, tunnels, mines, or natural cavities; or on active leafhoppers, etc.; or in nests of other Hymenoptera.   Or rarely food is brought to larvae in the nest from day to day, by the parent wasps.

Family 10.   Digger wasps, mud-daubers, and thread-waisted wasps, Family Sphecidae.

Family 11.   Dryinid wasps, Family Dryinidae.

Family 12.   Spider wasps, Family Pompilidae.

Family 13.   Cuckoo wasps, Family Chrysididae.

Family 14.   Velvet "ants," Family Mutillidae.

Family 15.   Vespoid digger wasps, Family Scoliidae.

Family 16.   Mud wasps, Family Eumenidae.

b. *Social wasps with a sterile worker caste in addition to both males and females*, the workers taking most care of the females' young.   Eggs laid in cells of nests composed of paper, which the adults make of wood.   Larvae are fed from day to day on juices of insects or sweets.   Wings folded lengthwise when at rest.   Trochanters one-segmented.

Family 17.   Hornets, yellow-jackets and paper-nest wasps, Family Vespidae.

Ants

Hairs of body not branched.   Petiole of abdomen with one or two swellings or tubercles (Fig. 161).   Often wingless.   Social insects with a sterile worker caste, in addition to males and females.   Nests in soil, wood, or stems of plants, without well-defined cells for each larva to live in.   Larvae fed on regurgitated food from adults, or on bits of insects, seeds, fungi, etc.   Trochanters one-segmented.

Family 18.   Ants, Family Formicidae.

Bees

Broader bodied and more hairy than ants or wasps.   The hairs of the thorax branched or plumose (Fig. 163).   First segment of hind tarsus often broad, flattened, and brush-like for assembling pollen, and hind tibia often specialized as a "pollen basket" (Fig. 47,C).   The pronotum does not extend backward on the sides to the tegulae at base of wings.   The mouth parts have the glossae well developed to form chewing-lapping type of mouth parts.   Trochanters one-segmented.   All individuals winged.   Nests provisioned with nectar and pollen from flowers, as food for the young.

a. *Solitary bees consisting of only males and females.*

1. Tongues (labia) short and broad.   Eggs laid in nests burrowed in ground, in pithy plants, or in crevices of walls or buildings.   Cells separated by a silky secretion.

Family 19.   Bifid-tongued bees, Family Prosopidae.

Family 20.   Colletids, Family Colletidae.

2. Tongues long and slender (Fig. 150).

i. Nests in burrows in plant stems, in cavities about buildings or in the soil.   Eggs laid in cells made of pieces of leaves cut from growing plants, or of plant fibers or of clay mixed with saliva.   Pollen-collecting brushes on underside of abdomen of females.

Family 21.   Leaf-cutting bees, Family Megachilidae.

ii. Nests in tunnels cut in the solid wood of buildings or trees.   Eggs laid in cells separated by cemented sawdust.

Family 22.   Large carpenter bees, Family Xylocopidae.

iii. Nests in tunnels in pithy plants.   Cells separated by plugs of plant fiber.

Family 23.   Small carpenter bees, Family Ceratinidae.

b. *Parasitic or "cuckoo" bees* that lay their eggs in nests of other bees and so steal the food or parasitize the rightful owners. Tongues long. Legs not adapted for collecting pollen. Males and females only, no worker caste.

Family 24. Cuckoo bees, Family Nomadidae.

c. *Gregarious bees consisting of only males and females.* Nests placed near together in the soil or in the face of cliffs, often with a common entrance or corridor. Tongues short or long and pointed.

Family 25. Mining bees, Family Andrenidae (Halictidae).

Family 26. Anthophorids, Family Anthophoridae.

Fig. 162.—Nest of a bumblebee, *Bombus auricomus* Robertson, in a deserted mouse nest. Note *A*, the wax-pollen lining material for protection of the nest; *B*, four empty cocoons from which adult bees have emerged and which are used later for the storage of honey and pollen; *C*, a wax-pollen mass enclosing eggs; *D*, cocoons containing larvae, pupae, or adult bees not yet emerged; and *E*, the remnants of the old mouse nest below. (*From T. H. Frison.*)

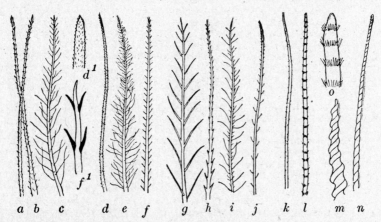

Fig. 163.—Branched, plumose, and threaded hairs, showing a characteristic of bees. *a* to *f* are from bumblebees; *g* to *j* of *Melissodes* sp.; *k* to *o* of a leaf-cutting bee, *Megachile* sp. (*From Comstock's "Introduction to Entomology," after J. B. Smith.*)

d. *Social bees, with a worker caste in addition to males and functional females.* Eggs laid in cells made of wax secreted by worker bees (Fig. 22).
1. Nests commonly in deserted mouse nests on the ground (Fig. 162). A number of eggs usually laid together in one waxen cell.
   Family 27. Bumblebees, Family Bombidae (Bremidae).
2. Nests built in trees or in hives provided by man. A single egg laid in each cell (Fig. 22).
   Family 28. Honeybees, Family Apidae.

As a group, the Hymenoptera may be considered more beneficial than injurious. There are, to be sure, a number of serious pests. But the work of the very many insect parasites and predators, the activities of the bees in pollinizing plants, and the production of honey and wax, undoubtedly offset many fold the injury inflicted by members of this order. As in the Coleoptera and Lepidoptera, there are no parasites of the larger animals.

ORDER DIPTERA

Flies, Mosquitoes, Gnats, Midges

The Diptera are a well-marked group in respect to the condition of the wings, and fairly homogeneous in general appearance; but in habits and in most other characteristics the order presents great diversity.

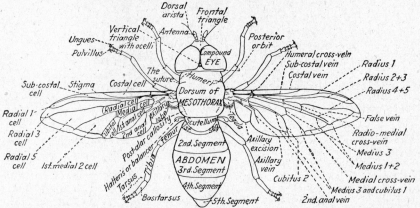

Fig. 164.—Dorsal view of a male syrphid fly to show names of the principal parts of the body.

The following names of veins are used by some authors instead of the ones shown in the figure; *auxiliary* instead of subcostal; *1st longitudinal* instead of radius 1; *2d longitudinal* instead of radius 2 + 3; *3d longitudinal* instead of radius 4 + 5; *4th longitudinal* instead of medius 1 + 2; *anterior cross-vein* instead of mediocubital cross-vein; *lower cross-vein* instead of medius 3; *5th longitudinal vein* instead of medius 3 + cubitus 1; *posterior cross-vein* instead of medial cross-vein; *anal cross-vein* instead of cubitus 2; *6th longitudinal vein* instead of 2d anal. Also *alula* or *squama* instead of tegula.

The following names of cells are used by some authors instead of the ones shown in the figure: *marginal* instead of radial 1; *submarginal* instead of radial 3; *1st basal* instead of radial; *1st posterior* instead of radial 5; *2d basal* instead of medial; *discal* instead of 1st medial 2; *3d posterior* instead of cubital; *anal* instead of 1st anal; *axillary* instead of 2d anal. (*From Metcalf, Ohio Biol. Surv. Bul.1.*)

They are set apart from all other orders of insects by having a single pair of wings (the front pair) developed for flight, and each of the hind wings reduced to a short, slender thread, with a knob at the end of it

(Fig. 164). These rudimentary second wings are called *halteres* or balancers. There is some evidence that they are orienting organs, serving to keep the insects balanced, something like the semicircular canals in the skull of the vertebrates. A few other insects have a single pair of wings, such as certain beetles and May-flies, but these never possess halteres. Many of the Diptera are wingless, having lost the first pair of wings; but even then the halteres usually remain, so that the possession of halteres is perhaps the most distinctive thing about this order. The front pair of wings is similar to those of bees and wasps in texture, *i.e.*, transparent and with comparatively few veins, as a rule. They are small in comparison with the size of the insect, a condition associated with very swift flight. While usually clear, they may have a color pattern (Fig. 171), and sometimes the veins are bordered with scales.

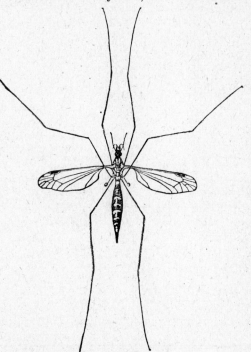

The three body regions are very distinct in Diptera. The head is large, often hemispherical, and attached to the thorax by a very slender stem or neck. By reason of the fact that only the front

FIG. 165.—A crane fly or tipulid, male, adult. (*From Sanderson, "Insect Pests," after Weed.*)

wings are functional, the thoracic mass is largely made up of the mesothorax. A small, distinct, semicircular part of the mesothorax overhanging the base of the abdomen is called the *scutellum*. The abdomen is of varied shape, usually shows from four to nine segments, and the cerci, ovipositor, and male genitalia are normally introverted or retracted, so as to be invisible without special preparation.

The mouth parts of adult Diptera are rather varied in form. Two distinct types are represented; the piercing-sucking type and the sponging type (see pages 120, 128). There are several varieties of the former, called subtypes (see Figs. 66, 67, and 74). So far as known, no adult fly masticates solid foods; and few, if any, pierce plants to suck the sap. The majority probably feed upon the nectar and pollen of flowers; many others depend upon liquid organic matter such as that from decomposing plant or animal bodies, flowing sap, and honeydew; or they dissolve solid substances in their saliva, *e.g.*, sugar, and sponge up the solution. A number of species are predaceous on other insects, sucking the juices from their bodies. The females of hundreds of species, representing at least eight families, suck the blood of warm-blooded animals, and, in the Muscidae and Hippoboscidae, the males also have this bad habit. There are many adults that take no food whatever, this life stage generally being short and occupied almost exclusively with the business of getting the eggs developed, fertilized, and laid.

The metamorphosis is complete or complex. The larvae are well separated from the adults both structurally and in habits and specialized to a more extreme degree than the larvae of any other order. There are very few cases where the larvae and adults live together and partake of the same kind of food, as is so common among the beetles. The larvae are always legless and in the larger part of the order have no distinct head (Figs. 36,C and 77). In those species where the head is distinct, mosquitoes, for example (Figs. 88,E,G and 570), the mouth parts of

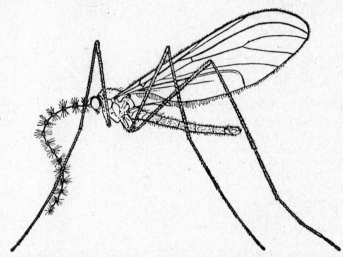

Fig. 166.—A gall gnat, *Hormosomyia oregonensis* Felt, male, side view, much enlarged.
(*From Cole and Lovett, "Oregon Diptera."*)

the larvae are of the chewing type; but in the great majority of species the body tapers gradually to the front end and terminates in a small conical segment which can be protruded or retracted. This head segment bears no eyes, and no true mouth parts. There is a pair of minute rudimentary sense organs and a pair of prominent mouth hooks which work vertically to tear the tissues upon which the larva feeds or into which it tunnels. The larvae typically have a complex pair of spiracles on the truncate last segment of the abdomen (Fig. 547,C), and, sometimes at least, another pair near the front end of the body; but commonly none along the sides of the body on the other abdominal segments. Such larvae are called *maggots*. They live mostly buried or hidden in decaying animal or vegetable matter, in water or mud, or inside the bodies of plants, insects, and other animals. There are strikingly few that feed externally upon plants and comparatively few crop pests of any kind. The most serious are certain gall flies, *e.g.*, the Hessian fly (page 401), the many fruit flies of the family Trypetidae, some leaf miners, root maggots, and borers in the stems of plants. The attacks of the larvae and adults upon animals are much more serious. This is the most dangerous order for the carrying of human and animal diseases (see pages 25–29). Many species suck the blood of animals as adults or live as larvae in their bodies. On the other hand, we must note the great benefit that accrues to us from the work of scavenger larvae and from those that are predaceous or parasitic on various insects and from the employment of Drosophila in genetic experiments (see page 67).

Fig. 167.—A black fly, *Simulium* sp., female, side view, enlarged. (*From Osborn, "Agricultural Entomology," after U.S.D.A.*)

Fig. 168.—A robber fly, side view, natural size. (*From Kellogg's "American Insects."*)

The pupae of Diptera (Fig. 89,*G,H*) have the appendages free from the body wall and can be distinguished from the pupae of all other orders by having a single pair of wing pads. In the higher families the pupa is protected in a unique manner. Instead of shedding the skin of the last active larval instar, when the pupa is formed, this dried larval skin is retained about the pupa and serves as a cocoon. It is often inflated like a large seed, and its walls become very hard and thickened to form an airtight and watertight case. Such a case about a larva or pupa is called a *puparium* (Figs. 36,*D* and 90,*C*). All flies of the suborder Cyclorrhapha spend the pupal stage in a puparium; most of the suborder Orthorrhapha do not. In the latter group the pupa commonly has no particular protective covering; rarely a cocoon is formed.

Fig. 169.—A bee fly, *Bombylius major* Linné, twice natural size. (*From Kellogg's "American Insects."*)

A good many of the flies are ovoviviparous, and in this order we have a few groups in which a viviparous reproduction occurs, *e.g.*, the Hippoboscidae and the tsetse flies of the family Muscidae. Another remarkable method of reproduction is the production of young by larvae and pupae, to which the term *paedogenesis* is applied. This occurs in certain of the gall midges.

It is easy to mistake many of the flies for other kinds of insects. If the observer is not careful to note the number of wings and especially the presence of halteres, he will be likely to place many of his flies as Hymenoptera, because they are commonly banded with yellow and black, like the wasps, or densely covered with hairs like bees. A few species are louse-like or tick-like in form but can be distinguished from ticks by the number of legs and from lice by the nature of the mouth parts, which are exposed piercing organs (Figs. 560, 561).

Small- to medium-sized, soft-bodied insects. Adults mostly diurnal, with only one pair of wings, the front pair, which are narrow, membranous, with few veins. The hind wings are modified into halteres. Head, thorax and abdomen very distinct. Head vertical, very free-moving, subhemispherical. Compound eyes large; three ocelli usually present. Mouth parts sponging or piercing-sucking; labial palps always wanting. Prothorax and metathorax fused with the mesothorax, but scutellum distinct. Tarsi generally five-segmented. Metamorphosis complete. Larvae legless, usually maggot-like with head greatly reduced; the mouth parts in these cases replaced by a pair of mouth hooks articulating to an internal cephalopharyngeal skeleton. Larval spiracles generally restricted to a small pair on prothorax and a large group on last segment of abdomen. Pupa generally with free appendages; sometimes obtect in the Orthorrhapha; very rarely in a cocoon, but often enclosed in a puparium.

Fig. 170.—A thick-headed fly, *Physocephala affinis* Williston, enlarged ½. (*From Kellogg's "American Insects."*)

Fig. 171.—A trypetid fly, the white-banded cherry fruit fly, *Rhagoletis cingulata* Loew, female, enlarged. (*From Lochhead, "Economic Entomology," after Caesar.*)

There are comparatively few crop pests, but many that attack animals. The following are the most important and the student should refer to them for additional figures and descriptions: Hessian fly (page 401), cabbage maggot (page 539), onion maggot (page 528), apple maggot (page 603), clover seed midge (page 436), mosquitoes (page 875), black flies (page 881), horse flies (page 814), horse bots (page 823), ox-warble flies (page 838), screwworm fly (page 843), horn fly (page 829), sheep "tick" (page 853), and house fly (page 906).

Specialists are fairly well agreed about the classification of the Diptera, at least in its major aspects.

A SYNOPSIS OF THE ORDER DIPTERA

A. **Suborder** *Orthorrhapha.*—Straight-seamed Flies. The adult insects escape from the pupal skin or pupal case through a T-shaped or straight split down the back or a transverse split between the seventh and eighth segments of the abdomen. Pupa usually naked, sometimes in a cocoon, rarely in a puparium. Adults do not have a small, lunate-shaped sclerite above the antennae known as the frontal lunule. Larvae often with a distinct head.

1. *Nemocera. The Long-horned Flies.*—Antennae usually long and slender, of 6 to 39 similar segments. Palps usually four- or five-segmented, generally pendulous. Larvae have a distinct head, eyes, and true mandibles working transversely. First anal cell of wings almost never narrowed toward the wing margin. Discal cell generally absent. Mostly very slender flies.

Family 1. Crane flies, Family Tipulidae (Fig. 165).
Family 2. Moth flies and sand flies, Family Psychodidae.
Family 3. Mosquitoes, Family Culicidae.
Family 4. Midges, Family Chironomidae.
Family 5. Gall gnats, Family Cecidomyiidae (Itonididae) (Fig. 166).
Family 6. Fungus gnats, Family Mycetophilidae.
Family 7. Buffalo gnats or black flies, Family Simuliidae (Fig. 167).

2. *Brachycera. The Short-horned Flies.*—Antennae usually short, of three segments, last segment sometimes annulate or with a style in addition, like a small whip of withered segments at the end. Palps one- or two-segmented, porrect. Larvae often have the head reduced in size and invaginated, and mouth hooks working vertically, instead of mandibles. Discal cell usually present. First anal cell always closed or narrowed toward the wing margin.

Family 8. Net-winged midges, Family Blepharoceridae.
Family 9. Horse flies, Family Tabanidae.
Family 10. Soldier flies, Family Stratiomyiidae.
Family 11. Snipe flies, Family Rhagionidae (Leptidae).
Family 12. Robber flies, Family Asilidae (Fig. 168).
Family 13. Bee flies, Family Bombyliidae (Fig. 169).
Family 14. Long-legged flies, Family Dolichopodidae.

B. **Suborder** *Cyclorrhapha.*—Circular-seamed Flies. The adults escape from the pupal case through a split that runs round the end of the case and releases a circular lid that is pushed off or aside. Pupa always enclosed by the skin of the last active larval stage, which hardens to form a puparium. Adults with a frontal lunule and antennae generally of three segments, the third bearing an arista or style. Head of larvae always greatly reduced and invaginated into the pharynx. First anal cell always closed.

3. *Aschiza. Flies without a Frontal Suture.*—Cap of puparium pushed off by expansion of the face of the adult, when it is ready to emerge; therefore there is no frontal lunule.

Family 15. Humpbacked flies, Family Phoridae.
Family 16. Flower flies or hover flies, Family Syrphidae (Fig. 36).

4. *Schizophora. Flies with a Frontal Suture.*—A line or seam circles round above the base of the antennae and sometimes extends down nearly to the mouth on either side of the face. This is the vestige of a crack in the head through which a membranous, expansible, bladder-like structure, known as the *ptilinum*, is forced out when the adult is ready to emerge. By inflating the ptilinum with body fluids, the cap of the puparium is forced off. The bladder is then withdrawn into the head and is seen only if one catches the adult very shortly after its emergence.

  a. Acalyptratae. Flies with small *tegulae; i.e.*, small, flat membranous expansions connecting the base of the wing, behind, to the thorax. They do not have a complete transverse suture across near the middle of the thorax. They are all small flies, some very small. The eyes of males do not come together on top of the head.

Family 17. Thick-headed flies, Family Conopidae (Fig. 170).
Family 18. Ortalid flies, Family Ortalidae.
Family 19. Fruit flies, Family Trypetidae (Fig. 171).
Family 20. The frit fly and others, Family Oscinidae.
Family 21. Ephydrid flies, Family Ephydridae.
Family 22. The pomace fly and others, Family Drosophilidae.

  b. Calyptratae. Flies with well-developed *tegulae, i.e.*, thin, subcircular membranes just behind the base of the wing close against the thorax (see Fig. 164). The thorax has a complete transverse suture near mid-length, above. This division includes our commonest and best-known flies. They are all medium to large in size. The males can often be distinguished from the females, by having the eyes contiguous, at least for a short distance, at the top of the head.

Family 23. Anthomyid flies, Family Anthomyiidae.
Family 24. House fly family, Family Muscidae.
Family 25. Flesh flies, Family Sarcophagidae.
Family 26. Tachina flies, Family Tachinidae.
Family 27. Botflies, Family Oestridae (see Figs. 546 and 554).

5. *Pupipara.*—Louse-like, often wingless flies, with a very tough skin, indistinctly-segmented abdomen, and legs inserted far apart on the sternum. External parasites on mammals (including bats), on birds, or on insects. The larvae develop viviparously until full grown and are born shortly before pupation, all growth taking place at the expense of the mother fly which nourishes the larva from special uterine glands (see Fig. 560). Antennae one- or two-segmented.

Family 28. The sheep tick and louse flies, Family Hippoboscidae (Fig. 560).
Family 29. Bee lice, Family Braulidae.

ORDER SIPHONAPTERA (APHANIPTERA OR SUCTORIA)

THE FLEAS

This is a very small order, very well defined, and not closely related to any other group of insects. All the species are wingless, and all are, in the adult stage, external parasites on warm-blooded vertebrates. They are facultative, rather than obligatory parasites (see page 55). They are almost unique among insects in being flattened, or thin, from side to side, like a sunfish. The legs are long, adapted for jumping, and the coxae are abnormally large, often being actually the largest segment of the leg (Fig. 578).

The body wall is hard or tough, polished, and provided with many backwardly directed hairs and with short, stout spines, often arranged so regularly as to resemble combs.

There are no compound eyes and often the simple eyes also are wanting. The three-segmented antennae are concealed in grooves just behind the eyes (see Fig. 68,*A*).

The mouth parts (Fig. 68) are piercing-sucking in type, but differ from any of the other subtypes in having two pairs of palps and the maxillae not stylet-like. The mandibles and labrum-epipharynx are adapted as stylets for piercing and forming the salivary duct and food channel; the maxillae are broad, triangular plates that do not enter the wound but do bear segmented palps. Another unusual feature of the fleas is that the three segments of the thorax are very distinct and free from each other.

These are almost the only external insect parasites of our larger animals that have a complete metamorphosis. The adult stage is the only stage that is known to people generally. The eggs are not fastened to the host, like the eggs of lice, and so are not noticed. They drop off the host or are laid on the floor of the nest, kennel, or dwelling of the host. The larvae that hatch from the eggs (Fig. 88,*L*) are very slender, cylindrical, whitish maggots with a distinct head, antennae, and mouth parts, but no ocelli and no legs. There are 12 well-marked body segments in addition to the head. There are usually minute paired spiracles on the thoracic and abdominal segments, and each segment bears a transverse band of very long, stiff hairs. The larvae, which are very active, live on such dead animal and vegetable matter as they find in the cracks of floors or in the dirt about the sleeping quarters of their hosts. When full-grown they spin a cocoon of silk, covered with particles of dirt, inside of which the pupa is formed. The pupa may be recognized by its compressed body, absence of wings, inconspicuous eyes, antennae which are shorter than the head, and mandibles elongated for piercing. The appendages of the pupa are free from the body wall (Fig. 89,*J*).

Small, laterally compressed, jumping insects, entirely wingless, usually spiny, and in the adult stage living as external parasites on warm-blooded animals. Compound eyes small or absent; antennae short, and concealed in grooves; no neck. Mouth parts piercing-sucking, with two pairs of palps. Thoracic segments distinct. Coxae very large, those of each pair nearly contiguous; hind legs fitted for jumping; tarsal segments five; claws strong. Cerci minute. Metamorphosis complete. Larvae scavengers; slender, cylindrical, without legs or eyes, but with well-developed head, chewing mouth parts, and spiracles on the principal body segments. Larvae scarcely ⅕ inch long; with transverse rows of very prominent setae. Pupae without wings, enclosed in a cocoon.

The only families of much importance are:

The common fleas, Family Pulicidae.

The sticktights or chigoes, Family Echidnophagidae.

The fleas of the first family are important, not only as very universal pests of cats, dogs, and hogs, often infesting houses where pets are kept and seriously annoying the persons that live there, by biting them; but also especially as the known carriers of the dread bubonic plague or "black death" (see page 889). The second family includes the sticktight flea that attaches to the head of chickens, the females remaining fixed in this position sucking the blood; and the true chigoe or jigger,[1] the females of which, after mating, burrow into the flesh of man, especially about the feet, and, as the eggs develop, increase enormously in size, causing bad sores.

[1] Not to be confused with the chigger mite (see p. 897).

CHAPTER IX

INSECT CONTROL

To a large extent the value of entomology is based on insect control. To a still greater extent the support given to this branch of science by the public is in direct proportion to the efficiency of measures for insect control which have been developed by entomological workers. While the lessening of insect damage or the control of insect outbreaks is not the only end and aim of insect study, it is the most important. The study of insects is also of the greatest value in helping to solve questions arising in the field of general biology and in aiding in the understanding of the natural laws governing the development and abundance of plants and animals.

Insect control in its broadest sense includes everything that makes life hard for insects and tends to kill them and to prevent their increase or spread over the world. The control of insects can be accomplished in many ways, which may be classified as follows:

AN OUTLINE OF CONTROL MEASURES FOR INSECTS

A. **APPLIED CONTROL**: Measures to destroy insects that depend upon man for their application or success and can be influenced by him to a considerable degree.

    I. **Chemical Control**: Preventive or remedial operations which depend for their effectiveness upon the *chemical* action or *chemical* properties of the substances used.

        a. Insecticides: Substances that *kill* insects by their chemical action.

            1. *Stomach Poisons:* sprays, dusts, or dips that kill the insect when they are swallowed.

            Examples: Lead arsenate, calcium arsenate, paris green; sodium fluoride, cryolite, fluosilicates; borax, thallium, phosphorus, and mercury compounds; phenothiazine, hellebore, fixed nicotines, and other organic toxicants.

            2. *Contact Poisons:* sprays, dusts, or dips that kill the insect without being swallowed.

            Examples: nicotine preparations, lime-sulphur, oil emulsions, pyrethrins, rotenone, synthetic thiocyanates, and cyclohexylamines.

            3. *Fumigants:* chemicals used in the form of a gas to kill insects; usually applied in an enclosure of some kind.

            Examples: hydrocyanic acid gas, carbon bisulphide, nicotine, sulphur dioxide, paradichlorobenzene, naphthalene, chloropicrin, ethylene oxide and dichloride, methyl bromide and formate.

        b. Auxiliary, Synergistic, or Supplemental Substances: Materials added to sprays or dusts, which are not primarily toxicant but which make the insecticide cover more economically; stick or adhere better to plants; spread over foliage, fruit, bark, or the bodies of the insects more quickly or completely; bring insecticidal substances into solution or emulsion; mask distasteful or repellent properties of insecticides; or activate the chemical action of the toxicant.

        Examples: (*a*) Carriers, such as water in a spray, talc or gypsum in a dust, or bran in a poison bait; (*b*) emulsifiers, such as soaps, flours, calcium and potassium caseinate, cresylic acid, blood albumin; (c) stickers, such as lime, rosin, glues; (d) spreaders, wetting agents and detergents, such as, caseinates, oleates, stearates, and a variety of sulphated and sulphonated organic compounds—sodium sulphate of lauryl, oleyl, or stearyl alcohols or the sulphonic acid of an aromatic hydrocarbon; (e) stabilizers, such as caustic soda, glue, gelatin.

    c. *Attractants or Attrahents:* Substances used in poisoned baits, sprays or dusts to induce insects to eat the poisoned material, or to lure insects into traps.

    d. *Repellants:* Substances that *keep insects away* from crops and animals, because of their offensive appearance, odor, or taste.

    Examples: Bordeaux mixture, creosote, naphthalene, oil of citronella, tars, cinchona alkaloids,

II. Mechanical and Physical Control: Special operations against insects that kill them by their *physical* or *mechanical* action.

    Mechanical Measures: The operation of machinery or application of manual operations.

    a. *Hand Destruction:* hand-picking, jarring, swatting, worming.

    b. *Mechanical Exclusion:* screening, tree banding, linear barriers, fly nets.

    c. *Use of Traps, Suction, and Other Collecting Machines.*

    d. *Use of Crushing, Dragging, or Grinding Machines.*

    Physical Measures: The application or manipulation of physical (as opposed to chemical) phenomena.

    a. *Manipulations of Water or Humidity:* drainage and dehydration of breeding media or flooding and syringing.

    b. *Manipulations of Temperature:* superheating, burning, use of low temperatures.

    c. *Use of Electric Shock, Electrostatic Field.*

    d. *Use of Light and Other Radiant Energy.*

    e. *Use of Sound Waves.*

III. Cultural Control or Use of Farm Practices: Regular farm operations performed so as to destroy insects or prevent their injuries.

    a. *Crop Rotations.*

    b. *Tilling of the Soil.*

    c. *Variations in the Time or Method of Planting or Harvesting.*

    d. *Destruction of Crop Residues, Weeds, Volunteer Plants, Trash.*

    e. *Use of Resistant Varieties.*

    f. *Pruning, Thinning.*

    g. *Fertilizing and Stimulating Vigorous Growth.*

IV. Biological Control:[1] the introduction, encouragement, and artificial increase of predaceous and parasitic insects, other animals, and diseases.

    a. *Protection and Encouragement of Insectivorous Wild Birds and Other Animals.*

    b. *Use of Domesticated Fowls and Mammals.*

    c. *Introduction, Artificial Increase, and Colonization of Parasitic and Predaceous Insects, Protozoa, Mites, Worms, Birds, Fishes, Toads, and Other Animals.*

    d. *The Spread and Increase of Fungous, Bacterial, and Protozoal Diseases of Insects and the Liberation of Infected Insects.*

    e. *The Use of Growing Plants to Destroy, Repel, or Prevent Damage by Insects.*

V. Legal Control: the control of insects by controlling human activities.

    a. *Inspection and Quarantine Laws to Prevent the Introduction of new pests from foreign countries or their spread within a country.*

    b. *Laws to Enforce the Application of Control Measures, such as spraying, the cleaning up of crop residues, fumigation, and eradication measures.*

---

[1] It will be noted that biological control is related to natural control, being that phase of natural control in which man plays an important part. When man does something to make the work of the natural enemies of insects more effective, we say that is biological control.

    c. *Insecticide Laws to govern the manufacture and sale and to prevent the adultera-tion and misbranding of insecticides.*

    d. *Poison Residue Laws which fix a tolerance of various insecticides upon food products offered for sale or transportation.*

B. NATURAL CONTROL:[1] All of the measures that destroy or check insects which do not depend upon man for their continuance or success, and cannot be greatly influenced by man.

   I. Climatic Factors such as rainfall, sunshine, cold, heat, and wind.

  II. Topographic Features, such as rivers, lakes, mountains, type of soil, and other characteristics of the country that serve as barriers.

 III. Predators and Parasites naturally present in the region, including insects, birds, reptiles, mammals.

 IV. Insect Diseases naturally present in the region, such as entomophagous fungi, bacteria, and other microorganisms.

A. APPLIED CONTROL

Applied control includes those methods, under the control of man, which it is necessary to use when harmful insects have not been held in check by natural factors. Under this heading we have (*a*) chemical control by the use of insecticides, repellants, attractants, and auxiliary substances; (*b*) physical and mechanical control by specially designed machines or other devices, and the special manipulation of physical factors of the environment; (*c*) cultural control by variations in the usual farm operations; (*d*) biological control by the introduction and establishment of insect enemies; and (*e*) legal control, by regulating commerce, farming, and other human activities that affect the prevalence and distribution of dangerously destructive insects, the success of insect-control operations, or the health of man.

Applied control of insects is, as a rule, expensive, and the amount which one can reasonably expect to save by the control applied must be weighed against the expense involved. When insects are present in numbers sufficient to cause a heavy loss of property, there is usually a strong desire on the part of the property owner to stop this loss. The cost of killing or checking the insects causing the trouble must be carefully considered, and the most economical means of efficient control employed.

A practical applied method of control has not been worked out for many of the insects attacking field and forest crops, and certain classes of those attacking livestock and man. Applied-control measures have been developed for most of the orchard, truck-crop, greenhouse, stored-grain, shade-tree, and household insects. In general, crops or articles of high value can be protected by applied control measures, while those of low value per acre will not justify the expense of this type of control.

INSECTICIDES

Insecticides are those substances which kill insects *by their chemical action.* Insecticides may be grouped into three general classes: (*a*) stomach poisons, (*b*) contact poisons, and (*c*) fumigants. Fumigants or poison gases are generally the most effective insecticides to use when the insects and the products they are damaging are in a tight enclosure such

---

[1] It will be noted that biological control is related to natural control, being that phase of natural control in which man plays an important part. When man does something to make the work of the natural enemies of insects more effective, we say that is biological control.

as a house, storeroom, or greenhouse. Sometimes fumigants are used to destroy insects in their burrows in the soil or in wood, and sometimes portable enclosures are placed over plants out-of-doors to fumigate them. Generally when plants, animals, or products in the open are to be treated, a spray or dust is applied. These are of two fundamentally different kinds known as *stomach poisons* and *contact poisons*.

If the insect to be destroyed has chewing mouth parts (see Figs. 62 and 63) or the plants are found to be riddled with visible holes (Fig. 1,*B*), a stomach poison should be used. If the insect has piercing-sucking mouth parts (see Figs. 64 and 65) or the plants are wilted, yellowed, browned, or white-spotted or the leaves curled, but without visible holes (see Fig. 2), a contact poison should be used. *Contact poisons may be used to kill any insect that can be reached with a spray or dust, regardless of the type of mouth parts,* but stomach poisons have not been successfully used to kill insects with piercing-sucking mouth parts.

Stomach Poisons

(See Tables X and XI)

Insecticides of this class are generally applied against chewing insects but may also be used for insects with sponging, siphoning, or lapping mouth parts under certain conditions. There are four principal ways of using stomach poisons:

a. The natural food of the insect, *e.g.*, the foliage of plants or the feathers of birds, is covered with the poison so thoroughly that the insect cannot feed without getting some of the poison.

b. The poison is mixed with a substance (an *attractant*, see page 254) that is very attractive or tasty to the insect—if possible more attractive than its normal food—and the poison-bait mixture is placed where the insects can easily find it.

c. Certain poisons may be sprinkled over the runways of insects, so that they get it upon their feet or antennae. In cleaning their appendages with their mouth parts, some of the poison will be swallowed, especially if it is a substance irritating to the feet or antennae.

d. To a limited extent, the tissues of plants and animals have been impregnated with poisons so that insects eating these tissues are killed by the poison. Thus, animals may be dipped in arsenical solutions, so that ticks, sucking the blood from beneath the skin, get a fatal dose of absorbed poison; plants sprayed with bordeaux mixture may absorb enough of the copper in the spray to kill sap-sucking insects that subsequently feed upon them; and the sap stream of dying trees may be poisoned with zinc chloride or copper sulphate so that bark beetles feeding in these trees will be killed and prevented from multiplying and spreading to other trees. No practical method has yet been discovered, however, for poisoning the sap of trees to kill insects attacking them, without killing the trees.

A satisfactory stomach poison must be sufficiently active to kill quickly. It must be inexpensive. It must be available in large quantities. It must not be distasteful to the insects against which it is used, so as to repel them. It must be a sufficiently stable chemical so that it will not undergo chemical changes destroying its toxicity or making it harmful to plants, during shipment and storage or when it is mixed with other chemicals as a spray or dust, or simply from being exposed to the moisture and gases of the atmosphere. If the poison is to be applied to living plants, it must also meet the following requirements: It must not kill or burn the plant to which it is applied. In general this means that stomach poisons applied to foliage *must be insoluble in water*, for soluble substances will generally be taken into leaf or root in solution and so poison the plant. The margin of safety between a dose of poison necessary to kill the insect and the slightly larger dose that will burn or damage the plant is, with most insecticides, very slight. It must spread uniformly and adhere well to the plant surfaces to which it is applied. Since most stomach poisons are not soluble in water, they should be of such a physical nature that they will remain in suspension well in the spray liquid. Great fineness of particles is important, because such substances will cover the plant better, either as a spray or dust, and also because in general the finer the particles the greater their toxicity to the insect. Finally, the ideal insecticide would not leave any residue dangerous to the health of man or animals on the food parts of plants treated: very few stomach poisons meet this requirement.

Some idea of the relative toxicity to insects of the various stomach poisons may be gained from the following table, adapted from the *Journal of Economic Entomology*, **29**: 1160–1166, 1936. It should be clearly understood, however, that there are astonishing variations in the toxicity of the same poison to different insects; and that only a test of each poison against each pest under consideration can be depended upon to determine its usefulness.

TABLE IX.—MEDIAN LETHAL DOSAGE OF CERTAIN STOMACH POISONS, IN MILLIGRAMS PER KILOGRAM OF BODY WEIGHT, FOR THE SILKWORM LARVA

Poison	Median Lethal Dose, Milligrams	
Rotenone (p. 265)	3	
Deguelin (p. 265)	10 to 12	
Arsenic trioxide (p. 249)	15 to 20	
2-4-dinitro-6-cyclohexyl-phenol	16	
Calcium 2-4-dinitro-6-cyclohexylphenate	20	73[1]
Arsenic pentoxide (p. 249)	20 to 40	
Malachite green	25	
Paris green (p. 248)		40[1]
Tephrosin (p. 265)	30 to 60	
Cuprous cyanide (p. 253)	37	
Cryolite (p. 251)	50 to 180	
Acid lead arsenate (p. 244)	62 to 90	90[1]
Sodium fluosilicate (p. 251)	90 to 130	
Barium fluosilicate (p. 251)	90 to 170	
Sodium fluoride (p. 251)	110 to 150	
Calcium arsenate (p. 245)	260 to 780+	
Basic lead arsenate (p. 244)	900	
Phenothiazine (p. 252)		1,120+[1]
Toxicarol (p. 265)	1,540	
Zinc arsenite (p. 250)		1,990+[1]

[1] These figures were from tests against the imported cabbageworm.

The Arsenicals.—The various compounds of the element arsenic, collectively called arsenicals, comprise the most important and most used stomach poisons for insects that have ever been discovered. Certain of these arsenicals meet nearly all the requirements for an ideal stomach poison, listed above, except that they leave residues upon the treated plants that may be dangerous to man and animals. Until about 1926, the amount of poison remaining upon fruits and vegetables that had been sprayed with arsenicals was not considered dangerous to health. Following the lead of Great Britain, the United States government, through the Food and Drug Administration, has, since 1926, fixed the tolerance of poisons remaining upon fruits or vegetables placed upon the market. At present the tolerances are:

> 0.01 gr. of arsenic trioxide per pound
> 0.025 gr. of lead per pound
> 0.02 gr. of fluorine per pound.

0.01 gr. per pound is equivalent to 1 lb. of the poison in 350 tons of fruits or vegetables.

Fruits or vegetables carrying toxic materials in excess of these amounts cannot legally be marketed if shipped interstate. Several states have also established tolerances, in general conforming with the federal regulations.

The residue tolerance law has somewhat curtailed the use of arsenicals which, during the twenties, was at times so great that the demands could not be supplied by manufacturers. More than 45,000,000 pounds of calcium arsenate, 40,000,000 pounds of lead arsenate, and 3,000,000 pounds of paris green were used in the United States in 1936 and a total of 58,000,000 pounds in 1929. These laws have stimulated a tremendous amount of research to find satisfactory substitutes for the arsenicals; but, up to 1938, the general practice was still to apply as much arsenical as necessary to control the pests on fruits, such as apples and pears, and then, by washing, to remove enough of the residue to meet the legal requirements before marketing the produce.

The element arsenic is apparently not poisonous, but many of its compounds are very toxic. The element occurs uncombined in the earth in scattered particles but is more commonly found in combination with metals such as silver, antimony, copper, and iron as arsenides. The arsenic used in insecticides is secured largely from the flue dust and fumes from the smelting of various metal ores.

The element forms two oxides: As_2O_3 known as "white arsenic," arsenious oxide, or arsenic trioxide; and As_2O_5, known as arsenic oxide, or arsenic pentoxide. When these two oxides are dissolved in water, they form, respectively, arsenious acid and arsenic acid; and when these acids act upon metals or bases, the salts so formed are known, respectively, as arsenites and arsenates. In most cases the arsenites are less stable and more toxic to both insects and plants, while the arsenates, although less violently poisonous to insects, are more stable and safer to use on plants. Consequently, the arsenates, such as lead arsenate and calcium arsenate, are chiefly used for spraying upon plants but are not toxic enough to be effective in most poison baits. The arsenites, such as paris green and sodium arsenites, however, are used in poison baits but cannot be sprayed upon foliage for they "burn" it severely.

The two points of greatest significance about the arsenicals are (a) the percentage of total arsenic in the particular insecticide and (b) the proportion of the arsenic which is soluble in water.

Generally speaking, the killing power of an arsenical is in direct ratio to the percentage of metallic arsenic it contains, although the metal cation with which the arsenic is combined may have toxicity of its own, as in the case of lead or copper arsenate and others. The danger of "burning" or injury to plants is generally in direct ratio to the percentage of its arsenic that is present *in water-soluble form,* since such water-soluble arsenic can enter the living parts of the plant foliage and poison them. The ideal arsenical would be one having a very high arsenical content, none of which should be soluble in water but all of it readily soluble in the digestive juices of the insect.

Lead Arsenate.—Lead arsenate was first used as an insecticide in Massachusetts, in 1892, and was developed by the federal Bureau of Entomology in connection with their work of controlling the gypsy moth, in response to the need of a stomach poison that could be applied at greater strength than paris green without burning. Its popularity and usefulness expanded rapidly, and at the present time dry powdered lead arsenate is the most widely used and important stomach poison. It may be used to destroy or prevent the attacks of most kinds of chewing insects which leave definite holes in the leaves except (*a*) on very tender plants or crops such as citrus where it affects the quality of the fruit; (*b*) for a few insects that are repelled by it or are very hard to kill; or (*c*) upon parts of plants to be used as human food or feed for animals. It is also used to destroy soil-infesting insects by working it into the top soil at the rate of 1 or more pounds per 100 square feet.

Of the numerous forms of lead arsenate known to chemists, the acid orthoarsenate,[1] $PbHAsO_4$, and the basic orthoarsenate, $Pb_4(PbOH)(AsO_4)_3$, are the ones chiefly used as insecticides. Most commercial products are mixtures of the two, in which the acid form predominates. The acid lead arsenate contains about 20 per cent metallic arsenic equivalent, much less than $\frac{1}{4}$ per cent of which is usually soluble in water.

The basic lead arsenate is not generally on the market in all areas, but may be obtained for special purposes. It has been used for the control of chewing insects upon tobacco, Persian walnut, and in regions of very high humidity where the acid lead arsenate may break down and cause burning. It is more stable than the acid form and on this account will not burn even very tender foliage. Because it contains only about two-thirds as much arsenic (about 14 per cent metallic arsenic equivalent) and is not so easily acted upon by the digestive fluids of insects, it requires a larger dosage and a longer time to kill insects.

Compared with the other arsenicals, lead arsenate is a relatively weak poison; but it is as nearly insoluble in water and as stable as any arsenical spray material. This high degree of safety to foliage is its greatest merit. Once the spray has dried upon foliage it adheres, like paint, for a long time. As manufactured it is very finely divided, light and fluffy, commonly having a volume of 50 to 100 cubic inches per pound (from 350 to 400 grams per 1,000 cubic centimeters). It consequently handles nicely as a dust and also remains in suspension in spray liquids fairly well. Spreaders or deflocculators, such as calcium caseinate,

---

[1] According to the California Department of Agriculture: "Acid or standard arsenate of lead is a chemical combination of arsenic pentoxide and lead monoxide containing *not less than* 30 *per cent arsenic pentoxide* and *not less than one part of arsenic pentoxide to* 2.14 *parts of lead monoxide.* Basic arsenate of lead is a chemical combination of arsenic pentoxide and lead monoxide containing *not less than* 22 *per cent of arsenic pentoxide and not more than one part of arsenic pentoxide to* 3.10 *parts of lead monoxide.*"

soybean flour, sulphated alcohols, or sulphonated compounds are frequently added to improve wetting, spreading, and adhesive properties; or the lead arsenate may be made to absorb gelatin or gums to form colloidal lead arsenate. It may be combined with such insecticides as, nicotine, nicotine sulphate, bordeaux, wettable sulphurs, self-boiled lime and sulphur, certain oil emulsions, and fluosilicates and cryolite. The alkalies in some soaps and very hard alkaline waters may decompose the acid lead arsenate forming water-soluble arsenic, but the basic lead arsenate is not affected in this way.

Lead arsenate may be applied either as a dust or as a spray. As a dust it is usually diluted with from 5 to 20 parts of some inert carrier such as dusting sulphur, talc, hydrated lime, or gypsum. As a spray lead arsenate powder is mixed in water or other spray solutions such as bordeaux and is kept from settling out by agitation while it is applied. A standard dosage is about $\frac{1}{3}$ to $\frac{1}{2}$ ounce (2 or 3 level teaspoonfuls) to each gallon of water or 2 to 3 pounds to 100 gallons, but for some insects it is necessary to use it twice that strong. It is very extensively used to protect deciduous fruits, garden and truck crops, ornamental plants, and forest and shade trees from injury by chewing insects. The great advantage of lead arsenate is its high degree of safety to plants, its stability in storage and after it has been applied, and its spreading and adhesive qualities.

Its disadvantages are its relatively low arsenical content and consequent weakness compared with other arsenicals and the fact that the residue remains upon food products and feeds as a poison dangerous to man and animals if taken in large quantities. Lead arsenate sprays will spot or discolor white painted surfaces.

Calcium Arsenate or Arsenate of Lime.—Although calcium arsenates were used as insecticides at least as early as 1907, the earlier compounds contained or produced so much water-soluble arsenic that they burned foliage badly. The high cost of lead arsenate during the World War led to renewed efforts to produce a satisfactory calcium arsenate, and by 1915 the federal Bureau of Entomology demonstrated that, when properly manufactured, this insecticide may be satisfactory as a spray for many purposes. The discovery about 1920 that undiluted calcium arsenate applied as a dust to cotton will control the cotton boll weevil, for which other arsenicals had been useless, greatly stimulated interest in the perfection of the product and led to the use of as much as 20,000 tons of it a year. It is also used to some extent on potatoes, tomatoes, and some shade trees. It is considerably cheaper per lethal unit than lead arsenate.

The commercial calcium arsenates are a mixture of tricalcium meta-arsenate, $Ca_3(AsO_4)_2$, and acid calcium metaarsenate, $CaHAsO_4$ (in which the former predominates), with a considerable excess of lime. As found on the market, they are very fluffy, white powders, having a volume of 75 to 100 cubic inches per pound (about 400 grams per 1,000 cubic centimeters) and indistinguishable in appearance from lead arsenates. These insecticides contain from 25 to 30 per cent metallic arsenic equivalent and are therefore about a third stronger poisons than lead arsenates. When freshly manufactured, they have as small a percentage of water-soluble arsenic as lead arsenates, but the action of carbonic acid and ammonia in the atmosphere and the dilute alkaline solutions like rain water or dew decompose the material, liberating water-soluble

TABLE X.—POISONS TO BE USED FOR CHEWING INSECTS

Stomach poison	Time of killing	Adherence	Injury to plants	Suspension	Price	How applied	Best uses
Lead arsenate	Quickly	Very well	Very slight tendency to burn	Settles moderately	Moderate; 15 to 60¢ a pound	Dusts, sprays, and in combinations	Best spray for most chewing insects
Calcium arsenate	Rather slowly	Well	Tendency to burn	Settles moderately	Moderate; 10 to 50¢ a pound	Dusts, sprays, and in combinations	Use on resistant plants for insects hard to kill
Paris green	Very quickly	Poorly	Burns tender foliage	Settles rapidly	Moderate; 25¢ to $1 a pound	Dusts, sprays, baits, and in combinations	For potato and other hardy plants
Zinc arsenate	Slowly	Well	Slight tendency to burn	Settles moderately	Moderate; 15 to 25¢ a pound	Sprays and in combinations	Control of orchard insects
Cryolite	Rather slowly	Poorly	Slight tendency to burn	Settles moderately	Moderate; 15 to 25¢ a pound	Dusts, sprays, and in combinations	Control of orchard and truck crop insects, especially for beetles
Barium fluosilicate	Rather slowly	Poorly	Tendency to burn	Settles moderately	Moderate; 20 to 40¢ a pound	Dusts, sprays, and in combinations	Control of orchard and truck crop insects, especially for beetles
Magnesium arsenate	Quickly	Well	Slight tendency to burn	Settles very slowly	Moderate; 20 to 75¢ a pound	Dusts, sprays, and in combinations	Effective for most chewing insects.
Sodium arsenite	Rather quickly	Poorly	Burns severely	Soluble in water	Cheap; 10 to 40¢ a pound	Poison baits	For grasshoppers, cutworms and armyworms
White arsenic	Very quickly	Moderately well	Burns severely	Soluble in water	Cheap; 12 to 50¢ a pound	Poison baits	For grasshoppers, cutworms and armyworms
Sodium fluoride	Slowly	Very poorly	Burns severely	Soluble in water	Moderate; 30 to 75¢ a pound	Dip or dust on animals or in houses	For household insects and chewing lice
Derris, cubé, or timbo (rotenone)	Slowly	Moderately well	Does not burn	Settles slowly	High; 25¢ to $1 a pound	Dust and spray	Wherever arsenicals would endanger health
Pyrethrum	Quickly	Poorly	Does not burn	Settles slowly	High; 50¢ to $1 a pound	Dust and spray	Tender foliage; nearly ripe fruits
Hellebore	Slowly	Very poorly	Does not burn	Settles slowly	High; 20¢ to $1 a pound	Dust and spray	On tender plants and wherever arsenicals would endanger human health
Fixed nicotine	Quickly	Well	Does not burn	Good; settles slowly	Rather high; 20 to 40¢ a pound	Spray	Orchard and truck insects
Phenothiazine	Slowly	Poorly	Some plant injury	Good; settles slowly	High; 50 to 80¢ per pound	Spray	Orchard and truck crops; in dry years

TABLE XI.—POISONS TO BE USED FOR CHEWING INSECTS

Kind of spray	Amount of poison required for following amounts of spray						
	200 gal.	150 gal.	100 gal.	50 gal.	25 gal.	5 gal.	1 gal.
Lead arsenate, in powdered form..........	4 lb.	3 lb.	2 lb.	1 lb.	8 oz.	1.6 oz.	1/3 oz. or 3 tsp.
Calcium arsenate in powdered form..........	3 lb.	2.25 lb.	1.5 lb.	0.75 lb.	6 oz.	1.2 oz.	1/4 oz. or 2 tsp.
Paris green..........	2 lb.	1.5 lb.	1 lb.	0.5 lb.	4 oz.	0.8 oz.	1/6 oz. or 1.5 tsp.
Zinc arsenate, in powdered form..........	4 lb.	3 lb.	2 lb.	1 lb.	8 oz.	1.6 oz.	1/3 oz. or 3 tsp.
Cryolite, in powdered form..........	8 lb.	6 lb.	4 lb.	2 lb.	1 lb.	3.2 oz.	2/3 oz. or 6 tsp.
Sodium arsenite..........	4 lb.	3 lb.	2 lb.	1 lb.	8 oz.	1.6 oz.	1/3 oz. or 3 tsp.
Hellebore..........	Use for garden spraying only				3 lb.	10 oz.	2 oz.
Barium fluosilicate..........	4 lb.	3 lb.	2 lb.	1 lb.	8 oz.	1.6 oz.	1/3 oz. or 3 tsp.
Derris, cubé, or timbo (rotenone)..........	For garden use			1 lb.	1 lb.	3.2 oz.	2/3 oz. or 6 tsp.

arsenic. Buyers should insist upon material manufactured the year in which it is to be used. The suspension is a little better, adhesiveness slightly less, and dusting qualities better than that of lead arsenate.

Calcium arsenates are compatible with lime-sulphur, nicotine, oil emulsions, bordeaux, and sulphides. They are incompatible with soaps (though less affected than lead arsenate), with nicotine sulphate in sprays but not in dusts, flotation sulphur, all the fluosilicates except calcium fluosilicate, and to some extent with fish oils and vegetable oils.

Calcium arsenate may be applied in the same ways as lead arsenate (see page 245). A standard dosage for a spray is $\frac{1}{4}$ ounce ($1\frac{3}{4}$ to 2 level teaspoonfuls) to each gallon or $1\frac{1}{2}$ to 3 pounds to 100 gallons of water or other spray liquid. An equal amount of freshly slaked lime or a third more of hydrated lime should be added to the water unless it contains an excess of lime as in the case of bordeaux.

It is distinctly more poisonous to insects than lead arsenate but also more toxic to plants, especially when not fresh. Only the more resistant plants, such as cotton, potato, and apple will stand it, and it must not be applied to stone fruits or many vegetables and flowers. It is likely to cause very severe burning when used upon plants attacked by diseases. Its advantages are its moderately high arsenical content, its physical fineness, its cheapness, and the fact that it is compatible with lime-sulphur. Its disadvantages are its instability, the danger of burning foliage, its toxicity to higher animals, and its incompatibility with nicotine sulphate in sprays.

Paris Green.—The control of insects by applying stomach poisons to their food plants was unknown and not practiced anywhere in the world prior to about 1860. London purple (a mixture of calcium arsenite and calcium arsenate with other by-products from the manufacture of certain dyes and pigments, now entirely obsolete as an insecticide) and paris green, an important pigment in former times, were the first stomach poisons to come into general use for dusting and spraying plants. Their use originated in the west-central United States about 1865 against the Colorado potato beetle. At first applied as dusts, because there were no spray pumps, their use slowly expanded in spite of fear and prejudice, and paris green remained the leading insecticide for chewing insects until the discovery of lead arsenate in 1892. The use of paris green has rapidly declined during the present century, and it is now recommended for use only (*a*) in poison baits; (*b*) against a few insects which are very hard to kill, and when attacking very hardy plants; and (*c*) in combination with bordeaux mixture. It should not be applied to stone fruits or tender vegetables and flowers.

The chemical name of paris green is copper acetoarsenite. The formula usually given is $Cu(C_2H_3O_2)_2 \cdot 3Cu(AsO_2)_2$. It usually contains arsenic equivalent to about 33 to 39 per cent metallic arsenic, from 2 to 3 per cent of which is generally in water-soluble form. It also contains about 30 per cent copper oxide and 10 per cent acetic anhydride, which are expensive but add nothing to the value of the spray. It is generally manufactured as a brilliant green, coarse powder, composed of minute spheres of various sizes. Impure grades are commonly marketed. A simple test for purity is that the paris green is completely soluble in ammonia leaving no residue visible under a microscope. On account of its coarseness it settles out of the spray solution rapidly and does not adhere well. It has a large percentage of water-soluble arsenic and breaks down easily, causing severe burning of foliage. On account of its high arsenic content and its ready solubility in the digestive tract of the insect, it is one of the quickest killing insecticides. Its distinctive color is also an advantage, preventing it being confused with other common substances. However it is more expensive than other arsenicals, and lead arsenate and calcium arsenate have largely taken its place for application to foliage. An average strength spray is made by mixing $\frac{1}{6}$ ounce ($1\frac{1}{2}$ level teaspoonfuls) and $\frac{1}{3}$ ounce of hydrated lime to each gallon of water or 1 pound of paris green and 2 pounds of lime to 100 gallons of spray. It cannot be combined with lime-sulphur, soaps, sulphides, or any fungicide containing ammonia.

White Arsenic, or Arsenic Trioxide, or Arsenious Oxide, As_2O_3.—This is a very active poison and is one of the cheapest arsenicals. It consists of about 75 per cent arsenic. It is the form from which all other arsenicals are derived. Because it is completely soluble in water, it is so likely to burn foliage that it is not safe to use in spraying plants, but it is used in poison baits for grasshoppers, armyworms, ants, and cockroaches. It is sold in the form of a white powder. When dissolved in water, it forms arsenious acids. These acids acting upon bases produce the series of salts known as *arsenites, e.g.,* paris green and sodium arsenite.

Arsenic Pentoxide or Arsenic Oxide, As_2O_5.—When arsenic trioxide is heated in the presence of nitric acid, it is readily converted into arsenic pentoxide, a salt containing about 65 per cent arsenic. Its only importance in entomology is that it is the basis for the manufacture of the very valuable arsenates. With water, arsenic pentoxide forms arsenic acids, and the action of these acids on bases produces the series of salts known as *arsenates.* The arsenates are, in general, less toxic than the arsenites but, because of greater stability, are less likely to burn plants and consequently are of greater value as insecticides.

Magnesium Arsenate.[1]—The use of this material as an insecticide has developed since about 1918, when greatly improved methods of manufacture began to produce a highly insoluble form. It has been recommended chiefly against the Mexican bean beetle, because it does not burn the foliage of beans. Although it has a high arsenical content equivalent to 42 to 46 per cent metallic arsenic, it has given a poor kill of most insects and is more expensive than lead or calcium arsenate. Both the trimagnesium orthoarsenate, $Mg_3(AsO_4)_2$, and the dimagnesium or acid orthoarsenate, $MgHAsO_4$, have been used. As a spray it is used $\frac{1}{3}$ to $\frac{2}{3}$ ounce to the gallon or 2 to 4 pounds to 100 gallons; as a dust 1 part is diluted with 5 to 10 parts of hydrated lime or other carrier.

Zinc Arsenate, $Zn_3(AsO_4)_2$.—This compound is used in codling moth control in place of lead arsenate, principally because it does away with the lead residue. In general, the results with this material have been good. It has proved nearly as effective as lead arsenate. It has only a slight tendency to burn the foliage. In this respect it is about equal to lead arsenate. The metallic arsenic equivalent is about 32 per cent. Zinc arsenate is not compatible with flotation sulphur and with certain soaps. It has been used most extensively in the arid sections of the Northwest.

[1] See *U.S.D.A., Bur. Ent., Plant Quar., Cir.* E-451, 1938.

Manganese Arsenates.[1]—Various compounds of manganese with arsenic have been tried as insecticides. The trimanganoarsenate, $Mn_3(AsO_4)_2$, and the dimanganoarsenate, $MnHAsO_4$, are perhaps the most important. The total arsenic content is about 34 to 39 per cent, and it is a somewhat unstable compound. Manganese arsenates are compatible with lime-sulphur, flotation sulphur, and wettable sulphurs. They have sometimes been found effective but frequently cause burning of foliage and have not come into general use.

Zinc Arsenite, $Zn(AsO_2)_2$.—This arsenical ranks next to paris green in killing efficiency, close to lead arsenate in adhesiveness and suspensibility, and is cheaper than either. It contains arsenic equivalent to about 54 per cent. It has been used to some extent for spraying potatoes, elm, and cabbage but is not safe on orchard or bush fruits or other tender foliage. It is incompatible with soaps, lime-sulphur, or glucose.

Sodium Arsenate, Na_3AsO_4 and Na_2HAsO_4.—This is a highly soluble form of arsenical which cannot be used for spraying the foliage of plants. The metallic arsenic equivalent is from 36 to 40 per cent. It is used in poison baits and for making other, more stable, insecticides.

Sodium Arsenite, $NaAsO_2$ and Na_2HAsO_3.—This is a very highly soluble form of arsenical used mainly in poison baits and stock dips and generally sold in liquid form. It is extremely toxic to plants and is extensively used as a weed killer. The metallic arsenic equivalent is 44 to 57 per cent.

There are several other arsenical compounds that have been used to a limited extent, but which have not proved of sufficient value to warrant their general adoption, mainly because of their liability to burn foliage. Among these may be mentioned arsenite of lime, arsenite of lead, and london purple.

Fluorine Compounds.—Compounds of this highly active and poisonous element were practically unknown as insecticides until the second decade of the present century. Its first important use was in the form of dry sodium fluoride for cockroaches, poultry lice, and ants; then fluorine compounds were discovered to have value in poison baits; and then, stimulated in part by the arsenical·residue laws, the highly insoluble fluosilicates and fluoaluminates were found to be very effective stomach poison sprays and dusts for use on plants. However, fluorine has been made subject to a residue law similar to that for arsenic.

Our supply of fluorine is derived from the widely distributed minerals, fluorspar, CaF_2, and cryolite, Na_3AlF_6. The insecticides are derivatives of hydrofluoric acid, especially sodium fluoride; of fluosilicic acid, especially sodium and barium fluosilicates; and of fluoaluminic acid, especially sodium fluoaluminate, commonly known as natural or synthetic cryolite. These are all on the market as fine white powders.

The question of the toxicity of fluorine compounds to man and domestic animals is one of great moment. Marcovitch has claimed that the minimum lethal dose for animals is about as follows, in grams per hundred pounds of body weight:

Calcium arsenate	0.0019 g.	Barium fluosilicate	0.0083 g.
Lead arsenate	0.0035	Sodium fluoride	0.0095
Sodium fluosilicate	0.0081	Cryolite	0.0237

He considers cryolites nonpoisonous and claims that natural foods contain 8 or 9 times as much fluorine as the average fruits and vegetables that have been sprayed with cryolite. The fear that the use of fluorine compounds as sprays for fruits and vegetables might result in mottling of teeth appears to have no foundation: this trouble results from the use

---

[1] See *U.S.D.A., Bur. Ento. Plant Quar., Cir.* E-408, 1937.

of water containing toxic amounts of fluorine, for a considerable time for drinking and especially for cooking, which concentrates the fluorine in the food.

Sodium Fluoride,[1] NaF.—This poison has come into general use since 1915 as a means of combating chewing lice on animals and poultry and for the control of household pests, particularly cockroaches. While it is primarily a stomach poison, it also acts to a slight extent as a contact poison when the powder or solution is taken into the tracheae or directly through minute pores in membranous areas of the body wall as at the junction of head and thorax or the attachment of the legs to the thorax. It is believed, however, that sodium fluoride kills cockroaches, ants, and possibly chewing lice, largely in the manner described under (c) on page 241; consequently, it is necessary that the powder be kept dry and finely pulverized in order to be effective. This insecticide has also been highly recommended as a poison for baits for grasshoppers, cutworms, earwigs. The commercial form of sodium fluoride is used undiluted as a powder or mixed with other, nontoxic, dusts and in solution. It should not be sprayed on the foliage of plants. Sodium fluoride is poisonous to man and should not be sprinkled over foods. It is also somewhat irritating to the skin and to the respiratory passages.

Fluosilicates or Silicofluorides and the Fluoaluminates (Cryolites).— Sodium fluosilicate (Na_2SiF_6) can be used for the same purposes as sodium fluoride. It has also been recommended *as a dust* for the cotton boll weevil and many other beetle pests of crops. Barium fluosilicate, $BaSiF_6$, is a much more insoluble salt and has been used *as both a dust and a spray* for many fruit and garden pests. It is more popular than sodium fluosilicate. Perhaps the best of the fluorine compounds for spraying or dusting upon plants are the cryolites (sodium fluoaluminate, Na_3AlF_6), but they are not satisfactory for poison baits. Both the natural mineral, and synthetic forms are available. They are about twice as soluble as barium fluosilicate but still only 1 part in over 1,600 of water. Barium fluosilicate and cryolite are used at 3 to 8 pounds to 100 gallons of water, or as a dust (diluted 1:3 with tobacco dust, cheap flour, clay, or talc) using 5 to 15 pounds per acre, for crops such as beans, against the Mexican bean beetle; or for blister beetles on flowers and vegetables. For codling moth and plum curculio on apple, these materials, combined with fish oil or summer spray oils, have been reported as very satisfactory especially in dry regions or dry seasons. The residues are somewhat harder to wash off than arsenical residues and require an alkaline instead of an acid wash. Potassium hexafluoaluminate (K_3AlF_6) has been found successful for the control of the Mexican bean beetle. Cryolites, fluosilicates, and fluoaluminates are compatible with lime-sulphur, wettable sulphur, neutral fish-oil soap, summer spray oils, gypsum, and lead arsenate. They are incompatible with alkaline soaps, calcium carbonate, or other calcium and magnesium salts occurring in hard waters. Barium fluosilicate is not compatible with nicotine sulphate. Lime is not a satisfactory diluent or carrier for these dusts. .

The principal difficulty with this group of stomach poisons has been to manufacture them in a sufficiently fine, light, and flocculent

---

[1] This should not be confused with *sodium chloride*, common salt, as the names sound much alike.

form to cover foliage as well as the arsenicals do.   Barium fluosilicate may be injuriously corrosive to parts of spraying machinery.

The fluorine compounds in general kill more quickly than the arsenicals; they are often cheaper; they are less toxic to the higher animals; in certain tests they have proved to be safer to use on plants; they act both as stomach poisons and as contact poisons and are irritating to the appendages of beetles and other insects causing them to take the poison into their mouths even if they do not feed upon the treated foliage; in some cases they have a valuable repellent effect upon the pests.

Organic Stomach Poisons.—Several of the substances whose most important action is as contact poisons have also an effect as stomach poisons when they are eaten by insects.   These have received exhaustive consideration as possible substitutes for the standard stomach poisons, since the Food and Drug Administration has ruled against the presence of arsenic, lead, and fluorine on human food products.   *Nicotine*, in a fixed or nonvolatile form, such as nicotine bentonite, nicotine tannate, nicotine humate, and combinations of nicotine and oil are known to be toxic in this way.   Preparations of derris, cubé, and pyrethrum have been effective against many chewing insects and have been used as arsenical substitutes for the control of cabbageworms, celery insects, and certain pests of melons, peppers, small fruits, and tobacco.   Freshly prepared commercial or home-mixed dusts containing 0.5 to 1.5 per cent rotenone (2 to 6 per cent total extractives of derris or cubé roots) in nonalkaline carriers such as fine clay, talc, tobacco dust, or sulphur, applied at the rate of 15 to 40 pounds per acre, will control the above mentioned garden pests.   Sprays containing from 0.02 to 0.025 per cent rotenone (0.07 to 0.1 per cent total extractives of derris or cubé roots) in water, with a suitable spreader, are also effective for most leaf-eating insects but soon lose their toxicity upon exposure to outdoor weather.

Pyrethrum dusts made of pure, fresh, ground pyrethrum flowers containing approximately 1 per cent total pyrethrins, diluted with 5 parts of dusting sulphur, talc, or tobacco, are also effective for a short time as stomach poisons.   Sprays made of pyrethrum extracts have some stomach poison properties also.

The principal weakness of these organic substances is that they lose their strength and effectiveness very soon after they are applied to plants.   It is therefore recommended that they be applied late in the day whenever possible.   Some success has attended the use of antioxidants and pigments such as lampblack to prevent the deterioration of the fugitive poisonous principles after they are exposed to light and air. It is also necessary to add some material as a spreader and sticker to make the sprays effective.   For a full discussion of derris and pyrethrum and nicotine see under Contact Poisons, pages 257–266.

Phenothiazine (*Thiodiphenylamine*, $C_6H_4NHC_6H_4S$).—Among the hundreds of organic compounds prepared synthetically in the search for better and safer insecticides is this material, first tried in a practical way in 1934.   It is a light-brown crystalline compound, melting at 356°F., with a neutral reaction, insoluble in water, and only slightly soluble in mineral oils and other organic solvents.   It is said to have very high toxicity to mosquito larvae, to equal lead arsenate in toxicity to codling moth, and to be less than one-seventy-fifth as toxic to man as lead or calcium arsenates.   An interesting use for it has been the demonstration that feeding it to cattle at the rate of 0.1 gram per kilogram of body weight will prevent the development of horn fly larvae in the dung of the animals. As a spray the material does not weather well, and it has given indifferent results in most sections, against codling moth.

Hellebore.—At one time *hellebore* was considered an important stomach poison, but it is little used now. It is made by drying and pulverizing the roots of certain plants of the lily family, known as the white hellebore (*Veratrum album*) and the American hellebore or Indian poke (*Veratrum viride*). The poisonous alkaloid in these roots is called veratrine. It is a fugitive poison, and hellebore is, when fresh, apparently about one-third as toxic to insects as lead arsenate. Because it is only mildly poisonous to man and soon loses its toxicity after it is applied, it may be used on food parts of fruits and vegetables without danger to the consumer. It must be stored in a tightly closed receptable. It may be combined with any fungicide or with contact insecticides. It is used as a dust diluted with flour, talc, or lime or as a tea at the rate of 1 or 2 ounces to the gallon of hot water. A special use for it is as a larvicide for house flies in manure. One half pound in 10 gallons of water is recommended to 8 bushels of manure. Unlike borax there is no limit to the amount of such treated manure that may be applied to soil. It should be noted that the black hellebore from the roots of *Helleborus niger*, which has some medicinal uses, is not employed as an insecticide.

Minor Stomach Poisons.—Of the many substances that have been tested as stomach poisons for insects, the following have proved their usefulness in certain limited fields. A form of *phosphorus*, known as white or yellow phosphorus, incorporated in a sweet sirup, forms a valuable bait for cockroaches, mice, and rats. The phosphorus must be handled with great caution, not only because of its toxicity, but also because it ignites spontaneously at temperatures above 95°F. The compound *thallium sulphate*, Tl_2SO_4, is combined with a sweet or fatty carrier to form a valuable poison for ants. *Borax* or sodium tetraborate ($Na_2B_4O_7$) has been used to kill house fly maggots in manure or other refuse; to prevent the breeding of mosquitoes in water to be used only for laundering purposes; and as an emulsion with oil, or as glyceroboric acid to treat maggot-infested wounds of animals and repel the flies. *Formaldehyde* (formalin, 40 per cent aqueous solution) is valuable as a house fly poison in baits, as a larvicide for mosquitoes and to destroy the potato scab gnat in seed potatoes. Metallic *mercury* incorporated in a heavy oil, such as vaseline, makes an effective ointment for lice on poultry or on man. The salts of mercury, "bichloride of mercury" or "corrosive sublimate," $HgCl_2$, and "calomel," Hg_2Cl_2, are specific remedies for certain root-infesting insects, such as the cabbage maggot (see page 539), the onion maggot (page 528) and the larvae of flea beetles and fungous gnats. The $HgCl_2$ may also be used in book bindings and ant tapes as a repellent to ants, cockroaches, and termites.

There are many other substances under test as stomach poisons, some of which give much promise but have not passed the experimental stage. Among the latter may be mentioned sulphur nitride, cuprous cyanide, diphenyl amino arsenious oxide and certain nitrophenols.

Poison Baits.—In certain cases where spraying is not practicable, as for certain household insects, insects living inside of fruits or vegetables or underground, and chewing insects attacking areas of crops too great to be protected by spraying, stomach poisons may be mixed with materials known to be attractive to the particular pest, and such *poisoned bait* exposed where the insects may get it. Poison baits have been extensively used for grasshoppers and armyworms, especially since about 1914. The success with these pests has stimulated efforts to control many other pests by this method. Success with a poison bait will depend upon many considerations such as attractivity, palatability, toxicity, stability, physical condition, and time, place, and method of exposure. Baits with a dry carrier have been used extensively for grasshoppers, crickets, cockroaches, silverfish, earwigs, cutworms, armyworms, white grubs and wireworms. Baits in a liquid carrier have been widely employed for ants, house flies, fruit flies, root-maggot flies, and adult Oriental fruit moth, greenhouse leaf tyer, and corn earworm.

Poison Baits for Grasshoppers, Cutworms, Armyworms, Crickets, and Earwigs.—For these insects a very strong and soluble form of arsenical is generally used. Lead arsenate should never be used, since it is too weak a poison; white arsenic, paris green, and arsenite of soda are commonly used. Sodium fluoride has been recommended as a poison in baits for the European earwig and grasshoppers and some investigators have found sodium fluosilicate less repellent and more effective for cutworms, than the arsenicals. The most common carrier is wheat bran, moistened with enough water to make it thoroughly wet but not enough to drip without being squeezed. Sometimes hardwood sawdust, fresh horse droppings, ground corncobs, or the hulls of seeds are used as the carrier. "Black strap" molasses is usually added, and sometimes other attractants such as chopped whole oranges or other fruits, oils, amyl acetate, or salt, are recommended. At the present time, after a great deal of careful experimentation to test the effectiveness of different combinations, the following formulas are most widely recommended:

FORMULA A

The Water or Kansas Bait

	Amount for 10 *Acres*	Amount for a Lot 50 × 100 *Ft.*
Bran	100 lb.	1½ lb.
Poison	4 lb.	1 oz.
Water	7 to 10 gal.	1 pt.

FORMULA B

The Oil Bait

	Amount for 10 *Acres*	Amount for a Lot 50 × 100 *Ft.*
Bran	100 lb.	1½ lb.
Poison	4 lb.	1 oz.
Lubricating oil (20 to 30 S.A.E.)	2 gal.	¼ pt.

Instead of all bran, up to 50 per cent hardwood sawdust, middlings, cheap flour, whey, or corncobs, ground in a hammer mill using a ¼-inch screen, may be used. The bran should be coarse-flaked and free from shorts which cause lumpiness. For poison one may choose white arsenic, paris green, sodium pyroarsenate, dry sodium arsenite, sodium fluosilicate or barium fluosilicate, or liquid sodium arsenite (2 quarts). Lead arsenate is not effective in poison baits. If a dry poison is used, it should be mixed thoroughly with the bran, then the water and other ingredients or the oil added and mixed until every flake of bran is moistened. Liquid sodium arsenite may be used, by mixing with the water, and the poisoned water added to the bran; but liquid poison has not proved to be satisfactory with the oil baits. The poisoned bran is then scattered broadcast over fields to be protected and adjacent waste land at the rate of 8 to 10 pounds per acre. It may be scattered by hand from pails or bags or by the use of an oat seeder, limestone seeder, or special poison-bait spreaders, and it has been distributed over vast areas from airplanes. Every effort should be made to scatter the poison thinly so that no lumps or piles of the bait are left on the plants or the ground. If this is done there is no danger of poisoning livestock or wild birds. For grasshoppers, the bait should be distributed in the morning so as to be fresh when the hoppers begin feeding. For cutworms, armyworms, and earwigs the bait is best distributed in the evening, since these insects are nocturnal. See also baits recommended for cockroaches (page 781) and for silverfish (page 779).

Poison Baits for Ants.—These have been extensively used for the Argentine ant and to a less extent for other species. For species that prefer fatty foods the carrier must be a fat, such as ground meat or grease; while for the sweet-eating ants a sirup, sugar, or honey is used. For the latter species at least, it has been found desirable to use sirup very weakly poisoned with chemically pure sodium arsenite, in order that the worker ants, which alone forage in houses and orchards, may have time after eating the poison bait to carry it back to the nest and feed some of it to the queens and the larvae. In this way the entire colony can be killed. Several formulas found effective for ants are given on page 772. These sirups are put in soda-fountain straws cut in short pieces, or in aluminum or tin boxes or cans or paperoid cups, having a piece of sponge or blotting paper in the bottom as a footing for the insects, and the sides or covers having holes punched in them or fitting on so as to leave passages through which the ants can enter the boxes. These containers are placed along the runways of the ants out of the reach of children and kept filled so that a part of the sponge is above the sirup. From 75 to 150 such cans are used in each city block or 5 to 10 around each dwelling, when fighting the Argentine ant, and entire municipalities have been freed of this pest by this method. Fall and early spring are the best times to use this control.

Poison Baits for Flies and Moths.—Commercial poisoned fly papers to be soaked in sweetened water have been used for years against the house fly. A mixture of 1 tablespoonful of formalin in 1 pint of mixed water and milk is also commonly recommended for this pest. For a bait nonpoisonous to man, a 1 per cent solution of sodium salicylate has been recommended. The fruit, vegetable, and root-maggot flies, whose maggots develop internally in a great variety of fruits and vegetables, or in the soil, have been killed in many parts of the world with mixtures of sodium arsenite, potassium arsenate, lead arsenate, or other arsenical, borax, or tartar emetic in molasses and water. These mixtures are frequently sprayed over the foliage of the plants on which the maggots are feeding, and the adult flies lap up the tiny droplets of sweetened poison from the leaves. Poisoned baits consisting of molasses or saccharine in water have also been tried for moths whose larvae are destructive to crops. These are usually exposed in cans or pails, suspended among the crops to be protected. Poisons used include sodium arsenite, barium chloride, sodium fluosilicate, zinc phosphide, and tartar emetic. Recently extensive attempts to bait the Oriental fruit moth have been carried on in which a great variety of attractants, such as various acids, oils, and alcohols have been added to the poisoned sirup or produced in it by the fermentation caused by yeasts and bacteria, but so far without leading to practical recommendations for control.

Poison Baits for Subterranean Crop Pests.—Sometimes poison baits are buried in the soil to destroy wireworms, white grubs, and similar root-feeding insects. Fresh vegetables, green foliage, germinating seeds, or balls of dough made from scorched flour have been used as carriers for the poison. The soil must be free of growing or rotting vegetation and the baits must be planted about every 3 feet.

CONTACT POISONS

(See Tables XII and XIII)

In order to kill an insect with stomach poisons, the insect must swallow the poison. Insects with piercing mouth parts take their food from beneath the surface and consequently get none of the poisons applied to the surface of foliage or fruits; no effective method of poisoning the sap of plants, without injuring the plants, has ever been discovered. Consequently, for the piercing-sucking plant pests we must use a contact poison.

Insecticides of this class are used in liquid and dust form to kill insects by coming in contact with or entering their bodies, *other than by the mouth,* and especially through the spiracles. There are many kinds, of which the following are the more important: (*a*) organic substances such as nicotine, pyrethrum, rotenone, and quassia; (*b*) caustic inorganic compounds like lime-sulphur; (*c*) oils, soaps and emulsions; (*d*) synthetic organic preparations such as the thiocyanates and cyclohexylamines; and (*e*) sulphur and other dusts. They kill insects by entering the tracheae as liquids or gases or by penetrating through membranous areas of the body wall and producing a chemical action on the vital tissues, such as blood or nerves; or, rarely, by their physical action in clogging the breathing tubes and smothering the insects.

Contact poisons are used chiefly for (*a*) insects (Fig. 64) that suck the sap of plants (horticultural sprays and dusts); (*b*) for household insects and other pests living in dwellings, barns, and storerooms, such as clothes moths, bedbugs, flies, chicken mites, and the like; (*c*) for bloodsucking insects attacking domestic animals—stable flies, black flies, mosquitoes (livestock sprays); and (*d*) for certain kinds of chewing insects that cannot be readily or safely killed by applying stomach poisons to their food.

In spraying or dusting with contact poisons, the aim should be to hit every insect present with the poison, since only those which are hit by it will be killed. Unlike stomach poisons, contact poisons cannot be applied in advance of an expected outbreak as a preventative.

In this class of insecticides we have a greater variety and diversity of substances in general use than is the case with the stomach poisons. Many substances that act as contact insecticides are also toxic to plants, and the margin between the point of injury to the insect and injury to the plant is often small. For this reason, one must use caution in applying these insecticides. For some of the most resistant insects it is best to apply the spray in winter or before the plants have started active growth, because they will stand stronger applications at that time. Such sprays are called *dormant sprays*, to distinguish them from the *summer sprays* and *dusts* which are applied while the trees and other plants are in foliage. The principal dormant sprays are oil emulsions and lime-sulphur; the most important summer contact sprays are nicotine, pyrethrum, rotenone, synthetic organic chemicals, and sulphur, while the oils have important uses in summer, as well as in winter.

Nicotine, $C_{10}H_{14}N_2$.[1]—Tobacco is one of the first materials to be used as an insecticide. As early as 1763, a French paper recommended the use of finely powdered tobacco mixed in water, with the addition of some lime, to destroy plant lice without injury to the foliage. It was not until 1809, however, that the presence of a volatile poisonous substance in tobacco was discovered, and in 1828 this was recognized as an alkaloid and named nicotine. Tobacco belongs in the same family of plants (Solanaceae) as the potato and tomato. The most important species are *Nicotiana tabacum* and *Nicotiana rustica*.

Nicotine ($C_{10}H_{14}N_2$) is the principal alkaloid in tobacco, in which it occurs at greatly varying concentrations depending upon the species and variety of plant, the part of the plant, the climate, soil, culture, time of harvest, and method of curing. It commonly occurs to the extent of 2 to 5 per cent in the leaves, while stalks, pods, and roots have rapidly decreasing amounts, and the seeds only a trace. Pure nicotine is a colorless oily liquid which becomes viscous and rapidly changes through red and brown to black upon exposure to air. It is readily soluble in water and alcohol and is slightly soluble in oils. The characteristic odor of tobacco is not due to the nicotine, which has a disagreeable odor like pyridine, a bitter burning taste, and a density slightly greater than water. It boils at 477°F. (247°C.). On account of its high boiling point, nicotine does not form a true gas under ordinary conditions but the vapor is used as a fumigant. Nicotine is one of the most violent poisons for animals but, fortunately, is not very toxic to plants.

Since its early discovery the principal use of nicotine has been for various piercing-sucking insects which cannot be destroyed by the stomach poisons. It has been used in a variety of ways which may be summarized as follows:

1. As a liquid extract: (*a*) homemade decoctions, (*b*) standardized concentrated free nicotine, (*c*) standardized concentrated nicotine sulphate, (*d*) extracts mixed with activators, (*e*) fixed nicotines.

2. As a dust (page 259): (*a*) finely powdered dried tobacco, (*b*) nicotine extracts mixed with dust carriers.

3. As a fumigant (page 284): (*a*) by burning dried tobacco or boiling it in water; (*b*) by vaporizing the concentrated extracts with heat, in buildings, or in portable machines, sometimes under canvas hoods; (*c*) by burning paper, punks, or powders soaked in the extracts.

Teas or decoctions of tobacco can be made at home by soaking tobacco in warm limewater. The difficulties are that the nicotine content of

[1] See "A Bibliography of Nicotine, Parts I and II," *U.S.D.A., Bur. Ento. Plant Quar., Cirs.* E-384, E-392, 1936.

different lots of tobacco varies so greatly that to make a spray strong enough to kill aphids might require anywhere from 11 to 145 pounds to the hundred gallons of water, depending upon the nicotine content of the tobacco used. There is no simple test that a farmer can use to tell how strong a given sample of tobacco is.

Accordingly, manufacturers have placed on the market extracts made from the waste of smoking tobaccos and from diseased leaves and stems, concentrating them greatly and putting them up in handy containers of known nicotine strength. In America the most extensively used brands have been standardized at 40 per cent nicotine alkaloid by weight. There are two, fundamentally different preparations: (*a*) those containing *free nicotine*, which may consist of 40 per cent actual nicotine, *e.g.* Nicofume, or 50 or 95 per cent actual nicotine; and (*b*) those in which the nicotine has been treated with sulphuric acid so as to convert it into *nicotine sulphate*, $(C_{10}H_{14}N_2)_2 \cdot (H_2SO_4)$. Black Leaf 40, one of the common trade brands, also contains 40 per cent nicotine, as nicotine sulphate. Nicofume is only slightly heavier than water (sp. gr. 1.013) but Black Leaf 40 is nearly one-fifth heavier (sp. gr. 1.188). The important differences between these two kinds of products are that free nicotine is volatile, more active than nicotine sulphate, and consequently kills more quickly; but its action does not extend over so long a period. It is used chiefly for fumigating and for spraying flowers that might be stained by the nicotine sulphate sprays. Nicotine sulphate is the most extensively used contact insecticide for aphids and other soft-bodied sucking insects on plants. It does not volatilize so rapidly as free nicotine and therefore remains as a poisonous residue for a short time upon the objects to which it is applied. In order to cause a more rapid liberation of the nicotine from nicotine sulphate, soap, hydrated lime, lime-sulphur, or ammonium sulphate may be added to the spray or dust.

To determine how much any commercial nicotine spray may be diluted for general use, the percentage of actual nicotine guaranteed on the label should be divided by 0.06, which will give the number of parts of water that can safely be added to 1 part of the concentrated spray. Thus 1 pint of a 95 per cent actual nicotine preparation would make (95 ÷ 0.06 or 1,583 pints) nearly 200 gallons of standard aphid spray. To be effective against most insects, the diluted spray should ordinarily contain from 0.06 to 0.05 per cent of nicotine, or 1 part of 40 per cent nicotine sulphate or free nicotine to 800 to 1,000 parts of water, or 1 pint (13.5 ounces avoirdupois) to the 100 gallons of spray. For small quantities, use 1 to 2 teaspoonfuls to each gallon of water. Unless an arsenical, lime-sulphur, bordeaux mixture, or oil emulsion is to be used in the same mixture, soap should always be dissolved in hot water and added to the nicotine spray to aid in liberating the nicotine and to serve as a spreader. Some liquid fish-oil soaps are excellent for this purpose. Calcium caseinate and ammonium sulphate are also used. A cubic inch of soap to each gallon of spray, or 4 to 6 pounds to the 100 gallons, is about the right amount. Soap-oil emulsions are also excellent spreaders, wetters, and carriers for nicotine. By using certain oil emulsions or oxidized oils as activators and carriers with nicotine, the amount of nicotine required to kill certain insects may be reduced to 1 part of the 40 per cent concentrate to 2,000 or 3,000 of the spray. Such activators are sold under many trade names, such as Penetrol.

The above sprays of nicotine soon disappear from the plants. Considerable attention has been directed toward the perfection of relatively insoluble and nonvolatile nicotine preparations that might remain effective as spray deposits over a longer period of time. These act as stomach poisons as well as contact poisons. Promising results have been obtained with such so-called "fixed nicotines" as nicotine-oil combinations, nicotine tannate, and other nicotine soaps, nicotine bentonite with sulphur and soaps or oils, and nicotine silicotungstate, which are coming into more extensive use as substitutes for lead arsenate and other stomach poisons.

Nicotine is an effective dip for external parasites of animals such as sheep ticks, lice, and scab mites. The nicotine content of the dip should be about the same as for plant sprays with the addition of 2 per cent flowers of sulphur.

Dusts containing nicotine are of two very different kinds. The dry roots, stems, and especially the waste from tobacco factories are finely powdered and used especially for soil-infesting insects. For this purpose the fumigating type of tobacco dust should be secured, for tobacco refuse from which the nicotine has been extracted is sold for fertilizer and is, of course, valueless as an insecticide. Such powdered tobacco is of variable and unknown nicotine content and should be called "tobacco dust" to distinguish it from the more extensively employed "nicotine dusts."

Nicotine dusts are made by mixing the concentrated nicotine extracts with some light, very finely divided material as a carrier or diluent. It is generally recommended that the nicotine sulphate be used in making dust except in extremely dry atmospheres, when free nicotine should be used. The kind of carrier used has an important effect upon the dust. Carriers have been classified as (*a*) adsorbent carriers such as bentonite, talc, kaolin, and other colloidal materials, which prevent the volatilization of the nicotine; (*b*) inert carriers such as gypsum, clays, sulphur, slate dust, and similar materials, which have little or no effect upon the volatilization of the nicotine; and (*c*) active carriers such as hydrated lime and carbonate of lime, which have the effect of changing nicotine sulphate into free nicotine in the presence of moisture.

In general the finer the carriers are, the more effective these dusts will be. They are not generally so effective if applied when the plants are wet with dew and rain. A hot, dry, still day is preferable since nicotine is more effective at high temperatures and low humidity. The efficiency of the dusts is increased by the use of trailers (see page 492). Although there has been much discussion of the kind of electrical charge carried by particles of dust from different carriers, it appears that this is not a matter of great practical importance. It is generally recommended that the user mix his own dusts because the freshly mixed dusts have a much higher killing power, can be prepared at less cost than the commercial dusts, and can be prepared on short notice from the separate ingredients which are stable in storage.

For mixing small quantities of dust, place the hydrated lime, dusting sulphur, or other dust to be used as a diluent, in a can, churn, or small keg, which has a tight-fitting lid. Pour the correct amount of 40 per cent nicotine sulphate, or 40 per cent free nicotine, over the dust in the container; add 10 to 15 small stones about the size of hens' eggs, close the lid tightly and roll the container back and forth for 10 minutes.

TABLE XII.—POISONS TO BE USED FOR SOFT-BODIED AND SUCKING INSECTS

Contact poisons are used for killing soft-bodied piercing-sucking insects, and are effective only when applied directly to the insect.

Contact poison	Time of killing	Injury to plants	Price	How applied	Best uses
Nicotine sulphate	Quickly	Does not burn	Moderate; $1.50 to $2.50 a pound	Dust, spray, and in combinations	Best general insecticide for all soft-bodied sucking insects
Laundry soaps	Moderate to slowly	Slight to severe	Cheap; 8 to 10¢ a pound	Spray	Some soft-bodied insects
Fish-oil soaps	Quickly	Burn tender foliage	Moderate; 5 to 15¢ a pound	Spray	Certain sucking insects, as squash bug, harlequin cabbage bug
Kerosene emulsion	Quickly	Likely to burn foliage	Cheap; 20 to 50¢ a gallon	Spray	Sucking insects on plants with hardy foliage
Lubricating-oil emulsions	Moderately to slowly	None in dormant stage	Cheap; 20¢ to $1 a gallon	Dormant spray or delayed dormant	Scale insects, mites, aphids
White-oil emulsions	Moderately	None on most trees	High; 60¢ to $2 a gallon	Summer spray	Combined with nicotine sulphate for aphids, codling moth
Miscible oils	Quickly	None in dormant stage. Very likely to burn foliage	Moderate; 40¢ to $1.50 a gallon	Dormant spray or delayed dormant	Scale insects and a few other sucking insects and mites
Liquid lime-sulphur	Slowly	None in dormant stage. Burns foliage[1]	Moderate; 12 to 75¢ a gallon	Dormant spray or delayed dormant; summer fungicide	Scale insects and mites
Dry lime-sulphur	Slowly	None in dormant stage. Burns foliage[1]	Moderate; 10 to 50¢ a pound	Dormant spray or delayed dormant	Scale insects and mites
Tobacco dust (powdered tobacco plant)	Moderately	Does not burn	Moderate; 6 to 20¢ a pound	Dust on plants or soil	Kills soft-bodied insects and repels some others
Derris, cubé, timbo (rotenone) extracts	Very slowly	Does not burn	Moderate; $1.50 to $2.50 a pound	Spray and dust	General insecticide for soft-bodied and sucking insects
Pyrethrum extracts	Quickly	Does not burn	High; 75¢ to $2.50 a pound	Spray and dust	For household and garden insects and animal parasites
Thiocyanogen compounds (Lethane, Loro, etc.)	Very quickly	Use with caution	High; $8.50 a gallon	Spray	Red spider mite, thrips, and other very resistant pests of plants; and household and livestock pests
Organic amines	Very quickly	Use with caution	High; $8.50 a gallon	Spray	The same

[1] Does not burn apple foliage if applied at summer strength.

TABLE XIII.—POISONS TO BE USED FOR SOFT-BODIED AND SUCKING INSECTS

Amount of poison required for the following amounts of spray

Kind of spray	200 gal.	150 gal.	100 gal.	50 gal.	25 gal.	5 gal.	1 gal.
40 per cent nicotine sulphate (1 part to 800 parts water)	1 qt.	1.5 pt.	1 pt.	8 fl. oz.	4 fl. oz.	0.8 fl. oz.	1 tsp.
Lubricating-oil emulsion[1] (2 % oil)	6 gal.	4.5 gal.	3 gal.	1.5 gal.	3 qt.	0.6 qt.	4 fl. oz.
Kerosene emulsion[1] (10 % oil)	30 gal.	22.5 gal.	15 gal.	7.5 gal.	3.75 gal.	3 qt.	1.2 pt.
Kerosene emulsion[1] (5 % oil)	15 gal.	11.25 gal.	7.5 gal.	3.75 gal.	1.87 gal.	1.5 qt.	0.6 pt.
Fish-oil or laundry soaps	12 lb.	9 lb.	6 lb.	3 lb.	1.5 lb.	5 oz.	1 oz.
Lime-sulphur solution[2,3] (winter strength)	25 gal.	18.75 gal.	12.5 gal.	6.25 gal.	3.12 gal.	2.5 qt.	1 pt.
Lime-sulphur solution[2,3] (summer strength)	4 gal.	3 gal.	2 gal.	1 gal.	2 qt.	0.4 qt.	3 fl. oz.
Wettable or flotation sulphur	10 lb.	7.5 lb.	5 lb.	2.5 lb.	1.25 lb.	4 oz.	1 oz.
Bordeaux mixture[3] (3–5–50 mixture):							
Copper sulphate	12 lb.	9 lb.	6 lb.	3 lb.	1.5 lb.	5 oz.	1 oz.
Hydrated lime	20 lb.	15 lb.	10 lb.	5 lb.	2.5 lb.	8 oz.	2 oz.
Derris or other rotenone extracts (standard, 1 % rotenone)	2.5 pt.	1.8 pt.	1.25 pt.	10 fl. oz.	5 fl. oz.	1 fl. oz.	1.5 tsp. about 6 cc.
Ground derris or cubé root (4 to 5 % rotenone)	5 lb.	3.75 lb.	2.5 lb.	1.25 lb.	10 oz.	2 oz.	0.5 oz.
Pyrethrum extracts (0.9 % pyrethrins)	1 pt.	0.75 pt.	0.5 pt.	0.25 pt.	2 fl. oz.	3 tsp. 12 cc.	2 cc.
Thiocyanogen compounds (Lethane, Loro, etc.)	3 qt.	4.5 pt.	3 pt.	1.5 pt.	0.75 pt.	2.5 fl. oz.	3 tsp. or 15 cc.

[1] From 66⅔ per cent oil in the stock emulsion.

[2] From 33° Bé. material.

[3] These sprays are important fungicides, useful for the control of plant diseases.

Do not roll rapidly, as the dust will mix better if the container is rotated at about 30 to 40 revolutions per minute. Be sure that the dust and nicotine are thoroughly mixed. If it is to be stored, it should be placed in a tight container, and kept in a dry, cool place.

For a 1 per cent dust use 1¼ pounds 40 per cent free nicotine (or nicotine sulphate) to 48¾ pounds of the carrier.

For a 2 per cent dust use 2½ pounds 40 per cent free nicotine (or nicotine sulphate) to 47½ pounds of the carrier.

For a 3 per cent dust use 3¾ pounds 40 per cent free nicotine (or nicotine sulphate) to 46¼ pounds of the carrier.

For a 5 per cent dust use 6¼ pounds 40 per cent free nicotine (or nicotine sulphate) to 43¾ pounds of the carrier.

The mixer should never be more than one-third full while it is being used. As the materials for making the dust will keep much better than the prepared dust, it is not advisable to make it in large quantities.

Nicotine is incompatible with calcium cyanide dusts and with barium fluosilicate. It may be incorporated with most other insecticides and fungicides.

As a contact insecticide, nicotine combines the valuable features of great toxicity to insects, great safety to plants, and a high degree of safety to the operator. It may be made to serve, in the form of fixed nicotine (see page 259), also as a stomach poison.

New Forms of Nicotine.—A synthetic preparation having the same empirical formula as nicotine, but somewhat less toxic, has been prepared and identified as β-pyridyl-α-piperidine. *Anabasine* is the name given to a levorotary form of nicotine found in the plant *Anabasis aphylla*, which is about as toxic to most insects as nicotine.

Pyrethrum.[1]—The use of pyrethrum as an insecticide originated in the Transcaucasus region of Asia about 1800. For many years, its nature was kept a secret by these Asiatic countries and the product was sold very extensively, as a flea and louse powder, at exorbitant prices. Following the publication of the nature of the product about 1850, its use became world-wide, and it was the most important export of Dalmatia and adjacent countries until the time of the World War. Since that time, the world market has been seized by Japan and that country now supplies the great bulk of the more than 10,000 tons of pyrethrum consumed annually, although some is now being grown in the United States. Until about 1919–1925 it was used almost exclusively in the form of dry, powdered flower heads, for household pests and human parasites. Then kerosene extracts of the flowers began to replace the powder. About 1926 it began to be used as a spray on plants; and by 1929 chemical studies of the nature of the poison had made it possible to produce a standardized extract, of constant and reliable insecticidal value.

Pyrethrum is the dried flowers or buds of plants of any one of three species[2] of the genus Chrysanthemum (Pyrethrum). The determination

---

[1] GNADINGER, C. B., "Pyrethrum Flowers," McLaughlin, Gormley, King & Co., 1936.

[2] Especially *Chrysanthemum cinerariaefolium;* but also *C. roseum* and *C. marshallii.* The so-called Persian insect powder was made from the last two species. About 1840

of the constituents of these flowers that kill insects baffled chemists for more than 60 years. Finally in 1924 it was announced that their toxic properties are due entirely to two ketone esters—thick oily substances called Pyrethrin I, $C_{21}H_{30}O_3$, and Pyrethrin II, $C_{22}H_{30}O_5$, which are usually present in amounts varying from 0.2 or 0.4 to 1.2 or even 2 per cent of the dried flowers. The best flowers will average 0.9 per cent pyrethrins. These pyrethrins are highly fugitive and temperamental substances, and the best grade pyrethrum products are secured when only the mature, fully opened flowers, from the best localities, are used and are grown, harvested, cured, and assayed in the most careful manner. Since it is impossible for the consumer to distinguish inferior from high-grade pyrethrum, it should be required that all pyrethrum products state the actual per cent of pyrethrins contained. Unless this is done, the consumers' only protection lies in buying stock known to be fresh and manufactured by companies of established reputations.

Light, moisture, and air all play an important part in bringing about the deterioration of pyrethrum. Decomposition takes place most rapidly when the ground flowers are exposed to high temperatures, in sunlight and in the open air. Whole flowers decompose more slowly, and deterioration can be retarded for a long time by storage in tightly sealed, tin containers and especially at temperatures below freezing.

The pyrethrins vary greatly in their effect upon different animals. They appear to be harmless to man when smeared upon the skin or taken in small quantities by the mouth or into the respiratory tract. About the only way they are harmful to the higher animals is by intravenous injection. Upon insects and most invertebrate animals they act slowly but fatally through the skin, and especially by way of the tracheae, as contact poisons. Many mites, mealy bugs, and scale insects, however, are highly resistant. Since they are also quite harmless to plants, these poisons offer great possibilities in the control of insects upon plant products to be used as human food.

There are three principal fields of usefulness of the pyrethrum products:

a. As household insecticides; both the powdered flowers and the extracts are widely used as the principal killing agent in thousands of commercial insecticides used for flies, bedbugs, ants, clothes moths, mosquitoes, and silverfish. For this purpose the extracts are incorporated in an oil that will evaporate without leaving a stain upon furniture, draperies, clothing, or wall paper; and the odor of the oil is masked with various perfumes. There is a tendency to standardize the strength of household sprays at a content of 0.10 to 0.12 gram of pyrethrins per 100 cubic centimeters, which will result from the proper extraction of 1 pound of high-grade flowers to a gallon of spray.

b. A second important use of pyrethrum is as dusts or sprays for both sap-sucking and chewing insects attacking vegetables, ornamentals, and fruits. This field has developed almost entirely since 1930. For this purpose extraction is not made by oils, because of the toxicity of oils to plants. If, however, the finely ground pyrethrum is treated with ethylene dichloride and then evaporated until about 1 pound of concentrate is left for each 15 pounds of ground flowers treated, a product

---

the value of the first-named species was discovered, and it is now the only species of commercial importance.

called *oleoresin of pyrethrum* is secured. This may then be emulsified with an alkaline-free, potash, cocoanut oil soap and is commonly marketed at a strength of 0.9 gram of pyrethrins per 100 cubic centimeters. Such pyrethrum soaps will kill most aphids at dilutions of 1 part to 400 to 800 parts of water and the more resistant beetles and caterpillars at 1 to 100 to 200. The oleoresin may also be dissolved in acetone or alcohol, or the original extraction made directly in these solvents and all soap omitted. A suitable wetting or spreading agent may be added at the time of dilution. This method has the advantage that the correct amount of spreader can be added regardless of the dilution of pyrethrins desired.

All pyrethrum extracts should be purchased on the basis of guaranteed pyrethrin content. Most insects are killed by sprays containing from 0.002 to 0.004 per cent pyrethrins. If, therefore, the user will divide the per cent pyrethrins I and II contained in the concentrate purchased, by 0.004, he will have the number of parts water that can be added to 1 part of the concentrate to make an effective spray for the more resistant pests; twice that amount of water may be used for pests that are easier to kill. For example, a pyrethrum extract containing 0.9 per cent pyrethrins I and II may be diluted 1 part to 225 (0.9 ÷ 0.004) for cabbageworms or leaf tyers and 1 to 450 (0.9 ÷ 0.002) for most aphids and thrips.

The pyrethrin extracts may also be mixed with fine talcs, powdered charcoal, tobacco dust, sulphur, and various earths to make insecticides to be applied in the dust form. These should not be prepared until ready to be used, as they deteriorate rapidly. They should be called *pyrethrum dusts* to avoid confusion with the powdered flowers or *pyrethrum powders*. The pure fresh pyrethrum powder (0.9 per cent pyrethrins) may be diluted with from 2 to 5 parts of one of the inert carriers to make a good dust. All pyrethrum dusts must be applied in the late afternoon or early evening for best results.

c. The third great use of pyrethrum is in livestock or cattle sprays. In this case, the requirements of the carrier are that they shall have no physiological effect upon the animal, that they shall not stain the coat, or taint milk from cows. In addition to killing flies which are hit by the spray, it should have a repellent effect for at least a day. They are usually made with a less volatile oil than household sprays and about the same pyrethrin content plus 5 per cent pine oil, oil of camphor, cloves, safrol, or other aromatic.

d. Among miscellaneous uses for pyrethrum may be mentioned its use for protection of valuable turf, such as golf greens, to destroy sod webworms without injuring plants, and its use to destroy mosquito larvae in pools without injuring fish, vegetation, or waterfowl.

Pyrethrum extracts containing soaps must not be mixed with calcium arsenate, lime-sulphur, or hydrated lime. The acetone or alcoholic extracts are compatible with lead arsenate, bordeaux, and oil emulsions. As already indicated, the principal advantages of this group of insecticides are their high toxicity to insects and their very slight toxicity to both plants and the higher animals. In general the pyrethrins are somewhat more toxic to insects than nicotine. They are, however, expensive. They deteriorate very rapidly after being applied to plants or otherwise exposed to air, and it is difficult to keep the stored concentrates for more than one season.

Derris, Cubé, Timbo, and Other Rotenone-bearing Insecticides.[1]—
Explorers have recorded for many years the use, by the natives of the
Dutch East Indies and other tropical countries, of the macerated roots of
certain tropical plants to poison fish and as a poison for arrows used
in warfare. When these materials are thrown into the water, fish rise
to the surface and dart about wildly, giving the natives good sport in
spearing them, and then become temporarily stupefied. Such poisoned
fish cause no ill effects when eaten because the flesh of the fish is not
affected and the poison is relatively harmless to man as a stomach
poison, though very poisonous when taken intravenously or hypodermically. It does, however, paralyze the tongue and organs of speech
when taken into the mouth. Upon insects, protozoa, tadpoles, and
some of the other lower animals these substances are extremely toxic,
slowly causing paralysis by acting upon the respiratory center of the brain.
Although used to kill leaf-eating caterpillars as early as 1848, these
rotenone-bearing roots had little use as insecticides, in America, until
the federal ruling against residues of arsenic, lead, and fluorine upon food
products stimulated an intensive search for insecticides that do not
contain cumulative poisons. The most important of these rotenone-
bearing plants are various kinds of derris, tuba or deguelia, especially
Deguelia malacensis, *D. chinensis*, *D. elliptica*, and *D. uliginosa*, from the
Old World tropics; cubé or timbo, *Lonchocarpus nicou*, from South and
Central America; *Cracca* (= *Tephrosia*) *virginiana*, *C. toxicaria*, and *C.
vogelii*; and certain species of Millettia and Ormocarpum. Probably the
most important poison in these plants is *rotenone* ($C_{23}H_{22}O_6$). Other
toxic ingredients have been isolated from most of the plants, and these
accompanying poisons are believed to have especial value when the
material is used as a stomach poison. Some of the other poisons, with
their empirical formulas, are *deguelin*, $C_{23}H_{22}O_6$; *tephrosin*, $C_{23}H_{22}O_7$; and
toxicarol, $C_{23}H_{22}O_7$.

The amount of these poisons in the roots of the plants varies greatly:
rotenone commonly ranges from 2.5 to 7.5 per cent; the other poisons
combined commonly vary from 7 to 20 per cent more. These fish poisons
may be prepared as insecticides simply by grinding the roots very finely
and applying them in the dust form, or by extracting the poisons with
solvents such as acetone, carbon tetrachloride, sulphuric ether, or a light
oil and incorporating the extracts in water as a spray. Commercial
preparations should state the *per cent of rotenone* contained and also the
per cent of total acetone or ether extracts. Good grades of derris powder
commonly contain 4 or 5 per cent rotenone and about three and a half
times that amount of total extractives and are ground to a fineness that
will pass a 250- to 325-mesh sieve. For use against common garden
and fruit insects, such dusts may be diluted with from 3 to 7 parts by
weight of very finely ground talc, sulphur, tobacco, nutshells, wood,
or inert clays, making a dust ready to apply which contains from 0.5
to 1 per cent rotenone and from 1.75 to 3.5 per cent total extractives.
It should be applied 15 to 25 pounds per acre on such crops as cabbage
and celery. *Alkaline carriers, such as lime, must not be used with rotenone.*

[1] See "A Digest of the Literature of Derris," *U.S.D.A.*, *Misc. Pub.* 120, 1932;
"Tephrosia as an Insecticide—A Review of the Literature," *U.S.D.A.*, *Bur. Ento.
Plant Quar.*, *Cir.* E-402, 1937; "Lonchocarpus Species Used as Insecticides," *U.S.D.A.*,
Bur. Ento. Plant Quar., *Cir.* E-367, 1936; and "Lonchocarpus—A Review of Recent
Literature," *U.S.D.A.*, *Bur. Ento. Plant Quar.*, *Cir.* E-453, 1938.

To make an equally effective spray, 2 to 5 pounds of derris powder may be suspended in water or 2 to 5 pounds of the roots may be cut in small pieces and macerated; 2 pounds of a neutral soap or the equivalent of a sulphonated oil dissolved in water added; and the mixture diluted to make 100 gallons of spray. The stickers or spreaders used must be alkali-free. For small amounts of spray, an ounce of a 4 or 5 per cent rotenone dust and a teaspoonful of spreader should be used in 2 gallons of water. Commercial derris-extract spray concentrates containing 1 per cent rotenone and $3\frac{1}{2}$ to 4 per cent total extractives may be diluted 1 part to 600 or 800 parts of water, for aphids; the rotenone content in the diluted spray being about 0.00125 per cent.

Very many uses have been found for these poisons during the past decade. In addition to their effectiveness for both piercing-sucking insects, such as aphids and red bugs, and chewing insects, especially caterpillars, upon plants, they make excellent dusts for external parasites of animals, such as fleas and lice. They have come to be the most widely used remedy for ox warbles (equal parts derris root powder and vaseline or olive oil). They make effective mothproofing liquids (0.05 per cent rotenone in acetone) and have valuable repellent properties for certain insects. They are, however, ineffective for many kinds of pests such as mealy bugs, squash bugs, scale insects, and many beetles and caterpillars. They act very slowly, often requiring several days to kill. The poisonous principles all deteriorate rapidly when exposed to sunlight and air; sprays and dusts usually lose their effectiveness within a week after application. They are more effective when they can be used during periods of cloudy weather, and the dusts should be applied to foliage moist with dew. The rotenone products are in general intermediate in cost between nicotine and pyrethrum. The outstanding advantages of this group of poisons are that they are harmless to plants, relatively nontoxic to man, and act as both contact poisons and stomach poisons. At least some of the preparations are compatible with oils, bordeaux, and arsenicals. They are incompatible with lime or other alkaline materials.

Synthetic Organic Insecticides.—The most recent development in the long search for the ideal insecticide is the synthesis or making to order of compounds having the desired properties. Some remarkable results have been secured by this method. In one case it is claimed that 1,047 different compounds were made and tested before the satisfactory one was found and put on the market. In most cases the synthetic, organic insecticides are specific in their action. They are frequently very toxic to one or two groups of insects, but of very little value against others. The development of this type of insecticide is expensive, because of the large number of materials which must usually be compounded and tested, before a satisfactory one is developed. The tests must include a large number of insect and plant species and a great variety of conditions, often requiring several years to complete.

Thiocyanogen Compounds.—Certain aliphatic thiocyanates, alkyl thiocyanates, or allyl thiocyanates, while relatively harmless to man, have recently been shown to have rather specific toxicity for certain insects and will probably become established contact poisons for certain pests of greenhouse plants and the household. In general they are more dangerous to plants than the natural organic insecticides. These synthetic compounds are made by introducing the thiocyanate anion (SCN) into organic compounds. For example, a bromine-substituted compound, such as γ-bromopropyl phenyl ether, will react with potassium thiocyanate in alcoholic solution, yielding γ-thiocyanopropyl phenyl ether. This material, dissolved in Penetrol and diluted with water to a strength of 0.1 per cent of the thiocyanogen compound and 0.5 per cent of Penetrol, gave excellent control of the bean aphid, a mealy bug, and the red spider mite and was noninjurious to most plants tested. Among other promising thiocyanates are lauryl thiocyanate, on the market as Loro, a spray for use on certain kinds of plants; and β-butoxy β'-thiocyanodiethyl ether, on the market as Lethane 440, as a general-purpose contact spray for plants and under various trade names as a household and livestock spray for household and stored-product pests, as well as the lice and flies that attack domestic animals. Lethane kills insects so quickly that it is possible to use it for very resistant insects at strengths beyond the tolerance of certain plants, syringing it off after it has overcome the insects but before the plants have become injured. It is compatible with all standard spray mixtures and with hard water. Further experimentation is needed to determine its effectiveness for various insects and under varied conditions. The thiocyanogen compounds appear to kill insects by paralysis resulting from injury to the central nervous system.

Organic Amines.—An insecticide containing N,N-amyl-benzyl cyclohexylamine, sold under the trade name of Cyclonox is effective for red spider mites, thrips, and leafhoppers. It is a colorless, oily compound, miscible in oils and, by means of a special emulsifier, in water.

Sulphur, Lime-sulphur and Other Sulphur Compounds.—Sulphur and its compounds are among the most important contact poisons. The following list will indicate some of its important uses:

A. *Sulfur Dusts:*
  1. Especially toxic to mites, such as chiggers and red spider, for thrips, for newly hatched scale insects, and as a stomach poison for some caterpillars. ·
  2. Important as a fungicide carrier for other dust insecticides, such as nicotine and lead arsenate. The following mixture or some modification of it is extensively used on peach.
     Dusting sulphur... 60 lb.
     Dusting talc... 35 lb.
     Light lubricating oil....................................... 5 lb.
B. *Mechanical Spray Mixtures of Sulphur with other Substances:*
  1. Wettable Sulphur: Sulphur so treated as to be readily miscible in water is widely used in spray mixtures, especially for its fungicidal effect.
  2. Flotation sulphur is an extremely finely powdered sulphur which is more active chemically than flowers of sulphur. It is compatible with lead arsenate and magnesium arsenate, but not with calcium, manganese, or zinc arsenate.
  3. Self-boiled Lime and Sulphur:
     Sulphur.. 16 lb.
     Unslaked lime (high-grade quicklime)...................... 16 lb.
     Water... 100 gal.
     The sulphur is intimately mixed with the lime by adding it as the lime begins to slake. The mixture is cooled before a chemical combination of the lime and sulphur takes place. Used as a summer fungicide for stone fruits and may aid in checking San Jose scale.

4. Colloidal Sulphur: Made by passing sulphur fumes into soap and water or glue and water, or by passing hydrogen sulphide (H_2S) gas into a solution of sulphur dioxide (SO_2). Recommended, in combination with lead arsenate, for apple scab, codling moth, and leaf rollers.

C. *Chemical Compounds of Sulphur with Calcium, Ammonia, Potassium, and Barium:* Used as a dormant spray for many scale insects, and fungous diseases, and as a foliage spray especially for fungous diseases, thrips, and various mites.

1. Liquid, Fire-boiled Lime-sulphurs:
   a. Homemade concentrated lime-sulphur
   b. Commercial Lime-sulphur: Dilute to a strength of 5° Bé, before applying as a winter scalecide (1 part to 7 or 8 of water).
2. Dry Lime-sulphur: Use 30 or more pounds to 100 gallons of water as a winter scalecide.
3. Potassium Sulphide: Dissolved 1 ounce in 2 gallons of water forms an effective spray for red spider.
4. Ammonium Polysulphide: Ammonium hydroxide (28 per cent NH_3) is saturated with hydrogen sulphide and about 10 per cent of sulphur is dissolved in the resulting ammonium sulphide solution, to make an effective spray (at 1 to 200 parts of water) for red spider mite on out-of-door crops.

Sulphur and lime are both practically insoluble in water, but, when boiled together, they combine to form a series of salts, some of which are soluble and others insoluble. Lime-sulphur is a mixture of the following compounds:

CaS_5—Calcium pentasulphide
CaS_4—Calcium tetrasulphide
CaS_2O_3—Calcium thiosulphate
$CaSO_3$—Calcium sulphite

The first three are soluble in water; the calcium sulphite is comparatively insoluble. It, together with any uncombined sulphur or excess lime, magnesium, or iron from impure lime, settles out and forms the sediment or "sludge" in home-boiled lime-sulphurs. In the commercial lime-sulphurs the sludge is removed before shipping. Of the first three compounds the pentasulphide and tetrasulphide, together called polysulphides, are believed to be the valuable ingredients, and it is to them that the lime-sulphur owes its killing power as well as its characteristic red color.

Solutions of lime-sulphur were first used as stock dips; they were tried as insecticides in California in 1886; and have, since about 1902, been a standard remedy for controlling certain scale insects, particularly the San Jose scale. These solutions are also efficient fungicides, and are used in a large number of combinations. They are very disagreeable to use because of their caustic properties, which cause them to burn the face, eyes, and hands of the operator, and because of their odor. They should never be used with soaps or with soap-oil emulsions without stabilizers.

Lime-sulphur solutions can be made at home by cooking together fresh or stone lime and sulphur. A number of formulas are in use, of which the following is typical:

Lump or stone lime.. 100 lb.
Commercial ground sulphur............................... 200 lb.
Water, enough to make.................................... 100 gal.

Heat about one-third of the required amount of water and to this add the lime. As soon as the lime starts slaking, add the sulphur, which should have been previously mixed thoroughly with enough water to make a thick paste. Then add the remainder of the water and boil for from 45 minutes to 1 hour, adding more water as necessary to keep up to the original level. If cooked by steam, as is usually done, a mechanical agitator should be provided. When the free sulphur has all disappeared, the mixture should be strained and may be stored in barrels, tanks, or cisterns.[1]

In order to use lime-sulphur solutions intelligently, their strength, or concentration, must be tested. This is done by means of a hydrometer, an instrument that works on the principle that any object will sink in any liquid until it displaces its own weight of that liquid. The lighter the liquid, the deeper it will sink; the heavier the liquid, the higher it will float. The stem of the hydrometer is graduated to make measurements easy. Two standard scales of graduation are used, *the specific gravity scale*, which begins at 1, for water at 70°C., and the *Baumé (Bé.) scale* which begins at 0 for water at 70°C. The correspondence between these two scales and the method of testing lime-sulphurs are shown in Fig. 172.

The concentrated home-boiled and commercial lime-sulphurs must always be diluted. Different batches of the home-boiled preparations will vary somewhat in strength. The many brands of commercial lime-sulphur on

[1] See *Ill. Agr. Exp. Sta., Cir.* 492, 1939.

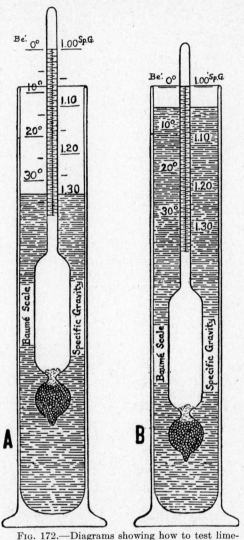

Fig. 172.—Diagrams showing how to test lime-sulphur solution with the hydrometer. *A* shows the way in which the hydrometer should float in good, full-strength, commercial lime-sulphur; *B*, the way it should float in lime-sulphur properly diluted for winter or dormant spraying. The type of hydrometer in common use gives the Baumé reading in a scale on the left side and the specific gravity reading in a scale on the right side of the spindle. Note that the smallest units on the Baumé scale are ½° while the smallest units on the specific gravity scale are 0.005. The density of other spray solutions may be measured in the same way, if a hydrometer of the correct scale is used. (*Original; drawing by R. D. Glasgow.*)

the market usually test about 33°Bé. The 33° material should be diluted, 1 part to 7 or 8 of water, for use as a dormant or winter spray on fruit trees; or about 1 to 49 of water for a summer spray. Lime-sulphur which has been drawn from different levels in storage tanks or barrels will give varying readings on the hydrometer. It is best to test the diluted spray mixture when ready to be applied rather than to depend on a dilution table. When the application is to be made to dormant trees, *the diluted lime-sulphur mixture should give a reading of about 5° on the Baumé hydrometer, or* 1.035 *specific gravity* (see Fig. 172). *For summer spraying, the reading should be about* 1°*Bé., or about* 1.005 *specific gravity.* However, since conditions vary so widely in different localities, the state experiment station should be consulted regarding the best dilution to use for various insects and diseases.

Dry Lime-sulphur.—There are a number of dry powdered materials on the market as substitutes for the liquid lime-sulphur. Dry lime-sulphur is liquid lime-sulphur with nearly all of the water removed. The chemical content of the material is slightly changed during the drying. Tests by several experiment stations indicate that these materials are about as effective as liquid lime-sulphur when mixed so as to give the same content of sulphur in the dilute spray that occurs in liquid lime-sulphur. If made to contain the same amount of sulphur as liquid lime-sulphur, the materials are much more expensive. It is also troublesome because of the amount of sludge in the dilute material. On the other hand, they are much more convenient to ship, store, and handle than liquid lime-sulphur. A typical analysis of a dry lime-sulphur shows active ingredients: CaS_5 and CaS_4 about 65 per cent, CaS_2O_3, about 5 per cent, free sulphur, about 10 per cent; and inert ingredients about 20 per cent.

Selenium Compounds.—The rather rare element selenium resembles sulphur very closely in chemical properties. Potassium-ammonium selenosulphide, $(KNH_4S)_5Se$, at a strength of 1 to 600 in water, with a suitable spreader, is a specific poison for red spider mites. Selenium is, however, a very dangerous, cumulative poison which may be taken up from the soil by plants, rendering them poisonous to man and animals. Consequently selenium sprays should never be used on vegetables or fruits.

Oils and Emulsions.—Although recommended as insecticides as early as 1763, oils were probably little used until well along in the nineteenth century. At first, petroleum, turpentine, and kerosene were applied, unmodified, and, while very toxic to the insects, they also sickened or killed the plants. The next attempt was to mix the kerosene mechanically with water, but such mixtures were very unstable and often resulted in disastrous destruction of trees and other foliage. Consequently oils did not become satisfactorily useful until entomologists learned how to overcome the severe effects upon plants by emulsifying the oils in water. The first of such emulsions were made from kerosene, about 1870. In 1874 a good formula for a kerosene, soap, and water emulsion was discovered; from that time on, oils have become increasingly important and formulas and uses have multiplied continually. In 1904 the first commercial emulsion or miscible oil was placed on the market. In 1919–1923 lubricating-oil emulsions were found to have great efficiency in killing San Jose scale. About 1930 certain highly refined neutral or white oils, free from unsaturated hydrocarbons, acids, and highly volatile elements, were found safe to use upon plants in foliage, and thus the field for oil sprays was greatly enlarged.

The more important fields of usefulness for oil insecticides are

a. As dormant (winter) sprays for scale insects, mites, insect eggs, and some hibernating caterpillars.

b. As summer or foliage sprays for aphids, mealy bugs, and scale insects.

c. As parasiticides for lice, fleas, and mites on animals.

d. As carrier for other insecticides particularly for pyrethrum, rotenone, nicotine, paradichlorobenzene, and sulphur.

e. Mixed with lead arsenate and other stomach poisons to increase their effectiveness.

f. As attractants in poison baits (see page 254).

The term "oil" is a generic expression for a great variety of chemical substances which are composed principally, if not exclusively, of carbon and hydrogen. At ordinary temperatures they are greasy fluids, insoluble in water but readily soluble in ether, chloroform, carbon disulphide, carbon tetrachloride, and the like. They are readily inflammable, are slightly lighter than water, and have a peculiar property of penetrating readily into the pores of dry substances.

There are three great classes of these oils: (*a*) The *fixed oils*, such as linseed, soybean, castor, neat's-foot, fish oil, and many others, derived from both plants and animals, which are essentially glycerides that saponify or form soaps with alkaline bases, setting free, glycerin. Fish oil, which is important in making insecticidal soaps, and soybean oil are those which have been most used for insecticides. (*b*) The *volatile* or *ethereal oils*, which differ from the fixed oils in not being greasy or viscous and not being capable of saponification. They are derived from special glands of plants and usually have a very pungent odor characteristic of the plant. Common examples are menthol, camphor, eugenol, oil of peppermint, wintergreen, and citronella. Their chief use in entomology is as attractants in baits and as repellents. (*c*) The *petroleum oils*, which are derived from sedimentary rocks, are, for the entomologist, by far the most important class. They are complex solutions of hundreds of hydrocarbons from which are derived natural gas, kerosene, gasoline, lubricating oils, asphalt, tar, and many other mixtures.

In order to understand the effects of different oils upon insects and upon plants, the following properties must be considered:

a. Volatility or distillation range, which is measured by the temperature at which the particular fraction distills. Spray oils are referred to as *light* (boiling points between 160 and 300°F., such as gasoline), *medium* (boiling points between 300 and 575°F., such as kerosene), or *heavy* (boiling points above 575°F., such as the lubricating oils). The lower the volatility or "heavier" the oil, *i.e.*, the higher the boiling point, other things being equal, the more effective oils are in killing insects, up to the point where they are so slightly volatile that not enough will evaporate to penetrate the tracheal system and so kill the insect. For dormant spraying, the oil should distill, at an even rate, about 90 per cent of its volume between 590 and 700°F., and not over 2 per cent at 230°F., for 4 hours. Unfortunately, the heavier oils are also the more toxic to plants so that the practical problem is to find the lightest oil that will kill the insect or the heaviest that can be used with safety on the particular plant.

b. The second important property to consider concerning spray oils is their *viscosity* which is defined as resistance to flowing—the opposite of fluidity. It is measured in arbitrary units, which are the number of seconds required for a given volume (60 cubic centimeters) of the oil to flow through a standard orifice at a definite temperature (100°F.) (Saybolt test). While it is customary to speak of oils of low viscosity as "thin" and those of high viscosity as "heavy," it should be understood that viscosity is a very different property from specific gravity. Other properties being equal, oils of low viscosity are safer to use upon foliage than those of high viscosity. For dormant sprays on deciduous trees, an oil having a viscosity between 100 and 200 seconds (Saybolt), at 100°F., is considered satisfactory, a lower range often being used in more northern areas and a higher range in warmer territory.

c. The third property of great importance in spray oils is their *purity* or *degree of refinement*. An oil of the proper volatility and the right viscosity for spraying may still be quite unsuitable because of the presence of impurities, especially unsaturated and aromatic hydrocarbons.[1] These compounds are chemically active and easily oxidized, causing the oils to become turbid and acid in reaction. It is generally believed necessary to remove 85 to 90 per cent of these unsaturated hydrocarbons, if the oil is to be used as a foliage spray; whereas, for dormant spraying, the removal of 65 to 75 per cent of the unsaturated hydrocarbons is satisfactory. The refining is brought about chiefly by treating with sulphuric acid and washing out the resulting sludge. For any oil of unknown purity, the degree of refinement may be determined by treating a sample with sulphuric acid. If the unsaturated hydrocarbons have already been removed, this treatment will not remove any portion of the oil. Such an oil is said to be "100 per cent unsulphonatable" or to have a "purity of 100 per cent." The ordinary lubricating oils are, as stated above, about 65 to 75 per cent unsulphonatable, and, for dormant spraying, a higher point of purification is considered unnecessary.

For certain purposes, oils of the proper grade may be applied straight or unmixed. Crude petroleum and raw linseed oil are excellent remedies for the lice upon animals. Oils of the proper grade are applied to standing water to kill mosquito larvae. Gasoline and similar light oils are sometimes applied to upholstered furniture to destroy clothes moths and bedbugs. Recently, machines have been perfected which atomize oils, dilute them with air, and apply them as a fog to insect-infested plants. In this way a very thin film of oil, in a highly active condition has been used to destroy leafhoppers, aphids, thrips, scale insects, and the eggs of mites, without prohibitive injury to plants. Although the attempts about 1888–1894 to use mechanical mixtures of oils and water upon plants resulted disastrously, recent renewal of interest in this method of dispersing oils has resulted in the perfection of the tank-mixture method by which very effective agitation in the spray tank assures a uniform mixture of the oil in water, without the use of highly stable emulsions. Thus an ounce of powdered blood albumin, 3 ounces of fuller's earth, and 100 gallons of a refined petroleum oil are effectively emulsified in the spray

---

[1] An unsaturated hydrocarbon is one in which the valence of carbon (4 in organic compounds) is not fully satisfied.

tank to make a quick-breaking emulsion. Devices known as injectors are sometimes attached to the spray pump to produce a satisfactory mixture of oil with the other spray ingredients.

Oil Emulsions.—The most extensive use of oils against insects is in the form of emulsions. Spray oil emulsions are mechanical mixtures of oil and water in which the oil is very evenly dispersed as small globules in the water and kept so, at least until it has been applied to the plant, by the use of a third substance known as the *emulsifier*. The emulsifier collects at the interface between the oil and water, forming an extremely thin layer about each minute oil droplet and thus, by reducing the interfacial tension between the oil and water, prevents the oil droplets from coalescing. In such tiny particles they are deadly to insects, yet they can be applied to many plants without injuring them. The proper making of emulsions is an exacting and difficult science. Variations in the behavior of different emulsions may be brought about by the use of different oils, different emulsifiers, or different waters used in dilution; by method of mixing and duration of agitation; by the amount of emulsifier; by temperature; and in other ways. Emulsifiers are of two principal classes: (*a*) *chemically-active emulsifiers,* such as soaps, saponin, bordeaux, various sulphated alcohols, sulphonated oils, organic acids, oleates, oxidized petroleum hydrocarbons, and vegetable oils; (*b*) *chemically inert emulsifiers,* such as soybean flour, calcium or ammonium caseinate, blood albumin, chalk, talc, kaolin, glue, and vegetable gums. Some emulsions are made simply by agitating or stirring the ingredients together properly, others require both heating and agitation. Almost any degree of permanency or quick-breaking desired can be produced by using oils of different viscosity, by varying the nature or amount of the emulsifier used, or by varying the size of the droplets of oil produced at the time of dispersal. The nature and amount of emulsifier also determine the amount of oil retained or "built up" on the plant surface and hence must be gauged carefully to give just enough build up to kill the insect but to avoid the excess which will kill the bark or spot the foliage or fruit. Emulsions are usually first made with a minimum of water, in which condition they are stored and sold. Before use they are diluted with water to the strength of oil desired. The oil content of an emulsion is always expressed in per cent by volume. The following rule may be used to calculate readily how to dilute a stock emulsion with water to give any desired amount at any desired concentration. *Multiply the amount of spray desired by the per cent of oil desired and divide the product by the per cent oil in the stock or concentrate.* Thus, if one desires to make 50 gallons of a 3 per cent oil spray from a 90 per cent oil stock, multiply 50 by 3 and divide by 90, which gives a quotient of $1\frac{2}{3}$, the number of gallons of the concentrated stock needed to make the 50 gallons of spray. Or to make 100 gallons of a 2 per cent spray from a $66\frac{2}{3}$ per cent stock: $100 \times 2 \div 66\frac{2}{3} = 3$, the number of gallons of the stock to use in 100 gallons of spray. If hard water is to be used for diluting the stock emulsion, a stabilizer such as caustic soda, casein, glue, or gelatin is often added to the stock emulsion before dilution to prevent the breaking of the emulsion.

The advantages of oils may be stated briefly as: high toxicity to most insects, their "creeping" or covering capacity, ease of mixing, and pleasantness of handling. Their disadvantages are chiefly their lack of stability in storage, danger of injuring plants, injury to spray hose, and the lack of fungicidal value as compared with lime-sulphur. Soap-oil emulsions are generally not compatible with lime-sulphur, the arsenicals, or fluosilicates; while those made with inert emulsifiers are compatible with nearly all standard fungicidal and insecticidal materials. Among the many kinds of emulsions recommended for various purposes, some of the better known are the lubricating-oil emulsions, the white- or summer-oil emulsions, coal-tar-oil emulsions, distillate-oil emulsions, kerosene emulsions, carbon disulphide emulsions, and the miscible oils. All of them can be purchased already emulsified, under various trade names.

Soaps.—A soap is a salt of a fatty acid, such as oleic, stearic, or palmitic acid derived from animal or vegetable oils, and an alkali-metal base such as sodium or potassium hydroxide. The sodium base forms hard soaps, the potassium base, soft soaps. Most soaps when dissolved in water at sufficient strengths have value as contact insecticides. They have been used against insects since 1787. Potash fish-oil soaps are the most widely used for spray purposes. Potash vegetable-oil soaps of cocoanut, linseed, soybean, or corn oil are of equal merit and do not have such a disagreeable odor. Rosin fish-oil soap at 1 ounce to the gallon or 6 pounds to 100 gallons, or a good commercial insecticide soap, applied on calm humid days or during a light rain or mist, when the evaporating power of the atmosphere is very low, is surprisingly effective against many kinds of pests. Sodium soybean-oil soap has been found of value in the control of insects. Soaps are used chiefly in the preparation of emulsions and as spreaders, wetting agents, and stabilizers for nicotine and pyrethrum sprays. The potash soaps are readily soluble in alcohol or cresol and are used in solution in oils to form the miscible oils. Soaps may be used in sprays containing lead arsenate, especially if hard water has to be used, in which case the soap may free the hard water of soluble calcium and magnesium salts.

FUMIGANTS

(See Table XIV)

Gaseous poisons used to kill insects are called fumigants. Their application is generally limited to plants or products in tight enclosures. Fumigants are used to combat all kinds of insects, regardless of the type of mouth parts. They are employed for a great variety of insects and under many different conditions.

A. *Not Involving Living Plants:*
  1. Mills, Factories, Storerooms, Packing Plants, Warehouses, Groceries, Museums, Insect Collections, and Herbariums:
      a. For pests of food products, including seeds, cereal products, candies, dried fruits, meats, cheese, nuts, and tobaccos.
      b. For pests of carpets, rugs, upholstery, woolens, furs, skins, dried plants, wood, paper, and leather goods.
  2. Human Habitations, Hotels, Prisons, Barracks, Hospitals, Theaters, and Camps:
      a. For any of the pests mentioned in 1.
      b. For bedbugs, lice, fleas, and mosquitoes.
  3. Ships, Railway Passenger, Freight, and Bunk Cars, and Automobiles:
      a. For any of the pests in 1 or 2.

　　b. For plant pests which are under quarantine and likely to be transported by rolling stock.

　　c. For rats which are hosts of fleas, the carriers of bubonic plague.

　4. Vacuum and Industrial Fumigation in Especially Constructed Chambers or Vaults, Usually in Partial Vacuum:

　　a. For bales of cotton, human baggage, bags of flour, and other materials requiring great penetration.

　　b. For packages of cereals, tobaccos, candies, and the like, to make sure they leave the factory free from living insects.

　5. Fumigation of Soil: for ants, grubs, wireworms, and many other subterranean pests.

　6. Fumigation of the Intestinal Tracts of Animals: for bots and intestinal worms.

B. *Involving Living Plants:*

　7. Fumigation of Nursery Stock:

　　a. For subterranean pests.

　　b. For scale insects.

　　c. For plants being imported from foreign states or countries.

　8. Fumigation of Cavities in Trees or Buildings: for borers, nests of bees and the like.

　9. Fumigation of Citrus Trees and Other Plants Involving the Use of Portable Tents or Boxes: for scale insects and other pests not controlled by sprays or dusts.

　10. Greenhouse Fumigation: for all kinds of greenhouse pests.

　11. Open Field Fumigation: rarely used for chinch bugs, grasshoppers, and other pests.

General Directions for Fumigating.—The first law of fumigation is to safeguard human lives. No one should undertake the use of these dangerous chemicals unless he has been thoroughly trained in the procedure and has adequate gas masks to protect the operators themselves. It is not safe to fumigate one room or apartment while other parts of the building are being occupied. Buildings being fumigated should be plainly placarded and securely locked or guarded, so that no one may enter until they have been thoroughly ventilated. In general it will be well to avoid windy or cold weather. Regardless of the kind of fumigant to be used, great care is required to make the enclosure to be fumigated as nearly airtight as possible. In dwellings, greenhouses, mills, and storerooms all cracks, broken glass, chimney holes, ventilators, and other openings must be very tightly closed. Strips of gummed paper or even 4-inch strips of newspaper, soaked in water or flour paste or smeared with heavy grease will be found useful to paste over cracks; pieces of wallboard, cut to fit, for large openings; and rags or cotton waste for small holes. For large cracks up to 5 or 6 inches wide, a paste made by mixing ground asbestos, 3 parts, with calcium chloride, 1 part, and enough water to make a thick "putty," is excellent, as the hygroscopic calcium chloride keeps the paste from drying out and cracking away. Masking or scotch tape, plastic clays, and heavy sheets of kraft paper sealed with Plastic Elastic are also excellent aids in sealing a building. More failures in fumigation are due to insufficient preparation of the building than to any other cause.

Where living plants are being treated, especially greenhouse and citrus plants, the dosage used will have to be very carefully determined; the duration of exposure is rarely over 1 hour; and ventilation must be carefully timed. The temperature must be between 40 and 80°F. and for greenhouse plants, between 55 and 68°F., and it should *rise slowly* to

TABLE XIV.—FUMIGANTS TO BE USED FOR INSECTS IN ENCLOSURES

Fumigant	Dosage per 1,000 cu. ft.	Killing power	Boiling point, °F.	Specific gravity of gas	Penetration	Danger from fire	Effect on plants	Special cautions	Cost of materials per 1,000 cu. ft.	Best uses
Hydrocyanic acid gas, HCN (for stored products)	12 oz. of 88% calcium cyanide	Very high	79	0.9483	Slight	Little	Kills	Poisonous to moist foods. Do not breathe	75¢ to $2	For mill and household insects
Hydrocyanic acid gas, HCN (for greenhouse)	¼ oz. of 48% calcium cyanide	High	79	0.9483	Slight	None	Use with caution	Must not be breathed; use at night	2¢	For greenhouse insects; citrus trees
Carbon bisulphide, CS₂	10 lb.	High	115	2.63[1]	Very good	Very inflammable	Kills	Use caution with seeds. Inflammability	$1 to $2.50	For stored grains, clothes moths and carpet beetles
Carbon tetrachloride, CCl₄	40 lb.	Very weak	169	5.311	Good	None	Kills	Use caution with seeds	$2 to $5	Where there is danger from fire
Nicotine 40 per cent, C₁₀H₁₄N₂ (for greenhouse)	½ to 1 oz.	Good	448	......	Slight	None	Safe	None. Have house dry	2¢	For greenhouse aphids
Sulphur dioxide, SO₂ (burning sulphur)	4 to 8 lb.	High	14	1.433	Good	Safeguard generator in which sulphur is burned	Kills	Tarnishes, bleaches, taints food, prevents germination	10 to 75¢	Few
Naphthalene, C₁₀H₈	100 lb.[2]	Fair	424	4.414[1]	Slight	None	Kills	Taints food; kills seeds	$5 to $25	For household and greenhouse[2]
Paradichlorobenzene, C₆H₄Cl₂ (for household)	100 lb.	Fair	342	5.07[1]	Good	None	Kills	Taints food; use caution with seeds	$25 to $50	For household insects; peach borers
Chloropicrin, CCl₃NO₂	1 to 3 lb.	Very high	233	5.7[1]	Very good	None	Kills	Corrodes. Persists. Tear gas	85¢ to $2.50	Mills, elevators, stored grains, warehouses
Ethylene oxide, C₂H₄O	2 or 3 lb.	High	51	1.5	Good	Inflammable[3] and explosive[3] above 3½ lb. per 1,000	Kills	Use caution with seeds	$1 to $4	Industrial, in vaults
Ethylene dichloride, CH₂Cl·CH₂Cl	3 to 5 lb.	Fair Slow	182	3.5	Good	Inflammable[4] and explosive[4]	Kills	Don't use on high fat foods	90¢ to $1.40	Stored grains. Industrial in vaults
Methyl bromide, CH₃Br	2 lb.	High Slow	40	3.5	Good	None	Use with caution	Avoid breathing	$1.60 to $2	Fruits, plants, railroad cars. Industrial in vaults

[1] The figures given are for the undiluted form of these gases. It should be understood that the undiluted form of these gases is nonexistent, except at their boiling points or above. At ordinary temperatures (70°F.) the heaviest mixture of paradichlorobenzene and air that could exist is about 1.267 times as heavy as air; of carbon bisulphide and air, about 1.06 times as heavy as air; and of carbon tetrachloride and air, about 1.583 times as heavy as air.

[2] For greenhouse plants, ⅛ to 2 oz. may be used with caution at high humidity.

[3] One part ethylene oxide mixed with 9 parts carbon dioxide makes it entirely noninflammable. Use 10 to 30 lb. of the mixture per 1,000 cu. ft.

[4] Three parts ethylene dichloride mixed with 1 part carbon tetrachloride makes it entirely noninflammable. Use 14 to 18 lb. of the mixture per 1,000 cu. ft.

prevent condensation of moisture upon the leaves and resultant burning by the gas. Generally treatment must be made at night or in darkness, since sunlight, during or within an hour before or after fumigation, is likely to cause injury. Where no living plants are involved, the aim will be to use a slight excess of the fumigant; to leave the materials exposed as long as convenient; or, if the building does not need to be entered, to neglect ventilation altogether; to have the temperature high— between 70 and 100°F.; and to neglect the trend of temperature and the amount of light as of little importance. Dry food substances may be used after treatment with the common fumigants, but those containing moisture may be poisoned.

Before fumigating any building, provision should be made for ventilating it after the fumigation is over. If a gas mask, correctly charged for the particular gas being used, is available, it is exceedingly useful in setting off the charge and for entering the building afterward to open

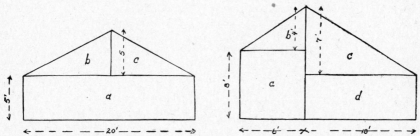

Fig. 173.—Diagrams showing how to compute the cubical contents of a greenhouse. At the left an even-span greenhouse; at the right a three-quarter-span greenhouse. (*From U.S.D.A., Farmers' Bul.* 1362.)

doors and windows. Otherwise several doors or windows must be arranged so they can be opened *from the outside*.

The dosage for fumigants is proportioned to the number of cubic feet in the enclosure. Consequently the first step in fumigating is to determine the cubical contents of the enclosure. If separate rooms or several floors are to be treated, it is best to generate the exact amount of gas needed in each room, and separate measurements and calculations should be made for each. The following directions for determining the cubical contents of a greenhouse are taken from Weigel and Sasscer, *U. S. Department Agriculture, Farmers' Bulletin* 1362, 1923.

"To determine the cubical contents of an even-span greenhouse (Fig. 173, *left*), compute the number of square feet in the rectangle, a, and in the right-angle triangles, b and c, and multiply the sum of the three by the length of the greenhouse.

"For example: $a = 5 \times 20 = 100$ square feet; $b = 5 \times 10 \div 2 = 25$ square feet; $c = 5 \times 10 \div 2 = 25$ square feet; $a + b + c = 150$ square feet; 150 square feet $\times$ 100 feet (length of house) = 15,000 cubic feet, the cubical contents of the greenhouse.

"To determine the cubical contents of the three-quarter-span greenhouse (Fig. 173, *right*), multiply the sum of the rectangles a and d and the right-angle triangles b and c by the length of the house.

"For example: $a = 6 \times 8 = 48$ square feet; $d = 18 \times 5 = 90$ square feet; $b = 6 \times 4 \div 2 = 12$ square feet; $c = 18 \times 7 \div 2 = 63$ square feet; $a + d + b + c = 213$ square feet; 213 square feet $\times$ 100 feet (length of house) = 21,300 cubic feet, the cubical contents of the greenhouse.

"In estimating the cubical contents of a greenhouse, it is not necessary to make allowance for the space occupied by the elevated benches and pots.

"To determine the total quantity of fumigant to be used, multiply the number of thousand cubic feet contained in the greenhouse [or other enclosure] by the quantity of fumigant to be used per 1,000 cubic feet; for example, if ½ ounce cyanide is to be employed per 1,000 cubic feet, and the greenhouse contains 15,000 cubic feet, the total amount of cyanide necessary would be 7½ ounces."

Knowing the number of cubic feet in the enclosure, the amount of materials to use will depend further (a) on the dosage which experimentation has shown to be necessary to kill the particular insects under treatment and (b) *where living plants are involved*, upon the dosage which experience has shown can be used without burning those particular kinds of plants. Of the gases discussed here, only hydrocyanic acid, nicotine, naphthalene, and methyl bromide can be used on growing plants.

Hydrocyanic Acid, HCN.[1]—Of the chemicals employed as fumigants for insects, hydrocyanic acid gas is by far the most extensively used. Hydrocyanic acid (while originally used to protect insect collections from destruction by museum beetles) was first extensively employed against the cottony cushion scale (see page 302) on citrus trees in California in 1886–1887. It was used for nursery stock in 1890 or 1894, and for greenhouse insects in 1894. Its first use for household insects was in 1898 and for mill insects in 1899, while vacuum fumigation was perfected in 1913–1914; liquid HCN was first recommended in 1915; and the calcium cyanides became available about 1916.

Hydrocyanic acid is a very volatile, colorless liquid with a specific gravity of 0.697 at 65°F. and weighing about 6 pounds to the gallon. In its physical properties it is much like water, but boils at 80°F. (26°C.) to form a gas slightly lighter that air (sp. gr. 0.9483) and extremely poisonous to insects, animals, and plants. One pound of the gas occupies about 14 cubic feet at atmospheric pressure. It diffuses rapidly, but, because it is lighter than air, the diffusion is mainly upward and outward. Under normal atmospheric pressure, the gas does not readily penetrate closely packed materials, such as piles of grain, sacks of flour, rolls of cloth, or bales of cotton. This lack of penetration may be overcome by using the gas in a partial vacuum.

Hydrocyanic acid is one of the most deadly gases known, quickly killing animals and also killing plants when used in too large doses or for too long exposures. Its action on animals is twofold, as it affects both the nerve centers and the respiratory organs. Extreme care is necessary in using this gas to prevent injury to the operator or others who may be exposed to this action. Its use should never be undertaken by inexperienced persons. Operators working with it should be protected by gas masks especially charged for HCN. All persons must be out of the building treated: it is not safe to treat certain rooms or apartments while other parts of the building are occupied. Occupants of adjoining buildings should be notified to keep their windows closed. The building being treated should be securely locked and placarded. Liquid foodstuffs in unsealed containers must be removed before treating. The building and contents must be thoroughly aired before it is reoccupied.

[1] See "A Bibliography of Cyanide Compounds Used as Insecticides, 1930," *U.S.D.A., Bur. Ento. Plant Quar., Cir.* E-354, 1935; R. J. Wilmot, "A Bibliography on the Use of Hydrocyanic Acid Gas as a Fumigant," *Fla. Agr. Exp. Sta.*, April, 1935.

This will require especial care in cold weather, including a thorough airing after being heated.

While the gas burns freely in the air, it is not considered dangerously inflammable or explosive unless used at about 4 pounds actual HCN, or more per 1,000 cubic feet or a volume concentration of above 5.6 per cent. For complete safety, however, all fires should be extinguished in buildings being treated. The draft of a fire will also exhaust the gas from a building rapidly. It is a very weak acid having a slight stinging odor somewhat like that of sparks from metal, described by some as resembling the odor of peach stones or bitter almonds. One of the warnings of the presence of this gas is a somewhat salty, metallic feeling in one's mouth. There are also very effective chemical detectors which indicate minute traces of the gas. It is readily soluble in water; hence any standing water must be removed from the enclosure before fumigating. It will tarnish such metals as brass, gold, or nickel, but the tarnish is easily removed by rubbing with a polishing cloth, or it may be prevented by coating the metal with grease. It has no effect on the colors of most fabrics or papers.

Hydrocyanic acid is formed from the cyanides of alkali metals especially sodium cyanide, calcium cyanide, and potassium cyanide. Calcium cyanide is sufficiently unstable that it yields its gas simply upon exposure to moist atmosphere, but the other salts require treatment with an acid such as sulphuric acid in order to liberate the gas rapidly. The reaction when sodium cyanide, sulphuric, acid, and water are mixed together is:

$$NaCN + H_2SO_4 \rightarrow NaHSO_4 + HCN\uparrow$$

The principal present uses of this powerful insecticide cover nearly all of the conditions listed on pages 274 and 275.

There are four principal methods of applying this gas in any enclosure:

a. By the pot method where the gas is generated from earthenware jars or paraffined barrels placed within the enclosure to be fumigated. This is done by mixing water and sulphuric acid in the container and, while this mixture is still hot, as a result of the chemical action, adding the proper amount of sodium cyanide. This method, which at one time was used almost exclusively, has been largely supplanted by the other methods, but it is still often the best and cheapest method to use, especially where great accuracy of dosage and small amounts of the gas are needed. A number of containers, of sufficient capacity that the contents do not come within 4 inches of the top, must be provided as generators. For greenhouse work half-gallon glazed earthenware jars are excellent; for household fumigation 3- or 4-gallon jars are generally used; while for very large jobs paraffined wooden barrels have been used. Not more than 5 pounds of cyanide should be used in a 3- or 4-gallon generator, and there must be at least one for each room. The generators should be placed at intervals over the enclosure, depending on its size, and the proper amount of water should be placed in them. The correct amount of acid should then be carefully poured into the water. Considerable heat is generated by the reaction of the acid and water, and, unless the generators are of good grade, they may break, which may result in the burning of floors or injury to the operator. To protect rugs and floors, the generators may be set in boxes or tubs of soil or ashes. The cyanide should previously have been accurately weighed into paper bags and placed beside the containers. Having everything in readiness, the operator should quickly drop each bag into its generator, starting with those farthest from the exit, and at once leave the room. If more than one floor of a building is to be treated, the fumigation should start on the upper floors. The gas soon forms an equal concentration in all parts of a building. Following fumigation and immediately after ventilating, the residue remaining in the containers should be removed and buried or poured into the sewer. The cyanide used should be of C.P. grade. Cyanide containing more than a trace of sodium chloride, or sodium nitrate, is not suitable for fumigation purposes. The dosage of hydrocyanic acid gas required for different insects will vary with the insect and with the conditions of treatment. For fumigation where the treatment of living plants is not involved, the usual proportions are *for each* 1,000 to 1,500 *cubic feet of space:*

Water.. 3   pt.
Commercial sulphuric acid (sp. gr. 1.83)................... 1½ pt.
Sodium cyanide[1] (98% pure)............................. 1   lb.

In the treatment of living plants the proportions used are about the same, but the dosage is greatly reduced. For greenhouse plants the following formula is generally safe to use, *for 1,000 cubic feet.* This dosage will need to be increased for the more resistant insects or may have to be reduced if very tender plants are to be treated.

Water.. ⅜ to ¾ fl. oz. (11 to 22 cc.)
Commercial sulphuric acid (sp. gr. 1.83)......... ³⁄₁₆ to ⅜ fl. oz. (5.5 to 11 cc.)
Sodium cyanide[1] (98% pure)................... ⅛ to ¼ oz. (3.5 to 7 g.)

b. In *the machine method*[2] a carefully measured quantity of sulphuric acid is introduced to a solution of sodium cyanide in a special chamber or machine, or sodium cyanide in solution may be admitted to a larger quantity of sulphuric acid; thus successive charges of the gas are generated and led from the machine into the enclosure to be fumigated through pipes leading from the machine. This method has been used chiefly in fumigating citrus trees covered with portable tents.

c. In *the liquid method*[3] hydrocyanic acid, stored in the liquid form in steel containers, is pumped out in the desired quantity and liberated through pipes and nozzles to the area to be treated. Twenty cubic centimeters (13 grams) of liquid hydrocyanic acid is equivalent to 1 ounce of sodium cyanide by the pot method, or 1 pound is considered equivalent to 2¼ pounds of sodium cyanide, and sufficient for about 3,000 cubic feet in well-built mills or dwellings. This method is limited to cases where large amounts of the gas are needed at one time. It has been used extensively in sterilizing warehouses and citrus trees.

[1] Potassium cyanide is now little used in fumigation. About one-fourth less hydrocyanic acid gas is given off where this form of cyanide is used.

[2] See *U.S.D.A., Farmers' Bul.* 1321, 1923.

[3] See *Calif. Agr. Exp. Sta. Bul.* 308, June 1919; and *U.S.D.A., Farmers' Bul.* 1321, 1923.

d. In *the dry method* calcium cyanide, or liquid hydrocyanic acid absorbed in inert earths or crude paper disks, is hermetically sealed in cans, and the gas is liberated by opening the tins and scattering the dry material over the floor or blowing it into the atmosphere. This, the newest method, is also the simplest. When these calcium cyanides are exposed to the atmosphere, the following reaction takes place:

$$Ca(CN)_2 + 2H_2O \rightarrow Ca(OH)_2 + 2HCN\uparrow$$

or

$$CaH_2(CN)_4 + 2H_2O \rightarrow Ca(OH)_2 + 4HCN\uparrow$$

It will therefore be evident why the atmosphere must be moist and also that the residue remaining after the reaction has gone to completion is harmless calcium hydroxide. There are two principal types of calcium cyanide on the market. One form in common use (Cyanogas) contains from 40 to 50 per cent calcium cyanide and gives off its gas very slowly. It will give off about half as much hydrocyanic acid gas as an equal amount by weight of sodium cyanide. Another form of dust, such as Calcyanide, contains 88 per cent calcium cyanide and will yield about 90 per cent as much gas as an equal weight of sodium cyanide. In this form of dust the gas is generated very rapidly upon exposure to the moist air. *No acid is required to generate the gas from calcium cyanide dusts, the necessary reaction taking place when it is simply exposed to the atmosphere.* Calcium cyanide has been used for killing scale insects on citrus trees by spreading it on the ground or by blowing the calcium cyanide dust into tents placed over the trees. For household insects, mill insects, and greenhouse insects the proper amount of the dust is spread out, not over ⅛ inch thick, on papers, on the floor, or on the walks in the greenhouse. It is used very extensively for the control of rodents by placing or blowing a small amount of material into their burrows. Calcium cyanide in flake form has also been used in combating chinch bugs, to kill the bugs trapped by barriers. The dosage to be used will depend upon the nature of the materials to be treated, the tightness of the building or other enclosure, and the kind of cyanide dust used. With an 88 per cent calcium cyanide dust ¾ pound to each 1,000 cubic feet of space is recommended for dwellings, mills, etc., and ⅛ ounce per 1,000 cubic feet is about a minimum dosage for ordinary greenhouse fumigation. With a 48 per cent dust the dosage should be about twice the above recommendation. In greenhouse fumigating it is recommended that the above dosage be tried in an experimental way. If a satisfactory kill is not secured, the dosage should be gradually increased, provided no burning results. If the plants are injured, the dosage must be reduced. The correct dosage for particular plants and insects can be determined by consultation with a trained entomologist. In any case, the plants should not be watered for some hours previous to the fumigation.

The other type of dry cyanides, such as the Zyklon products, undergo no chemical change when exposed, but the absorbed liquid hydrocyanic acid, when exposed to the air, undergoes a change of state, simply evaporating to form the deadly gas, leaving the harmless, inert, carrier material. Some commercial cyanide products contain a small per cent of another gas, such as chloropicrin, as a warning gas.

It has recently been discovered that for some insects the effectiveness of hydrocyanic acid gas can be increased about 30 per cent by applying the required amount in fractional or tandem doses, rather than all at once. For example, the required dose of liquid hydrocyanic acid may be divided into four equal parts and one such part applied every 2 hours, so as to keep the concentration of the gas at about 3.5 milligrams per liter of air in the enclosure.

Hydrocyanic acid fumigation appears not to conflict with the use of insecticides except bordeaux mixture which should not be used before or after fumigating with this gas. The principal merits of hydrocyanic acid as a fumigant are its extreme toxicity to insects, the fact that there is a small margin of safety between the dosage required to kill many plant-infesting insects and the slightly higher dosage that will kill the plants themselves, that it can be used safely to treat seeds to be used for planting, that it is not detrimental to foods except such as have a very high water content, that it does not seriously bleach or tarnish household articles. Its principal disadvantages are its lightness and consequent lack of penetrating power, the very great danger to human lives when it is not used with extreme care, and that it must be used in darkness or subdued light when treating living plants, many of which will not stand a concentration sufficient to kill some of their insect parasites.

Vacuum Fumigation.—The lack of penetration of hydrocyanic acid and other gases under atmospheric pressure is overcome and the time required for effective fumigation is greatly reduced, in a very ingenious way, by using the gas in a partial vacuum. This method is used particularly with hydrocyanic acid, carbon disulphide, ethylene oxide, methyl bromide, and methyl formate in industrial fumigation. The products to be treated are placed in a tight-sealing steel chamber or vault. The air is pumped out until a 27- to 29-inch mercurial vacuum, or 1 to 3 inches absolute pressure, is reached. The poison gas, thoroughly mixed and heated to about 120°F., is then introduced until atmospheric pressure is attained, penetrating very completely through the bales, bundles, sacks, or packages to effect complete destruction of insect life in 1 to 2 hours. The gas is then pumped out, the treated materials are "air washed" by admitting air, and they are ready immediately for storage or shipment in a clean sanitary condition. The reduction in oxygen content resulting from the removal of the air also makes the insects much more susceptible to the effects of the gas; carbon dioxide is frequently mixed with the toxic gas (except in the case of hydrocyanic acid) and this further stimulates the pests to respire the poison rapidly. This method is used especially for imported products likely to be infested with dangerous foreign insects; for packaged foods, confections, and tobaccos; and for eradicating human lice from clothing, bedding, and baggage of soldiers, hobos, and the inmates of public institutions. All cotton and cotton waste must be vacuum fumigated to meet quarantine regulations for entry into the United States to prevent the introduction of the pink bollworm. Potatoes shipped out of certain states are vacuum fumigated to prevent spread of the potato tuber moth. Nursery stock is often so treated to assure its freedom from scale insects.[1] Thousands of tons of dried fruits, nuts,

[1] MACKIE, D. B., "Vacuum Fumigation and Pest Control," Vacufume Company, Inc., Los Angeles.

candies, coffee, cereal, seeds, tobaccos, vegetables, furniture, books, furs, mattresses, and other articles of commerce, packed in cellophane or other cartons, boxes, bags, barrels, or bales, are vacuum fumigated every year. This process does not leave any distasteful or poisonous residue, nor does it bleach, stain, or corrode containers. In this way the great loss of prestige and patronage, that inevitably result when such products are received infested with worms, moths, or weevils, is avoided. The special chambers are available in sizes ranging from a few cubic feet to fumigators large enough to treat an entire railway car. The cost of treatment is generally less than $1/4$ cent a pound.

Carbon Bisulphide, CS_2.[1]—Although the discovery of the insecticidal value of carbon bisulphide is attributed to a Frenchman, who in 1858 found the vapor to be effective for grain-infesting insects, its early development was in connection with the fight against the grape phylloxera (see page 651).

The commercial grade of carbon bisulphide which is generally employed for insecticide work is a nearly colorless, ill-smelling liquid[2] that changes to a gas very rapidly when exposed to the air. The liquid is about one-fourth heavier than water and each volume of liquid forms about 375 volumes of gas, 1 pound of the gas occupying about 5 cubic feet at atmospheric pressure. The gas is 2.63 times as heavy as air and therefore sinks to the bottom of any container in which it is used. The gas does not discolor fabrics, but the impure commercial liquid leaves a tenacious, yellowish residue. It is highly inflammable and explosive when mixed with air, and for this reason it must be used with caution. A flame of any kind, a lighted cigar, or even the spark from hitting metal against metal or from an electric switch may cause an explosion of the gas. Because of the fire hazard this chemical cannot be shipped by mail or express. In some cases, its use in buildings invalidates insurance policies. Its boiling point is 115°F. (46°C.), and its flash point is about 297°F. (147°C.). Carbon bisulphide is only slightly soluble in water but is itself a good solvent for rubber, gums, waxes, varnishes, and oils. The gas is deadly to all forms of insect life, if used at sufficient strengths and at temperatures in which the insects are active (70°F. or above). It is, on the whole, the best fumigant where much penetration is required, as for treating bins of grain or seed, or piles of feed that can be treated in a tight room or container. Carbon bisulphide cannot be used to treat growing plants. If breathed by man, the gas causes dizziness and nausea, and produces somewhat the effect of intoxication. It is deadly if inhaled in large amounts.

The chief fields of usefulness for carbon bisulphide, at present, are (*a*) for insect pests in stored grains and seeds, in granaries, mills, and storerooms, where there is no fire hazard; (*b*) for insect pests of clothing, fabrics, furniture, etc., in houses, warehouses, etc.; (*c*) as emulsions, for soil insects and other vermin in the soil; (*d*) for borers in the wood of trees, and insects such as honeybees in buildings; (*e*) for bots and intestinal worms in the stomachs of animals.

To use this gas, the sides and bottoms of the bins, rooms, or containers to be fumigated must be made as nearly airtight as possible. Where there is sure to be some leakage, the amount of carbon bisulphide

---

[1] Also called *carbon disulphide*.
[2] The odor is not that of carbon bisulphide but is due to traces of hydrogen sulphide.

must be increased above that given below. One should never attempt to fumigate a room or bin with large cracks or openings in the bottom or sides. The best results will be obtained at temperatures from 75 to 90°F. Do not fumigate when the temperature is below 60°F. One pound of carbon bisulphide to each 100 cubic feet of space or 80 bushels of grain, or 2 to 3 gallons per 1,000 bushels of grain, is usually sufficient. It cannot be depended upon to give a satisfactory kill to a depth of more than 6 feet in a bin of grain. The liquid may be applied directly to grain or seeds, but this may injure germination; better results will be obtained by pouring the carbon bisulphide on gunny sacks, rags, or cotton waste. This gives a rapid evaporation which is more effective than when the liquid is exposed in shallow pans. Another method sometimes used is to spray the liquid into the air through a nozzle or to hasten evaporation with hot water conducted through a pan of carbon bisulphide in coils of pipe. For soil fumigation it is often made into an emulsion (see page 392), and for internal animal parasites it is given in gelatine capsules (see page 827). While not necessary, it is better to cover the tops of open bins with a tarpaulin or blanket. The room or bin should be kept closed from 36 to 60 hours. Such exposure will not injure the milling qualities of grain or the germination of most seeds and will leave no poisonous residues on feeds.

The outstanding merit of carbon bisulphide is its superior penetrating powers which make it effective in masses of grain or fabrics where hydrocyanic acid and other light gases cannot reach. It is also simple to use, and because of the foul odor of the commercial form it is not likely to overcome anyone unawares. Seeds can generally be treated without injury to their germination if they are fairly dry when treated, and the use of carbon bisulphide as a soil fumigant stimulates the growth of many kinds of crops. The great limitation of carbon disulphide is the fire hazard which attends its use. Commercial mixtures of 20 per cent carbon bisulphide, 80 per cent carbon tetrachloride, and a trace of sulphur dioxide, which are said to have a greatly reduced fire hazard, have been placed on the market, but the mixture cannot be made at home. It is recommended as effective at $1\frac{1}{2}$ to 2 pounds per 100 cubic feet of space or 3 to 5 gallons per 1,000 bushels of grain. It is also exceedingly toxic to plants, so that it cannot be used for greenhouse fumigation. There is also some difficulty at moderate temperatures in getting the liquid to evaporate rapidly enough to yield a toxic concentration.

Nicotine, $C_{10}H_{14}N_2$.—Nicotine, whose history and chemistry have been discussed under contact insecticides (see page 257), is also used extensively as a greenhouse fumigant, where its high degree of safety both to the plants and to the operator have made it a favorite, and to some extent also against orchard and garden pests and poultry lice.

For fumigating, free nicotine and not nicotine sulphate must be used. It may be driven off from the dried parts of tobacco plants by burning them, but, because of the difficulty in gauging the strength in that way, commercial free-nicotine preparations, such as the Nicofume products, are generally used. The nicotine may be volatilized by painting or dropping the liquid over hot steam pipes, by heating it in shallow pans, or by forcing it through heated tubes. The nicotine is also absorbed in known amounts in paper punk, powders, and other combustible materials, from which it is liberated by burning. Such materials should smoulder;

if they burn actively, most of the nicotine may be destroyed. Such a preparation is now marketed to be burned in the container. Small holes are punched near the top of the container and the nicotine-impregnated powder is ignited with a "sparkler." The entire house may be fumigated or the burning, pressure fumigators may be attached to poles and carried along rows of plants, thus releasing the toxic gas in close contact with insects to be destroyed. With such a preparation it is possible to do "spot fumigating," *i.e.*, a few plants or a single bench in a large greenhouse may be fumigated without treating the whole house. Nicotine has been used to fumigate trees for the destruction of *adult* codling moths by forcing the liquid, with compressed air, through hot tubes (about 665°F.) into the intake of a power duster. Ten cubic centimeters for an average tree is said to be adequate. Its effectiveness can be greatly increased by applying it under a gastight tent or canvas placed over the tree for $\frac{1}{2}$ minute or longer. For treating canning and truck crops, a device known as a Nicofumer, mounted on a truck, pumps liquid nicotine through heated pipes and discharges the fumes through pipes that are enclosed by the apron of a canvas trailer from 25 to 100 feet long, which holds the fumes about the plants for a brief time as it passes over them. This has proved very effective for the control of aphids on peas, cabbage, beans, and other crops.

The dosage generally required to kill greenhouse aphids is from $\frac{1}{2}$ to 1 ounce of 40 per cent free nicotine per 1,000 cubic feet, with an overnight exposure. This is equivalent to from 80 to 175 grains of actual nicotine per 1,000 cubic feet. Other commercial preparations should be used at the same strength of actual nicotine. Since the actual amount of nicotine contained is usually stated in grains, the correct dosage can easily be computed, or the manufacturer's directions should be followed. Contrary to directions often given, fumigation of plants with nicotine is best done under dry conditions and at temperatures between 50 and 70°F. It should never be done in bright sunshine. Violets and ferns are among the few kinds of plants likely to be severely injured by nicotine. Poultry may be fumigated while they are roosting by painting the 40 per cent nicotine sulphate upon the perches (see page 866). This will free them of lice and certain mites and ticks but does not kill the eggs.

The advantages of nicotine as a fumigant are its safety to plants and the fact that it is not seriously dangerous to human life because it is so irritating as to drive anyone away before overcoming him. It is, however, a rather weak fumigant; the fact that it does not exist as a true gas at working temperatures makes it somewhat erratic; it does not penetrate well and, as usually generated, has a persistent odor that makes it unsatisfactory for the fumigation of houses or food products.

Sulphur Dioxide, SO_2.—This gas is used to some extent as an insecticide but is not generally so effective as those already discussed. It is usually generated by burning sulphur that is dampened with alcohol; or a mixture of 58 per cent sulphur, 38 per cent saltpeter, (KNO_3), and 4 per cent potassium chlorate to cause it to burn more readily; or sulphur candles which may be purchased ready prepared for burning. A concentration of from 5 to 10 per cent of sulphur dioxide in the atmosphere (4 to 8 pounds per 1,000 cubic feet of space) is effective for most insects. The gas is very irritating to animals. Sulphur dioxide is noninflammable, but corrodes metals, ruins paint, and discolors some fabrics. It is slightly heavier than air, having a specific gravity of 1.433. This gas destroys the germinating power of most grains and affects the baking qualities of flour. It cannot be used on living plants, and the release of the gas from a large enclosure may kill surrounding vegetation. It has a very strong bleaching effect, which may be partly overcome by having the space to be treated very dry. The gas can now be obtained in liquid form, shipped in steel cylinders; this is the most convenient way to use it where any large-scale fumigating operations are to be undertaken.

Paradichlorobenzene, $C_6H_4Cl_2$.—This chemical, which may be purchased in the form of white, readily volatile crystals, is now widely used as a soil fumigant for killing the larvae of the peach tree borer (see page 631) and to some extent for root aphids, wireworms, and sweetpotato weevil. It vaporizes very slowly (boiling point, 344°F.) to form a noninflammable gas with an ether-like odor. The undiluted vapor is 5.1 times as heavy as air and so will readily penetrate the air spaces in soils or the crevices in the underground parts of trees. It is employed to some extent as a repellent for pests of fabrics and, at about 100 pounds per 1,000 cubic feet, in household fumigation, but imparts an odor to grains or foods. The gas is noninflammable, and does not bleach or discolor metals or fabrics.

Naphthalene, $C_{10}H_8$.—Like paradichlorobenzene, this chemical is widely sold as a white crystalline salt and as the familiar "moth balls." It vaporizes very slowly (boiling point, 424°F.) to form a noninflammable gas with a pungent, tarry odor and a specific gravity of about 4.4. In the control of clothes moths and carpet beetles it is about as effective as paradichlorobenzene. It is recommended to destroy gladiolus thrips (see page 738) and has been used as a fumigant in greenhouses, especially for red spider mites and thrips, which are very resistant to hydrocyanic acid. The material is vaporized by heat or by driving a current of air over the chemical using from $\frac{1}{8}$ to 2 ounces per 1,000 cubic feet for 8 to 15 hours at temperature of 70 to 85°F. and relative humidity of 60 to 80 per cent.

Chloropicrin,[1] CCl_3NO_2.—This is one of the tear gases which, since the World War, has been employed to some extent as an insecticide. At ordinary temperatures it is a colorless, yellowish, very stable liquid, with a specific gravity of 1.654, and weighs $13\frac{3}{4}$ pounds per gallon. It volatilizes very slowly, boiling at 233°F. (112°C.). The specific gravity of the vapor is 5.7 giving it remarkable penetrating power. It is noninflammable and not explosive unless heated very rapidly. It has been employed chiefly in flour mills, grain elevators, and farmers' bins, by spraying it into the atmosphere or pouring it upon sacks laid over the materials being fumigated. Sacked seeds may be fumigated by injecting a small amount into each sack. It may also be used to destroy subterranean insects, garden centipedes, and nematodes in the soil. For most grain and household insects it requires about twice as many pounds per 1,000 cubic feet as of hydrocyanic acid but only $\frac{1}{5}$ to $\frac{1}{8}$ as much as of carbon bisulphide. One to three pounds per 1,000 cubic feet are recommended. It is said to be effective at lower temperatures than hydrocyanic acid or carbon bisulphide, and it is so irritating to insects that it will drive worms out of nuts and other insects out of cracks or upholstery to die in the open. It has no bleaching or tarnishing effects

[1] See "Bibliography of Chloropicrin, 1848–1932," *U.S.D.A.*, *Misc. Publ.* 176, 1934.

but corrodes some metals slightly. Its great advantages are its unusual weight and penetrating property which make it possible to fumigate mills without cleaning out the machinery and to "spot fumigate" a part of an enclosure without treating the entire space; the quickness with which it kills; its effectiveness over a wide range of temperatures; and the fact that it is not inflammable and not highly toxic to man. Its powerfully irritating effect upon the eyes and respiratory passages is valuable as a warning of its presence, but greatly limits its usefulness. It is also difficult to remove from an enclosure or from materials fumigated, by ventilation. It cannot be used at all for growing plants and will injure the germination of certain kinds of seeds. It also prevents fermentation and cannot be used for brewers' malt.

Ethylene Oxide, C_2H_4O.—At ordinary temperatures this chemical is a colorless gas, with a faint sweetish odor, the boiling point being 51°F. (10.5°C.). The gas is about $1\frac{1}{2}$ times as heavy as air. At low temperatures it is a colorless mobile liquid with a specific gravity of 0.887, readily miscible with water, alcohol, acids, and ammonia. It is stored and transported in steel cylinders. In toxicity to insects it stands between chloropicrin and carbon bisulphide. Three pounds per 1,000 cubic feet is a standard dosage for tightly sealed buildings. Owing to its low boiling point, it is unusually effective at low temperatures and is recommended where fumigation must be done at temperatures below 70°F. It is not considered dangerously toxic to man. It leaves no residual taste or odor on such sensitive products as tobacco, coffee, or nut meats. It has no effect upon the baking or milling qualities of grains but may destroy their germinating power. It is not corrosive to metals or injurious to fabrics or finishes. It is said to kill the egg stage before the active stages, so that products freed of visible stages will not be likely to contain viable eggs to reestablish the infestation. The gas cannot be used for fumigating living plants, and the germination of seeds is likely to be seriously affected. At concentrations slightly above the effective dosage the vapors are inflammable, so that this gas is always used in combination with carbon dioxide. The ethylene oxide from pressure cylinders (1 part) may be poured over finely crushed, solid carbon dioxide or "dry ice" (9 parts), and the mixture shoveled into bins of grain or other enclosures and allowed to vaporize. Three pounds of ethylene oxide and 27 pounds of solid carbon dioxide are recommended for each 1,000 bushels of grain. Commercial products such as Carboxide, consisting of 1 part by weight of ethylene oxide and 9 parts of carbon dioxide, are sold in cylinders of various sizes, under pressure, as a homogeneous liquid. Since the two materials have about the same vapor pressure, the gases remain intimately mixed when released from the cylinder. The carbon dioxide makes the material noninflammable; reduces the absorption of the toxicant gas by the materials being fumigated, so that much smaller dosages are effective; and, by stimulating the insects to much more rapid respiration, causes them to be killed much more quickly. The principal use of this gas mixture is in the sterilizing of packaged food products, tobaccos, candies, nuts, dried fruits, furs, and the like, in specially built, vacuum-fumigating chambers. Ten to thirty pounds of the mixture per 1,000 cubic feet at 27-inch vacuum effectively destroys all insect life in from 1 to 3 hours. In airtight vaults at atmospheric pressure the same dosage is effective in from 12 to 48 hours.

Ethylene Dichloride,[1] $CH_2Cl \cdot CH_2Cl$.—At ordinary temperatures this material is a colorless liquid with an odor resembling ether and very slightly soluble in water. The specific gravity of the liquid is 1.27, 1 gallon weighing 10.4 pounds. The boiling point is 182°F. (83.7°C.), and the vapor is approximately 3.5 times as heavy as air. It is less toxic than ethylene oxide, being intermediate between carbon bisulphide and carbon tetrachloride in efficiency. It is also very slow to kill insects, death sometimes occurring from 24 to 72 hours after exposure. Alone it is dangerously inflammable, but the addition of 1 volume of carbon tetrachloride to 3 volumes of ethylene dichloride makes it perfectly free from fire hazard either as a liquid or as a gas. The mixture is stored and transported in ordinary tins or cans and is used simply by pouring it over cloths or directly upon the material being fumigated and allowing it to vaporize. In unfinished rooms it may be sprayed about freely, even upon rugs, clothing, or upholstery; but it ruins paints and varnishes. It is rather slow to vaporize and a forced circulation of air over the liquid is advisable. Dosages of the mixture recommended range from 10 to 18 pounds per 1,000 cubic feet in airtight vaults at atmospheric pressure for 24 hours. In trunks, clothes chests, or tightly sealed, plastered closets, 3 fluid ounces to 5 cubic feet or 3 to 4 gallons to 1,000 cubic feet will be required and at least a 24-hour exposure. The temperature should be above 75°F. It has been found to be effective for the peach tree borer when applied as an emulsion in water and potash fish-oil soap or other emulsifier (see page 632). This gas is not considered dangerous to man unless exposure continues for a considerable time. It is clean, safe, and pleasant to use and keeps indefinitely in tight containers. It cannot be used on growing plants but is not injurious to the germination of most seeds. Food substances rich in fat absorb appreciable amounts of the gas and require long periods of aeration to remove the taint, and the gas is said to leave a disagreeable flavor in tobaccos.

Carbon Tetrachloride, CCl_4.—This is a thin, transparent, colorless, oily fluid, which boils at 168.8°F. (76°C.) to form a gas with a specific gravity of 5.31 and a pungent, not disagreeable odor. It is much less toxic to most insects than hydrocyanic acid (requiring about forty times as much) or carbon bisulphide (requiring about ten times as much). Its only claim to consideration as an insect fumigant is that it is not inflammable or explosive. At present it is employed chiefly in mixtures with ethylene dichloride to reduce the fire hazard.

Methyl Bromide,[2] CH_3Br.—This compound has the lowest boiling point of all fumigants in general use at the present time. The boiling point is about 40°F. (4.5°C.); consequently it is a gas at ordinary temperatures. The colorless liquid has a specific gravity of 1.732. It is readily soluble in mineral oils, slightly soluble in water, and somewhat caustic to the skin. One pound of the gas occupies about 4 cubic feet at atmospheric pressure. It has a specific gravity of 3.29 to 3.5, is noninflammable, and has such a slight, chloroform-like odor that toxic amounts of it may be respired unawares. It is shipped and stored in sealed tins and applied by sprinkling or spraying; or in mixture with carbon dioxide under the name of Guardite Gas. It is about as toxic to many insects as chloropicrin and is used at dosages from 1 to 2½

---

[1] See "A Bibliography of Ethylene Dichloride," *U.S.D.A., Misc. Publ.* 117, 1932.
[2] See *Jour. Econ. Ento.,* **31**: 70–84, 1938.

pounds per 1,000 cubic feet for 12 to 16 hours in buildings. It is not narcotic to the insects and kills very slowly. It is recommended for use in ordinary atmospheric fumigation, as in mills and greenhouses, and in vacuum chambers, and also for such products as fresh fruits and green vegetables, dried fruits, and other stored food products, household goods, nursery stock and ornamental flowers, and for rodent control. It does not injure the germination of seeds if they are dry. High temperature and humidity increase the effectiveness of this gas. It appears to be a very promising fumigant but has been under investigation only since 1932; much is still to be learned about it.

Other Fumigants.—Methyl formate, $HCOOCH_3$, is used for industrial fumigation in vacuum fumigators as described on page 282. Proxate is a mixture of 60 ounces of methyl formate in 50 pounds of carbon dioxide. Fifteen to thirty pounds of the mixture are used per 1,000 cubic feet. Propylene dichloride, $CH_2ClCHClCH_3$, has been used much like ethylene dichloride in mixtures with carbon tetrachloride. It is slightly less dense and somewhat less toxic than ethylene dichloride. Methyl thiocyanate, CH_3SCN; cyanogen chloride, $CNCl$; nitrobenzene, $C_6H_5NO_2$; tetrachloroethane, $CHCl_2CHCl_2$, and many other gases have shown promise as insect fumigants, but their use is still in the experimental stage. Formaldehyde has been used as a fumigant to kill germs in buildings following contagious diseases but *is useless as a fumigant for insects.*

REPELLENTS

Substances which are only mildly poisonous, or which may not be active poisons, but which prevent damage to plants or animals by making the food or living conditions of the insects unattractive or offensive to them, are called repellents. These substances are rarely, if ever, repellent to all kinds of insects. Such chemicals can sometimes be employed to advantage where it is impossible to use an insecticide and may afford a greater or less degree of protection to manufactured products, growing plants, or the bodies of animals. Among the many examples may be noted the following: (*a*) *Repellents against crawling insects.* Examples are the creosote lines used as barriers to the migration of chinch bugs (page 356); orthodichlorobenzene and other chemicals used to protect buildings from termites; ant tapes, usually containing bichloride of mercury, which are placed about table legs and the like to keep ants from crossing; heavy oils at the base of poultry roosts as a barrier to poultry mites; and certain chemical bands about tree trunks. (*b*) *Repellents against the feeding of insects.* Examples are the application of bordeaux, lime, and similar washes to plants to ward off leafhoppers and some chewing insects; the dusting of cucurbits to protect them from cucumber beetles; mosquito "dopes" and fly sprays to lessen the attacks of bloodsucking flies and mosquitoes; the application of sulphur to the body to keep chiggers from attacking; the use of smoke and smudges to repel biting flies; the chemical treatment of logs to keep beetle borers from destroying log cabins and other rustic work; moth balls, oil of cedar, and mothproofing treatments to protect materials from attack by clothes moths and carpet beetles. (*c*) *Repellents against the egg-laying of insects.* Examples are the use of pine tar oil to keep screwworm flies from laying eggs about wounds of animals (page 845).

Bordeaux Mixture.—Bordeaux mixture originated in France as a spray to control the downy mildew disease, *Peronospora viticola*, of grapes, about 1882, and was first used in the United States in 1887. While primarily a fungicide, bordeaux mixture is very repellent to many insects, such as flea beetles, leafhoppers, and potato psyllid when sprayed over the leaves of plants. By the addition of lead arsenate, a poisoned bordeaux is made that is widely used as a preventive of, and a remedy for, insects and plant diseases on many crops. It is to some extent an ovicide and has some residual toxic effect upon the sap, which kills leafhoppers and psyllids for some days.

Bordeaux mixture is made at different strengths for different purposes. The strength is generally indicated by numbers, 4–8–100 or 6–10–100, of which the first number designates the number of pounds of copper sulphate and the second the number of pounds of lime to be mixed with 100 gallons of water. The formula for the widely used 6–10–100 mixture is as follows (any other strength desired can be made by varying the amounts of lime and copper sulphate:)

	Field Formula	*Garden Formula*
Water (cold)	100 gal.	3 gal.
Hydrated lime	10 lb.	5 oz.
Pure copper sulphate (bluestone)	6 lb.	3 oz.

For controlling chewing insects, the above mixture is generally poisoned by adding:

Lead arsenate	2 to 4 lb.	1 to 2 oz.

Where it can be secured, unslaked stone or rock lime can be used instead of the hydrated lime; 6 pounds to 100 gallons of water, or 1 ounce to 1 gallon, is sufficient. *Air-slaked lime should never be used.* Only wooden, earthenware, or glass containers should be used for mixing and storing bordeaux. For making small amounts, it is best to suspend, the evening before spraying is to be done, the proper amount of copper sulphate in a sack near the top of a tub, barrel, or spray tank containing about three-fourths of the water to be used. If this has not been done, the sulphate can be dissolved quickly in a little hot water and then diluted with cold water. Quickly dissolving powdered copper sulphates can be purchased from most dealers for use instead of the crystal form. Slake the lime, or mix the hydrated lime, in a small quantity of water, dilute it, and then pour the limewater into the copper sulphate solution and stir vigorously. When properly mixed, a beautiful blue, voluminous, colloidal precipitate of tetracupric sulphate, $4CuO \cdot SO_3$, and pentacupric sulphate, $5CuO \cdot SO_3$, is formed, which will remain in suspension for several hours. The precipitated particles are exceedingly thin and, when sprayed upon foliage, dry upon the leaves with remarkable covering power and tenacity. A gallon of good bordeaux has sufficient precipitate to cover 1,000 square feet of leaf surface. Its efficiency is said to depend upon the gradual formation upon the leaf surface of small quantities of soluble copper over a long period of time. The lead arsenate may then be stirred up with a little of the bordeaux and added to the mixture, together with enough water to make the full amount. For small gardens it is more convenient to buy one of the commercial bordeaux, but these are not generally so effective as the freshly mixed material.

If much spraying is to be done, stock mixtures of limewater and copper sulphate solution should be prepared. These can be kept indefinitely in covered jars or, if not tightly covered, by adding water to keep up to the original level. These stock solutions are best prepared by dissolving 1 pound of copper sulphate to each gallon of water and by mixing or slaking lime and then diluting it so that each gallon contains $1\frac{2}{3}$ pounds of hydrated lime (or 1 pound of stone lime). Do not mix the two solutions until ready to spray. Never mix them without first diluting one or both. When ready to spray, fill the spray tank about three-fourths full of cold water. If you want a 6–10–100 bordeaux, add 6 gallons of the limewater stock mixture (representing 10 pounds of lime) to each 100 gallons that the spray tank will hold. Start the agitator of the pump going, or stir vigorously, and pour in 6 gallons (representing 6 pounds) of the copper sulphate stock solution for each 100 gallons the spray tank will hold. Continue agitation or vigorous stirring for several minutes. If the bordeaux is to be poisoned, add the lead arsenate, reduced to a thin paste, and more water as necessary to make a full tank. To make good bordeaux observe the following precautions: (a) Do not allow the bluestone solution or the bordeaux to stand in contact with any metal; (b) use only pure lime and bluestone; (c) have one or both of the chemicals diluted with nearly all of the water before the other is added; (d) use water as cold as possible; (e) agitate thoroughly after mixing; and (f) use within a few hours after mixing. If spraying is interrupted, the mixture can be preserved for a day or two by adding an ounce of sugar to each 25 gallons and stirring thoroughly. Bordeaux is somewhat incompatible with soaps and must not be used before or shortly after fumigating plants with hydrocyanic acid. It is often combined with an arsenical or other stomach poison, with nicotine sulphate, and with oils. It may, however, reduce the efficiency of oil emulsions for scale insects or whiteflies, either by the chemical reaction or because the bordeaux kills the fungi that parasitize the insects.

Creosote and Coal Tar.—These complex chemical substances and their derivatives, such as naphthalene and carbolic acid, are very repellent to certain insects, particularly to the chinch bug, adult clothes moths, and certain flies and beetles. Thin lines of creosote poured on the ground or creosoted paper strips have been used extensively in the Middle West for diverting the migrating hordes of chinch bugs when they leave the small-grain fields for the corn. These substances are also used in treating wood to render it distasteful to certain boring insects, especially termites.

Strong soap solutions mixed with creosote or carbolic acid are repellent to the adults of certain wood borers. Nearly any very fine dust, such as gypsum or tobacco, acts as a repellent to leaf-feeding insects. Certain oils, such as lubricating oil, kerosene, castor oil, fish oil, oil of tar, pine oil, rosin oil, oil of pennyroyal, oil of citronella, lemon oil, oil of tansy, pyrethrum extracts or fine sulphur dust are used, alone or in combination with other chemicals, to protect the bodies of man and animals from insects such as biting flies, mosquitoes, chiggers, ticks, and fleas. Powdered alum, camphor, and turpentine are also employed for this purpose. Repellents do not kill insects unless used in combination with traps, and have to be renewed frequently, as the insect attack recurs.

Livestock Sprays.—The manufacture and sale of livestock sprays have become a considerable industry during the past quarter century. The earliest sprays used were made as repellent as possible and consisted especially of pine tar oil, creosote, naphthalene, and other strong smelling chemicals. These materials often contaminated milk and stained the animal's coat objectionably. About 1930 pyrethrum extracts came to be used extensively, and attention was directed toward killing the flies, instead of merely repelling them, and toward avoiding staining properties and all odors offensive to man. One pound of pyrethrum powder or the equivalent extract to a gallon of moderately heavy mineral oil gives sufficient toxicity, and aromatics such as clove, citronella, sassafras, camphor, and pine oil may be added to repel the flies for 5 to 10 hours. Since about 1937 some newer and apparently very superior synthetic chemicals, such as Lethane, have become available. They give, not only a much quicker knockdown of the flies, but also a higher per cent kill. The effectiveness of livestock sprays varies greatly with weather conditions and method of application. Wind and sun reduce the period of effectiveness. The sprays should be applied indoors or out of the wind with a sprayer that creates a mist or fog about the animal, and application should be from the rear of the animal. A pint of spray is enough for a dozen cows. Probably the best time to apply to dairy cattle is after the morning milking which should give protection for most of the day. Sprays with a high oil content should be used very cautiously on cattle.

Mothproofing.—In recent years great interest has been taken by entomologists and manufacturers in the treatment of fabrics, such as clothing, rugs, upholstery, and the bindings of books, with chemicals which render them either poisonous or distasteful to insects that might otherwise feed upon them. Various materials are used some of which are effective for some months (see page 790).

PHYSICAL AND MECHANICAL CONTROL

Aside from the destruction of insects that may be accomplished by *ordinary* farm practices (see below) there are certain *special* physical and mechanical measures that are of value. The more important of these are outlined and defined on page 239. These differ from chemical control measures in the nature of their effect upon the insects, which is a physical action not involving a chemical action upon the insect. They may be arbitrarily distinguished from the cultural control measures in that they involve the use of special equipment or operations, which would not be performed at all were it not for the insects, and they generally give immediate, tangible results. On this account they are psychologically good and generally popular. They are in general costly in time and labor, often do not destroy the pest until much damage has been done, and rarely give adequate or commercial control. There are two subdivisions, the *mechanical measures*, which involve the operation of machinery or manual operations; and the *physical measures*, which employ in a destructive way certain physical properties of the environment (see outline, page 239).

The destruction of insects or their egg masses by hand is sometimes the most practical method to employ in areas where labor is very cheap and where the insects or their eggs are large or conspicuous, not too active, or occur in relatively restricted areas. In other cases it is possible to prevent the invasion of a crop by migrating insects, through the use of physical barriers. The crawling hordes of chinch bugs and armyworms may often be stopped by constructing deep, dusty-sided furrows around the fields toward which they are traveling. The same result may be achieved by using barrier lines of certain heavy-bodied oils, poured along on the ground. Low fences of sheet metal are employed against mormon and coulee crickets (page 467). Such linear barriers are practicable against the nonflying insects, for crops of limited area, or for a very limited period of attack. Screening of houses has come to be a regular practice in all civilized countries as a protection from flies and mosquitoes and the diseases they carry. Sometimes individual plants, storage houses, seedbeds, fields, or vineyards are protected by covering them with screens of thin cloth or wire screen. Because of the expense involved, this measure can be resorted to only in cases where the crop has a high money value. Sticky bands around tree trunks are often employed against insects that infest trees by climbing up the trunks; they are of no value for insects that fly into the trees to feed or to lay eggs. Collars about individual plants to protect from cutworms, bags over clusters of fruits to protect from fruitworms, chips under melons to protect from the melonworm, or the complete wrapping of tree trunks with paper against the flatheaded borers are other examples. Fly nets, muzzles, and other similar devices are of some value in protecting animals from certain insect parasites.

Mechanical devices, such as hopperdozers, hoppercatchers, aphidozers, fly traps, moth traps, maggot traps, light traps, electric traps, and others have been used successfully for catching and killing a variety of insects. Since insects for the most part lack cunning or intelligence, insect traps are often surprisingly simple. They generally take advantage of some dominating, fixed tropism or instinct which the insect species has been observed to follow rigidly. Insect traps are sometimes *merely mechanical*, such as window traps for flies, the maggot trap for the house fly (page 909), or boards for squash bugs (page 502); sometimes they employ a *bait* as in the cone traps for flies or Japanese beetle traps (page 623). Sometimes they are *stationary*, as the sticky bands about tree trunks or the electrified screens or light traps to which many insects are attracted; and sometimes *moving*, as the hopperdozers (page 341). Rarely, crushing, grinding, suction, or dragging machinery may be successfully employed to destroy some kind of pest. These will be discussed further under the insects for the control of which they have been found most useful.

Physical measures involve especially manipulations or changes in temperature or humidity or employ radiant energy in some way to destroy a pest. Cranberries are protected from some of their insect enemies by flooding the bogs at the proper time, and some other crops, in areas where irrigation is possible, are protected in the same way. The draining of swamps, marshes, and other standing water is the most effective method of destroying mosquitoes and horse flies.

The Use of Low Temperatures.—Artificial heating or cooling of stored products, or the mills or factories where such products are processed, is a common method of preventing insect damage. Nearly all insects become inactive at temperatures between 60 and 40°F. Few insects are killed at these temperatures unless exposed to them for a considerable length of time. Insects in hibernation frequently withstand temperatures of −20 to −30°F., or lower. It is not certain that exposure to such temperatures will kill the eggs of such species as the grain weevils. But practically no damage from insects will occur at temperatures below 40°F. Low temperatures are not so effective as high temperatures in killing insects, but storage of food products or clothing at points below or near freezing will prevent all insect damage. Changes from low to high temperatures and back to cold are more effective in killing insects than constant low temperatures.

Superheating.—Abnormally high temperatures are employed against (a) insects in cereals, coffee beans, and other seeds, and their processed derivatives; (b) insects, such as the Mediterranean fruit fly, in oranges or grapes; (c) insects in clothing, bedding, baggage, bales of cotton, and other fibers; (d) insects, mites and eelworms infesting bulbs; (e) insects infesting soil; and (f) insects infesting logs. Sometimes exposure to the sun's rays is sufficient especially in tropical regions. For insects and mites infesting bulbs, a temperature of 110 to 111.5°F. for a few hours is used. Careful experiments by a number of entomologists have shown that no insect can long survive when exposed to temperatures of 140 to 150°F. Most insects, including those which attack stored grains, are killed by 3 hours' exposure to temperatures from 125 to 130°F., and this, or a slightly longer exposure, will destroy all stages of these insects. Many mills and large elevators have equipped their buildings with enough heating pipes to enable them to raise the temperature to 125 to 150°F. for several hours during periods of warm weather, and thus kill all insects in the buildings at a much less expense than by fumigating. In general, the higher the humidity, the more effective superheating will be.

There are a number of heat-treating machines now on the market which raise grain to a high temperature while it is being passed through them. Such machines are fairly effective in cleaning infested grain and other seeds, but the exposure of the grain to high temperature must be of sufficient duration to kill all stages of the insects. It should be understood that the heat-treatment of grain will cause a certain shrinkage due to loss of moisture. In applying heat to piles of clothing, bins of grain, or bales of goods, it must be borne in mind that it requires a long time for the heat to penetrate and that the temperature on the surface will have reached the killing point long before the insects within the material have been affected.

Recently new investigations of the possibilities of utilizing light to lure and destroy insects have shown promise by very carefully fixing the candle power or the exact color or portion of the spectrum used. It is noteworthy that many insects are much more strongly attracted by the blue portion of the spectrum and by ultraviolet than by the red rays. By exposing insects to an electrostatic field of high frequency and great current volume at high voltage, between two electrodes, sufficient heat may be quickly generated within the bodies of insect pests to kill them before the product which they are infesting is damaged. In one test the insects in grain were burned to a crisp in a few seconds while the temperature of the grain was raised to only 125°F.

CULTURAL CONTROL OR THE USE OF FARM PRACTICES

The cultural control measures differ from physical and mechanical control in generally involving the use of *ordinary* farm practices and farming machinery and in being usually preventive, indirect, or intangible, so that the farmer has much difficulty in being sure how effective they are. They must usually be employed far in advance of the time when damage by the pest becomes apparent, and they often do not make a strong appeal to the farmer. However, they are the cheapest of all control measures, once research has revealed an effective and practicable procedure; in fact, they often cost the farmer nothing at all because they are merely variations in the time or manner of performing operations which are necessary in the production of a crop. Often, with crops of great acreage and low unit value, they are the only control measures that can be employed profitably. The opportunity for cultural control of insects usually results from the interplay of the complicated metamorphoses of insects and the change of the seasons. The result is often some particularly *weak point* in the life cycle or adaptation of the insect pest to its environment, at which point it may be attacked by a cultural control measure.

In order to control insects by cultural practices, it is necessary that one understand the life history and habits of the insect with which one is dealing. A control that would be effective against one kind of insect might be useless against a closely related kind because of a difference in habits. These operations, to be effective, must also be used at the proper stage of the development of the insect. It is useless to try to destroy white grubs by late fall or winter plowing, after they have gone down a foot or more below the surface of the soil, or to kill insects by burning their hibernating places, before they have entered them in the fall or after they have left them in the spring.

Crop Rotations.—In a state of nature, the plants growing on the land in any of the great agricultural areas of the world are quite different from those which are grown after such lands have been placed under cultivation. There was in most of these areas a predominance of grasses but with a mixture of legumes and plants of many other botanical families. Such plants grew from year to year with little change in the proportion of one over the other. The insects depending on these wild plants were always assured of a food supply sufficient to maintain them but, with the exception of a few general feeders, the food plants were not abundant enough to permit a great increase of any one species.

Under farming conditions great changes take place in the character of the plants grown on the land. There are no longer a great number of species, generally intermixed, but a few species occupying the land in nearly pure stands of thousands and hundreds of thousands of acres. This affects the insect population of the land in two general ways. Many of those which depend on the plants of one family, or even on one species of plant, find their food supply cut off, except in the small uncultivated areas, and may nearly, or quite, disappear from the region, as certain species of billbugs in drained bottom lands. Others take to the cultivated crop closely related to their wild food plant and find it, perhaps, more palatable. Such insects may, and generally do, increase enormously and become very destructive, as have the chinch bug and the cotton boll weevil.

Among insects that injure cultivated crops, the number of general feeders is very small. Those which feed on the plants of one family are numerous, and there are many that feed on only a few very closely related species. The following lists of the principal insect enemies of corn, wheat, and clover in Illinois include only those insects that are of sufficient importance to be considered as doing commercial damage to these crops. Under each crop, the pests are listed in the order of their destructiveness to this crop in Illinois.

TABLE XV.—A COMPARISON OF THE MOST IMPORTANT INSECTS ATTACKING THREE MAJOR FIELD CROPS IN ILLINOIS

Corn Insects	Wheat Insects	Red Clover Insects
Northern corn rootworm	Hessian fly	Clover bud weevil
White grubs[1]	Chinch bug[1]	Clover leaf weevil
Wireworms[1]	Wheat jointworm	Grasshoppers[2]
Chinch bug[1]	Wheat stem maggot	Clover root curculio
Corn earworm	Wireworms[1]	Clover seed chalcid
Southern corn rootworm[1]	Grasshoppers[2]	Pea aphid
Corn root aphid	Armyworm[2]	Variegated cutworm[2]
Armyworm[2]	Wheat head midge	Clover seed caterpillar
Grasshoppers[2]	Wheat stem sawfly	Clover seed midge
Black cutworm	Wheat sawfly	Green cloverworm
Seed-corn maggot	Billbugs[1]	Clover leaf tyer
Common stalk borer	Frit fly	Armyworm[2]
Sod webworm	Wheat head armyworm	Clover root borer
Billbugs[1]	English grain louse	Clover stem borer
Morning-glory flea beetles	Green bug	Leafhoppers
Corn-seed beetles	Sorghum webworm	
Carrot beetle	White grubs[1]	
Corn prionus	False wireworms	
Clover rootworm	Southern corn rootworm[1]	
Corn leaf aphid	Variegated cutworm[2]	
Imbricated snout beetle		
Pale-striped flea beetle		
Thief ant		
Green June beetle		
Fall armyworm		
Variegated cutworm[2]		

[1] Species of importance to corn and wheat only.
[2] Species of importance on clover, wheat, and corn.

In the above list, it will be noted that 8 of the insects which are listed as pests of corn are also listed as pests of wheat, but that only 3 of the 50 insects are serious pests of all three of these crops. Wheat and corn are grasses, clover is a legume; and it will be seen from a study of these lists that much can be accomplished in preventing insects from becoming seriously abundant in our fields, if a good rotation is practiced where a crop of one plant family follows that of a different family. It is not possible in many cases in grain-farming areas to put such a rotation into effect in all fields each year; but a large part of the increased yields obtained from a rotation where grains follow legumes, and legumes grains, is due to the reduction in insect damage. Crops of the same group, such as corn, oats, and wheat, grown on the same land, year after year, give a condition favorable to the insects that attack the grass crops; and the same is true of a number of years' cropping of ground with plants of any one family.

Crop rotations will be most effective for insects that are restricted feeders, that have limited powers of migration or sluggish habits, and that are slow breeders spending a relatively long time in the feeding stage. Because of the ease with which most insects move about, many of the species which feed on any crop will be found in the fields the first year they are planted to such a crop; these insects may occur in numbers sufficient to cause severe damage. For this reason, rotation of crops cannot be depended upon for combating all insects attacking field crops. Generally, however, infestations in such fields will be later and lighter than in fields continued in the same crop, and crop rotations are by far the best, and in some cases almost the only, means we have of controlling certain insects. The application of rotations to the control of different species will be discussed in more detail under the insects attacking certain crops.

Tilling or Cultivating the Soil.—Insects are greatly affected, directly, by the texture of soils, their chemical composition, the percentage of soil moisture, the temperature, and other soil organisms; and, indirectly, by the influence of these things upon their food plants. Consequently various methods of stirring and managing the soil have a profound effect upon many insects. When the exact effects are understood, much can be accomplished in the control of some crop pests by cultivating the soil at a certain time of the year or in some special manner. The best method to employ will depend on the life history and habits of the species to be controlled. Deep, thorough, and frequent cultivation of fields infested by the corn root aphid and its attendant ant is the best method of freeing the soil of these insects. Some species of insects that go through a part of their development in the ground can be easily killed if the soil is cultivated while they are in their pupal cells; the plum curculio and certain wireworms are examples. Others may be killed in the same way in the hibernating shelters in which they pass the winter, *e.g.*, the Oriental fruit moth. Some degree of control may be obtained over certain insects by planting infested land to row crops which require frequent cultivation; white grubs and certain flea beetles are examples. Caking of soil usually works a hardship on subterranean insects, and the pale western cutworm and certain thrips never become abundant in caking soil. Tillage may, therefore, favor certain pests and under some conditions should be avoided at particular seasons. Rolling or packing the soil tends to raise the water level and may drive certain subterranean insects above the surface where their natural enemies can get at them.

Infestation may sometimes be entirely prevented, if the ground is kept in a state of clean cultivation during the egg-laying period of some of the crop-infesting insects, as the southern corn rootworm, that will not deposit their eggs on the bare soil. Certain other species, as the pale western cutworm, prefer the bare ground, and with these, cultivation should be avoided until after the eggs have been laid. With some of the soil-infesting insects, such as white grubs, plowing at a certain time in the year will destroy large numbers of the larvae, or aid in their destruction, by exposing them to birds and other animals that feed upon them; while plowing at other times will be of no value in reducing their numbers.

Destruction of Crop Residues, Weeds, and Trash.—The destruction of crop residues is often of great importance in insect control. In some sections of North America where the European corn borer is well established, it has become necessary to practice rotations and cultural methods that permit the utilizing, plowing under, or destruction by burning during the fall, winter or early spring, of all crop residues and weeds remaining in the fields.   In some of the areas infested by the European corn borer such a cleanup of all corn refuse has been made compulsory.   Insects often have a much longer season of activity than the annual crops they attack.   They are often supported by, and increase their numbers upon, weeds and volunteer plants growing earlier in spring and later in fall than the planted crop.   Much can be accomplished in the control of flea beetles, common stalk borer, corn root aphid, the green bug, hornworms, southern corn rootworms, and many others by eliminating weeds, especially those closely related to the planted crop, from the field and field margins.   During the winter many crop insects hide under surface trash such as boards, boxes, sacks, brush heaps, stone piles, dense grass, fallen leaves, and other dead vegetation.   Many codling moths winter in such shelter, asparagus beetles often gather in great numbers in cracks of wooden posts, and squash bugs are frequently plentiful in board piles. Neat husbandry, especially over the winter, reduces the insect population that will have to be fought the following season.

Variations in the Time of Planting and Harvesting.—The time of planting a crop has a very great influence on the infestation of the crop by some insects.   By changing or carefully selecting the time when a crop is planted, we may avoid the egg-laying period of a particular pest; get young plants well established before the attack comes; allow a shorter period of susceptibility during which the insect will attack, as in the case of the seed-corn maggot in a cool, wet spring; or even get a crop matured before a certain pest becomes abundant, as with the cotton boll weevil or an early radish crop and the cabbage root maggot.   Early-planted corn will largely escape injury by the corn earworm in most sections of the country.   On the other hand, early-planted corn may be heavily infested by the southern corn rootworm or the European corn borer. In places where both kinds of these insects are present, the best time of planting corn will depend to a considerable extent on which of these insects is the more destructive.

There is no better example of the importance of farm practices in insect control than the effect on the infestation by the Hessian fly of early and late seeding of wheat.   During most seasons early-sown wheat will be moderately to heavily infested, and medium-late-sown wheat will not be seriously infested.   Indeed, there are many years when a difference of a few days in the time of seeding will make the difference between a good crop and a very poor one, all because of the difference in the amount of infestation by the Hessian fly (see page 404).

The following table shows the average yields of wheat and the average infestation by Hessian fly obtained in an 8-year experiment from fields at eight different points in Illinois.   The first column shows the average yield obtained from all seedings made before the normal safe-seeding date; the second column the average yield from approximately the same number of seedings made after the normal safe-seeding date.   The third and fourth columns show the average percentage of plants which were infested by Hessian fly, when examined on Nov. 10 each year; the third column giving the infestation in all seedings made before the safe-seeding date for the particular locality; and the fourth column the infestation in all seedings made after the safe-seeding date.   In each locality each year there were from three to five seedings at intervals of about 5 days before the normal safe-seeding date and a corresponding number at similar intervals after that date. It will be noted that there was an average gain of 5.8 bushels per acre attributable solely to the difference in date of planting.

TABLE XVI.—AVERAGE YIELDS OF WHEAT AND PERCENTAGE OF INFESTATION BY THE HESSIAN FLY FOR EIGHT YEARS

	Average yield		Average per cent of infestation	
Location of field	From wheat sown *before* the safe-seeding date, bushels	From wheat sown *after* the safe-seeding date, bushels	In wheat sown *before* the safe-seeding date, per cent	In wheat sown *after* the safe-seeding date, per cent
Rockford, Ill.....	21.8	28.1	24.5	1.7
Bureau, Ill......	27.4	32.9	45.5	5.2
La Harpe, Ill....	30.8	36.5	38.0	1.8
Urbana, Ill......	29.5	37.1	32.6	5.4
Virden, Ill.......	23.6	28.4	48.0	6.3
Centralia, Ill.....	14.5	21.9	81.0	8.0
Carbondale, Ill...	21.5	23.9	16.0	1.0
Grand Chain, Ill.	15.5	21.4	32.3	1.0
Average.......	23.1	28.8	39.7	3.8

With crops of indeterminate growth such as clover, alfalfa, strawberries, and the like it should be possible to do much to destroy populations of such insects as the clover seed midge and chalcid and the clover head caterpillar by clipping or harvesting the crop at carefully chosen times before a particular brood of the pests has completed that part of their development which is dependent upon the growing crop. But recommendations must be worked out for each locality and with reference to conditions in different years.

The Use of Resistant Varieties.—Some strains or varieties of cultivated plants are more or less resistant to certain of the insects that attack them. It is only by taking advantage of this resistance of the American grape roots to the grape phylloxera, which is native to North America, that we are able to grow European grapes in this country or in most of the large vineyard areas of Europe. By grafting these varieties upon American rootstocks, the injury is avoided. There is some very good evidence to show that a marked difference exists in the resistance of different varieties of grain to attacks by insects. Some varieties of corn and sorghums have been found to resist the attacks of chinch bugs. Extensive research along this line suggests the possibility that resistant varieties of corn may help to solve the problem of European corn borer control. Resistance or tolerance of plants to insect attack is due to a great variety of conditions, either physical, chemical, or physiological, and in most cases it is not clearly understood. Northern Spy and other varieties of apple have so much hard tissue (sclerenchyma) in the circumference of the roots that woolly apple aphids cannot penetrate it with their mouth stylets. Roughness of surface or hairiness of some varieties of soybeans, cotton, and red clover gives them a high degree of immunity from attack by tiny leafhoppers which are very destructive to smooth-leaved varieties. Certain kinds of cattle have skins so thick and tough that cattle ticks and horn flies are much less troublesome. Others have a very oily, thin coat of hair unfavorable to lice. The rind of many citrus fruits, while green, contains an oil which kills the young maggots of the Mediterranean fruit fly. Pistillate varieties of strawberries and figs are not attacked by the strawberry weevil and the fig wasp, respectively. Acidity or distastefulness of sap, thickness of husks, vigor, early maturity, and unusual recuperative ability are other qualities which may give plants and animals a valuable degree of tolerance to insect attack. By hybridization, grafting, and pureline selection, the desirable resistance factors may be combined or intensified in crops or animals. The work now in progress with hybrid corn is adding greatly to our knowledge of resistance to insects. Special strains of corn, resistant to certain insect pests, are being made to order, for areas where these pests are usually abundant. This field offers possibilities that warrant much careful investigation.

Other Cultural Control Measures.—Finally, the importance of good husbandry, as defined by agronomists, horticulturists, entomologists, and plant pathologists, working in cooperation, cannot be overstressed. The use of good seed, excellent preparation of seedbeds, conservation and regulation of soil moisture, proper pruning and thinning when necessary, and the judicious use of fertilizers, all offer possibilities of stimulating plant growth in such way as to make possible the growing of profitable crops, where the neglect of one or more of these factors may result in loss or disaster.

Biological Control: The Introduction and Encouragement of Natural Enemies

Among the many adverse factors which continually affect every insect species in the struggle for existence are the other living things that feed upon it. These are collectively known as its natural enemies. The fact that man has, during the last few centuries, learned something about the habits, ecology, and interrelations of insects now enables him to take sides in the constant warfare that insects are carrying on against each other. The facts at hand concerning the insect population of a given area of the earth show that more than two-thirds of the species present are feeders on plants, or plant products, or in other words, are competing with man for the products of the soil. From one-fourth to one-third of the insects present in any given area feed on other insects, and many of these are of great benefit to man in reducing the plant-feeding species. Others feed on those that attack the plant feeders, and so become the enemies of man. This complicated relationship of man and insects has already been discussed in Chapter II.

As long ago as the seventeenth century man first conceived the idea of taking advantage of the food preferences of these natural enemies of insects to destroy or suppress pest species. *Biological control* may be defined as the destruction or suppression of undesirable insects, other animals, or plants, by the introduction, encouragement, or artificial increase of their natural enemies.

Among the natural enemies of insects which may be used in this way are

1. Predaceous and parasitic insects (see pages 54–64).
2. Predatory vertebrates.
3. Nematode parasites.
4. Protozoan diseases.
5. Parasitic fungi.
6. Bacterial diseases.
7. Virus diseases.

Of the many possible applications of biological control only two seem to have been used with success, and the entomologist is vitally interested in both of them. *First,* the control in this way of insect pests of growing crops, living animals, or stored products. The plant- or animal-feeding insect is the pest; the attacking organism may be of any one of the groups listed as natural enemies above. *Secondly,* the control of weeds or plant pests by the insects that feed upon them. In this case the insect is the attacking organism, a benefactor, not a pest. Four principal methods have been employed: (*a*) Collecting parasites or predators in places where they have naturally developed or assembled in great numbers and releasing them (perhaps àfter storing them over winter), in places where they may do the most good: either concentrating the beneficial organism on a small area or dispersing it more widely from a center of great abundance. (*b*) Collecting and storing or handling the host insects in such a way as to kill them but permit any parasites or predators among them to escape. (*c*) Rearing under favorable conditions, great numbers of parasites or predators and releasing them, whenever and wherever needed, especially at the time when the normal fluctuations of the pest insect have reached their point of greatest abundance. (*d*) Importing parasites, predators, or diseases from a foreign country. The last-named has been the method most extensively used. It has been used especially in cases where the damage is being caused by a species of insect that was unintentionally imported, whose natural enemies have not been brought with it to that part of. the earth where it has become established. The beneficial organism is usually sought in the original home of the pest. There is no hope of *exterminating* insects by this method, and it is seldom possible so to reduce them that no other control measures need be used. When any parasite becomes so abundant that it nearly wipes out a pest insect, it too must suffer a marked decline in numbers because of the reduction of its food supply. A scarcity of parasites permits the pest insect again to increase until the parasite once more overtakes it. Thus a single active parasite of a pest species will tend to cause more or less regular periods of abundance and scarcity of the parasite and its host. Consequently it is often of advantage if a parasite can be introduced which will find several hosts in the region where it is liberated. During the periods of the pest abundance, it will be necessary to depend on artificial measures for controlling the plant-feeding insect. However, if biological control results in a pronounced reduction in the abundance of a pest, so as to decrease the amount of damage it would have done or the extent of other control measures required, it should be considered a success. This is the only type of control that is self-perpetuating. Since it may be expected to continue indefinitely into the future, a tremendous initial expense may prove to be a very low total cost. Clausen[1] stated that at least 24 pest species have been adequately controlled in one or more countries by the biological method. Sweetman[2] asserts that in a dozen of these cases no other control measures are now required. There are many other examples where partial control has been achieved or where the final result is still in doubt.

Parasitic and Predaceous Insects.—The classical example in this type of biological control was the introduction of the Australian lady beetle or vedalia (*Rodolia cardinalis* Muls.) into California to destroy the cottony cushion scale (*Icerya purchasi* Maskell). The cottony cushion scale was unwittingly introduced into California about 1868. It soon became a most serious pest of the orange, spreading rapidly over the state, and by 1890 had killed hundreds of thousands of trees, threatening to wipe out the orange industry over the entire state. It was traced to Australia and New Zealand; in New Zealand it was very destructive, also, but in Australia little injury resulted from it. Accordingly the United States government sent an entomologist to Australia to search for the natural enemies that it was felt must be holding it in check there. The lady beetle was found. About 140 of them were carefully shipped to California. They were turned loose on screened orange trees and allowed to feed on the cottony cushion scale. Within a year and a half the progeny of these few beetles had increased to such numbers that they had checked the cottony cushion scale over the whole state. They have since nearly eliminated this worst of orange insects as a pest in California. Since that time this lady beetle has been shipped to 40 different countries in at least 32 of which it has become established (see Clausen, *l. c.*). It has never failed to bring the cottony cushion scale under control, wherever it has been established. A few of the other outstanding examples of success with imported entomophagous insects are the control of the woolly apple aphid (page 605) by a minute internal wasp parasite,[3] which is native to the United States and which has been established in 28 different countries; the control of a mealy bug,[4] affecting avocado, fig, guava, and cocoanut, by a minute chalcid wasp[5] in the Hawaiian Islands; the control of the sugarcane leafhopper[6] in the Hawaiian Islands, by a predaceous bug[7] that devours the eggs of the leafhopper; the control of the cane borer weevil[8] by a parasitic tachinid fly[9] in the Hawaiian Islands. Of the 24 most successful cases of biological control of pest insects, about 75 per cent have been by parasites and 25 per cent by predators. More than 50 per cent have been used against Homoptera, especially scale insects. More than 75 per cent of the cases have been achieved on islands where a temperate climate prevails. And the most

[1] *Ann. Ento. Soc. Amer.*, **29**: 201–223, 1936.
[2] "Biological Control of Insects," Comstock Publishing Company, Inc., 1936.
[3] *Aphelinus mali* (Haldeman), Order Hymenoptera, Family Aphelinidae.
[4] *Pseudococcus nipae* (Maskell).
[5] *Pseudoaphycus utilis* Timberlake, Order Hymenoptera, Family Encyrtidae.
[6] *Perkinsiella saccharicida* Kirkaldy, Order Homoptera, Family Cicadellidae.
[7] *Cyrtorhinus mundulus* (Bredd.), Order Hemiptera, Family Miridae.
[8] *Rhabdocnemis obscura* (Boisduval), Order Coleoptera.
[9] *Ceromasia sphenophori* Vill., Order Diptera, Family Tachinidae.

successful cases have resulted from the use of a single natural enemy, although in a number of somewhat successful projects in the United States, the attempt has been to introduce all possible natural enemies of the pest species, with the hope that the combined attack of many predators and parasites upon all stages of the pest and at all seasons would keep its numbers down to an abundance of little importance. In the United States, in addition to the cottony cushion scale, a vast program of biological control has been attempted against the gypsy moth (page 701), brown-tail moth (page 704), European corn borer (page 360), Japanese beetle (page 621), Oriental fruit moth (page 636), alfalfa weevil (page 420), and others. In each case exhaustive studies of the natural enemies have been made in the countries from which the pest was imported and literally millions of these harmless insects have been released on American soil to wage war against the pests. During the past 50 years about 75 foreign parasites and predators have been established in the United States, 95 in the Hawaiian Islands, 18 in the Fiji Islands, and smaller numbers in many other areas.

The most elaborate attempts to bring about biological control, by rearing great numbers of a native entomophagous insect and releasing them, have been with the minute egg parasites called Trichogramma.[1] By rearing another insect such as the Angoumois grain moth, in cages, great numbers of its eggs are provided as food for the Trichogramma and hundreds of millions of the tiny egg parasites have been so produced, for release in greenhouses, orchards, and gardens. At times it has been possible to place orders for a million Trichogramma at 15 to 20 cents a thousand from a commercial entomologist and have them delivered by air mail within a few days. In introducing a parasite into a locality where it has not been previously known to occur, great care must be taken. It must be ascertained without question (1) that the insect to be introduced is a parasite on the particular insect it is desired to control, or on other plant-feeding species; (2) that it is never by any chance a plant feeder; and, (3) that it will not attack some of the other primary parasites already present in the locality and so do more harm than good. These, and many other points, must be carefully investigated before the introduction of the parasite is attempted. The establishment of the parasite is often attended with great difficulty, especially if it has to be brought from a distant part of the world, or from the southern to the northern hemisphere. A very high degree of skill in the handling of parasites has been developed by some entomologists in the laboratories where this work is under way. The introduction of parasites has been one of the significant contributions of entomology to the advancement of agriculture.

Predatory Vertebrates.—Among the vertebrates that are predatory upon insects, the birds are doubtless most effective, because their ecological activities are nearly identical with those of insects. More than half of the food of this most abundant and most mobile group of terrestrial vertebrates is insects. They stand supreme among the vertebrate enemies of insects. Much has been done by man to attract them about his habitations and encourage their abundance, but there have been no spectacular cases of the transportation of birds from one country to another to combat insects (see also page 53). Fishes of certain species

[1] *Trichogramma evanescens* Westwood and *T. minutum* Ashmead, Order Hymenoptera, Family Trichogrammatidae.

have been employed to destroy mosquito larvae cheaply and effectively. The giant toad of Mexico, Central America, and South America has been introduced into the Hawaiian Islands, the Philippines, and West Indies and is said to have brought under control the white grubs destructive to sugarcane in Puerto Rico.

Nematode Parasites.—Many species of the phylum Nemathelminthes (page 158) are parasitic in the bodies of beetles, grasshoppers, cockroaches, moths, and other insects. Species attacking the striped cucumber beetle and the Japanese beetle are considered important in the control of these pests, and some effort has been made to use the latter[1] in biological control. Much more study of this subject is needed.

The Use of Insect Diseases.—Insects, like other animals, suffer from the attacks of diseases. At times, under favorable conditions, a disease may become epidemic on a species of insect and within a few days or weeks reduce the species from a point of great abundance to one of scarcity. Insect diseases may be caused by protozoa, fungi, bacteria, or viruses. All of these minute organisms live abundantly on and in the bodies of insects. Among such diseases, which have been employed in biological control, are the brown,[2] red,[3] and yellow[4] fungi that live on the bodies of whiteflies which attack citrus in Florida. The spores of these fungi are sometimes mixed with water and sprayed over trees infested with the whiteflies, much as a chemical spray would be applied. The Empusa diseases of grasshoppers, house flies, and other pests; *Beauveria* (= *Sporotrichum*) *globuliferum* attacking chinch bugs; and Metarrhizium and Cordyceps on white grubs kill millions of their hosts, in certain areas and certain seasons, but their utilization in biological control has not yet been demonstrated as practicable. The protozoan, *Coccobacillus acridiorum*, has been used against grasshoppers in parts of Africa; *Bacterium thuringiensis* and other species against the European corn borer in Europe and the pink bollworm in Egypt; and *Perezia pyraustae* against the European corn borer in the United States. Anyone who has seen an epidemic of wilt disease or polyhedral disease among gypsy moth or cabbage looper larvae or sac brood among honeybees need not be told that they are terrible and virulent diseases. They are easily contracted by eating contaminated food and many efforts have been made to spread such infections by spraying trees or crop plants with solutions containing crushed, diseased larvae, but it appears that environmental factors hold the whip hand and no extensive epidemics have been produced artificially. The spread of an insect disease is so dependent on weather conditions, abundance and habits of the host insect, and other factors not controllable by man that little progress has been made in the artificial spread of diseases for controlling insects. The facts now known, however, warrant further experimental work along this line.

Biological Control of Weeds.—The classical example of biological control of weeds concerns the fight against the cacti in Australia. About 1840 a doctor immigrating to Australia carried with him a single potted plant of the prickly pear, *Opuntia inermis*. This plant was a curiosity in that country and cuttings from it were spread far and wide. They

---

[1] *Neoplectana glaseri.*
[2] *Aegerita webberi.*
[3] *Aschersonia aleyrodis.*
[4] *Aschersonia goldiana.*

grew and thrived beyond belief, and within 30 years it was realized that the innocent curio had developed into a terrible weed pest. Whereas in America this cactus grows to heights of 6 to 10 inches; under Australian conditions it reaches a height of 6 to 10 feet, with prickly branches so dense that no one can penetrate through it. It spread rapidly over farms and grazing land, crowding out and smothering all other crops. By 1910 it had claimed 10 million acres. By 1916, 23 million acres were overrun with it, and it was spreading at the rate of about 1 million acres a year. Eventually 50 to 60 million acres were rendered absolutely useless by the prickly pear—a gigantic jungle, made impenetrable by the riotous growth of spiny cactus. Every conceivable method of destruction was tried. Mechanical cutters and rollers and poison sprays and gases proved to be either inefficient or too costly. In 1913 the Australian government undertook to control this weed by the use of its insect enemies, introduced from Texas, Mexico, India, Ceylon, Uruguay, and South Africa where the cactus was native. Of the many insects introduced, the most promising is a moth borer, *Cactoblastis cactorum*. From the original importation in 1925 many thousand millions of caterpillars have developed. With as many as a million per acre eating the interior of the plant and opening it to further destruction by rots, the cactus is soon reduced to dry skin and fiber. Other enemies of the plant including certain kinds of mealy bugs, true bugs, and red spiders have contributed to the good work; and it is hoped that the entire infested area may be eventually reclaimed by the careful manipulation of these insect benefactors. In every case elaborate tests are made to insure that the insects so introduced will feed on nothing but cactus. Many other attempts have been made in various countries to control plant pests with their insect enemies and with sufficient success to warrant the belief that under certain conditions plant eating insects may be among man's greatest allies.

LEGISLATION FOR INSECT CONTROL[1]

In the early days of agricultural development in this and other countries, plants and plant products were brought into, or sent out of, the country with little or no thought concerning the insect pests that might be transported along with them. In fact, it is only since the middle of the past century that any serious attempt at legislation to restrict the spread of insect pests has been attempted by any country. The introduction of the grape phylloxera from America to the vineyards of France, some time about 1860, caused such serious destruction to the French vineyards and to those in other European countries to which it was carried that it became apparent something must be done to prevent the unrestricted movement of infested vines to all parts of the world. In 1881 representatives of many of the European countries in which grapes were extensively grown met and agreed on regulations restricting the movement of infested grape stalks. The spread of certain plant pests in the United States in the latter part of the past century, particularly the San Jose scale, stimulated the passage of insect legislation in this country. By the close of the century nearly every state had passed laws restricting the shipment of infested nursery stock. Although an act was passed

[1] See STRONG, LEE A., "Legislation in the United States for the Control of Insects and Other Plant Pests," *U.S.D.A., Bur. Ento. Plant Quar., Cir.* E-455, 1936.

by the federal government in 1905 "to prohibit importation or interstate transportation of insect pests," it was not until 1912 that the United States had adequate federal laws to control the menace of foreign plant pests.

At present, there are four classes of insect legislation: (a) legislation to prevent the introduction of new pests from foreign countries; (b) legislation to prevent the spread of established pests within the country, or within the state; (c) legislation to enforce the application of control measures that have been found effective in preventing damage by established pests; and (d) legislation to prevent the adulteration and misbranding of insecticides.

Quarantine and Inspection Laws.—The Federal Horticultural Board, composed of representatives of the federal Bureaus of Plant Industry and Entomology and the Forest Service, was created by the Plant Quarantine Act of 1912 and had charge of federal regulatory measures dealing with insects and plant diseases until July 1, 1928, when these functions were transferred to the Plant Quarantine and Control Administration, and later (1932) to the Bureau of Plant Quarantine, which on July 1, 1934, was merged with the Bureau of Entomology to form the Bureau of Entomology and Plant Quarantine. This bureau now has charge of the administration of measures which provide: (a) for the regulation by inspection or treatment, or prohibition by quarantine, of movements of nursery stock from foreign countries; (b) the examination at ports of entry of all restricted plant materials, ships, passengers, baggage, airplanes, and foreign mail packages; (c) in cooperation with the several states, the administration of domestic quarantines of a part of the United States or a part of a state against the interstate or intrastate movement of any article or commodity which is likely to cause wider dissemination of a plant pest new to, or not widely distributed within, the United States; (d) the inspection of, and certification that, domestic nursery stock, fresh fruits, vegetables, seeds, and other plant products exported for propagation are apparently free from dangerous insects and plant diseases, so as to meet the requirements of foreign countries; (e) the control or eradication of specific insect pests and plant diseases which are not widely or generally distributed, such as the pink bollworm, the European corn borer, or the Mediterranean fruit fly; or which threaten outbreaks of such widespread destructiveness as to constitute a general menace, such as chinch bugs and grasshoppers. Every state in the union now has in force regulatory measures forbidding the movements of certain plants into, or within the state, at least until such plants have been inspected or fumigated, or both; and in many cases reinspection is required at the point of destination. It is illegal to ship nursery stock anywhere in the United States unless it is accompanied by a certificate of inspection, stating that it has been found apparently free from certain seriously destructive insect pests and plant diseases.

Nursery stock may be defined in a general way as including: (a) all woody plants and parts thereof and (b) florists' stock and plant products for propagation which are customarily grown in the out-of-doors for at least a part of the year. It, therefore, includes all fruit, nut, forest, and shade trees, either deciduous or evergreen; all shrubs, vines, cuttings, grafts, scions, buds, and seedlings; fruit pits and seeds of trees and shrubs; and herbaceous plants grown outside of greenhouses. It does not include soft, succulent plants that die down to, or including, the roots in winter; bulbs, roots, and bedding plants grown indoors; or seeds of field crops, vegetables, or flowers. However, Federal Plant Quarantine 37 included the plant products just mentioned unless specifically excepted. Field, vegetable, and flower seeds; and fruits, vegetables, cereals, and other plant products *imported for food, medicine,* or *manufacturing purposes* were at once excepted. That is, no permit is required to import the last-named products unless a special quarantine is in effect against them.

The Plant Quarantine Act of 1912, also provided that whenever it seems necessary, in order to prevent the introduction or spread of any dangerous insect or plant disease, the federal government has the power, after a public hearing, to prohibit the importation or shipment, interstate, of any class of plants or plant products from any foreign country or locality and from any state or portion of a state or territory in this country. Such specific prohibitions are called *quarantines.* Many of the states, also, have quarantine laws.

There are thus in the eyes of the law three great classes of plant materials:

1. Materials which can be brought in without any restriction. This includes the two groups mentioned near the end of the second paragraph above.

2. Those things which can be imported by any responsible person, but only under federal permit.[1]

3. Materials which cannot be imported at all except under special permit in limited quantities for experimental purposes by the United States Department of Agriculture.

At present foreign plant quarantines are in force restricting the importation of certain fruits and vegetables; cotton lint and cottonseed products; certain cereals, plants, and plant products for propagation or as packing materials. To indicate the need for such legislation it should be pointed out that during a 7-year period federal authorities intercepted:

From Germany, 12 infested shipments containing 15 kinds of insect pests.
From England, 154 infested shipments containing 62 kinds of insect pests.
From Japan, 291 infested shipments containing 108 kinds of insect pests.
From France, 347 infested shipments containing 89 kinds of insect pests.
From Holland, 1,051 infested shipments containing 148 kinds of insect pests.
From Belgium, 1,306 infested shipments containing 64 kinds of insect pests.

At present domestic plant quarantines are in force against the unrestricted movement of certain commodities from the known infested areas on account of the gypsy moth, the Japanese beetle, the pink bollworm of cotton, the Mexican fruit fly, the Dutch elm disease, the white pine blister rust, and the black stem rust of grains; and certain state quarantines against the European corn borer, the alfalfa weevil, the phony peach disease, and others are in force.

In general, it may be said of quarantine measures that the enforcement of such measures will check the spread of certain insects, but cannot be depended upon to stop such spread. A considerable expense is justified if the quarantine checks the spread of a new pest long enough

[1] See "Service and Regulatory Announcements," *U.S.D.A., Bur. Ento. Plant Quar., Cirs.* 126, 1936, and 384, Oct. 19, 1935.

to permit the development of control measures, the introduction of parasites, or changes in agricultural practices best suited to prevent loss by the newly established pest. It should be borne in mind, however, that regulatory measures are always expensive measures and that it is impossible to keep a strong-flying insect out of adjacent territory from that at present occupied by it, merely by passing laws. It is certainly desirable that all reasonable restrictions be placed on the importation of new insect pests from foreign countries and that every precaution possible be taken to inspect plants and plant products entering this country, to see that they are free from foreign insects.

The history of an insect in its native country cannot always be relied on as a criterion of what the insect will do when established in some locality in another part of the world, where it is not held in check by the natural enemies found in its native home. For this reason it is best from a legal standpoint to consider that all foreign plant-feeding insects and zoophagous parasites are dangerous, until they have been proved otherwise. At the present time practically all civilized nations have in effect regulatory measures restricting the movement of plant products from other countries, and it is probable that these measures will become more strict in the future, rather than more lenient. Unfortunately, there have been a few attempts to use quarantines as trade barriers. It is hardly necessary to state that this should never be tolerated.

Compulsory Clean-up Measures.—While the regulations governing the control of pests in nurseries are, in most states, broad enough to allow for the enforcement of clean-up measures against orchard or field-crop insects, as well, it is only in a comparatively few states that such measures are generally enforced. Indeed, it usually requires serious loss of property or personal injury before the public will wholeheartedly back up such measures. Some of the Eastern states, notably Massachusetts, have rigidly enforced the control of such pests as the gypsy and brown-tail moths. If premises are infested in some states, the property owner is given legal notice that he must, before a given date, take measures to control certain insect pests. If such measures have not been taken before this date, the work is done by a force of men employed by the city or town in which the property is located. The cost of this work is assessed against the property in the form of a tax. If unpaid, it constitutes a lien against the property as much as any other unpaid tax and is collected in the same way. This measure has had a thorough test in the courts, and has now been in force for a number of years.

Insecticide Laws.—Aside from the legislation having to do directly with combating insect pests or controlling their spread, the federal government and many of the states have laws standardizing the grades of insecticides and fungicides which are sold within the state. The Federal Insecticide Act of 1910 makes it illegal to manufacture, sell, transport interstate, import, or export insecticides or fungicides which are adulterated or misbranded. An insecticide or fungicide is considered *adulterated* if its strength or purity fall below the professed standard under which it is sold, if any substance has been substituted wholly or in part for the substance named, if any valuable constituent has been abstracted, or, if intended for use on vegetation, it should contain any substance injurious to such vegetation. An insecticide or fungicide is considered *misbranded* if the label or circular accompanying the package is false or misleading in any particular regarding the weight or measure of contents, the place of manufacture, or the ingredients contained. For example, a substance offered for sale as an insecticide must bear a label stating the amount of arsenic it contains and what percentage of the arsenic is water soluble; and either the name and percentage of each active ingredient and the total percentage of inert substances, or the name and percentage of each inert substance. The U. S. Department of Agriculture through the Food, Drug and Insecticide Administration collects samples, examines them, analyzes them chemically, and tests them by actual application to the pests they are recommended to control. If they are found to be misbranded, if they fail to produce the results claimed, or if they are injurious to plants or animals upon which they are recommended to be used, this administration, after a hearing to determine that the law has indeed been violated, may cause the manufacturer, dealer, or importer, to be tried and fined or imprisoned or both. This administration also is charged with enforcing the poison residue regulations (see page 243).

B. NATURAL CONTROL

Natural control includes control (a) by climatic factors such as rainfall, sunshine, cold, heat, and wind movement; (b) by the physical character of the country, such as large bodies of water, mountain ranges, streams, the character of the stream flow, and the type of soil; (c) by the *natural* presence and abundance of predaceous and parasitic insects, birds, fishes, reptiles, and mammals and by cannibalism; and (d) by the presence of diseases which attack insects and conditions favorable to the spread of such diseases.

CONTROL BY CLIMATIC FACTORS

Under natural control, climatic factors are perhaps the most important. A few species of insects have become adapted to variations in climate to such an extent that they occur throughout the world, in temperate and, in some cases, in temperate and tropical zones. Few, if any, species of insects occur in all three zones—the arctic, temperate, and tropical—except such species as infest stored products, the dwellings of man, or the bodies of animals, which are therefore not subjected to a very marked degree to the climatic changes of any region.

As a general statement, it may be said that the insect life of a region is dependent, directly or indirectly, on the temperature of the region, the soil, and the amount of moisture. There are many species of insects which have become adapted to certain climatic conditions and thrive under these conditions even though they may seem at first to be unfavorable. Great numbers of mosquitoes and certain species of flies occur in the arctic regions during the brief summers. At the height of warm weather, the total number of insects in such regions is very large, but never, on the whole, as large as the numbers found in the tropics. Winter temperatures control the distribution of many insects; the harlequin cabbage bug, which cannot survive our northern winters, is an example.

A very warm, moderately humid climate and fertile soil offer conditions favorable for the greatest development of insect life. A poor soil can support only a limited amount of plant growth and, therefore, a limited insect population. A warm, wet climate is unfavorable to many insects. Such a climate creates a condition where insect diseases will flourish and also presents many physical factors unfavorable to insect life. A hot and very dry climate also is unfavorable, only a comparatively few species of insects having become adapted to life under desert conditions.

The amount of sunshine occurring in a given region also is important. Many species of insects are influenced to a marked degree by the rays of the sun. Some apparently seldom fly except during periods when the sun is shining; and, as flight is the chief means of dissemination of most species over any area, this is an important factor influencing the general abundance of insects; the chinch bug is one of the best examples.

Wind movement is also of great importance. Many of the smaller and frailer species of insects which normally fly for considerable distances are unable to leave the ground during strong winds or, if they do take flight, are so buffeted and beaten by the wind that they soon die; the Hessian fly is thus affected. Some species of insects, as certain mosquitoes, habitually fly against the wind while many others fly with the wind; their dispersal, or the direction of their dispersal, is to an even greater extent dependent upon wind movement at the time of year when they are in their winged or adult stage.

The brown-tail moth has spread very slowly in a westerly direction from the original point of establishment, but its spread to the east and north has been rapid, due chiefly to southwesterly winds during the time when the adults are flying.

Control by Topographic Factors

Large bodies of water, such as oceans, offer effective barriers to the natural spread of nearly all species of insects. Certain species that do not possess the power of flight are affected in their spread by smaller bodies of water, such as lakes or large streams, and are largely dependent upon man or other animals for the passage of such barriers. The granary weevil is unable to fly but has been carried by man over much of the earth. Mountain ranges also are effective barriers to the spread of insects and offer varying conditions of climate through which many insects cannot pass unaided. The Colorado potato beetle began spreading from Colorado to the East about 1859 and reached the Atlantic Coast in 15 years, while it was more than 50 years in crossing the Rocky Mountains. The character of the streams and the number of ponds and lakes control to a great extent the insect life of a country. Certain flies, mosquitoes, and beetles live in their immature stages in slowly moving streams or still water, while others, such as black flies and certain caddice-flies, live only in swift-flowing streams.

The character of the soil of any region exerts a marked influence over the insect inhabitants of that region. This is true not only because the soil has a direct influence on the plant growth of the region and thus indirectly affects the insects that live on plants, but also because many insects spend the whole or a part of their life in the soil. Certain soils are very favorable to their growth, while they would be unable to live in different soils, or in the same type of soil under different conditions. Certain species of wireworms live only in poorly drained soils, and certain species of tiger beetle larvae, which live in sandy soils, are unable to exist in clay soils.

CONTROL BY NATURAL ENEMIES

Predaceous and Parasitic Insects.—Of the natural factors that tend to reduce the plant-feeding insects, the number of the insects feeding on other insects in a locality is sometimes as important as the climatic factors. It has been demonstrated many times that a plant-feeding insect removed to a part of the world where its insect enemies are not present, and with suitable food plants present, is able to increase to far greater numbers than was the case in its native home. Indeed it would probably be very difficult to produce crops in most regions of the earth, if the predaceous and parasitic insects were not present to keep down the plant-feeding species.

Weiss, in classifying the insects of New Jersey according to their food habits, found that 28 per cent of the approximately 10,000 kinds of insects known to occur in that state were feeders on other insects.

Cannibalism, the devouring of individuals by their own kind, is an important factor in reducing the numbers of certain insects. A conspicuous example is the larvae of the corn earworm.

Birds.—From an insect standpoint a birdless country would be a highly desirable place in which to live, as in such a country they would be safe from the attacks of many of their most persistent enemies. A proportion of the food of most birds is made up of insects, and the food of many species is largely of insect origin. The actual number of insects eaten in a day by certain birds is surprising, in some cases being almost or quite equal to the weight of the bird itself. This is especially true of the nestlings during the most rapid period of their growth. Thanks to the studies which have been carried on by the United States Bureau of Biological Survey, we now have sufficient data on the food of nearly all the common species of birds to enable us to know their value as insect destroyers.[1] Some of the common species, such as robins and catbirds, eat insects mainly during the summer when this kind of food is most abundant. Others such as some of the woodpeckers and creepers subsist largely on an insect diet throughout the year. Even those species that are largely grain feeders, such as the blackbird and English sparrow, will often congregate in large numbers in areas where insect outbreaks are occurring and will feed mainly on insects during the period when they can be easily obtained (see also page 53).

While birds cannot be expected to become sufficiently abundant in any thickly settled farming area for us to depend upon them alone to prevent insect damage, they are of great value. Most birds earn, many times over, the fruit and berries they take from our orchards and gardens. The value of most birds as insect destroyers alone will warrant all the protection we can afford them. This protection is needed, not only against killing by the use of firearms and nest robbing, but also against the cat, which is the birds' worst enemy in many sections. The number of birds in a local area may be increased by providing food during the winter months and water and suitable nesting materials and places during spring and summer, and by planting seed- and fruit-bearing shrubs. The

---

[1] Henderson, Junius, "The Practical Value of Birds," Macmillan, 1927.

setting aside of tracts of land as game preserves also will aid in increasing the numbers of song birds valuable as insect destroyers.

Mammals and Other Animals That Feed on Insects.—Many of the small mammals feed to a great extent on insects. Some of the ground squirrels eat white grubs and other soil-infesting insects, but these make up only a small part of their food. Moles, shrews, and skunks depend largely on insects for their food and destroy very large numbers of the soil-infesting kinds. Some species of snakes, newts, and salamanders also subsist largely on an insect diet. The toad is one of the most useful of our common small animals, its food consisting almost entirely of insects, more than 60 per cent of which are of injurious species. These are eaten in very great numbers, the toad devouring in 24 hours an amount of insect food equal to about four times its stomach's capacity.

Value of a Knowledge of Natural Control.—As the factors of natural control cannot be greatly influenced by man, it might seem that a knowledge of such factors would not be of much value. The opposite is true, however. Knowing the climate, soil, and topography of a region, we may, to a certain extent, be able to tell the kinds of insects that will be most common in that region, and, with a knowledge of their food-plant preferences, the crops that will be most subject to injury. This is of the greatest value in estimating the amount of injury that may be expected from foreign insects newly established in a country.

This knowledge of natural control is also of great practical value in enabling one to tell the effect of the weather of a season on insects and from this to predict the relative abundance of injurious species the next year. For example, a winter period of very low temperature with no snow will kill most of the eggs of the gypsy moth. Long periods of dry, hot weather, during the summer will prevent the apple maggot from ever becoming a serious pest in areas where such weather is the rule. Heavy rains during the time when the eggs of the chinch bug are hatching will often terminate a period of several years of serious destruction by this insect. In the case of the pale western cutworm, in the prairie provinces of Canada, it has been found possible, according to Seamans, to predict the irregular and very serious outbreaks of this cutworm by noting the number of wet days in the preceding May and June. If there are fewer than 10 days in these 2 months when it is too wet to work in the soil, there will be an increase and probably an outbreak of the cutworm the following spring. If there are more than 15 such wet days in May and June, little trouble may be expected from this insect the following season. The explanation of this correlation between weather and cutworm abundance is as follows: The pale western cutworms work below ground except when the soil is wet. When driven above ground by heavy rains or by irrigation, they are attacked by several kinds of parasites, which reduce their numbers so that no outbreak is possible the following year. If the soil is dry enough so they may remain below ground most of their period of larval activity in May and June, the parasites do not reach them, the cutworms remain healthy and may increase their numbers to epidemic proportions by the following year.

Taking into account what we know of the effect of various natural factors on insect abundance, it is now possible to warn growers of threatening outbreaks of certain of the more carefully studied species of insects in time to apply measures of control. In Illinois, for example, during the past 20 years fruit growers have been informed, about 14 days in advance, the time when the eggs of the codling moth will start hatching. Field checks have shown that the predictions have usually been accurate to within 24 hours and that there has never been an error of more than 3 days. The importance of a knowledge of the effect of these natural factors on insect life is just beginning to be realized, and a more thorough study of them is badly needed.

CHAPTER X

APPARATUS FOR APPLYING INSECTICIDES

Probably the first apparatus for applying insecticides in liquid form was a bundle of twigs or feathers, or a brush broom. For applying dry insecticides, a bellows or blowing tube was used. Such crude apparatus was the only kind in general use up to less than 100 years ago. The development of spraying and dusting machinery has been largely an American achievement, of which we may be justly proud.

The two general methods of applying insecticides are as *sprays*, in which water or oil is the carrier for the poison; and as *dusts*, in which some fine dry powder is the carrier.

SPRAYING MACHINERY

A good sprayer is one that will apply a liquid to the surface of the object treated in such a manner as to give a uniform covering with the

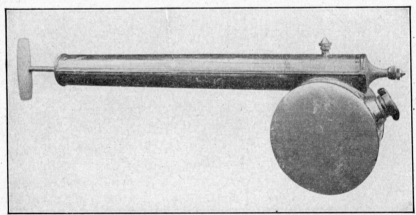

Fig. 174.—A hand atomizer, useful for spraying small plants. (*From U.S.D.A., Farmers' Bul.* 908.)

least expenditure of power, labor, and material possible to produce this result. In addition, a sprayer should have sufficient capacity to cover the objects to be treated in a short space of time. The period of time available for any one application depends on the kind of plants or insects being treated but is not usually over 5 days. The sprayer must be strongly built and of the proper material to withstand wear and the chemical action of the spray mixtures. All working parts should be of noncorrosive metals such as brass, bronze, or stainless steel. It must be so constructed that all parts are readily accessible for cleaning or making repairs. It must be as simple as possible in construction. A supply of spare parts should be available at some near-by point. It is best to

314

buy of a reliable company whose machines have stood the test of practical work.

General Suggestions about Spraying.—No agricultural operation requires greater care and thoroughness than spraying. When stomach poisons are being applied, the aim should be to cover every bit of leaf and fruit surface and succulent buds and stems. If a contact insecticide is being used, it should be remembered that only those insects will be killed that are actually hit by the spray. While thoroughness is required, the application should be stopped as soon as dripping begins.

The application of sprays must ordinarily be very accurately timed to coincide with the period in the development of the insects when they are most easily killed, or with the stage in seasonal development of the plant when it will best withstand the treatment. Since these times will vary widely in different sections of the country, the grower should secure the advice of a trained entomologist, who knows the local conditions, regarding spray programs for different crops and insects. In general, the earlier the spray is given after the insects appear, the easier it is to destroy them.

Spraying should not be done in rainy weather, although arsenical sprays will adhere satisfactorily to plants if the spray has time to dry before rain falls. Winter sprays should not be applied when the temperature is below freezing or when the trees are wet with snow or rain.

Good pressure is essential to the best results in spraying and the spray outfit should be sufficiently powerful to maintain a pressure of at least 125 pounds. For orchard and vineyard spraying, best results are secured at pressures of at least 250 to 500 pounds per square inch.

When using poisons, which do not go into solution, good agitation of the spray material is very important to maintain an even mixture.

The best pump can easily be ruined by failure to give it proper care. Clean water should always be pumped through the sprayer after using, to flush out all the corrosive and abrasive spray materials. Wooden tanks are best stored in damp places. The metal parts should be kept oiled to prevent rust.

Since most spraying materials are violent poisons, they should be plainly labeled and, together with mixing vessels, kept out of reach of children. Animals must be kept away from liquid sprays and not allowed to pasture under sprayed trees. Fruits and leafy vegetables must not be sprayed immediately before marketing, and should always be cleaned of any spray residue whether or not it is considered poisonous.

Kinds of Sprayers.—The size and type of sprayer to be used will depend on the amount and kind of work to be done, but the sprayer should always be large enough. The engine for operating the sprayer should be able to maintain sufficient pressure without strain. Spraying, at best, is a disagreeable task, and the use of a machine which does not have sufficient capacity is very likely to mean that the work will be poorly done.

Some of the more important types of sprayers are (1) hand atomizers; (2) compressed-air sprayers; (3) bucket pumps; (4) wheelbarrow or "estate" sprayers, hand- or motor driven; (5) barrel pumps; (6) double-acting tank pumps; and (7) power sprayers.

For a few plants in a backyard garden or flower bed a small sprayer of the atomizer type (Fig. 174) may be used. This type of sprayer is made in several styles. Those giving a continuous fine spray are to be preferred.

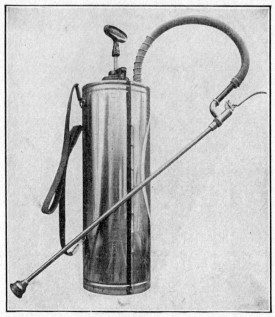

FIG. 175.—A compressed-air sprayer, useful for a few plants or shrubs.   (*From U.S.D.A.,* *Farmers' Bul.* 908.)

FIG. 176.—A bucket pump. (*From the Deming Manufactur-ing Co.*)

FIG. 177.—A wheelbarrow sprayer. (*From the Hardie* *Manufacturing Co.*)

For the average garden a good compressed-air sprayer of from 1- to 3-gallon capacity is fairly satisfactory. This type of sprayer (Fig. 175) should be equipped with a short extension rod and an angle nozzle. Such a sprayer may be used for covering a few shrubs, or bush fruits, but is not suitable for trees.   There are several other types of hand sprayers suitable for use in small gardens such as the bucket pumps (Fig. 176), knapsack sprayers, and wheelbarrow sprayers (Fig. 177).

For large gardens of ½ to 2 acres, or for a small orchard for family use, a barrel sprayer (Fig. 178) with an extension rod at least 8, and not over 12 feet long is fairly effective.   This sprayer should be so constructed that a man can maintain from 125 to 175 pounds pressure, with one discharge pipe in operation.   It should have a good agitator.   The working parts should be of brass or bronze.   The air chamber should be in the

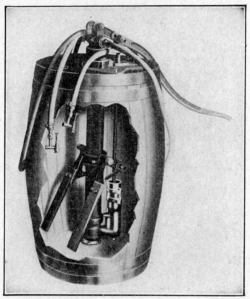

Fig. 178.—A barrel pump suitable for spraying a garden or small orchard.   (*From U.S.D.A.*, *Farmers' Bul*. 908.)

barrel, not projecting much above the top, and sufficiently large to maintain an even pressure.

A tank outfit equipped with a double-acting two-cylinder hand pump (Fig. 179) will be much more satisfactory for an orchard of from 20 to 40 trees.   With this type of sprayer a pressure of from 100 to 125 pounds may be maintained, and, while this is not sufficient for the best results, it will give fairly clean fruit where the spray material is properly mixed and thoroughly applied at the proper time.   A lead of at least 25 feet of hose should be supplied.

For orchards of more than 50 large trees, a power sprayer (Fig. 181) is the only type that can be depended upon to give good results.   There are many styles and makes of well-built power spray outfits on the market. In selecting the one best suited for any particular orchard, the points mentioned in regard to the essentials of a good sprayer should be kept in mind.   The engine should, in all cases, be large enough to give an excess of power for the kind of work to be done.   This means that it should be capable of maintaining a pressure of 250 to 500 pounds.   If extension rods are to be used, one should have at least 2 horsepower to each discharge pipe and, if the spray gun is used, at least 3 horsepower for each gun. Most of the larger-type sprayers are now sold with engines of from 8 to 25 horsepower as standard equipment.

Parts of a Sprayer.—Some of the more important parts of a spray outfit are as follows:

The *cylinders* (Fig. 180) are the part of the pump in which the pressure is developed.   They must be of noncorrosive material.   Other things

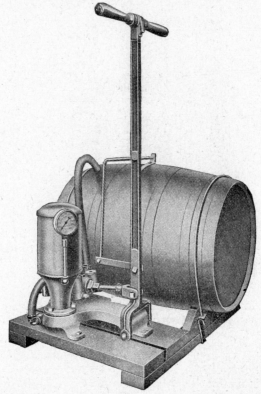

Fig. 179.—A double-acting, horizontal tank pump.   (*From the Hardie Manufacturing Co.*)

being equal, a cylinder of large diameter is to be preferred.   The larger power pumps have 2, 3, or 4 cylinders.

The *plungers* or *pistons* (Fig. 180) force the spray liquid through the cylinders.   The plunger is made to fit tightly inside the cylinder by means of *packing*.   Some arrangement should always be provided for easily tightening or renewing the packing.

Valves, usually of the ball type, and *valve seats* (Fig. 180) direct the flow of the spray.   They are very likely to become clogged with particles of spray materials or dirt and *must* be readily accessible for cleaning.

An *air chamber* (Fig. 180) to equalize the pressure and remove the excessive strain should be provided on every pump.   Its volume should be equal to at least half the capacity of the pump per minute. The discharge opening leading to the nozzles is located at the bottom of the air chamber.

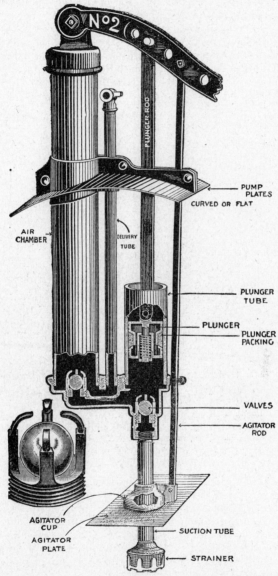

FIG. 180.—Sectional view of a hand pump to show arrangement of the valves, plunger, and packing, and the course of the spray; with the principal parts named. The part labeled plunger tube is better known as the *cylinder*.   (*From the Spramotor Company.*)

The *tank* (Figs. 178 and 179), in which the spray ingredients are mixed and held, is of metal or wood. Brass is resistant to all ordinary spray mixtures.

Since many spray ingredients are not soluble in water, an *agitator* (Fig. 180) is of great importance to keep the finely divided particles evenly distributed in the water.

A *pressure gage* (Fig. 179) and pressure regulator are necessary to show the pressure being maintained and to provide a means of adjusting to the pressure desired.

Strainers, over the intake to the tank, and over the suction or intake pipe leading to the cylinders (Fig. 180), are of great importance in keeping the spray liquid free from troublesome foreign matter.

The *discharge pipe* should be as free as possible from abrupt angles that cut down the pressure between pump and nozzles.

FIG. 181.—Spraying an apple orchard with tractor-drawn sprayer, operated by a power take-off from the tractor. Such an outfit will maintain 500 pounds pressure for 14 nozzles. The caterpillar tractor and large tires permit operations on very soft, wet ground. The spray tower, to enable one operator to cover the tops of the trees, was made from an empty steel drum. (*From John Bean Manufacturing Co.*)

The *hose* (Figs. 177 and 181) for orchard spraying should be at least 25 feet long, of ½- to ¾-inch inside diameter and four- to seven-ply strength. For solid-stream forest-tree spraying, hose of much greater strength and size and sometimes leads ½ mile long are required.

Each lead of hose is provided either with an *extension rod* and *nozzles*, which may number from 1 to 16, or with some type of *spray gun* which may be regulated to discharge various types of spray, from a solid stream carrying to great heights to a fine mist.

Nozzles and Spray Guns.—The actual application of the spray is usually made from an extension rod equipped with from 1 to 16 nozzles of one of several types, or with the spray gun.

The spray gun (Fig. 182) is made of a short, heavy rod about ¾ inch in diameter and 2 to 3 feet long with one or two large nozzles at the tip

The opening in the nozzles is rather large, and the gun throws a large volume of spray. A rotating handle at the base enables the operator to regulate the stream, from a coarse spray that will carry to the tops of the trees, to a fine mist for the lower or near-by branches.

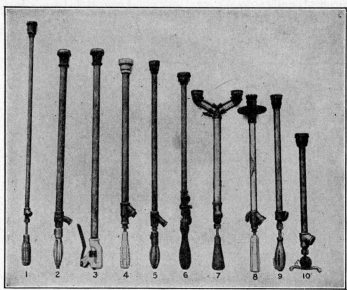

Fig. 182.—Ten different designs of spray guns. 1, Friend Universal; 2, Wardel (new design); 3, Wardelo; 4, Hardie; 5, Bean; 6, Myers; 7, Boyce; 8, Comet; 9, Hayes; 10, Friend. (*From Anderson and Roth, "Insecticides, Fungicides and Appliances," Wiley.*)

According to Anderson and Roth,[1] the following amounts of spray will be discharged from a spray gun with disks having different-sized apertures, and with a pressure of 250 pounds:

Size of Aperture, Inches	Rate of Discharge, Gallons per Minute
$\frac{3}{32}$	5
$\frac{7}{64}$	6
$\frac{1}{8}$	9
$\frac{9}{64}$	12
$\frac{3}{16}$	20

Instead of a spray gun, an extension rod may be used to support the nozzles in spraying trees. Spray rods consist of light metal pipes, from 4 to 12 feet in length, with a cutoff at the base where they are attached to the hose, and with one or more nozzles attached to the tips. The rods are generally made of light brass, enclosed in bamboo. Iron or steel rods are sometimes used but are heavy and hard to hold.

Spray nozzles are of many styles but may be grouped into four general types, the bordeaux (Fig. 184), the vermorel (Fig. 183), the disk type (Fig. 184), and the solid-stream type (Fig. 185). In the bordeaux type the stream is broken by a beveled obstruction, which may be set at any

[1] ANDERSON and ROTH, "Insecticides, Fungicides and Appliances," Wiley, 1923.

angle to the direction of the stream.   This type of nozzle gives a driving
fan-shaped spray of rather coarse spray particles.   It is very wasteful
of spray material, is heavy, and wears rapidly where high pressures are
used.   It will drive the spray with considerable force against the object
to be sprayed, and this may sometimes be an advantage.   This type of
nozzle is now little used.

The vermorel and disk nozzles consist of a cap enclosing a small
chamber, called the *whorl chamber*.   The spray enters this chamber at

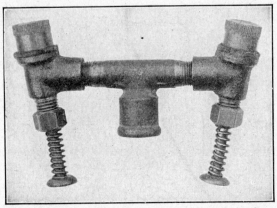

Fig. 183.—A double vermorel nozzle with disgorgers; each nozzle throws a misty cone-
shaped spray.   (*From U.S.D.A., Farmers' Bul.* 908.)

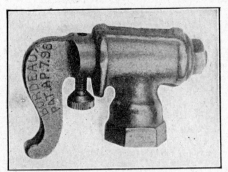 

Fig. 184.—*Left*, a bordeaux nozzle, which throws a flat, fan-shaped, driving spray.
Right, an angle disk nozzle, which throws a cone-shaped spray; of larger capacity than the
vermorel nozzle.   (*From U.S.D.A., Farmers' Bul.* 908.)

an angle through one or more openings and is discharged through a
small aperture in the center of the thin disk that covers the outer end
of the whorl chamber.   The angle at which the liquid enters the chamber,
the depth of the chamber, the size of the discharge aperture, and the
pressure determine the fineness of the spray particles.   The shallower
the chamber, the smaller the particles, if the other factors are equal.
The same effect is produced by increasing the angle at which the liquid
enters the chamber, or, up to a certain point, by increasing the pressure
or decreasing the size of the discharge aperture.   Increasing the pressure
with the other factors equal will increase the fineness of the spray par-

ticles.   The spray from this type of nozzle is thrown out in the form of a hollow cone, the size and shape of the cone also depending on the factors just mentioned.   Enlarging the aperture increases the width of the cone and the amount of liquid discharged.   Increasing the pressure has the same effect.   In the earlier types of vermorel nozzles, a disgorging pin was placed in the center of the chamber, the point of the pin being held so that it could be forced through the aperture to dislodge particles that had clogged the opening (see Fig. 183).

The disk nozzle as now constructed will, with proper adjustments, give a fine spray with enough driving power to cover the foliage.   The larger disk nozzles, such as used on the spray gun, will deliver a large volume of liquid, enabling the operator to work rapidly.   The fact that the spray is delivered in the form of a hollow cone has some disadvantages, but the fact that the nozzle is, or should be, moved continuously while the spray is being discharged should insure a thorough distribution.   The disk nozzles and Vermorel nozzles are generally used for orchard spraying in clusters of from two to eight attached to a Y-shaped extension on the end of the spray rod, or singly on 3- or 4-foot rods for garden work.

Solid-stream Spraying.—This method of spraying has been developed in the eastern section of the United States, where, because of the presence of the gypsy moth, elm leaf beetle, and other insects, it has become necessary to spray large street and park trees or good-sized tracts of woodland.   To accomplish such work requires a nozzle that will throw a stream from 25 to 125 feet (Fig. 185) and give a thorough covering of the foliage at such heights.

Fig. 185.—Spray gun for use on shade and forest trees, which will enable operator to cover trees 100, or more feet tall, from the ground.   (*From John Bean Manufacturing Co.*)

"Shade-tree nozzles" of several types are made by the manufacturers of spraying equipment.   They are so designed that, by the adjustment of the nozzle or changing the tips, the spray stream may be caused to break at from 10 to 100 feet from the nozzle (Fig. 186). With a pressure of from 400 to 800 pounds it is possible with these nozzles to spray 100- to 125-foot trees from the ground and get a very satisfactory coverage of the foliage.   In the work in woodlands and large parks it is often necessary to use long leads of hose, 1 to $1\frac{1}{4}$ inches in diameter and, at times, $\frac{1}{2}$ to 1 mile in length.   Where such length of hose is used, a pump pressure of from 500 to 800 pounds is necessary to maintain a pressure of 250 pounds at the nozzle.   For this type of spraying, an engine of 20 or more horsepower is needed.   Most of these large sprayers are mounted on trucks with double transmission or are drawn by a tractor equipped with a power "take-off," so that the full power of the engine can be used for the pump when not moving or can be used to drive both, when in motion[1] (see Figs. 181 and 186).

Spray Booms.—For the spraying of such crops as grapes, small fruits, nursery trees, and field crops in general, many special types of sprayers have been developed.   These usually consist of an arrangement of nozzles of the disk type, attached to frames of metal called *booms*, in

---

[1] "Solid Stream Spraying against the Gypsy and Brown-tail Moths in New England," *U.S.D.A., Dept. Bul.* 480, 1917.

such a manner and in sufficient numbers to insure a thorough covering of the foliage, often on both the upper- and undersurfaces. For use on field crops such as peas or beans, booms are often used with from 30 to 70 nozzles. With such sprayers care should be taken that hose connections from the pump are adequate to supply each nozzle with the spray mixture, that the nozzles are readily detachable for cleaning, that the pump and motive power to drive it are large enough to maintain the

Fig. 186.—Spraying shade and woodland trees with a modern, high-pressure, motor-truck outfit. The same motor drives the truck and operates the sprayer. (*From The Hardie Manufacturing Co.*)

pressure necessary to efficient operation, and that they will maintain the capacity rate of discharge of the nozzles. Where traction-driven outfits are operated, a triplex pump should be used. With such outfits it is seldom possible to develop more than 150 pounds pressure if 6 to 10 nozzles are attached to the boom. Because of the difficulty of maintaining even this low pressure and because of the slipping of the wheels in soft ground, an engine-driven outfit is much more satisfactory. For such outfits about 1 horsepower is necessary for each 4-gallons-per-minute discharge capacity of the nozzles.

Stationary or Central Spray Plants.—The idea of using a central spray plant, from which the spray material could be pumped through pipes to all parts of the orchard, is supposed to have originated in California and to have first been used in that state about 1910. In this type of orchard spray equipment a large powerful pump and central mixing and storage tanks are located in some part of the orchard, usually near the center, convenient to a supply of water. From these central tanks, which are equipped with powerful agitators, large pipes run out over the orchard to connect in turn with lateral pipes running between the rows of trees at regular intervals. At intervals along these lateral pipes are attached hose connections. The distance apart of the lateral pipes and the distance between the hose connections on the pipes vary in different orchards with the topography of the ground, the type of planting, and

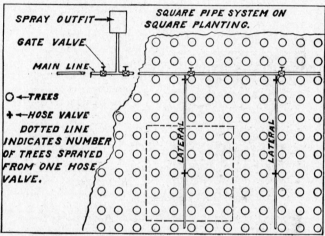

Fig. 187.—Diagram showing practical plan of piping an orchard, using the "square system" of piping from a stationary spray plant. (*From John Bean Manufacturing Co.*)

the planting distance used in the orchard. The two chief kinds of systems are known as the square system (Fig. 187), which is perhaps most commonly used, and the long system, which is the type used in all orchards planted on the diagonal. In another type of system the pipes are connected with a return line to the central plant. This type, however, is not in very general use. The hose connections and the lateral pipes are usually spaced for the use of 75 to 100 feet of hose. This length of hose enables a man to attach it to the pipe and, with the one attachment, spray from three to four rows of trees in each direction. The lateral pipes are sometimes run along the surface of the ground or just under the surface. In other orchards they are run as an overhead system. The pipes have to be arranged so that they can be drained after using. This is accomplished by opening stopcocks at the ends of the pipes and blowing the pipes dry of spray material from the central pump.

Some of the advantages of a central spray plant are that the orchard can be sprayed regardless of the condition of the soil. No time is lost in stopping to fill spray tanks and mix the materials, as the operator at the central plant mixes the material and fills one of the central tanks;

then, while pumping from this tank, he mixes and fills the other so that he can alternate in pumping the material from one or the other of the tanks without loss of time. It makes possible the spraying of closely planted trees without knocking off the fruit by the passage of a spray rig between the trees. It is easy to spray orchards on steep hillsides. There is a saving of from 35 to 50 per cent in time and labor required in the actual spraying. The original cost of a central spray system is considerably more than that of equipping with a sufficient number of movable spray rigs, to take care of the same acreage.

DUSTING MACHINERY

Although the original method of poisoning plants for insect control was by dusting, spraying has been much more extensively used during the past 50 years. Dusting is easier, lighter, pleasanter, and faster work than spraying. Two pounds of dust mixture have ordinarily as much covering power as 10 gallons (80 pounds) of spray. The dust outfit is also much lighter than the spray rig, and an equal force of men can, with a dust outfit, cover from several to many times the acreage possible by liquid applications.

Fig. 188.—A useful shaker for applying dusts, made from a gallon tin pail and a wooden extension handle. The pail should have a tight-fitting lid. (*From Ill. Agr. Exp. Sta. Bul.* 249.)

There is probably not an orchardist, truck gardener, or grower of field crops, who would not, by preference, apply insecticides and fungicides in the dust form if he felt that he could secure as good, or very nearly as good, results with dusting as he can with spraying. Certain chewing insects have been fairly well controlled for many years by applications of dust to the leaves of the plants. Other chewing insects and sucking insects have not been controlled at all in this way. In most cases it is desirable to make applications for the control of diseases and insects at the same time, and for this reason a dust which would come into general use must be one that will contain a stomach poison, a contact insecticide, and a fungicide. Until such a satisfactory combination dust is discovered, general reliance for the protection of many crops will be placed upon liquid-spray mixtures.

A good duster, like a good sprayer, is one that will spread the insecticide—in this case in dry form—in such a manner as to give the most uniform coating possible to the plants to be treated. For the reason that spraying has proved somewhat more effective than dusting, there has been a wider range of development in spraying machinery, and, because of the difference between the machinery required for the application of a liquid and that for a dust, sprayers have always been, and will continue to be, somewhat more complicated than dusters. The large dusters now in use for field crops and orchard and shade tree dusting are very efficient.

The only dusters in use, up to about 1900, consisted of modified bellows, or small hand machines, from which the dust was discharged by the air blast created by a revolving fan.   At present, there are many kinds and types of dusters on the market.   These consist, in general, first, of small *hand dusters* adapted to small gardens, having a metal chamber with a plunger which is pulled back and forth by means of a handle, and which discharges a blast of air through a second chamber containing dust.   Many different forms of these plunger dusters are made; in fact, there are almost as many different kinds as there are companies making them.   Some of them are well constructed and, with reasonable care, will last for several years.   Next in size are the *blower dusters* (Fig. 189), which consist of an enclosed fan rotated by a hand crank which sends a continuous blast of air through a small chamber

Fig. 189.—Dusting potatoes with a hand-power blower duster suitable for garden work. The figure shows the use of the Y attachment to cover two rows.   (*From Niagara Sprayer Co.*)

into which the dust is fed from a hopper.   Most of this type of dusters do, and all of them should, contain some sort of an attachment for keeping the dust stirred in the hopper and for providing an even feed through the discharge chamber.   These dusters can be used for a considerably larger area of truck crops than can be covered by the same sized sprayer.   In fact, 1 or 2 acres may be protected with a good blower type of hand duster.   Dusting several acres with a hand duster is tiresome work, and, if several applications of dust have to be made to such acreage, a larger, power-driven machine should be used.

Another type of duster of about the same size is the *bellows duster*. With this type the air is forced through the discharge chamber by the extension and compression of a bellows attached to the back of the duster.   This gives a discharge of the dust in puffs instead of a continuous stream as with the blower type.   The bellows-duster, or "puff" duster as it is sometimes called, is well adapted to hill crops but is not suitable for row crops.   It is also fairly satisfactory to use on small trees and shrubbery.

There are a number of dusters which have been developed for field-crop dusting, in which the fan is geared to the wheel or wheels on which the duster runs. Such dusters are called *traction dusters*. A type which is generally used in the South for small acreages of cotton, has much the same appearance and is built in much the same way as a one-horse cultivator. The operator guides the machine by two handles, which extend back from the frame. The discharge pipes extend back parallel with the handles but are slightly longer and can be adjusted to deliver the dust at an angle on each side of the operator, exposing him to the dust most of the time while the work is being done. Such dusters will do fairly well for 5 to 10 acres of field crops.

There are a number of larger-type traction dusters for work on field crops, which consist, in general, of a two-wheeled frame with some arrangement for driving the fan from the wheels. These machines will develop a fair amount of power and are capable of turning a 12- to 16-inch

Fig. 190.—Dusting potatoes with a two-horse field-crop traction duster, with discharge pipes arranged to cover four rows. (*From Niagara Sprayer Co.*)

fan at 2,500 to 3,000 revolutions per minute. The hopper of such a machine holds from 50 to 100 pounds of dust, or enough to cover several acres. Such machines are drawn by at least two mules and, if properly constructed, will do fairly effective work. There are various modifications of the discharge from these machines depending on the type of crop to be treated (Figs. 190 and 191). Some of these large dusters (Fig. 190) adapted for field crops have four main discharge pipes, each pipe branching again to permit the use of eight nozzles, and so making it possible to apply the dust to each side of the four rows covered by the machine on each trip through the field. The greatest difficulty with these machines is to get an even distribution of air into all the pipes. With certain types of field crops, particularly those that are broadcasted, it has not been possible to secure an even distribution with the branched pipes. Some machines are now made with a single discharge pipe which conveys the dust into a triangular trough-shaped boom. This type of discharge boom has been particularly developed for work on the pea aphid and has proved more satisfactory than the numerous discharge pipes. Owing to the fact that the speed of the machines will vary with the rate at which they are drawn through the field and that the wheels may slip in soft ground, it is always difficult to maintain an even discharge from traction-type dusters.

Most of the dusters now in use for orchard and field-crop work, are engine-driven. With a 4- to 10-horsepower engine operating the fan of such a duster, it is possible to get a very uniform distribution over the plants treated. The orchard dusters (Fig. 192) are built on the same principle as the traction field-crop dusters, but, instead of having many discharge pipes, the dust is blown out over the trees from one large flexible pipe. The rate of discharge of the dust is controlled by the size of the opening from the hopper into the discharge chamber. This rate can be maintained more uniformly than the discharge of liquid from a sprayer. In the larger-type dusters the hopper discharges directly on the fan instead of into a separate chamber. There is some advantage in this method of discharge; the dust is given a somewhat higher velocity, and the force of discharge can be more easily regulated by increasing or diminishing the speed of the fan.

Fig. 191.—Dusting grapes with a vineyard traction duster. (*From Niagara Sprayer Co.*)

Another type of duster has a mixing attachment in the hopper by which it is possible to mix the ingredients of the dust immediately before it is applied. This is of importance where volatile substances, such as nicotine, are to be used in the dust, as it insures an absolutely fresh material. There are many other styles of dusters made to meet the particular needs of certain crops, such as vineyard dusters, nursery dusters, corn dusters, cotton dusters, and dusters for large shade trees.

With dusting machines, as with spraying machines, one should carefully study the problem before selecting a machine. Find out first the rate per tree or per acre at which the dust should be applied for the control of the particular insects to be combated and get a machine capable of applying dust somewhat in excess of this rate. For extensive work a power-driven machine is highly desirable. The machine should have provision in the hopper for keeping the dust stirred, so that it will not cake and clog, and some method of insuring a positive feed at a uniform rate into the discharge chamber. With some of the volatile dusts, particularly with nicotine where it is used for rather resistant insects, it has been necessary to attach a trailer behind the duster. This trailer consists of a piece of canvas from 10 to 30 feet wide, attached across the back of the duster and dragging behind for a distance of from 10 to as much as 100 feet (Fig. 306).   The purpose of these trailers is to confine the fumes given off from the dust and expose the insects to them for a longer period.

Fig. 192.—Dusting apple trees with an engine-driven orchard duster.   (*From Niagara Sprayer Co.*)

The best trailers are constructed of special gasproof canvas to prevent the insecticide vapors from escaping through the fabric.

AIRPLANE DUSTING AND SPRAYING

One of the most recent developments has been that of airplane dusting (Fig. 193).   The first use of the airplane in the control of insects was made in Ohio, in 1921, against the catalpa sphinx.   By this method large quantities of the dust are discharged from hoppers built into the airplanes, which fly over the fields or woodlands.   The dust so discharged is distributed by the air from the propeller of the plane and settles in a very uniform manner over the plants or trees beneath.   In airplane dusting, planes specially designed for this work fly at speeds of 70 to 120 miles an hour and at heights of 10 to 35 feet from the ground.   Swaths from 100 to 200 feet wide are dusted, and the plants are covered at the rate of 4 to 12 acres a minute of actual dusting time or 200 to 500 acres an hour by a single plane.

The airplane has been successfully used in the control of the cotton boll weevil, gypsy moth, sugarcane borer, and spruce budworm; in dusting peaches for insects and plant diseases; in destroying mosquito larvae in impenetrable swamps; and for certain shade- and forest-tree insects.   It has also been of great assistance in scouting for insect outbreaks in remote areas of forest lands.   Certain commercial companies have aided greatly in the perfection of planes and in placing this work on a practical basis.   Where large areas of crops occur, sufficiently concentrated to justify the employment of a plane, commercial companies will contract to do dusting at prices below the cost of ground dusting. The work is, in general, quite as well done, requiring little, if any, more materials and with little more danger to surrounding fields and homesteads than results from ground dusting. The speed with which the application can be made is a point of very great advantage in the control of many pests. Successful airplane dusting is impossible on many crops if there is a wind movement above 8 to 10 miles per hour. This makes it impractical to attempt this form of insecticide application in some sections. The autogyro has

Fig. 193.—Dusting cotton with an airplane. The cloud of dust discharged at a height of about 10 feet above the ground covers a swath 150 feet or more in width, at a speed of 75 to 100 miles an hour, or from 200 to 500 acres an hour. (*From U.S.D.A.*)

been used for both dusting and spraying and has some advantages, particularly in the matter of safety.

Another point in regard to dusting is the fact that the dust on leaving the machine is given a certain electrical charge. This charge may be the same as that held by the leaf surfaces over which it is blown; if this is the case, the dust is repelled; as objects with the same electrical charge are mutually repellent. If the charge of the dust particles is different from that of the leaf, they will be attracted, and stick to the leaf surface. There is apparently a difference in the charge given to dusts of different kinds by passage through a duster. It has not been found possible to control this charge in a practical way.

CHAPTER XI

INSECTS INJURIOUS TO CORN

Corn is the most important farm crop in the United States, having an average value for the past 10 years of more than $2,000,000,000. Insects attack every part of the plant throughout its growth, and year after year destroy about 9 per cent of the crop. The portion of this one crop lost because of insect damage would, at current market prices, be sufficient to support 20 large universities, or build 2,000 miles of modern hard-surfaced road each year. Control is difficult, on the whole, because of the extensive areas and the relatively low value of the crop per acre, which make it necessary to depend mostly upon farm practices and other indirect control measures. Nevertheless, many of the insect pests of corn can be very effectively controlled if the best known remedies are properly applied.

FIELD KEY FOR IDENTIFICATION OF INSECTS INJURING CORN

A. *Insects chewing leaves or stalk above ground leaving visible holes:*

   1. Leaves of plant ragged and sometimes entirely eaten off, the injury beginning about the margins of the fields. Tips of ears gnawed. Fields presenting a ragged appearance, bare stalks standing without leaves.....................
   ....*Grasshoppers*, page 335.
   2. Leaves of young corn with green portion eaten out, presenting a whitened bleached appearance. Plant growth is retarded and leaves wilt and hang limp. Leaves covered with dull or shining-black or greenish-black, small, jumping beetles from a little smaller than a pinhead, to several times as large. Some kinds transmit a bacterial wilt disease causing death of plants..............
   ....*Flea beetles*, page 386.
   3. Unfolding corn leaves with rows of holes running across the leaves. Cavities eaten in the sides of the plant by black, tan, or brown, often mud-covered beetles, with slender snouts nearly ⅓ as long as rest of body. Beetles from ¼ to ⅞ inch long. Injury most severe on reclaimed or sod land............
   ....*Billbugs*, page 357.
   4. Grayish long-headed beetles, ½ inch long, with faint white stripes along their sides, eat the margins of the leaves............*White-fringed beetle*, page 472.
   5. Corn leaves with irregular holes with ragged edges eaten out, giving plant a ragged appearance. Often the plant is entirely stripped of leaves. Dark-green worms up to nearly 2 inches in length with light stripes on the sides and down the middle of the back, feed usually at night and hide under clods or in the heart of the plant during the day. Skin when seen through a lens appears smooth. Worms often crawling into the corn in large numbers from near-by fields of grass or small grains.........................*Armyworm*, page 341.
   6. Injury similar to A, 5, by worms which often crawl in great numbers from field to field. Holes eaten in leaves have smooth edges. Distinguished from true armyworms, under a lens, by the greater length of the hairs and their more prominent black bases on the smooth skin, by the prominent λ on the front of the head, and by the somewhat different striping..*Fall armyworm*, page 344.
   7. Corn plants under 1 foot in height cut off, mostly at night, below, at, or slightly above the surface of the ground. Fat, well-fed, smooth-appearing worms with

3 pairs of slender legs and 5 pairs of prolegs, of varying sizes up to 2 inches in length and of several shades and markings, hide under the ground or clods near the injured plants. Many kinds curl the body and "play 'possum" when disturbed...*Cutworms*, page 346.

8. Young corn plants up to 8 or 12 inches in height eaten into or cut off near the surface of the ground. A loose silken web containing many bits of dirt, leading to a short, silk-lined tunnel in the ground. Short, dirty-colored, brown-spotted, coarse-haired worms, from ¼ to ½ inch long, usually hidden in these silken tunnels, try to escape when disturbed.............*Webworms*, page 383.

B. *Small insects sucking sap from leaves or stems, not leaving visible holes, but causing wilting, spotting, discoloration, or death of the plants:*

1. Leaves of the corn plant covered with masses of soft-bodied bluish-green plant lice or aphids, of the size of pinheads, most abundant in the curl of the plant and on the developing tassel. Infested plants scattered over the field........ ....*Corn leaf aphid*, page 350.

2. Corn wilting, drying out, and falling down, on the side of the field next to small grains. Small, active, reddish or black-and-white, somewhat flattened, sucking bugs clustered behind the lower leaves and over the entire lower part of the stalk; when crushed, giving off a vile odor. Fields invaded by hordes of these insects at the time of small-grain harvest..............*Chinch bug*, page 351.

C. *Insects boring or tunneling in the stalks or stems:*

1. White, chunky, legless, humpbacked grubs, with a distinct, hard, brown or yellow head, tunnel in the pith of the stalk, dwarfing and often killing the plants. Most severe on old corn ground.................................. ....*Maize billbug or curlew bug*, page 359.

2. Flesh-colored caterpillars from ½ to 1 inch long, with inconspicuous, small, round, brown spots scattered over the body, boring in all parts of the stalk, shank of ear and ear. Tassels frequently broken off from injury at the base. Many caterpillars often found in one stalk.....*European corn borer*, page 360.

3. Cornstalks, especially those around margins of fields, bored during June and July, by very active, dark-brown worms or caterpillars, with two white stripes on each side, which are broken for about ¼ their length near the middle of the body, and a continuous white stripe running down the center of the back. The caterpillars reach a length of 1½ inches........*Common stalk borer*, page 364.

4. Stalks bored throughout their length by white worms, up to 1 inch in length, conspicuously marked with rounded brown spots; not common in the North.. ....*Southern cornstalk borer*, page 366.

5. Stalks bored by slender, greenish worms with fine longitudinal brownish markings, about ¾ inch long when full-grown. Stalks become much distorted and curled by the early-season injury. Most abundant in the south............ ....*Lesser cornstalk borer*, page 367.

D. *Insects attacking the ear:*

1. Dark-green and coppery-red beetles about ½ inch long by ¼ inch across, feeding on silks and husks, especially at the tip of the ear. Wing covers of the beetles shining. Four prominent white spots on the tip of the abdomen, which projects from under the greenish-brown wing covers..*Japanese beetle*, page 621.

2. Silk fouled with moist masses of excreta, many of the silks eaten off; kernels at the tip of the ear eaten by worms up to 2 inches long, varying in color from very dark green to light green; live beneath the husks. Bodies of the worms sparsely haired, skin rough-appearing under a lens. Worms nearly 1¾ inches long when full-grown and usually only one to the ear. Occasionally the worms enter at the butt of the ear, but usually at the tip. They do not tunnel into the cob...*Corn earworm*, page 368.

3. The ear, its cob and shank tunneled throughout and many of the kernels and the silk eaten by flesh-colored, inconspicuously spotted caterpillars up to 1 inch long. Many worms often found in one ear. Husks perforated with small holes with exuding frass.................*European corn borer*, pages 360 and 371.

4. Ripening kernels of grain sorghum heads and broomcorn eaten out or entirely consumed by sluggish caterpillars, somewhat flattened and thickly clothed with spines and hairs, the body greenish with four, red or brown, longitudinal stripes above..........

Sorghum webworm, Celama sorghiella Riley (see *Tex. Agr. Exp. Sta. Bul.* 559).

E. *Insects attacking the roots or underground stem:*

1. Plants weak, leaves reddish or yellowish. Bluish-green plant lice or aphids, about the size of pinheads, sucking the sap from the roots; always attended by ants...................................*Corn root aphid*, page 371.

2. Plant makes slow growth, dies, and sometimes falls over. Often distinct areas in field showing injury. Roots eaten off clean, not tuneled, by white, curved-bodied or U-shaped, six-legged grubs, from ½ to over 1 inch long, with large brown heads and distinct jaws.................*White grubs*, page 374.

3. Slick, shining, smooth, reddish-brown, very tough-skinned, six-legged worms, 1 to 1½ inches long, with the last segment of the abdomen often curiously ornamented, bore through the soil and chew off small roots or tunnel larger roots.......................................*Wireworms*, page 377.

4. Corn on land which has been in this crop for one or more years falls over about the time the tassels appear, frequently after a rain has softened the ground. Roots eaten off or containing many, small, brown tunnels in which will sometimes be found slender, whitish worms not more than ½ inch long with distinct brown heads and 6 short legs.............*Northern corn rootworm*, page 380.

5. Corn falls as in *E*, 4, but the injury is not confined to old corn ground. Underground parts of stalk as well as roots show tunnels. Slender yellowish-white worms having much the same appearance as those described under *E*, 4, but slightly larger and more robust...........*Southern corn rootworm*, page 381.

6. Corn plants following clover sod are stunted, seldom reaching a height of over 8 or 10 inches. Plants wilt during hot, dry days. Small, short, white, fat-bodied grubs, not over ⅛ inch long with light-brown heads, gnaw on the roots. Grubs hold body in a curved position. Cease feeding during June.........

....*Grape colaspis*, page 383.

7. Corn wilts and dies when from 1 to 4 feet high. Sometimes falls over. Injury most common to corn in old sod ground. Large, fleshy, legless, whitish grubs from ½ to 2 inches long, with the body enlarged just behind the head, not curved, and with strong brown jaws.....................................

....*Corn prionus* (see *Eighteenth Rept. Ill. State Ento.*, page 128, 1891–1892).

8. Plants wilt and die from attacks of white, legless, but not curved-bodied grubs, up to ½ inch long, which eat off the underground stem and taproot........

....*White-fringed beetle*, page 472.

9. Roots or stem eaten off. Plump, smooth-appearing, grayish or greenish-white worms, up to 2 inches long, unmarked except for brownish head and shield behind head, with 3 pairs of slender legs and 5 pairs of prolegs, remain in the soil day and night; curl the body when dug out or disturbed...............

....*Cutworms*, page 346.

10. Roots and stem eaten by dirt-colored, brown-spotted, coarse-haired caterpillars, from ¼ to ½ inch long, which are hidden in a loose-woven dirt-covered silk-lined tunnel in the soil...................*Webworms*, page 383.

11. Plantlet weak, reddish to yellowish in color. Small brown ants make conical mounds of dirt and tunnel the ground along the corn roots, sometimes feeding on the starchy parts of the kernel. Carrying about and caring for small bluish-green aphids, on the corn roots..............*Cornfield ant*, page 371.

F. *Insects attacking planted seed, eating into it or devouring the plantlet as it germinates, so that plants often fail to come up:*

1. Seeds fail to sprout or plantlet is weak. White, very slender worms or larvae, slightly over ¼ inch long when full-grown, with yellowish-brown head and 6 short legs burrowing in the kernels and sometimes in the sprout.............

....*Pale-striped flea beetle*, page 387.

2. Seeds fail to sprout, or plants die when small. Slick, shining, brown to reddish-brown, smooth, hard, six-legged worms, 1 to 1½ inches long, boring through the kernels and young plants...........................*Wireworms,* page 377.
3. Seed does not sprout. Brownish or blackish-brown beetles, about ⅓ inch in length, feeding in kernels of corn in the ground............................
...............*Corn-seed beetles:* the darker, brown-striped one, *Agonoderus pallipes;* the uniform chestnut-brown one, *Clivina impressifrons,* page 388.
4. Seed produces a weak sickly sprout. Starchy parts of kernel eaten out and scattered through the ground by very small orange-colored ants............
....*Thief ant* or *fire ant,* page 389.
5. Seeds fail to sprout or make a weak sprout. Dirty yellowish-white legless maggots about ¼ inch long, blunt at the posterior end, tapering sharply to the head, may be found burrowing in the kernels of corn in the ground, or in the earth around the kernels.......................*Seed-corn maggot,* page 389.
References.—Eighteenth and Twenty-third Repts., Ill. State Ento., 1891, 1892, 1905.

1 (27, 75, 146). Grasshoppers or Locusts[1]

Importance and Type of Injury.—Few, if any, other species of insects have caused greater direct loss of crops than have grasshoppers. From ancient to modern times they have caused the death through famine of millions of human beings. In any section of the world when these insects are abundant, man has to make a determined fight to save his crops. Damage is most severe in those parts of the world where the annual rainfall is 25 inches or less. Corn is seldom attacked by grasshoppers until the plant has reached a height of 20 or more inches. Plants that are attacked have the tips of the ears, the tassel, and the leaves eaten, and the stalks present a general ragged or bare appearance. Grasshopper injury usually starts on the sides of the field, as the insects seldom originate in the cornfield.

Plants Attacked.—Various species of grasshoppers attack nearly all cultivated and wild plants.

Distribution.—Grasshoppers occur over the entire world.

Life History, Appearance, and Habits.—Grasshoppers that attack corn practically all pass the winter in the egg stage. These eggs are laid in packet-like masses nearly 1 inch long and from ½ to 1½ inches below the surface of the soil (Fig. 194). Each egg mass consists of from 20 to 120 elongate eggs, securely cemented together, the whole mass somewhat egg-shaped and dirt-covered (Fig. 195). They are mainly deposited in uncultivated ground such as field margins, pasture land, and roadsides. In the middle western and western states they are frequently laid in considerable numbers in clover, alfalfa, and stubble fields. The pellucid grasshopper lays its eggs chiefly in sod land and almost entirely in heavy soils. The Rocky Mountain grasshopper and *Melanoplus packardi* chiefly in fields planted to crops, while the two-striped and the differential grasshoppers deposit their eggs mostly at edges of fields, along roadsides, or in drift soil. Most of the important species pass the winter in the egg stage. A few species winter as partly grown nymphs or as adults. In the latitude of central Illinois, hatching of the more common and typical species begins in mid-May and continues until July. The

[1] Many species of the Order Orthoptera, Family Locustidae.

Fig. 194.—Differential grasshopper, *Melanoplus differentialis* Thomas, laying eggs in soil; enlarged. A part of the soil has been removed to expose the abdomen and the egg mass. (*From U.S.D.A., Farmers' Bul. 691.*)

Fig. 195.—A clump of grama grass, showing a number of grasshopper egg masses among the stems and roots (indicated by arrows). (*From S. D. Agr. Exp. Sta. Bul. 172.*)

young hoppers (Fig. 84,*E*) differ but little from the adult, except in size and the fact that they lack wings.   They change their skins several times in the course of their growth and at the last molt acquire full-sized usable wings.   With most of the species which injure corn, growth is completed from the middle of August to the first of September.   The adults, however, continue to feed until the first heavy frost.   The eggs are mainly deposited during the latter part of September and October.   Of the many species of grasshoppers occurring in the United States, the following are the most destructive.   The species destructive in any given year will vary with the locality, altitude, latitude, weather conditions, and kinds of crops that are prevalent there.

The red-legged grasshopper[1] is one of the smallest of the more destructive species, being less than 1 inch long when full-grown.   It is very destructive in fields of legumes and is common along roadsides.   In the Middle West this and the Rocky Mountain grasshopper[2] have been very destructive to soybean crops by cutting through the

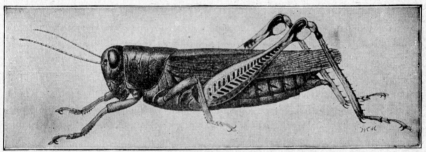

Fig. 196.—Adult of the differential grasshopper, *Melanoplus differentialis* Thomas, side view, enlarged about ½.   (*From Ill. State Natural History Surv.*)

pods and causing the seeds to mold.   Its general color is brownish red, the hind tibiae being usually pinkish red with black spines.   It can be distinguished from the Rocky Mountain grasshopper by the shape of the cerci (Fig. 46) which in this species are twice as long as wide, widest at the base, and incurved on both upper and lower margins; the tip of the abdomen in the male is not notched.

The Rocky Mountain grasshopper[2] is one of the most widespread and generally destructive species, able to survive well on dry native grasses and waste land, as well as in nearly all cultivated crops.   It shows some preference for light sandy soils.   It lays its eggs over a wide area and in diverse situations.   This species is very similar in size and appearance to the red-legged grasshopper, being scarcely 1 inch long when full-grown.   The hind tibae are generally not so bright pink as in the red-legged grasshopper.   The cerci at the tip of the abdomen (Fig. 46) are about two-thirds as broad as long, widest beyond the middle and incurved on the upper margin only; the tip of the abdomen in the males has a distinct median notch or incision.   This species is called migratory because the larger nymphs usually migrate from their breeding grounds to more succulent vegetation, and, when abundant, the adult often flies for many miles to new feeding grounds.

The clear-winged grasshopper[3] (Fig. 197) is perhaps second in importance only to the Rocky Mountain grasshopper.   It is one of the most common western species, especially at higher elevations and in more northern latitudes, but occurs also in the

---

[1] *Melanoplus femur-rubrum* (De Geer).

[2] *Melanoplus mexicanus spretus* (Walsh), formerly called also the lesser migratory grasshopper, *M. atlanis* Riley.

[3] *Camnula pellucida* (Scudder).

eastern part of the country.   Like the Rocky Mountain grasshopper it is adapted to survive upon a great diversity of vegetation, and it survives drought well.   It prefers to lay its eggs in sod land and in heavy soils.   Its hind wings are nearly colorless and transparent.

The differential and the two-striped grasshoppers are better adapted to cultivated crops, such as vast areas of lush grain, and do not survive drought conditions.   In dry years they persist only in irrigated districts and along streams.   The differential grasshopper[1] (Fig. 196) is, next to the Carolina grasshopper, the largest of the destructive species, reaching a length of 1½ to 1¾ inches.   It is brownish or olive green in color, with a good deal of yellow on the under parts and with the chevron-like black markings on the hind femora very prominent.   The cerci of the male have a short rounded thumb-like projection on the lower margin.   This species is especially destructive to corn.   The two-striped grasshopper[2] ranges from 1 to 1½ inches in length.   It is a robust species, the upper part of the body olive with a yellow stripe on each side, extending from the head to the tip of the wing.   There is a dark stripe on the upper half of the hind femur.   The species is very common in clovers.   It

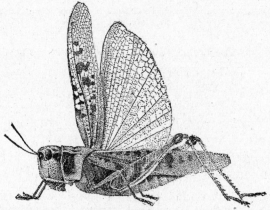

Fig. 197.—Adult of the clear-winged grasshopper, *Camnula pellucida* (Scudder), about twice natural size.   (*From Ill. State Natural History Surv.*)

frequently matures by late June, and there may be a partial second generation of both this species and the Rocky Mountain grasshopper.

The Carolina grasshopper[3] is a very large species, reaching a length of 2 inches. While generally less destructive than the others described, it is so common along roadsides, railroads, and paths that it is one of the most commonly observed species. It is brown, mottled with fine specks of gray and red on the wing covers.   The hind wings are black with distinct yellow margins.   It flies readily when disturbed.

Control Measures.—Grasshopper control consists of three distinct measures: (*a*) to destroy the eggs in the fall and winter and (*b*) and (*c*) to combat the grasshoppers at the time they are attacking crops.   The first of these measures consists of fall plowing or disking of areas in which the grasshoppers have laid their eggs, thus exposing the eggs to the action of the weather and to birds during the winter and early spring.   Plowing or disking to a depth of 4 inches, followed by packing or firming of the soil, is usually sufficient to destroy the eggs.   This is

[1] *Melanoplus differentialis* Thomas.
[2] *Melanoplus bivittatus* Say.
[3] *Dissosteira carolina* (Linné).

often quite effective, especially in the West, where there are areas of uncultivated land interspersed with the cultivated areas, if wind erosion and drying out of the soil are not too serious.

The most effective and generally accepted method of grasshopper control is by the use of poisoned baits. Hundreds of different baits have been tried. Practically all of them consist of (a) a *base* or *carrier* of bran diluted with either sawdust, cottonseed hulls, ground corncobs, or other similar materials; (b) a strong *arsenical* or other *stomach poison;* and (c) *water* or *oil* to make the carrier moist. Some of the baits which have been used most generally and which give the highest kill under average conditions are:

	Amount for 10 Acres	Amount for a Lot or Garden 50 × 100 Ft.
Bran	100 lb.	1½ lb.
⎰Sodium arsenite, paris green, or white arsenic	4 to 6 lb.	1 to 1½ oz.
or		
⎱Liquid sodium arsenite	2 qt.	⅟₁₆ pt. (30 cc.)
Sufficient water to make a stiff mash, usually	7 or 8 gal.	1 pt.
Bran	100 lb.	1½ lb.
Sodium arsenite (liquid)	2 qt.	⅟₁₆ pt. (30 cc.)
Black strap molasses	2 gal.	¼ pt.
Sufficient water to make a good mash, usually	7 or 8 gal.	1 pt.
Bran	100 lb.	1½ lb.
Lubricating oil, 20 to 30 viscosity	2 gal.	¼ pt.
Sodium arsenite, paris green, or white arsenic	4 to 6 lb.	1 to 1½ oz.

In any of the above formulas hardwood sawdust may be substituted for one-half the bran, or corncobs which have been ground in a hammer mill to the same particle size as bran will also make a very satisfactory substitute. Cottonseed hulls, oat hulls, and several other materials have been tested. On the whole, the sawdust or ground corncobs make the best bran substitute. Many experiments have been tried with the use of fruit extracts, ground fruits, amyl acetate, and other chemicals to make grasshopper bait more attractive. It is doubtful if these will pay a profit in additional kills of grasshoppers.

The best results are obtained by spreading the bait at the rate of about 10 pounds per acre. This may be done by hand from pails or bags carried on the arm. According to Parker:[1]

"Spreading by hand from a horse-drawn wagon or motor truck is the common method of application. One man scattering from a wagon can cover 10 acres per hour; two men in a truck can cover 20 acres. During the last two years an inexpensive home-made bait-spreading machine has been developed with which 40 to 50 acres can be treated per hour.

Special mechanical spreaders are described in the *Canadian Department of Agriculture Publication* 606, March, 1938.

[1] PARKER, R. L., Mimeographed Report of the Fourth International Conference for Anti-locust Research, Cairo, Egypt, April, 1936.

"Airplanes fitted with special equipment for bait spreading can cover from 100 to 150 acres per hour.   Their use in the United States has not been extensive because all available funds for grasshopper control are generally expended for materials, and the labor for spreading is provided by the land owners."

The bait should be applied early in the morning as this is the time of heaviest feeding of the grasshoppers.   They do not feed to any extent at temperatures below 65 or above 90°F.   It is of the utmost importance to poison the small grasshoppers as soon after they hatch as possible.

Fig. 198.—A hopper catching machine or hopperdozer designed to be pushed in front of a tractor, truck, or automobile, over heavily infested fields containing crops that will not be ruined by it.   The upper figure shows the front of the machine, consisting of a back, against which the jumping hoppers strike, and pans below to hold a mixture of oil and water which kills them as they drop into it.   The lower figure shows the means of attachment to tractor or truck.   (*From Ill. State Natural History Surv.*)

Sometimes the adults can be poisoned as they congregate for egg-laying. The oil used should be a fresh, lubricating, mineral oil such as is used for automobile crankcases of 20 to 30 S.A.E. rating, the latter being preferable in very hot dry weather.   Advantages claimed for the oil baits are that they remain moist and attractive to grasshoppers for a week or more, whereas the standard bait dries out in a few hours and then is not attractive to hoppers.   Grasshoppers take water baits only while they are fresh and moist.   Water baits applied during the middle of the day or evening are a complete loss, but in a rush the oil baits may be applied any time of the day.   Water baits ferment, cake, and mold in storage and cannot be made up in advance or saved for later application; oil baits can be kept for months if necessary. Water baits will burn crops if applied at a rate of more than 10 or 12 pounds per acre and heavy rains follow; oil baits do not burn plants to which they cling. With the oil baits, the flakes of bran separate better, which helps to avoid excessive applications and prevents danger to birds and livestock. The oil bait is not so hard on the hands of men scattering it. The cost of materials for these baits will run from 10 to 40 cents per acre, depending upon kinds of ingredients used and quantities purchased. The oil baits cost slightly more than the water bait. One application usually checks damage promptly and kills from 50 to 90 per cent of the hoppers within 5 days; a second and third application may be needed. During the last several years organized grasshopper control campaigns have been carried on, covering the upper Mississippi and Missouri river valleys. From the period of 1925 to 1934 it is estimated that grasshoppers destroyed crops to the value of $249,000,000 in the most heavily infested states. During this same period more than $4,500,000 was spent on grasshopper control, and it is believed that every dollar expended resulted in a saving of about $50 worth of crops, on the average.

Where for any reason it is impossible to control grasshoppers by the use of baits, as it is in some of the legume crops, hopper catchers or hopperdozers may be used to lessen the hopper damage. The hopper-dozers (Fig. 198) are merely devices placed on the front end of an automobile, truck, or tractor and pushed slowly over the field at from 5 to 7 miles per hour, or drawn by teams, causing the hoppers to jump or fly up and fall back in the catching device. There are many homemade types of these catchers. The hopperdozer consists of a shallow pan partly filled with water with a little kerosene over the top. On the whole, hopperdozers are much more expensive to operate and less efficient than poisoning. It is rarely possible to catch 50 per cent of the hoppers present. In certain crops such as soybeans they may be used to advantage but cannot be used in corn, small grains, and other crops that would be seriously broken down by the machine. As high as 4 to 8 bushels of hoppers per acre have been caught with these machines. There are about 200,000 grasshoppers in a bushel. It has been estimated that where grasshoppers are present at the rate of 15 to 20 per square yard, they will eat 1 ton of alfalfa, per day, in each 40-acre field.

References.—*Minn. Agr. Exp. Sta. Tech. Bul.* 141, 1914; *Colo. Agr. Exp. Sta. Bul.* 280, 1923, and *Ext. Cir.*, Ser. 1, 180 *A*, 1921; *S. D. Agr. Exp. Sta. Bul.* 172, 1917; *U.S.D.A.*, *Farmers' Bul.* 1691, 1938; *Jour. Econ. Ento.*, **7**: 67, 1914, **10**: 524, 1917, **11**: 175, 1918, **12**: 337, 1919, **13**: 232, 237, 1920, and **14**: 138, 1921; *U.S.D.A., Tech. Bul.* 190, 1930; *Iowa Agr. Coll. Ext. Bul.* 182, 1932; *Ill. Agr. Exp. Sta. Bul.* 442, 1938.

2 (28, 48). ARMYWORM[1]

Importance and Type of Injury.—This insect fluctuates greatly in abundance, undergoing cycles which reach destructive peaks at greatly varying periods of years. During epidemics it often destroys much of the vegetation over many hundreds of square miles. Corn under 8 inches in height that is attacked by armyworms will usually have the leaves eaten off entirely. With larger corn the midrib of the leaves will sometimes be left, but the center of the young stalk is so eaten out that

---

[1] *Cirphis unipuncta* (Haworth), Order Lepidoptera, Family Noctuidae.

it dies. The dark-green worms (Fig. 199), up to 2 inches in length, with white strips on the sides and down the middle of the back, will be found hiding under clods and stones or in the center leaves of the plant during the day. The damage usually starts at the sides of the field, where the worms have moved in from some other crop.

Plants Attacked.—All grass crops especially corn, timothy, millet, bluegrass, small grains, and some legumes; and many other plants under stress of hunger.

Distribution.—United States and Canada, east of the Rockies, and many other parts of the world.

Life History, Appearance, and Habits.—The winter is passed mainly in the partly grown larval stage; but the fact that the moths are abroad very early in the spring in the northern states would indicate that some of the insects winter as adults, or as pupae, or that there is a spring flight northward from the southern part of the range of the insect. The partly grown worms shelter in the soil about clumps of grasses, or under litter on the ground. They begin feeding early in the spring, become full-grown by the latter part of April in the latitude of central Illinois, and pupate just below the surface of the soil. The pupae are dark brown, about ¾ inch long, tapering sharply at the tail, and blunt at the head end. They remain in this stage for 2 weeks, or longer if the weather is cool, and then transform to uniform, pale-brown or brownish-gray moths with a wing expanse of about 1½ inches (Fig. 200). There is a single, small, but prominent, white dot in the center of each front wing. The moths are strong fliers, but remain hidden during the day,

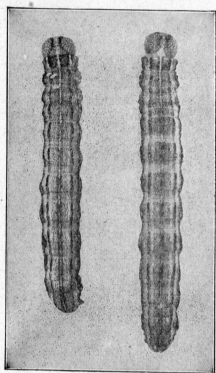

FIG. 199.—Full-grown armyworms, *Cirphus unipuncta* (Haworth), the left one showing eggs of a tachinid fly parasite attached to the skin. Twice natural size. (*From Ill. State Natural History Surv.*)

becoming active at night. They are attracted to lights, and are strongly so to sweets or decaying fruit. The females lay their greenish-white eggs in long rows or clusters on the lower leaves of grasses to the number of 500 or more. The leaf is generally folded lengthwise, and fastened about the eggs with a sticky secretion. The young worms are pale green in color and have the looping habit of crawling until about half grown. They may often be found by thousands in fields of grass or small grains, and because of their habit of feeding at night, their presence is generally not suspected until the crop is nearly destroyed. When the food supply becomes exhausted in the fields where they have hatched, these caterpillars move out in hordes or armies and attack crops in near-by fields. These crawling masses of worms have given them their common name. On becoming full-grown, the worms are nearly 1½ inches long, of a general greenish-brown color, with longitudinal stripes as follows: a narrow broken stripe down the center of the back, bordered by a wide, somewhat darker, mottled one reaching halfway to the side; as seen from the side there are three stripes of about equal width: next to the wide mottled one on the upper side a pale orange, white-bordered stripe, next a dark-brown, light-mottled one just reaching to the spiracles, and just below the spiracles, a pale-orange, unmottled one edged with white. The head is honeycombed with dark lines, and each proleg has a dark band on its outer side and a dark tip on the inner side. They then enter the ground and change to the pupal stage, emerging as moths in from 14 to 20 days. There are from two to three generations each year. The larvae of the first generation do most of the damage, in June in the latitude of central Illinois. The larvae of the last generation are abundant in late August and September.

Fig. 200.— Armyworm. Adult moth, natural size. (*From Ill. State Natural History Surv.*)

Control Measures.—One of the most effective methods of controlling an outbreak of armyworms is to poison them by scattering a poison-bran mixture in the fields where they are feeding, or across the line of march of the worms when they are leaving fields where food is scarce. Directions for making and applying poisoned bait are given under Grasshoppers, page 339 and on page 254. For armyworms, as for cutworms, the bait should be spread in the late afternoon or early evening.

Where the worms are advancing from one field to another, they may be stopped by plowing deep furrows in front of their line of advance and dragging a log or keg of water back and forth in the furrow until a very fine dust mulch has been worked up. The worms tumbling into this furrow will be unable to crawl up the steep dusty side, and may be crushed by the continued passage of the log, or by spraying with kerosene or other contact poison.

Armyworm outbreaks usually originate in fields of small grain or grasses, especially where there is a very rank growth of vegetation, or where the grain has fallen down and lodged. Such situations should be watched, especially during May, and if the young worms are found, the poison bran bait should be applied immediately.

The armyworm is preyed upon by a number of insects, especially certain parasitic flies,[1] which lay their eggs on the backs of the worms, mostly on the fore part of the body (see Fig. 199, left-hand worm and Fig. 37). The young maggots hatching from these eggs bore into the worms and kill them. They are also preyed upon by several ground beetles and certain parasitic wasps. Perhaps the most efficient insect enemy of the armyworm is an extremely small, black, wasp-like insect[2] that deposits its eggs inside the eggs of the armyworm. The other parasites attack

---

[1] *Winthemia quadripustulata* (Fabricius) and other Tachinid flies, Order Diptera, Family Tachinidae.

[2] *Telenomus minimus* Ashmead, Order Hymenoptera, Family Scelionidae.

the worms when they are partly, to nearly fully grown, and thus prevent an excessive increase in the next generation, but do not kill the worms until after most of their feeding has been done. The egg parasite, on the other hand, by preventing the eggs from hatching, stops all damage by these insects.

References.—*N. Y. (Cornell) Agr. Exp. Sta. Bul.* 376, 1916; *Jour. Agr. Res.*, **6**: 799, 1916; *U.S.D.A., Farmers' Bul.* 731, 1916; *Ill. Natural History Surv., Ento. Ser., Cir.* 7, 1921; *Conn. Agr. Exp. Sta. Bul.* 408, pp. 191–200, 1938.

3. FALL ARMYWORM[1]

Importance and Type of Injury.—Besides the army cutworms and the true armyworm (see page 341), this insect, which is a member of the same family, often develops the marching habit, the caterpillars crawling in great droves, which may be very injurious to field and vegetable crops. They eat the foliage and tender stems of many plants, often taking everything clean as they go, and then disappear suddenly. The larvae often attack the ears of corn in a manner identical with the corn earworm. They are especially bad in the South in seasons following a cold wet spring.

Plants Attacked.—Corn, sorghums, and other plants of the grass family are probably the preferred food, and the insect is often called the "grassworm"; but it attacks also alfalfa, beans, peanuts, potato, sweetpotato, turnip, spinach, tomato, cabbage, cucumber, cotton, tobacco, all grain crops, clover, and cowpeas.

Distribution.—This insect is a continuous resident of the Gulf states and the tropics of North, Central, and South America and some of the West Indies. It often migrates northward as far as Montana, Michigan, and New Hampshire.

Life History, Appearance, and Habits.—This tropical insect is apparently unable to live through the winter in any section where the ground freezes hard. In southern Florida and along the Gulf Coast, several stages may be present and more or less active during the winter months. In the spring, as they increase in numbers, swarms of moths are produced that fly northward, sometimes covering hundreds of miles before they alight to lay their eggs. About 1,000 eggs are laid by each female, in masses averaging about 150, usually on green plants, and covered with hairs from the moth body. The small larvae feed down near the ground gregariously at first, especially in the heart of the plant, and are not generally noticed until they have reached a length of 1 or 1½ inches, by which time, if abundant, they are consuming so much grass or grain that they create alarm. They do not leave the plant to hide in the soil during the daytime, as do the armyworm and climbing cutworms. The full-grown larvae (Fig. 201) vary in color from light tan or green to nearly black. They have three yellowish-white hair lines down the back from head to tail; on the sides next to the yellow lines is a wider dark stripe and next to it an equally wide, somewhat wavy, yellow stripe, splotched with red. These worms are very similar to the true armyworm in appearance but can be distinguished by the more prominent white inverted Y

---

[1] *Laphygma frugiperda* (Smith and Abbott), Order Lepidoptera, Family Noctuidae. A closely related species, *L. exigua* Hübner is known as the sugarbeet armyworm and the asparagus fern caterpillar. It attacks those plants and corn, cotton, peas, and peppers (see *Fla. Agr. Exp. Sta. Bul.* 271, 1934).

on the front of the head and by the more prominent black tubercles from which the fine scattered hairs on the body arise. The corresponding hairs on the true armyworm are much shorter and the tubercles smaller. The fall armyworm can be distinguished from the armyworm, also, by the fact that it feeds on cotton, tobacco, legumes, and many vegetables as well as grasses, while the armyworm feeds chiefly on grains and grasses and attacks other plants only when driven by hunger.

Fig. 201.—Larva of the fall armyworm, side view, about twice natural size. (*From U.S.D.A., Tech. Bul.* 34.)

When abundant, the caterpillars eat all the food at hand and then start to crawl in great armies into adjoining fields. While these "forced marches" may come in the fall in the North, in the extreme South they occur in midsummer or even in early spring. Gardens may be invaded and consumed in a few nights. Suddenly, when full-grown, all the caterpillars disappear almost as if by magic, having dug into the ground

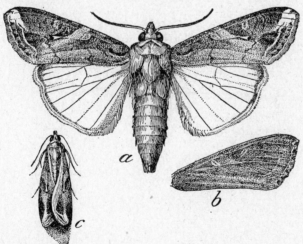

Fig. 202.—Adult of the fall armyworm. *a*, male; *b*, right front wing of female, about twice natural size; *c*, moth in resting position, about natural size. (*From U.S.D.A., Tech. Bul.* 34.)

about an inch to pupate. Within 2 weeks a new swarm of moths emerges from the ground, which generally flies far before laying eggs, and so the entire country may be invaded during the summer. The adult moth (Fig. 202) is similar to many cutworm moths, about 1½ inches across the wings, the hind wings grayish white and the front pair dark gray, mottled with lighter and darker splotches and having a noticeable whitish spot near the extreme tip. They are active mainly at night and not much noticed. Only one generation of larvae is usually abundant in any one community in the North, but in the South there may be 5 to 10 generations in the same locality in one year.

Control Measures.—In favorable seasons a number of parasitic enemies keep the fall armyworm caterpillars down to moderate numbers. As is generally the case, however, cold, wet springs check the parasites more than they do the insects that they feed upon. In such seasons, especially, watch should be kept of grassy fields for the appearance of the young worms. If they appear, they may be controlled by the measures recommended for the armyworm (page 343). After the worms have disappeared, fields in which they have been feeding should be disked or otherwise lightly cultivated, if practicable, to break up the pupae and throw them out on the surface, where natural enemies and weather conditions will destroy many of them. Keeping fields of cotton and corn free from grass will do much to prevent injury by this insect, since the infestations almost always start among grasses.

References.—U.S.D.A., *Farmers' Bul., Tech. Bul.* 34, 1928, and 138, 1929.

4 (49, 67, 80). CUTWORMS[1]

Importance and Type of Injury.—There are a great many species of cutworms, and they vary greatly in numbers from year to year. They frequently make it necessary to replant corn, and they destroy from 5 to 50 per cent of the stand of other crops. They injure plants in four principal ways: (*a*) The solitary, surface cutworms eat off plants just above, at, or a short distance below, the surface of the soil and sometimes drag them to their burrows in the soil. Most of the plant is not consumed, merely being eaten enough to cause it to fall over. Consequently these caterpillars have great capacity for doing damage. Among the important surface cutworms are the black cutworm, the bronzed cutworm, the clay-backed cutworm, and the dingy cutworm. (*b*) The climbing cutworms climb the stems of herbaceous plants, vines, shrubs, and trees and eat buds, leaves, and fruits of vegetables and orchard and vineyard crops. The variegated and the spotted cutworms are important species that sometimes assume the climbing as well as the army habit. (*c*) The army cutworms are those which occur in great numbers and, after consuming nearly all the vegetation in one area, crawl along on the ground by the thousands to adjacent fields. They feed largely from the tops of the plants, without cutting them off, but, when abundant, will consume succulent plants clean to the ground. The armyworm (page 341) and the fall armyworm (page 344) are army cutworms that are separately discussed because of their great importance. The army cutworm[2] also has this habit developed to an unusual degree. (*d*) The subterranean cutworms, unlike all the others, remain in the soil to feed upon roots and underground parts of the stems. Consequently they cannot be controlled by the use of poisoned baits. The pale western and the glassy cutworms are important subterranean forms. In most cases the smooth, brownish, greenish, or nearly white well-fed-appearing worms (Figs. 203, 204, 205, 206) will be found during the daytime hiding in the soil close to the stems of the plants which they have cut off or fed upon.

[1] Many species of Order Lepidoptera, Family Noctuidae.
[2] *Chorizagrotis auxiliaris* (Grote).

Plants Attacked.—Nearly all plants, except those with hard, woody stems are fed upon by cutworms. Some of the crops most seriously injured are corn, beans, cabbage, cotton, tomatoes, tobacco, and clover.

Distribution.—Cutworms of various species are of world-wide distribution. Certain species are confined largely to southern, and others to northern, climates. Some prefer dry conditions, while others are most abundant in wet areas or overflowed land.

Life History, Appearance, and Habits.—The majority of cutworms pass the winter in the partly grown to fully grown larval stage. Some, however, hibernate as adults, and others as pupae, in the soil. In typical

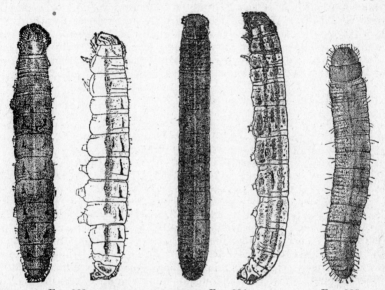

FIG. 203.   FIG. 204.   FIG. 205.

FIG. 203.—Larva of the spotted cutworm, *Agrotis c-nigrum* (Linné), dorsal and side view, somewhat enlarged. (*From Ill. State Natural History Surv.*)

FIG. 204.—Larva of the black cutworm, *Agrotis ypsilon* (Rottemburg), dorsal and side view, somewhat enlarged. (*From Ill. State Natural History Surv.*)

FIG. 205.—Larva of the glassy cutworm, *Sidemia devastatrix* (Brace), dorsal view, somewhat enlarged. (*From Ill. State Natural History Surv.*)

cases, the worms remain as small larvae in cells in the soil, under trash or in clumps of grasses during the winter. They start feeding in the spring and continue growth until early summer, when they change in the soil to a brown pupal stage and later to the adult or moth stage (Fig. 207). With most of our common species, there is but one generation a year; a few species have two to four generations; and, in others, the generations are so broken up that adults may be found at almost any time from late spring to midautumn. The eggs of most species are laid on the stems of grasses and weeds or behind the leaf sheath of such plants. Certain of the moths, notably the black cutworm,[1] lay their eggs on low spots in the field, or land that has been subject to overflow. Certain others lay their eggs on the bare ground, or on ground that has been somewhat packed by the passage of vehicles or animals. The egg stage commonly

[1] *Agrotis ypsilon* (Rottemburg).

lasts from 2 days to 2 weeks.   The larvae in most cases remain below the surface of the ground, under clods, or other shelters during the day and feed at night.   The time required to grow from newly hatched caterpillars, about $\frac{1}{25}$ inch long, to nearly 2 inches long, varies from 2 weeks to 5 months.   They then dig down several inches in the soil where they make cells in which they pupate from 1 to 8 weeks or over winter. The adults upon emerging crawl out of the ground through the tunnel made by the larva going down.   So far as their life cycles are concerned the commoner cutworms fall into two distinct groups.   (*a*) Those with a single generation a year are mostly northern in distribution and winter as larvae, regardless of the latitude or length of season where they live. They are held to a single generation by the remarkable fact that, while all other stages are accelerated by higher temperatures, the *prepupal stage* (from the time the full-grown larvae cease feeding until they pupate) is increased or delayed by higher temperatures thus compensating for the longer season in the South and preventing additional generations. (*b*) Those which undergo several generations a year are most abundant in the South and winter as pupae.

FIG. 206.—Larva of the bronzed cutworm, *Nephelodes emmedonia* (Cramer), dorsal view, somewhat enlarged.   (*From Ill. State Natural History Surv.*)

The abundance of a given species from year to year is greatly affected by rainfall, which may prevent the moths from laying their eggs or, by flooding the soil, may force the larvae up to the surface during the daytime so that their parasites destroy nearly all of them.   From the several dozen destructive species a few of the worst are briefly characterized to show something of the range of appearance and habits in the family. There are many other species of nearly equal importance.

The black cutworm[1] is a cosmopolitan species with a restless, pernicious habit of cutting off many plants while satisfying its appetite.   It lays its eggs singly or a few together on the leaves or stems of plants often in low or overflowed land.   The winter is spent in the larval or pupal stage, and there are two generations a year in Canada, and four in Tennessee.   The larva is greasy gray to brown above, with faint lighter stripes.   The skin has strongly convex, rounded, isolated granules of large and small size.

The dingy cutworms[2] are northern species which winter as partly grown larvae and have but one generation a year.   The larvae are dull dingy brown with a broad buff-gray dorsal stripe, subdivided into triangular areas on each segment and margined by a narrow dark stripe on each side.   The skin granules are round, coarse, isolated, and slightly convex.   These species are said to be very resistant to drought, and they sometimes assume the climbing habit.   The eggs are laid singly or a few together.

The bronzed cutworm[3] (Fig. 206) is one of the most strikingly marked of all species, the larva being dark bronzy brown, striped from head to tail with five, clear-cut pale lines about half as wide as the brown between them.   One stripe runs down the middle of the back and one below the spiracles on either side of the body.   The

[1] *Agrotis ypsilon* (Rottemburg).
[2] *Feltia subgothica* (Haworth) and *Feltia ducens* Walker.
[3] *Nephelodes emmedonia* (Cramer).

skin is granulate. This is a northern species, especially troublesome, on corn, grains, and grasses. It winters as partly grown larvae and has a single generation a year.

The variegated cutworm[1] is found throughout most of the cultivated parts of the earth. It is said to have destroyed $2,500,000 worth of crops in the United States in a single year. It shows some preference for garden crops and the foliage, buds and fruits of trees and vines, tobacco, ornamentals, and greenhouse plants. The eggs are laid in bare patches of 60 or more on stems or leaves of low plants, on twigs or branches of trees, or on fences and buildings. It may complete 3 or 4 generations a year, wintering mostly as pupae. The larva has a distinct pale-yellow dot on the mid-dorsal line of most of the segments and frequently a dark W on the eighth abdominal segment. The skin is smooth, and the general color ashy or light dirty brown, lightly mottled with darker brown.

The spotted cutworm[2] is a troublesome species of general distribution throughout North America, Europe, and Asia, though scarce in the South. It shows a preference for garden crops. It winters as large larvae and undergoes two or three generations a year. Eggs are laid either singly or in rows or patches of a hundred or more mostly on leaves. On the posterior half of the larva, each segment has a pair of elongate wedge-shaped black dashes on the upper side, which increase in size and lie closer together posteriorly. There is also a dark stripe through the spiracles. The skin is smooth.

The army cutworm[3] is a western species, adapted to arid conditions. It is a surface feeder, burrowing but little, and often assumes the army habit. It is said to have destroyed 100,000 acres of winter wheat in 1 year in Montana alone. It has one generation a year, wintering as half-grown larvae. The eggs are laid singly in or upon the soil. The larvae are pale greenish gray to brown with the back pale-striped and finely splotched with white and brown but without prominent marks. The skin is covered with fine close-set pavement-like granules.

The pale western cutworm[4] is an underground feeder which has destroyed millions of dollars worth of small grains, beets, and alfalfa in the western half of the United States and Canada. The body is gray-ish, unmarked by spots or stripes; the skin has fine, flat, pavement-like granules. This species lays its whitish, spherical eggs, singly or a few together on the soil. The larvae hatch during warm periods in winter or very early spring, and the single generation of larvae have usually completed feeding by the end of June.

The glassy cutworm[5] is widespread except in the more southern states. It is a strictly subterranean species which prefers sod and is most troublesome to crops following sod in low ground. There is one generation a year, the species wintering as small larvae.

FIG. 207.—Adult of the clay-backed cutworm, *Feltia gladiaria* Morrison, slightly enlarged. (*From Ill. State ·Natural History Surv.*)

The body is greenish white like a grubworm, uncolored except for the reddish head and cervical shield, and has a somewhat translucent or glassy appearance. The skin is not granulated.

The yellow-striped armyworm or cotton cutworm[6] is a day-feeding species that has been very destructive to cotton by devouring the young plants as well as by boring into squares and bolls, but it is a general feeder on many crops. The female lays her eggs in masses on foliage, trees, or buildings and covers them with scales from

[1] *Lycophotia* (= *Peridroma*) *saucia* Hübner (= *P. margaritosa* Haworth).
[2] *Agrotis c-nigrum* (Linné).
[3] *Chorizagrotis auxiliaris* (Grote).
[4] *Porosagrotis orthogonia* (Morrison).
[5] *Sidemia devastatrix* (Brace).
[6] *Prodenia ornithogalli* Guenee.

her body. The species winters in the pupal stage, has several generations a year, and is most abundant in the South. The larva has a pair of dorsal, triangular, black spots on most of the segments and commonly a bright orange stripe just outside these spots on each side.

Control Measures.—The species of cutworms which attack the plant above or at the surface of the ground, including the climbing cutworms, may be controlled very effectively by the use of the poison-bran bait described for the control of grasshoppers (see page 339), but it should be applied in the evening as for the armyworm. Those species that feed mostly below the surface cannot be controlled by the use of poisons. The eggs of many species of cutworms are laid very largely in grass lands. One of the best methods of avoiding damage by these insects is to rotate the crops in such a manner that corn is not planted on sod ground unless such sod has been broken early in the fall or during late summer. Summer plowing, before the eggs are laid, and fallowing until frosts occur are of value against all species which lay their eggs upon low-growing vegetation. Ditches and dusty furrows with postholes are of value in checking the advance of the army species. For the climbing cutworms, bands of tanglefoot about the trunks of trees and grapevines will give protection. Tobacco and garden crops which are started in seedbeds may have the tops and stems, but not the roots, dipped as they are transplanted in bordeaux mixture (page 290) poisoned with 6 pounds of lead arsenate or 3 pounds of calcium arsenate to 100 gallons. Where the underground species are abundant, a special study of conditions will have to be made, as no general recommendations will apply.

Cutworms are subject to attacks by other insects, especially by certain flies which lay their eggs on the backs of the worms, and by ground beetles. They are readily fed upon by many species of birds, and the eggs are attacked by certain small wasp-like parasites.

References.—CROSBY and LEONARD, "Manual of Vegetable-garden Insects," pp. 260–301, Macmillan, 1918; *Can. Dept. Agr., Div. Ento. Bul.* 3, 1912; *Jour. Agr. Res.*, **46**: 517–530, 1933; *U.S.D.A., Tech. Bul.* 88, 1929; *Mont. Agr. Exp. Sta. Bul.* 225, 1930.

5. CORN LEAF APHID[1]

Importance and Type of Injury.—Corn infested by this insect shows numerous greenish or greenish-blue aphids in the curl of the leaves and upper parts of the stalk. Leaves are sometimes entirely covered with these aphids. Winged and wingless individuals will be found during the summer. Infested corn leaves are frequently mottled with yellowish or reddish-yellow patches. It is more destructive in the South. By feeding on the tassel and silk and coating them with honeydew, it may seriously interfere with pollinization of corn. This honeydew may attract great numbers of corn earworm moths to the ears and result in increasing the infestation by that pest. The feeding of the aphids causes a discoloration of the brush of broomcorn. It is a serious pest of fall and winter barley in the southwestern states, sometimes weakening these crops to such an extent that very little grain is produced. It is said to have reduced the yield of grain sorghums in western Kansas by one-third. It is the disseminator of a serious mosaic disease of sugarcane.

Plants Attacked.—The insect has been found on corn, barley, sugarcane, millet, broomcorn, sorghums, Sudan grass, and many other wild and cultivated plants of the grass family. It shows a preference for sorghums. It winters on barley in the Southern states and California.

[1] *Aphis maidis* Fitch, Order Homoptera, Family Aphididae.

Distribution.—The insect is common throughout the corn-growing areas of the United States and Canada, being more abundant in the South. Its range extends throughout the tropical and temperate regions of the world.

Life History, Appearance, and Habits.—Our knowledge of the life history of this insect is incomplete. In the North Central states, it appears in cornfields about midsummer. In the South the insect multiplies rapidly and does its greatest damage in the winter months. Of the females, only the winged and wingless ovoviviparous forms are known. Males have been noted only very rarely and the egg-laying true females have never been found. No observations have been made on the winter stages in the northern states, and it is not known whether this species passes the winter in the egg stage in this section or whether it migrates up from the South during the spring and early summer. The female aphids cluster in large numbers on the plants, sometimes almost entirely covering the leaves. Some winged females are present throughout the summer. The number of ovoviviparous generations produced in a year varies from about 9 in central Illinois to as many as 50 in southern Texas. The insects feed until they are killed by a heavy frost, or the drying up of their food plants.

Control Measures.—The damage which this insect causes can be largely prevented by early planting of the crops and by proper tillage and fertilization to hasten their growth and maturing. Pasturing infested winter barley is recommended to free the crop of this pest.

References.—*Twenty-third Rept. Ill. State Ento.*, p. 123, 1905; *U.S.D.A., Tech. Bul.* 306, 1932.

6 (31). CHINCH BUG[1]

Importance and Type of Injury.—For more than 150 years chinch bugs have caused serious losses to American agriculture. Webster estimated that the insect caused a total damage in this country, from 1850 to 1900, of $350,000,000. In 1934 they caused an estimated loss of over $40,000,-000 in Illinois, alone. The first indication of the presence of chinch bugs in the field will often be the wilting and drying out of the infested corn. Usually this occurs on the side of the field next to small grains. Occasionally injured plants may appear in any part of the field. Small black-and-white to gray-and-white or red insects will be found behind the sheaths of leaves or in the soil about the base of the plant, often with slender beaks inserted in the plants from which they suck the sap. At small-grain harvest, hordes of these insects will be found crawling over the ground from cut grain into fields of corn or other growing grass crops.

Plants Attacked.—The insect feeds only on the plants belonging to the grass family. This includes all of our cultivated and wild grasses, corn, and small grains.

Distribution.—The chinch bug has been found throughout the United States, in southern Canada, in Mexico and in Central America. Its areas of greatest destructiveness are in the Mississippi, Ohio, and Missouri river valleys.

Life History, Appearance, and Habits.—The chinch bug (Fig. 208) hibernates only in the adult stage, the full-grown insect being about $\frac{1}{6}$

---

[1] *Blissus leucopterus* (Say), Order Hemiptera, Family Lygaeidae. The hairy chinch bug, *Blissus hirtus* Montandon, predominantly a short-winged form, is sometimes abundant in turf in the northeastern United States, killing the grass in spots. Short cutting, frequent watering, and top dressing help to prevent destruction of the grass, and rotenone or nicotine dusts or sprays will kill the bugs.

Andre has also described what he believes to be a distinct species, *Blissus iowensis* Andre, which is also predominantly short-winged and shorter haired than *B. hirtus*, and requires much longer to develop than *B. leucopterus*.

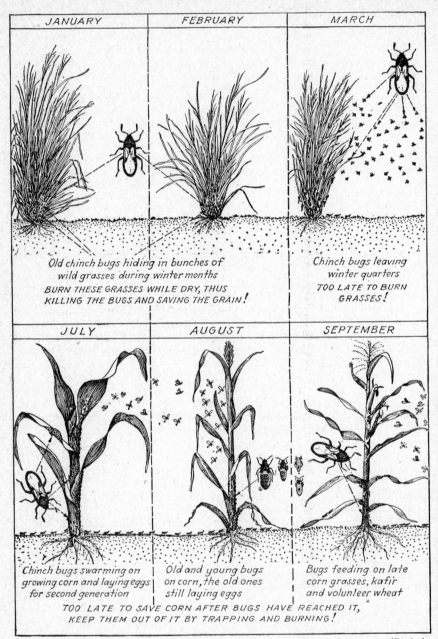

FIG. 208.—Chart showing seasonal history of chinch bug, *Blissus leucopterus* (Say), in the Central States.   During the winter from December to February, bugs are hibernating at the edges of woodland, under fallen leaves, and in bunches of grass and other shelters. In February or March, flight to young grain begins and continues until about the middle of May.   In June and July the bugs crawl in great numbers from ripening wheat to young

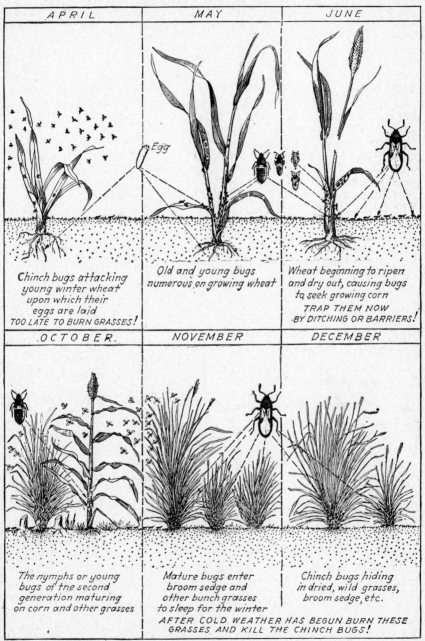

APRIL	MAY	JUNE
Chinch bugs attacking young winter wheat upon which their eggs are laid *TOO LATE TO BURN GRASSES!*	Egg Old and young bugs numerous on growing wheat	Wheat beginning to ripen and dry out, causing bugs to seek growing corn *TRAP THEM NOW BY DITCHING OR BARRIERS!*

OCTOBER	NOVEMBER	DECEMBER
The nymphs or young bugs of the second generation maturing on corn and other grasses	Mature bugs enter broom sedge and other bunch grasses to sleep for the winter *AFTER COLD WEATHER HAS BEGUN BURN THESE GRASSES AND KILL THE CHINCH BUGS!*	Chinch bugs hiding in dried, wild grasses, broom sedge, etc.

succulent corn and at that time may be trapped in ditch barriers or by lines of creosote. Becoming adult about August 1, they scatter among the corn plants, where eggs are laid and the second generation is begun. When these become adult from September 23 to the early part of November, they find hibernating quarters, where they remain until spring. (*After U.S.D.A., Farmers' Bul.* 1498.)

to ⅕ inch in length, with a black body. The white wing covers are each marked with a triangular black patch at the middle of their outer margins. The legs are reddish to reddish yellow. The insect gives off a vile odor, when crushed, that is somewhat distinctive and always remembered by one who has smelled it. They hide away in almost any kind of shelter during the winter; but in the Middle West few of them remain in corn-fields, and they are found chiefly along the south side of hedgerows, bushy and grassy fence rows or roadsides, and the south and west edges of wood-lands. They have been found in large numbers in soybean stubble, where the beans have been harvested with a combine, and they occur in con-siderable numbers in the underground nests of field mice. Where the clump-forming native prairie grasses are present, the bugs seem to prefer such clumps for winter quarters. They have been taken, however, in a great variety of different shelters. As many as 5,000 bugs may frequently be found on a square foot of surface in favorable hibernating places. The adult chinch bugs remain in their hibernating quarters until the temperature reaches a point above 70°F. for several hours during which the sun is shining. They may move about on warm days earlier in the spring, and occasionally mating takes place before they leave their hiber-nating quarters. When the above-mentioned temperature is reached on sunny days, they crawl up the stems of grasses or other plants about their hibernating quarters and take flight, usually going to fields of small grain. Here they feed by sucking the sap from wheat, rye, oats, or barley, and the females deposit their eggs behind the "boots" of the lower leaves or, if the ground is loose, upon the roots. The insects mate repeatedly, laying a few eggs each day for 3 weeks or 1 month, an average of about 200 eggs being laid by each female. The eggs are nearly cylindrical, are three or four times as long as broad, are yellow, and have four short nip-ple-like projections on the cap at the head end. These eggs hatch into small, very active, reddish bugs with a band of white on the back just behind the wing pads. They become dark as they grow older and, at the last molt, acquire full-sized wings. The bugs require about 30 to 40 days to complete their development. This usually does not occur until after small-grain harvest, especially that of wheat. They are dependent for their food supply on the sap of growing grass plants. It is, therefore, necessary for them to leave the dried stubble field when grain is cut. As they are still wingless, they usually migrate on foot to fields of corn, oats, or grasses, where they complete their growth. The adult stage is reached during the early part of the summer to midsummer. The adults may remain in the situations where they are feeding but usually fly for a few days after reaching the adult stage. Mating again takes place, and the eggs of a second generation are deposited on corn or grasses. The first-generation adults die by mid-September, and the second-generation nymphs complete their growth by the approach of cold weather. In parts of the Southwest, as in Oklahoma and Texas, a partial third genera-tion often occurs, the first generation becoming adult about the time small grains mature. Migration to corn and sorghums, therefore, takes place on the wing and creosote barriers are not practicable. During the warm sunny afternoons of early fall, they fly from the cornfields to their winter quarters. As they are seeking warm, sheltered places at this time, most of them will congregate on the south and west sides of the situations which afford them shelter during the winter.

Two forms of the chinch bug occur. In the more common form the adult has black-and-white well-developed usable wings. The less common form is sometimes considered a distinct species, the hairy chinch bug.[1] It occurs more generally in the northeastern states and at more northern latitudes, feeds more on grasses, often becoming a pest in lawns and greens, and does not have the pronounced migration from one food plant to another that occurs in the case of the long-winged form in the large grain-growing sections.

Control Measures.—As the chinch bug feeds only on plants of the grass family, the growing of nongrass crops is of great value during years of chinch-bug outbreaks, not only because these crops will not be injured in

Fig. 209.—Creosote barrier for protecting corn from migrating chinch bugs. The creosote is poured on the ground, at the brow of a ridge made by throwing a furrow toward the field to be protected. Note that the bugs, which came out of a grain field at the left, had already killed several rows of corn; while the corn on the right has been protected. (*From Ill. State Natural History Surv.*)

the least by the bugs, but also because the larger the area in such crops, the less will be that in which the chinch bug will find feeding and breeding places. Neighbors should cooperate in planning rotations so that corn is not planted adjacent to fields of winter or spring wheat, barley, or oats.

It has also been found possible to reduce the injury by chinch bugs to corn by planting a strong-growing legume crop such as soybeans or cowpeas in the field of corn. These plants are not of themselves repellent to the chinch bugs, but, by producing a dense shade around the base of the corn plants, they give a condition which is unfavorable to the bugs and which they avoid. Chinch bugs are primarily sun-loving insects and always seek the thinner parts of fields or poorer stands of any of the crops on which they feed. Certain varieties of corn and grain sorghums have been found very resistant to the attacks of second-generation chinch bugs, but none chinch-bug proof.

[1] *Blissus hirtus* Montandon.

Winter burning over the hibernating quarters has been of value in the areas west of the Mississippi River, where the insects winter mostly in bunch grasses. This practice is of little value in other areas. In many cases it should be discouraged as causing more harm than good. Probably this measure never destroys more than 25 to 50 per cent of the bugs over any large area. However, it is the last opportunity to prevent damage to spring wheat, barley, and oats, and it is often recommended that waste land, roadsides, ditch banks, and the margins of woodlands in which a dozen or more bugs are found per square foot, be burned over with a backfire, if this can be done without endangering property.

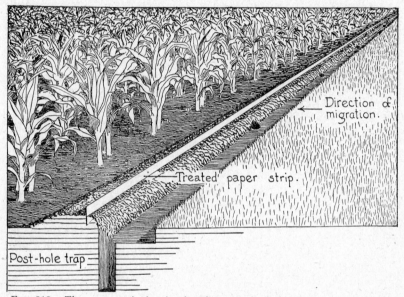

Fig. 210.—The paper strip barrier for the control of chinch bugs. Tarred paper, saturated with creosote and cut into strips about 4 inches wide, is set on edge and buried half in the soil at the summit of a furrow thrown toward the corn. This presents a barrier wall in addition to the repellent odor of the creosote and prevents bugs being blown across the barrier. (*From Ill. State Natural History Surv.*)

One of the most effective methods of combating the chinch bugs is trapping them at the time of small-grain harvest when they are traveling on foot from small-grain fields to fields of corn or other growing grass crops. This may be done by constructing a barrier line around the margin of the small-grain field along which the bugs can be stopped and killed or trapped. One of the effective barriers is made by pouring a narrow line of crude creosote (Fig. 209) along the brow of a smooth ridge thrown up with the plow around the margin of the infested small-grain field. The creosote should be poured on the side of the ridge next the small grain so the bugs will be climbing the ridge as they approach it. A strip of creosote making a line 1 inch wide on the soil is sufficient to turn the bugs, as they are strongly repelled by the odor of this chemical. Daily applications of the creosote are necessary for a period of 10 days to 2 weeks; 50 gallons of creosote are usually sufficient to maintain 1/4 mile of barrier for a season. Postholes 18 inches to 2 feet deep may be dug on the inner or small-grain side of this line, the tops of these holes flared and dusted. The bugs may be caught in such holes by the bushel as they travel along the creosote line seeking a place where they can escape from the field. Dusting of the top of the holes makes it impossible for the bugs to obtain a foothold upon it, and they roll into the holes. A small amount of kerosene poured into the holes will kill all of the bugs wet by it. Since 1933 paper barriers have come into use especially in gravelly or gumbo-soil areas. These are constructed by plowing a shallow furrow and digging postholes as for the creosote line. Tarred felt (not asphalt) paper strips about 4 or 5 inches wide and thoroughly soaked in creosote are placed against the steep edge of the furrow and the soil banked against the paper so as to hold it erect and projecting 2 or 3 inches above the soil (Fig. 210). Additional creosote is applied against the side of the paper strips every 2 or 3 days. Advantages claimed for the paper strips are that they require only about half as much creosote, that bugs are not blown across them by wind as sometimes happens with the creosote line on the soil, and that in certain soils they are much easier to make bugproof than the earlier type of barrier.

References.—Kans. Agr. Exp. Sta. Bul. 191, 1913; *U.S.D.A., Farmers' Buls.* 1498, 1934, and 1780, 1937; *Iowa Agr. Ext. Cir.* 199, 1934; *Ill. Agr. Exp. Cir.* 431, 1935; *Ill. Natural History Surv. Bul.* 19, Art. 6, 1932.

7 (40). BILLBUGS[1]

Importance and Type of Injury.—Billbugs injure plants in two important ways: (*a*) The adult snout beetles eat small holes into the stems, through which they consume a quantity of tissue from beneath the surface. They also eat out holes into which to insert their eggs. These punctures, made while the developing leaves are curled in the heart of the plant, show up, after the leaves expand, as transverse rows of punctures or holes across the leaves (Fig. 211). The perforated leaves often fall or twist into a curl so as to interfere with the growth of following leaves. When the punctures are made low down on the stalk, the plants send out excessive suckers or sprouts. (*b*) The second phase of injury, and the most harmful, is done by the larvae or grubs. A number of species feed upon the fibrous roots of small grains and cultivated grasses, in timothy corms, and in the stems of small grains, causing the heads to bleach and the straws to fall or lodge. The maize billbug and the curlew-bug also cause serious losses to corn by tunneling and feeding inside the stalk.

Plants Attacked.—While corn is attacked most conspicuously by most of the adult billbugs, nearly all the cultivated and wild grasses, the small grains, rice, peanuts, reeds, rushes, and cattails are also hosts for the grubs and adults of various species.

Distribution.—Taken together, the billbugs may be said to be present throughout the grasslands and cultivated areas of the United States and Canada. Their destructiveness has been most apparent from the Great Plains eastward and that of the maize billbug and curlew-bug in the southern states.

---

[1] A number of species of the genus Calendra, formerly Sphenophorus, Order Coleoptera, Family Curculionidae.

Life-history, Appearance, and Habits.—Billbugs usually winter in the adult stage. The several species range from ⅕ to ¾ inch in length. The head is prolonged into a cylindrical curved snout, about ⅓ to ¼ as long as the rest of the body, at the end of which are small chewing mouth parts. The body wall and wing covers are very hard. The beetles "play 'possum" when disturbed and are often so covered with mud that they are practically invisible on the soil, so long as they remain motionless. Upon coming out of hibernation, the insects feed, mate, and lay eggs for about 2 months. For each whitish kidney-shaped egg laid, a small hole is gouged out in the stem of the host plant with the mouth parts. The tiny grubs hatch in 4 to 15 days. They are white, short, chunky, humpbacked grubs, without legs, and with a distinct, harder, brown

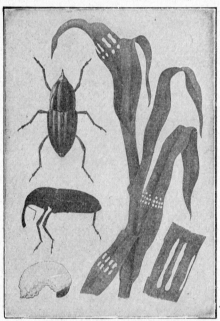

Fig. 211.—Billbugs: adults, larva, and characteristic injury, from adults feeding on corn. (*From Ill. State Natural History Surv.*)

Fig. 212.—Adults of the maize billbug, *Calendra maidis* (Chittenden), feeding on corn. (*From Kan. Agr. Exp. Sta., Tech. Bul.* 6.)

or yellow head. The larvae feed and grow for several weeks, eating out the pith of the stem and, if this becomes exhausted, descending to the soil to complete growth by feeding upon the fibrous roots. The larvae pupate either in the stems of plants or in the soil among the roots. The adults transform in the fall and may remain within the pupal cell over winter, or they may emerge and feed for some time before entering hibernation. In general there is one generation a year, but in the warmer parts of the country they winter in various stages.

Among the most important species are the following:

The maize billbug[1] is $\frac{2}{5}$ to $\frac{3}{5}$ inch long, rather broad-bodied, reddish brown or black, but often so covered with mud that the ground color is invisible. The raised longitudinal lines on the wing covers run about two-thirds of their length. The attacks of the adults upon young corn plants cripple them severely, causing excessive suckering and killing small plants outright. The larvae feed in the pith of the corn stalk for 40 to 50 days from early June to September. They pupate in August and early September, always in the larval tunnel in the upper part of the taproot. Adults that emerge early leave the pupal cells and find winter shelter in coarse grasses or other litter about the fields, but those transforming later winter in the base of the stubble. The larval injury also stunts and sometimes kills the plants. Corn is the principal food of both larvae and adults, although the insect breeds also in some of the large swamp grasses, mainly in grama grass.

The curlew-bug[2] averages $\frac{3}{8}$ inch long and is naturally brown with golden reflections, the more elevated bumps being polished black. Each wing cover has, besides the longitudinal lines, a prominent dent near the base and an elevation or hump near the tip. Like the maize billbug, this species feeds in the larval stage in the base of the stalk and the taproot of corn. It sometimes destroys entire fields of corn and, in certain years, in the South Atlantic states, may do more damage than all other corn insects combined. Besides corn, it injures rice and peanuts and, after the corn is older, lays its eggs in nut grass or chufa. When pupating in corn, the pupae are so low down in the root that they usually remain in the soil when stalks are pulled out by hand.

The remaining species do not develop as larvae in the stalks of corn but often attack it destructively in the adult stage. The bluegrass billbug[3] is about $\frac{1}{4}$ inch long, the body heavily and evenly marked with rounded punctures on the upper surface and tapering strongly to the tip of the abdomen. The grubs excavate the stems and corms or eat the rootlets of timothy, bluegrass, redtop, and all the small grains. Satterthwait considers this the most destructive billbug, because of the reduction it causes in hay and pasture crops and the premature failure of sod lands. It often dwarfs corn crops planted on spring-plowed old sod. The larval period is only a little over 3 weeks, and pupation occupies about 8 days more, in the corms or in the soil. Adults may emerge 45 days after the eggs were laid.

The bluegrass billbug prefers upland, while the timothy billbug[4] works in timothy in the larval stage, in a similar way, but prefers lower ground. The adult is larger than the bluegrass billbug, nearly $\frac{3}{16}$ inch long and the punctures on the prothorax form a distinct pattern. This species is believed responsible for many of the early failures of stands of grass in meadows. The adults attack corn, causing excessive suckering and failure to form ears.

The clay-colored billbug[5] is more than $\frac{1}{2}$ inch long, buff-colored, with fine punctures on the upper surface. The adults usually kill corn plants when they feed upon them. Sometimes they feed upon the kernels of wheat in the unripened heads, as they do also on millet and wild grasses. The grubs normally develop in the "nuts" at the ends of the roots of rushes and on sedges and reeds. (See Fig. 128.)

The cattail billbug[6] is a large species, about $\frac{5}{8}$ inch long, with pale yellow velvet-like hairs in the pits on its back. Normally breeding in cattail flags and bur reed, this species may be disastrous to corn, when it is planted in low, infested, waste land.

---

[1] *Calendra maidis* (Chittenden), Order Coleoptera, Family Curculionidae.
[2] *Calendra callosa* (Olivier).
[3] *Calendra parvula* (Gyllenhal).
[4] *Calendra zeae* (Walsh).
[5] *Calendra aequalis* (Gyllenhal).
[6] *Calendra pertinax* (Olivier).

The adults attack the stalks below the surface of the ground, causing the plants to be dwarfed and unproductive.

Control.—Crop rotations in which corn does not follow corn are of great help in eradicating the maize billbug and the curlew-bug. Chufas and other sedges must also be eliminated in order to check the curlew-bug. If cornstalks are plowed out and dragged with a spring-tooth harrow to remove the dirt from the taproots, after cold weather has begun, the hibernating beetles of these species will die. Merely raking and burning the stalks is only partly effective, especially for the curlew-bug, since the stalks break off, leaving the taproot containing the beetles in the soil. For the bluegrass, timothy, and clay-colored billbugs, crop rotations in which corn is not planted on ground that has been in sod or on reclaimed swampland where reeds, rushes, or cattails grew, during the preceding season, are of chief importance. Such land should be planted to soybeans, cowpeas, cotton, potatoes, flax, and other crops not known to be attacked by billbugs, the first year after breaking. Proper drainage is generally of great benefit. Fall plowing, followed by clean cultivation to keep the seedbed in perfect condition until the soil is warm and there is sufficient moisture to insure quick germination before planting, will go far to prevent serious losses from these pests.

References.—*Sixteenth Rept. Ill. State Ento.*, 1890; *Twenty-second Rept. Ill. State Ento.*, 1903; *U.S.D.A., Farmers' Bul.* 1003, 1932, and *Bur. Ento. Bul.* 95, Parts II and IV, 1911, 1912; *Kans. Agr. Exp. Sta., Tech. Bul.* 6, 1920; *N.C. Agr. Exp. Sta., Tech. Bul.* 13, 1917.

8. European Corn Borer[1]

Importance and Type of Injury.—The presence of the European corn borer is often indicated by cornstalks with the tassels broken or bending over. Sometimes the stalks are so heavily infested that they break over at various points and collapse (Fig. 213). Other indications of its attack are small areas of surface feeding on the leaf blades, with fine, sawdust-like castings on the upper sides of the leaves or stalks; small holes in the stalks, often with slimy borings protruding from the holes; worms boring through the stem and along the entire length of the ear and cob; numerous flesh-colored, inconspicuously spotted caterpillars, from $\frac{1}{2}$ to 1 inch long, boring in all parts of the stalks. This insect has caused the complete loss of crops of early sweet corn.

Plants Attacked.—Nearly all herbaceous plants large enough for the worms to enter. In the single-generation areas of the Middle West and in Ontario the corn borers feed almost entirely on corn, although some of the common weeds, vegetables, flowers, and field crops are often found infested when such plants are grown in close proximity to badly infested cornfields. In the area in New England where the corn borer has two generations a year, the insect feeds extensively on potatoes, beans, beets, celery, dahlias, asters, chrysanthemums, gladioli, and many weeds, as well as corn, even when such plants are not associated with corn. It has been found feeding on more than 200 kinds of plants.

Distribution.—This insect is distributed over the greater part of Europe and parts of Asia. While the insect has been known to be present

[1] *Pyrausta nubilalis* Hübner, Order Lepidoptera, Family Pyralididae. *U.S.D.A., Tech. Bul.* 59, 1928.

in North America only since 1917 and was probably introduced into this country in shipments of broomcorn from Italy or Hungary some time about 1908 or 1909, it has shown itself capable of being one of the most

Fig. 213.—A field of dent corn totally ruined by the European corn borer; Ontario, Canada, 1925. (*From Ill. Agr. Exp. Sta. Cir.* 321.)

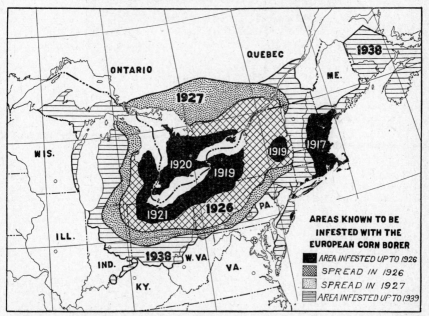

Fig. 214.—Map showing the known distribution of the European corn borer in North America, Jan. 1, 1939. The figures show dates of earliest discovery in various states as follows: 1917, first discovered in America near Boston, Mass.; 1919, in eastern New York, near Schenectady and in western New York, near Silver Creek; 1920, in southern Ontario, near Port Stanley; 1921, in northern Ohio and southern Michigan. The areas marked 1927 and 1938 indicate maximum known distribution at those dates. (*Original*.)

destructive insect pests of corn. By 1938, the insect had spread over practically all New England, southward along the coast to Delaware and Maryland, over most of Pennsylvania, Ohio, and Indiana, along the eastern border of Quebec, New Brunswick, and Nova Scotia. It will probably in time include the whole of the corn-growing area of North America.

Life History, Appearance, and Habits.—This insect passes the winter in the form of a full-grown worm or caterpillar (Fig. 215), in the stems of the food plants on which it has been feeding. These worms are from ¾ inch to nearly 1 inch in length. The body is flesh-colored, rather inconspicuously marked with small, round, brown spots. The two spots nearest the middle of the back on any segment are farther apart than their own diameter. They may be found in all parts of the stem and ear; but, especially in cornstalks, are most abundant, in winter, just above the ground surface. In the spring of the year, the caterpillar constructs a flimsy cocoon in its burrow, and in this transforms to a smooth, brown pupal stage. The moths (Fig. 216) begin emerging during June and

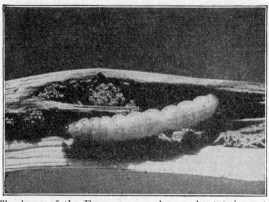

Fig. 215.—The larva of the European corn borer, about twice natural size. It is easily confused with the smartweed borer which is native to the United States. In the European corn borer the two spots on the back of each abdominal segment are usually much smaller than in the smartweed borer, and more widely separated, their distance apart usually exceeding the width of one spot. In the native species, the distance between the spots is usually less than the width of one spot. A faint stripe can usually be seen on the middorsal line of the European corn borer. (*From Ill. State Natural History Surv.*)

continue to come out in the northern states until August. The adult female moths are a pale yellowish brown with irregular darker bands running in wavy lines across the wings. The male moth is distinctly darker, having the wings heavily marked with olive brown. The moths have a wing expanse of about 1 inch. They are strong fliers but move about mainly at night. The females lay their eggs (Fig. 217) in groups of 5 to 50 on the undersides of the leaves of their food plants, especially on the lower leaves of young corn plants. Each female will lay, on the average, from 500 to 600 eggs, sometimes many more. The eggs ordinarily hatch in a week or less, depending upon the temperature; the young larvae feed until nearly half grown in spaces between closely appressed leaves, in the tassel, beneath the husks, or between the ear and the stalk. When about half-grown, they begin to eat into the stalk, the ear, or the thicker parts of the leaf stem and become borers. They continue to feed in this way until they are full-grown. Dry summers, extremely cold winters, and heavy rains at the time of hatching are very unfavorable to this insect. In dry summers many of the larvae perish before they can bore into the plant.

The large numbers of them which frequently occur in a single plant, often cause the plant stem to collapse. Three hundred and eleven borers have been taken in a hill of corn containing four stalks, and forty-two have been taken from a single ear of field corn. Entire fields sometimes average 10 borers per plant.

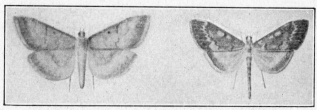

Fig. 216.—Adults of the European corn borer; female at left, male at right; slightly enlarged.
(From Mich. Agr. Exp. Sta.)

The infestations in the United States apparently did not originate from a single importation. The corn borers along the Atlantic seaboard are of a two-generation strain or variety, maturing adults in June and again in August and September; whereas those in western New York to Indiana and Wisconsin appear to have been derived from a one-generation strain. Here the adults emerge in June and July, and the larvae from their eggs become practically full-grown by September but remain as larvae in their tunnels over the winter. In the southern part of the western area, however, the insect is beginning to change over to a two-generation cycle especially when the larvae feed upon

Fig. 217.—Egg masses of European corn borer on under side of corn leaves. About natural size. *(From Spencer and Crawford, Ontario Dept. Agr. Bul. 295.)*

sweet corn. The two-generation corn borers are much more destructive, since they feed on many vegetables and flowers as well as corn. At Norfolk, Virginia, three generations and on the Island of Guam five generations have been observed in a year. In both areas, hibernation takes place mainly in the full grown larval stage.

Control Measures.—About the only effective control measure thus far developed consists in the destruction or utilization, during the fall, winter, or early spring, of all crop residues and plant refuse in which the borers may pass the winter. The most effective methods of control are *clean* plowing under of all crop and weed refuse or raking together this plant refuse and burning. The borers are killed in corn that is cut and shredded or placed in a silo. Disking the fields or *ordinary* plowing under of cornstalks is of little value. Late planting of corn is of some value in preventing injury in the one-generation areas.

Other measures which may prove of considerable importance in controlling this insect are rotations in which a maximum acreage will be in crops not seriously affected. In general the legume crops have suffered very slightly, although some damage has been done to cowpeas. Soybeans, red clover, and alfalfa have had almost no damage from this insect. If shock fodder is fed to cattle in feed lots, all the refuse stalks should be cleaned up from the lots and burned by early spring, as it has been shown that many borers may come through the winter in such stalks and spread from them to fields planted in corn in the vicinity the next season. Where corn is cut for silage or shredded for fodder, the stubble should not be left over 2 inches high. Machines for cutting the corn close to the ground have been developed to leave only a 2-inch stubble. As corn is the favorite food plant among the cultivated crops, considerable damage to other crops may be avoided, in the two-generation area, if corn is planted at some distance from them. This holds particularly true for beets, beans, spinach, rhubarb, celery, and outdoor flowering plants. All weeds should be kept down around roadsides and field margins or burned during the winter. For the protection of early sweet corn, dahlias, and other plants, the value of which warrants the expense, sprays of ground derris (4 pounds to 100 gallons of water with proper spreading agents), applied so as to thoroughly wet the whorl, the developing tassel, the ear, the junctures of leaf blades with stem, and the tillers, have been recommended. The first application should be made as soon as the eggs begin hatching and should be followed by at least three succeeding applications at 5-day intervals.[1] Valuable or valued plants, such as dahlias and other flowers, may also be largely protected by growing them beneath tobacco cloth. Millions of parasites introduced from the borer's native home in Europe have been released in American cornfields and at least a half dozen beneficial species have become established. It is hoped that they may aid greatly in control. Much research has been conducted to produce borer-resistant or borer-tolerant corn with encouraging results, but the matter is at this date in the experimental stage.

References.—Mass. Agr. Exp. Sta. Bul. 189, 1919; N. Y. (Cornell) Agr. Exp. Sta. Ext. Bul. 31, 1919; U.S.D.A., Farmers' Bul. 1548, 1935, and Dept. Bul. 1476, 1927, and Tech. Buls. 53, 1927, and 77, 1928; Ill. Agr. Exp. Sta. Cir. 321, 1928; Ohio Agr. Exp. Sta. Bul. 429, 1928; Conn. Agr. Exp. Sta. Bul. 395, 1937; U.S.D.A., Cir. 132, 1930.

9. Common Stalk Borer[2]

Importance and Type of Injury.—Injury by the common stalk borer is usually confined to the margins of the field for 2 or 3 to 20 rows into the field. Occasionally when fields are very weedy the previous year, the injury may extend over the entire field. Stalks of corn or other plants attacked by this insect will show irregular rows of holes through the unfolding leaves. Plants will often show an unnatural growth, twisting or bending over, presenting a stunted appearance, and often not producing ears. Holes will be found in the sides of the stalks with moist castings being thrown out. Inside the stalks will be found very active dark brown worms ranging from ¾

[1] See U.S.D.A., Bur. Ent. Plant Quar., Cir. E-410, June, 1937; Conn. Agr. Exp. Sta. Bul. 395, June, 1937.

[2] Papaipema nebris Guenee, Order Lepidoptera, Family Noctuidae.

inch to nearly 2 inches in length. All but the larger of these worms (Fig. 218,*b*) have a single continuous white stripe down the back, with broken white stripes on the sides extending from the head about one-sixth the length of the body, interrupted for one-fourth the length and then starting again and extending over the posterior half of the body.

Plants Attacked.—This insect is almost a universal plant feeder, attacking and working in stems of any plants large enough to shelter it and soft enough so that it can bore into them. It apparently prefers giant ragweed and corn in its later stages. It is often a pest in flower and vegetable gardens, where it tunnels in the stalks of dahlias, hollyhocks, tiger lilies, asters, rhubarb, peppers, potatoes, tomatoes and tobacco.

Distribution.—It is distributed generally throughout the United States east of the Rocky Mountains

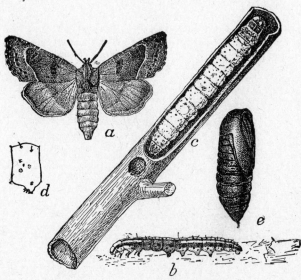

Fig. 218.—The common stalk borer, *Papaipema nebris* Guenee. *a*, adult; *b*, half-grown larva, side view; *c*, full-grown larva in burrow; *d*, side of one segment of larva; *e*, pupa. All slightly enlarged. (*From Chittenden, U.S.D.A.*)

Life History, Appearance, and Habits—So far as known, the insect hibernates only in the egg stage, the subglobular, whitish to grayish, ridged eggs being laid during the late summer and fall on grasses and weeds, often in the creases of rolled or folded leaves, sometimes over 2,000 from a single female. They hatch very early in the spring, mostly in May, into the small, brown, white-striped caterpillars which frequently bore first into the stems of grasses, particularly bluegrass, on which the eggs are often laid. They may enter corn and other plants at the side of the stem and burrow upward, destroying the heart of the plant while the outer leaves remain green, or they may enter near the top of a stem and work downward, causing that part of the plant to wilt and die. As the caterpillars grow and grass stems become too small for them, they migrate, about July 1, to attack larger-stemmed plants. They are extremely uneasy individuals, never seeming contented with their location, and frequently changing from the stem of one plant to that of another. This habit is responsible for their causing a much greater amount of injury than would be the case if they stayed in one plant. While the giant ragweed seems to be the favorite food of this insect, it will frequently leave these weeds for corn or garden crops, if the two are growing close together. The worms become full-grown during the latter part of July and the first of August, at which time they may lose their striping, and become a plain dirty-grayish color (Fig. 218,*c*). They are then about 1 to 1½ inches in length. They transform just below the soil surface or, rarely, inside the stems of their food plants into a brown pupal stage (Fig. 218,*e*) and after 2 to 6 weeks, emerge, in late August and September as grayish moths (Fig. 218,*a*) with a wing expanse of a little over 1 inch. The front wings are usually a dark grayish brown, with a number of small white spots on the disk and along costal and apical margins, or (in the variety *nitela*) without spots and the apical third with diagonal lines or bands of lighter and darker color. The hind wings are a pale gray-brown. There is considerable variation in the appearance of the adults. They mate and the females lay their eggs on grasses and weeds as above mentioned. Eggs have been taken on ragweed, dock, pigweed, burdock, and several grasses. There is only one generation a year.

Control Measures.—A thorough clean-up preferably by burning over the margins of the fields during the fall and a cleanup and burning of all crop refuse is the best means of fighting this insect. Mowing fence rows about mid-August, just before the adults begin laying eggs, will greatly reduce the number of eggs laid about those fields; but mowing in the early part of the growing season may drive great numbers of the borers into adjacent crops. Individual plants may be saved by splitting the infested stem very carefully lengthwise, removing the borer and binding the stem together; or by injecting, from an oil can, a half teaspoonful of carbon disulphide through the entrance hole and plugging it with clay or gum.

References.—*Twenty-third Rept. Ill. State Ento.*, p. 44, 1905; *Rept. N. J. State Ento.* for 1905, p. 584, 1906; *Iowa Agr. Exp. Sta. Res. Bul.* 143, 1931.

10. SOUTHERN CORNSTALK BORER[1]

Importance and Type of Injury.—This is one of the most destructive corn insects in many parts of the South, often responsible for reduction in yields of 15 to 50 per cent; but, because of the insidious method of attack, the damage it does is not generally appreciated. Corn infested by the Southern cornstalk borer is usually twisted and stunted, often with an enlargement of the stalk at the surface of the ground. The leaves will sometimes be ragged, broken, and dangling, showing many holes along the leaf which have been eaten out while it was still curled in the heart of the plant. Inside of the stalk, usually well above the ground, will be found dirty grayish-white worms about 1 inch in length when full-grown, conspicuously marked with many dark-brown spots (Fig. 219).

Plants Attacked.—The southern cornstalk borer feeds principally on corn. It has also been taken on sorghum and Johnson grass.

Distribution.—This is a southern insect, damage being limited mainly to states from Maryland and Kansas on the north to, and including, the southern and southwestern states. The insect is found also in Mexico, and southward to South America.

Life History, Appearance, and Habits.—The winter is passed in the full-grown larval stage, mostly in the taproots of the old cornstalks. The larvae, which are about 1 inch long, are yellowish, with very pale spots during the winter; but during their feeding period in the summer they are conspicuously spotted with eight, rounded, brown or black spots in a transverse row on the front of each body segment and two others behind these. The insect will be found in the lower part of the stalk, just above the roots. It remains in the larval condition until early spring when it changes inside the stalk to a naked brown pupa, the larva having first made a silk-lined exit tunnel to the outside of the stalk, the cover to which is not completely eaten away. The adult moths emerge from the

[1] *Diatraea crambidoides* Grote (= *D. zeacolella* Dyar), Order Lepidoptera, Family Pyralididae.

larval burrows in midspring. These moths are of a general light-straw color, with a wing expanse of 1¼ inches. The labial palps extend in front of the head like a short beak. They are active at night only, unless disturbed. They lay their flattened, whitish or yellow, oval eggs, in small groups, overlapping shingle-fashion. Each female commonly lays 300 to 400 eggs on the undersides of the leaves. The worms hatching from these eggs feed at first on the leaves but soon enter the stalk, boring up and down in the pith. They change from one plant to another, as does the common stalk borer. The first generation worms become full-grown a little before midsummer and pupate inside the stalks. Those of the second generation reach maturity in the early fall and may remain as larvae during the winter. There are from one to three genera-

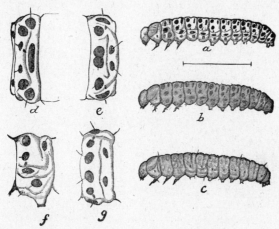

Fig. 219.—The southern corn-stalk borer, *Diatraea crambidoides* Grote. *a, b, c,* color varieties of the larvae; *d,* third thoracic segment; *e,* eighth abdominal segment; *f,* and *g,* a middle segment in side and dorsal views. The line shows natural size. (*From Howard, U.S.D.A.*)

tions annually. The shortest recorded development from egg to adult is 36 days.

Control Measures.—As the insects hibernate in the stalks of corn and other food plants, a thorough cleaning up and burning of cornstalk refuse and corn stubble, immediately after harvest, is the most effective control measure. Rotation in which corn follows some other crop, at as great distance as practicable, is also helpful in keeping down this insect. Late-fall and winter plowing followed by thorough harrowing or breaking will destroy most of the hibernating larvae in the stalks.

References.—*Va. Agr. Exp. Sta., Tech. Bul.* 22, 1921; *N. C. Dept. Agr. Bul.* 274, 1920; *S. C. Exp. Sta. Bul.* 294, 1934; *U.S.D.A., Tech. Bul.* 41, 1928.

11. LESSER CORNSTALK BORER[1]

Importance and Type of Injury.—In the southern part of the United States, corn is sometimes injured by slender greenish worms boring into the lower part of the stalk, usually within 2 inches of the soil surface. Corn under 18 or 20 inches high attacked by this insect becomes much distorted and curled, and frequently fails to

---

[1] *Elasmopalpus lignosellus* (Zeller), Order Lepidoptera, Family Pyralididae.

produce ears or good stalks.   A dirt-covered silken tube usually leads away from the tunnel in the plant.   Mostly troublesome on dry sandy soils.

Plants Attacked.—Besides corn, the insect feeds on peas, beans, peanut, cowpeas, crab grass, Johnson grass, wheat, turnips, and several other crops.

Distribution.—The range of the insect is from Maine to southern California but it is largely confined, so far as injury goes, to the more southern states.   It is also found in Mexico, Central America, and South America.

Life History, Appearance, and Habits.—The insect hibernates in the larval, pupal, and adult stages, but usually in the southern states as a larva, which transforms to a pupa before spring.   The moths (Fig. 220,*a,c*) emerge from the pupae, or become active, early in the spring and lay their greenish-white eggs on the leaves or stem of the plants on which the larvae feed.   The eggs hatch in about 1 week; the bluish-green brown-striped caterpillars (Fig. 220,*d*) feed at first on the leaves or roots but later burrow into the stems of corn and other plants.   They eat into the heart of the

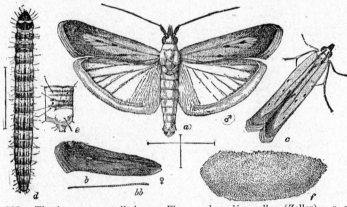

FIG. 220.—The lesser cornstalk borer, *Elasmopalpus lignosellus* (Zeller).   *a*, male; *b*, forewing of dark female; *bb*, antenna of female; *c*, male at rest; *d*, larva, dorsal view; *e*, side view of a middle segment; *f*, cocoon.   Lines indicate natural size.   (*From Chittenden, U.S.D.A.*)

unfolding leaves of the corn, sometimes killing it.   The larvae become full-grown in about 2 to 3 weeks.   Their presence in the corn is indicated by masses of borings which are pushed out through the holes in the stalk.   The insects leave their burrows when full-grown and spin silken cocoons (Fig. 220,*f*) under trash on the surface of the ground, in which they change to brownish pupae about $\frac{1}{3}$ inch long.   From these the moths emerge in from 2 to 3 weeks.   They have a wing expanse of nearly 1 inch. The front wings are brownish yellow with grayish margins with several dark spots. In the female the front wings are nearly black.   A second generation is produced in all the southern states where this insect is injurious.

Control Measures.—As is the case with the common stalk borer, practically the only effective control is fall or winter cleanup of the fields and field margin and rotations with crops not attacked by this pest.   Winter plowing has given control in some cases or is at least helpful in reducing the numbers of the insects in the field.   Early planting also is recommended in some sections in the southern states as a control for this pest.

Reference.—*U.S.D.A., Dept. Bul.* 539, 1917.

12 (63, 71, 111, 260).   CORN EARWORM[1]

Importance and Type of Injury.—This insect has been called the worst pest of corn, considering the United States as a whole.   It is claimed that

[1] *Heliothis obsoleta* Fabricius, Order Lepidoptera, Family Noctuidae.

the American farmers grow on the average two million acres of corn each year to feed the corn earworm. Corn attacked by the corn earworm will show the ears with masses of moist castings at the end, and the kernels, especially about the tip of the ear, eaten down to the cob by large brownish to greenish, striped worms, which are nearly 2 inches long when full-grown (Fig. 221). In the worst years 70 to 98 per cent of the ears of field corn the country over are attacked and as high as 5 to 7 per cent of the kernels of field corn and 10 to 15 per cent of canning corn may be consumed. By cutting off silk, pollination is prevented and nubbins

Fig. 221.—Nearly full-grown larva of corn earworm, feeding in tip of ear of corn. (*From Britton, Twenty-first Rept. State Ento., Conn., 1897.*)

result. Molds may be carried in through the husks resulting in diseased ears that may cause death among livestock to which they are fed. The presence of the worms in ears of sweet corn is most repulsive to consumers and very troublesome to commercial canners. These worms vary greatly in color from a light green or pink to brown or nearly black and are lighter on the under parts. They are marked with alternating light and dark stripes running lengthwise of the body. The stripes are not always the same on different individuals, but there is usually a double middorsal dark line the length of the body. The head is yellow and unspotted, and the legs are dark or nearly black. The skin of the insect is somewhat coarse and, when looked at under a magnifying glass, shows many small thorn-like projections. Usually the injury starts at the tip of the ear. Occasionally the worms enter through the side or at the butt.

Plants Attacked.—The corn earworm is a very general feeder, attacking many cultivated crops and weeds. It is seriously injurious to the tomato, tobacco, cotton, and vetch, as well as to corn, and has been given the names of tobacco budworm, tomato fruitworm, cotton bollworm and vetchworm.

Distribution.—This insect is of world-wide distribution. Its damage is most severe, however, in the South.

Life History, Appearance, and Habits.—In the southern United States, at least, the insect passes the winter in the form of a brown pupa, which will be found from 2 to 6 inches below the surface of the soil. In the spring and early summer moths (Fig. 222) emerge from these pupae and crawl up exit holes which the larvae prepared before pupating. The moths have a wing expanse of about 1½ inches. They vary in color, the average having the front wings of a light grayish brown, marked

Fig. 222.—Adult corn earworm with wings spread. Somewhat enlarged. (*From Ill. State Natural History Surv.*)

with dark-gray irregular lines and with a dark area near the tip of the wing. The irregular lines often shade into an olive-green. The hind wings are white with some dark spots or irregular dark markings.

The moths fly during warm cloudy days, but mainly at dusk of the evening. They feed on the nectar of many flowers and, during the warm evenings, deposit their eggs on the plants in which the larvae feed. Each moth will lay from 500 to as many as 3,000 eggs, the average being probably over 1,000. These eggs are laid singly and are of a hemispherical shape with ridges along their sides resembling very much a minute sea urchin. They are yellowish, about half as large as a common pinhead. The early generations feed in the curl of young corn plants and on tomatoes, cotton, tobacco, beans, and legumes. Fresh corn silk is one of the favorite places for egg laying of the moths of the later generations. The eggs hatch in from 2 to 10 days, and the worms feed at first on the leaves, or bore directly into the corn silk. They feed on the silk until it becomes dry and then on the kernels at the tip of the ear for 2 to 4 weeks, molting 5 times. On becoming full-grown the larvae crawl down the stalk or drop to the ground, into which they burrow and excavate a small smooth-walled cell, commonly 3 to 5 inches deep, where they pupate, coming out as moths again after another period of 10 to 25 days, although this period may be prolonged during cold weather. The worms do not always remain in the first ear which they entered but frequently go from one ear to another. They are cannibalistic and usually only one full-grown worm is found in each ear. There are from two to three generations of the insect each season, the number depending on the latitude. While the winter is passed in the pupal stage, it is very doubtful in view of recent work on the life history of this insect if the pupae ever survive the winter north of about 40° north latitude. The infestations in the northern states are, in all probability, caused by a migration of the adults from points farther south. Some moths may also develop from worms in green corn or string beans, shipped to northern markets in the early spring.

Control Measures.—No effective control of the corn earworm in field corn has been devised. Covering the ends of the ears shortly after the silks appear with paper bags, held with a rubber band; clipping off the silks and tips of the husks as soon as pollination is completed to remove eggs and young larvae; and inserting small quantities of hexachlorethane (0.5-gram tablets) inserted into the tip of the ear beneath the husks and held with a wire clip to destroy the larvae as a fumigant; or treating the fresh silks with a highly refined mineral oil of 100 to 200 viscosity are the measures proposed to protect corn intended for table use.

The time of planting will have a marked effect on injury by this insect but will not always be the same in different years, *i.e.*, in some years early-planted corn will be injured, while in most years, the latest corn suffers the worst damage. The moths prefer to lay their eggs on fresh corn silk, so that corn which silks before or after the greatest abundance of moths will largely escape infestation (see also pages 452, 463, and 525). Fall plowing to disturb the overwintering pupal stage may be of importance in reducing the numbers of moths that emerge in the spring, in areas where the insect winters in the pupal stage.

References.—Ky. Agr. Exp. Sta. Bul. 187, 1914; *U.S.D.A., Farmers' Bul.* 1310, 1923; *Jour. Econ. Ento.,* **9**: 395, 1916; and **13**: 242, 1920; *U.S.D.A., Tech. Bul.* 561, 1937; *Md. Agr. Exp. Sta. Bul.* 348, 1933; *Ark. Agr. Exp. Sta. Bul.* 320, 1935; *Jour. Econ. Ento.,* **31**: 459, 1938.

OTHER INSECTS FEEDING ON THE EAR

A large number of different species of beetles are sometimes found feeding on the tip of corn ears where the kernels have been exposed by the feeding of the corn earworm or by birds. These beetles may be regarded as accidental feeders on corn, and practically never attack uninjured ears. The European corn borer described on page 360 is the most serious insect attacking the ears of corn, in sections where it is abundant. The Japanese beetle also causes severe damage to corn, in areas where it is abundant, by feeding on the husks, kernels, and silk at the tip of the ear (see page 621).

13 (285). CORN ROOT APHID[1] AND CORNFIELD ANT[2]

Importance and Type of Injury.—Corn infested by the corn root aphid, and its ever-attendant ant, germinates normally, and the plants reach a height of from 3 or 4 to 9 or 10 inches, when growth becomes greatly retarded, especially during dry years. The plants often take on a yellowish or reddish tinge to the leaves. An examination of the field will show numerous small anthills around the injured corn plants, and small brownish ants tunneling along the corn roots. Clinging to the corn roots will be found many bluish-green aphids, about the size of pinheads, when fullgrown, the younger aphids being much smaller.

Plants Attacked.—This species of aphid is known to infest the roots of corn, cotton (on which also it is a serious pest), a number of different grasses, and several weeds, particularly smartweed. A very similar species[3] occurs on the roots of aster.

[1] *Anuraphis maidi-radicis* (Forbes), Order Homoptera, Family Aphididae.
[2] *Lasius niger americanus* Emery, Order Hymenoptera, Family Formicidae.
[3] *Anuraphis middletoni* (Thomas).

Distribution.—The insect is common throughout the corn- and cotton-growing areas east of the Rocky Mountains.

Life History, Appearance, and Habits.—The winter is passed, at least in the northern part of the country, only in the egg stage. These eggs (Fig. 223,*E*) are collected in the fall by the small brown cornfield ants and stored in their nests over winter. The ants (Fig. 223,*F*) pile the

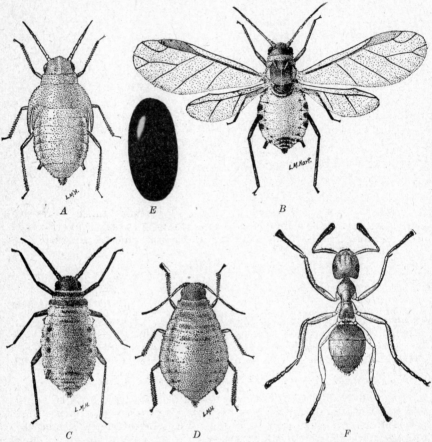

Fig. 223.—The corn root aphid and the corn field ant. *A*, full grown nymph of the winged female aphid, × 20; *B*, winged ovoviviparous female, which produces young without mating, × 14; *C*, the male aphid, which appears only in autumn, × 25; *D*, his mate, the oviparous female, which also occurs only in autumn, × 14; *E*, one of the overwintering eggs produced by the pair, × 25; *F*, worker of the corn field ant, which cares for the aphids, × 8. (*From Ill. State Natural History Surv.*)

eggs in their nests and move them about according to moisture and temperature conditions in the soil. In the early spring, about the time the young smartweed plants begin to appear in the field, the aphid eggs begin hatching. The ants seem instinctively to know that the young aphids must have something to feed on and carry them in their jaws to the roots of the smartweed and some of the grasses on which the aphids feed. Here the young aphids insert their beaks and suck the sap, growing rather rapidly, and in about 2 or 3 weeks all become full-grown females and begin giving birth to living female young. These young become mature females and begin giving birth to others after a period of from 8 days to 2 weeks, reproduction being by parthenogenesis. Only female aphids appear during the summer months. During July and August winged individuals (Fig. 223,*B*) frequently make their appearance on the roots and will sometimes crawl to the surface of the ground and fly to other fields, thus spreading the infestation. Local distribution of the aphid in the field is almost entirely dependent on the cornfield ants. This aphid is rarely found on roots except where attended by the ants and, if placed on the surface of the ground, is apparently helpless so far as finding a place to feed is concerned. An ant finding one of these aphids, however, immediately picks it up, carries it underground, and places it on the roots of one of its food plants. The aphids apparently have been dependent on the ants so long that they have entirely lost the faculty of taking care of themselves. This interrelation between the ant and the aphid is one of mutual benefit, as the aphid is protected and kept by the ant where a supply of food is accessible; while the ant, in turn, derives a large part of its food from the sweet sticky exudation known as *honeydew* given off from the anal opening of the aphid. Honeydew is slightly modified sap which the aphids suck from the plant in excess of their needs, so that the ants, in effect, make use of the piercing-sucking mouth parts of the aphids to get their food. The ants act as though they knew the importance of preserving the eggs of the aphids over winter and that corn is the favorite food of these aphids. They will carry the aphids for some distance in order to pasture them on the roots of corn. In one instance under observation, the ants moved 156 feet from a timothy meadow into the third row of a field of corn, carrying with them not only their own young, but also a large number of aphids which they began pasturing on the corn roots.

While only the female aphids are present in the fields during the summer, on the approach of cold weather these females give birth to different forms which consist of wingless true males and females. These (Fig. 223,*C,D*) mate, and the females, instead of giving birth to young, lay dark-green shiny eggs, dying shortly afterward. The eggs are gathered up by the ants and stored in their nests during the winter.

Control Measures.—The best and most effective method of combating these aphids is a measure directed, not against the aphids, but against their attendant ants. This consists of thorough deep cultivation of the soil early in the spring before planting. The land should be plowed $6\frac{1}{2}$ to 7 inches deep, followed by two or three deep diskings at about 3-day intervals. The object of this heavy cultivation is to break up, scatter, and destroy the ant nests in the soil. If the field is plowed to a depth of only $4\frac{1}{2}$ or 5 inches, many of the lower chambers of the ant nests will not be thrown out in the furrow; 6-inch plowing, however, throws out about 95 per cent of the ant nests, and the disking, following this, breaks up and scatters the young of the ants and their aphids, so that they are not able to reestablish their nests in time to cause injury to the young corn. This treatment will also drive many of the ants out of the fields, as has been shown by watching at night the margins of fields where such treatments have been given.

Extensive experiments have been carried on to find a treatment which could be applied to seed corn that would repel the cornfield ants and thus prevent their placing the corn root aphids on the roots of young corn. Owing to the fact that soil is one of the best deodorants, most of the chemicals which have been used for treating the seed have not proved effective. One of the most effective treatments of this sort has been to moisten the seed thoroughly with a solution made by stirring 3 to 4 fluid ounces of oil of tansy into 1 gallon of wood alcohol, and moistening, but not wetting, the corn seed with this solution before planting. Some injury is likely to occur if cool, wet weather follows planting of such treated seed.

References.—*Ill. Agr. Exp. Sta. Buls*, 178, 1915, and 130, 1908; *U.S.D.A., Farmers' Bul.* 891, 1917; *Forty-sixth Rept. Ill. Agr. Exp. Sta.*, 1933.

14 (39, 87, 216, 273). WHITE GRUBS OR JUNE BEETLES[1]

Importance and Type of Injury.—White grubs are among the most destructive and troublesome of soil insects. When fields infested with white grubs are planted to corn, the corn usually comes up but the plants cease growing after reaching a height of from 8 inches to 2 feet. The corn will show a patchy growth with varying sized areas in the field where the plants are dead or dying. If the injured plants are pulled up, the roots will be found to have been eaten off, and from 1 to as many as 200 white curved-bodied grubs, from ½ to over 1 inch in length, will be found in the soil about the roots. The grubs (Fig. 224,*lower figure*) are white with brown heads and six prominent legs. The hind part of the body is smooth and shiny, with dark body contents showing through the skin. There are two rows of minute hairs on the underside of the last segment which will distinguish the true white grubs from similar-looking larvae (see Fig. 402,*E*). Injury is usually most severe to crops following sod.

Plants Attacked.—All grasses and grain crops; potatoes, beans, strawberries, roses, nursery stock, and nearly all cultivated crops.

Distribution.—Throughout North America.

Life History, Appearance, and Habits.—The winter is passed in the soil both as adults and as larvae of several distinct sizes. In the spring after the trees have put forth leaves, the adults (Fig. 224,*center*) become active, fly about during the night, and feed on the foliage of trees and the leaves of some other plants. They leave the soil just at dusk and remain on the trees during the night, mating and feeding. At the first streaks of dawn, they return very promptly to the soil, where the females lay their pearly white eggs (Fig. 224,*left*) from one to several inches below the surface. The eggs are generally laid in grasslands or patches of grassy weeds in cultivated fields. Clean stands of clover or alfalfa and clean-cultivated row crops are not likely to be infested by the egg-laying females. The eggs hatch in 2 or 3 weeks, and the young grubs feed on the roots and underground parts of plants until early fall, when they are about ½ inch long. They then work their way down in the soil usually below the frost line and have been taken 5 feet below the surface.

As the soil warms in the spring, they work upward and, by the time plant growth is well started, they are feeding a few inches below the surface. Feeding continues throughout the season and, on the approach of cold weather, they again go deep into the soil, where the second winter is passed, the grubs then being about 1 inch long. The third season they

[1] *Phyllophaga* or *Lachnosterna* spp., Order Coleoptera, Family Scarabaeidae.

come up near the surface of the ground and feed until late spring or early summer; they then change to the pupal stage (Fig. 224,*upper figure*) in cells in the earth about 6 or 8 inches below the surface.   During the latter part of the summer, they change to the adult beetle but do not leave the soil until the following spring.   There may be some movement of the

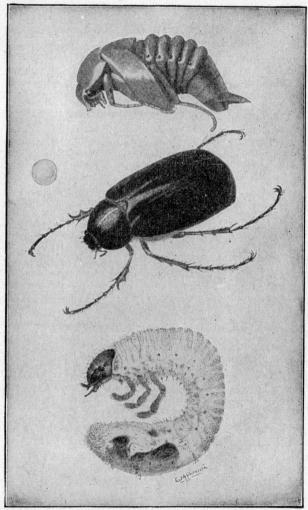

Fig. 224.—Life stages of a white grub, *Phyllophaga rugosa* (Melsheimer).   Above, pupa; at left, egg; at center, adult; below, larva.   (*From Ill. State Natural History Surv.*)

beetles downward to below the line of severe freezing.   The overwintering population of white grubs, therefore, consists of adults that have not yet taken flight from the soil and of larvae usually of two distinct sizes, the smaller about 9 months, and the larger about 1 year and 9 months old.

The adults are the well-known, brown, or brownish-black June beetles, May beetles, or "daw bugs."   About 200 species are known.   They vary somewhat in their life history, some completing their growth in 1 year, while others require as much as 4 years. The 3-year life cycle is by far the most common. There are several other closely related beetles that attack corn, the grubs differing somewhat in structure but having the same general appearance. There are also a number of grubs, very similar in appearance, which occur in the soil and manure, but which eat only decaying vegetable matter and do not feed on the living roots of plants. These lack the double row of spines on the underside of the last body segment which is characteristic of the true white grubs.

Control Measures.—The most important control measures for white grubs are based upon three observations regarding the life cycle. (*a*) The larvae prefer to feed upon crops of the grass family, such as corn and other cereals, upon potatoes, or upon strawberries; while legumes such as clover,

Fig. 225.—A badly injured pasture sod; the bluegrass roots have been entirely eaten off by white grubs, permitting the rolling back of the sod, like a carpet, to expose the grubs. (*From Ill. State Natural History Surv.*)

alfalfa, and soybeans are much less severely damaged. Consequently, land in which numerous white grubs are found while plowing the soil should not be planted to corn, potatoes, or other seriously injured crops, but to soybeans, clovers, or one of the small grains. (*b*) The beetles prefer to lay their eggs in grassy, weedy fields and will not as a rule deposit eggs in fields of clover or alfalfa, unless there is a considerable admixture of grass or weeds in such fields. (*c*) While white grubs are troublesome every year, the most severe injury occurs in regular 3-year cycles. This is because most of the insects reach the adult stage in the years 1938, 1941, 1944 and each third year thereafter. Severe damage occurs the year after the adults are abundant and lay their eggs, or the second year of the life cycle. Throughout the central and eastern United States, the years of most severe damage, as shown by Davis, will be 1939, 1942, 1945, and each third year thereafter. During the years when heavy flights of beetles are expected, farmers should make every effort to keep corn and other cultivated fields free of grass and weedy growth during April, May, and June and would do well to have as much of their farms in legume crops as possible. This will reduce the number of eggs laid on their farms. The year following heavy flights of May beetles, it is well to avoid planting corn and potatoes on fields that were in sod or covered with a grassy, weedy growth the preceding spring.

One of the best ways of cleaning grubs out of fields is to pasture the land with hogs during summer and early fall. When hogs are allowed to run on heavily infested land, they will usually root out and eat the grubs, fattening themselves and nearly freeing the land of grubs. If this is done during a summer in which most of the grubs hatch from eggs (1941, 1944, etc.), most of the damage by these pests may be prevented. Hogs intended for breeding stock should not be pastured on grub-infested land, however, as the white grubs are the intermediate host of the giant, thorn-headed worm,[1] one of the intestinal parasites of hogs.

Plowing infested fields between mid-July and mid-August, especially in the year when most of the grubs are pupating (1940, 1943, etc.), will kill many of the pupae and newly transformed adults. Plowing until about October first will also crush many larvae and expose them to birds, if done before cool weather when they work their way down below the plow line.

White grubs in the soil are attacked by several insect parasites, especially the larvae of certain wasps[2] which often greatly reduce their numbers, and whose cocoons are often abundant in grub-infested fields. Birds, especially crows and blackbirds often follow the plow, picking up the grubs as they are turned out in the furrow. On lawns, parks, and golf courses, where the expense is not prohibitive, the use of soil insecticides, such as lead arsenate and carbon bisulphide emulsion, will kill the grubs in the soil. Directions for their use are given on page 392.

References.—Ill. Agr. Exp. Sta. Buls. 116, 1907, 186, 1916, and 187, 1916; *U.S.D.A., Farmers' Bul.* 1798, 1938; *Jour. Econ. Ento.*, **24**: 450–452, 1931, and **30**: 615–618, 1937.

15 (37, 68, 86, 274). WIREWORMS[3]

Importance and Type of Injury.—Wireworms are among the most difficult insects to control, the most destructive and most widespread pests of corn, small grains, grasses, potatoes, and other root crops, vegetables, and flowers. Crops that are attacked by wireworms often fail to germinate, as the insects eat the germ of the seeds or hollow them out completely, leaving only the seed coat. The crop may not come up well, or it may start well and later become thin and patchy, because the worms bore into the underground part of the stem causing the plantlet to wither and die, though they do not cut it off completely. Later in the season the worms continue to feed upon the small roots of many plants. The wireworm larvae (Fig. 226) are usually hard, dark-brown, smooth, wire-like worms, varying from $\frac{1}{2}$ to $1\frac{1}{2}$ inches in length when grown. Some species are soft, and white or yellowish in color. Their injuries are usually most severe to crops planted on sod ground, or the second year from sod.

[1] *Macrocanthorhynchus hirudinaceus*, Phylum Nemathelminthes.
[2] *Tiphia* spp. and *Elis* spp., Order Hymenoptera, Family Scoliidae.
[3] Many species of the Order Coleoptera, Family Elateridae.

Plants Attacked.—Wireworms are especially destructive to corn and grasses, but all the small grains and nearly all cultivated and wild grasses are attacked.   Among the garden crops severely damaged are potatoes, beets, sugarbeets, cabbage, lettuce, radish, carrot, beans, peas, onions, asters, phlox, gladioli, and dahlias.   Leguminous plants, such as velvet

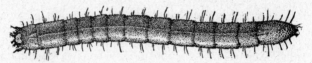

Fig. 226.—Larva of the corn wireworm, *Melanotus cribulosus* (Leconte), about 3 times natural size.   (*From Ill. State Natural History Surv.*)

beans, and certain small grains, such as oats, are more resistant to wireworms than other crops, but clovers, alfalfa, peas, and beans may suffer considerable damage.

Distribution.—Throughout North America and most of the world.

Life History, Appearance, and Habits.—There are many different species of wireworms that attack our cultivated crops, including corn.

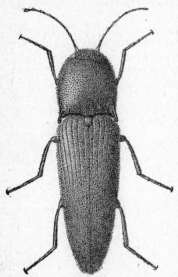

Fig. 227.—A click beetle, adult of the wireworm, *Melanotus fissilis* (Say), about 5 times natural size. (*From Ill. State Natural History Surv.*)

The winter is passed mainly in the larval and adult stages in the ground.   In the early spring the adults (Fig. 227) become active and fly about, some species being strongly attracted to sweets; these can be taken in large numbers by placing a few drops of sirup on the tops of fence posts or other exposed places out-of-doors. They are "hard-shelled," usually brownish, grayish, or nearly black in color, somewhat elongated, "streamlined" beetles, with the body tapering more or less toward each end.   The head and thorax fit closely against the wing covers, which protect the back of the abdomen.   The joint just in front of the wing covers is loose and flexible, and, when the beetles are placed or fall on their backs, they flip the middle part of the body against the ground in such a manner as to throw themselves several inches into the air.   The chances of their alighting on their feet seem to be about fifty-fifty; but they will generally keep trying until they come down right side up, when they make use of their legs to escape.   This habit has afforded amusement to most country boys and girls and has given the insects such names as click-beetles, snapping beetles, and skipjacks.   The females of the species that are most injurious to corn burrow into the soil and lay their eggs mainly around the roots of grasses.   The adults live 10 to 12 months, most of which time, and all that of the other stages, is spent in the soil. The egg stage requires a few days to a few weeks.   The larvae hatching from these eggs spend from 2 to 6 years in the soil feeding on the roots of grasses and other plants. As the soil becomes hot and dry, the larvae migrate downward so that it is often hard to find them in dry summer weather, even in severely infested fields. The last segment of the larva is usually characteristically ornamented and serves to distinguish different species during this stage (Fig. 228). Most species change to a naked, soft pupa, and a few weeks later to the adult stage, in cells in the ground, during the late summer or fall of the year in which they become full-grown. The adults, which are commonly about ½ inch long, remain in the soil until the following spring. There is much overlapping of the generations so that all stages and nearly all sizes of larvae may be found in the soil at one time. The larvae move only a few yards, at most, during their long lifetime, and the adults often remain and lay their eggs near where they developed so that marked differences in infestation often occur in near-by fields.

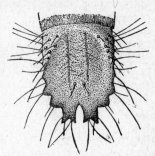

Fig. 228.—Terminal segment of larva of a wireworm, *Drasterius elegans* Fabricius, much enlarged. (*From Ill. State Natural History Surv.*)

Control Measures.—On large acreages devoted to field crops, chief dependence for avoiding wireworm damage will have to be placed upon cultural control measures. The nature of these measures will vary with the species of wireworm present and much more research is needed to develop the best control for the various sections of the country. In the northwestern United States and western Canada, clean summer fallowing or resting of the land every second or third year; shallow tillage to prevent the growth of all vegetation, especially large weeds, in the early part of the summer; and the avoidance of deep plowing which allows the wireworms to penetrate into the soil more successfully are recommended. On the other hand plowing to a depth of 9 inches about the first of August and allowing the dry lumpy soil to lie undisturbed for a few weeks is said to kill great numbers of the pupae and adults, by breaking up their cells in the soil. Hawkins claims that from 1 to 4 years of clean cultivation of infested fields will reduce the wireworm population to harmless numbers, in Maine, and that new generations do not start to any extent in cleanly cultivated fields. In parts of the West, however, the heaviest infestations of wireworms are in fields that have been under continuous cultivation. The growing of small grains or legumes on second-year sod also helps in lessening the damage by these insects, as the percentage of such plants killed by them is not large enough seriously to affect the yield of these crops. Hay crops or long-standing pasture or grain crops are favorable to wireworms and should be omitted as much as possible from the rotation. Certain species of wireworms are abundant only in poorly drained soils. The proper draining of such soils will entirely prevent damage by these species. In irrigated districts all stages of wireworms can be killed by flooding the land so that the water stands a few inches deep for a week during warm weather when the temperature of the soil at a depth of 6 inches averages 70°F. or higher. Allowing the top 18 inches of soil to become very dry for several weeks during the summer at least once in 6 years is also recommended in the Pacific Northwest. For controls practicable on smaller acreages or high-priced crops see page 483. The most effective rotations, methods of tillage, planting dates, and other farm practices will have to be worked out for each farm area, with especial reference to the kinds of crops desired and the particular species of wireworms present. Extensive experiments carried on at a number of different points to develop a treatment for seed corn that would prevent wireworm injury have failed in showing any practical method by which this can be done.

References.—U.S.D.A., Farmers' Buls. 725, 1916, and 733, 1916; Can. Dept. Agr., Pamphlet 33 (n.s.), 1923; Maine Agr. Exp. Sta. Bul. 381, 1936; Clemson, S. C., Agr. Coll. Cir. 163, 1938; Pa. State Coll. Exp. Sta. Bul. 259, 1930; Can. Dept. Agr., Ent. Branch, Saskatoon Leaflet 35, 1933.

16. NORTHERN CORN ROOTWORM[1]

Importance and Type of Injury.—This is perhaps the most important corn pest in the upper Mississippi Valley. Corn makes a slow growth, this checking of growth being most noticeable about the time the tassel appears. Plants are undersized and frequently fall over after a heavy rain. Small roots are eaten off and larger roots tunneled by thread-like,

FIG. 229.—Northern corn rootworm, Diabrotica longicornis (Say), larva, 5 times natural size. (From Ill. State Natural History Surv.)

white worms about ½ inch long (Fig. 229) with yellowish-brown heads and six small legs on the fore part of the body. The skin of the body is somewhat wrinkled. The larvae of this insect are also able to transmit bacterial wilt of corn (see page 386). The insects are so numerous that fields are practically sure to be heavily infested by the end of the second year in corn.

Plants Attacked.—The larvae attack only corn, so far as known. The adults feed on a large number of plants that flower in summer and early fall.

Distribution.—The insect occurs in greatest abundance in the north part of the Mississippi Valley. It is not injurious in the southern states, and occurs in small numbers east of New York and west of Kansas and Nebraska.

Life History, Appearance, and Habits.—The winter is passed only in the egg state. The eggs are deposited in the fall in the ground around the roots of corn, and in no other known situation. They hatch rather late in the spring, and the larvae work through the ground until they encounter the roots of corn. Although the larvae feed to a slight extent on some of the native grasses, most of them will die if corn is not planted in the field where the eggs were laid. Larvae have been found in the roots of corn growing in ground following 1 year in oats. The worms burrow through the roots, making small brown tunnels. They become full-grown during July, leave the roots, and pupate in cells in the soil, the pupa being pure white and very soft. The adult stage (Fig. 230) is reached during the latter part of July and August. The beetles leave the soil and feed on the silk of corn and the pollen of this and many other plants. The eggs

[1] Diabrotica longicornis (Say), Order Coleoptera, Family Chrysomelidae.

are deposited in cornfields during September and October, and nearly all the beetles die at the time of the first heavy frost. They are about ⅙ to ¼ inch long, of a uniform greenish to yellowish green, and are very active, tumbling off the flowers or out of the corn silk when disturbed.

Control Measures.—As the eggs are laid only in cornfields and the larvae feed mainly on the roots of this plant and cannot migrate to other fields, a rotation that will put any crop other than corn on the land for a period of 1 year, between corn crops, will effectively prevent damage.

References.—Eighteenth Rept. Ill. State Ento., p. 135, 1891–1892; *Ill. Agr. Exp. Sta. Bul.* 44, 1896; *Jour. Econ. Ento.*, **25**: 196–199, 1932.

17 (95). SPOTTED CUCUMBER BEETLE OR SOUTHERN CORN ROOTWORM[1]

Importance and Type of Injury. Cornfields infested with this insect start growth in a normal way, and the plant begins to show the effect of the infestation when from 8 to 20 inches tall in the North or much earlier in the South. From then on, the plant makes a very poor growth, or none at all, and frequently dies. Sometimes the heart of the plant is killed by the larvae, the lower leaves remaining green. As when attacked by the northern corn rootworm, the larger plants will fall after heavy rains. Examination of the plants will show the roots tunneled and eaten off by larvae about ½ to ¾ inch in length, with yellowish-white, somewhat wrinkled bodies, six very small legs, and brownish heads (Fig. 231).

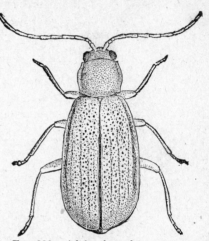

Fɪɢ. 230.—Adult of northern corn rootworm, *Diabrotica longicornis* (Say), about 10 times natural size. (*From Ill. State Natural History Surv.*)

Fɪɢ. 231.—Larva of the spotted cucumber beetle or southern corn rootworm, *Diabrotica duodecimpunctata* (Fabricius), about 5 times natural size. (*From Ill. State Natural History Surv.*)

The last segment of the abdomen has a nearly circular margin and is brownish in color. In addition to the injury to the roots, the lower part of the stalk will usually be bored through by the grubs. This insect has also been shown to have a part in the dissemination of bacterial wilt of corn (see page 386). As a garden pest, both larvae and adults do a large total damage, which is seldom fully appreciated because distributed over so many different crops (see page 497).

---

[1] *Diabrotica duodecimpunctata* (Fabricius), Order Coleoptera, Family Chrysomelidae.

Plants Attacked.—This insect has been taken from a very large number of plants, including more than 200 of the common weeds, grasses, and cultivated crops. It goes by several names, according to the food plants, and is perhaps best known as the twelve-spotted cucumber beetle. It is also frequently called the overflow worm, and the budworm.

Distribution.—The insect is widely distributed, occurring over the greater part of the United States east of the Rocky Mountains, in Southern Canada, and in Mexico. It is more abundant and destructive in the southern part of its range. The variety *tenella* extends into New Mexico, Arizona and California.

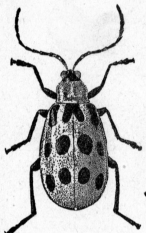

Fig. 232.—Adult, spotted cucumber beetle, *Diabrotica duodecimpunctata* (Fabricius), about 6 times natural size. (*From Ill. State Natural History Surv.*)

Life History, Appearance, and Habits.—This insect passes the winter in the form of a yellowish or yellowish-green beetle about ¼ inch long, with 12 conspicuous black spots on the wing covers (Fig. 232). The head is black, and the antennae, which are about one-half to two-thirds as long as the body, are dark or nearly black. The beetles hibernate in nearly any kind of shelter but seem to prefer the bases of plants which are not entirely killed down by the frost. They become active very early in the spring, flying about during the first days when the temperature reaches 70°F. or above. There is some evidence that the adults migrate northward in the spring. The females deposit their eggs in the ground around the bases of plants. The young larvae, on hatching, bore in the roots of plants and the underground parts of the stem. They become full-grown during July. The insect has two generations in the southern part of its range and at least a partial second generation is produced in the North.

Control Measures.—It is extremely difficult to prevent damage to corn by these insects, as the eggs are frequently laid in the fields after the corn is up and there is no method by which infested soil can be cleaned of the larvae. About the most effective method is late planting on land which has been plowed early in the spring or in the fall and cultivated frequently before planting, so that all vegetation has been kept down. In certain seasons when the beetles have been very abundant, fields handled in this way have been practically the only ones to escape injury. Rotation of crops is of no value in controlling this species. The injury is usually most severe during wet years, or the first season following wet years, and is also often serious on land that has been overflowed. Damage is often most severe on land of high fertility that produces a heavy early growth of vegetation. This may be due to preference of the female beetles for such soil in which to lay their eggs or to the fact that they are attracted to the rank vegetation which generally follows an overflow.

References.—*U.S.D.A. Farmers' Bul.* 950, 1918; *S.C. Agr. Exp. Sta. Bul.* 161, 1912; *Eighteenth Rept. Ill. State Ento.*, p. 129, 1891–1892; *Ark. Agr. Exp. Sta. Bul.* 232, 1929; *Ala. Agr. Exp. Sta. Bul.* 230, 1929, and *Cir.* 65, 1934.

18 (58). GRAPE COLASPIS[1]

Importance and Type of Injury.—Corn that has been planted on clover sod will sometimes wilt when the plants are about 6 to 10 inches high. The plants may die, or merely be greatly retarded in growth. An examination of the roots and soil about them will show numerous, curved, fat-bodied, very short-legged grubs from $\frac{1}{8}$ to $\frac{1}{6}$ inch in length (Fig. 233). Injury is most severe during late May and June, and usually on corn following clover sod.

Plants Attacked.—Adults of this insect have been found on a number of crops including timothy, June grass, grapes, strawberries, beans, clover, buckwheat, potatoes, c o w p e a s, muskmelons, and apples. Its habit of feeding on grape has given it the common name of grape colaspis. It is also known as the clover rootworm.

Distribution.—T h i s b e e t l e r a n g e s throughout eastern North America and into Arizona and New Mexico.

Life History, Appearance, and Habits.— So far as known, the winter is passed in the young larval stage. The larvae are active early in the spring and generally become full-grown during the first part of the summer; in central Illinois, by about June 15. They pupate in the earthen cells in the soil and emerge during July as pale brown elliptical beetles (Figs. 234 , 277). The body of the beetle is about $\frac{1}{6}$ inch long and covered with rows of evenly spaced punctures. The adults fly about freely in

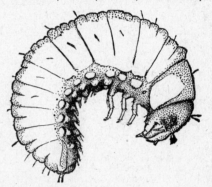

FIG. 233.—Larva of the grape colaspis, *Colaspis brunnea* (Fabricius), about 15 times natural size. (*From Ill. State Natural History Surv.*)

the field and, as above stated, are very general feeders. Mating takes place and eggs are deposited in midsummer about the roots of several of the above mentioned food plants, but particularly on those of timothy, grape, and clover. There is only one generation of the insect each year.

Control Measures.—Injury by this insect frequently occurs on spring-plowed red clover sod or spring-plowed timothy, but seldom where such ground is broken in the fall. Injury to corn has rarely been recorded following crops other than clover or timothy. A rotation that will avoid putting corn on spring-plowed clover or timothy sod will nearly always prevent injury by this insect.

References.—*Twenty-third Rept. Ill. State Ento.*, p. 129, 1891–1892; *Trans. Ill. State Acad. Sci.*, **24**: 235–240, 1931.

19 (47, 69, 79). WEBWORMS[2]

Importance and Type of Injury.—Corn on spring-plowed sod land will sometimes be cut off near the surface of the ground in much the same manner as where attacked by cutworms. An examination of the surface of the ground will show a loose silken web containing bits of dirt leading to a short silk-lined tunnel in the ground, usually at the base of the plant (Fig. 235). Short, rather thick-bodied, usually spotted and coarsely haired, active worms, from $\frac{1}{4}$ to $\frac{3}{4}$ inch long, will be found in these silk-lined tunnels (Fig. 236). Frequently the cutoff corn plants are dragged to the tunnel. In years

---

[1] *Colaspis brunnea* (Fabricius), Order Coleoptera, Family Chrysomelidae.
[2] Species of the subfamily Crambinae, Order Lepidoptera, Family Pyralididae. Among the dozen or more species of importance are the corn root webworm, *Crambus caliginosellus* Clemens, the bluegrass webworm, *Crambus teterrellus* (Zincken), and the striped or black-headed webworm, *Crambus mutabilis* Clemens.

when any one of several species is abundant they may destroy from 10 to 80 per cent of a stand of corn. Several species of webworms cause serious and widespread injury to golf greens and lawns, causing ragged, patchy appearance or sometimes large areas killed completely. The presence of many blackbirds, robins, flickers, and other birds pecking holes in the turf is an almost sure sign of the presence of these or other insect pests.

Plants Attacked.—The webworms feed on the grass plants, including corn, more particularly on bluegrass, timothy, and other pasture and field grasses, and on tobacco and certain weeds. Certain closely related species attack clovers.

Distribution.—There are a number of species of webworms occurring in different parts of the United States and Canada.

Fig. 234.—Adult of the grape colaspis, *Colaspis brunnea* (Fabricius), about 15 times natural size. Line shows natural size. (*From Ill. State Natural History Surv.*)

Life History, Appearance, and Habits.—The webworms, which are injurious to corn, pass the winter in the larval stage in silk-lined nests in grass and sod lands. They become active early in the spring, feed in much the same way as cutworms, although eating the leaves to a greater extent, and become full-grown during late June and July. Pupation takes place in a silken cocoon a short distance underground. The moths (Fig. 237) are, in most cases, pale brown in color, and have a pronounced projection from the front of the head. They vary in size from ½ to nearly 1 inch in length. The projection from the head, which is formed by the labial palps held close together, has given them the name of snout moths. One will frequently stir them up when walking across grasslands. They have a very quick, jerky, zigzag flight, usually going only a rod or two, when they alight, roll their wings close about the body, and hide by crawling down into the grass. The moths lay their eggs around the lower parts of grass stems or drop them at random as they fly about over the grass, soon after emerging, and the young larvae hatch and feed for a short time before going into hibernation. With most species, there are two or three generations each season. The corn webworm has only one generation a year, so far as known.

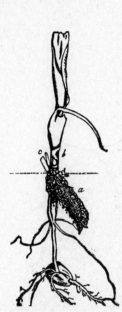

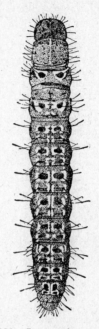

FIG. 235.—Corn plant injured by a sod webworm, *Crambus luteolellus* Clemens: *a*, silk-lined tunnel in the soil; *b*, gnawed surface of stalk; *c*, tip of severed leaf drawn into mouth of nest. (*From Ill. State Natural History Surv.*)

FIG. 236.—Larva of a sod webworm, *Crambus mutabilis* Clemens, dorsal view, about 4 times natural size. (*From Ill. State Natural History Surv.*)

Control Measures.—The most effective control measure for these insects, is early fall plowing of sod land which is to be used for corn the next season. Adults of the webworm will not lay their eggs in plowed ground, so that this measure will effectively prevent damage. If it is impossible to plow early, the land should be plowed in the fall and harrowed to expose the hibernating worms to the weather and their natural enemies. If fields are damaged early in the spring to such an extent that they should be replanted, the replanting should be made between the rows and the first planting of corn left as long as possible, undisturbed, to serve as food for the worms until the

FIG. 237.—Adult of a webworm, *Crambus mutabilis* Clemens, about twice natural size. (*From Ill. State Natural History Surv.*)

second planting becomes too large for them to cut off. Poison-bran bait, as recommended for cutworms and armyworms, is of no value against the webworms. For the control of these pests on lawns and golf greens see page 392.

References.—*U.S.D.A.*, *Farmers' Bul.* 1258, 1922; *Twenty-third Rept. Ill. State Ento.*, p. 36, 1905; *Jour. Agr. Res.*, **24**: 399–425, 1913; *Rept. Ento. Soc. Ontario* for 1934, pp. 98–107.

20 (21, 65, 76, 108). FLEA BEETLES[1]

Importance and Type of Injury.—The injury by these beetles is most severe in cold or wet years. They may reduce the stand or yield from 5 to 25 per cent. The adults eat very small holes in the green portion of the leaves, giving the whole plant a bleached appearance; growth is retarded, and the leaves wilt even during wet weather. Small to very small, shining, roundish, black, brown, or grayish-black beetles will be found feeding on the leaves. These beetles jump readily when approached. The hind legs are distinctly enlarged and thickened. Occasionally the injury will be caused by somewhat elongated dark-green beetles with white stripes on the back (Fig. 241). The most serious phase of injury by these insects is the dissemination of a bacterial wilt of corn, known as Stewart's disease[2] which causes great damage, frequently amounting to the loss of practically the entire crop, especially in the early-

FIG. 238.—Adult of the corn flea beetle, *Chaetocnema pulicaria* Melsheimer, about 20 times natural size. (*From Ill. State Natural History Surv.*)

FIG. 239.—Adult of the sweetpotato flea beetle, *Chaetocnema confinis* Crotch, about 25 times natural size. (*From Ill. State Natural History Surv.*)

maturing varieties of sweet and field corn in the middle and southern states. The adults of the brassy or corn flea beetle[3] are not only responsible for most of the spread of bacterial wilt throughout the summer, by directly infecting the leaves as the beetles feed, but are largely responsible for the successful wintering of the bacteria which occur in enormous numbers in the alimentary canal of these beetles in hibernation. The toothed flea beetle[4] has also been shown to be capable of spreading the disease, and the larvae of the northern and southern corn rootworms (pages 380 and 381) start infections at the base of the plants as they bore through the roots and crown.

Plants Attacked.—Nearly all kinds of plants are attacked by adult flea beetles. The species that are most injurious to corn also feed on millet, sorghum, broomcorn, sweetpotato, sugarbeet, oats, morning-glory, bull nettle, and cabbage.

Distribution.—Flea beetles of various species are of world-wide distribution. The corn flea beetle and sweetpotato flea beetle,[5] perhaps the two most injurious species on corn, occur generally over the eastern part of the United States. Their place is taken in the West by the western black flea beetle.[6]

Life History, Appearance, and Habits.—Practically all the flea beetles injurious to corn, with the probable exception of the pale-striped flea beetle (see page 387),

[1] Order Coleoptera, Family Chrysomelidae.
[2] Caused by *Aplanobacter stewarti* (Smith) McCulloch. See *U.S.D.A., Tech. Bul.* 362, October, 1937.
[3] *Chaetocnema pulicaria* Melsheimer.
[4] *Chaetocnema denticulata* Illiger.
[5] *Chaetocnema confinis* Crotch.
[6] *Phyllotreta pusilla* Horn.

pass the winter in the full-grown beetle stage.   They shelter largely along bushy fence rows, roadsides, or the edges of woodlands.   Some of the species prefer shelters afforded under trees.   In the spring the insects become active as soon as vegetation is well started.   The beetles after mating, lay their eggs on the leaves of plants or in the ground about the roots or underground stems.   The larval habits of most species are not well known.   Nearly all the damage to corn is caused by the adults feeding on the leaves and disseminating wilt disease.   This occurs during the first 2 or 3 weeks after the corn has come up, and the injury is usually most severe during cold seasons, when the growth of the corn plant is slow, thus giving the flea beetles a long period during which they are attacking the small plants.   In seasons when the weather is favorable for growth, the plant usually outgrows the attack of these insects. Bacterial wilt disease, on the other hand, is serious only after mild winters, because cold winters are deadly to the flea beetles which harbor the bacteria in their bodies over the winter.   Several of the species attacking corn have a single generation, while others produce two generations each season.   Only the overwintering adults are of importance, however, so far as their injury to corn goes.   The species which most commonly cause injury to corn are the western black flea beetle, the corn flea beetle (Fig. 238), the toothed flea beetle, the sweetpotato flea beetle (Fig. 239), and the smartweed flea beetle.[1]

Control Measures.—In combating flea beetles on corn, keeping the fields free from weeds is probably of first importance.   Larvae of some of the species injurious to corn feed on the roots of weeds and grasses, and the adults on the leaves of these plants as well as on corn.   Fields that have been kept clean the previous season, both in the field and around the margins, are very seldom injured by flea beetles.   The next important control measure is planting sufficiently late so that the corn will make a quick growth.   If the beetles are working on the corn, frequent cultivation will drive them temporarily from the plants and prevent continued feeding, thus lessening damage.   In cases of severe infestation it is possible to catch large numbers of the beetles by attaching vertical sheets covered with sticky tanglefoot to the cultivator frames in such a manner that the sheets pass between the corn rows and just above the surface of the ground.   Such measures will seldom be necessary if frequent cultivation can be given.   According to McColloch, lead arsenate dusts are effective and practical when the infestation is local in the field.

References.—*Twenty-third Rept. Ill., State Ento.*, pp. 109–111, 1905; *U.S.D.A., Dept. Bul.* 436, 1917, and *Farmers' Bull.* 1371, 1934, and *Tech. Bul.* 362, 1937; *Phytopathology*, **25**: 32, 1935; *Jour. Agr. Res.*, **52**: 585–608, 1936.

21. Pale-striped Flea Beetle[2]

Importance and Type of Injury.—Corn seed attacked by the larva of this flea beetle often fails to sprout or produces a pale, weak plant.   If the seed is examined, it will

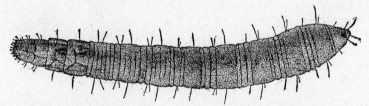

Fig. 240.—Larva of the pale-striped flea beetle, *Systena blanda* Melsheimer, dorsal view, about 15 times natural size.   (*From Ill. State Natural History Surv.*)

be found to be injured by very slender white worms, a little over ¼ inch long, with light-brown heads, six very short legs, and the body tapering slightly toward the head (Fig. 240).   Injury is usually most serious during periods of cool weather, which retards the germination of the seed after planting.

[1] *Systena hudsonias* Forster.
[2] *Systena blanda* Melsheimer, Order Coleoptera, Family Chrysomelidae.

Plants Attacked.—The adults of this species have been found feeding on a great variety of cultivated plants and weeds, including watermelon, pumpkin, pea, bean, eggplant, potato, sweetpotato, mint, pigweed, lamb's-quarters, purslane, ragweed, cocklebur, wild sunflower, alfalfa, and many others. Lamb's-quarters and shepherd's-purse seem to be preferred by the larvae.

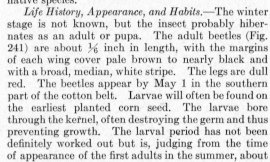

Distribution.—This species is distributed generally over temperate North America. It is probably a native species.

Life History, Appearance, and Habits.—The winter stage is not known, but the insect probably hibernates as an adult or pupa. The adult beetles (Fig. 241) are about 1/6 inch in length, with the margins of each wing cover pale brown to nearly black and with a broad, median, white stripe. The legs are dull red. The beetles appear by May 1 in the southern part of the cotton belt. Larvae will often be found on the earliest planted corn seed. The larvae bore through the kernel, often destroying the germ and thus preventing growth. The larval period has not been definitely worked out but is, judging from the time of appearance of the first adults in the summer, about

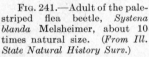

Fig. 241.—Adult of the pale-striped flea beetle, *Systena blanda* Melsheimer, about 10 times natural size. (*From Ill. State Natural History Surv.*)

1 month. There is probably one complete generation a year in the latitude of central Illinois.

Control Measures.—The damage by this insect has been most severe in fields which were weedy the previous season. Keeping down weeds will help in preventing damage the next season. Early plowing and late planting of cornfields are also of value. These measures will starve out many of the larvae in the soil before the corn is planted. The most effective measure of control is planting good seed sufficiently late so that the corn will make a quick, strong growth, and the seed will not lie in the soil long enough to be seriously damaged by the larvae.

References.—*Twenty-third Rept. Ill. State Ento.*, p. 107, 1905; *N. Y. (Cornell) Agr. Exp. Sta. Memoir* 55, 1922; *Mich. State Board Agr. Rept. for* 1933–1934.

22. CORN-SEED BEETLES[1,2]

Importance and Type of Injury.—Sometimes when corn seed fails to sprout, an examination of the kernels will show dark-brown, striped,[1] or nearly chestnut-brown[2] beetles about 1/4 to 1/3 inch long eating out the contents (Fig. 242). This injury seldom occurs except where seed of low vitality has been used or when cold weather has greatly delayed germination.

Plants Attacked.—These two beetles are mainly feeders on insects, or insect remains, and only rarely attack seeds.

Distribution.—The greater part of the United States and Canada.

Life History, Appearance, and Habits.—The life history of these beetles is not known. They probably pass the winter in the pupal or adult stage, as they are abroad very early in the spring. They may often be seen in large numbers at electric lights.

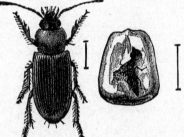

Fig. 242.—Adult of the corn-seed beetle, *Agonoderus pallipes* Fabricius, and injury to kernel. Line indicates natural size. (*From Ill. State Natural History Surv.*)

Control Measures.—Planting sufficiently late to insure quick germination of the corn seed, and using seed of good vitality is the best method of overcoming the damage caused by these insects.

[1] *Agonoderus pallipes* Fabricius, Order Coleoptera, Family Carabidae.
[2] *Clivina impressifrons* Leconte, Order Coleoptera, Family Carabidae.

Reference.—*Eighteenth Rept. Ill., State Ento.*, p. 11, 1891–1892; *Trans. Ill. State Acad. Sci. for* 1934, pp. 138, 139.

23 (285). THIEF ANT[1]

Importance and Type of Injury.—An examination of a corn kernel that has failed to sprout, or that has produced a weak plant, will sometimes show many little starch grains scattered through the soil about the kernels and the entire inside of the seed hollowed out. Frequently very small orange-red ants will be found actively working in the kernels (Fig. 243).

Plants Attacked.—Seeds of corn, sorghum, millet, and probably other plants.

Distribution.—General throughout North America.

Life History, Appearance, and Habits.—The winter is passed in all stages of development in nests in the soil. These nests are often made in the walls of the nests of larger species of ants. The workers of this small species obtain a part of their food by preying upon the helpless larvae and pupae of the larger species. This has given them their common name of thief ant. The workers of this species are about $\frac{1}{20}$ inch long, of an orange-yellow color. The males are slightly larger and black. The females, or queens, are brown and very much larger, being about $\frac{1}{4}$ inch long. Certain other species of ants rarely cause similar injury.

FIG. 243.—Worker of the thief ant, *Solenopsis molesta* Say, about 25 times natural size. (*From Ill. State Natural History Surv.*)

Control Measures.—The seed treatments that have been used for preventing injuries by these ants have not proved of much value. The best method of preventing their injuries is thorough cultivation of the soil before planting. This will break up their nests, scatter the young, and greatly reduce their numbers in the field. Surface planting aids in control.

References.—*Ill. Agr. Exp. Sta. Bul.*, 44, p. 214, 1896; *Ill. Agr. Exp. Sta. Cir.* 456, 1936.

24. SEED-CORN MAGGOT[2]

Importance and Type of Injury.—The seed attacked by the seed-corn maggot usually fails to sprout, or, if it does sprout, the plant is weak and sickly. The pale or dirty-colored, yellowish-white maggots (Fig. 244) will be found burrowing in the seed. Injury is usually most severe in wet, cold seasons and on land rich in organic matter.

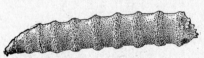

FIG. 244.—Larva of the seed-corn maggot, *Hylemyia cilicrura* Rondani, side view, 8 times natural size. (*From Ill. State Natural History Surv.*)

Plants Attacked.—Corn, beans, peas, cabbage, turnip, beets, radish, seed potatoes, and several others.

Distribution.—This species is widely distributed in Europe. It was first found in this country in 1856, in New York. It has now spread over nearly the entire United States and southern Canada.

Life History, Appearance, and Habits.—The winter is probably passed in the soil of infested fields in the maggot stage inside of a dark-brown capsule-like puparium about $\frac{1}{5}$ inch long, or as free maggots in manure or about the roots of clovers. The flies (Fig. 245), which are grayish brown in color and about $\frac{1}{5}$ inch long, are abroad in the fields early in

[1] *Solenopsis molesta* Say, Order Hymenoptera, Family Formicidae.
[2] *Hylemyia cilicrura* Rondani, Order Diptera, Family Anthomyiidae.

May in the latitude of central Illinois.   They deposit their eggs in the
soil where there is an abundance of decaying vegetable matter or on the
seed or plantlet.   The eggs hatch readily at temperatures as low as 50°F.,
and the larvae and pupae may develop at any temperature from 52 to
92°F.   The maggots burrow in the seed, often destroying the germ.
When full-grown, they are of a yellowish-white color, about ¼ inch long,
sharply pointed at the head end, legless, and very tough-skinned.   They
change to the pupal stage inside the brown puparium in the soil and in
from 12 to 15 days emerge as adults.   Since the insect has been carried
through its entire life cycle in 3 weeks, there are probably from 3 to 5
generations each year throughout the corn belt.

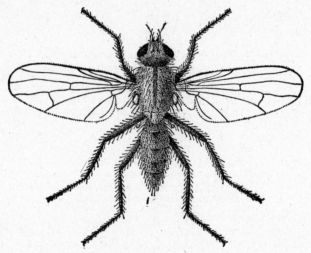

Fig. 245.—Adult of the seed-corn maggot, *Hylemyia cilicrura* Rondani, 8 times natural size.
(*From Ill. State Natural History Surv.*)

Control Measures.—No effective chemical control is known.   Shallow
planting in a well-prepared seedbed, sufficiently late to get a quick germination of the seed, is probably the best means of preventing injury.
Land that is heavily manured, or where a cover crop is turned under,
should be plowed early in the fall if possible, so it will be less attractive to
the egg-laying flies, the following spring.   Prompt resetting or replanting
of the damaged crops will usually give a stand.

References.—*U.S.D.A., Div. Ento., Bul.* 33, (n.s.), pp. 84–92, 1902; *N. Y.* (*Cornell*)
Agr. Exp. Sta. Memoir 55, 1922.

INSECTS THAT ATTACK SORGHUM, GRAIN SORGHUMS, BROOMCORN, AND SUDAN GRASS

The insects attacking these grasses and forage crops are largely the same as those
attacking corn.   The corn leaf aphid occasionally causes considerable damage to
broomcorn because of the discoloration of the broomcorn head.   Chinch bugs are
very serious pests of millet and Sudan grass, making it almost impossible to grow
either of these crops in areas where the bugs are very abundant.   They also find
Sudan grass an especially favorable place in which to pass the winter, and many of
them hibernate in this grass in areas where it is grown.   In general the sorgo and
Kafir varieties of grain sorghums are more resistant to chinch bug injury than the

milos and feteritas.[1]   Millet is one of the favorite places for second-generation army-worm moths to lay eggs.   For the general corn insects, control measures are the same as given in preceding pages.

. The ripening kernels of grain sorghum heads and broomcorn are often eaten out or entirely consumed by sluggish caterpillars, somewhat flattened and thickly clothed with spines and hairs, the body greenish, with four red or brown, longitudinal stripes on the upper part, which are the larval stage of the sorghum webworm.[2]   Early planting and other measures to hasten maturity of the crop and a thorough cleanup of the crop residue are suggested for control.   Grain sorghums, sweet sorghums, Sudan grass, and broomcorn are severely damaged by the tiny, grayish to red, headless maggots of the sorghum midge,[3] which extract the plant juices of the developing seeds, blasting or blighting the grain spikelets.   Planting sorghums as far as possible from, and to leeward of, sources of infestation such as Johnson grass, and old fields of sorghum, broomcorn, and Sudan grass; the use of a single resistant variety in the field; and the burning of crop refuse during the winter may aid in control.

INSECTS THAT ATTACK SUGARCANE

Most of the insects that attack corn also attack sugarcane, but there are some species especially destructive to this crop.   The sugarcane moth borer[4] is a very close relative of the southern corn stalk borer.   The larvae bore into the interior of the stalks of sugarcane as well as corn, broomcorn, sorghums, and Sudan grass, killing the young canes and weakening the larger ones, so as to lower both quantity and quality of the juice produced.   Material control has been claimed by the propagation and release of great numbers of the egg parasite, *Trichogramma minutum;* by clean culture, low cutting of the cane, and submerging seedcane in water to kill the larvae.

INSECTS THAT ATTACK LAWNS, GOLF GREENS, PASTURES, AND HAY GRASSES

Among the most serious pests of turf are the webworms (page 383), cutworms (page 346), white grubs (page 374), annual white grub (page 394), wireworms (page 377), green June beetle (page 623), billbugs (page 357), Japanese beetle (page 621), Asiatic garden beetle (page 471), ants (page 767), chinch bugs (page 351), and earth worms.   Occasionally large, dark-colored, thick-skinned maggots will be found around the roots of grasses in pastures.   These maggots are the young of the crane flies.[5] They are often called "leather jackets."   In most cases these insects are entirely harmless, feeding only on the decaying vegetable matter in the soil.   One species, the range crane fly,[6] has occasionally been destructive in the West.   These species all work more or less in the soil.   Grasses are often attacked above ground by the armyworm (page 341), the fall armyworm (page 344), chinch bugs (page 351), grasshoppers (page 335), crickets (page 466), and leafhoppers.[7]   In some of the western States

[1] *U.S.D.A., Tech. Bul.* 585, 1937.

[2] *Celama sorghiella* Riley, Order Lepidoptera, Family Nolidae.

[3] *Contarinia sorghicola* Coquillet, Order Diptera, Family Itonididae.   See *U.S.D.A., Farmers' Bul.* 1566, 1928.

[4] *Diatraea saccharalis* (Fabricius), Order Lepidoptera, Family Pyralididae.   See *U.S.D.A., Tech. Bul.* 41, 1928.

[5] Many genera and species, Order Diptera, Family Tipulidae (see *U.S.D.A., Bur. Ento., Bul.* 85, 1910).

[6] *Tipula simplex* Doane, Order Diptera, Family Tipulidae (see *U.S.D.A., Dept. Cir.* 172, 1921).

[7] See *U.S.D.A., Bur. Ento. Bul.* 108, 1912.

the range caterpillar[1] has caused serious losses of range grasses.   On hay and pasture crops about the only control measures, the expense of which can be justified, are rotations with crops not of the grass family, and pasturing with hogs as recommended on page 377.   In some cases sowing pastures with mixtures of grass and legume seeds has reduced the damage by these pests.

General control measures for the soil-infesting species on golf greens, lawns, and parkways, include the following.   Any of them should first be tried on a small scale, in an experimental way, since their effectiveness and safety will vary with different climatic, soil, and turf conditions.

Lead arsenate.—The mixing of lead arsenate with each top dressing, so as to give 5 to 15 pounds of this poison per 1,000 square feet, during the entire season, is a valuable protection against webworms, cutworms, many of the beetle grubs, and earth worms.   New lawns or greens being constructed may be made highly insectproof by mixing lead arsenate uniformly with the upper 3 inches of top soil at the rate of 35 pounds per 1,000 square feet.   If neither of the above measures has been used, a spray of 1 or 2 pounds of lead arsenate to 20 gallons of water applied per 1,000 square feet will check attacks of webworms, armyworms, beetle grubs, grasshoppers, and other leaf eaters.   Very heavy applications of lead arsenate are sometimes made to the soil to check infestations of Japanese beetles, but these excessive applications kill all vegetation and are warranted only as emergency measures.

Pyrethrum.—All insects in the superficial layer of the sod and soil can be killed by a high-grade, freshly prepared, horticultural pyrethrum extract, using 1 fluid ounce to 5 gallons of water, sprinkled upon the turf at the rate of 1 gallon to the square yard.   This treatment, which will cost about 3 cents a square yard for materials, is especially valuable, applied to small sample areas, to determine whether webworms and other pests are present, because it causes the worms to come up to the top of the turf and wriggle about within a few minutes after it is applied.

Carbon Disulphide.—A dilute mixture of carbon disulphide poured about the roots of grasses at the rate of from 1 quart to 2 or 3 gallons per square foot, depending upon the per cent of carbon disulphide in the diluted emulsion, will kill all insect life to a depth of 2 to 6 inches.   The emulsion can be homemade or purchased ready prepared.   The strength of the solution should vary with soil temperature.   A proportioning machine has been invented to feed any desired amount of a stock solution into a stream of water flowing through a hose.   Anyone contemplating using this method should consult *U.S. Department of Agriculture, Circular* 238, for directions and cautions regarding its use.   It will cost $2 to $3 per 1,000 square feet, but, when properly used, it is promptly and highly effective against the worst of the soil pests.

Ants may usually be eradicated from turf by sprinkling a little brown sugar thoroughly mixed, dry, with paris green, ½ ounce to each pound of sugar, over the grass where the ants are noticed.   Definite nests or mounds should be destroyed by pouring a little carbon disulphide into a hole punched into the center of the nest and closing the hole with soil.

---

[1] *Hemileuca oliviae* Cockerell, Order Lepidoptera, Family Saturniidae (see *U.S.D.A.*, *Bur. Ento. Bul.* 85, Part V, p. 59, 1910).

Short cutting and frequent watering and top dressing help to prevent destruction of lawns by chinch bugs (page 351); and rotenone or nicotine dusts or sprays will kill the bugs.

If earth worms become objectionable, they may be killed by a solution of mercuric chloride, $HgCl_2$, 3 ounces in 50 gallons of water, sprinkled evenly over 1,000 square feet of turf. This material should not be mixed or handled in metal containers.

References.—*Mich. State Coll. Ext. Bul.*, 125, 1932; *U.S.D.A.*, *Cir.* 238, 1932; *Bul. U.S. Golf Assoc., Green Sec.*, **12**: 14–17, 1932; *Nat. Greenkeeper*, **6**: 1932; and **7**: 8–13, 1933.

25. ASIATIC BEETLE[1]

Importance and Type of Injury.—This insect is principally a pest of lawns and turfs and a great nuisance in nurseries. It was first found in Connecticut in 1920. Although the adult beetles do some damage by chewing the blossoms of flowers, the larvae, which can be distinguished from the native white grubs only by careful exam-

FIG. 246.—The Asiatic beetle, *Anomala orientalis* Waterhouse, *A*; and the Asiatic garden beetle, *Autoserica castanea* Arrow, *B*. Each enlarged about three times. (*From New Jersey Dept. Agr.*)

ination (Fig. 402), kill grasses especially of lawns, by eating off the roots close to the soil surface.

Distribution.—Parts of Connecticut, New York, and New Jersey have been infested by this insect. It had previously been found in Japan and Hawaii.

Plants Attacked.—Various grasses, nursery stock, sugarcane in Hawaii.

Life History, Appearance, and Habits.—In Connecticut most of the insects winter as half- to nearly full-grown larvae, about 1 inch long, in the soil to a depth of 5 to 12 inches. In late April the larvae come up near the surface and resume feeding until early June, when they form pupal cells at a depth of about 6 inches. The adults emerge from late June to late August. They are broad-bodied spiny-legged convex-backed beetles (Fig. 246), about $5/8$ inch long, varying in color from straw yellow to black, but mostly straw-colored with a varying extent of dark markings. They are active during the daytime. They dig into the ground to lay their round, white eggs, singly, in the upper 5 or 6 inches of the soil. Eggs may be found all summer long. The newly hatched larvae feed until cold weather and are of various sizes when they go into hibernation. Most of them complete their life cycle in 1 year, but some pass two winters as larvae. They can best be distinguished from our native white grubs by the transverse anal opening as contrasted with the V- or Y-shaped vent of the white grubs (see Fig. 402).

Control Measures.—Grub-proofing the soil with lead arsenate or fumigating with carbon disulphide emulsion are recommended. Attempts have been made to suppress

---

[1] *Anomala orientalis* Waterhouse, Order Coleoptera, Family Scarabaeidae.

the insect by imported parasites and by quarantines to prevent its spread, especially in shipments of plants with a ball of earth about the roots.

References.—Conn. Agr. Exp. Sta. Bul. 304, 1929; *U.S.D.A., Cirs.* 117, and 238, 1930.

26. ANNUAL WHITE GRUB[1]

Importance and Type of Injury.—Grubs of the general appearance and habits of the white grubs, but without the double row of small spines on the underside of the last abdominal segment and with the anal slit transverse instead of V-shaped (Fig. 402), feed on the roots of grasses, burrowing extensively through the soil just below the turf. The adults have occasionally caused personal injury by burrowing in the external ear of sleeping persons.

Life History, Appearance, and Habits.—The nearly full-grown larvae winter at depths of a foot or more and ascend to feed for a short time by May first. They then pupate for 2 or 3 weeks at an average depth of about 6 inches, after mid-June. In certain years the adults are present in enormous numbers from late June to August. They are attracted to lights, lay eggs in the soil, and in most respects resemble the common white grubs, except that they complete a generation in 1 year, instead of the usual 3 years for the Phyllophaga.

Control Measures.—The only known methods of control are by soil poisoning or fumigation, as recommended on pages 283, 392.

References.—Jour. Econ. Ento., **11**: 136–144, 1918, and **31**: 340–344, 1938.

[1] *Ochrosidia villosa* Burmeister, Order Coleoptera, Family Scarabaeidae.

CHAPTER XII

INSECTS INJURIOUS TO SMALL GRAINS

FIELD KEY FOR THE IDENTIFICATION OF INSECTS INJURING WHEAT AND OTHER SMALL GRAINS

A. Insects chewing leaves, heads, or stalks above ground, leaving visible holes:

1. Early-sown wheat in the fall sometimes eaten to the ground. In the spring, the heads and leaves are eaten off................... *Grasshoppers*, page 396.
2. Fields of wheat stripped of leaves and the awns of the heads eaten off by large green and dark-green striped worms up to 2 inches in length, with 3 pairs of legs and 5 pairs of prolegs. Worms feed mainly at night and remain at base of plants during the day.............................. *Armyworm*, page 396.
3. Part of wheat head eaten off by grayish or greenish-gray worms, with clear-cut brown and yellow stripes on the body; worms hide about the base of the plant during the day. Body slenderer and head relatively bigger than in the armyworm, and with two straight dark bands over top of head.................. *Wheat-head armyworm, Neleucania albilinea* (Hübner) (see *Iowa Agr. Exp. Sta. Bul.* 122, 1911.)
4. Pale-green larvae up to ⅔ inch long, with distinct heads and at least 10 pairs of legs and prolegs, feeding on the edges of the wheat leaves; often hold the hind part of the body curled against the leaf............. *Wheat sawflies*, page 397.

B. Small insects sucking sap from leaves, stems, or developing grain, not leaving visible holes, but causing wilting, discoloration, falling over, or death of the plants:

1. Areas of dead and whitened plants appearing in fields in early spring, especially during periods of cool weather. Plants in and about such areas covered with very small, winged and wingless, green, sucking aphids... *Green bug*, page 397.
2. Injury much the same as in *B*, 1, but with the insects more generally distributed over the fields. Later large numbers of the aphids are found in the heads.... .... *English grain aphid*, page 399.
3. Patches of wheat dying in the late spring in parts of the field where the stand is the poorest, or on thin ground. Great numbers of small, red, brown, or black-and-white bugs, the largest only ⅕ inch long, clustered on the lower parts of the stems and on the lower leaves.................. *Chinch bug*, page 400.
4. Plants in the fall with stiff, bluish-green leaves; center shoot often missing; or plants dead. Whitish, legless and headless maggots up to ³⁄₁₆ inch long; or brown, capsule-like cases of about the same length, behind the lower leaf sheaths. In the spring, many of the straws fall when the heads are beginning to fill; brown seed-like cases about ⅛ inch long found behind the lower leaf sheaths but not inside the straw...................... *Hessian fly*, page 401.
5. Heads of wheat full of very small pink or reddish maggots, ¹⁄₁₂ inch long or less, that lie among the bracts and feed on the kernels. Kernels shriveled.......: *Wheat midge, Contarinia tritici* (Kirby) = *Thecodiplosis mosellana* Gehin (see *Purdue Agr. Exp. Sta. Cir.* 82, 1918.)

C. Insects boring or tunneling in the stem or straw:

1. Stalks damaged by a brown- and white-striped caterpillar working inside the stem. The stem always with a hole in the side from which castings are pushed out... *Common stalk borer*, page 364.
2. Plants in fall having much the appearance of *B*, 4, but with very slender, greenish, footless maggots in the enlarged part at the base of the stem. In the spring, scattered wheat heads are blasted, turn white just after forming, but do not fall over. The stalk is eaten off inside the leaf sheath at first or second

joint below the head, by a pale-green very slender maggot, about ¼ inch long
....*Wheat stem maggot*, page 410.

3. Straw falls as the heads fill, as in *B*, 4, but the inside of the straw is eaten out
   down to the ground by a pale-yellow, wrinkled-bodied larva about ½ inch long
   ....*Wheat stem sawfly*, page 405.

4. Wheat in early spring with crown of plant eaten out by small, yellowish,
   maggot-like larva less than ¼ inch long, with distinct brown jaws, stunting or
   killing the plant. Later in the spring, after the plant has jointed, it may be
   injured at the joint by a similar worm eating inside the stem, and the straw
   weakened so that the head falls.................*Wheat strawworm*, page 408.

5. Straw fallen as in *B*, 4, the weakening of the straw caused by small, yellow,
   legless, brown-jawed larvae, working in hard, knotty galls, usually just above
   one of the lower joints. Many bits of the galls and hardened straws coming
   through the thresher in the grain.................*Wheat jointworm*, page 406.

6. Stems neatly severed with a concave cut from the inside, at or slightly above
   the surface of the ground, the adjacent stem packed with frass. A light-yellow
   larva, up to ½ inch long found inside the straw near the base...............
   Black grain-stem sawfly, Trachelus tabidus (Fabricius) (see *Jour. Econ. Ento.*,
   28: 457–458, 1935).

D. *Insects that attack the plant under ground:*

1. Tender portions of grain plants eaten off just below the ground, or in wet
   weather aboveground, especially in the spring..............................
   ....*Pale western and other cutworms*, page 349.

2. Young wheat plants killed out in the fall and sometimes in the spring, over
   irregular areas in the field. Slender, shining, smooth, tough, brown or yellowish-brown larvae, the largest 1 to 2 inches long, with 6 slender legs, found
   around the wheat roots.............................*Wireworms*, page 411.

3. Germinating seeds in the ground and young plants eaten by slender, pale-brown
   or yellowish-brown or nearly black, 6-legged larvae, from ½ to 1 inch in length,
   the body prominently jointed. Legs and antennae longer than in true wireworms. Injury occurs in drier parts of wheat belt.*False wireworms*, page 412.

4. Wheat plants, especially in early-sown fields, killed out in the fall over irregular
   areas in the field. Plants eaten off just above the roots by slender, white,
   6-legged, brown-headed larvae which are less than half an inch long and very
   difficult to find.......................*Spotted cucumber beetle*, page 381.

5. Roots of plants eaten off in the late spring and sometimes in the fall in early-
   sown fields, by white, curved-bodied grubs, from ½ to over 1 inch in length,
   with reddish-brown heads and 6 long legs. Grubs feed just below the surface
   of the ground.................................*White grubs*, page 413.

6. Wheat stalks die at the ground just before harvest. Lower part of stalk
   hollowed out by white, short, very fat, legless larvae about ¼ inch in length.
   Stalks sometimes fall.................................*Billbugs*, page 413.

A. INSECTS THAT ATTACK WHEAT

27 (1, 75, 146). GRASSHOPPERS

Grasshoppers, when abundant, especially in the spring-wheat growing sections,
frequently cause severe injury to the wheat by eating off the bracts or sometimes
cutting off newly formed heads. As a rule, severe damage by grasshoppers to wheat
occurs only in the western states. Early-sown wheat is sometimes killed by grasshoppers feeding upon it in the fall. Grasshoppers may be controlled on wheat by
the methods given for controlling these insects on corn.

For a full description of grasshoppers, see Corn Insects, page 335.

28 (2, 48). ARMYWORM[1]

Importance and Type of Injury.—Armyworms are most apt to be destructive to
wheat and other small grains when wet weather has caused a rank growth in the fields.

[1] *Cirphis unipuncta* Haworth, Order Lepidoptera, Family Noctuidae.

The female moths lay their eggs in large numbers among the rank or lodged grain in the fields, and the worms hatching from these eggs may suddenly appear in such numbers as completely to strip the leaves from the grain; they then crawl out into other near-by fields. For full description of this insect, see Corn Insects, page 341.

Control Measures.—In wheat and other small-grain fields the only method of controlling the armyworm is to watch the fields closely in years when the moths are known to be abundant and, if the worms are found in the fields, to apply poison bran at once (see page 339).

SAWFLIES THAT FEED ON THE WHEAT LEAVES[1]

There are several species of sawflies that feed on the leaves of wheat. In all these the larvae may be recognized by their wrinkled bodies, in most cases of a pale green color, and by the number of abdominal prolegs of which there are always six pairs or more. They occur throughout the wheat-growing areas of the country but are seldom if ever of economic importance. Occasionally, some of these species will become abundant enough partially to strip the plants of leaves, but, as this stripping does not occur until the wheat head is partially filled, it has but little effect on the yield of the grain. They are, in general, heavily parasitized, and no special control measures are necessary to keep down their numbers.

29. GREEN BUG[2]

Importance and Type of Injury.—This grain aphid, because of its general distribution and great prolificacy, causes a loss of from 1 to possibly 3 per cent of the wheat crop of the entire world. In some years the damage has amounted to 25 per cent of the wheat crop in the southwestern part of the United States. Wheat or other small-grain fields infested by the green bug usually show small deadened areas appearing in the field during the late winter or early spring. An examination of such areas will show the plants swarming with numbers of tiny green aphids or plant lice (Fig. 247,*A*), which are sucking the sap. During periods of spring weather, favorable to the aphids, these deadened spots may spread rapidly and the entire field be killed out.

Plants Attacked.—The green bug feeds on all the small grains and many of the wild and cultivated grasses. It has also been found on rice, corn, sorghum, and several other cultivated plants.

Distribution.—The insect is of European origin and was first recorded in the United States in 1882 in Virginia. It has now spread so it is generally distributed over practically all the United States, its range extending on the north into Canada. The green bug is not common in the New England states. Its greatest damage has been done in the large grain-growing states west of the Mississippi, including Oklahoma, Texas, Kansas, and Nebraska.

Life History, Appearance, and Habits.—In the southern states this insect passes the winter in the active nymphal and adult stages, feeding on the stems of plants and giving birth to living young during the warmer periods of weather. In the more northern states the winter is passed in the form of black shiny eggs, deposited on the leaves of plants on which the insect feeds. These eggs hatch during the winter or early spring, producing numbers of pale green, wingless female insects, about $\frac{1}{16}$ inch in length when full-grown, and having a dark-green stripe down the back

---

[1] *Dolerus arvensis* Say, *Dolerus collaris* Say, and *Pachynematus extensicornis* (Norton), Order Hymenoptera, Family Tenthredinidae.

[2] *Toxoptera graminum* Rondani, Order Homoptera, Family Aphididae.

(Fig. 247,*A*).   In from 7 to 18 days after hatching, these females begin giving birth to living young.   These may become either winged or wingless.   The winged individuals differ slightly in appearance from those hatching from the winter eggs, being slightly larger and with filmy wings having an expanse of about ¼ inch.   The head is brownish yellow and there are blackish lobes on the back of the thorax.   In about 15 days these females, in turn, begin giving birth to living young, and the insects continue thus, generation following generation.   Each female begins reproducing when from 7 to 18 days old and continues to reproduce for about 20 to 30 days, giving birth during this time to an average of from 50 to 60 young.   It will be seen from these figures, that the rate of reproduction is enormous and almost beyond comprehension.   If unattacked by natural enemies, in the course of a single season the mass of aphids which could

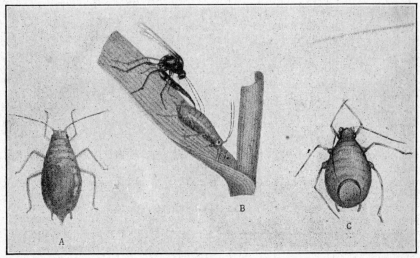

Fig. 247.—The green bug.   *A*, a parasitized specimen, the larva of the parasite showing through its body wall.   *B*, a female of the parasite, *Lysiphlebus testaceipes* Cresson, ovipositing in a green bug.   *C*, the dead "shell" of a green bug showing the circular hole through which an adult parasite has emerged.   All much enlarged; actual size about the same as the letters in this figure.   (*Modified from Kan. Univ. Bul., vol. 9, no. 2.*)

be produced would be so great as to destroy all vegetation upon which they could feed.   On the approach of cold weather, the female aphids give rise to winged males and females.   These mate, and the mated females produce eggs, in which stage the insect passes the winter in the colder parts of the country.   A single female may at times both lay eggs and give birth to living young.   There are from 5 to 14 generations each season, all except the last being composed entirely of females.

Control Measures.—The abundance of this insect and resultant injury are very largely dependent on weather conditions and can be but little influenced by man.   No serious outbreak of the green bug has occurred except when the preceding summer was comparatively cool and moist.   The worst outbreaks have always occurred in seasons of mild winters followed by cool, late springs.   The reason for this is that the green bug will start reproducing at temperatures a little above 40° F. and can reproduce at a fairly rapid rate at temperatures between 55 and 65° F.

It is preyed upon by a number of insect enemies, but particularly by a little wasp[1] (Fig. 247,*B*). This tiny wasp stings the aphids and deposits its eggs within their bodies. These eggs hatch into little maggots which eat out the body substance of the aphid, and finally emerge as adults, through a hole cut in the back of the aphid (Fig. 247,*C*). These wasps are practically always present in areas where the green bug is abundant. When the temperature is below 65°F., the wasp will reproduce very slowly or hardly at all. Long periods of cool wet weather thus permit the green bug to increase in enormous numbers, while its most effective natural enemy can increase only very slowly. This relationship between the two insects, and the effect of the weather upon them, is apparently responsible for the abundance of the aphids during the years of mild winters and cool springs. In normal years the wasp-like parasites reproduce at a sufficiently rapid rate to hold down excessive abundance of the aphids.

Of the control measures that can be exercised by man, the most effective is the destruction of volunteer grain, especially oats. The heavy masses of volunteer oats in the spring, with their thick growth, are particularly favorable to the aphids. The aphids will often start in fields having an abundance of such volunteer grain and, with favorable weather, spread from these centers over adjoining areas. It has been suggested that the green bug can be killed by covering the first brown spots in the field with straw and burning the straw. This method, however, is not practical under most conditions.

References.—*Kan. Univ. Bul.*, vol. 9, no. 2, 1909; *U.S.D.A.*, *Farmers' Bul.* 1217, 1921, and *Bur. Ento. Bul.* 110, 1912.

30. ENGLISH GRAIN APHID[2]

Importance and Type of Injury.—Grain infested by the English grain aphid will show somewhat the same appearance as that infested by the green bug. However, the infestation is not usually confined to small spots in the field. After the wheat or small grain begins to head, very large numbers of these aphids will often be found clustered in the bracts of the wheat heads, or in the heads of other grain (Figs. 248 and 249). Their feeding may shrivel the growing wheat kernels and in the early spring cause the death of the wheat plants.

Plants Attacked.—This insect feeds on all the small grains and many of the wild and cultivated grasses. It has been found in small numbers on corn but is not an important pest of this plant.

Distribution.—The English grain aphid is generally distributed throughout the United States and southern Canada wherever small grains are grown.

Life History, Appearance, and Habits.—This insect passes the winter mainly in the fully or partly grown stages. A few individuals go through the winter in the egg stage. It will be found in the heaviest growth of grain and especially in clumps of volunteer oats, rye, or wheat that has made a rank growth. The overwintering forms are all females, and, as the weather becomes warm in the spring, they begin giving birth to living young which may become winged or wingless. They feed during the early spring on the growing grains, sucking the sap from the leaves and stems. As the heads begin to form, many of the aphids will gather in the heads, causing shriveling and shrinking of the newly formed grain. After the harvest of small grains the insects migrate to wild or cultivated grasses where they spend the summer. In the fall, after the winter grains are planted, they go back to them or, as above stated, gather in large

[1] *Lysiphlebus testaceipes* Cresson, Order Hymenoptera, Family Braconidae.
[2] *Macrosiphum granarium* (Kirby), Order Homoptera, Family Aphididae.

numbers in clumps of volunteer grain. The males appear during the fall and early winter and mate with the true females, which lay eggs on the grains where they have been feeding. Only a comparatively small number of eggs are laid, the average number being about eight. The wingless females are of a pale-green color with long black antennae and have a long black cornicle extending backward from each side of the abdomen. The winged individuals are about the same size and of the same general color. The lobes on the thorax, however, are brown or blackish. The wing expanse is a little over $\frac{1}{4}$ inch.

Control Measures.—As is the case with the green bug, we are largely dependent upon natural enemies for the control of this insect. It is usually held in check by several parasites, by lady beetles, and other aphid eaters. During cool springs, it may become sufficiently abundant to cause damage. Destruction of volunteer grain in the fall is of value in controlling this species.

Reference.—*Jour. Agr. Res.*, **7**: 463–480, 1916.

31 (6). CHINCH BUG[1]

Importance and Type of Injury.—The chinch bug is as destructive to small grains as it is to corn. Wheat fields infested with the chinch bug show a deadening and drying out of the plants early in the spring. As the wheat begins to head, deadened areas will appear over the fields. Usually these areas are in the spots where the soil is the poorest or where the stand of wheat has been partly killed out by the winter or by heavy rains early in the spring. Thus in dry years these spots will usually appear in the higher parts of the field, while following wet springs they will usually be noticed in the low, wet spots. For full description of the chinch bug, see Corn Insects, page 351.

FIG. 248.—A head of wheat infested by the English grain aphid, *Macrosiphum granarium* (Kirby). (*From Ill. State Natural History Surv.*)

Control Measures.—No effective practical control has been developed which will rid infested wheat fields of the chinch bug. The sowing of wheat on fertile soil, which will promote a heavy growth, is of value, as chinch bugs avoid shade and dampness and will not be found in large numbers in fields having good

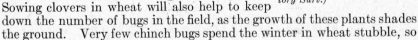

FIG. 249.—The English grain aphid clustered about the bracts of a wheat head, enlarged about $3\frac{1}{2}$ times. (*From Ill. State Natural History Surv.*)

stands of wheat with a strong uniform growth. Sowing clovers in wheat will also help to keep down the number of bugs in the field, as the growth of these plants shades the ground. Very few chinch bugs spend the winter in wheat stubble, so

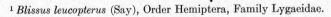

[1] *Blissus leucopterus* (Say), Order Hemiptera, Family Lygaeidae.

that burning over the infested stubble during the latter part of the summer is practically of no value for the control of these pests.

References.—U.S.D.A., Farmers' Bul. 1498, 1936; *Kan. Agr. Exp. Sta. Bul.* 191, 1913; *Ill. Agr. Exp. Sta. Cir.* 268, 1923, and *Buls.* 243, 1923, and 249, 1924; *Ill. Agr. Exp. Sta. Cir.* 431, 1935.

32. HESSIAN FLY[1]

Importance and Type of Injury.—The type of injury caused by the Hessian fly is not conspicuous. Wheat infested in the fall is stunted in growth; the leaves of the plants take on a dark bluish-green color, become distinctly thickened, and stand more erect and stiff than those of uninfested plants. The central growing shoot is often lacking. Small white or greenish-white, shiny, legless and headless maggots about $\frac{3}{16}$ inch in length; or brown, elongated, capsule-like cases (*puparia*), about $\frac{1}{8}$ inch long, containing white maggots, will be found behind the sheaths of the lower leaves of the plant, usually below the surface of the ground (Fig. 250, *December to March*). The injury is caused entirely by the larvae, which withdraw the sap from the lower parts of the stem. Heavily infested plants generally die during the winter. In the early spring the appearance of the injured plants is much the same as that in the fall. Later in the spring the capsule-like "flaxseeds" will be found behind the leaf sheath above the surface of the ground, sometimes as high as the second or third joints. Infested straws usually break over when the heads begin to fill (Fig. 250, *June*). Heavily infested fields will frequently have 50 to 75 per cent or more of the straws fallen. The yield of infested grains is seriously reduced.

Plants Attacked.—The principal food plants of the Hessian fly are wheat, barley, and rye and are preferred in about the order named, wheat being by far the favorite food of this insect. Oats are never injured by this insect. It has been taken rarely and in very small numbers on certain species of wild grasses. It has also been taken in very small numbers on emmer and spelt. Eggs have been found on oats, foxtail, and einkorn, but no larvae have ever been known to develop on these plants.[2]

Distribution.—The original home of the Hessian fly was possibly in the southern Caucausas region of Russia. It was probably introduced into North America in straw bedding used by the Hessian troops during the Revolutionary War, as it was first noted on Long Island about 1779. The insect has now spread to all the principal wheat-growing areas of the world. It is not found in a few of the arid wheat-growing sections in the plains states, but occurs on the Pacific coast.

Life History, Appearance, and Habits.—The Hessian fly passes the winter in the full-grown maggot stage. Occasionally a partly grown maggot will survive. In most cases, however, the maggot is within the brown puparium, or what is commonly called the "flaxseed." These overwintering stages will be found hidden away behind the leaves of the volunteer or early-sown wheat (Fig. 250, *November to March*), between leaf sheath and stem, or in some cases among the stubble of the previous season's crop. The insect is inactive during the winter, all those in the flaxseed stage having finished feeding. In the spring, shortly after the wheat plant starts its active period of growth, the maggots change inside

---

[1] *Phytophaga destructor* (Say), Order Diptera, Family Cecidomyiidae.
[2] *Kan. Agr. Exp. Sta. Tech. Bul.* 11, 1923.

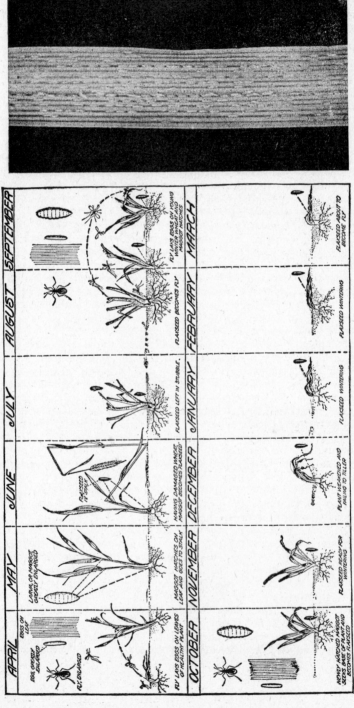

Fig. 250.—Chart showing the relation of the Hessian fly to the wheat plant, each month of the year. *(From Lochhead, "Economic Entomology," Blakiston, after U.S.D.A.)* At the right, photograph of Hessian fly eggs on leaf of wheat, enlarged about four diameters, showing how the eggs are laid in the natural grooves of the wheat leaf on the upper side. *(Original.)*

the puparia to the pupal stage and in a week or two emerge as small two-winged flies a little less than $\frac{1}{8}$ inch in length. These flies are all a sooty-black color, much smaller than the common house mosquito, and are very frail creatures, never feeding so far as known. During windy weather they remain clinging to the leaves or clustering about the base of the wheat plants. On warm days they fly about over the fields and mate, and the females, whose abdomens are of an orange-red color, lay eggs of the same reddish color in the grooves on the upper sides of the wheat leaves. The eggs (Fig. 250, *April and October*) are very small and can just be seen without the use of a magnifier. They are slender, often laid 2 to 15 in a string, end to end, and, when viewed through a lens, they have somewhat the appearance of a string of "wieners." They are likely to be overlooked by any but the most careful observers. The adult flies probably never live more than 4 days and most records show that they usually die in 3 days or less. While incapable of flying long distances by their own efforts, they may be carried by a moderate wind for a distance of several miles. Adult Hessian flies may be found fairly abundant at a height of 25 feet above the ground, and they may occur in considerable numbers at least 2 miles from any of their known food plants. The female flies lay from 250 to 300 eggs, the average number of the fall generation being 285, and for the spring generation slightly less. Many eggs are often deposited on a single plant, as many as 319 having been counted in the fall on one plant. The eggs hatch in from 3 to 10 days, depending on the temperature. The young maggots, which are reddish when they first emerge from the egg, soon turn white. They work their way down the grooves of the leaves as far as they can go behind the leaf sheath, without cutting through the sheath or the stem. Here they start feeding by rasping on the straw and sucking up the sap which oozes out from the irritated surface. They do not move about after feeding has begun. The maggots never enter the straw of the plant as do those of some other insects which are sometimes mistaken for the Hessian fly. With favorable weather the maggots will become full-grown in about 2 weeks (Fig. 250, *May and October*). The outer skin then loosens from an inner skin and forms the brown protective case known as the flaxseed or puparium. Most of the maggots developing from eggs laid in the spring reach the flaxseed stage some time before the wheat begins to head. Normally, there is one spring generation. Under certain weather conditions, particularly those of an early wet spring, a second or supplementary spring generation may be produced. Flies of the supplementary generation emerge and lay their eggs on the late tillers, normally causing but little injury. Nearly all the flies will be in the flaxseed stage 2 weeks or more before wheat harvest. They remain in this stage in the dry stubble during the summer (Fig. 250, *July, August*) and emerge again as flies as soon as sufficient rain has fallen to cause a growth of volunteer wheat in the fields. The stimulus that causes the estivating larvae in the puparia to pupate is the soaking of the straw and the puparia by late summer rains with the average temperature above 45°F. The same conditions start the growth of volunteer wheat. Having pupated, the time before the flies emerge is dependent upon temperature. At 40°F. the pupal period averaged 30 days; at 50°F., 15 days; at 60°F., about 11 days; while at 66°F., it averaged a little over 7 days. It is said that above 75°F. the larvae will not pupate. During wet seasons, adult flies may begin coming out by the middle of July and a nearly full, summer generation may be produced in volunteer wheat. In normal years the flies do not start emerging before late summer or early fall and lay their eggs in such volunteer wheat as may be present in the fields or in early-sown wheat. In central Illinois, emergence starts usually about September 1. If no green wheat, rye, or barley is available, the flies will die without depositing many of their eggs. Emergence ceases on the approach of cold weather. The maggots developing from the fall generation of flies will nearly all become full-grown before the first hard frost or at least before the ground freezes in the fall. Maggots that are less than half-grown, are likely to be killed by freezing weather.

It will be seen from this description that there are normally two full generations of flies each year, but under exceptionally favorable weather conditions there may be three, four, or even five. In California there is

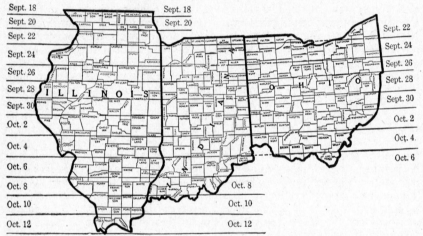

Fig. 251.—Map showing normal safe dates for sowing wheat in several North Central States to escape injury by the Hessian fly. (*Original.*)

generally a single generation, the insects remaining in the puparia from June to January, and infestation of the new crop comes mostly from the stubble of the preceding year.

Control Measures.—The three measures which have been found of most value in preventing damage by this insect, throughout most of the wheat belt, are:

a. Sowing sufficiently late in the fall so that the wheat will not come up until after the adult flies have emerged, laid their eggs, and died. The accompanying map (Fig. 251) shows the normal safe dates for sowing wheat to escape injury by the Hessian fly and to make the largest yields, in some of the North Central states.

b. Keeping down all growth of volunteer wheat in the fields, or about stacks, on which the fall generation of flies might deposit their eggs, and thus carry the insect through until the following spring.

c. Plowing under infested wheat stubble as soon as possible after harvest. The adult Hessian fly is such a weak insect that it cannot work its way up through the soil, and, where wheat stubble is thoroughly plowed under and firmed, this will prevent the Hessian fly from emerging. This is a most effective control measure but frequently cannot be practiced because wheat is often used as a nurse crop for grasses or clovers which remain on the ground for a year or more after the wheat is harvested.

The proper date of seeding wheat to escape infestation by the fall generation of Hessian fly has now been worked out by the entomologists in the experiment stations of all the principal wheat-growing states. There will be some variation in the time of emergence of the fly during different seasons, as its development is so largely dependent on weather conditions. However, long-continued experimental seedings in many states have shown that it is nearly always possible to sow late enough to avoid any but a very light infestation by the Hessian fly and still to secure a sufficient plant growth to permit the wheat to withstand the cold weather of the winter. The results secured from experimental seedings conducted for 8 years, in Illinois, are summarized on page 299. Under California conditions, where the adults emerge from February to April, *early* planting and the stimulation of rapid growth are recommended.

Many experiments have been carried on to develop other methods of control. Those sometimes recommended, but which have been found of practically no value, are pasturing wheat with cattle or sheep, rolling wheat to crush the maggots or puparia, mowing wheat in the spring after the spring generation of flies has emerged, and early planting of strips of wheat as traps. Under certain conditions rotation of crops is of some value in preventing injury. Where no grass or clover is sown in the wheat, and the weather is sufficiently dry, summer burning of stubble may be of slight value. In most cases, however, the flaxseeds are sufficiently low in the wheat so that they are not killed by the fire. Some varieties of wheat have been found resistant to Hessian fly, but up to the present time no varieties have been found sufficiently resistant to attacks of this insect, and having other desirable qualities, to warrant recommending them generally in any of the large wheat-growing areas. Maintaining the wheat ground in a good state of fertility is of considerable value, as a wheat plant on rich ground will overcome the attack of one or two maggots of the Hessian fly and still produce a fairly good yield, whereas on poor ground such a plant would be killed. Where late seeding is practiced, the seedbed should be put in the best condition possible so that a vigorous growing plant may be obtained. No method has yet been developed for controlling Hessian fly by direct application of insecticides.

References.—*Kan. Agr. Exp. Sta. Tech. Bul.* 11, 1923; *Ohio Agr. Exp. Sta. Bul.* 177, 1906; *Jour. Agr. Res.*, **12**: 519–527, 1918; *U.S.D.A., Tech. Bul.* 81, 1928.

33. Wheat Stem Sawfly[1]

Importance and Type of Injury.—This native grass-feeding sawfly has acquired an appetite for small grains and become a menace to the small-grain crops in the more northern wheat belt. In some years it destroys 50 per cent of the crop. Wheat infested by this sawfly will show fallen straw in much the same manner as fields infested by Hessian fly or joint-

[1] *Cephus cinctus* Norton, Order Hymenoptera, Family Cephidae.

worm.   Examination of the straw will show the inside filled with fine
sawdust-like cuttings, among which will be found a wrinkled-bodied,
nearly legless, brown-headed larva ⅓ to ½ inch long, of a pale-yellow
color and with a short, pointed projection at the "tail" end.

Plants Attacked.—Wheat, spring rye, barley, spelt, timothy, quack
grass, and some of the native grasses.

Distribution.—The insect is native to North America and is most
destructive in the northern wheat-growing states west of the Mississippi
River.   A closely related species, the European wheat sawfly,[1] has been
found in New York and Pennsylvania.

Life History, Appearance, and Habits.—The winter is passed as a
mature larva in the base of the wheat straw near the surface of the soil.
In the spring the larva transforms to a pupa inside the wheat straw and
the adult emerges during June.   The adult is wasp-like in appearance,
black, with yellow rings on the abdominal segments.   The females lay
their eggs by thrusting them into the plant tissues on the upper parts of
the wheat stem.   The larva feeds within the stem, boring down through
the joints, and, by late summer, has reached the lower parts of the plant
close to the surface of the ground.   Here it cuts a V-shaped groove
entirely around and inside the stem, which causes the stem to break off,
and plugs itself in the base of the plant with its frass, thus forming a
chamber in which it hibernates and later pupates.

Control Measures.—Plowing under infested stubble in the fall, making
sure that it is thoroughly turned under to a depth of 5 or 6 inches is the
best method of control.   The overwintering larvae remain so close to the
surface of the soil that it is impossible to kill many of them by burning.
Cutting grains as early as possible without seriously affecting the yield or
grade of wheat does much to reduce damage by this insect.   Rotation of
crops which will put some immune crops, as corn, winter rye, flax, oats,
alfalfa, or sweet clover, on the wheat-stubble land, is also a help in combating this insect.   Some varieties of wheat, especially the solid-stemmed
varieties, are resistant.

References.—*U.S.D.A., Tech. Bul.* 157, 1928; *Sci. Agr.*, **15**: 30–38, 1934; *Bul. Ento.
Res.*, **22**: 547–550, 1931.

34. Wheat Jointworm[2]

Importance and Type of Injury.—In the large wheat-growing areas
east of the Mississippi the wheat jointworm is probably second in importance only to the Hessian fly as an insect pest of wheat.   Infested fields
just before harvesttime will show many of the straws broken off and bent
over in a manner similar to fields infested by the Hessian fly.   An examination of the fallen straws will show numerous, hard, gall-like swellings
filling the entire straw, often for an inch or more, usually just above a
joint (Fig. 252).   Inside these swellings, in oval cavities, are small, yellowish maggots about ⅐ to ⅙ inch in length.   When infested fields are
threshed, many small bits of broken straw containing these galls will come
through into the grain or be thrown out in large numbers around the
separator.

[1] *Cephus pygmaeus* (Linné).
[2] *Harmolita tritici* (Fitch), Order Hymenoptera, Family Chalcididae.

Plants Attacked.—Wheat is the only host known in the East, but it is reported from grass in California.

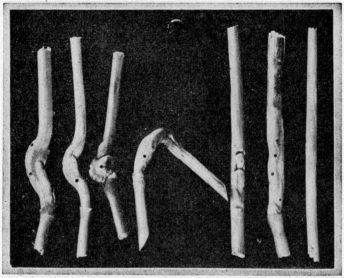

F IG . 252.—Characteristic galls in wheat straws caused by the wheat jointworm, showing exit holes of adults or parasites; about natural size. (*From U.S.D.A., Dept. Bul.* 808.)

Distribution.—The wheat jointworm is a native insect and is generally distributed in the states east of the Mississippi. Its range extends, however, into some of the states west of the Mississippi, but it has caused little damage in this area except in Missouri. The absence of the insect west of the Mississippi appears to be due to the harder stems of the wheat in that area.

Life History, Appearance, and Habits. The winter is passed in the gall-like hardened swellings inside the wheat straw, the insect being in the pupal or larval stage, mainly in the former. Those that have passed the winter in the larval stage change to pupae early in the spring, and all emerge as adults about the time the active growth period of the wheat starts, or the plants are beginning to form joints. The adult insects are about $\frac{1}{10}$ to $\frac{1}{8}$ inch in length and are jet black with the exception of the joints of the legs and two spots on the shoulders which are yellow. The females after mating insert their eggs

F IG . 253.—Adult female jointworm with her ovipositor thrust through the leaf sheath into the straw. Enlarged about 6 times. (*From Ohio Agr. Exp. Sta.*)

just above the wheat joints and inside the straw (Fig. 253). They drill a tiny hole into the wall of the straw by means of a stiff hair-like ovipositor attached to the underside of the abdomen. Usually a number of eggs are

laid in one place, sometimes as many as 25.   Occasionally, only one egg
will be deposited in a plant.   The larvae feed within the walls of the straw,
and the irritation set up by this feeding causes the straw to thicken, each
larva being separated from the others in a little cavity of its own.   Often
the swellings around the larvae cause the straw to twist or the formation
of the galls makes the straw so brittle that it is broken over in the field.
The larvae complete their growth about the time the wheat matures but
remain in the larval condition inside the straw until fall, when most of
them pupate.   The height of the jointworms inside the straw will vary
in different seasons.   Sometimes the galls will be just above the first
joint, within 8 or 10 inches of the surface of the ground, and at other
times the galls may be as high as the third joint and will, therefore, be
cut off with the straw when the wheat is harvested.   The variation in
the height of the galls is due to the difference in the development of the
plant at the time the eggs are deposited, the females tending to lay their
eggs in the uppermost parts of the plant.

   Control Measures.—If wheat stubble can be thoroughly burned,
practically all the overwintering jointworms remaining in the stubble
may be killed.   Plowing the infested stubble under shortly after harvest,
being sure that all stubble is turned under to a depth of 5 or 6 inches, is
a good control measure.   These measures are especially advisable where
combine harvesters and threshers are used.   In seasons when the larvae
are well up in the straw, it is possible by cutting the wheat low to remove
nearly all the insects in the straw.   Such straw may then be baled and
sold for use in cities or, if it cannot be disposed of in this way, it should be
burned.   Rotation of crops, putting any other crop than wheat on the
land, also may be of some help in reducing the numbers of this insect.   In
ordinary years, most of the infestation originates from not burning, or not
plowing under, the stubble of the previous season, so that every effort
should be made to burn or turn under infested stubble as soon after harvest
as possible.

   References.—*U.S.D.A.*, *Farmers' Bul.* 1006, 1918; *Ohio Agr. Exp. Sta. Bul.* 226,
1911; *U.S.D.A.*, *Dept. Bul.* 808, 1920; *Utah Agr. Exp. Sta. Bul.* 243, 1933.

35. Wheat Strawworm[1]

   Importance and Type of Injury.—Injury by this insect is of two distinct kinds: that caused by the first-generation larvae in young plants
in early spring and the later injury by the second generation in the maturing straw.   In the spring plants attacked by the wheat strawworm show
a stunted appearance; the crown of the plant is usually eaten out, including the developing head; and the plant killed or so injured that no head is
produced.   The later injury, after the plant has started to form joints,
has a somewhat stunting effect, weakening the straw, although a head
may be produced.

   Plants Attacked.—Wheat.   A closely related species attacks rye.

   Distribution.—The insect is generally distributed in the wheat-growing
regions west of the Mississippi River and is found in small numbers, but
is rarely destructive, in the states east of the Mississippi River.

   Life History, Appearance, and Habits.—The insect passes the winter in
the larval and pupal stages in the stubble or in stacks of straw.   In the

   [1] *Harmolita grandis* (Riley), Order Hymenoptera, Family Chalcididae.

early spring the adult insects gnaw small round holes in the straw, through which they emerge. Usually emergence begins during April or, in early springs, in March. These adults from the overwintering pupae are wingless and about ⅙ inch in length. The general color of most of the insects is brownish. They have much the appearance of ants. Upon examination with a lens their bodies will be seen to be quite hairy. They deposit their eggs in the stem walls at the base of the young wheat plants a little above the ground. The larvae, on hatching, work their way into the stem, eating off the developing head and preventing the formation of any tillers on the plant. The larvae, which are yellowish, legless, with small heads, and only about ⅕ inch long, become full-grown the latter part of April or first of May. They pupate within the plant and emerge as adults

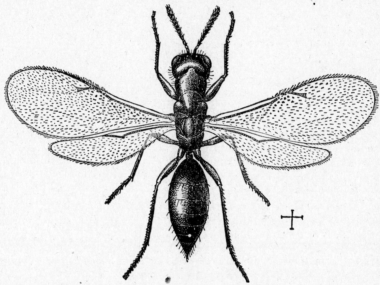

Fig. 254.—Adult female of the summer generation of wheat strawworm, greatly enlarged. Lines show natural size. (*From U.S.D.A., Dept. Bul.* 808.)

the latter part of May. These adults (Fig. 254) are larger than the early spring adults, less hairy, somewhat black in color, and are practically all winged. The females deposit their eggs inside the straw during the late spring, usually about the time the wheat is heading. In general, only one egg is laid in a wheat stem. The larva feeds within the stem and remains in this stage during the summer, either in the cut straw or, if it was working in the lower part of the stem, in the standing stubble. They change to the pupal stage in midautumn and pass the winter in this stage. There are thus two generations of the insect a year, the adult females of the two generations being strikingly different in size and appearance.

Control Measures.—As the majority of the adults, of the spring generation, are wingless, and those of the summer generation are not strong fliers, one of the best methods of controlling this insect is by rotation of crops. Wheat should not be sown within 300 feet of any straw of the previous season. Volunteer wheat should be destroyed by early spring. In dry-farm areas, allowing the land to lie fallow every other year is recommended. Baling and shipping infested straw to cities or towns is also a help in reducing the numbers of this pest. Many of the overwintering pupae may be destroyed where the stubble can be closely burned.

References.—U.S.D.A., Dept. Buls. 808, 1920, and 1137, 1923; and Farmers' Bul. 1323, 1923; Utah Agr. Exp. Sta. Bul. 243, 1933; Jour. Econ. Ento., **24**: 414–416, 1931.

36. Wheat Stem Maggot[1]

Importance and Type of Injury.—Wheat attacked by this insect in the fall of the year appears much the same as that infested by the Hessian fly, the plants taking on a darker appearance and remaining stunted, with

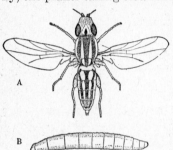

FIG. 255.—The wheat stem maggot, *Meromyza americana* Fitch. *A,* adult, 6 times natural size. *B,* larva, 6 times natural size. (*From Ill. State Natural History Surv.*)

stiff, s o m e w h a t thickened leaves. An examination of these plants will reveal slender, pale-green maggots (Fig. 255,*B*) working inside the lower part of the stem or crown of the plant. These maggots are about ¼ inch long.

The summer type of injury differs from that of the winter. The first indication of the presence of the insects is usually the dying out and whitening of the wheat heads and upper parts of the straw shortly after the head begins to fill while the lower stem and leaves are still green. The maggots will be found at this time of the year inside the straw just above the last or next to the last joint. The whitened heads are very conspicuous in the green fields of wheat often giving an exaggerated idea of the importance of the insect. Its injury rarely amounts to 1 or 2 per cent, but occasionally may be much higher.

Food Plants.—The principal food plants among the cultivated crops are wheat, rye, barley, and oats. It also feeds on bluegrass, timothy, and a number of the wild and introduced grasses.

Distribution.—The insect is a native species, occurring over practically the entire United States, in Mexico, and in the principal agricultural regions of Canada.

Life History, Appearance, and Habits.—The insect, so far as known, passes the winter only in the larval stage, the maggot being hidden away inside the lower parts of the stem of the wheat or the other plants on which it feeds. In the spring these larvae change inside a green puparium to the pupal stage and emerge mostly in June as yellowish-white flies, about ⅕ inch long, with three conspicuous black stripes on the thorax and abdomen and with conspicuous bright-green eyes (Fig. 255,*A*). The females, after mating, deposit their eggs on the leaves or stems of wheat and grasses on which the larvae feed. The young maggots crawl down behind the leaf sheaths to the tender soft part of the stems and tunnel into them, in the case of wheat feeding along the stem for a distance of 2 or 3 inches. The injured stem is partly severed and the head turns white and dies. When the larva becomes full-grown, the outer skin loosens from an inner skin (the integument of the last instar

[1] *Meromyza americana* Fitch, Order Diptera, Family Oscinidae.

larva) and forms a pale-green slender puparium in which the maggot changes to the pupal stage and later to the adult stage. These adults emerge about midsummer and lay their eggs on wild grasses or volunteer grain. The larvae of this summer generation become full-grown by the last of August or during September and transform to adults that emerge and lay eggs for the fall generation which develops on winter wheat as above described.

Control Measures.—Rotations with other crops and the destruction of volunteer grains will tend to reduce the numbers of this pest. Destruction of straw that is heavily infested by these maggots, or the baling and selling of straw off the farm, if done soon after wheat harvest, will help somewhat in reducing the number of insects.

References.—*U.S.D.A., Bur. Ento. Bul.* 42, p. 43, 1903, and *Dept. Bul.* 1137, 1923; *S.D. Agr. Exp. Sta. Bul.* 217, 1925.

<h3 style="text-align:center">37 (15, 68). WIREWORMS[1]</h3>

Importance and Type of Injury.—The different species of wireworms doing the greatest amount of injury to wheat are much the same as those

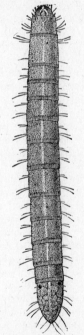

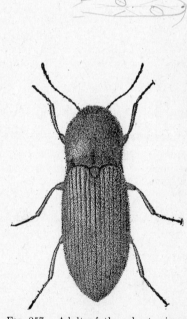

FIG. 256.—The wheat wireworm, larva, 5 times natural size. (*From Ill. State Natural History Surv.*)

FIG. 257.—Adult of the wheat wireworm, *Agriotes mancus* (Say), 7 times natural size. (*From Ill. State Natural History Surv.*)

which injure corn. So far as is known, all the wheat wireworms (Fig. 256) have a 4-year, or longer, life cycle. The adult beetles (Fig. 257) have the same appearance and character as those attacking corn. For full description of these insects, see Corn Insects, page 377.

[1] *Agriotes mancus* (Say), and others, Order Coleoptera, Family Elateridae.

Control Measures.—In general, the control measures for combating these insects given under corn will apply to the species attacking wheat. In the grain-growing sections of the Northwest, the following recommendations have been made for reducing the great losses caused by these pests: (*a*) summer fallowing of all fields where wireworms have recently caused damage; (*b*) shallow tillage, especially in spring and early summer, leaving the soil below 3 inches as firm as possible and destroying absolutely all vegetation; (*c*) seeding only when there is sufficient soil moisture to produce prompt germination; (*d*) the use of 5 to 10 pounds of extra seed per acre to allow for thinning by wireworms; (*e*) the application of commercial fertilizers.

References.—*U.S.D.A.*, *Farmers' Bul.* 1657, 1931; *Can. Ento. Branch, Saskatoon Leaflet* 35, 1933; *Jour. Econ. Ento.*, **27**, 308–314, 1934, and **18**: 90–95, 1925; *Idaho Agr. Exp. Sta. Res. Bul.* 6, 1926.

38. False Wireworms[1]

Importance and Type of Injury.—In the drier wheat-growing sections of the West, wheat is often seriously injured by the larvae of this and several closely related species of beetles. The germinating seed in the ground and the young plants are fed upon in the fall, and the small plants are often destroyed in the spring. The seed grain in the ground is nibbled, and the germ often eaten out. The young wheat plants are killed over irregular areas in the field, frequently in the vicinity of straw stacks or weed patches. Injury is most severe during dry years.

Plants Attacked.—Native, dry prairie grass, oats, millet, corn, kafir, alfalfa, cotton, beans, sugarbeets, garden crops, and other plants. Wheat is much the preferred food plant.

Distribution.—States between the Mississippi River and the Pacific Coast, ranging northward into British Columbia and southward to Texas.

Life History, Appearance, and Habits.—These insects pass the winter in the form of partly grown larvae or as adults. The adults (Fig. 259) are dark to black beetles a little under 1 inch in length. The wing covers of some species are distinctly ridged, others are smooth or granulate, and they are grown fast together, making it impossible for the insects to fly. When disturbed, they have a peculiar habit of placing their heads on the ground and elevating the hind part of their bodies as though standing on their heads. The larvae (Fig. 258) closely resemble wireworms in appearance, but have longer legs and antennae. Their bodies are brown or yellowish brown in color and prominently jointed. Some species of larvae are nearly black. The adult insects become active early in the spring and lay their eggs in the soil, from 10 to as many as 60 being deposited in a place. The larvae feed on the seeds, roots, and underground parts of the stems of the plants which they attack and under favorable conditions complete their growth in from 110 to 130 days.

Fig. 258.— A false wireworm, *Eleodes letcheri vandykei* Blaisdell, dorsal view of larva, much enlarged. (*From U.S.D.A., Bur. Ento. Bul.* 95, *Part V.*)

[1] *Eleodes opaca* (Say), *Eleodes suturalis* (Say), and others, Order Coleoptera, Family Tenebrionidae.

They pupate in earthen cells in the soil, the pupal period lasting from 10 to 25 days. Most of the damage is caused by the larvae. The adult beetles are general feeders. They are very long-lived, in some cases having been known to live for as much as 3 years in the adult stage. A second period of egg-laying usually occurs in the late summer or early fall, the larvae from these eggs hibernating in the soil in a partly grown condition.

Control Measures.—The most effective control measure for these insects is rotation of crops that will bring corn or other cultivated crops on the ground for at least 2 years between crops of wheat. General, thorough cleanup of straw and wheat refuse in the fields is also of much help in preventing damage. A dust furrow with post holes, as for army-worms, has been used to check the crawling adults as they migrate into

Fig. 259.—Adults of a false wireworm in characteristic attitudes. Somewhat enlarged.
(*From U.S.D.A., Bur. Ento. Bul. 95, Part V.*)

new fields. Exhaustive experiments to discover a method of poisoning seed wheat to prevent injury by the larvae have failed to reveal a practical method.

References.—*Jour. Agr. Res.*, vol. 22, no. 6, 1921, and vol. 26, no. 11, 1923; *U.S. D.A., Bur. Ento. Bul.* 95, Part V, pp. 73–87, 1912; *Tex. Agr. Exp. Sta. Ann. Rept.*, **56**: 52–53, 1933.

39 (14, 87, 216, 273). WHITE GRUBS[1]

Importance and Type of Injury.—Where wheat is sown early on ground heavily infested with white grubs, the young plants may be damaged in the fall by having their roots eaten off by these insects. The most serious damage to wheat occurs in the spring from grubs feeding on the roots. Heavily infested fields may be nearly destroyed. Feeding usually starts during the latter part of May or the first of June, and continues until wheat harvest. The brown-headed, curved-bodied, six-legged grubs, 1 inch or more in length, will be found just below the surface of the soil about the wheat roots. For a full description of white grubs see Corn Insects, page 374.

Control Measures.—The control measures given for preventing damage by this insect on corn apply equally well to wheat. In addition, most of the fall damage may be prevented by seeding sufficiently late so that the grubs will not feed on the wheat roots before they start their winter migration to below the frost line. Seeding late enough to avoid damage by Hessian fly will prevent most of the fall damage by grubs. Wheat is not so severely damaged as corn but, where possible, should not be planted on ground known to be infested by white grubs.

40 (7). BILLBUGS[2]

Wheat infested by billbugs will often show deadened stalks shortly before harvest, or the stalks may fall in much the same manner as where wheat is heavily infested

[1] *Phyllophaga* or *Lachnosterna* spp., Order Coleoptera, Family Scarabaeidae.
[2] *Calendra* spp., Order Coleoptera, Family Curculionidae.

with Hessian fly or jointworm. An examination of the lower parts of the plant will show short, white, curved-bodied, legless grubs with brown heads, feeding in the lower part of the stem and crown of the plant. The burrows made by the insect in the plant are filled with fine, sawdust-like castings. The grubs are rather small, seldom attaining ¼ inch in length. Damage is usually more severe on sod ground, in lowlands, or about the margins of fields. Damage by billbugs to wheat is not usually severe, but has apparently been increasing somewhat during the last few years in the midwestern states. For a full description of billbugs and their control, see Corn Insects, page 357.

B. INSECTS THAT ATTACK RYE

Rye is not so subject to injury by insects as is wheat. It is fed upon to a limited extent by the Hessian fly, but the injury to rye is not nearly so severe. In one case where comparisons could be made of the number of eggs laid on wheat and rye, in adjoining strips alternating through a field, it was found that only about one-sixth as many eggs of the Hessian fly were laid on the rye as on the wheat. The Hessian fly maggots seem to have greater difficulty in developing on rye than on wheat. The same control measures apply as on wheat.

Rye is especially subject to attack by chinch bugs, unless the stand is heavy so that the bugs are repelled by the shady condition. It is also subject to the attack of the sawflies and jointworm. None of these insects are serious pests on rye.

C. INSECTS THAT ATTACK OATS

Oats are comparatively free from serious insect injury. They suffer more severely perhaps than other small grains during outbreaks of armyworms. Oats are never attacked by the Hessian fly. In most years oats have not made sufficient growth to be attractive to chinch bugs when they leave their winter quarters and, therefore, escape injury, except such as may occur from migrating bugs at wheat harvest. In years when the English grain aphid is abundant, it sometimes severely injures oats.

D. INSECTS THAT ATTACK BARLEY

Barley is somewhat more subject to insect injury than oats. It is one of the favorite foods of the chinch bug, and fields of barley growing in areas generally infested by the chinch bug are usually more severely injured and contain greater numbers of these insects than adjoining or near-by fields of rye, oats, or wheat. Barley is also fed upon to some extent by the Hessian fly, although it does not damage this crop to nearly the extent that it does wheat. Armyworms, grasshoppers, stem maggots, and aphids also attack barley. For the control of these insects, see the discussions under Wheat.

E. INSECTS THAT ATTACK RICE

In the United States the principal insect pests of rice are the rice stink bug,[1] which sucks the sap from the developing rice grains; and rice water weevil,[2] the legless, small-headed, white grubs of which at first tunnel and later chew at the roots from the outside; the larvae of the sugarcane moth borer (see page 391); the rice stalk borer,[3] the caterpillars of which burrow through the rice stalks causing them to break over

---

[1] *Solubea pugnax* Fabricius, Order Hemiptera, Family Pentatomidae.
[2] *Lissorhoptrus simplex* Say, Order Coleoptera, Family Curculionidae.
[3] *Chilo plejadeilus* Zincken, Order Lepidoptera, Family Pyralididae.

and reducing the yield of grain; the sugarcane beetle,[1] the adult of which gnaws at the stems, between the surface of the ground and the roots of the plant; the southern corn rootworm (page 381); the fall armyworm (page 344); and the chinch bug (page 351).

General control measures include the keeping down of grasses near rice; winter plowing, burning over, or pasturing; rotating rice with crops not attacked by these insects; and submersion of fields when young rice is attacked by the sugarcane beetle, rootworms, or chinch bugs.

Reference.—U.S.D.A., Farmers' Bul. 1543, 1927.

[1] *Eutheola rugiceps* Le Conte, Order Coleoptera, Family Scarabaeidae.

CHAPTER XIII

INSECTS INJURIOUS TO LEGUMES

FIELD KEY FOR THE IDENTIFICATION OF INSECTS INJURING CLOVERS, ALFALFA, SWEET CLOVER, COWPEAS, AND SOYBEANS

A. *Chewing insects that eat away portions of leaves, buds, or stems above ground.*

1. Irregular holes, usually extending from the margin of the leaf inward, made by brownish or grayish jumping insects up to 1½ inches long.................. ....*Grasshoppers*, page 335.

2. Leaves of clover and alfalfa eaten in early spring by green, legless, curled, narrow-headed grubs with a whitish stripe down the middle of the back, up to ½ inch in length. The grubs hide about crown of plant during day and feed at night. Robust, brown, gray-mottled, oval-bodied beetles, ⅓ inch long, with the thorax and head narrowed to a short snout, feed on leaves in late spring and early fall.....:......................*Clover leaf weevil*, page 418.

3. Alfalfa plants have upper leaves shredded and growing tips eaten off in spring by dark-green larvae about ⅓ inch long. Badly infested plants have a whitened, bleached appearance. Small, dark-brown, grayish-mottled, oval, snout beetles about ³⁄₁₆ inch long feed on the leaves. Confined to Rocky Mountain region................................*Alfalfa weevil*, page 420.

4. Red clover buds dying and the growth of the plant stunted. Small, pale, brown or green, legless grubs, up to ¼ inch long, feed in the heads and inside the lateral buds. Green or bluish-green snout beetles, about ⅛ inch long, with black heads and snouts, found on leaves and stems. Injury occurs in spring months only and is most severe during dry seasons.................. ....*Clover bud weevil*, page 422.

5. Long, gray, black, or striped beetles, from ½ to ¾ inch long, feed on leaves of alfalfa during late summer and early fall. The growth of the plants is stunted and the leaves present a ragged appearance.........*Blister beetles*, page 424.

6. Oval beetles about ⅕ inch long, green or greenish yellow in color and with 6 conspicuous black spots on each wing cover, often abundant on cowpeas and soybeans eating the leaves................*Spotted cucumber beetle*, page 381.

7. Reddish or yellowish beetles, about ⅙ inch long, with 3 black spots near the inner edge of each wing cover, eat holes in the leaves of cowpeas. Slender, white larvae feed on the roots and root nodules.....*Bean leaf beetle*, page 489.

8. Soybeans and cowpeas grown near garden beans in certain sections have the leaves skeletonized by coppery-brown beetles, ¼ inch long, with 8 small black spots on each wing cover; and by oval, yellow, very spiny larvae up to ⅓ inch long which feed from underside of leaves......*Mexican bean beetle*, page 487.

9. Grayish long-headed beetles, ½ inch long, with white stripes along their sides, eat the margins of the leaves...............*White-fringed beetle*, page 472.

10. Short-snouted, dark-gray or brownish beetles, about ⅙ inch long, eat off small soybean plants on spring-broken clover sod as fast as they appear through the ground....................................*Clover root curculio*, page 438.

11. Alfalfa and, more rarely, clover leaves are eaten by dark-green caterpillars, about 1 inch long, with a light stripe containing a crimson hair line along each side of the body. Caterpillars with 5 pairs of prolegs and 3 pairs of thoracic legs. Sulphur-yellow butterflies with black borders on the wings fly about over the fields.............................*Alfalfa caterpillar*, page 425.

12. Leaves of clover, alfalfa, soybeans, and cowpeas eaten off by light-green worms, about 1¼ inch long, with a narrow white stripe and a second faint white line

on each side. These caterpillars have only 4 pairs of prolegs in addition to the 6 slender thoracic legs. They drop off the plants when disturbed. Injury most common in southern part of the United States.......................... ....*Green cloverworm*, page 426.

13. Light webs of silk cover alfalfa plants or surface of ground about the base of the plants in newly sown fields in early fall. Yellowish-green worms, up to 1 inch long, with scattered hairs and conspicuous black spots, feed on the leaves and new growth within these webs................................. ....*Garden, sugarbeet, and alfalfa webworms*, pages 427 and 383.

14. Leaves and stems of cowpeas, alfalfa, and vetch, and pods of cowpeas and soybeans eaten by greenish, white-striped caterpillars, up to 1¾ inches long, sparsely haired, and very variable in color; the skin rough-appearing under a lens.......................................*Corn earworm*, page 368.

15. Leaves of clover or alfalfa eaten, or plant stripped of foliage and tender shoots, by dark-green worms up to 2 inches in length with light stripes on the sides and down the middle of the back. These caterpillars, which feed at night and hide under clods, stones, or in heart of the plant during the day, have a skin that appears smooth under a lens. Worms often crawl over the soil in great armies. ....*Armyworm*, page 429.

16. Plants of clover, alfalfa, and other legumes are cut off at the surface of the ground, or leaves eaten, by plump, cylindrical worms of several shades and markings up to 1½ or 2 inches long, with 6 short slender legs near the head and 5 pairs of prolegs.................................*Cutworms*, page 429.

B. *Piercing-sucking insects that take the sap only, causing wilting, whitening, browning, reddening, and dying of the leaves and stems:*

1. Stems and leaves of clover and alfalfa covered with small, green plant lice or aphids. Plants wilt and die; leaves and stems coated with sticky fluid from the aphids...*Pea aphid*, page 430.

2. Attack similar to *B*,1 on cowpeas...................*Cowpea aphid*, page 430.

3. In the Rocky Mountain states, plants are stunted and seeds stuck together in pellets when threshed, by an insect similar to *B*,1......*Clover aphid*, page 430.

4. Leaves of clover and alfalfa have a somewhat mottled, whitened appearance due to many very fine, white spots, or become dwarfed and yellowish or reddish in color. Numerous, elongate, active, wedge-shaped bugs, mostly less than ¼ inch long, feeding on underside of leaves. Fields swarming with these small, variously colored, flying and jumping insects...........*Leafhoppers*, page 430.

5. Greenish to yellowish-brown, flat bugs, about ¼ inch long, cause blasted buds, flower drop, and shriveled seeds, by puncturing the terminal growth of alfalfa.. ....*Lygus bugs*, page 431.

6. Large, flattened, shield-shaped, bright-green, stinking bugs about ⅔ inch long, and various sized nymphs with reddish markings, suck sap from and poison soybeans and cowpeas, causing pods to drop or to form hardened, knotty areas, or to produce stunted and distorted seeds.............................. *Green stink bug, Acrosternum (Nezara) hilaris* Say and *pumpkin bug, Nezara viridula* Linné (see *Va. Agr. Exp. Sta. Bul.* 294, 1934).

C. *Insects that bore in the stems:*

1. Stems of red and sweet clover are swollen or cracked open, with the pith eaten out. Stems sometimes break off. Yellowish, smooth-sided, cylindrical worms about ½ inch long with two curved hooks at end of body, feeding on the pith in these tunnels. Parent beetle a smooth, narrow, hard-shelled insect ⅓ inch long with blue wing covers and bright red head and prothorax..................... *Clover stem borer, Languaria mozardi* Latrielle (see *U.S.D.A., Dept. Bul.* 889, 1920).

D. *Insects that eat into, suck sap from, or live within the flowers, heads or seeds:*

1. Cowpeas in the pod, or in storage, contain white footless grubs or short, chunky, brownish beetles, about 1/10 inch long. Both beetles and grubs feed inside seed, later leaving the seed through a small round hole.......................... ....*Cowpea weevil or four-spotted bean weevil*, page 441.

2. Legless, whitish, grubs, up to ¼ inch long, with small yellowish heads, bearing a white Y on the front, feed inside the seeds within the growing pods of cowpeas. Bronzy-black snout beetles, ⅕ inch long, the thorax and convex elytra having very conspicuous round punctures, feed on the plants and deposit eggs in holes eaten through the pods. They rarely fly and "play 'possum" at the slightest disturbance...
   Cowpea curculio or pod weevil, Chalcodermis aeneus Boheman (see Ala. Poly. Inst. Agr. Exp. Sta. Bul. 246, 1938).

3. Red-clover heads with much the same appearance as in D,5, but with a small, somewhat hairy caterpillar, about ¼ inch long and with distinct head and legs, feeding on the developing seeds and destroying many of the florets at the base..
   ....Clover head caterpillar, page 432.

4. Seeds of red and other clovers and alfalfa each completely occupied and later broken and cracked open, by a very small, fat, white, legless, maggot-like larva that reaches full growth inside the seeds. Infested seeds are often dull-colored. Eggs laid by very small, black, four-winged, wasp-like insects, ⅟₁₆ inch long, that fly about fields and crawl over the heads..........Clover seed chalcid, page 434.

5. Red clover heads fail to develop evenly, only a part of the pink florets opening, the rest of the head remaining green. Very small, pinkish, legless maggots, ⅟₁₂ inch long, feeding on the outside of the green seeds causing them to shrivel and dry up...................................Clover seed midge, page 436.

E. *Insects that attack the plant underground:*

1. Plants wilt and die during periods of dry weather. Plants are scored along the roots and often girdled near the crown, by short-snouted, dark-gray or brownish beetles or grayish-white, legless, brown-headed grubs, about ⅙ inch long. Injury most severe in late spring and early fall. .Clover root curculio, page 438.

2. Clover plants turn brown, wilt, and die. No feeding is apparent above ground but roots are found scored on the surface and tunneled through, by very small, black or dark-brown, cylindrical beetles or very small, legless, curved, brown-headed grubs about ⅟₁₀ inch long. Injury is most common in old stands of clover..Clover root borer, page 439.

3. White, curved-bodied grubs, about ⅛ inch long, with 6 slender legs and brown head and prothoracic shield, gnaw at the roots; chunky light-brown beetles, ⅟₁₀ inch long, feed on the leaves...................Grape colaspis, page 440.

4. Plants wilt and die from the attacks of white, legless, but not curved-bodied grubs, up to ½ inch long, which eat into the underground stem and taproot but do not eat the smaller roots...............White-fringed beetle, page 472.

A. CLOVER AND ALFALFA INSECTS

41. Clover Leaf Weevil[1]

Importance and Type of Injury.—The damage by this insect is most apparent in clover fields during the early spring. In late, cool, dry springs, red clover and alfalfa plants are frequently totally destroyed. The insect is never of great importance in wet springs. At this time of the year, the leaves of the clover or alfalfa plants will be found with smooth-edged notches eaten out of their sides or occasionally with whole leaves eaten off. Small, green, fat-bodied, legless larvae will be found during the daytime hidden away around the base of the clover plants, occasionally feeding on the leaves. Their bodies are nearly always curved so that the head and tail nearly touch. There is a pale yellowish-white stripe, edged with red, down the center of the back. The full-grown larvae are approximately ½ inch in length (Fig. 260).

Plants Attacked.—Red clover, sweet clover, and alfalfa are the main plants fed upon by the larvae of this beetle. Beans have been very little

[1] *Hypera punctata* (Fabricius), Order Coleoptera, Family Curculionidae.

injured.   The adult insects feed on a great variety of flowers and have been observed in large numbers on goldenrod and also on the heads of wheat, as well as many of the common weeds and flowering plants.

Fig. 260.—Larva of the clover leaf weevil, *Hypera punctata* (Fabricius), about 3 times natural size.   (*From Ill. State Natural History Surv.*)

Distribution.—This insect is probably a native of southern Europe. It was not known in this country until 1880, when it was first reported as injuring clover in New York.   It has now spread over most of the United States and into Canada.

Life History, Appearance, and Habits.—This insect passes the winter mainly in the form of partly grown larvae, around the crowns of the plant; to a lesser extent in the egg stage; and to a still lesser extent in the full-grown or adult beetle stage.   All eggs have hatched by early spring and the young larvae feed mainly at night on the leaves.   During the day, they remain well hidden in trash on the surface of the ground, or about the crown of the plants.   The larvae become full-grown during late spring and spin coarse brown or greenish-brown cocoons just beneath the ground surface, about the crowns of the plants, or occasionally on the stems of the leaves.   These cocoons are thin, about ⅓ inch in length, and have the appearance of a coarse network of somewhat stiff threads.   The pupal

Fig. 261.—Adult of the clover leaf weevil, *Hypera punctata* (Fabricius), about 6 times natural size.   (*From Ill. State Natural History Surv.*)

stage is passed in this cocoon, the adult beetle coming out in early summer.   The beetles are dark brown, flecked with black on the back and paler brown underneath.   A strong, robust snout projects from the head (Fig. 261).   The beetles feed very actively for a short time after emerging from the cocoon, and then become rather sluggish and feed but little until fall; they again become active in the fall, mating takes place, and the females lay their eggs mainly during September and October.   The eggs are pale yellow in color and are deposited in the stems of the leaves, on the stalks, or about the crown of the food plant. Some of these eggs hatch during the fall, while others do not hatch until the following spring.

Control Measures.—It is difficult to apply any direct control measure to fields infested by these insects that will kill the beetles without injuring the plants. In most seasons, the insects are held in check by a fungus disease which attacks the larvae. The larvae infested with this disease turn yellowish and later brownish in color, and usually remain curled around the tips of the leaves during the daytime, not hiding away as do the healthy larvae. Where numbers of the insect are found in this condition, one may be sure that little damage to the crop will result, although the weevils may be very abundant in the field. Clover fields that are found very heavily infested early in the spring, so that growth has practically stopped and the plants are being killed, should be plowed up and planted to some of the grass crops, or, where possible, to small grain. The insects are very difficult to poison.

References.—*Ill. Agr. Exp. Sta. Bul.* 134, 1909; *U.S.D.A., Dept. Bul.* 922, 1920; *N. Y. (Cornell) Agr. Exp. Sta. Bul.* 411, 1922; *Okla. Agr. Exp. Sta. Rept.* for 1932–1934, pp. 268, 269, 1935.

42. ALFALFA WEEVIL[1]

Importance and Type of Injury.—This is the most important insect enemy of alfalfa. During the past few years its injuries have been much less severe than during the period from 1920 to 1930. It may, and many times it does, destroy at least one cutting of hay during the season. The plants attacked show a skeletonizing or shredding of the tips of the new growth, this injury increasing from early spring until shortly before the time of the first cutting of alfalfa. In heavily infested fields, the growing tips are eaten off, the growth of the plants stunted, and the green part of the leaves eaten out to such an extent that the fields appear to be suffering from severe frost injury, presenting a bleached-out appearance. The plants are covered with green larvae about $\frac{1}{4}$ inch long when full-grown. They are plump-bodied, legless, but with well-developed ridges on the underside of the body which take the place of legs (Fig. 262,*b*). The adult snout beetles (*d*) will also be found in numbers during the spring and early-summer months in the injured fields.

Plants Attacked.—Alfalfa is the most important food plant. Bur clover, yellow sweet clover, and rarely vetch and a few of the true clovers are also attacked.

Distribution.—The insect was probably imported from southern Europe some time about 1900. The first noticeable injury occurred in the vicinity of Salt Lake City, Utah, in 1904. Since then, the insect has spread to include within its range the large alfalfa-growing districts of the West, reaching to Colorado, Wyoming, Montana, Idaho, Nevada, California, and Oregon. It has spread at the rate of 20 miles or more per year, the rate of spread depending on the type of farming, and the natural factors in the different areas where it is working.

Life History, Appearance, and Habits.—This weevil (Fig. 262,*d*) winters only in the adult stage. The beetles are grayish brown, or nearly black, with short grayish hairs giving them a somewhat spotted appear-

[1] *Hypera postica* (Gyllenhal) [= *Phytonomus variabilis* (Herbst)], Order Coleoptera, Family Curculionidae.

ance. They are about ⅛ inch long with a medium-sized beak about one-half the length of the thorax, projecting downward from the front of the head. On leaving their hibernating quarters about the crowns of the alfalfa plants, or under leaves and rubbish, the beetles feed for a few days, and mate, and the females lay their shining, oval, yellowish eggs (*a*) in the stems of the alfalfa. They first make cavities in the stems with their beaks, and in these insert from 1 to as many as 40 eggs. Each female beetle lays from 600 to 800 eggs during the spring. The larvae (*b*) on hatching are nearly white but soon become green with a prominent, white, middorsal stripe. They feed in the interior of the stalk for 3 or 4 days then make their way to the opening leaf buds at the tips of the

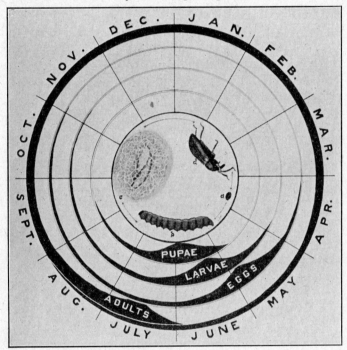

Fig. 262.—Diagram of seasonal history of the alfalfa weevil in Colorado. *a*, egg; *b*, larva; *c*, pupa in cocoon; *d*, adult. Enlarged 3 times. (*From Eleventh Ann. Rept. Colo. State Ento.*, 1919.)

plants, where they feed for some time concealed. This stunts the plant and produces a new growth below the tip which in turn is eaten off by the weevil larvae, leaving nothing but the woody fibers. In heavily infested fields, practically the entire first cutting is ruined, and the growth of the second crop is delayed. Most of the larvae become full-grown about the time of cutting the first crop. They then go to the ground, and spin a net-like, nearly spherical cocoon (*c*) in which they change to the pupal stage. After about 10 days in this stage, they emerge as adults. The new adults tend to appear in two divisions, coinciding roughly with the first harvest in early June and the second harvest in late July. These adults feed on the alfalfa plants the remainder of the summer and go into winter quarters early in the fall.

Control Measures.—The most effective control measures against this insect in most of the areas where it is now present are either the careful timing of cuttings or spraying or dusting with a stomach poison. The best time for cutting to check the insect should be learned from authorities in each state. In general, whenever the feeding of the weevils has checked the growth of the first crop and most of the eggs have been laid, alfalfa should be cut and hay removed promptly to allow the hot sun to kill the larvae in the stubble. The best time to poison the insects is when the young larvae become sufficiently abundant almost to stop the growing of the alfalfa plants. As the eggs hatch over a considerable period, it is necessary to wait until this time to get all the insects with one spray. The best time to apply the spray is usually about 2 weeks before the first-crop alfalfa is normally ready for cutting. Spraying at this time with lead arsenate or zinc arsenate, 2 pounds in 100 gallons of water, per acre; or dusting with equal parts of calcium arsenate and sulphur, 5 pounds per acre, has given almost complete control. Other methods which have not been found so effective consist of spraying the alfalfa stubble after the first crop has been cut, pasturing, disking, or dragging the field with a brush drag which knocks the larvae from the stubble and causes their death by contact with the hot dry ground. Dragging the field with a brush or wire drag after it has been harrowed, to create a dust mulch around the crowns of the plant has given fairly good results, but timely clipping and poisoning are by far the best methods of control. Proper handling of the crop with reference to the time of cutting will often prevent the necessity of using insecticides. In some sections a fair degree of control has been brought about by certain imported parasites[1] but they cannot be depended upon to prevent damage by the weevil. Quarantines against the importation of hay, straw, alfalfa meal, and some other farm products from the infested areas have been passed to prevent the transportation of the weevil into other areas.

References.—*U.S.D.A., Farmers' Bul.* 1528, 1927; *Calif. Agr. Exp. Sta. Bul.* 567, 1933; *Jour. Econ. Ento.*, **25**: 681–693, 1932, and **27**: 960–966, 1934; *Colo. Agr. Exp. Sta. Bul.* 567, 1933; *Cal. Dept. Agr. Spcl. Pub.* 166, 1939.

43. CLOVER BUD WEEVIL[2] OR LESSER CLOVER LEAF WEEVIL[2]

Importance and Type of Injury.—The clover bud weevil since about 1910 has become one of the most important and destructive pests of red clover in the middle western states. The injury is most severe during dry seasons and frequently amounts to an infestation of 90 to 98 per cent of all plants in the field. Infested clover plants show a deadening of the leaves and a general checking of the growth, which is particularly noticeable during dry seasons. If such plants are examined, small slits will be found cut in the stem, usually just above an axil of the stem or at the lateral buds, and the buds eaten into both at the terminal and on the sides of the stems. The heads of the plant become stunted and misshapen. Examinations made during May will show small pale-green larvae which are feeding in the stems, the newly forming buds, or the florets in the head.

[1] Especially *Bathyplectes curculionis* (Thomson), Order Hymenoptera, Family Ichneumonidae.

[2] *Hypera* (= *Phytonomus*) *nigrirostris* (Fabricius), Order Coleoptera, Family Curculionidae.

Plants Attacked.—The insect seems to prefer red clover, but feeds on all the common species of clovers, and also on sweet clover and alfalfa. On the last two plants it has been of little importance.

Distribution.—The insect was probably introduced into this country in the eastern part of Canada some time about 1875 or 1880. It has now spread over the entire eastern half of the United States and into the Pacific Northwest. In the midwestern states it has increased very rapidly in abundance since about 1915.

Life History, Appearance, and Habits.—The winter is passed only in the adult stage, the overwintering beetles being about ⅛ inch long, of a beautiful deep green, or blue-green color, with small black heads and a glossy-black, slender beak approximately as long as the thorax (Fig. 263). They shelter to some extent around the crowns of the clover plants in the field but in Illinois have been found in greatest abundance in woodland areas. A number of hibernating adults have been taken from around the base of a single tree or stump in the center of large areas of heavy woodlands. They are also found along bushy hedges, fences, roadsides, and other areas where trash occurs to give protection from the winter weather. They fly from their hibernating quarters about the time clover growth starts in the spring. The adults feed for a few days on the clover leaves; the females then begin laying their eggs. The eggs are deposited in small slits, cut in the stem of the clover plant or in the bud at the axil of the leaf or in the terminal bud of the plant. Usually but one egg is laid in a place, although two or three have occasionally been found.

Fig. 263.—Adult clover bud weevil, dorsal and side views. About 5 times natural size. (*From Ill. State Natural History Surv.*)

Egg-laying extends over nearly a month, from early April to early May, each female laying a total of from 200 to 300 eggs. The eggs hatch in from 2 to 3 weeks, and the young larvae, which are at first white, but later change to a brownish-white, begin feeding on the plant tissue. Where feeding starts at the buds, these may be entirely killed. If they start feeding in the head, it is destroyed entirely or in part. They occasionally tunnel in the stems, causing the stem above the point where they are working to wilt and die. The larval period extends from 20 to 25 days, or possibly somewhat longer, depending on the weather conditions of the season. Full-grown larvae are legless whitish grubs, with a black head and a dark line across the body just behind the head; they usually lie in a curve. They may spin their cocoons on the ground around the base of the plant, but more commonly in the part of the plant where they have been feeding. These cocoons are elliptical or nearly round, about ⅐ inch across. They are transparent, of a somewhat whitish or yellowish-white color. The insects remain in the pupal stage for from 5 to 12 days, emerging as full-grown beetles about 10 days before the time for cutting the first crop of red clover. Newly emerged beetles are brown in color but begin to show green by the end of 1 or 2 days, and by the third or fourth day after emergence are a pronounced grass green.

They feed in the clover field for 2 to 3 weeks, gradually becoming less abundant as they fly from the field to seek shelter in which to pass the winter.   In Illinois most of the beetles have left the fields by mid-July. There is probably only one generation of the insect each season.

Control Measures.—No satisfactory method of control for this insect has yet been developed.   Spraying, dusting, and cultural practices that have thus far been tried have not proved of much practical value.   Clover on fertile soil during a moderately wet or wet season will not suffer severely from the attacks of this insect; but on poor soil in a normal year, or in a dry season, it may be severely damaged.   Pasturing clover lightly until about May first may avoid much of the damage to the seed crop, by delaying the maturing of the crop.

Reference.—*Ohio State Univ. Ext. Bul,* vol. 16, no. 10, 1920–1921; *Okla. Agr. Exp. Sta. Rept.* for 1932–1934, pp. 268, 269, 1935.

44 (102). BLISTER BEETLES[1]

Importance and Type of Injury.—Clover, alfalfa, and soybeans are rarely seriously damaged by these insects.   Frequently, however, they are present in such numbers in alfalfa fields as to cause some alarm and a little damage.   The long, black, gray, spotted or striped black-and-yellow beetles, with very conspicuous heads and necks, long legs, and rather soft wing covers which do not completely cover the tip of the abdomen, (Fig. 264), will be found clustered on the tips of alfalfa and clover plants, feeding on the flowers and leaves.   Growth of the plants is somewhat stunted, and the field presents a ragged appearance where the beetles are numerous.

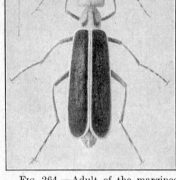

FIG. 264.—Adult of the margined blister beetle, *Epicauta marginata* Fabricius.   Line shows natural length. (*From Ill. State Natural History Surv.*)

Plants Attacked.—Blister beetles are general feeders attacking many flowering plants, [field and garden crops, clover, alfalfa, soybeans, and weeds.

Distribution.—Throughout the United States and arable parts of Canada.

Life History, Appearance, and Habits.—The blister beetles, whose life histories are known, pass the winter in the larval stage.   The larvae of some species feed on the eggs of grasshoppers and others in the cells of certain burrowing bees.   They have a very interesting and complicated life history.   Those which feed on the eggs of grasshoppers are, of course, beneficial in their larval stage.   Some of the blister beetles which feed on clover and alfalfa, notably the gray blister beetle, have two generations a year. Nearly all the other species have only one. The adult beetles vary in appearance, the gray blister beetle and the black blister beetle being of a uniform color and from ½ inch to nearly 1 inch in length.   Several other species are spotted, striped, or marked with different colors (see also page 509).

Control Measures.—Probably the best control for blister beetles on alfalfa or clover is to go over the field with a hopperdozer as described for catching grasshoppers. Once over the field with this machine will probably clean out the blister beetles. Dusting with a mixture of 25 per cent cryolite and 75 per cent dusting gypsum is a very effective control on crops of sufficient value to make such control profitable.

[1] *Epicauta* spp., *Cantharis* spp., *Macrobasis* spp., Order Coleoptera, Family Meloidae.

In most cases it is not necessary to take any active steps to combat these insects on clover and alfalfa.

Reference.—U.S.D.A., Dept. Bul. 967, 1921.

45. ALFALFA CATERPILLAR[1]

Importance and Type of Injury.—This insect is particularly a pest of alfalfa in the southwestern states. Infested fields show part of the leaves eaten out or entirely consumed by a green worm, which in the southern states appears during late February or March. Previous to the appearance of the worms, sulphur-yellow butterflies with black

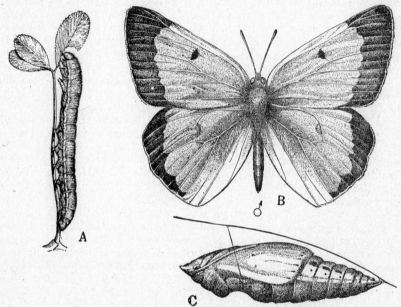

FIG. 265.—The alfalfa caterpillar. *A*, full-grown larva feeding on leaf; *B*, adult male butterfly; *C*, chrysalid or pupa, showing how it is suspended from the plant. Slightly enlarged. (*From U.S.D.A., Dept. Bul.* 124.)

margins on the upper surface of the wings will be found in large numbers hovering over the alfalfa plants, on which they are depositing their eggs.

Plants Attacked.—The insects feed mainly on alfalfa, although they are occasionally taken on clover and several other legumes.

Distribution.—The alfalfa caterpillar is found throughout the United States and southern Canada, with the possible exception of the extreme southeastern United States. Its heaviest damage, however, occurs in the Southwest, and the alfalfa-growing regions along the Pacific slope.

Life History, Appearance, and Habits.—Throughout most of its range the insect passes the winter in the pupal stage (Fig. 265,*C*) on the plants. In the extreme southern part of its range, larvae may be found during the winter months, and, occasionally, adults will be seen on the wing in these sections; so it probably hibernates in all stages. In the spring the overwintering pupae change to yellow butterflies (Fig. 265,*B*) having a

[1] *Eurymus* (= *Colias*) *eurytheme* Boisduval, Order Lepidoptera, Family Pieridae.

wing expanse of about 2 inches. The entire under surface of the wings is solid sulphur yellow. The upper surface of the wings is bordered with black. The butterflies, on emerging, lay their eggs to the number of from 200 to 500, on the under sides of the alfalfa leaves. These eggs hatch in a few days into small dark-brown worms which soon change to a green color. The worms grow very rapidly, becoming full-grown in from 12 to 15 days, at which time they are nearly 1½ inches in length, of a dark grass-green color, with a fine white stripe on each side of the body through which runs a very fine red line (Fig. 265,A). These worms change to the pupal stage without spinning a cocoon. They attach the narrow tail end of the pupa to the alfalfa stalk and throw a loop of silk about their bodies a little above the middle, which holds the head upright. The pupal stage lasts from 5 to 7 days before the emergence of the adult butterflies. In the extreme southwestern states, there are from five to seven generations a year, while probably at least two generations always occur in the northern part of the insect's range.

Control Measures.—Spraying or dusting the crop, as recommended for the alfalfa weevil, will serve as an effective control for the alfalfa caterpillar. Such measures should be taken only when the insect is very abundant and serious damage to the crop is threatened. Cutting the infested alfalfa as low as possible and removing the hay from the field is also an effective control measure. This cuts off the food supply of the young caterpillars and exposes them to their insect enemies, of which several kinds are nearly always present in the infested fields. Many caterpillars are also killed by exposure to the heat of the sun. In cutting, care should be taken that the field is left clean, without a sufficient amount of leaf growth to support the growing caterpillars until they mature. Keeping down weed growth around the field and pasturing, where it can be properly done, so that the animals are not left too long on the field, also are recommended. It is of the greatest importance that alfalfa growers know the parents of this caterpillar and, when swarms of the yellow butterflies (Fig. 265,B) are noted about the field, keep close watch of the growing alfalfa crop and if necessary take the control measures above recommended before the worms have caused serious damage.

References.—U.S.D.A., *Farmers' Bul.* 1094, 1920, and *Dept. Bul.* 124, 1914; *Bul. Brooklyn Ento. Soc.*, **28**: 108–115, 1933.

46. Green Cloverworm[1]

Importance and Type of Injury.—The green cloverworm is only occasionally of importance as a pest of clover and alfalfa. It is nearly always present in fields of these crops grown in the eastern United States. When it is abundant, the infested fields present a ragged appearance, many of the leaves having been eaten off by green worms with two narrow white stripes down each side of the body (Fig. 266). The insect is more important in the southern than in the northern states.

Plants Attacked.—Alfalfa, clover, garden and field beans, soybeans, cowpeas, vetch, strawberry, raspberry, and many of the common weeds and other legumes.

Distribution.—Eastern United States to the plains states and southeastern Canada.

Life History, Appearance, and Habits.—The green cloverworm passes the winter in the pupal or adult stage. The adults are dark-brown, black-spotted or mottled moths with a wing expanse of about 1¼ inches (Fig. 267). When at rest, the wings are held in such a position as to give the insect a triangular appearance. The moths shelter around barns, haystacks, and other protected places. They become active

[1] *Plathypena scabra* (Fabricius), Order Lepidoptera, Family Noctuidae.

about the time clover growth is well started, and, after mating, the females lay their eggs singly on the undersides of the leaves of the plants which they attack. The eggs hatch into small worms which are nearly the shade of the alfalfa leaf. They feed on the leaves, completing their growth under normal conditions in about 4 weeks. They then crawl down the plant and work their way under litter or just below the surface of the soil, where they change, inside a light silken cocoon, to a brown pupal stage. They remain in this stage for from 10 days to 3 weeks and then emerge as moths.

There are from three to four generations in the southern part of the United States, and probably two in the northern part of the insect's range.

Control Measures.—In most years these insects are not sufficiently abundant to necessitate taking active measures of control. When fields are being injured, the alfalfa or clover

FIG. 266.—Larva of the green cloverworm, *Plathypena scabra* Fabricius, and its injury on clover. About natural size. (*From Ill. State Natural History Surv.*)

FIG. 267.—Adult of the green cloverworm, enlarged about ½. (*From Ill. State Natural History Surv.*)

should be cut as soon as possible and removed from the field. This exposes the worms to the bright sunlight and to their insect and bird enemies and is usually the only control measure necessary. Hopperdozers have been used with a fair degree of success in combating the insects.

References.—*U.S.D.A., Farmers' Bul.* 982, 1918, and *Dept. Bul.* 1336, 1925; *Calif. Mon. Bul.*, **22**, 156–160, 1933; *Rept. Ont. Ento. Soc.*, **62**: 75–82, 1931.

47 (19, 69). GARDEN WEBWORM[1] AND ALFALFA WEBWORM[2]

Importance and Type of Injury.—These insects are of some importance on clover but are more particularly pests of alfalfa and some other crops. During the spring and summer months, fields that are infested will show light webs over the leaves, in which will be found greenish to yellowish-green, somewhat hairy worms with black dots over their bodies (Fig. 268). Under protection of the webs, the larvae may consume all the green tissue of the leaves. When full-grown, they are an inch or a little over in length, greenish to nearly black, with a light stripe extending down the middle of the back and with three dark spots on the side of each segment, from each of which projects one to three bristle-like hairs. When disturbed, the worms drop to the ground or crawl down into tubular webs which they have spun. Fields that are heavily infested will show a considerable amount of webbing over the alfalfa, and the leaves inside these webs will be nearly all eaten off. In the fall, the worms work in somewhat the same manner as cutworms on the newly sown alfalfa. At

[1] *Loxostege similalis* (Guenee), Order Lepidoptera, Family Pyralididae.
[2] *Loxostege commixtalis* (Walker).

this time, they hide in silken-lined burrows on or in the ground to which they retreat when disturbed.

Plants Attacked.—The garden webworm is a general feeder. Among the plants attacked are alfalfa, clover, beans, soybeans, cowpeas, sugar-beets, peas, strawberries, wild sunflower, thistles, pigweed, ragweed, sweet clover, lamb's-quarters, and a number of others. The alfalfa webworm will eat almost any succulent growth, except small grains and grasses. It has been recorded as a very serious pest of alfalfa and sugar-beets, especially in seasons following a dry year.

Distribution.—The insect is native to both North and South America and occurs generally in the farming areas of the United States, Canada, and Mexico.

Life History, Appearance, and Habits.—The winter is passed in the pupal stage in the soil about the plants on which the fall generation of larvae fed. In the extreme South it is possible that some of the insects live through the winter in the larval stage. Adult moths emerge from the pupae early in the spring and deposit their eggs on the leaves of their food plants. The eggs are laid in masses of from 2 or 3 to 20 or 50.

FIG. 268.—The garden webworm, *Loxostege similalis* (Guenee), caterpillar or "worm," about twice natural size. (*From U.S.D.A., Farmers' Bul.* 944.)

The garden webworm moths are about ¾ inch across the wings, buff-colored with shadings and irregular markings of light and dark gray; the alfalfa webworm is similar in appearance, but measuring 1 to 1¼ inches from tip to tip of wings. They are rather weak insects, probably living but a short time. When one is going through infested fields, they will frequently fly up, going a short distance, but usually alighting within a rod or two of the point where they were first disturbed. They are not active during the daytime, unless disturbed, but often assemble about lights on warm nights. The larvae hatch from the eggs in 3 days to 1 week. They begin feeding on the under side of leaves, protecting themselves inside the light webs above described. It usually requires about 3 to 5 weeks for the worms to reach the full-grown stage and go into the ground for pupation. When short of food, the alfalfa web-worm caterpillars sometimes migrate like armyworms. The pupal period is ordinarily from 7 days to 3 weeks. There are several generations each season. There are said to be five full generations each year in Oklahoma. In the northern part of the range, there are probably two or three generations each season.

Control Measures.—Fields which become infested during the summer months can usually be cleaned of the webworms by cutting the alfalfa. This cuts off the food supply of the worms, as they are unable to feed on the dried alfalfa hay, and also by destroying their webs exposes many of them to bird and insect enemies. Fields of newly seeded alfalfa infested in the early fall may be protected by dusting with calcium arsenate at the rate of about 10 pounds per acre, or spraying with calcium arsenate or lead arsenate, 4 pounds to 100 gallons of water, per acre. Either one of these methods is very effective in killing the fall generation of larvae provided a thorough application of the poison is made. On sugarbeets, the alfalfa webworm is controlled by a fine spray of 4 pounds of paris green or 8 pounds lead arsenate in 50 gallons of water per acre. To be most effective the sprays should be applied as soon as the eggs hatch and before there is much webbing over of the plants. The migrating alfalfa webworms may be stopped by dusty or vertical-sided furrows, constructed across the path of their migrations. Keeping field margins closely cut to keep down all growth of the weeds on which this insect feeds is also of considerable help in preventing damage. Insect parasites usually prevent damage for more than 1 year successively.

References.—Okla. Agr. Exp. Sta. Bul. 109, 1916; *U.S.D.A., Farmers' Bul.* 944, 1918; *Calif. Mon. Bul.* **23**: 236, 237, 1934; *Colo. Agr. Coll. State Ento. Cir.* 58, 1933.

48 (2, 28). ARMYWORMS *

The true armyworm and the western army cutworm[1] often feed on clover and alfalfa to a limited extent. They prefer the grass crops, however, and normally cause but little damage to legumes. Occasionally armyworm outbreaks are associated with outbreaks of the variegated cutworm, and these latter insects, which prefer legumes, may cause serious damage to clover fields, which is mistaken for damage by the armyworm.

Control Measures.—The control is the same as that given on page 343.

49 (4, 67, 80). CUTWORMS

Several species of cutworms are troublesome in clover and alfalfa fields. In the midwestern states the variegated cutworm[2] (Fig. 269) is the one most generally

FIG. 269.—Larva of the variegated cutworm, dorsal view. Enlarged about ½. (*From Ill. State Natural History Surv.*)

abundant in clover and alfalfa during the spring and summer months. The dingy cutworm,[3] clay-backed cutworm,[4] and several others are also common in fields of clover and alfalfa at this time of the year. In the fall, newly sown alfalfa fields are often seriously damaged by the yellow-striped armyworm[5] and the fall armyworm.[6] Both these insects migrate up from the South, and the moths lay their eggs in the newly sown alfalfa. The worms hatching from these eggs feed on the alfalfa in the general manner of cutworms, not infrequently destroying the entire field. Several other species of cutworms are abundant in clover and alfalfa in different parts of the country.

Control Measures.—Cutworms may be controlled in these crops by the use of the poison-bran bait, made and applied in the manner described for the control of these insects on corn (see page 350).

References.—U.S.D.A., Farmers' Bul. 739, 1916; *Can. Dept. Agr. Bul.* 22, 1923.

[1] *Chorizagrotis auxiliaris* (Grote), Order Lepidoptera, Family Noctuidae.
[2] *Lycophotia saucia* Hübner, Order Lepidoptera, Family Noctuidae.
[3] *Feltia subgothica* Haworth.
[4] *Feltia gladiaria* Morrison.
[5] *Prodenia ornithogalli* Guenee.
[6] *Laphygma frugiperda* Smith and Abbott.

50 (92). Aphids

The pea aphid, which is fully discussed on page 490, is often a serious pest on clover and alfalfa, which are its winter hosts. In the northwestern United States the clover aphid[1] has in some instances caused the total loss of seed crops of red and alsike clovers. Chief losses have been due to the presence of honeydew on the seeds, causing them to stick together in lumps or solid cakes in storage. The winter hosts of this aphid are pome fruit trees. The cowpea aphid[2] attacks cowpeas much as the pea aphid injures peas. Cutting clover and alfalfa as early as possible, or the use of a brush drag on hot days, after the first crop has been cut for hay, has given good results. Where these crops are being grown for seed and become heavily infested, closely pasturing with sheep, or clipping as close to the ground as possible, promptly removing the hay, and allowing the second crop to produce seed, will often prevent a heavy loss of the crop.

References.—Idaho Agr. Exp. Sta. Res. Bul. 3, 1923.

51. Leafhoppers[3]

Importance and Type of Injury.—Where leafhoppers are abundant in any crop, the plants show a lack of vigor, growth is retarded, and in most cases the leaves have a somewhat whitened, mottled appearance, or turn yellow, red, or brown, due to sucking out of the sap by the hoppers, which feed mainly on the undersides of the leaves. In walking through infested fields, large numbers of tiny mottled or speckled, green, yellow, brown-gray, or various-colored insects (Fig. 270) will hop or fly for short distances ahead of one. With certain species of leafhoppers the feeding produces a burning effect on the plants, and causes the tips to wither and die somewhat as though scorched by bright sunshine or injured by drought. Leafhoppers are especially destructive to clovers during the seedling stage and just after a cutting has been made. Some species are carriers of plant diseases. See Potato Leafhopper (page 510) and Beet Leafhopper (page 545).

Plants Attacked.—Nearly all cultivated and wild plants are attacked by various species of leafhoppers. All the clovers, alfalfa, and many of the grasses and small grains, also orchards, vineyards, and forest and shade trees are infested and damaged to some extent.

Distribution.—Leafhoppers occur throughout the world.

Life History, Appearance, and Habits.—The winter is passed in various stages, according to the different species. Some go through the winter in the egg stage in the stems of various plants, a large number of species pass the winter in the form of full-grown insects, which hide away in shelters around and in the field crops which they attack, while a few pass the winter in the partly developed, or nymphal stages. In the Gulf states, the females may continue to reproduce through the winter and the winged forms migrate northward in spring. In most cases that are known, the insects lay their eggs in the stems, buds, or leaves of their food plants. These hatch into wingless but very active nymphs. The

---

[1] *Anuraphis bakeri* Cowen, Order Homoptera, Family Aphididae.

[2] *Aphis medicaginis* Koch.

[3] Many species of the Order Homoptera, Family Cicadellidae.

nymphs feed by sucking the sap and sometimes inject a substance which is distinctly poisonous to the plant tissue, killing the areas around their feeding punctures. They develop from small nymphs to adults, molting their skins several times during this process, but not passing through any distinct pupal stage and never spinning cocoons or forming a chrysalis. The adults vary in size from ½₀ to ¼, and rarely ½, inch in length. They are all good jumpers or hoppers, as their common name implies. The adults are winged but use their legs to a large extent in jumping from one part of the plant to another. The general outline of their bodies is long and slender, and, as above stated, they vary greatly both in color and in shape.

Control Measures.—Leafhoppers are extremely difficult insects to control. Cutting and removing the crops from heavily infested clover

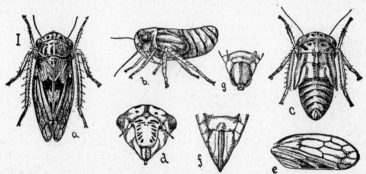

Fig. 270.—The clover leafhopper, *Agallia sanguinolenta* Provancher. *a*, adult; *b*, nymph, side view; *c*, nymph, dorsal view; *d*, face; *e*, front wing; *f* and *g*, genitalia and last segment of female and male. The line shows natural size. (*From Osborn and Ball.*)

and alfalfa fields, especially just after the majority of the eggs have been laid, is about the most effective method of control, as this removes the eggs, drives out some of the leafhoppers, and starves some of the young nymphs. Immediately after cutting, large numbers of leafhoppers can be caught by running a hopperdozer over the infested field; or this machine may be used on the growing crops. It is most effective when the back and sides of the machine have been thoroughly coated with some sticky material such as tree tanglefoot that will catch the hoppers as they fly against it. Certain pubescent native strains of clover are noticably resistant to leafhopper attack and should be used when they can be obtained.

References.—*Jour. Agr. Res.*, **17:** (no. 6): 399–404, 1927; *Ann. Ento. Soc. Amer.*, **16:** 363, 1923; *U.S.D.A.*, *Farmers' Bul.* 737, 1916, and *Bur. Ento. Bul.* 108, 1912; *Ky. Agr. Exp. Sta. Cir.*, 44, 1936.

52. Bugs of the Genus Lygus Affecting Alfalfa[1]

Importance and Type of Injury.—Several species of Lygus injure the tender growing or fruiting parts of alfalfa by puncturing the tissues with their piercing-sucking mouth parts to obtain food. In addition to the

[1] This article condensed from a paper prepared for this purpose by Lloyd L. Stitt of the Bureau of Entomology and Plant Quarantine.

physical injury a toxic reaction upon the plant cells near the puncture apparently results from their feeding. In alfalfa, the feeding of Lygus bugs[1] causes "blasted" buds, excessive flower fall, and brown shriveled worthless seeds. In many western areas the species *Lygus hesperus* constitutes approximately 80 per cent of the Lygus population on alfalfa-producing seed. Alfalfa seed fields often fail to produce a profitable crop due to the damage caused by these bugs.

Plants Attacked.—These species of Lygus are general feeders and are found on many herbaceous plants and also on trees. The cultivated crops damaged in the western and southwestern states are alfalfa, cotton, beans, and probably sugarbeets grown for seed. Weed hosts include winter mustard, lamb's-quarter, mare's-tail, and slim aster.

Distribution.—The species of Lygus considered here occur throughout the Rocky Mountain and Pacific states.

Life History, Appearance, and Habits.—In most regions these species of Lygus pass the winter as adults in hibernation, but in warmer climates the adults can be swept from plants all winter. Eggs and nymphs have been found in southern Arizona in all months except December. At an average mean temperature of 85.5°F. the egg stage requires about 8 days, the five nymphal instars about 11 days. A generation requires from 20 to 30 days in the summer under the climatic conditions of Arizona. The adults (Fig. 297) are about $\frac{1}{4}$ inch long by $\frac{3}{32}$ inch broad, flattened, oval, and show a variation in color, from pale greenish to yellowish-brown. Eggs are found in the flowers, buds, bracts, nodes, and internodes. The egg is slightly curved, elongated, and bears a lid on its truncated apex. When the nymphs first hatch they are very pale green and have an orange spot on the middle of the abdomen. Shortly after feeding begins, they become darker green in color and the third, fourth, and fifth instars have four noticeable black spots on the thorax.

Control.—As yet no fully satisfactory control has been developed for the species of Lygus attacking alfalfa. However, recent investigations indicate that the starting of the alfalfa seed crop at the same time in all fields throughout a neighborhood keeps the infestation low during the development of that crop; whereas, fields producing seed at other times, in many cases receive migrations of adults resulting in extremely high infestations and heavy damage. Clean cultural practices, such as clean cutting of the hay crop preceding the seed crop, result in a great reduction of the nymphal population.

References.—*Idaho Agr. Exp. Sta. Res. Bul.* 11, 1933; *U.S.D.A., Tech. Bul.* 5, 1927; *Jour. Econ. Ento.,* **29**: 454–457, 1936.

53. CLOVER HEAD CATERPILLAR[2]

Importance and Type of Injury.—Clover heads infested by the clover head caterpillar present somewhat the same appearance as those infested by the clover seed midge. The head is usually irregular in appearance and will have the flowers opening on only one side of the head. Such heads are often pink on one side and green on the other. An examination

---

[1] *Lygus hesperus* Knight, *L. elisus* Van D., *L. pratensis oblineatus* Say, Order Hemiptera, Family Miridae.

[2] *Laspeyresia* (= *Grapholitha*) *interstinctana* Clemens, Order Lepidoptera, Family Tortricidae.

of the infested heads will show a small, somewhat hairy caterpillar feeding on the seeds at the base of the clover florets, preventing them from opening (Fig. 271,*A*). These caterpillars are very small, being only about ¼ inch long when full-grown.

Plants Attacked.—The insect feeds mainly in the heads of red clover but has also been found on white clover, alsike clover, and mammoth clover. It also attacks the leaves of these plants when no heads are present.

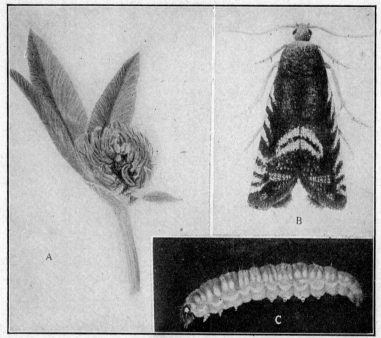

Fig. 271.—Clover head caterpillar, *Laspeyresia interstinctana* Clemens. *A*, larva at work in head of red clover, slightly enlarged; *B*, adult, with wings folded, about 8 times natural size; *C*, larva, side view, about 8 times natural size. (*From Ill. State Natural History Surv.*)

Distribution.—The clover head caterpillar is distributed generally over the eastern part of the United States and southern Canada.

Life History, Appearance, and Habits.—The insect passes the winter in both the larval and pupal stages, both being found under trash and around clover fields. Pupation occurs inside of small silken cocoons about ⅓ inch in length. These are very hard to find because of the dirt which is usually attached to the outside of them. The overwintering larvae pupate in the spring, and, from these and the overwintering pupae, the moths begin emerging just about the time that red clover is starting to come into bloom. They are small dark-brown moths about ¼ inch long. The wing margins are marked with six or seven short dashes, silvery white in color. The marks on the inner margins form a double crescent when the wings are folded (Fig. 271,*B*). They lay their eggs on the leaves, stems, and heads of clover. The young larvae hatching from these eggs feed on the clover leaves, or more usually work their way into the green clover heads, feeding at the base of the florets and destroying from half to all the florets in a head, so that seed is not produced. The larvae are said not to destroy the hardened seed, but eat the green, soft, newly formed seeds. They complete their growth in 4 or 5 weeks, being about ¼ inch long and of a greenish to greenish-white color (Fig. 271,C). They spin their cocoons either in the head or about the base of the clover plants and change inside the cocoons to the brown pupal stage, which lasts about 15 to 20 days. The moths of the second generation emerge about midsummer in time to lay eggs upon the normal second crop of blooms. A third generation occurs throughout the entire southern part of the insect's range, the larvae feeding in the crowns of the plant.

Control Measures.—The only measures that have been found at all practical for reducing the numbers of these insects are the same as those given for the control of the clover seed midge. Fall clipping will help to destroy the overwintering stages. The first crop may best be used for seed when this insect is abundant. Rotation of crops is also of some benefit in keeping down the numbers of this insect.

References.—*N. Y. (Cornell) Agr. Exp. Sta. Bul.* 428, 1923; *Ill. Agr. Exp. Sta. Bul.* 134, 1909.

54. CLOVER SEED CHALCID[1]

Importance and Type of Injury.—This is one of the most important insect pests of alfalfa and clover seed but has no effect on the production of hay from these crops. The worst damage from this insect occurs when both first and second crops of alfalfa are used to produce seed in the same community. Infested plants have little to distinguish them from those that are uninfested. A close examination of the seeds, however, will show many of them broken or cracked open. Threshed seed will show many empty shells of the seed, or parts of such shells. Very small, white, maggot-like larvae develop in the infested seeds, eating the contents (Fig. 272,e). In some sections this insect has made the growing of clover and alfalfa seed unprofitable. Examinations of alfalfa seed in Utah over a 4-year period showed that from 8 to 14 per cent of the seed was destroyed by this insect.

Plants Attacked.—The insect attacks nearly all the clovers and alfalfa.

Distribution.—General over the United States and southern Canada.

Life History, Appearance, and Habits.—The insect passes the winter in the full-grown larval stage (Fig. 272,b) inside the infested seeds on the surface of the ground. Occasionally the insect may pupate in the fall and remain inside the seed in this stage. The adult insects begin emerging during the late spring, usually the latter part of May in the latitude of central Illinois. Adults of the first generation occur in great numbers during the first part of June. The adult insect (Fig. 272,a) is a very active little wasp-like creature, jet metallic black in color, with legs of a dark-brownish color, and tarsal claws of a light yellowish brown. They are so small, being only about 1/15 inch in length, that one has great difficulty in seeing them in the field and will never suspect their numbers

[1] *Bruchophagus gibbus* (Boheman) (= *B. funebris* Howard), Order Hymenoptera, Family Chalcididae.

unless they are collected in a fine-mesh net. The female insects lay their eggs in clover in which the seed has formed but has not yet hardened. The egg is pushed inside the soft seed and hatches into a white, footless, maggot-like larva that consumes the entire inside of the seed, leaving only a thin outer shell. During threshing, most of the light, infested seeds are blown out with the straw. The larvae complete their growth in 2 weeks or more, depending on the weather. As the adult insects continue to emerge over a long period, the first-generation larvae occur over a correspondingly long period. Upon becoming full-grown, they pupate inside the seed, either in the head or in seed which has dropped to the surface of the ground, and emerge about midsummer as adults. These adults lay the eggs for a second generation, and in the South a third gen-

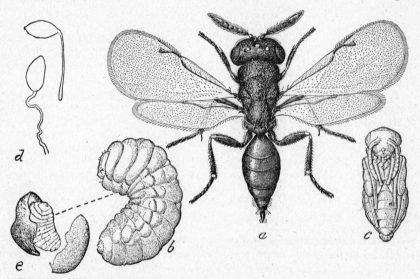

Fig. 272.—Clover seed chalcid. *a*, adult; *b*, larva; *c*, pupa; *d*, egg; *e*, larva in broken seed; all greatly enlarged. (*a, b, c, from U.S.D.A., Farmers' Bul.* 636; *d, from Ill. State Natural History Surv.; e, from Ohio State Univ.*)

eration occurs late in the season. The generations overlap, however, so that the insects are present in all stages in the field practically at all times from the first of June until September.

Control Measures.—This insect has been found extremely difficult to control. In certain years the use of the first crop of clover, instead of the second crop, for seed, will greatly increase seed yields. When the seed crop is known to be heavily infested, as indicated by the presence of many adult wasps about the heads, many of the insects can be killed by early cutting of the crop and removing the hay from the field before the seed has had time to mature. If this is generally practiced in a community and volunteer red clover, alfalfa, and bur clover are kept down, the numbers of the insect may be somewhat reduced. Destruction of chaff and screenings left in the hulling operation is important in preventing the emergence of adults the following spring. However, at present, we are nearly dependent on the natural enemies of the insect to hold it in check, as no really effective control measure has been developed.

References.—*U.S.D.A., Farmers' Bul.* 1642, 1931; and *Dept. Bul.* 812, 1920; *Ill. Agr. Exp. Sta. Bul.* 134, 1909; *Proc. Utah Acad. Sci.,* **11** : 241–244, 1934.

55. Clover Seed Midge[1]

Importance and Type of Injury.—The clover seed midge, if very abundant, is capable of practically destroying the red clover seed crop. Its presence is indicated by the failure of many of the clover florets to open (Fig. 273, *May and June*). Where many maggots are present in the head, it will present a stunted appearance with irregular bloom, most of the head remaining green and the florets never opening. Infested seed presents a shriveled appearance or scarcely forms within the base of the clover head. The insect has no effect on the production of clover for a hay crop.

Plants Attacked.—Red clover seed is apparently the favorite food of this insect. It has been found in small numbers in alsike, mammoth, crimson, white, and sweet clover but is of no importance on these crops.

Distribution.—The insect occurs generally throughout the United States from the East to the Pacific Coast; also in the southern part of Canada.

Life History, Appearance, and Habits.—The clover seed midge passes the winter in the larval stage inside a frail silken cocoon which is spun on, or shortly below, the surface of the ground (Fig. 273, *September*). Occasionally the larvae will crawl under trash and not spin a cocoon. In this stage the insect is about $\frac{1}{10}$ inch in length and of a reddish-pink color. In early spring the larvae change inside the cocoons to the pupal stage and emerge during April and May as very small, fragile, delicate-winged flies. These flies are shaped like mosquitoes but much smaller, dusty gray to black on the body, with a bright-red abdomen which is especially noticeable in the female. The female is equipped with a long ovipositor equal in length to her body. With this, she deposits her eggs in the young clover heads just as they are appearing from the buds (Fig. 273, *May*). The pale-yellow eggs are attached singly or in clusters to the hairs about the calyx of the clover blossoms, each female laying about 100 eggs. The eggs hatch in from 3 to 5 days, and the young maggots work their way to the top of the flowers, and down inside the unopened petals. Here they feed by sucking the sap and thus destroy the ovules and prevent the formation of seed. Upon becoming full-grown, in about a month, the larvae drop to the ground, generally during periods of rain (Fig. 273, *June*). They work their way below the surface, and there change inside their cocoons to the pupal stage. The first of the summer-generation adults usually appear during the first part of July. They lay their eggs in the second-crop clover, and from these come the overwintering maggots. In the southern part of the country, a third generation is produced.

Control Measures.—When this insect alone is causing serious damage to clover seed, most of the injury may be prevented by cutting the clover a little before the uninjured heads have come into full bloom and removing the hay from the field as soon as it is dry. This is a most effective control measure, as the cutting at this time will kill practically all the young midge larvae within the clover heads. The insect may also be controlled to some extent by clipping clover about 2 weeks before any of the heads show the bloom, and again about 1 month later. This brings the first-

[1] *Dasyneura leguminicola* (Lintner), Order Diptera, Family Cecidomyiidae.

crop clover into bloom before the summer-generation midges have emerged. These measures apply mainly to the control of the clover seed midge and also to the clover head caterpillar. Where several other clover

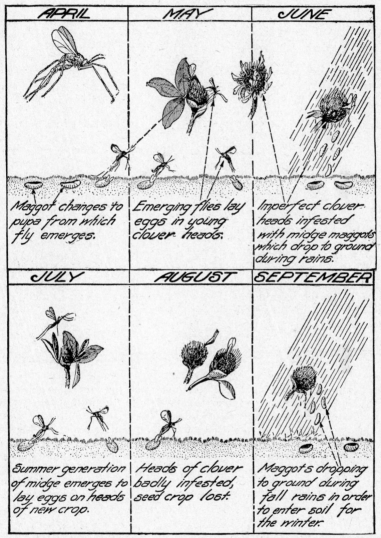

APRIL	MAY	JUNE
Maggot changes to pupa from which fly emerges.	Emerging flies lay eggs in young clover heads.	Imperfect clover heads infested with midge maggots which drop to ground during rains.

JULY	AUGUST	SEPTEMBER
Summer generation of midge emerges to lay eggs on heads of new crop.	Heads of clover badly infested, seed crop lost.	Maggots dropping to ground during fall rains in order to enter soil for the winter.

Fig. 273.—Diagram showing seasonal history of the clover seed midge, *Dasyneura leguminicola* (Lintner). (*From U.S.D.A., Farmers' Bul. 971.*)

insects are present, however, this practice may not be found to give the best production of seed.

References.—*Ore. Agr. Exp. Sta. Bul.* 203, 1917; *Ill. Agr. Exp. Sta. Bul.* 134 1909 *U.S.D.A., Farmers' Bul.* 971, 1932; *Rept. Ento. Soc. Ont.*, **65**: 22–28, 1935.

56. Clover Root Curculio[1]

There are several species of clover root curculios, or clover Sitonas, that attack red, sweet, and alsike clover and alfalfa. Our knowledge of the different species is incomplete. Probably the most important is *Sitona hispidula* (Fabricius), but there are two other common species.[2]

Importance and Type of Injury.—These insects work in such a way that much of their injury is not noticed. They may destroy from 60 to

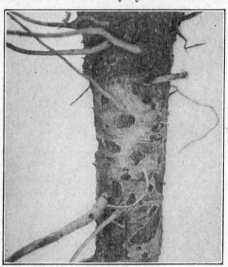

Fig. 274.—Work of clover root curculio on alfalfa root, about 3 times natural size. (*From Ill. State Natural History Surv.*)

Fig. 275.—Adult clover root curculio, *Sitona flavescens* Marsham, enlarged about 12 times. (*From Ill. State Natural History Surv.*)

80 per cent of the plants in young stands of alfalfa. In most years they are not of much importance. Clover or alfalfa plants infested by these insects wilt and often die, especially during periods of dry weather. If the plant is dug and examined, the roots will be found scored and furrowed on the outside with numerous burrows, oftentimes nearly girdled (Fig. 274). Small, grayish-white, footless, brown-headed grubs, about ⅙ inch long, will be found on the roots. Small, grayish or brownish beetles, of about the same length, with blunt short snouts (Fig. 275), may be found eating rounded areas from the leaves or gnawing stems and leaf buds during the day, or hidden away among the trash on the ground or around the crown of the plant.

Plants Attacked.—The insects feed on all the common clovers, alfalfa, soybeans, and cowpeas, and doubtless on some other legumes. White clover seems to be slightly preferred by the insects, although they are often very abundant on the other varieties of clover as well as on alfalfa.

Distribution.—These insects are quite generally distributed over the United States and southern Canada. These species are probably all of

[1] *Sitona* spp., Order Coleoptera, Family Curculionidae.
[2] *Sitona flavescens* Marsham, and *S. crinitus* Gyllenhal.

European origin but have been known in North America for at least 60 years.

Life History, Appearance, and Habits.—The winter is passed in the egg, adult, and larval stages. Most of the insects pass the winter as young larvae. In the spring these larvae develop by feeding on the clover roots and crown of the plant, pupate during late March and April, and emerge during May and June as beetles. These beetles feed actively for about 1 month or 6 weeks, often when abundant doing considerable damage to clover fields. They become less active during the middle of

the summer, and, although they remain about the clover fields, they feed but little. In the early fall they again become active, feed and mate, and the females deposit their eggs about the crowns of the plants. The eggs are laid at intervals extending over several weeks, often as late as the middle of November. Nearly all the eggs hatch in the fall but some hatch the following spring. A considerable number of beetles survive the winter, and some eggs are laid in the spring, at least in the midwestern states.

Control Measures.—Practically no effective control measures for combating these insects have been developed. Rotation which will put infested fields in a grass or cultivated crop will drive out the beetles. If the land is plowed late in the fall or early in the spring, practically all the insects in the field will be destroyed. If, on the other hand, the land is plowed early in the fall, many of the adult beetles will migrate out of the field and crawl considerable distances to other fields containing crops on which they feed and survive.

References.—*Ill. Agr. Exp. Sta. Bul.* 134, 1909; *U.S.D.A., Farmers' Bul.* 649, 1915; *Ky. Agr. Exp. Sta. Cir.* 42, 1934; *Jour. Econ. Ento.,* **27**: 807–814, 1934.

57. CLOVER ROOT BORER[1]

Importance and Type of Injury.—This insect is apparently of much less importance than formerly. This is due, in part at least, to better systems of crop rotation. Infested clover plants turn brown, wilt, and die, generally having the appearance of suffering from the attack of some disease. An examination of the roots, however, will show numerous burrows running through them and also grooves on the surface of the roots (Fig. 276). These burrows cut off the circulation of the plant and frequently kill it. Small, white, brown-headed, footless grubs, about $\frac{1}{10}$ inch long, will be found boring in the roots. Dull-black or dark-brown, somewhat hairy, cylindrical, hard-bodied beetles, about $\frac{1}{12}$ to $\frac{1}{10}$ inch in length, may be found in the burrows in the roots

Fig. 276.—Tunnels of the clover root borer, *Hylastinus obscurus* (Marsham), in clover root. About natural size. (*From Webster, U.S.D.A.*)

or around the crown of the plant.

Plants Attacked.—This insect prefers red and mammoth clover. It has also been taken on white and sweet clover and alfalfa, peas, and vetch but is apparently of little importance on these crops.

[1] *Hylastinus obscurus* (Marsham), Order Coleoptera, Family Scolytidae.

Distribution.—The clover root borer is distributed over the northern part of the United States and eastern Canada, with the possible exception of the Great Plains. It is of European origin and was probably brought into this country about 1870.

Life History, Appearance, and Habits.—The insect passes the winter in the larval and adult stages, nearly all the insects being in the adult stage. The winter is passed in the ground in the clover roots. In the spring the female beetles deposit a few eggs in cavities eaten out in the crown of the clover plant, on the sides of the roots, or in burrows inside the clover roots. Egg-laying extends over the spring months. The larvae, upon hatching, tunnel through the roots, making irregular branched burrows. Most of them become full-grown during late summer and transform to pupae during midfall. All stages of the insect can be found during the late summer and fall months. There is but one generation each year.

Control Measures.—The clover root borer is seldom seriously destructive except in fields where clover has been allowed to stand more than two seasons. Where clover is allowed to stand for only two seasons, very little trouble from this insect is experienced.

References.—*U.S.D.A., Bur. Ento. Cir.* 67, 1905; *U.S.D.A., Dept. Bul.,* 1426, 1926.

58 (18). GRAPE COLASPIS

This insect has been fully covered as a pest of corn, which is the crop most seriously damaged by it (see page 383). It sometimes causes con-

FIG. 277.—Injury to clover foliage by beetles of the grape colaspis, and an adult, natural size on leaf. (*From Ill. State Natural History Surv.*)

siderable injury to clover plants by the feeding of the grubs on the clover roots. The adults eat the leaves (Fig. 277). No method of control on the clover plant has been developed.

B. SOYBEAN AND COWPEA INSECTS[1]

Soybeans and cowpeas have been, on the whole, more free from serious insect injury than most of the field crops. Grasshoppers are probably the worst insect pests of these crops and, in years when they are abundant, may seriously damage, or nearly destroy, soybean fields. They may be

[1] See *Ohio Agr. Exp. Sta. Bul.* 366, 1923.

controlled by the same methods as those given for the control of these pests on corn; see page 338. The poison-bran bait, however, should be applied to adjoining fields to kill the hoppers before they enter the soybeans, as they seldom originate in these fields, but migrate into them from adjoining fields. If the poison bait is applied in the soybean fields, it falls between the plants where the ground is heavily shaded, and very little of it is eaten.

The green cloverworm (see page 426) has, during certain years proved to be a very destructive pest of soybeans and cowpeas in the southern states. It may be controlled on these crops by dusting with calcium arsenate, at the rate of 5 to 7 pounds per acre, applying the material with a cotton duster or a field-crop duster. In most years, the insect is of no importance on these crops (see *North Carolina Agricultural Extension Circular* 105, 1920).

The clover root curculio may completely destroy soybeans if they are planted on spring-broken clover sod. The adult beetles of the curculio attack the soybeans as they first come up and eat off the plants as fast as they appear through the ground. No injury has been reported to these crops when planted on clover-sod ground, plowed in the fall or early winter. Where soybeans are to follow spring-broken clover sod, it should be well harrowed and planting delayed as late as possible (for full description of this insect, see page 438).

In the southern states the cowpea is sometimes badly damaged by the caterpillars of the corn earworm or cotton bollworm. There is no effective method of controlling this insect on cowpeas. Damage occurs only in scattered seasons during periods of their greatest abundance.

Throughout the Gulf states cowpeas, string beans, Lima beans, and to some extent, cotton and strawberry plants are severely injured by the cowpea curculio,[1] which causes wormy peas and beans within the growing pods. Dusting with fluosilicates and the frequent picking of the ripe pods and storing on a clean, tight, dry floor are helpful in control. The cowpea weevil[2] and some other weevils often are very destructive to stored cowpea seed. The insects deposit their eggs in the cowpeas in the field, the infestation originating from infested seed or from scattered cowpeas remaining in the field from the previous year. (The control of these insects is given under Stored Grain Insects, page 811.) (See U.S. Department Agriculture, Technical Bulletin 599, 1938.)

The striped blister beetle (see page 510) is sometimes a serious pest in soybeans (see *U.S. Department Agriculture, Leaflet* 12, 1927). The spotted cucumber beetle is sometimes abundant on soybeans and cowpeas but never sufficiently injurious to warrant taking special control measures. Several aphids also attack these crops but are not of special importance. Leafhoppers are often abundant, especially in fields of soybeans, and occasionally cause some injury to the leaves.

The Mexican bean beetle[3] (see Garden Insects, page 487) has caused some injury to soybeans and cowpeas where these crops were grown in close proximity to heavily infested garden beans. It has not been of importance in the areas where soybeans or cowpeas are grown alone as field crops.

[1] *Chalcodermis aeneus* Boheman, Order Coleoptera, Family Curculionidae (see *Ala. Poly. Agr. Exp. Sta. Bul.* 246, 1938).

[2] *Callosobruchus maculatus* (Fabricius), Order Coleoptera, Family Mylabridae.

[3] *Epilachna varivestis* Mulsant, Order Coleoptera, Family Coccinellidae.

CHAPTER XIV

COTTON INSECTS

FIELD KEY FOR THE IDENTIFICATION OF INSECTS INJURING COTTON

A. *Insects that chew the foliage, often stripping the plants of leaves, or cutting off and devouring young seedling plants:*
  1. Large and small, lubberly grasshoppers often invade cotton from near-by waste lands and defoliate the plants.....................................
    ....*Lubber grasshoppers, differential grasshoppers,* and others, page 335.
  2. Seedlings cut off, gnawed, or rasped, about 1 inch above the ground, or on larger plants just below the apical bud, by brownish to black, jumping insects, with long antennae..
    Field cricket, Gryllus assimilis Fabricius, page 466 (see also *U.S.D.A., Cir.* 75, 1929).
  3. Cotton in newly broken fields, or fields that were very grassy the year preceding, attacked while small by swarms of robust, long-legged, grayish or brown beetles, $\frac{1}{2}$ to 1 inch long, that lack the underwings and cannot fly. They feed in late afternoon and at dusk, and may eat off the plants over a large field..
    ..*Wingless May beetles* or *June bugs* (see *U.S.D.A., Farmers' Bul.* 223, 1905).
  4. Dark-brown to black, humpbacked beetles, nearly $\frac{1}{4}$ inch long, with a slender snout $\frac{1}{3}$ as long as the body, bent down between the legs, and prominent, round punctures on the back, appear during early summer in fields that were in cowpeas the previous year, and eat small holes in leaves or tender parts of stem.........*Cowpea pod weevil* (see *N.C. Dept. Agr. Bul.*, vol. 29, no. 6).
  5. Cotton sometimes seriously defoliated, especially in wet seasons, by slender, green, black-and-white-striped caterpillars, with four black dots on each segment above, and ranging up to $1\frac{1}{2}$ inches long. They crawl with a slightly looping movement and feed only on cotton.......*Cotton leafworm*, page 444.
  6. Cotton, along with most other plants, defoliated by tan or green to nearly black caterpillars with three yellow hair lines down the back and a wider one on the side, that is splotched with red. Body covered with fine scattered hairs arising from black tubercles, and a prominent inverted Y on the front of the head...*Fall armyworm*, page 344.
  7. Young plants cut in two at or near the surface of the soil, or the leaves eaten, at night. Plump dirty-looking caterpillars, of various shades and markings of green, brown, and black, up to $1\frac{1}{2}$ inches long, found in daytime in the soil. Often curl the body when disturbed...................*Cutworms*, page 346.
  8. Foliage devoured by very hairy or woolly caterpillars, up to 2 inches long, black-bodied and covered with long black and red hairs...................
    ...*Salt-marsh caterpillars*, page 476 (see *U.S.D.A., Farmers' Bul.* 223, 1905).
  9. Pieces of the leaves cut off and carried to their nests in the soil by reddish-brown ants varying from $\frac{1}{12}$ to $\frac{1}{3}$ inch in length........................
    ....*Leaf-cutting ant*, page 771 (see *U.S.D.A., Bur. Ento. Cir.* 148, 1912).
  10. Many-legged hard-shelled crayfish, with a pair of pincher legs in front, crawl out of burrows in wet soil and devour the plants. Not insects but Crustacea
    ....*Crayfish* or *crawfish*.
B. *Insects that suck the sap from leaves, stem, squares, or bolls:*
  1. Very slender, almost microscopic, yellowish, active nymphs and brown-bodied, bristly winged adults, only $\frac{1}{25}$ inch long, suck sap from the leaves and buds, causing retarded growth, defoliation, and malformed seedlings...............
    Thrips, especially *the tobacco thrips, Frankliniella fusca* (Hinds), *the flower*

442

thrips, F. tritici (Fitch), and *the bean thrips, Heliothrips fasciatus* Pergande (see *S.C. Agr. Exp. Sta. Bul.* 306, 1936; *Calif. Agr. Exp. Sta. Bul.* 609, 1937).

2. Leaves of cotton curled or dwarfed; the undersides thickly dotted with small, winged or wingless, tan, brown, green, or black plant lice of various sizes, which suck the sap..........................*Melon aphid* or *cotton aphid*, page 446.

3. The lint, especially of long-staple cotton, is stained an indelible yellow color by flattened, rather narrow, long-legged bugs of various sizes up to ⅗ inch long, with head and prothorax bright-red and the rest of the body dark-brown, crossed with light-yellow lines. The bugs puncture the seeds in the developing bolls and cause a juice to exude that stains the lint.......................
Cotton stainer, Dysdercus suturellus Herrich-Schaeffer and other species (see *U.S.D.A., Bur. Ento. Cir.* 149, 1912; and *Jour. Agr. Res.,* **31**: 1137–1147, 1925).

4. Squares and young bolls drop, shrivel, or decay; small, round, blackened spots appearing on their surface, and stems often split lengthwise where they have been punctured by sap-sucking bugs......................................
Cotton-leaf bug, stink bugs, and leaf-footed plant bugs (see *U.S.D.A., Farmers' Bul.* 223, 1905, and *Tech. Bul.* 296, 1932).

5. Shield-shaped, bright-green, bad-smelling, flattened bugs, the winged adults about ⅝ inch long, suck sap from the buds and bolls, causing them to shed...
Green stink bug, Acrosternum hilaris Say (see *Va. Agr. Exp. Sta. Bul.* 294, 1934).

6. Small squares shed, buds blasted, and growth stunted; resulting in shortening of internodes and development of an excessive growth of branches close together, with deformed leaves, on young plants; and an excessively tall, whip-like growth of the main stem and a suppression of fruiting branches on older cotton. Small greenish bugs from ⅟₂₅ to ⅛ inch in length, with scarlet eyes, cluster on the growing tips..............................*Cotton fleahopper*, page 446.

7. During hot weather the leaves of cotton become blotched with large irregular areas of reddish brown so that entire patches of varying size in the fields look red. Undersides of such leaves, when examined under a lens, are seen to be swarming with minute, pale-greenish to reddish, six- or eight-legged mites....
....*Red spider mite*, page 482.

C. Insects that bore or tunnel in the stalks or puncture and split them:

1. Caterpillars, up to 1½ inches long, prominently striped with brown and white in front and behind, but with a large grayish-brown "bruised-looking" area in front of the middle, burrow through the heart of the stem, killing the terminals. Especially bad around weedy margins of fields in early summer. Stripes disappear when caterpillars become full grown. . . . *Common stalk borer*, page 364.

D. Insects that eat into, or feed inside of the squares and bolls:

1. Squares flaring, and either falling to the ground, or hanging withered and dry on the plant. Squares and bolls with wart-like scars, covering punctures about the diameter of a pin. Inside of bolls decayed and soiled. White, curved-bodied, brown-headed, legless grubs, up to ½ inch long, feed in the squares and bolls destroying the unopened buds, lint, and seed. Hard-shelled, grayish to brown, long-legged beetles, averaging ¼ inch long, with a slender snout half as long as the body, puncture squares and bolls to feed and lay eggs in them.
....*Cotton boll weevil*, page 447.

2. Flowers fail to open, squares fall, and bolls are eaten out but not decayed-looking. A pinkish-white, brown-headed caterpillar up to ½ inch long, found inside the fruits or seeds; easily distinguished from *D*,1 by having 8 pairs of legs and prolegs. Larvae often bore inside of the seeds, and web 2 seeds together, so that "double seeds" are found when the cotton is ginned. Distinguished from other caterpillars by having 4 teeth on the mandibles and the crochets on the prolegs horseshoe-shaped. Restricted to western Texas, New Mexico, and Arizona......................................*Pink bollworm*, page 452.

3. Light- to dark-green, pinkish or nearly black, more or less striped caterpillars, up to 1¾ inches long, bore into squares and bolls and eat out the interior. The worms have 3 pairs of slender legs behind the head, 5 pairs of fleshy prolegs, and

the skin is rough and thorny appearing under a lens......................
...........................*Cotton bollworm* or *corn earworm*, page 452.

4. Squares eaten out as in *D,3* by bright-green oval, distinctly flattened caterpillars, covered with short velvety hairs, and with the head drawn under the
front of the body; not exceeding ¾ inch in length........................
Cotton square borer, Strymon (= *Uranotes*) *melinus* (Hübner) (see *Tex. Agr. Exp.
Sta. Bul.* 401, 1929).

E. *Insects that attack the roots, below ground:*

1. Seeds fail to sprout or plant dies when small. Slick, whitish, slender, cylindrical worms up to 1 inch long, with 6 small legs behind the head, are found in
the soil or eating into the seeds and roots.................................
Corn and *cotton wireworm* (*Horistonotus uhleri* Horn) (see *S.C. Agr. Exp. Sta.
Bul.* 180, 1914).

2. Seedlings weak, yellowish in color. Many small brown ants tunneling in the
soil about the roots and caring for bluish-green, wingless plant lice or aphids
about the size of pinheads, that cluster on the roots and suck the sap..........
...........................*Corn root aphid* and *cornfield ant*, page 371.

F. *Insects that lay eggs in the stalks of the plant:*

1. Cotton stalks in the fall of the year, are punctured and may be split by rows of
round holes of the diameter of a pin and with an elongate curved white egg
within. No injury to cotton....................*Tree crickets*, page 662.

The following very important cotton insects are not discussed in this
chapter, but accounts of their damage, habits, and control are given in
connection with other crops, which they also attack:

Fall armyworm, page 344.
Cutworms, page 346.
Red spider mite, page 482.
Wireworms, page 377.

59. Cotton Leafworm[1]

Importance and Type of Injury.—Slender, greenish, looping worms,
with black-and-white stripes and a number of small rounded black spots
scattered over the body (Fig. 278,*B*), rag or strip the leaves of cotton,
especially late in the season. Four of the black dots form a square on the
back of each segment, and the last pair of prolegs stands out conspicuously behind the body. The adult moth sometimes causes injury to ripe
peaches, grapes, and other fruits by lacerating them with spines on the
end of its tongue. In this respect it is a striking exception to the rule
that adults of Lepidoptera are not injurious.

Plants Attacked.—The larva feeds only on cotton. Adults sometimes
feed destructively on ripe peaches, grapes, and other fruits.

Distribution.—Eastern United States from the Great Lakes southward
into the tropics. More destructive in the Gulf states than farther north.

Life History, Appearance, and Habits.—This insect is not known to
winter in the United States. It is a tropical insect, but frequently the
adult moths fly northward in great numbers as far as the Great Lakes,
and even into Canada. The eggs are deposited on the underside of
cotton leaves. The larvae reach full size (1½ inches in length) in 2 to 3
weeks. They pupate on the plant, inside a fold of leaf or fastened by
their tails to the foliage (Fig. 278,*A*). A generation is completed in
about a month. There are from three to seven generations in the cotton

---

[1] *Alabama argillacea* (Hübner), Order Lepidoptera, Family Noctuidae.

belt, and the moths of each successive generation fly farther and farther to the north. The moths (Fig. 278,*C*), which measure about 1¼ inches

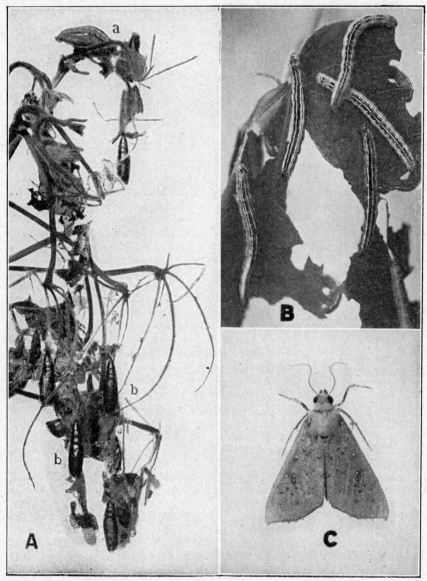

Fig. 278.—The cotton leafworm, *Alabama argillacea* (Hübner). *A*, larva, pupae, and injury: note how the leaves have been stripped, the larva at *a*, and the pupae attached by their posterior ends at *b,b* (*from U.S.D.A.*). *B*, larvae natural size, devouring foliage. *C*, adult female moth, enlarged ½, in resting position. (*From Ala. Poly. Inst.*)

from tip to tip of wing, are of an olive-tan color with three more or less prominent, wavy, transverse bars on each front wing. All stages die out during the winter in the United States; the species surviving only in the tropics.

Control Measures.—Where poisoning for the boll weevil is practiced, this insect will not give trouble. If it appears early in the season, an application of calcium arsenate should be made. Late in the season the feeding of the leafworm is considered a benefit in boll-weevil territory because it destroys the top growth which could not make cotton and which would sustain the boll weevil late in the fall. When the adults are troublesome in orchards it has been found possible to prevent much injury by applying a sulphur dust with a power duster. In small orchards or vineyards many of the moths may be killed by setting around the trees a few pans of crushed fruits poisoned with sodium arsenite, 2 teaspoonfuls to the gallon.

60 (13, 96). APHIDS[1,2,3]

Almost as soon as cotton has put out leaves, small, soft-bodied, pale-green plant lice fly to them and start their families. In cool, wet seasons, when their own enemies cannot work against them so well, they may become abundant enough to stunt and deform the plants. Often, when the hot weather of summer comes on, they practically disappear.

The most important species, feeding above ground, is the cotton aphid[1] or melon aphid. The cowpea aphid[2] often migrates into cotton from fields of soybeans and cowpeas. Aphids are often especially abundant where cotton is regularly dusted for boll weevil, since this kills the predators and parasites of the aphids. Spraying or dusting with nicotine may be used to destroy them, but the expense will seldom be justified (see also page 497). The corn root aphid[3] and its attendant ant are pests of cotton, the roots of which they attack just as they do corn in some parts of the cotton belt, especially in the Carolinas. The life history, descriptions, and control are discussed under Corn Insects (see page 371).

61. COTTON FLEAHOPPER[4]

Importance and Type of Injury.—In scattered areas throughout the cotton belt this small green sucking bug has sporadically caused serious losses by sucking the sap from the very small squares and other terminal growth, resulting in excessive shedding and an abnormal whip-like growth of the plant as described in the key.

Plants Attacked.—In addition to cotton, croton, evening primrose, goatweed, sage weed, horsemint, orach, wild sunflower and at least 30 other species of weeds.

Distribution.—General throughout the cotton belt and over much of the United States.

Life History, Appearance, and Habits.—The cotton fleahopper hibernates in the egg stage on croton, sage weed, goatweed, Atriplex, and other weeds. In southern Texas it appeared on horsemint early in March, migrated to cotton late in April, and deserted the cotton by the end of July, feeding for the remainder of the season on croton, evening primrose,

---

[1] *Aphis gossypii* Glover, Order Homoptera, Family Aphididae.
[2] *Aphis medicaginis* Koch.
[3] *Anuraphis maidi-radicis* (Forbes), Order Homoptera, Family Aphididae.
[4] *Psallus seriatus* (Reuter), Order Hemiptera, Family Miridae.

snap beans, and potatoes. The eggs, which are yellowish white, about ⅟₃₀ inch long by a fourth as wide, are inserted beneath the bark, especially just below the growing tips. In a little over a week they hatch, and the greenish nymphs begin sucking the sap from the terminal bud cluster, including young leaves, stems, and squares. The extent of the injury suggests that a toxin may be inserted in the plant as the insect feeds. The nymphs molt five times, and within 10 to 30 days are mature bugs. The adult is about ⅛ inch long, flattened, elongate ovate in outline, with prominent antennae, the body pale yellowish-green in color, with minute black hairs and black specks over the upper surface. The generations are not distinctly separated, as reproduction is continuous throughout the warm season.

Control.—Dusting at 5- to 7-day intervals, with a mixture of 1 part calcium arsenate to 2 parts of dusting sulphur (97 per cent pure and 98 per cent passing a 325-mesh screen), or 10 parts paris green and 90 parts sulphur, using 12 pounds per acre, gives the best control under conditions of heavy infestation and also gives a high control of boll weevil and leaf-worm. Sulphur, alone, is more effective against the nymphs and should be used when the fleahoppers are mostly in the nymphal stage and other cotton insects are not threatening. The eradication of weeds and the destruction of cotton stalks during fall and winter are recommended.

References.—*Tex. Agr. Exp. Sta. Bul.* 339, 1926, and 380, 1928; *S.C. Agr. Exp. Sta. Bul.* 235, 1927; *U.S.D.A., Dept. Cir.* 361, 1926, and *Yearbook* pp. 282–285, 1935, and *Cir.* E-430, 1938; *Jour. Agr. Res.*, **40**: 485–516, 1930; *Jour. Econ. Ento.*, **30**: 125–134, 1937; *U.S.D.A., Tech. Bul.* 296, 1932.

62. Cotton Boll Weevil[1]

Importance and Type of Injury.—No insect pest in all the world has gained greater notoriety than the cotton boll weevil. Very careful data have been collected as to the direct loss in production of seed and lint due to this pest. The most recent estimates place the current loss at 3,000,000 to 5,000,000 bales of cotton a year, or from 20 to 40 per cent of the normal production. This loss amounts to from $100,000,000 to $200,000,000 a year. It seems safe to say, therefore, that the total loss inflicted upon the cotton interests of the United States during the first 45 years of the boll weevil invasion has been between $4,000,000,000 and $5,000,000,000. In addition to this direct loss, there have been enormous financial losses due to depreciated land values, the abandonment of cotton growing, the closing down of cotton gins and oil mills, the interruption of railroad, bank, and mercantile business, and many other economic disturbances in territory where the boll weevil has appeared. It should not be supposed that the cotton farmer sustains all the loss. On the contrary, it is borne chiefly by the users of cotton goods. Dr. W. E. Hinds estimated that each person in the United States pays $10 a year more for cotton goods than he would if the boll weevil were not present.

The injury is caused by the adult weevils and their young or grubs. The adults (Fig. 279,*c*) puncture the squares and bolls by chewing into them with their long slender bills to feed on the tissues inside and to lay their eggs in the holes. This causes the squares to flare (*b*) and either drop off or hang withered and dry (*a*). The grubs that hatch from the eggs feed inside the squares and bolls, destroying the developing flower

---

[1] *Anthonomus grandis* Boheman, Order Coleoptera, Family Curculionidae.

so that it fails to bloom or else develops seeds with little fiber.   A good indication of the importance of this feeding is the fact that where it is prevented by poisoning the weevils, gains of from 500 to 1,000 pounds of seed cotton per acre are often secured, the average gain being 300 to 400 pounds per acre.

Food Plants.—The larvae of this pest feed exclusively on cotton, but the adults have been known to feed also on okra, hollyhock, and hibiscus.

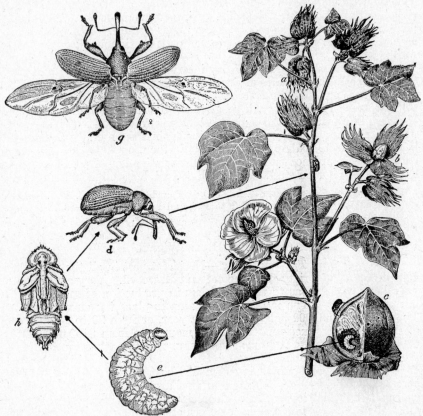

Fig. 279.—The cotton boll weevil, *Anthonomus grandis* Boheman.   On the right a cotton plant attacked by the boll weevil, showing at *a*, a hanging dry infested square; at *b*, a flared square with weevil punctures; at *c*, a cotton boll sectioned to show attacking weevil and larva in its cell; *g*, adult female with wings spread as in flight; *d*, adult from the side; *h*, pupa ventral view and *e*, larva.   (*Rearranged from U.S.D.A.*)

A variety of the weevil[1] in Arizona feeds on a wild cotton plant.[2]   This variety also attacks cultivated cotton.

Distribution.—The cotton boll weevil is not a native of the United States.   Its original home was in Mexico or Central America.   Little was known about it previous to 1892, when it first appeared in southern-most Texas.   From that date it spread to the north and east, an average distance of about 60 miles a year (Fig. 280).   An average of more than

[1] *Anthonomus grandis thurberiae* Pierce, Order Coleoptera, Family Curculionidae.
[2] *Thurberia thespesioides.*

20,000 square miles of new territory was infested by the weevil in each of the first 35 years after it crossed the Rio Grande River. It now occupies practically all the important cotton-growing area in the United States. The weevil does not seem to be able to maintain itself in the extreme northern edge of the cotton-growing area. It also ranges southward throughout Mexico to Guatemala and Costa Rica and is known to occur in Cuba.

Life History, Appearance, and Habits.—The boll weevil winters chiefly in the adult stage in all kinds of shelter; under dead leaves and piled cotton stalks, among Spanish moss on trees, under loose bark, about gins and barns, in woods, along fences, and in many other protected places. The adult beetle (Fig. 279,*d*,*g*) is a small, hard-shelled weevil, averaging about ¼ inch long, of a yellowish, grayish, or brownish color, becoming nearly black with age. It has a slender snout, half as long as the body, and close-fitting smooth wing covers with fine parallel lines and covered with short gray down or fuzz. The most characteristic feature about the boll weevil is the two spurs or teeth near the end of the front femur, the inner one being much longer than the other, and a single tooth on the middle femur.

From about the time early cotton is up (March), until early or even late June, the adults straggle out of hibernation, feeding at first on the tender terminal growth of the plants. They concentrate at first on the older or earlier planted fields. The weevils prefer to attack the blossom buds or squares after they are about 6 days old. They eat cavities into them and lay a single egg in each hole and usually one egg to a square. Each female may lay from 100 to 300 eggs. Later in the season eggs are laid in the bolls, but the squares are always preferred. The egg hatches in 3 days to the grub or larva. The larva feeds and grows 1 or 2 weeks inside the square or boll, molting two or three times and becoming about ½ inch long. It is white, legless, curved-bodied, and much wrinkled, with brown head and mouth parts (Fig. 279,*e*). It never leaves the square or boll in which the egg was laid but transforms to the pupal stage in the hollowed-out dirty cavity formed by its feeding. The pupal stage (*h*) lasts 3 to 5 days or more, when the final change is made to the adult. The adults eat their way out of the squares or bolls, mate, and within 4 or 5 days may be laying eggs for a new generation. Since the life stages may be passed in from 15 to 25 days, it is apparent that there may be from 2 or 3 to as many as 8 or 10 generations in a single year. Late in the season, as cotton becomes mature (mid-August to early September), the weevils spread extensively, taking flights by short stages that may total 20 to 50 miles or more. The chief spread of the insect comes at this time. At the approach of colder weather, the insects go into hibernation as already described.

Control Measures.—Because all the developing stages of the weevil are spent inside the squares and bolls of the cotton plant and the adults insert their slender snouts below the surface to feed, this insect is one of the most difficult of all known pests to control. Generally speaking the weevils will be least destructive following low winter temperatures and when hot dry weather prevails during June, July, and August. The bolls or fruits of the crop are produced over a period of about 2 months and protection from weevils during this heavy fruiting period is essential to making a crop. Successful control can be achieved only by a combination of measures, of which the following are the most important.

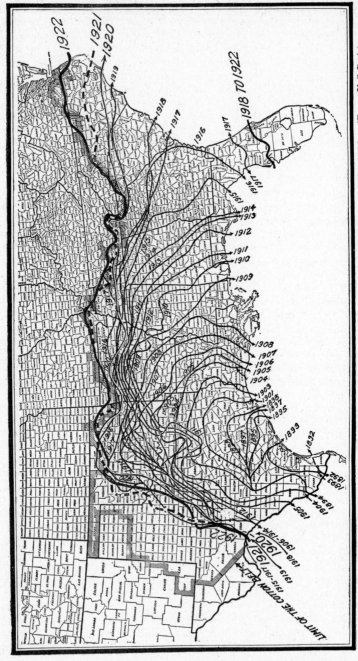

FIG. 280.—Map showing spread of cotton boll weevil over the Southern States from 1892 to 1922. (From U.S.D.A.)

1. Fall Destruction of Plants in the Field.—When examination shows that nearly all the squares in the field are being punctured by weevils, the cotton should be harvested as soon as possible and the stalks cut or plowed out, immediately collected together, and burned or plowed under deeply. No more cotton can be made from punctured squares, and the prompt destruction of the plants prevents the maturing of thousands of late weevils, which are the chief ones to winter over. The destruction of the plants before the first killing frost, and fall plowing, will remove hibernating places and make possible the early planting of the next crop.

2. Making an Early Crop.—Since the weevils prefer squares in which to lay their eggs and such squares are doomed to make no cotton, it is of the greatest importance to get the blossoms past the square stage to the boll stage before weevils become abundant enough to damage many of them. The use of early-maturing varieties; planting as early as the seed-bed can be thoroughly prepared; enriching the soil by the use of legumes and commercial fertilizers; and frequent shallow cultivation not too close to the plants are important in making the early crop.

3. Dust Poisoning.—When the crop has been made by the best farming possible, it should be protected and further increased by poisoning the weevils with calcium arsenate. This will generally pay where the land normally yields at least one-third to one-half bale per acre in the absence of the boll weevil or when poisoned. An average gain of 300 to 400 pounds of seed cotton per acre is commonly secured where proper dusting is done. Proper dusting requires (*a*) the use of good machinery,[1] (*b*) the use of calcium arsenate dust containing not less than 40 per cent total arsenic pentoxide, of which not more than ¾ per cent is in water-soluble form, and of a density between 80 and 100 cubic inches per pound (see also control of cotton fleahopper, page 447); (*c*) dusting at night or on calm humid days if ground machines are used; (*d*) making three or four applications at intervals of 4 or 5 days and using 4 to 8 pounds per acre; (*e*) delaying the first application until 10 out of 100 squares on the average are punctured by weevils, unless overwintering weevils are unusually abundant, when one dusting of 3 or 4 pounds per acre may be warranted just as the squares begin to form. Dusting may be discontinued when infestation has been reduced below 10 per cent of the squares; but later applications may be profitable when there is a good crop of young bolls subject to attack. To avoid the possibility of soil injury, especially on light sandy soils, growers should not use heavier applications of calcium arsenate than are recommended and may dilute the arsenate with equal parts of hydrated lime.

Since about 1923, applying the poison by airplane has become very successful. Specially designed planes fly from 5 to 25 feet above the cotton tops at a speed of about 80 to 100 miles an hour. They discharge the calcium arsenate dust from special hoppers in a swath 200 to 250 feet wide and dust the cotton at a rate of 400 to 500 acres per hour. The work is done during the daytime, can be made more thorough than hand- or ground-machine dusting, and requires only about two-thirds as much poison. The dust delivered from airplanes sticks to the foliage better because of the electrical charge it carries. The total cost for airplane dusting averages about $1 per acre for each application.

---

[1] See "Machinery for Dusting Cotton," *U.S.D.A., Farmers' Bul.* 1729, p. 19.

4. **Presquare Poisoning.**—When the infestation as the very first squares appear on the plants amounts to more than 20 weevils per acre, a poison sirup or calcium arsenate dust should be applied promptly to the tips of every plant. The poison sirup is made by mixing, for each acre to be treated on a given day, 1 pound of calcium arsenate powder in 1 gallon of water, stirring to a white paste. Add 1 gallon of good-grade table sirup and stir thoroughly. It may be applied with homemade mops or swabs to the growing tips of the plant. In the Carolinas 2 or 3 such applications and fewer subsequent dustings are preferred. Cotton which has been poisoned must not be used later for pasture.

References.—U.S.D.A., Bur. Ento. Bul. 114 (U.S. Senate Document 305) 1912, and *Farmers' Bul.* 1329, 1923; *Quar. Bul. State Plant Board Fla.,* **8**: 2, 1924; *U.S.D.A., Tech. Buls.,* 112, 1929, and 487, 1935; and *Cirs.,* E-430 and E-431 (mimeographed), 1938; *Ark. Agr. Exp. Sta. Bul.* 271, 1932; *Jour. Econ. Ento.,* **25**: 772–776, 1932; How-ard, L. O., "The Insect Menace," pp. 312–322, 1931; the publications of local experiment stations.

63 (12, 71, 111, 260). COTTON BOLLWORM OR CORN EARWORM[1]

This caterpillar, which also attacks corn (see page 368), tobacco, tomatoes (see page 525), beans, vetch, alfalfa, and many garden plants and flowers, destroys the squares and bolls by eating into them. Unlike the boll weevil the parent stage is a moth that does no harm except to lay the eggs. The eggs are laid on the leaves and outside of squares. The caterpillars do not remain in one boll but bore out again and crawl from square to square, a single worm often destroying all the fruits on a branch of the plant. The pupal stage is passed in the soil at a depth of 1 to 4 inches. The first two generations are usually passed on corn, tobacco, and other plants, and it is not until the third generation (August and later) that the bollworm becomes destructive to cotton. On this account every effort should be put forth to make an early crop, as suggested for controlling the boll weevil. Since the insect winters in a helpless pupal stage in the soil, deep plowing and cultivating the soil in fall and winter is effective by turning up the pupae and exposing them to natural enemies and to the weather. Dusting as for the boll weevil gives effective control for this insect also.

References.—U.S.D.A., Bur. Ento. Bul., 50, 1905, and *Farmers' Bul.* 1595, 1929; *Ark. Agr. Exp. Sta. Bul.,* 320, 1935.

64. PINK BOLLWORM[2]

Importance and Type of Injury.—The pink bollworm is considered one of the six most destructive insects in the world. Its attack causes the squares to fall or blossoms fail to open and, as the season advances, the bolls are rendered worthless by the feeding within them of a number of pinkish-white caterpillars, which very often eat into the seeds. The caterpillars range up to ½ inch in length, are cylindrical, and are pinkish on the upper part. The lint fails to develop, or is so stunted as to be unpickable, and the yield of oil from the seed is greatly decreased. Double seeds (Fig. 282) are found when the cotton is ginned, two partly eaten seeds being fastened together. A loss of from one-fifth to one-half of

[1] *Heliothis obsoleta* Fabricius, Order Lepidoptera, Family Noctuidae.
[2] *Pectinophora gossypiella* (Saunders), Order Lepidoptera, Family Gelechiidae.

the value of the crop is regularly suffered in infested areas, and the establishment of this pest throughout our cotton belt would doubtless result in "an annual loss of 20 to 40 per cent of the crop, or a money loss of several hundred million dollars annually" (Hunter).

Plants Attacked.—Cotton and rarely okra; and possibly hollyhock and other malvaceous plants.

Distribution.—The pest is believed to be a native of India, whence it has been carried by commerce to most of the other cotton-producing

Fig. 281.—Larva of the pink bollworm. About 6 times natural size. (*From U.S.D.A., Dept. Bul.* 1397.)

countries of the world. It is well established and a very destructive pest in nearly all the cotton-growing countries of Asia, and adjacent islands, including Hawaii and the Philippines; in north, east, and west Africa; in Australia; in Brazil; in the West Indies; and in Mexico. It was brought to Mexico in 1911 in seed imported from Egypt, and in 1915 or 1916 was carried across the border into Texas in cotton lint and seed. Since its first discovery in Texas in 1917, it has been found at many

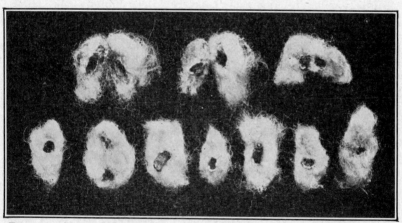

Fig. 282.—Cotton seeds containing pink bollworms opened to show the cells. Both single- and double-seeded cells are shown, the double-seeded ones being broken apart. (*From U.S.D.A., Dept. Bul.* 1397.)

points in Texas and near the border in New Mexico and Arizona, and also in Louisiana, Florida, and Georgia. A number of outlying infestations have been eradicated at great cost.

Life History, Appearance, and Habits.—The winter is passed in the larval stage (Fig. 281), the insect being curled up in a small cocoon in stored seed (Fig. 282), in the soil, or in bolls in the field. Some of these larvae are long-cycle, or resting, caterpillars that may remain in this stage as long as 2½ years. These are frequently found inside the seeds, and there is the greatest danger of spreading the insect to great distances by shipping seed.   Pupation takes place in the upper 2 inches of the soil or on the ground under trash or in the seeds or bolls.   The pupa is not quite smooth but is covered with very short fine pubescence.   In 1 to 3 weeks the small dark-brown moths emerge.   They measure about ¾ inch from tip to tip of wings, and somewhat resemble the common clothes moth. The wings are narrow, with a wide fringe, and are peculiarly pointed at the tip.   The first segment of the antenna has five or six long stiff hairs, and the palps are long and curved.   The moths are seldom seen, as they hide during the daytime.   They have been taken at altitudes upward of half a mile in the air, and are doubtless spread far in strong winds.

The greenish-white eggs are scattered all over the cotton plant, but mostly on the bolls, from 1 to nearly 100 in a place, and hatch in 4 to 12 days.   The larvae promptly drill into the squares and eat the developing flowers or into the bolls, where they consume both lint and seeds. After growing for 3 or 4 weeks, they may change to pupae and form adults for a new generation, or they may hang over as the resting larvae. There are from four to six generations in a year.

Control Measures.—The worms can be killed in cottonseed by heating to 145°F., without injury to the seed.   The cleaning and fumigating of gins, oil mills, warehouses and seeds with carbon bisulphide are important. Early maturity of the crop, followed by the burning of old stalks and bolls, is the best farm practice.   Careful application of arsenicals to the plants reduces the infestation materially, but no really satisfactory control for this serious pest has been discovered.   Every effort is being made to prevent its spread over the cotton belt.

Strict quarantines are in force to prevent continuous reinfestations from cottonseed carried across the Mexican border or from other quarantined areas.   A very effective X-ray device for detecting the larvae in infested seed has been perfected.

Plowing the soil to a depth of 6 inches, followed within a month by a standard 7-inch irrigation effectively destroyed the insects on the fields in Texas.

References.—*U.S.D.A.*, *Dept. Buls.* 918, 1921, and 1374 and 1397, 1926; *Jour. Agr. Res.*, **9**: 343–370, 1917; *Jour. Econ. Ento.*, **24**: 795–807, 1931.

CHAPTER XV

TOBACCO INSECTS

FIELD KEY FOR THE IDENTIFICATION OF INSECTS INJURING TOBACCO

A. Insects chewing leaves or stalk above ground, leaving visible holes:

1. Very young plants in seedbeds partially to completely defoliated by tiny dark-purple yellow-spotted bugs, only $\frac{1}{25}$ inch long, with a spherical soft body and very distinct head, that jump actively by the use of a forked tail-like appendage...
   Garden springtail, Sminthurus hortensis Fitch (see *Conn. Agr. Exp. Sta. Bul.* 379, 1935).

2. Large, grayish to brownish, actively jumping hoppers often invade tobacco fields in great numbers from surrounding grass and clover, when these crops are harvested, and rag the leaves severely...............*Grasshoppers*, page 335.

3. Very small, oval, active, jumping beetles, about $\frac{1}{16}$ inch long, dark-brown in color, with more or less black across the wing covers, or entirely black eat tiny "shot holes" in the leaves, especially of newly set plants or plants in seed beds....................*Tobacco flea beetle* or *potato flea beetle*, page 457.

4. Large green caterpillars, up to 4 inches long, with diagonal white bars on the sides and a slender horn at the tip of the body, cling to the vines and rapidly strip off the foliage, including the veins of the leaves. Numerous pellets of excrement on the ground............................*Hornworms*, page 458.

5. Large holes, not including the veins, are eaten in the leaves or whole seedlings devoured by plump, smooth, greenish or brownish, more or less striped or spotted caterpillars, up to 1½ inches long, which usually lie curled up in the soil during the daytime................................*Cutworms*, page 460.

B. Insects sucking sap from leaves or stems, not leaving visible holes, but causing wilting, yellowing, spotting, or death of the leaves:

1. Very slender, almost microscopic, yellowish, active nymphs and the brown-bodied bristly-winged adults, only $\frac{1}{25}$ inch long, suck sap from the leaves, causing the veins to become outlined with silver and the leaves peppered with minute black spots...
   Tobacco thrips, Frankliniella fusca Hinds (see *Conn. Agr. Exp. Sta. Bul.* 379, 1935; *S.C. Agr. Exp. Sta. Bul.* 271, 1931).

2. Small plants or the leaves or tops of older plants wilt suddenly. Broad, flat, shield-shaped bugs, about ½ inch long and of a buff to greenish color, suck at stems, petioles, or veins..
   Spined tobacco bug, Euschistus servus (Say) and *E. variolarius* Beauv. (*N.C. Dept. Agr. Spcl. Bul.* (Supplement, October, 1909), pp. 58–59).

3. Second-crop and late tobacco leaves yellow and later split and become ragged-looking, due to the feeding of a greenish-black slender bug, the largest only $\frac{1}{8}$ inch long and about $\frac{1}{4}$ as broad, with long, slender legs and antennae.....
   Tobacco suck fly, Dicyphus minimus Uhler (see *Fla. Agr. Exp. Sta. Bul.* 48, 1898).

4. Flattened, coppery-brown, somewhat mottled, winged bugs, $\frac{3}{8}$ inch long, and their greenish soft-bodied nymphs suck sap from the buds before the leaves unfold, causing leaves to be distorted and curly. Winged ones fly actively when disturbed................................*Tarnished plant bug*, page 479.

C. Insects boring in stalk or stem:

1. Smooth, tough, yellowish to reddish-brown, six-legged slender worms, up to 1½ inches long, burrow into the plant at the surface of the soil and hollow out

455

the stalk or taproot above and below the entrance hole, causing the plants to wilt. The worms may also eat off the roots.............................

Wireworms, or *"pith worms"* (see *N.C. Dept. Agr. Spcl. Bull.*, (Supplement, October, 1909), pp. 46–53).

2. Yellow to pinkish-white, swift-crawling caterpillars, up to ⅗ inch long, live in silken tubes covered with grains of soil, at the surface of the ground, near the plant. They burrow into the plant below ground and then tunnel upward, causing the plant to wilt and sometimes cutting off the terminal bud.........
....*Corn root or tobacco webworm*, page 461.

3. Young plants are tunneled up and down the stem for an inch or so from a small entrance hole just below the ground by small dirty-white maggots, up to ¼ inch long, broad at the posterior end and tapering to a pointed head, causing them to wilt and die................................*Seed corn maggot*, page 389.

D. *Insects eating into the buds or seed pods:*

1. Greenish caterpillars with pale longitudinal stripes, up to 1½ inches long, tunnel holes into the leaf buds as the plants begin to top, causing the leaves to be misshapen and ragged with large holes. Later the caterpillars bore into the seed pods....................................*Tobacco budworm*, page 461.

2. Green, tan, or blackish caterpillars, nearly 1¾ inches long when full-grown, attack the plant as in *D*,1, but mostly later in the season. Bodies of the worms sparsely haired, the skin rough and thorny-looking under a lens, often heavily striped and with more or less reddish markings.............................
...............................*Corn earworm* or *false budworm*, page 463.

E. *Insects mining between upper and lower surface of the leaves:*

1. Grayish, irregular, blotch mines (later turning brown), especially in the older leaves; grayish-white caterpillars, up to ⅛ or ½ inch long, with a pinkish or greenish tinge and brown at each end, may be found in the mines between upper and under leaf surface..
.............*Tobacco leaf miner* or *splitworm*, or *potato tuberworm*, page 463.

F. *Insects attacking the plant underground, or uprooting and eating off the young plants near the surface of the ground:*

1. Garden crops, peanuts, tobacco, and others grown in moist, light soils have the seedlings uprooted, the roots cut off, and pits eaten into the underground parts by a weird-looking brown cricket, up to 1½ inches long, covered with fine velvety hairs; with long hind wings, half-length front wings, large hind legs, and broad, rake- and shovel-like front legs, which they use to make small, mole-like burrows just beneath the soil..
Mole crickets, *Scapteriscus vicinus* and *Gryllotalpa hexadactyla* Perty, Order Orthoptera, Family Gryllidae (see *U.S.D.A.*, *Farmers' Bul.* 1561, 1928).

2. Seedbeds in use for more than 1 year become infested with large, white, six-legged grubs, that crawl on their backs over the surface of the ground at night and burrow and work up the soil to such an extent that the small plants are covered or uprooted and seriously damaged.......*Green June beetle*, page 623.

G. *Insects attacking tobacco and tobacco products in storage:*

1. Cigars, cigarettes, and package tobaccos, infested with small, white, curved, very hairy grubs and very small light-brown beetles (1/16 inch long) which eat holes through the wrappers or feed within the packages. The beetles have smooth wing covers and antennae not enlarged at end......................
....*Cigarette beetle*, page 785.

2. Small, white, curved grubs, similar to *G*,1, but not hairy; and small, rather narrow, oval, reddish-brown beetles, with parallel lines on the wing covers and last 3 segments of the antennae enlarged, tunnel through tobacco and many other substances...............................*Drug store beetle*, page 786.

3. Very active, whitish caterpillars, peppered with many small scattered brown spots and sometimes tinged yellow, brown, or pink, up to ⅝ inch long, eat holes much larger than those made by the cigarette beetle in tender parts of leaves and contaminate stored tobacco and other dried vegetable products with silk and excrement...........................*Tobacco moth*, page 787.

65 (76). Tobacco Flea Beetles[1]

Importance and Type of Injury.—This is considered the most injurious insect of tobacco, at least in some sections. The adult flea beetles chew small rounded holes into or through the leaves, especially from the underside (Fig. 283). They attack the young plants in the seedbeds almost as soon as they come up, and often ruin entire beds. After the plants are transplanted, they weaken or kill them in the field. Damage continues until the crop is harvested, the mature leaves often being spotted with holes which greatly lessen the quantity and the quality, especially of cigar-wrapper tobacco. It has been estimated that a flea beetle eats ten times its own weight in a day. The larvae feed on the roots of tobacco and other plants of the same family, cutting off the small roots and sometimes tunneling into the stalk. As a result of the feeding of the beetles, fungus diseases are encouraged and spread among the tobacco. The tobacco flea beetle is most destructive in the southern tobacco areas, while in Connecticut the adults of the closely related potato flea beetle[2] (page 468), by migrating into tobacco from potato and tomato fields, cause more damage than the tobacco flea beetle. Morgan estimated the loss in Kentucky and Tennessee in 1 year (1907) as close to $2,000,000.

Fig. 283.—Tobacco leaf badly damaged by tobacco flea beetle. (*From Ky. Agr. Exp. Sta. Bul.* 266.)

Plants Attacked.—Tobacco, tomato, potato, eggplant, pepper, nightshade, Jimson weed, ground cherry, and horse nettle are attacked regularly, and many other plants to a less extent.

Distribution.—Probably in all tobacco-growing sections of the United States and Canada.

Life History, Appearance, and Habits.—The brownish black-clouded hard-shelled little beetles, only ⅟₁₆ inch long (Fig. 284), hide during the winter in great numbers under leaves, grass, and trash on the ground about tobacco fields, especially along the margins of woods. Here they hibernate or remain more or less active, depending on the temperature. As soon as the plants come up in the seedbeds the beetles attack them, increasing in numbers from then until harvesttime, when they begin to disappear.

The eggs are laid mostly on the surface of the soil under the plant. They hatch in a week, and the delicate, slender, white larvae burrow into the soil and feed on the fibrous rootlets mostly within a few inches of the surface. They become full-grown in a couple of weeks. They are then

[1] *Epitrix parvula* (Fabricius), Order Coleoptera, Family Chrysomelidae.
[2] *Epitrix cucumeris* Harris.

about ⅙ inch long by ⅟₆₀ inch thick and dirty white except for the brownish mouth parts. After 4 or 5 days spent as a whitish pupa in the soil, the new adult emerges. There are probably three or four generations a year in the South.

Control Measures.—The ground surrounding the seedbed should be burned over in late winter. Less injury will occur if the seedbeds are located as far from good hibernating quarters as possible and at a considerable distance from the new fields. Seedbeds should be boarded or planked up on the sides and covered with tobacco cloth having at least 25 strands to the inch, making the covering beetle-tight. Surrounding the seedbed should be a 2- or 3-foot strip of early-seeded tobacco, protected with poles and cloth and kept dusted with a mixture of 1 pound paris green in 5 pounds lead arsenate. The top soil of seedbeds should be sterilized with steam or other heat before planting. If the beetles get into the seedbeds or attack the plants later in the season, treat them every 4 days to 1 week with a dust containing 1 per cent rotenone; with paris green, 1 part and lead arsenate 5 parts; with barium fluosilicate 4 parts and 1 part of sterilized tobacco dust or clay; or with a dust containing 36 per cent cryolite. In the seedbeds the dust should be used at the rate of ½ pound per 100 square yards. In the field the amount should vary from 3 to 10 pounds per acre, depending upon the size of the plants. Instead of the dust an 8-8-100 bordeaux mixture with 4 pounds lead arsenate and 2 pints 40 per cent nicotine sulphate in each 100 gallons has been recommended. After the crop is harvested, much can be done to check flea beetles by destroying the stalks and the "suckers" that grow up in the fields.

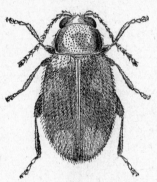

Fig. 284.—Tobacco flea beetle. Adult, about 20 times natural size. (*From U.S.D.A., Farmers' Bul.* 1425.)

References.—*N.C. Agr. Exp. Sta. Bul.* 239, 1919; *U.S.D.A., Farmers' Buls.* 1352, 1923, and 1425, 1924; *Conn. Agr. Exp. Sta. Bul.* 364, 1934; *Ky. Agr. Exp. Sta. Res. Bul.* 266, 1926; *U.S.D.A., Bur. Ent. Plant Quar. Cir.* E-373, 1937; *Jour. Econ. Ento.*, **25**: 1187–1190, 1932, and **26**: 233–236, 1933.

66 (110). Tobacco Hornworms[1]

The best known of tobacco insects, and among the most injurious, are the large green tobaccoworms with white bars on the sides and a slender horn at the end of the body and the parent "tobacco flies," or hawk moths, that lay the eggs of the hornworm. These worms infest the plants all summer long and are such ravenous feeders as to ruin many leaves and, where abundant, defoliate the plants. Because of their concealing coloration, they can often be located most easily by the pellets of excrement which they drop upon leaves or the ground, above which they are feeding. The life history and descriptions of these pests are given on page 523.

[1] *Protoparce sexta* (Johanssen) and *P. quinquemaculata* Haworth, Order Lepidoptera, Family Sphingidae.

Control.—Where labor is cheap and plentiful, hand-picking of these worms is fairly effective and laborers should always be instructed to destroy the worms wherever they are encountered while working among the plants (Fig. 285). Dusting the plants with paris green, 1 part, in hydrated lime or sterilized tobacco dust, 6 parts, or a dust containing 36 per cent cryolite, applied very evenly, at the rate of 7 or 8 pounds per acre (4 pounds per acre for shade-grown tobacco) is recommended. The first application should be given as soon as the young worms become numerous on the plants and later applications should be given as needed. Tobacco

Fig. 285.—Hornworms on tobacco plants, killed by the application of lead arsenate. (*From U.S.D.A.*)

grown under shade cloth will be largely free of these worms, if the cloth is kept intact throughout the growing season. Fall and winter plowing to destroy the pupae, where it is not otherwise objectionable, reduces the population of hornworms the following season. Stalks and stubble should be destroyed immediately after harvest. It has been found possible to poison the adult moths by using isoamyl salicylate as an attractant and a 5 per cent solution of tartar emetic as a poison, exposed in special bait feeders (see *Jour. Econ Ento.*, **26**: 227–233, 1933).

References.—*U.S.D.A., Farmers' Bul.* 1356, 1923; *Conn. Agr. Exp. Sta. Bul.* 379, 1935; *Tenn. Agr. Exp. Sta. Bul.* 120, 1918; *N.C. Agr. Exp. Sta. Ext. Cirs.* 174, 1929, and 207, 1936.

67 (4, 49, 80). Cutworms

Importance and Type of Injury.—Cutworms of a score of species often cut off newly transplanted tobacco at the surface of the ground in the manner discussed under Corn Insects (page 346) and garden crops (page 475) (Fig. 286). Some climbing cutworms attack the leaves of older plants.

Life History, Appearance, and Habits.—The cutworms injurious to tobacco are divided by Crumb (*l.c.*) into two groups. One group, including the variegated cutworm[1] and the black cutworm,[2] pass the winter as naked brown pupae in the soil, transform to moths very early in spring, and produce during the season three or four generations. The species

Fig. 286.—Tobacco plant ruined by a cutworm with the larva in feeding position. (*From U.S.D.A., Farmers' Bul.* 1494.)

of this group are destructive throughout the tobacco-growing regions of the United States and Canada. The other group consists of species that lay their eggs chiefly in weedy or grassy fields in late summer or fall. The eggs hatch and the larvae spend the winter partly grown and feed destructively upon the newly set plants in the spring. These species include the dingy cutworm,[3] the clay-backed cutworm,[4] the dark-sided cutworm,[5] and the spotted cutworm.[6] Species of this group are rarely destructive south of Virginia and Tennessee and all but the last-named have a single generation a year.

Control Measures.—If cutworms are present, poisoned-bran bait consisting of:

[1] *Lycophotia saucia* Hübner, Order Lepidoptera, Family Noctuidae.
[2] *Agrotis ypsilon* Rottemburg, Order Lepidoptera, Family Noctuidae.
[3] *Feltia subgothica* (Haworth) and *F. ducens* Walker.
[4] *Feltia gladiaria* Morrison.
[5] *Euxoa messoria* (Harris).
[6] *Agrotis c-nigrum* (Linné).

	For 5 to 10 *Acres*	For 75 *Sq. Yd.* *of Seedbed*
Wheat bran..	100 lb.	3 lb.
Paris green.......................................	2 lb.	1 oz.
Water..	enough to moisten	
Molasses, may be added...........................	2 qt.	⅛ pt.

should be applied several days before the tobacco plants are set. To determine whether the bait should be applied, Crumb suggests placing rather large compact bunches of freshly cut clover, dock, or chickweed on well-plowed soil. If cutworms are present in the soil they will collect under such vegetation, and an examination in 2 or 3 days will indicate whether the bait should be applied. Poison-bran bait may be broadcast late in the afternoon at the rate of 10 to 20 pounds dry weight per acre, or 4 pounds to 100 square yards of seedbed. Or a small handful may be placed about each hill as the plants are set. It should not be scattered upon the leaves. Fall or late spring plowing is recommended to reduce the cutworms.

References.—U.S.D.A., Farmers' Bul. 1494, 1926; *Fla. Agr. Exp. Sta. Bul.* 48, 1898; *Tenn. Agr. Exp. Sta. Bul.* 159, 1936.

68 (15, 37). Wireworms or Pithworms[1]

Wireworms are among the most destructive tobacco insects. Newly transplanted tobacco often wilts and dies within a few days. If the stems are split open, the larvae are found tunneling near the ground level. Other larvae will be found in the adjacent soil, where they may feed throughout the season, by chewing off the smaller roots. These pests are discussed more fully on pages 377, 411 and their control on pages 379, 483.

69 (19, 47). Corn-Root or Tobacco Webworm[2]

Importance and Type of Injury.—This small caterpillar, also known as the corn webworm, causes much damage to tobacco in Virginia and adjoining states by cutting into the young plants at the surface of the ground and then boring up or down in the stem. Usually the attack comes after the plants have become established, and it may necessitate one or more replantings.

Plants Attacked.—Tobacco, corn, and certain weeds such as buckhorn, wild carrot, plantain, asters, and daisies.

Distribution.—Throughout the eastern states, especially destructive in Virginia.

Life History, Appearance, and Habits.—Similar to that given for sod webworms (page 384).

Control.—No practical control measures are known except the indirect ones described on page 385.

Reference.—U.S.D.A., Dept. Bul. 78, 1914.

70. Tobacco Budworm[3]

Importance and Type of Injury.—Tiny, rust-colored to pale-green, striped caterpillars eat into the buds or unfolded leaves of tobacco as

---

[1] Order Coleoptera, Family Elateridae.

[2] *Crambus caliginosellus* Clemens, Order Lepidoptera, Family Pyralididae.

[3] *Heliothis* (= *Chloridea*) *virescens* (Fabricius), Order Lepidoptera, Family Noctuidae.

the plants begin to top.   If the holes are made in the tips of the buds, the leaves that expand from the buds are often ragged and distorted (Fig. 287).   If the tiny holes penetrate the unfolded leaves, these leaves will have large unsightly holes when they are fully expanded.   Only a single larva is commonly found on a plant, and the larvae migrate but little. The attack on the buds renders the leaves unfit for cigar wrappers and greatly cuts the price.   The larvae of the second generation often feed on the "suckers" and eat into the seed pods.

Plants Attacked.—Tobacco, cotton, ground cherry, and other solanaceous plants and also geranium and ageratum.

Distribution.—From Missouri, Ohio, and Connecticut, southward. Most injurious in the Gulf states; rarely in Kentucky and Tennessee.

Fig. 287.—Injury to the tobacco plant by the tobacco budworm.   (*From U.S.D.A., Farmers' Bul.* 819.)

Life History, Appearance, and Habits.—The budworm winters in the soil as a mahogany-colored spindle-shaped pupa, about ¾ inch long. The moths emerge in the spring and deposit their eggs singly on the underside of leaves.   They are about 1½ inches across the spread wings. The front wings are light-green and crossed by four oblique light bands, the inner three of which are edged with black.   Eggs are laid singly on the underside of the leaves.   The tiny larvae that hatch from the eggs make their way to the buds and crawl down among the unfolded leaves, where they may cause great damage before they are detected. The full-grown larva is 1½ inches long, of a pale-green color, and marked with several longitudinal pale stripes.   Pupation occurs in the soil, and there are two generations of the insect each season.

Control Measures.—A dose of poison should be placed in each bud before the young larva gets into it.   For this purpose a mixture of 1 pound lead arsenate in 75 pounds of corn meal (6 heaping teaspoonfuls to the peck) is recommended, applying a small quantity directly upon the bud of each plant by hand or by means of a sifter can, using 8 to 12 pounds of the mixture per acre, once or twice a week. As the buds expand, two applications a week from the time the plants are set out until topping of the plants is completed will be needed to give entire protection. If workmen will kill all worms seen while working among the tobacco, much of the damage may be prevented. As soon as the plants are harvested, the remnants should be cut and burned and the soil plowed. Infestation the following year may be reduced if the leftover plants in seedbeds are destroyed as soon as planting is finished and if the suckers are removed from plants and destroyed. The seedbeds should be carefully covered as recommended for the tobacco flea beetle.

References.—U.S.D.A., Farmers' Bul. 819, 1917, and 1531, 1927; *Jour. Econ. Ento.* **29**: 282–285, 1936.

71 (12, 63, 111, 260). Corn Earworm or False Budworm[1]

This very destructive pest of corn and cotton (see pages 368 and 452) attacks tobacco in a manner very similar to the tobacco budworm. It is usually not so destructive to tobacco as the true tobacco budworm. Very early in the season in the extreme south, and after corn has become mature in most sections, the moths, of which there are several generations, may lay their eggs on tobacco. The caterpillars may be destroyed by the same measures recommended for the tobacco budworm.

72 (106). Tobacco Splitworm[2] or Potato Tuberworm[2]

This insect is discussed as the potato tuberworm on page 516. When attacking the tobacco, the pinkish-white caterpillars, about ⅓ inch long, mine between the upper and lower surfaces, especially of older leaves, causing unsightly gray to brown blotches that render the product unfit for cigar wrappers.

Control Measures.—It is advisable to transplant as early as possible and make every effort to mature an early crop. Early leaves that are badly infested should be pruned off and the stubble and sucker growth destroyed. Cleaning up trash about the field and barns will help to prevent injury. Potatoes should not be followed with tobacco or grown near tobacco fields.

Reference.—U.S.D.A., Dept. Bul. 59, 1914.

[1] *Heliothis obsoleta* Fabricius, Order Lepidoptera, Family Noctuidae.

[2] *Gnorimoschema* (= *Phthorimoea*) *operculella* (Zeller), Order Lepidoptera, Family Gelechiidae.

CHAPTER XVI

INSECTS INJURIOUS TO VEGETABLE GARDENS AND TRUCK CROPS

An abundance of fresh vegetables in the diet is known to be very important to human health. The use of vegetables, now negligible in many families, would be much more extensive if the control of the many insects that discourage the home gardener were better understood. The quality of commercially grown vegetables would be better and the price could be lowered, also, if insects were not permitted to add so much to the expense of production. A very slight amount of feeding upon vegetables such as cucumbers, melons, tomatoes, and lettuce render them unsalable. This makes the insect losses to truck crops high in proportion to the amount of the plant consumed.

A knowledge of the habits, life cycles, and control measures effective against these pests will enable any grower to reduce very much the extensive damage he suffers from garden insects.

The more important garden insects are discussed under the following groups:

A. General Garden Pests, page 464.
B. Insects Injurious to Peas and Beans, page 485.
C. Insects Injurious to Cucurbits, page 494.
D. Insects Injurious to Potatoes, page 506.
E. Insects Injurious to Sweetpotatoes, page 517.
F. Insects Injurious to Tomatoes, Peppers, and Eggplant, page 522.
G. Insects Injurious to Onions, page 526.
H. Insects Injurious to Cabbage and Related Vegetables, page 529.
I. Insects Injurious to Beets, Spinach, Lettuce, Carrots, Parsnips, Celery and Related Vegetables, page 542.
J. Insects Injurious to Asparagus, page 549.
K. Insects Injurious to Sweet Corn, page 552.

General References.—Crosby and Leonard, "Manual of Vegetable-garden Insects," Macmillan, 1918; *U.S.D.A., Farmers' Bul.* 1371, 1934; *Can. Dept. Agr. Bul.* 161 (n.s.), 1932; *Cornell Ext. Bul.* 206, 1931; *Mich. Agr. Exp. Sta. Spcl. Bul.* 183, 1929; *Mich. State Coll. Ext. Bul.* 180, 1937; *Purdue Univ. Ext. Bul.* 186, 1932; *N.D. Agr. Exp. Sta. Cir.* 42, 1933; *Fla. Agr. Exp. Sta. Bul.* 232, 1931; *Calif. Agr. Ext. Serv. Cir.* 87, 1934; *U.S.D.A., Bur. Ento. Plant Quar. Cir.* E-376, 1937; *Ill. Agr. Exp. Sta. Cir.* 437, 1937.

A. GENERAL GARDEN PESTS

While many garden insects feed on only one kind of crop or a few closely related ones, there are a number of others that feed on a variety of plants. Some of these are discussed in this section before the pests of special crops are taken up. The following species are general garden pests. They are also included in the field keys under the crops to which they are most destructive.

FIELD KEY FOR THE DETERMINATION OF SOME GENERAL GARDEN PESTS ATTACKING MANY CROPS

A. *Insects chewing leaves, buds, or stems of plants:*

1. Black, jumping, somewhat square-backed insects, with antennae longer than the body and the females with a needle-like ovipositor almost as long, eat seedling plants or the fruits of vegetables mostly at night and hide among surface trash or in soil cracks in the daytime........*Field cricket*, page 466.

2. Insects somewhat similar to *A*,1, but heavier and clumsier, with very small wings and long, sword-shaped ovipositors. Migrate from waste land often attacking crops in great migrating swarms, feeding during warm sunny days. Only in the Rocky Mountain states.*Mormon-cricket* or *coulee-cricket*, page 467.

3. Brownish, grayish, jumping insects up to 1½ inches long, often migrate into gardens from surrounding fields and strip leaves, eat tender stems and devour fruits of vegetables of all kinds....................*Grasshoppers*, page 468.

4. Many minute holes eaten in leaves by very small, hard-shelled black beetles from ¹⁄₁₆ to ⅛ inch long, which rest and feed chiefly on the underside of leaves and jump readily when disturbed...................*Flea beetles*, page 468.

5. Cinnamon brown velvety-looking beetles, something like small May beetles, ⅜ inch long, eat irregular holes from margins of leaves of nearly all garden and flower plants at night. Insects congregate about artificial lights. Only along Atlantic seaboard..........................*Asiatic garden beetle*, page 471.

6. Grayish long-headed beetles, ½ inch long, with faint white stripes along their sides, eat the margins of the leaves............*White-fringed beetle*, page 472.

7. Yellowish-green caterpillars, up to 1 inch long, with scattered fine hairs and conspicuous small dark spots, skeletonize and devour the leaves, spinning considerable silk over the foliage, webbing the leaves together..............
....*Beet* or *alfalfa webworm*, page 474.

8. Plant stems eaten off at night and left lying on the ground, or leaves devoured, or fruits eaten into by plump, smooth, greasy-looking, greenish, brownish, or grayish, spotted or striped caterpillars, up to 1½ inches long, which dig into the soil to hide during the daytime...................*Cutworms*, page 475.

9. Very hairy, yellowish, brown, or reddish-brown- and black-banded caterpillars up to 2 inches long, eat large holes in leaves...........*Woolly-bears*, page 476.

10. Large smooth-edged holes eaten in leaves, or plants stripped of leaves, by caterpillars up to 1½ inches long, striped in several colors, with a white λ on front of head, often migrating in "armies."........*Fall armyworm*, page 344.

B. *Insects sucking sap from leaves, buds, fruits, or stems, not leaving visible holes but causing wilting, rolling of leaves, spotting and discoloration of foliage, and discoloration and distortion of fruits:*

1. Small, soft-bodied, sluggish, green or blackish bugs of various sizes, the largest scarcely larger than a pinhead, either winged or wingless, cluster on underside of leaves and terminal shoots. Leaves often sticky with honeydew...........
....*Aphids*, page 477.

2. Olive-green bugs, flecked with yellow and dark-brown, with flattened, oval bodies, somewhat triangular in front, up to ¼ inch long by half as wide, cause spotted and deformed leaves, discolored stems, and dwarfed or dying buds, by sucking sap and poisoning tissues. Bugs run or fly readily when disturbed....
....*Tarnished plant bug*, page 479.

3. Blackish, rather soft-winged, jumping bugs, only ¹⁄₁₀ inch long, and their greenish nymphs cause spotting and dying of leaves. Resemble flea beetles but have piercing-sucking mouth parts..........*Garden fleahopper*, page 481.

4. Leaves turn yellow, become blotched with reddish-brown or white spots and die. Underside of leaf with inconspicuous white webs among which are many microscopic, 8-legged, reddish or greenish mites and their round pearly eggs, looking, to the naked eye, like fine white dust............*Red spider mite*, page 482.

C. *Insects living in the soil and feeding on the roots under ground:*

1. Young plants may be rooted out and the roots of others eaten off or scarred by weird-looking brown crickets, up to 1½ inches long, and covered with fine

velvety hairs. The rake- and shovel-like front legs are used to make tunnels beneath the soil in which they live..
..................*Mole crickets* (see *U.S.D.A., Farmers' Bul.* 1561, 1928.)

2. Slender, tough-skinned, brownish, polished-looking worms, up to 1¼ inches long, with chewing mouth parts and 6 short legs, eat into planted seeds or eat off rootlets of plants.....................................*Wireworms*, page 483.

3. White thick U-shaped grubs with brownish heads and 6 slender legs, up to 1½ inches long, eat off roots or gnaw into tubers........*White grubs*, page 484.

4. Plants wilt and die from the attacks of white legless, not curved-bodied grubs, up to ½ inch long, which eat into underground stems and larger roots. Only in Gulf states.....................................*White-fringed beetle*, page 472.

5. Roots of vegetables chewed off by a somewhat slender white grub with a Y-shaped anal opening and a transverse row of short setae in front of it.......
....*Asiatic garden beetle*, page 471.

6. Very delicate, slender, cylindrical, white larvae, not over ⅓ inch long, with distinct heads and 6 very short legs, eat tiny rootlets....*Flea beetles*, page 468.

7. Caterpillars similar to *A,8,* but usually paler, feed on roots or stems below ground...*Cutworms*, page 475.

8. Cylindrical, many-segmented worms, with 2 pairs of legs on each segment, ranging up to 1 or 2 inches long, eat roots, tunnel into tubers and attack leaves and fruits lying on the soil..........................*Millipedes*, page 485.

9. Rootlets eaten off and underground parts of plant scarred by small, nearly white, very active creatures with a pair of legs on each principal segment of the body, of all sizes up to ¼ inch long...............*Garden centipede*, page 759.

73. Field Cricket[1]

Importance, Type of Injury, and Plants Attacked.—Everywhere present in small numbers, this familiar insect (Fig. 105) sometimes becomes abundant enough to cause serious losses. In California and some Gulf states they have damaged cotton by cutting off seedling plants just above the ground or just below the apical bud and cutting off many leaves that drop to the ground. In the Great Plains, they destroy the seeds of alfalfa and all the cereals by devouring them or cutting off the heads especially in the "dough" stage, but also after the grain is shocked or stacked. The fruits of tomatoes, all the cucurbits, peas, beans, strawberries, and others may be seriously damaged or entirely ruined especially in the Gulf states. These pests also frequently eat holes in paper and rubber and in cotton, linen, woolen, or fur garments, either out-of-doors or indoors, especially when soiled with perspiration or foods.

Distribution.—Throughout the United States, southern Canada, and much of South America.

Life History, Appearance, and Habits.—The life history varies greatly throughout the range of the insect, but in the northern states the majority hibernate in the egg stage and complete a single generation in a year, while in the Imperial Valley of California and the Gulf Coast they winter as nymphs or remain more or less active the year round and may produce as many as three generations. Hatching of overwintered eggs takes place during late May and June. There are 8 to 12 nymphal instars, in all but the first of which the nymphs feed like the adults. The total developing period occupies from 9 to 14 weeks. The body of the adults varies from ⅗ to 1 inch in length, the antennae are a half longer than the body, the hind legs are very heavy for jumping, the cerci are about ⅓ as

[1] *Gryllus assimilis* Fabricius, Order Orthoptera, Family Gryllidae.

long as the body, and the ovipositor of the female is $\frac{2}{3}$ to nearly 1 inch long. The adults vary so much in color, length of wings, and other structural features that no fewer than 45 different scientific names have been proposed. Throughout most of the northern states, adults are present from late July until heavy frosts and lay their eggs mostly in August and early September. The chirping of the males during this part of the year is one of the most familiar out-of-door sounds. During sunshiny days crickets remain most of the time under shelter of vegetation, under surface trash, or in cracks or excavations that they make in the soil. Feeding, mating, and egg-laying take place from late afternoon until late morning. Damp soil is preferred in which to lay the eggs: this may be found along ditches, beneath cracks in baked soil, or by digging pits the size of the end of a finger with the four front legs. The long ovipositor is then forced down into the soil and a small number of eggs laid in each hole beneath the soil but not enclosed in a pod-like secretion as grasshopper eggs are. The eggs are banana-shaped, $\frac{1}{12}$ inch long, and straw- to cream-colored. Females commonly produce about 300 eggs.

Control Measures.—The use of poisoned bran baits as for grasshoppers (page 339) is effective against this pest. Deep fall plowing to bury the eggs, the maintenance of a fine dust mulch to kill and drive away the active stages, rotations in which cotton does not follow small grains, and, in houses, placing bran bait (page 254) on papers on the basement floor, out of the way of pets and children, or dusting with pyrethrum or sodium fluoride have been found useful control measures.

References.—*S.D. Agr. Exp. Sta. Bul.* 295, 1935; *U.S.D.A., Tech. Bul.* 642, 1939.

74. MORMON-CRICKET[1] AND COULEE-CRICKET[2]

Importance and Type of Injury.—These insects, which are not true crickets but are related to the katydids and longhorned grasshoppers, have somewhat the appearance and habits of the field cricket. For nearly a century the mormon-cricket has periodically overwhelmed settlers and homesteaders in the Rocky Mountain area by migrating from its native breeding grounds in the high rugged hills into the arable lands of the valleys and eating garden crops, small fruits, legumes, and especially the heads of grain in the milk or dough stage. A very serious outbreak of these pests in the Great Salt Lake basin in 1848 was terminated so spectacularly by great flocks of gulls that the grateful settlers erected an imposing monument in commemoration.

Distribution.—The mormon-cricket occurs in large areas of Utah, Colorado, Wyoming, Montana, Idaho, Nevada, Oregon, and Washington. The coulee-cricket has been very destructive in Montana and Washington.

Plants Attacked.—Nearly all field and garden crops, many fruits, shrubs, trees, and weeds, especially bitterroot and the wild mustards. The coulee-cricket is especially fond of sagebrush, *Artemisia rigida*, and also eats dung and dead animals. Both species are very cannibalistic.

Life History, Appearance, and Habits.—The winter is passed in the egg stage, $\frac{1}{4}$ to 1 inch deep in the soil, the mormon-cricket choosing

[1] *Anabrus simplex* Haldeman, Order Orthoptera, Family Tettigoniidae.
[2] *Paranabrus scabricollis* Thomas, Order Orthoptera, Family Tettigoniidae.

especially barren soil in sunny locations; the coulee-cricket choosing the base of tiny grass stools. The elongate eggs are deposited singly by thrusts of the long ovipositor. These eggs hatch in the first warm days of spring, February to April, usually a full month earlier than grasshoppers. They pass through their seven nymphal stages in about 75 to 100 days and become adult from early June to mid-July. Coulee-crickets nearly all disappear by July but the mormon-cricket continues egg-laying all summer long, and the embryos become well developed before winter but do not hatch until the following spring. The adults are about 1 inch long, heavy-bodied and clumsy. The wings are very small and useless except that the male uses them to produce his mating chirp. The tarsi are four-segmented, the antennae as long as the body, and the female has a sword-shaped ovipositor also as long as the body. Unlike the field cricket, the mormon- and coulee-crickets are active only during the warm sunny part of the day and seek shelter at night and in cloudy or rainy weather. From the time they are about half-grown, they begin migrating from $\frac{1}{8}$ mile to over a mile a day, going in a straight line, in no predictable direction, over all kinds of obstacles, often in bands covering a square mile in area and often 100 to 500 individuals per square foot.

Control Measures.—Since the insect cannot fly in any stage of its life, linear barriers consisting of 10-inch strips of sheet metal or 8-inch boards with the upper 4 inches covered with metal or oilcloth, set on edge; or, for the coulee-cricket, vertical walled ditches have been used extensively to stop migrating swarms. Soil pits or water traps are made at intervals to catch the crickets halted by the barrier. The application of sodium or calcium arsenite dusts 1 part in 3 parts of hydrated lime directly upon the swarms of insects gives a good kill. Poisoned baits have not been very effective against the mormon-cricket.

References.—*U.S.D.A., Tech. Bul.* 161, 1929; *Wash. Agr. Exp. Sta. Bul.* 137, 1937; *Univ. Idaho Ext. Bul.* 100, 1936.

75 (1, 27, 146). GRASSHOPPERS[1]

In years when grasshoppers are abundant, they may become serious on garden crops, especially where the gardens are surrounded by fields of clover or by waste lands in which the grasshoppers have not been controlled. The ravenous grasshoppers strip the leaves, devour fruits, and eat the tender stems from almost all kinds of vegetable crops. These insects are more fully treated under Corn Insects, page 335. To protect gardens from attack by grasshoppers, the poisoned baits described on pages 254 and 339 are recommended. The poisoning must be done on land around the garden as well as in the garden itself.

76 (20, 65, 108, 121). FLEA BEETLES[2]

Importance and Type of Injury.—The name *flea beetle* is applied to a variety of small beetles which have the hind legs enlarged and jump vigorously when disturbed. When flea beetles are abundant, the foliage of garden plants may be so badly eaten that it can no longer function and the plant dies. Since they are small, and also rather active, they do not take much food in one spot; their injury consists of very small, rounded or irregular holes eaten through or into the leaf, so that leaves

[1] Many species of Order Orthoptera, Family Locustidae.
[2] Order Coleoptera, Family Chrysomelidae.

look as though they had been peppered with fine shot (Fig. 62,*I*). These small holes give an opportunity for the entrance of destructive plant diseases, and the beetles may carry the disease organisms from one plant to another and spread them as they feed. The potato flea beetle is a means of spreading early potato blight in this way, and the corn flea beetle spreads the bacterial wilt of corn. Besides the injury by the adults, the larvae of some flea beetles commonly feed on the roots of the same plants, riddling them with tunnels or eating off small rootlets. The larvae of other kinds feed, along with the adults, on the foliage, or mine in the leaves or tunnel in the stems.

Plants Attacked.—Some flea beetles are rather general feeders, but perhaps the majority attack only one plant or the closely related crops

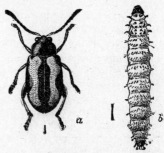

Fig. 288.—The potato flea beetle, *Epitrix cucumeris* Harris, adult. Line shows natural size. (*From Chittenden, U.S.D.A.*)

Fig. 289.—The striped cabbage flea beetle, *Phyllotreta vittata* Fabricius, *a*, adult, *b*, larva. Lines show natural size. (*From Riley, U.S.D.A.*)

of a single plant family. Among the most destructive garden species are the potato flea beetle,[1] eggplant flea beetle,[2] spinach flea beetle,[3] horse-radish flea beetle,[4] sinuate-striped flea beetle,[5] striped cabbage flea beetle,[6] sweetpotato flea beetle,[7] and the mint flea beetle.[8] Tomatoes, peppers, and cucumbers are also subject to severe injury.

Life History, Appearance, and Habits.—The life history varies greatly with the different species. Usually the winter is passed in the adult stage, the beetles hibernating under leaves, grass, or trash about the margins of fields, along ditch banks, fence rows, margins of woods, and similar protected places. The potato flea beetle (Fig. 288) and eggplant flea beetle are about $\frac{1}{16}$ inch long and nearly a uniform black in color. The equally small, tobacco flea beetle[9] is yellowish-brown with a dark cloud across the wings, and the sweetpotato flea beetle[7] (Fig. 324) is of about the same size but with a bronzy reflection. The striped cabbage flea beetle (Fig. 289) and the sinuate-striped flea beetle are about

[1] *Epitrix cucumeris* Harris.
[2] *Epitrix fuscula* Crotch.
[3] *Disonycha xanthomelaena* Dalman.
[4] *Phyllotreta armoraciae* (Koch).
[5] *Phyllotreta zimmermanni* (Crotch).
[6] *Phyllotreta vittata* Fabricius.
[7] *Chaetocnema confinis* Crotch.
[8] *Longitarsus menthaphagus* Gentner (see *Mich. Agr. Exp. Sta. Spcl. Bul.* 155, 1926).
[9] *Epitrix parvula* Fabricius.

$\frac{1}{12}$ inch long, with a curious, crooked, yellowish stripe on each wing cover. The pale-striped[1] and horse-radish flea beetles[2] are $\frac{1}{8}$ inch

long, with a broader, nearly straight, yellowish stripe on each wing cover; the smartweed flea beetle,[3] of the same size, is bluish-black, without stripes, and very straight-sided. Among the largest of our common species is the spinach flea beetle[4] (Fig. 292), which is fully $\frac{1}{5}$ inch long, with greenish-black wing covers, a yellow pro-thorax, and a dark head. Nearly all these species are elongate oval in outline, with narrowed prothorax and narrower head. The antennae are one-half to two-thirds as long as the body, and the hind femurs are distinctly thickened, enabling the beetles to jump away quickly when they are disturbed. Many of the species emerge from hibernation, beginning in late May, and feed on

FIG. 290.—Eggs of the spinach flea beetle, *Disonycha xanthomelaena* Dalman, about 10 times natural size. (*From Ill. State Natural History Surv.*)

FIG. 291.—Larva of spinach flea beetle, dorsal view, about 10 times natural size. (*From Ill. State Natural History Surv.*)

FIG. 292.—Adult of spinach flea beetle, about 10 times natural size. (*From Ill. State Natural History Surv.*)

weeds and the foliage of trees until the garden plants are available, when they migrate to them. They are frequently serious pests in seedbeds and on newly transplanted vegetables. The eggs, which are so small

[1] *Systena taeniata* Say.
[2] *Phyllotreta armoraciae* (Koch).
[3] *Systena hudsonias* Forster.
[4] *Disonycha xanthomelaena* Dalman.

as never to be seen by the grower, are scattered in the soil about the plants by the potato flea beetle and the tobacco flea beetle and laid in clusters upon the leaves, by the spinach flea beetle (Fig. 290). They require about 10 days to hatch. The striped cabbage flea beetle deposits her eggs in tiny cavities gnawed in the stem of the plant, the horse-radish flea beetle deposits them in clusters on the leaf petioles, and the sinuate-striped flea beetle lays her eggs singly on the leaves of cabbage, turnip, and radish. The larvae of flea beetles are mostly whitish, slender, delicate, cylindrical worms from ⅛ to ⅓ inch long when full-grown, with tiny legs and brownish heads. They feed on the roots, underground stems, or tubers of vegetables or of weeds, for 3 or 4 weeks. The larva of the sinuate-striped species, however, mines in the leaves, the horse-radish flea beetle burrows in the leaf petiole, and the spinach flea beetle feeds exposed on the underside of the leaves. The last-named species (Fig. 291) is a short leaden-gray wrinkled grub about ¼ inch long in the full-grown larval stage. Pupation usually occurs in the soil for 7 to 10 days. There are generally one or two generations a year.

Control Measures.—Flea beetles are hard to control with arsenicals. Arsenicals are apparently distasteful to them, and they do not readily eat enough sprayed foliage to be killed by the poison. Best results have generally been secured by applying bordeaux mixture or bordeaux with lead arsenate or calcium arsenate 4 pounds to 100 gallons, as a repellent (see page 290) very thoroughly to the plants at 7- to 10-day intervals, since these insects detect any unsprayed portions of the foliage and feed upon them. Plants such as tomatoes and eggplant may be dipped in poisoned bordeaux when they are transplanted. Fluorine compounds have, in general, proved much more toxic to flea beetles than arsenicals. Dusts or sprays of barium fluosilicate, or natural or synthetic cryolite, have given good results.

Seedbeds may be covered with strips of gauze or tobacco cloth, 25 strands to the inch, to exclude these pests from delicate young plants. Sticky shields or specially made boxes having the inner walls covered with tree tanglefoot may be passed over infested plants and catch thousands of the beetles as they jump off the disturbed plants. Keeping down weeds in and around the garden is often the most important method of holding these pests in check, since the adults often feed on weeds in early spring and late in fall, and the larvae may develop in great numbers on the roots of certain weeds. In the case of the mint flea beetle care should be taken not to introduce the eggs in soil about the roots used for planting new fields.

References.—Crosby and Leonard, "Manual of Vegetable-garden Insects," 1918; *U.S.D.A.*, *Dept. Buls.* 535, 1917, and 902, 1920; *N. Y. (Geneva) Agr. Exp. Sta. Bul.* 442, 1917; *N. Y. (Cornell) Agr. Exp. Sta. Memoirs* 55, 1922; *Can. Dept. Agr. Ento. Cir.* 2, and *Pamphlet* 80 (n.s.), 1927; *U.S.D.A.*, *Farmers' Bul.* 1371, 1934; *Jour. Econ. Ento.*, **29:** 586–589, 1936; *Ohio Agr. Exp. Sta. Bul.* 595, 1938.

77. Asiatic Garden Beetle[1]

Importance and Type of Injury.—Irregular holes are eaten, especially from the margins of the leaves on warm nights, by beetles which may be found during the day in loose soil at the base of the injured plants. The

---

[1] *Autoserica castanea* Arrow, Order Coleoptera, Family Scarabaeidae.

adults occur in such numbers around bright lights that they are a serious nuisance about residences, and patrons are driven away from amusement parks, swimming pools, and restaurants.   Their whitish grubs feed especially on the roots of grasses but do not cut them off so destructively as do the Japanese beetle larvae.

Plants Attacked.—More than 100 different plants, including vegetables such as beets, carrots, corn, eggplant, kohlrabi, parsnips, peppers, and turnips; ornamentals such as asters, dahlias, chrysanthemums, roses, zinnias, delphiniums, azaleas, rhododendrons, and viburnum; and peach, cherry, and strawberry.

Distribution.—First found in America in New Jersey, in 1922, the insect has since been discovered at widely scattered points along the Atlantic seaboard from Massachusetts and eastern New York to South Carolina.   It occupies a zone of continuous infestation of about 4,000 square miles centering about New York City.   It is believed that this pest cannot thrive in areas of low summer rainfall.

Life History, Appearance, and Habits.—The winter is passed as small grubs, 6 to 10 inches deep in the soil.   They resume feeding near the soil surface in early spring, pupate in late May and June, and emerge as adults in June and July.   The adults (Fig. 246) are shaped like May beetles, but are only $\frac{3}{8}$ inch long.   They are cinnamon-brown in color, with fine longitudinal striae on the wing covers and a fine pubescence which gives them a velvety appearance.   They are troublesome all summer.   Eggs are laid in greatest numbers in grassy areas, overgrown with such weeds as orange hawkweed, goldenrod, sorrel, and asters, which keep the soil moist and cool. There is a single generation a year.   The larvae may be distinguished from related white grubs, by their more slender bodies, the Y-shaped anal opening, and the transverse row of short setae in front of it, as shown in Fig. 402.

Control Measures.—In lawns and golf courses, complete protection from the larvae has been secured by treating with lead arsenate, 10 pounds to each 1,000 square feet (see page 392).   Strawberry plots have been protected by mulching with hay.   When infested overgrown land is plowed up, it may be planted the first year to potatoes or some other crop that is not injured.   Spraying the preferred food plants with 6 pounds lead arsenate and 4 pounds of flour per 100 gallons of water is effective in killing the beetles, but in areas of great abundance successive hordes of beetles may attack the plants each warm night and destroy them in spite of the sprays.   Funnel-shaped light traps have captured as many as 2,000 beetles per hour per trap.

References.—*U.S.D.A.*, *Cirs.* 238, and 246, 1932; *Jour. Econ. Ento.*, **27**: 476–481, 1934, and **29**: 348–356, 1936.

78. The White-fringed Beetle[1]

Importance and Type of Injury.—One of the most recent of the many imported insect pests is this beetle which was first found in Florida in the summer of 1936.   As many as 200 to 300 grayish long-headed beetles, $\frac{1}{2}$ inch long, with faint white stripes on their sides, are found per plant feeding upon the margins of the leaves or crawling upon the ground. The lower part of the stem and taproot or the planted seeds of many

---

[1] *Naupactus leucoloma* Boheman, Order Coleoptera, Family Curculionidae.

field and garden crops are attacked in the spring by vast numbers of small white legless grubs, up to ½ inch long, which cause the plants to yellow, wilt, and die. Although the insect cannot fly, the eggs are deposited upon many parts of plants that are moved in commerce, the larvae may be transported in small quantities of soil, and the adults, which have a pronounced tendency to climb upward, readily cling to objects being transported. Entomologists who have studied the insect feel that it may become a serious pest in many regions of the United States.

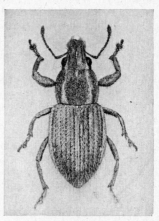

Plants Attacked.—These insects are very general feeders. The adults have been observed feeding upon more than 50 different kinds of plants, ranging from herbaceous weeds and crops to vines and trees. The adults are not voracious feeders and most of the damage is done by the larvae which feed on the roots of many plants, including the following crops: peanuts, corn, sugarcane, cotton, cowpeas, velvet beans, cabbages, collards, sweetpotatoes, c h u f a, l u c e r n e, Mexican clover, and blackberries.

Fig. 293.—White-fringed beetle, *Naupactus leucoloma* Boheman, about twice natural size. (*From U.S.D.A., Bur. Ento. and P.Q.*)

Distribution.—The white-fringed beetle is a native of South America, where it is widely distributed over the southern part of the continent. It has also recently been reported from Australia. In the United States it has been found from western Florida to eastern Louisiana near the Gulf Coast.

Life-history, Appearance, and Habits.—The insect passes the winter months as large larvae, mostly in the upper 9 to 12 inches of the soil; but some have been taken 24 inches deep in the soil. The whitish naked pupae have been found mostly in the ground 3 to 6 inches deep, from late May to the end of July. From the middle of June until the end of July the adult beetles emerge and crawl to a near-by favorite food plant. The elytra of the beetles are grown together so that they cannot fly. They are dark-gray beetles, slightly less than ½ inch long, less than ⅓ as broad over the basal half of the wing, whence the body narrows gradually to the end of the short, broad snout and tapers rapidly to the tips of the wing covers. The margins of the elytra are banded with white, and there are two pale lines extending along the sides of the head and prothorax, one above the eye and one below it. The body is covered with dense, short, pale hairs, longer toward the tip of the elytra. The adults feed for some days upon the margins of older leaves toward their base but eat relatively little and are active chiefly in the afternoon. No males of this species have been found. The females, which reproduce parthenogenetically, begin to lay eggs in 10 to 12 days after emerging from the pupal stage. A female may lay many hundreds of eggs, the maximum observed being 1,847. The average life time of the females is probably several months, during which time, they frequently crawl from ¼ to ¾ mile. The whitish oval eggs are deposited in gelatinous masses, as many as 60 eggs per cluster. They adhere to sticks, stones, the base of plant stems, and other objects on or near the ground. The soil adheres to the egg masses, making them very difficult to see. In midsummer the eggs hatch in about 2 weeks, but in cooler weather, the egg stage may last 1 or 2 months. It is claimed that moist conditions are necessary for hatching. The larvae feed entirely below ground down to a depth of 6 inches or more, beginning in late July and continuing until cold weather. They chew away the lower part of the stem and taproot of many kinds of plants. When full-grown the larvae measure about ½ inch long. They are yellowish-white, the back evenly rounded, upward; they are legless, and sparsely covered with short hairs. From late May to the end of July, the larvae pupate at depths of 3 to 6 inches in the soil. Apparently there is normally one generation a year.

Control.—The known infested areas have been placed under rigid quarantine to restrict the movement of all products likely to convey the white-fringed beetle in any stage. Since the insect does not fly, ditches about a foot wide and a foot deep, with well-packed vertical sides, are very effective in stopping the migrations of the insect. Post holes 8 to 10 inches deep with several inches of loose soil in the bottom, dug at intervals of 15 to 20 feet along the bottom of the trench, serve to trap the beetles. The loose soil absorbs sufficient heat to kill the beetles in the bottom of the holes, or, if rain falls, the resulting mud smothers them. Kerosene may be used to make certain the destruction of all trapped beetles. In cities and about industrial plants the pests may be combated by cleaning the soil of all vegetation.

References.—*U.S.D.A., Bur. Ent. Plant Quar., Cirs.* E-420, January, 1938, and E-422, March, 1938 (mimeographed).

79 (19, 47, 69). Beet Webworm[1] and Alfalfa Webworm[2]

Importance and Type of Injury.—These small webbing caterpillars, by devouring the foliage, have caused the abandonment of thousands of acres of sugarbeets, often destroy fields of alfalfa, and take from 10 to 100 per cent of truck crops. When they have eaten everything in a field, they often migrate like armyworms.

Plants Attacked.—The larvae are general feeders and destructive to almost any succulent garden or field crop, but grains and grasses are not preferred. Among the garden crops seriously attacked are cabbage, carrot, beets, beans, peas, potatoes, spinach, and cucurbits.

Distribution.—From the Mississippi Valley westward to the Continental Divide.

Life-history, Appearance, and Habits.—These pests winter as larvae in silk-lined cells or tubes in the soil, within which they pupate in late spring. The adults from this overwintering generation are present in the fields from late March to late June. They are night-active moths, with a wing expanse of 1 to 1¼ inches. The alfalfa webworm is buff-colored, irregularly marked with light and dark gray and with a row of spots on the underside of the hind wings near the apical margin. The adult beet webworm is smoky brown, mottled with dusky and straw-colored spots and lines. It can be distinguished from the alfalfa webworm by the dark markings on the underside of the hind wings, being in the form of a con-

---

[1] *Loxostege sticticalis* Linné, Order Lepidoptera, Family Pyralididae.
[2] *Loxostege commixtalis* Walker.

tinuous dark line near the margin. The whitish to yellow or green, oval eggs are laid in groups of 2 to 20, mostly on the underside of the leaves. The sugarbeet webworm eggs are laid in a single row, end to end, while the alfalfa webworm eggs are in overlapping groups. As the larvae skeletonize and devour the leaves, they spin a web, drawing leaves together, and also forming a tube several inches long which leads to a hiding place, as under a clod of earth, into which they retreat when disturbed. The full-grown larvae are yellowish or greenish to nearly black, 1 to 1¼ inches long. The beet webworm larva has a black stripe down the middle of the back, while the alfalfa webworm has a broad light-colored stripe down the middle of the back. On each side of each segment there are three small dark spots, from which arise one to three setae. The insects pupate 2 to 3 weeks in earthen silk-lined cells, an inch or so under ground. The pupa of the beet webworm has eight, bristle-like appendages at the end of the abdomen, while in the alfalfa webworm these appendages are spoon-shaped. There are three partial generations over much of the range.

Control Measures.—On vegetable crops the control measures are the same as for cabbageworms (page 532). Sugarbeets should be sprayed with paris green, 8 pounds to 100 gallons of water, or lead arsenate, 16 pounds to 100, using 50 gallons to the acre, before the small larvae have spun much web. Spraying young alfalfa with lead arsenate, 5 pounds, soybean flour, 4 ounces, and water 100 gallons, applied 75 to 100 gallons per acre, has given satisfactory control under midwestern conditions. Dusting with lead arsenate in a carrier such as 300-mesh talc or gypsum is slightly less effective. Growers should watch their fields and gardens for the appearance of the moths, which fly up as the plants are disturbed, and for the egg masses on the underside of the leaves and should begin spraying just as soon as the eggs start hatching. The migrating caterpillars can be stopped by trap furrows as recommended for armyworms and the white-fringed beetle (pages 343, 474). Moths lay fewer eggs in fields kept free from weeds, especially pigweed and lamb's-quarters. Late-seeded alfalfa may escape serious damage.

References.—*Colo. State Ento. Cir.* 58, 1933; *U.S.D.A., Bur. Ent. Bul.* 109, 1912.

80 (4, 49, 67, 259). CUTWORMS[1]

Importance and Type of Injury.—Recently set or young seedling plants are often cut off during the night at the surface of the soil (Fig. 286) and left lying to wilt on the ground near by. Plump, soft-skinned, greasy-looking caterpillars, varying in length up to 1½ inches or more, which generally roll the body tightly when disturbed and "play 'possum," are found in shallow holes in the soil about the base of the plants. Some species climb up the plants and eat at the leaves or chew their way into fruits, such as tomatoes, while still others feed entirely below ground. Several species, when they become abundant, crawl from the fields where they developed, in great numbers like marching armies, and may invade gardens and rapidly devour all kinds of garden crops.

Plants Attacked.—Nearly all garden vegetables, as well as flowers, field crops, and fruit trees are attacked by cutworms of one kind or another.

[1] Various species of Order Lepidoptera, Family Noctuidae.

The distribution, life history, and control of cutworms are discussed under Corn Insects, page 347, and Tobacco Insects, page 460. Sod or weedy land intended for vegetables should be plowed in late summer and kept fallow until late in the fall, since it is during this period that the cutworm moths lay their eggs, chiefly on rough, grassy, or weedy land. Some species of cutworms, however, deposit their eggs on fence posts and the stems of large weeds along fences and ditches. In small gardens certain measures are practicable that cannot be used on field crops. The worms may be concentrated by laying small boards about the garden or they may be dug out and destroyed by hand. Some gardeners place cylinders of tin (such as tin cans with bottoms cut out) partly sunken in the soil about choice plants when they are transplanted, to keep the cutworms away. The poisoned baits recommended under Corn Insects are effective for many species.

81. Woolly-bear,[1] Yellow-bear,[2] and Salt-marsh Caterpillars[3]

Importance and Type of Injury.—Very hairy or woolly, yellowish and brown caterpillars (Fig. 294), ranging up to 2 inches in length, riddle the foliage of many

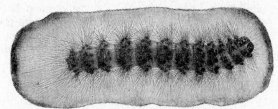

Fig. 294.—The salt-marsh caterpillar, *Estigmene acraea* Drury, larva, natural size. (*From Ill. State Natural History Surv.*)

garden crops in summer and autumn. The smaller worms are usually found feeding together on the underside of leaves. The larger ones feed exposed and, when full-grown, are often seen in the autumn scurrying over the surface of the ground, apparently in great haste to get somewhere.

Plants Attacked.—One or more of these species attack practically all the garden and field crops.

Distribution.—General throughout the United States.

Life History, Appearance, and Habits.—The woolly-bear winters in the larval stage in some protected place and feeds briefly in the spring before changing to a pupa. The other two species hibernate in the pupal stage inside thin silken cocoons heavily covered with interwoven hairs from the body of the caterpillar. The adults are snow-white or yellowish-winged moths, from 1½ to 2 inches across the wings, with yellowish black-spotted abdomens. The yellow bear

Fig. 295.—The salt-marsh caterpillar, male moth, natural size. (*From Ill. State Natural History Surv.*)

adult is nearly pure white except for the abdomen, each wing having a few small black spots. The salt-marsh caterpillar adult has the wings peppered with a number of small black spots, white above and yellow below, in the female. In the male (Fig. 295) the hind wings are yellow both above and below. The adult of the woolly-bear is entirely yellowish and

[1] *Isia isabella* Smith and Abbott, Order Lepidoptera, Family Arctiidae.
[2] *Diacrisia virginica* Fabricius.
[3] *Estigmene acraea* Drury.

with a few black spots on the wings. The adults appear in spring and lay their spherical eggs in patches on the leaves. The larvae attain full size in a month or two. The woolly-bear larva is black at each end, with a median band of brown of variable extent, and, as pointed out by Comstock, has an evenly clipped appearance. The other two species have hairs of different lengths, yellowish, tawny, or grayish in color, and not so dense as completely to hide the skin. There are generally two generations a year in the North.

Control Measures.—These caterpillars may be readily killed by applications of lead arsenate, barium fluosilicate, or cryolite, 4 pounds to 100 gallons or 2 ounces to 3 gallons, or a dust consisting of 1 part of any of these poisons in 9 parts of gypsum or talc.

Reference.—*Can. Dept. Agr. Bul.* 99 (n.s.), 1934.

82 (50, 60, 96, 117, 126, 267). Aphids or Plant Lice[1]

Importance and Type of Injury.—Every vegetable crop grown may be destroyed, or its commercial value ruined, by aphids. These insects are soft-bodied "plant lice," on the average about as big as pinheads, and usually green in color, although some are brown, yellowish, pinkish, or black (Fig. 296). They all feed by thrusting sharp hollow stylets, from their beaks, in among the cells of the plant and sucking out the sap. This causes the blighting of buds, dimpling of fruits, curling of leaves or appearance of discolored spots on the foliage. As the aphids become more abundant, the plants gradually wilt from loss of sap and possibly by being poisoned by the saliva of the aphids; the leaves may become yellowish or brown, and the plant usually dies.

The presence of aphids makes vegetables unattractive and detracts from their flavor and market value, and when they occur even in small numbers may occasion much work to remove them in preparing the vegetables for use. They excrete a honeydew from their intestines that gums up the plants and often serves as a medium on which a fungus may grow that further spoils vegetables. Recently it has been shown that aphids are important natural agencies in spreading certain plant diseases, such as mosaics and blights; and this phase of damage may exceed in seriousness their direct injury by feeding.

Plants Attacked.—All vegetables including those of the cabbage family, cucumbers, melons, beans, peas, potatoes, tomatoes, lettuce, turnips, spinach, and other garden crops have serious aphid pests.

Distribution.—As a group, world-wide; and many species are nearly cosmopolitan.

Life History, Appearance, and Habits.—Aphids (Fig. 296) typically winter as fertilized eggs on some perennial plant; some winter on the dead remnants of annual vegetables; while the overwintering condition of some species is not known. The eggs are small, ovate, blackish objects, glued on their sides generally to the stems of plants or in crevices about the buds (Fig. 367). When the weather becomes warm enough, small nymphs hatch from the eggs, which grow quickly to full size but never get wings. Since each of these is the start for a great colony of aphids that may be produced during the season, they are called *stem-mothers*. They are all females which have the remarkable ability to reproduce young like themselves, without mating. These young are born ovoviviparously, *i.e.*,

---

[1] Many species of the Order Homoptera, Family Aphididae.

already hatched from the egg; and differ from their stem-mothers, in having only one parent and in not passing through an exposed egg state. They are like the stem-mothers in being wingless and in producing young

FIG. 296.—Garden aphids.   (*From U.S.D.A.*)

ovoviviparously, beginning when they themselves are only a week or so old and producing from a dozen to 50 or 100 active nymphs within the next week or two.   In this way a succession of generations is produced, the young clustering about their mothers until patches on the plant may be crowded with them. At some time during this period, either all or a part of certain generations of these females may develop wings. These may fly to other plants of the same kind, or in some species they habitually fly to a different kind of plant (usually an annual), known as the *summer host.* Such winged ones are called *spring migrants.* They settle down on the new host plant and start a succession of generations there; all produced, as before, from unfertilized eggs that hatch in the body of the mother.

As shortening days forecast the end of the season, and before the summer-host plant dies, a generation is usually produced that is all winged but is often of two kinds. Some of them are winged males, the first appearance of males in the aphid colonies being at the approach of cold weather. The others are winged females which are called *fall migrants* and which may serve to return the species to the kind of perennial plant from which their distant ancestors flew away in the spring. These fall migrants give birth to nymphs in the normal manner, but the nymphs, when grown, are wingless true females that cannot reproduce unless they mate with the males which are of the preceding generation. After mating, the true female lays from one to four or more large fertilized eggs in a sheltered place about the plant, and dies; or, sometimes, simply dries up about the single egg she is capable of maturing. From these eggs arise the stem-mothers of the next spring, which differ from all the hosts of other aphids produced during the year in having both male and female parents. In some species the males and true females have no mouth parts.

This is a kind of standard life cycle. Variations from it will be noted in the discussion of particular species, but the essential features of this life cycle should be fixed in mind because the details are not repeated for the aphids discussed under particular crops.

Control Measures.—In general the most satisfactory control of aphids is obtained by the use of one of the plant poisons—nicotine, derris, cubé or timbo—applied as a spray or dust (see pages 257, 265). For certain low-growing dense crops such as melons, turnips, and the like, dusting is superior because the dust will circulate among the plants and even penetrate into curled leaves and reach the aphids in positions where a spray could not be driven. Perhaps the most effective control is by application of nicotine vaporized by heat and air blasts. Special machinery for such applications is in extensive use for control of these pests on canning crops.

References.—"The Life-cycle of Aphids," *Ann. Ento. Soc. Amer.,* **13**: 156–162, 1920; *Ann. Appl. Biol.,* **22**: 578–605, 1935; *Jour. Econ. Ento.,* **31**: 60–64, 1938.

83 (182). TARNISHED PLANT BUG[1]

Importance and Type of Injury.—This small, brownish, flattened bug is provided with piercing-sucking mouth parts with which it takes the sap of a great variety of plants. As it feeds it appears to introduce some poisonous substance into the plant. Its feeding causes various sorts of injuries; the leaves may become deformed, as in beets and chard; the stems or leaf petioles scarred and discolored, as in the "black joint" of

---

[1] *Lygus pratensis* Linné, Order Hemiptera, Family Miridae.

celery; or the buds and developing fruit dwarfed and pitted, as in the case of beans, strawberries, peaches (see Fig. 407), and pears.

Plants Attacked.—Beet, chard, celery, bean, potato, cabbage, cauliflower, turnip, salsify, cucumber, cotton, tobacco, alfalfa, many flowering plants, and most deciduous and small fruits—more than 50 economic plants, besides many weeds and grasses.

Distribution.—Throughout the United States, and in many other parts of the world.

Life History, Appearance, and Habits.—The adult bugs, and probably also the nymphs, hibernate under leaf mold, stones, bark of trees, among the leaves of such plants as clover, alfalfa, and mullein, and in many other protected places. They are about ¼ inch long by less than half as broad, flattened, oval in outline, with the small head projecting in front, and of a general brown color much mottled with small, irregular splotches of white,

Fig. 297.—Adult of the tarnished plant bug, *Lygus pratensis* Linné, about 3 times natural size. (*From Ill. State Natural History Surv.*)

Fig. 298.—Tarnished plant bug, nymph, last instar, about 3 times natural size. (*From Ill. State Natural History Surv.*)

yellow, reddish brown, and black. Along the side of the body at the posterior third is a clear-yellow triangle (the cuneus) tipped with a small, round, intensely black dot (Fig. 297). They become active very early in spring, when they attack the buds of fruit trees, causing serious injury to the terminal shoots and fruits (see page 627). They do not appear to lay their eggs on these plants to any great extent, but migrate to various herbaceous weeds, vegetables, and flowers, where the eggs are either inserted full length into the stems, petioles, or midribs of leaves or into buds, or are tucked in among the florets of the flower head. The egg is elongate, slightly curved, and the outer end is cut off squarely, the lid which covers this end being usually flush with the stem. After about 10 days, a small yellowish-green nymph ⅟₂₅ inch long, oval in outline and provided with long legs, antennae, and piercing-sucking mouth parts, emerges from the egg and begins feeding on the sap. It grows rapidly, molting five times, the larger nymphs gradually taking on the appearance of the adult and being marked with four rounded black dots on the thorax and one on the base of the abdomen (Fig. 298). The life cycle is completed in 3 or 4 weeks, so that probably three to five generations occur each season. By late summer they occur everywhere in profusion but, because of their obscure and protective coloration and shy and hiding habits, are not much noticed.

Control Measures.—No generally effective control for this pest is known. Finely ground sulphur as a spray and hydrated lime and sulphur as a dust have given fair control on celery. Dusting with pyrethrum,

3 parts, in sulphur, 7 parts, 25 pounds per acre; or hydrated lime 75 pounds per acre at 10-day intervals, beginning when vegetable plants are small; or 5 per cent nicotine applied under a canvas to confine the fumes have been recommended.   Cleaning up of weeds and destruction of favorable hibernating places may help to keep their numbers down.   Haseman recommends the use of sticky shields.   On potatoes, bordeaux mixture has been found to be somewhat repellent.

References.—*Mo. Agr. Exp. Sta. Res. Bul. 29, 1918; N.Y. (Cornell) Agr. Exp. Sta. Bul. 346, 1914; Jour. Econ. Ento.,* **25**: 671–678, 1932, and **26**: 148–150, 1933.

84. GARDEN FLEAHOPPER[1]

Importance and Type of Injury.—These small bugs look somewhat like black aphids, and they suck the sap from leaves and stems in a manner like aphids.   Small

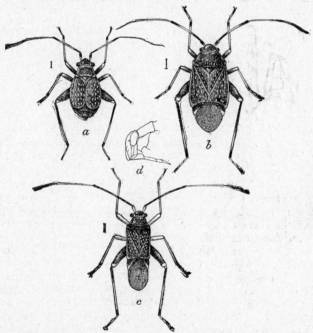

FIG. 299.—The garden fleahopper, *Halticus citri* Ashmead.   *a*, short-winged female; *b*, long-winged female; *c*, male; *d*, side view of head showing piercing-sucking mouth parts, Lines indicate natural size.   (*From Chittenden, U.S.D.A.*)

pale spots often appear on the leaves where the fleahoppers have sucked out the sap, and badly infested leaves are killed.   The insect occurs sporadically.

Plants Attacked.—Bean, beet, cabbage, celery, corn, cowpea, cucumber, eggplant, lettuce, pea, pepper, potato, pumpkin, squash, sweetpotato, tomato, various legumes, ornamentals, and many weeds.

Distribution.—General throughout the United States, except in the western part.

Life History, Appearance, and Habits.—The insect winters in the adult stage. In the spring the females insert their eggs in the leaves or stems of the plants on which they feed, the eggs being laid in punctures made by the mouth parts.   The greenish nymphs of various sizes appear on the underside of the leaves in early spring

[1] *Halticus citri* Ashmead, Order Hemiptera, Family Miridae.

and grow rapidly to blackish adults. The females (Fig. 299) are of two kinds: a long-winged form (*b*), which is $\frac{1}{12}$ inch long by about one-third as wide, nearly straight-sided, and with the overlapping tips of the wings transparent; and a short-winged form (*a*), about $\frac{1}{10}$ inch long, oval-bodied, and more than half as wide as long. The latter form lacks the transparent tips of the wings, and this, together with the jumping habit of the species, gives it a strong resemblance to a flea beetle. The legs are long and the antennae are longer than the body and very slender. Five generations have been recorded in South Carolina.

Control Measures.—A spray of 40 per cent nicotine sulphate, 1 to 700, with soap, or a 3 per cent nicotine dust, applied so as to hit all the insects, is an effective control for this pest. Destruction of the weeds on which the insect lives will do much to keep it from reaching serious numbers.

References.—*U.S.D.A., Dept. Bul.* 964, 1921; *Conn. Agr. Exp. Sta. Bul.* 344, pp. 160, 161, 1933.

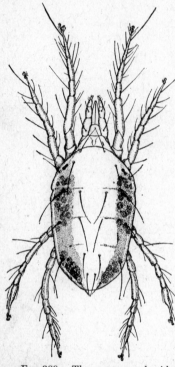

FIG. 300.—The common red spider, *Tetranychus telarius* Linné, adult female, highly magnified. (*From U.S.D.A., Farmers' Bul.* 831.)

85 (158, 269). RED SPIDER MITE[1]

Importance and Type of Injury.—In periods of dry hot weather the leaves of beans and other plants become blotched with pale yellow and reddish brown, in spots ranging from small specks to large areas, on both upper and under surfaces; the leaves have a pale sickly appearance and gradually die and drop. The under surfaces of such leaves look as though they had been very lightly dusted with fine white powder. When examined under a lens, the fine white specks are seen to consist of empty wrinkled skins and minute spherical eggs and to be suspended on almost invisible strands of silk. Upon this silk, and beneath it on the surface of the leaf, are resting or running about numerous minute, whitish, greenish, or reddish, eight-legged mites of several sizes, up to about $\frac{1}{60}$ inch long (Fig. 300). These mites live upon the sap of the plant, which is drawn by piercing the leaf with two sharp slender lances attached to the mouth. Besides the loss of sap it seems possible that the leaves are poisoned by the feeding of the mites.

Plants Attacked.—The vegetable crops most seriously injured by the red spider are beans, corn, tomato, eggplant, celery, and onion. Additional food plants and the life history are given in the discussion of this mite as a greenhouse pest, page 748.

Distribution.—World-wide.

Control Measures.—The overwintering mites may be reduced in numbers by the destruction of such weeds as pokeweed, Jerusalem oak, Jimson

---

[1] *Tetranychus telarius* Linné, Class Arachnida, Order Acarina, Family Tetranychidae. *Tetranychus pacificus* MacGregor does similar damage on the Pacific Coast.

weed, wild blackberry, and wild geranium, about the garden, and by the
destruction or spraying of violets, berry bushes, and other host plants,
wherever they retain green foliage over the winter. Once a garden is
infested, a very thorough spraying, so as to cover the underside of all
infested leaves with one of the following sprays, will check them. The
application should be repeated a week later to destroy such mites as were
in the egg stage at the first spraying. One pound of glue dissolved in 8
gallons of water makes a very cheap and effective spray. Organic thiocyanates such as Lethane are very effective. All mites covered will be
killed by a spray of potassium sulphide 1 ounce in 2 gallons of water; or by
1 to 1½ gallons, 33° Bé., lime-sulphur in 100 gallons bordeaux; or by summer-oil emulsions at a strength of 1 or 2 per cent oil; or by dusting with
superfine sulphur.

References.—U.S.D.A., Dept. Bul. 416, 1917; *Ore. Agr. Exp. Sta. Bul.* 121, 1914;
Proc. Ohio Vegetable Growers' Assoc., **19:** 80–91, 1934.

86 (15, 37, 68). WIREWORMS[1]

Importance and Type of Injury.—When garden or truck crops are
planted on land that has been in sod for several years, the wireworms
that were attracted to the grass plants remain in the soil until their
life cycle is completed and may damage the vegetables on such plots for
from 1 to 5 years. Other species breed and thrive in the intensely cultivated truck-crop areas and are not bound up with sod conditions in any
way. They feed below ground, chewing off small roots and scoring the
surface or tunneling through larger roots, tubers, and underground parts
of stems. Another serious injury is their habit of feeding on newly
planted seeds, which prevents the plants from coming up.

Plants Attacked.—The seeds of corn, beans, peas, the tubers of potatoes, and the roots of turnips, sweetpotatoes, carrots, rutabagas, radish,
sweet corn, cabbage, cucumber, tomato, onion, watermelon, many other
vegetables, and nursery stock suffer from the attacks of these pests.

Distribution.—See Corn Insects, page 377.

Life History, Appearance, and Habits.—These points are discussed
under Corn Insects (page 378). Some of the species injurious in gardens
have a 2-year life cycle, while others require at least 6 years.

Control Measures.—If a long-standing sod or other field known to be
infested is to be planted to vegetables, it should be plowed in midsummer
and thoroughly cultivated until cold weather to destroy the pupae and
adults. Some crop not seriously injured by wireworms, such as rape,
buckwheat, or clover should be planted in the soil for 2 years following
the sod. In the West certain species of wireworms deposit their eggs
by preference in cultivated land. Rotations are not effective for such
species. Cultural practices, adapted to each region and crop, are the
most practical means of control.

In small gardens, and in other cases where the value of the crop and
the labor costs do not make it prohibitive, wireworms may be trapped
by baits made of germinating seeds, such as peas, beans, or corn; of
graham or rice flour made into a stiff dough; or of slices of fresh vegetables
or bunches of fresh clover or cabbage leaves. These are placed under
boards or tiles or buried in the soil 2 to 4 inches deep, at intervals of

[1] Several species of the Order Coleoptera, Family Elateridae.

3 to 10 feet over the field, the spots being marked with wires or stakes. Cull potatoes cut in two and pressed into furrows 2 or 3 inches deep, at distances of 3 feet apart each way, attract great numbers of wireworms. They should be dug up and the worms destroyed a week after planting and a second set of baits may be set. ·Baits may be drilled into the soil to a depth of 3 inches and the wireworms that collect about them killed by drilling calcium cyanide into the same rows 10 days to 2 weeks later, to a depth of 5 inches. These treatments should be made about the time early garden vegetables are germinating, on soil plowed the preceding summer or fall and kept free of vegetation. Such fields should not be planted to any crop until the wireworms have been destroyed. Wireworms have been trapped by these means at the rate of from 6,000 to 80,000 per acre. Poisoning wireworms in the soil can be accomplished also with one of the following fumigants, used when the temperature is 70°F. or above and the soil is dry and not hard packed. The ground to be treated should be plowed deeply, smoothed, and marked off into 2-foot squares. At the intersection of the marks, holes should be punched with a stick or augur to a depth of 3 or 4 inches, and 1 fluid ounce of carbon bisulphide should be poured into each hole and the hole immediately closed by pressing the soil together. After 5 days the soil should again be plowed deeply and prepared for planting. Materials for this treatment will cost $1.50 or more per 1,000 square feet.

Crude flake naphthalene used at the rate of 800 pounds per acre or 20 pounds per 1,000 square feet, evenly mixed with soil as it is plowed to a depth of 7 to 12 inches, and subsequently worked over, will kill a large percentage of the wireworms. The cost for materials will be about one-half or one-third as much as for carbon bisulphide. The soil can be planted a week later. A combination of the above treatments has been recommended, in which paradichlorobenzene or crude chipped naphthalene, 13 ounces, is dissolved in carbon bisulphide, 1 fluid ounce, and the latter emulsified with a neutral wetting agent to make 1 gallon in water; the mixture should be applied at approximately 1 pint per large plant.

References.—U.S.D.A., Bur. Ento. Bul. 123, 1914, and Farmers' Buls. 725, and 733, 1916; and Dept. Bul. 156, 1915; Can. Dept. Agr. Pamphlet 33 (n.s.), 1923; N.Y. (Cornell) Agr. Exp. Sta. Buls. 33, 1891, and 107, 1896; Jour. Econ. Ento., 17: 562, 1924, and 19: 636, 1926; Jour. Econ. Ento., 30: 332–336, 1937; Can. Dept. Agr. Leaflet 35, 1935.

87 (14, 39, 216, 273). WHITE GRUBS[1]

Like wireworms, white grubs develop from eggs laid chiefly in sod land. When such soil is broken and planted to vegetables, the grubs seriously injure the latter and remain in the same fields 2 to 3 years before they are full-grown. If gardens are allowed to grow up to grass and weeds, the parent May beetles may lay their eggs directly in the garden. All kinds of roots may be eaten, and tubers and fleshy roots like potatoes and beets have the surface scarred and gouged with broad shallow cavities $\frac{1}{4}$ to $\frac{1}{2}$ inch deep.

Control Measures.—Vegetables should not be planted in soil that has been covered with a grass sod for years, nor in fields where white grubs are abundant. Legumes may be grown in such soil for a year or two

[1] Phyllophaga spp., Order Coleoptera, Family Scarabaeidae.

until the grubs have disappeared. In general, garden crops should not be planted in soil which was covered with grasses or small grains during a spring in which a heavy flight of May beetles occurred, since such soil is nearly certain to be infested with eggs and grubs for the next 3 years. Clover sod, or fields bearing clean-cultivated row crops during beetle flights, should be given preference for trucking during the years that immediately follow. Gardens that are surrounded by such trees as poplars, oaks, lindens, willows, ash, maple, and walnut are usually infested with white grubs every year, because the May beetles are attracted to such trees to feed and lay their eggs in the soil near by.

The distribution, life cycle, additional control measures and references for white grubs are given under Corn Insects, page 374. See also control for Wireworms, page 483.

88 (280). MILLIPEDES OR THOUSAND-LEGGED-WORMS[1]

Importance and Type of Injury.—Although not insects, these pests, which are often wrongly called wireworms, injure plants in much the same way as wireworms and white grubs. They eat the roots of various plants, tunnel into the root vegetables and tubers, eat planted seeds, and also devour leaves and bore into fruits that lie in contact with the soil.

Plants Attacked.—Corn, potatoes, parsnips, carrots, beets, turnips, radishes, lettuce, cabbage, cauliflower, beans, peas, tomatoes, muskmelons, cucumbers, squash, and others.

Distribution.—Although these pests are widely distributed, chief complaints of injury in the field have been from New York, Ohio, and neighboring states and the Pacific Coast states.

Life History, Appearance, and Habits.—The life cycles of millipedes are not well known. They have a simple development. The eggs are laid in the soil, sometimes in masses, and hatch into small worms differing from the full-grown ones in having fewer segments and legs. The adults (Fig. 301) are 1 or 2 inches long and are at once distinguished from insect larvae by the large number of legs that they possess; two pairs to each apparent body ring, the total number often being more than 100.

FIG. 301.—A typical millipede, *Julus impressus,* enlarged. (*From Ill. State Natural History Surv.*)

Control Measures.—The only control measure known to be effective is to poison or trap the worms with baits similar to those recommended for wireworms. The bait, however, should be placed on the surface of the soil in balls or small piles. Care must be taken to keep poultry away from the piles of poisoned bait.

B. INSECTS INJURIOUS TO PEAS AND BEANS

FIELD KEY FOR THE DETERMINATION OF INSECTS INJURING PEAS AND BEANS

A. *Insects eating holes in the leaves:*
 1. Irregular holes, usually extending from the margin of leaf inward, eaten by robust, brown or grayish, elongate wedge-shaped insects up to $1\frac{1}{2}$ inches long, with short or long wings and the third pair of legs very long. Jump vigorously when disturbed....................................*Grasshoppers,* page 335.
 2. A coppery-brown oval beetle, $\frac{1}{4}$ inch long, with 8 small black spots on each wing cover, eats irregular holes from underside of leaves; or oval, yellow, very spiny larvae up to $\frac{1}{3}$ inch long eat rectangular holes from underside of leaves, the holes separated by slender parallel strips of leaf that are untouched. Both

[1] *Julus impressus, Julus hortensis* Wood, *Julus hesperus* Cham., Class Diplopoda.

the adult and the larva leave the transparent upper epidermis uneaten......
....*Mexican bean beetle*, page 487.

3. Reddish to yellowish beetles, ⅙ inch long, usually with three black spots in a row along inner edge of each wing cover and costal margin also edged with black, eat rounded holes through the leaf and chew at the stems..................
....*Bean leaf beetle*, page 489.

4. A greenish or yellowish beetle ¼ inch long, with 6 black spots on each wing cover, eats irregular holes in the leaves; especially around their margins......
....*Spotted cucumber beetle*, page 490.

5. A uniformly tan-colored beetle, about ⅙ inch long, the wing covers with regular rows of small punctures, eats irregular holes in the leaves..................
....*Grape colaspis*, page 383.

6. Cinnamon-brown velvety-looking beetles, about ⅜ inch long, eat irregular holes in the leaves, especially at the margin. Much attracted to lights. Found in the eastern states.........................*Asiatic garden beetle*, page 471.

7. Very hairy, yellowish, brown or reddish-brown, and black-banded caterpillars eat the foliage.....................................*Woolly-bears*, page 476.

8. Very large holes or entire leaves eaten at night by fleshy grayish or brownish, slimy worm-like creatures, without legs or segmentation, up to 3 inches long. Hide under trash about base of plants during the day......................
....*Garden slugs* (not insects), page 760.

B. *Insects sucking sap from leaves and stems:*
   1. Foliage of beans, peas, and many other crops are silvered, bleached, wilted, and covered with tiny black dots of excrement by very slender, reddish-yellow active nymphs and grayish-black adult bugs, only $\frac{1}{25}$ inch long, with 2 white bars across the bristly front wings...
      Bean thrips, Hercothrips fasciatus (Pergande) (see *Calif. Agr. Exp. Sta. Bul.* 609, 1937).
   2. Small, greenish, long-legged, winged or wingless aphids up to ⅙ inch long with 2 slender tubes projecting from body near tip of abdomen, cluster on underside of leaves and along stems of peas and suck sap, causing plants to wither and bronzy-looking patches to appear in the field............*Pea aphid*, page 490.
   3. Similar to the above on terminal growth of beans, but not over half so large, and nearly black in color. Plants often become covered with a black "soot."
      Bean aphid, Aphis rumicis Linné (see *Ann. Appl. Biol.*, **8**: 51; **9**: 135; and **10**: 35).
   4. Shield-shaped bright-green bad-smelling flattened bugs, the winged adults about ⅝ inch long, suck sap from pods of beans and inoculate them with disease causing pods to shed and the seeds to be distorted..................
      Green stinkbug, Acrosternum hilaris Say (see *Va. Ag. Exp. Sta. Bul.* 294, 1934).
   5. Similar to B, 3, but $\frac{1}{10}$ inch long and lacking the tubes on abdomen. Jumps readily when disturbed. Wings thick, at least at base...................
      ....*Garden fleahopper*, page 481.
   6. The foliage of beans is dwarfed, crinkled, and curled, and rosettes form, or small triangular brown areas appear at tips of leaves, gradually spreading around entire margin. Few pods and few beans per pod are produced. One or more winged or wingless, elongate, wedge-shaped green bugs up to ⅛ inch long by ¼ as broad, found on lower side of leaves; run or jump quickly when disturbed.....................................*Potato leafhopper*, page 510.
   7. Leaves become blotched with red and yellow, and die. Underside of leaf with inconspicuous white webs among which are many microscopic 8-legged reddish or greenish mites and their round pearly eggs.......*Red spider mite*, page 482.

C. *Insects attacking the seeds within the growing pods in the field or living inside of the seeds in the field or in storage:*
   1. Interior of peas in the field and in storage devoured by short, white, footless grubs and chunky, brownish, white-flecked beetles, ⅕ inch long, with 2 small black spots at tip of abdomen, beyond wing covers, and a single sharp tooth near apex of hind femur. Only 1 insect to a seed...........*Pea weevil*, page 808.

2. Interior of beans and peas eaten out in the field and in storage by grubs similar to *C*,1, and by smaller brownish beetles, ⅛ inch long, with 1 large and 2 small teeth near apex of hind femur. Often several insects in 1 seed..............
....*Bean weevil*, page 810.
3. Interior of cowpeas eaten out as in *C*,1 and 2, by a more cylindrical, larger-headed grub, which is found in seeds in the field only. A black humpbacked beetle, nearly ¼ inch long, with a slender snout ⅓ as long as the body, and prominent round punctures on the back, makes punctures in the pods and lays eggs in the seeds..
Cowpea curculio, Chalcodermus aeneus Boheman (see *Fla. Agr. Exp. Sta. Bul.* 232, 1931).
4. Peas within the pod partly eaten, showing irregular superficial cavities, web-covered, and surrounded by granular pellets of excrement, among which is a white caterpillar, up to ½ inch long....................*Pea moth*, page 493.
5. Seeds of peas, beans, vetch, and locust trees attacked in the green pods by a caterpillar up to 1 inch in length, and varying from white or greenish to reddish in color..
Legume-pod moth (Etiella zinckenella Treit) (see *U.S.D.A., Bur. Ento. Bul.* 95, Part VI, p. 82, 1912).
D. *Insects cut off plants near surface of ground or feed on underground parts:*
1. Plants chewed just below surface of soil, or roots eaten off by a slender white worm, up to ⅓ inch long, with 6 short legs, and dark head and tip of body. Plants wilt and die.....................*Larvae of bean leaf beetle*, page 489.
2. Plants chewed off near the surface of the ground and sometimes dragged to burrows in the soil. Dull-green, brown, gray, or black, cylindrical, greasy-looking worms, variously striped or spotted, with distinct head, 6 short slender legs on the thorax, and 5 pairs of prolegs, up to 1½ or 2 inches long, are found in the soil about plants during the daytime..................*Cutworms*, page 475.
3. Elongate, brownish, cylindrical or somewhat flattened worms, with many legs, up to 1 or 1½ inches long, eat the roots and the foliage that rests on the soil. Distinguished from wireworms by having 1 or 2 pairs of legs on each body segment..........*Millipedes* or *thousand-legged-worms* (not insects), page 485.
E. *Plants fail to come up from seed:*
1. Seeds have been eaten into. Slender tough-skinned brownish polished-looking worms about 1 inch long, with chewing mouth parts and 6 short legs, are found in seeds or in soil near by............................*Wireworms*, page 483.
2. The germ of the softened kernel has been eaten out by one or more yellowish-white footless maggots, pointed at the head end; ¼ inch long................
....*Seed corn maggot*, page 389.

89. Mexican Bean Beetle[1]

Importance and Type of Injury.—Where it occurs, this insect is a serious enemy of all kinds of snap beans and Lima beans. It varies greatly in the same area according to weather and other factors affecting it. It may cause heavy damage one season and be very difficult to find the next year. Both larvae and adults feed on the leaves, usually on the under surface, leaving the upper surface more or less intact except as it breaks through upon drying out. The larvae eat out somewhat regular areas, leaving slender parallel strips of untouched leaf between them, giving the plants a characteristic lace-like skeletonized appearance. When abundant, the insects also attack the pods and stems, and the plants may be shredded and dried out so that they die within a month after the attack begins, often before any crop is matured.

[1] *Epilachna varivestis* Mulsant (= *E. corrupta* Mulsant), Order Coleoptera, Family Coccinellidae.

Distribution.—For three-quarters of a century the Mexican bean beetle, also known as the bean lady beetle, was a more or less serious pest only in the western part of the country, from Colorado southward. In 1920 it was discovered in northern Alabama, having been shipped there, it is believed, in alfalfa hay a year or two earlier. By 1938 it had spread to all the United States east of the Mississippi River, with the exception of Wisconsin, and into southern Ontario. Further spread is sure to occur, and in time its range will probably extend over almost the entire country.

Plants Attacked.—All kinds of garden beans and cowpeas, soybeans, and beggar-tick. It has been most injurious to garden and field beans and occasionally to soybeans and cowpeas. In cases of extreme infestation, alfalfa, clovers, vetch, and some grasses and weeds are fed upon.

Life History, Appearance, and Habits.—This insect passes the winter only in the adult stage, occurring on the ground among leaves and other rubbish, sometimes in groups, especially in woodland and hedgerows,

Fig. 302.—Mexican bean beetle; showing eggs, larvae, pupa (above), and adults (right) on bean leaves, about natural size. (*Modified from Tenn. State Board of Ento.*)

near the fields where beans were grown. The beetles (Fig. 302) are from ¼ to ⅓ inch long, very convex, short ovate in outline, and from yellow to coppery-brown in color. Each wing cover has eight small black spots that form three rows across the body when the wings are at rest. The general appearance is similar to that of our common beneficial lady beetles (see page 59), but it is larger than most of them. Some of the beetles appear in gardens and fields of beans about the time the earliest garden beans are coming up from the seed, while others continue to straggle out of winter quarters for nearly 2 months. After feeding a week or two on the beans, the adults deposit eggs on the underside of the leaves. These are about ¹⁄₂₀ inch long, orange-yellow in color, and fastened on end in close groups of 40 to 50 or more. The eggs hatch in 5 to 14 days, depending on the temperature, and the larvae feed as already described from 2 to 5 weeks. When full-grown, they are ⅓ inch or more in length by half as broad, are oval, are yellow, and have their backs protected by six rows of long branching black-tipped spines (Fig. 302). When growth is completed, the larvae cement the hinder part of their bodies to the underside of uninjured leaves of beans or other plants, often gathering in groups. The pupa (Fig. 302) pushes out of the larval skin crowding it back to the tip of the abdomen, which remains covered with this spiny wrinkled skin. The exposed part of the pupa is nearly bare, smooth, orange-yellow and rounded in front. The adult emerges in about 10 days and may lay eggs for the second generation within 2 weeks more. From egg to adult occupies a month, on the average, and in the southeastern states there are three or even four partial generations. In the western and northern states there is one generation, with a partial second. Most injury occurs in July and August. Dispersal takes place chiefly in late summer by the flight of the adults and during the autumn months, when the beetles gradually leave the plants and seek hibernating places.

Control Measures.—Any one of the following sprays or dusts, made of freshly manufactured high-grade insecticides and applied to the foliage of beans *so as to cover the underside of the leaves thoroughly* will protect the crop if applied every 10 days, beginning as soon as the beetles or their eggs are found on the plants. The dusts should be applied 12 to 15 pounds per acre when the foliage is moist; the sprays 80 to 100 gallons per acre. Recommended dusts: (*a*) cryolite, 1 part to dusting gypsum, talc, or sulphur, 3 parts; (*b*) magnesium arsenate, 1 part, hydrated lime 4 parts. Recommended sprays: (*a*) cryolite, 6 pounds to 100 gallons water or 1 ounce to each gallon; (*b*) magnesium arsenate, 3 pounds, and Kayso or skim-milk powder, 3 pounds, to 100 gallons of water; (*c*) calcium arsenate, 3 pounds in bordeaux mixture 100 gallons. All the above sprays and dusts leave a residue on the plants that may be poisonous to man and *for pod or green beans should not be used after the pods begin to form.* If the insects have not been brought under control by that time, derris or cubé root powders, containing at least 4 per cent rotenone, should be used as a dust, 1 part to 7 parts of dusting talc, clay, sulphur or tobacco; or as a spray 3 to 5 pounds in 100 gallons of water with a wetting agent. Bean foliage sprayed or dusted with the poisonous insecticides should not be used as hay.

Where the beetles are abundant, it is recommended that bush beans be grown instead of pole beans, that no larger acreage be planted than can be properly sprayed or dusted, that pods be picked promptly, and that the remnants from crops of green beans be plowed under as soon as the crop is harvested. Since damage occurs mostly during July and August, very early and late summer, quick-maturing varieties of green beans may be grown with less injury.

References.—*U.S.D.A., Farmers' Bul.* 1407, 1924, and *Dept. Bul.* 1243, 1924; *Jour. Econ. Ento.,* **21**: 178, 1928.

90. Bean Leaf Beetle[1]

Importance and Type of Injury.—Injury by the bean leaf beetle is twofold: the reddish to yellowish, dark-spotted adult beetles feed on the underside of the leaves, eating rounded holes in them and upon the stems of seedling plants at or below the soil level. The slender white larvae chew the roots and nodules, and feed on the stem just below the surface of the soil, more or less completely girdling the plant. This insect, which is said to disseminate cowpea mosaic, may cause a loss of from 10 to 50 per cent of the crop.

Plants Attacked.—Beans, peas, cowpeas, soybeans, corn, beggar-tick, bush clover, tick trefoil, quail pea, hog peanut, and other weeds.

[1] *Cerotoma trifurcata* (Förster), Order Coleoptera, Family Chrysomelidae.

Distribution.—Abundant in the southeastern states and ranging north to Kansas, Minnesota, Canada, and New York, and westward to Texas and New Mexico.

Life History, Appearance, and Habits.—Winter is passed in the adult stage, in or near bean fields of the preceding season, and the beetles are ready to attack the plants as soon as they appear above ground. The adults (Fig. 303,*a*) vary much in color and markings, but are typically reddish to yellowish in color, about $\frac{1}{5}$ to $\frac{1}{4}$ inch long, with three or four black spots in a row along the inner edge of each wing cover and a black band all around near the outer margin of the wing covers. They are found on the underside of the leaves and when disturbed generally drop. The females descend to the ground to lay their eggs in the soil about the bases of the plants. The lemon-shaped orange-colored eggs are found in small clusters of a dozen or two and each female may lay 40 or more such clusters during a period of about a month. The eggs hatch in from 1 to 3 weeks, depending upon the season, and the slender white larvae find their way to the base of the stem or roots and feed, as described above, for 3 to 6 weeks. The larvae (Fig. 303,*c*) are whitish, dark brown at both ends, conspicuously segmented, and have six very small legs near the head. When full-grown, the larva forms an earthen cell within which the white soft-bodied pupal stage (*b*) is completed in about a week. In the North these adults constitute the overwintering population. In the South there are one or two partial generations in addition.

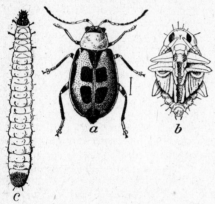

Fig. 303.—Bean leaf beetle. *a*, adult; *b*, pupa; *c*, larva. Line indicates natural size of beetle. (*From U.S.D.A., Farmers' Bul.* 856.)

Control Measures.—Where this insect has been destructive, applications of insecticides recommended for the control of the Mexican bean beetle (see page 489) should be made to the vines as soon as the beetles appear on them, to catch the females before they have laid their eggs. Especial care should be taken to cover the undersides of the leaves. The destruction of the food plants, by plowing so as to expose the roots as soon as possible after harvest, and of the wild host plants, such as beggar-tick, hog peanut, tick trefoil, and bush clover, will tend to prevent abundance of the pests. Timing the planting of the crops so that they germinate between the times of appearance of the generations of adults is recommended.

References.—U.S.D.A., Bur. Ento. Bul. 9, 1897; Jour. Econ. Ento., **8**: 261, 1915; S.C. Agr. Exp. Sta. Bul., 265, 1930; Ark. Agr. Exp. Sta. Bul., 248, 1930.

91 (17, 95). Spotted Cucumber Beetle

This general pest often eats large irregular holes in the foliage of beans and cuts off the growing tips. When the leaves are parted, the yellowish-green, black-spotted adults (Fig. 232) drop or fly away. The stems of the plants are often girdled by the feeding of many beetles at or near the surface of the ground so that the plant gradually dies. The insect is further discussed under Corn Insects, as the southern corn rootworm (page 381), and under Cucurbit Crops (page 497). A thorough application of the sprays recommended for the Mexican bean beetle will control this insect on beans.

92 (50). Pea Aphid[1]

Importance and Type of Injury.—This insect is not of special importance in small gardens, but it is most feared of all the pea insects by the

---

[1] *Illinoia pisi* (Kaltenbach), Order Homoptera, Family Aphididae.

large growers and canners.   Infested peas wilt.   Bronzy patches appear in the fields. . Upon brushing or shaking of the plants, small green insects, the largest of them only ⅙ inch long, fall to the ground.   If the plants are examined, the terminal shoot and leaves and the stems may be found crowded or lined with myriads of these green, long-legged, either winged or wingless aphids (Fig. 304), which suck the sap and possibly also poison the plant.   When the aphids are abundant, their shed skins give the plants and the ground a whitish appearance.   When the insects are not abundant enough to kill the plants or even to cut the yield greatly, the quality of the peas is often affected. The aphids are often overlooked and the injury laid to the more conspicuous insects, especially lady beetles, that are feeding upon the aphids, or to disease, such as root rot.

Distribution.—Throughout the pea- and alfalfa-growing sections of the United States and Canada.

Plants Attacked.—Garden and field peas, sweet peas, clover, sweet clover, alfalfa, and weeds of the legume family.

Life History, Appearance, and Habits.—This aphid winters on alfalfa, clovers, especially red and crimson clover, and other perennial plants.   It may go through the winter either in the egg stage or as ovoviviparous

Fig. 304.—Pea aphids on stems of clover, natural size. (*From Ill. State Natural History Surv.*)

females in the northern states and, farther south, may continue to breed all winter on these and other plants.   In the spring it increases on the winter-host plants and in the latitude of central Illinois winged migrants begin to spread to other plants, including peas, about May 1. These winged females (Fig. 305) start colonies on new plants by giving birth to active young nymphs which molt four times, reach the adult stage, and begin reproducing in about 12 days.   Each female commonly produces 6 or 7 young a day until from 50 to 100 or more have been born.   There are from 7 to 20 or more such generations of females in the course of a year.   When unchecked by natural enemies their increase is phenomenal, but they are much subject to fluctuations due to weather conditions, the attacks of many predators and parasites, and fungous diseases.   They usually become most abundant and injurious to peas during June, in the latitude of Illinois.   Most of the adults are wingless, but when they become crowded on the plant some winged ones appear, and these spread the insect widely. During midsummer they are found

chiefly on other leguminous plants but in early fall again become abundant on peas wherever these are available.

As the amount of daylight diminishes in the fall and the temperature drops, the ovoviviparous females give birth to young, some of which become sexually mature, egg-laying, wingless females and others winged males. The eggs are laid chiefly on the leaves and stems of alfalfa and red clover and are about $\frac{1}{30}$ inch long, light green in color when newly laid but turning a shiny black. These fertilized

Fig. 305.—Winged, ovoviviparous female adult of the pea aphid. (*From Ill. State Natural History Surv.*)

eggs live over winter and give rise in the following spring to the ovoviviparous stem-mothers which start the next season's infestations.

Control Measures.—The most satisfactory methods for the control of pea aphids are (*a*) dusting with ground derris, cubé, or timbo root in a carrier of talc, gypsum, or other suitable fine dust. For large fields the

Fig. 306.—Four-row traction duster equipped with "trailer" to hold the fumes from nicotine dust. (*From U.S.D.A., Farmers' Bul.* 1499.)

dust should be applied under at least a 25-foot canvas trailer, at the rate of 35 to 40 pounds per acre, with the dusting machine moving at not over 3 miles per hour. (*b*) Spraying with 3 pounds ground derris, cubé, or timbo root, containing not less than 4 per cent rotenone, in 100 gallons of water, and applied at from 125 to 200 gallons per acre. A nicotine vaporizer has been developed which is very effective, but at present is mainly adapted for large field operations. For truck gardens a nicotine dust may be used. The following formula is very effective:

Hydrated lime... 25 lb.
Monohydrated copper sulphate........................... 1½ lb.
40 per cent nicotine sulphate........................... 1 qt.

References.—U.S.D.A., Dept. Bul. 276, 1915; *Md. Agr. Exp. Sta. Bul.* 261, 1924.

93. Pea Moth[1]

Importance and Type of Injury.—This insect is a serious pest in the more northern areas of canning peas, dried peas, and seed peas industries. Growing peas within the pod have irregular cavities eaten out of the side, seldom exceeding half of their substance, by a yellowish-white caterpillar up to ½ inch long with both extremities of the body darker, and small dark spots and pale short hairs scattered over the body. The presence of the worms is not easily detected except by opening the pods, when the seeds are seen to be spoiled and the pod partially filled with the pellets of excrement and silk of the caterpillars (Fig. 307). Affected pods yellow, or ripen prematurely.

Plants Attacked.—This insect attacks all varieties of field and garden peas, sweet peas, and vetch.

Distribution.—The insect has been present in America only since about 1900, and has been recorded from New York, eastern Canada, Michigan, Wisconsin, Washington, and British Columbia.

Life History, Appearance, and Habits.— Winter is passed as inactive larvae enclosed by strong cocoons of fine silk, about ⅜ inch long and covered with soil particles. These are found a short distance below the surface of the soil in the fields. Others winter in similar cocoons in cracks and crevices about the barns where the peas were stored before threshing. They change to brownish pupae in late spring.

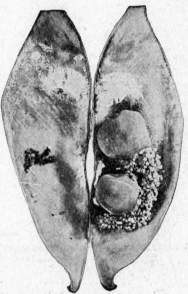

Fig. 307.—Work of the pea moth. (*From Wis. Agr. Exp. Sta. Bul.* 310.)

About the time peas come into blossom (latter half of July in northern Wisconsin) the adults appear from the cocoons. They are frail, very small, day-active moths, about ½ inch across the spread wings, of a general brown color with short black-and-white oblique lines along the front margin of the forewings. They may be found zigzagging about the plants in late afternoon, mating and laying eggs, but not themselves injuring the peas. The eggs are white, flattened, and a little smaller than an ordinary pinhead. They are deposited one in a place upon the pods, leaves, flowers, or stems of the peas, and on other plants growing in the pea fields. Promptly upon hatching, the minute caterpillars drill into the pods, casting out a little frass, which however soon blows or wears away, leaving almost no indication of their entrance. The larvae devour parts of several seeds, which are further contaminated with their excrement and silk. Within 2 to 4 weeks they are full-grown, when they eat out of the pods and enter the soil or other protected places mostly during August. Here they form cocoons in which some of the larvae may transform to moths and emerge

---

[1] *Laspeyresia* (= *Grapholitha*) *nigricana* Stephens, Order Lepidoptera, Family Tortricidae.

2 to 12 weeks later; but the majority of them remain in the cocoons as larvae during the next 10 or 11 months.

Control Measures.—Seed and dried peas or vetch should not be grown in the same areas where canning peas are an important industry. Other measures recommended include: threshing promptly within a day or two of harvesting, so that larvae still in the pods might be killed by the thresher; burning all remnants of the crop immediately; thorough disking of the soil after harvesting; deep plowing in the fall to destroy the larvae in their winter nests; and the use of early varieties, maturing before mid-July, in Wisconsin.

References.—*Wis. Agr. Exp. Sta. Bul.* 310, 1920; *Proc. Ento. Soc. Nova Scotia for* 1919, pp. 11–20, 1920; *Mich. Agr. Exp. Sta. Rept.* 44, pp. 266–268, 1931; *Wash. Agr. Exp. Sta. Bul.* 327, 1936.

C. INSECTS INJURIOUS TO CUCURBITS

FIELD KEY FOR THE DETERMINATION OF INSECTS INJURING CUCUMBER, SQUASH, MELONS, AND OTHER CUCURBITS

A. *Insects eating holes in the leaves or chewing at the stems:*
1. Small beetles, about ⅕ inch long, with three black stripes down the back, separated by wider stripes of bright-yellow, eat irregular holes in the leaves and feed especially about the base of the stem, often girdling the plant at or near the surface of the ground.................*Striped cucumber beetle*, page 495.
2. Beetles of the same general shape but slightly larger than *A*,1, green or greenish-yellow in color and with 6 conspicuous black spots on each wing cover, attack especially the leaves and flowers of the plants..............................
....*Spotted cucumber beetle*, page 497.
3. Leaves have small rounded holes eaten into them so that they look as though peppered with fine shot. Very small black beetles about 1⁄16 inch long, or larger ones ⅛ inch long, with a broad yellow longitudinal stripe on each wing cover, are found mostly on underside of leaves, jumping vigorously when disturbed...*Flea beetles*, page 468.
4. Stems of young plants are chewed or cut off near or below the ground, by dull-greenish, brownish, or grayish, cylindrical, "greasy-looking" worms, variously marked with spots, stripes, or oblique bands; the largest ones 1½ inches long; found in the soil during the day; often coil up when disturbed..............
....*Cutworms*, page 475.
B. *Insects sucking sap from the leaves or stems, causing wilting, curling, and dying of leaves:*
1. Elongate, flattened, oval, blackish-brown bugs, ⅔ inch long, and powdery-white black-legged nymphs, from 3⁄16 to ½ inch long, hiding under wilted leaves, clods, etc., or shying about the vines, suck the sap and poison the plants. Have a stinking odor when crushed...................*Squash bug*, page 499.
2. Developing leaves at tips of vines are curled, wilted, and shriveled, the undersides thickly dotted with small, winged or wingless, tan, brown, green, or black aphids of various sizes, which suck out the sap........*Melon aphid*, page 497.
C. *Insects boring through the center of the vines causing extensive wilting and rotting of the stem:*
1. Yellowish sawdust-like excrement accumulates in small masses from holes in the vines. Within the vine are found whitish grub-like caterpillars up to 1 inch long and ¼ inch in diameter, with brown heads and short legs........
....*Squash borer*, page 502.
2. Injury similar to *C*,1, especially late in the season, caused by a smaller and slenderer green worm, not over ¾ inch long, the smaller ones with a row of black spots across each segment......................*Pickleworm*, page 504.
D. *Insects boring into the fruits:*
1. Greenish to yellowish, brown-headed caterpillars, up to ¾ inch long, those under ½ inch with conspicuous cross rows of blackish spots, bore into fruits of

muskmelon, cucumber, and squash, pushing out a small mass of excrement behind them and causing the fruits to rot and sour.....*Pickleworm*, page 504.

2. Injury very similar to *D*,1 is caused by a larger, more slender worm, ranging up to 1¼ inches long, greenish-yellow in color, with light-brown head and 2 long, whitish well-separated stripes down the back. They wriggle very actively when disturbed...................................*Melonworm*, page 505.

3. Plain green or greenish black-spotted, or yellow and black-striped beetles, about ⅕ inch long, gnaw into the rind of squashes, pumpkins, melons, and related fruits in the fall of the year, often working together in great numbers. They disfigure the fruits but seldom penetrate deeply enough to cause rotting or fermentation..
   Striped and spotted cucumber beetles and northern corn rootworm, pages 495 and 380.

E. *Insects attacking the roots and boring in the underground stem, or devouring the planted seeds:*

1. Very slender whitish worms, the largest ⅓ inch long, with the two extremities brown, with 6 short legs behind the head and a pair of blunt prolegs on the last segment, devour the smaller roots and tunnel through the underground stems and larger roots.................Larvae of *striped cucumber beetle*, page 495.

2. Thick white soft-bodied grubs, up to an inch or more long, with reddish-brown heads, 6 slender legs, and body somewhat slenderer near the middle, chew the roots. Lie curled in a "U" or semicircle.............*White grubs*, page 484.

3. Slender, tough, smooth, brown worms up to 1 inch long, with chewing jaws and 6 small legs, eat into seed before or as it germinates and later chew at the roots and underground parts of the stem..............*Wireworms*, page 483.

94. STRIPED CUCUMBER BEETLE[1]

Importance and Type of Injury.—This is the most serious pest of cucurbits throughout America east of the Rocky Mountains. On the Pacific Coast, a distinct but very similar species, the western striped cucumber beetle,[2] is the cause of a similar injury and may be controlled in the same way. The beetles feed on the plants from the moment they appear above ground in the spring until the last remnants of the crop are removed or destroyed by frost. They work down to meet the germinating plants before they reach the surface of the soil. They chew the leaves and tender shoots, and especially the stem near or below the surface, partially or completely girdling it. They feed in the blossoms and, in autumn especially, gnaw holes in the rind of the fruits. They are known carriers of the bacterial wilt of cucurbits, the bacillus[3] causing this disease living over winter in the intestines of this and the related spotted cucumber beetle. In the spring the beetles inoculate the disease into the interior tissues of the new plants as they feed, and spread it from plant to plant and field to field wherever infective beetles go and feed. This insect is also one of the most important agencies in the spread of cucumber mosaic. Furthermore, the larvae of the beetle injure the vines during the summer by devouring the roots and tunneling through the underground parts of the stems. Many plants are killed early in the season by this beetle and the wilt disease it spreads, and such vines as survive the first attack have the yield greatly reduced by the work of the adults and larvae.

---

[1] *Diabrotica vittata* (Fabricius), Order Coleoptera, Family Chrysomelidae.
[2] *Diabrotica trivittata* (Mannerheim).
[3] *Bacillus tracheiphilus* Erw. Sm. (see *U.S.D.A., Dept. Bul.* 828, 1920).

Plants Attacked.—Cucumbers, muskmelons, winter squashes, pumpkins, gourds, summer squashes, and watermelons appear to be injured about in the order named. So far as known the larvae can develop only on these and related cucurbits. The beetles, however, also feed on beans, peas, and corn and the blossoms of many other cultivated and wild plants. This pest is not troublesome in sandy soils.

Distribution.—This native insect ranges from Mexico into Canada, east of the Rocky Mountains.

Life History, Appearance, and Habits.—Only the unmated adults live through the winter. They are ordinarily not found in old cucurbit patches, but in neighboring woodlands, under fallen leaves, strips of bark, or rotten logs, or under protecting trash in lowlands, hedgerows, or weedy fence rows, practically always in direct contact with the soil, near the previous crop or about their wild food plants, such as goldenrod and asters. The beetles emerge from hibernation from early April on, in central Illinois, becoming active at temperatures above 55°F., but not taking flight below 60°F. Before cucurbits are available for them to feed upon, they subsist on the pollen, petals, and leaves of buckeye, willow, wild plum, hawthorn, thorn apple,

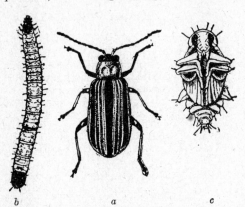

Fig. 308.—The striped cucumber beetle. *a*, adult; *b*, larva; *c*, pupa; about 5 times natural size. (*After Chittenden, U.S.D.A.*)

apple, elm, syringa, and related plants for several weeks. As soon as cucumber, squash, or melon vines appear above ground, they come to these plants, settling upon the young vines and devouring the seed leaves and the stems. The beetles (Fig. 308,*a*) are familiar to nearly every grower. They measure about ⅕ inch long by 1/10 inch wide. The upper surface is about equally black and yellow, the folded wing covers forming three longitudinal black stripes. Mating and egg-laying take place soon after the beetles migrate to the vine crops. The orange-yellow eggs are laid about the base of these plants, often below the surface of the soil or in cracks in the ground. The larvae (*b*) that hatch from them work their way into the soil and feed for from 2 to 6 weeks on the roots and underground parts of the stem, frequently entirely destroying the root system. When full-grown, they are about ⅓ inch long by one-tenth as broad. The last segment of the abdomen is more flattened than in the related northern and southern corn rootworms. The whitish pupal stage (*c*) is also found in the soil and lasts for about a week. Adults from this first generation appear in midsummer over a period of 6 weeks or more and feed extensively on stems, leaves, and blossoms of cucurbits and also on legumes. In the warmer latitudes these beetles soon mate and produce another generation during late summer and early fall, and the adults from these and even from partial third or fourth generations are the ones found in late fall on goldenrod, sunflower, and aster blossoms and on the fruits of cucurbits. In the northern part of its range there is only one generation.

Control Measures.—This pest is hard to control because of the under-ground position of the larvae, the habit of the adults of attacking the plant low down or even below ground, and the fact that the adults detect arsenicals on the plants and seek out places that are not covered by these poisons, on which to feed. A dust of 1 part calcium arsenate and 9 parts burned gypsum, $CaSO_4 \cdot 2H_2O$, has given very good results in control of this insect. The dust must be thoroughly mixed and applied, once or twice a week during the first month of growth of the crop, to keep the plants covered from the time they first appear above ground until the attack by the first-generation beetles is over. Other insecticides that have given excellent control are: cryolite, 1 part, dusting gypsum, 3 parts; talc or gypsum containing 1 per cent rotenone; or sprays of 40 per cent nicotine sulphate, 1 to 800 of water containing Penetrol, 1 to 200, as an activator.

Where the acreage is small enough to make it practicable, the plants may be covered, from the moment they appear through the ground, with wire- or cloth-screen protectors made in the form of cones or hemispheres. Such protectors keep the beetles off until the plants get a good start, when they must be removed. It is well to plant an excess of seed, and thin out after the plants are started.

References.—Ohio Agr. Exp. Sta. Bul. 388, 1925; *U.S.D.A., Farmers' Bul.* 1322, 1923; *Wis. Agr. Exp. Sta. Bul.* 355, 1923.

95 (17, 91). SPOTTED CUCUMBER BEETLE[1]

This species, belonging in the same genus as the striped cucumber beetle and the northern corn rootworm, is a much more general feeder than either. While the total damage done by it, because it also injures field crops, is probably greater than either of the others, its injury to cucurbits is much less noticeable. This injury is similar to that of the striped cucumber beetle. The larva of this insect is the well-known southern corn rootworm. In addition, the adult is a constant pest of string and Lima beans, peas, potato, beet, asparagus, eggplant, tomato, cabbage, and many other garden plants, and the larvae develop on the roots of corn, beans, small grains, alfalfa, and many wild grasses. A fuller description of the insect is given under Corn Insects, Southern Corn Rootworm, page 381. On cucurbits it may be controlled by the same methods as recommended for the striped cucumber beetle.

96 (60). MELON APHID[2]

Importance and Type of Injury.—After the vines of cucurbits begin to "run," a single hill here and there will often be found to have the edges of the leaves curled downward or some of them wilted, shriveled, and browning (Fig. 309). The undersides of such leaves, inside the curl, are generally crowded with very small, yellow, green, and black aphids, some winged, others wingless, and of several sizes. These insects

---

[1] *Diabrotica duodecimpunctata* Olivier, Order Coleoptera, Family Chrysomelidae. A similar species, the Western spotted cucumber beetle, *D. soror* LeConte, replaces this species west of the Rocky Mountains.

[2] *Aphis gossypii* Glover, Order Homoptera, Family Aphididae.

suck the sap from the leaves, weakening the plants and reducing both the quantity and quality of the fruit.   In years of abundance they kill the plants and ruin the crops over extensive areas.   In the North they are especially destructive in hot dry summers, following cool wet springs which have reduced the efficiency of their natural enemies.   This has been considered the most destructive aphid occurring in this country.

Fig. 309.—Tip of a melon vine badly infested with melon aphids.   (*From Chittenden, U.S.D.A.*)

Plants Attacked.—This insect feeds on a wide variety of plants but is of economic importance chiefly on the cucurbits, cotton (on which it is a very serious pest), okra, and citrus fruits.   It also feeds on strawberry, bean, beet, spinach, eggplant, asparagus, a number of ornamental plants, and many weeds, especially shepherd's-purse, peppergrass, pigweed, and dock.   It also occurs on certain of these plants in the greenhouse.

Distribution.—Throughout the United States, southward to Central and South America; most destructive in the South and Southwest.

Life History, Appearance, and Habits.—In the North the insect winters on live-forever, in the egg stage, where this plant is found.   These eggs are fertilized in the fall by males that developed on the same plant. In the extreme South, at least, the insect continues breeding ovoviviparously throughout the winter. It appears on cucurbits in late spring or early summer, and, if the weather is favorable and it is not checked by sprays or natural enemies, increases and spreads with astonishing rapidity. Fifty-one generations were recorded in 12 months in Texas, the average young per female being more than 80. Many winged individuals are produced, and wherever they fly or are blown by the wind a new colony is soon started.

Control Measures.—Since the attack on cucurbits commonly begins in small scattered spots over the field, such spots should be watched for and the aphids destroyed upon them before they become generally established over the whole crop. Besides the wilting and curled leaves, attention may be attracted to the aphids early by the visits of ants, bees, wasps, and flies to the colonies to get the honeydew and by the white cast skins of the aphids sticking to the leaves. Every effort should be made to apply control measures before the leaves on the terminal shoots curl too badly. A 3 per cent nicotine dust, made according to the directions given on page 259 is a very effective insecticide for these aphids on the vine crops. Dusting is much more effective if the plants are covered with a piece of canvas or tarpaulin beneath which the dust is applied. The tarpaulin should be left in place for 3 to 5 minutes. Nicotine sulphate 40 per cent, used 1 to 800 in soapy water, kills all the aphids wetted by it, but it does not cover so well as the dust, which may penetrate even to the interior of curled leaves. Whenever a bordeaux mixture is being applied for fungus diseases, 1 pint of nicotine sulphate may be added to each 100 gallons of this spray to destroy these aphids if they are present.

References.—*U.S.D.A., Farmers' Bul.* 1499, 1926; *Tex. Agr. Exp. Sta. Bul.* 257, 1919; *Ill. Agr. Exp. Sta. Cir.*, 297, 1928; *U.S.D.A., Farmers' Bul.* 1282, 1922; *Maine Agr. Exp. Sta. Bul.* 326, 1925.

97. SQUASH BUG[1]

Importance and Type of Injury.—There is no more vexatious pest of the garden than the squash bug. Leaves of plants attacked by it rapidly wilt as though the sap flow had been cut off or poisoned, and soon become blackened, crisp, and dead. Small plants may be killed entirely; larger plants usually show certain leaves or runners that are affected. In many localities it is practically impossible to grow certain varieties of squashes, as the plants are killed by these bugs before any fruits are matured. The bugs possess a remarkable vitality and tenacity of life, to the great annoyance of the gardener. Plants in farm or city gardens suffer most severely although large commercial plantings also are badly injured.

The attack of this pest may be identified by finding the brownish-black, flat-backed adult bugs, about ⅝ inch long (Fig. 311,*a*), and their numerous, whitish, black-legged nymphs (*c,d,e*), which range in size from ³⁄₁₆ to ½ inch long. They are usually more or less hidden about the base of the plant under the deadened leaves or under clods, and, when approached, they shy around the vines or walk rapidly to cover.

Plants Attacked.—All the cucurbits or vine crops are attacked, but the bugs show a marked preference for squashes and pumpkins. Among

[1] *Anasa tristis* De Geer, Order Hemiptera, Family Coreidae.

the squashes, the winter varieties, such as hubbards and marrows, suffer most severely.

Distribution.—Throughout the whole United States from Central America to Canada.

Life History, Appearance, and Habits.—Only the unmated adult bugs pass the winter. They hibernate in all kinds of shelter under the protection of dead vines, leaves, clods, stones, piles of boards, outbuildings, and dwellings which they enter in the fall. They are rather slow to appear in spring and feed only on cucurbits. By the time vines begin to "run," the adults fly into the fields and gardens and apparently locate their food by the odor of the plants. Mating occurs in the spring and egg-laying begins soon afterward. The egg clusters (Fig. 311,*b*) will usually be found on

Fig. 310.—Squash bug adults and nymphs massed upon cucurbit fruit in fall after foliage has been killed by frost; about ⅓ natural size. (*Original.*)

the underside of leaves in the angle formed by the veins. The eggs are yellowish-brown to very dark bronzy-brown according to their age, elliptical in outline, and about 1/16 inch long. They are laid in groups commonly numbering a few dozen. The eggs lie on their sides, sometimes close together, sometimes separated by more than their own diameter. They are usually placed in rows in two directions, the rows meeting each other at an acute angle. The overwintering adults live and continue laying eggs until about midsummer.

A week or two after the eggs are laid, the small nymphs hatch. They are at first strikingly colored, the abdomen being green, the head, thorax, antennae and legs crimson, soon darkening to reddish-brown. The older nymphs are grayish-white in color, with nearly black legs and antennae.

There are five nymphal instars. When the insects are crushed, a disagreeable odor is given off from two oval spots on the middle of the upper side of the abdomen; or, in the adults, from near the base of the legs. This gives rise to the name "stink bug" often wrongly applied to these insects.[1] Nymphs and adults all feed by sucking the sap of the plant. The youngest nymphs are gregarious, those from one egg cluster feeding close together.

From 1½ to 2 months after the eggs were laid, the new bugs begin to transform to adults, there being usually a period of scarcity of mature bugs following the disappearance of the parents, before the young have become adult. The new adults do not mate or lay eggs until the following spring, at least in the North, and there is probably only one generation a year throughout the country. These adults and the nymphs from later

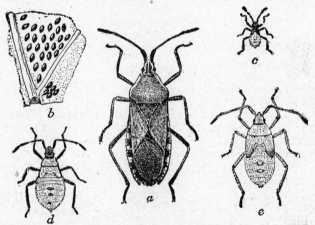

Fig. 311.—Life stages of the squash bug. *a*, adult; *b*, eggs (about natural size); *c*, *d*, and *e*, nymphs in different stages. *a*, *c*, *d*, and *e* twice natural size. (*From Chittenden, U.S.D.A.*)

laid eggs are usually present in great numbers when the first frosts kill the leaves and vines. The bugs then collect in dense groups upon the protected or sunny faces of unripe fruits, where they continue to suck the sap. They gradually fly and crawl away in search of winter shelter, and such nymphs as do not succeed in reaching the adult condition die during the early winter.

Control Measures.—No very satisfactory control measure for squash bugs has been discovered. The adult bugs are so resistant to contact sprays and poison gases that their destruction by insecticides is very difficult. A spray of kerosene extract of pyrethrum containing 2.15 per cent pyrethrins, 1 pint, carefully emulsified with potash fish-oil soap, 1 quart, and diluted with water to make 100 gallons will kill most of the nymphs hit by it. Cloudy murky days when the evaporating power of the air is very low should be chosen for spraying whenever possible as the insecticide is more effective under such conditions. Since the striped cucumber beetle will generally be present about the same plants, a dust of pyrethrum, 2 parts, calcium arsenate, 1 part, and gypsum, 9 parts, applied

[1] The true stink bugs are the shield-shaped bugs of the related family Pentatomidae.

5 to 25 pounds per acre, depending upon the size of the plants, gives fair control of both pests. If these treatments are repeated three times at 10-day intervals, nearly all the insects can be caught as the susceptible early-instar nymphs. One of the most effective controls known is to collect the bugs by hand and to crush the egg masses as fast as they appear on the young plants in spring. Pieces of boards, shingles, and similar flat objects are often placed out among the plants. The bugs collect under them at night and if they are examined each morning many may be destroyed. In the fall great numbers of the bugs can be caught and killed on the fruits after the vines have died. As soon as the crop is harvested, the vines should be removed from the field and burned or worked into a compost heap. The margins of fields and the grounds surrounding the garden should be as free as possible of rubbish, piles of leaves, boards, and other shelter for the bugs during the winter.

References.—N.H. Agr. Exp. Sta. Bul. 89, 1902; Jour. Econ. Ento., **13**: 416–425, 1919, and **16**: 73, 1923; Thirty-fourth Rept. Conn. State Ento. Bul. 368, pp. 224–231, 1935; Okla. Agr. Exp. Sta. Bul. 34, pp. 269–271, 1935.

98. SQUASH BORER[1]

Importance and Type of Injury.—This pernicious borer is in many localities a very destructive pest of squashes and pumpkins, sometimes destroying 25 per cent or more of the crop. Attention will often first be called to it by the sudden wilting of a long runner or an entire plant. Examination of the wilted vine will reveal masses of coarse greenish-yellow excrement which the borer has pushed out of the vine through holes in the side. If such a vine is split open, it will be found hollowed out and partly filled with moist slimy frass similar to that cast out on the ground. In the midst of this material is a thick, white, wrinkled, brown-headed caterpillar, the largest ones 1 inch long and almost $\frac{1}{4}$ inch thick (Fig. 312,d). Plants injured to the point of wilting usually contain a number of borers. Their excrement should be watched for, especially about the base of the plant close to the roots. Infested vines are often completely girdled and usually become rotten and die beyond the point of attack.

Plants Attacked.—Squashes, such as hubbards, marrows, cymblings, and other late varieties, summer squash, pumpkins, gourds, cucumbers, and muskmelons are injured, about in the order named. The wild cucumber also is infested.

Distribution.—East of the Rocky Mountains from Canada to South America.

Life History, Appearance, and Habits.—The insect winters an inch or two below the surface of the soil inside of a tough dirt-covered silk-lined black cocoon (Fig. 312,e) about $\frac{3}{4}$ inch long, in either the larval or pupal stage. If as a larva, the change to the mahogany-brown pupa (f) occurs in the spring. Two or three weeks later, about the time the vine crops are beginning to run, the pupa tears open the end of its cocoon and wriggles up through the soil to the surface. Its skin then splits down the back and reveals a beautiful wasplike moth, 1 to $1\frac{1}{2}$ inches across the wings (Fig. 312,a,b). The front wings are covered with metallic-shining olive-brown scales, but the hind wings are transparent. The abdomen is ringed with red, black, and copper; and, like its relative the peach borer

[1] *Melittia satyriniformis* Hübner, Order Lepidoptera, Family Aegeriidae.

(page 629), the moth flies swiftly and noisily about the plants during the daytime, seeming more like a wasp than a moth. Small, oval, somewhat flattened, brownish eggs, $\frac{1}{25}$ inch long, are glued one in a place on the stems and leaf stalks, especially toward the base of the plant. The small borers enter the stem a week or two later and tunnel along, eating out the inner tissues for about a month. They have a brownish head, six short slender legs on the thorax, and five pairs of short prolegs. Each proleg bears two transverse rows of crochets. Small borers may often be found in the leaf stems, but most of them occur toward the base of the plant. Later in the season they are often found throughout the stem, and even in the fruits. The larvae are full fed in 4 to 6 weeks. They then desert

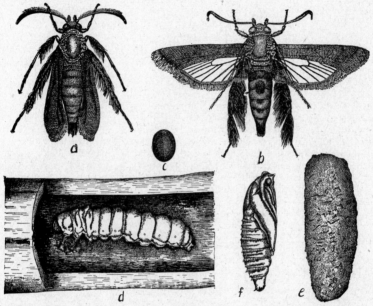

FIG. 312.—The squash vine borer. *a*, male; *b*, female; *c*, egg; *d*, larva in stem; *e*, earth-covered cocoon; *f*, pupa; about twice natural size. (*From Iowa Agr. Exp. Sta. Cir.* 90.)

their burrows and make cocoons in the soil. In the more northern states the larvae remain in these cocoons until spring before pupating, there being only one generation a year. Farther South, at least a part of the larvae pupate promptly after leaving the plants, and some of these change to adults, giving rise to a second generation of larvae during August and September. In the Gulf states, two generations are believed to be the rule.

Control Measures.—Like most borers, this insect presents great difficulties in control since no insecticides can reach the point of feeding. The following sprays have been recommended as killing many of the young insects before they get into the stems: (*a*) 40 per cent nicotine sulphate 1 part in 100 parts of water with $\frac{1}{2}$ part potash fish-oil soap; (*b*) lead arsenate, 3 pounds, and fish oil, 1 quart, in 100 gallons of water; (*c*) poisoned bordeaux, three or four sprays at weekly intervals, directed chiefly upon the vines within 4 feet of the roots, may be required to give control; (*d*) a mixture of ground derris 20 per cent, sulphur 25 per cent, and dusting clay 55 per cent, has given a 74 per cent increase in yield under New Jersey conditions. It was used at the rate of 30 to 65 pounds per acre for three applications. The best thing to do with infested plants is to slit the stems lengthwise at the point of attack, crushing or removing the borers and immediately covering the stems with moist earth. Since girdling is most likely to occur at the base of the plant, it is well to cover the vines with soil a few feet from the base so that they may form supplementary roots and save the vine in case it is cut off at its base. To reduce injury the following year, all vines should be raked together and burned as soon as the crop can be harvested. The soil should be harrowed in the fall and turned under deeply in the spring to prevent the emergence of adults from the cocoons.

References.—*U.S.D.A., Farmers' Bul.* 668, 1915; *Mass. Agr. Exp. Sta. Bul.* 218, pp. 70–80, 1923; *Conn. Agr. Exp. Sta. Bul.* 328, 1931; *Jour. Econ. Ento.*, **28**: 229–331, 1935.

99. PICKLEWORM[1]

Importance and Type of Injury.—Throughout the Gulf states, ripening fruits of muskmelons, cucumbers, and squashes are bored into, from the side next the ground, by a whitish to greenish caterpillar, up to ¾ inch long, brownish at the head end and the smaller ones with a transverse row of black spots on each segment. The larvae push out small masses of green sawdust-like excrement from their holes in the fruit. The fruits soon rot, sour, and mold after the interior has been exposed to the air by these burrows. Earlier in the season larvae work in the stems, terminal buds, and especially in the blossoms of squash. Late maturing crops are frequently almost totally destroyed by these worms.

Plants Attacked.—Muskmelon, cucumber, and squash, are seriously injured; watermelon rarely; and pumpkin not at all.

Distribution.—Especially destructive in the southern states but occasionally so, as far north as Missouri, Illinois, Michigan, and New York. The insect is found from Canada to South America.

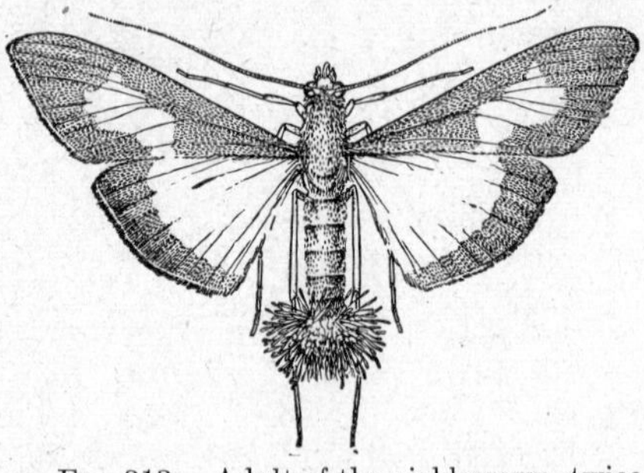

FIG. 313.—Adult of the pickleworm, twice natural size. (*From Crosby and Leonard, "Manual of Vegetable-garden Insects," copyright,* 1918, *by Macmillan. Reprinted by permission.*)

Life History, Appearance, and Habits. The insect hibernates in the pupal stage surrounded by a thin cocoon of silk and usually in a roll of leaf from the food plant. The cocoons generally lie on the ground in or near the old food plants but are sometimes suspended on weeds and other plants nearby. A very small percentage of them survive the winter and the spring cultivation of the soil; these give rise in late spring (early June in North Carolina) to striking-looking moths a little more than 1 inch wide across the spread wings (Fig. 313). The two pairs of wings are margined with a band of yellowish-brown, about ⅛ inch wide, and the body above is of the same color, with purplish reflections in certain lights; while a median spot on the front wings and the basal two-thirds of the hind wings are transparent yellowish-white. The tip of the abdomen has a prominent rounded brush of long hair-like scales. The moths fly late at night and eggs are deposited in clusters of two to seven on the tender buds, new

[1] *Diaphania nitidalis* (Stoll), Order Lepidoptera, Family Pyralididae.

leaves, underside of fruits, or stems. The larvae work at first in the buds, blossoms, and tender terminals, and some of them complete their growth in the vegetative part of the plant. Many, however, begin to wander about when they are partly grown and find their way to the fruits. Several fruits, especially of muskmelon, may be entered and ruined by a single caterpillar before its growth is completed. When full-grown they are greenish or coppery all over, except for the brown head and brown area just behind the head, although younger larvae are conspicuously marked with about 100 black spots evenly scattered over the body. After feeding for about 2 weeks, they desert their burrows, roll a leaf about themselves and spend the next week or 10 days in the pupal stage inside of a thin cocoon. The first generation is few in numbers, but the moths emerging from these pupae in early July lay sufficient eggs so that the second generation may do some damage to early muskmelons. It is, however, during August and September that the really severe injury by this insect occurs, probably by the third- or fourth-generation larvae. Muskmelons harvested before early July or mid-July are rarely injured. Four or five weeks are required for a generation, and there are four full generations and a partial fifth in North Carolina.

Control Measures.—After the worms begin to attack a crop, there is no known means of saving it. A few fruits can be enclosed in paper bags or placed on chips to avoid the worms, but for the commercial crop, protective measures must be taken during the preceding fall and spring. As soon as a crop is harvested, the vines and unused fruits and adjoining weeds and trash should be collected together and burned or converted into compost, to destroy the worms in them. Early in the fall the fields should be plowed in order to bury the pupae that have fallen to the ground. Every effort should be made to get the crop matured early, since the early cantaloupes and cucumbers almost always escape. To protect muskmelons and cucumbers, early squash may be used as a trap crop. Four to eight rows of squash to the acre, should be planted at two-week intervals, the first when the main crop is planted, in order that there may be an abundance of fresh squash blossoms throughout the season. The moths prefer to lay their eggs on squash, and the worms feeding in squash do not wander from fruit to fruit as those on other crops. Before the worms become full-grown in the squash blossoms, *either the infested blossoms must be picked off and destroyed or else the entire vines removed and destroyed as the later planted ones come into bloom.*

Muskmelons may be completely protected by this means if the trapped worms in the squash plants are destroyed regularly. The flavor of the fruits is not affected by growing squash among them, but the seeds from such fruit cannot be used.

Reference.—*N.C. Agr. Exp. Sta. Bul.* 214, 1911.

100. MELONWORM[1]

The melonworm is a close relative of the pickleworm, and its life cycle and habits are very similar, except that this species feeds much more extensively on the foliage than does the pickleworm, rarely enters the vine or leaf petioles, appears a little later, and attacks pumpkins as well as the other cucurbits, but it is rare in watermelons. It is rarely

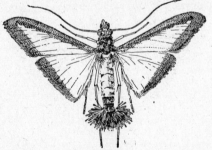

Fig. 314.—Adult of the melonworm, twice natural size. (*From Crosby and Leonard, "Manual of Vegetable-garden Insects," copyright, 1918, by Macmillan. Reprinted by permission.*)

injurious north of the Gulf states, although the adults at least are found from South America to Canada. The adult of the melonworm (Fig. 314) is a beautiful moth with a wing expanse of about $1\frac{3}{4}$ inch. The wings are pearly white, with a narrow dark-brown band about $\frac{1}{16}$ inch wide all around the front and outer margins. The body in front of the wings is dark-brown, while the hinder part of the thorax and abdomen is silvery white with a large bushy tuft of darker hair-like scales at the tip

[1] *Diaphania hyalinata* (Linné), Order Lepidoptera, Family Pyralididae.

of the body. The greenish caterpillar may be distinguished from the pickleworm in all but the smallest and largest stages by having two white well-separated slender stripes the full length of the body on the upper side and by lacking the dark spots. They are somewhat more slender than pickleworms, and more active in their movements.

Control Measures.—Since the larvae of this species feed a good deal on the foliage, where the expense can be justified, melonworms may be controlled by repeated weekly dustings with either derris powder (4 per cent rotenone), cryolite, or lead arsenate, 1 part, in talc or sulphur, 3 parts, begun within 10 days after the plants appear above ground. The application should be made while the worms are still small, and should be directed against the undersides of the leaves. Dusting is best because the cloud of dust will cover the undersides of the low-growing leaves more successfully. In general, the control measures recommended for the pickleworm will also help to control this species.

Reference.—*N.C. Agr. Exp. Sta. Bul.* 214, 1911.

D. INSECTS INJURIOUS TO POTATOES

FIELD KEY FOR THE DETERMINATION OF INSECTS INJURING THE POTATO

A. *Insects chewing holes in the leaves:*

1. Very convex, nearly hemispherical beetles, about 3/8 inch long, yellow in color, with black spots on the prothorax and 5 black stripes on each wing cover; and brick-red humpbacked soft-bodied larvae of various sizes up to 3/5 inch in length, with 2 rows of black spots along each side of the body; eat the leaves and tender shoots.............................. *Colorado potato beetle,* page 507.
2. Elongate, nearly parallel-sided, not very hard, active, long-legged beetles, either black, grayish, or black with narrow gray or yellow stripes or margins on the wing covers, swarm upon the plants, devouring blossoms and leaves...... ....*Blister beetles,* page 509.
3. Very small, elongate oval, black beetles from 1/16 to 1/8 inch in length rest on the leaves and jump readily when disturbed. Eat small rounded holes into the leaves. Leaves look as though peppered with fine shot. Very delicate slender white larvae, up to 1/5 inch long, with brown heads and 6 very short legs near the head, sometimes found feeding on roots or tubers...*Flea beetles,* page 468.

B. *Insects sucking sap from leaves or stems:*

1. Small, very active, greenish, slender, wedge-shaped jumping bugs, from 1/8 inch down in length, suck sap from underside of leaves, causing tips of leaves to turn brown, followed by the browning and curling of the entire margin, the tissue along the midrib dying last of all................*Potato leafhopper,* page 510.
2. Small, soft-bodied, green, or sometimes pink aphids, the largest from 1/6 to 1/8 inch long, some winged and others wingless, cluster on underside of leaves and terminal shoots causing them to wilt, curl, and die. Vines become covered with sticky "honeydew."..........................*Potato aphid,* page 513.
3. Tiny scale-like flat nymphs, margined with a white fringe, suck sap from shaded parts of foliage, producing rolled or cupped, yellow or reddish leaves ("purple-top" or "psyllid yellows"), killing or stunting the plants, and causing tiny malformed unmarketable potatoes. The adults are small "jumping plant-lice." *Potato psyllid, Paratrioza cockerelli* (Sulc.) (see *Jour. Econ. Ento.,* **30**: 377–379, 891–898, 1937).
4. Large, slender, shield-shaped, flat-backed bugs suck sap, causing tops to wilt; one, an inch long with hind femora and tibiae greatly swollen............... ....*Big-footed plant bug, Acanthocephala femorata.* Another, with the hind tibiae expanded like a leaf, 3/4 inch long and with a transverse yellow line across the wing covers.............................. *Leaf-footed plant bug, Leptoglossus phyllopus* (see *Fla. Agr. Exp. Sta. Bul.* 232, pp. 80, 81, 1931).

C. Insects boring in the stem and leaves:

1. Legless white grubs, about ¼ inch long, with brown heads, bore up and down the stalks, causing the leaves and stem to wilt and die. Small dark-gray straight-sided snout beetles, about ⅕ inch long, with 3 small black spots at the junction of prothorax and wing covers, feed by gouging out slender deep holes in the stems.....................................*Potato stalk borer*, page 515.

2. Caterpillars, up to 1½ inches long, prominently striped with brown and white in front and behind, but with a large grayish-brown "bruised-looking" area about the middle, mine in the leaves or burrow through the heart of the stems, killing the terminals. Especially troublesome around weedy margins of fields in early summer. Stripes become fainter and disappear when caterpillars are full-grown.....................................*Common stalk borer*, page 364.

3. White caterpillars, up to ½ inch long, with a pinkish or greenish tinge, and brown at each end, form blotch mines in the leaves or bore through petioles and terminal stems, causing the shoots to wilt and die..*Potato tuberworm*, page 516.

D. Insects attacking the tubers under ground:

1. Tubers tunneled with deep, more or less cylindrical burrows, about the diameter of a match, by shining, slick, reddish-brown, tough, 6-legged worms up to 1½ inches long by ⅛ inch thick...........................*Wireworms*, page 377.

2. Tubers gnawed and eaten away in irregular, broad, scabby areas over the surface, by white curved-bodied grubs with reddish-brown heads and 6 long slender legs..*White grubs*, page 374.

3. White caterpillars, up to ½ inch long, with a pinkish or greenish tinge, and brown at both ends, tunnel through tubers in the field or in storage..........
....*Potato tuberworm*, page 516.

4. Tubers in low ground or when stored in damp places occasionally attacked by a slender white black-headed maggot, only ⅙ inch long when full-grown, that bores through the flesh, causing superficial wounds that resemble potato scab.. *Potato scab gnat, Pnyxia scabiei* Hopk. (see *Ohio Agr. Exp. Sta. Bul.* 524, 1933).

5. Yellowish-white legless maggots, about ¼ inch long, burrowing over the surface and through the tubers of seed potatoes..........*Seed corn maggot*, page 389.

101. COLORADO POTATO BEETLE[1]

The common yellow and black-striped "potato bug" is perhaps the best known beetle in all America. When first known to man, it occupied the eastern slopes of the Rocky Mountains from Canada to Texas, and its food was the weed known as buffalo bur or sand bur.[2] It was described and named by Thomas Say, one of the earliest American entomologist-explorers, in 1824, and for 30 years longer it continued to live as an obscure beetle of no importance to man. The pioneer settlers, pushing westward across the continent, finally brought to this insect a new food, the potato. The insect soon largely deserted the weeds for the cultivated plant and began spreading eastward from potato patch to potato patch, often destroying the entire crop wherever it appeared. It was recorded in Nebraska in 1859, in Illinois in 1864, in Ohio in 1869, and reached the Atlantic coast in 1874. Its average annual spread, was about 85 miles a year. Nothing was known about spraying in those days, and the insect multiplied and spread almost unchecked until about 1865, when it was discovered that paris green could be used to poison it.

Importance and Type of Injury.—Both the yellow- and black-striped hard-shelled beetles and their brick-red black-spotted soft-skinned young or larvae feed by chewing the leaves and terminal growth of the

---

[1] *Leptinotarsa decemlineata* (Say), Order Coleoptera, Family Chrysomelidae.
[2] *Solanum rostratum.*

potato. Unless killed by stomach poisons, they soon devour so much of the vines that the plants die and the development of tubers is prevented or the yield greatly reduced.

Distribution.—Throughout the United States except parts of Florida, Nevada, California, and in eastern Canada.

Plants Attacked.—Potato is the favorite food. When this cannot be found, the insect may survive on tomato, eggplant, tobacco, pepper, ground cherry, thorn apple, Jimson weed, henbane, horse nettle, bella-donna, petunia, cabbage, thistle, mullein, and perhaps other plants. It is a pest chiefly of potato.

Life History, Appearance, and Habits.—The adult stage goes through the winter buried in the soil to a depth of several inches, seldom more than 8 or 10. This is the only stage that survives the winter. The beetles come out of the ground in spring in time to meet the first shoots of volunteer or early-planted potatoes. These adults (Fig. 315,*a*) are familiar to nearly everyone and may be recognized by the alternate black and yellow stripes that run lengthwise of the wing covers, five of

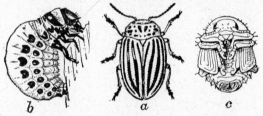

Fig. 315.—Colorado potato beetle, *Leptinotarsa decemlineata* (Say). *a*, adult; *b*, larva or slug, side view; *c*, pupa; about twice natural size. (*From U.S.D.A., Farmers' Bul.* 1349.)

each color on each wing cover. They are about ⅜ inch long by ¼ inch wide and very convex above. The orange-yellow eggs are deposited on the underside of the leaves, in close-standing groups averaging a couple of dozen each. A number of batches of eggs are matured by each female until an average of about 500 are deposited in the course of 4 or 5 weeks. The overwintering adults then die, and, from 4 to 9 days after the eggs were laid, small, humpbacked, reddish, chewing larvae hatch and likewise attack the leaves. They grow very fast, passing through four instars, similar except for size, and become full-grown in 2 or 3 weeks. The largest (Fig. 315,*b*) are a little more than ½ inch long, the back arched in almost a semicircle, with a swollen head and two rows of black spots on each side of the body. They usually feed in groups, completely consuming the leaves. These slug-like, reddish larvae are frequently supposed to be a different kind of potato pest from their parents.

If the full-grown larvae are watched, however, they will be found to descend into the soil, make a spherical cell and transform to a yellowish, motionless, pupal stage (*c*) which lasts 5 to 10 days. Then the adult beetles appear from the pupae, crawl up out of the ground and, after feeding for some days, may lay eggs for a second generation. Two generations appear to be the rule, although there may be only one in the North and there is a partial third in the more southern part of their range.

Control Measures.—Both adults and larvae are easily controlled by thorough spraying or dusting with any good stomach poison, as lead arsenate, calcium arsenate, or cryolite, 5 pounds to 100 gallons of water; or paris green, 4 pounds, and hydrated lime, 4 pounds, to 100 gallons of water; used about 100 gallons per acre for large plants.   Dusts of the above stomach poisons, diluted with about six parts of hydrated lime or talc are also effective.   These treatments should be made at any time that beetles or larvae appear on the vines.   All parts of the vines must be covered with the spray.   When potatoes are being sprayed with bordeaux mixture for flea beetles, leafhoppers, or potato blights, the arsenical may be added to the bordeaux when needed to destroy the "bugs," thus saving the expense of a separate application.   For small patches of potatoes the stomach poisons may be dusted lightly over the foliage from an open-meshed bag, sifter can, or hand duster.

References.—*Iowa Agr. Exp. Sta. Bul.* 155, 1915; *U. S. Bur. Ento. Cir.* 87, 1907; *Jour. Agr. Res.*, **5**: 917–925, 1916; *Jour. Econ. Ento.*, **26**: 1068–1075, 1933; *Boyce Thompson Inst. Contrib.*, **3**: 1–12, 1931.

102 (44). BLISTER BEETLES[1]

Importance and Type of Injury.—Older people often call these "old-fashioned potato bugs," because they were much more noticeable on potato before the Colorado potato beetle invaded the central and eastern states. They are slender beetles, about four times as long as wide, rather soft, with the head distinctly set off from the prothorax, and the tip of the abdomen exposed beyond the tip of the wing covers (Fig. 316).   They are black or grayish colored, or black with narrow yellowish or gray stripes or margins on the wings.   Only the adults feed on foliage, but they are very ravenous and may destroy many plants.

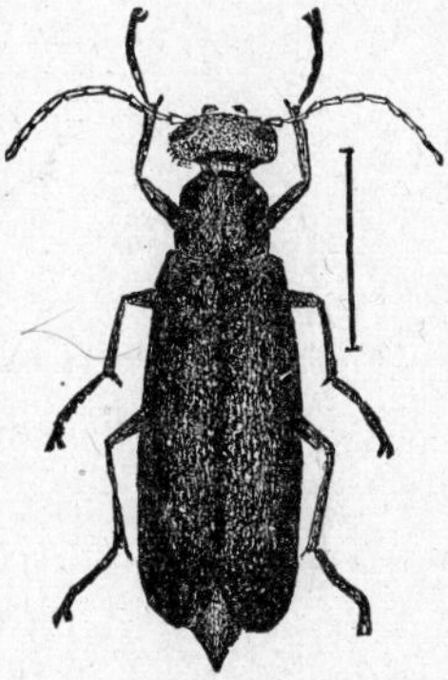

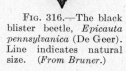

Plants Attacked.—The several species of blister beetles attack potato, tomato, eggplant, sweetpotato, bean, pea, soybean, cowpea, melon, pumpkin onion, spinach, beet, carrot, peppers, swiss chard, radish, cabbage, corn, oats, barley, clover, cotton, clematis, aster, chrysanthemum, zinnia, and other crops, and weeds.

Distribution.—Blister beetles will be found in all parts of the United States and Southern Canada.

FIG. 316.—The black blister beetle, *Epicauta pennsylvanica* (De Geer). Line indicates natural size.   (*From Bruner.*)

Life History, Appearance, and Habits.—The life cycles of only a few American species have been worked out. That of the striped blister beetle[2] is briefly as follows: The insect winters as a full-fed, but not fully transformed, larva, known as a pseudopupa (Fig. 317,*g*), in an earthen cell in the soil.   It is about ⅖ inch long, yellow, tough-skinned and with much reduced legs and mouth parts.   According to Crosby and Leonard, it may pass through several years in this condition, but usually in the spring it molts, acquiring functional legs and moving about a while before it pupates.   The true pupal stage lasts about 2 weeks, and then the adults come out suddenly in great numbers in June and July.   They are very restless, active beetles that tend to feed together in swarms.   Their bodies contain an oil known as cantharidin, which will blister the tender skin if the beetles are crushed on it (see page 21).   The females lay their eggs in clusters of 100, more or less, in holes that they make in the soil.   The eggs are elongate, cylindrical, yellow objects which hatch in 10 days to 3 weeks into very active, strong-jawed little larvae (*c*) that burrow through the soil until they find an egg

[1] *Epicauta vittata* Fabricius, *E. pennsylvanica* (De Geer) and other species, Order Coleoptera, Family Meloidae.

[2] *Epicauta vittata* Fabricius.

mass of a grasshopper.   They gnaw into the egg pod and begin eating the eggs.   During the next 4 or 5 weeks, the larva molts four times, undergoing a remarkable series of changes in form and appearance known as a *hypermetamorphosis*, during which its legs, mouth parts, and other appendages become progressively smaller and smaller (Fig. 317,*c,d,f,g*), until it reaches the pseudopupa stage, already described, in which it winters.   This species apparently has one generation or less a year.   Some other species appear to have two generations a year.   Several of them are known to feed as larvae in the egg capsules of grasshoppers.   So we have come to believe that the larvae of blister beetles are very beneficial, although the adults may be very injurious.   In Europe, some species feed in the nests of solitary bees, eating the eggs and honey.   More work upon the life cycles of our species is much needed.

Control Measures.—Blister beetles are hard to control because they are very active and are repelled by, or resistant to, arsenicals.   Barium fluosilicate or cryolite, 1 part to 3 parts of talc or dusting gypsum, applied as a dust at 25 to 30 pounds per acre, will give quick control.   Applications should be made promptly at the first appearance of the beetles.   Arsenicals may be combined with bordeaux mixture, which is very

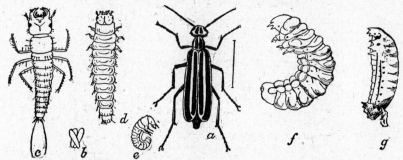

Fig. 317.—Life stages of the striped blister beetle, *Epicauta vittata* Fabricius. *a*, adult; *b*, eggs; *c*, first larval (*triungulin*) stage; *d*, second (*caraboid*) stage; *e*, same as *f*, as doubled up in pod; *f*, third (*scarabaeoid*) stage; *g*, pseudopupa or *coarctate* larval stage.   All but *e* enlarged.   (*From U.S.D.A.*)

repellent to the beetles, and bordeaux used alone will give fair protection to the vines. Plants that will not stand a heavy application of stomach poisons may be protected by knocking the beetles off into pans of kerosene, or if very valuable the plants may be covered with mosquito netting.

References.—Iowa Agr. Exp. Sta. Bul. 155, 377–380, 1915; Crosby and Leonard, *"Manual of Vegetable-garden Insects,"* pp. 302–312, Macmillan, 1918; *Jour. Econ. Ento.*, **27**: 73–79, 1934.

103 (155). Potato Leafhopper[1]

Importance and Type of Injury.—This little wedge-shaped green leafhopper, only ⅛ inch long, is the most injurious pest of potatoes in the eastern half of the United States.   The insects feed on the underside of the leaves, sucking out the sap from the veins and, in some manner not fully explained, cause the trouble known as tipburn or hopperburn (Fig. 318).   It appears that only this species of leafhopper produces hopperburn and only upon certain plants, such as potato, eggplant, rhubarb, dahlia, and horsebean.   The first symptom of this trouble is the appearance of a triangular brown spot at the tip of the leaf.   Similar triangles may appear at the end of each lateral veinlet, or the entire margin may roll upward and turn brown at one time, as though scorched

[1] *Empoasca fabae* Harris, Order Homoptera, Family Cicadellidae.

by fire or drought.   These brown margins increase in width until only a narrow strip of the leaf along the midrib remains green, the rest is shriveled and dead, with the leaf veins much distorted.   Older leaves below the growing tips usually burn first, but in cases of heavy infestations every leaf rapidly succumbs and the vines die long before the normal development of the tubers has been completed, thus greatly cutting the yield by producing mostly tubers so small that they are often not worth harvesting.   Many other agencies cause a browning of the leaf area of potatoes but the typical relation of the spots to the veins, just described, is characteristic of this particular trouble.   The diagnosis may be confirmed by finding the leafhoppers on the underside of the leaf.   The number of hoppers to an acre of potatoes may run between five and six millions.[1]   Upon bean and apple, stunting, dwarfing, crinkling, and tight

FIG. 318.—Work of the potato leafhopper, showing typical hopperburn at left, compared with a normal leaf at right.   (*From Parrot.*)

curling of the leaves are characteristic symptoms.   Alfalfa leaves become yellowed and clover leaves reddened when attacked, with a great reduction of the vitamin A content.   It has been shown that this species feeds on the phloem cells of the veins, which become torn and distorted and the xylem tubes plugged, so that food substances in the leaves are not properly translocated.

Three other species, so similar to *E. fabae* that only microscopic examination of the internal genitalia can be depended upon to distinguish them, have been found occurring as major pests of truck crops in various parts of the United States.   *E. filamenta* DeLong occurs in arid regions at high elevations in the Rocky Mountains, where the annual rainfall is under 10 inches, as a pest on potatoes, sugarbeets and beans.   *E. abrupta* DeLong occurs at low altitudes and low relative humidity from Texas to Oregon, along the Mexican border and Pacific Coast.   It causes a speckled white-stippled appearance of leaves of potato and bean, which is caused by withdrawing of sap from the mesophyll tissues.

[1] *Jour. Econ. Ento.*, **14**: 62–68, February, 1921.

E. arida DeLong also occurs in low altitude and low relative humidity areas in California.

Plants Attacked.—For many years this insect has been known as the apple leafhopper because its most extensive known injury was to apple nursery stock. It also feeds on a large number of cultivated and wild plants, including beans, potatoes, eggplant, rhubarb, celery, dahlia, alfalfa, soybeans, clovers, and sweetclover. On peanuts it causes a diseased condition known as "peanut pouts."

Distribution.—The potato leafhopper occurs over the eastern half of the United States, westward to Colorado and Wyoming, at elevations below 4,600 feet and where the average annual rainfall is 25 inches or more, with resultant high average humidity.

Fig. 319.—Adult of the potato leafhopper, *Empoasca fabae* Harris, greatly enlarged. Line at left indicates natural size. (*From Ill. State Natural History Surv.*)

Life History, Appearance, and Habits.—Although many kinds of leafhoppers may be found under surface vegetation during the winter in the northern states, it has been impossible to find the potato leafhopper in hibernation, or to keep them alive through the winter when placed in promising hibernating quarters. No wild host plant has been found in the North, upon which it breeds in early spring, before cultivated crops are up. The adults apparently migrate into the northern states from areas of milder climate, instead of wintering over in the North. In Florida and other Gulf states it breeds on alfalfa and other legumes or on castor bean and other weeds during the entire winter. The adults are about ⅛ inch long, by one-fourth as broad, of a general greenish color and somewhat wedge-shaped (Fig. 319). They are broadest at the head end, which is rounded in outline, and taper evenly to the tips of the wings. There are a number of faint white spots on the head and thorax, and one of the characteristic marks of this species is a row of six rounded white spots along the anterior margin of the pro-thorax, which can be seen with a hand lens. The hind legs are long and enable the insect to jump a considerable distance.

Large numbers of flying adults often appear suddenly in fields of beans as soon as these plants come up in early June. Except in cool wet seasons they are not attracted to potatoes until the plants are considerably larger. The possible reason is that bean plants are much higher in sugar content when they first break through the ground than potato plants are. Potatoes become much sweeter as they grow older, and it is then that the leafhoppers swarm upon them. Beginning from 3 to 10 days after mating, the very small, whitish, elongate eggs, only ¹⁄₂₄ inch long, are thrust into the main veins or petioles of the leaves on the underside, by use of the female's sharp ovipositor. An average of two or three eggs is laid daily, and the females live about a month or more. The eggs hatch in about 10 days, and the nymphs become full-grown in about 2 weeks. The nymphs are similar in shape to the adults but lack the wings and are very small and pale colored so they are really hard to see on a leaf. They usually complete their growth on the leaf where they hatched, feeding from the underside and increasing in size, greenness, and activity as they shed their skins; at the fifth molt they appear as adults. Both nymphs and adults are very active, the adults flying or jumping when disturbed, while the nymphs more characteristically run sideways over the edge of the leaf to the side which is turned downward. Two nearly complete generations and a partial third and fourth are produced at the latitude of central Illinois, so that there are individuals to infest both early and late potatoes and other truck crops.

Control Measures.—Fortunately bordeaux mixture, which is the most important potato spray for fungus diseases and flea beetles is also effective for leafhoppers, causing the young leafhoppers to die shortly after hatching. This effect persists for several days after the spray is applied. It should be applied as soon as the plants are 4 to 8 inches high and at 1-week or 10-day intervals as long as the vines can be kept green. This usually requires four or five sprayings. An 8-12-100 homemade bordeaux is recommended (see page 290). Experience in several states indicates that where potatoes are properly sprayed with bordeaux, the yield has been increased about one-third. On the average, the total cost of sufficient sprays to control this insect would be from $8 to $10 per acre. Sprays consisting of dry wettable sulphur, 10 pounds in 100 gallons of water, also give good control of leafhoppers and have a residual effect similar to bordeaux. Pyrethrum sprays kill the leafhoppers present at the time, but have no residual value, and the plants soon become heavily infested again. On beans 1 part pyrethrum dust in 10 to 20 parts of 300-mesh dusting sulphur has given excellent control. This dust also controls red spider mites, which are often very troublesome on beans at the same time.

References.—*U.S.D.A., Farmers' Bul.* 1225, 1921; *Fla. Agr. Exp. Sta. Bul.* 164, 1922; *Jour. Econ. Ento.*, **21**: 183–188, 261–267, 1928, **24**: 361–367, 475–479, 1931; **27**: 525–533, 1934, and **28**: 442–444, 1935; *Iowa Agr. Exp. Sta. Res. Bul.* 78, 1923; *U.S.D.A., Tech. Bul.* 231, 1931; *Jour. Agr. Res.* **43**: 267–285, 1931; *U.S.D.A., Tech. Bul.* 618, 1938.

104. POTATO APHID[1]

Importance and Type of Injury.—The epidemics of this insect are extremely sporadic. Severe outbreaks (Fig. 320) result in the complete browning and killing of the vines of the potato by curling and distorting the leaves from the top of the plant downward. On tomato the most noticeable injury is the devitalizing of the blossom clusters so that the blossoms fall and no tomatoes set. Green or pinkish, winged or wingless aphids cluster in shaded places on the leaves, stems, and blossoms. Winged migrants spread from field to field, so that the epidemic may sweep over a district in an alarming manner. Following such an outbreak, the insect may not visit a community in conspicuous numbers again for many years. The transmission of tomato and potato diseases, such

---

[1] *Illinoia solanifolii* (Ashmead), Order Homoptera, Family Aphididae.

as mosaics, leaf roll, and spindling tuber, by the feeding of these aphids, causes more injury to the plants than sucking the sap.

Plants Attacked.—Potato, tomato, eggplant, pepper, ground cherry, sunflower, pea, bean, apple, turnip, buckwheat, aster, gladiolus, iris, corn, sweetpotato, ragweed, lamb's-quarters, shepherd's-purse, and many other weeds and crops.

Distribution.—Maine to California and Florida to Canada; probably in every state.

Life History, Appearance, and Habits.—The winter eggs (see page 477) are deposited chiefly on rose, and the aphids may be found regularly on the succulent parts of rose bushes in the spring. During the first half of July, in Maine, winged aphids develop that fly to potatoes, and some of

Fig. 320.—A field of potatoes killed by the potato aphid.   (*From Ohio Agr. Exp. Sta. Bul.* 317.)

the others crawl to their summer host. A generation may be developed on potato every 2 or 3 weeks, and each unmated female may give birth to 50 or more active nymphs within 2 weeks' time. Thus the vines rapidly become covered with aphids, which blight the stems and wither the leaves. On a single large tomato plant, 24,688 aphids have been counted. By the middle of September, in Maine, and usually by early October, in Ohio, the aphids have all deserted potatoes and dispersed to other plants, of which the rose is the favorite. Here wingless, egg-laying females develop as the last generation of the season, and, after mating with males that fly over from the summer-host plants, the winter eggs are laid on stems and leaves of the rose.

The potato aphid when full-grown is nearly ⅛ inch long, of a clear green or pink, glistening color and with long slender cornicles (see Fig. 64,*B*). The wingless ones tend to drop from the plant when disturbed.

Control Measures.—Nicotine sulphate, 1 part to 500 parts of soapy water, thoroughly applied from one to three times, at intervals of 2 or 3 days, will control the severest outbreaks. Nicotine dusts may also be used (see page 259). Because of the many host plants on which this aphid may develop, clean cultivation in the truck patch is important in preventing outbreaks. Rosebushes should not be permitted to grow in abundance near potato fields, since they afford a place for an abundance of overwintering eggs.

References.—*Maine Agr. Exp. Sta. Bul.* 242, 1915, and 323, 1925; *Ohio Agr. Exp. Sta. Bul.* 317, 1917; *Jour. Econ. Ento.,* **25**: 634–639, 1932.

105. POTATO STALK BORER[1]

Importance and Type of Injury.—In some sections, this stalk borer is abundant enough in certain years to destroy entire fields of potatoes. Throughout much of its range it is of little importance. Chief injury is due to the larvae eating out the interior of the stalks, causing the entire plant to wilt and die. The adults eat slender deep holes in the stems.

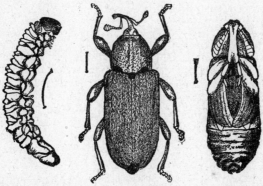

FIG. 321.—Potato stalk borer; larva, adult, and pupa; the lines show natural size. (*From Sanderson and Peairs, "Insect Pests of Farm, Garden and Orchard," Wiley, after J. B. Smith.*)

Plants Attacked.—Potato, eggplant, and related weeds, such as Jimson weed, horse nettle, and ground cherry. Most injurious to early potatoes.

Distribution.—Over the United States, except the northernmost states.

Life History, Appearance, and Habits.—The insect goes through the winter in or among the old vines in the adult stage. These adults are blackish snout beetles, about 1/5 inch long, covered with flattened gray hairs or scales that give them a frosted appearance. There are three distinct black dots at the base of the wing covers (Fig. 321). In late spring they emerge and feed by eating deep holes in the stems of the new plants. The eggs are placed singly in similar cavities in the stem or leaf petioles and hatch in a week or 10 days. The larvae eat up and down in the stems, completely hollowing them out for several inches. They are yellowish-white, legless, wrinkled grubs with brownish heads, and range up to about 1/3 inch long (Fig. 321). Before pupating, the larva packs its burrow with excelsior-like scrapings from the stem and chews an exit hole nearly through the stem for the escape of the adult. A week or two later the pupae transform to adults which may be found in the stalks, from late July on, in the northern states. They do not come out of the larval burrows until the following spring unless the stalks are broken open.

Control Measures.—The only practical control is to collect and burn all potato vines in infested fields, as soon as the crop is harvested. As with most potato insects, the destruction of Jimson weed, horse nettle, and ground cherry will help to keep down their numbers.

[1] *Trichobaris trinotata* (Say), Order Coleoptera, Family Curculionidae.

References.—*U.S.D.A., Bur. Ento. Bul.* 33, 1902; *Kan. Agr. Exp. Sta. Bul.* 82, 1899.

106 (72). POTATO TUBERWORM[1]

Importance and Type of Injury.—The tubers of potatoes in the field and in storage are riddled with slender, dirty-looking, silk-lined burrows of pinkish-white or greenish caterpillars (Fig. 322), that range up to ¾ inch in length, with dark-brown heads. Some of them burrow in the stems and petioles or mine in the leaves. When working on tobacco, the insect is known as the split worm. Brown blotches caused by this worm, between the upper and lower epidermis of tobacco leaves, make the leaves

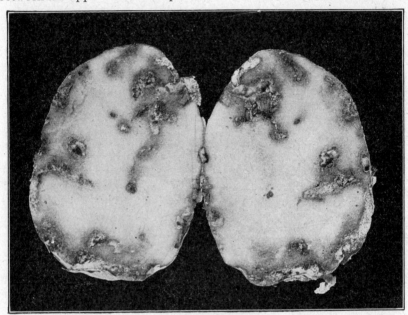

Fig. 322.—Potato cut open to show injury by larvae of potato tuberworm, *Gnorimoschema operculella* (Zeller). (*From U.S.D.A., Dept. Bul.* 427.)

unfit for wrappers. It is very destructive to potatoes in warm dry regions where it occurs.

Plants Attacked.—Potato, tobacco, tomato, eggplant, and weeds of the same family.

Distribution.—Southern United States from California to Florida and northward to Washington, Colorado, Virginia, and Maryland.

Life History, Appearance, and Habits.—In warm storage the insect may continue to reproduce as long as the potatoes contain enough food to mature the larvae. The moths escape from storehouses in early spring, or winter in all stages in sheltered culls out-of-doors. They lay eggs, one in a place, chiefly on the underside of the leaves, or upon exposed tubers. Most of the larvae first produce blotch mines on the leaves but subsequently work down into the stems. They become mature in 2 or 3 weeks,

[1] *Gnorimoschema* (= *Phthorimoea*) *opercullella* (Zeller), Order Lepidoptera, Family Gelechiidae.

pupate in a grayish, silken, dirt-covered cocoon, about ½ inch in length, entangled in the dead leaves or among trash on the ground, and emerge as adults in a week or 10 days. The adults are very small narrow-winged night-active moths, ½ inch from tip to tip of the wings, and grayish-brown, mottled with darker brown. An entire generation may develop in a month of warm weather, and five or six in a year. The damage is most severe in years of low rainfall and high temperatures. The later generations infest the tubers in the field by working down through cracks in the soil to lay eggs upon them, or the larvae may migrate from the stems to the tubers, or, especially at digging time, the eggs may be laid upon the exposed tubers. The caterpillars at first work just under the skin of the potato but later tunnel through the flesh. Before pupating, they come out of the potatoes and spin up among them or in cracks about the storehouse.

Control Measures.—Potatoes should be kept well cultivated and deeply hilled during their growth. At harvesttime, to prevent the caterpillars migrating from the wilting vines to the tubers, infested vines should be cut and burned or removed from the field a few days before digging and never piled over dug potatoes. The tubers should not be left exposed to the egg-laying moths during late afternoon or overnight. Harvest should be very thorough and all culls should be destroyed. Tubers infested in storage may be saved by fumigating with carbon bisulphide (see page 283), when the temperature is above 65°F., using 5 pounds to each 1,000 cubic feet of space, for 48 hours. Several fumigations are generally required to give effective control. Potatoes and potato bags should be fumigated before being transported.

References.—U.S.D.A., Dept. Bul. 427, 1917; *Va. Truck Exp. Sta. Bul.* 61, 1927; *Calif. Agr. Ext. Ser. Cir.* 99, 1936; *Jour. Econ. Ento.,* **25** : 625–634, 1932.

E. INSECTS INJURIOUS TO SWEETPOTATOES

FIELD KEY FOR THE IDENTIFICATION OF INSECTS INJURING SWEETPOTATOES

A. *Insects chewing holes in the leaves or vines:*
  1. Oval beetles, about ¼ inch long, that look like a drop of molten gold when alive, nearly hemispherical, with the margins of the body extended so as to hide the head and most of the legs; and elongate oval, brown larvae, up to ⅜ inch long, the margin of their bodies surrounded by about 30 thorny spines, but their backs completely covered with a dirty mass of excrement and shed skins that conceal the body, eat rounded holes in the leaves or devour them completely..............................*Golden sweetpotato beetle,* page 518.
  2. Similar to *A,* 1, but the beetle is dull-yellow in color, with five longitudinal black stripes on the wing covers. The larva with shorter marginal spines and yellowish-white in color with a median gray line. A thick tail-like projection, carried at an angle of 45 degrees from the leaf, bears the shed skins, but no excrement..............................*Striped sweetpotato beetle,* page 518.
  3. Similar to *A,*1, but 5⁄16 inch long, golden yellow, with three small black spots on each wing cover, arranged in a triangle. The larva straw-yellow with two dark spots behind the head; the marginal spines not very long, black at their tips; the back covered with excrement carried on two long spines and drawn out sideways into long shreds........*Black-legged sweetpotato beetle,* page 518.
  4. Beetle similar to the above, golden around the margin, the disk mottled with black and yellow, the black extending out as a slender tooth to each shoulder.

The larva is dull-green, bluish along the back, and covered with broad branching masses of excrement supported on the anal spines.....................
....*Mottled sweetpotato beetle*, page 518.

5. Beetle larger than the above, ⅓ inch long, yellow to brick-red in color, with 15 to 20 small rounded black spots on the forward two-thirds of the back; very convex, and the margins of the body not extended. Larva light-yellow, with many small brown spots, ½ inch long, marginal spines long, black-tipped. An irregular mass of excrement held slanting back from the body or vertically over it.....................................*Argus tortoise beetle*, page 518.

6. Very small, black, jumping beetles only 1/16 inch long, with a bronzy reflection, eat long narrow channels in the leaves, especially on the upper surface, parallel with the veins, during May and early June...*Sweetpotato flea beetle*, page 520.

7. Metallic, bluish-green, oblong-oval beetles, a little over ¼ inch long by two-thirds as wide, eat the tender leaves about the crown of the plant, completely devouring them from the margin inward (see also *C*, 2)....................
Sweetpotato leaf beetle, *Typophorus viridicyaneus* (Crotch) (see *U.S.D.A.*, *Cir.* 495, 1938).

8. Leaves and vines eaten by shiny, slender, ant-like snout beetles, ¼ inch long, with blue-black head, wing covers, and abdomen, but the middle region of the body and the legs bright-red (see also *C*,1, below).....................
....*Sweetpotato weevil*, page 520.

9. Holes eaten in leaves or leaves skeletonized by bluish-green caterpillars with yellow heads, up to 1 inch in length, which feed inside of folded leaves held together by silk. Injurious in the Gulf states........................
Sweetpotato leaf roller, *Pilocrocis tripunctata* Fabricius (see *U.S.D.A. Dept. Bul.* 609, 1917).

10. Plants or separate leaves cut off and the heart of the plant eaten out during the night. Plump, greenish-gray or blackish, more or less mottled or striped caterpillars found during the day just below the surface of the soil..........
....*Cutworms*, page 475.

B. Insects sucking sap from the plants:

1. Greenish, motionless, oval, flat nymphs with white waxy spines radiating from their bodies, the largest only 1/16 inch long, suck sap from the underside of the leaves. The small but conspicuous four-winged white bugs, like tiny moths, fly up when the plants are disturbed...................................
Sweetpotato whitefly, *Bemisia inconspicua* Quaintance (see *Fla. Agr. Exp. Sta. Bul.* 134, 1917).

C. Insects burrowing in the fleshy roots or "tubers" in the field or in storage:

1. Legless, white, fat grubs with pale-brown heads, up to ⅓ inch long, make winding, excrement-filled tunnels through the tubers, causing them to decay and become bitter and unfit for use. The red-and-blue adult beetles also found in the tunnels (see also *A*, 8, above).............*Sweetpotato weevil*, page 520.

2. Pale-yellow, rather plump larvae, up to nearly ½ inch long, with 6 small legs and a distinct brownish head, at first burrow through the vine under ground and later tunnel through the fleshy roots during the growing season. Usually lie curled when in the soil (see also *A*, 7, above).............................
Sweetpotato leaf beetle, *Typophorus viridicyaneus* (Crotch) (see *U.S.D.A.*, *Cir.* 495, 1938).

107. SWEETPOTATO OR TORTOISE BEETLES, OR "GOLD BUGS"[1]

Importance and Type of Injury.—The foliage of sweetpotatoes is very commonly cut full of holes, or entire leaves may be eaten by beautiful oval beetles a little squared at the shoulders, of a golden color, sometimes with black stripes or spots, and about ¼ inch long. Similar injury is performed by the spiny dirt-laden larvae found on the underside of the leaves. When they attack newly set plants the injury may be severe.

[1] Order Coleoptera, Family Chrysomelidae.

Plants Attacked.—These beetles restrict their feeding largely to the plants of a single family, the morning-glory family, of which the best known kinds are sweet-potato, morning-glory, and bindweed.

Distribution.—As a group, the tortoise beetles occur over nearly all the United States and arable Canada.

Life History, Appearance, and Habits.—The tortoise beetles live through the winter in the beetle stage in dry sheltered places, under bark or trash. They come out of hibernation rather late and are found feeding and mating on the plants during May

Fig. 323.—The argus tortoise beetle. Sweetpotato leaf showing: *a*, adult; *b*, larva; and *c*, pupa. About 3 times natural size. (*From Jour. Agr. Res., vol. 27, no. 1.*)

and June. They are turtle-shaped, flat below, and the sides of the prothorax and wings are extended beyond the sides of the body so as to hide the head and much of the legs. The recognition marks of the more important species[1] are given in the key on pages 517 and 518.

The females of the striped sweetpotato beetle[2] lay their eggs one in a place, on the leaf stems or veins of the underside of the leaf, covering each white egg with a little daub of black pitchy material that hides and protects it. The argus tortoise beetle[3] (Fig. 323,*a*) lays her eggs in clusters of 15 to 30, each egg attached to the leaf by a slender pedicel. In about a week or 10 days the eggs hatch, and during June and

---

[1] Golden sweetpotato beetle, *Metriona bicolor* (Fabricius). Striped sweetpotato beetle, *Cassida bivittata* Say. Black-legged sweetpotato beetle, *Jonthonota nigripes* (Olivier). Mottled sweetpotato beetle, *Chirida guttata* (Olivier). Argus tortoise beetle, *Chelymorpha cassidea* (Fabricius).

[2] *Cassida bivittata* Say.

[3] *Chelymorpha cassidea* (Fabricius.)

July the curious larvae of these beetles (Fig. 323,*b*) are found feeding mostly on the underside of the leaves. They are about ⅜ inch long, provided with conspicuous thorny spines all around the margin, two of which at the posterior end are nearly as long as the body. On these long spines, which may be turned up over the back like a squirrel's tail, the larva packs all its excrement and the skin it sheds at each molt, tying the dirty mass together with silk. The larvae thus come to look like moving bits of dirt or excrement and are often called "peddlers." When growth is completed, the larvae fasten themselves to leaves. The somewhat different-looking pupae (*c*) are exposed by molting the skin, and a week or so later the new adults emerge. They feed a little during August and then go into winter quarters.

Control Measures.—Spraying with lead arsenate, 3 pounds to 100 gallons of water, or cryolite, 4 pounds to 100 gallons of water, with 3 ounces of soybean flour, gives good control. These sprays may be applied in the seedbed or in the field.

References.—*N.J. Agr. Exp. Sta. Bul.* 229, 1910; *Jour. Agr. Res.*, vol. 27, no. 1, Jan. 5, 1924; *U.S.D.A., Farmers' Bul.* 1371, 1934.

108 (76). Sweetpotato Flea Beetle[1]

Importance and Type of Injury.—Long narrow grooves are eaten in the leaves, especially on the upper surface, along the veins, during May and early June by very

small, chunky, black, jumping beetles about 1/16 inch long, with bronzy reflections and reddish-yellow appendages (Fig. 324). When these channels are numerous, the leaf may wilt and turn brown and the plant be killed or badly stunted.

Plants Attacked.—Besides sweetpotato, bindweed, and morning-glory, corn, wheat, oats, rye and other grasses, red clover, sugar-beets, raspberry, and box elder.

Distribution.—General east of the Rocky Mountains.

Life History, Appearance, and Habits.—The small beetles winter under protecting trash in fence rows, the margins of wood lots and other sheltered places. They come out of hibernation and attack the plants about the time they are set out from the seedbeds. By the end of June, all have usually left the sweetpotato, migrating especially to bindweed, about which they lay their eggs, and then die. Their white larvae feed on the small roots of bindweed, becoming full-grown and producing the new generation of

Fig. 324.—Sweetpotato flea beetle, adult, about 24 times natural size. (*From Ill. State Natural History Surv.*)

beetles in late July and August. These beetles make their characteristic feeding channels on bindweed and morning-glory in the fall, but rarely attack sweet-potato until they come out of hibernation the following spring. They sometimes injure small grains in the fall.

Control Measures.—Plants in the seedbed should be thoroughly sprayed with lead arsenate, 2 pounds to 100 gallons of water. Plants set out late, after the beetles have migrated to weeds, are less liable to injury.

Reference.—*N.J. Agr. Exp. Sta. Bul.* 229, 1910.

109. Sweetpotato Weevil[2]

Importance and Type of Injury.—Also known as the sweetpotato root borer, this insect is most injurious to the roots or "tubers" which may be

---

[1] *Chaetocnema confinis* Crotch, Order Coleoptera, Family Chrysomelidae.

[2] *Cylas formicarius* (Fabricius), Order Coleoptera, Family Curculionidae.

honeycombed by numerous, fat, legless, white grubs, with pale-brown heads, ranging up to ⅓ inch in length. Their tunnels are tortuous and filled with excrement, and, when badly infested (Fig. 325), the roots are unfit even for stock feed. From 25 to 75 per cent of the crop is often destroyed.

Plants Attacked.—Sweetpotato, morning-glory, and other plants of the same family.

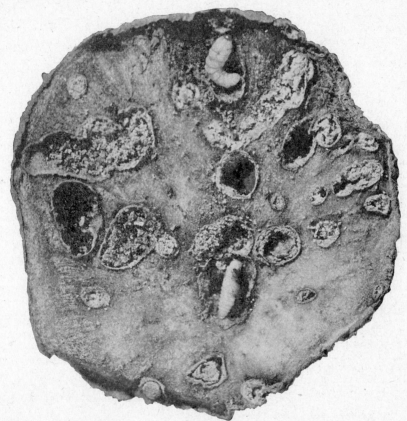

Fig. 325.—Sweetpotato cut open to show injury by sweetpotato weevil. Larva in burrow at top; pupa below. Three times natural size. (*From U.S.D.A., Farmers' Bul.* 1020.)

Distribution.—Confined in this country to the Gulf states. Probably imported from Asia.

Life History, Appearance, and Habits.—Breeding is continuous throughout the winter months, especially in potatoes in storage; and all stages may be found practically every week of the year. The eggs are deposited singly, in small cavities eaten out of the stem, or by preference in the "tuber." They hatch in less than a week and the grubs eat down through the stem or into the potato, feeding for 2 or 3 weeks and causing the potato to develop a bad odor and a bitter taste. The larvae may also be found in the slips, draws, vines, or smaller roots, before the potato develops.   About a week more is spent in the pupal stage in a cavity in the tuber, and then the beetle eats its way out.   The adult is the only stage generally seen.   It is a shiny, slender-bodied, ant-like snout beetle, about ¼ inch long, with blue-black head, wing covers, and abdomen, but the middle region of the body (prothorax) and the legs are bright-red.   The adults feed on the stems and leaves, and soon deposit eggs for another generation.   The generations require 1 month to 6 weeks each and follow each other as long as growing plants or stored potatoes are available. The adults do not fly far.   Spread to new fields or territory usually results from the planting of infested slips or draws.

Control Measures.—Sweetpotatoes or slips from infested territory should never be used for planting.   Other measures that will prevent serious damage by this weevil are cleaning up the vines and all infested tubers from the fields promptly after harvest and feeding, burning, or burying them deeply; turning hogs into the fields after the potatoes have been dug; observing quarantine and certification regulations designed to protect the growers; destroying volunteer sweetpotatoes and related weeds; putting sweetpotato plantings as far away from those of the preceding year as possible; dipping the plants, before setting, in lead arsenate, 1 pound in 10 gallons of water; spraying the vines at 10-day intervals with lead arsenate, 2½ pounds in 100 gallons of water; and cultivating or mulching the soil to prevent exposure of the roots by cracking of the soil. Federal and state quarantines prohibit the movement from the known infested area, of plant material which is likely to carry the insect.

Reference.—U.S.D.A., Farmers' Bul. 1371, 1934.

F. INSECTS INJURIOUS TO TOMATOES, PEPPERS, AND EGGPLANT

FIELD KEY FOR THE DETERMINATION OF INSECTS INJURING THE TOMATO, PEPPERS, OR EGGPLANT

A. Insects cutting off the newly set plants close to the ground:
   1. Plants chewed off at night and left lying on the soil.   Plump greasy-looking green tan or blackish caterpillars up to 2 inches in length, some of them spotted or striped, found in the soil about plants during the daytime..............
   ....*Cutworms,* page 475.
B. Insects devouring or eating holes in the foliage:
   1. Very small, oval, black, brassy or pale-striped beetles, from ⅟₁₆ to ⅛ inch long, eat small round "shot holes" in the leaves, and jump vigorously when disturbed.
   ....*Flea beetles,* page 468.
   2. Elongate, nearly parallel-sided, not very hard, long-legged beetles, black, brown, grayish, or black with light margins on the wing covers, swarm actively over the plants, devouring leaves and blossoms.......*Blister beetles,* page 509.
   3. Large green caterpillars, up to 4 inches long, with diagonal white bars on the sides and a slender horn at the tip of the body, cling to the vines and strip off the foliage.........................:............*Tomato* and *tobacco hornworms,* page 523.
C. Insects that suck sap from the stems, buds or leaves:
   1. Small, soft-bodied, green or pinkish plant lice, the largest ⅙ inch long, winged or wingless, cluster on underside of leaves and on terminal shoots, sucking their sap and causing them to wilt, curl, and die.   Vines become covered with sticky honeydew..*Potato aphid,* page 513.
   2. Slender, wedge-shaped, greenish or yellowish leafhoppers, under ⅛ inch long, suck sap of tomatoes and transmit a virus which causes tomato yellows or curly-top; the leaves droop, thicken, become crisp, and later develop a yellow color with purple veins or, in the greenhouse, transparent veinlets..........
   ....*Beet leafhopper,* page 545.

3. Tiny scale-like flat nymphs, margined with a white fringe, suck sap from shaded parts of foliage, producing rolled or cupped, yellow or reddish leaves ("psyllid yellows"), killing or stunting the plants, and causing tiny malformed unmarketable fruits..

Potato psyllid, Paratrioza cockerelli (Sulc.) (see *Jour. Econ. Ento.*, **30**: 377–379, 891–898, 1937).

4. Stems and leaves develop a white fuzzy appearance almost like a white mold, owing to the development of hair-like outgrowths under which many minute mites live...

Tomato erinose, Eriophyes cladophthirus Nalepa (see *Fla. Agr. Exp. Sta. Bul.* 76).

D. *Insects that bore into the fruit or buds, or mine into the leaves:*

1. Fruits tunneled, soured, and decayed. Plump striped light-green or tan to nearly black worms, $1\frac{3}{4}$ inches long when full-grown, found partly or wholly buried in the fruits which they are eating, or hiding about the ground in daytime......*Corn earworm* or *tomato fruitworm* and *climbing cutworms*, page 525.

2. Leaves of tomato, in the greenhouse or out-of-doors, have serpentine mines or large white blotches adjacent to folded leaves, which are held together by light webs. Developing buds and ripening fruits have pinholes bored into them, especially near the stem, by tiny, yellowish, gray, or green purple-spotted brown-headed caterpillars, only $\frac{1}{4}$ inch long................................

Tomato pinworm, Gnorimoschema lycopersicella (Busck), Order Lepidoptera, Family Gelechiidae (see *U.S.D.A., Cir.* 440, 1937; *Mon. Bul. Calif. Dept. Agr.*, **24**: 301–309, 1935; *Jour. Econ. Ento.*, **26**: 137–143, 1933).

3. Damage similar to *D*, 2, especially on the fruits, is caused by a whitish caterpillar, up to $\frac{1}{2}$ inch long, with a pinkish or greenish tinge and brown head.. ....*Potato tuberworm*, page 516.

4. Legless white brown-headed grubs, less than $\frac{1}{4}$ inch long, are found tunneling in the seed mass in the center of the pods of peppers. Shining, brownish-black to grayish snout beetles, averaging $\frac{1}{8}$ inch in length, with a single stout spine at the middle of the front femur, lay eggs in buds or fruits.................

Pepper weevil, Anthonomus eugenii Cano (= *A. aenotinctus* Champ) (see *Fla. Agr. Exp. Sta. Bul.* 310, 1937).

5. Fruits of peppers and eggplant are stung by barred-winged flies laying eggs in them. Legless, nearly headless maggots, up to $\frac{1}{2}$ inch long, devour the core causing the fruits to drop or decay.......................................

Pepper maggot, Spilographa electa Say. (see *N.J. Agr. Exp. Sta. Bul.* 373, 1923).

General References.—Cornell Univ. Ext. Bul. 206, 1931; *Calif. Agr. Ext. Serv. Cir.* 99, 1936.

110 (66). Tomato and Tobacco Hornworms

Importance and Type of Injury.—The best-known tomato insects are the large, green, white-barred worms, up to 3 or 4 inches long, with a slender horn projecting from near the rear end (Fig. 326,*b*). They eat the foliage ravenously. They are more seriously injurious to tobacco and are known among tobacco farmers as tobaccoworms and "tobacco flies." Some people suppose that they can sting with their horns, but they are entirely unable to hurt a person in any way.

Plants Attacked.—Tomato, tobacco, eggplant, pepper, potato, and related weeds.

Distribution.—Both species occur throughout most of the United States, often in the same garden. The southern or tomato hornworm[1] ranges from the northern states southward far into South America. The northern or tobacco hornworm[2] ranges from the southernmost United States into Canada.

[1] *Protoparce sexta* (Johanssen), Order Lepidoptera, Family Sphingidae.

[2] *Protoparce quinquemaculata* (Haworth), Order Lepidoptera, Family Sphingidae.

Life History, Appearance, and Habits.—The winter stage of the hornworms is very often spaded up or plowed out in the spring. It is a mahogany-brown, hard-shelled, spindle-shaped pupa about 2 inches long, with a slender tongue case projecting from the front and bent around like a pitcher handle (Fig. 326,*c*). From these cases appear in May or June large swift-flying hawk moths or hummingbird moths, 4 or 5 inches from tip to tip of wings. They fly at dusk and hover about beds of petunias or patches of Jimson weed and other flowers with deep tubular corollas, sipping the nectar with their very long tongues. When not in use, the tongue is coiled up like a watch spring under the head. The moths are grayish or brownish in color, with white and dark mottlings. The adult of the tobacco hornworm[1] may be distinguished from the adult of the other species by the two clear-cut, narrow, zigzag, dark stripes that extend diagonally across each hind wing; these stripes are

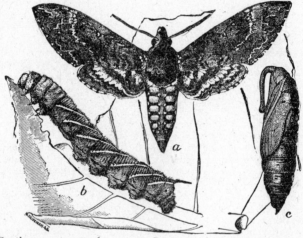

FIG. 326.—Southern or tomato hornworm, *Protoparce sexta* (Johanssen), showing adult moth, worm or larva, and pupa; about ⅔ natural size. (*From U.S.D.A.*)

indefinite and obscured in the tomato hornworm (Fig. 326,*A*). The tobacco hornworm, adult, has five pairs of orange-yellow spots on the abdomen, while the tomato species has six pairs of such spots.

The moths themselves do no injury; but deposit spherical greenish-yellow eggs, one in a place, on the lower side of the leaves. The small larvae which hatch from the eggs in about a week, feed ravenously for about 3 or 4 weeks, during which they shed their skins five times and increase to a length of 3 or 4 inches. The two species are similar, but may easily be distinguished by the diagonal white stripes on each side of the body. In the tobacco hornworm larvae the horn is black and there are eight stripes, each hooking backward from its lower end, forming an L or V; while in the tomato hornworm (Fig. 326,*b*) the horn is red and there are seven oblique stripes which do not turn backward.

When full-grown, the larvae, by using their mouth parts and legs, dig into the soil 3 or 4 inches and change to pupae, which may go through the following winter before they transform to moths. In the southern

[1] *Protoparce quinquemaculata* (Haworth), Order Lepidoptera, Family Sphingidae.

half of their range, however, the pupal stage lasts only about 3 weeks, and at least a part of the adults emerge and produce a second generation late in the season.

Control Measures.—Since the large caterpillars are hard to poison and the small ones hard to see, gardeners will do well as a routine practice to dust their tomato plants with equal parts of calcium arsenate and hydrated lime applied as a dust, or to spray them with poisoned bordeaux, at 10-day intervals from the time they start growing until the first fruits are about half-grown. Dusting the plants with powdered lead arsenate, using 5 pounds per acre on large plants; or spraying with the same material using 3 or 4 pounds in 100 gallons of water is also effective. On tomatoes spraying or dusting with arsenicals should be discontinued at least 10 days before the tomatoes ripen. In small gardens the worms may often be destroyed by hand more easily than by spraying the plants, locating the well-concealed worms by their droppings. Fall plowing destroys many of the pupae. A 5 per cent solution of tartar emetic in isoamyl salicylate as an attractant, put out in exposed pans or feeders, one per acre, attracts and kills many of the adult moths.

These caterpillars would be much more destructive if it were not for their natural enemies. The most commonly noticed is a braconid wasp.[1] The caterpillars are often found with small white objects covering their backs (Fig. 41), which are generally thought to be eggs. They are, however, cocoons enclosing the pupal stage of the parasite. The eggs of the parasite had been previously thrust through the skin of the hornworm, and the larvae, after feeding within the worm body, ate out through the skin and spun the cocoons. The adult wasps later cut out circular lids and escape from the cocoons to attack other worms. Worms with cocoons on their backs should not be destroyed.

References.—*Tenn. Agr. Exp. Sta. Bul.* 93, 1911; *Ky. Agr. Exp. Sta. Bul.* 225, 1920; *U.S.D.A., Farmers' Bul.* 1356, 1923; *Jour. Econ. Ento.*, **26**: 227–233, 1933.

111 (12, 63, 71, 260). Tomato Fruitworms

Importance and Type of Injury.—Among the most serious enemies of the tomato are the plump, greasy, greenish or brownish, striped caterpillars that eat into the fruits from the time they are formed until they are ripe (Fig. 327). These insects frequently make it necessary to discard from 5 to 25 per cent of the tomatoes in commercial tomato-growing areas. Slightly damaged fruits, containing larvae or their shed skins, often contaminate the products from tomato canneries, causing great loss of prestige. The worms are rather restless, and shift from one fruit to another so that a single caterpillar may spoil many fruits without eating the equivalent of a single one. Fifty to eighty per cent of the fruits are sometimes destroyed by these caterpillars. The most serious of these tomato fruitworms is the corn earworm[2] or tobacco false budworm, which has been discussed on page 368. It also attacks the fruits of beans, peppers, okra, and eggplant but is not injurious to any of the cucurbit or cruciferous vegetables. Certain cutworms[3]

[1] *Apanteles congregatus* (Say), Order Hymenoptera, Family Braconidae.

[2] *Heliothis obsoleta* Fabricius, Order Lepidoptera, Family Noctuidae.

[3] Especially the variegated cutworm, *Lycophotia saucia* Hübner, Order Lepidoptera, Family Noctuidae.

attack the fruit in the same way (see page 346). One of the worst infestations by cutworms that the writers have seen was brought about by the owner's cutting a quantity of wild grasses and spreading over the tomato fields as a mulch.

Control Measures.—Tomatoes can be protected from injury by these fruitworms by spraying the vines with lead arsenate, 2 to 4 pounds in 100 gallons of bordeaux mixture, or water, about the time the earliest fruits are the size of small marbles and again in 10 days. Dusting with equal parts of cryolite and talc, beginning when the plants are about a foot in diameter and repeating twice at 2-week intervals, has given good control. Cryolite and arsenicals must not be used after the fruits are approximately half-grown. Dusting with ground derris, cubé or timbo, containing not less than 0.5 per cent rotenone, at the rate of 25 to 30

Fig. 327.—Tomatoes with larvae of the tomato fruitworm feeding in them, about ½ natural size. (*From Tenn. Agr. Exp. Sta. Bul.* 133.)

pounds per acre is effective if used while the worms are less than half-grown. The infested fruits should be picked and destroyed by burning or burying a foot or more deep.

References.—Tenn. Agr. Exp. Sta. Bul. 133, 1925; *Ill. Agr. Exp. Sta. Cir.* 428, 1935; *Calif. Agr. Ext. Serv. Cir.* 99, 1936.

G. INSECTS INJURIOUS TO ONIONS

There are two very serious pests of the onion widely distributed in America: the onion maggot, which feeds in the bulbs; and the onion thrips, a very small, slender bug that draws the sap from the leaves.

112. ONION THRIPS[1]

Importance and Type of Injury.—The onion thrips is a very minute insect that punctures the leaves or stems and sucks up the exuding sap, causing the appearance of whitish blotches and dashes on the leaves. As the attack increases in severity, the tips of the leaves first become blasted and distorted and later whole plants may wither, brown, and fall over on

[1] *Thrips tabaci* Lindeman, Order Thysanoptera, Family Thripidae.

the ground. The insects may be found in greatest numbers between the leaf sheaths and the stem. The bulbs become distorted and remain undersized. Entire fields are often destroyed by this pest, especially in dry seasons.

Plants Attacked.—Nearly all garden plants, many weeds, and some field crops. Seriously injurious to onion, cauliflower, cabbage, bean, cucumber, squash, melon, tomato, turnip, beet, sweet clover, and other plants.

Distribution.—In all onion-growing sections of the United States and Canada.

Life History, Appearance, and Habits.—The adults and nymphs both winter on plants or rubbish in the fields or about weedy margins. They are slender, yellow, active bugs, pointed at both ends, the largest of them only $\frac{1}{25}$ inch long (Fig. 328). The males are wingless and very scarce—the females regularly reproducing without mating. The females have four extremely slender wings which could hardly serve for flight except for the fringe of very long hairs on their hinder margins. The feet also are remarkable in these insects, the tarsus ending in a small bladder, without claws. These bugs squirm in between the leaves and feed, mostly out of reach of insecticides. They rasp and puncture the surface of the leaf with their stabber-like mouth parts (see Fig. 71) and swallow the sap, together with bits of leaf tissue. White bean-shaped eggs are thrust into the leaves or stems nearly full length and hatch in 5 to 10 days. The nymphs are very similar to the adults, but paler in color. They become full-grown in from 15 to 30 days, passing through four instars, two of which are passed in the soil and without taking food. After the fourth molt the adult females return to the plants and soon lay eggs for another generation. Eggs, nymphs, and adults are found together throughout the summer. It is thought that there are usually five or six generations a year.

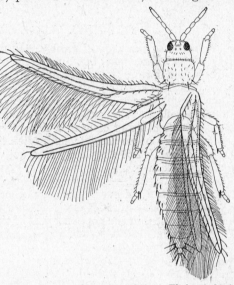

FIG. 328.—The onion thrips, *Thrips tabaci* Lindeman, adult, enlarged about 50 times, left wings spread to show the fringe of bristles. Note also tarsi without claws. (*Drawn by Kathryn Sommerman.*)

Control Measures.—No satisfactory practical remedy has been discovered. Spraying with nicotine sulphate, 1 part to 400 or 500 parts of soapy water, using high pressure, holding the nozzle close to the plants, and using enough spray to wet them thoroughly, will destroy many of the thrips. Nicotine sulphate, 1 quart, and 6 pounds of soap should be dissolved in each 100 gallons of spray. Nicotine may also be used in the dust form. In either case, a second application should be made a week later to catch those that were in the egg stage, or in the soil, at the first application. A dust of crude naphthalene flakes, 40 pounds, and hydrated lime, 60 pounds, has given increased yields of nearly 100 bushels per acre under New York conditions. Five applications were necessary. Crude chipped naphthalene, 2 pounds, and carbon bisulphide, 3 pints, emulsified in 100 gallons of water has also been found effective. Set onions should not be grown near seed onions. After the crop is harvested, the tops should be raked together and burned. The margins of fields should be burned over where practicable to destroy the weeds on which the thrips develop. Certain varieties of sweet Spanish onions possess considerable resistance to injury.

References.—U.S.D.A., Farmers' Bul. 1007, 1919; Jour. Econ. Ento., **27**: 109–112 1934, **29**: 335–339, 1936, **30**: 332–336, 1937.

113. Onion Maggot[1]

Importance and Type of Injury.—In dry years the onion maggot is of little or no importance, but during the second, third, or later years of a series of wet springs, it may destroy 80 to 90 per cent of the crop. Small white maggots up to 1/3 inch in length (Fig. 329,*d*) bore through the underground stem and into the bulbs, causing the plants to become flabby and turn yellow. They mine out the small bulbs completely, leaving only the outer sheath and causing a thin stand that is often blamed to poor seed. Larger bulbs are attacked, often by several maggots that eat out cavities which, if not completely destructive to the bulbs, cause subsequent rotting in storage.

Plants Attacked.—This insect is of no importance to any crop except the onion, rarely, if ever, attacking other plants.

Distribution.—Like its close relative the cabbage maggot, the onion maggot is a pest in the northern part of the United States and in Canada, rarely injurious in the South.

Life History, Appearance, and Habits.—The insect winters mostly as larvae or pupae in chestnut-brown puparia (Fig. 329,*h*) which resemble grains of wheat, often buried several inches in the soil. They are frequently found very abundantly in piles of cull onions. Some adults may survive under the protection of sheds and trash. Those in puparia transform to adults and emerge from the soil over a period of several months in late spring. The adults are slender, grayish-bodied, large-winged, rather bristly flies, only about 1/4 inch long (*a*, *b*). The females lay elongate white eggs about the base of the plant or in cracks in the soil. The eggs hatch in 2 to 7 days, varying with temperature and humidity. The maggots crawl down the plant, mostly behind the leaf sheaths, and enter the bulbs, consuming and spoiling them as stated above. They feed for 2 or 3 weeks. When full-grown, they are about 1/3 inch long and can be distinguished from the closely related cabbage maggot by the middle lower pair of tubercles at the rear end, which are single- and not double-pointed as in the cabbage maggot. Pupation occurs in the soil about the plant and after 2 or 3 weeks the adults emerge and lay eggs for another generation. A third generation often attacks the onions shortly before harvest, causing them to rot very badly in storage.

---

[1] *Hylemyia antiqua* Meigen, Order Diptera, Family Anthomyiidae.

Control Measures.—This insect can be controlled by the chemicals recommended as soil treatments for the cabbage maggot (page 541). Where large acreages of onions are grown for "sets," spraying with bordeaux-oil emulsion gives effective control. This is made from an 8-12-100 bordeaux mixture (see page 290) by mixing with it 3 gallons of lubricating-oil stock emulsion (see page 577) to 97 gallons of the bordeaux. Three to five applications of the spray at weekly intervals, the first when the plants are an inch high, will give practically 100 per cent control. 100 to 125 gallons per acre will be required. Applications can be made by hand sprayers, or, in large fields, by sprayers attached to tractor cultivators. Cull onions should be burned, buried, or hauled far away from onion land, immediately after harvest. The first-generation adults can be attracted to rows of cull onions, planted around the margins of the

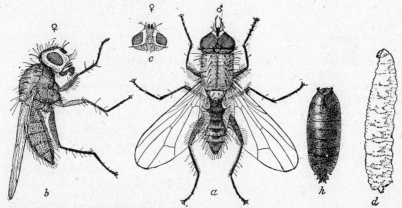

Fig. 329.—The onion maggot. *a*, adult male from above; *b*, adult female from the side; *c*, head of female; *d*, larva or maggot; *h*, puparium; about 5 times natural size. (*From Ill. State Natural History Surv.*)

field and at intervals through the field. These come up earlier and grow faster than the seeded onions and are attractive to the egg-laying flies, thus giving partial protection to the main crop. The maggots in the culls must later be destroyed by spraying with oil, using about 1 gallon to 25 feet of row.

References.—*Jour. Econ. Ento.*, **18**: 111, 1925, and **11**: 82, 1918; *Penn. Agr. Exp. Sta. Bul.* 171, 1922; *Ill. Agr. Exp. Sta. Cir.* 337, 1936.

H. INSECTS INJURIOUS TO CABBAGE AND RELATED VEGETABLES

FIELD KEY FOR THE DETERMINATION OF INSECTS INJURING CRUCIFERAE

A. Insects that eat holes in the leaves and into the heads:
  1. Young plants showing many small holes with yellow margins made by the feeding of very small, jumping, black beetles with a crooked yellow stripe on each wing cover...........................*Cabbage flea beetle*, page 469.
  2. Chunky, ash-gray weevils, about ⅛ inch long, with short snouts, gouge out leaves and stems, and their white brown-headed grubs hollow out the stems..
  Cabbage curculio (*Ceutorhynchus rapae* Gyllenhal) (see *U.S.D.A., Div. Ento. Bul.* 23, pp. 39–50, 1900).

3. Velvety green caterpillars with a very slender orange stripe down the middle of the back; of all sizes up to 1¼ inches long, and with 8 pairs of legs and prolegs, rag the leaves and eat their way beneath the outer leaves, leaving accumulations of dirty pellets in the leaf axils. White butterflies, nearly 2 inches from tip to tip of the wings, each of which has a few black spots on it, are usually found about the patch......................*Imported cabbageworm*, page 530.

4. Caterpillars of similar size and habits to *A*, 3, but more striped, the body smooth, and with only 6 pairs of legs and prolegs, loop over the plants, humping the back high at each "step." Eggs laid at night by somber-brown moths with a silvery spot at the middle of each front wing...*Cabbage looper*, page 533.

5. Small pale-green caterpillars, not over ⅓ inch long, eat small rounded holes in the leaves from the underside. Wriggle actively when disturbed............ ....*Diamond-back moth*, page 534.

B. *Insects that suck the sap from leaves and stems:*

1. Very small, whitish-green bugs or aphids rest in groups, or great soggy patches, in the heart of the plant or on the undersurface of leaves, causing leaves to cup and curl, wilt, and turn yellow..........*Cabbage and turnip aphids*, page 536.

2. Showy, red- and black-spotted, stinking bugs, flat and shield-shaped, up to ⅜ inch long, cause the plants to wilt and die. Not injurious north of fortieth parallel.......................................*Harlequin bug*, page 537.

C. *Insects that eat out the interior of the leaf, forming blotches but not holes through the leaf:*

1. Winding white trails or broad whitish spots appear on leaves, made by small white maggots feeding between the two surfaces of the leaf................. ....*Leaf miners*, page 539.

D. *Insects that attack the roots:*

1. Plants become stunted and frequently wilt down suddenly during the day and die without apparent external cause. Roots scarred and tunneled by white maggots up to ⅓ inch long, without legs or distinct head................... ....*Cabbage maggot*, page 539.

General Reference.—Ill. Agr. Exp. Sta. Cir. 454, 1936.

114. IMPORTED CABBAGEWORM[1]

Importance and Type of Injury.—The first-formed outer leaves of cabbage, cauliflower, and related plants, unless sprayed or dusted, are usually riddled with large holes of irregular shape and size, and the outer layers of the cabbage heads are eaten into by velvety-green worms of all sizes up to 1¼ inches long (Fig. 330). If the leaves are parted, masses of greenish to brown pellets (the excrement of the worms) are found caught in the angles of the leaves. So much of the leaf tissue is generally devoured by these worms that the growth of the plants is seriously interfered with (see Fig. 1), the heads of cabbage and cauliflower are stunted or do not form at all; other leafy vegetables are rendered unfit for consumption.

Plants Attacked.—All the vegetables of the cabbage or mustard family are attacked by these worms—cabbage, cauliflower, kale, collards, kohlrabi, Brussels sprouts, mustard, radish, turnip, horse-radish, and many related weeds. The worms also feed on nasturtium, sweet alyssum, mignonette, and lettuce.

Distribution.—This very common pest was unknown in the new world previous to 1860, when the butterflies were first taken at Quebec, Canada. Within 20 years it had spread over all the United States east of the Mississippi River. It now occurs as a pest throughout the United States and most of Canada.

[1] *Ascia* (= *Pieris*) *rapae* (Linné), Order Lepidoptera, Family Pieridae.

Life History, Appearance, and Habits.—Neither the greenish worms nor their well-known white butterfly parents persist through the winter;

Fig. 330.—Imported cabbageworms, *Ascia rapae* (Linné), and their injury to cabbage leaf. Natural size. (*From Conn. Agr. Exp. Sta. Bul.* 190).

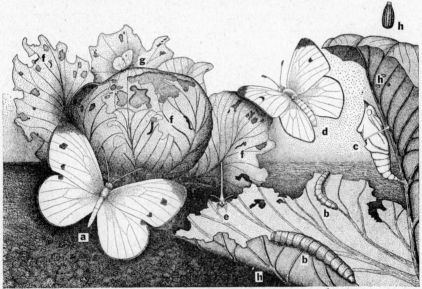

Fig. 331.—The imported cabbageworm, *Ascia rapae* (Linné). *a*, adult female butterfly; *b*, caterpillars or larvae; *c*, the pupa or chrysalis, showing how it is suspended by a girdle of silk about the middle of the body and a pad at the tip of the abdomen; *d*, adult male; *e*, front view of an adult with wings folded, in the manner characteristic of butterflies; *f*, larvae and their destructive work on cabbage plant; *g*, adult with folded wings, side view; *h*, eggs, natural size on the leaves and much enlarged at upper right. (*Drawing by Kathryn Sommerman, in part after N. Y. Exp. Sta. Cir.*)

but only the pupal stage. This is a naked, grayish, greenish, or tan-colored chrysalid with some sharp, angular projections over its back and

in front.   It is suspended from some part of the plant or on a building or
other object near the cabbage patch.   The tail end of the pupa is fastened
with a button of silk and kept from hanging head downward by a single
loop of silk that encircles the body near the middle, like a girdle (compare
Fig. 265,C).   Early in spring the familiar white butterflies (Fig. 331),
with three or four black spots on the wings, split out of the chrysalids and
fly about the gardens, alighting frequently to glue an egg to the underside
of a leaf of cabbage or related plant.   In all, several hundred eggs are laid
by a female.   They are just big enough to be seen, are shaped like a short,
very thick bullet, are deep yellow in color, and have ridges running both
lengthwise and crosswise.   Each egg gives rise in about a week to a very
small greenish caterpillar which feeds voraciously on the leaves and
reaches a length of an inch or a little more in about 2 weeks.   These
caterpillars (Fig. 330) are an intense leaf green, except for a very slender
orange stripe down the middle of the back, and another broken stripe
along each side of the body, which is formed by a pair of elongate yellowish spots near each spiracle.   The worm has a velvety appearance, due
to numerous, close-set, short, white and black hairs that form a kind of
white bloom over the body.   The crawling of these caterpillars is slow
and even, the body being supported by three pairs of slender legs and
five pairs of fleshy prolegs.   When full-grown, they frequently crawl
some distance away, fasten their tails with silk to some support, spin a
silken girdle about the middle of the body and change to the pupal
stage.   In summer this stage lasts 1 to 2 weeks and other generations
succeed until from three to six are completed.

   Control Measures.—The imported cabbageworm is easily controlled
by the use of derris or cubé sprays, without danger of poisoning the consumer.   A spray of 6 pounds ground derris or cubé root (containing
4.5 to 5 per cent rotenone), 4 ounces soybean flour or other good wetting
and sticking agent, and 100 gallons of water makes a very effective
treatment.   Thorough spraying must be done.   The spray should be
applied at the rate of 75 to 150 gallons per acre, depending on the size
of the plants.   Derris powder, containing at least 4 per cent rotenone,
mixed with 7 parts of dusting clay, talc, sulphur or tobacco (not lime),
using 15 to 40 pounds per acre, has also given excellent control.   This
worm is also easily controlled by arsenicals, the only difficulty being to
make the spray adhere to the very waxy leaves.   For this purpose a
spreader or sticker should be added.   Arsenicals should not be used on
cauliflower, broccoli, collards, Brussels sprouts, and other vegetables of
which the sprayed parts are eaten, except while the plants are small.
On cabbage, arsenicals must be discontinued at least 2 weeks before
harvest and all the loose leaves trimmed from the heads when they are
harvested.   On this account and because of the difficulty of covering
the heavy foliage of the larger plants, every effort should be made to
destroy these insects while the plants are small.   Several dustings of
lead arsenate, 1 part, in hydrated lime, 3 parts; or paris green, 1 part, and
calcium arsenate, 4 parts, in hydrated lime, 20 parts, applied 20 to 35
pounds per acre; or a spray of lead arsenate, 4 to 8 pounds, with a good
spreading and sticking agent, in 100 gallons of water will give good control; 35 to 50 pounds of the dusts should be used per acre when the
diamond-back moth is prevalent.   The old stalks should be destroyed
and the field plowed soon after the crop is harvested.   Weeds such as

wild mustard, peppergrass, and shepherd's-purse, on which the first generation of worms develop, should be destroyed.

A number of natural enemies prey on these caterpillars and in certain seasons and sections reduce them to a point of no importance. One of these is a small wasp[1] that was purposely brought to America in 1883 from England and has since spread widely over the country. It lays its eggs in the bodies of the caterpillars, and its young, feeding internally, devour the worm so that it dies. The larvae of the wasp then spin their small yellowish cocoons, about as big as grains of wheat, in masses on the leaves, often in contact with the cabbageworm on which they fed. A still smaller wasp[2] attacks the worms in a similar way but its young remain inside the worms until the little wasps pupate and then the adult parasites eat out of the dead cabbageworm chrysalids. More than 3,000 of these parasites have been reared from a single cabbageworm.

References.—U.S.D.A., Farmers' Bul. 766, 1916; Wis. Agr. Exp. Sta. Res. Bul. 45, 1919; N.Y. (Geneva) Agr. Exp. Sta. Bul. 640, 1934; Jour. Econ. Ento., **27**: 440–445, 1934.

115. Cabbage Looper[3]

Importance and Type of Injury.—This species attacks the plant in the same manner as the imported cabbageworm, and the two species are commonly found on the same plant. In certain seasons or sections the cabbage looper is more destructive than the imported cabbageworm.

Plants Attacked.—In addition to all of the plants of the cabbage family, this species also attacks lettuce, spinach, beet, pea, celery, parsley, potato, tomato, carnation, nasturtium, and mignonette.

Distribution.—Throughout the United States, from Canada into Mexico; a native species.

Life History, Appearance, and Habits.—The cabbage looper winters as a greenish to brownish pupa, nearly ¾ inch long, wrapped in a delicate cocoon of white tangled threads attached by one side usually to a leaf of the plant on which the larva fed. The cocoon is so thin that the outline of the pupa can be seen inside.

Fig. 332.—Moth of the cabbage looper. One and one-third times natural size. (From Crosby and Leonard, "Manual of Vegetable-garden Insects," copyright, 1918, by Macmillan. Reprinted by permission.)

These pupae transform in the spring to moths of a general grayish-brown color, about an inch long, with a wing spread of nearly 1½ inches (Fig. 332). The mottled brownish front wings have a small silvery spot near the middle, somewhat resembling the figure 8; the hind wings are paler brown to bronzy. They are nocturnal and much less conspicuous about the fields than the cabbage butterflies, but nevertheless manage to lay many small round greenish-white eggs, singly, on the upper surface of the leaves.

[1] *Apanteles glomeratus* Linné, Order Hymenoptera, Family Braconidae.
[2] *Pteromalus puparum* Linné, Order Hymenoptera, Family Chalcididae.
[3] *Autographa brassicae* (Riley), Order Lepidoptera, Family Noctuidae.

All injury is by the greenish larvae (Fig. 333) which are similar in size and habits to the imported cabbageworms.   The body tapers to the head. There is a thin but conspicuous white line along each side of the body just above the spiracles and two others near the middle line of the back.   The

FIG. 333.—Full-grown cabbage looper. About 1½ times natural size. (From Crosby and Leonard, "Manual of Vegetable-garden Insects," copyright, 1918, by Macmillan. Reprinted by permission.)

larva has three pairs of slender legs near the head and three pairs of thicker club-shaped prolegs behind the middle. The median half of the body is without legs, and this region is generally humped up when the insect rests or moves. From this looping habit the common name is derived. Two to four weeks of feeding bring the small looper to full size. It then spins a cocoon similar to that in which the winter is passed, and, in the summer months, appears as an adult again within 2 weeks.   There may be three, four, or more generations in a year, the number of worms usually increasing with each generation.

Control Measures.—The same measures are recommended as for the imported cabbageworm, but very thorough dusting or spraying must be done, because the worms crawl very actively and will migrate to parts of a plant that have not been covered by the poison.   It is very important to kill these worms while they are small, as the larger ones are more difficult to poison.   Derris and cryolite dusts (pages 265, 251) have given good results against this pest.   The looper caterpillars are often almost completely destroyed, usually late in the season, by a wilt disease which causes their bodies to rot.

References.—U.S.D.A., Bur. Ento. Bul. 33, 1902; Ill. Agr. Exp. Sta. Cir. 437, 1939.

116. DIAMOND-BACK MOTH[1]

Importance and Type of Injury.—This is one of the minor cabbageworms, seldom devouring more than a small percentage of the leaves.   The very small caterpillars work on the underside of the leaves, eating many small holes, giving a shot-hole effect all over the leaves.   In dry seasons they sometimes become abundant enough to cause appreciable injury to young cabbage.   On the vegetables, of which the outer leaves are eaten, and on greenhouse plants they are more serious.

Plants Attacked.—In addition to practically all the Cruciferae, the diamond-back moth attacks some ornamental and greenhouse plants such as sweet alyssum, stocks, candytuft, and wallflower.

Distribution.—Introduced into the United States from Europe sometime before the middle of the nineteenth century, it now occurs wherever its host plants are grown.

Life History, Appearance, and Habits.—The small grayish moths spend the winter hidden under the remnants of the cabbage crop left in the field.   They are about ⅓ inch long, the folded wings flaring outward and upward toward their tips, and, in the male, forming a row of three, diamond-shaped yellow spots where they meet down the middle of the back.   The hind wings have a fringe of long hairs.   The minute yellowish-white eggs are glued to the leaves, one, two, or three in a place, and in a few days the very small greenish larvae are at work on the underside of the leaves.   They become full-grown in from 10 days to a month.   They rarely exceed ⅓ inch in length, are pale-yellowish-green in color with fine scattered erect black hairs over the body (Fig. 334,A), and can be distinguished from small cabbageworms of other kinds by

[1] *Plutella maculipennis* (Curtis), Order Lepidoptera, Family Plutellidae.

their nervous habit of wriggling actively when disturbed or dropping on a silken thread. The cocoon (*B*) within which the full-grown caterpillar changes to the moth, is a beautiful gauzy sack ⅜ inch long, but so thin and loosely spun that it hardly conceals the pupa. It is usually fastened to the underside of a leaf. The little moth emerges

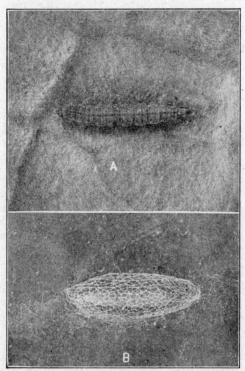

Fig. 334.—Diamond-back moth: larva and silken cocoon. Four times natural size. (*From Mont. Agr. Exp. Sta. Cir.* 28.)

from it within a week or two and promptly starts another generation, of which there may be from two to six or more a year in temperate regions.

Control Measures.—The same as for the imported cabbageworm, but the rotenone-bearing dusts are not effective against the larger larvae.

References.—*Jour. Econ. Ento.*, **30**: 443–448, 1937; *Ill. Agr. Exp. Sta. Cir.* 437, 1939.

OTHER CABBAGEWORMS

There are several other kinds of caterpillars that often feed on cabbage and related plants and some of them may be locally more abundant than the ones described above. The potherb butterfly,[1] the southern cabbage butterfly,[2] and the gulf white butterfly[3] are very closely related to the imported cabbageworm; while the cross-striped cabbageworm[4] has numerous black transverse bands across the body and is the young of a small yellowish-brown moth, and the zebra caterpillar[5] is light-yellow in color with

[1] *Pieris oleracea* Harris, Order Lepidoptera, Family Pieridae.
[2] *Pieris protodice* Boisduval and Leconte, Order Lepidoptera, Family Pieridae.
[3] *Pieris monuste* Linné.
[4] *Evergestis rimosalis* Guenee, Order Lepidoptera, Family Pyralididae.
[5] *Mamestra picta* Linné, Order Lepidoptera, Family Noctuidae.

three broad, longitudinal, black stripes, across which run many fine, yellow, transverse lines.   The cabbage webworm[1] and the purple-backed cabbageworm[2] usually feed beneath a protecting silken web or burrow into the leaves.   All these worms may be controlled by the same methods given for the imported cabbageworm, but for the last two named the applications must be made early, before the young worms gain protection under their webs or in their tunnels.

Reference.—CROSBY and LEONARD, "Manual of Vegetable-garden Insects," 1918.

117 (82). CABBAGE[3] AND TURNIP[4] APHIDS

Importance and Type of Injury.—These two plant lice, while easily distinguished by specialists, are very similar in general appearance, and, indeed, were not recognized as separate species until 1914.   The nature of

Fig. 335.—Head of cabbage ruined by the cabbage aphid.   Note the numerous small lice on the leaves.   (*From Ill. State Natural History Surv.*)

attack is similar, and they may be considered together.   Plants in seedbeds and at all subsequent stages of their growth are frequently covered with dense clusters of whitish-green plant lice about the size of the smallest bird shot, which suck the sap from the leaf (Fig. 335).   The affected leaves curl and crinkle or form cups, completely lined with the aphids, and, in severe infestations, wilt and die.   The plants, if not killed, are dwarfed, grow slowly, and form small light heads not suitable for marketing.   In cases of bad infestation the entire plants become covered with a disgusting mass of the small, soggy lice, and the dying leaves and plants rapidly decay.

Plants Attacked.—The cabbage aphid is recorded from cabbage, cauliflower, Brussels sprouts, kohlrabi, collards, kale, turnip, and radish.   The

[1] *Hellula undalis* Fabricius, Order Lepidoptera, Family Pyralididae.
[2] *Evergestis straminalis* Hübner, Order Lepidoptera, Family Pyralididae.
[3] *Brevicoryne brassicae* Linné, Order Homoptera, Family Aphididae.
[4] *Rhopalosiphum pseudobrassicae* (Davis), Order Homoptera, Family Aphididae.

turnip aphid from cabbage, collards, kale, rape, mustard, rutabaga, lettuce, wild mustard, and shepherd's-purse. Doubtless both occur on other plants of this family.

Distribution.—Both species probably occur throughout North America wherever their host plants grow.

Life History, Appearance, and Habits.—The life cycle is, in general, like that given as typical for aphids (see page 477). The cabbage aphid winters in the northern states as small black fertilized eggs laid in depressions upon the petioles and underside of leaves of cabbage. The turnip aphid probably winters in a similar way, although the sexual individuals and eggs of this species have not been described. Farther south the species continues to reproduce ovoviviparously throughout the winter. In cage experiments Paddock carried the turnip aphid through 25 generations in 12 months in Texas, while 16 generations of the cabbage aphid have been observed from April to October. When their food becomes unsatisfactory from any cause, winged females are developed which spread the species from plant to plant and start new families wherever they alight. Each female commonly produces from 80 to 100 young during her lifetime of about a month.

Control Measures.—The control measures recommended for aphids on page 479 are effective for this species. Dusts containing 2.4 per cent nicotine (see page 259) will penetrate into and under the leaves very effectively; and in some severe outbreaks have been found to be the most satisfactory means of control. These dusts should be applied when the plants are dry and the temperature above 70°F. On account of the waxy powder that covers the bodies of the lice, and the tendency of the leaves to form pockets or cups in which the lice are protected, it is essential, when nicotine sprays are used, to add a good spreader (see page 258), to use a spray pressure of 200 pounds or more, to apply 100 to 125 gallons per acre, and to repeat 2 or 3 days later. A spray of $1\frac{1}{3}$ pints nicotine sulphate, 6 pounds of soap, and 100 gallons of water is very effective. Plants may be dipped in such a spray when transplanting. The destruction of the old stalks of cabbage and other crops as soon as the crop is harvested will help to prevent destructive outbreaks of these aphids.

References.—*U.S.D.A., Dept. Cir.* 154, 1921; *Purdue Agr. Exp. Sta. Bul.* 185, 1916; *Tex. Agr. Exp. Sta. Bul.* 180, 1915.

118. Harlequin Bug[1]

Importance and Type of Injury.—The harlequin bug, "fire bug," or "calico back," is the most important insect enemy of cabbage and related crops in the southern half of the United States. It often destroys the entire crop where it is not controlled. It sucks the sap of the plants, taking its food entirely from beneath the surface, sapping them so that they wilt, brown, and die. The gaudy, red-and-black-spotted, stinking bugs (Fig. 336), about $\frac{3}{8}$ inch long, flat and shield-shaped, and the smaller similar-looking nymphs have a very characteristic pattern. They may be found in all stages of development from early spring to winter, and dozens to the plant in severe cases.

Plants Attacked.—Horse-radish, cabbage, cauliflower, collards, mustard, Brussels sprouts, turnip, kohlrabi, radish, and, in the absence of

[1] *Murgantia histrionica* (Hahn), Order Hemiptera, Family Pentatomidae.

these favorite foods, tomato, potato, eggplant, okra, bean, asparagus, beet, and many other garden crops, weeds, fruit trees, and field crops.

Distribution.—This is a southern insect ranging from the Atlantic to the Pacific, and rarely if ever injurious north of about the fortieth parallel. It first spread over the South from Mexico, shortly after the Civil War, and it was believed by many that it was brought in by the Yankee troops.

Life History, Appearance, and Habits.—Throughout most of its range the insect continues to feed and breed during the entire year.   Farther

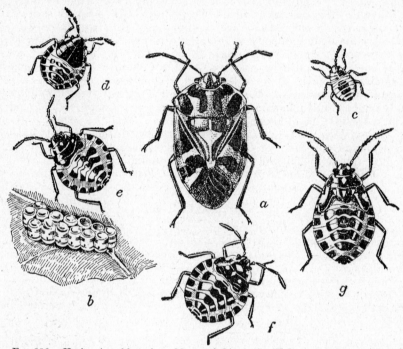

Fig. 336.—Harlequin cabbage bug, *Murgantia histrionica* (Hahn).   *a*, adult; *b*, egg mass; *c*, first stage nymph; *d*, second stage nymph; *e*, third stage nymph; *f*, fourth stage nymph; *g*, fifth stage nymph—all enlarged about 3 times.   (*From U.S.D.A., Farmers' Bul.* 1061.)

north the approach of winter drives the bugs into the shelter of recumbent cabbage stalks, bunches of grass, and other rubbish, and only the adults survive severe winter weather.   The first warm days of spring tempt them out of hiding, and they begin feeding on weeds, being ready to lay eggs by the time the earliest garden plants are set out.   The eggs are laid mostly on the underside of the leaves.   They are like tiny white kegs standing on end in double rows, about a dozen glued together, each "keg" bound with two broad black hoops and with "round black spots set in the proper place for bungholes" (Fig. 336,*b*).   The eggs hatch in from 4 to 29 days, the time varying with the temperature, and the very young bugs begin the business of destroying the plants.   They feed and grow for 4 to 9 weeks, passing through five distinct instars before they are capable of mating and laying eggs for a second generation.   Three generations, and a partial fourth, may succeed each other before cold weather puts a stop to their rapid increase.

Control Measures.—Much time and quantities of arsenicals are wasted every year by persons who do not understand the way in which the harlequin bugs feed and who try to kill them with applications of paris green and other stomach poisons. Strong contact sprays and dusts such as Lethane, summer-oil emulsions, or soap, 1 pound in 6 gallons of water, if applied during cloudy, humid days will kill the bugs wet by the spray. The most important control is the destruction of the adults in the fall and in the spring as they come out of hibernation and before they have begun egg-laying. Hand destruction may be facilitated by the use of trap crops of mustard, kale, turnip, or radish, planted very early in spring or late in fall after the main crop is harvested. When the bugs have concentrated on these small patches, they should be killed by spraying with kerosene, or by covering the trap crop with straw and burning. Trap crops should never be used unless they can be given careful attention to destroy the bugs attracted to them. Weeds such as wild mustard, Amaranthus, and others of the mustard family should be kept down. After the bugs have been reduced by the diligent use of the above methods, the stragglers that remain must be cleaned up by hand-picking early in the day and destroying the bugs and their egg masses wherever found.

Reference.—U.S.D.A., Farmers' Bul. 1712, 1933.

119. Leaf Miners

This group of plants is often disfigured and damaged by several species of small flies that live in the maggot stage by eating the tissue of the leaves, between the upper and lower surfaces. Their feeding causes the production of large whitish blotches or blasted areas, or, in the case of the serpentine leaf miner,[1] slender, white, winding trails through the interior of the leaf. The leaves are greatly weakened and the mines serve as points where disease and decay may start, but the chief loss is to those vegetables of which the green leaves are eaten and which are rendered unattractive and unsalable by these flies. No practical control measures are known.

Reference.—Jour. Agr. Res., **1**: 59–87, 1913.

120. Cabbage Maggot[2]

Importance and Type of Injury.—Plants attacked by the cabbage maggot appear sickly, off color, and runty, and, if the attack is severe, they wilt suddenly during the heat of the day and die. Roots of cabbage, cauliflower, rape, and the fleshy parts of turnips and radishes show brownish grooves over their surface and slimy winding channels running through the flesh, while many of the small fibrous roots are eaten off. Legless white maggots, from ¼ to ⅓ inch long, blunt at the rear end and pointed in front, are often found in these burrows.

Fig. 337.—Female of the cabbage maggot, *Hylemyia brassicae* (Bouché). About twice natural size. (*From Can. Dept. Agr., Pamphlet 32, n.s.*)

In most sections early cabbage after transplanting, late cabbage while still in the seedbed, early turnips, and late spring radishes are most severely injured. Very early radish and late cabbage, after being set

---

[1] *Agromyza pusilla* Meigen, Order Diptera, Family Agromyzidae.
[2] *Hylemyia brassicae* (Bouché), Order Diptera, Family Anthomyiidae.

out, usually escape injury.   The pest fluctuates much in abundance in different sections and years, but frequently 40 to 80 per cent of the plants are destroyed, resulting in the loss of thousands of dollars annually in many of the states where it occurs.

Plants Attacked.—This fly is chiefly injurious to plants of the mustard family or Cruciferae, such as cabbage, cauliflower, broccoli, Brussels sprouts, radish, and turnip, but also attacks beet, cress, celery, and some other vegetables to a slight extent.

Distribution.—Introduced from Europe early in the nineteenth century, it has spread widely over North America.   It is a serious pest in Canada and the northern part of the United States but is seldom injurious south of the fortieth degree north latitude.

FIG. 338.—Root of cabbage showing cabbage maggots and their destructive work.   (*From Can. Dept. Agr., Pamphlet 32, n.s.*)

Life History, Appearance, and Habits.—This insect goes through the winter chiefly as a pupa in a hard brown egg-shaped puparium, about ¼ inch long, and buried from 1 to 5 inches in the soil.   In spring, about the time early cabbage plants are set out (mid-May in the latitude of Chicago) the end of the puparium is broken open and a small gray fly emerges and crawls out of the soil.   These flies (Fig. 337) are similar in general appearance to the common house fly but only about half as long (¼ inch long), dark ashy gray with black stripes on the thorax and many black bristles over the body.   The cells of the wing that open nearest to its tip are both wide open at the margin.   They fly about close to the ground and deposit their small, white, finely-ridged eggs on the plants near where the stem meets the ground or in cracks and crevices in the soil.   Three to seven days later, the eggs hatch and the very small maggots promptly seek the roots and eat into them.   Each larva feeds for 3 to 4 weeks, and the roots often become riddled with their tunnels (Fig. 338).   The larva has at the blunt rear end 12 short, pointed, fleshy processes arranged in a circle around the two button-like spiracles.   The two processes nearest the middle line below are double-pointed.

When the maggots are abundant, the underground parts of the plants soon become honeycombed and rotten.   Over 125 maggots have been taken from the roots of a single plant.   Upon completing its growth the larva may pupate in its burrow, but more generally crawls away from the root into the soil a short distance and there forms its puparium.   Two or three weeks later, on the average, the adults break out of the puparium and may push up through the soil from a depth of 6 inches or more. Undoubtedly some of these puparia of the first generation remain until the following spring, but most of them transform to adults in late June and July, and lay eggs upon late cabbage and other plants. In most sections the injury from this second generation during dry midsummer weather is not severe, since the insect requires cool moist weather and succulent plants in which to thrive. Enough transform, however, to produce a partial third generation in autumn when they are sometimes very destructive to fall radishes and turnips. In some sections a partial fourth generation has been reported.

Control Measures.—The most successful control is to treat the base of the plants and the adjoining soil with either (*a*) bichloride of mercury (corrosive sublimate, $HgCl_2$), 1 ounce, in 10 gallons of water; (*b*) calomel (mercurous chloride, Hg_2Cl_2), 3 or 4 ounces, and 1 ounce of gum arabic in 10 gallons of water; or (*c*) a dust of 1 part calomel in 25 parts of gypsum, talc, or hydrated lime. Two gallons of the solution or 6 to 10 pounds of the dust are enough for 100 plants. For an acre, 20 to 30 gallons of the solution or about 75 pounds of the dust will be needed. The first application should be made 4 or 5 days after transplanting and repeated 10 and 20 days later. If bichloride of mercury is used, 2 to 4 ounces ($\frac{1}{2}$ teacupful) is poured close around the base of each plant, wetting the stem and surrounding soil but not the leaves. The chemical should be dissolved in a small quantity of hot water in wooden, glass, or earthenware vessels and then diluted. The solution may be applied to a few plants by pouring from a sprinkling can with the "rose" removed or by means of a pail and dipper. More rapid application can be made from a knapsack-sprayer tank equipped with a short hose, a 2-foot extension rod, and a pinch cock, allowing the liquid to run down by gravity to the base of the plant. For large areas, a barrel pump tank with two leads of hose and a man for each hose will be most efficient. *All metal containers should be thoroughly rinsed after using this solution in them, to prevent corrosion.* About 2 gallons of solution will be needed to treat a hundred plants. In transplanting, the above solution should be used instead of water to puddle the plants; and it may be used instead of water in transplanting machines. In seedbeds the material may be applied evenly all along the rows, using about 1 gallon to each 35 feet of row, wetting the soil to a depth of $\frac{1}{4}$ inch. Three applications at 10-day intervals will be needed in the seedbed. On radishes the row treatment should be applied shortly after the plants come up, and only one treatment will usually be needed. The newer calomel treatment is used exactly like the bichloride of mercury treatment, except that 3 or 4 ounces to 10 gallons may be used, without injury to the plants. Advantages claimed for the calomel treatment are that it is not corrosive to metal containers, it is not poisonous to man, one application at the greater strength is sufficient for the seedbed and the field crop, and it does not retard growth or maturity of the crop, as the bichloride of mercury may do.

Seedbeds may be protected from these flies and from flea beetles and other pests by covering them with thin cloth, such as hospital gauze, having from 20 to 30 threads to the inch, securely tacked to the framework around the bed and supported by wires across the beds at intervals of 5 or 6 feet.

References.—N.Y. (Geneva) Agr. Exp. Sta. Buls. 442, 1917, and 419, 1916; Proc. Ento. Soc. Nova Scotia for 1919, p. 41, 1920; Can. Agr. Dept., Ento. Bul. 12, pp. 9–29, 1916; Conn. Agr. Exp. Sta. Bul. 338, 1932; Jour. Econ. Ento., **25**: 709–712, 1932; Farm Res., N.Y. (Geneva) Agr. Exp. Sta., April, 1936.

I. INSECTS INJURIOUS TO BEET, SPINACH, LETTUCE, CARROT, PARSNIP, CELERY, AND RELATED VEGETABLES

FIELD KEY FOR IDENTIFICATION

A. Insects chewing the foliage, tender stem or seed heads:
  a. Insects feeding exposed upon the foliage:
1. Small, greenish-black, hopping beetles, $\frac{1}{4}$ inch long, with a broad yellow collar behind the head, and grayish to purple, short, warty grubs (up to $\frac{1}{3}$ inch long) eat small holes in the leaves or skeletonize them............ ....*Spinach flea beetle*, page 543.
2. Elongate, nearly cylindrical, rather soft-bodied long-legged beetles, with head and prothorax well marked off and tip of abdomen exposed beyond the wing covers, and either black, gray, or striped yellow and black, swarm upon beets and rapidly devour the tops................*Blister beetles*, page 509.
3. Yellowish-brown hopping beetles, $\frac{1}{8}$ inch long, with a yellow stripe lengthwise of each wing cover, which divide the back into 5 stripes of equal width, eat tiny "shot holes" in the leaves of beets and beans from the underside.. ....*Pale-striped flea beetle*, page 387.
4. Small cinnamon-brown velvety May beetle-like pests, only $\frac{3}{8}$ inch long, eat irregular holes from the margins of leaves on warm nights............ ....*Asiatic garden beetle*, page 471.
5. Large green caterpillars, up to 2 inches long, with a black crossband and 6 yellow spots on each segment, eat the leaves of celery, parsnip, and carrot.. ....*Black swallowtail butterfly*, page 545.
6. Lettuce, celery, and beets are frequently attacked by a pale-green caterpillar, up to $1\frac{1}{4}$ inches long, narrowing gradually to the head, and with light and dark stripes. They have only 3 pairs of prolegs and consequently crawl with a looping movement.. *Celery looper, Autographa falcifera* Kirby (see *Eleventh Rept. Ill. State Ento.*, pp. 38–43, 1882; *Fla. Agr. Exp. Sta. Bul.* 250, 1932).
  b. Caterpillars working under the protection of a silken web:
1. Lettuce, beet, cabbage, cucurbits, pea, beans, potato, and tomato are attacked by yellowish-green worms, up to 1 inch long, with scattered hairs and conspicuous black spots, feeding within light webs of silk that they spin over the plants, especially near the ground or in the soil................. ....*Garden webworm*, page 427.
2. Sugarbeets, garden beets, carrots, cabbage, peas, spinach, beans, potatoes, squash, and other foliage are ragged and webbed by pale-green caterpillars, up to an inch in length, with one middorsal and two lateral dark stripes. Worm retreats into a tubular, silken tunnel when disturbed.............. ....*Beet webworm*, page 474.
3. Caterpillars similar to *b*, 2, but with a light, middorsal stripe............. ....*Alfalfa webworm*, page 474.
4. The flower heads of parsnip and celery are webbed together and devoured by small yellowish to grayish-green black-spotted caterpillars, the largest about $\frac{3}{5}$ inch long, which also mine in the stems.....*Parsnip webworm*, page 544.
5. Celery, beet, and spinach foliage is ragged by greenish watery-looking whitestriped caterpillars, the largest $\frac{3}{4}$ inch long, which work on the underside of leaves or fold and web the leaves together and feed within. Squirm actively when disturbed...................*Greenhouse* or *celery leaf tyer*, page 544.
B. Insects sucking the sap from leaves and stems:
1. Slender wedge-shaped greenish or yellowish leafhoppers, under $\frac{1}{8}$ inch long, suck sap of beets and poison them, causing the rolling and shriveling of leaves

and the appearance of warts along the veins........*Beet leafhopper*, page 545.

2. Pale-green or pinkish, winged or wingless aphids, only $\frac{1}{12}$ inch long, suck the sap of spinach, beets, celery and about 100 other plants, causing stunting, wilting, and unmarketable condition of the plants.............................
.............................*Green peach aphid* or *spinach aphid*, page 546.

3. Broad, oval, black bugs, up to $\frac{1}{10}$ inch long, convex above, suck the sap of celery and other plants causing the leaves to wilt and die.   They give off a vile odor when crushed.....................................*Negro bug*, page 546.

4. Very active, shy, flat-backed, sucking bugs, less than $\frac{1}{4}$ inch long, cause wilted leaves or large brown dead spots on the stem of celery and related vegetables, known as "black joint".....................*Tarnished plant bug*, page 479.

C. *Insects mining in the leaves:*

1. Blister-like or blasted spots on the leaves of spinach, chard, beets and related plants are made by small maggots, not over $\frac{1}{3}$ inch long, eating the interior of the leaf without consuming either surface.......*Spinach leaf miner*, page 547.

D. *Insects attacking the roots:*

1. Aphids cluster on the roots of beets and other vegetables, sucking the sap and forming moldy white-looking clumps.................................
.................................*Sugarbeet root aphid* and others, page 548.

2. Stout, broad, reddish-brown, spiny-legged beetles, $\frac{1}{2}$ inch long by $\frac{1}{4}$ inch broad, gouge out unsightly holes in the roots of celery, carrots, parsnips, sugarbeets, sunflowers, and other vegetables and field crops..*Carrot beetle*, page 548.

3. Irregular zigzag dark grooves over the surface, or burrows through the roots of carrots and tunnels in the stalks and heart of celery, parsley, and dill, are made by fat, white, legless grubs.....................................*.
Carrot weevil, Listronotus latiusculus Boheman (see *Jour. Econ. Ento.*, **20:** 814–818, 1927, and **31:** 262–266, 1938; *Cornell Ext. Bul.* 206, 1931).

4. Roots of plants chewed off by a somewhat slender, whitish grub, with a Y-shaped anal opening and a transverse row of short setae in front of it........
....*Asiatic garden beetle*, page 471.

5. Yellowish-white maggots, about $\frac{1}{3}$ inch long, chew off the small roots of celery and the bottom of the taproot of carrots and parsnips causing plants to yellow and make stunted growth.   Roots may be riddled with rust-red burrows and surface scars.....................................*Carrot rust fly*, page 549.

121 (76). Spinach Flea Beetle[1]

Importance and Type of Injury.—Small holes are eaten in the leaves, or the leaves are skeletonized from beneath, by small, jumping, greenish-black beetles, with a yellow collar behind the head, and by grayish to purple, warty, short, cylindrical worms, all under $\frac{1}{4}$ to $\frac{1}{3}$ inch long.

Plants Attacked.—Spinach, beet, and pigweed, chickweed, lamb's-quarters, and other weeds.

Distribution.—General east of the Rocky Mountains.

Life History, Appearance, and Habits.—The insect winters as a greenish-black oval beetle, $\frac{1}{5}$ to $\frac{1}{4}$ inch long, with a yellow prothorax (Fig. 292).   In April and May the beetles appear on the plants and lay small clusters of orange eggs (Fig. 290) placed on end at the base of the plant or on the soil near by.   The dirty-gray to purplish young or larvae (Fig. 291) feed on the underside of the leaves, becoming $\frac{1}{4}$ to $\frac{1}{3}$ inch long within 2 to 4 weeks.   They are very warty, cylindrical grubs, each wart terminating in a short black hair.   When disturbed, the larvae and beetles "play 'possum"·and drop to the ground.   A pupal stage of a week or 10 days is passed in the soil, and the new adults appear in July,

[1] *Disonycha xanthomelaena* Dalman, Order Coleoptera, Family Chrysomelidae.

lay eggs over a period of nearly 2 months, and the second generation matures before winter sets in.

Control Measures.—Dusting or spraying with 1 per cent rotenone made from ground derris, cubé, or timbo root will give control if thoroughly applied.

Reference.—U.S.D.A., Farmers' Bul. 1371, pp. 12, 13, 1934.

122. PARSNIP WEBWORM[1]

Importance and Type of Injury.—The flower heads of parsnip and celery are webbed together with silk and devoured by small, yellow, greenish, or grayish caterpillars covered with small black spots and short hairs. They interfere seriously with the production of celery and parsnip seed. After consuming the unripe seed the caterpillars mine in the stems, and when full-grown are about ⅗ inch long.

Plants Attacked.—Parsnip, celery, wild parsnip, wild carrot, and related weeds.

Distribution.—Southern Canada and the northern states east of the Mississippi.

Life History, Appearance, and Habits.—The grayish moth, an inch across the wings, winters under loose bark and in other protection and lays its eggs in late spring on the developing flower head and other parts of the plant. After destroying the flower buds and seeds, the caterpillars pupate in their mines, emerging as adults in late summer, when they seek hibernating places. The insect is especially abundant on the heads of wild parsnips.

Control Measures.—Spraying or dusting with calcium arsenate at the rate of 6 to 8 pounds to the acre is an effective remedy, but much of the damage has usually been done before the caterpillars are noticed. Injured flower heads can be cut and burned in August, before the moths emerge. Wild host plants should be destroyed about the farm.

Reference.—Ont. Dept. Agr. Bul. 359, 1931.

123 (256). CELERY OR GREENHOUSE LEAF TYER[2]

Importance and Type of Injury.—This insect, which is more fully discussed as the greenhouse leaf tyer (page 733), is a major pest of celery culture and also destructive to spinach, beans, and beets grown in the field, especially following mild winters. It usually appears in injurious numbers within a few weeks of harvesting the crop, feeding upon the tenderest growth just above the heart of the plant, covering the leaves with its webs and excrement and greatly lowering the value of the crop. For descriptions of the insect see pages 733. There are 3 or 4 generations a year in the South.

Control Measures.—Early harvesting of infested crops and plowing under all crop refuse immediately after harvest aid greatly in control. Fresh pyrethrum dust diluted with an equal weight of dusting tobacco, sulphur, or hydrated lime and applied against the dry hearts of the celery plants, using 25 pounds per acre, kills many of the larvae and drives

[1] *Depressaria heracliana* (De Geer), Order Lepidoptera, Family Oecophoridae.

[2] *Phlyctaenia rubigalis* (Guenee) (= *P. ferrugalis* Hübner), Order Lepidoptera, Family Pyralididae.

others out of their webs.   A second, similar application ½ hour later to destroy these exposed worms gives a very effective control.

Reference.—Fla. Agr. Exp. Sta. Buls. 250 and 251, 1932.

124. Black Swallowtail Butterfly[1]

Importance and Type of Injury.—Large green caterpillars, with a black crossband on each segment, which is indented by six yellow spots on its front margin, ranging up to 2 inches in length, eat the foliage of celery and related plants, stripping the leaves clean as they go (Fig. 148,*a,b*).   They sometimes seriously injure young plants but are usually more noticeable because of their gaudy appearance than for the injury they do.

Plants Attacked.—Celery, dill, parsnip, carrot, parsley, caraway, and many other plants of the same family.

Distribution.—Throughout North America east of the Rocky Mountains.   In the West it is replaced by the western parsley caterpillar.[2]

Life History, Appearance, and Habits.—In the northern states the winter is passed as a dirty tan-colored chrysalid suspended from the host plants and other objects by a silk button and girdle, as described for the imported cabbageworm (see page 531). The adults are large black swallowtail butterflies, expanding nearly 4 inches, with numerous yellow spots on the outer part of the wings and also a row of blue patches on the hind wing (Fig. 148,*c*).   The eggs are scattered about on the leaves, in May and June, and hatch into the curious caterpillars.   When disturbed, the caterpillars protrude from the head end two, soft, orange-colored horns (known as *osmeteria*), that give off a sickening sweet odor which is probably a protection from some enemies. The larval stage lasts about a month, and the pupal stage 9 to 15 days in summer. There are two or three generations a year.

Control Measures.—Hand-picking of the caterpillars, or dusting young plants with lead arsenate or calcium arsenate, will easily destroy the worms.   Poisons must not be applied to foliage vegetables ready to be put on the market, but derris or other rotenone sprays or dusts may be used at that time.

Reference.—Scudder, "Butterflies of the Eastern United States," p. 1353, 1889.

125. Beet Leafhopper[3]

Importance and Type of Injury.—Beets are attacked by a disease known as "curly top" or blight, which is caused by the feeding of a small leafhopper.   This stunts the plants, kills them, or greatly reduces the sugar content of the beets and the crop of seed.   The leaf veins become warty, the veinlets transparent, the petioles kinked, and the leaves rolled upward at the edges, brittle and shriveled.   In many localities the growing of sugarbeets has been abandoned because of this pest.   The attack is generally sporadic, and there is no way of predicting when a destructive outbreak will occur.   This insect is also the carrier of tomato yellows.

Plants Attacked.—Sugarbeets, table beets, mangels, tomatoes, and certain weeds, a total of more than 100 kinds of plants.

Distribution.—Western United States from Canada to Mexico and east to Missouri, Illinois, and Texas.

Life History, Appearance, and Habits.—The cause of the trouble known as curly-top is a small wedge-shaped leafhopper, of a pale-greenish or yellow color, often with some darker blotches, about ⅛ inch long,

---

[1] *Papilio polyxenes* Fabricius, Order Lepidoptera, Family Papilionidae.

[2] *Papilio zelicaon* Lucas, Order Lepidoptera, Family Papilionidae.

[3] *Eutettix tenellus* (Baker), Order Homoptera, Family Cicadellidae.

with long slender hind legs that enable it to jump quickly into the air. It also flies readily, and when flying looks like a tiny white fly. The adult females winter on and about their wild food plants such as salt bush, Russian thistle, greasewood, wild mustard, filaree, and sea blite. Two generations usually develop on sugarbeets. Egg-laying in these desert host plants begins by early or mid-March, and the first generation normally matures on these plants. From early May to early June the first-generation adults may fly for hundreds of miles in great swarms, alighting in beet fields wherever the crop is up. The adults, and later the nymphs, feed by inserting the slender mouth parts into the plant, introducing a poisonous or disease-producing substance (filterable virus) that causes the curly top condition to develop. The eggs are inserted full length into the veins, leaf petioles, or stems, and hatch in 2 weeks to tiny, pale-colored wingless nymphs that settle in the center of the plant. In from 3 weeks to 2 months the bugs are full-grown. There may be from one to three or more generations.

Control Measures.—Curly top is caused only by the feeding of the leafhopper, but no practical control measure for the bug has been discovered. It flies so readily that it is hard to destroy by contact sprays or dusts. Early planting and thorough cultivation are recommended to produce a crop in spite of the presence of the leafhoppers. The insect is generally less destructive in seasons following severe winters and in areas of high rainfall. Some progress has been made in developing resistant varieties of sugarbeets.

References.—*Utah Agr. Exp. Sta. Bul.* 234, 1932; *U.S.D.A., Tech. Bul.*, 206, 1930; *Hilgardia*, **7**: 281–350, 1933; *Jour. Econ. Ento.*, **27**: 945–959, 1934.

126 (82, 181). Green Peach Aphid (Spinach Aphid)[1]

This insect, which is further discussed as the green peach aphid, has been very destructive to spinach in the large trucking sections of the Atlantic Coast. It has been estimated that $750,000 worth of damage was done to the spinach crop in Virginia in a single year (1907). It also attacks celery, lettuce, beets, tomato, eggplant, potato, the Cruciferae, cucurbits, and other vegetables. Its life cycle on the peach is discussed on page 625. On spinach and other vegetables the best control is the use of a 3 per cent nicotine dust, at the rate of 30 or more pounds per acre.

127. Negro Bug[2]

Fig. 339.—The negro bug, *Allocoris pulicaria* (Germar): adult, about 10 times natural size. (*From Ill. State Natural History Surv.*)

Importance and Type of Injury.—Short, oval, black bugs (Fig. 339), about $\frac{1}{10}$ inch long, together with small black and reddish nymphs, sometimes congregate on celery, corn, wheat, and other crops and suck out the sap, stunting the plants and causing the leaves to wilt and die. Outbreaks of the bug in Michigan and Ohio have destroyed thousands of dollars worth of celery, but the pest occurs only sporadically. Secretions of the bugs give a foul taste to raspberries and blackberries over which they crawl.

Plants Attacked.—Celery, corn, wheat, other grasses, some ornamental flowers, and many weeds, such as beggar-ticks, Lobelia, and Veronica.

[1] *Myzus persicae* (Sulzer), Order Homoptera, Family Aphididae.
[2] *Allocoris pulicaria* (Germar), Order Hemiptera, Family Cydnidae.

Distribution.—Throughout the United States and Canada, east of the Rocky Mountains.

Life History, Appearance, and Habits.—The insects winter as adult bugs which are often mistaken for beetles because of the hard shell over the back. This is not formed of the two wing covers but is a greatly enlarged thoracic shield, under the edges of which the wings slip when the bug comes to rest. The eggs are laid singly on the leaves and hatch in about 2 weeks into reddish nymphs which gradually grow into the form of the adult as they feed on the plants. They become adult by midsummer, and, after feeding for a few weeks, seek hibernating places long before cold weather.

Control Measures.—Weeds on which the bugs feed should be destroyed. When they attack cultivated crops a spray of nicotine sulphate, 1 quart, water 100 gallons, and soap, 8 pounds, will kill all of the bugs hit by it.

Reference.—*Mich. Agr. Exp. Sta. Bul.* 102, 1893.

128. Spinach Leaf Miner[1]

Importance and Type of Injury.—This insect is of less importance in the United States and Canada than in Europe. Blasted spots or blister-like blotches appear

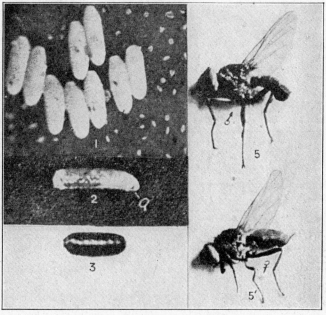

Fig. 340.—Spinach leaf miner, *Pegomyia hyoscyami* Panzer: 1, eggs on leaf, greatly magnified; 2, maggot or larva, the mouth hooks at *a*; 3, puparium; 5, male adult above; 5' female adult, about 4 times natural size. (*From N.Y. (Geneva) Agr. Exp. Sta. Bul.* 99.)

on the leaves of spinach, chard, and related plants, where small maggots have eaten out the tissue of the leaf between upper and lower surfaces. The leaf vegetables are rendered unfit for greens, and the development of seeds and roots, such as beets, is decreased by the partial defoliation.

Plants Attacked.—Spinach, beet, sugarbeet, chard, mango, and many weeds, including chickweed, lamb's-quarters, and nightshade.

Distribution.—Probably introduced from Europe previous to 1880, it is now generally distributed over the United States and Canada.

[1] *Pegomyia hyoscyami* Panzer, Order Diptera, Family Anthomyiidae.

Life History, Appearance, and Habits.—The winter is probably passed mostly in puparia (Fig. 340,*3*) in the soil.   In April and May the slender-bodied, grayish, black-haired, two-winged flies (*5*), about ¼ inch long, appear in the fields, and the females deposit small white eggs (Fig. 340,*1*), one to five side by side, on the underside of the leaves. Upon hatching, the tiny maggot at first eats a slender winding mine in the leaf, but as it increases in size the mine is widened to form a blotch that often joins the mines of other maggots in the same leaf.   The maggots (*2*) may migrate from leaf to leaf, and become full-grown in 1 to 3 weeks.   Pupation takes place chiefly in the upper 2 or 3 inches of the soil, but some transform among trash on the ground or even in the larval mines.   Within 2 to 4 weeks the adults appear from the pupae and start a new generation.   Three or four generations may be completed during the season.

Control Measures.—The increase of the flies may be checked by destroying their host weeds.   Screening the beds with cheesecloth or tobacco cloth, as recommended on page 458, will keep the plants free of this pest.   The screen should be removed a week before harvesting.   Spinach grown very late in fall, in winter, or very early in spring, may escape injury.

References.—*N.Y. (Geneva) Agr. Exp. Sta. Bul.* 99, 1896; *Ann. Appl. Biol.*, **1**: 43–76, 1914.

129. SUGARBEET ROOT APHID[1]

Aphids are frequently found on the roots of the vegetables in this group.   One of the most destructive is the sugarbeet root aphid, which is found in the western half of the United States on the roots of sugarbeets, beets, mangels, and many weeds, such as lamb's-quarters, yarrow, dock, goldenrod, and grasses.   It reduces both the size and the quality of the beets by sucking the sap from the roots.

The insects winter in part as fertilized eggs on the bark of poplar trees, and in part as wingless females on the roots of herbaceous plants.   The insects are yellow in color and have a mass of fine cottony-looking waxy threads toward the end of the body, so as to appear like white mold on the roots.   A migration of winged aphids from the poplars to beets takes place in July, and a return migration to poplars in September and October.

Control Measures.—Where beets are grown on irrigated land, aphids can be kept in control and the yields increased by giving five or more irrigations at 10-day intervals, during July and August.

Reference.—*Utah Agr. Exp. Sta. Leaflet* 41, 1934.

130. CARROT BEETLE[2]

Importance and Type of Injury.—The roots of celery, carrot, and parsnip are gouged by the feeding of broad reddish-brown stout-legged beetles, about ½ inch long, and slightly over half as wide (Fig. 341).

FIG. 341.—The carrot beetle, *Ligyrus gibbosus* (DeGeer), adult. The line indicates natural size. (*From Ill. State Natural History Surv.*)

Plants Attacked.—Carrot, parsnip, celery, beet, potato, cabbage, corn, cotton, sunflower, dahlia, and other crops and weeds, especially Amaranthus.

Distribution.—Over much of the United States except the most northern states.

Life History, Appearance, and Habits.—The adult beetles winter in the soil to a depth of 4 feet, emerge in spring, and lay eggs at night in the soil.   The eggs increase greatly in size before hatching.   The larvae resemble the common white grubs, being curved and white with a bluish cast and red-brown heads.   They feed largely on

[1] *Pemphigus betae* Doane, Order Homoptera, Family Aphididae.

[2] *Ligyrus gibbosus* (DeGeer), Order Coleoptera, Family Scarabaeidae.

grasses and decaying vegetation in the soil but often attack the roots of crops, a dozen or more beetles sometimes being found around a single plant. A generation is completed in a year, the adults being present and injurious from late April to August.

Control Measures.—No successful control measure is known.

References.—*Jour. Econ. Ento.*, **10**: 253, 1917; *Ill. Agr. Exp. Sta. Cir.* 437, 1939.

131. CARROT RUST FLY[1]

Importance and Type of Injury.—Celery plants, after getting a good start, wilt, and the outer leaves turn yellow, on account of the eating off of most of the fibrous roots by a very slender yellowish-white legless maggot, about ⅓ inch long when full-grown. Carrots and parsnips become stunted, and the lower end of the taproot is found to be eaten off. In severe attacks the entire root becomes scarred and riddled with the burrows of the larvae, the burrows taking on a rust-red color. Injury may continue in stored carrots if the temperature is favorable.

Plants Attacked.—Carrots, parsnips, celery, parsley, celeriac, wild carrots.

Distribution.—Starting near Ottawa in 1885, as an importation from Europe, the insect has spread over much of eastern Canada and the United States as far west as Oregon.

Life History, Appearance, and Habits.—The winter is passed in the slender brown puparia, about ⅕ inch long, buried in the soil, or as maggots in the roots. The shining-green, yellow-headed flies are abroad in May and deposit eggs about the base of the plants from which the maggots issue and work down in the soil to attack the smaller tender tips of the roots. As they increase in size and numbers, nearly the entire root system may be destroyed during the month or more while the larvae are growing. The pupal stage is passed in the soil near the roots, and a second generation of flies emerges in August and a partial third generation in September to attack late carrots and celery.

Control Measures.—Carrots planted after June 1 and harvested early in September may escape injury. The screening and insecticide measures recommended for the cabbage root maggot are effective for this pest. Naphthalene and derris dust, applied at the time the flies are abundant, have given 90 to 98 per cent control. Mercurous chloride seed treatment has been 90 per cent effective.

Reference.—*Jour. Econ. Ento.*, **24**: 189–196, 1931; *Mass. Agr. Exp. Sta. Bul.* 305, pp. 28–36, 1934.

J. INSECTS INJURIOUS TO ASPARAGUS

FIELD KEY FOR THE DETERMINATION OF INSECTS INJURING ASPARAGUS

1. Young shoots are gouged and scarred and, later in the season, the foliage is devoured and the stems scarred, by hordes of brilliant, blue-, red-, and yellow-spotted beetles, about ¼ inch long, and by their slug-like dull-gray larvae. Eggs, 1/16 inch long and dark-brown, are found standing on end in rows like the teeth of a comb, along the stems and leaves.......*Asparagus beetle*, page 550.
2. New shoots, foliage, and stems scarred and chewed by a light-brown to reddish-orange, rather straight-sided beetle, a little over ¼ inch long, with 6 small black spots scattered over each wing cover. Orange to brownish larvae ⅓ inch long feed mostly in the berries. Eggs laid flat on their sides....................
................................*Twelve-spotted asparagus beetle*, page 551.
3. Whitish, legless and headless maggots, up to ⅕ inch long, mine beneath the epidermis of the stems, sometimes girdling the plants and causing them to yellow and die. Adults are small, black, shiny two-winged flies, ⅙ inch long......
....*Asparagus miner*, page 552.

General Reference.—*Iowa Agr. Exp. Sta. Cir.* 134, 1932.

[1] *Psila rosae* Fabricius, Order Diptera, Family Psilidae.

132. Asparagus Beetle[1]

Importance and Type of Injury.—Wherever asparagus is grown, in the area infested by this insect, the voracious beetles (Fig. 342) make their appearance as soon as the shoots push above the soil in the spring.   They gnaw out the tender buds at the tips and cause the tips to be scarred and browned.   The eggs are also deposited in great numbers on the tips and make the plants unfit for sale.   After the leaves come out, the beetles

Fig. 342.—Asparagus beetle, *Crioceris asparagi* (Linné).   At left, eggs on asparagus shoots, natural size; at right, adult and larva, about 4 times natural size.   (*From Conn. Agr. Exp. Sta. Forty-fifth Rept.*, 1921.)

and their grayish slug-like larvae gnaw the surface of the stems and devour the leaves, thus robbing the root system of the food materials needed to form a good crop of shoots the following season.   The larvae also excrete a black fluid that stains the plants.   The injury is especially serious in new beds.

Plants Attacked.—Asparagus only, so far as known.

Distribution.—The northeastern fourth of the United States, from Missouri eastward and from Tennessee and North Carolina northward into Canada; also throughout much of California and Colorado and Oregon.   The insect is an importation from Europe, first found in

[1] *Crioceris asparagi* (Linné), Order Coleoptera, Family Chrysomelidae.

America on Long Island in 1860. In Illinois it first became destructive from 1910 to 1915.

Life History, Appearance, and Habits.—The beetles winter in sheltered places such as decayed or split fence posts, under loose bark of trees, or in the hollow stems of old asparagus plants. The adult is a brilliantly colored, active beetle, about ¼ inch long, with bluish-black straight-sided wing covers, each with three large, yellowish, squarish spots along each side and reddish margins. The reddish prothorax and the head are much narrower. Egg-laying begins soon after they appear in the field in April or May. Shoots of asparagus are often seen which are literally blackened by hundreds of eggs, all standing on end in rows of from three to eight (Fig. 342). The eggs hatch within a week, and the very small grubs migrate to the tips of the leaves and begin feeding upon them. The later stages of the larvae are similar to the young except for a gradual increase in size up to ⅓ inch long. The color is dull-gray, with black head and legs. The body is plump, soft, wrinkled, and not hairy. After feeding for 10 days to 2 weeks, the larvae disappear into the soil and form a yellowish pupa. A week or two later the new adults emerge from the soil and promptly start another generation. Each generation requires from 3 to 7 or 8 weeks, and there are from two to five generations in the course of a year. Cold weather kills the eggs and larvae and drives the beetles into hibernation. The insect appears to be much less destructive in wet seasons.

Control Measures.—During the cutting season the tips should be kept sprayed or dusted with derris as recommended for cabbageworms (page 532). The spray will require a good sticker, and dusts should be applied when the tips are wet with dew, to attain satisfactory coverage. Newly set beds should be sprayed or dusted with lead arsenate as soon as the beetles appear and again after the foliage is fully formed. For a spray, use 6 pounds lead arsenate and 6 pounds of soap, or ½ gallon of summer oil, in 100 gallons of water. Keeping down volunteer plants and cutting the shoots very clean and deep every day or two will tend to remove the eggs before the larvae can establish themselves in the patch. A few plants here and there may be left to grow and kept covered with the spray. After cutting season the plants should be dusted with lead arsenate, 1 part to 8 or 10 parts of lime; or sprayed as recommended above, not only to lessen the injury to the development of the plants, but also to reduce the number of beetles the following spring. The spray may be applied with a potato sprayer by raising the booms high enough to avoid injuring the plants.

References.—*U.S.D.A., Farmers' Bul.* 837, 1917; *N.Y. Agr. Exp. Sta. Bul.* 278, pp. 70–71, 1934.

133. TWELVE-SPOTTED ASPARAGUS BEETLE[1]

Importance and Type of Injury.—Injury by the adult is the same as by the asparagus beetle, and the gnawing is especially noticeable on the young shoots. The larvae however, do little damage, since they feed almost entirely in the fruits or berries. In some sections this species is more abundant and destructive than the former.

Plants Attacked.—Asparagus only.

[1] *Crioceris duodecimpunctata* (Linné), Order Coleoptera, Family Chrysomelidae.

Distribution.—Imported from Europe some time previous to 1881, when it was reported in Maryland, the insect has now spread over much the same territory as the asparagus beetle. It was first taken in Illinois in 1925; it has not been recorded from the states west of the Mississippi.

Life History, Appearance, and Habits.—The life history is similar to that of its close relative, described above. The adults, which are reddish-orange or tan, with 6 prominent black spots scattered over each wing cover, appear a little later in the spring and do not lay eggs until shortly before the berries form. The greenish eggs are glued, singly, on their sides to the leaves and hatch in 1 to 2 weeks. The orange-colored larvae, upon hatching, crawl to a developing berry, bore into it, and eat out the pulp. Before they become full-grown, they usually destroy three or four berries, by eating out the seeds. Pupation takes place in the soil. The new adults (Fig. 343) appear in late July and produce another generation in early September.

Fig. 343.—Twelve-spotted asparagus beetle, *Crioceris duodecimpunctata* (Linné). Four times natural size. (*From Conn. Agr. Exp. Sta. Forty-fifth Rept.*, 1921.)

Control Measures.—The same as for the asparagus beetle. In small gardens, gathering and burning the berries will give control.

References.—U.S.D.A., *Farmers' Bul.* 837, 1917; *N.Y.* (*Geneva*) *Agr. Exp. Sta. Bul.* 331, 1913; *Ill. Agr. Exp. Sta. Cir.* 437, p. 45, 1939.

134. ASPARAGUS MINER[1]

Importance and Type of Injury.—The maggots of this small fly mine up and down the stems of asparagus just beneath the surface, near the base of the plant. Where abundant, they may girdle the plants, causing the foliage to yellow and die prematurely.

Plants Attacked.—Asparagus.

Distribution.—Northeastern United States and southern Canada and also California.

Life History, Appearance, and Habits.—The insect winters in puparia in the larval tunnels under the epidermis of the stems and from 1 to 6 inches below the surface of the soil. The flies appear in the fields the latter half of May and thrust their eggs beneath the epidermis of the stem near or below the surface of the soil. The egg stage lasts 2 or 3 weeks, and the larvae feed for about the same period before pupating. In 3 weeks more, chiefly in July, the new adults are abroad, and the second-generation larvae mine in the stems during late summer. There are two generations a year.

Control Measures.—The rust-resistant strains of asparagus are said to be less injured by this insect. If old stalks are pulled up and burned, the puparia in the stems will largely be destroyed. Spraying with nicotine sulphate, 1 part to 500 parts of soapy water, will destroy many of the larvae in their galleries. In irrigated districts, winter flooding is recommended.

References.—U.S.D.A., *Bur. Ento. Cir.* 135, 1911; *N.Y.* (*Cornell*) *Agr. Exp. Sta. Bul.* 331, 1913.

K. INSECTS INJURIOUS TO SWEET CORN

The insects of sweet corn are, without exception, the same as those of field corn, already discussed in Chapter XI. One of the most destructive to sweet corn is the corn earworm, which eats the kernels at the end of the ears (see page 368). In seasons when this insect is abundant, nearly every ear will be attacked by the worms, rendering the ears repulsive to the consumer and causing great additional expense to canners. Experiments

[1] *Agromyza simplex* Loew, Order Diptera, Family Agromyzidae.

in cutting off the fresh corn silk the second day after its appearance or enclosing the ends of the ears in paper bags the second day after the appearance of the silk have given a fair degree of control. These methods are not practical for use in large fields (see also page 371). If corn can be produced to be past the fresh-silk stage before the moths become abundant, they will pass it by and lay eggs elsewhere. In the latitude of northern Illinois, early-maturing or early-planted corn is injured least, while in the latitude of central and southern Illinois medium plantings suffer least.

In the area where it is abundant, the European corn borer is by far the most destructive pest of sweet corn. Sweet corn is more severely injured than field corn because of the smaller stalks. The control of this pest, so far as it has been developed, is given on page 363.

The soil-infesting larvae, such as white grubs, wireworms, and cut-worms, are especially destructive to gardens when planted on sod land. This practice should never be followed. Sod land that is to be put to trucking should be seeded to clover, alfalfa, or small grains for at least one year. Gardens surrounded by the favored food trees of the May beetles are especially liable to injury by white grubs every year. Since corn insects of many kinds, notably the northern corn rootworm and corn root aphid, accumulate and increase in numbers when corn is planted year after year on the same soil, sweet corn should not be planted in fields that have been in corn for several years. Sweet corn will never be injured severely by the northern corn rootworm unless grown in soil that has been in corn for one or more years previously. For the corn root aphid, the cultivation of the soil suggested on page 373 will be most effective.

Armyworms and grasshoppers are periodically very destructive, but can be brought under control by the use of poisoned-bran bait. It may be necessary in the case of small gardens, to apply the poisoned bait to surrounding fields. Keeping down the growth of weeds about gardens will help in the control of many pests of corn, especially the corn root aphid and the common stalk borer.

For further information regarding the special control of insects on sweet corn, see pages 332 to 394.

CHAPTER XVII

INSECTS INJURIOUS TO DECIDUOUS FRUITS AND BUSH FRUITS

A large number of insects find their favorite or only food supply on the roots, trunk, branches, leaves, or fruit of our deciduous tree and bush fruits. Many of the insects which formerly fed on the fruits or foliage of uncultivated plants have found the abundant food supply furnished by large orchards or plantations of the bush fruits much to their liking, and now feed almost exclusively on these cultivated crops. The world-wide commerce in fruits and fruit-producing plants has led to the importation and general spread of many serious fruit pests in this and other countries. In most parts of this country it is now impossible to produce marketable fruit, regardless of the fertility of the soil, favorable climate or varieties grown, unless the insects are controlled. In other words, insects have come to be one of the limiting factors in fruit production.

In the United States and Canada fruit trees are subject to attack by many insects, among which are several that, in a single season, may ruin the crop or even destroy the trees of a well-established mature orchard. The most important of these fruit insects are discussed in the following order.

A. Apple Insects, page 554.
B. Pear Insects, page 613.
C. Quince Insects, page 619.
D. Peach Insects, page 620.
E. Plum Insects, page 638.
F. Cherry Insects, page 640.
G. Apricot Insects, page 644.
H. Grape Insects, page 644.
I. Currant Insects, page 655.
J. Gooseberry Insects, page 660.
K. Raspberry Insects, page 660.
L. Blackberry Insects, page 664.
M. Strawberry Insects, page 664.
N. Blueberry and Huckleberry Insects, page 671.

General Reference.—SLINGERLAND and CROSBY, "Manual of Fruit Insects," Macmillan, 1915.

A. APPLE INSECTS

FIELD KEY FOR THE IDENTIFICATION OF INSECTS INJURING THE APPLE

A. External chewing insects that eat holes in leaves, buds, bark, or fruit:
   1. Foliage of apple skeletonized and eaten off by grayish or brownish measuring-worms about 1 inch in length, with 3 pairs of legs near the head and 2 pairs of prolegs near the end of the body. Damage occurs in the spring about the time the trees have come into full foliage. Injured trees have the appearance of having been scorched by fire. The caterpillars causing the injury spin down from the foliage on silken threads when the trees are jarred or the insects are disturbed. Grayish, wingless, soft-bodied, spider-like moths, crawling up

the trunks of the trees in the spring and mating with grayish-brown males, with wings about 1½ inches from tip to tip. . . . *Spring cankerworm*, page 559.

2. Injury and appearance of insects the same as in *A*, 1, except the moths appear on trunks of trees during the fall and the caterpillars have 3 pairs of prolegs instead of 2. Injury to the trees occurs at the same time in the spring. . . . . .
. . . . *Fall cankerworm*, page 561.

3. Loosely woven white webs, enclosing the leaves of several branches, within which very hairy, pale-yellow, black-spotted caterpillars may be found feeding on the protected foliage. Caterpillars are about 1¼ inches long when full-grown. The webs contain black pellets of excrement, giving them an unsightly appearance. Injury most conspicuous in the late summer and early fall. . . .
. . . . *Fall webworm*, page 562.

4. Large, thick, white webs in the forks or crotches of trees in the early spring; leaves stripped from the branches within a considerable distance of these nests. Brownish-black, hairy caterpillars, ½ inch to 2 inches in length, with a light stripe down the back, sheltering within the webs during the daytime or feeding on the new foliage. Silken roadways or paths lead from the web to other parts of the tree. Small shiny varnished-appearing black egg masses, about ½ inch long, encircle small twigs during the winter. . *Eastern tent caterpillar*, page 563.

5. Apple leaves skeletonized by yellow-striped, hairy caterpillars, up to 1½ inches in length, with 2 pencil-like tufts of long black hairs projecting in front, and a single similar tuft projecting backward from near the tail. Frothy, white, egg masses, about ½ inch across, attached to grayish, hairy, empty cocoons on folded dead leaves or bark will be found on the twigs and branches of infested trees during the winter. . . . . . *White-marked tussock moth*, page 698.

6. During the winter season several leaves, tightly webbed together and firmly attached to a twig near the tip of a branch, enclose from 25 to several hundred small dark-brown, very hairy caterpillars about ¼ inch long. If caterpillars or their hairs come in contact with the skin, a distinct burning or itching results, which may persist for several days. White moths, with brown abdomens, flying in great numbers at night, during July. The moths are strongly attracted to lights. . . . . *Brown-tail moth*, page 704.

7. Pale-yellow masses of eggs about 1 inch across, with a felt-like covering of tawny hairs, attached to the bark of trees, or on buildings, walls and other objects during the winter. Somewhat flattened, bluish, stiff-haired caterpillars, feed on the foliage of all kinds of deciduous trees and also upon evergreens. Caterpillars with 5 pairs of conspicuous bluish tubercles on the back, behind the head, followed by 6 pairs of somewhat larger, red tubercles on the following segments. Foliage often completely stripped from trees in infested areas. . . . . *Gypsy moth*, page 701.

8. Yellow- and black-striped caterpillars, with a distinct ridged yellow ring about the neck, or with a red hump just back of the head with a row of black spines projecting from it, feed, many together, on the foliage during the summer and early fall. Caterpillars, when full-grown, are 1½ to 2 inches in length. They rear head and tail when disturbed. Small trees, or single branches on large trees, completely stripped of leaves by these caterpillars. . .
. . . . *Yellow-necked* and *red-humped caterpillars*, page 564.

9. Leaves, especially those at tips of new growth near the top of the tree, skeletonized or eaten out in part, by dark-green active caterpillars with four shining black tubercles on the back just behind the head. Caterpillars web 2 or 3 leaves together and feed within the enclosed area. Injury is most severe during the late summer and early fall. . . . . *Apple leaf skeletonizer*, page 565.

10. Very tough, blackish, conspicuously curled cases, about ¾ inch long, usually partly enclosed in 1 or 2 dead leaves, attached to the ends of twigs and in the forks of small twigs during the winter. Cases enclose dark-brown hairy caterpillars, about ⅓ inch long, which eat out the opening buds in the spring. Fruit sometimes specked with small feeding punctures during August. . . . .
. . . . :. . . *Leaf crumpler*, page 566.

11. Tiny brownish-gray tough silken cases, about $\frac{1}{6}$ inch long, bent at the top, so that they roughly resemble a pistol in outline, standing out at right angles to twigs and branches during the winter. Brown worms, about $\frac{1}{8}$ inch long, sheltering in these cases, feed on the newly opening buds in the spring and on the fruit and foliage during the late summer...*Pistol casebearer*, page 567.

12. Cases attached at right angles on the twigs and branches as in A, 11, but cases not bent over at the top. In the spring the insect moves about over the leaves eating or mining out small areas.................*Cigar casebearer*, page 567.

13. The newly opening buds and leaves of the apple are fed upon by a small brown worm, about $\frac{1}{4}$ inch long, with a black head; leaves and buds webbed with silk where the worm is feeding. In the winter small silken cases containing brown worms are attached to the axil of the bud or twig. Fruit on infested twigs often shows small areas of feeding, somewhat resembling injury by late codling-moth larvae....................*Budworms* or *bud moths*, page 569.

14. Large, irregular cavities eaten in the sides of small apples, from 1 to 3 weeks after the fruit has set. Green worms about $\frac{3}{4}$ inch long, sometimes with fine stripes on the sides, are found on the twig or branch adjacent to the injured apple or actively feeding on the apple. Apples at picking time deformed or misshapen, with brown, healed-over cavities in the sides.................
....*Green fruitworms*, page 570.

15. Injury resembling A, 14, but injured areas much more shallow, often merely the surface of the apple eaten away. Injury most common where the leaves rest against the fruit. Very active, translucent, greenish worms, with 4 white stripes on the back, the outer stripes broader than the inner ones, feeding on the fruit, or skeletonizing the leaves during early summer. Worms about $\frac{1}{2}$ inch long when full-grown.....................................
Palmer worm, Dichomeris ligulella Hübner (see *N.Y. (Cornell) Agr. Exp. Sta. Bul.* 187, 1901).

16. Injury similar to A, 14, but usually including nearly all the apples on a branch. Foliage ragged and buds eaten off by small, greenish-brown, very active worms about $\frac{3}{4}$ inch long when full-grown. Injury occurs from the time the buds separate until about 4 weeks thereafter. Injured leaves rolled, folded, or drawn together about the fruit with light silken threads. Apples at picking time with brown, scarred areas in the sides...*Fruit tree leaf roller*, page 571.

17. Fruits of apple scarred by shallow feeding, most common where a leaf touches the fruit. Skin of the fruit is eaten. Injury occurs late in the season......
Red-banded leaf roller, Eulia velutinana Walker (see *Penn. Agr. Exp. Sta. Bul.* 169, 1921; *Va. Agr. Exp. Sta. Bul.* 259, 1927).

18. Brownish or grayish, winged, vigorously jumping insects, from 1 to 2 inches in length, with conspicuous heads and eyes, feed on the foliage of apple during the late summer, sometimes stripping the trees and scoring the bark............
....*Grasshoppers*, page 573.

19. Very small, dull-black, snout beetles, about $\frac{1}{10}$ inch long, which hop vigorously when disturbed, eat small holes in the underside of leaves from the time when the foliage first starts until midsummer. Foliage of heavily infested trees has the appearance of having been riddled with fine bird shot. Yellowish-brown mines start near the center of the leaf and run to conspicuous blister-like cells in the outer leaf margin........................*Apple flea weevil*, page 573.

20. Twigs and small branches of apple, up to 2 feet in length, are pruned off, especially on young trees. Injury occurs during May and the first of June. At this time large grayish snout beetles nearly $\frac{3}{4}$ inch long, with bodies spotted with black, may be found resting on the trees, or cutting off the twigs and buds. Injury occurs only in the vicinity of woodlands or on newly cleared ground..
New York weevil, Ithycerus noveboracencis (Förster) (see Slingerland and Crosby, "Manual of Fruit Insects," page 210).

21. Buds of apple and the newly set fruit cut off by grayish plump-bodied snout beetles, slightly smaller than A, 20. Beetles are about $\frac{1}{2}$ inch in length, of a

greenish-gray color, with 2 irregular light bands across the wing covers. The wing covers are bent down at the end and terminate in an acute angle. Body covered with overlapping scales............*Imbricated snout beetle*, page 574.

B. *Insects that suck sap from leaves, buds, twigs, branches, trunk, or fruits:*

1. Small gray or brownish-gray spots about $\frac{1}{16}$ inch across, each with a central nipple, on twigs and fruit, often surrounded by reddish or pinkish, inflamed-appearing areas. Twigs covered with a gray coating, having the appearance of ashes. If the larger grayish dots on the bark are lifted, a lemon-yellow, soft-bodied creature, nearly the size of a pinhead, will be found beneath. Trees where the grayish covering occurs on the trunk and branches have thin, yellow, spotted foliage, often in a dying condition...*San Jose scale*, page 575.

2. Grayish-white, pear-shaped, flat scales, about $\frac{1}{8}$ inch long, adhering tightly to the bark of branches and trunk. Smaller three-ridged, straight-sided scales scattered among the larger pear-shaped individuals. In the winter, the larger grayish-white scales will be found to cover a number of small reddish-purple eggs. In the summer, a yellowish, soft-bodied sucking insect will be found under the scale.....................................*Scurfy scale*, page 578.

3. Brownish to grayish-brown, hard, polished scales, very closely resembling a half oyster shell in appearance, and about $\frac{1}{8}$ inch in length, by one-third as wide, adhering tightly to the bark of the apple, so thick as to overlap. In the winter, from 40 to 60 pearly white eggs will be found under most of the larger scales. Scales often in patches on the bark. Heavily infested trees lacking in vigor or dying. Fruit sometimes specked with scales, but usually lacking in reddened areas such as occur around San Jose scale...................... ....*Apple oyster-shell scale*, page 579.

4. Greenish-black shiny eggs about the tips and buds of twigs. Many small soft-bodied sucking aphids, clustered on the buds or curling the newly formed leaves of apple trees in the spring. The feeding causes the dwarfing of the fruit, or practically stops growth, although the apples remain on the tree. Winged and wingless aphids will be found after the time of bloom. Bodies of the insects with a distinct waxy coating, some of them with a pinkish tinge. These aphids migrate from the tree by early summer...................... ....*Rosy apple aphid*, page 581.

5. Dark, greenish-black, shiny eggs about the buds and in crevices in the bark during the winter. Soft green-bodied aphids appear on the buds as soon as the green begins to show in the spring. These aphids continue to feed on the buds, new growth, and water sprouts throughout the summer, often causing a curling of the new growth. Some dwarfing of apples, as in *B*, 4.............. ....*Green apple aphid*, page 583.

6. Shiny greenish-black eggs on twigs and bark as in *B*, 5. Large numbers of green-bodied aphids clustered on the opening buds. They nearly all leave the apple before the time of bloom. Injury by this species is very slight......... ....*Oat aphid* or *apple-grain aphid*, page 584.

7. White cottony masses of wax-like material covering the backs of dark purplish-brown aphids, about $\frac{1}{20}$ inch long, which cluster on wounds along the trunk and branches of apple and other trees (see also *E*, 1)...................... ....*Woolly apple aphid*, page 605.

8. Small greenish, active, slender, winged insects, about $\frac{1}{8}$ inch long, accompanied by pale, greenish-white, wingless, active nymphs, sucking the sap from the underside of leaves of apple. Foliage of infested trees pale in color, with small white dots showing on the leaf; new foliage curled, leaf margins slightly burned; and very small black specks of excrement on the fruit.............. ....*Apple leafhoppers*, page 584.

9. Fruit with irregular russeted spots, misshapen, pitted, or dimpled in appearance, but without cavities, and dwarfed in size; flesh often hard or woody under the injury. Elongate-oval reddish bugs, about $\frac{1}{4}$ inch long, somewhat plump of body, with head of orange color, feeding on the new foliage, and later on the fruit, by sucking the sap through their slender beaks............ ....*Apple red bugs*, page 585.

10. Large numbers of reddish, minute, rounded eggs on the ends of twigs of apple during the winter. Very small, yellowish-green or reddish, 8-legged mites, about $\frac{1}{50}$ inch long, feeding on the underside of foliage of apple and other trees. Infested foliage has a pale sickly appearance. Numerous delicate webs over the foliage and other parts of the tree....*Red spider mite*, page 482.

11. Leaves of apple, plum, peach, and other fruits show a pale, sickly appearance, or with brownish-red specks. At a distance, foliage has the appearance of being dusty. Very light webs over the surface of the leaves, upon which are crawling minute, reddish 8-legged mites.......*European red mite*, page 586.

12. Similar to *B*, 10, but eggs somewhat larger and more reddish in color, often occurring in such numbers as to give a reddish tinge to the twigs. Infested trees with pale foliage which sometimes drops off during late summer. Small 8-legged reddish mites about $\frac{1}{30}$ inch long, with very long front legs, working on the underside of the foliage......................*Clover mite*, page 588.

13. Very small, elongated, 4-legged whitish mites, about $\frac{1}{125}$ inch in length, under bark of trees during the winter. Dark-brown, blister-like, scabby areas on the leaves and occasionally on the fruit. Badly infested trees with distinct yellowish tinge to the foliage................*Pear leaf blister mite*, page 616.

C. *Insects that bore into the trunk, branches, or twigs:*

1. Shallow, irregular burrows in the inner bark, mostly on the sunny sides of apple and other trees. During the early summer months, dark olive-gray flattened beetles, about $\frac{1}{2}$ inch long, blunt at the head end and tapering to a rounded point at the tail, and with metallic-colored, roughened, wing covers will be found on the sunny parts of the trunk and branches. White, flattened, legless grubs from $\frac{1}{2}$ to 1 inch long, with the body greatly enlarged just behind the head, are found in the shallow burrows under the bark................
....*Flatheaded apple borer*, page 589.

2. Nearly round holes, a little larger than a lead pencil, in the lower parts of the trunks of apple trees. Bits of yellowish-brown frass or sawdust forced from small holes in the bark at the base of apple trees. Yellowish-white legless grubs up to 1 inch long, with strong brown jaws and body enlarged but not flattened just behind the head, bore in the inner bark and wood of trees about the base. Velvety-brown cylindrical beetles, about 1 inch long, with two conspicuous white stripes on the back, crawling over the trunk and feeding on the foliage of apple from June to September............................
....*Roundheaded apple borer*, page 590.

3. Many small round holes, about the size of a No. 6 shot, in the bark of the twigs and branches of apple; seldom occur except on trees lacking in vigor. Small black beetles about $\frac{1}{10}$ inch long, the body blunt at either end, crawling over the bark and excavating small holes, usually starting at the base of a bud or twig. Underneath the bark are numerous, fine, sawdust-filled burrows up to 4 inches in length, radiating from a short parent gallery, and often containing small white grubs...............................*Shot-hole borer*, page 592.

4. Small holes through the bark of apple somewhat smaller than *C*, 3, but extending directly into the wood, where they branch several times. Burrows in the wood usually stained a dark color. Brownish or brownish-black blunt-ended minute beetles, working in these holes and occasionally crawling over the bark of the tree...*Pinhole borers*, page 718.

D. *Caterpillars, worms, maggots, or weevil grubs burrowing into, or feeding inside, the fruits:*

1. Newly set apples with crescent-shaped punctures in the side of the fruit, often with a small hole cut on the inside of the apex of the crescent. Early-injured apples often containing small white fat-bodied footless grubs. Small round cavities or pits eaten in the surface of the apple from midsummer until midfall. Dark-brown snout beetles, about $\frac{1}{3}$ inch in length, with grayish-white patches on the back and four slightly raised humps on the wing covers, feeding on the surface of the fruit...............................*Plum curculio*, page 594.

2. Injury similar to *D*, 1, but lacking the crescent-shaped cuts. The cavities are eaten deeper into the fruit and are often close together, so that the side of the fruit may be peppered with small holes about ⅛ inch deep. Apples knotty, misshapen, and undersized. Dead areas of skin often present around the cavities. A brown to light-brown beetle, with slender snout ½ the length of the body, and with 4 very distinct humps on the wing covers, but lacking the white markings of *D*, 1, may be found feeding on the fruit. Injured fruit seldom drops unless heavily fed upon early in the season...................
....*Apple curculio*, page 597.
3. Apples with holes eaten in the flesh of the fruit, the burrow generally running to the core, and seeds more or less injured. Masses of dark castings often protruding from these holes especially at the calyx. Pinkish-white, brown-headed worms, about ¾ inch long, feeding inside the apple, or resting in tough cocoons of white silk, spun under the bark on the trunk or in other shelters about the tree. Minute, flat, white, shiny eggs, ¾ the size of a pinhead, on the leaves adjacent to the growing apples or on the skin of the apple............
....*Codling moth*, page 599.
4. Injury similar to *D*, 3, but holes in the apple about half the size, seldom reaching the core, usually going in from the side of the fruit. Areas of varying size where the larva has eaten under the skin, without feeding deeply on the flesh.
.........*Lesser appleworm* (see *N.Y. (Cornell) Agr. Exp. Sta. Bul.* 410, 1922).
5. Numerous, brown, twisting mines, containing yellowish-white maggots, up to ⅜ inch in length, run through the flesh of the fruit. These mines are most abundant in early varieties of apples, particularly in sweet apples. Mines in the fruit indicated by dark lines on the surface of the apple. Flies a little smaller than the house fly and of a general black color, marked with white bands on the abdomen and black bands on the wings, rest on the surface of the fruit, and deposit their eggs in the flesh........*Apple maggot*, page 603.
E. Insects that attack the roots of the tree:
1. Trees stunted or unthrifty. Large knots on the roots and base of the trunk, often with many short, fibrous roots extending from the knots or tumors. Numerous water sprouts about the crown of the trunk. Purplish soft-bodied aphids on the affected parts, more or less heavily covered with a white powdery or cottony mass of secretion..................*Woolly apple aphid*, page 605.
F. Insects that injure twigs and branches by laying their eggs beneath the bark:
1. Twigs and small branches of apple and many other trees split and scored where bark of twigs has been pushed back and the wood at short intervals raised in small bundles of splinters, with double row of egg punctures beneath. During May and June wedge-shaped, transparent-winged, black-bodied insects, about 1½ inches long, blunt at the head end and with conspicuous eyes at each corner of the head, resting on the foliage, or flying about the trees. Many of them utter a shrill high-pitched song.......................................
........................*Periodical cicada*, or *seventeen-year locust*, page 607.
2. Double rows of punctures or slits, about ¼ inch long, in the bark of twigs and branches, usually most abundant in trees standing in sod or surrounded by weeds and grasses. Slits are crescent-shaped and farthest apart at the middle. The scars remain on the injured twigs for several years. In the late summer and early fall, small, triangular, green, very active, jumping insects, about the size and shape of a beech nut, make these punctures in the twigs, in which their eggs are deposited............................*Buffalo treehopper*, page 610.

135. Spring Cankerworm[1]

Importance and Type of Injury.—For more than two centuries out-breaks of cankerworms have periodically defoliated shade and fruit trees in the eastern United States. In unsprayed or poorly sprayed orchards

[1] *Paleacrita vernata* Peck, Order Lepidoptera, Family Geometridae.

they may cause complete defoliation and loss of the crop; but they are of no importance in well-sprayed orchards. The foliage of the trees is eaten and skeletonized by measuring-worms. The injury occurs just about the time the trees have come into full foliage. Silken threads are spun from branch to branch on the tree and from the branches to the ground. Brown and brownish-green measuring-worms about 1 inch long spin down from the tree when it is jarred or shaken. Heavily infested orchards have much the appearance of having been scorched by fire.

Trees Attacked.—Apple, elm, and many other fruit and shade trees.

Distribution.—General, east of the Rocky Mountains; southeastern Canada; also in California and Colorado.

Life History, Appearance, and Habits.—The winter is passed in the form of naked brown pupae about ½ inch long by ⅛ inch thick. These pupae are found in the soil from 1 to 4 inches below the surface, and in greatest numbers close to the base of the trees. The moths begin emerging during warm periods in February and continue coming out until the end of April. The male moth is strongly winged and is of a dull-gray appearance, being much the color of a well-weathered piece of board. These moths may be seen flitting about from tree to tree at dusk and after dark in the spring evenings. The female moth is wingless, with a gray spidery body. She differs from the fall cankerworm female (Fig. 345)

FIG. 344.—Larvae of the spring cankerworm, *Paleacrita vernata* Peck; side view, showing the two pairs of prolegs near tip of abdomen. About twice natural size. (*From U.S.D.A., Farmers' Bul.* 1270.)

by having a dark stripe down the middle of the back and two transverse rows of small reddish spines across each abdominal segment on the upper side. On emerging from the ground, the female crawls to a tree and up the trunk, or onto the branches, where she mates with the male and deposits her oval dark-brown eggs in irregular masses under the loose scales of bark. These eggs hatch in about a month into small greenish or brownish measuring-worms, which at once begin to feed on the foliage. These worms (Fig. 344) can be distinguished from the fall cankerworm by having only two pairs of prolegs, near the end of the body. They vary from light-brown to nearly black and usually have a yellowish stripe below the spiracles and the under parts partly black. When not feeding, the larvae tend to rest upon the twigs more than upon the leaves. They feed for 3 weeks to 1 month and, if abundant, may completely strip the foliage from the trees. At the end of the feeding period they crawl or spin down to the ground where they excavate the small cells in which they change to the pupal stage and pass the remainder of the summer and the following winter.

Control Measures.—As the wingless female moths crawl up the tree to deposit their eggs, they may be stopped and trapped by placing a band of sticky material, such as tanglefoot, around the trunk of the tree at from 2 to 4 feet from the ground. This should be put on before the first warm weather in February or March. When the moths are abundant, the bands should be watched very closely from February to May, as the trapped moths will often become so numerous as to bridge the bands with their bodies, allowing those emerging later to cross without being caught. All trees within 200 feet of others should be banded, since the moths and small larvae may be blown into banded trees by strong winds, from infested trees near by. As the caterpillars are easily poisoned and do their feeding when the heaviest spray applications of the year are being made, they are never abundant in well-sprayed orchards. They may be easily killed by a spray of lead arsenate at the rate of 3 pounds to 100 gallons of water. Under forest conditions, lead arsenate, 4 pounds in 10 gallons of water per acre, applied from an autogyro gave good control. In unsprayed orchards late spring and summer cultivation, with care to work close to the trunks of the trees, will help to some extent in reducing their injuries, by destroying the pupae.

References.—*Ohio Agr. Exp. Sta. Cir.* 65, 1907; *U.S.D.A., Farmers' Bul.* 1270, 1922, and *Dept. Bul.* 1238, 1924.

136. FALL CANKERWORM[1]

This insect has practically the same life history as that of the spring cankerworm and is controlled by the same measures. It differs from the other species in that the moths emerge late in the fall, some time after the ground has been frozen. The insect passes the winter in the egg stage instead of in the pupal stage, as is the case with the spring species. The ashy-gray eggs are laid in single-layer groups, on branches and trunks of trees, each egg having a flowerpot-like shape. The worms hatch and

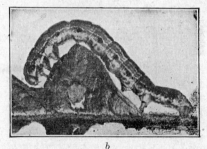

a   *b*

FIG. 345.—At left, female moths of fall cankerworm, *Alsophila pometaria* (Harris); about twice natural size. The female of the spring cankerworm is similar but has a black stripe down the middle of the back and transverse rows of reddish spines on the upper side of the abdomen. At right, larva of the fall cankerworm, side view, showing the three pairs of prolegs near the tip of abdomen. About twice natural size. (*From U.S.D.A., Farmers' Bul.* 1270.)

feed on the leaves in the same manner and at the same time as the spring cankerworm. In this species the larva has three pairs of prolegs near the tip of the body (Fig. 345). They are brownish above, with three narrow whitish stripes along the sides above the spiracles and a yellow stripe below the spiracles, and the underparts light apple-green. They remain more upon the leaves and spin cocoons in the soil before pupating. The

[1] *Alsophila pometaria* (Harris), Order Lepidoptera, Family Geometridae.

females may be distinguished from those of the spring cankerworm by lacking the dark middorsal stripe, and the double rows of reddish spines across the upper side of the abdominal segments.

Trees Attacked.—Apple, elm, and many other fruit and shade trees.

Distribution.—Much of the northern United States and southern Canada; also Colorado, New Mexico, and California.

Control Measures.—The control measures are the same as for the spring cankerworm, except that the trees should be banded in the late fall and not in the spring. Cultivation is of little value for this species, since it forms a tough cocoon.

137. FALL WEBWORM[1]

Importance and Type of Injury.—This insect is of no importance in orchards that are sprayed regularly for the control of codling moth and other pests, but in neglected orchards and upon shade trees and shrubs, it often makes very unsightly webs. Its presence is indicated by loosely woven, dirty white webs (Fig. 346) enclosing the foliage on the ends of the branches. Several branches are sometimes covered by one of these webs. The webs enclose many pale-yellow black-spotted very hairy caterpillars, about 1 inch long, which feed upon the surface of the leaves. These webs contain a quantity of black pellets of excrement from the worms, making them very unsightly.

Fig. 346.—Nest or web of the fall webworm showing worms inside the web. Much reduced. (*From U.S.D.A., Farmers' Bul.* 1270.)

Trees Attacked.—The fall webworm has been found feeding on more than 100 fruit, shade, and woodland trees. It does not attack evergreens.

Distribution.—General over the United States and southern Canada.

Life History, Appearance, and Habits.—This insect passes the winter in the form of brown pupae, enclosed in lightly woven, silken cocoons. These cocoons will be found under trash on the ground or sometimes under the bark of trees. The moths begin emerging during the spring and continue to come out over a long period. Both sexes are winged, satiny white, sometimes with brown or black spots. They lay their eggs on the leaves, in masses, partly covered with white hairs, and the caterpillars hatching from these eggs construct webs over the leaves inside of which they feed. They continue feeding for about 1 month to 6 weeks, and upon becoming full-grown, crawl down the tree and construct the cocoons in which they pupate. The adults emerge late in the summer and lay eggs for a second generation of the worms in early fall, which, upon becoming full-grown, spin the cocoons in which they pass the winter as pupae.

Control Measures.—Sprays for the second and third generations of codling moth will usually control the webworm. In young orchards

---

[1] *Hyphantria cunea* (Drury), Order Lepidoptera, Family Arctiidae.

and on shade trees and shrubs the webs may easily be removed by hand at less expense than will be incurred for spraying.

References.—Ann. Rept. Smithsonian Inst. for 1921, pp. 395–414, 1923; Can. Dept. Agr. Bul. 3(n.s.), 1922; U.S.D.A., Farmers' Bul. 1270, 1931.

138. EASTERN TENT CATERPILLAR[1]

Importance and Type of Injury.—This insect, which has been observed in America since 1646, periodically at intervals of 10 years or so, becomes so abundant as to defoliate unsprayed orchards and many kinds of shade trees. Large thick webs, containing many hairy brown caterpillars, are constructed in the forks and crotches of trees (Fig. 348) in the spring.

FIG. 347.—Eastern tent caterpillar, egg masses on twig. About natural size. (*From U.S.D.A., Farmers' Bul.* 1270.)

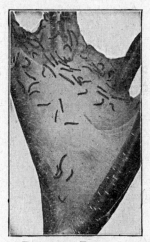

FIG. 348.—Eastern tent caterpillar. Larvae and nest in crotch of wild cherry tree. Greatly reduced. (*From U.S.D.A., Farmers' Bul.* 1270.)

The leaves may be stripped from all the branches within a yard or more of these nests. These caterpillars do not feed within their webs, but congregate there during the night and in rainy weather.

Trees Attacked.—Wild cherry, apple, peach, plum, and more rarely witch hazel, rose, beech, birch, barberry, oak, willow, and poplar.

Distribution.—This species occurs throughout the eastern United States and Canada, westward to the Rocky Mountains, and with closely related forms (see page 698) covers the United States.

Life History, Appearance, and Habits.—This insect passes the winter as a dark-brown collar-like mass of eggs securely attached to, and often encircling, small twigs. These egg masses (Fig. 347) are about ¾ inch long by ½ inch in diameter and contain several hundred eggs. They have a shiny, varnished appearance. The eggs hatch early in the spring as soon as the apple leaves begin to unfold or a little earlier. The caterpillars (Fig. 349) gather in a near fork of the limbs, a colony often being

[1] *Malacosoma americana* (Fabricius), Order Lepidoptera, Family Lasiocampidae.

made up of all the caterpillars hatching from several egg masses.   Here they construct their webs and sally forth to attack the newly opening leaves.   They spin a fine thread of silk wherever they crawl and, in the course of a few days, well-defined silken pathways lead from the nest to the favored feeding spots on the tree.   As the caterpillars grow, their nest webs are enlarged.   They become full-grown in 4 to 6 weeks.   They are then about 2 inches long, thinly covered with long soft light-brown hairs.   The general color is black.   There is a white stripe down the back bordered with reddish-brown, and along each side there is a row of oval blue spots and brown and yellow lines.   They now scatter to some distance from the nest and spin white cocoons, usually on the tree trunk or some near-by object, in which they change to brown pupae and later emerge as light reddish-brown moths with two whitish stripes running obliquely across each forewing.   The females deposit their eggs on the

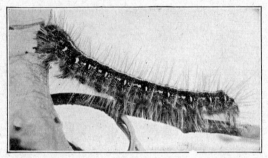

Fig. 349.—Larva of eastern tent caterpillar, *Malacosoma americana* (Fabricius); slightly enlarged.   (*From U.S.D.A., Farmers' Bul.* 1270.)

twigs for the next season's caterpillars early in the summer.   There is only one generation each year, about 9 months being spent in the egg stage.

Control Measures.—Spraying with lead arsenate, 3 pounds in 100 gallons of water, will give good control.   Dusting with 15 per cent lead arsenate in 85 per cent of a suitable carrier is fairly effective.   Arsenicals should be applied when the webs are first noticed.   Spraying for the control of codling moth will control this pest.   Wild cherries and plums should not be allowed to grow within a quarter mile of an apple orchard. The removal of such trees will aid greatly in keeping down several other orchard pests.   The winter egg clusters may be readily seen and should be pruned out and burned in winter.   The nests may be removed by winding them upon the end of a pole bearing a conical brush, or having nails driven into the end of it.   They should then be burned.

Reference.—Conn. Agr. Exp. Sta. Bul. 378, 1935.

139. Yellow-necked[1] and Red-humped Caterpillars[2]

Importance and Type of Injury.—Black- and yellow-striped caterpillars, up to 2 inches long, with yellow rings around their necks, or, in the case of the red-humped caterpillar, with a pronounced red hump just back of the head, with a row of spines projecting from it.   These caterpillars will be found feeding in colonies on the leaves of the apple, pear, and some forest trees during July and August, completely defoliating small trees, or single branches on large trees.

[1] *Datana ministra* (Drury), Order Lepidoptera, Family Notodontidae.
[2] *Schizura concinna* (Smith and Abbott), Order Lepidoptera, Family Notodontidae.

Trees Attacked.—Apple, pear, cherry, quince, and many shade and forest trees.

Distribution.—General over the United States and Canada.

Life History, Appearance, and Habits.—The yellow-necked caterpillar passes the winter in the form of a brown naked pupa 2 or 3 inches below the surface of the ground. The red-humped caterpillar passes the winter as a full-grown larva in a cocoon on the ground, pupating early in the summer. Both emerge as brown moths, about 2 inches across the wings, which fly to the apple and related trees, where the females lay their eggs on the undersides of the leaves in masses of 50 to 100 (Fig. 350). The young caterpillars hatching from the eggs feed at first on a single leaf with their heads all pointing toward the outer edge of the leaf. At first they skeletonize the leaf, but within a few days they increase in size and of necessity spread over a number of leaves, sometimes all the leaves on a single twig or small branch, and begin consuming the entire leaf. During the course of their feeding, they sometimes migrate from one part of the

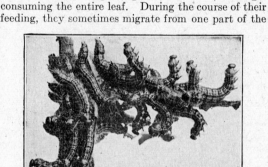

Fig. 350.—Egg mass of yellow-necked caterpillar, slightly enlarged. (*From U.S.D.A., Farmers' Bul.* 1270.)

Fig. 351.—Cluster of larvae of yellow-necked caterpillar, *Datana ministra* (Drury); showing position assumed when alarmed. About natural size. (*From U.S.D.A., Farmers' Bul.* 1270.)

tree to another. When disturbed, these caterpillars raise both ends of their bodies in the air, clinging to the plant with the prolegs near the middle (Fig. 351). They become full-grown in about 3 weeks, at which time the yellow-necked caterpillars enter the ground and change to the pupal stage, and the red-humped caterpillars spin their cocoons on the ground.

Control Measures.—In young orchards where the trees are 3 years old or under, a careful inspection of the trees during late July or early August will show where these insects are starting to feed, and they may be easily removed by hand and killed. A spray of lead arsenate or cryolite, 4 pounds to 100 gallons of water, will kill these caterpillars.

References.—*N.H. Agr. Exp. Sta. Bul.* 139, 1908; *Conn. Agr. Exp. Sta. Bul.* 344, pp. 78, 79, 1933.

140. APPLE LEAF SKELETONIZER[1]

Importance and Type of Injury.—This insect is not troublesome every year, but fluctuates greatly in numbers. It has been most injurious in the states of the upper Mississippi Valley. The leaves at the ends of branches, and particularly in the top of the tree, are loosely folded and covered with a very light web or in some cases two or three leaves are webbed together. The leaves have the green fleshy part of the upper side eaten off entirely or in part, giving them a brownish, deadened appearance. In severe infestations the foliage of the tree has the appearance of injury by severe

---

[1] *Psorosina hammondi* (Riley), Order Lepidoptera, Family Pyralididae.

drought or fire.   Injury first becomes apparent in July and usually is most severe during the first part of September.

Trees Attacked.—Apple and sometimes plum and quince.

Distribution.—Most abundant in the central part of the United States.

Life History, Appearance, and Habits.—The insect passes the winter in the form of a brown pupa, about ⅓ inch long, usually in fallen leaves on the ground about the orchard.   In late spring it changes into a dark-brown moth, about ⅓ inch long, with

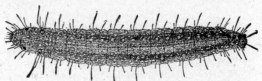

Fig. 352.—Apple leaf skeletonizer, larva, about 5 times natural size.   (*From Ill. State Natural History Surv.*)

wings mottled with silver bands (Fig. 353).   These moths deposit their eggs on the leaves, and the first-generation caterpillars appear during June or July.   These caterpillars (Fig. 352) are brownish-green, about ½ inch long, with four shining black tubercles on the back just behind the head.   They feed for about 3 weeks and, on becoming full-grown, pupate in their webs on the leaves, emerging again as moths in August and depositing their eggs for the second generation of caterpillars, which appear during the latter part of that month and September.   These caterpillars become full-grown by the approach of cold weather, and pupate in the leaves, which later fall to the ground.

Fig. 353.—Apple leaf skeletonizer, *Psorosina hammondi* (Riley); adult with wings spread. About 5 times natural size. (*From Ill. State Natural History Surv.*)

Control Measures.—The ordinary spray schedule, where second- and third-generation codling moth sprays are regularly applied, will usually control this insect.   Occasionally in orchards that have been neglected for a year or two, or in young orchards that have not come into bearing, it will be necessary to give a spray for this insect alone.   Such a spray should be applied when the first appearance of injury occurs, in July, using lead arsenate at the rate of 4 pounds to each 100 gallons of water, being careful to coat the upper surface of the leaves and drive the spray into the webs.

References.—*U.S.D.A., Farmers' Bul.* 1270, 1922; *Fifteenth Rept. Ill. State Ento.,* pp. 58–64, 1889; *Trans. Ky. State Hort. Soc.,* **77**: 82, 1932.

141. Leaf Crumpler[1]

Importance and Type of Injury.—The tough curled cocoons, from ¾ to 1½ inches long, are surrounded by dead crumpled leaves and are tightly fastened to the twigs of the apple (Fig. 354).   The buds are eaten off when they are just beginning to unfold.   The fruit is sometimes scored and pitted by the feeding of the worms.

Trees Attacked.—Apple, plum, crab apple, quince, cherry, wild cherry, wild plum, and pear.

Distribution.—This insect, like the preceding one, is abundant in the upper Mississippi Valley and other northern states.

Life History, Appearance, and Habits.—The winter is always passed as a dark-brown, somewhat hairy caterpillar, about ⅓ to ½ inch long, enclosed in the tough, curled, grayish cases above mentioned.   In the spring of the year, about the time that the apple buds open, these worms become active, loosen their cases from the points

[1] *Mineola indigenella* (Zeller), Order Lepidoptera, Family Pyralididae.

where they have been attached to the tree during the winter and, carrying the cases with them, begin feeding on the newly opening buds, later fastening several leaves together with silken threads. The latter part of May and during June, they change to the pupal stage, and later emerge as moths having a wing expanse of about ¾ inch. The wings are brownish in color, with white mottlings. The moths deposit their eggs on the new apple leaves, the young caterpillars of the next generation appearing in about 2 to 3 weeks. These begin feeding at once on the shoots and leaves, and they construct the silken, curved cornucopia-shaped cases in which they feed for the remainder of the summer. In the early fall they attach these cases securely to the

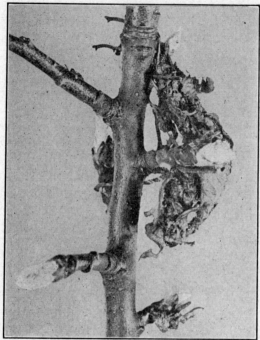

Fig. 354.—Winter nest of leaf crumpler enclosing larva, on apple twigs; slightly enlarged. (*From Ill. State Natural History Surv.*)

twigs and there pass the winter. This insect is never destructively abundant in well-sprayed orchards.

Control Measures.—In orchards where this insect is causing injury, particular attention should be given to applying lead arsenate in the cluster bud, calyx, and 3-weeks' applications. No special sprays are required in commercial orchards other than those regularly given for the control of codling moth, curculio, and other leaf-feeding insects. If young orchards become infested, they should be sprayed about Aug. 1 to 10, with lead arsenate, 6 pounds, and soybean flour, 4 ounces, to 100 gallons of water.

References.—*U.S.D.A., Farmers' Bul.* 1270, 1931; *Fourth Rept. Ill. State Ento.*, pp. 65–74, 1889.

142. Pistol Casebearer[1] and Cigar Casebearer[2]

Importance and Type of Injury.—Tiny, brownish-gray, tough, silken cases, about ¼ inch long, either slightly curved[2] or curled or bent over at the top, so that they roughly resemble a pistol in outline[1] are attached to the leaves, twigs, fruits, and

[1] *Coleophora malivorella* Riley, Order Lepidoptera, Family Coleophoridae.
[2] *Coleophora fletcherella* Fernald, Order Lepidoptera, Family Coleophoridae.

branches, and stand at right angles to them (Fig. 355,b). Small brown worms, enclosed in these cases, which they drag about with them, feed on the leaves, buds, and fruit and eat numerous small holes over the surface. (See also Fig. 90,I.)

Trees Attacked.—Apple, pear, quince, plum, cherry, and haw. The Jonathan apple is apparently immune.

Distribution.—General in apple-growing sections, Virginia to Kansas and north-ward into Canada. The cigar casebearer ranges westward to New Mexico, Montana, and British Columbia.

Life History, Appearance, and Habits.—The winter is passed in the larval or worm stage inside the little pistol- or cigar-shaped cases, the cases being about ⅛ inch long and firmly attached to the bark of twigs or branches. The partly grown larvae

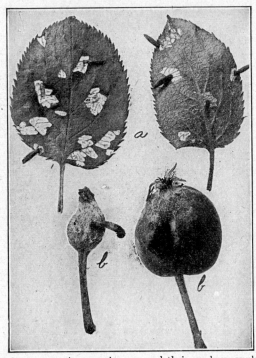

Fig. 355.—Casebearers: *a*, cigar casebearers and their work on apple leaves; *b*, pistol casebearer and its work on young fruits. Natural size. (*From Lochhead, "Economic Entomology," Blakiston, after Caesar.*)

within these cases are light-brown with dark-brown heads. The insects begin feeding about the time the buds unfold in the spring, and reach the full-grown stage about June 1. The cigar casebearer, by protruding the head end from the case, eats in between leaf surfaces, making small blotch mines (Fig. 355,a). This species also makes complete new cases in the spring when the larvae start to feed. About June 1, they change to the pupal stage, and a little later to small mottled gray moths, with lanceolate wings fringed with long hairs and expanding about ½ inch, which deposit their eggs on the underside of the leaves. The larvae of the next generation appear during the late summer. These worms feed on the leaves until early fall, when they migrate to the twigs and branches, where they pass the winter. These little insects are usually not abundant enough to cause commercial loss.

Control Measures.—The ordinary spray schedule for bearing trees will usually control these insects. Where this schedule is not followed, and the casebearers are abundant, a spray given in the cluster-bud stage, using lead arsenate, 3 pounds to 100 gallons of water, or nicotine sulphate, 1 to 800, with an activator, will kill most of them.

References.—N.Y. (Cornell) Agr. Exp. Sta. Buls. 93, 1895, and 124, 1897; *U.S.D.A., Farmers' Bul.* 1270, 1922; *W. Va. Agr. Exp. Sta. Bul.* 246, 1931.

143. BUDWORMS[1]

There are several species of bud moths which cause injury to the apple. They vary somewhat in their life history, but the means of control are the same for all species.

Importance and Type of Injury.—The larvae of these insects eat out the newly opening buds (Fig. 356). The worms form small silken cases on the leaves or twigs, webbing together bits of the foliage. Small pits are eaten in the fruit, usually where a leaf lies in contact with it.

Trees Attacked.—Apple, pear, cherry, wild plum, haw, and possibly others.

Distribution.—Most important in the northern apple-growing sections.

Life History, Appearance, and Habits.—The bud moths or budworms pass the winter as small brown worms with black heads, about ¼ inch long, in small silken cases attached to the axil of a twig or bud. These cases are inconspicuous and not easy to find. In the spring when the leaves begin to open, the little dark-brown caterpillars emerge from their hibernating cases and attack the buds and leaves. They eat out the buds or cut off the stem of a leaf and, folding the edges together, attach

FIG. 356.—Spring foliage injury by budworm on apple. Reduced ½. (*From U.S.-D.A., Farmers' Bul.* 1273.)

it by silken threads to other leaves on the tip of the twig. They live in these cases for from 5 to 7 weeks, feeding at night. In early summer the worms, which are then about ½ inch long, change within their silken homes into a brown pupal stage, and later emerge as grayish or brownish moths. The moths deposit eggs for the next generation, which appears from mid-June to mid-July. These little worms feed in much the same manner as those of the spring generation, although one of the species, the lesser bud moth,[2] mines the leaves to some extent.

Control Measures.—Where abundant, a strong lead arsenate spray, 4 pounds to 100 gallons of water, should be applied just as the buds are opening, and again in the cluster-bud and calyx sprays.

References.—U.S.D.A., Dept. Bul. 1273, 1924; *Can. Ento.*, **65**: 160–168, 1933.

[1] *Spilonota ocellana* Denis and Schiffermüller, and others, Order Lepidoptera, Family Olethreutidae.

[2] *Recurvaria nanella* (Hübner), Order Lepidoptera, Family Olethreutidae.

144. Green Fruitworms[1]

Importance and Type of Injury.—It frequently happens that an orchardist will find large holes eaten in the young apples, from 3 to 4 weeks after the petals fall. This injury is often caused by the larvae of several species of moths which are known as green fruitworms. When the apples are about the size of marbles, one may find the entire side or end of the fruit eaten out (Fig. 357). A close examination will often show that the majority of the young apples on a single branch or twig have been injured in this way. Usually one will be able to find a greenish or greenish-white worm, from ¼ to 1¼ inches long, somewhere upon the twig or branch, possibly feeding actively on the side of the apple. The injury is sometimes severe enough entirely to destroy the apple; more often the fruit will continue to grow but will be worthless. These worms are seldom very abundant in an orchard.

Plants Attacked.—Nearly all the deciduous forest trees, all common tree fruits and some field crops.

Distribution.—General over the eastern United States and Canada.

Life History, Appearance, and Habits.—The winter is passed in two stages, but most commonly as a grayish moth with a wing expanse of a little over 1 inch. These moths hibernate in woodlands or sheltered places about orchards. A small percentage of the insects pass the winter in the pupal stage. This stage will be found in a closely woven, silken cocoon 2 or 3 inches beneath the surface of the soil. Those that have gone through the winter as pupae emerge as moths early in the spring. The moths fly to the orchards as soon as growth starts, and deposit their eggs, one in a place, on the twigs and branches of trees, the moths dying after they have completed egg-laying. There

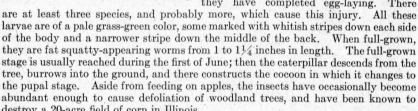

Fig. 357.—Green fruitworm hollowing out a small apple. About natural size. *From Ill. State Natural History Surv.*)

are at least three species, and probably more, which cause this injury. All these larvae are of a pale grass-green color, some marked with whitish stripes down each side of the body and a narrower stripe down the middle of the back. When full-grown, they are fat squatty-appearing worms from 1 to 1¼ inches in length. The full-grown stage is usually reached during the first of June; then the caterpillar descends from the tree, burrows into the ground, and there constructs the cocoon in which it changes to the pupal stage. Aside from feeding on apples, the insects have occasionally become abundant enough to cause defoliation of woodland trees, and have been known to destroy a 20-acre field of corn in Illinois.

Control Measures.—Green fruitworms are sometimes plentiful and destructive in orchards receiving the regular codling moth sprays. This is due mainly to the fact that these worms are so large by the time the calyx spray is applied that they can withstand a dose of poison sufficient to kill smaller insects such as the codling moth, lesser appleworm, and curculio. The most effective measure for controlling green fruitworms is by heavy applications of poison early in the season, when the worms are still small. In orchards where these insects are abundant, the trees should be thoroughly sprayed with 4 pounds of lead arsenate to each 100 gallons of the spray mixture, at the cluster-bud stage of the apple, *i.e.*, the spray applied when the fruit

---

[1] *Graptolitha bethunei* Grote and Robinson, *Graptolitha laticinerea* (Grote); *G. antennata* Walker, and others, Order Lepidoptera, Family Noctuidae.

buds are just beginning to separate, but before the petals have opened. This is practically the only effective method of controlling these insects.

References.—U.S.D.A., Farmers' Bul. 1270, 1931; N.Y. (Geneva) Agr. Exp. Sta. Bul. 423, 1916; Can. Dept. Agr. Ento. Branch, Tech. Bul. 17, 1919; U.S.D.A., Bur. Ento. Plant Quar. Cir. 270, pp. 31, 32, 1933.

145. Fruit Tree Leaf Roller[1]

Importance and Type of Injury.—This insect varies greatly in numbers from year to year. In years of abundance it may ruin 80 to 90 per cent of the apple crop. From shortly after the buds open to about 3 weeks after the petals fall, small greenish to greenish-brown worms feed on the leaves, buds, and small fruits of the apple. In most cases, a light web is spun about several leaves, and these are rolled or drawn together, often enclosing a cluster of newly formed apples. Small apples have cavities eaten out of the side or center, somewhat like those made by the green fruitworm. The trees may be partially to completely defoliated, with numerous fine, white, silken webs over the bark and trunk. At picking time the apples have deep, russeted, elongate scars in the side (Fig. 358).

Fig. 358.—Mature apple showing injury caused by fruit tree leaf roller feeding on the young fruit. (*From Ill. State Natural History Surv.*)

Plants Attacked.—Nearly all kinds of deciduous fruits, many forest trees, some bush fruits, and some herbaceous plants.

Distribution.—General in the apple-growing sections of the United States and Canada.

Life History, Appearance, and Habits.—The fruit tree leaf roller passes the winter in the egg stage. These eggs are laid in masses of 30 to 100, and are closely plastered on the twigs, branches, and occasionally on the trunk of the tree. They are covered with a smooth coating of dull-brown or gray varnish-like material, which protects them from the weather and prevents the individual eggs from showing in the cluster (Fig. 359). The egg masses blend almost perfectly with the bark on which they are deposited, making it extremely difficult to see them. The eggs begin hatching about the time the apple fruit buds are beginning to separate in the spring. The young worms crawl to the leaves, and feed for about 1 month. They vary somewhat in color but in general are pale-green with a brown head and a brown plate just back of the head. When about ¾ inch in length, the larvae pupate within the folded or rolled leaves, or crawl to the trunk or branches of the tree and there construct a somewhat flimsy cocoon. From this cocoon the moths, which have brownish front wings,

[1] *Cacoecia* (= *Archips*) *argyrospila* (Walker), Order Lepidoptera, Family Tortricidae.

expanding ¾ to 1 inch, mottled with pale gold and usually forming two large patches on the front margin (Fig. 360), emerge during late June or July.

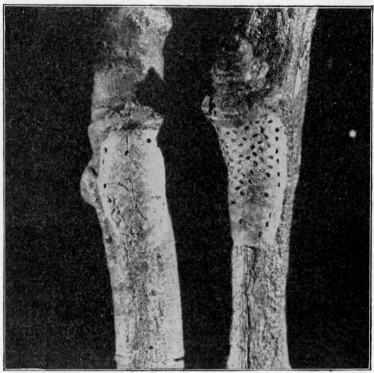

Fig. 359.—Egg masses of the fruit tree leaf roller on twigs of apple about 5 times natural size. Egg mass on the right shows exit holes of newly hatched larvae. (*From Ill. State Natural History Surv.*)

Within a few days they mate and lay their eggs in the situations above described. The insect remains in the egg stage until the following spring. There is but one generation each year.

Fig. 360.—Adult female of the fruit tree leaf roller, about twice natural size. (*From Ill. State Natural History Surv.*)

Control Measures.—While it would appear that the fruit tree leaf roller might be controlled by poison sprays, actual experience in the orchard has shown that it is very difficult to prevent damage by this insect, even with heavy applications of poison. Very effective control has been obtained by thoroughly spraying the trees, during the dormant stage, with a boiled or cold-mixed quick-breaking lubricating-oil emulsion, or other good dormant oil spray, at a strength of 5 to 6 per cent oil. Commercial miscible oils and oil emulsions of several kinds have been found effective for this work. They should be used at the strength recommended by the manufacturers. In spraying for the control of the fruit tree leaf roller, very thorough work must be done, as it is necessary, in order to prevent hatching, to hit each egg mass and the egg masses are often laid on the smaller twigs and in the axils of the twigs.

References.—Colo. State Ento. Cir. 5, 1912; *Mont. Agr. Exp. Sta. Bul.* 154, 1923; *Idaho Agr. Exp. Sta. Bul.* 157, 1928; *Ill. State Natural History Surv., Ento. Ser. Cir.* 9, 1926; *N.Y. (Geneva) Agr. Exp. Sta. Bul.* 583, 1930.

146 (1, 27, 75). GRASSHOPPERS[1]

Grasshoppers are not generally considered as orchard pests, but occasionally during years of abundance when they have eaten most of the green growth in fields, or during periods of drought, they will migrate to orchards and sometimes completely strip the foliage from trees during July and August. Sometimes the bark is scarred and roughened by the feeding of these insects. Outbreaks of grasshoppers in the vicinity of orchards should be controlled before the insects migrate to the trees. This may best be done by the use of the poisoned bran bait (see page 254). If the grasshoppers have already invaded the orchard, the trees should be sprayed with lead arsenate at the rate of 4 pounds to 100 gallons of spray, adding 8 pounds of hydrated lime to prevent burning the foliage.

References.—Mont. Agr. Exp. Sta. Bul. 148, 1922; *Iowa Agr. Coll. Ext. Bul.* 182, 1932; *Ill. Agr. Exp. Sta. Bul.* 442, 1938.

147. APPLE FLEA WEEVIL[2]

Importance and Type of Injury.—This insect has been more destructive in Ohio and Illinois than in other states. Very small dull-black snout beetles about $\frac{1}{10}$ inch long puncture the newly opening leaves and buds of apple trees early in the spring. In the summer the leaves have numerous small holes eaten out from the underside, in cases of severe injury appearing as if riddled by very fine bird shot (Fig. 361). Yellowish-brown larval mines start from near the center of the leaf and run to small blister-like cells at the margin of the leaf.

Trees Attacked.—Apple, haw, winged elm, hazelnut, quince, wild crab, blackberry.

Distribution.—From Missouri and Illinois to eastern New York and southward to the Ohio River.

Life History, Appearance, and Habits.—The adults pass the winter in trash, grass, and leaves just at the surface of the ground, usually under apple trees. They become active as soon as the buds begin to swell in the spring, some crawling up the trunk of the trees, while others mount the stems of weeds and grasses and fly to the newly opening buds. They feed for 1 or 2 weeks on the newly forming leaves, dropping to the ground during periods of cold weather and storms. When the leaves are about two-thirds grown, the females begin depositing their eggs along the midribs. These eggs hatch into small grubs, which feed between the upper and under surface of the leaf, eating out a little mine which is extended to the margin of the leaf. Here they hollow out a cell about $\frac{1}{4}$ inch across and, in this, complete their growth and change to pupae, emerging as full-grown beetles during the latter part of May and June. The newly emerged beetles feed for from 2 weeks to 1 month on the leaves and then seek shelter in the trash about the base of the trees, remaining there until the following spring. There is but one generation each year.

Control Measures.—Spraying with any good fluorine spray will give control. In case of moderate to heavy infestations two sprays should be applied—in the prepink

[1] Various species of Order Orthoptera, Family Locustidae.
[2] *Orchestes pallicornis* Say, Order Coleoptera, Family Curculionidae.

and pink stage of bloom development. Cryolite or barium fluosilicate, 5 pounds, and flotation or other wettable sulphur, 8 pounds, in water 100 gallons make a very effective spray. Care should be taken to apply the spray to the undersides of the leaves. This spray is to be given only when the recommended spray schedule for the control of other apple insects (see page 612) is not effective in controlling this insect. Spraying beneath the tree in late fall with a 15 per cent kerosene emulsion applied in

Fig. 361.—Injury to apple foliage by feeding of apple flea weevil adults, in midsummer.
(*From Ill. State Natural History Surv.*)

sufficient amounts thoroughly to wet down the surface of the ground is another effective measure of control.

Reference.—*Ill. State Natural History Surv. Bul.*, vol. 15, art. 1, 1924; *Jour. Econ. Ento.*, **29**: 381, 382, 1936.

148.—Imbricated Snout Beetle[1]

Importance and Type of Injury.—Occasionally, buds of apple and newly forming fruits are injured by a grayish snout beetle, about ½ inch long. This insect has a body which is very plump and well rounded (Fig. 362,*a,b*). It injures the apple by

---

[1] *Epicaerus imbricatus* (Say), Order Coleoptera, Family Curculionidae.

eating out the buds, or cutting off the young fruit and leaves. It is a very general feeder and is found on many other plants, causing great damage to strawberries. It has seldom been abundant enough to cause serious injury in well-cared-for orchards.

Life History, Appearance, and Habits.—The eggs of this insect (Fig. 362,*e,f*) are laid on the leaves of many plants, and the immature stages (*c,d*) are passed in the stems or on the roots of legumes or some other field crops. The adult insects make their appearance on the trees during the latter part of May and June, and continue feeding for about 1 month. There is one generation a year.

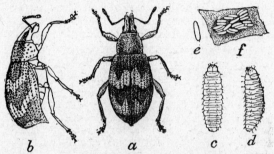

Fig. 362.—Imbricated snout beetle, *Epicaerus imbricatus* (Say). *a, b*, adult, dorsal and side views, 3 times natural size; *c, d*, larva, dorsal and side views, enlarged; *f*, eggs on leaf, twice natural size; *e*, egg more enlarged. (*From Ill. State Natural History Surv.*)

Control Measures.—If the beetles are causing damage to young trees, many of them may be caught by jarring them upon sheets spread under the trees. A strong lead arsenate or cryolite spray, used at the rate of 6 pounds of powder to 100 gallons of water, will kill many of the insects.

149 (172, 179, 189). San Jose Scale[1]

Importance and Type of Injury.—If infestation by this insect is allowed to develop unchecked, it will result in the death of the trees in home or commercial orchards. In 1922, more than 1,000 acres of mature apple trees were killed in southern Illinois by the San Jose scale. Trees lightly infested show small grayish specks on the surface of the bark which are disk-shaped and just discernible to the eye (Fig. 363). Under the hand lens these specks show a raised nipple-shaped spot at the center. Frequently the bark is reddened for a short distance around each of the scales, this being especially true on young trees and on new growth of old trees. On heavily infested trees the entire surface of the bark is covered with a gray layer of overlapping scales, appearing as if the twig or branch had been sprinkled with wood ashes when wet. Infested trees show a general decrease in vigor and thin foliage, which is usually more or less yellowed and spotted because of the presence of scales upon it. Terminal twigs characteristically die first. Sometimes there is an abundance of water sprouts along the larger branches. The fruit on infested trees is also attacked by the scale, the insects being most abundant around the blossom and stem ends, often forming a gray patch about the calyx of the apple. The fruit on infested trees often has a spotted or mottled appearance because of a small red inflamed area surrounding each of the scales (Fig. 64,*E*).

[1] *Aspidiotus perniciosus* Comstock, Order Homoptera, Family Coccidae.

Trees Attacked.—Apple, pear, quince, peach, plum, prune, apricot, nectarine, sweet cherry, currant, gooseberry, and osage orange, besides many bush fruits and shrubs and some shade and forest trees.

Distribution.—Throughout the commercial fruit-growing sections of the United States and Canada and in many other parts of the world.

Life History, Appearance, and Habits.—The insect passes the winter in a partly grown condition. Nearly all the scales surviving a temperature of 10°F. above zero are in the second nymphal instar, often known as the sooty-black stage, and are about one-third grown. The insect remains dormant, tightly fastened to the bark of the tree until the sap starts flowing in the spring. It then begins to grow and usually becomes

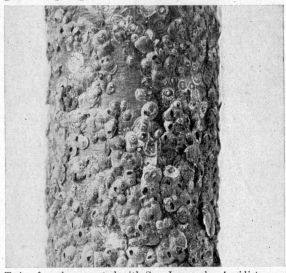

Fig. 363.—Twig of apple encrusted with San Jose scale, *Aspidiotus perniciosus* Comstock. Many of the scales show exit holes of a hymenopterous parasite. Both male and female scales can be recognized. Enlarged several times. (*From Ill., State Natural History Surv.*)

full-grown about the time the apple trees come into bloom. At this time two forms of scales will be noted, one nearly round about $\frac{1}{12}$ inch across, with a raised nipple in the center, and the other oval, about $\frac{1}{25}$ inch long by half as broad, with a raised dot nearer the larger end of the scale. The latter scales are the waxy covering of the males, which emerge as small, yellow, two-winged insects (Fig. 364) at about this time. The larger round scales cover the bodies of the females, which remain under the scales throughout their lives, and, after mating, begin to give birth to living young. They continue to reproduce for a month or more, depending on the temperature. These young have the appearance of very small yellow mites or lice (Fig. 64,*F*). They have six well-developed legs and two antennae; they crawl about over the surface of the bark for a short time. On finding a place which is attractive to them, they insert their slender, thread-like mouth parts through the bark and begin sucking the sap. Very shortly after this they molt or shed their skins and with the old skin lose their legs and feelers, becoming mere flattened yellow sacks attached to the bark of the tree by their sucking mouth parts. As the insect grows, a waxy secretion is given off from the body and this hardens into the protective scale under which the insect lives. There are from two to possibly six generations of the scale each year; the smaller number occurring in the northern part of the country, and the larger in the extreme southern. The insect increases most rapidly in dry hot seasons, it having been determined that the progeny of a single female insect could be well over 30,000,000 in a single year.

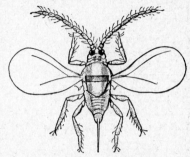

The insect is carried accidentally from orchard to orchard on the bodies of birds and larger insects and also to a greater extent by being blown through the air by the wind. It has been spread throughout the entire country on shipments of infested nursery stock, the original infestation in this country

Fig. 364.—San Jose scale, adult male greatly enlarged. (*From Ill. State Natural History Surv.*)

having been started from an importation of Chinese plants which were set out on the grounds of a large estate at San Jose, Calif., about 1870; from that city the scale gets the name by which it is known in North America.

Control Measures.—The best method of controlling this insect is by the use of a dormant spray which will kill the insect by the action of the spray on its body. Spraying trees during the winter period with one of the many forms of oil emulsions is the best means of control. There are many excellent brands of emulsion now on the market. They should be used according to the directions of the manufacturers. Good formulas for homemade emulsions follow:

Bordeaux cold-mixed: Pump together equal parts of oil (see page 270) and 8-8-100 bordeaux mixture, sending the material at least twice through the pump. For a 2 per cent strength use 4 gallons of this stock emulsion in 100 gallons of water, or 1 part to 25 of water.

Calcium caseinate cold-mixed: Pump together 2 gallons of the right grade oil (see page 271) and 1 gallon of water in which has been dissolved 4 ounces of calcium caseinate. For a 2 per cent strength use 3 gallons of the stock emulsion in 100 gallons of water, or 1 part to 33 parts of water.

These emulsions should be used as soon as made, since the oil will separate upon standing. Oil emulsions must not be mixed or stored in tanks which have contained lime-sulphur without a thorough cleaning. From the time of the general spread of the scale in the east, about 1905, up to 1921, lime-sulphur was the material generally used for this purpose. Lime-sulphur is effective in the cooler parts of the country. It should be used at a dilution of 1 part to 7 or 8 parts of water where the concentrated lime-sulphur solution tests 33° Bé. (see page 269 and Fig. 172). Fall spraying with lime-sulphur may be practiced where for any reason it is not easy to spray in the spring, but the results obtained have not been so good as those from the spring applications. With the oil sprays, the application may be made at any time when the tree is in a dormant condition and the temperature is above freezing. In some of the western states damage to the trees has resulted where oils have been used just before a drop in temperature to 20°F. below zero or lower. The spring applications are most effective, if made just before the leaves appear on the trees. In order to control San Jose scale, it is necessary to do the most thorough and careful spraying. Bear in mind when spraying, that only those insects which are actually hit by the spray will be killed, that the insect to be hit varies from the size of a pencil point to that of a pinhead, and that they are generally distributed all over the bark of the tree on the twigs, branches, and trunk.

References.—Ill. Agr. Exp. Sta. Cir. 180, 1915; *Bul. Ill. State Nat. Hist. Surv.,* vol. 17, art. V, 1928; *U.S.D.A., Farmers' Bul.* 650, 1915, and *Bur. Ento. Bul.* 62, 1906. Experiment station publications of nearly all fruit-growing states.

150 (237). SCURFY SCALE[1]

Importance and Type of Injury.—This insect is found on the bark of the tree in the form of a grayish-white scale, pear-shaped in outline, about

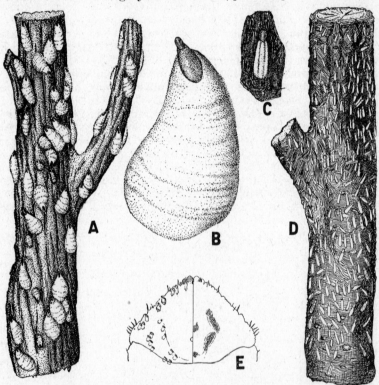

FIG. 365.—Scurfy scale, *Chionaspis furfura* Fitch. *A*, female scales on twig, about twice natural size; *B*, a single female shell about 16 times natural size, showing exuviae of first and second nymphal stages at the pointed end; *C*, a single male shell about 16 times natural size, showing the three longitudinal ridges; *D*, twig bearing male scales, which are usually assembled on a separate twig from the females; *E*, pygidium or terminal abdominal segment from body of female, showing microscopic features used in classifying Coccidae: the dorsal surface is shown on the left, the ventral surface on the right. (*Original.*)

⅛ inch long, rounded at one end and tapering to a rather sharp point at the other (Fig. 365). The scales are usually more abundant on the shaded parts of the tree and in orchards where trees have not been prop-

[1] *Chionaspis furfura* Fitch, Order Homoptera, Family Coccidae.

erly pruned or the foliage is too dense. The fruit is sometimes spotted by this insect. However, it is not of much importance in vigorous well-sprayed orchards.

Plants Attacked.—Pear, apple, gooseberry, currant, black raspberry, Japan quince, mountain ash, and other common deciduous trees and bush fruits.

Distribution.—This is a native insect which is found in the United States, from Idaho and Utah eastward.

Life History, Appearance, and Habits.—The scurfy scale passes the winter as reddish-purple eggs, averaging about 40 under each female scale. The eggs hatch rather late in the spring (late May or June) after the apple trees have come into full leaf. The minute, purplish, young nymphs crawl about for a few hours and then settle down on the bark for the remainder of their lives. The female scales are, when full-grown, about 1/8 inch in length, the males, which are usually concentrated on certain twigs by themselves, are only about one-fourth as long, and are covered by a narrow, straight-sided, intensely white scale with three distinct ridges along the back of it. From these scales the minute two-winged males emerge and fly about to seek the females. The males die soon after mating and the females after they have laid their eggs. There are two generations of the insect each year throughout most of the country, the eggs of the summer generation being deposited in July and hatching the latter half of that month. Some of these nymphs usually settle upon the fruit. Eggs of the overwintering generation are usually laid in late August and September; but only one generation occurs in the North.

Control Measures.—Spraying during the dormant period with a tar distillate oil emulsion, about 4½ per cent, is a very effective control. Dormant mineral-oil emulsions, activated with an ovicidal chemical, such as 2 to 3 per cent dinitro-o-cyclohexylphenol, are also very effective. Liquid lime-sulphur, about 5°Bé. will also give good results.

References.—N.Y. (Geneva) Agr. Exp. Sta., Farm Res., January, 1938; Jour. Econ. Ento., **26**: 912, 913, 1933.

151 (238). APPLE OYSTER-SHELL SCALE[1]

Importance and Type of Injury.—Small dark-brown scales are found adhering closely to the bark of the tree, appearing very much like half of a minute oyster shell (Fig. 366). The scales are about 1/8 inch long by one-third as wide. They usually are more or less clustered on the bark and a heavily infested tree may have the bark entirely covered. The bark of the injured tree usually becomes cracked and scaly. The trees lose vigor, the foliage is undersized and specked with yellow, and in severe infestations the death of the tree results.

Plants Attacked.—Apple, pear, mountain ash, quince, plum, raspberry, currant, almonds, apricots, grapes, figs, and Persian walnuts. A very closely related species, but not the same, is very destructive to shade trees (see page 707). That species has only one generation and does not attack the apple or other fruit trees.

Distribution.—This scale is largely confined to the northern two-thirds of the United States and southern Canada.

[1] *Lepidosaphes ulmi* (Linné), Order Homoptera, Family Coccidae. A very similar species, *Lepidosaphes ficus* (Signoret), is a pest of figs in California.

Life History, Appearance, and Habits.—The insect passes the winter in the form of grayish-white minute eggs, tightly enclosed under the wax of the parent scale.   From 40 to 150 or more of these eggs will be found under each female scale.   The eggs hatch late in the spring after the apple trees have bloomed.   The young nymphs, which are very small and whitish in color, crawl about over the bark for from a few hours to 1 or 2 days.   Once having inserted their beaks into the bark, they begin the

FIG. 366.—Twig infested with apple oystershell scale, *Lepidosaphes ulmi* (Linné); about 3 times natural size. (*From Ill. State Natural History Surv.*)

formation of a waxy scale coating, which covers their bodies, and soon shed their skins and their antennae and legs along with them.   The scale is white at first but later changes to polished brown. The insects become full-grown about the middle of July; the males then emerge as winged insects, and the females, after mating, deposit their eggs under their scales, the body gradually shrinking toward the pointed end of the scale as the eggs are deposited. The females die shortly after the last eggs are laid. These eggs hatch in about 2 weeks, and the second generation of scales becomes full-grown during the early fall, there being two generations of this insect each season over much of its range.   In the northern part of its range, probably only one generation occurs.

Control Measures.—The oyster-shell scale does not yield readily to dormant treatments of ordinary oil emulsion or lime-sulphur.   Sprays of 3 to 5 per cent tar distillate oil emulsions, applied in the dormant stage of the tree, are very effective.   They should be applied according to the manufacturer's directions.   A 2 per cent summer-oil emulsion applied just at the time that the young scales are hatching, in late May or early August, and crawling about over the bark, unprotected by their waxy covering, is very effective.   No fixed date can be given for applying this spray and it will be necessary, in order to accomplish the best results, to watch the trees carefully and spray as soon as the crawling young appear in numbers.   In the latitude of central Illinois hatching occurs about the first of June.

References.—*Ohio Agr. Exp. Sta. Cir.* 143, 1914; *U.S.D.A., Farmers' Bul.* 1270, 1922; *Proc. N.Y. Hort. Soc.*, **79**: 15–24, 1934; *N.Y. (Cornell) Agr. Exp. Sta. Memoir* 93, 1925.

APPLE APHIDS

Importance and Type of Injury.—Aphids of three different species are common on the foliage, fruit, and twigs of apple trees throughout the country.   These are known as the rosy apple aphid, the green apple aphid, and the apple-grain aphid.   The last named is of little importance, although usually the most abundant early in the season.   The relative abundance varies greatly with different seasons, sometimes one species being extremely abundant and the others scarce, while in other years all three may be very numerous. In many years the injury from aphids is slight. When they are abundant, their feeding causes the leaves to curl and the stems and twigs to become stunted and unhealthy in appearance. The new twig growth sometimes assumes a curled and twisted appearance, even forming a loop. In many cases, particularly when attacked by the green and rosy aphids, the apples remain very small. Many of these fruits are also somewhat misshapen, hard, and knotty in texture and character- istically puckered around the calyx end (Fig. 368). The rosy apple aphid is the most injurious species of aphid occurring on the foliage of the apple. A fourth species, the woolly apple aphid, does not injure the fruits but is common on trunks, branches, and roots.

152. Rosy Apple Aphid[1]

Plants Attacked.—Apple is the favor- ite host, although the insect feeds also on pear, thorn, and Sorbus and, during the summer, on the narrow-leaved plantain.

Distribution.—General throughout the United States and apple-growing sections of Canada.

Life History, Appearance, and Habits. The dark-green shiny ovate eggs, in which stage the insect passes the winter, are attached to the bark of the twigs and branches on all parts of the tree and usually hidden away in crevices in the bark or the depressions and wrinkles formed around the buds, twigs, and old wounds (Fig. 367). The eggs begin to hatch when the buds start opening in the spring. With this species, the eggs do not all hatch at once but continue for 2 weeks or sometimes longer. The young aphids (Fig. 369) make their way to the newly opening buds and feed on the out- side of the leaf-bud and fruit-bud clusters, until the leaves have begun to unfold. They then work their way down the in- side of the clusters and begin sucking the sap from the stems and newly formed

Fig. 367.—Eggs of the three com- mon species of apple aphids, enlarged. (*From N.Y. (Cornell) Agr. Exp. Sta. Memoir* 24, *June*, 1919.)

fruits. Their feeding causes the leaves to curl, and this affords the aphids protection from some of their natural enemies and from sprays or dusts applied to the tree for their control. The aphids which hatch from the eggs are all females and are called the stem-mothers, as they are the mothers of the season's brood. In about 2 weeks or a little longer, depending on the weather, these stem-mothers, without mating, begin

[1] *Anuraphis roseus* (Baker), Order Homoptera, Family Aphididae.

giving birth to young, and these young in turn begin reproducing in about 2 weeks.   The body of the aphid has a somewhat waxy coating and usually a slight purplish or rosy tinge.   They continue on the apple during May, and in smaller numbers through June and July.   During the early summer they migrate to stems and stalks of the narrow-leaved plantain, where they feed and reproduce until fall.   In the fall winged ovovivi-

Fig. 368.—Apples injured by rosy apple aphid. (*From N.Y.* (*Cornell*) *Agr. Exp. Sta. Memoir 24, June,* 1919.)

parous females fly back to apple trees and give birth to the true egg-laying females.   The males develop a little later and fly to mate with the true females, which then deposit their eggs in the situations above mentioned. The eggs hatch the following spring.

Control Measures.—Damage by this and the other species of aphids that winter in the egg stage on apple can be prevented by dormant or delayed dormant sprays of one of the tar distillate emulsions, at concentrations of 2 to 5 per cent of the oil.   The commercial tar-oil emulsions are in most cases cheaper than the homemade emulsions.   They should be applied according to the manufacturer's directions. All aphids are killed which are thoroughly wet with nicotine sulphate used at the rate of 1 part to 800, in lime-sulphur solution, or in water with 4 pounds of potash soap to each 100 gallons. After the leaves are about half-grown, nicotine dusts (see page 259) have given better results than sprays.

It is only during an occasional season that these insects become destructively abundant. They are preyed upon by many natural enemies, including the lady beetles, syrphid flies, and aphid-lions. In seasons when the spring is warm, the natural enemies usually become sufficiently abundant to control the aphids, and it is rarely necessary in such years to resort to artificial measures of control. When the spring is

Fig. 369.—Newly hatched nymphs of the three common apple aphids. 1, the green apple aphid; 2, the rosy apple aphid; and 3, the apple-grain aphid, much enlarged. (*From N.Y. (Geneva) Agr. Exp. Sta. Bul.* 431.)

cold and backward, the aphids usually increase more rapidly than their enemies, and it is in such seasons that the greatest damage occurs.

References.—Maine Agr. Exp. Sta. Bul. 233, 1914; *N.Y. (Cornell) Agr. Exp. Sta. Memoirs* 24, 1919; *N.Y. (Geneva) Agr. Exp. Sta. Bul.* 636, 1933; *U.S.D.A., Farmers' Bul.* 1128, 1920.

153. Green Apple Aphid[1]

The green apple aphid has much the same life history as the rosy apple aphid. It passes the winter in the same manner in the egg stage on the bark of wood of the previous season's growth. Unlike the rosy aphid, it remains on the apple trees during the entire summer but does not cause curling of the leaves. It is difficult for the orchardist to distinguish the nymphs and eggs of this species from those of the rosy aphid and apple-grain aphid (see Figs. 367, 369). The adults are yellowish-green, with the head, tips of antennae, legs, and cornicles dark.

Trees Attacked.—Apple, pear, wild crab, hawthorn, and possibly others.

Distribution.—General in the apple-growing sections of North America.

Control Measures.—The control measures are the same as for the rosy aphid, but somewhat more effective because of the fact that the eggs of this species nearly all hatch within a few days and the young will all be clustered on the buds at the time when the most effective spraying can

[1] *Aphis pomi* De Geer, Order Homoptera, Family Aphididae.

be done.   Since the green apple aphid remains on the apple during the summer, and there may be some migration from one orchard to another, a summer application of nicotine may be necessary.   This species is also subject to wide fluctuations in abundance, being controlled by the same natural enemies as those attacking the more injurious rosy aphid.

References.—See under Rosy Apple Aphid.

154. Apple-grain Aphid[1]

This aphid spends practically its entire feeding period on various grains and grasses, being particularly abundant on the small grains commonly grown in the United States.   It is of very little importance as an apple pest, as it does not remain on the apple tree long enough to cause serious deforming of the fruit, twigs, or foliage.   Its eggs are laid on the apple twigs and cannot readily be distinguished from the two species previously discussed.   In some states its eggs are by far the most abundant of the species found on apple, and for this reason one cannot tell from the number of aphid eggs found on twigs during the winter whether or not serious injury from these insects is likely to occur the following spring. The appearance of great numbers of these aphids on the buds in early spring may cause alarm, but they soon disappear from the apple and do not curl the leaves.   The adults are yellowish-green, with darker bands across the abdomen.   The antennae, legs, and cornicles are yellow except at the extreme tips.

Plants Attacked.—Apple, pear, wild crabs, and haws; and grains and grasses during the summer.

Distribution.—General throughout the country.

Control Measures.—No control measures are necessary for this species on apple.

References.—See under Rosy Apple Aphid.

155 (103). Apple Leafhoppers[2]

Importance and Type of Injury.—The importance of these insects varies greatly in different years.   Damage results from both the devitalizing effects of their feeding upon the tree and the spotting of the fruit by their excrement.   During late summer and fall apple foliage becomes pale in color, with little specks of greenish-white showing through from the under surface of the leaves.   The new foliage becomes lightly curled, the margins of leaves sometimes burned, and the fruits speckled with minute deposits of excrement.   Many flimsy, white, shed skins of the leafhoppers are left on the undersides of the leaves.   Severely injured leaves fall from the trees.   One of the species[3] is also a pest of first importance on potatoes (see page 510).

Plants Attacked.—Apple, rose, currant, gooseberry, raspberry, potato, sugarbeet, bean, celery, many other trees, shrubs, and some herbaceous flowers, grains, grasses, and weeds.

Distribution.—General in the United States.

Life History, Appearance, and Habits.—The apple and potato leafhopper[3] apparently does not winter in the northern states (see page 512).

---

[1] *Rhopalosiphum prunifoliae* (Fitch), Order Homoptera, Family Aphididae.
[2] Several species of the Order Homoptera, Family Cicadellidae.
[3] *Empoasca fabae* Harris.

The rose leafhopper,[1] the white apple leafhopper,[2] and another common apple leafhopper,[3] pass the winter in the egg stage in the bark of apple, rose, and other plants. Those hibernating as adults become active very early in the spring, the common species flying about and mating before the buds of the trees have begun to show green. When the leaves appear, the insects begin laying their eggs, which are pushed into the midrib or the larger veins and stems of the leaves. The first-generation nymphs appear about the time the leaves become full-grown, or shortly thereafter. They, as well as the adults, feed by sucking the sap from the underside of the leaves. The nymphs are a pale-green to greenish-white in color, are wingless, but are very active, running forward, backward or sidewise with equal ease. They reach maturity about midsummer, change to the adult stage, and deposit eggs for the second generation, which becomes full-grown during the early fall. Some of the species found in orchards differ somewhat in their life history from the above, but not to an extent to be of any significance in the application of control measures.

Control Measures.—Where the leafhopper nymphs average 50 or more per 100 leaves in the spring (May in central Illinois), spraying is advised. Forty per cent nicotine sulphate, 1 pint, and soap, 2 pounds, per 100 gallons of water containing ½ per cent summer oil, has given good control. Lead arsenate, 3 pounds per 100 gallons, may be used in this combination for codling moth control. Sprays containing 1 per cent rotenone have given fair control. If there are fewer than 50 nymphs per 100 leaves, the expense of spraying for these insects is not generally warranted. They are best controlled in the spring.

FIG. 370.—Apples deformed by apple red bugs. Note the dimpled appearance. (*From U.S.D.A., Farmers' Bul.* 1270.)

References.—U.S.D.A., Dept. Bul. 805, 1919; *N.Y. (Geneva) Agr. Exp. Sta. Bul.* 541, 1918; *Jour. Econ. Ento.,* **17:** 594, 1924; *Conn. Agr. Exp. Sta. Cir.* 111, 1936.

156. Apple Red Bugs

Importance and Type of Injury.— The apple-growing states in the East often suffer considerable losses from certain sucking insects known as apple red bugs. Infested fruit (Fig. 370) has a pitted or dimpled appearance, is dwarfed in size, is somewhat hard or woody in texture, and is sometimes russeted in spots. The general appearance is somewhat like that of the injury by the rosy apple aphid except for the dimpling or pitting and russeting, which do not usually occur in the case of injury by aphids.

Trees Attacked.—Apple, pear, haw, and probably others.

Distribution.—States east of the Mississippi River. Most destructive in New York, New England, and southeastern Canada.

---

[1] *Empoa rosae* (Linné).
[2] *Typhlocyba pomaria* McAtee.
[3] *Empoasca maligna* Walsh.

Life History, Appearance, and Habits.—Red bugs pass the winter in the egg stage. These eggs are laid in the bark of branches of the trees in the case of the dark apple red bug,[1] and in the bark lenticels in the case of the light or false apple red bug.[2]   They hatch early in the spring.   The young nymphs feed at first on the foliage and later on the young apples.   They are piercing-sucking insects and feed entirely on the sap of the leaves or juice of the fruit.   Wherever they insert their beaks in the apple, the surrounding tissue becomes hardened and ceases to grow, and the entire fruit is stunted.   These two species have essentially the same habits.   Upon becoming

full-grown, the adults of the dark apple red bug are about $\frac{1}{4}$ inch long, reddish black in color and covered on the upper surface with white flattened hairs.   In the false red bug (Fig. 371) the head and front part of the body are of an orange color. There is but one generation annually, nearly all the injury being caused by the young or nymphs, during the first month after the fall of the petals.

Control Measures.—Red bugs cannot be controlled by the application of any stomach poison, as they do not feed on the surface of the leaves or the fruit.   Where abundant they may be controlled by the use of a contact poison such as 40 per cent nicotine sulphate 1 part to 800 parts of water, or 1 pint to 100 gallons.   This should be applied in the cluster-bud spray and again in the calyx spray (see Spray Schedule, page 612). Nicotine sulphate may be combined with the ordinary-strength lime-sulphur and lead arsenate, or may be applied alone.   In the latter case, it is much more effective if 1 pound of potash fish-oil soap is dissolved in each 25 gallons of the spray mixture.

Fig.   371.—Adult of false apple red bug, *Lygidea mendax* Reuter.   About $3\frac{1}{2}$ times natural size.   (*From U.S.D.A., Farmers' Bul.* 1270.)

The spray should be applied on warm days by two men working from opposite sides of the tree, as the red bugs are so active that they may escape being hit by a spray applied to one side of the tree at a time.   On cool days, the insects will not be feeding in the trees.

References.—*N.Y.* (*Cornell*) *Agr. Exp. Sta. Buls.* 291, 1911, and 396, 1918; *N.Y.* (*Geneva*) *Agr. Exp. Sta. Bul.* 490, 1921, and *Tech. Bul.* 52, pp. 66–69, 1933.

157. European Red Mite[3]

Importance and Type of Injury.—This insect has become one of the most important fruit pests of the eastern United States and Canada. The injured trees, if the infestation is slight, show specking of the foliage; if the infestation is heavy, the foliage is pallid and sickly or bronzed in appearance, and from a little distance it has the appearance of being covered with dust.   Many of the injured leaves drop.   The fruit is undersized and of poor quality and color, and fruit buds are greatly weakened or prevented from forming.

Trees Attacked.—The mite occurs on many deciduous fruit and shade trees and shrubs but is most injurious to plum, prune, apple, and pear.

Distribution.—This pest, which is widely distributed over continental Europe, was first found in the United States in 1911 and has rapidly appeared in many localities in the northeastern and northwestern quarters of the United States, north of latitude 37°N.

Life History, Appearance, and Habits.—The mite passes the winter as somewhat spherical eggs (Fig. 372,*A*), of a bright-red to orange color, on

---

[1] *Heterocordylus malinus* Reuter, Order Hemiptera, Family Miridae.

[2] *Lygidea mendax* Reuter, Order Hemiptera, Family Miridae.

[3] *Paratetranychus pilosus* (Canestrini and Fanzago), Order Acarina, Family Tetranychidae.

twigs and smaller branches of the trees. The egg has a distinct style or stalk about as long as the diameter of the egg and without guy lines. The eggs hatch in the spring just before the time of apple bloom. The young mites crawl to the leaves and suck the sap from the unfolding leaves, becoming full-grown in 2 to 3 weeks if the weather is warm, or after a longer period if cool. They do not spin much silk as the common red spider does. After mating, the females lay their eggs to the average number of about 30 or 35. Eggs from unmated females develop into males, those from fertilized eggs into both males and females. Generation succeeds generation, all stages of the mites being found on the trees

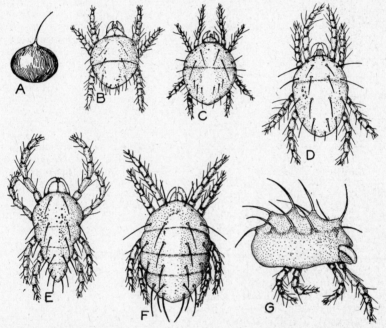

FIG. 372.—European red mite, *Paratetranychus pilosus* (Canestrini and Fanzago). *A*, egg; *B*, first instar or six-legged stage; *C*, protonymph; *D*, deutonymph; *E*, adult male; *F*, adult female; and *G*, adult female in side view. All enlarged about 75 times. (*From U.S.D.A.*)

during the summer months. The average length of the life cycle is about 21 days, and there are from four to eight generations a year. The adult females are bright to brownish red, unspotted, the body quite elliptical in outline, about 1/75 inch in length. There are four rows of long curved spines down the back, each borne on a whitish tubercle. Hot, dry weather is favorable to the increase of these mites.

Control.—Thorough spraying, while the trees are completely dormant, with a good miscible oil or homemade soap-lubricating-oil emulsion has proved very effective in the control of this species. Such sprays should contain not less than 2½ per cent actual oil. Where these sprays are applied, a satisfactory commercial control can be obtained. Spraying during the summer with bordeaux containing 6 quarts of summer oil per 100 gallons of water will aid in controlling these mites. The same schedule should be followed as that recommended for the control of apple scab and blotch.   The dormant spray, however, is the most satisfactory method of control.

References.—Conn. Agr. Exp. Sta. Bul. 252, 1923; Can. Dept. Agr., Ento. Branch, Cir. 39, 1925; U.S.D.A., Tech. Bul. 25, 1927; Jour. Agr. Res., vol. 36, no. 2, 1928; U.S.D.A., Tech. Bul. 89, 1929; Can. Jour. Res., 13: 19–38, 1935; Mass. Agr. Exp. Sta. Bul. 305, pp. 28–36, 1934; Jour. Econ. Ento., 29: 546–550, 1936.

158. The Pacific Red Spider Mite[1]

This species, the females of which are almost indistinguishable from the common red spider, but the males distinct, is rated by certain writers as the worst crop pest on the Pacific Coast.   It is especially destructive to deciduous orchard crops, vineyards, and ornamentals, often causing complete defoliation.   It ranges from Oregon to California.   Neutral oil emulsions and dry lime-sulphur sprays, applied early in an infestation before the mites have spun much silk, have given the best control.   Well-watered and fertilized vigorous trees, by producing more shade and humidity, generally escape serious damage.

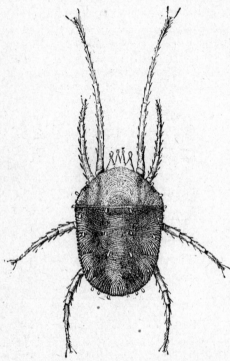

Reference.—U.S.D.A., Cir. 157, 1931.

159. Clover Mite[2]

Importance and Type of Injury. In dry seasons these mites sometimes do considerable harm to apples, by sucking the sap of the buds and leaves.   Infested foliage takes on a thin yellowish appearance and, during prolonged drought, many of the leaves fall from the trees.   Occasionally during the dormant stage of the tree one will find small reddish or pinkish eggs attached to the surface of the bark around the buds and tips of the twigs.   They are some-

Fig. 373.—Adult clover mite, *Bryobia praetiosa* Koch; greatly enlarged.   (*From U.S.D.A., Farmers' Bul.* 1270.)

times so numerous that the twigs have a reddish appearance.   These are the eggs of a little creature related to spiders and ticks, and known as the clover mite.   The eggs of some other species of mites are found in the same situations.   The most characteristic and easily noticed indication of the presence of this mite is that of the eggs during the winter.

Plants Attacked.—Apple, peach, prune, plum, pear, cherry, almond, raspberry, and many forest trees and herbaceous plants.

Distribution.—Northern United States and Canada.

[1] *Tetranychus pacificus* MacGregor, Order Acarina, Family Tetranychidae.
[2] *Bryobia praetiosa* Koch, Order Acarina, Family Tetranychidae.

Life History, Appearance, and Habits.—The eggs are deposited by the adult females during the late fall. The mites hatch in the early spring and feed on the foliage of trees to some extent, though they depend mainly on the leaves of many herbaceous plants. They are eight-legged reddish creatures, smaller than a pinhead, with the front legs much longer than the others (Fig. 373). Their injuries are more apt to cause damage to the stone fruits, particularly the plum. There are a number of generations throughout the growing season. Occasionally in the fall the mites invade houses in large numbers, and, while they cause no damage inside the houses, their presence is extremely annoying.

Control Measures.—A dormant spray of one of the tar distillate oil emulsions, at 2½ per cent oil, is effective in controlling the clover mite by killing the overwintering eggs. If the mites become destructively abundant during the summer, spraying with 1½ to 2 per cent summer-oil emulsion has given good control.

References.—Colo. Agr. Exp. Sta. Bul. 152, 1909; Conn. Agr. Exp. Sta. Bul. 327, pp. 574, 575, 1931.

160. FLATHEADED APPLE BORER[1] AND PACIFIC FLATHEADED BORER[2]

Importance and Type of Injury.—This is one of the worst enemies of deciduous trees and shrubs. It kills many trees and shrubs in the nursery and many more after they have been set in orchards, parks, city streets and lots, and along highways. It is especially destructive during the first 2 or 3 years after the trees are planted, or in very dry seasons, or where parts of trees, that have been shaded, are exposed to the sun by pruning. The presence of this insect is indicated by shallow, broad, irregular mines or burrows on the main trunk or large branches, just under the bark and in the wood, the larger ones going into the wood a distance of an inch or two. Above these burrows are dark-colored dead areas of bark, often with sap exuding. These burrows are packed tightly with fine sawdust, except where they go into the wood; here they are usually packed with coarse, excelsior-like fibers. They are nearly always on the sunny side of the tree but may extend completely around the tree. Injuries usually result in killing large areas of

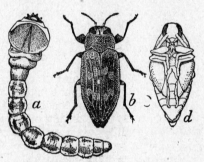

FIG. 374.—Flat-headed apple borer, *Chrysobothris femorata* (Olivier). *a*, larva; *b*, adult beetle; *d*, pupa, about twice natural size. (*From Chittenden, U.S.-D.A.*)

bark and sometimes in girdling and killing the tree or infested branches. There are several other closely related species that also sometimes attack apple.

Trees Attacked.—Nearly all fruit, woodland, and shade trees.

Distribution.—Generally distributed throughout the United States and the fruit-growing sections of Canada. The Pacific flatheaded borer occurs throughout western North America from Canada to Arizona and Texas.

Life History, Appearance, and Habits.—The winter is passed in the grub or borer stage. These are of different sizes, from ½ to 1 inch in length. The larger, nearly full-grown borers will be found from 1 to

---

[1] *Chrysobothris femorata* (Olivier), Order Coleoptera, Family Buprestidae.
[2] *Chrysobothris mali* Horn.

2 inches deep in the wood of the tree, usually to a less depth in the southern states. In the spring they change to yellow pupae (Fig. 374,*d*) and later to beetles. The full-grown grub is about $1\frac{1}{4}$ inches in length, legless, of a yellow to yellowish-white color with a broad flat enlargement of the body just back of the head (Fig. 374,*a*). It usually lies with the body curved to one side. The adult beetle (*b*) is about $\frac{1}{2}$ inch long by $\frac{1}{5}$ inch wide. It is of a dark olive-gray to brown color with a metallic luster on the irregularly corrugated wing covers. The body is very blunt at the head, and tapers to a rounded point at the posterior end. They are decidedly sun-loving insects, and will be found in greatest numbers on the sunny sides of trees or logs. The female beetle lays her yellow disk-like wrinkled eggs in cracks in the bark of trees, nearly always selecting a tree that is unhealthy, or a spot on a healthy tree where the bark has been injured, as by sunscald or a bruise. The eggs are laid from May to August, and most of the borers complete their growth by fall, the life cycle occupying only 1 year.

Control Measures.—Nearly all damage by this insect may be prevented by wrapping the trunks of the trees the first year they are set. Wrappings should be placed before the middle of May and should extend from the ground level to the lower branches. Any good grade of paper, even several thicknesses of old newspapers, may be used. The paper should be held in place with twine and should remain on the trees through the second year. The only effective remedy, once a tree is infested, is to cut out the grubs with a sharp-pointed knife. This should be done during the late summer or early fall. The presence of the insects is indicated by darkened areas of bark and fine bits of sawdust protruding through the bark. Tree borer paints, such as paradichlorobenzene, 2 pounds, in cottonseed oil or dormant oil emulsion, 1 gallon, applied when growth is not rapid, will kill the small larvae but are of little value. Care should be taken to avoid planting of nursery stock that is infested and to keep trees in a healthy, vigorous condition by proper planting, watering, pruning, cultivation, and avoidance of wounds. Wounds should be promptly covered with a good tree paint. The shading of the trunks of young trees by pruning to head the trees low, or by wide stakes, tends to keep the adults away.

References.—*U.S.D.A., Farmers' Buls.* 1065, 1919, and 1270, 1931, and *Tech. Bul.* 83, 1929.

161. ROUNDHEADED APPLE BORER[1]

Importance and Type of Injury.—This insect may cause a loss of 10 to 20 per cent of the apple trees in an orchard in years when it is abundant. This damage is caused by the borer, or grub, which feeds on the inner bark and sapwood of the tree. The burrows of this insect are usually made in the base of the trunk, from 1 or 2 inches below the surface of the ground to a foot or more above ground. They extend through the sapwood and heartwood, often seriously weakening young trees, and sometimes girdle the tree. The presence of the insect is indicated by coils or piles of sawdust-like particles adhering to the bark, or on the ground about the base of the trunk (Fig. 375), and by darkened areas in the bark

[1] *Saperda candida* Fabricius, Order Coleoptera, Family Cerambycidae.

about the base of the tree. To make a thorough examination for the presence of this insect, it is necessary to remove the earth about the base of the trunk to a depth of 2 or 3 inches and examine the bark carefully.

Trees Attacked.—Apple, pear, quince, haw, mountain ash, serviceberry, and occasionally some others.

Distribution.—United States and Canada, from the Dakotas and Texas, eastward; also New Mexico and British Columbia.

Life History, Appearance, and Habits.—The winter is passed only in the larval or borer stage. The borers are of two sizes; those which have hatched from eggs laid in the past season, and those from eggs laid a year earlier, the latter being full-grown and an inch or over in length. They are of a creamy-yellow color, with a brown head and a rounded thickening of the body just behind the head (Fig. 376). The two-year-old larvae

Fig. 375.—Castings of roundheaded apple borer at base of apple tree. Trees showing such condition should receive treatment by hand or by fumigation to destroy the borers beneath the bark. (*From U.S.D.A., Farmers' Bul.* 1270.)

Fig. 376.—Larva of roundheaded apple borer, nearly full-grown. About natural size. (*From U.S.D.A., Farmers' Bul.* 1270.)

are usually found in the tree to a depth of 1 to 2 inches. In the spring of the year these larger larvae change to the pupal stage and, after 2 weeks' to a month's time, emerge as robust velvety-brown cylindrical beetles with a conspicuous white stripe on each side of the body above, and the underside of body, legs, and head, except the eyes, also white (Fig. 377). The beetles are very striking in appearance. They are abroad from June to the first of September. The adults crawl over the surface of the tree and feed to some extent on the foliage and on the new twig growth. They have well-developed wings, but are rather sluggish creatures and usually fly only short distances. They lay their eggs during the summer months, these being deposited in the bark of the trunks of apple trees from just below the surface of the ground to 18 or 20 inches up on the trunk, or occasionally even higher than this. They are usually laid in cracks of the bark, which are sometimes enlarged by the beetles with their strong jaws. The borers, upon hatching 2 or 3 weeks later, feed at first on the outer bark and, as they increase in size, work into the wood of the tree.   The older, or nearly full-grown ones, are found several inches within the wood and are usually a foot or more above the surface of the ground, while the young borers may work several inches below the surface.   When full-grown, the larva hollows out a cell in the wood in which to pupate.   In most cases, the insect requires 2 years in which to complete its growth but may require 3 under unfavorable conditions.

Control Measures.—Hand worming in late August or early September and again in late April is advised.   The base of the tree should be carefully examined, 1 or 2 inches of the soil around the trunk removed, and the young borers dug out or probed out. So far as possible, workmen should avoid cutting across the grain of the bark.   A flexible wire is used to reach the borers deep in the wood.   The larger borers may be killed by injecting carbon disulphide or nicotine paste material into the burrows and plugging the opening with putty.   A mixture of calcium arsenate and raw linseed oil applied to the openings of the burrows has been used with some success.   The ordinary spray schedule applied for controlling other orchard insects aids to some extent in controlling the round-headed apple borer, as the adults will frequently get enough poison to kill them when feeding on the new bark or the leaves.   All growth of crabs, haws, and particularly serviceberry and mountain ash should be kept down within ¼ mile of the apple orchard.

Fig. 377.—Adult male and female roundheaded apple borer, *Saperda candida* Fabricius.   Male on right, female on left, about natural size.   (*From U.S.D.A., Farmers' Bul.* 847.)

References.—*U.S.D.A., Farmers' Bul.* 1270, 1931, and *Dept. Bul.* 847, 1920; *Ark. Agr. Exp. Sta. Bul.* 146, 1918; *Can. Dept. Agr. Ent. Branch Cir.* 73, 1930.

162 (183).   SHOT-HOLE BORER OR FRUIT TREE BARK BEETLE[1]

Importance and Type of Injury.—Small holes, about the size of a pencil lead, are often eaten through the bark on the twigs of healthy fruit trees, especially above a bud or other projection.   The holes are sometimes indicated by a small amount of sawdust or borings on the bark of the tree.   On peach, cherry, and other stone fruits these holes are usually covered and sealed in by dried droplets of gum, which hang from the twigs like tear drops.   When the insects become very abundant, wilting and yellowing of the foliage occur and are usually followed by the death of the tree.   Bark of twigs, branches, and trunks of weakened, infested trees is perforated with numerous small shot-hole-like openings, from which the beetle gets the name "shot-hole borer."   The removal of the bark exposes many, small, winding, sawdust-filled, gradually enlarging

[1] *Scolytus rugulosus* Ratzeburg, Order Coleoptera, Family Scolytidae.

galleries leading out from a shorter central gallery (Fig. 378,*C*). In nearly all cases, the attack of the borers results in the death of the tree or branches where they are numerous, but these beetles are very rarely the *primary* cause of the death of the tree.

Trees Attacked.—Apple, peach, plum, cherry, quince, serviceberry, choke cherry, and many other trees.

Distribution.—This insect is a native of Europe but is now generally distributed over the United States.

Life History, Appearance, and Habits.—During the winter, this insect is in the grub or larval stage in the inner bark. At this time, the legless

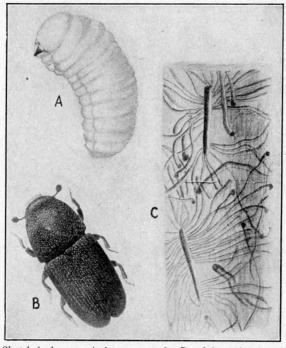

Fig. 378.—Shot-hole borer. *A*, larva or grub; *B*, adult or beetle, about 15 times natural size; *C*, parent and larval galleries in sapwood of injured twig from which the bark has been removed; about natural size. (*From Ill. State Natural History Surv.*)

grubs are about ⅛ inch long, of a pinkish-white color, with a slight enlargement of the body just behind the head (Fig. 378,*A*). In the early spring they change to the pupae and emerge as full-grown insects during June and July. The adult insects (*B*) are about ¹⁄₁₀ inch long, by nearly half as wide, black in color, the body very blunt at either end. They have well developed wings and are capable of flying considerable distances. The insects mate, and females seek out trees that are in a somewhat unhealthy condition. They enter the bark at a point along the branch or twig, usually just above a slight projection, or at a lenticel, and excavate a gallery about 1¼ to 2 inches long, usually running in the same direction as the length of the trunk or branch. They deposit their white spherical eggs at short intervals on either side of this parent gallery.

The female usually dies with the tip of her body blocking the entrance to her egg gallery. The young grubs hatching from these eggs start burrowing in the inner bark, in general at a sharp angle from the parent gallery. They continue their burrows until they become full-grown, about 6 to 8 weeks. The larval burrows are from 2 to 4 inches long, enlarging slightly throughout their length and diverging gradually outward from the parent gallery. They are packed with frass, while the parent gallery is clean. When full-grown, the larva changes to a pupa and later to an adult beetle at the end of the larval burrow beneath the bark. The beetles emerge from holes bored directly outward through the bark. There are from one to three generations of the insect each year, the larger number occurring in the South.

Control Measures.—Many experiments have been carried out to test the effect of treating the bark of infested trees with some substance strong enough to kill the young larvae in their burrows. Some fairly favorable results have been obtained by using a solution of carbolineum. In many cases, however, severe injury to the trees has resulted from the use of this substance. Probably the best method of preventing and overcoming the attack of this insect is to provide adequate water and food for the trees. Trees that are in a backward, somewhat sickly, condition should be given a heavy treatment in the spring with some strong nitrogenous fertilizer, such as nitrate of soda or sulphate of ammonia, applying the material to the surface of the soil above the roots of the tree. The dosage should vary with the size of the tree. Ten-year-old trees may safely receive from 3 to 5 pounds of either of these fertilizers, and the amount may be doubled on 20-year-old trees. Painting with a light distillate type of oil in which flake naphthalene has been dissolved, $\frac{3}{4}$ pound per gallon, will kill a large proportion of the insects.

During the winter all badly diseased trees or branches in or near the orchard should be cut out and all prunings promptly burned. Infested firewood should be burned before the spring following cutting. This applies, not only to the apple, but to the other favored food plants of this insect (see also pages 627 and 718).

References.—*U.S.D.A. Farmers' Bul.* 1270, 1922; *Ohio Agr. Exp. Sta. Bul.* 264, 1913; *Rept. Neb. State Ento.*, 1909; *Univ. Colo. Agr. Ext. Serv. Cir.* 64, 1932; *U.S.D.A., Farmers' Bul.* 1666, p. 131, 1931.

163 (187, 191, 194). Plum Curculio[1]

Importance and Type of Injury.—This snout beetle, which is primarily a pest of peach, plum, cherry, and other stone fruits, is sufficiently fond of the apple to make it second to the codling moth in importance as a pest of this fruit. Injury on apple is shown by small crescent-shaped cuts in the skin of small fruits, some of them with a little round hole opposite the concave side of the crescent, into which an egg is usually deposited (Fig. 380). Later these injuries develop into swellings or knots, protruding from the surface of the fruits, each with a small puncture in the skin at its apex. Apples will sometimes show depressions instead of swellings, with the curculio injury at the center of the depression. An examination of such apples will sometimes reveal a grayish-white curved worm inside. Many of the infested fruits drop during late May and June. During late

[1] *Conotrachelus nenuphar* Herbst, Order Coleoptera, Family Curculionidae.

summer, numerous, round, feeding holes or punctures are made through the skin of the apple and other fruits, and the flesh is eaten out beneath these punctures (Fig. 381). The infested apples are often hard, knotty, and misshapen (Fig. 379). In peaches, plums, and cherries serious losses result from the feeding of the grubs in the fruit and their presence in ripened marketed fruits (see pages 635, 639, and 642).

Trees Attacked.—Plum, pear, apple, peach, cherry, apricot, prune, nectarine, quince, and other cultivated and wild fruits.

Fig. 379.—Apples deformed by the plum curculio. (*From U.S.D.A., Farmers' Bul.* 1270.)

Distribution.—East of the Rocky Mountains in the United States and Canada.

Life History, Appearance, and Habits.—This insect passes the winter as a dark-brown snout beetle, about ¼ inch long, with grayish or whitish patches on its back and four humps on the wing covers (see Fig. 382). A strong curved snout, about one-third the length of the body, projects forward and downward from the head of the insect. These beetles seek protection in and around orchards or near-by woodlands, where they find

Fig. 380.—Egg punctures made by the plum curculio on apples. (*From Ill. State Natural History Surv.*)

shelter during the winter, under fallen leaves, in piles of stone, and about rock outcroppings and fences. They become active about the time the apples and peaches bloom, or possibly in some years a little earlier than this. They fly to the trees, feed on the newly forming apples or buds, petals, shucks, and newly set fruits. They then mate, and the females begin laying their eggs. In this operation the female first eats a small round hole into the skin of the fruit, then turns around and deposits a shiny white egg in the hole, and finally cuts a crescent-shaped slit beneath the egg with her mouth parts so as to leave the egg in a gradually dying flap of the fruit (Fig. 380). Over 500 eggs have been deposited by single females, but the average is believed to be less than 100. Upon hatching, 2 to 12 days later, the young grub eats into the flesh of the apple to the core and seeds. The curculio larva is grayish-white, legless, curved-bodied, with a small brown head (Fig. 382). It is about ⅓ inch long when full-grown. Infested apples nearly always fall to the ground before the curculio has completed its growth. When the apple remains on the tree, the larva does not develop, but peaches infested after the stone starts

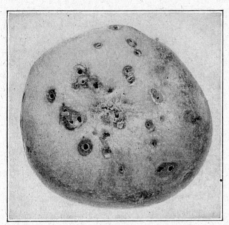

FIG. 381.—Late summer feeding punctures of the plum and apple curculios, on apple. (*From Ill. State Natural History Surv.*)

to harden, and nearly all cherries, remain on the tree until ripe. Two or three weeks are generally spent in the fruit in the larval stage. Upon becoming full-grown the insect leaves the apple and works its way into the ground, an inch or two, excavating a little cavity, in which, after about 2 weeks, it changes to the pupal stage (Fig. 382). About 1 month after the larva enters the soil, it changes to the adult insect, and the summer generation of beetles begins to appear in the orchard. In the latitude of central Illinois, most of the beetles come out during July. In the southern states, there is, in some seasons, a partial second and possibly sometimes a partial third generation. North of the 39° north latitude there is probably only one generation annually. The adult beetles on emerging fly to the fruit and, during the remainder of the summer, feed on the apples,

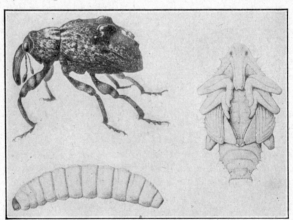

FIG. 382.—The plum curculio, *Conotrachelus nenuphar* Herbst. Adult above at left in side view; larva below at left, side view; pupa at right, ventral view; about 5 or 6 times natural size. (*From Ill. State Natural History Surv.*)

making small holes through the skin and, with the aid of their curved snouts, eating out a cavity in the flesh beneath these holes. Some of them begin seeking winter quarters in August, while others remain on the trees as late as mid-October or, in the South, the first of November.

Control Measures.—To control curculio in badly infested areas, it may be necessary to combat the insect by spraying, jarring, and the removal of infested dropped and cull fruits. Sprays applied during the maximum feeding and egg-laying period of the beetles are the quickest method of control. In most cases only two sprays are necessary—one at the time of petal fall and another 1 week later. These should consist of lead arsenate, 4 pounds, and soybean flour, $\frac{1}{4}$ pound, in 100 gallons of the recommended fungicide. In some sections cryolite may be used instead of the lead arsenate. Thorough application is necessary. Jarring the beetles from the trees in the early morning, upon sheets placed on the ground, especially under trees bordering woodlands, ditches, and hedgerows, enables the grower to collect and destroy great numbers of the beetles and prevent them migrating into the interior of the orchard. If jarring is not practiced, it is well to apply an extra spray to the margins of orchards near favorable hibernating places.

In badly infested orchards it will be of advantage to pick up and destroy the dropped fruits during June and July, if this can be done at small expense. These may be disposed of by soaking them with waste oil or by enclosing in paper-lined sacks and placing in the hot sun for several days or burying. Once well established in an orchard, the curculio may become such a serious pest that it will warrant taking the most vigorous measures possible to secure effective control (see also page 635).

References.—*U.S.D.A., Dept. Bul.* 1205, 1924; *Del. Agr. Exp. Sta. Buls.* 175, 1932, and 193, 1935; *Conn. Agr. Exp. Sta. Bul.* 301, 1929, and *Cir.* 99, 1934; *N.Y. (Geneva) Agr. Exp. Sta. Bul.* 606, 1937; *Va. Agr. Exp. Sta. Bul.* 297, 1935; *U.S.D.A., Tech. Bul.* 188, 1930.

164. APPLE CURCULIOS[1,2,3,]

Importance and Type of Injury.—There are several forms of these insects: (*a*) the apple curculio;[1] (*b*) its variety, the western or larger apple curculio;[2] (*c*) a distinct species,[3] which also attacks apple; and (*d*) a variety of the latter, the cherry curculio.[4] Either may cause very serious damage, amounting to the loss of 50 per cent or more of a crop. This insect is somewhat like the plum curculio in general appearance but differs from it in its habits and manner of injury to the apple. It is not so important as an apple pest. The apples attacked are misshapen, knotty, and undersized. Small holes are eaten in the sides or ends of the apple, many holes often being made close together causing a deadened area on the skin of the apple (Fig. 381). There are sunken pits or sharp-pointed protuberances on the apple marked at their center by a small puncture in the skin. Infested apples sometimes drop as is the case with the plum curculio. The injury may be distinguished, on mature fruits, from that of the plum curculio by the larger number of punctures close together through the skin, by the larger deadened areas on the fruit surface as above mentioned, and by the absence of crescents.

Trees Attacked.—Apple, haw, wild crab, quince, pear, shadbush, cherry,[4] and some others.

[1] *Tachypterellus* (= *Anthonomus*) *quadrigibbus* (Say), Order Coleoptera, Family Curculionidae.

[2] *Tachypterellus quadrigibbus magnus* List.

[3] *Tachypterellus consors* Dietz.

[4] *Tachypterellus consors cerasi* List.

Distribution.—The apple curculio is found east of the Continental Divide; the larger apple curculio from Illinois and Missouri to Nebraska and Texas; *Tachypterellus consors* from the Rocky Mountains to the Pacific Coast; and the cherry curculio is an enemy of cherries in Colorado and New Mexico.

Life History, Appearance, and Habits.—This insect, like the plum curculio, passes the winter in the full-grown or beetle stage in leaves and rubbish in dry places on the ground, especially under trees on which they developed. It is of a brown to light-brown color, with four very distinct humps on the back (Fig. 383). The snout is longer and more slender than that of the plum curculio, being nearly as long as the insect body. The head is small, and the body enlarges toward the base of the abdomen, giving the insect a distinct triangular outline when viewed from above. Before fruit sets, the adults feed on buds, fruit spurs, and terminal twigs,

Fig. 383.—Adult apple curculio, *Tachypterellus quadrigibbus* (Say), 12 times natural size. (*From N.Y. Agr. Exp. Sta.*)

blighting the tender shoots. The fruit may be attacked as soon as it is set. The females eat similar cavities in the fruit, in which they lay their eggs, but do not make the crescent-shaped slits characteristic of the egg-laying scars of the plum curculio. The eggs are deposited from May to mid-July, each insect laying up to 125, with an average of several dozen, one in a place. Many of the larvae develop in "June drops," and many in mummied apples on the trees. This insect also differs in its habits from its near relative, the plum curculio, in that the larvae pupate within the apple. A total of 5 to 6 weeks are spent in the fruit as egg, larva, and pupa. The new adults emerge from the fruits from mid-July to early September and feed until they go into hibernation, sometime in September, upon the maturing fruits. So far as we know, there is only one generation of this insect annually.

Control Measures.—General measures of orchard sanitation are fully as important in the control of this insect as for the plum curculio. It is difficult to kill by means of poison sprays, as the adult beetles feed but little on the surface of the fruit, taking most of their food through their long beaks from the flesh of the apple beneath the skin. Heavy spraying with lead arsenate has reduced the infestation only about 60 per cent. On the whole, it may be said that keeping the orchard and vicinity free from grass, bushes, or other places offering hibernating quarters for the adult insects is the most effective means of controlling the apple curculio.

References.—*N.Y.* (*Geneva*) *Agr. Exp. Sta. Tech. Bul.* 240, 1936; *Jour. Agr. Res.*, **36**: 3–249, 1928; *Jour. Econ. Ento.*, **26**: 420–424, 1933, **29**: 697–701, 1936; *Colo. Agr. Exp. Sta. Bul.* 385, 1932.

165 (175). CODLING MOTH[1]

Importance and Type of Injury.—This is the most persistent, destructive, and difficult to control of all the insect pests of the fruit of the apple. If left to itself, it will usually infest from 20 to 95 per cent of the apples in an orchard and, in spite of man's best efforts to control it, it usually injures from 5 to 10 per cent of the fruit. It has forced the abandonment

FIG. 384.—Codling moth larva and pupa within cocoons from beneath bark of apple tree, about twice natural size. (*From U.S.D.A., Farmers' Bul.* 1270.)

of apple growing as an industry in certain large sections of the country. Apples attacked by this insect have holes eaten into the side, or from the blossom end, to the core. The seeds and core are tunneled and eaten by pinkish-white brown-headed worms about ¾ inch long when full-grown. Dark masses of frass or castings often protrude, especially at the blossom end, from the holes eaten in the apples. Even poisoned larvae may eat far enough into the fruits before they die to lower the grade of the fruit by their "stings." These small holes of the size of a pin prick, and less than ¼ inch deep, with a little dead tissue around them, lower the grade of the fruit.

Plants Attacked.—Apple, pear, quince, wild haw, crab, English walnut, and several other fruits.

Distribution.—Throughout the apple-growing sections of the world.

Life History, Appearance, and Habits.—The codling moth passes the winter in the full-grown larval stage in a thick silken cocoon (Fig. 384). The larvae are pinkish-white caterpillars with brown heads and are about ¾ inch long. These cocoons are generally spun under loose scales of the bark on the trunks of apple trees, under other shelters about the base of the trees, or on the ground near by. Many of the larvae winter in or around packing sheds. They remain dormant and are able to

[1] *Carpocapsa pomonella* Linné, Order Lepidoptera, Family Tortricidae. The lesser appleworm, *Laspeyresia* (= *Grapholitha*) *prunivora* Walsh, causes injury similar to that of the codling moth and may generally be controlled by the same measures.

withstand low temperatures. A drop in temperature to $-25°F$. or below, however, will kill many of the larvae. During the winter, birds, especially woodpeckers, find and eat large numbers of the larvae. In mid-spring the worms change inside their cocoons to a brownish pupal stage (Fig. 384), and, after a period of from 2 to 4 weeks or more, they emerge from the cocoons, beginning in early May, as grayish moths with somewhat iridescent, chocolate-brown patches on the back part or tip of the front wings and faint wavy crossbands of brown on the rest of the wings. The moths (Fig. 386) have a wing expanse of from 1/2 to 3/4 inch. During

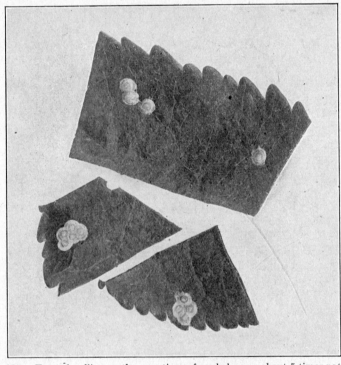

FIG. 385.—Eggs of codling moth on sections of apple leaves; about 5 times natural size. From eggs laid in cages. In the orchard eggs are usually laid singly. (*From Ill. State Natural History Surv.*)

the day the moths remain quiet, usually resting on the branches or trunk of the tree. The coloring of the wings is such that it blends with that of the bark, making the insect very inconspicuous. About dusk of the evening, if the temperature is above 55 to $60°F$., they become active, mate, and the females lay their eggs. If the temperature is low, they remain quiet and few eggs will be deposited. Consequently, if the temperature is high and the weather dry during the period of egg-laying and hatching, the codling moth is likely to be very destructive that year. Each female usually deposits more than 50 eggs during her lifetime. The eggs are white, flattened, pancake-shaped, and about 1/25 inch in diameter (Fig. 385). The eggs of the first generation are laid, one in a place, almost entirely on the upper side of the leaves, the twigs, and the fruit spurs, usually a short distance from a cluster of apples. Most of the eggs are laid 2 to 6 weeks after the apples have bloomed and hatch in from 6 to 20 days, depending on the temperature and to some extent on the rainfall. The worms feed slightly on the leaves but in a few hours crawl to the young apples and chew their way into the fruit, usually entering by way of the calyx cup at the blossom end. After entering the fruit, they work their way into the core, often feeding on the seeds (Fig. 387). Some of the infested fruits drop from the tree and the larvae complete their growth on the ground. Upon becoming full-grown, in 3 to 5 weeks, they burrow to the outside of the apple and either crawl to, or down, the trunk of the tree; or drop to the ground and crawl back to the trunk or to some other object. Under loose bits of bark or other shelter on the trees or on the

Fig. 386.—Adults of the codling moth, *Carpocapsa pomonella* Linné; about 5 times natural size. (*From Ill. State Natural History Surv.*)

ground, such as discarded sacks, tree prunings, weed stems, and other litter, they spin their cocoons, and change as before to the pupal, and later to the adult, stage.

In the latitude of southern Illinois, there is nearly a full first, nearly a full second, and a partial third generation of this insect each season. In the latitude of northern Illinois, there is nearly a full first generation, a partial second, but no third generation. The emergence of the moths of the second generation extends over about 6 weeks, and eggs of this generation may be deposited in the northern part of the United States from early July to mid-September. In the South, eggs may be laid as late as October. The hibernating larvae consist of individuals from all generations.

Control Measures.—The measures necessary to control this insect vary greatly, depending upon the age of the trees, the varieties of apples grown, the location of the orchard, proximity to other orchards and woodland, location and care of the packing and storage sheds, the type of pruning, and many other factors.   In sections where the climate is favorable to the development of this insect, the task of control is an exacting and continuous one.

Spraying with lead arsenate, fixed nicotine, or cryolite is at present the most effective known method.   These materials are used in combination with various fungicides, stickers, and spreaders.   It is highly important that the sprays for this insect be applied at the proper time.   The first and most important spray for codling moth control is that known as the *petal fall* or *calyx spray*.   This is applied when about three-fourths of the petals have fallen from the apple blossoms.   The spray should not be applied when the trees are in full bloom, because of the danger of poisoning honeybees.   Following the calyx spray, additional sprays, known as *cover sprays*, are applied at about weekly intervals, in order to poison the newly hatched larvae as they feed briefly upon the leaves and eat into the skin of the fruits.   In the North Central states a typical schedule

Fig. 387.—Apple injured by codling moth, showing larva in fruit.   Slightly enlarged.
(*From U.S.D.A., Farmers' Bul.* 1270.)

consists of (*a*) a calyx spray of lead arsenate, 2 pounds, and flotation or other mild sulphur, 8 pounds in 100 gallons of water; (*b*) 1 week later a spray of lead arsenate, 4 pounds, and flotation or other mild sulphur, 8 pounds per 100 gallons; (*c*) 1 week later, lead arsenate, 4 pounds, and soybean flour, 4 ounces in weak bordeaux, 100 gallons; (*d*) 1 week later with the same materials.   In severe infestations or especially favorable years, (*e*) a fourth, first brood, cover spray may be necessary.   (*f*) The sprays for second- and third-brood worms should be timed according to the seasonal development of the insect.   Information concerning the best schedule and combination of materials to use in each locality should be obtained from the experiment station entomologist.   In most cases, if apples or other fruits receive more than two sprays of lead arsenate or cryolite, the fruit will have to be washed in a solution of weak nitric acid or other chemical to reduce the residue of lead, arsenic, or fluorine below the legal tolerance (see page 243).

It has been found that the development of the codling moth is largely dependent on the temperature.   Development is nearly at a standstill at temperatures below 50°F. and is retarded above 86°F.   Temperatures above 50°F. and below 86°F. have been called *effective temperatures*. It has been found that approximately 550 day-degrees of effective temperature are required to bring about the hatching of the earliest larvae of the first generation, and 1,000 day-degrees additional for hatching of the first larvae of the second generation, and so on for each additional generation. A spray should always be applied just before the time of the hatching of the first larvae of any generation. For a fuller discussion of this method of forecasting the time of appearance of any generation of the codling moth, see the first reference on page 603.

Aside from spraying, there are several other measures which help in keeping down the codling moth. These consist of a thorough cleanup of the orchard, scraping the loose bark from old trees, and removing rubbish from the ground. In cases of abundance, the trees should be banded during the summer with chemically treated bands, which will kill the codling moth larvae that seek shelter under them to spin their cocoons. The chemical most generally used for this purpose is beta-naphthol in an oil carrier. Bands can be purchased treated ready to apply or may be prepared by the fruit grower. Directions for making them can be obtained from the state experiment stations or the United States Bureau of Entomology and Plant Quarantine. The bands should be in place not later than 5 weeks after petal fall; June 1 in the latitude of southern Illinois and June 15 in the latitude of northern Illinois. Removing cull apples from the orchard and a thorough cleanup of refuse and rubbish around the packing shed also will help in keeping down the numbers of this insect. The codling moth is preyed upon by many insect enemies, but these are never sufficient to reduce its numbers so that artificial control measures may be omitted.

References.—*Ill. State Natural History Surv., Bul.*, vol. 14, art. VII, 1922; *N.M. Agr. Exp. Sta. Tech. Bull.* 127, 1921; *U.S.D.A., Dept. Bul.* 932, 1921, and *Farmers' Bul.* 1326, 1931; *Ill. State Hort. Soc. Trans.*, **67**: 184–190, 1934; *Jour. Econ. Ento.*, **30**: 404–427, 1937; publications of nearly all experiment stations.

166. Apple Maggot[1]

Importance and Type of Injury.—Apples in the colder sections of the United States are often badly injured by the maggots of medium-sized black, white, and yellow flies. These maggots bore through the flesh of the apple and are known as the apple maggot, or more commonly as the "railroad-worm." Where this insect is abundant, it is one of the most serious pests of apples, especially early varieties. Infested apples have brown winding galleries running through the superficial part of the flesh and minute egg punctures and distorted, pitted areas on the surface. Heavily infested early varieties of fruit will be reduced to a brown rotten mass filled with yellowish legless maggots, about 1/4 inch in length and tapering toward the head. In later varieties the injury consists of corky streaks. When the fruit is slightly infested, there is no external indication of the presence of the maggots, but, when the fruit becomes ripe, the burrows show as dark lines under the skin (Fig. 389). There is a marked difference in the susceptibility to attack by different varieties, the thin-skinned early maturing varieties being most severely injured.

[1] *Rhagoletis pomonella* (Walsh), Order Diptera, Family Trypetidae.

Plants Attacked.—The apple maggot is a native insect, probably feeding originally on haws. It has been found in wild crabs, is a serious pest of blueberries, and feeds to some extent in huckleberries, European plums and cherries. A closely related form is a serious pest of blueberries.

Distribution.—The apple maggot is a northern insect occurring as far west as North and South Dakota, southward and eastward to Arkansas and Ohio, and throughout the northeastern states and southeastern Canada. A small variety of this species, breeding in snowberry, has been taken in the western states.

Life History, Appearance, and Habits.—The winter is passed in the pupal stage within a brown puparium about ¼ inch long. These puparia are buried in the soil to a depth of from 1 to 6 inches or more. The adult flies emerge over a period of a month or two in summer. They are black

Fig. 388.   Fig. 389.

Fig. 388.—Apple maggot. *Rhagoletis pomonella* (Walsh); adult on fruit, natural size. (*From Ont. Dept. Agr. Bul.* 271.)

Fig. 389.—Section through an apple infested with the apple maggot, showing a full-grown larva, natural size. (*From Ont. Dept. Agr. Bul.* 271.)

in color, with white bands on the abdomen, four on the female and three on the male, and are a little smaller than the house fly. The wings are conspicuously marked with four oblique black bands (Fig. 388). They drink drops of water that have accumulated on the fruit and leaves. The females lay their eggs singly in punctures in the skin of the apple, made by a sharp ovipositor attached to the tip of the abdomen. Egg-laying does not usually take place until 2 or 3 weeks after the flies have emerged. The eggs hatch in from 5 to 10 days, and the maggots develop slowly in the green fruit and do not usually complete their growth until the infested apples have dropped from the tree. After the fruit has fallen, growth is rapidly completed, and the larvae leave the apple and enter the ground, where the puparia are formed within which they pupate. In the southern part of the range of the insect there is a partial second generation, the adults emerging in the early fall. Some of the insects remain in the puparium for a year or two before emerging. These long-term pupae emerge later than the others and complicate control.

Control Measures.—The most effective control for the apple maggot is to spray all the trees in the orchard at the time the adults make their appearance in midsummer, using lead arsenate at the rate of 3 pounds to 100 gallons of water. Throughout most of the insect's range, this spray should first be given during the last week in June and should be followed by a second, and possibly a third application, at intervals of two or three weeks. The date of emergence of the flies can be determined by placing a screen or cheesecloth cage over an area at least a yard square, where maggots pupated in the soil the previous year. The flies feeding on the surface of the fruit and foliage are poisoned by the spray. The insects in picked fruits may be destroyed by placing the fruits in cold storage for 4 or 5 weeks. Picking up all early-dropped fruit every few days and feeding it to hogs will destroy many of the larvae before they have left the apples.

References.—Jour. Agr. Res., vol. 28, Apr. 5, 1924; *Nova Scotia Dept. Agr. Bul.* 9, 1917; *N.H. Agr. Exp. Sta. Bul.* 171, 1914; *U.S.D.A., Tech. Bul.* 66, 1928; *N.Y. (Geneva) Agr. Exp. Sta. Bul.* 644, 1934; *Vt. Agr. Exp. Sta. Bul.* 43, pp. 8–14, 1935; *Jour. Econ. Ento.*, **29**: 542–544, 1936.

167 (243). WOOLLY APPLE APHID[1]

Importance and Type of Injury.—White cottony masses cover purplish aphids, clustered in wounds on the trunk and branches of apple, quince, elm, pear, and mountain ash, or on large knots on the roots and underground parts of the trunk (Fig. 390). Infested trees often have many short fibrous roots. These injuries sometimes cause the death of the tree, stunting, or serious retardation of growth. The injury on elm causes the formation of close clusters of stunted leaves or rosettes, at the tips of the twigs, the leaves being lined with purplish masses of aphids, covered with white powdery secretion.

Trees Attacked.—Apple, pear, hawthorn, mountain ash, elm.

Distribution.—World-wide.

Life History, Appearance, and Habits.—In the North the winter is passed in the two forms, the eggs and the immature nymphs. The nymphs hibernate under ground on the roots of apple. In the warmer parts of the country the egg-laying females may winter over on apple trunks and branches. Wherever apple and elms are grown in the same community, eggs are normally deposited in cracks or protected places on the bark of elm trees in the fall. The eggs hatch early in the spring. The aphids that emerge from these eggs are wingless and feed on the elm buds and leaves for two generations during May and June. They then produce a winged form which migrates to the apple, hawthorn, and mountain ash. They feed to some extent in wounds on the trunk and branches, and many work their way down the trunk below the surface of the ground. Their most severe injury is caused by feeding on the roots. During the summer the aphids reproduce by giving birth to living young. In the fall the wingless males appear and mate with wingless females, each female laying a single egg in the situations above described. Some winged females are present during the entire summer. The body of this aphid is really of a reddish or purplish color but is nearly hidden under masses of bluish-white cottony wax that is exuded by the insect.

[1] *Eriosoma lanigerum* (Hausman), Order Homoptera, Family Aphididae.

Control Measures.—Infested nursery stock should never be planted. Some varieties of apples, particularly the Northern Spy, show resistance

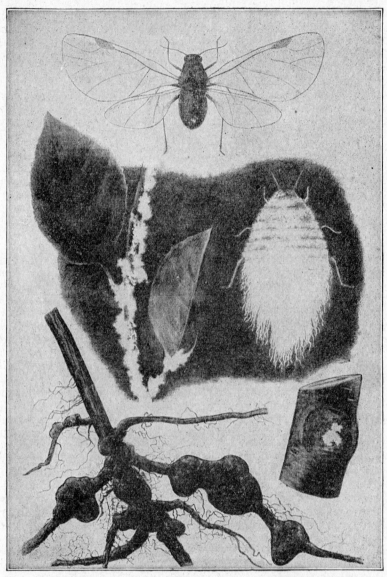

Fig. 390.—Woolly apple aphid, *Eriosoma lanigerum* (Hausman). Winged female above, greatly magnified.   At center, cluster of wax-covered aphids on twig, a single wingless female much enlarged at right.   Roots showing characteristic galls produced by root-infesting form, below.   A cluster of aphids in pruning scar at lower right.   The last two reduced in size.   (*From Ill. State Natural History Surv.*)

to attack.   The most important insect enemy of this aphid, a wasp-like parasite, *Aphelinus mali*, has been transported to 28 different foreign countries in attempts to control this pest. In the Pacific Northwest and some other areas it has been so successful that no other control measures are necessary. The forms of the woolly apple aphid living on the trunk and branches of the tree may be killed by thorough spraying with the spray solutions already recommended for the green and rosy apple aphids. The spray will have to be applied with strong pressure in order to hit the bodies of the aphids, which are well protected by their waxy covering. The root-infesting forms may be killed on nursery stock by dipping the roots in a strong nicotine solution. In California paradichlorobenzene, applied from September to November, has been successful against the root-infesting forms, using $\frac{3}{4}$ to 1 ounce for 4- to 6-year-old trees and removing the residue after 2 weeks. Various other methods of soil fumigation and applications of liquids have been tried against these insects but, up to the present time, without much success. Cultivation and fertilization that will keep the trees in a vigorous growing condition will help in lessening the damage by these insects.

References.—*Va. Agr. Exp. Sta. Tech. Bul.* 57, 1935; *U.S.D.A., Farmers' Bul.* 1270, 1922; *Maine Agr. Exp. Sta. Buls.* 217 and 220, 1913, and 256, 1916.

168. PERIODICAL CICADA OR SEVENTEEN-YEAR LOCUST[1]

This insect (Figs. 393 and 394) is so well known that a description is hardly needed. The body is wedge-shaped, nearly black and from 1 inch to $1\frac{1}{2}$ inches long, including the wings. It is, however, frequently confused with the common "harvestmen" or dog-day cicadas,[2] which are believed to have 2-year life cycles. There are two races of the periodical cicada which cause damage to fruit trees. One of these has a life cycle of 13 years and is abundant only in the southeastern part of the United States (Fig. 391). The other race, which is the true seventeen-year cicada, or seventeen-year "locust," is more abundant in the northeastern part of the United States and has a 17-year life cycle (Fig. 392). There are a number of broods which overlap, the adults appearing in different years. Adults of both races are smaller than the dog-day cicadas and appear during May, June, and very early July, while the dog-day cicadas are present every year during July, August, and September. The dog-day cicadas never cause extensive injury. In both races of the periodical cicada, the body is brownish black, unspotted above, the margins of the wings have a distinct reddish tinge, and a black "W" is present near the lower margin of the front wing. The dog-day cicadas are much larger, have a greenish margin to the wings, and numerous lighter markings on the thorax and abdomen.

Importance and Type of Injury.—Roughened punctures in the twigs and small branches of apple and many other trees, usually from 1 to 4 inches in length. The bark is pushed from the wood and the wood of the twig cut and raised so that a series of small bundles of splinters protrude from the surface (Fig. 5,*A*). This injury is caused, not by the feeding of the insect, but by the female cicada depositing her eggs in the twigs (Fig. 394).

[1] *Magicicada septendecim* (Linné), Order Homoptera, Family Cicadidae.
[2] *Tibicen linnei* (Smith and Grossbeck), and other species, Order Homoptera, Family Cicadidae.

Distribution.—The thirteen-year form ranges from Virginia and southern Iowa and Oklahoma on the north and west, to the Atlantic Coast and Gulf of Mexico. The seventeen-year form ranges from Massachusetts, Vermont, Michigan, and Wisconsin on the north, Kansas on the west, to Texas, northern Alabama, and northern Georgia on the south. Both are most abundant east of the Mississippi River. This insect is not known to occur outside of eastern North America.

Life History, Appearance, and Habits.—This species has the longest developmental period of any known insect. The eggs are laid during late

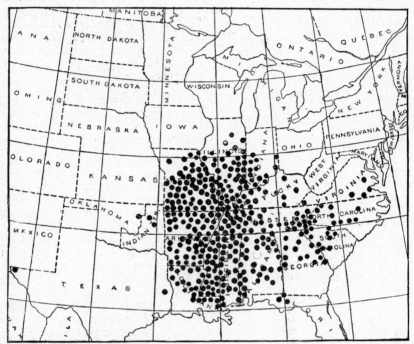

Fig. 391.—Periodical cicada. Map showing distribution of the combined broods of the 13-year race, broods of which will appear in some part of the dotted area during 1945, 1946, 1948, 1949, 1950, and each thirteenth year after each of these years. (*From U.S.D.A., Bur. Ento. Bul.* 71.)

May, June, and early July, in the above-described punctures, in the twigs and small branches of trees. The females lay from 400 to 600 eggs, depositing from 12 to 20 in each puncture beneath the bark. Many egg punctures may occur in a single line, as many as 50 having been found along one branch. The eggs hatch in about 6 or 7 weeks (see Fig. 82,*F* to *J*). The somewhat ant-like young drop to the ground and enter the soil through cracks, or at the base of plants. Here they excavate a small cell about a tree rootlet from which they suck the sap. They grow very slowly, and their feeding usually has no noticeable effect on the trees, on the roots of which thousands of the young cicadas occur. After 13 or 17 years of this underground existence, the nymphs have become full-grown. The insects are now about an inch in length and somewhat resemble a small crayfish (Fig. 84,*L*). The full-grown nymphs burrow to the surface of the soil and emerge through small holes about ½ inch in diameter. They sometimes construct mud cones or "chimneys" about

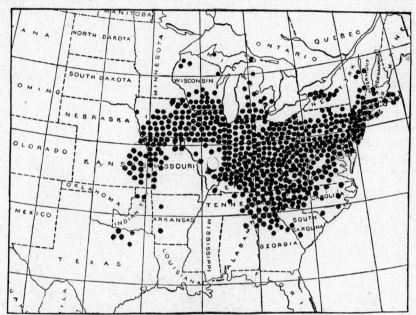

Fig. 392.—Periodical cicada. Map showing distribution of the combined broods of the 17-year race, broods of which will appear in some part of the dotted area in 1940, 1944, 1945, 1946, 1947, 1948, 1949, 1950, 1951, 1952, 1953, 1956, and each seventeenth year after each of these years. (*From U.S.D.A., Bur. Ento. Bul.* 71.)

these holes to a height of 2 to 3 inches, with the opening near the ground. They appear in large numbers at about the same time, emergence usually starting soon after sunset. They crawl upon the trunk of some tree or the stem of a weed, which they grasp firmly. The skin splits down the middle of the back, and the adult insect gradually works its way out (Fig. 393). They remain on the support until their wings harden and the bodies dry, but by the following day they are ready to take flight, the empty skins often remaining clinging to tree trunks and other supports for months. As many as 20,000 to 40,000 may emerge from the ground under one large tree. They fly about during the day, mate, and feed by sucking the sap from the twigs of trees. The injury caused in this way is very slight. Four or five days after emergence the males start "singing." This song is a very high-pitched, shrill call, produced by two drum-like membranes on the

Fig. 393.—Adult periodical cicada beginning to issue from nymphal shell. About natural size. (*From U.S.D.A., Bur. Ento. Bul.* 71.)

sides of the first abdominal segment, and not by rubbing wings or legs together, as many people wrongly believe. The adults live from 30 to 40 days and have all disappeared by the second week in July, in Illinois.

Control Measures.—No spray or dust has been found which can be applied to the trees that will prevent the egg-laying of the female cicadas. Young trees, or particularly valuable fruit trees, may be covered with mosquito bar or other cheap cloth, during the period when the adult cicadas are on the wing. It has been found possible to reduce their numbers greatly, in valuable groves or parks, by banding the trees with sticky "tanglefoot" and raking off the trapped cicada nymphs each morning. Spraying with strong nicotine or pyrethrum solutions will kill all the insects hit by it, but will rarely be worth while. Orchards set on soil where trees were standing during the last appearance of the adults, will be more heavily infested than those set on prairie soil, or at some distance from woodland. Young fruit trees should not be heavily pruned the winter or spring before the appearance of the adult cicadas. After

Fig. 394.—Periodical cicada. Female depositing eggs in apple tree and characteristic egg punctures. (*From Wellhouse, "How Insects Live," after Snodgrass. Copyright, 1926, by Macmillan. Reprinted by permission.*)

the cicadas have disappeared, the seriously weakened twigs should be pruned off and burned. The years in which adult cicadas will appear in any locality can be obtained from the entomologists of the state experiment stations.

References.—U.S.D.A., Bur. Ento. Bul. 71, 1907; Ohio Agr. Exp. Sta. Bul. 311, 1917, and Cir. 142, 1914; Mo. Agr. Exp. Sta. Bul. 137, 1915; Jour. Econ. Ento., **29**: 190–192, 1936, and **30**: 281–294, 1937.

169. Buffalo Treehopper[1]

Importance and Type of Injury.—Double rows of curved slits are found in the bark of small branches and twigs. If these slits (Figs. 5,*C* and 395) are carefully cut open, a row of from 6 to 12, small, elongated, yellowish eggs will be found embedded in the inner bark just under each slit. The bark of infested trees presents a roughened, somewhat scaly and cracked appearance and never makes a very vigorous growth.

[1] *Ceresa bubalus* (Fabricius), Order Homoptera, Family Membracidae.

Plants Attacked.—Apple, pear, peach, quince, cherry, elm, locust, cottonwood, and many other trees. The nymphs feed on weeds, grasses, corn, and legumes.

Distribution.—General throughout the United States and southern Canada.

Life History, Appearance, and Habits.—The insect remains in the egg stage during the winter, hatching rather late in the spring into tiny, pale-green, very spiny nymphs (Fig. 84,*B*), which drop from the tree and feed on the sap of various grasses and weeds, reaching the adult stage during August. The new adults then deposit their eggs in the bark for the next year's generation, after which the adults die. The full-grown buffalo treehopper (Fig. 396) is a peculiarly shaped little insect of a light-green color, triangular in outline, very blunt at the head end with a short horn at each upper corner, and pointed behind. They are about ¼ inch long by two-thirds as wide at the head end. The female treehopper has a very sharp, knife-like ovipositor with which she cuts the slits in the twigs and through these forces her eggs into the inner bark. A single slit is made through the bark, eggs are thrust to the right and left beneath the bark and the scar later separates to form the characteristic double crescent.

Fig. 395.—Twigs of apple showing injury by the egg punctures of the buffalo tree-hopper, slightly enlarged. (*From Ill. State Natural History Surv.*)

Several other species of treehoppers[1] attack apple and other fruit trees, most of them having practically the same life history as that of the buffalo

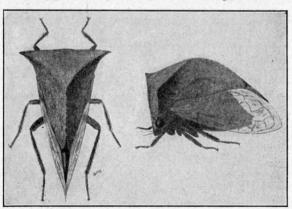

Fig. 396.—Buffalo treehopper, adults in dorsal and side views about 6 times natural size. (*From Ill. State Natural History Surv.*)

treehopper. Some species differ in their manner of laying eggs, and produce a single row of slits in the bark of the tree; others lay their eggs in gummy covered masses on the surface of the bark.

[1] The green-clover treehopper, *Ceresa bubalus* Fabricius; the dark-colored tree-hopper, *Ceresa basalis* Walker; *Stictocephala gillettei* Godg.; and others.

Control Measures.—The best method of controlling this insect is clean cultivation of the orchard, keeping down all weeds or grassy growth and avoiding cover crops of alfalfa.   When the insect is abundant, the growing of summer cover crops, such as clover or cowpeas, should be discontinued for one or two seasons, and the orchard kept clean of all vegetation until the latter part of July.   Dormant sprays of 4 to 6 per cent of oil have been found to kill 70 to 100 per cent of the overwintering eggs.

References.—U.S.D.A., *Farmers' Bul.* 1270, 1922; *N.Y.* (Geneva) *Agr. Exp. Sta. Tech. Bul.* 17, 1907; *U.S.D.A., Cir.* 106, 1930, and *Tech. Bul.* 402, pp. 30–32, 1934.

Spray Schedule for Apples

In most of the apple-growing sections of the north central United States the following spray schedule is followed as a part of the regular orchard practice.   The grower should follow the spray schedule recommended by the agricultural experiment station of his own state:

I. *Dormant Spray.*—Applied before the trees come into leaf, or as a delayed dormant spray, when the buds are showing tip green.   Applied generally for the control of scale insects and leaf roller.
Materials to be used: Commercial or homemade oil emulsions, at a strength of 2 per cent oil; or liquid lime-sulphur (33° Bé.) 1 part to 7 parts of water to make a 5.5° Bé., dormant spray.

II. *The Prepink and Cluster-bud Sprays.*—Mainly for disease control, but of some value in the control of casebearers and budworms.
Materials to be used:

> Water.. 100 gal.
> Sulphur (dry or wettable sulphur or liquid lime-sulphur,
>   according to season and locality)....................... 8 lb.
> Lead arsenate or cryolite................................ 3 lb.[1]

III. *The Petal-fall or Calyx Spray.*—The application should be begun when the petals are three-fourths fallen.   This is the most important spray for the control of the codling moth.
Materials to be used: Microfine or other mild sulphur, 8 pounds; lead arsenate, 2 pounds; water, 100 gallons.

IV. *The 1-week or 10-day Spray.*—Applied 1 week to 10 days after *III.*   Mainly for disease control but is of importance in the control of the plum curculio, codling moth, and some other insects.
Materials to be used: Lead arsenate, 3 to 4 pounds; soybean flour, $\frac{1}{4}$ pound; and lime, 4 pounds in 100 gallons of water; or weak bordeaux ($\frac{3}{4}$–$1\frac{1}{2}$–100), 100 gallons.

V. *The Three-weeks Spray and Later Cover Sprays for First-brood Codling Moth and Certain Apple Diseases.*—From 2 to 4 additional sprays will be needed for the first-brood codling moth.
Materials to be used: Same as in the 1-week or 10-day spray, except if codling moth infestation is heavy, use 2 quarts of summer oil in the second and third cover sprays, in addition to the materials in *IV.*

VI. *Later Sprays.*—For time to apply and materials to use in the sprays for the second and later broods of codling moth, the grower should consult his state experiment station.

---

[1] If needed for defoliating insects.

B. PEAR INSECTS

FIELD KEY FOR THE IDENTIFICATION OF INSECTS INJURING THE PEAR

A. *External chewing insects that eat holes in leaves, buds, bark, or fruit:*
  1. Soft, fleshy, dark-green to orange, slimy, slug-like larvae, up to ½ inch in length, feed on the surface of pear and cherry leaves, skeletonizing the leaves. Most abundant during late spring and again in late summer. In late spring black and yellow wasp-like insects, about ⅕ inch in length, lay their eggs on the leaves..*Pear slug,* page 641.

B. *Insects that suck sap from leaves, buds, twigs, branches, trunk, or fruit:*
  1. Buds and young flowers of pear, prune, and some other fruits fail to open, and become brown and blasted in appearance. Very small, black, slender-winged insects feed within the buds or opening flowers. Young fruits shrivel and drop...*Pear thrips,* page 614.
  2. Pear trees with foliage showing a brownish color, blackening or drying up about midsummer. Very small, shining, cicada-like insects, about ⅒ inch long, under the bark of infested trees during the winter. Minute orange-yellow eggs on bark at base of fruit spurs in spring. Translucent, yellow, olive, or black, very small, but broad, wingless nymphs, sucking the sap from the stems of leaves, and sometimes from the undersurface of the leaf. Sticky drops of nearly colorless liquid on the leaves and fruit, sometimes covered by a black growth of soot-like fungus...........................*Pear psylla,* page 614.
  3. Fruit and branches specked with small blackish or grayish-brown scales, circular in outline and from very small to a little larger than a pinhead, with a raised, dark-gray, nipple-shaped area in the center. Bright lemon-yellow, soft-bodied insects lying under the protecting scales. In heavy infestations, bark of twigs and branches completely coated with a gray covering of scales..*San Jose scale,* page 616.
  4. Very minute, elongated, nearly white-bodied, four-legged mites, about 1⁄125 inch long, sheltering under the bud scales of trees in winter and forming reddish-brown blister-like galls on the undersides of the leaves during the growing season. Galls often so thick as entirely to coat the undersurface of the leaves giving them much the appearance of being infected with some fungus..*Pear leaf blister mite,* page 616.

C. *Insects that bore into the trunk and branches of the tree:*
  1. Slender, bronzy, shining beetles, about ⅓ inch in length, on the bark of the sunny sides of the pear trees during May and June. Twisting or winding, brown burrows running through the inner bark, indicated by swelling or cracking of the outer bark. Where burrows are numerous, trees are sometimes girdled and killed...........................*Sinuate pear borer,* page 618.

D. *Caterpillars, worms, or maggots that burrow into or feed inside the fruits:*
  1. Dark masses of wet frass protrude from the sides of the fruits or from the blossom end, being forced out from holes which extend through the flesh of the fruit usually to the core. Pinkish-white brown-headed worms feeding in these holes. Flat, shining, white eggs on the leaves in the vicinity of fruits or on the skin of the fruit....................................*Codling moth,* page 619.
  2. Caterpillars similar to *D*, 1, but more pinkish, less conspicuous, and not over ½ inch long, boring inside the fruit, often without external evidence of their presence....................................*Oriental fruit moth,* page 636.
  3. Small fruits misshapen, bloated, lopsided, and with dark blotches; usually drop a few weeks after setting. White to orange maggots, not over ¼ inch long, from a few to over 100 per fruit, may consume entire interior of the small fruits...
    Pear midge, Contarinia pyrivora Riley (see *N.Y. (Geneva) Agr. Exp. Sta. Tech. Bul.* 247, 1937).

Many other apple insects may be found injuring the pear. See the Key on pages 554–559.

170. Pear Thrips[1] and Bean Thrips[2]

Importance and Type of Injury.—Pear thrips attack the buds of fruit trees very early in the spring, before the buds open, causing them to shrivel and turn brown. The female thrips also injure young fruits by depositing their eggs in the stems of the blossoms. These egg punctures cause the fruit to drop. Heavily infested orchards appear as though injured by fire. In some sections the insect is very important.

Distribution.—This insect is an imported species and was first found in California in 1904. It now occurs along the Pacific Coast, in California, Oregon, and British Columbia. In the eastern part of the country, it is recorded from New York, Pennsylvania, and Maryland.

Plants Attacked.—The insect is primarily a pest of pear, but it attacks also apple, apricot, cherry, grape, peach, plum, prune, and several other fruit trees, as well as poplar, maple, shadberry, willow, currant, and several shrubs and weeds.

Life History, Appearance, and Habits.—The winter is passed from 5 to 7 inches deep in the soil, in small cells. The insects are in the newly formed adult stage. They remain in these cells until early spring, appearing on the trees in New York about the first of April and in California in February. The adults are very active, working their slender bodies in between the bud scales and feeding upon the swelling buds. These adults are about $\frac{1}{20}$ inch long. The females soon begin laying their eggs in the fruit stems, midribs, and stems of leaves. The egg-laying period extends over about 3 weeks. The young nymphs begin hatching in 2 weeks and feed in large numbers within the opening fruit buds. The young are white in color in contrast with the black adults. The young become full-grown in about 4 weeks, and, still in the nymphal stage, they drop to the ground, which they enter. There they form the cells in which they pass the summer and hibernate during the winter (see Fig. 116, page 197).

Control.—The best method of controlling these insects is to spray thoroughly with a miscible-oil or lubricating-oil emulsion at a strength of 2 per cent oil, plus nicotine at the rate of 1 pint to 100 gallons of water. This spray should be applied as soon as the adult thrips appear on the buds. The time of their appearance will vary according to the locality and season, so that one will have to keep watching the trees in order to know the best time for applying this spray. Where thrips are abundant, a second application of the same materials should be made at the cluster-bud stage of the fruit. In applying an oil spray at this stage, care must be taken to use an oil which is not injurious to the foliage. Serious infestations can be greatly reduced by thorough, deep cultivation of the orchard in the late summer and early fall, if this can be done without injury to the trees, or the forms in the soil may be killed by irrigation.

References.—*U.S.D.A., Bur. Ento. Bul.* 68, pp. 1–16, 1909; *N.Y. (Geneva) Agr. Exp. Sta. Bul.* 484, 1921; *Jour. Econ. Ento.*, **27**: 879–884, 1934.

171. Pear Psylla[3]

Importance and Type of Injury.—In sections where this insect has become established, *i.e.*, in the northeastern part of the country, it is

---

[1] *Taeniothrips inconsequens* Uzel, Order Thysanoptera, Family Thripidae.

[2] *Hercothrips fasciatus* (Pergande).

[3] *Psylla* (= *Psyllia*) *pyricola* Förster, Order Homoptera, Family Chermidae.

one of the most important pests of the pear. The leaves on heavily infested trees turn brown and often drop, the fruit drops prematurely,

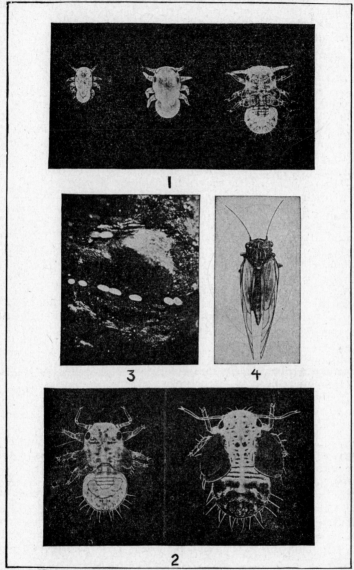

FIG. 397.—The pear psylla, *Psylla pyricola* Förster. 1, first, second, and third instar nymphs; 2, fourth and fifth instar nymphs; 3, eggs; 4, winter adult. All much enlarged. (*From N.Y.* (*Geneva*) *Agr. Exp. Sta. Bul.* 527.)

or is undersized and of poor quality. Dark, reddish-brown, four-winged, cicada-like insects, about $\frac{1}{10}$ inch in length (Fig. 397,*4*), will be found under the bark during the winter; and much smaller, very broad, active, yellow nymphs (Fig. 397,*1,2*) will be found on the fruit and leaves during the growing season. The leaves and fruit of badly infested trees will be covered with honeydew, which in turn is generally coated with a black fungus, later in the season.

Trees Attacked.—Pear.

Distribution.—This insect is of European origin, was brought to Connecticut about 1832, and has spread over the eastern states.

Life History, Appearance, and Habits.—The pear psylla passes the winter as an adult under the bark of the trees or in other sheltered places about the orchard. The adults come out of hibernation early in the spring, and the females deposit their pear-shaped, orange-yellow eggs (Fig. 397,*3*) in cracks in the bark or about the buds. The eggs are attached by a short stalk and have a thread-like filament projecting from the unattached end. The eggs hatch in 2 weeks to a month and by the time the trees are in full bloom many very small, yellow, wingless nymphs, about $\frac{1}{80}$ inch long, may be found on the stems and undersides of the leaves, from which they are sucking the sap. These nymphs complete their growth in 1 month, or slightly less. There are from three to five generations each season, the eggs of the later generations being laid on the leaves and stems.

Control Measures.—One of the most effective methods of controlling this insect is to spray the trees very thoroughly with oil emulsions or miscible-oil sprays, applied at the strength found effective for the control of San Jose scale. Tar distillate oils at 3 per cent strength are also effective. The spraying should be done as soon as the trees lose their leaves in the fall or just before growth starts in the spring. As many of the psylla are on the small branches and twigs, very thorough spraying is necessary to hit the bodies of all the insects. The trees may be thoroughly sprayed with dormant-strength lime-sulphur solution (1 to 9), applied just before the blossom buds open. This spray will prevent the hatching of any eggs that are thoroughly coated with it. Here again the most thorough application is necessary. Very good control has been obtained by summer spraying with bordeaux mixture (4-6-100) and 1 pint of 40 per cent nicotine sulphate to each 100 gallons. Dusting with strong nicotine dusts has given fair control.

Reference.—N.Y. (*Geneva*)*Agr. Exp. Sta. Buls.* 387, 1914, and 527, 1925, and *Cir.* 129, 1932.

172 (149, 179, 189, 204). San Jose Scale

Varieties of pears, such as Duchess, Seckel, Bartlett, and Bosc, are subject to attack by the San Jose scale and should be given the same dormant treatment as the apple. Mature Kieffer and Garber pears are seldom seriously injured by San Jose scale and rarely need to be sprayed for the control of this insect (see page 575).

173. Pear Leaf Blister Mite[1]

Importance and Type of Injury.—This insect is of moderate importance in most pear-growing sections, but control measures are generally necessary in commercial orchards. Brownish blisters appear on the undersides of the pear and apple leaves. The blisters are commonly $\frac{1}{8}$ inch across, or massed together in such a way as nearly to cover the underside of the leaf surface (Fig. 398). Upon examination with a

[1] *Eriophyes pyri* Pagenstecher, Order Acarina, Family Eriophyidae.

lens, these blisters will be found swarming with very small, whitish or pinkish mites, with an elongate, tapering, ringed abdomen and only two pairs of legs located near the head (Fig. 399). Fruit buds turn brown and flare open during the winter, or produce weak flowers, and russeted globular or misshapen fruits, due to the work of this mite under the fruit bud scales.

Fig. 398.—Pear leaves infested with the pear leaf blister mite. Reduced. (*From Slingerland and Crosby, "Manual of Fruit Insects," copyright, 1915, by Macmillan. Reprinted by permission.*)

Trees Attacked.—Pear, apple, and related wild trees.

Distribution.—General in fruit-growing sections of North America. It was introduced into this country about 1870.

Life History, Appearance, and Habits.—The adult mites, which are only about $\frac{1}{125}$ inch long (Fig. 399), enter the bud scales in August and September where they

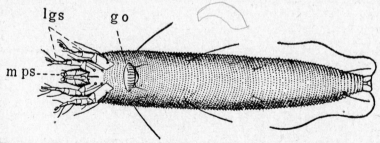

Fig. 399.—Pear leaf blister mite, *Eriophyes pyri* Pagenstecher, ventral view, magnified about 450 diameters. *m ps*, mouth parts; *lgs*, the two pairs of legs; *g o*, opening of the reproductive system. (*From U.S.D.A., after Nalepa.*)

spend the winter. In warmer regions eggs are deposited within the buds, where they hatch and develop during the winter, destroying the bud tissues as already described. In the Northwest the mites winter in the egg stage and hatch early in the spring. As soon as the foliage has started to come out in the spring, they become active and start feeding on the undersides of the leaves, causing the brownish blisters in which the eggs are laid and the young develop. There are a number of generations each year. The creatures may also attack the fruit, causing it to be dwarfed or to drop when the mites are numerous.

Control Measures.—Where leaf blisters are serious, pear leaf blister mites are readily controlled by winter spraying with lime-sulphur, or miscible oil, at the same strength as for San Jose scale. Where the injury to buds is serious, a spray of 2 or 3 per cent light oil emulsion plus 5 to 7 per cent lime-sulphur should be applied when the mites are migrating to the buds in fall and before they have worked down beneath the bud scales.

References.—*N.Y. (Geneva) Agr. Exp. Sta. Bul.* 306, 1908; *Calif. Agr. Exp. Sta. Cir.* 324, 1932; *U.S.D.A., Farmers' Bul.* 1666, pp. 64, 65, 1931; *Jour. Econ. Ento.,* **25**: 985–988, 1932.

BORERS

Pears are subject to attack by the flatheaded apple borer and some species are also injured by the roundheaded apple borer. The control methods for these insects are the same as those given for their control on apple (see pages 589–592).

174. SINUATE PEAR BORER[1]

Importance and Type of Injury.—Trees infested by this insect have narrow winding burrows in the inner bark and sapwood and discolored canker-like areas from base of trunk to small branches, the presence of these burrows being indicated by splitting and dark, dead lines in the outer bark. If the burrows are numerous, the bark will be killed. Entire plantings of young trees are sometimes girdled and killed where the insects are abundant.

Trees Attacked.—Pear, hawthorn, mountain ash, and cotoneaster.

Distribution.—This is a European insect first found in New Jersey in 1894. Its known range in North America includes several of the eastern states, principally New York and New Jersey.

Life History, Appearance, and Habits.—The winter is passed in the larval stage in burrows in the tree. The larvae, which are very slender, are of two sizes in winter; the smaller, about ½ inch long, in the inner bark, and the larger, about 1½ inches long, in small cells in the sapwood which are plugged at each end with coarse sawdust. The larger larvae pupate in the early spring and emerge as beetles during the last of May and June. These beetles are very slender, about ⅓ inch long and about one-fifth as wide, and of a purplish-bronze color. They are somewhat flattened and boat-shaped in outline; tapering to a blunt point at the tail end. The beetles feed on the foliage of the pear and related wild and cultivated fruit trees; the females, after mating, lay their eggs in cracks in the bark. The grubs on hatching burrow in the bark as above described. The larvae grow rather slowly, changing to pupae the second spring after the eggs are laid. There is probably one generation each two years.

Control Measures.—The dead or dying trees, or branches that are very heavily infested, should be cut off and burned during the winter. If the foliage of the pear trees is sprayed heavily with lead arsenate, 6 pounds to 100 gallons of water, just about the time the adult beetles emerge and again 2 weeks later, most of them will be killed when they feed on the poisoned leaves. This is the most effective control.

Reference.—*Conn. Agr. Exp. Sta. Bul.* 266, 1921; *N.Y. Agr. Exp. Sta. Bul.* 648, 1934.

[1] *Agrilus sinuatus* (Olivier), Order Coleoptera, Family Buprestidae.

175 (165). Codling Moth

The fruit of the pear is subject to attack by the codling moth, but not to so great an extent as is the apple. The control measures are the same as for the apple (see page 599), but fewer sprays are required.

C. QUINCE INSECTS

FIELD KEY FOR THE IDENTIFICATION OF INSECTS INJURING THE QUINCE

1. Grayish scales adhering to the bark and fruit (see Field Key for the Identification of Insects Injuring the Apple, *B*, 1)............*San Jose scale*, page 575.
2. Worms feeding in, and boring holes through, the fruit (see Field Key for the Identification of Insects Injuring the Apple, *D*, 3)....*Codling moth*, page 599.
3. Pinkish to creamy-white, short-legged worms with brown heads, up to ½ inch long, boring in the twigs, causing them to wilt, and also in the fruit. Entrance holes of worms often through the stem, not showing on the outside of the fruit..*Oriental fruit moth*, page 636.
4. Irregular cavities eaten in the flesh of the quince, with very small openings through the skin. White legless grubs feeding in the fruit of the quince during early summer but seldom causing the fruit to drop. Grayish-brown, rather broad-shouldered snout beetles, the snout about one-third as long as the body, feeding on the flesh of the fruit. Injured fruit becomes deformed and knotty..*Quince curculio*, page 619

Apple Insects That Injure the Quince

The quince is often infested with San Jose scale (see page 575). It is particularly subject to injury by the roundheaded apple borer (see page 590), which seems to prefer the quince above any of the other fruit trees. The oriental fruit moth (see page 636) prefers the quince to most other fruits. In sections where it is abundant, it is very difficult to produce quinces that can be sold on commercial markets. The fruit is attacked by the codling moth (see page 599), and the leaves by several of the insects common on apple.

176. Quince Curculio[1]

The most important insect which confines its attack to the quince is the quince curculio. This curculio differs slightly from the apple and plum curculios in that it goes through the winter in the grub stage in the soil. The adult (Fig. 400) is a broad-shouldered snout beetle, grayish brown in color, without humps on the back.

Fig. 400.—The quince curculio, *Conotrachelus crataegi* Walsh; adult, natural size and enlarged. (*From Slingerland, N.Y. (Cornell) Agr. Exp. Sta. Bul.* 148.)

Control Measures.—Summer cultivation of the orchard is of no value in combating this insect, since it does not enter the soil until late summer. As the grubs leave the fruit before it drops, picking up the dropped quinces cannot be practiced as a control measure.

[1] *Conotrachelus crataegi* Walsh, Order Coleoptera, Family Curculionidae.

Thorough spraying with lead arsenate and lime, 4 pounds of each to 100 gallons of water, is the most effective method of keeping down this insect. The application should be made at approximately the same stage of development of the quince as for the plum curculio on apple (see page 597).

References.—*N.Y. (Cornell) Agr. Exp. Sta. Bul.* 148, 1898; *N.Y. State Dept. Agr. Bul.* 116, 1919; *Conn. Agr. Exp. Sta. Bul.* 344, pp. 174, 175, 1933.

D. PEACH INSECTS

FIELD KEY FOR THE IDENTIFICATION OF INSECTS INJURING THE PEACH

A. *External chewing insects that eat holes in the foliage or fruit; or moths sucking juices from the ripening fruits:*
   1. Robust metallic-green to greenish-bronze, or copper-colored beetles, from ⅓ to ⅝ inch long, feeding on the fruit of the peach. The abdomen of the insect projects back from the wing covers and is marked by two conspicuous white spots. The injured fruit appears to have been gouged or partly peeled; sometimes where the beetles are numerous, the entire skin is eaten off............ ....*Japanese beetle*, page 621.
   2. Large, somewhat flattened, dark-green beetles up to 1 inch in length, with a brownish-yellow tinge to the wing covers and thorax, feeding on the fruit of the peach, and also on the foliage, during July and August. Fruit scored somewhat as in *A*, 1. Beetles drop to the ground when the trees are shaken in the daytime. Feed mainly at night................*Green June beetle*, page 623.
   3. Tan-colored moths, with a purplish tinge and a small, oval, dark spot near the middle of each front wing, distinctly triangular in outline and about ⅞ inch long, suck the juice from cracks or from small punctures that they make in ripening fruits. Feed on peach, grape, apple, and other fruits............... ....Adult of *cotton leafworm*, page 444.
B. *Insects that suck sap from buds, leaves, fruits, and terminal twigs, sometimes killing the latter:*
   1. Black shiny eggs on the tips of twigs and in cracks in the bark of larger branches of peach during the winter. Pale-greenish or black aphids sucking the sap of the newly formed fruit, leaves, and young twigs. Leaves on infested trees curl and turn yellow. Insects sometimes so numerous as to cover completely the twigs and foliage............*Green peach aphid*, or *black peach aphid*, page 625.
   2. Young peach trees in the orchards, and particularly trees in the nursery row, with the terminal growth and some of the laterals dying back, but not containing borers. Coppery-brown bugs, with oval bodies, somewhat triangular in front and about ¼ inch long by half as wide, feeding on the terminals, twigs, foliage, and peach fruits. Body of the bug is somewhat spotted and flecked with yellow and dark-brown. Bugs fly readily when approached............ ....*Tarnished plant bug*, page 627.
C. *Motionless insects found on the bark of trunk and branches and on leaves, sucking the sap:*
   1. Round, flattened, grayish scale insects, about 1/16 inch across, closely adhering to the bark of trees, often with reddened areas of bark about the point where the insects are attached. Lemon-yellow, sack-like, sucking insects sheltered under the waxy scales...............................*San Jose scale*, page 624.
   2. Peach twigs, particularly on the undersides, nearly covered with sharply convex, brownish scales, about 1/12 inch in diameter, of a general dark-brown color, with some black lines, particularly about the margins. In summer, fruit and leaves of infested branches coated with honeydew or masses of sooty, black fungus.................................*Terrapin scale*, page 625.
   3. Brown flattened scales, less than ⅛ inch long, adhere tightly to the bark on the underside of the branches, in winter. Whitish masses of cottony material,

nearly ⅓ inch across, beneath and behind these scales from May to July......
....*Cottony maple scale,* page **709.**
D. *Insects that burrow in the trunk, branches, or twigs:*
 1. Small bits of gum exuding from many points along the trunk and branches of peach. Galleries about the size of a pencil lead and a little over 1 inch long, with sawdust-filled burrows radiating from them, in the inner bark of the tree. Brownish, or brownish-black, blunt-ended beetles, 1/16 inch long, are found boring the galleries in the inner bark, or crawling over the bark of the tree.....
 ....*Shot-hole borer* and *peach bark beetle,* page **627.**
 2. Masses of gum, mixed with brown sawdust-like bits of frass and bark, exuding from around the base of the trunk. White caterpillars, with brown heads, up to 1 inch long, burrowing in the bark from 8 or 10 inches above, or 3 or 4 inches below, the surface of the soil. Brown tough-skinned pupal cases protruding from the gummy masses about the base of the trunk during the latter part of the summer and early fall. Black and yellow wasplike moths flying rapidly, in daytime, about the base of the trunk, in late summer or early fall..........
 ....*Peach borer,* page **629.**
 3. Masses of gum containing brown sawdust, exuding from the upper part of the trunk and branches of the peach, most frequently from forks of limbs. White caterpillars with brown heads, up to a little over ¾ inch in length, working in the bark under masses of gum. Metallic, blue-black, yellow-marked, wasplike moths flying rapidly about the trees after midsummer.................
 ....*Lesser peach borer,* page **632.**
 4. Very small, inconspicuous, silk cocoons closely attached to the larger branches and the trunk of the tree during the winter. Brown worms with black heads, from 1/16 to ⅛ inch in length, inside these cocoons. Small cavities eaten into the ends of the new growth of peach shoots in the spring, causing them to wilt and die back. Small masses of gum exude from injured twigs. Occasionally the small brown worms will be found in the peach fruits....................
 ....*Peach twig borer,* page **634.**
 5. Pinkish to creamy-white, short-legged worms with brown heads, the largest ½ inch long, bore in the twigs of peach, causing them to wilt. Injury is similar to D, 4..*Oriental fruit moth,* page **636.**
E. *Worms—caterpillars or weevil grubs—burrowing into and feeding inside the fruits, or weevils scarring the surface of the fruits:*
 1. Crescent-shaped punctures, or slits, with a small hole often cut at the inside of apex of the crescent, on the young peach fruits for the first 2 or 3 weeks after the blooming period. Fat-bodied, brown-headed, white, footless grubs feeding in the flesh of the peach, causing irregular greenish areas on the surface of the fruit and black cavities within the fruit. Many of the grub-infested fruits drop when less than one-fourth grown. Dark-brown snout beetles about ⅓ inch long, with grayish-white patches on the back of the wing covers, feeding and laying eggs on peach fruits for a few weeks after the fruit has set and feeding again during the latter part of the summer...............*Plum curculio,* page **635.**
 2. Pinkish to creamy-white short-legged worms, with brown heads, up to ½ inch long, bore in the fruits of the peach. Entrance holes of worms often through the stem, not showing on the outside of the fruit. Gray moths with chocolate-brown markings on the wings, about ¼ inch long, flying about the trees at dusk or resting on the trunk........................*Oriental fruit moth,* page **636.**

177 (274). JAPANESE BEETLE[1]

Importance and Type of Injury.—Metallic green or greenish-bronze beetles, ⅓ to ⅝ inch long, with reddish wing covers and with two prominent, and several smaller, white spots near the tip of the abdomen and along the sides, feed on the surface of the fruit and leaves of deciduous

---

[1] *Popillia japonica* Newman, Order Coleoptera, Family Scarabaeidae.

fruits. The fruit may be partly peeled and gouged in irregular shallow patches, or nearly devoured. The leaves are skeletonized on many trees and plants. Grass is sometimes killed by the feeding of the larvae on the roots. Where the beetle has become established in the eastern United States, it is so abundant as to cause serious injury to tree fruits and also to many field crops and some truck crops on which it feeds.

Plants Attacked.—The adult beetles feed on the foliage and fruits of about 250 kinds of plants, including nearly all the deciduous fruits and small fruits, shade trees, shrubs, corn, soybeans, garden flowers, vegetables, and weeds. The larvae are serious pests of lawns and grass roots, vegetables and nursery stock.

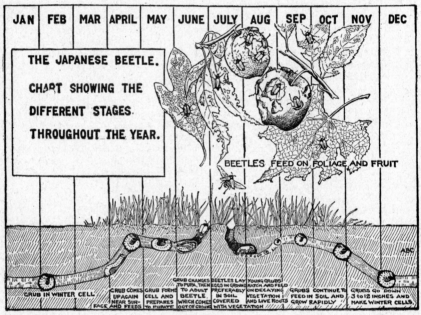

Fig. 401.—Diagram of the life cycle of the Japanese beetle, *Popillia japonica* Newman. (*From Penn. Dept. Agr. Bul. 390.*)

Distribution.—This insect was imported into New Jersey from Japan, on the roots of nursery stock, about 1916. It is well established in a large area in the eastern United States, including parts of the states of Pennsylvania, New Jersey, Delaware, New York, Connecticut, Maryland, Virginia, Massachusetts, and the District of Columbia. In most of the other states east of the Mississippi River, except those bordering on the Gulf of Mexico, it occurs in restricted areas, especially in the large cities.

Life History, Appearance, and Habits.—The winter is passed as a grub, about ½ to ¾ inch long, buried in the soil (Figs. 401 and 402,*B*). Growth is completed during June, and the adults emerge in greatest numbers in July. The female beetles lay their white spherical eggs, in groups, 2 to 6 inches deep in the soil. The grubs feed mainly on decaying vegetation, at first, but later on the fine roots of grasses and other plants. They frequently cause serious damage to lawns and golf greens. The grub resembles the common white grub but is only ¾ to 1 inch long, when full-grown, and can be distinguished from its near relatives by the appearance of the last ventral segment, as shown in Fig. 402,*B*. The nature of the life cycle is shown by Fig. 401. The adults are most abundant during July and August, flying and feeding actively on warm sunny days. They prefer to feed on parts of the plant exposed to the sun. As a rule there is one generation annually; but the grubs may take 2 years to develop in wet cold soils.

Control Measures.—Spraying with arsenicals as recommended for the codling moth will, to a large measure, protect fruit from injury by these insects, not by poisoning the beetles, but by preventing their feeding on the fruit, as they are strongly repelled by such sprays. The following are recommended as special, protective sprays: (*a*) aluminum sulphate, 3 pounds, hydrated lime, 20 pounds, in water 100 gallons (nonpoisonous); (*b*) lead arsenate, 6 pounds, and a suitable sticker, such as flour or fish

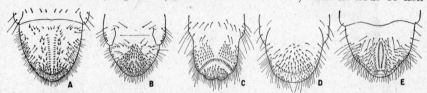

Fig. 402.—Terminal abdominal segments of the larvae of five species of "white grubs," ventral surface; showing the features by which the different species may be distinguished from each other. *A*, the Asiatic beetle, *Anomola orientalis; B*, the Japanese beetle, *Popillia japonica; C*, the Asiatic garden beetle, *Autoserica castanea; D*, the annual white grub, *Ochrosidia villosa;* and *E*, one of the native white grubs with a 3-year life cycle, *Phyllophaga hirticula;* all enlarged about three diameters. (*Redrawn from Conn. and N. J. Agr. Exp. Sta. publications.*)

oil, in 100 gallons of water (poisonous); (*c*) the standard derris or pyrethrum spray (page 266) (nonpoisonous and does not leave a noticeable residue). Thousands of beetles can be caught by jarring them from trees and shrubs upon sheets or canvas in the early morning. The larvae may be killed by drenching the sod of lawns and golf courses with carbon bisulphide emulsion or by applying lead arsenate, 400 to 1,500 pounds per acre, as recommended on page 392. About the roots of nursery stock paradichlorobenzene is used. Home grounds, parks, and estates can be protected to some extent in the less heavily infested districts, by the use of properly constructed traps, baited with geraniol and eugenol, suspended about every 50 feet, in sunny spots near trees or shrubs throughout the summer. Directions for securing and using such traps should be secured from the U. S. Department of Agriculture.

References.—*N.J. Dept. Agr. Bur. Statistics and Inspection, Cir.* 46, 1922; *Pa. Dept. Agr. Bul.* 390 (vol. 7, no. 11), June, 1924; *U.S.D.A., Dept. Bul.* 1154, 1923; *Jour. Econ. Ento.,* **19**: 786–790, 1926; *Conn. Agr. Exp. Sta. Bul.* 411, 1938; *U.S.D.A., Cirs.* 237, 1936, and 332, 1934, and *Tech. Bul.* 478, 1935.

178. Green June Beetle[1]

Importance and Type of Injury.—Serious damage is caused by this insect to lawns, golf courses, vegetable gardens, and ripening fruits. Large, somewhat flattened, green beetles, with the margins of the body bronze to yellow, nearly 1 inch long and half as broad (Fig. 403), feed on

[1] *Cotinis nitida* (Linné), Order Coleoptera, Family Scarabaeidae.

the foliage of the peach and also on the fruit just before ripening. Thick-bodied dirty-white grubs, always crawling on their backs (Fig. 403), feed on grass roots, tobacco plants, and decaying vegetation in the soil.

Plants Attacked.—The adult beetles feed on the foliage of a number of

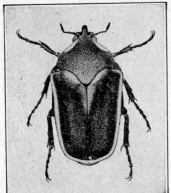

different trees and plants; also occasionally attacking ears of corn, fruits, and vegetables of the garden. The larvae also do considerable injury to the roots of grasses in lawns and golf courses, to a number of vegetables and ornamental plants, and in the tobacco seedbeds.

Distribution.—This insect is confined to

a   b

Fig. 403.—Left, adult of green June beetle, *Cotinis nitida* (Linné). Right, full-grown larva of green June beetle in natural position when crawling on its back. Twice natural size. (*From U.S.D.A., Dept. Bul. 891.*)

the southern United States, extending north into Long Island and southern Illinois.

Life History, Appearance, and Habits.—The winter is passed as a grub deep in the soil. In the spring the grubs burrow close to the surface and feed mainly on decaying vegetable matter. After a heavy rain they occasionally come out of the soil and crawl on the surface of the ground. When so crawling, they generally lie on their backs and work themselves forward by a peculiar movement of the body ridges. The grubs become full-grown by midspring, change in an earthen cell in the ground to the pupal stage, emerge as beetles, and feed on the foliage of trees during July and August. Their eggs are laid in soil rich with decaying vegetable matter, on which the grubs feed until cold weather. There is one generation each year.

Control Measures.—Piles of grass clippings or manure should not be left near lawns or orchards, as they attract the beetles as a place for egg-laying. Grubs from these piles will migrate to the lawns, and the adult beetles will injure the foliage and fruit in the orchards. Thorough wetting of the lawns in the evening will bring the grubs to the surface so that they can be destroyed. "V"-shaped troughs or flower pots sunk in the ground have proved very effective in trapping the larvae. In tobacco seedbeds they have been controlled by spreading poison-bran bait (see page 339), at the rate of 1 pound to each 10 square yards.

References.—*N.C. Agr. Exp. Sta. Bul. 242, 1921; U.S.D.A., Dept. Bul. 891, 1922, and Farmers' Bul. 1489, 1926.*

179 (149, 172, 189, 204). SAN JOSE SCALE

The San Jose scale is fully as serious a pest of the peach as of the apple. The life history of this insect on the peach is practically the same as on the apple, and the control measures are the same (see page 575). Oil sprays should be used with caution on peach during the active growing season.

180 (241). Terrapin Scale[1]

Importance and Type of Injury.—During the winter the undersides of peach twigs are sometimes found nearly covered with shiny, convex, brownish scales about $\frac{1}{12}$ inch in diameter (Fig. 405). In the summer, the fruit will be covered with masses of honeydew on which a sooty black fungus grows.

Plants Attacked.—All the common fruits and many shade trees and shrubs.

Distribution.—Eastern and southern United States.

Life History, Appearance, and Habits.—The scale passes the winter as fertilized females closely adhering to the bark of the twigs (Fig. 404). These vary somewhat in color but are of a general dark brown, with some black lines, marking off squarish areas about the margins of the scale. The scale margins are also generally somewhat crinkled. They resume feeding early in the spring, and by the last of May or early June the female, which is then about $\frac{1}{7}$ inch in diameter, begins giving birth to living young. These accumulate in a dome-shaped cavity underneath her body, and a day or two after birth the tiny nymphs, which are oval, white and with well-developed legs and antennae, crawl from under the dead female's body and out to the undersides of the leaves. Here they insert their beaks along the larger veins and feed by sucking the sap. After about a month on the leaves, the females migrate back to the twigs, and a week or two later the winged males find them and mating occurs. The males die in midsummer, but the females continue to feed until cold weather, hibernate through the winter, feed again in spring as stated above, and finally die during the following summer. There is but one generation a year.

Fig. 404.—Mature female of the terrapin scale, *Lecanium nigrofasciatum* Pergande; (*a*) ventral, (*b*) dorsal, and (*c*) lateral views. About 8 times natural size. (*From U.S.D.A., Dept. Bul.* 351.)

Control Measures.—This scale cannot be satisfactorily controlled by applications of lime-sulphur. Control has been obtained with dormant sprays of a 2 per cent lubricating-oil emulsion; or the commercial miscible oils or oil emulsions, used at the strengths recommended by the manufacturers for the control of San Jose scale.

Reference.—*U.S.D.A., Dept. Bul.* 351, 1916, and *Farmers' Bul.* 1557, p. 32, 1931.

181 (126). Green Peach Aphid[2]

Importance and Type of Injury.—Greenish smooth-looking aphids or plant lice, suck the sap from the new fruits and twigs of peach and a number of other trees and herbaceous plants.

Plants Attacked.—The food plants of this insect include peach, plum, apricot, cherry, many ornamental shrubs, and garden and flowering plants (see also page 546).

Distribution.—This insect is a native of Europe but is now generally distributed over North America.

Life History, Appearance, and Habits.—The winter is passed as black shining eggs on the bark of the peach, plum, apricot, and cherry. The young aphids, which are

---

[1] *Lecanium nigrofasciatum* Pergande, Order Homoptera, Family Coccidae. A related species, *Lecanium coryli* (Linné), has been introduced into the Northwest (see *Can. Dept. Agr. Ento. Cir.* 77, 1931).

[2] *Myzus persicae* (Sulzer), Order Homoptera, Family Aphididae.

pale yellowish-green in color with three dark lines on the back of the abdomen, begin to hatch about the time the peach blooms.   On becoming full-grown, they begin giving birth to living young, generally remaining on the peach for two or three generations, after which most of the individuals acquire wings and migrate to garden plants in late spring.   On the approach of cold weather in the fall, the females fly to the peach, where they give birth to the true sexual females.   These mate with males that fly over from the summer host plants, and the fertilized females deposit their eggs in the above mentioned situations.

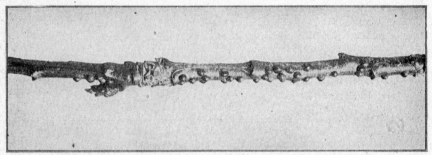

FIG. 405.—Terrapin scale on peach twigs during winter, about natural size.   (*From U.S.-D.A., Dept. Bul.* 351.)

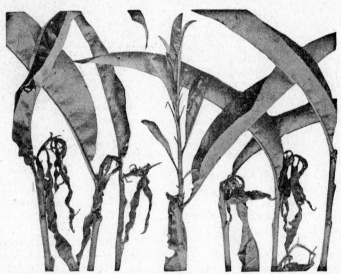

FIG. 406.—Tips of peach nursery trees injured by the feeding of the tarnished plant bug.   (*From Slingerland and Crosby, "Manual of Fruit Insects," copyright 1915, by Macmillan.   Reprinted by permission.*)

Control Measures.—The control measures for this aphid are the same as for other aphids, *i.e.*, spraying or dusting with a strong contact insecticide.   A 40 per cent nicotine-sulphate solution, used at 1 pint to 100 gallons of water, with 1 pound of potash fish-oil soap dissolved in each 25 gallons of water, will kill all aphids hit by it. A 3 per cent nicotine dust also is effective.

References.—*Pa. Agr. Exp. Sta. Buls.* 185 and 186, 1924; *Colo. Agr. Exp. Sta. Bul.* 133, 1908; *N.J. Agr. Exp. Sta. Cir.* 107, p. 5, 1919; *Va. Agr. Truck Exp. Sta. Bul.* 71, pp. 812–815, 1930.

182 (83). Tarnished Plant Bug[1]

Importance and Type of Injury.—This insect, which feeds on many different kinds of plants, is particularly injurious to the peach. In the nursery rows peach trees will be found with the main terminal and a number of laterals wilting and dying back (Fig. 406), but not containing borers. Young trees thus injured become scrubby or brushy. The insect also causes sunken areas on the sides of the fruit which are free from down, looking as though the peach had been gouged or partly peeled, when small. This injury is commonly called "cat-facing" (Fig. 407).

Plants Attacked.—Many kinds of trees and herbaceous plants; a very general feeder.

Distribution.—Throughout the world.

Life History, Appearance, and Habits.—The tarnished plant bug passes the winter as a full-grown insect in many different kinds of shelter but seems to prefer plants which do not entirely die down. Large numbers of them have been found in winter between the leaves of mullein, wild parsnip, alfalfa, and clover. It is of a coppery-brown color, flecked with darker brown and yellow. The body is oval in shape, somewhat triangular in front, about $\frac{1}{4}$ inch long, and about half as wide (Fig. 297). The insects become active during the first warm days of spring. They fly to trees, feed by sucking the sap and, while feeding, apparently inject some substance that is highly injurious to the plant tissue. The feeding causes the peach to die back, resulting in a condition usually referred to by nurserymen as "stopback" (Fig. 406). There are

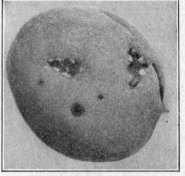

Fig. 407.—Typical "cat-facing" of peach, caused by feeding of the tarnished plant bug. (*From Ill. State Natural History Surv.*)

several generations of the insect each season, the eggs being laid mainly in the stems of herbaceous plants, including some cultivated plants and weeds. A few eggs are laid in the peach terminals and also occasionally in the peach fruits (see also page 479).

Control Measures.—No satisfactory control measures have as yet been found for this insect. Keeping down weeds about the orchard is of some value in reducing its numbers, but they should not be allowed to grow in early spring and cut later, as this practice may drive the bugs to cultivated plants. Dusting with 300-mesh sulphur has been of value on some truck crops.

References.—Mo. Agr. Exp. Sta. Res. Bul. 29, 1918; N.Y. (Cornell) Agr. Exp. Sta. Bul. 346, 1914; Ill. State Natural History Surv. Bul., vol. 17, art VI, 1928.

183 (162). Shot-hole Borer[2] and Peach Bark Beetle[3]

Importance and Type of Injury.—These two important species of bark beetles attack the trunk and branches of the peach but are of little impor-

---

[1] *Lygus pratensis* Linné, Order Hemiptera, Family Miridae.

[2] *Scolytus rugulosis* Ratzeburg, Order Coleoptera, Family Scolytidae.

[3] *Phthorophloeus liminaris* (Harris), Order Coleoptera, Family Scolytidae.

tance on vigorously growing trees. They are so similar in their habits that a separate description will not be given. The shot-hole borer, or fruit tree bark beetle, is further discussed on page 592. Small beads of gum will be found exuding from points on the bark of the trunk and branches of the peach; also small holes in the bark, about the size of a No. 6 shot. Galleries of borers are found in the inner bark radiating from a main or parent gallery and terminating in small round exit holes through the bark.

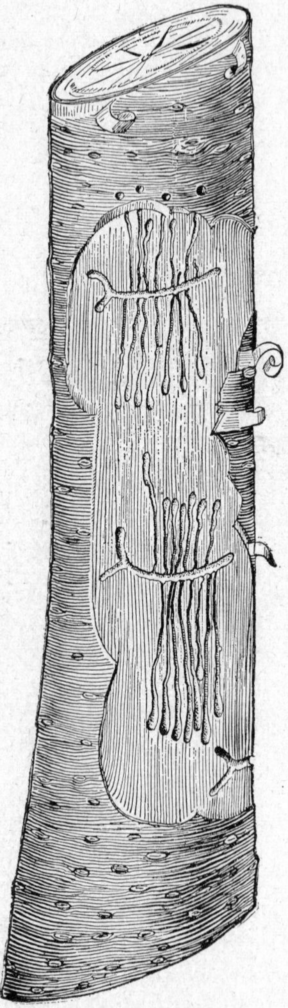

Trees Attacked.—The shot-hole borer attacks peach, cherry, plum, apple, and other fruit and shade trees. The peach bark beetle attacks cherry and other stone fruits, but *not* the pome fruits.

Distribution.—The peach bark beetle is most abundant in the eastern half of the United States. The shot-hole borer occurs throughout the country.

Life History, Appearance, and Habits.— These insects are small brownish or blackish beetles about $\frac{1}{10}$ inch or less in length. The peach bark beetle winters in the adult stage, either in its pupal cells in dead or dying wood or in special hibernating cells cut in the bark of healthy trees. The shot-hole borer passes the winter in the form of a small pinkish-white legless grub in the inner bark of the tree. These grubs transform to beetles in their burrows and emerge in the late spring. Both species are good fliers and are seemingly attracted to trees that are in a poor condition or those having dying branches. The parent beetles excavate a burrow in the inner bark of such trees, and along the sides of this the female lays her eggs. The parent gallery of the shot-hole borer (Fig. 378) generally runs the long way of the branch. That of the peach bark beetle (Fig. 408) generally runs crosswise and forks to form a "Y" at one end of the gallery. The grubs hatching from these eggs work out through the inner bark for a distance of 2 to 3 inches, gradually enlarging their burrows as they go. The larval burrows of the peach bark beetle follow mostly the grain of the wood, while the larval burrows of the shot-hole borer run chiefly across the grain. Upon becoming full-grown, they change to beetles

Fig. 408.—Work of the peach bark beetle, *Phthoro-phloeus liminaris* (Harris), in wood of peach tree, showing parent galleries, larval galleries, and exit holes. (*From U.S.D.A., Farmers' Bul.* 763.)

and emerge through small holes which they cut in the bark. There are at least two generations of these insects in the latitude of central Illinois, and a partial third farther south.

Control Measures.—As the insects are attracted mainly to unhealthy or injured trees, the best control measure is to keep the trees in a vigorous condition during the summer, either by cultivation, or by fertilizing with nitrogenous fertilizers. All peach prunings and dying or diseased trees in or around the orchard should be removed and burned during the winter. There is no method of spraying which has been found effective for these insects, which will not also injure the tree.

References.—*U.S.D.A.,* *Farmers' Bul.* 763, 1916; *Ohio Agr. Exp. Sta. Bul.* 264, 1913; *Mich. Agr. Exp. Sta. Bul.* 284, pp. 30, 32, 1933.

184. PEACH BORER[1]

Importance and Type of Injury.—The most important insect enemy of the peach is the peach borer. Nearly everyone who has tried to grow peaches, in a commercial orchard, a farm orchard, or a back yard is familiar with the work of this insect. Masses of gum exude from around the base of the trunk from about 1 foot above, to 2 or 3 inches below the surface of the soil, with bits of brownish frass or sawdust, mixed with the gum (Fig. 409). White worms with brown heads will be found burrowing in the bark of the peach trunk from 2 or 3 inches below the surface of the ground to 10 inches above. Deadened areas in the bark, where these

FIG. 409.—Base of peach tree showing evidence of attack by the peach borer; the break in the bark and exuding gum, containing frass, indicate the presence of borers beneath the bark. (*From Ill. State Natural History Surv.*)

worms have eaten out the living tissue, in many cases cause the death of the tree.

Trees Attacked.—Peach, wild and cultivated cherry, plum, prune, nectarine, apricot, and certain ornamental shrubs of the genus Prunus.

Distribution.—All sections of the United States and Canada westward to Arizona, Utah, Oregon, and British Columbia.

Life History, Appearance, and Habits.—This insect always passes the winter in the worm or larval stage. These worms (Fig. 410) vary greatly in size, some being over ½ inch in length, while others are very small, in some cases not over ⅛ inch long. The difference in the size of the worms comes from the fact that the eggs are laid over a considerable period of time. In the spring of the year the worm becomes active as soon as the soil is sufficiently warm. The larger ones will complete their growth by the middle or latter part of May and will be found under the bark close to the ground. They are then about 1 inch in length, whitish, with a dark-brown head and plate behind the head. They change, in closely spun, silken, dirt-and-gum-covered cocoons on the surface of their burrows or in the soil, to the brown pupal stage. Just before the moth

---

[1] *Conopia* (= *Aegeria*) *exitiosa* (Say), Order Lepidoptera, Family Aegeriidae.

emerges, the pupa forces its way out of the cocoon, and the empty pupal skin generally remains protruding from the cocoon. In the latitude of southern Illinois the first adults begin to appear in July and emergence continues until the latter part of September or possibly in a few cases into October, the greatest number coming out during August. The female insect is a blue-black moth having clear hind wings, with an orange cross-band on the abdomen (Fig. 411). When in flight the moths closely resemble some of the larger wasps, for which they are frequently mistaken. Unlike most moths they are active in the daytime. The male has both wings nearly clear and several narrow yellow bands across the abdomen.

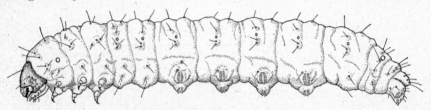

Fig. 410.—The peach borer, *Conopia exitiosa* (Say). *Upper figure*, the larva about four times natural size, slightly from beneath to show the prolegs, each with the characteristic double transverse row of crochets (*Drawn by Kathryn Sommerman*). *Lower figure*, the frass-covered silken cocoon of the pupal stage, from which the pupa has projected its body before the adult moth emerged, leaving the empty exuviae clinging to the end of the cocoon. (*From Ill. State Natural History Surv.*)

The females lay from 200 to 800 eggs mostly on the trunks of the trees or in cracks in the soil within a few inches of the trunk. They are seemingly attracted to trees which previously have been infested by the borer or to those to which some mechanical injury has occurred. Upon hatching from the eggs, in about 10 days, the worms work their way into the bark of the tree and feed during the late summer or early fall in the outer layers of the bark, penetrating deeper into the bark as they become larger. Where the eggs are laid over a period of several months, it is easy to understand the great variation in the size of the larvae found in the bark in the fall. There is, as a rule, one generation each year.

Control Measures.—While this insect, if not combated, is almost sure to kill peach trees within a few seasons, we are fortunate in having a control measure which, if properly applied, is nearly 100 per cent effective. This control method consists of placing a small amount of the crystals of paradichlorobenzene, generally called "PDB" ("painless death to borers"), on the surface of the soil around the trunk of the tree during the fall (Fig. 412). This method, which has come into general use since about 1920, is now practiced in most of the commercial peach orchards in the United States. The amount of PDB to be applied will vary with the size of the tree. The following rules should be strictly observed in using this chemical. For trees under 3 years of age, use ½ ounce per tree, for trees from 3 to 6 years old, ¾ ounce, and for older trees from 1 to 1½ ounces, depending on the diameter of the trunk at the surface of the ground. Apply the crystals in a ring completely encircling the trunk, not

Fig. 411.—Peach borer, *Conopia exitiosa* (Say); adults, natural size. The ones at center and lower left are males, the other two females. Natural size. (*From Sanderson and Jackson, "Elementary Entomology," after Slingerland.*)

closer to the bark than 1 inch, nor at a greater distance than 3 inches. Cover the crystals with several shovelfuls of earth to confine the PDB gas. Do not apply the treatment during the summer, as the borers are not then in the tree, nor should it be applied in the late fall, when the soil temperature is likely to be below 60°F. for the first 2 weeks after the material is applied. The crystals applied in warm soil quickly volatilize into a heavy gas which penetrates through the soil crevices and into the burrows of the borers and kills them. The best dates for applying paradichlorobenzene can be obtained from entomologists of the state experiment stations. In the latitude of southern Georgia the best dates for making these applications will usually be between Oct. 15 and 20; in central Georgia, between Oct. 10 and 15; in northern Georgia, between Sept. 25 and Oct. 5; in the latitude of southern New Jersey, Oct. 1 to 10; in northern New Jersey, Sept. 20 to Oct. 1; in southern Illinois, Sept. 25 to Oct. 15; and in northern Illinois, Sept. 20 to Oct. 20.

Instead of the "death-ring" treatment the base of the tree may be sprayed with one of the following emulsions: (a) Dissolve 2 pounds of paradichlorobenzene in 1 gallon of dormant miscible oil or cottonseed oil, warmed to about 90°F.; emulsify this solution in 1 pint potash fish-oil soap and stir in enough water to make a total of 4 gallons of spray. Each pint then contains 1 ounce of PDB. This should be sprayed about the base of the trunk, using ¼ to ½ pint per tree for trees under 3 years old, 1 pint for 4- or 5-year old trees, and 1½ pints for older trees. Mound as when the PDB ring is used. (b) Stir 9 parts by volume of ethylene dichloride into 1 part of potash fish-oil soap, then slowly stir into this mixture 8 volumes of water. This stock emulsion should be diluted further before applying, according to the age of the trees to be treated. Use 3 volumes of the stock and 7 volumes of water, ¼ to ½ pint per tree for 2- or 3-year old trees; 2 volumes of the stock to 3 of water, ½ pint for 4- or 5-year old trees; and a half and half mixture, ½ pint per tree for older trees. Mound as in the other treatments. The ethylene dichloride treatment may be used in colder weather than the paradichlorobenzene.

FIG. 412.—A ring of paradichlorobenzene properly placed about the trunk of a peach tree to kill peach borers. The ground has been cleared of trash before applying the chemical. Soil will be placed over the ring to confine the fumes. (*From Ill. State Natural History Surv.*)

Spring treatments are not recommended, although, if no treatment has been given in the fall, it is in some cases advisable to apply the material in the spring. The treatment will not be effective if given before the soil temperature has reached an average of about 60°F.; by this time the young borers have been actively at work in the bark for several weeks, and will have caused very serious damage. In the East and South, there has been some injury from this treatment applied to young trees, and there it is recommended that the soil be removed from around the tree 4 to 6 weeks after treatment. In any case, the mounds of earth should be leveled before July in order not to force the moths to lay their eggs high up on the trunk.

The borers may be dug out by the use of a sharp knife, or probed out with a flexible wire, first removing the soil from around the base of the tree, to a depth of about 3 inches. This method is slow and laborious, and it is not possible in most cases to get all the borers.

References.—*N.J. Agr. Exp. Sta. Bul.* 391, 1922; *U.S.D.A.*, *Dept. Buls.* 796, 1919, and 1169, 1923; *Ohio Agr. Exp. Sta. Bul.* 329, 1928; *U.S.D.A.*, *Tech. Bul.*, 58, 1928; *N.Y.* (Geneva) *Agr. Exp. Sta. Cir.* 172, 1937; *U.S.D.A.*, *Cir.* E-424 (mimeographed), 1938; *Jour. Econ. Ento.*, **25:** 786–799, 1932, and **29:** 754–756, 1088–1092, 1936.

185. Lesser Peach Borer[1]

Importance and Type of Injury.—This, too, is one of the most destructive insect enemies of stone fruits. It causes injury similar to, and only slightly less important than, that of the peach borer. Masses of gum

---

[1] *Synanthedon pictipes* (Grote and Robinson), Order Lepidoptera, Family Aegeriidae.

exude from the trunk and branches of the peach tree where injury has occurred. Frequently these masses of exuding gum will be found in the forks, especially where the forks have split, and around wounds in the bark (Fig. 413). White worms with brown heads will be found working in the inner bark. Bits of brownish castings and sawdust-like material are mixed with the masses of gum. The larvae very closely resemble those of the peach borer, but their attacks may usually be distinguished in the orchard by the fact that they attack the upper part of the trunk and the larger branches.

Trees Attacked.—Peach, plum, cherry, wild plum, wild cherry, Juneberry, and possibly some others.

Distribution.—Peach-growing sections of the United States, except the western states; most abundant in the South.

Life History, Appearance, and Habits.—The lesser peach borer passes the winter in the inner bark of the peach tree in the form of partly grown larvae. These larvae vary in size from $\frac{1}{4}$ inch to nearly 1 inch in length. They start to feed very early in the spring, the larger individuals completing their growth by late April and the smaller ones by the last of June. They pupate in cocoons inside the burrows, but always come close to the surface of the bark, leaving the exit of the burrow covered by a thin silken web.

FIG. 413.—Work of the lesser peach borer, *Synanthedon pictipes* (Grote and Robinson). (*From Ohio Agr. Exp. Sta. Bul.* 307.)

Just before the moths emerge, the pupae push themselves out of the cocoon and partly out of their burrows. Emergence of the moths takes place during May and June. The moths are clear-winged, of a metallic blue-black color with pale-yellow markings on the abdomen. They are extremely active and dart about with great rapidity on warm sunny days. Their general appearance is much like that of some of the larger wasps. The female lays her eggs in cracks and crevices in the bark, usually around the crotch of the tree or wounds in the bark. The eggs hatch in about 10 days after they are deposited, and the larvae work their way into the bark. A partial second generation occurs over most of the North Central states, only one in South Dakota, and probably nearly a full second generation in the South.

Control Measures.—Experiments in painting thoroughly over all areas of the tree showing damage by this borer, with a mixture of 2 pounds of paradichlorobenzene dissolved in 1 gallon of dormant miscible oil and then further diluted with 2 gallons of water, giving a mixture containing $\frac{2}{3}$ pound in each gallon, has given a very good kill. The treatment is best made in September or early October, after the eggs have hatched, but can be made in the spring, about April 1.

The lesser peach borer is of little importance in well-sprayed, cultivated, and properly fertilized orchards. The best method of preventing this insect is to keep the trees in a vigorous, healthy condition and to avoid, so far as possible, injuries to the bark and crotches of the tree. Trees with wide-angled crotches should be chosen for planting. The worms may be dug out by hand, as recommended under the peach borer.

References.—Ohio Agr. Exp. Sta. Bul. 307, 1917; *S.D. Agr. Exp. Sta. Bul.* 288, 1934; *U.S.D.A., Cir.* 172, 1931; *Jour. Econ. Ento.,* **29:** 754–756, 1936; *Ill. Natural History Surv. Cir.* 31, 1938.

186. Peach Twig Borer[1]

Importance and Type of Injury.—In unsprayed or neglected orchards, the peach twig borer is of considerable importance. Small brown worms may be found working

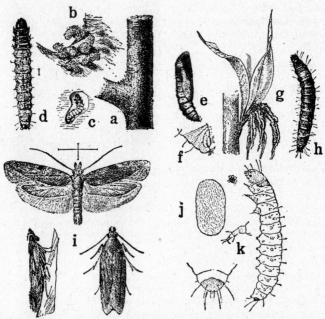

Fig. 414.—The peach twig borer, *Anarsia lineatella* Zeller. *a, b, c, d,* larva and its winter nests; *e, f,* pupa with detail of posterior end; *g,* new shoot of peach withering from attack of larvae; *h,* larva enlarged; *i,* adult moths; *j,* egg greatly enlarged; *k,* larva, side view and details of structure. The lines show natural size. (*From U.S.D.A.*)

inside the twigs and new growth, as well as in the fruits. The twigs die back and small masses of gum exude from them.

Trees Attacked.—Peach, plum, apricot and almond.

Distribution.—General in peach-growing sections, most serious on the Pacific Coast.

Life History, Appearance, and Habits.—This insect passes the winter as a partly grown, chocolate-brown caterpillar, with a black head, from ⅛ to ¹⁄₁₆ inch in length, hidden in a silken cocoon closely attached to the bark of trunk and branches (Fig. 414*a,b,c*). About the time the peach leaves appear, the larvae leave their winter

[1] *Anarsia lineatella* Zeller, Order Lepidoptera, Family Gelechiidae.

nests and start boring in the ends of the tender new growth, causing a wilting and dying back of the twigs (g) in which they are working. A larva may feed in more than one twig before it completes its growth. Upon becoming full-grown, the larvae are about ½ inch in length. They then spin cocoons on the larger branches or trunk. Here they change into small moths, gray in color, with a wing-spread of about ⅓ to ½ inch (i). The moths mate, and the female lays her eggs on the twigs. Another generation of the worms soon hatches and feeds almost entirely in the fruits. There are probably from one to four generations of the insect in the United States each season, the larger number occurring in the South.

Control Measures.—Damage by this insect is confined, as previously stated, largely to unsprayed orchards. The most effective control measure in the central states is a delayed dormant spray with lime-sulphur or oil emulsion, such as is used for controlling the San Jose scale. This spray kills the overwintering larvae in their cocoons. Where such treatments are regularly given, practically no damage from the peach twig borer occurs. In some sections the oil emulsions have not been successful. In some of the western states it has been found that the addition of 3 pounds of lead arsenate to 50 gallons of lime-sulphur, 1 to 10, applied as near the pink stage of the buds as possible, gives good control. Basic lead arsenate, 3 pounds in 100 gallons of water, applied in the spring has given good control in California. For apricots, poisoned bordeaux is preferred.

References.—*Jour. Econ. Ento.*, **15**: 395, 1922; *Colo. Agr. Exp. Sta. Bul.* 119, 1907; *Calif. Agr. Exp. Sta. Bul.* 355, 1923; *Jour. Econ. Ento.*, **29**: 156–160, 1936.

187 (163, 191, 194). PLUM CURCULIO[1]

The plum curculio, which has been described under apple insects (see page 594), is one of the most serious insect pests of the fruit of the peach. The adults damage the fruits by their feeding and egg-laying punctures and the larvae tunnel through the fruit, feeding on the pulp. Infested fruits generally drop. The plum curculio is also one of the main agencies in spreading the brown rot of the peach.

Control.—The method of control is, in general, the same as that given for the insect on the apple. In peach orchards which have been badly infested, it is often advisable to gather the peach "drops" during May and the first part of June. The orchard should be gone over every few days, if possible, and the dropped peaches gathered and burned or sacked and buried to a depth of at least 2 feet. For the control of this insect, general orchard sanitation is as important in peach orchards as in apple orchards. Cultivating the soil during the late spring and early summer will destroy the larvae and pupae of the curculio in their cells in the earth. Jarring the tree early in the morning and catching the beetles on sheets spread under the tree is of help in control.

The peach tree is more sensitive to arsenical injury than apple, and care must be used in applying these poisons. Two pounds lead arsenate, 3 pounds zinc sulphate, and 3 pounds lime in 100 gallons of water makes an effective spray. A dust of 5 pounds lead arsenate, 90 pounds gypsum, and 5 pounds of light oil will give nearly as good control as the spray and is not so likely to injure the foliage. Cryolite, 3 pounds, and summer oil stock emulsion, 2 quarts, in 100 gallons of water may be used in place of lead arsenate. This is not so likely to cause damage to the foliage. The first spray should be given about 10 days after the peach has bloomed, or as soon as the brown skin or shuck is pushed off by the forming fruit. The second spray should be given 3 weeks later. A third and fourth

[1] *Conotrachelus nenuphar* (Herbst), Order Coleoptera, Family Curculionidae.

spray are usually necessary on the later maturing varieties. The last spray for curculio should be applied about a month before the fruit is ripe. In badly infested peach orchards it may be of advantage to apply an additional spray immediately after the fall of the petals, but this is not generally recommended. The airplane is being used to apply the dusts in sections where there are many large peach orchards (see Fig. 193).

Reference.—U.S.D.A., Tech. Bul. 188, 1930; Del. Agr. Exp. Sta. Buls. 175, 1932, and 193, 1935; Ill. Natural History Surv. Cir. 37, 1939.

188. Oriental Fruit Moth[1]

Importance and Type of Injury.—This is, in many parts of the eastern United States, the most important insect pest of the fruits of the peach. The earliest indication of injury by this insect is similar to that of the peach twig borer and consists of a dying back of the new growth of twigs in the spring (Fig. 416). The worms found burrowing in the twigs, however, are not brown like the peach twig borer, but pinkish or creamy white, with brown heads. The fruit shows injury similar to that of the codling moth in apples. A pinkish-white, short-legged larva, $\frac{1}{2}$ inch long, with a five-toothed, comb-shaped plate on the last segment of the body, may be found inside the fruit (Fig. 417). In many cases the fruit will not show any blemish through the skin, the young worms having entered through the stem. The fruit may look perfect at the time of picking but breaks down shortly after packing and shows numerous feeding burrows of the larvae. Their attacks cause an increase in the amount of brown rot.

Trees Attacked.—Peach, quince, apple, pear, apricot, plum, and several other fruits.

Distribution.—The oriental fruit moth is an imported insect, having been brought in from the Orient previous to 1915, on imported nursery stock. It is now well established over the eastern United States in areas where peaches are grown and has also been found in the Southwest.

Life History, Appearance, and Habits.—The winter is passed as full-grown larvae about $\frac{1}{2}$ inch in length, closely resembling the codling moth. The larvae enclose themselves in silken cocoons which they spin on the bark of the trunk or in many cases in rubbish, weeds, and mummied fruits, on the ground around the peach orchard. In the spring the insect changes to the pupal, and then to the moth stage. The moths (Fig. 415) are gray, with chocolate-brown markings on the wings. They very closely resemble the codling moth but are somewhat smaller in size, being about $\frac{1}{4}$ inch long, with a wing spread of only $\frac{1}{2}$ inch. The females lay their flat whitish eggs on the leaves, and, in some cases, on the twigs, generally shortly after the peaches bloom. The first-generation worms, upon hatching, bore in the tender twigs and, upon becoming full-grown, spin cocoons in which they transform to moths. There are from one to seven generations of the insect each year. The later generations attack the fruit in much the same manner as the codling moth attacks the apple, except for the fact that the larvae do practically no external feeding upon the leaves and in working their way through the skin of the peach appar-

[1] *Laspeyresia* (= *Grapholitha*) *molesta* (Busck), Order Lepidoptera, Family Tortricidae.

ently swallow little or none of the particles which they bite off. As already mentioned, many of the larvae enter the fruit through the stems.

Fig. 415.—Adult of the Oriental fruit moth, *Laspeyresia molesta* (Busck). At left, with wings folded, resting on a leaf; at right, with wings spread; about 5½ times natural size. (*From U.S.D.A.*)

Fig. 416.—Typical injury to peach twigs by larvae of the Oriental fruit moth. *Left*, early summer injury; *right*, a type of injury, consisting of a mass of gum, leaves, and frass, found in fall and winter. (*From Purdue Agr. Exp. Sta. Cir. 122.*)

These feeding habits prevent the effective use of poison sprays for the control of this insect.

Control Measures.—In many areas the most important control is to plant only early-maturing varieties, which may often be picked before this pest attacks the fruits, even where the insect is abundant. Applications of dusts impregnated with light-grade mineral oil (100 viscosity by the Saybolt test) have given about 80 per cent control under Illinois conditions. A formula that has given good results is sulphur 60 per cent, 300-mesh talc 35 per cent, mineral oil 5 per cent, all by weight. This or similar dusts should be applied at 5-day intervals, beginning about 20 days before the peaches are picked. These oil dusts act as irritants, not as poisons.

Fig. 417.—Peach injured by oriental fruit moth, showing larva and result of its feeding. About natural size. (*From Purdue Agr. Exp. Sta. Cir. 122.*)

The application of paradichlorobenzene, as used for the control of the peach borer (see page 631) will kill the overwintering larvae near the base of the trunk. Cultivating the soil of infested orchards to a depth of 4 inches, 1 to 3 weeks before blooming time, will kill many of the over-wintering larvae in the soil. Prompt destruction of cull fruits, the screening of packing sheds, and cleaning of used baskets are advised in the northern peach areas. The small wasp parasite, *Macrocentrus ancylivorus* Rohwer, and others have accomplished effective control of this pest in many areas and in many years.

References.—*Va. Agr. Exp. Sta. Tech. Bul.* 21, 1921; *Pa. Dept. Agr. Bul.*, vol. 8, no. 9, June, 1925; *Purdue Agr. Exp. Sta. Cir.* 122, 1925; *U.S.D.A., Tech. Buls.* 152, 183, and 215, 1930; *Conn. Agr. Exp. Sta. Bul.* 313, 1930; *Ill. Nat. History Survey Cir.* 26, 1934; *Ohio Agr. Exp. Sta. Bul.* 569, 1936.

E. PLUM INSECTS

The insects that attack the plum are very much the same as those that attack the peach; in fact, there are only one or two insects of special importance on the plum.

FIELD KEY FOR THE IDENTIFICATION OF INSECTS INJURING THE PLUM

1. About the time the plum trees come into full foliage, silken webs are found enclosing the leaves at the ends of branches. Feeding on the leaves within the webs are smooth-bodied, grayish-yellow, many-legged caterpillars, up to ¾ inch long. Black wasp-like insects deposit eggs on the midribs of the plum leaves just previous to the appearance of the webs..................................
   Web-spinning sawfly, Neurotoma inconspicua (Norton) (see *S.D. Agr. Exp. Sta. Bul.* 190, 1920).
2. Tips of new growth on plum trees literally covered with brownish, blackish, or light-green aphids. Leaves badly curled, growth stunted, fruit sometimes misshapen and shriveled..........*Rusty plum aphid* or *hop aphid*, page 639.

3. Plums of the Japanese and European varieties have the branches coated with grayish scales, having much the color of wet ashes. Mature trees of the Americana varieties are seldom injured (see *C*, 1 under Key for Insects Injuring the Peach)..*San Jose scale*, page 639.
4. Leaves of plum and other fruits show a pale sickly appearance as though covered with dust or brownish-red specks. Very light webs over the surface of the leaves, upon which are crawling minute, reddish, eight-legged mites.......
....*European red mite*, page 586.
5. Partly grown plums are "stung" and drop from the tree (see *E*,1 under Key for Insects that Attack the Fruit of the Peach)........*Plum curculio*, page 639.

189 (149, 172, 179, 204). San Jose Scale[1]

This scale is very destructive to plums belonging to the Japanese or European varieties, *i.e.*, such varieties as Lombard, Green Gage, Fallenberg, Burbank, and many others. The scale is not, however, of great importance on mature American plum trees, such as the Wild Goose or other varieties of *Americanas*. The San Jose scale may cause injury to young plum trees of any variety (see page 575). Control is the same as on peach.

190. Plum Aphids

There are several species of aphids which attack the plum. They often appear in very large numbers on the tips of the plum twigs (Fig. 418) and by their feeding cause splitting of the fruits, curling of the leaves, stunting of the growth, and injury of the tree and fruit by the discharge of honeydew. The hop aphid,[2] the rusty plum aphid,[3] and the mealy plum aphids[4] are common species.

Fig. 418.—Rusty plum aphids, *Hysteroneura setariae* (Thomas), clustered on plum. About natural size. (*From U.S.D.A., Farmers' Bul.* 908.)

Control Measures.—While several species of aphids are involved, the control measures are essentially the same. As soon as the aphids appear, the trees should be thoroughly sprayed with 40 per cent nicotine sulphate, 1 pint, fish-oil or other soap, 4 pounds (or 1 gallon summer-oil emulsion), and water 100 gallons. Nicotine may also be applied as a dust, using 2½ per cent actual nicotine. Dormant sprays containing 2 per cent coal-tar distillate are effective in killing the overwintering eggs.

References.—*Pa. Agr. Exp. Sta. Bul.* 182, 1923; *U.S.D.A., Dept. Bul.* **774**, 1913; *Okla. Agr. Exp. Sta. Bul.* 88, 1910; *Calif. Agr. Exp. Sta. Bul.* 606, 1937.

191 (163, 187, 194). Plum Curculio[5]

The plum curculio has already been discussed under both apple and peach (see pages 594 and 635). It is perhaps more destructive to plums

---

[1] *Aspidiotus perniciosus* Comstock, Order Homoptera, Family Coccidae.
[2] *Phorodon humuli* (Schrank), Order Homoptera, Family Aphididae.
[3] *Hysteroneura setariae* (Thomas), Order Homoptera, Family Aphididae.
[4] *Hyalopterus pruni* (Geoffroy), and *H. arundinus* (Fabricius).
[5] *Conotrachelus nenuphar* (Herbst), Order Coleoptera, Family Curculionidae.

than to any other fruit. The curculio is one of the main agencies in spreading brown rot. If this very destructive disease of the fruit is to be controlled, it is necessary to control the plum curculio.

Control Measures.—The methods of control which are given for this insect on the peach will apply equally well for the plum. Particular attention should be given to clean cultivation of the plum orchard and to picking up the early-dropped plums. To control this pest on plums, the first spray should be applied just before the blossoms open, using lead arsenate 4 pounds, the recommended sulphur fungicide, soybean flour 4 ounces, and water 100 gallons. Apply again as soon as the blossom skins or shucks have fallen from the plums. A third spray is usually needed about 10 days after the second. On some varieties and in case of severe infestations an additional spray is needed 3 weeks after the third.

F. CHERRY INSECTS

The cherry, like the plum, is attacked by many of the insects which affect the peach. Sour cherries are not subject to severe injury by San Jose scale, but it is a serious pest of sweet cherries. Cherries are sometimes attacked by the peach borer, lesser peach borer, and bark beetles; the same control measures may be used as on peach. Some of the aphids that attack the plum attack also the cherry, and there are several other species which confine their work largely to this tree. The methods of control are the same as for these insects on plum.

The sweet cherry is attacked by practically the same insects as those feeding on the sour cherry and is also subject to injury by the San Jose scale. Wherever sweet cherries are grown, especially while the trees are young, they should receive a dormant treatment for San Jose scale, the same as that given the apple or peach.

FIELD KEY FOR THE IDENTIFICATION OF INSECTS INJURING THE CHERRY

1. Fawn-colored very long-legged slender-bodied beetles, about $\frac{1}{2}$ inch long, attack ripening fruits often in swarms. Most abundant in sandy areas and disappear within a month............................*Rose chafer*, page 645.
2. Soft, fleshy, dark-green to orange, slimy slug-like larvae, up to $\frac{1}{2}$ inch in length, feed on the surface of cherry leaves and skeletonize the leaf. Most abundant during late spring and again in late summer. In the late spring black-and-yellow wasp-like insects, about $\frac{1}{5}$ inch long, lay their eggs on the leaves of cherry......................................*Pear* or *cherry slug*, page 641.
3. Grayish, very thin, flaky scales, up to $\frac{1}{8}$ inch across, and with a raised reddish nipple-like area in the center of the scale are found clustered or massed on the bark of cherry. The scales covering the reddish-yellow bodies of the insects are thinner and more translucent than those of the San Jose scale.............. ....*Forbes' scale*, page 642.
4. Grayish scales on the bark of sweet cherry, rarely on sour cherry (see *C*, 1, under Key for Insects Injuring the Peach)...........*San Jose scale*, page 575.
5. Cherries are "stung" but seldom drop. White footless grubs, up to $\frac{1}{3}$ inch long, feed inside the fruit (see *E*, 1, under Key to Insects that Attack the Fruit of the Peach)................................*Plum curculio*, page 642.
6. Injury similar to *5*, but the egg-punctures in the fruits lack the crescent. Blossoms as well as fruits are eaten into by a brown snout beetle with four humps on the wing-covers................................. ....*Cherry curculio* (see page 597; also *Colo. Agr. Exp. Sta. Bul.* 385, 1932).

7. Cherries misshapen, undersized, ripening prematurely, or the fruit partly decayed or shrunken and wrinkled on one side.   Yellowish-white maggots, up to ¼ inch long, burrow in the injured fruit.   Black flies, smaller than a house fly, with the abdomen marked with white crossbands and the thorax margined with yellow, and with four black bands on the wings, are found resting on the foliage or on the fruit..........................*Cherry fruit flies*, page 643.

192. Pear Slug[1]

Importance and Type of Injury.—Small, fleshy, dark-green to orange, slug-like, slime-covered larvae, up to ½ inch in length, with the front part of the body enlarged, feed on the surface of the cherry leaves and skeletonize the leaves, leaving only a framework of veins (Figs. 154, 419).

Trees Attacked.—Cherry, pear, and plum.

Distribution.—Throughout the United States.

Life History, Appearance, and Habits.—This insect, like many other sawflies, passes the winter in a cocoon formed in an earthen cell 2 or 3 inches below the surface of the ground.   In the late spring, shortly after the cherries have come into full leaf, black-and-yellow sawflies, ⅕ inch long, emerge from these cocoons. The insects are a little larger than the common house fly. They have four wings. The female sawfly inserts her eggs in the leaves, and after a few days these eggs hatch into the soft-bodied worms or slugs which feed on the leaf as above described.   The feeding period varies from 2 to 3 weeks.   As the slugs grow in size, they become somewhat lighter in color, until when full-grown they are nearly orange-yellow.   They

Fig. 419.—Larvae of the pear slug, *Eriocampoides limacina* Retzius, and characteristic injury on leaf.   About twice natural size. (*From Slingerland and Crosby, "Manual of Fruit Insects," copyright 1915, by Macmillan. Reprinted by permission.*)

then crawl or drop to the ground, into which they burrow, and there change to the pupal stage.   Adults emerge during late July and August and lay eggs for the second generation of slugs.   It is this generation that usually causes the greatest amount of injury, especially on young trees, which they may completely defoliate.   When this second generation becomes full-grown, they go into the ground and remain there during the winter.

Control Measures.—The pear slug is very easily poisoned.   A spray of lead arsenate, 2 pounds, soybean flour, 4 ounces, and water, 100 gallons, is very effective.   Dusting with lead arsenate, 1 pound, and talc, gypsum, or hydrated lime, 5 pounds, will also give good control.   The spray schedule recommended for the control of the curculio and fruit flies is effective in controlling the slug.

References.—*State Ento. S.D. Cir.* 27, 1918; *Mich. Agr. Exp. Sta. Spcl. Bul.* 244, 1933.

[1] *Eriocampoides limacina* Retzius, Order Hymenoptera, Family Tenthredinidae.

193. FORBES' SCALE[1]

Importance and Type of Injury.—As already stated, the San Jose scale is of little or no importance on sour cherries. The Forbes' scale, which resembles the San Jose scale in general appearance, is often very abundant, although not seriously destructive to the tree. Grayish, thin, flaky scales are massed on the bark of the trunk and branches of the tree, sometimes completely covering the bark (Fig. 420). Examination with a microscope will show a raised reddish area in the center of each scale which will distinguish it from the San Jose scale.

Plants Attacked.—Cherry, apple, apricot, pear, plum, quince, and currant.

Distribution.—Throughout the United States east of the Rocky Mountains.

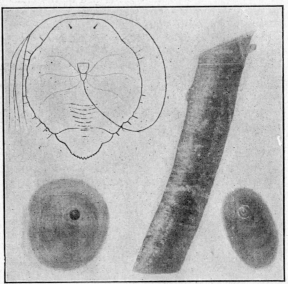

FIG. 420.—Forbes' scale, *Aspidiotus forbesi* Johnson. Mature female removed from scale, upper left, greatly enlarged; female scale, lower left, much enlarged; male scale, lower right, much enlarged; infested twig, somewhat reduced. (*From Ill. State Natural History Surv.*)

Life History, Appearance, and Habits.—The winter is passed as a partly grown scale, somewhat resembling the San Jose scale, but the raised area in the center of the scale has a reddish or orange color. In Illinois, the first young appear in May, and are produced both from eggs and by the birth of nymphs. There are from one to three generations each season.

Control Measures.—This insect can be controlled by dormant applications of oil sprays or lime-sulphur, as given for the control of the San Jose scale (page 577).

Reference.—*Twenty-second Rept. Ill. State Ento.*, p. 115, 1902; *Mich. Agr. Exp. Sta. Spcl. Bul.* 239, 1933.

194 (163, 187, 191). PLUM CURCULIO

Most of the wormy cherries in many parts of the country are caused by the plum curculio. This insect is especially abundant on cherry trees grown in small orchards or in cities. The control measures for this insect are the same as those given for its control on the plum (see page 640).

[1] *Aspidiotus forbesi* Johnson, Order Homoptera, Family Coccidae.

195. Cherry Fruit Flies[1,2]

Importance and Type of Injury.—There are two closely related flies which attack the cherry and cause wormy fruits. Cherries are somewhat misshapen and undersized, turn- ing red ahead of the main crop, oftentimes with one side of the fruit partly decayed and shrunk- en or wrinkled and closely attached to the pit. Yellowish- white footless maggots, up to ¼ inch long and pointed at the head end, will be found in the flesh of the fruit (Fig. 421). The small ones are very hard to detect in raw fruits but become more visible and tend to sink to the bottom when broken fruits are boiled for a minute. Brown burrows extend through the fruit.

Trees Attacked.—Cherry, pear, plum and wild cherries. The blackbodied species[2] prefers sour to sweet cherries.

Distribution.—Northern United States and Canada.

Fig. 421.—Section of ripe cherry showing maggot of the cherry fruit fly at work. About 3 times natural size. (*From N.Y. (Cornell) Agr. Exp. Sta. Bul.* 325.)

Life History, Appearance, and Habits.—The winter is passed in brown, capsule-like cases (puparia) in the soil. The adult flies begin to emerge during late spring, and fly to cherry trees, where the females feed by scraping the surface of leaves and fruit and sucking up solutions from them. They lay their eggs through small slits cut in the flesh of the fruit. These flies (Fig. 422) are a little smaller than a house fly, of a general black color, with yellow margins on the thorax. The white-banded cherry fruit fly[1] has four white crossbands on the abdomen, which are not found on the black-bodied cherry fruit fly.[2] There are blackish bands on the wings. The maggots on hatching from the eggs feed in the flesh of the fruit, causing decay and the injuries already described. On becoming full-grown, they drop to the ground and work their way below the surface, where they change into the pupal stage in which they pass the winter.

Fig. 422.—Adult of the cherry fruit fly, *Rhagoletis cingu- lata* (Loew), about 4 times natu- ral size. (*From N.Y. (Cornell) Agr. Exp. Sta. Bul.* 325.)

Control Measures.—The spray schedule used for the control of the plum curculio will give a very satisfactory control of these insects. In this schedule the spray consists of lead arsenate, 3 pounds, the recom-

[1] *Rhagoletis cingulata* (Loew), Order Diptera, Family Trypetidae.
[2] *Rhagoletis fausta* Osten Sacken.

mended sulphur fungicide, and water 100 gallons.   This should be applied immediately after the adult flies appear in the orchard, just after the blossom buds open.   A second spray, using the same ingredients, should be given as soon as the shucks have fallen from the young cherries.   A third spray of the same materials should be given 10 days later.   These sprays protect the fruit by poisoning the adult flies.   Cherries sprayed in this way must be very thoroughly washed before being eaten or canned, to remove poisonous residue.   Rotenone, ground derris, or timbo root (containing 4 to 4.5 per cent rotenone), in a spray at the rate of 2 pounds to 100 gallons of water, has given good control and will not leave objectionable residue.   Cull cherries about canneries and picking stands should be immediately destroyed by heat or burying deeply.

References.—N.Y. (Cornell) Agr. Exp. Sta. Bul. 325, 1912; *Ont. Dept. Agr. Bul.* 227, 1915; *Mich. Agr. Exp. Sta. Cir. Bul.* 131, 1930; *Jour. Econ. Ento.*, **26**: 431–438, 1933, and **28**: 205–207, 1935.

G. APRICOT INSECTS

The apricot is injured by practically the same insects as the plum, and the same control measures will apply on this tree.   Plum curculio is often very destructive to apricot, and, where this fruit is grown, special attention should be given to this insect.   San Jose scale is also often very serious, especially on the younger trees.

H. GRAPE INSECTS

FIELD KEY FOR THE IDENTIFICATION OF INSECTS INJURING THE GRAPE

A. External chewing insects that eat holes in the leaves, buds, or fruit:
  1. Fawn-colored, long-legged, slender-bodied beetles, about ½ inch long, feed on the blossoms, newly set grapes, and leaves.   Most abundant in sandy areas and during the first 2 weeks after blooming...............*Rose chafer*, page 645.
  2. Small, active, jumping beetles of a metallic greenish-blue color, feed on the unfolding leaves of grape in the spring.   Light-brown black-spotted grubs, the largest about ⅓ inch in length, feed in company with the beetles and, later, on the opened leaves.............................*Grape flea beetle*, page 647.
  3. Small, chain-like holes are eaten in the leaves by brown to grayish beetles about ⅓ inch long; especially abundant about 2 weeks after grapes bloom..........
    ....*Grape rootworm*, page 653.
  4. Leaves skeletonized; all the green eaten and buds and fruits devoured by metallic-green to coppery-red beetles, ⅓ to ⅝ inch long, with white spots at the tip of the abdomen beyond the wing covers...........*Japanese beetle*, page 621.
  5. Greenish-gray to purplish caterpillars up to ½ inch in length, with brown heads, feed on or in the grape berries, under protecting webs, and often attach bits of leaves to the berries.............................*Grape berry moth*, page 648.
  6. Buds and new growth, especially of vines growing in or near grass sod, eaten by plump, nearly bare, caterpillars, with 6 slender legs and 5 pairs of prolegs, somewhat ornamented with spots or stripes, which climb into the vines to feed at night, and hide on or in the soil during the day.......................
    Climbing cutworms, Lampra alternata, and others (see *Mich. Agr. Exp. Sta. Spcl. Bul.* 239, 1933).
B. Insects that suck sap from the leaves:
  1. Vines lacking in vigor, sickly in appearance, the foliage pale, with many tiny whitish spots over the leaves.   Very slender yellowish bugs, with red marks on the wings, about ⅛ inch long by a fourth as wide, and their active light-colored

nymphs, suck the sap from the underside of leaves. The nymphs run sidewise as readily as forward..........................*Grape leafhopper*, page 650.
C. *Insects that produce small galls or swellings on the leaves:*
  1. Vines lack vigor, often dying. Rounded, irregular galls, about half the size of a pea, on the leaves, sometimes covering them. The galls open on the underside of the leaves, and the inside is lined with many, small, wingless, pale-yellow aphids (see also *D*, 1).........................*Grape phylloxera*, page 651.
D. *Subterranean insects that attack the roots of the vines:*
  1. Vines lack vigor. Knot- or gall-like swellings on the roots are covered with aphids similar to *C*, 1, causing the roots to rot off. .*Grape phylloxera*, page 651.
  2. Vines show a lack of vigor, the foliage yellowing, and little growth being made. White grubs, about ½ inch long, with brown heads, eat off the small, fibrous roots and score the bark of the larger roots during late summer (see also *A*, 3) ....*Grape rootworm*, page 653.
  3. Whitish caterpillars with brown heads, up to 1¾ inches long with crochets on the prolegs in 2 transverse rows, burrow in the roots eating out the wood and inner bark, most of them a foot or more from the base of the vine, killing the roots and disastrously weakening the vines................................
  Grapevine rootborer, Memythrus polistiformis Harris (see *W. Va. Agr. Exp. Sta. Bul.* 110, 1907).

196. Rose Chafer[1]

Importance and Type of Injury.—The leaves of grape, and especially the blossoms, are eaten by gray or fawn-colored, long-legged, slender beetles about ½ inch long (Fig. 423,*a*). Newly set grapes are eaten and bunches of grapes nearly ruined. The insects are most abundant on the grape for the first 2 or 3 weeks after bloom, but will be found in smaller numbers for the next week or 10 days. The beetles also eat ripening cherries and riddle the buds and leaves of roses. Poultry eating these beetles may be poisoned and killed.

Plants Attacked.—Grape, apple, peach, cherry, pear, strawberry, rose, hydrangea, peony, blackberry, raspberry, Virginia creeper, corn, bean, beet, pepper, cabbage, poppy, hollyhock, mullein, clover, small grains and grasses, and a great variety of other plants, trees, and shrubs.

Distribution.—The insect is generally distributed over the eastern United States and southeastern Canada, and westward to Colorado and Texas. Its injuries are more severe in sandy areas.

Life History, Appearance, and Habits.—The winter is passed in the larval or grub stage. The larva (Fig. 423,*b*) closely resembles that of the common white grub but is somewhat more slender and much smaller. When full-grown, it is about ¾ inch in length. The larvae are found in uncultivated land, especially in sandy areas, to a depth of 10 to 16 inches. In the spring the nearly full-grown larvae work their way toward the surface of the ground and feed for a short time on the roots of grasses, grains, weeds, and other plants. They pupate during May, and remain in this stage (*e*) for about 3 weeks, emerging as adults at about the time when the grapes come into bloom. The adult (*a*,*f*) is a very ungainly beetle, nearly ½ inch in length, with reddish-brown head and thorax and the undersurface of the body blackish in color. The whole body is covered with small yellow hairs, which give the beetle a fawn-colored appearance. These beetles feed chiefly on the surface of the plants, and the females, after mating, deposit their eggs in groups of from 6 to 25, at a depth of about 6 inches in the soil. While the eggs are grouped, each egg

---

[1] *Macrodactylus subspinosus* Fabricius, Order Coleoptera, Family Scarabaeidae.

is laid in a separate pocket in the soil. The eggs hatch in 1 to 2 weeks, and the young larvae feed on the roots of grasses and other plants for the remainder of the summer, going down in the soil on the approach of cold weather.

Control Measures.—The most important method of controlling this insect is thoroughly to cultivate the areas about the vineyard where the eggs may have been deposited. Very few grubs of the rose chafer have been found in land where cultivated crops, such as potatoes and corn, are being grown. Cultivation is most effective in killing the insects if carried on during May and early June, when the rose chafer is in the pupal stage. In order to secure the maximum protection from cultivation, it is neces-

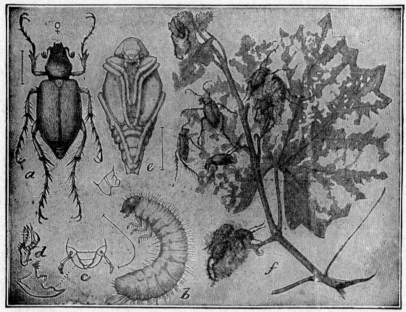

Fig. 423.—The rose chafer, *Macrodactylus subspinosus* Fabricius. *a*, adult; *b*, larva; *c* and *d*, mouth parts; *e*, pupa; *f*, injury to leaves and blossoms of grape, with beetles at work. *a*, *b*, *e*, enlarged, the lines show natural size; *c* and *d*, more enlarged; *f*, slightly reduced. (*From U.S.D.A., Farmers' Bul.* 721.)

sary that all grape growers unite in keeping the areas, surrounding vineyards, under cultivation. A spray consisting of 6 pounds of lead arsenate, or 4 pounds calcium arsenate, and 1 or 2 gallons of molasses to 100 gallons of water, applied as soon as the beetles appear, and at intervals of about 1 week if the infestation continues severe, has proved very effective in controlling the beetles. Cryolite at the same strength should give control. Not more than two such sprays should be applied because of the residue and danger of injuring the vines. Choice flowers and fruits may be protected by covering with cheesecloth during the 4 or 5 weeks when the adults are present.

References.—*Fifty-third Rept. Ento. Soc. Ont.*, p. 60, 1922; *U.S.D.A., Bur. Ento. Bul.* **97**, Part III, 1911; *Boyce Thompson Inst. Contrib.* 4, 1932; *U.S. D.A., Farmers' Bul.* **1666**, 1931.

197. Grape Flea Beetle[1]

Importance and Type of Injury.—Small jumping beetles, about ⅕ inch long, having a dark metallic greenish-blue color (Fig. 424), feed on the buds of the grape just as they are starting to unfold in the spring. The buds are eaten off, and the newly opening foliage presents a ragged, tattered appearance. Light-b r o w n, black-spotted grubs about ⅓ inch long will be found feeding on the newly opening leaves, along with the beetles.

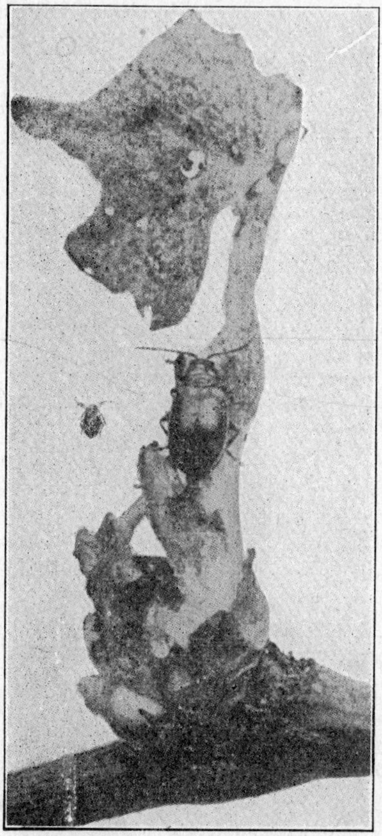

Plants Attacked.—Grape, plum, apple, quince, beech, elm, and Virginia creeper.

Distribution.—Eastern two-thirds of the United States.

Life History, Appearance, and Habits.—The adult grape flea beetles pass the winter in hibernation in or near the grape vineyards under any shelter which they can find. They become active fairly early in the spring and at once start feeding on the opening grape leaves. The females soon start laying their eggs, which are deposited in masses underneath the loose bark of canes, or occasionally on the upper surface of the leaves. The eggs are light yellow in appearance and are fairly conspicuous. Small dark-brown to blackish, spotted grubs hatch from

Fig. 424.—Grape flea beetle, *Altica chalybea* Illiger, natural size and enlarged. (*From N.Y. (Cornell) Agr. Exp. Sta. Bul.* 157.)

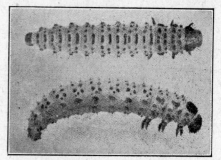

Fig. 425.—Larvae of the grape flea beetle, dorsal and lateral views. About 4 times natural size. (*From N.Y. (Cornell) Agr. Exp. Sta. Bul.* 157.)

these eggs and feed on the surface of the leaves and on the clusters of the blossom buds. On becoming full-grown, the larvae, which are then lighter brown, with regular rows of blackish spots (Fig. 425), drop or crawl to the ground, enter the soil, and there change to the beetle stage. The adult beetles of the summer generation make their appearance during July and August, and feed on the grape leaves until the approach of cold weather. They rarely cause much injury during this part of the season.

Control Measures.—A strong lead arsenate spray, applied as soon as the beetles make their appearance on the vines in the spring, is somewhat effective as a control measure, but, because of the limited amount of foliage which can be covered with poison at this time, it is often hard to protect the vines from injury by this insect. When the grubs are present on the leaves, using lead arsenate at the rate of 4 pounds

---

[1] *Altica chalybea* Illiger, Order Coleoptera, Family Chrysomelidae.

to 100 gallons of water, or bordeaux mixture, is effective in killing the grubs.   On a few backyard vines, the insect may be controlled by spreading a piece of cloth, dipped in oil, under the vines and jarring the beetles upon it.   Fall clean-up of the vineyards, as recommended for the control of the grape berry moth, will also help in keeping down the numbers of this insect.

References.—*U.S.D.A., Dept. Bul.* 901, 1920; *N.Y. (Geneva) Agr. Exp. Sta. Bul.* 331, 1910; *Mich. Agr. Exp. Sta. Spcl. Bul.* 139, 1933.

198. GRAPE BERRY MOTH[1]

Importance and Type of Injury.—This insect is almost universally present wherever grapes are grown, either as a few vines in a back yard or in extensive vineyards.   The grape berries are webbed together, turning dark purple in color, and drop from the stems when the grapes are about the size of garden peas.   Small holes are eaten in the nearly ripened grapes, the sides of which are attached by a light web to a bit of leaf, or to adjoining berries.   Small silken cocoons lie in small, semicircular flaps

Fig. 426.—Grape berry moth, *Polychrosis viteana* (Clemens).   Adult, natural size and enlarged.   (*From N.Y. (Cornell) Agr. Exp. Sta. Bul.* 223.)

cut in the grape leaves, folded over, and held together by a light web. When this insect is abundant, it will often destroy as much as 60 to 90 per cent of the fruit in unsprayed vineyards.

Plants Attacked.—Cultivated and wild grapes.

Distribution.—Troublesome chiefly in the northeastern fourth of the United States and southeastern Canada, but extending westward to Wisconsin and Nebraska, and southward to Louisiana and Alabama.

Life History, Appearance, and Habits.—This insect passes the winter in grayish silken cocoons, nearly always folded in fallen grape leaves. Occasionally the cocoons may be attached to the loose scales of bark or in rubbish on the ground.   In the late spring, shortly before or just after the grape has bloomed, a grayish or grayish-purple moth, about ½ inch across the wings (Fig. 426), emerges from the cocoon, and the female lays her flattened, circular, cream-colored eggs at dusk, upon the fruit, stems, flower clusters, or newly forming grape berries.   The little worms hatching from these eggs spin a silken web wherever they go and thus web together the fruit clusters, or parts of clusters (Fig. 428).   They feed on the grape berries and each worm of this generation will usually destroy a number of grapes.   On becoming full-grown, the worms are ⅓ to ½ inch

[1] *Polychrosis viteana* (Clemens), Order Lepidoptera, Family Olethreutidae.

long, greenish gray with brown heads (Fig. 427). Each cuts out and folds over a little flap of leaf and within this spins a cocoon. The moths

of the second generation emerge during July and deposit their eggs on the grape berries as did those of the first generation. While a single worm may find sufficient food in one berry to complete its development, it often goes from one grape to another, especially where the berries are touching, and thus may destroy three or four grapes before it completes its growth. Development, from the deposition of the eggs to the emergence of the adults, commonly averages about 5 weeks. When full-grown, they again go to the leaves where they spin their cocoons, pupate, and remain in this stage during the winter. There are two generations normally in the northern states and three farther south.

Control Measures.—Spraying is the most effective control for this insect. Spraying with lead or calcium arsenate, 4 pounds, fish-oil soap or other spreader, 1 pint, in 100 gallons of

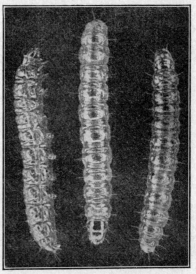

Fig. 427.—Grape berry moth. Larvae or caterpillars, about 4 times natural size. (*From N.Y.* (*Cornell*) *Agr. Exp. Sta. Bul.* 223.)

8-8-100 bordeaux mixture will give satisfactory control. When either of these arsenicals is used the residue is likely to be above the legal tolerance.

Fig. 428.—Work of spring generation of grape berry moth larvae among blossoms and young fruits in June. Natural size. (*From N.Y.* (*Cornell*) *Agr. Exp. Sta. Bul.* 223.)

In the Middle West three sprays should be applied: one shortly after the fruit has set, the second about 10 days later, and a third when the grapes are approximately half-grown. In some sections a spray just before blossoming is recommended. On account of objectionable arsenical residue on the marketed fruits the third spray should contain only 2 pounds lead arsenate with 2 pounds rosin fish-oil soap, or 3 quarts of summer-oil emulsion, to each 100 gallons of spray. To cover the berries of the grape thoroughly, it is necessary to use a high-pressure sprayer that will break the spray into very fine particles. In addition to the spraying, a partial control of the insect may be accomplished by thoroughly cleaning up around the grape vineyards and raking up and burning the fallen leaves during the fall or winter, or by plowing the vineyards and adjacent land as soon as frost is out of the ground in spring.

References.—U.S.D.A., Dept. Bul. 550, 1917; *Del. Agr. Exp. Sta. Buls.* 176, 1932, and 198, 1936; *Mich. Agr. Exp. Sta. Spcl. Bul.* 239, 1933.

199. Grape Leafhopper[1]

Importance and Type of Injury.—There is no insect which attacks the grape that is so universally present, year in and year out, as the grape

Fig. 429.—The grape leafhopper, *Erythroneura comes* (Say), about 7 times natural size.
(*From U.S.D.A., Dept. Bul.* 19.)

leafhopper. The quantity of fruit harvested may be reduced as much as 30 per cent and the quality greatly lowered. Very small whitish spots appear over the grape leaves (Fig. 2,*B*), and they later become brown and shriveled. The entire leaf becomes pale greenish yellow, the vines show very little vigor, and the foliage presents a general sickly appearance. Numerous, small, pale, red-flecked, very active insects are found sucking sap from the undersides of the leaves.

Plants Attacked.—Grape, Virginia creeper, apple, and many other plants.

Distribution.—Throughout the United States.

Life History, Appearance, and Habits.—The adult grape leafhoppers (Fig. 429, *right*) are of a pale yellowish color, with red markings on the wings. They pass the winter among the fallen grape leaves, grasses, or other shelters in the vicinity of the vineyard. The hoppers are about ⅛

[1] *Erythroneura comes* (Say), Order Homoptera, Family Cicadellidae. In northwestern Ohio and southwestern Michigan, at least, the three-banded grape leafhopper, *Erythroneura tricincta* var. *cymbium* McAtee, is more injurious than *E. comes*.

inch or less in length by one-fourth as wide. They become active about the time the grape leaves are half-grown, and fly to the leaves, on which they feed by sucking the sap from the underside. After feeding 2 or 3 weeks, the female hoppers begin to lay their eggs. These eggs, which are very small, are pushed into the tissue of the leaf. They hatch into pale-green or greenish-white nymphs (Fig. 429, *left*), which are wingless, but extremely active, and which feed by sucking the sap, remaining almost entirely on the undersides of the leaves. They become full-grown in from 18 days to 5 weeks, depending largely on the temperature of the season. There are probably two generations of the insect in most of its range, with a third generation in the South. The nymphs of the last generation become full-grown during September, and the adults seek the shelters above described about the time of the first frost.

Control Measures.—Because of the fact that these insects are sap feeders and because of the activity of the adults and nymphs, they are very difficult pests to control. A spray combination that has given entirely satisfactory results, under midwestern conditions, consists of 40 per cent nicotine sulphate, 1 pint, and soybean flour, 4 ounces, in 8-8-100 bordeaux, 100 gallons, applied while the nymphs are small. Derris and pyrethrum sprays and dusts have also been found effective. Great care should be used in making the application to be sure that the undersides of the leaves are thoroughly covered. All grape leafhoppers, whether adults or nymphs, which are hit with the above mixtures, will be killed; but it is frequently the case that poor control of the insect is obtained because the spray is not thoroughly applied. To prevent the leafhoppers from escaping the spray, a boom should be used that will apply the spray to both sides of the leaves at once. Calcium cyanide dusts, applied beneath tents, and nicotine dusts, such as the following, have been recommended in California: 40 per cent nicotine sulphate, 10 per cent; sodium carbonate, 10 per cent; and hydrated lime, 80 per cent. As the adult leafhoppers, like many of the other grape insects, pass the winter in shelters afforded by trash, weeds, or grasses, thoroughly cleaning up around the vineyards is of value in controlling these insects and should be done as a general practice in the vicinity of grape vineyards.

References.—*Calif. Agr. Ext. Serv. Cir.* 72, 1937; *Ariz. Agr. Exp. Sta. Cir.* 146, 1924; *Rept. Ento. Soc. Ont. for* 1922, p. 48, 1922; *N.Y. (Geneva) Agr. Exp. Sta. Bul.* 344, 1922; *U.S.D.A., Dept. Bul.* 19, 1914; *Del. Agr. Exp. Sta. Bul.* 198, 1936.

200. Grape Phylloxera[1]

Importance and Type of Injury.—This aphid is the most destructive grape pest known in the western United States and Europe. Fortunately for the grape growers of the eastern United States, it practically never causes serious damage in this section of the country. Within 25 years after this insect was introduced into France from America, about 1860, it had destroyed nearly one-third of the vineyards in that country— more than 2,500,000 acres. Small galls about the size of half a pea form on the leaf surface (Fig. 430), sometimes so numerous as practically to cover the entire leaf. The galls are open on the underside of the leaf. They contain many small, wingless, yellowish aphids. This form rarely

[1] *Phylloxera vitifoliae* Fitch, Order Homoptera, Family Phylloxeridae.

occurs in the west.   Numerous knots or galls form on the grape roots and
rotting of the roots, yellowing of the grape foliage, and general decrease in
vigor, or the death of the vines, result from injury by this insect.

Plants Attacked.—Grape.

Distribution.—General in North America where grapes are grown.

Life History, Appearance, and Habits.—The life history of this insect
is extremely complicated, as there are four distinct forms of adults, besides
the immature stages.   Only a brief outline of its life history can be given.
The winter is passed both as eggs attached to the canes of the grape plants
and in the form of yellowish aphids on the nodules or galls on the grape
roots (Fig. 431).   The root-infesting forms become active, feeding on the
roots as soon as growth starts in the spring.   The eggs on the canes hatch
in the spring after the foliage of the grape has come out, and the yellow

Fig. 430.—Phylloxera galls on wild grape leaf.   About natural size.   (*From Slingerland
and Crosby, "Manual of Fruit Insects," copyright, 1915, by Macmillan.   Reprinted by
permission.*)

aphids developing from these eggs migrate to the leaves where they begin
feeding.   This injury to the leaf causes the formation of galls.   As soon
as the aphids have become full-grown, they give birth to living young
inside the galls, and these young shortly begin forming other galls, several
generations being passed on the leaf.   Some of these leaf-inhabiting
aphids drop to the ground and burrow beneath the soil to the roots, where
they cause the formation of the root galls and where they can live for a
number of generations.   Toward the fall of the year, winged forms are
produced on the grape roots which leave the ground and lay eggs on the
vines.   These eggs hatch into males and true females, mating takes place,
and each fertilized female lays a single egg, which remains on the cane
during the winter.

Control Measures.—In the eastern United States, the form of phylloxera which causes the galls on the leaves is very abundant, especially
on some of the native grapes.   The root-infesting form, while present,

is very rare in the East and does not cause any serious damage but is the only form found in the Pacific Coast states. The insect is a native of eastern United States, and the grapes growing in that section have acquired practical immunity to its attack. The best-known and most effective remedy for combating this insect is the grafting of European grapes on resistant rootstocks native to eastern United States. This practically does away with any injury by the insect. Nearly all the grapes sold in nurseries are grafted on native rootstocks. In certain parts of California and in Europe, where the European rootstocks are used, the insect is controlled by flooding the vineyards at certain times during the season, also by soil fumigation with carbon bisulphide. In Europe, to

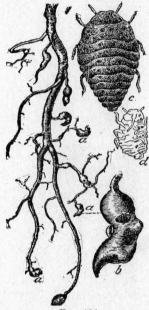

Fig. 431.   Fig. 432.

Fig. 431.—The grape phylloxera, *Phylloxera vitifoliae* Fitch. *a*, galls on grape roots caused by feeding of the insects; *b*, gall, much enlarged, with aphids feeding; *c*, adult aphid, greatly enlarged; *d*, shed skin of the same. (*From Herrick, "Manual of Injurious Insects."*)

Fig. 432.—Adult of the grape rootworm, *Fidia viticida* Walsh, on leaf, about 3 times natural size. (*From N.Y. (Geneva) Agr. Exp. Sta. Bul. 453.*)

kill all of the phylloxera in the soil of an infested vineyard, before setting new vines, chemicals are sometimes injected into the vines, killing vines and insects together.

References.—U.S.D.A., Dept. Bul. 903, 1921; *U.S.D.A., Tech. Bul.* 20, 1928; Essig, "Insects of Western North America," pp. 225–227, Macmillan, 1926.

201. Grape Rootworm[1]

Importance and Type of Injury.—Vines show a lack of vigor, the leaves turn yellow, and little new growth is made. An examination of the roots will sometimes show small whitish grubs eating off the small feeding roots and gouging out numerous channels in the bark of the larger roots. Grayish-tan beetles feed upon the leaves, throughout the summer. Not the

---

[1] *Fidia viticida* Walsh, Order Coleoptera, Family Chrysomelidae. A very similar species, the western grape rootworm, *Adoxus obscurus* (Linné), occurs in California.

entire leaf is eaten, but series of small holes are made through the leaf in chain-like rows (Fig. 433).

Plants Attacked.—Grape and related wild plants.

Distribution.—Eastern United States, except extreme north and south.

Life History, Appearance, and Habits.—The grape rootworm passes the winter as small curved-bodied brown-headed grubs of various sizes, up to ½ inch long. During the winter these worms make their way deep into the soil and will be found 2 feet or more below the surface of the

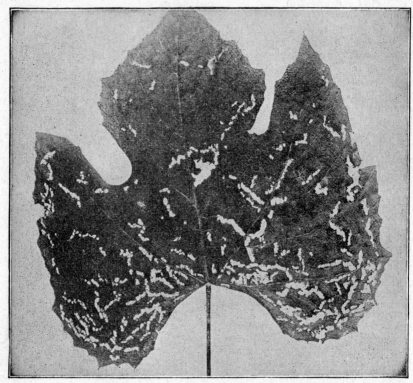

Fig. 433.—Typical injury of adult grape rootworms on leaves. About natural size. (*From N.Y. (Geneva) Agr. Exp. Sta. Bul.* 453.)

ground. As the soil warms in the spring, the grubs migrate back to within 1 or 2 inches of the surface and excavate small cells, in which they change to very soft white pupae in late May and June. About 2 weeks after the grapes bloom, the insects emerge from these cells as small brown or grayish beetles. These beetles are about ¼ inch long and are rather chunky in appearance (Fig. 432). They feed on the upper surface of the grape leaf, producing the effect on the foliage above described. Shortly after commencing to feed, the females deposit their yellowish eggs, in masses averaging about 20 or 30 in a cluster. The eggs are attached to the grape canes, usually under the loose bark scales, although sometimes not so protected. The beetles do not all come out at once, and the period of their feeding and egg-laying may extend over a month or 6 weeks.

The eggs hatch after a short time, and the young grubs drop to the ground, where they burrow into the soil until they encounter grape roots. They at once start feeding on these roots, cutting off the slender feeding roots and channeling or gouging the bark of the larger roots. On the approach of cold weather they work their way deeper into the soil, where they remain during the winter. There is one generation each year.

Control Measures.—As the beetles of the grape rootworm feed extensively on the upper surface of the leaves, they may be poisoned by lead arsenate, or cryolite, applied at the rate of 4 pounds to 100 gallons of water or of bordeaux mixture. The spray for this insect should be given as soon as any of the feeding punctures are noticed on the leaves. Thorough spraying of all grape foliage is essential in the control of this insect. The beetles are somewhat repelled by the lead arsenate, and, if a part of the foliage is not sprayed, they will be nearly sure to feed upon it. Bordeaux mixture alone has no marked repellent effect on this beetle, and vineyards sprayed with bordeaux, without the addition of the lead arsenate, have in a number of cases been very seriously damaged by the grape rootworm. Sprays for the grape berry moth will also destroy these adults. Intensive shallow cultivation of the vineyard soil, up to the time of emergence of adults in late June, will destroy many of the pupae.

References.—*U.S.D.A., Bur. Ento. Bul.* 68, Part VI, 1908; *N.Y. (Geneva) Agr. Exp. Sta. Bul.* 453, 1918; *N.Y. (Geneva) Agr. Exp. Sta. Bul.* 519, 1924; *Penn. Agr. Exp. Sta. Bul.* 433, 1926; *Mich. Agr. Exp. Sta. Spcl. Bul.* 239, 1933.

I. CURRANT INSECTS

FIELD KEY FOR THE IDENTIFICATION OF INSECTS INJURING CURRANTS AND GOOSEBERRIES

1. Greenish many-legged worms with black-spotted bodies, up to 1 inch in length, feed on the edges of the currant leaves. Injury usually starts about the time the plants come into full foliage and is first noticeable in the center of the bushes. Black-bodied, wasp-like sawflies, about ¼ inch in length, deposit pearly-white elongated eggs, in rows, on the vines and midrib on the underside of the currant leaves.................................*Imported currantworm*, page 656.

2. Currant leaves show bright-red, cupped, or wrinkled areas. Numerous, small, greenish-yellow, somewhat flat-bodied aphids on the undersides of the curled and distorted leaves. Where injury is severe the leaves drop from the plants..*Currant aphid*, page 657.

3. Scales up to ⅟₈ inch across, rounded in outline, with a grayish, nipple-shaped projection in the center, are clustered or distributed singly over the bark of the currant. Lemon-yellow, sacklike, sucking insects on the bark under the scale. Infested plants often so heavily covered as to entirely coat the bark and kill the plants...................................*San Jose scale*, page 658.

4. Small whitish areas on the upper sides of the currant leaves. Leaves turn brown and drop. New growth of the currant wilts and dies. During late spring, bright-red wingless bugs, with black dots on the thorax and yellow stripes on the sides, suck the sap from the leaf and new twig growth. Yellowish adult bugs, about ⅓ inch long by one-half as wide, with 2 distinct black stripes on each wing cover, feed on the currant leaves and new growth during late summer..............................*Four-lined plant bug*, page 658.

5. Currant canes put forth undersized foliage in the spring, the tips of the canes often being dead and sometimes broken off. Burrows running the entire length of the cane, mainly in the pith, and containing yellowish grub-like larvae, about ½ inch long. Small round holes bored through the sides of the cane and in the early spring covered with a silken web. Yellow clear-winged moths fly

very actively about the currant bushes in the early summer................
....*Currant borer*, page 659.
6. Fruits of currants and gooseberries turn red and drop from the plants. A small white maggot, up to $\frac{1}{4}$ inch long, feeds inside of the berry.................
Currant fruitfly, Epochra canadensis Loew (see *Wash. Agr. Exp. Sta. Bul.* 155, 1938).

202. IMPORTED CURRANTWORM[1]

Importance and Type of Injury.—This insect is present on practically every currant bush each season, and control measures should be a part of the regular routine of currant growing. Many-legged, smooth, greenish worms, with numerous black spots over their bodies, feed on the edges of the currant leaf (Fig. 434). When disturbed, the worms raise the front and hind part of their bodies from the leaf. When they are abundant, the leaves are stripped from the currant, the injury usually starting in the thick foliage near the center of the plant.

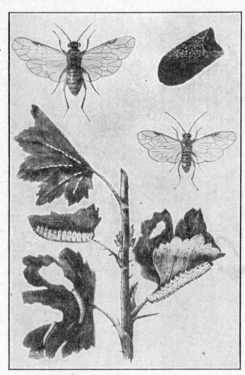

Plants Attacked.—Currant and gooseberry.

Distribution.—General over the United States and southern Canada. Imported from Europe about 1857.

Life History, Appearance, and Habits.—The winter is passed in a small capsule-like cocoon on or near the surface of the ground in the larval or pupal stage. The adult saw-flies (Fig. 434) are black, about $\frac{1}{3}$ inch long, with the abdomen marked with light yellow. They deposit their white, flattened, shining eggs in rows on the veins and midribs of the underside of currant leaves shortly before the plant comes into full

FIG. 434.—Imported currantworm. *Pteronidea ribesii* (Scopoli). Adult female at upper left, empty cocoon at upper right, male below it. Characteristic strings of eggs on veins on underside of upper leaf. Larvae of several instars feeding below. Enlarged about $\frac{1}{2}$. (*From Ill. State Natural History Surv.*)

foliage (Fig. 434, *upper leaf*). The worms hatch just about the time that the currant leaves are full-grown and feed along the margins of the leaves, consuming the entire leaf as they go. They feed for from 2 to 3 weeks, when they go to the ground and transform within their cocoons to the pupal stage. The second generation of the insect is usually not so numerous as the first and appears on the vines during late June or July.

[1] *Pteronidea ribesii* (Scopoli), Order Hymenoptera, Family Tenthredinidae.

A partial third generation occurs in the South. The larvae of the later generations construct cocoons in the soil in which they pass the winter.

Control Measures.—The currantworm is so easily poisoned that there is no necessity of using strong insecticides for its control. Ground derris, timbo, or cubé root, containing 4.5 per cent rotenone, used at 2 pounds to 100 gallons of water, with a sticker such as 4 ounces of soybean flour, should give excellent control. Dusts containing 1 per cent rotenone should also give good control. For a few bushes early in the season, before the currants are ripening, 1 ounce of lead arsenate in 3 gallons of water is satisfactory. Pyrethrum sprays or dusts may also be used successfully.

References.—*S.D. Tenth Rept. Ento.*, p. 26, 1919; *Can. Ento.*, **52**: 106, 1920; *Cornell Univ. Ext. Bul.* 306, 1934.

203. CURRANT APHID[1]

Importance and Type of Injury.—Currant leaves become crinkled or cupped (Fig. 435). The leaf surface becomes bright red in color, just above the points where the cupping occurs. Numerous, small, greenish-

FIG. 435.—Leaves curled and cupped by currant aphids. (*From Slingerland and Crosby,* "*Manual of Fruit Insects,*" *copyright,* 1915, *by Macmillan. Reprinted by permission.*)

yellow, somewhat flat-bodied aphids are found on the undersides of the curled and distorted leaves, most numerous inside the cups.

Plants Attacked.—Currants and sometimes gooseberry.

Distribution.—Throughout the United States and Canada.

Life History, Appearance, and Habits.—The currant aphid goes through the winter in the form of shining black eggs, which are found on the canes of the currant plants, particularly on the new growth. The young aphids hatch from these eggs soon after the leaves appear in the spring. They crawl to the leaves and suck the sap from the underside. Their feeding soon causes the distortion of the leaf described above. If very numerous, the infested leaves will drop from the plants. There are

[1] *Capitophorus* (= *Myzus*) *ribis* (Linné), Order Homoptera, Family Aphididae. *Aphis varians* Patch also attacks currants.

a number of generations each season, all those appearing during the summer consisting of females which give birth to living young as soon as they have become full-grown. Winged females migrate to certain weeds and reproduce during the summer, their offspring returning to the currants in fall to join those wingless strains which remained upon the currant all summer. In the fall of the year these females give birth to males and females, which mate and deposit their eggs on the canes where the insect passes the winter.

Control Measures.—The aphids, being sucking insects, cannot be controlled by any stomach poisons. The best control consists in thoroughly spraying the undersides of the leaves early in the season with a nicotine solution, such as a 40 per cent nicotine sulphate, used 1 part to 800 parts of water, or 1 pint to 100 gallons of water. Where used with water alone, soap should be added at the rate of 6 to 8 pounds of potash fish-oil soap to each 100 gallons of water. Lead arsenate can be combined with this spray, controlling the sawfly and other leaf-feeding insects as well as the aphids. Dusting with a 2 per cent nicotine dust also is very effective for the control of this insect. In spraying or dusting, care must be used to get the material on the underside of the leaves and up into the ridges and cup-shaped depressions caused by the feeding of the aphids.

References.—*N.Y. (Geneva) Agr. Exp. Sta. Bul. 517, 1924; N.Y. (Geneva) Agr. Exp. Sta. Bul. 139, 1897.*

204 (149, 172, 179, 189). SAN JOSE SCALE

The San Jose scale is a serious pest of currants and will soon kill out currant bushes if a plantation becomes infested and no treatment is given. It is easily controlled by the application of lime-sulphur or oil emulsions as recommended for the control of this insect on the apple (see page 577).

FIG. 436.—Four-lined plant bug, *Poecilocapsus lineatus* Fabricius. Adult, natural size and enlarged. (*Reprinted from Sanderson and Peairs, "Insect Pests of Farm, Garden and Orchard," Wiley.*)

205. FOUR-LINED PLANT BUG[1]

Importance and Type of Injury.— Where this insect is feeding on the leaves of currants, distinct white or dark-colored spots, $\frac{1}{16}$ to $\frac{1}{8}$ inch in diameter, looking much like fungus disease spots, appear on the upper sides of the leaves. If the insect is abundant, these areas come together and the leaf turns brown and usually drops. New growth of the currant will sometimes wilt where this bug has fed upon it.

Plants Attacked.—General; gooseberry, rose, and many other shrubs, garden flowers, and herbaceous plants.

Distribution.—General east of the Rocky Mountains.

Life History, Appearance, and Habits.—This insect passes the winter in the form of slender white eggs, about $\frac{1}{6}$ inch long, which are inserted in slits in the canes of currants and other plants. Several eggs are generally laid at one place. They are forced into the cane at right angles to the long axis of growth, the tips of the eggs

[1] *Poecilocapsus lineatus* Fabricius, Order Hemiptera, Family Miridae.

usually protruding from the cane. The eggs hatch from May to the latter part of June. The bright-red to orange nymphs have black dots on the thorax and, in the last stage, a yellow stripe on each side of the wing pads. They feed by inserting their beaks in the leaves and new twig growth and sucking the sap. They become full-grown in about a month to 6 weeks. The adult insects (Fig. 436) are ¼ to ⅓ inch long, of a general greenish-yellow color, with four distinct black stripes down the wing covers on the back. In the fall the female insects lay their eggs in the canes as above described. There is but one generation of the insect each year.

Control Measures.—As these insects are exceedingly active, they are somewhat hard to control. The best method is to spray with a contact poison such as 40 per cent nicotine sulphate, ½ pint, soap, 1 pound, in 25 gallons water, or derris or pyrethrum with a good wetting agent, in May or early June before the adults appear, or to dust with a 4 per cent nicotine.

Reference.—*N.Y.* (*Cornell*) *Agr. Exp. Sta. Bul.* 58, 1893; *Mo. Agr. Exp. Sta. Bul.* 342, pp. 19, 20, 1934.

206. Currant Borer[1]

Importance and Type of Injury.—This insect is of European origin but is now well distributed in America. It is almost sure to cause trouble in both commercial and garden currant plantations. Canes have yellowish under-sized foliage appearing on them in the spring. Such canes usually die within 2 or 3 weeks. Dead canes usually have the tips broken off showing a burrow running nearly the entire length of the cane, partly in the pith and partly in the wood.

Plants Attacked.—Gooseberry, black elder, sumach, and currant. The insect is more destructive to the black than to the red currant.

Distribution.—Throughout North America.

Life History, Appearance, and Habits.—This insect passes the winter as nearly full-grown yellowish borers, or larvae, about ½ inch long, which will be found inside the canes usually a short distance above the ground (Fig. 437). The larvae feed a little in the spring and eat an exit hole through the side of the cane which they cover with a silken web. They transform to the pupal stage inside the burrow a short distance from this exit hole. The adult insects emerge during June or July, in central Illinois. They are black-and-yellow clear-winged moths, about ½ inch long, which one would readily mistake for small wasps. They are extremely active, flying with great rapidity. They deposit their eggs on the bark of the currant canes, and the grubs hatching from these eggs bore into the canes, feeding on the pith and the wood. They are nearly full-grown by midfall and remain in the canes during the winter as above stated.

Fig. 437.— Imported currant borer, *Synanthedon tipuliformis* (Linné), in its burrow ready to pupate. Enlarged ½. (*From Slingerland and Crosby,* "*Manual of Fruit Insects,*" copyright, 1915, *Macmillan. Reprinted by permission.*)

Control Measures.—As practically no external feeding is done by this insect and as it cannot be reached with contact sprays, the only effective method of control is to cut out and burn the infested canes. This can best be done shortly after the currant leaves appear in the spring, as

[1] *Synanthedon tipuliformis* (Linné), Order Lepidoptera, Family Aegeriidae.

it will then be very easy to distinguish the infested canes because of their weak, sickly appearance.   The canes should be cut close to the ground and removed and burned before the last week in May, in the latitude of southern Illinois.

References.—Wash. Agr. Exp. Sta. Bul. 36, 1898; *U.S.D.A., Farmers' Bul.* 1398, 1933.

J. GOOSEBERRY INSECTS

The gooseberry is injured by much the same insects as the currant. San Jose scale is often very destructive on gooseberry but is not nearly so conspicuous on this plant as on the currant, the heaviest infestation often occurring close to the ground.   If the scale is present in the vicinity of gooseberry plantations, they should be thoroughly sprayed, even though a casual examination does not show much scale on the canes. The imported currant sawfly is fully as destructive to the gooseberry as to the currant and can be controlled by the same measures.   The gooseberry fruitworms[1] feed on the pulp of the fruits, causing browning at the blossom end, premature coloring and often the drying up of the berries.

K. RASPBERRY INSECTS

FIELD KEY FOR THE IDENTIFICATION OF INSECTS INJURING THE RASPBERRY AND BLACKBERRY

1. Raspberry leaves with pale-green, spiny, many-legged worms feeding on the edges of the leaf.   Insects more abundant about the time raspberry comes into full foliage, sometimes nearly stripping the plants of leaves.................. *Raspberry sawflies, Monophadnoides rubi* Harris and *Priophorus rubivorus* Rohwer (see *N.Y. (Geneva) Agr. Exp. Sta. Bull.* 150, 1898).
2. Bases of raspberry plants, particularly on the more heavily shaded parts of the stem, covered with white, scurfy-appearing scales.   Reddened areas on the bark around the points where the scale is attached.   These scales are about $\frac{1}{5}$ inch across when full-grown, rounded in outline, and flattened.   In the winter many reddish eggs will be found beneath the protecting scale....................... *Rose scale, Aulacaspis rosae* Bouché (see Essig, "Insects of Western North America," p. 307, 1926).
3. Leaves of raspberry and blackberry become spotted with white or brownish flecks, due to feeding on the undersurface of minute 8-legged reddish or greenish mites, causing the leaves to fall and the fruits to dry up.................... ....*Red spider mites,* page 482.
4. Fleshy enlargements or swellings of the raspberry canes, up to several inches long, and $\frac{1}{2}$ inch or more in diameter.   Slender, white-bodied grubs, up to $\frac{1}{2}$ inch in length, feed on the pith of the canes in, or near, the swollen areas. Dull, bluish-black beetles, $\frac{1}{3}$ inch in length by $\frac{1}{4}$ as wide, with a dull coppery-red area on the back just behind the head, are found on the canes during late spring and summer........................*Red-necked cane borer,* page 661.
5. Buds and blossoms of the raspberry and loganberry with numerous holes eaten in them; leaves partly skeletonized; whitish grubs about $\frac{1}{4}$ inch long, feed inside the ripening fruit.   Small light-brown beetles about $\frac{1}{8}$ inch in length feeding on the tender leaves and newly opening buds........................... ....*Raspberry fruitworm,* page 661.
6. Raspberries and blackberries make very little growth.   Many of the canes wither and die.   Fruit, if any, is small.   Roots and lower parts of the canes

---

[1] *Zophodia grossulariae* Riley and *Zophodia franconiella* Hulst, Order Lepidoptera, Family Pyralididae.

girdled and burrowed by whitish, brown-headed caterpillars, up to ⅔ inch long, which enter the canes near the ground and tunnel into crown of plant, just beneath the bark.....................................

Raspberry root borer, Bembecia marginata (Harris), Order Lepidoptera, Family Aegeriidae (see *Wash. Agr. Exp. Sta. Bul.* 155, 1938).

7. Tips of shoots of young canes wilted, with a purple discoloration at the base of the wilted part, or broken off clean as though cut with a knife; caused by the feeding of a small white maggot which tunnels the pith and girdles the canes from the inside, then tunnels on down below the break................................

Raspberry cane maggot, Pegomyia (= *Hylemyia*) *rubivora* (Coquillet), Order D i p t e r a, Family Anthomyiidae (see *Ill. Agr. Exp. Sta. Cir.* 427, 1935).

8. Irregular rows of minute round punctures, each about the diameter of a pin and extending into the pith of the raspberry. The raspberry cane frequently dies above these punctures, or splits and breaks off during the winter. Cricket-like pale-green insects, about 1 inch in length, with very long feelers and long hind legs crawl about over the plants and insert their eggs in the canes during the late summer.....*Tree crickets*, page 662.

FIG. 438.—Section of blackberry cane showing gall or swelling caused by larva of red-necked cane borer. (*From U.S.D.A., Farmers' Bul.* 1286.)

207. RED-NECKED CANE BORER[1]

Importance and Type of Injury.—The injury caused by this beetle is shown by enlargements or swellings of the raspberry or blackberry canes (Fig. 438). The canes frequently die and sometimes break off at the point where the swelling occurs.

Plants Attacked.—Raspberry, blackberry, and dewberry.

Distribution.—Eastern United States.

Life History, Appearance, and Habits.—The winter is passed in the form of a slender white grub inside the pith of the raspberry cane. These grubs are about ½ inch in length. In the spring they complete their growth, change inside the stem to the pupal stage, and emerge during May and June as dull bluish-black beetles, with a metallic luster, about ⅓ inch in length by one-fourth as wide, with a coppery-red or brassy prothorax. These beetles lay their eggs in the bark of the cane, usually near the base of the leaf. The young larvae burrow upward in the sapwood and also around the cane, sometimes making several girdles, and causing the formation of the galls or swellings. There is but one generation of the insect each year.

Control Measures.—The only method of control is to cut out the infested canes during the winter, burning the galls, thus destroying the young larvae within them.

References.—*U.S.D.A., Farmers' Bul.* 1286, 1922; *Mich. Agr. Exp. Sta. Quar. Bul.* 14, pp. 267–269, 1932.

208. RASPBERRY FRUITWORM[2]

Importance, Type of Injury, and Life History.—This is one of the most destructive pests of raspberry and loganberry in the northern part of the

[1] *Agrilus ruficollis* (Fabricius), Order Coleoptera, Family Buprestidae.
[2] *Byturus unicolor* Say, Order Coleoptera, Family Dermestidae.

United States. Light-brown beetles, ⅛ to ⅙ inch long, feed on buds, blossoms, and tender leaves and lay eggs on blossoms and young fruits from which slender white grubs, up to ⅓ inch long, with a brown area on the upper side of each segment and prominent prolegs at the tip of the abdomen, bore into the fruit, making it unfit for food; the grubs feed especially in the receptacle. The grubs become full-grown as the fruit ripens, drop to the ground, and pupate in the soil. They winter in the adult stage in the soil, emerge about mid-April, and begin mating and egg-laying as the berries blossom.

Control.—Derris or cubé sprays, containing at least 0.01 per cent rotenone, (2 pints of 4.5 per cent rotenone in 100 gallons of water); or dusts containing ½ per cent rotenone in talc, three applications at weekly intervals, beginning 10 days after the first blossoms appear, are recommended for control. Thorough cultivation during late summer will destroy many of the insects in the soil.

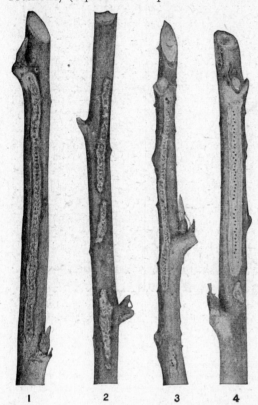

1   2   3   4

Fig. 439.—Egg punctures of a tree cricket in raspberry canes. (*From N.Y. (Geneva) Agr. Exp. Sta. Bul.* 388.)

References.—Conn. Agr. Exp. Sta. Bul. 251, 1923; *U.S.D.A., Bur. Ento. Cir.* E-433, 1938; *Wash. Agr. Exp. Sta. Bul.* 155, 1938.

209. Tree Crickets[1]

Importance and Type of Injury.—The insects known as tree crickets (Fig. 440) sometimes cause injury to tree and bush fruits. Small round holes about the diameter of a pin (Fig. 439) are drilled singly in the twigs or brambles. In each hole, which extends into the cambium or sapwood, the insect deposits a single pale-yellow egg, about ⅛ inch long. The punctures are usually made in a single row along one side of the cane or stem, sometimes as many as 50 to 75 in a row, and about 25 to the inch. These egg punctures often serve as the entrance points for tree and bramble diseases. Canes frequently die above these punctures, and, where they are numerous, the canes split and break off. The adult insects are sometimes very injurious by eating holes in ripe fruits.

[1] *Oecanthus niveus* (DeGeer), *O. nigricornis* Walker, and others, Order Orthoptera, Family Gryllidae.

Plants Attacked.—Apple, prune, plum, peach, cherry, raspberry, loganberry, blackberry, and others.

Life History, Appearance, and Habits.—The insects winter in the egg stage in twigs or brambles, the young hatching in the spring. The pale-green slender nymphs (Fig. 85) feed on the foliage of various plants or on small, sluggish insects, fungi, pollen, or ripe fruits. They grow rather slowly, reaching the adult stage in late summer. The adult has somewhat the appearance of a cricket but is pale green in color and has a longer, more slender body and smaller head. The antennae or

Fig. 440.—The snowy tree cricket, *Oecanthus niveus* (DeGeer); characteristic posture of female in act of laying eggs. (*From N.Y. (Geneva) Agr. Exp. Sta. Bul.* 388.)

feelers are much longer than the body. The males have stiff veins in the flat wings, which are adapted to making sounds. The songs of the males in late summer are described as a series of short, clear, musical whistling notes, indefinitely repeated, often synchronized, and varying in frequency with the temperature. The females deposit their eggs in the fall, and there is only one generation each year.

Control Measures.—Spraying or dusting with arsenicals in early summer, while the crickets are young, has given complete control. Pruning out and burning the parts of canes containing the eggs, after the last crop has been removed and before spring, aids in control.

References.—*N.Y.* (*Geneva*) *Agr. Exp. Sta. Bul.* 388, 1914; *Ore. Agr. Exp. Sta. Bul.* 223, 1926; *U.S.D.A., Cir.* 270, pp. 56–58, 1933.

L. BLACKBERRY INSECTS

The blackberry is comparatively free from insect injuries. It is infested to some extent by the rose scale, reference to which has been made under Raspberry Insects, page 660. The canes are occasionally punctured by tree crickets (page 662). The raspberry sawfly also attacks the blackberry and can be controlled by the same methods as recommended for other sawflies on page 657. The canes of the blackberry are occasionally injured by the larvae of the red-necked cane borer (page 661), which causes irregular swellings or galls along the canes. The insect is rarely numerous enough to be considered of any economic importance. Leafhoppers (page 650) and mites (page 482) sometimes become thick enough on the foliage to cause a whitening of the leaves. The blackberry psyllid[1] curls and stunts the new foliage in some sections. Certain leafhoppers suck the sap of the leaves, causing white mottling. Whitish, sawfly larvae,[2] up to $\frac{1}{3}$ or $\frac{1}{2}$ inch long with brown heads and plates on the thorax, mine between upper and lower epidermis of the leaves, causing large blotch mines so that the plants appear as though singed by fire. On the Pacific Coast the blackberry mite[3] feeds near the base of the drupelets around the core preventing the fruits from ripening in part or in whole— the so-called "redberry disease." The affected fruit becomes very bright colored, in parts varying from a single drupelet to an entire berry, hard, and clings to the canes until winter. Control is effected by a spray of lime-sulphur, 5.5° Bé. in March or April, followed by a spray of 5 pounds wettable sulphur in 100 gallons of water, in early May.

M. STRAWBERRY INSECTS

FIELD KEY FOR THE DETERMINATION OF INSECTS INJURING THE STRAWBERRY

A. *External chewing insects that eat holes in the leaves or chew off the lower surface of the leaves:*
   1. Very small, green or metallic-blue, jumping beetles, less than $\frac{1}{6}$ inch across, feeding on the strawberry plants early in the spring. Leaves riddled with small round holes, often drying up and turning brown.....................
      ....*Strawberry flea beetle*, page 665.
   2. Grayish plump-bodied snout beetles, about $\frac{1}{2}$ inch long, the wing covers crossed by 2 light bands, bent down at the end and terminating in an acute angle, and most of the body covered with small flattened scales, sometimes defoliate the plants.....................*Imbricated snout beetle*, page 574.
   3. Small greenish or bronze caterpillars folding or rolling together the strawberry leaves and feeding within the rolled portion. Heavily infested plants have a whitish appearing foliage. Injury occurs from early spring until early fall. Small grayish-brown moths, about $\frac{1}{4}$ inch long, the wings marked with wavy bands of light brown, fly about strawberry beds from April to September.....
      ....*Strawberry leaf rollers*, page 666.

[1] *Trioza tripunctata* Fitch, Order Homoptera, Family Chermidae (*N.J. Agr. Exp. Sta. Bul.* 378, 1923).

[2] *Metallus rubi* Forbes, Order Hymenoptera, Family Tenthredinidae (see *N.Y. (Geneva) Agr. Exp. Sta. Tech. Bul.* 133, 1928).

[3] *Eriophyes essigi* Hassan, Order Acarina, Family Eriophyidae (see *Wash. Agr. Exp. Sta. Bul.* 279, 1933; *Calif. Agr. Exp. Sta. Bul.* 399, 1925).

B. *Insects that injure the leaves by sucking the sap:*

1. Almost microscopic, white to light-brown mites, feed on young unfolded leaves, in crown of plant, stunting and dwarfing them and causing a rosette of leaves due to failure of stems to elongate... *Strawberry crown mite*, or *cyclamen mite*, page 750 (see also *Wash. Agr. Exp. Sta. Bul.* 155, 1938).

2. Leaves curled, spotted, and dying, with minute white and dark specks and some silk on underside, among which almost invisible, 6- and 8-legged greenish or reddish mites crawl and feed.................... *Red spider mites*, page 482.

C. *Insects injuring the crown of the plant:*

1. White, thick-bodied grubs, about ⅕ inch in length, boring in the crowns of the strawberry plant. Crown of the plant eaten out so that it dies or is so weakened that very few runners or new growth is produced. Reddish-brown snout beetles, about ⅙ inch in length, shelter under litter or about the plants in strawberry fields during the winter......... *Strawberry crown borer*, page 669.

2. Strawberry plants eaten off close to the ground by curved-bodied, light brown-headed grubs, about ⅕ inch long. Grubs most abundant during late spring and again during the late summer. Black beetles, ⅙ inch long, with a short blunt snout protruding from the front of the head, clustered about the bases of strawberry plants and sometimes feeding on the leaves...................
.... *Strawberry root weevil*, page 669.

D. *Insects injuring the roots:*

1. Strawberry plants lacking in vigor, foliage of a pale color, the fruit drying up or failing to mature properly. Roots of the injured plants covered with dark bluish-green aphids, their slender beaks inserted in the roots, from which they suck the sap. Usually attended by small brown ants. Aphids of the same description feeding on the leaves during early spring.........................
.... *Strawberry root aphid*, page 667.

2. Large white grubs, up to 1 inch in length, with brown heads, usually holding the body in a curved position, and with distinct, well-developed, slender legs, feed on the roots of strawberries from early spring to early fall.............
.... *White grubs*, page 670.

3. Small, white, brown-spotted grubs, about ⅛ inch in length, feeding on the roots of strawberries during May and June. Infested plants are weakened and have poorly colored foliage. Brownish or coppery-colored beetles, about ⅛ inch long, feeding on the foliage of strawberries during the early fall......
.... *Strawberry rootworms*, page 668.

E. *Insects injuring the buds and fruits:*

1. Buds and newly formed fruit of the strawberry dried up on the partly severed stems, or entirely eaten off. This injury is caused by dark reddish-brown snout beetles, from 1/12 to ⅛ inch long, which make punctures in the buds, in which they insert eggs. Very small, legless, white soft-bodied grubs, feed within the strawberry buds............................... *Strawberry weevil*, page 670.

210. STRAWBERRY FLEA BEETLE[1]

Importance and Type of Injury.—Very small metallic-blue beetles feed on the foliage of the strawberry plants. The leaves are riddled with large numbers of small round holes, often drying up and browning around these holes.

Plants Attacked.—The beetles lay their eggs on strawberry, evening primrose, and other plants of the same family. They also feed on some greenhouse plants.

Distribution.—Throughout the United States and Canada.

Life History, Appearance, and Habits.—These beetles, in common with many other kinds of flea beetles, hibernate as adults. They emerge early in the spring, and their principal damage is done before the strawberry plants bloom. There are from one to two generations a year, the larger number occurring in the South.

[1] *Altica ignita* Illiger, Order Coleoptera, Family Chrysomelidae.

Control Measures.—Thorough spraying of the strawberry plants with 8-8-100 bordeaux mixture, applied whenever the insects are abundant, is very effective. Usually this application should be made about a week before the strawberries bloom. Spraying or dusting with cryolite will also give good control.

Reference.—*Conn. Agr. Exp. Sta. Bul.* 344, p. 165, 1933.

211. Strawberry Leaf Rollers[1]

Importance and Type of Injury.—Small greenish or bronze caterpillars, up to ½ inch long, fold or roll the strawberry leaves (Fig. 442), fastening them together and feeding within. The leaves have a brown appearance and much of the foliage is killed. Heavily infested beds have a whitened

Fig. 441.   Fig. 442.

Fig. 441.—Larva of the obsolete-banded strawberry leaf roller, *Archips obsoletana* Walker, on leaf. Three times natural size. (*From Slingerland and Crosby, "Manual of Fruit Insects," copyright, 1915, Macmillan. Reprinted by permission.*)

Fig. 442.—Strawberry leaf roller, *Ancylis comptana* Frölich. Leaves folded by larvae, slightly reduced. (*From Iowa Agr. Exp. Sta. Bul.* 179.)

or grayish appearance instead of the usual green of healthy strawberry plants, and the fruits are withered and deformed.

Plants Attacked.—Strawberry, blackberry, dewberry, and raspberry.

Distribution.—Northern United States, Louisiana, Arkansas, and southern Canada.

Life History, Appearance, and Habits.—The winter is passed in the larval and pupal stages, the pupae inside of silken cocoons in the folded leaves and the larvae in silken shelters under trash on the surface of the ground. In the spring of the year the larvae transform to pupae and a little later emerge as small grayish or brownish moths, with a wing expanse of about ⅖ inch. The wings are marked with wavy bands of light and dark color. In southern Illinois the moths are abroad in April, and in the northern part of the state by the middle of May or a little earlier. They lay their eggs singly, 20 to 120, on the undersides of the strawberry leaves. These eggs hatch in about 1 week and the caterpillars coming

[1] *Ancylis comptana* Frölich and *Archips obsoletana* Walker, Order Lepidoptera, Family Tortricidae.

from them (Fig. 441) feed at first on the undersurface of the leaves, under a silken cover, migrating when about half-grown to the upper side of the leaves, where they fold a leaf about themselves in the characteristic manner, holding it with fine silken threads. They feed for 35 to 50 days and then transform into brown pupae inside the folded leaves. There are two, and a partial third, generations throughout most of the range of the insect. The hatching of the second generation begins about the middle of July.

Control Measures.—Dusting or spraying the plants with cryolite or fluosilicates, in water or bordeaux mixture, as the first blossoms appear, and twice more at weekly intervals, gives effective control. The treatment may need to be repeated for the second generation. Because of the poisonous residue, resulting from these sprays, derris or timbo root, $2\frac{1}{2}$ pounds, and soybean flour, 4 ounces, in water, 100 gallons, should be substituted in any sprays applied after the fruits are formed and may be used at any stage in the development of the plants and fruits. Any of the sprays should be applied, however, before the leaves have become folded to any extent, as the poison will not be effective against the leaf rollers after they have constructed their silken cases. The numbers of the insect may be kept down if the strawberry beds are mowed close to the ground and burned over shortly after the fruit is picked.

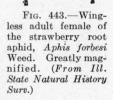

References.—*Neb. State Ento. Cir.* 7, 1908; *Iowa Agr. Exp. Sta. Bul.* 179, 1918; *Iowa Agr. Exp. Sta. Cir.* 110, 1928; *Jour. Agr. Res.*, **44**: 541–558, 1932; *Jour. Econ. Ento.*, **28**: 388–390, 1935.

212. STRAWBERRY ROOT APHID[1]

Importance and Type of Injury.—Strawberry plants are lacking in vigor, the foliage of a pale color. The fruit dries up or fails to mature properly.

Plants Attacked.—Strawberry.

Distribution.—United States east of the Rocky Mountains.

FIG. 443.—Wingless adult female of the strawberry root aphid, *Aphis forbesi* Weed. Greatly magnified. (*From Ill. State Natural History Surv.*)

Life History, Appearance, and Habits.—These insects pass the winter in the form of black shining eggs which are attached to the leaves and stems of the strawberry plant. The eggs hatch early in the spring into dark bluish-green aphids (Fig. 443), which feed on the new leaves of the strawberry. Wherever these aphids become abundant, they are soon found by colonies of the brown cornfield ant and are carried by these ants to the strawberry roots, where they feed by sucking the sap from the roots. There are a number of generations of the insect during the year. All those occurring in the field during the summer are females and reproduce by giving birth to living young. On the approach of cold weather in the fall winged females are produced. These make their way to the leaves of the plant and there give birth to the sexual females and males. After mating, these true females lay the winter eggs above described.

[1] *Aphis forbesi* Weed, Order Homoptera, Family Aphididae.

Control Measures.—No very effective control measure is known.   In setting new beds, uninfested plants should be selected, and the ground, wherever beds are set, should be given a thorough and deep cultivation in the early spring, to break up and drive out the ants which may be in the soil.

References.—*N.J. Agr. Exp. Sta. Bul.* 225, 1909; *Trans. Ky. State Hort. Soc.,* **77**: 120, 121, 1932.

213 (255). STRAWBERRY ROOTWORMS[1]

Importance and Type of Injury.—There are several species of small beetles, the young of which are injurious to strawberries. Small, white, brown-spotted grubs feed on the roots of the strawberries during May, June, and July. These grubs (Fig. 444) are about ⅛ inch in length and are very much smaller than the common white grub. The foliage of strawberries is destroyed by small bronze-brown or copper-colored adult beetles, about ⅛ inch long (see also pages 383 and 732).

Plants Attacked.—Strawberry, raspberry, grape, rose, and many other plants.

Distribution.—Over most of the United States.

Life History, Appearance, and Habits.— These insects pass the winter as full-grown

| FIG. 444.—Strawberry rootworm, *Paria canella* (Fabricius). Larva or grub, about 15 times natural size. (*From U.S.D.A., Farmers' Bul.* 1344.) | FIG. 445.—Adult of the strawberry rootworm *Paria canella* (Fabricius). About 10 times natural size. (*From U.S.D.A., Farmers' Bul.* 1344.) |

beetles (Fig. 445). They come out of their winter shelters during April or May. The females deposit their eggs in the ground near the plants. The young grubs hatching from them feed on the roots for about 3 months, becoming full-grown during August and transforming to a pupal stage and later to the adult beetle. There are two generations annually in California. The beetles do considerable feeding on the foliage of the strawberry during the early fall.

Control Measures.—Dusting with a mixture of talc or sulphur and ground derris or timbo root, containing 1 per cent rotenone as applied, may be used in the spring or fall when the beetles are feeding on the leaves. Lead arsenate, 4 pounds, soybean flour, 4 ounces, and water, 100 gallons, gives good control. This spray should not be used close to picking time.

References.—*U.S.D.A., Farmers' Buls.* 1344, 1923, and 1362, p. 59, 1922; *Thirteenth Rept. Ill. State Ento.,* p. 159, 1883. *U.S.D.A., Dept. Bul.* 1357, 1926; *Mich. Agr. Exp. Sta. Spcl. Bul.* 214, pp. 94, 95, 1931.

[1] *Paria* (= *Typophorus*) *canella* (Fabricius), *Colaspis brunnea* Fabricius, and *Graphops pubescens* Melsheimer, Order Coleoptera, Family Chrysomelidae.

214. Strawberry Root Weevil[1] or Crown Girdler

Importance and Type of Injury.—Strawberry plants stunted, the leaves closely bunched together and very deep colored, or dying, the fine roots and the crown having been eaten off close to the ground, by small, fat, curved grubs with light-brown heads.

Plants Attacked.—Strawberry, the related small fruits, and many other plants. Nursery evergreens have been seriously attacked by the larvae.

Distribution.—Northern United States and Canada.

Life History, Appearance, and Habits.—The insect passes the winter in both the larval and adult stages, under surface trash and in old strawberry crowns. The adults are nearly black beetles, about ¼ inch long or a little less, having short blunt snouts protruding from the front of their heads. They feed upon the foliage and berries. No males have ever been found and the females reproduce partheno-genetically. The grubs, or larvae, are white, legless, up to ⅜ inch long, with light-brown heads, and hold their bodies in more or less of a curved position. The grubs begin feeding as soon as the weather becomes warm in the spring and about the same time the beetles leave the shelters in which they have passed the winter and gather in the strawberry beds. The wing covers of the beetles are tightly grown together and the insect is unable to fly. In southern Illinois the insect is most abundant in the adult stage during May and June, and again in late July and August. The female beetles lay small, white, spherical eggs among the roots or in the crown of the strawberry plants, these eggs hatching into the grubs above described. There are probably two generations a year in the latitude of Illinois.

Control Measures.—Spraying with an arsenical very early in spring is of some value in control. Destroying old beds promptly after the last picking and rotating new beds to fields that have been in cultivation, but had not had strawberries on or near them for at least 1 year, is of value. This insect may be kept out of the strawberry fields by constructing a barrier of tarred boards around the new fields. Various poison baits of bran and molasses—or chopped raisins, dried apples, or other fruits and shorts poisoned with sodium fluosilicate—have given good control. About a tablespoonful is dropped on the ground about each plant, just before the overwintering weevils attack them and again just before the new adults appear in midsummer. A formula that has given good control is:

Ingredients	For 1½ Acres	For 30 Acres
a. Bran	5 lb.	100 lb.
b. Molasses	1 pt.	2½ gal.
or brown sugar	1 lb.	20 lb.
c. Calcium arsenate or sodium fluosilicate	¼ to ½ lb.	5 to 10 lb.
d. Water	2 qt.	10 gal.

References.—*Maine Agr. Exp. Sta. Bul.* 123, 1905; *Can. Dept. Agr., Pamphlet* 5 (n.s.), 1931; *Wash. Agr. Exp. Sta. Bul.* 199, 1926; *Ore. Agr. Exp. Sta. Bul.* 330, 1934.

215. The Strawberry Crown Borer[2]

This pest damages strawberries in much the same manner as the strawberry root weevil. It lives upon cultivated and wild strawberries and cinquefoil. It has been especially destructive in the bluegrass region of Kentucky and Tennessee but ranges also to Illinois, Missouri, and Arkansas. Most of the insects winter as reddish-brown snout beetles, about ⅙ inch long, under trash in strawberry patches. Most of the

---

[1] *Brachyrhinus ovatus* (Linné), Order Coleoptera, Family Curculionidae.

The black vine weevil, *B. sulcatus* (Fabricius), has very similar habits. The adults are ⅓ to ⅖ inch long and have patches of golden hairs upon the wings when freshly emerged (see *Wash. Agr. Exp. Sta. Bul.* 199, 1926; *Ore. Agr. Exp. Sta. Cir.* 79, 1923; *U.S.D.A., Tech. Bul.* 325, 1932). The rough strawberry weevil, *B. rugostriatus* Goeze, is a very similar species, destructive in the West.

[2] *Tyloderma fragariae* (Riley).

eggs of the first generation are laid after mid-April in small holes eaten in the leaf bases of the plants.   The white thick-bodied grubs, up to ⅕ inch long, tunnel through the strawberry crowns, killing or stunting the plants and greatly reducing the yield. The control measures are much the same as for the root weevil and consist (a) of setting only plants or crowns dug before the first beetles are active in the spring (Mar. 1, in Kentucky) and certified as free from infestation; (b) the destruction of old beds promptly after the last picking; (c) setting new beds on soil in cultivation for at least 1 year, and at least 1,000 feet from a source of infestation; and (d) the use of barriers of boards set on end with an L-shaped iron turned outward at the top.

References.—Tenn. Agr. Exp. Sta. Bul. 128, 1923; Jour. Econ. Ento. **31**: 385–387, 1938.

216 (14, 39, 87, 273). White Grubs[1]

White grubs or grubworms are among the insects most commonly injurious to strawberries.   Large white grubs with brown heads feed on the roots of the strawberry plants, causing the plants to die over areas of varying size.   It is practically impossible to clean a strawberry bed, once infested with white grubs, without plowing up the bed.   In setting new patches, ground should be selected that has been in some clean-cultivated crop for 1 or 2 years previously; as the June beetles prefer to lay their eggs in grassland or land with a heavy growth of grassy weeds.   If possible, strawberry beds should be located at some distance from the trees on which the beetles feed and near which they usually lay their eggs.   Thorough cultivation of the soil in the spring before the plants are set will be of some benefit but will not clean all of the grubs out of the soil (see further discussion of white grubs on page 374).

217. Strawberry Weevil[2]

Importance and Type of Injury.—The work of this beetle kills buds and fruits, leaving them hanging on partly severed stems (see Fig. 5,*E*).

Plants Attacked.—Strawberry, wild blackberry, raspberry, dewberry, and cinquefoil.

Distribution.—Eastern United States.

Life History, Appearance, and Habits.—This insect passes the winter in the form of a dark reddish-brown snout beetle, from 1/12 to ⅛ inch long (Fig. 446), sheltered under trash.   These adults become active early in the spring, about the time the strawberries are coming into bloom.   The adult beetle makes a puncture in the strawberry bud with her long beak, and in this inserts an egg.   She then crawls down and girdles the stem of the bud.   The young grubs, which are legless,

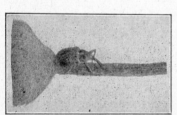

Fig. 446.—Strawberry weevil, *Anthonomus signatus* Say, on stem of strawberry.   About 4 times natural size.   (*From Ark. Agr. Exp. Sta. Bul. 185.*)

white, and soft-bodied, feed within buds and, after about 4 weeks, change within the bud to a pupal stage.   The adults emerge a little before midsummer. They feed for a short time and then go into hibernating quarters about midsummer, remaining there until the following spring.

[1] Many species of Phyllophaga, Order Coleoptera, Family Scarabaeidae.
[2] *Anthonomus signatus* Say, Order Coleoptera, Family Curculionidae.

Control Measures.—Dusting with lead or calcium arsenate, 1 part, in sulphur, 4 or 5 parts, using two applications of 10 to 30 pounds per acre, the first as the plants are coming into bloom, has given good control of both the weevil and the common red spider. These poisons must not be used within 3 weeks of ripening. Clean cultivation and resistant varieties help in control. The planting of one row of staminate variety to each five rows of pistillate varieties has also given good results.

References.—Tenn. State Bd. Ento. Bul. 30, 1919; N.J. Agr. Exp. Sta. Bul. 324, 1918; Ark. Agr. Exp. Sta. Bul. 185, 1923; U.S.D.A., Cir. E-346, 1935; Jour. Econ. Ento., 30: 437–438, 1937; Rept. Ento. Ark. Agr. Exp. Sta., 1930–1931, pp. 45–49; Ore. Agr. Exp. Sta. Bul. 330, 1934.

N. BLUEBERRY AND HUCKLEBERRY INSECTS

The most injurious insect attacking this group of fruits is the blueberry maggot, a variety of the apple maggot which is discussed on page 603. Eggs are laid in the ripe berries and the maggots eat the pulp of the fruits, causing many fruits to drop, spoiling the sale of others, and occasioning great difficulties in sorting the fruits for canning. Dusting blueberry land with calcium arsenate, 6 to 7 pounds per acre, making two applications 7 to 10 days apart (in mid-July in Maine), has given excellent control by killing the adult flies. Recently derris dusts applied from airplanes or autogyros, using 10 to 15 pounds per acre, have been recommended. Many other pests of these fruits are recorded in *Maine Agricultural Experiment Station Bulletin* 356, 1930. The blueberry thrips and certain caterpillars and sawflies are recorded as of economic importance.

References.—U.S.D.A. Cir. 196, 1931; Jour. Econ. Ento., 30: 294–297, 1937.

CHAPTER XVIII

CITRUS INSECTS

Only the more important pests of citrus are discussed here. For more complete information about these species and for the other minor species, the reader is referred to the various publications put out by the states where citrus is extensively grown.

General References.—U.S.D.A., Dept. Bul. 907, 1920; *Fla. Agr. Exp. Sta. Buls.* 126, 1915, and 148, 1918; *U.S.D.A., Farmers' Buls.* 1321, 1923, 933, 1918, 862, 1917, and 674, 1915; *Calif. Agr. Exp. Sta. Bul.* 214, 1911; *Calif. Fruit Growers' Exchange, Bur. Pest Control, Buls.* 5, 1928, and 6, 1929; Essig, "Insects of Western North America," pp. 269–322, Macmillan, 1926; *Fla. Agr. Ext. Serv. Bul.* 67, 1932; Quayle, H. J., "Insects of Citrus and Other Subtropical Fruits," Comstock Publishing Co., 1938.

FIELD KEY FOR THE DETERMINATION OF INSECTS INJURING CITRUS FRUITS

A. *Motionless insects, covered by a firm waxy shell, sucking sap from fruits, leaves, or twigs:*
  1. Foliage of trees showing yellow leaves or brown areas which break out of the leaves. Fruits uneven in color, with elongate, oyster-shell-shaped scales attached to the skin of the fruit. Many dead twigs.......................
  ....*Long scale* or *purple scale*, page 675.
  2. Leaves entirely yellow or spotted with yellow. No honeydew on the leaves or fruit:
    a. Many yellow spots on the fruits. Entire branches of trees covered with a thin coating of circular yellowish or reddish scales, about $\frac{1}{12}$ inch in diameter...................................*California red scale*, page 676.
    b. A sprinkling of purplish-red round scales on leaves and fruit only.........
    ....*Florida red scale*, page 676.
  3. Foliage of trees somewhat discolored. Fruits and leaves covered with honeydew or a black, sooty growth of mold. Bodies of insects not covered with a separable shell:
    a. Nearly round scales, about $\frac{1}{8}$ inch across, of a brown or blackish color, often with a raised H on the back, are found over the twigs, leaves, and fruits...*Black scale*, page 678.
    b. Oval, brownish or grayish scales on twigs and leaves, up to $\frac{1}{6}$ inch long and very flat....................*Soft brown scale*, or *citricola scale*, page 679.
    c. Large white, cottony, fluted objects, nearly $\frac{1}{2}$ inch long, on the twigs. Smaller oval yellowish-brown scales on the leaves.....................
    ....*Cottony cushion scale*, page 302.
B. *Insects usually more or less active; not covered with a separable shell; sucking sap from fruits, leaves, or twigs:*
  1. Leaves of trees showing pale spots and with light webs on their undersurface. Minute red or yellow mites on the leaves or fruits. Fruits with a grayish or silvery sheen to the skin. Fruit drops............*Red spider mites*, page 682.
  2. Flowers and buds show retarded growth and the new growth distorted. Many tiny, slender, dark or yellowish insects among buds and flowers. Skins of fruits show erect shallow scabs or smooth scars............*Thrips*, page 673.
  3. Masses of cottony white material covering flat, oval, purplish, soft-bodied bugs on leaves or at the angles where fruits touch. Fruits coated with very sticky honeydew or black sooty mold......................*Mealy bugs*, page 679.

4. Foliage of trees covered with honeydew or blackened with a heavy growth of sooty mold. Fruits undersized or of poor color. At certain seasons, very small white insects fly from the tender growth of trees when disturbed. Small pale-green short-oval flat scale-like nymphs fixed to the underside of the leaves. Brown, bright-red, or yellow fungi often growing on the bodies of the nymphs.
....*Whiteflies*, page 681.

5. Foliage more or less tightly curled and more or less covered beneath with soft-bodied plant lice...
............*Melon aphid* (page 497), *green citrus aphid*, and others, page 447.

6. Fruits russet-brown or, on grape fruit, chamois color, but smooth............
....*Rust mite*, page 683.

218. Thrips

Importance and Type of Injury.—Thrips injure the citrus fruits by attacking the flowers and buds, causing them to fall. The growth of the

Fig. 447.—Oranges injured by thrips, *Scirtothrips citri* (Moulton), showing characteristic rings. (*From Essig, "Insects of Western North America," after Quayle. Copyright, 1926, Macmillan.*)

young trees is often retarded and new growth distorted by their feeding on the foliage. The fruit also is attacked and is reduced in size; the skin is scarred, often with a definite ring around the stem end where the thrips feed beneath the sepals while the fruit is small (Fig. 447).

Plants Attacked.—All citrus, many deciduous fruits, and many other plants.

Distribution.—Thrips are world-wide in their distribution. The citrus thrips[1] is the species causing the greatest amount of injury in the warmer, more arid regions of California; it does not occur in Florida. The flower thrips[2] is most important in Florida. The greenhouse thrips (page 736) and the bean thrips (page 486) also do some damage to citrus fruits.

Life History, Appearance, and Habits.—The winter is passed either as adults or in the egg stage on the stems and leaves of the infested trees. The eggs are inserted into the tissue at the base of the flowers, in the new foliage, or in the stems bearing leaves or fruits. The young thrips are yellowish, very small, slender, active creatures. The nymphs, during the first two instars, feed by rasping the plant surface and sucking the sap that flows from these injured spots. They complete the feeding period in from 3 days to 3 weeks. There follow two nonfeeding nymphal stages, which are passed among the trash on the surface of the soil or in soil crevices. From the last of these the adults emerge to start new generations every 2 or 3 weeks.

Control Measures.—Dusting with superfine sulphur, 2 or 3 applications at monthly intervals, one of the applications being made when the trees are in full bloom, is an effective control. Spraying with lime-sulphur, $1\frac{1}{2}$ to 2 gallons, and calcium caseinate, $\frac{1}{2}$ pound, in 100 gallons of water, when part of the petals have fallen and again 2 or 3 weeks later, is also recommended.

CITRUS SCALE INSECTS

The scale insects of citrus are probably the most destructive of any group of insects which attack these trees. As with the greenhouse scales, the citrus scales can be divided into three classes, armored scales, unarmored scales, and mealy bugs.

ARMORED SCALES

In this class of scales, a protective covering of wax is secreted from the body of the insect to form two protective scales, one above and the other beneath the body of the scale. The upper covering is thick and hard; the lower plate, or scale, is very thin and delicate, fitting closely to the surface of the plant where the insect is feeding. In this class of scales the eggs are laid under the protective scale, or in some cases the young are born alive. In either case the young scales move about for a short time, select a favorable location on their food plant, and there insert their beaks; in the case of the females they do not move for the remainder of their lives. On starting to feed, they begin to secrete fibers of wax, which form the covering or scale over the body. The insects molt or shed their skins very shortly after beginning to suck the sap, and at this first molt lose their legs. The females later molt a second time and become adults, while the males, after a further metamorphosis, develop into small two-winged adults, which are incapable of feeding. They mate with the females and die very shortly thereafter. The female after being fertilized increases very rapidly in size and produces her eggs

---

[1] *Scirtothrips citri* (Moulton), Order Thysanoptera, Family Thripidae.

[2] *Frankliniella cephalica bispinosa* Morgan [= *F. tritici* (Fitch)], Order Thysanoptera, Family Thripidae.

or begins giving birth to living young. The scales that attack citrus develop more slowly during the colder periods of the year, but all stages of these insects can usually be found on trees at any season. Some of the most destructive of the armored scales are the purple scale, the long scale, the yellow scale, and the orange or red scale.

219. PURPLE SCALE[1]

Importance and Type of Injury.—This is one of the most widely distributed and destructive of citrus pests. The foliage of trees infested by the purple scale turns yellow about the areas where the scales are feeding. In Florida these yellow areas may turn brown and break out, making holes through the leaves. Fruit attacked by this scale is stunted, ripen-

FIG. 448.—The purple scale, *Lepidosaphes beckii* (Newman), male and female scales on orange leaf, about 5 times natural size. (*From Essig, "Insects of Western North America," copyright, 1926, Macmillan. Reprinted by permission.*)

ing is delayed, the coloring of the fruit is very uneven, and flavor is affected. The scales are difficult to remove from the fruit before marketing, requiring a vigorous scrubbing to detach. Their feeding also permits the entrance of various fungi.

Distribution.—This insect is the most important pest of orange groves in Florida and other Gulf states; and, although mainly confined to coastal regions, it is rated as third in importance in California, being outranked only by the California red scale and the black scale.

Plants Attacked.—The purple scale is primarily a pest of citrus fruit, and particularly of the orange and grapefruit. It also occurs on avocado, croton, eucalyptus, fig, olive, yew, and other plants.

---

[1] *Lepidosaphes beckii* (Newman), Order Homoptera, Family Coccidae. The long scale, *L. gloveri* (Packard), is very similar to the purple scale but is much narrower and straighter.

Life History, Appearance, and Habits.—The female scale deposits from 40 to 80 eggs beneath her scale, dying soon after her full quota of eggs is laid.   These eggs hatch in from 2 weeks to 2 months.   The young very pale scales crawl about the bark for a short time.   They are said to be strongly repelled by light and to seek the shaded part of the tree and fruit as a place for inserting their beaks before starting to feed.   They are, however, repelled by total darkness, seeking a position somewhat intermediate between that of exposure to sunlight and the most heavily shaded parts of the tree.   The first molt occurs in about 18 or 20 days, after which the insect forms a thicker scale which is of a reddish or purplish-brown color.   The full-grown scale is about $\frac{1}{8}$ inch long, and shaped like an oyster shell (Fig. 448).   After the insects have fed for about 2 months, the males change to two-winged active insects, which fly about, and mate with the females, which shortly begin to deposit their eggs. Each female lays from 40 to 80 eggs underneath her protective scale. There are three main generations a year.

Control Measures.—In Florida, it has been found possible to control the purple scale and the whitefly by the same applications.   The more thorough spraying, however, is usually required for the control of the purple scale.   The Florida Experiment Station recommends the following control measures.   Spraying with an oil emulsion made after the formula:

Fish-oil soap	8 lb.
Lubricating oil, 24 to 28° Bé	2 gal.
Water	1 gal.

This is a cold-mixed emulsion.   A similar boiled emulsion is composed of:

Fish-oil soap	2 lb.
Light-grade lubricating oil	2 gal.
Water	1 gal.

The latter spray should be mixed and prepared according to the recommendations given for the preparation of oil emulsions on page 273.   One gallon of either of the stock mixtures is diluted to make 50 gallons with soft water; or the entire amount makes 200 gallons of spray.   Under average conditions the spray should be applied in May and again in September or early October.   Two sprays are all that are generally required to keep down this scale, if they are applied with sufficient thoroughness.   In California fumigation with either liquid hydrocyanic acid or calcium cyanide, from July to October, sometimes preceded by a $1\frac{2}{3}$ per cent oil spray, is used.

Reference.—Calif. Agr. Exp. Sta. Bul. 226, 1912.

220 (263). California and Florida Red Scales[1,2]

Importance and Type of Injury.—The California red scale[1] is said to be the worst pest of citrus in California.   It is also important in Texas, but it is of little importance in Florida.   A similar scale known as the Florida red scale[2] is of much importance in the Gulf states, where it causes

[1] *Aonidiella* (= *Chrysomphalus*) *aurantii* (Maskell), Order Homoptera, Family Coccidae.

[2] *Chrysomphalus aonidum* (Linné), Order Homoptera, Family Coccidae (Fig. 489).

a similar injury except that it attacks the leaves and fruit only. The Florida red scale deposits eggs which hatch in a few hours, while the California red scale is ovoviviparous. No injury from the secretion of honeydew occurs to trees where this scale is present, but as the California red scale infests all parts of the trees, including the leaves, twigs, and fruit, its feeding is often a serious matter. Infested trees have the leaves spotted with yellow or entire leaves turning yellow and yellow spots on the fruit, but not the marked discolorations that often appear with other scales. The entire bark of twigs and branches may be covered by round, or nearly round, distinctly reddish scales up to about $\frac{1}{12}$ inch in diameter, with central exuviae (Fig. 449).

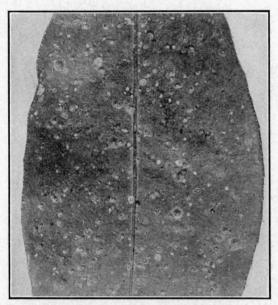

Fig. 449.—The California red scale, *Aonidiella aurantii* (Maskell), showing mature males and females and young; about natural size. (*From E. O. Essig, Univ. of Calif.*)

Plants Attacked.—Primarily citrus, but it also infests acacia, eucalyptus, fig, grape, privet, quince, rose, English walnut, willow, and many other plants.

Distribution.—Generally abundant in the coastal citrus-growing regions of California, and in a number of the citrus districts in the Gulf states. It also occurs in many other parts of the world where citrus is grown.

Life History, Appearance, and Habits.—The life history of this scale somewhat resembles that of the San Jose scale on deciduous fruits. The young scales are born at the average rate of two or three a day for a period of about 2 months during the summer. Small numbers of young are born during warmer periods in winter months, but very few, if any, are produced during the colder weather. The young insects feed for from $2\frac{1}{2}$ to $3\frac{1}{2}$ months before reaching maturity. The males develop into yellowish, minute, two-winged insects, which mate with the females and die, the mated females living for about 2 months longer and during this time reproducing.   There may be as many as four generations a year in southern California.

Control Measures.—The control of this scale is the same as that of the other armored scales, and consists of spraying with oil emulsions or fumigating with hydrocyanic acid gas.   The control will vary with the conditions in the orchard.   For the best method to follow, refer to the special publication on this subject.

Reference.—Calif. Agr. Exp. Sta. Bul. 222, 1911.

Unarmored Scales

The unarmored scales are often called also soft scales and are, as a class, larger than the armored scales.   No true scale of wax is formed separate from the body wall.   The protective covering is the body wall, which is heavily sclerotized.   Both sexes move about during the early part of their lives, but the females cannot crawl after the eggs have formed.   These scales discharge quantities of honeydew from their bodies, and the sooty fungus that grows in the accumulations of this honeydew often causes nearly as much injury as the feeding of the scales.   The black scale, the soft brown scale (page 742), the citricola scale, and the hemispherical scale are typical examples of this class.

221 (264). Black Scale[1]

Importance and Type of Injury.—This scale is considered one of the most destructive insect pests of citrus in California.   The principal damage is caused, in addition to the feeding of the insect, by the sooty mold fungus which grows on the honeydew given off by this scale.   This fungus makes it necessary to wash all fruits from badly infested groves and interferes with the physiological functions of the leaves.

Trees Attacked.—Orange, grapefruit, lemon, plum, almond, apple, pear, apricot, beech, fig, grape, pepper tree, oleander (the most common host in Florida), rose, English walnut, and a number of other plants.

Distribution.—All the principal citrus-growing regions of the world; and in greenhouses in colder areas.   It is not important in the Gulf states.

Life History, Appearance, and Habits.—The winter is passed in all stages of growth, but mainly as partly grown females.   Most of the over-wintering scales become full-grown early in the spring.   The full-grown females are nearly hemispherical in shape, being about ⅕ inch across and from 1/25 to ⅛ inch thick, dark brown to black in color, with a median longitudinal ridge and two transverse elevations on the back forming a letter H (Fig. 450).   They deposit an average of 2,000 eggs.

Fig. 450.—The black scale, *Saissetia oleae* (Bernard), on twig, about twice natural size. (*From E. O. Essig, Univ. of Calif.*)

[1] *Saissetia oleae* (Bernard), Order Homoptera, Family Coccidae.

The eggs, which are about $\frac{1}{80}$ inch in length, are white at first, later changing to orange. Most of the eggs are laid during the spring months. The eggs hatch in about 20 days. The young remain beneath the parent scale for some hours and then emerge and crawl about, but always start feeding within 3 days. Most of the young settle on the leaves or new growth. When partly grown, most of the scales migrate to the twigs and branches. From 8 to 10 months are required for the scales to complete their growth. The males go through a pupal stage, and in the adult stage are active, two-winged insects. They are, however, very rare and reproduction is generally by parthenogenesis. There is one generation a year over most of the area but two generations occur in some coastal sections.

Control Measures.—Under most conditions, black scale has been controlled by fumigation with hydrocyanic acid, at a time when no eggs are present, the dosage varying with the infestation and the condition of the orchard. In most of the areas of southern California, where the black scale has apparently developed resistance to fumigation, spraying has given much better control. Good control may be obtained by spraying with an emulsion of light-medium oil, having a distillation of 50 to 65 per cent at 636°F., a viscosity of 52 to 70 seconds, and 80 to 95 per cent unsulphonatable. This is made into an emulsion with potassium fish-oil soap, ammonium caseinate, or fine clay, and applied at a strength of $1\frac{1}{2}$ to 2 per cent oil, during the summer months. Citricola scale and red spider also are controlled by this spray if the infestations are not very severe.

Reference.—*Calif. Agr. Exp. Sta. Bul.* 223, 1911.

222. CITRICOLA SCALE[1]

This species, which is very similar to the soft brown scale (page 742), is grayish when mature and deposits over 1,000 eggs during a period of 1 to 2 months in spring. The young nymphs are very flat and transparent. They feed on the underside of the leaves and on the smaller twigs during the summer but migrate to the twigs during the winter, becoming adults in the spring. Fumigation with hydrocyanic acid, spraying with oil emulsions, and dusting with sulphur are all used in its control on citrus.

223 (265). MEALY BUGS[2,3,4,5]

Importance and Type of Injury.—Citrus trees infested with mealy bugs will have masses of white cottony-appearing insects clustered on the leaves and twigs and at the angles where fruits touch (Fig. 451). Infested fruit is generally coated with a very sticky honeydew, which makes washing of the fruit necessary. The insects sometimes become sufficiently abundant to kill the trees.

Plants Attacked.—All citrus and many ornamental and greenhouse plants.

[1] *Coccus pseudomagnoliarum* (Kuwana).
[2] Long-tailed mealy bug, *Pseudococcus longispinus* (Targioni), Order Homoptera, Family Coccidae.
[3] Citrophilus mealy bug, *Pseudococcus gahani* Green.
[4] Grape mealy bug, *Pseudococcus maritimus* (Ehrhorn).
[5] Citrus mealy bug, *Pseudococcus citri* (Risso).

Distribution.—World wide. In this country, the insects are more destructive in California, although they are found in the other citrus-growing states.

Life History, Appearance, and Habits.—The insects can be found in all stages of development throughout the year. The adult long-tailed mealy bug has four waxy filaments at the tip of the abdomen which are nearly as long as the rest of the body; the citrophilus and grape mealy bugs have these filaments about one-third as long as the body; whereas in the citrus mealy bug they are scarcely longer at the tail end than elsewhere. The citrophilus mealy bug has a much darker body fluid than the other species. On the citrus mealy bug the white powder over the back is very dense. Except in the case of the long-tailed mealy bug, which is ovoviviparous, the mature females, which are about ⅛ to ⅓ inch long (Fig. 451), deposit from 300 to 600 eggs in a cottony mass of

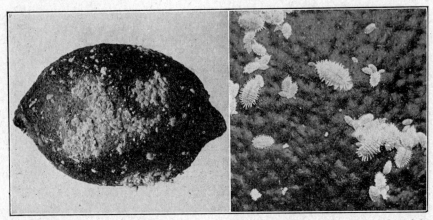

Fig. 451.—Citrus mealy bugs; at left, an infested lemon, slightly reduced; at right, adult females and nymphs of various sizes on fruit of orange, about twice natural size. (*From E. O. Essig, Univ. of Calif.*)

wax secreted from their bodies. The eggs hatch in from 6 to 20 days, and the young mealy bugs feed by sucking the sap or juices from the leaves or fruit. They move about but little and require from 1 to 4 months to complete their growth. The adult males have two wings, the females are wingless throughout life. There are generally two to four generations a year.

Control Measures.—Oil emulsions are only partially successful in the control of mealy bugs. Clear water, if applied in amounts up to 1,000 gallons per tree, and with considerable force, will control the pests. Fumigation is of little value, except for the long-tailed mealy bug. Biological control, by means of lady beetles and hymenopterous parasites introduced from Australia and artificially reared in insectaries for liberation in the orchards, has been so successful that, at the present time, other control measures are not needed. Up to 20,000,000 beetles a year have been produced in this way in California.

Reference.—*U.S.D.A., Farmers' Bul.* 1309, 1923.

224. WHITEFLIES[1,2,3,4]

Appearance and Type of Injury.—Infested trees have a blackened appearance due to a sooty mold which grows in the honeydew given off by the whitefly nymphs. This sweetish, sticky honeydew is discharged in large quantities from the alimentary tract. Trees are stunted, through loss of sap, and the fruit is undersized and of poor color.

Distribution.—Whiteflies are widely distributed throughout the world. They are serious pests of citrus in Florida and the Gulf states but are, at present, of little importance in California. The common citrus whitefly[1] (Fig. 452) is the only one found in California. The cloudy-winged whitefly[2] and the woolly whitefly[3] are also serious pests in Florida. The citrus blackfly,[4] which occurs at Key West and in the West Indies, in the nymphal stages has a black body with a white fringe; the adult has a dark-brown body with the wings partly clouded.

Food Plants.—All species of citrus and many other plants. In Florida these insects breed in large numbers on chinaberry.

Fig. 452.—Citrus whitefly, *Dialeurodes citri* (Riley and Howard); adults and eggs on leaf, slightly enlarged. (*From Florida Agr. Exp. Sta. Bul.* 183.)

Life History, Appearance, and Habits.—The life histories of all species of whiteflies which are of importance on citrus are very much alike. All stages of the insects may be found throughout the year, but little breeding takes place during cold periods. The oval eggs, less than $\frac{1}{100}$ inch in length, are attached to the under sides of the leaves by a short stalk. The eggs of the citrus whitefly are pale yellow, those of the cloudy-winged whitefly are black, and those of the woolly whitefly are brown and curved like miniature, fat sausages. They hatch in from 4 to 12 days into active pale-yellow flattened six-legged "crawlers," or nymphs. They move about for a short time, mainly on the lower sides of the leaves, as they avoid strong light. These crawlers soon insert their beaks into the leaves and begin sucking the sap. They soon molt, losing their legs in the process, and then have the appearance of very minute, flattened, oval bodies, attached to the undersides of the leaves by their sucking beaks (see Fig. 498), and with a marginal fringe of short, white, waxy filaments somewhat like a mealy bug. After two more molts the adults (Fig. 452) emerge. They are small four-winged insects about $\frac{1}{12}$ inch long. Both sexes are winged and both feed by sucking sap. They have a white appearance due to the fine white powder which completely covers the

---

[1] *Dialeurodes citri* (Riley and Howard), Order Homoptera, Family Aleyrodidae.
[2] *Dialeurodes citrifolii* (Morgan).
[3] *Aleurothrixus howardi* (Quaintance).
[4] *Aleurocanthus woglumi* Ashby (see *U.S.D.A., Dept. Bul.* 885, 1920).

wings and body.   In Florida there are mainly three generations a year of
the common citrus whitefly.

Control.—Spraying with oil emulsions as recommended for the purple
scale has given the best control.   Sprays should be applied according
to conditions in the orchard.   At least two sprays a year, applied in
May and September, should be used in badly infested orchards.   Keeping
down water sprouts is of some value in controlling this insect.   Certain
fungi that grow on the bodies of the whiteflies are important natural
enemies of these insects in Florida but are of no importance in control
in California.   In Florida they generally control the insects during the
summer rainy season.   Cultures of these fungi may be secured from state
officials and sprayed over the trees to start the destructive parasitism of
the whiteflies (see also page 304).

Reference.—Fla. Agr. Ext. Serv. Bul. 67, 1932.

225. Red Spider Mites[1,2,3]

Importance and Type of Injury.—These are among the most destructive of all pests of citrus fruits.   The citrus red mite[2] is most injurious in

Fig. 453.—The citrus red mite, *Paratetranychus citri* McGregor; mites and one of their
eggs showing the silken threads, like guy lines, that stretch from the leaf to the top of the
egg stalk.   *(From Quayle, Univ. of Calif.)*

the more arid, coastal regions of California, while the six-spotted mite[3]
is much more destructive in the humid Gulf states.   Infested foliage
shows many pale spots where the mites have sucked out the green part of
the leaf.   The leaves are webbed on the undersides, and those heavily
infested turn silvery or brown and may drop.   The fruits also become
gray or yellow and may fall.   The feeding of the six-spotted mite is
restricted to limited areas on the underside of the leaves.   These areas
become depressed, yellow, and webbed beneath and shiny and yellow on
the upper side.

Plants Attacked.—These red spider mites do little damage to plants
other than citrus.

[1] Order Acarina, Family Tetranychidae.
[2] *Paratetranychus citri* McGregor.
[3] *Tetranychus sexmaculatus* Riley.

Distribution.—Although both species occur in both of the citrus belts of the United States, the citrus red mite does not thrive in the more humid Gulf states or in the drier areas in California more than 60 miles from the ocean. The six-spotted mite is the most common mite on citrus in the Gulf states and in California is limited to the coastal citrus belt.

Life History, Appearance, and Habits.—Throughout most of the citrus-growing area of the United States all stages of the mites may occur at all times of the year. Their numbers are greatly reduced during the winter months and often also in midsummer. The eggs are fastened to the leaves or to the silk spun by the mites. Those of the citrus red mite have a vertical stalk like a mast, from the top of which about a dozen threads extend, like guy lines, to the leaf. There are three growing stages, the six-legged "larva" and two eight-legged nymphal stages, differing chiefly in size. The adult citrus red mite is velvety red or purplish in color with about 20 prominent bristles over the body each arising from a conspicuous tubercle (Fig. 453). The six-spotted mite is pinkish, greenish, or yellowish, sometimes with six dark spots. The bristles do not arise from conspicuous tubercles. A generation may be completed in from 3 to 5 weeks.

Control Measures.—Oil sprays (as for the black scale) or wettable sulphur, 5 to 10 pounds, with lime-sulphur, 2 gallons in 98 gallons of water, have been chiefly used in control.

Reference.—*Jour. Agr. Res.*, vol. 36, no. 2, 1928.

226. CITRUS RUST MITE[1]

Importance and Type of Injury.—Second only to the purple scale and the citrus whitefly as a citrus pest in the Gulf states, this mite is said

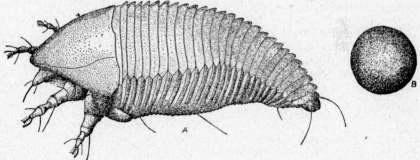

Fig. 454.—Citrus rust mite, *Phyllocoptes oleivorus* (Ashmead). *A*, adult, side view, × 700; *B*, egg, × 825. (*From U.S.D.A.*)

to reduce the value of half the oranges and grapefruit in Florida from 25 to 50 cents a box. By rupturing the cells and sucking sap from the skin of the fruit and the leaves, it causes a serious russeting or silvering of the fruit and a powdery appearance lowering quality and attractiveness of the fruit and reducing the vigor of the trees.

Plants Attacked.—Lemon, grapefruit, and orange are injured in the order named. All other citrus fruits and some ornamental plants are attacked.

[1] *Phyllocoptes oleivorus* (Ashmead), Order Acarina, Family Eriophyidae.

Distribution.—In the United States most important in the Gulf states, though it occurs also in California.

Life History, Appearance, and Habits.—The mites occur on the trees throughout the year but are least abundant in January and February. Males of this species are unknown. The females deposit their pale-yellow, smooth, spherical eggs in depressions on the fruits and leaves. After 2 to 8 days the nymphs hatch. They feed like the adults, molt twice at from 1- to 6-day intervals, and are then the slender elongate females which are only $\frac{1}{200}$ inch long, with two pairs of legs at the head end and a tapering ringed abdomen (Fig. 454). Generations succeed each other at intervals as short as 1 or 2 weeks, and the mites become most abundant about July 1.

Control.—Dusting with sulphur, when the foliage is wet with dew, or spraying with wettable sulphur, 5 to 10 pounds, and lime-sulphur, 1 to 2 gallons, in 98 gallons of water are recommended as most effective. Oil sprays and bone glue, 15 pounds in 100 gallons of water, are also used.

Reference.—U.S.D.A., Tech. Bul. 176, 1930.

227. MEDITERRANEAN FRUIT FLY[1]

Importance and Type of Injury.—This pest is capable of preventing successful growing of a number of important fruits in areas of mild winters where there is a year-round succession of cultivated or wild fruits upon which it can develop. The maggots develop in the pulp of the fruit, devouring it and favoring the development of bacterial and fungous diseases; the egg punctures made by the adults may affect shipping qualities of the fruit.

Plants Attacked.—Many deciduous and citrus fruits including especially peaches, nectarines, plums, grapefruit, oranges, and other citrus except lemons. Apples, pears, quinces, coffee, and nearly 100 other cultivated and wild fruits are more or less attacked.

Distribution.—This insect is not now known to be present in the continental United States. It was discovered in Florida in 1929, in scattered locations over an area of about 10,000,000 acres. It was never found in wild host plants away from cultivated land. A most vigorous and remarkably successful campaign of eradication, involving the expenditure of over $7,000,000 of state and federal funds, resulted in its complete extermination, and none of the insects have been found there since July, 1930. It occurs in Bermuda and Hawaii and in nearly all subtropical countries except North America. Constant vigilance is required to prevent its reintroduction to this country in shipments of horticultural products or in the baggage of travelers.

Life History, Appearance, and Habits.—In cool regions the insect winters as pupae or adults, while in warmer regions, where fruits are available, reproductive activity may be continuous throughout the year. From 2 to 10 eggs are deposited through a small hole, the size of a pinprick, made in the rind of the fruit by the ovipositor, but many additional eggs may be laid in the same hole by other females. A single female may produce 800 eggs. The eggs may hatch in 2 to 20 days, and the larvae burrow in the pulp for 10 days to 6 weeks before completing growth, when they are a little more than $\frac{1}{4}$ inch long. They then desert the fruit and form a

[1] *Ceratitis capitata* (Wiedeman), Order Diptera, Family Trypetidae.

puparium in the soil, within an inch or two of the surface or under other protection, or even exposed in boxes or wrappers. Pupation may be completed in 10 to 50 days. The adult (Fig. 455) is about the size of the house fly. The thorax is glistening black with a characteristic mosaic pattern of yellowish-white lines. The abdomen is yellowish with two silvery crossbands, and the wings are banded and blotched with yellow, brown, and black. The adults have sponging mouth parts and take only liquid foods. The females may oviposit successfully after as much as 10 months of inactivity. Under most favorable conditions the life cycle may be passed in 17 days, commonly in 3 months; and there may be from 1 to 12 or more generations a year in various parts of the world.

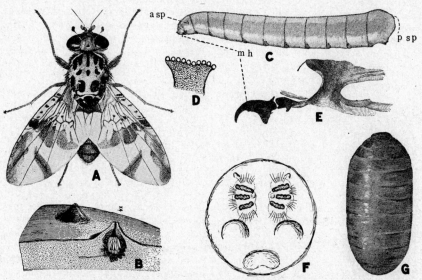

Fig. 455.—Mediterranean fruit fly, *Ceratitis capitata* Wiedeman. *A*, adult male fly; *B*, two egg cavities in rind of grapefruit, showing conical elevation left by withering of the rind, the right one in section to show mass of eggs and a single tunnel made by a newly hatched larva; *C*, a larva in side view, *a sp*, anterior spiracle, *p sp*, posterior spiracles, *m h*, mouth hooks; *D*, anterior spiracle, more enlarged; *E*, mouth hooks and cephalo-pharyngeal skeleton of larva; *F*, posterior view of last segment of larva, showing posterior spiracles of full grown larva; *G*, puparium. (*Rearranged from U.S.D.A.*)

Control Measures.—The most effective control is a bait spray consisting of an arsenical or copper carbonate in sweetened water. One or two pints are applied to several areas on each tree every week while the crop is in a susceptible condition. Traps baited with sweetened bran, fruit juices, and the like are extensively used to catch the adults. Fallen fruit should be collected daily and buried to a depth of several feet or burned or cooked to destroy the larvae. If fruits are held at 31°F. for 11 days all stages of the fly are killed. Introduced parasites have greatly reduced the pest in Hawaii.

References.—*U.S.D.A., Dept. Buls.* 536, 1918, and 640, 1918; *Calif. Fruit Growers' Exchange Bul.* 6, 1929.

CHAPTER XIX

INSECTS ATTACKING SHADE TREES AND SHRUBS

Nearly all our forest trees, shade trees, and shrubs are attacked by a large number of insects. No attempt is made in this book to treat all of the insects of forest trees. More than a thousand different species are known to feed upon the oaks, and correspondingly large numbers attack various other shade trees. A few shade trees are relatively free from injury. Shrubs are subject to attack by scale insects, especially the oyster-shell scale and San Jose scale. Poplar, linden, dogwood, willow, rose, and lilac are some of the trees and shrubs most subject to insect attack.

Where an especially destructive insect pest is well established, one should avoid planting trees or shrubs particularly likely to be attacked by this pest. In most cases such trees or shrubs can be grown if frequently sprayed, but other species which are not subject to attack by the same insect and which are nearly as attractive in appearance could well be substituted. As an example, one should avoid planting white, green, black, or red ash in any of the cities of central Indiana, Illinois, Ohio, and neighboring states, as such trees are certain to be infested with the oyster-

TABLE XVII.—Showing the Relative Likelihood of Injury to Trees by Insect Pests. Trees Having the Higher Numbers Should Be Given Preference for Planting

Sweet gum	6	European linden	3
Tree of Heaven	6	Horse chestnut	3
Ginkgo	5	Buckeye	3
Red oak	5	American elm	3
Scarlet oak	5	Hackberry	3
Oriental plane	5	Water or red elm	3
Tulip, or tulip poplar	5	Soft or silver maple	2
Sycamore	5	European elm	2
Sugar maple	5	Scotch elm	2
Norway maple	5	American linden	2
White oak	5	Cottonwood	1
Burr oak	5	Carolina poplar	1
American plane	4	Lombardy poplar	1
Red maple	4	Balm of gilead	1
Honey locust	4	Black locust	1
Spruces	4	Boxelder	1
Blue ash	4	American mountain ash	1
European mountain ash	4	Green ash	1
White pine	3	Black ash	1
Catalpa	3	White ash	1

This table follows a plan originated by Dr. E. P. Felt, formerly State Entomologist of New York.

shell scale. The blue ash, sycamore, various maples, and many other trees can be planted in these localities, and will never be injured in the least by this scale.

The table on p. 686 shows the relative resistance to insect attack of the different varieties of shade trees commonly grown in the latitude of central Ohio and Illinois. The figure 6 has been placed opposite trees which are practically immune from insect injury; 5 indicates some damage; trees having one somewhat serious enemy are rated at 4; and those having at least one notorious insect pest at 3. Greater likelihood of injuries is indicated by 2, and still more by 1. The species of trees are arranged according to the comparative injury by insects.

General References.—DOANE, VAN DYKE, CHAMBERLIN, and BURKE, "Forest Insects," McGraw, 1936; HERRICK, "Insect Enemies of Shade Trees," Comstock Publishing Co., 1935; FELT, "A Manual of Tree and Shrub Insects," Macmillan, 1924; GRAHAM, "Principles of Forest Entomology," McGraw, 1939; *U.S.D.A., Misc. Pub.* 273, 1938; *N.Y. State Museum, Mem.* 8, Vols. 1 and 2, 1905; *Ohio Agr. Exp. Sta. Bul.* 332, 1918; *U.S.D.A., Farmers' Bul.* 1169, 1921.

FIELD KEY FOR THE DETERMINATION OF INSECTS INJURING SHADE TREES AND SHRUBS

A. External chewing insects that eat holes in leaves, buds, bark, or fruit:

1. Yellow-and-black beetles and grubs, about ¼ inch long, eating out the green parts of elm leaves. Large numbers of grubs collect on the bark about the base of the trunk or on the ground near by........*Elm leaf beetle,* page 691.

2. Leaves of willows, poplars, and cottonwood partly or completely eaten. Convex, oval, yellow or reddish beetles, spotted or striped with black, about ¼ to ½ inch long, feed, along with their soft-bodied, more elongate, blackish young or larvae, on underside of leaves.................................... *Cottonwood, poplar,* and *willow leaf beetles, Chrysomela scripta* Fabricius, *Chrysomela lapponica* Linné, *Chrysomela tremulae* Fabricius, *Chrysomela interrupta* Fabricius, and others.

3. Spruce or balsam trees appearing as though scorched by fire and slowly dying. Thick dark-brown caterpillars, ¾ inch or less in length, covered with pale-yellow warts, eat off the needles and spin silken threads over the terminal shoots, webbing them together, in early spring.....*Spruce budworm,* page 692.

4. Shade trees, particularly evergreens, stripped of their leaves by brownish, rather fat-bodied worms, which live within tough silken bags. Bags up to 2 inches in length hanging in large numbers from the leaves and twigs of infested trees. Bags remaining on the trees during the winter months...... ....*Bagworm,* page 693.

5. Catalpa trees completely stripped of their leaves by dark-colored caterpillars marked with varying amounts of green and yellow. Caterpillars up to 3 inches in length, with a curved black horn projecting from the last segment of their bodies....................................*Catalpa sphinx,* page 695.

6. Large colonies of black caterpillars stripping the leaves from hickory, walnut, and related trees. All foliage completely stripped from single branches. Caterpillars coming down the trunks of the trees in large numbers and hanging in masses on the trunks while shedding their skins........................ ....*Walnut caterpillar,* page 696.

7. Loosely woven, flimsy, white webs of silk, enclosing the leaves of branches of varying size up to several feet across, contain very hairy pale-yellow black-spotted caterpillars, up to 1¼ inches long, which eat the foliage without leaving their tent. Injury most noticeable in late summer or fall in the north...... ....*Fall webworm,* page 562.

8. Tent-like webs of silk in the forks of wild cherry, wild plum, crab apple, and other trees in spring. Brownish, very hairy caterpillars, 2 inches or less in

length, with a light stripe down the back, hide in these tents during the day. At night they make silken pathways along the branches to feed upon the foliage...................................*Eastern tent caterpillar*, page 563.

9. Pale-blue hairy caterpillars, with a row of keyhole-shaped white spots down the middle of the back and pale-yellow stripes on the sides, defoliate poplars, oaks, maples, and many other trees, without spinning tents; and often swarm into buildings............................*Forest tent caterpillar*, page 698.

10. Leaves of trees skeletonized or stripped by yellowish caterpillars with 2 long tufts or pencils of dark hairs protruding from near the head, and one tuft from near the tail. Trunk and branches of trees during winter months with dark-gray cocoons on the bark, cocoons often having frothy-white masses of eggs attached to them................*White-marked tussock moth*, page 698.

11. Buff-colored masses of eggs, up to 1 inch across, covered with felt-like hairs, on the trunks of shade trees and surrounding objects during the winter months. Large dark-gray, somewhat flattened caterpillars, with pairs of red and blue spots down the back, strip the leaves from trees during June and early July.. ....*Gypsy moth*, page 701.

12. During the winter season several leaves tightly webbed together and firmly attached to a twig near the tip of the branch, enclose 25 to several hundred small dark-brown very hairy caterpillars, about ¼ inch long. If the caterpillars or their hairs come in contact with the skin, a distinct burning or itching results, which may persist for several days. White moths, with brown abdomens, flying in great numbers at night during July, strongly attracted to lights......................................*Brown-tail moth*, page 704.

13. Poplar and willow leaves eaten off during midsummer by rather large black-bodied caterpillars. When full-grown, these caterpillars are about 2 inches long, with irregular whitish colorings on the sides of the back and a nearly square white patch of hairs on the middle of the back of each segment. The adult insect is a white moth with a wing spread of nearly 2 inches.......... ...*Satin moth, Stilpnotia salicis* (Linné) (see *U.S.D.A., Dept. Bul.* 1469, 1927).

14. Smooth greenish or brownish measuring-worms or looping caterpillars, which lack prolegs near the middle of the body, eat foliage of elm, hackberry, oak, and other deciduous trees, dropping on silk threads when disturbed.......... ....*Cankerworms*, pages 559 and 561.

15. Greenish, yellowish, whitish, or bluish worms, 1 inch or less in length, often spotted or striped with black or yellow, with dark heads and 9 to 11 pairs of legs and prolegs, defoliate elm, birch, poplar, willow, pine, larch, spruce, balsam, or other trees during the summer. Worms usually hold their bodies or tails coiled or curved over edge of leaves.............................. ....*Sawflies* (Order Hymenoptera, Family Tenthredinidae).

B. *Motionless scale insects on twigs, trunks, or leaves, sucking sap:*

1. The bark of shade trees, particularly elm and willow, with small grayish-white flat scales, often nearly covering the bark. From September to late spring, minute purplish eggs to the number of 50 or more will be found under each of these scales..*Scurfy scales*, page 707.

2. Surface of the bark of ash, poplar, and many other shade trees, covered with brownish-gray scales, which, upon close examination, will be found to resemble the half of a minute oyster shell. During the winter months, numerous pearly white, very small eggs to the number of 75 to 100 will be found packed beneath the larger of these scales......................*Oyster-shell scale*, page 707.

3. Small whitish scales on the needles or leaves of pine, spruce, hemlock, and various other evergreens. Often present in such numbers as to give the needles a grayish appearance. Very small, purplish eggs beneath these scales during the winter months.................................*Pine leaf scale*, page 708.

4. Underside of limbs and trunks of elm trees more or less completely covered with oval reddish-brown scales, up to ⅜ inch long, surrounded with a white fringe. Sticky honeydew over the trees and objects beneath, and many flies and wasps buzzing about....................*European elm scale*, page 708.

5. Whitish cottony-appearing masses of wax, nearly ⅓ inch across, on the undersides of the branches of maple, linden, and other shade trees. Such masses most conspicuous during June and July. Small, flat, brownish-appearing scales, adhering tightly to the bark of shade trees during the winter months...
....*Cottony maple scale*, page 709.

6. Plump, rounded, reddish-brown scales, about ⅛ inch across, sharply convex and oval in outline, clustered on the twigs and branches of various shade trees during the winter months. Scales often so thick as completely to cover the bark..*Terrapin scale*, page 709.

7. Very plump, almost hemispherical, brownish-gray hard scales up to ¼ inch or more in diameter, on the leaves and terminal twigs, especially of oaks of the burr oak group. Some of the twigs with leaves dead at the tips.............
............................*Oak Kermes* (see general references, page 687).

C. *Soft-bodied, more or less active, sap-sucking insects, on the leaves, branches, twigs, or trunks:*

1. Leaves of elm curled and bunched together in the spring, but no galls formed upon them. Numerous brownish aphids, with a purplish-white woolly wax covering their bodies, feeding within the curled leaves. Copious discharge of honeydew from such leaves...................*Woolly apple aphid*, page 711.

2. The trunks and undersides of the branches of pines and balsam become covered with white cottony flecks. The flocculent tufts of wax cover dark-brown aphids, that suck the sap and cause a sickly condition of the trees.....
.............................*Pine-bark aphid, Chermes pinicorticis* Fitch.

3. Large numbers of bright-red bugs, ½ inch or less in length, the larger ones with dark wings bordered with red and with 3 red lines on the thorax, suck sap from leaves and new growth of boxelder or ash. In the fall the bugs cluster in masses on the trunks or wander in great numbers up and down the tree trunks or crawl over walls and porches and enter houses.......*Boxelder bug*, page 711.

D. *Borers working in the trunks and branches:*

1. Shallow mines in the bark and sapwood of the trunk and larger branches of shade trees, these mines generally occurring on the south and southwest sides of the trees. Grayish-white grubs, with a pronounced flattened enlargement of the body just back of the head, working in these burrows. Adults are flattened beetles with the body tapering back from the shoulders, the antennae short, and their backs often irregularly roughened and metallic-colored.......
.........................*Flatheaded borers* or *metallic wood borers*, page 712.

   a. Birch trees with the leaves on the upper branches dying. Brown spots on the bark, and the inner bark of branches and trunk with many zigzag sawdust-filled burrows running through it. White slender grubs, up to 1 inch in length, in these burrows. Entire trees dying after 1 or 2 seasons.......
   ....*Bronze birch borer*, page 712.

2. Injury similar to *D*, 1, but mines more often extend through the solid wood of the tree, and the enlargement of the front end of the body is not flattened. Adults are generally more or less cylindrical beetles, with very long antennae, and often beautifully colored...
.......................*Long-horned borers* or *roundheaded borers*, page 713.

   a. Large amounts of excelsior-like sawdust accumulating about the base of poplar, particularly Carolina poplars, aspen, and cottonwood. Ragged, usually dark-colored holes in the trunk and branches, from which dark-colored sap is seeping and coarse sawdust is being forced out. Branches breaking, disclosing large burrows running through the wood, and greatly weakening the structure. Injury most severe on Carolina poplar, Lombardy poplar, and cottonwood...................*Poplar borer*, page 714.

   b. Black locust trees with swollen areas on the trunk and larger branches, the bark on these areas often cracking open, or trees breaking over at the point of injury. Wood of the trees honeycombed with burrows of rather large borers...*Locust borer*, page 715.

   c. Elms in a weakened, sickly condition with areas of the bark along the trunk becoming loose and easily detached from the tree. Such bark containing numerous burrows filled with brown sawdust. Yellowish-white grubs, up to a little over 1 inch in length, will be found working in these burrows. Injury occurring only on weakened or sickly trees............*Elm borer*, page 717

3. Terminals, or leaders, of pines, fir, or spruce turn brown and die, from mid-summer on, on account of the work of fat, white, legless grubs that tunnel through the wood and make "shot holes" through the bark of the twigs, which exude small drops of resin. Eggs laid by a reddish-brown snout beetle, ¼ to ⅓ inch long, irregularly blotched with white on the back...................
   White pine weevil, Pissodes strobi (Peck) (see *N.Y. (Cornell) Agr. Exp. Sta. Bul.* 449, 1926; *U.S.D.A., Cir.* 221, 1932).

4. Branches and limbs of willows or poplars in northeastern United States are bored with tunnels, deformed with knotty swellings, and splitting and breaking, with much sawdust and sap oozing at points of attack, and foliage wilted. Fat, white, legless grubs, ½ inch long or less, may be found tunneling in the trees from fall to midsummer. Eggs laid by blackish, white-flecked, chunky, snout beetles, about ⅓ inch long, with the rear third of the wing covers, the sides of the thorax and parts of the legs pink.....................................
   Mottled poplar and willow borer, Cryptorhynchus lapathi (Linné) (see *N.Y. (Cornell) Agr. Exp. Sta. Bul.* 388, 1917).

5. Fine sawdust deposited in small amounts over the trunk and branches of deciduous trees. Small, round holes in the bark, a little larger than a pinhead. Very small, blunt-headed black beetles crawling over the bark or boring through the bark. Numerous galleries in the inner bark and outer sapwood, branching out from a short parent gallery.......................................
   Elm bark beetles, page 720; *fruit tree bark beetle*, page 592; *hickory bark beetle*, page 719; *peach bark beetle*, page 627; and others.

6. Swarms of very small, chunky, cylindrical, brown or black beetles, usually ⅛ to ¼ inch long, attack the trunks of various conifers or evergreen trees, eating many holes into the bark and making galleries in the inner bark and outer sapwood. Small white grubs make similar tunnels radiating from the parent gallery. Sawdust sifts down, and pitch or resin accumulates, at the entrance holes. Affected trees show reddish tops and soon die..............
   ....*Bark beetles* and *ambrosia beetles*, page 718.

7. Large dark-colored burrows in the wood of locust, willow, chestnut, maple, and some other shade trees, but particularly abundant in black and red oaks. Dark-colored sap oozing from such burrows and discoloring the bark for some distance below the burrow. Large, pinkish, brown-headed borers, up to 3 inches in length, with well-developed prolegs, working in these burrows.............
   ....*Carpenter-worm*, page 722.

8. Shade trees along the eastern coast of the United States, especially elm and maple, with many dead branches in the tops of the trees, or with small branches broken over and hanging partly severed. Numerous holes along the infested branches from which sawdust is being forced out. Whitish brown-spotted caterpillars, up to 3 inches long, in these burrows.....*Leopard moth*, page 723.

9. The canes of lilacs dying or breaking over. Slightly enlarged, swollen areas on the canes, just above the surface of the ground. Bark cracked from infested canes, showing the presence of numerous burrows through the wood.........
   ....*Lilac borer*, page 725.

E. *Insects causing the development of galls on leaves and twigs:*
   1. Leaves of elm with raised, much wrinkled, greenish or pinkish galls on the upper surface. Such galls when broken open, found packed almost solid with greenish or brownish aphids. Small slits on the undersides of the leaves below the galls.....................................*Elm cockscomb gall*, page 725.
   2. See also Felt, "Key to American Insect Galls," *N.Y. State Mus. Bul.* 200, 1917.

F. *Twigs girdled, severed, or splintered so that they die and hang broken over or fall to the ground:*
   1. Many twigs, ⅓ to ¾ inch in diameter and several feet long, neatly cut off and lying on the ground beneath hickory, persimmon, oak, poplar, sour gum, honey

locust, and other deciduous shade, fruit, and nut trees.   Severed end of twig is convex and no borers in the new fallen twigs.   Grayish cylindrical hard-shelled beetles, nearly 1 inch long, with antennae longer than the body, girdle the twigs by cutting round and round them from the bark inward....................

..........*Twig girdler, Oncideres cingulata* Say, and other species (Fig. 5,*D*).

2. Twigs severed as in *F*, 1, from oak, maple, hickory, chestnut, locust, and other shade and fruit trees.   Severed end of twig is concave and with a central burrow leading from it up the twig, which is plugged with shavings and contains a white, cylindrical, conspicuously segmented grub, which cut off the twig from the inside........*Maple and oak twig pruner, Ellaphidion villosum* Fabricius.

3. Tips of small branches breaking off or hanging with dead leaves during the early summer, wood split at point of break, with tufts of splinters sticking up at short intervals along the twig.......................*Periodical cicada*, page 607.

228. Elm Leaf Beetle[1]

Importance and Type of Injury.—This beetle is one of the most destructive pests of the elm tree throughout the eastern United States.

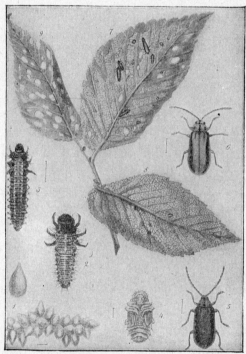

Fig. 456.—The elm leaf beetle, *Galerucella xanthomelaena* (Schrank).   *1*, egg mass, *2*, young larva, *3*, full-grown larva, *4*, pupa, *5* and *6* adult beetles, about twice natural size; *7*, *8* and *9*, injured leaves, reduced.   (*From Felt's "Manual of Tree and Shrub Insects," copyright, 1924, Macmillan.   Reprinted by permission.*)

Infested trees have a general yellow appearance of the foliage, with many leaves skeletonized.   Yellowish to dull-green beetles, about ¼ inch long, with an indistinct black stripe along each side (Fig. 456,*5,6*), or small yellow to black larvae (*2,3*) will be found skeletonizing the leaves or

[1] *Galerucella xanthomelaena* (Schrank), Order Coleoptera, Family Chrysomelidae.

crawling about on the bark of the trunk, sometimes clustered in great numbers about the base of the trunk.

Trees Attacked.—The different species of elm, the English elm and camperdown elm being most subject to attack.

Distribution.—The elm leaf beetle is of European origin and was probably brought into this country some time about 1834. It is now established over most of the eastern United States and has been found as far west as Indiana and Kentucky. It has never been taken in Illinois, Iowa, Missouri, or the other midwestern states but is destructive in Idaho, Washington, Oregon, and California.

Life History, Appearance, and Habits.—The adult beetles go through the winter hidden away in sheltered places which will afford them some protection from the weather, often in buildings, where they may be a nuisance in the fall or during warm winter weather. They are about ¼ inch long, of a light-yellow to brownish-green color, with several black spots on the head and thorax, and a somewhat indefinite, black or slate-colored stripe on the outer margin of each wing cover. The beetles fly to the elm trees shortly after they come into foliage in the spring and deposit double rows of yellowish eggs resembling minute lemons (Fig. 456,*1*) on the undersides of the elm leaves, usually about 25 in a place. The slug-like larvae hatching from these eggs are yellow in color, spotted and striped with black. They feed for about 3 weeks and, when full-grown, are ½ inch in length. They then crawl down the trunk of the tree, gathering in large masses about the base of the tree or in any shelter near by. Here they pupate, emerging as beetles in from 1 to 2 weeks. There are two, three, or more generations a year, depending on the locality.

Control Measures.—The elm leaf beetle may be held in check so that no damage from it will occur, if the trees are thoroughly sprayed with lead arsenate, 6 pounds per 100 gallons of water, with casein, lime, or soybean flour as a sticker. This should be applied when the leaves are nearly full-grown or as soon as feeding is noticed. The first spray should be applied about a week after the elms have come into full foliage. If only a part of the trees in the neighborhood have been sprayed, it will be necessary to make another application for the second generation of beetles, putting this application on about midsummer. Many of the larvae may be killed about the base of the tree, when they have come down to pupate, by spraying with derris, cubé, timbo, or other good contact sprays or even ordinary laundry soap, ¼ pound to the gallon.

References.—*Conn. Agr. Exp. Sta. Bul.* 155, 1907; *N.Y.* (*Cornell*) *State Museum Bul.* 156, 1912; *Ore. Agr. Exp. Sta. Cir.* 92, 1920; *N.Y.* (*Cornell*) *Agr. Exp. Sta. Bul.* 333, 1933; *Proc. Nat. Shade Tree Conf.*, **7**: 66–73, 1931.

229. SPRUCE BUDWORM[1]

Importance and Type of Injury.—In a list of the 20 most destructive insects in the United States, compiled by a group of federal entomologists a few years ago, this insect was given third place, being outranked only by the cotton boll weevil and the corn earworm. It is said to have destroyed 200,000,000 cords of balsam fir and red spruce in the forests of

---

[1] *Cacoecia* (= *Archips*) *fumiferana* (Clemens), Order Lepidoptera, Family Tortricidae.

the eastern United States and Canada between 1910 and 1925. The damage is all done by the caterpillars feeding on the foliage of the terminal shoots, which are webbed together to form shelters in which the larvae pupate and hibernate. The tops of the trees first appear as though scorched by fire and, if heavily infested, the trees may die over large areas.

Trees Attacked.—Firs, spruces, larch, hemlock, and pines.

Distribution.—The insect occurs over the coniferous forests of the northern United States and southern Canada.

Life History, Appearance, and Habits.—Winter is passed as tiny caterpillars in very small silken cases attached in crevices on the twigs near the buds. As the balsam buds burst in the spring the caterpillars emerge and feed for 3 to 5 weeks on the tender needles, becoming full-grown in June or July. The mature caterpillars are about 1 inch long, dark brown, with a yellowish stripe along each side and covered with yellowish tubercles. They pupate for 1 week or 10 days within loose cocoons of silk among the damaged foliage and then appear as moths from early July to early August. The moths which are reddish brown to yellowish gray, with the forewings splotched with a number of gold-enbrown spots and expanding about 1 inch, lay their greenish scale-like eggs in overlapping masses of about a dozen, along the underside of the balsam and spruce needles. After 10 days as eggs, the caterpillars hatch but feed only a short time before going into hibernation. There is only one generation a year.

Control.—The insect can be controlled on shade trees and ornamentals by spraying with lead arsenate, 3 pounds to 100 gallons of water, just after the buds open. Under forest conditions airplane dusting with calcium arsenate is effective. It is recommended that the oldest stands of fir should be cut first, since fast-growing trees are less severely injured.

References.—*Can. Dept. Agr., Div. Forest Insects, Spcl. Cir.*, 1931; *Maine Agr. Exp. Sta. Bul.* 210, 1913; *Can. Dept. Agr. Tech. Bul.* 37 (n.s.), 1924.

230. BAGWORM[1]

Importance and Type of Injury.—Infested trees have the foliage stripped or very much ragged. The stripped trees usually die. Numerous spindle-shaped sacks or bags (Fig. 457) from ¼ to 1½ inches in length hang down from the twigs, leaves, branches, and sometimes on the bark of the trunk. During the summer these bags contain dark-brown shiny-bodied worms.

Trees Attacked.—This insect is a very general feeder, attacking practically all deciduous and evergreen trees.

Distribution.—The bagworm is found from the Atlantic states to the Mississippi Valley. Related species extend westward and southward to Texas.

Life History, Appearance, and Habits.—The winter is passed as pale, whitish eggs, enclosed by the pupal skin inside the bag in which the female worm lived during the summer. These eggs hatch rather late in the spring, after the trees have come into full foliage. In the latitude of central Illinois they hatch about June 10 to 15. The young worms, on hatching, almost immediately spin a silken sack or bag about themselves and then begin feeding on the foliage. As they feed, they attach to the

[1] *Thyridopteryx ephemeraeformis* Haworth, Order Lepidoptera, Family Psychidae.

bag bits of the leaves on which they are feeding. The bag is carried about by the insect wherever it goes, the larva merely protruding the front end of its body from the bag. It is almost impossible to draw the larva out of the bag without crushing its body. This bag offers almost complete protection from birds, but the worms are parasitized by several species of flies and wasp-like parasites. The worms are of a brown color over the head and thorax, the parts enclosed by the bag being lighter and softer. When full-grown, the insect measures about 1 to 1¼ inches in length, the bags at that time being from 1½ to 2½ inches in length. They pupate within the bags during early September, attaching the bag to the twigs with numerous threads of silk, forming a strong thick loop (Fig. 457). The male changes to a black-winged moth, with a wing expanse of about

FIG. 457.—Cases of the bagworm, *Thyridopteryx ephemeraeformis* Haworth, fastened to twig of cedar, as found in winter. Natural size. (*From Ill. State Natural History Surv.*)

1 inch. These moths emerge from the bags and fly actively. The female moths, however, are wingless and merely protrude their abdomens from the tip of the bag during mating and then retreat within the bag, where they deposit their eggs and shortly afterward wriggle out of the bag, drop to the ground, and die.

Control Measures.—On small trees, or those that can be readily reached, a thorough cleanup of the bags during the winter by hand-picking and burning will effectively control the insects. This method is one of the best to use when infestations are first starting, but for a general infestation over shade trees in a neighborhood, the most effective control is spraying. Lead arsenate, 5 pounds, soybean flour, 4 ounces, and water, 100 gallons, makes a very effective spray, if applied when the caterpillars are not over half-grown. This should be applied during the latter half of June in the latitude of central Illinois. For spraying later in the season, the lead arsenate should be increased to 8 pounds. A weaker spray is not effective, as these insects are very hard to poison. Where the bags are collected, care should be taken that they are removed from the trees and burned for, if they are merely thrown about the ground, the eggs will hatch, and many of the worms will find their way to the trees.

References.—Mo. Agr. Exp. Sta. Bul. 104, 1912; *N.J. Dept. Agr. Cir.* 243, 1934.

231. CATALPA SPHINX[1]

Importance and Type of Injury.—This is one of the most serious insect pests of the catalpa tree.   Infested trees will have the leaves eaten off by

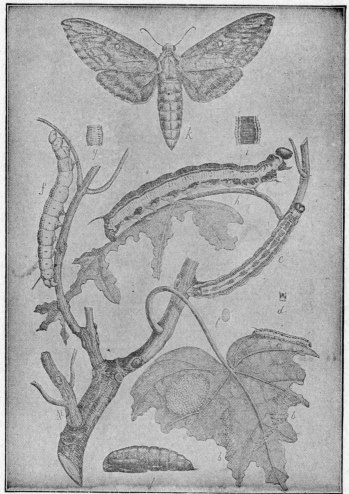

FIG. 458.—Catalpa sphinx, *Ceratomia catalpae* Boisduval.   Above at *k*, adult female moth; at center (*e, f, h*) three caterpillars stripping leaves; below at *a*, an egg mass on leaf; and at *j* the pupa; *b* and *c*, partly grown larvae; *l*, a single egg enlarged; *d, g*, and *i* show the variation in color pattern of single segments of larvae from above.   Slightly reduced. (*From U.S.D.A.*)

dark or black caterpillars from 1 to 3 inches in length with dark-green markings on their bodies, and a sharp horn at the tip of the abdomen (Fig. 458).   The markings vary greatly in different individuals.

[1] *Ceratomia catalpae* Boisduval, Order Lepidoptera, Family Sphingidae.

Trees Attacked.—Catalpa.

Distribution.—This insect is widely distributed in the United States and is of greatest importance in a zone from New York to Colorado.

Life History, Appearance, and Habits.—The winter is passed as brown naked pupae in the soil, 2 or 3 inches below the surface. These pupae (Fig. 458,*j*) will be found under, and in close proximity to, catalpa trees. The moths (*k*) emerge shortly after the catalpas have come into full leaf, and deposit their white eggs in masses on the undersides of the leaves. Sometimes as many as 1,000 eggs have been found in a single mass. The moths are of a general gray color with a wing expanse of from 2½ to 3 inches. They fly mainly at night and are seldom seen. The eggs hatch in 10 days to 2 weeks, and the young caterpillars at once begin feeding on the foliage. At first, they feed in groups (*b,b*) but later separately. The full-grown caterpillars are about 3 inches in length, with a moderately large black horn protruding from the tip of the body. The backs of the worms are often almost completely covered with the whitish cocoons of a wasp-like parasite.[1] Upon becoming full-grown, they go down the trunk of the tree, enter the ground, and there change to the pupal stage. There are two generations of the insect each season, at 40° north latitude, the second generation of worms appearing on the trees during late August and early September.

Control Measures.—While these worms are large and very ravenous feeders, they may be very easily controlled by spraying or dusting. Spraying with lead arsenate, at the rate of 4 pounds to 100 gallons of water, applying the spray as soon as any of the caterpillars are noticed on the leaves, will afford almost complete protection. Dusting with lead arsenate, or calcium arsenate, at the rate of 8 to 10 pounds per acre is also very effective. The insects are heavily parasitized, and severe outbreaks usually last for only 1 or 2 seasons, with a break of about the same period before the worms will again become numerous enough to strip the trees completely. If discovered while small, on small trees, hundreds of them can be killed by burning single leaves.

Reference.—*Ohio. Agr. Exp. Sta. Bul.* 332, pp. 238–241, 1918.

232. WALNUT CATERPILLAR[2]

Importance and Type of Injury.—

FIG. 459.—Walnut caterpillar, egg mass, about twice natural size. (*From Ohio Agr. Exp. Sta. Bul.* 332.)

Walnut, hickory, and other trees, have the branches or entire trees stripped of their leaves during July and August. Masses of large dark-bodied white-haired caterpillars cluster on the trunk of the tree (Fig. 460) or feed together on the leaves.

---

[1] *Apanteles congregatus* Say, Order Hymenoptera, Family Braconidae.

[2] *Datana integerrima* Grote and Robinson, Order Lepidoptera, Family Notodontidae.

Trees Attacked.—Walnut, including English and Japanese walnuts, butternut, pecan, hickory, and occasionally peach, willow, honey locust, beech, sumac, apple, and oak.

Distribution.—The insect is common throughout the eastern and southern United States and westward to Kansas.

Life History, Appearance, and Habits.—The winter is passed in the form of a brown naked pupa about 1 inch or a little over in length. The pupae will be found from 2 to 6 inches beneath the surface of the soil in the vicinity of the trees on which the insect feeds. The adult moths emerge from these pupae during late June and July. They measure 1½ to nearly

Fig. 460.—A cluster of walnut caterpillars, *Datana integerrima* Grote and Robinson, molting on the trunk of a walnut tree. (*From Ohio Agr. Exp. Sta. Bul.* 332.)

2 inches across the expanded wings. The wings are of a general light brown, with dark-brown wavy lines running across them. The hind wings are lighter brown without the crosslines. A dark-brown tuft of hair covers the back of the thorax. The moths are strong fliers and deposit their eggs in masses of 200 to 300 (Fig. 459) on the underside of the leaves of their food plants. These hatch in about 2 weeks into small reddish white-striped worms with black heads. As they grow, the color of the body changes to brown and later, in the full-grown stage, to black. They are covered with rather soft, long, frowzy white hairs. The full-grown caterpillar is 2 inches or a little over in length. They feed together, several hundred in a place, and a single colony will often strip one or two branches on the tree. As the worms grow, they have a peculiar habit of coming down the tree to change their skins. The entire colony crawls

down the trunk at the same time, so that one will often find masses of worms, from 4 to 8 inches across, clinging to the trunk of the tree (Fig. 460). The shed skins remain on the tree trunk, having much the appearance of dead worms. On becoming full-grown, the caterpillars leave the tree, crawl away for a short distance, and enter the ground, where they later change into the brown pupal stage. There is but a single generation over most of the range of the insect, although possibly two in the southern areas where it occurs.

Control Measures.—The most effective method of controlling this insect is to spray the infested trees with lead arsenate at the rate of 4 pounds to 100 gallons of water. This spray should be applied as soon as the caterpillars are seen feeding upon the foliage. On small trees the colonies may be removed before much of the foliage has been destroyed, using a pole pruner, and cutting off and burning the few leaves on which the caterpillars are starting to feed. If the trees are watched every day, it is possible to crush many of the caterpillars when they come down the trunk to shed their skins.

Spraying, however, is the only really effective remedy that can be depended upon to protect trees from all damage by these insects. Banding trees with sticky material is of no benefit, as the moths do not crawl up the trunk to deposit their eggs.

References.—*U.S.D.A., Farmers' Bul.* 1169, 1921; *N.Y. State Museum Memoir* 8, **1**: 303, 1905; *Ark. Agr. Exp. Sta. Bul.* 224, 1928.

233. FOREST TENT CATERPILLAR[1]

This "forest armyworm" is a serious defoliator of many shade and forest trees, and sometimes, as it migrates by millions to new food plants, it swarms over and into cabins and other buildings to the annoyance and dismay of farmers and tourists, who may be driven away from the woods. It appears to prefer aspen or poplar, especially where they are growing in pure stands, but it is seriously destructive to oaks and maples and feeds on basswood, ash, elm, birch, conifers, and, during its migrations, upon field and vegetable crops.

It is a close relative of the tent caterpillar (page 563) and has a similar life history, but it makes no tent. Eggs are laid in a similar manner and carry the insect over winter, hatching the first half of May. The pale-blue larvae, which have a row of keyhole-shaped, white spots down the middle of the back and pale-yellow stripes on the sides, are troublesome from mid-May to the end of June. After 10 to 12 days in cocoons, the adults appear in swarms about mid-July. Spraying with lead arsenate, 3 pounds to 100 gallons of water, at a cost of about $15 per acre gave good control over areas accessible to power sprayers. Spraying protective zones 100 to 300 feet wide around farm premises and summer resorts effectively protected them.

234. WHITE-MARKED TUSSOCK MOTH[2]

Importance and Type of Injury.—This insect is usually considered more of a shade tree than orchard pest, but sometimes, especially in the

---

[1] *Malacosoma disstria* Hübner, Order Lepidoptera, Family Lasiocampidae.

[2] *Hemerocampa leucostigma* Smith and Abbott, Order Lepidoptera, Family Lymantriidae.

North, it becomes destructive in orchards. The foliage is skeletonized by yellowish-black, hairy, striped caterpillars, up to 1¾ inches long. They are easily recognized by the three pencil-like tufts of long black hairs that project, one on each side of the red head and the third from the tail; by the four tufts of short, white, erect hairs on the back; and by the two bright-red spots on the back toward the hind end. Fruits are sometimes scarred by the shallow feeding of the caterpillars on the surface.

Trees Attacked.—Apple, pear, quince, plum, and other deciduous fruits and almost all shade trees except conifers. Especially destructive to shade trees in cities.

Distribution.—Eastern United States and Canada and westward to British Columbia and Colorado. Less troublesome in the South.

Life History, Appearance, and Habits.—The winter is passed in the egg stage. The eggs are laid in conspicuous masses, nearly an inch long, and 50 to 100 eggs each (Fig. 463), attached to the trunk, branches, or dead leaves of the tree or near-by objects, usually on top of the dirty-grayish cocoon from which the female moth emerged. The eggs are covered with a mass of white, stiff substance, having the appearance of hardened lather. They hatch in the late spring

Fig. 461.—Larvae of the white-marked tussock moth, *Hemerocampa leucostigma* Smith and Abbott, slightly enlarged. (*From U.S.D.A., Farmers' Bul.* 1270.)

into light-brown hairy caterpillars, marked as described above (Fig. 461), which feed on the surface of the leaves, skeletonizing them. They become full-grown during July, spin their cocoons on the trunk and branches, and within these transform to the pupal, and later to the adult, moth stage. As is the case with cankerworms, the male moths, which are gray, with dark wavy bands across the front wings, are well provided with wings and are strong fliers, while the wings of the female are mere stubs and cannot be used at all for flight (Fig. 462). Except in the most northern states the moths emerging during midsummer soon deposit eggs for a second generation of caterpillars, which feed upon the trees during the latter part of August and early September. They transform to moths again during September and October. The females lay eggs on their cocoons, and these carry the insect through the succeeding winter. In the more southern part of its range the tussock moth may produce a third generation.

Control Measures.—On shade trees it is often possible to hold this insect in check by daubing the overwintering egg masses with creosote, as described for the gypsy moth (page 704). On fruit trees an application

Fig. 462.—The tussock moth, *Hemerocampa leucostigma* Smith and Abbott, adult female clinging to the cocoon from which she has emerged. Natural size. (*From Ohio Agr. Exp. Sta. Bul. 332.*)

Fig. 463.—Eggs of the white-marked tussock moth as laid by the wingless female on top of her cocoon. (*From Ill. State Natural History Surv.*)

of lead arsenate, at the rate of 2 pounds to 100 gallons of water, should be given as soon as the caterpillars appear in numbers upon the trees. Outbreaks of the tussock moth, while often causing considerable damage for 1 or 2 seasons, are usually short-lived because of the presence of numerous insect parasites that always reduce outbreaks.

Reference.—N.Y. (Geneva) Agr. Exp. Sta. Bul. 312, 1909.

235. GYPSY MOTH[1]

Importance and Type of Injury.—This imported insect is one of the most serious pests of shade trees, both evergreen and deciduous trees, stripping the foliage and often causing the death of the trees. Millions of dollars have been spent in fighting this pest in the New England states.

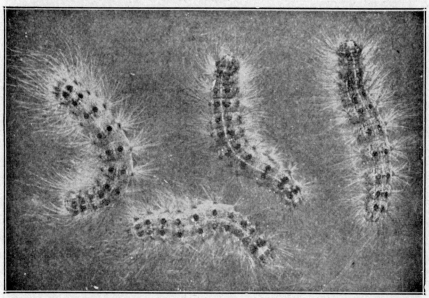

Fig. 464.—Full-grown larvae of the gypsy moth, *Porthetria dispar* (Linné). The tubercles on the back of the segments are blue on the front half of the body, and red on the rear half. Natural size. (*From U.S.D.A.*)

Shade, fruit, and woodland trees are stripped of their leaves during May and June by small flattened pale-brown caterpillars with long tufts of rather stiff brown and yellow hairs projecting from the sides of the body. The caterpillars (Fig. 464) are about 2 inches long when full-grown. There are five pairs of blue tubercles followed by six pairs of red tubercles, arranged in two rows down the back.

Plants Attacked.—This caterpillar feeds on nearly all deciduous and evergreen trees and shrubs. It also attacks the foliage of some garden plants and is frequently a very serious pest of cranberries. More than 500 different species of plants are included in the list of those fed upon by this insect.

Distribution.—So far as known, the insect is confined to the New England states, a small area in southeastern Canada, and small outlying

[1] *Porthetria dispar* (Linné), Order Lepidoptera, Family Lymantriidae.

infestations in New York, Pennsylvania, and New Jersey. It was originally brought into this country from Europe about 1869.

Life History, Appearance, and Habits.—The winter is passed in the egg stage. These eggs are laid in masses up to an inch long and averaging about 400 eggs. They are covered with a coating of hair and are about the color of chamois skin. In the latitude of Boston the eggs hatch during late April and early May. The young caterpillars are very voracious feeders. During the first two instars, they do not feed on evergreens but, as they grow larger, readily attack them. They become full-grown during the first half of July and spin a very loose light cocoon on the trunks of trees and other near-by objects. Within this they change to a dark-brown pupal stage. The moths begin emerging during the latter part of

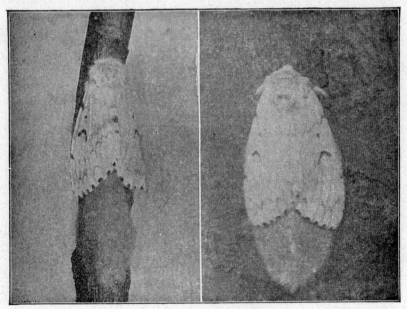

FIG. 465.—Adult females of the gypsy moth, *Porthetria dispar* (Linné), depositing their egg-masses. Enlarged about ½. (*From U.S.D.A.*)

July. The males are dark brown in color, with small bodies well equipped with wings, and are strong fliers. The female moth (Fig. 465) is a light-buff color, with irregular darker markings across the wings. She is a very heavy-bodied insect and is able only to flutter along the ground but cannot travel by sustained flight through the air. After mating, she deposits her eggs on various objects, frequently on the undersides of stones, on buildings, on the trunks of trees, or in many other places. Eggs are laid mostly in July and do not hatch until the following May. There is but one generation of the insect each year. Since the females are incapable of flight, most of the spread of the insect occurs when trees, lumber, stones, freight cars, automobiles, and other objects upon which egg masses have been deposited are shipped or moved to new locations, and when larvae spin down on silken threads and are transported by man or are blown by the wind, during the first and second instars of their

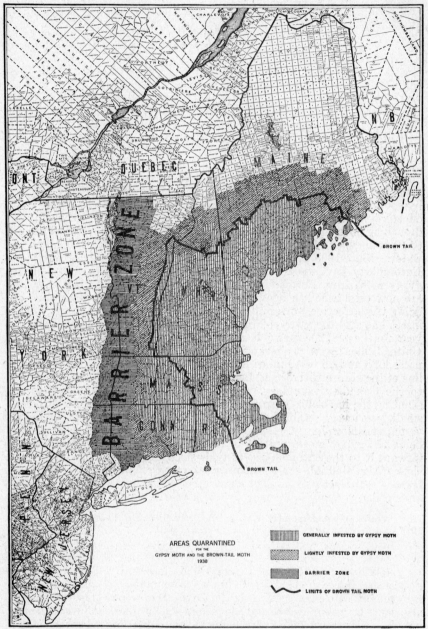

Fig. 466.—Map showing distribution of the gypsy and brown-tail moths in the New England states. Vertical shading indicates area generally infested by the gypsy moth at the end of 1938. Diagonal shading indicates area lightly infested by the gypsy moth at the end of 1938. The heavy black line shows the limits of brown-tail moth infested area at the end of 1938. The horizontal shading shows the location of the barrier zone where clean-up operations, to prevent a nation-wide spread of the gypsy moth to the westward, have been centered. (*After U.S.D.A., Bur. Ento. & P.Q.*)

existence. At this time they are carried long distances, having drifted up to 30 miles across Cape Cod Bay and to altitudes of over a third of a mile. The young larvae are equipped with hollow hairs, which greatly increase their buoyancy and enable them to drift with the wind as above described. The larvae are full-grown by late June, they pupate for about 2 weeks, and the moths lay their eggs during July to August. The eggs will not hatch until they have been exposed to low temperatures.

Control.—There are several effective control measures to use against this insect. During the winter, the egg masses may be killed by touching them with brushes or sponges wet with coal tar creosote. The larvae feed at night, coming down the trunk of the trees and hiding away in some sheltered place during the day. In the infested areas, bands of burlap are placed around the trees; these are examined each day, and the larvae which have sought shelter under them are killed. Bands of tree *tanglefoot* placed around the trunks of trees and freshened every 10 days by combing them, keep the larvae from ascending the trees. The most effective method of control is by spraying, using a high-power outfit and drenching both shade and woodland trees with a spray of lead arsenate, 5 to 10 pounds, and fish oil or raw linseed oil, 1 pint, to each 100 gallons of water. Since 1905, the state of Massachusetts and the federal Bureau of Entomology have been actively engaged in importing parasites of the gypsy moth from Europe and Asia. A large number of these parasites are now established in heavily infested areas in New England and are doing very effective work in keeping down this destructive insect. This insect has cost the state of Massachusetts more than a million dollars a year, for the past 20 years. It has been confined to the eastern part of the United States by strict quarantine measures. In 1923 a barrier zone, 250 miles long and 25 to 30 miles wide (Fig. 466), was established along the line of the Hudson River and Lake Champlain, from near New York city to the Canadian boundary. By cleanup operations and very careful scouting the insect has been prevented from crossing this zone, although small, previously established infestations occur in New York, Pennsylvania, and New Jersey. Anyone seeing an insect which they suspect of being the gypsy moth, outside of the known area of infestation, should forward it to the entomologist of their state at once, as it is possible to clean up isolated infestations at no great expense, but if the insect should become established over the country, it would certainly cause losses amounting to many millions of dollars a year.

References.—*Can. Dept. Agr. Bul.* 63 (n.s.), 1926; *U.S.D.A., Farmers' Bul.* 564, 1913, and *Cir.* 464, 1938, and *Bur. Ento. Bul.* 87, 1910.

236. Brown-tail Moth[1]

Importance and Type of Injury.—This insect is a very important pest of deciduous shade and fruit trees. The damage is caused by dark-brown hairy caterpillars stripping the leaves from trees. The caterpillars also cause great annoyance and sometimes serious illness to human beings because their hairy bodies are equipped with nettling hairs, which on entering the skin cause a rash that is very irritating. There have been some instances where the insect probably caused death from large numbers of the hairs being breathed into the lungs.

[1] *Nygmia phaeorrhoea* (Donovan), Order Lepidoptera, Family Lymantriidae.

Plants Attacked.—Apple, pear, cherry, oak, willow, and other deciduous trees and shrubs; rare on buckeye, ash, hickory, chestnut; never on evergreens.

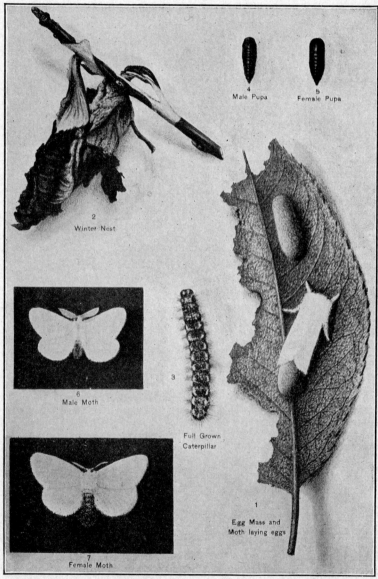

Fig. 467.—The brown-tail moth, *Nygmia phaeorrhoea* (Donovan), showing the various stages of the insect and a winter nest; about natural size. (*From Mass. State Forester.*)

Distribution.—The insect is a native of Europe and was introduced into eastern Massachusetts on imported nursery stock about 1897. Its known range is indicated by the heavy line in Fig. 466.   It has also been found in Nova Scotia and New Brunswick.

Life History, Appearance, and Habits.—The insect passes the winter in the form of very tiny caterpillars, living together in colonies of 25 to 500.   Each colony webs together several leaves and attaches them firmly to twigs by threads of very tough silk.   These winter nests (Fig. 467,*2*) are rather conspicuous on the trees.   The caterpillars are so well protected within these nests that they can withstand very low temperatures.   As soon as the leaves start coming out in the spring, the little caterpillars become active and crawl out from the nest to feed on the tender foliage.   They go back within the nest at night, for a time, but as they become larger remain on the foliage.   In the latitude of Boston they become full-grown about the last of June.   The full-grown caterpillars (Fig. 467,*3*) are about 1½ inches in length, dark brown in color, with a broken white stripe on each side of the body and a bright-red tubercle on the back of the eleventh and of the twelfth body segment.   They seek some sheltered place where they transform to the pupal stage (*4,5*) and remain in this stage for about 2 weeks; they then emerge as medium-sized moths during July.   These moths (Fig. 467,*6,7*) have a wing expanse of a little over 1½ inches.   The wings and thorax are pure white.   The abdomen is mostly brown, with a very conspicuous tuft of chestnut-brown hairs at the tip.   These moths are strong fliers, flying in large swarms at night.   They are attracted to strong lights, and sometimes in localities where they are abundant, they literally cover the sides of electric light and telephone poles and buildings, giving them the appearance of being covered with snow.   The female moth deposits from 200 to 400 eggs on the underside of a leaf of the food plant, mostly in July.   These eggs are laid in masses (Fig. 467,*1*) and are covered with brown hairs, giving them a dark chestnut-brown color.   The eggs hatch during August or early September, and the young caterpillars from each egg mass or from several near-by egg masses feed together for a short time on the terminal leaves of twigs and branches and then spin the web shelter in which they pass the winter.   The insect has been shipped to many parts of the world on nursery stock in these winter webs.   There is one generation a year.

Control.—The best and most effective method for controlling the brown-tail moth is to spray infested trees during the early spring when the caterpillars have started to feed or in early August as the new caterpillars begin hatching, using 6 pounds of lead arsenate in each 100 gallons of water.   In light infestations much can be done in keeping down the numbers of the insect by cutting off and burning the winter webs.   After a little practice these webs can be easily detected on the tips of the bare branches, and, with the aid of a long pole pruner and a ladder, nearly all the webs can be removed from the trees and burned.   With federal aid, about 24 million webs were cut and burned in the winter of 1933–1934 and 2 to 4 million in succeeding winters.   Since about 1915 natural enemies, especially a fungus disease[1] which kills the caterpillars, low winter temperatures, and applied control measures have brought about a remarkable decrease in the abundance of this insect.

References.—*Can. Dept. Agr. Bul.* 63 (n.s.), 1926; *U.S.D.A., Cir.* 464, 1938, and *Farmers' Bul.* 1335, 1923, and *Bur. Ento. Bul.* 87, 1910.

[1] *Entomophthora aulicae* Reich.

237 (150). Scurfy Scales[1]

Importance and Type of Injury.—These scales are often very injurious to smaller elm, willow, and dogwood. They are rarely a serious pest on large trees. Infested branches show small flattened dirty-white scales, about $\frac{1}{10}$ of an inch long, lying nearly flat on the bark. In the winter many reddish-purple eggs, just discernible with the naked eye, will be found beneath these scales. On heavily infested trees, the entire bark may be coated with the grayish scales.

The life history and control of these scales are similar to those of the apple scurfy scale given on page 578.

238 (151). Oyster-shell Scale[2]

Importance and Type of Injury.—Branches of trees or entire trees are dying, the bark cracking and having much the appearance of drying up on the branches. The bark is covered with small brownish-gray scales about $\frac{1}{8}$ inch long by $\frac{1}{16}$ inch wide, usually curved and closely resembling a minature oyster shell (see Fig. 366). The bark may be completely covered with these scales.

Trees Attacked.—All species of ash (with the exception of the blue ash), poplar, dogwood, elm, soft maple, linden, horse chestnut, lilac (with the exception of the white lilac), many species of rose, peonies, and many other shade trees and shrubs. The very closely related form attacking apple is described on page 579.

Distribution.—General throughout the United States.

Life History, Appearance, and Habits.—The winter is passed in the egg stage under the female scale. These eggs are elliptical, nearly white in color, and from 50 to 60 will be found under each female scale. They hatch late in the spring after the trees have come into full foliage. In the latitude of central Illinois, this is about June 1. The white six-legged young, just discernible to the naked eye, crawl about over the tree for a few hours and then insert their beaks into the bark and begin sucking the sap. They soon molt, and the females remain in this position for the rest of their lives. They grow rather rapidly, secreting the wax which forms into the brown protective scale over their bodies. About midsummer, or a little later, the males become full-grown and change, under their scales, to minute yellowish-white two-winged insects, which fly about for a short time, mate with the females, and die. After mating, the female deposits her eggs, her body gradually shrinking to the small end of the scale, where she finally dies. There is but one generation of this species of oyster-shell scale each year. Another kind of oyster-shell scale, somewhat closely resembling this one, occurs on certain species of dogwood. The dogwood form has two generations each year.

Control Measures.—These scale insects have several natural enemies that keep down their numbers. Certain mites feed on the eggs and may prevent damage by this scale for a number of years. A dormant spray of tar-distillate-oil emulsion, 5 per cent, plus dormant mineral-oil emulsion, 1 per cent, has given better than 98 per cent kill of the eggs. Mineral oil emulsion to which has been added 0.2 per cent dinitro-o-cyclohexylphenol

[1] Order Homoptera, Family Coccidae. The elm scurfy scale is *Chionaspis americana* Johnson; the one on dogwood is *C. corni* Cooley; the black-willow scale is *Chionaspis salicis-nigrae* (Walsh); and the one on hickory is *C. caryae* Cooley.

[2] *Lepidosaphes* sp., Order Homoptera, Family Coccidae.

has given about 99 per cent control under New York conditions. Spraying with 2 per cent summer-oil emulsion, just as the eggs are hatching (about June 1 in central Illinois) also gives satisfactory control. The planting of resistant shade trees, such as the blue ash, hackberry, hard maples, and oaks will keep down this scale, as it does not breed on these trees.

References.—*Ohio Agr. Exp. Sta. Cir.* 143, 1914; *Jour. Econ. Ento.*, **30**: 651–655, 1937.

239. PINE LEAF SCALE[1]

Importance and Type of Injury.—Infested trees have the foliage somewhat yellowed, with rather elongated, whitish scales up to $\frac{1}{8}$ inch in length, attached to the leaves. These white scales (Fig. 468), on the green leaves or needles of pine or other evergreens, often first attract notice to the presence of the insects.

Trees Attacked.—Pines, spruces, firs, cedars, and hemlocks.

Distribution.—Throughout the northern United States and southern Canada.

Life History, Appearance, and Habits.—The winter is passed in the form of very minute, purplish eggs underneath the gray parent scale. From 20 to 30 of these eggs will be found under each scale. The eggs hatch in midspring into crawling young, which move about for a short time and then settle down and secrete a scale about their bodies. They become full-grown by late summer, and a second generation is produced from eggs laid during August. These, in turn, become full-grown, and the females deposit eggs, by fall. In the latitude of Illinois and Ohio there are two generations annually, although probably only one farther north.

FIG. 468.—The pine leaf scale, *Chionaspis pinifoliae* (Fitch). One leaf shows the small male scales and the larger female scales. The width of the scales varies to some extent with the width of the leaf. About twice natural size. (*From Ohio Agr. Exp. Sta.*)

Control Measures.—Spraying just after the eggs hatch (about May 25 in central Illinois) with a summer-oil emulsion recommended for use on evergreens, at the strength recommended by the manufacturer, but not exceeding 2 per cent actual oil in most cases, is very effective in control. These sprays should be applied when the temperature is below 80°F. and the humidity high. Nicotine sulphate, 1 to 500, with a good spreader has given good control in Connecticut. A dormant fall or spring application of lime-sulphur, 1 to 9 of water, may be used, but oil sprays should not be applied in fall or winter or in very hot weather.

Reference.—*Ohio Agr. Exp. Sta. Bul.* 332, p. 291, 1918; *Jour. Econ. Ento.*, **24**: 115–119, 1931; *Conn. Agr. Exp. Sta. Bul.* 315, pp. 578–581, 1930.

240. EUROPEAN ELM SCALE[2]

Importance and Type of Injury.—This unarmored scale insect is often found in enormous numbers covering the underside of limbs and the trunks, weakening and killing the trees by sucking the sap, secreting sticky honeydew that gums and soots the leaves or smuts walks, benches, and grass, and attracting thousands of flies and wasps, making shade trees very unattractive.

---

[1] *Chionaspis pinifoliae* (Fitch), Order Homoptera, Family Coccidae.
[2] *Gossyparia spuria* (Modeer), Order Homoptera, Family Coccidae.

Trees Attacked.—All kinds of elms.

Life History, Appearance, and Habits.—The winter is passed as second-instar nymphs, motionless, about the color of the bark, and hidden in protected crevices on trunk or branches. The males form conspicuous white cocoons in very early spring in which they transform to minute, winged or wingless, reddish "gnats" in late April or May. The females also become adult early in May. They are from ⅙ to ⅜ inch long, oval in outline, reddish brown and surrounded by a white cottony fringe. From late May through June and July the females produce eggs which are deposited beneath them, like a hen sitting upon a nest. These eggs hatch within an hour and the lemon-yellow crawling nymphs swarm over twigs to the leaves, upon the underside of which they settle and feed until fall. Most of them migrate back to the limbs or trunk before the leaves fall. There is only one generation a year.

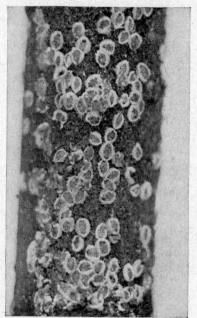

Fig. 469.—European elm scale, *Gossyparia spuria* (Modeer); scales of the female on twigs, twice natural size. Note the white fringe of secretion that surrounds the body of the insect. (*Original.*)

Control Measures.—Dormant sprays of 5 per cent miscible oil or 5 per cent oil and 5 per cent lime-sulphur combined are very effective. Summer white oil sprays, containing 2 per cent oil, with or without nicotine or derris extracts, applied during July and August are of value but less effective than the dormant sprays. Where spraying equipment cannot be procured, powerful streams of water from fire hydrants directed against the branches about the time the leaf buds are opening will dislodge great numbers of the females.

References.—U.S.D.A., Dept. Bul. 1223, 1924; Nev. Agr. Exp. Sta. Bul. 65, 1908.

241 (180). TERRAPIN SCALE[1]

Nearly hemispherical reddish-brown scales, about ⅛ inch across and very convex in outline, cluster on the bark of twigs and branches (see Fig. 405). The scale is generally somewhat mottled and streaked with black. The bark is often entirely covered for considerable distances along the branches.

Maple, elm, and sycamore are most severely injured. It also attacks osage orange, peach, plum, pear, quince, and some other shade trees. The distribution, life history, and control of this insect are discussed on page 625.

242. COTTONY MAPLE SCALE[2]

Importance and Type of Injury.—This is one of the most destructive scales on soft maple and also injures some other trees. Attention is

[1] *Eulecanium nigrofasciatum* Pergande, Order Homoptera, Family Coccidae.
[2] *Pulvinaria vitis* (Linné), Order Homoptera, Family Coccidae.

usually attracted to infested trees by the cottony-appearing masses of scale along the underside of the twigs and branches during May and June (Fig. 470). Branches of heavily infested trees die, and the foliage of the entire tree turns a sickly yellow. The reduction of the vigor of the tree by the scale often leads to attacks by bark beetles or other borers.

Plants Attacked.—Soft maple, linden, Norway maple, apple, pear, willow, poplar, grape, hackberry, sycamore, honey locust, beech, elm, plum, peach, gooseberry, Virginia creeper, currant, sumac, and some others.

Fig. 470.—The cottony maple scale, *Pulvinaria vitis* (Linné). The whitish masses, looking something like popped corn scattered along the twigs, are tufts of wax-like material secreted by the females and containing their eggs to the number of 1,500 or more per tuft. The brownish bodies of the females themselves stand somewhat upright at one end of the tufts they have secreted. (*Original.*)

Distribution.—The insect is distributed throughout the United States and Canada. It is most destructive in the northern part of the United States.

Life History, Appearance, and Habits.—The cottony maple scale passes the winter as a small, brown, flattened scale, a little less than ⅛ inch long, attached to the bark of twigs and small branches. These scales are all females. In the spring, as soon as the sap starts to flow, they grow very rapidly, and soon begin depositing their eggs, which are secreted in cotton-like masses of wax under the scale. This wax is secreted in such abundance that it forms a mass several times the size of the overwintering insect, and the body of the scale often becomes elevated at an angle from the twig. From 1,500 to 3,000 eggs are laid by each female. The eggs hatch during late June and July, in the latitude of central Illinois, and the young scales crawl from the twigs to the undersides of the leaves, where they suck the sap along the midrib or the veins. They become mature during August and September, mating takes place and the males die, the females crawling back to the twigs and small branches, where they pass the winter.

Control Measures.—Spraying during the early spring, just before the leaves put out on the maples, with any of the dependable miscible oils, diluted to give 2 per cent actual oil, or according to the manufacturer's directions, will give very good control. Care must be taken in spraying maples that the trees are dormant, since these trees, especially hard maples, are very sensitive to oil injury. Summer spraying is of little value.

References.—Twenty-sixth Rept. Ill. State Ento., p. 62, 1911; Ohio Agr. Exp. Sta. Bul. 332, 1918; Conn. Agr. Exp. Sta. Bul. 344, p. 127, 1934.

243 (167). WOOLLY APPLE APHID

This insect is described and the means of control are given under Apple Insects, page 605. It frequently becomes very abundant on elm where its feeding on the leaves in the spring causes them to curl and bunch together. No true galls are formed by this insect, so that it can be readily distinguished from the work of the cockscomb gall aphid.

OTHER APHIDS OF SHADE TREES AND SHRUBS

Nearly all shade trees and ornamental shrubs suffer to some extent from the attacks of aphids, especially *Spiraea vanhouttei*, roses of all varieties, snowball, poplar, sycamore, and maple. In general, the life history and control of these aphids are very similar to those described on pages 477 and 479.

244. BOXELDER BUG[1]

Importance and Type of Injury.—These strikingly marked, red-and-black sucking bugs (Fig. 471) often feed destructively upon the flowers, fruits, foliage, and tender twigs especially of boxelder and ash trees. They are however more important as a nuisance during fall and in warm days in winter when they swarm into houses or congregate in great numbers upon trunks of trees, porches, walls, and walks, often causing great alarm. They are not capable of biting or harming foods, clothing, or other household articles. The insects winter in buildings and other dry sheltered places in the adult stage. They are flat-backed rather narrow bugs, about ½ inch long, brownish black, with three longitudinal red stripes on the thorax and red veins on the wings. Nymphs of all sizes are bright red. There are two generations a year in warmer parts of the country.

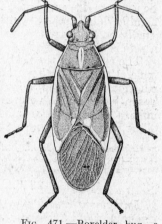

FIG. 471.—Boxelder bug, a pest of boxelder trees in summer and a household nuisance during fall, winter and spring. The light portions of the upper surface in the drawing are bright red. About 3½ times natural size. (*Drawn by Kathryn Sommerman.*)

Trees Attacked.—Boxelder, especially the flowers and fruit, is the favorite food. The insects also feed upon ash, maple, many fruit trees, and even house plants.

Distribution.—Especially troublesome in the Mississippi Valley; also in the eastern states.

Control.—Pyrethrum, derris, thiocyanate, or nicotine sprays at about twice the strength necessary to kill aphids will kill all insects wet by them. One tablespoonful of 40 per cent nicotine sulphate in 1 gallon of

[1] *Leptocoris trivittatus* Say, Order Hemiptera, Family Coreidae.

water in which 1 ounce of soap has been dissolved will kill all the bugs hit by the spray. When the adults invade houses, the household or fly sprays of the same ingredients are the best.

FLATHEADED BORERS[1]

These borers are the most important pests of newly set shade trees in the South and Middle West. There are several species of borers belonging to this group, which are very common, attacking practically all kinds of trees. The typical injury consists of rather shallow, long, winding, oval galleries packed with frass, beneath the bark, usually on the south or southwest side of the trees. Areas of the bark become entirely undermined. Some species kill trees by mining beneath the bark, while others are very destructive to lumber by mining into sapwood and heartwood. In these galleries are to be found medium to rather large, yellowish-white legless grubs, with a pronounced flattened enlargement of the body (thorax) just back of the head which bears a horny plate on both the upper and the underside. The adults are often beautifully colored or metallic, boat-shaped beetles, $\frac{1}{3}$ to 1 inch long, with the wing covers usually curiously roughened, like bark. A typical life history of these insects is given under Apple Insects, page 589.

The control of the flatheaded borers on shade trees is the same as on apple. Trees that have been taken from the nursery and are set in situations exposed to the sun should be protected at least during the first season by complete wrappings of paper from the ground to the lowest limbs. Under forest conditions, the only practical measure is prompt felling, peeling, and burning as for bark beetles (page 719).

References.—Jour. Econ. Ento., **11**: 334, 1918; *N.Y. State Museum, Memoir* 8, **2**: 653–658, 1905.

245. BRONZE BIRCH BORER[2]

Importance and Type of Injury.—This flatheaded borer is a very serious pest of birches and in northern states has destroyed nearly every white or paper birch in many localities. Infestation is first indicated by a browning of the tips of the upper branches, followed by the death of the entire tree. Infested branches will often appear somewhat swollen and brown, with ridges around the smaller ones. Small, slightly oval holes, about $\frac{1}{8}$ inch in diameter, will be found in the bark. An examination of the inner bark will show numerous burrows, tightly packed with sawdust, running in every direction, often extending into the wood, and containing slender white grubs about $\frac{3}{4}$ inch long, with a slight brownish enlargement of the body just back of the head.

Trees Attacked.—The white or paper birches are most severely injured. The insect has been found also in several other species of birch, poplar, cottonwood, and willow.

Distribution.—It is a native American insect, generally distributed throughout the northern United States and southern Canada, westward to Idaho, Colorado, and Utah.

---

[1] *Chrysobothris femorata* (Olivier) and many other species, Order Coleoptera, Family Buprestidae.
[2] *Agrilus anxius* Gory, Order Coleoptera, Family Buprestidae.

Life History, Appearance, and Habits.—The winter is passed as full-grown larvae in cells just within the sapwood.   These larvae are from ½ to ⅗ inch in length, white in color, very slender, with a slight enlargement of the thorax, and with two rather slender, brownish projections from the last segment of the body.   In the spring the larvae pupate within their cells in the wood and emerge during May, June, and July, as greenish-bronze beetles ¼ to ½ inch long, with rather blunt heads and slender, pointed bodies (Fig. 472).   The female beetles deposit their eggs singly in cracks in the bark or in crevices made by their jaws.   These soon hatch into tiny borers that work their way into the inner bark, becoming full-grown by fall and excavating cells in the sapwood.   The burrows of the insect may be 4 or 5 feet long, are very crooked, are filled with frass, and cross and recross, often completely cutting off the circulation of the sap.   There is but one generation of the insect each year.

Fig. 472.—Bronze birch borer, *Agrilus anxius* Gory, adult, about 6 times natural size. (*From Ill. State Natural History Surv.*)

Control Measures.—The only known methods of checking the insects are to set trees in well-selected sites, to keep them growing vigorously with plenty of water and fertilizer, to wrap the trunks during the period when adults are present, and to cut out and burn during the winter, not later than the first of April, all infested parts of trees.

References.—*Jour. Forestry,* **25**: 68–72, 1927, and **29**: 1134–1149, 1931.

Roundheaded Borers[1]

The family of beetles known as long-horned or roundheaded borers[1] are even more numerous and destructive than the flatheaded borers. The larvae, or grubs, work beneath the bark and also tunnel through the heartwood, often riddling the trunks of trees with holes that are as large as a pencil or much larger.   These grubs (Fig. 477) also have an enlargement of the body behind the head, but it is not so much flattened as in the flatheaded borers and has a horny plate on the upper side only.   They are generally legless.   The adults are usually cylindrical or straight-sided hard-shelled beetles, with antennae nearly as long as, or longer than the body and often gorgeously banded, spotted, or striped with contrasting colors.   A typical life history is given under Apple Insects, page 590.

Besides the locust borer, elm borer, and poplar borer, discussed below, there are a number of other species which attack various species of shade trees.   A few of these attack healthy trees, while others are attracted only to trees in a weakened or sickly condition.   The wood may be rendered worthless for lumber.   A minor annoyance often results when wood containing the larvae is stored in dwellings for firewood.   The adults which

[1] Order Coleoptera, Family Cerambycidae.

transform, though harmless, are alarming to persons who do not know what they are. The maintenance of healthy vigorous trees, and the avoidance or painting of wounds will go far to prevent attack. The only practical methods of control known are to use carbon bisulphide in the burrows as described for the poplar borer, or to cut and burn the badly infested trees during the winter when the insects are all within the trunk or bark in the larval stage. Linden, poplar, soft maple, oak, young hickory, and willow are particularly subject to attack by some of these borers.

246. Poplar Borer[1]

Importance and Type of Injury.—Damage by this insect frequently makes it impossible to grow certain species of poplars in some localities. It is the most destructive borer attacking poplar trees. Infested trees have many large burrows, with openings through the bark, on the trunk

and large branches. The entrance to these burrows is usually packed with coarse, excelsior-like wood fibers. An accumulation of these fibers or sawdust-like material will often be noticed around the base of the trunk. There is generally a discharge of sap from the opening to the burrow, which wets and discolors the bark of the tree for some distance below it. The wood of the tree is weakened, so that the branches or the main trunk break during periods of high wind.

Trees Attacked.—Carolina poplar, cottonwood, aspen, and Lombardy poplar are the most seriously injured. It has also been found in some other species of poplars and in willow.

Distribution.—This is a native insect, distributed generally over the northern United States and Canada and southward to Texas.

Fig. 473.—Poplar borer, *Saperda calcarata* Say. Adult female, slightly enlarged. (*From U.S.D.A., Farmers' Bul. 1154.*)

Life History, Appearance, and Habits.—The winter is passed as a yellowish, rather round-bodied grub, from 1 to nearly $1\frac{1}{2}$ inches in length. These grubs will be found in large burrows in the wood of the tree. They start feeding as soon as the weather becomes warm in the spring and continue throughout the season. They bore through the wood of the tree, cutting out large galleries, sometimes as much as an inch in diameter. Some of the borers become full-grown by late spring. They are then about 2 inches long. They pupate in cells in the wood, and emerge as beetles from July to September. The adult beetles (Fig. 473) are from 1 to $1\frac{1}{2}$ inches in length, with long antennae slightly darker than the body. The beetle is of a general light-gray color, with irregular, somewhat elongated, small yellowish spots, and the whole body is sprinkled with minute black dots. The female deposits her eggs in cracks in the bark. These eggs hatch in a few days into small grubs, which feed at first in the outer bark, but work their way rather quickly into the inner bark and sapwood. At least 2 years are required for the grubs to complete their growth. Under unfavorable conditions this period may extend over 3 years.

[1] *Saperda calcarata* Say, Order Coleoptera, Family Cerambycidae.

Control Measures.—Trees that are especially valued for shade, may be freed of most of the borers by carefully going over the tree and injecting a small amount of carbon bisulphide into all openings leading into the burrows. A machine oilcan is best suited for injecting the carbon bisulphide. The hole should be closed immediately after injecting the chemical, using a wad of wet clay or putty for this purpose. An examination of the infested trees two or three times from the first of June to the first of October, treating each time all burrows from which fresh sawdust is being forced out, will kill practically all insects in the trees. Experiments have shown that all grubs within 6 to 10 inches of the point where the carbon bisulphide is injected will be killed. The edges of the egg scars in burrow wounds may be painted with wood-preserving creosote during October. This will kill the young borers which are frequently found in these wounds. Badly infested trees should be cut during the winter or early spring, and all the large branches and trunk burned to prevent emergence of the adult beetles. Cutting and piling the wood will not be effective, as the larger grubs will complete their growth, and emerge as beetles to reinfest neighboring trees.

References.—*Ohio Agr. Exp. Sta. Bul.* 332, p. 319, 1918; *U.S.D.A.,* *Farmers' Bul.* 1154, 1920; *N.Y. State Museum Memoir* 8, **1**: 98, 1905.

247. LOCUST BORER[1]

Importance and Type of Injury.—Black locust trees have swollen areas on the trunk, often with the bark cracked open, exposing burrows in the tree sometimes ½ inch in diameter. The wood is nearly always discolored and blackened in and around these burrows. Young locust trees break over during the late summer, by reason of numerous burrows through the trees which have weakened them (Fig. 475). This insect is so numerous throughout most of the eastern United States as to prevent profitable growing of black locust. If this insect could be controlled, the black locust would become of much greater value as a farm wood-lot tree.

Trees Attacked.—Black locust.

Distribution.—The locust borer is found throughout the United States, westward to Arizona, New Mexico, Colorado, and Washington, and in southern Canada.

Life History, Appearance, and Habits.—The winter is always passed as a very small grub or larva within the inner bark or barely to the sapwood of the infested trees. Early in the spring these grubs begin feeding, and burrow their way into the wood of the tree, going through both the sapwood and the heartwood. These burrows are very irregular in direction and are frequently so numerous as to cause the death of the tree. The tree may break over during periods of high winds. The larvae become full-grown about midsummer. At this time they are ¾ to nearly 1 inch in length, legless, and tapering from the thorax backward, giving them a club-shaped appearance. These larvae excavate cells in the wood and here change to the pupal stage. The adult beetles begin coming out about the first of September in the latitude of Illinois. They are of a general black color, marked with bright-yellow crosslines (Fig. 474). The legs and antennae are dull red, the underside of the body jet

[1] *Cyllene robiniae* Förster, Order Coleoptera, Family Cerambycidae.

black. They are extremely active, flying readily from tree to tree and scuttling about over the bark. They feed on the flowers of goldenrod and a few other allied plants. Feeding in the adult stage is not necessary, as the females will lay their eggs without having fed. The elongate white eggs are tucked into crevices and cracks in the bark. These hatch in about 2 weeks, and the young work their way through the outer bark and into the inner bark, or just to the surface of the sapwood, before the approach of cold weather. There is one generation each year.

Control Measures.—Where black locust is interplanted with other trees, there is less damage from this borer. Dense stands are less injured than open ones. Rapidly growing trees escape severe injury. To

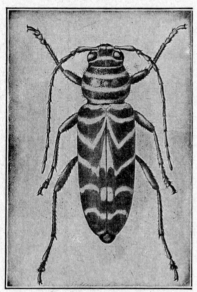

Fig. 474.—The locust borer, *Cyllene robiniae* Förster; adult female about twice natural size. (*From Ill. State Natural History Surv.*)

Fig. 475.—Young locust tree broken at point of injury by larvae of the locust borer. (*From Ohio Agr. Exp. Sta. Bul. 332.*)

accelerate growth, mulching with leaves to a depth of 6 inches has been recommended. If a plantation is growing very slowly or in cases where locust is planted for erosion control, the trees may be cut back very severely, after a few years' growth, thus causing the production of numerous root sprouts, greatly accelerated growth, and a shady condition unfavorable to the insects. For shade trees, a mixture of $\frac{1}{4}$ pound sodium arsenite dissolved in 5 gallons of water, to which is added 1 quart of dormant miscible oil, has given good control. This should be applied to the trunks when new growth starts in the spring.

References.—*U.S.D.A., Bur. Ento. Bul.* 58, Parts I and III, 1906; *Rept. Ky. State Forester*, 1915; *Ohio Agr. Exp. Sta. Bul.* 332, 1918; *Jour. Econ. Ento.*, **25**: 713–721, 1932.

248. Elm Borer[1]

Importance and Type of Injury.—Branches of elm dying, or entire trees with foliage undersized and of a general yellow color, is the symptom of attack by this round-headed borer. Numerous galleries run through the inner bark and sapwood. The outer bark is often darkened and loosened from the tree. These galleries are tightly packed with frass or brownish sawdust.

Fig. 476.—The elm borer, *Saperda tridentata* Olivier. Adult about 3 times natural size. The ground color is dark gray, while the light stripes on the sides of the body and the bands across the elytra are a beautiful bright red. (*From Ill. State Natural History Surv.*)

Fig. 477.—The elm borer, larva, about 3 times natural size. (*From Ill. State Natural History Surv.*)

Trees Attacked.—White elm and slippery elm. English and Scotch elms are not attacked.

Distribution.—Throughout the northeastern United States.

Life History, Appearance, and Habits.—The winter is passed as partly grown larvae in the bark and sapwood of the tree. In the spring these larvae begin feeding on the inner bark and wood, running their irregular galleries in all directions. They become full-grown late in the spring and change to a pupal stage in cells in the sapwood. They emerge during late spring and early summer as gray long-horned beetles, about ½ inch in length (Fig. 476). The wing covers are bordered on the outer margin with a narrow line of red, and there are three fine extensions of this red line across each wing cover. The eggs are laid in cracks in the bark, and the grubs, on hatching, work their way into the inner bark and sapwood. The galleries running through the inner bark cut off much of the sap flow. Most of the grubs (Fig. 477) are about two-thirds

[1] *Saperda tridentata* Olivier, Order Coleoptera, Family Cerambycidae.

grown by fall. There is probably one generation of the insect each year, except in cases where conditions are unfavorable to the growth of the grubs, when they may require two seasons to complete their growth.

Control Measures.—The most effective control for this beetle, as well as for many other shade-tree insects, is to keep the trees in as vigorous condition as possible. The elm borer seems to be attracted to trees which are in a sickly condition, or lacking in vigor because of insufficient plant food or moisture. Keeping the trees well supplied with water and well fertilized will do much to prevent injury by this insect. Sickly trees, or large branches which are dead or dying, should be cut out and burned during the winter months. Such trees or branches act as breeding places for the beetle, and as centers of infestation for all elms in the neighborhood.

References.—*Ill. Agr. Exp. Sta. Bul.* 151, 1911; *N.Y. State Museum, Memoir* 8, p. 67, 1905.

249. Bark Beetles and Ambrosia Beetles[1]

These short, cylindrical, reddish-brown to black beetles, ranging from $\frac{1}{25}$ to over $\frac{1}{3}$ inch long, are the most destructive group of insect pests attacking coniferous trees. Many species also attack deciduous trees.

Fig. 478.—Hickory bark beetle, *Scolytus quadrispinosus* Say; adult, about 7 times natural size. (*From Ill. State Natural History Surv.*)

Destruction of standing timber is estimated to exceed two billion board feet (valued at 60 million dollars) a year in the United States. The adults usually have a very large prothorax overhanging the head, the antennae are clubbed, the tibiae serrate, and the elytra often cut off obliquely at the end and provided with spines. The larvae are whitish, legless, stout, cylindrical, curved grubs. There are three main types of attack upon the trees: (1) The twig and cone beetles bore into the cones of trees or the pith of twigs and eat the wood. (2) The ambrosia or timber beetles tunnel into the sapwood and heartwood of unseasoned lumber, log cabins, and rustic work, making "pinholes" upon which they propagate fungi as food, often rendering lumber practically worthless. These adults do not eat the wood and their larvae do no damage. (3) The most destructive and best-known species, called bark beetles, mine just beneath the bark of standing or felled trees, feeding upon the cambium and adjacent tissues, and leaving characteristic engravings upon both the inner surface of the bark and the wood. Reddish, boring dust is pushed out of the tunnels and clings to the bark or accumulates about the base of the trees and often sap exudes and dries to form hard pitch or resin tubes about the entrance holes. These engravings can always be analyzed into (*a*) *feeding holes;* (*b*) *egg tunnels* or *parent galleries*, made by the female who lays her eggs in little niches along the sides; and (*c*) *larval tunnels* radiating from the parent gallery, in which the larvae feed and complete their growth, and at the end of which they pupate and transform to adults. In monogamic species the female starts the burrows, but there are polygamic species in which the male first eats out a nuptial chamber from which his several females each excavates an egg gallery, while the male remains throughout

[1] Order Coleoptera, Family Scolytidae.

his life in the nuptial chamber plugging the entrance hole with his body. Some attack perfectly healthy trees, others restrict themselves to trees already weakened from some other cause. Their feeding destroys the cambium, loosens the bark, and rapidly kills the tree usually from the top downward. Many foresters believe that these insects are largely responsible for forest fires by producing areas of dead and highly inflammable material. Among the most important groups are the genus Dendroctonus, including the mountain pine beetle (*D. monticolae* Hopkins), the southern pine beetle (*D. frontalis* Zimmerman), the western pine beetle (*D. brevicomis* Leconte), the black hills beetle (*D. ponderosae* Hopkins), and many others, which are especially destructive to mature trees. Species of the genus Ips, known as pine engravers, are primarily enemies of young trees and the tops of older ones. Species of the genus Phloeosinus attack cypress and the giant redwoods or Sequoia. In this family also are included the hickory bark beetle,[1] which has killed a large per cent of the hickory trees in the states of the Mississippi Valley and caused losses of about 15 million dollars a year, the shot hole borer (page 592), the peach bark beetle (page 627), the elm bark beetles (page

FIG. 479.—Work of the hickory bark beetle, *Scolytus quadrispinosus* Say, in a 12-inch hickory tree. The dead bark has been removed. Note the straight, vertical egg galleries made by the females and, radiating from each egg gallery, the numerous tunnels made by the growing larvae. (*From N.Y. State Coll. Forestry, Tech. Bul. 17.*)

720), and dozens of other species of only slightly less importance. Somewhat typical life histories of bark beetles are given on pages 593 and 628.

Control Measures.—No practical methods have been discovered to prevent the attack of bark beetles or to save the trees over large forest areas. To prevent further losses, the infested trees may be converted into lumber during the winter or early spring, burning all slabs and slash before the beetles emerge, or submerging the logs for at least 6 weeks to drown the insects. If milling is done 20 to 50 miles from the forest, the slabs need not be burned. Other methods used include felling the trees and burning the bark from them or peeling and exposing to the sun, even though the lumber is left to rot in the woods. For shade trees and farm wood lots the only effective means of control are to keep the trees in a vigorous growing condition by supplying enough water and plant food, and to keep all dying and injured wood pruned out. Such pruning should

[1] *Scolytus quadrispinosus* Say.

be done during the winter or early spring months, and all such wood should be carefully burned. If allowed to remain piled up or near the trees, it may act as a breeding place for bark beetles which will infest the growing trees in the neighborhood. No effective spray is known that can be used in combating these insects.

References.—Can. Dept. Agr. Tech. Bul. 14, Parts I and II, 1917–1918; *Miss. Agr. Sta. Tech. Bul.* 11, 1922; *Ore. Agr. Exp. Sta. Buls.* 147, 1918, and 172, 1920; *U.S.D.A., Farmers' Bul.* 1188, 1921, and *Bureau Ento. Buls.* 56, 1905, 58, 1906–1909, and 83, 1909.

250. ELM BARK BEETLES

Importance and Type of Injury.—The native, dark elm bark beetle[1] and the lesser European bark beetle[2] have taken on especial importance with the discovery that they are carriers of a virulent and destructive parasitic fungus[3] imported from Europe, and first recognized in America in 1930. The disease and the European bark beetle were both imported on Carpathian elm logs used for furniture veneer. By 1938 the disease had spread along the principal railroads as far west as Indianapolis. The heaviest infestation of the disease centers about New York City, but it is known to be present also as isolated infestations in New Jersey, Maryland, West Virginia, Virginia, Connecticut, Ohio, and Indiana (Fig. 480). In the Netherlands 60 to 70 per cent of the elms have died and been cut down due to this parasitic fungus. Infested trees show wilting leaves on certain branches, which shrivel, discolor, and fall, except at the end of the branch; the ends of the twigs curl in a peculiar manner; the larger limbs put out numerous trunk suckers, very evident in winter; and the dying twigs when examined in cross-section show a brown discolored ring beneath the cambium. Beetles flying from diseased trees bear spores in or upon their bodies and, when they make feeding or breeding cavities in healthy trees, inoculate the latter with the deadly disease.

Trees Attacked.—Elms are the favorite host of the beetles but the dark elm bark beetle also attacks basswood and ash.

Distribution.—The dark elm bark beetle probably occurs wherever the American elm is found throughout the United States and Canada. The lesser European bark beetle has been present in this country at least since 1909. It is known to occur throughout much of southern New England, southeastern New York, New Jersey, eastern Pennsylvania, Maryland, and along the Ohio river from Pennsylvania to Illinois (see Fig. 480). Another carrier in Europe, the larger European bark beetle[4] has been intercepted by American quarantine officials but so far is not known to be established in America. It is of interest to note that the disease is present in certain sections (Fig. 480), such as Ohio and Indiana, where the imported carrier is not known to occur and many sections, as about Boston, are, so far, free from the disease although the imported bark beetle occurs there. The disease spreads most rapidly where the bark beetle carriers are breeding, although it is readily disseminated also by transportation of diseased wood.

[1] *Hylurgopinus rufipes* Eich, Order Coleoptera, Family Scolytidae.
[2] *Scolytus multistriatus* (Marsham).
[3] *Ceratostomella* (= *Graphium*) *ulmi* Schwarz.
[4] *Scolytus scolytus.*

Life History, Appearance, and Habits.—The breeding habits of these insects is essentially like those described for bark beetles in general. The lesser European bark beetle larvae winter at the ends of their tunnels,

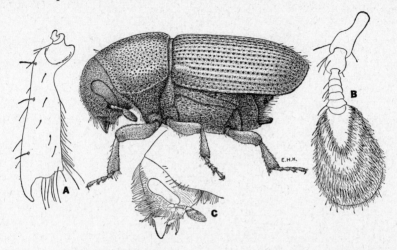

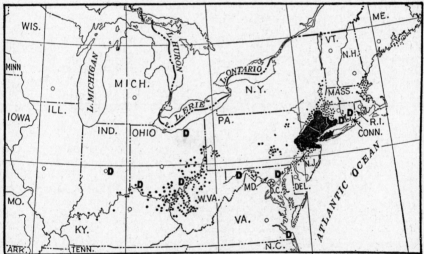

Fig. 480.—The smaller European elm bark beetle, *Scolytus multistriatus* (Marsham), about 20 times natural size. *A*, front tibia, *B*, antenna, and *C*, outline of the head from the side, more enlarged.

Below, map showing the known distribution of the beetle and of the Dutch elm disease which it disseminates, at the close of the year 1938. The solid black area in New Jersey, New York, and Connecticut shows the major zone of distribution of the fungous disease and the letters *D*, *D*, *D*, the known outlying areas where the disease occurs. The small round dots show the distribution of the imported bark beetle carrier. (*In part after Collins, Jour. Econ. Ento., April*, 1938.)

they pupate in the outer bark, and the adults begin to appear in May. They are stout reddish-brown beetles, about $\frac{1}{8}$ inch long, of the appearance shown in Fig. 480. They feed on the twigs, often in the crotches, of healthy elm branches especially in trees which contain, or are very near to other trees that contain, weak, injured or sickly portions. The females construct their brood chambers, however, only in the cambium of weakened, dying, or recently dead wood. The parent galleries or brood chambers run lengthwise of the branches and the larval galleries across the grain. The parent gallery is 1 to 2 inches long and from 80 to 140 eggs are laid in it. The larvae develop in a few weeks, and a second or summer generation of adults appears and lays eggs in August and early September. There are two generations a year, but egg-laying extends over a considerable time, and adults may be continuously present from May to September. The parent galleries of the small, black, native species are cut across the grain of the wood, and the larval tunnels run more or less lengthwise of the twigs.

Control.—A 10-mile protective zone has been established about the known infested area and is under constant surveillance; over 3,500,000 unhealthy trees have been removed and destroyed; and over 3,000 men have been employed, at certain times, in efforts to prevent the spread of the disease.

Control measures consist in keeping trees healthy, vigorous, and free from injuries and in the prompt removal and burning of diseased or infested trees or parts of trees likely to attract the beetles to feed or to lay their eggs.

Reference.—Cornell Agr. Exp. Sta. Serv. Bul. 290, 1934.

251. CARPENTER-WORM[1]

Importance and Type of Injury.—This insect is very common on many shade trees, especially on oak. Infested trees have large burrows running through the wood,

Fig. 481.—The carpenter-worm, *Prionoxystus robiniae* Peck, adult female, natural size. (*From Ohio Agr. Exp. Sta. Bul.* 332.)

with occasional openings through the bark of the trunk, from which sawdust may be forced out or from which a discharge of dark-colored sap will be oozing and discoloring the trunk. The burrows of the insect in the trunk will sometimes be as much as 1 to 1½ inches in diameter. These galleries occasionally are so numerous as to weaken

[1] *Prionoxystus robiniae* Peck, Order Lepidoptera, Family Cossidae. A very similar species, the lesser oak carpenter-worm, *Prionoxystus macmurtrei* Guer., occurs in eastern United States and Canada.

trees, and cause them to be broken off during high winds. The injury also greatly lessens the value of lumber or posts from infested trees.

Trees Attacked.—Oaks of the black and red oak groups, elm, locust, poplar, willow, maple, ash, chestnut, and others.

Distribution.—The insect is generally distributed throughout the United States and southern Canada.

Life History, Appearance, and Habits.—The winter is passed in the larval stage in the burrows in the trunk and large branches of the tree. The borers vary in size, according to their age, from 1 inch up to 2 inches or a little over. The presence of prolegs distinguishes them from beetle grubs. The general color is white tinged with pink, with a very dark-brown head and numerous dark-brown tubercles over the body. The insects pupate in large cells excavated in the wood of the tree and, just before emerging, the pupae work part way out of the burrow. The adult moth on emerging generally leaves the empty pupal skin protruding from the tree. The moths are abroad at night during the summer months. In this stage the insect is of very striking appearance (Fig. 481), having a wing expanse of as much as 3 inches, in the female. The moths are of a general gray color, mottled with black and lighter shadings, the hind wings being faintly tinged with yellow. In the male the hind wings have a distinct orange margin. The eggs are laid in crevices in the bark and the young borers on hatching immediately work their way into the wood of the tree. Three years are probably required for completion of the larval growth.

Control Measures.—These boring caterpillars may be killed by injecting carbon bisulphide or other fumigant into their burrows and closing the openings with mud or putty. This is practical only on valuable shade trees.

References.—*Ohio Agr. Exp. Sta. Bul.* 332, p. 329, 1918; *N.D. Agr. Exp. Sta. Bul.* 278, 1934.

252. LEOPARD MOTH[1]

Importance and Type of Injury.—Branches, especially in the tops of trees, are dead and smaller branches broken over or hanging partly cut off, during the latter part of the summer. The leaves wilt suddenly on small branches. There are holes along such branches, from which damp sawdust is being pushed out (Fig. 482).

Trees Attacked.—Elms and maples are the favored food plants, but the insect attacks also many other deciduous shade and fruit trees.

Distribution.—This moth, a native of Europe, was first found in the United States in New York, in 1879. It is distributed along the Atlantic Coast from Maine to Delaware.

Life History, Appearance, and Habits.—The

FIG. 482.—The leopard moth, *Zeuzera pyrina* (Linné); work in small twig, borings hanging from bark, nearly full-grown larva, and adult moth. One-half natural size. (*From Felt, "Manual of Tree and Shrub Insects," copyright, 1924, Macmillan. Reprinted by permission.*)

winter is passed as a partly grown larva, from 1 to 1½ inches in length, in burrows in the heartwood of infested trees. The worms (Fig. 482) are of a pinkish-white color, with many dark-brown spots distributed over the body, and have prolegs. In the spring the worms start feeding and boring through the wood, the smaller ones continuing to feed throughout the season, while the larger ones become full-grown in the late spring and change to a

---

[1] *Zeuzera pyrina* (Linné), Order Lepidoptera, Family Cossidae.

brown pupal stage within their burrows. When the moth is ready to emerge, the pupa forces itself partly out of the burrow, and the skin splits down the back, permitting the adult moth to escape. Emergence takes place from late May to early fall. The adults (Fig. 482) are very striking in appearance, being of a general white color, blotched and spotted with blue and black. They have a wing expanse of from 2 to 3 inches. The salmon-colored oval eggs, to the number of 400 to 800, are laid in crevices of the bark. They hatch in about 10 days, and the young borers

Fig. 483.—Ash tree injured by lilac borer, *Podosesia syringae* Harris. (*From Ill. State Natural History Surv.*)

rapidly work their way into the heartwood of the branch. It requires from 2 to 3 years for the insect to complete its growth.

Control Measures.—Where this insect is abundant, it is very important that all infested branches and badly infested trees be cut and burned during the fall and winter months. The presence of the insect is indicated by wilting of the leaves and by masses of wet frass thrown out from the burrows. In valuable shade or fruit trees the insect may be killed by digging or probing out the worms or by injecting commercial borer paste or carbon bisulphide into the burrows. The worms will attempt to clean out the burrows and so come in contact with the paste and be killed.

References.—Conn. Agr. Exp. Sta. Bul. 169, 1911; *Jour. Econ. Ento.*, **27**: 196, 197, 1934; *N.J. Agr. Exp. Sta. Rept.*, **50**: 146, 147, 1929; *U.S.D.A., Farmers' Bul.* 1169, 1925.

253. LILAC BORER[1]

Importance and Type of Injury.—Lilac canes are dying, the base of the canes exhibiting swollen areas where the bark is cracked and broken away from the wood. There are numerous holes through the bark and wood. Canes suddenly wilt and show fine, sawdust-like borings forced out from holes in the bark. Figure 483 shows typical injury on ash.

Plants Attacked.—Lilac and green, white, English, and mountain ash are most seriously infested, although the insect is occasionally taken in some other trees.

Distribution.—Eastern United States, north through southern Canada, and westward to Colorado.

Life History, Appearance, and Habits.—The winter is passed as a partly grown larva in the stems of lilac, usually near the surface of the soil, and in infested trees. The insect starts feeding in the spring and completes its growth by early summer. At this time it is a nearly pure-white worm, about 1½ inches long, with a brown head. It transforms in the burrow to a brown pupal stage and emerges in about 3 weeks, in June and July, as a clear-winged moth of a somewhat wasp-like appearance. The fore wings are of a general brown or chocolate color, the hind wings are clear, marked with a dark border. The body is mainly brown and the legs are marked with brown and yellow. The insect is about 1 inch in length with a wing expanse of 1½ inches. The moths are very active fliers. The females deposit their eggs on the bark about the base of lilac canes, or on ash. The worms, hatching from them, become about half-grown by cold weather. There is one generation a year.

Control Measures.—Painting infested areas on ash and lilac with a solution of paradichlorobenzene, 1 pound, in raw cottonseed oil, 2 quarts; or with paradichlorobenzene, 1 pound, in dormant miscible oil, 2 quarts, and water, 2 quarts, has given good control. Injecting carbon bisulphide into the burrows is effective if carefully done. Badly injured canes of lilac should be cut and burned during the winter.

References.—Twenty-sixth Rept. Ill. State Ento., 1911; *Mich. State Bd. Agr. Rept. for* 1934.

254. ELM COCKSCOMB GALL[2]

Importance and Type of Injury.—This aphid seldom causes the death of the elm trees but often becomes annoying because of the fact that its galls make the trees unsightly. Leaves of elms infested with this aphid have crinkled reddish galls ½ to ¾ inch long on their upper surfaces. The shape of these galls resembles a miniature rooster's comb (Fig. 484). During the early part of the growing season, the galls are green or reddish in color. During the late summer, they dry and become brown. An examination of the green galls will show that they are filled with small, rather smooth, greenish or brownish aphids. A narrow slit on the underside of the leaves serves as an entrance to the gall, although during the early part of the season this slit is so nearly closed that the aphids cannot get through it.

[1] *Podosesia syringae* Harris, Order Lepidoptera, Family Aegeriidae. *P. fraxini* (Luger) attacks ash and mountain ash in a similar manner.

[2] *Colopha ulmicola* (Fitch), Order Homoptera, Family Aphididae.

Trees Attacked.—Elm, especially red elm.

Distribution.—General over the United States.

Life History, Appearance, and Habits.—The winter is passed as dark-brown shiny eggs in cracks on the bark of elm trees. These eggs hatch in the spring when the leaves are partly grown. The aphids crawl to the leaves and begin feeding by sucking the sap. Where they start to feed, the peculiar gall formations start growing, enclosing the aphids. These aphids, which are all females, remain inside these galls, giving birth to living young, so that if the gall is broken open during the summer, it will be found swarming with aphids. As the aphids feed, they give off quantities of honeydew, which drops from the gall, as it begins to crack open late in the season, often

Fig. 484.—The elm cockscomb gall, *Colopha ulmicola* (Fitch). A number of galls on leaves of white elm. About natural size. (*From Washburn, "Injurious Insects and Useful Birds," Lippincott.*)

nearly coating the surfaces of walks and benches under the trees. This honeydew attracts flies and other insects, which add to the unsightly appearance of infested trees. During the summer, the galls crack open and the aphids make their way out. Winged migrants carry the species to the roots of various grasses, where the aphids live during the summer, returning to elms in the fall.

Control Measures.—There is no practical control measure for this insect other than cutting off the galls early in the season, as soon as they start showing on the leaves. This of course, can be done only on small trees, or those that are especially valuable.

References.—*Ohio Agr. Exp. Sta. Bul.* 332, p. 311, 1918; *Maine Agr. Exp. Sta. Bul.* 181, 1910; *Wis. Dept. Agr. Bul.* 123, 1931.

CHAPTER XX

INSECT PESTS OF GREENHOUSE AND HOUSE PLANTS AND THE FLOWER GARDEN

Few persons realize the importance of the annual crop of flowers and ornamentals and other hothouse products in this country. Approximately 4,000 acres of ground are enclosed in glass and devoted to the production of flowers and vegetables in the United States and the annual crop is valued at nearly $100,000,000. In addition to the commercial production there is an enormous interest in the growing of house plants and garden flowers by individuals who attach to their products a personal, but very real value, far beyond their market price.

The insects that attack greenhouse crops are the same as those that attack related crops in the field. Some of these insects are field pests only in the South but have adapted themselves to the semitropical conditions found in greenhouses in the North. It would seem at first thought that insects could be easily controlled in greenhouses, where climatic factors can be largely regulated by man; but this is far from being the case. Conditions must be maintained in the greenhouse to give a maximum rate of growth to the crop, and such conditions are frequently very favorable for the insects attacking the crops. The cost of producing crops under glass is high, and the loss of 10 to 25 per cent of the crop from attacks by insects generally means that the entire operation of producing and marketing the crop has been carried on at a loss to the grower. Crops grown under glass are more easily injured by heavy applications of insecticides, or other insect-control measures that affect the plants, than the same crops grown in the open. For this reason much greater care must be exercised in the control methods employed.

The best way to prevent damage by insects to greenhouse crops is to keep the insects out of the houses. This may be done by careful inspection of all plants brought into the houses and by thoroughly cleaning up the houses by heat or heavy fumigations in the intervals between crops. Open range houses are, on the whole, more difficult to keep free of insects than those built in sections, as some part of the open range house is nearly always occupied by a crop in some stage of growth. Under such conditions, strong fumigations which will kill nearly all plant and animal life cannot be used for cleaning up the house.

It will help greatly to avoid most insect injury, if the following precautions are taken: (1) Before a crop is put in, or the soil placed in the benches, the houses should be cleaned out as thoroughly as possible, and, if they have recently been infested with insects, a thorough fumigation with hydrocyanic acid gas should be given at a dosage equal to 1 ounce sodium cyanide to 100 cubic feet of space (see page 278). Such a fumigation will kill practically all insect life in any stage that may be present in the house. (2) A careful inspection should be made of the soil being brought into the benches, to make sure that this soil does not contain wireworms, white grubs, eelworms, cutworms, or other

727

insects that may be injurious to greenhouse crops. If the soil is infested, it should be sterilized with steam. (3) All plants brought into the greenhouse should be inspected to make sure that they are free from insect infestation. Many florists have suffered serious losses by bringing a few infested plants into their houses and thus starting an infestation which, before it was noticed, had spread throughout the entire range of their greenhouses.

Many of the best accounts of the life histories, habits, and control of greenhouse insects are contained in general publications dealing with this class of insect pests. Some of these publications are here listed and should be used as general references for further information on the insects treated in this chapter.

References.—Can. Dept. Agr., Bul. 7 (n.s.), 1922, and *Bul.* 99 (n.s.), 1934; *Cornell Univ. Ext. Bul.* 371, 1937; *Mich. Agr. Exp. Sta. Spcl. Bul.* 214, 1931; *Ill. State Natural History Surv. Cir.* 12, 1930; *N.J. Agr. Exp. Sta. Bul.* 296, 1916; *Fifth Rept. S.D. State Ento. for* 1924; *U.S.D.A., Farmers' Bul.* 1362, 1923; *Twenty-seventh Rept. Ill. State Ento. for* 1912.

KEY FOR THE IDENTIFICATION OF INSECTS INJURIOUS TO GREENHOUSE PLANTS AND THE FLOWER GARDEN

A. *External chewing insects that eat holes in leaves, buds, or stems:*

1. Brown, green, or gray, mottled, jumping insects of all sizes up to 1½ inches long, migrate into gardens from surrounding areas and strip leaves and eat tender stems of many kinds of flowers..............*Grasshoppers*, page 335.
2. Foliage of many kinds of flowers is eaten at night by dark reddish-brown beetle-like insects, up to ⅘ inch long, with a pair of sharp pinchers or forceps at the tip of the abdomen, ¼ as long as rest of body. Insects hide in soil during day...................
   European earwig, Forficula auricularia Linné, Order Dermaptera, Family Forficulidae.
3. Many, minute holes eaten in leaves by very small hard-shelled black beetles, from ¹⁄₁₆ to ⅛ inch long, which feed on the underside of leaves and jump readily when disturbed............................*Flea beetles*, page 468.
4. Fawn-colored long-legged cylindrical beetles, about ½ inch long, attack blossoms and buds of roses and other flowers and shrubs, during daytime, especially in regions of sandy soil, in June............*Rose chafer*, page 645.
5. Cinnamon-brown velvety beetles, shaped like a May beetle but only ⅜ inch long, eat irregular holes from margins of leaves of many garden flowers at night. Attracted to lights. Along Atlantic seaboard.....................
   ....*Asiatic garden beetle*, page 471.
6. Elongate, nearly cylindrical, not very hard, active long-legged beetles, either black, gray, or brown, sometimes with narrow, longitudinal stripes or with small spots, swarm upon plants in gardens, devouring blossoms and leaves. Cause blisters on skin when crushed...............*Blister beetles*, page 509.
7. Bright-red snout beetles, with the snout and undersurface black, about ¼ inch long, eat numerous holes in the buds of roses, so that blooms fail to develop. The whitish legless grubs feed in the "hips" or fruit....................
   Rose curculio, Rhynchites bicolor Fabricius, Order Coleoptera, Family Curculionidae.
8. Undersurface of the leaves eaten off by pale-green, very active caterpillars up to ¾ inch long. Such leaves are often covered with a light web enclosing several leaves or drawing parts of a single leaf together...................
   ....*Greenhouse leaf tyer*, page 733.
9. Injury appearing much the same as in *A*, 8. Small active greenish worms feeding at first as leaf miners but later eating off the surface of the underside of the leaves. Leaves rolled or drawn together with light webs............
   ....*Oblique-banded leaf roller*, page 735.

10. Ferns with the leaves partly stripped, presenting a very ragged appearance. Newly unfolding leaves gnawed or eaten away. Dark-green velvety caterpillars, with two wavy white lines down each side of the body, hiding in the soil about the base of the ferns, feeding very actively at night.............. ....*Florida fern caterpillar*, page 735.

11. Many kinds of plants cut off at the surface of the ground, or, in some cases, stripped of leaves, by fat, sleek caterpillars or worms of varying sizes and colors...*Cutworms*, page 736.

12. Leaves eaten from the new growth, or buds eaten into or partly eaten out, by greenish, dark-brown, or yellowish worms with slightly hairy bodies, differing from cutworms in their habit of feeding on the upper part of the plants. Injury common during the fall months.............*Corn earworm*, page 736.

13. Roses with the leaves riddled with small holes, bark of the new growth eaten off, and buds eaten out. Injured plants with most of the feeding roots eaten off by small curved-bodied whitish grubs. Brownish or brownish black, very active beetles, about ⅛ inch long, feeding on the leaves................... ....*Strawberry rootworm*, page 732.

14. Velvety-green caterpillars, with a very faint, orange stripe down the middle of the back, up to 1¼ inches long, rag the leaves of mignonette, alyssum, and nasturtium. White butterflies about 2 inches across the wings lay the eggs.. ....*Imported cabbageworm*, page 530.

15. Pale-green, white-striped, looping caterpillars, with only 3 pairs of prolegs, which hump the back high as they crawl, eat holes in leaves of mignonetet, geranium, chrysanthemum, and carnation.........*Cabbage looper*, page 533.

16. Pale-green caterpillars, not over ⅓ inch long, very active, eat small rounded holes in the leaves of stocks and wallflowers in midsummer. Wriggle actively when disturbed...........................*Diamond-back moth*, page 534.

17. Yellowish, brown, or reddish brown-and-black-banded caterpillars, up to 2 to 2½ inches long, very densely covered with short stiff hairs of even length, eat large holes in leaves of dahlias, verbenas, hydrangeas, and many other flowers. Smaller caterpillars often feed close together..................... ....*Woolly bears*, page 476.

18. Velvety-black caterpillars, with 2 conspicuous bright-yellow stripes on each side of the body, between which cross many narrow hair-lines of the same color, eat foliage of sweet peas, gladioli, lilies, and other flowers in late summer; full-grown worms about 2 inches long.................................... *Zebra caterpillar*, *Ceramica picta* Harris, Order Lepidoptera, Family Noctuidae.

19. Greenish or yellowish-green false caterpillars, up to about ½ inch long, some smooth, others bristly, skeletonize the leaves of roses, by eating the upper surface, leaving the lower epidermis and the veins to dry out and die. The worms, which have more than 5 pairs of prolegs on the abdomen, are sluggish, and leave a slimy secretion on the leaves which dries to form a glaze......... ....*Rose slugs*, Order Hymenoptera, Family Tenthredinidae.

20. Clean-cut, round or oval pieces of rose leaves, from ½ to 1 inch across, are cut off and carried away to a nesting site by large bumblebee-like insects.... *Leaf-cutter bees*, *Megachile* spp., Order Hymenoptera, Family Megachilidae.

21. Slender, active, wingless, very slender-waisted, hard-bodied insects crawl over plants infested with aphids or over the buds of peonies and other flowers to get sweet secretions of the aphids or the flowers. Others make mounds of soil thrown up from underground nests, especially about aster plants........ ....*Cornfield ant* and other *ants*, pages 371 and 767.

B. *Insects that suck sap from leaves, buds, or stems:*

1. Surface of the leaves whitened with many small flecks of light green or yellow. Tips of the leaves curling up and dying. Flower buds producing distorted blossoms which open only on one side. Undersides of the leaves with numerous black spots. General vigor of the plants greatly reduced by the feeding of many very small yellow or black, very slender, active insects............ ....*Greenhouse thrips*, page 736.

2. Greenhouse plants show a sickly appearance, the foliage turning yellow. Numerous brown or brownish-gray, flattened, motionless scales adhere to the undersides of the leaves or along the stems especially at the leaf axils. Scales usually of sufficient size to be easily seen, some up to $\frac{1}{8}$ inch in diameter. The scale covers the body of the insect but is easily lifted from it . . . .
. . . . *Armored scales*, page 731.

3. Canes of roses specked or nearly covered with snowy-white, thin, flat, rounded scales, with a nipple at the center, about $\frac{1}{12}$ inch in diameter; or smaller, slender scales with the exuviae at one end. Most abundant near the ground level. Beneath the scales, small legless insects suck the sap from the plants . . .
. . . *Rose scale, Aulacaspis rosae* (Bouché), Order Homoptera, Family Coccidae.

4. Foliage plants, such as oleander, fern, palm, Ficus, and Vinca having much the same appearance as in *B*, 2. Scales somewhat larger and adhering more tightly to the plants. Scales much more conspicuous than in *B*, 2, usually very convex. Scale covering cannot be lifted from the body of the insect, but forms a distinct part of the body wall . . . . *Tortoise scales*, page 742.

5. Plants, particularly foliage plants, such as coleus, orchids, poinsettias, ivy, and many others, with whitish clusters of soft-bodied insects, up to $\frac{1}{4}$ inch long, at the axils of the stems and leaves and along the stems. These insects have many whitish, waxy filaments protruding from the body. These filaments are so thick as to give the body a distinct bluish-white appearance. Insects seldom move unless disturbed . . . . *Mealy bugs*, page 743.

6. Plants infested in much the same way as in *B*, 5, by insects with rather small, dark-colored bodies, with rows of short waxy filaments radiating around the outer margin of the body, and forming a waxy tube extending back from the body several times its length . . . . *Greenhouse Orthezia*, page 745.

7. Greenhouse and garden plants having a weakened appearance; frequently with the leaves curled, the young growth distorted, and often coated to some extent with sticky honeydew. Small winged or wingless aphids, or plant lice, of various colors, sucking the sap from the more tender parts of the plants, buds, and blossoms . . . . *Greenhouse* and *garden aphids*, page 745.

8. Undersurface of the leaves, with small, oval, flat, pale-green motionless insects, less than $\frac{1}{30}$ inch in length, adhering to them. Many tiny four-winged snow-white flies on the underside of the leaves. These flies leave the plants in swarms when disturbed. Plants more or less covered with a coating of glazed, sticky material on which a sooty black fungus is frequently growing . .
. . . . *Greenhouse whitefly*, page 747.

9. Leaves of rose, dahlia, asters, and many other flowers are mottled with very fine whitish spots, or somewhat curled and withered. Tiny, elongate, active, somewhat wedge-shaped bugs, green or varicolored, suck sap from underside of leaves. Bugs jump, fly, or run sideways when disturbed . . . .
. . . . *Leafhoppers*, page 430.

10. Olive-green bugs, flecked with yellow and dark brown, with flattened, oval bodies, somewhat triangular in front, up to $\frac{1}{4}$ inch long by half as wide, by piercing and sucking the sap from buds of zinnias and dahlias, dwarf or deform them, and cause spotting and deforming of leaves and dead areas on stems of various flowers, sometimes in greenhouses . . . . . *Tarnished plant bug*, page 479,

11. Bright greenish-yellow bugs, of same shape as *B*, 10, but up to $\frac{1}{3}$ inch long. with 2 triangular, black spots on the prothorax, 2 black stripes on the fore part of each wing and a rounded black spot near the posterior end of the stripes, pierce leaves and buds of garden snapdragon, weigelia, zinnia, and dahlia, and sometimes in greenhouses, causing whitish to brown spots on upper side of leaves . . . . *Four-lined plant bug*, page 658.

12. Blackish, rather soft-winged, flattened or nearly globular, jumping bugs, only $\frac{1}{10}$ inch long, somewhat resembling flea beetles, and their greenish nymphs of various sizes, suck sap from primrose, ageratum, and other flowers in the garden, and sometimes in greenhouses, causing spotting and dying of leaves . . . . *Garden fleahopper*, page 481.

13. Plants with pale blotches or spots showing through the leaves, or with the entire leaf having a light color, often drying up or turning a reddish brown about the margins. Fine silk threads are spun on the undersides of the leaves, or formed into webs which entirely cover the surface of the plant. Minute, 6- or 8-legged, greenish or yellowish mites crawling about on the webs or underside of the leaves........................*Red spider mite*, page 748.

14. Cyclamen with the leaves much distorted, buds gnarled and knotty in appearance, usually failing to open; flowers on infested plants when open are streaked and blotched in appearance and quickly die. Infested foliage shows pockets or depressions in the leaf, or dark-purplish areas, often covered with small cracks, on the leaf surface. Tiny white or pale-brown mites, with 3 or 4 pairs of legs, working in the infested flowers, on the leaves, and about the base of the plants.......................................*Cyclamen mite*, page 750.

C. *Insects that bore in the stems of the flowers:*

1. Stems of dahlia, hollyhock, bleeding heart, larkspur, golden glow, and other flowers have the heart eaten out or filled with frass from a dark-brown caterpillar with two white stripes on each side of the body, which are interrupted for a space near the middle of the body, giving the appearance of a bruise....
....*Common stalk borer*, page 364.

2. Similar to C, 1, but the white stripes not interrupted......................
Burdock borer, *Papaipema cataphracta* Grote, Order Lepidoptera, Family Noctuidae.

3. Stems of iris bored by a smooth, cylindrical caterpillar, up to 1½ inches long, flesh-colored to pinkish, which tunnels from stems downward toward base of stalk...
Iris borer, *Macronoctua onusta* Grote, Order Lepidoptera, Family Noctuidae.

D. *Insects that make galls or mine in the leaves:*

1. Chrysanthemums with small blister-like cone-shaped galls on the upper surface, and occasionally on the under surface, of the leaves. Leaves curled and stems of heavily infested plants crooked and distorted. Flowers opening imperfectly if at all. Small gnat-like flies, about $\frac{1}{14}$ inch in length, and of a reddish-orange color, crawling about over the leaves. Minute yellowish-white maggots in the galls on the leaves and stems.........*Chrysanthemum midge*, page 751.

2. Flower buds and leaves of roses much distorted; terminal growth and buds dying and turning brown. An examination of such buds will show many small, white-to orange-colored maggots, up to $\frac{1}{12}$ inch long, feeding on the plant tissue within the buds. Small silken cocoons just under the surface of the soil about the plants...*Rose midge*, page 752.

3. Leaves of chrysanthemum or marguerite with irregular light-colored mines extending over their surface. When badly infested, the leaves may dry up, but usually remain attached to the plants. Heavily infested plants are stunted and produce small inferior flowers...
Chrysanthemum leaf miner or marguerite fly, *Phytomyza chrysanthemi* Kowarz, (see *U.S.D.A.*, *Farmers' Bul.*, 1362, 1922).

4. Columbine has conspicuous, slender, white, serpentine lines on the leaves due to the mining of a tiny maggot between upper and lower surface of the leaves. Larkspur leaves are similarly tunneled, the mines often coalescing to make yellowish blotches...
Columbine leaf miner, *Phytomyza aquilegiae* (Hardy), or *larkspur leaf miner*, *P. delphiniae* Frost, Order Diptera, Family Agromyzidae.

E. *Insects that attack the roots or bulbs:*

1. Greenhouse or garden plants sometimes cut off below the surface of the ground, or wilting and dying, without visible injury to the part of the plant above the ground. Examination of the roots will show large curved-bodied, brown-headed white grubs with six rather conspicuous legs, feeding on the underground parts of the plant...*White grubs*, page 753.

2. Appearance and injury much the same as in E, 1, but with underground parts of the plant eaten off, or bored through, by shining, hard-bodied, brown, slender worms up to 2 inches in length......................*Wireworms*, page 754.

3. Plants presenting a somewhat sickly appearance, with no visible injury to the part above the surface of the ground. Roots showing minute brownish scars or tunnels along their surface. Very small, somewhat thread-like, active, white maggots, not over ¼ inch long, embedded in the root tissues, or working in the soil about the roots..............*Fungus gnats* or *Sciara maggots*, page 754.
4. Narcissus and related bulbs with scars on the base of the bulbs, the bulbs soft and frequently rotting. An examination of the bulbs shows whitish or yellowish-white fat maggots inside the bulbs, eating out the plant tissue. Maggots, when full-grown, are from ½ to ¾ inch in length..*Narcissus bulb fly*, page 755.
5. Bulbs appearing as in *E*, 4, but containing smaller maggots of a grayish-yellow color, with bodies markedly wrinkled. Many maggots often occurring within one bulb. The maggots vary in size up to ½ inch in length................ ....*Lesser bulb fly*, page 756.
6. Plants of various species that are grown from bulbs turn a sickly yellow, failing to produce flowers; or, if producing flowers, with the flowers much distorted. Leaves stunted and plants generally of a very unhealthy appearance. Examination of the bulbs shows numerous pale-white 6- or 8-legged mites, sheltering and feeding behind the bulb scales. Infested bulbs with reddish-brown spots on the bulb scales....................................*Bulb mite*, page 756.

F. *Various small creatures, not true insects, attacking mostly the roots or leaves on the soil surface:*
1. Light-gray or slate-colored, flat-bodied, distinctly segmented creatures, usually about ½ inch long, with 7 pairs of legs, hiding under clods or bits of plant refuse on the surface of the benches, and feeding mainly at night on the roots and tender portions of nearly all greenhouse plants. When disturbed, they usually roll themselves into small tight balls.................................... ....*Sowbugs* or *pillbugs* (not true insects), page 757.
2. Hard-shelled, very active, many-legged creatures, up to 2 inches in length, usually with 2 pairs of legs on each body segment, crawling over the surface of greenhouse benches or hiding under any shelter on the surface of the soil. Bodies usually have a brown or pinkish-brown color. Creatures scuttle about very actively when disturbed. Are most abundant in damp parts of the bench.....................*Greenhouse millipedes* (not true insects), page 758.
3. Roots and underground parts of stems scarred or eaten off by small, nearly white, very active, many-legged creatures, ¼ inch long or less. They are very strongly repelled by light.. ....*Garden centipede, Scutigerella immaculata* (Newport), page 759.
4. Soft, gray or gray-and-brown-spotted, slimy, legless creatures, from ½ inch to as much as 4 inches in length, crawling about on the surface of the soil or on the plants. A sticky, viscid secretion is given off from the body and, on drying, forms a shiny trail where the creatures have crawled. Usually abundant only in the damper parts of the greenhouse, where they will be found under decaying wood, flower pots, and other shelters. They feed at night on the plant tissue of many greenhouse crops........*Slugs* or *snails* (not true insects), page 760.
5. Plants presenting a sickly appearance, often somewhat distorted, but with no visible injury to the parts above ground. An examination of the roots will show numerous knots or galls, or the roots distinctly swollen, enlarged, and of a gouty appearance. The cause of the injury cannot be seen, except when the roots are examined under high magnification, when numerous, very minute, nearly transparent, worm-like creatures will be found in the tissue of the swollen and distorted roots..........*Nematodes* or *eelworms* (not true insects), page 760.

255 (213). STRAWBERRY ROOTWORM[1]

Importance and Type of Injury.—Roses attacked by this beetle have the leaves riddled with small holes, often to such an extent that they appear to have been peppered with shot. The bark is eaten from the new growth, and the eyes of the buds

[1] *Paria canella* (Fabricius), Order Coleoptera, Family Chrysomelidae.

are frequently eaten out.   An examination of the roots will show small curved-bodied whitish grubs eating off the small feeding roots and scoring the bark of the larger roots (see also page 668 and Fig. 444).

Plants Attacked.—Rose, raspberry, blackberry, grape, oats, rye, peach, apple, walnut, butternut, wild crab, mountain ash, and several others.

Life History, Appearance, and Habits.—The females of the strawberry rootworm are about ⅛ inch long and two-thirds as wide (Fig. 445).   Most of the beetles are brownish, but some are brownish-black, usually with four black spots on the wing covers.   They lay their eggs in bunches of from 4 to 15 on dead leaves on the surface of the benches.   The eggs hatch in from 10 days to 2 weeks, and the small grubs work their way into the soil, and begin feeding on the rootlets.   They become full-grown in from 35 to 60 days.   They then change in the soil into the soft white pupal stage, in which they remain for about 2 weeks, emerging as adults at the end of this period. There are several generations a year under greenhouse conditions.   While this insect has been known for a number of years as a pest of strawberries, its injury to greenhouse plants has been comparatively recent.   Both sexes of the insect are found out-of-doors, but only females have been found in the greenhouse.

Control Measures.—Spraying the plants thoroughly, immediately after they have been cut back, with a mixture of 8 pounds lead arsenate and 6 pounds of fish-oil soap in 100 gallons of water, will give good control.   Dusting with sulphur, 85 parts, and lead arsenate, 15 parts, will help in keeping down the beetles on the new growth.   It is also advisable in badly infested houses to cover the soil at this time with a heavy application of finely ground tobacco dust.   As the soil will not be watered during the drying-out period of the plants, this layer of dust will remain effective for some time and will kill many of the beetles or young larvae hatching from the eggs.   All dried leaves on the benches should be collected and burned at short intervals.

Reference.—*U.S.D.A., Dept. Bul.* 1357, 1926.

256 (123). Greenhouse Leaf Tyer[1]

Importance and Type of Injury.—The undersurfaces of the leaves are eaten by slender, pale-green, active caterpillars (Fig. 486), which leave the upper epidermis intact.   The leaves are sometimes covered with a light web, enclosing several leaves or drawing the parts of a single leaf together.

Plants Attacked.—Chrysanthemum, snapdragon, cineraria, aster, sweet pea, ageratum, ivy, rose, and many other *soft-leaved* greenhouse plants, as well as many out-of-door weeds and cultivated crops such as celery, beets, spinach, beans.

Fig. 485.—Adult of the greenhouse leaf tyer, *Phlyctaenia rubigalis* (Guenee), in resting posture, about twice natural size. (*From Can. Dept. Agr. Bul.* 7, *n.s.*)

Life History, Appearance, and Habits.—The adult moths (Fig. 485) are of a brownish color, with the front wings crossed by dark wavy lines of a distinctive pattern, as shown in the figure.   The wings expand to about ¾ inch.   They remain quiet about the greenhouse most of the day, flying actively at night.   When disturbed during the day, they fly with a jerky motion, seeking shelter on the underside of the leaves or other objects after a very short flight.   The female moths lay their flattened, scale-like, watery-looking, shagreened eggs singly or in over-lapping groups on the undersides of the leaves, especially close to the soil.   These eggs hatch in about 5 to 12 days into small slender caterpillars which

---

[1] *Phlyctaenia rubigalis* (Guenee) (= *P. ferrugalis* Hübner), Order Lepidoptera, Family Pyralididae.

soon take on a pale-green color. When full-grown, the caterpillars are slightly less than ¾ inch long, of a pale-yellow color, with a broad white stripe running lengthwise over the back, and a dark-green band in the center of this white stripe. The younger ones have black heads. The larvae are almost invisible against a leaf. When ready to

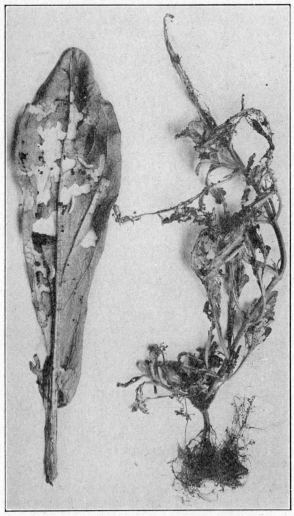

Fig. 486.—Destructive work of the greenhouse leaf tyer. Leaf of marigold at left, showing caterpillars at work; ageratum plant at right, destroyed by the caterpillars. (*From Can. Dept. Agr., Bul. 7, n.s.*)

pupate, the caterpillars usually form a shelter by rolling over the edge of the leaf and fastening it together with threads of silk. Inside this shelter, they spin thin silken cocoons, and then change to the pupal stage. After 10 or 12 days, the adult moths emerge from these cocoons. The length of the entire life cycle is about 40 days, and there may be 7 or 8 generations a year in greenhouses. The insects appear to be adversely affected by temperatures above 80 or 85°F.

Control Measures.—Good greenhouse sanitation is one of the most effective ways of preventing attacks of the greenhouse leaf tyer. Fumigation with calcium cyanide, as recommended for the control of greenhouse aphids, is quite effective in killing the adult leaf tyers. Such fumigations, however, kill only the adult stage of the insect. If the plants grown in the greenhouse are sufficiently hardy to withstand a fumigation with this material about once a week until four or five fumigations have been applied, practically all the leaf tyers can be cleaned out of the houses. Dusting infested plants with a dust consisting of 85 parts dusting sulphur and 15 parts lead arsenate is very effective in killing the larvae if thoroughly applied with a good dust gun. This treatment will burn some of the more tender plants and must be used with caution. Lead arsenate at the rate of 3 pounds to 100 gallons of water, plus a good spreader, is effective against the larvae and can be used on stock plants or when the plants are small. Because of the discoloration caused by the lead arsenate, pyrethrum-bearing insecticides are most satisfactory after the plants are half-grown. As a means of destroying the adults, fumigation with nicotine, applied as strong as the plants will bear and repeated at weekly intervals, is fairly effective. It aids in keeping down these insects if the adults are regularly caught in light traps. White 200-watt lamps should be hung at about 75-foot intervals along the benches, with a large pan containing a little water and kerosene beneath them. A light fumigation of nicotine will help to stir up the moths and bring them to the light traps.

References.—Jour. Agr. Res., vol. 29, no. 3, 1924, and **30:** 777–792, 1925; *Ill. Nat. History Surv. Cir.* 12, pp. 39–42, 1930.

257. OBLIQUE-BANDED LEAF ROLLER[1]

Importance and Type of Injury.—This leaf roller is common in greenhouses and also attacks the foliage of a great many shade trees, vegetables, and ornamental shrubs and flowers. The type of injury is much the same as that of the greenhouse leaf tyer. The pale-green black-headed young larvae, after hatching from the eggs, live for a time as leaf miners, and then feed for the remainder of their life on the undersides of the leaves. The adult moth is a little over 1 inch across the wings, of a reddish-brown color, with the front wings crossed by three distinct bands of dark brown. The control measures are the same as for the greenhouse leaf tyer.

Reference.—Jour. Econ. Ento., **2:** 391–403, 1909.

258. FLORIDA FERN CATERPILLAR[2]

Importance and Type of Injury.—Ferns attacked by this insect have the leaflets stripped from the old growth, and the new growth entirely eaten away. Large ferns may be completely stripped of leaves in 1 or 2 days. Ferns are disfigured to such an extent that they are useless for decorative purposes and cannot be sold.

Life History, Appearance, and Habits.—The adult moth is of a general brown color with a rather dark V-shaped patch near the center of the wings. The front wings, as a whole, are variegated. The wings expand about 1 inch. The eggs are laid on the undersides of the leaves and hatch in from 5 to 7 days into uniformly pale-green larvae. These larvae later change to a slightly darker green or to velvety black. There are two wavy white lines down each side of the body, and a central stripe of somewhat darker color. The full-grown larvae are about 1½ inches in length. These caterpillars feed

---

[1] *Cacoecia* (= *Archips*) *rosaceana* (Harris), Order Lepidoptera, Family Tortricidae.
[2] *Callopistria floridensis* Guenee, Order Lepidoptera, Family Noctuidae.

very actively at night, cutting off and devouring the fern leaves. On becoming full-grown, they work their way into the soil and there spin a cocoon in which they change to reddish-brown pupae. The pupal stage lasts about 2 weeks. There may be a complete generation of this insect every 7 or 8 weeks.

Control Measures.—Arsenical poisons have not been found particularly effective against this caterpillar. Recent experiments in Canada have shown that dusting ferns with pyrethrum powder, applying the dust twice each week, was very effective in controlling the fern caterpillar. A spray composed of:

Fresh pyrethrum powder.................................... ½ lb.
Common laundry soap....................................... ¼ lb.
Water.. 8 gal.

was also very effective in poisoning larvae of this insect. Pyrethrum or derris extracts used according to manufacturer's directions will give control.

259 (4, 67, 80). Cutworms[1]

The same cutworms that attack garden and field crops outdoors occur also in greenhouses. The liberal use of the poisoned-bran bait as recommended for the control of cutworms, on page 254, will be found effective in cleaning them out of greenhouse benches or beds.

260 (12, 63, 71, 111). Corn Earworm and Other Caterpillars

A full description of the corn earworm is given on pages 368 and 525. It frequently happens that the moths when abundant in the fall of the year will enter greenhouses in large numbers and lay their eggs on many of the greenhouse plants, particularly chrysanthemums. The injury caused by young earworms feeding on the chrysanthemum buds is often very severe. Dusting infested chrysanthemums with calcium arsenate, using the grade of calcium arsenate recommended for cotton dusting, has proved very effective in the control of these insects. They may also be killed by dusting the plants with calcium fluosilicate. There is a slight danger of burning with the calcium arsenate, but practically none with the calcium fluosilicate. The European corn borer (page 360), the cabbage looper (page 533), and the yellow woolly bear (page 476) also sometimes attack greenhouse plants.

261. Greenhouse Thrips[2]

Importance and Type of Injury.—Thrips occur very generally in greenhouses, and several species of the insect are found feeding on various plants. The surface of the leaves becomes whitened and somewhat flecked in appearance. The tips of the leaves wither, curl up, and die. Buds fail to open normally (Fig. 487). The underside of the leaves will be found spotted with small black specks. Where such spots are numerous, the appearance of the foliage becomes marred and, in many cases, plants are rendered so unsightly that they cannot be used for decorative purposes.

Plants Attacked.—Practically all plants found in greenhouses are attacked by thrips. Some of those which suffer most severely are roses, carnations, cucumbers, fuchsias, chrysanthemums, crotons, and cinerarias.

Life History, Appearance, and Habits.—There are slight variations in the life history of the species found in greenhouses. A typical life history is about as follows: The female insect deposits her eggs in slits in the leaf,

---

[1] Order Lepidoptera, Family Noctuidae.

[2] *Heliothrips haemorrhoidalis* (Bouché), Order Thysanoptera, Family Thripidae. The onion thrips (page 526), the flower thrips (page 674), the sugarbeet thrips, and others frequently do damage in greenhouses also.

inserting the minute white eggs in the leaf tissues. These eggs hatch in from 2 to 7 days, into active, very pale, white nymphs. The nymphs feed on the tissues of the leaf, rasping the leaf tissue with their mouth stylets and sucking up the sap which flows from the injured area. They pass through four instars in the course of their growth and, in the last two instars, they are inactive for a few days before changing to the adult. The adults of different species vary in color, some being yellowish, others nearly black, and others dark brown. They are less than $\frac{1}{10}$ inch long, slender-bodied, and are possessed of three pairs of legs, and four very slender wings with a fringe of long hairs around their margins (Fig. 487). The greenhouse thrips is dark brown with the appendages light-colored and $\frac{1}{20}$ to $\frac{1}{24}$ inch long; the antennae are eight-segmented; and the body surface is reticulated. Under greenhouse conditions generations follow each other throughout the year, the total time required for each genera-tion being from 20 to 35 days, depending on the climatic conditions and the species of thrips.

FIG. 487.—The greenhouse thrips, *Heliothrips haemorrhoidalis* (Bouché). Damaged rosebuds and adult thrips; the latter about 4 times natural size. (*From Can. Dept. Agr., Bul. 7, n.s.*)

Control Measures.—Spray-ing the undersides of the leaves with a sweetened poison mix-ture made by boiling together 5 pounds of soap, 1 pound of paris green, and 5 pounds dark-brown sugar in 2 quarts of water, until the sugar is all dissolved, is very effective when applied 1 quart of this stock solution in 25 gallons of water. Fumigation with naphthalene volatilized at the rate of 1 to $1\frac{1}{2}$ ounces per 1,000 cubic feet, or with nico-tine (page 284) or with hydro-cyanic acid (page 278), are effective. A spray of derris (0.0056 per cent rotenone with other extractives) and pyrethrum (0.01 per cent pyrethrins), with sulphonated castor oil (1 to 300) as a spreader, was found to be very effective. Rotenone dusts may also be used.

262. GLADIOLUS THRIPS[1]

Importance and Type of Injury.—Since 1929 this pest has spread with great rapidity over practically the entire United States, resulting in great commercial losses and incalculable disappointment to home gardeners and gladiolus fanciers. The injury is of three distinct types: (*a*) The foliage becomes blasted, with a characteristic silvery appearance, even-tually browning and dying, where the minute slender insects have punc-tured the surface cells and sucked up the exuding sap; (*b*) the flowers become flecked, spotted, and greatly deformed, and many spikes fail to

---

[1] *Taeniothrips gladioli* Moulton & Steinwender, Order Thysanoptera, Family Thripidae. Very similar injury to growing iris is caused by the iris thrips, *Bregmato-thrips iridis* Watson (see *U.S.D.A., Cir. 445*, 1937).

bloom at all; (c) the corms in storage are rendered sticky by sap exuding from punctured cells, become darker in color, and their surfaces corky, russeted, and greatly roughened. Infested corms may fail to germinate or may develop a weakened root system and plants that produce only small flowers or none at all.

Plants Attacked.—Gladioli, iris, lilies.

Distribution.—Probably over the entire United States where gladioli are grown.

Life History, Appearance, and Habits.—Apparently the insect does not winter out-of-doors except in very warm climates, but at digging time many of the thrips leave the foliage, collect on the corms, and hibernate where they are stored, or, if infested corms are stored at temperatures above 60°F., breeding and development may continue in storage. When infested corms are planted, the insects work along the developing shoot to the leaves and flowers. Although one of the largest of the thrips, the adults are only $\frac{1}{16}$ inch long and very slender. The basal third of the slender bristly wings and the third segment of the antennae are nearly white, in sharp contrast to the general brownish-black color of the body. The females may produce offspring either with or without mating, but the unfertilized eggs always develop into males. The tiny kidney-shaped eggs are thrust into the tissues of the growing plants or the corms. At favorable temperatures the eggs hatch in about a week, and the minute, yellowish, first- and second-instar nymphs begin feeding in a manner like the adults and become nearly as large in about a month. There follow two additional nymphal instars, during which the wing pads appear on the body and the nymphs are quiescent and take no food. A complete generation may require from 2 weeks (at 80°F.) to about a month (at 60°F.), and there may be six generations from June to September in the greenhouse. Effective temperatures for development are from 50 to 90°F.

Control Measures.—There is evidence that certain varieties of gladioli are resistant to injury by this pest. The most important control is to avoid planting infested corms and, in warmer areas, to avoid planting on soil infested the preceding year. The corms can be freed of the pests by placing them in paper bags or in trays or boxes covered with canvas or large sheets of paper and sprinkling over them flake naphthalene at the rate of 1 ounce per 100 corms or 1 pound for 2,000 corms. After 3 weeks, or at temperatures of 70°F. after 10 days, the excess naphthalene should be removed and the bulbs aired. Bichloride of mercury or corrosive sublimate (poison), dissolved at the rate of 1 ounce in $7\frac{1}{2}$ gallons of water in

	Small Quantity for Home Garden	Large Quantity for $\frac{1}{10}$ Acre
(A)		
Tartar emetic	2 oz.	4 lb.
Brown sugar	$\frac{1}{2}$ lb.	16 lb.
Water	3 gal.	100 gal.
(B)		
Paris green	1 rounded tbsp.	1 to 2 lb.
Brown sugar	2 lb.	65 lb.
Or		
Light grade molasses	$\frac{3}{4}$ pt.	3 gal.
Water	3 gal.	100 gal.

wooden or earthenware vessels, may be used by submerging the bulbs completely in the solution for from 8 to 12 hours and then planting or drying and placing in storage. The thrips can all be destroyed very effectively by submerging the corms in a hot-water bath between 112 and 120°F. for 20 to 30 minutes. After the plants start growing, they should be watched carefully, and, if the thrips or the characteristic injury appear upon them, the plants should be very carefully sprayed but not to the point of dripping, three to six times, at intervals of 2 to 6 days, with one of the formulas as shown on page 738.

References.—Ohio Agr. Exp. Sta. Bul. 537, 1934; *Colo. State Ento. Cir.* 64, 1935.

Scale Insects

Many greenhouse plants are attacked by some 20 to 25 species of scale insects which occur commonly in greenhouses. It is impossible to treat, in this book, all these different species separately. These species are figured and described by Dietz and Morrison in "The Coccidae or Scale Insects of Indiana," a bulletin from the office of the State Entomologist of Indiana, 1916.

The scale insects found in greenhouses may be divided into two large groups: (*a*) those having a distinct, hard, separable shell or scale over their delicate bodies, known as *the armored scales* or *Diaspidinae;* and (*b*) those in which the hard shell is not separable from the body, *the tortoise scales, soft scales,* or *Lecaniinae.* The tortoise scales have their bodies rounded and resemble somewhat the back of a turtle.

263. Armored Scale Insects[1]

In the first class, the armored scales, reproduction takes place by means of eggs in most cases, although in a few of these species the young are born alive. In cases where the eggs are laid, these eggs are protected by the scale of the mother insect until they hatch. In whichever manner the young are produced, they crawl from beneath the scale of the parent and move about actively for a short time, until, upon finding a location on the plants which seems favorable to them, they insert their thread-like mouth parts through the epidermis of the leaf or bark and begin feeding by sucking the sap. After feeding a short time, they molt and in this process lose their legs and antennae. The cast skin is incorporated into the scale, which now forms over the body of the insect and which is composed of fine threads of wax which have exuded from the body wall of the scale and have run together. The female scales molt twice during their life but always remain under the scale for their entire life. The males, after their second molt, have a more elongated body and, after a third and fourth molt, assume the adult form. In this stage they are very minute two-winged yellowish insects, with antennae,

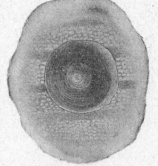

Fig. 488.—Ivy or oleander scale, *Aspidiotus hederae* (Vallot), on a bit of leaf. About 9 times natural size. (*From Ill. State Natural History Surv.*)

[1] Order Homoptera, Family Coccidae.

eyes, three pairs of legs, and a rather prominent long appendage projecting from the tip of the abdomen. They move about actively, seek out the female scales, and mate with them, but do not feed in this stage. After the female has mated, she continues to feed for some time and produces her eggs, or in the case of a few species as above mentioned, brings forth living young. The life history of all the armored scales is essentially the same.

Fig. 489.—The Florida red scale, *Chrysomphalus aonidum* (Linné). Scales of adults and nymphs on palm leaf, about 6 times natural size. (*From Mich. Agr. Exp. Sta.*)

A few of these scales, which are most common in greenhouses in the United States are:

Ivy or oleander scale, *Aspidiotus hederae* (Vallot) (Fig. 488).
Latania scale, *Aspidiotus lataniae* Signoret.
Greedy scale, *Aspidiotus camelliae* Signoret.
Florida red scale, *Chrysomphalus aonidum* (Linné) (Fig. 489).
Red or orange scale, *Aonidiella* (= *Chrysomphalus*) *aurantii* (Maskell) (Fig. 449).
Palm scale, *Chrysomphalus dictyospermi* (Morgan).
Boisduval's scale, *Diaspis boisduvalii* Signoret.
Cactus scale, *Diaspis echinocacti* (Bouché).
Rose scale, *Aulacaspis rosae* (Bouché).
Cyanophyllum scale, *Aspidiotus cyanophylli* Signoret.
Mining scale, *Howardia biclavis* (Comstock).
Camellia scale, *Fiorinia fioriniae* Targioni.
Fern scale, *Hemichionaspis aspidistrae* (Signoret) (Fig. 490).
Thread scale, *Ischnaspis longirostris* (Signoret) (Fig. 491).
Chaff scall, *Parlatoria pergandii* Comstock.

This class of scales attacks a wide variety of plants, including palms, Ficus, lantana, citrus plants, oleander, ivy, hibiscus, rose, and several other greenhouse plants.

Control Measures.—Spraying the plants or washing them with soft brushes dipped in a spray consisting of 1 pound of potash fish-oil soap in 3 gallons of water is effective in cleaning up these scales, if the treatment is persisted in until several applications have been made. Two hours after using this solution the plants should be washed with clear water in order to avoid any possible burning. Thiocyanate sprays, such as

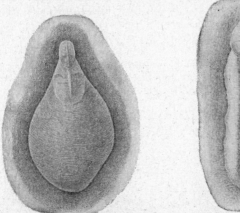
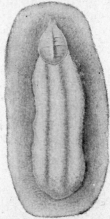

Fig. 490.—The fern scale, *Hemichionaspis aspidistrae* (Signoret). Female on a bit of leaf, about 20 times natural size, on the left. Male on a bit of leaf, about 60 times natural size, on the right. (*From Ill. State Natural History Surv.*)

Fig. 491.—The thread scale, *Ischnaspis longirostris* (Signoret), on a bit of palm leaf. About 15 times natural size. (*From Mich. Agr. Exp. Sta.*)

Lethane, and white-oil emulsions are also recommended. Repeated spraying with nicotine sulphate will aid in keeping the scales in check. Fumigation with hydrocyanic acid gas, used at the strengths recommended in the table on page 276 is effective in cleaning up infestations by this class of scale. Any of the above control measures should be repeated at 10- to 14-day intervals until all the scales have been killed. All incoming plants should be inspected and, if infested, should be freed of scale insects before they are placed in the greenhouse.

264. TORTOISE SCALE INSECTS[1]

This group of scales has a life history somewhat similar to that of the armored scales. The protective shell is, however, not formed of wax and the cast skins, but of the chitinous body wall, very little wax being

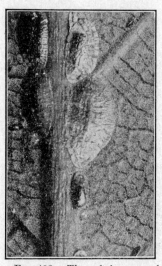

FIG. 492.—The soft brown scale, *Coccus hesperidum* Linné. Several scales on a bit of leaf, about 5 times natural size. (*From Can. Dept. Agr.*)

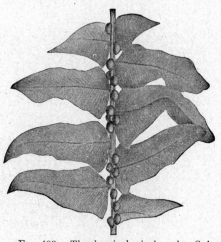

FIG. 493.—The hemispherical scale, *Saissetia hemisphaerica* (Targioni), on fern leaf. About natural size. (*From Ill. State Natural History Surv.*)

secreted. The body is generally smooth in outline, often very convex, and brown, black, or mottled in color. The young female scales may move about for a time after they have begun to feed. They retain their legs and antennae through adult life (Fig. 404,*a*). They reproduce by means of eggs and living young. The males in this class of scales, also, are winged and similar to those described above. Among the more common scales of this class are:

Soft brown scale, *Coccus hesperidum* Linné (Fig. 492).
Tessellated scale, *Eucalymnatus tessellatus* Signoret.
Hemispherical scale, *Saissetia hemisphaerica* (Targioni) (Fig. 493).
Black scale, *Saissetia oleae* (Bernard) (see Fig. 450).
Long scale, *Coccus elongatus* (Signoret) (Fig. 494).
The soft brown scale[2] (Fig. 492) differs from most of the tortoise scales in being very flat. It is oval in outline and greenish to brownish in color, often resembling very closely the part of the plant on which it is resting. It is also peculiar in being

[1] Order Homoptera, Family Coccidae.
[2] *Coccus hesperidum* Linné.

ovoviviparous, one or two young being born daily for a month or two. The young are sluggish, usually settle down near the female, and grow gradually to the adult female stage in about 2 months. It is a very general feeder in the greenhouse and also attacks tropical and subtropical fruits out-of-doors.

The hemispherical scale[1] (Fig. 493) is very convex, elliptical in outline, smooth and brown. They rarely move from the point where the young nymphs settle and insert their mouth parts. Growth is slow, there being only about two generations a year. Males are rare and reproduction is generally by parthenogenesis. From 500 to 1,000 eggs are laid beneath the female's body, the ventral wall shrinking dorsad to make a neat cup to protect them. It is a serious greenhouse pest and also attacks subtropical fruits.

The injury by these scales is very similar to that by the armored scales. The plants attacked include oleander, bay, Vinca, croton, cyclamen, fern, palm, Ficus, abutilon, and several other plants.

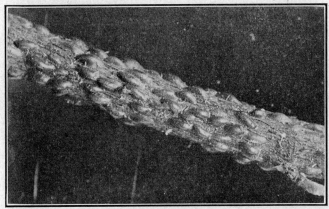

Fig. 494.—The long scale, *Coccus elongatus* (Signoret), adults on stem, about twice natural size. (*From Mich. Agr. Exp. Sta.*)

Control Measures.—Soft scales can be controlled by the same measures as those given for the control of the armored scales. One ounce of 40 per cent nicotine sulphate and 3 ounces of mild soap flakes in 3 gallons of water make an effective spray for this class of scales. Sprayed plants should be protected from direct sunlight for 2 hours, after which they should be washed with clear water.

265 (223). MEALY BUGS[2,3,4]

Mealy bugs are closely related to the scale insects; in fact they belong to the same family. There are several species which occur commonly in greenhouses and on flowering plants in houses. They are all very much alike in their life history, and differ but slightly in appearance. Mealy bugs may be placed in two groups: (*a*) *The short-tailed mealy bugs*, which reproduce by laying eggs; including the common or citrus mealy bug,[2] the palm mealy bug,[3] and the Mexican mealy bug.[4] In all these the filaments that surround the body are of about equal length and none of

---

[1] *Saissetia hemisphaerica* (Targioni).

[2] *Pseudococcus citri* (Risso), Order Homoptera, Family Coccidae.

[3] *Pseudococcus nipae* Maskell.

[4] *Phenacoccus gossypii* Townsend & Cockerell.

them more than one-fourth the length of the body.   (*b*) *The long-tailed mealy bug*[1] gets its name from the four filaments near the tip of the abdomen which may be as long as the body.   This species does not form an egg sac but gives birth to nymphs.

In general, the life history is as follows: The adult mealy bugs deposit their eggs in a compact, cottony, waxy sack, beneath the rear end of the body, to the number of 300 to 600.   Egg-laying continues for 1 or 2 weeks, and, as soon as it is completed, the female insect dies.   These sacks containing eggs will be found chiefly at the axils of branching stems

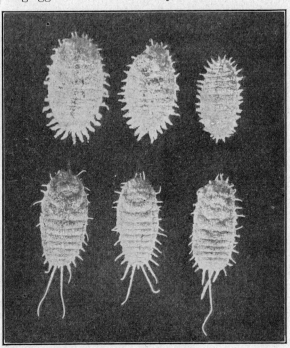

Fig. 495.—Mealy bugs.   Upper row, the citrus mealy bug; lower row, the long-tailed mealy bug.   All about 10 times natural size.   (*From Can. Dept. Agr., Bul. 7, n.s.*)

or leaves but occasionally on other parts of the plant.   Under greenhouse conditions the eggs hatch in about 10 days.   The young mealy bugs remain in the egg case for a short time and then crawl over the plants. They are flattened, oval, light-yellow, six-legged bugs, with smooth bodies.   They feed by inserting their slender mouth parts into the plant tissues and sucking the sap.   Shortly after they begin feeding, a white waxy material begins exuding from their bodies and forms a covering over the insect and about 36 prominent leg-like filaments, which radiate from the margin of the body on all sides.   They do not remain fixed but move about to some extent over the plant, although they are always very sluggish.   The female nymphs change but little in their appearance, except to increase in size, being about $\frac{1}{6}$ to $\frac{1}{4}$ inch long when full-grown. The males, when nearly grown, form a white case about themselves and,

[1] *Pseudococcus adonidum* (Linné) [= *P. longispinus* (Targioni)].

inside of this, transform into tiny, active, two-winged, fly-like insects. On emerging from the case, the males fly about actively and mate with the females, but very soon die.. The male is incapable of feeding in the adult stage. It takes about 1 month for the completion of a generation under greenhouse conditions.

Plants Attacked.—Many kinds of greenhouse plants are attacked by mealy bugs. They are especially troublesome on the soft-stemmed foliage plants, such as coleus, fuchsia, cactus, croton, fern, gardenia, and begonia. Among the many other plants attacked are citrus, heliotrope, geranium, oleander, orchids, poinsettias, umbrella plant, ivy, dracaena, and chrysanthemums.

Control Measures.—The best and most effective method of controlling mealy bugs is to syringe the plants with a white-oil emulsion or with nicotine sulphate with an activator. One method of controlling mealy bugs is to syringe the plants with water applied with as much force as the plants will stand without injury to the leaves. Fumigation with hydrocyanic acid gas is very effective in controlling the Mexican mealy bug and will kill many of the younger stages of other species. Relatively high concentrations of hydrocyanic acid gas ($\frac{1}{4}$ to $\frac{1}{2}$ ounce per 1,000 cubic feet, or more if the plants will stand it) for short periods of exposure are effective against all stages of the Mexican mealy bug and are less likely to injure plants than lower dosages for longer periods of time. Fumigations should be made at 5- to 7-day intervals for three to six times. Insecticides of the thiocyanate group, such as Lethane, and 40 per cent nicotine sulphate or free nicotine, 1 to 400, with a good spreader, are effective in killing the younger mealy bugs. Repeated applications are necessary to clean up the plants.

266. Greenhouse Orthezia[1]

This insect also is a close relative of the scales and mealy bugs. The small dark-green wingless bodies of the nymphs are about the size of pinheads and have rows of minute waxy plates extending back over their bodies. In the case of the female insect (Fig. 496) a white, waxy, fluted egg sac is attached to the body and extends backward for a distance of two or more times the diameter of the body, the total length being about $\frac{1}{16}$

Fig. 496.—Greenhouse orthezia, *Orthezia insignis* Douglas, nymph and adult females; the one on the right has the egg sack broken open to show the eggs and newly hatched nymphs. About 15 times natural size. (*From Mich. Agr. Exp. Sta.*)

inch. These insects resemble the mealy bugs quite closely in their habits and can be controlled by the same methods.

267 (82). Aphids or Plant Lice[2]

Aphids have already been described as pests of many of the outdoor crops (see pages 371, 477, 536). Everyone who has grown plants for a year or more in the greenhouse or flower garden is well aware that these

[1] *Orthezia insignis* Douglas, Order Homoptera, Family Coccidae.
[2] Order Homoptera, Family Aphididae.

little soft-bodied insects are also pests of ornamental crops.   A number of different species of aphids occur in greenhouses, and one or more of them attacks almost every species of plant that is grown under glass.   The aphids differ somewhat in size, appearance, and the color of their bodies. Some are green, others brown, reddish, or black in color.   Their life histories under greenhouse conditions, are very similar, except for some variation in the length of time which it takes the different species to develop.   So far as known, all species continue to go on, generation after

Fig. 497.—Rose aphids, *Macrosiphum rosae* (Linné), clustered on rosebuds.   Several times natural size.   Note the predaceous Syrphid fly larva eating an aphid on the lower part of the left-hand stem.   (*From U.S.D.A., Farmers' Bul.* 1362.)

generation, the year round, producing only the female forms; and these females, on becoming full-grown, give birth to living young without being mated.   Under greenhouse conditions, the true sexes do not appear, nor do the insects reproduce by means of eggs.   They feed by sucking the sap from the tender plants, often causing the plants to become deformed, the leaves curled and shriveled; and, in some cases, galls are formed on the leaves.   The insects are further injurious because of the fact that some species eject a sweetish honeydew, which is attractive to ants and on which certain sooty-appearing fungi grow.   This may render the plants so unsightly as to hinder their sale for any purpose.   Aphids are also the carriers of certain diseases of greenhouse plants. Some of the more common species occurring in the greenhouse are as follows:

Small green chrysanthemum aphid, *Myzus rosarum* Kaltenbach.
Black chrysanthemum aphid, *Macrosiphoniella sanborni* (Gillette).
Pea aphid, *Illinoia pisi* (Kaltenbach) (see Figs. 304 and 305).
Rose aphid, *Macrosiphum rosae* (Linné) (Fig. 497).
Corn root aphid, *Anuraphis maidi-radicis* (Forbes) (see Fig. 223).
Black violet aphid, *Neotoxoptera violae* Pergande.
Green peach aphid, *Myzus persicae* (Sulz).
Melon aphid, *Aphis gossypii* Glover.

Species of importance in the flower garden include:

Black bean aphid, *Aphis rumicis* Linné.
Green peach aphid, *Myzus persicae* (Sulz).
Columbine aphid, *Hyalopterus trirhoda* Walker.
Melon aphid, *Aphis gossypii* Glover.

Control Measures.—Aphids may be readily controlled under greenhouse conditions by fumigation, spraying, or dusting. Light applications of hydrocyanic acid gas, generated from calcium cyanide dust scattered upon the walks in the evening, are very effective. For most plants the dosage used should be $\frac{1}{4}$ ounce per 1,000 cubic feet of space. The same results may be secured by generating the gas from sodium cyanide by the pot method. One of the most effective and safest methods of destroying aphids is by the use of pressure fumigation with nicotine. A combination of nicotine and a combustible powder is prepared commercially. This mixture is sealed in cans of several sizes. To release the nicotine, two holes are punched in the can and the powder ignited with a slow match or sparkler. The pressure formed in the can forces out the nicotine vapor. Such pressure fumigators can be used for treating an entire house, or for "spot fumigation" of a part of a single bench. This product should be used according to the manufacturer's directions. For aphids in the greenhouse or flower garden, spraying with nicotine or ground derris or timbo root, or dusting with the same materials, is also a satisfactory method of control. Nicotine sprays are generally used at 1 pint nicotine sulphate and 1 pound fish-oil soap or other wetting agent to 100 gallons of water, or $\frac{1}{2}$ ounce of each to 3 gallons of water. Rotenone sprays are used at 0.75 per cent strength, with a good wetting agent. A dust containing 2 per cent nicotine or 1 per cent rotenone is also effective on some plants. Nicotine should not be used in any form on violets, ferns, or blooming orchids.

268. GREENHOUSE WHITEFLY[1]

Importance and Type of Injury.—Plants covered, especially on the undersides, with small, snow-white, four-winged flies and very small, oval, flat, pale-green nymphs, less than $\frac{1}{30}$ inch in length, which suck the sap (Fig. 498). Infested plants are lacking in vigor, wilt, turn yellow, and die. The leaves are covered with a coating of glazed, sticky material on which a sooty-colored fungus often grows, completely covering the foliage.

Plants Attacked.—Cucumber, tomato, lettuce, geranium, fuchsia, ageratum, hibiscus, coleus, begonia, solanum, and many other plants.

[1] *Trialeurodes vaporariorum* Westwood, and others, Order Homoptera, Family Aleyrodidae.

Life History, Appearance, and Habits.—The female whitefly deposits more than 100 minute yellowish eggs. These are attached to the underside of the leaves by a short stalk and are often laid in a small ring, as the female circles about with her mouth parts inserted in the leaf. On hatching, the nymphs are flat and nearly transparent. They settle upon the leaf, near the point where they hatch, and remain in this situation until they become adults. They suck the sap from the leaves, feeding greedily on the plant juices for about 4 weeks. In the course of this time, they pass through four instars. All the nymphs have fine, long and short, white, waxy threads radiating from their greenish bodies. The average duration of the nymphal periods is about 28 to 30 days. The

FIG. 498.—Greenhouse whitefly, *Trialeurodes vaporariorum* Westwood, adults and nymphs on underside of leaf. Four times natural size. (*From Can. Dept. Agr. Bul. 7, n.s.*)

adult whitefly is about ⅟₁₆ inch in length, very active, four-winged, with a yellowish body, and has the appearance of having been thoroughly dusted with some very fine white material (see Fig. 452). Both males and females fly, and they feed, like the nymphs, on the underside of the leaves, living from 30 to 40 days. Under greenhouse conditions the generations overlap, and all stages of the insect may be found on infested plants at any time.

Control Measures.—Fumigation with hydrocyanic acid gas is the best method of controlling the whitefly (page 278). The dosage used will depend on the plants infested and on the tightness of the greenhouse. One-eighth to one-fourth ounce of sodium cyanide per thousand cubic feet, or the equivalent in calcium cyanide, should be used. Three or four fumigations, at weekly intervals, will usually be necessary before the greenhouse is cleaned of these pests, because some of the nymphs and eggs will be likely to escape the first several fumigations.

If fumigation cannot be practiced, persistent spraying with nicotine sulphate or insecticides of the thiocyanate group, such as Lethane, as recommended for the control of mealy bugs (page 745), will be very effective. Such sprays will have to be applied four or five times before the plants are cleaned up, and it is advisable, on plants with tender foliage, to wash the plants with clear warm water 1 or 2 hours after spraying.

Reference.—*Twenty-seventh Rept. Ill. State Ento.*, p. 130, 1912.

269 (85). GREENHOUSE RED SPIDER MITE[1]

Importance and Type of Injury.—Leaves of plants infested by red spiders present a peculiar appearance. Those lightly infested have pale

[1] *Tetranychus telarius* (Linné), Order Acarina, Family Tetranychidae.

blotches or spots showing through the leaf. In heavy infestations the entire leaf appears light in color, dries up, often turning reddish-brown in blotches or around the edge (Fig. 499). Plants generally lose their vigor and die. The undersurface of lightly infested leaves will show silken threads spun across them. In heavy infestations these threads may form a web over the entire plant, upon which the mites crawl and to which they fasten their eggs. The underside of leaves, on close examination, will be found covered with minute eight-legged mites, showing as tiny, reddish, greenish, yellowish, or blackish, moving dots on the leaves. The color appears to vary in part with the kind of food.

Plants Attacked.—There are very few plants grown in greenhouses which are not subject to injury by red spiders. Some of the smooth hard-leaved plants, such as certain palms, are only moderately injured. Some of the worst injured are: cucumbers, tomatoes, carnations, chrysanthemums, melons, sweet peas, snapdragons, violets, roses, *Asparagus plumosus*, snapdragon, and fuchsia. Out-of-doors arborvitae, cedars and other evergreens, hydrangeas, roses, and clematis, suffer severely. Of all the plants infested, control is probably most difficult on roses.

Life History, Appearance, and Habits.—The adult female red spider is an eight-legged pale-yellow or greenish mite, about $\frac{1}{60}$ inch in length (Fig. 499). The male is slightly smaller, being only about $\frac{1}{80}$ inch long. Two dark spots, composed of the food contents, show through the transparent body wall. The body is oval in outline and sparsely covered

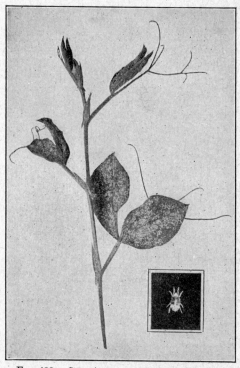

FIG. 499.—Greenhouse red spider mite, *Tetranychus telarius* (Linné). Injury to foliage of sweet pea, about natural size. Adult mite below, about 12 times natural size. (*From Can. Dept. Agr., Bul. 7, n.s.*)

with spines. The mite feeds through sucking mouth parts with which it pierces the epidermis of the leaf. After mating, the female red spiders begin laying eggs at the rate of from 2 to 6 a day, each depositing a total of 70 or more eggs during her lifetime. These eggs are spherical, shiny, and very minute; they are attached to the underside of the leaves, usually to the web which the mite spins wherever it goes over the plant. The eggs hatch in from 4 to 5 days into small, crawling young, which closely resemble the adults except that they have only three pairs of legs. The female mites molt three times in the course of their growth, the males molting only twice. A complete generation is produced every 20 to 40 days, but these generations overlap, so that all stages of the mites may be found in the greenhouse at any time.

Control Measures.—Red spiders cause the most damage in greenhouses where plants are grown that require a high temperature and rather dry atmosphere. Frequent syringing or spraying of plants with a stream of clear water, applied with sufficient force to tear up the webs of the mites and knock them from the plants, is one of the most effective methods of control. This treatment cannot be applied to many species of plants, which are injured by frequent watering or develop certain diseases if grown under moist conditions. If houses can be completely cleared of growing plants, red spiders may be cleaned out by heavy fumigations, using sodium cyanide at the rate of 1 pound to 1,000 cubic feet, or sulphur burned at the rate of 5 pounds per 1,000 cubic feet. *Such fumigations cannot be given when there are any valuable plants in the house.* The difficulty of control varies greatly on different kinds of plants. Spraying with certain organic thiocyanates, such as Lethane; or cyclohexylamine derivatives, such as Cyclonox, used 1 part to 400 of water, repeated in one week, then at monthly intervals; and derris or other rotenone sprays, with sulphonated castor oil as a spreader, have been found effective. It should be noted that neither nicotine nor hydrocyanic fumigations are effective for red spider mites, but naphthalene fumigation, as described on pages 286–737, has been used successfully. Frequent dusting of the beds and leaves of plants with superfine sulphur is also of value in red spider control. As the mites survive the winters outdoors and frequently become very abundant outside of the greenhouse, it is highly advisable to keep down all rank growth of weeds and grasses within the immediate vicinity, as the spiders often breed outside and later gain access to the greenhouses.

References.—*U.S.D.A., Farmers' Bull.* 1306, 1923; *Fifteenth Rept. S.D. State Entomol.,* 1925; *Mass. Agr. Exp. Sta. Bul.* 179, 1917; *Boyce Thompson Inst., Contrib.,* **2**: 512–522, 1930; *Jour. Econ. Ento.,* **30**: 512–522, 1937, and **31**: 211–216, 1938.

270. CYCLAMEN MITE[1]

Importance and Type of Injury.—Infested plants have the leaves distorted, with the buds failing to open or with small distorted flowers, presenting a streaked and blotchy appearance (Fig. 500). The foliage shows purplish areas. Small, white, green, or pale-brown mites, with six or eight legs, work about the base of the plants or in the buds or the injured areas on the leaves.

Plants Attacked.—Cyclamen is injured more than any other plant. This mite is also recorded as a pest of snapdragon, geranium, chrysanthemum, larkspur, begonia, fuchsia, and petunia; and of strawberry, out-of-doors.

Distribution.—It is an imported species. It was first noticed in New York in 1898 and in Canada in 1908. It is now generally distributed in greenhouses throughout the country.

Life History, Appearance, and Habits.—The adult mites are pale, shiny, brown creatures with four pairs of legs. The first instar is paler in color, glassy-looking, and has only three pairs of legs. The eggs are laid about the bases of cyclamen and other food plants and also in

---

[1] *Tarsonemus pallidus* Banks, Order Acarina, Family Tarsonemidae.

the injured areas in the leaves. They are only about $\frac{1}{50}$ inch long and can barely be seen with the naked eye. All stages of the mites will be found on the foliage of infested plants, usually in greatest numbers about the base of the plants and the flowers or buds. A generation may be completed in about 2 weeks.

Control Measures.—Great care should be used to avoid bringing infested cyclamen plants into a greenhouse. On geranium, snapdragon, and chrysanthemum the mites may be controlled by spraying with 40 per cent nicotine sulphate, 1 ounce, fish-oil soap, 4 ounces, and water, 4 gallons. This spray should be applied only in cloudy weather. Rotenone sprays are also effective. Immersing plants in water at 110°F. for 15 minutes kills the mites. The mites have been effectively destroyed on strawberry plants, by exposing the plants to a temperature of 110°F.

Fig. 500.—Cyclamen blooms destroyed by the cyclamen mite, *Tarsonemus pallidus* Banks; healthy bloom at right. About natural size. (*From Can. Dept. Agr., Bul. 7, n.s.*)

for 2 hours in a special chamber where the air was nearly saturated with water vapor. Infested empty houses may be cleaned up by thorough fumigation with sodium cyanide, 1 pound per 1,000 cubic feet. This treatment cannot be used in houses where there are growing plants.

References.—*Proc. Ento. Soc. Wash.*, **39**: 267–268, 1935; *Jour. Econ. Ento.*, **28**: 91–99, 1935.

271. Chrysanthemum Midge[1]

Importance and Type of Injury.—Infested plants have the leaves misshapen, in cases of light infestations, with small, somewhat blister-like cone-shaped galls on the upper surfaces. In severe infestations, the leaves are curled, the flowers are distorted, with crooked stems, and with numerous galls along the stems and leaves (Fig. 501).

Plants Attacked.—Chrysanthemums of all varieties.

Distribution.—This insect is a late importation from Europe, having first been found in this country in 1915. It is now generally distributed in greenhouses over the United States and Southern Canada.

[1] *Diarthronomyia hypogaea* Loew, Order Diptera, Family Cecidomyiidae.

Life History, Appearance, and Habits.—The adult insect is a very frail little long-legged orange-colored gnat, about $\frac{1}{14}$ inch in length. The female flies lay about 100 very minute orange-colored eggs on the surface and tips of the new growth of the chrysanthemums. The maggots hatching from these eggs, in 3 to 16 days, bore their way into the plant tissues. The irritation resulting from their feeding causes the growth of small cone-shaped galls. On the leaves these galls are usually on the upper side, but they may occur also along the stems, many galls developing so close together that they form masses or knots on the stems. Inside these galls the maggots develop, reach their full growth, and pupate. On emerging, the empty pupal skin will usually be left protruding from the gall. In nearly all cases, the adult flies emerge after midnight and before morning. The life cycle requires, on the average, about 35 days. There may be five or six generations a year under greenhouse conditions.

FIG. 501.—Injury caused by the chrysanthemum midge, *Diarthronomyia hypogaea* Loew. About natural size. (*From Mich. Agr. Exp. Sta. Spcl. Bul. 134.*)

Control Measures.—Careful inspection of all plants brought into the greenhouse, with the destruction of those found infested, is the best method of avoiding trouble from this insect. Galls are most likely to be found on varieties with light-green foliage. All infested foliage should be picked off and burned. Spraying every second day with nicotine sulphate or free nicotine, 1 fluid ounce in 6 gallons of water, is very effective. The spray should be applied in the early evening. When a house is generally infested, nightly fumigation with a weak charge of hydrocyanic acid, applied between midnight and daylight, until all the pests have disappeared, is the most effective method of control. These insecticides kill the adults, which emerge at night; consequently nightly treatments over a considerable period of time are necessary to eradicate the pest.

Reference.—*U.S.D.A., Dept. Bul.* 833, 1920.

272. Rose Midge[1]

Importance and Type of Injury.—Flower buds are distorted, turning brown and dying. Tender growth is sometimes curled and brown, the buds and young shoots failing to develop. An examination of buds will show whitish maggots clustered inside, mainly at the base, or on the upper sides of the tender leaves and leaf petioles. The maggots are

[1] *Dasyneura rhodophaga* (Coquillett), Order Diptera, Family Cecidomyiidae.

about $\frac{1}{20}$ inch in length when full-grown. They are at that time somewhat tinged with red.

Plants Attacked.—Roses.

Life History, Appearance, and Habits.—The adult midge is a very small two-winged fly, about $\frac{1}{20}$ inch in length, and of a reddish or yellowish-brown color (Fig. 502). They are usually most abundant in greenhouses during the summer and early fall. The females deposit their very minute yellow eggs, inserting them into the buds, usually just behind the sepals of the flower buds, or in the unfolding leaves. The whitish maggots issuing from these eggs feed on the tender tissue of the new growth and inside the buds, becoming mature in 5 or 6 days. They then drop to the ground, where they spin a cocoon in which the pupal stage is passed. The length of the life cycle varies with the temperature of the house, but, under favorable conditions, a complete generation may appear every 20 days. Usually the winter is passed in cocoons in the soil.

Control Measures.—Probably the cheapest and most effective control for this insect is to cover the surface of the beds with finely-ground tobacco dust, containing at least 1 per cent nicotine, putting on a sufficient amount of dust to make a layer $\frac{1}{4}$ inch deep over the surface of the beds. This not only kills the full-grown maggots when they drop from the plants but also kills the pupae in the cocoons and has been found entirely effective for cleaning up this insect. In

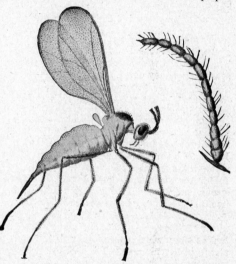

Fig. 502.—The rose midge, *Dasyneura rhodophaga* (Coquillett). Adult female, greatly enlarged, and antenna in detail. (*From Ill. State Natural History Surv.*)

applying the tobacco dust, it is necessary that care be used to make a complete covering over all parts of the beds and around the stems of the growing roses. In addition to covering the surface of the beds, all dirt walks and soil under the benches should be thoroughly sprayed with a 10 per cent oil emulsion.

Where the rose midge has become established in houses, a vigorous campaign should be waged against it, as it is possible for the insect to destroy the entire crop in a short time. Aside from the use of tobacco, all infested buds should be picked off and burned. Fumigation with nicotine every night for 2 or 3 months is fairly effective. Such fumigations kill the adults but do not injure the other stages of the insect. They are expensive, and too frequent fumigation may cause woody growth in the plants.

273 (14, 39, 87, 216). White Grubs

Importance and Type of Injury.—White grubs of the same species as those attacking the roots of plants outdoors (see Corn Insects, page 374) occasionally occur in numbers in greenhouses. They may be the grubs of the ordinary June beetles, or those of several other beetles, particularly of the genus Ochrosidia or the Japanese beetle (page 621) or cyclamen weevil.[1] In most cases, the insects are brought into the greenhouse with infested soil. In a few cases, such as the annual white grub (page 394), the beetles may gain access to the greenhouses at night, during the period of the year when the insects are abundant, and lay their eggs in the soil of pots or benches. The length of the life cycle under greenhouse conditions has never been carefully worked out. With the last-mentioned species, however, it does not occupy more than 1 year.

Control Measures.—The best method of controlling white grubs in greenhouse soils is to sterilize the soil with steam, hot water, or carbon bisulphide (see under Nematodes, page 761). A careful inspection should be made of all soil brought into the greenhouse, to make sure that it does not contain large numbers of these insects. The adult beetles may be kept out of the greenhouses by screening.

274. Other Greenhouse Beetles

Wireworms (see page 377) gain access to greenhouse soils in the same way as white grubs. The treatment for these insects under greenhouse conditions is the same. The cyclamen weevil[1] and the Japanese beetle (see page 621) are greenhouse pests in areas where they occur.

275. Fungus Gnats or Mushroom Flies[2]

Importance and Type of Injury.—Plants are lacking in vigor and leaves turning yellow, without visible injury to any part of the plant above the ground. Roots with small brown scars on the surface, or with small feeding roots and root hairs eaten off. Very small, thread-like, active, white maggots embedded in the root tissue, or working through the soil about the plant. Often injurious to potted plants. Root rots often follow the attacks of these insects. In mushroom houses the maggots destroy the spawn and tunnel through caps and stems of the mushrooms. These flies also transport the hypopi of mites (see page 757).

Plants Attacked.—Many species of plants grown in greenhouses.

Life History, Appearance, and Habits.—There are several species of so-called fungus gnats which cause injury to the roots and underground parts of the stems of plants in greenhouses. The adults are all very small, sooty-gray or nearly black, long-legged, slender flies or gnats, measuring from $\frac{1}{8}$ to $\frac{1}{10}$ inch in length. These flies deposit their eggs in clusters of 2 to 30 or more in the soil, each female laying from 100 to 300 eggs. The eggs are only about $\frac{1}{100}$ inch in length, almost too small to be seen, although the egg clusters may be readily found. After about 4 to 6 days, small legless maggots, with black heads and nearly transparent bodies, hatch and begin working their way through the soil. The maggots feed for 5 to 14 days and, on becoming full-grown, are about $\frac{1}{4}$ inch in length. They form a flimsy cocoon in or on the ground and there pupate. In 5 or 6 days the adult flies emerge from the cocoons and work their way to the surface of the soil. They live for about 1 week, while they mate and lay eggs for another generation.

Control Measures.—Thoroughly drenching the soil with a solution of mercuric bichloride, or corrosive sublimate, used at the rate of 1 ounce dissolved in 8 gallons of water, is very effective in killing these maggots and will not cause injury to any but the most tender-rooted plants. Mercuric bichloride is a corrosive and deadly poison, which should not be allowed to come in contact with the bare skin and should be used in glass or wooden containers. Allowing the soil to dry to as great a degree as possible without injury to the plants also is effective in killing many of the maggots; but it is seldom that all can be cleaned out by this method. Artificial light traps will destroy enormous numbers of the adults. Pyrethrum, derris, or timbo dusts, will kill the adults.

---

[1] *Brachyrhinus sulcatus* (Fabricius), Order Coleoptera, Family Curculionidae.
[2] *Sciara* spp. and others, Order Diptera, Family Mycetophilidae.

276. Narcissus Bulb Fly[1]

Importance and Type of Injury.—Bulbs of narcissus and other plants fail to grow. The bulbs become soft and the outer scales of the bulbs often have brown scars upon them. An examination of the bulbs will show a large whitish or yellowish-white maggot inside the bulb, feeding on the plant tissue (Fig. 503).

Plants Attacked.—Narcissus, hyacinth, amaryllis, galtonia, and several others.

Distribution.—The insect is of European origin and is now well established throughout the western part of Canada. It has frequently been found in shipments of foreign bulbs and has been reported from many points in the United States.

Life History, Appearance, and Habits.—The adult of the narcissus bulb fly is a shiny yellow-and-black hairy fly, about the size of a small bumblebee, which it somewhat resembles in appearance. The eggs are laid in the base of the leaves or in the necks of bulbs. The young maggots hatching from them bore into the bulb and rasp

Fig. 503.—Narcissus bulbs infested with larvae of the narcissus bulb fly, *Merodon equestris* (Fabricius), the bulb on the left opened to show the larva and its work; about natural size. (*From Can. Dept. Agr., Bul. 7, n.s.*)

or tear apart the plant tissues by means of their strong, somewhat hooked mouth parts. The maggots are whitish to yellowish-white in color, thick and fat in appearance, and reach a length of ¾ inch. The puparia are formed in the bulb or in the soil. There is probably not more than one generation a year.

Control Measures.—Vacuum fumigation of thoroughly cured and clean bulbs with 25 to 30 pounds of carbon bisulphide per 1,000 cubic feet for from 1¾ to 2 hours at a temperature between 70 and 80°F. is very effective. Treating bulbs with hot water has been found one of the most effective methods of controlling this insect. The bulbs should be submerged in water held at a temperature of 110 to 111.5°F. for 2½ hours. Treating in a chamber where the bulbs may be held at a temperature of 110 to 111°F. for a period of 2 hours has some advantages over the hot-water treatment. The air in the chamber should be held at near saturation by being drawn in through steam or water spray. When bulbs are taken up from the field, the infested ones can usually be sorted out by their lighter weight, softer condition, shrunken and corky basal plates, or partly decayed appearance. Bulbs showing indications of heavy infestation should be destroyed by burning, and the remainder of the lot sterilized in hot water.

References.—*Va. Truck Exp. Sta. Bul.* 60, 1927; *Jour. Econ. Ento.*, **21**: 342–357, 1928, and **25**: 1020–1026, 1932.

[1] *Merodon equestris* (Fabricius), Order Diptera, Family Syrphidae.

277. Lesser Bulb Fly[1]

Importance and Type of Injury.—The injury closely resembles that of the narcissus bulb fly. The maggots are grayish or yellowish gray, and the body is markedly wrinkled (Fig. 504). Many maggots will often be found in a single bulb. They reach a length of about ½ inch.

Plants Attacked.—Narcissus, hyacinth, amaryllis, onion, iris, shallot, and several other plants.

Distribution.—The distribution of this insect is about the same as that of the narcissus bulb fly. It is probably already established out-of-doors in some points in the United States.

Life History, Appearance, and Habits.—The adult fly is of a blackish-green color, ⅓ inch long, with the body nearly bare of hairs but with several white lunate markings on the sides of the abdomen. It somewhat resembles a small wasp. The general life history is the same as that of the narcissus bulb fly, but there are probably two generations a year.

Control Measures.—The control measures are the same as those suggested for the narcissus bulb fly.

Fig. 504.—The lesser bulb fly; *Eumerus strigatus* Fallen; bulb showing larvae, somewhat reduced. (*From U.S.D.A., Farmers' Bul.* 1362.)

278. Bulb Mite[2]

Importance and Type of Injury.—Bulbs of various species of plants rot and fail to produce growth (Fig. 505). Plants grown from bulbs turn yellow and present a general sickly appearance. The leaves of such plants are stunted and distorted, and the plants will generally fail to produce flowers or will produce only misshapen ones. Very small whitish mites, with six or eight brownish or pinkish legs, may be found in large numbers sheltering behind, or boring into, the bud scales.

Plants Attacked.—These mites attack practically all classes of bulbs, some of the more commonly infested being narcissus, lilac, hyacinth, crocus, gladiolus, amaryllis, and lily. They may also infest cereals.

Distribution.—These creatures are of European origin but have been now generally distributed throughout the United States and Canada in shipments of bulbs from Europe, Japan, and the Bermuda Islands.

Life History, Appearance, and Habits.—Infested bulbs will contain practically all stages of the bulb mite. These mites are ¹⁄₅₀ to ¹⁄₂₅ inch long, whitish, barely visible to the naked eye (Fig. 505). The eggs are laid behind the bud scales and soon hatch into the six-legged nymphs. After molting, these nymphs change to the eight-legged form and during this second instar of their life are most destructive to the bulbs. After two or more molts they become adults and begin to reproduce. Under certain conditions, probably unfavorable to the species, a remarkable, heavily chitinized, nonfeeding, but very active stage, known as the

---

[1] *Eumerus strigatus* Fallen, Order Diptera, Family Syrphidae.
[2] *Rhizoglyphus hyacinthi* Boisduval, Order Acarina, Family Tyroglyphidae.

hypopus, may intervene between the six-legged and eight-legged nymphal stages, lasting for 1 or 2 weeks before molting to the eight-legged nymph. The hypopi have a group of suckers on the ventral side, behind the anus. They readily attach to insects, mice, and other creatures and may be distributed in this manner to new and more favorable breeding places.

The mites apparently prefer healthy bulbs and migrate through the soil from the decaying bulbs to the more attractive food. They are readily transported from place to place in bulb shipments. The life of the female bulb mite is about 1 or 2 months, while that of the male is usually less. Each female deposits from 50 to over 100 eggs.

Control Measures.—All bulbs found infested by the mites should be destroyed by burning at the time of digging or when the bulbs are planted, either in the field or in the greenhouse. Infested bulbs can be recognized by their soft, mushy condition. They may be dipped for 10 minutes in a solution of 40 per cent nicotine sulphate, using 1 part of the nicotine to 400 parts of water, heated to 122°F. before dipping. If bulbs are submerged for 3 hours in water heated to 110 to 111.5°F., practically all the mites will be killed. Sound, healthy bulbs in which the root growth has not started will not be injured by this treatment. Prompt and rigorous destruction of all soft, rotten, and infested bulbs and the storage of bulbs at a temperature of 35°F. aid in control.

Fig. 505.—The bulb mite, *Rhizoglyphus hyacinthi* Boisduval; mites working in a rotten bulb, and their eggs; enlarged about 8 times. (*From Conn. Agr. Exp. Sta. Bul.* 225.)

Reference.—*Conn. Agr. Exp. Sta. Bul.* 402, 1937.

279. Sowbugs[1] or Pillbugs[2]

Importance and Type of Injury.—These creatures are not insects but are more closely related to crayfish. They frequently cause some damage in greenhouses. The plants infested show feeding about the roots and on the tender portions of the

[1] *Porcellio laevis* Koch, and others, Class Crustacea, Order Isopoda, Family Oniscidae.

[2] *Armadillidium vulgare* (Latrielle), Class Crustacea, Order Isopoda, Family Armadillididae.

stems near the ground. Light-gray to slate-colored fat-bodied and distinctly segmented creatures, with seven pairs of legs, and up to ½ inch long, will be found hiding
about the bases of plants or under clods or bits of manure on the surface of the benches
(Fig. 506).

Plants Attacked.—Roots and tender growth of nearly all greenhouse plants and
mushrooms are attacked.

Life History, Appearance, and Habits.—These creatures reproduce by means of
eggs which are retained in the marsupium of the female for about 2 months. Young
sowbugs, on hatching, do not leave the marsupium of the female for some time; 25 to
75 constitute a brood. The young are similar to adults except in size. About 1 year
is required for the young bugs to reach full growth. All stages of the bugs will be
found in infested greenhouses at the same time.

Control Measures.—One of the best methods of eliminating sowbugs is to poison
them, using a poison bran mash made of:

Dry bran...	1 pk.
Paris green..	¼ lb.
Molasses..	1 qt.
Water..	4 qt.

The ingredients should be very thoroughly mixed together, and the bran placed
over the benches in small piles, about a handful in a place. Many florists poison the

bugs by the use of a mixture of 1 part
paris green to 9 parts sugar. This is
either sprinkled over the benches or
placed in small piles under clods, bits of
manure, or boards, where the sowbugs
congregate. Hot water may often be
used to kill great numbers.

280 (88). MILLIPEDES[1]

Importance and Type of Injury.—
Young shoots of plants are chewed, and
the tender parts of the stem have areas
eaten out of the bark. Seedlings are
sometimes cut off. Infested houses have
many, long, slender, cylindrical, hard-
shelled, very active, crawling creatures
(Fig. 301), up to about 1 inch in length,
with two pairs of legs on nearly every
segment, hiding under clods on the
benches, or scuttling about among the
plants.

Fig. 506.—Sowbugs feeding in manure.
About natural size. (*From Can. Dept. Agr.,
Bul. 7, n.s.*)

Plants Attacked.—Many different spe-
cies of greenhouse plants. However, the main food of these creatures is decaying
vegetable matter, particularly manure, used for fertilizing in greenhouse benches.

Life History, Appearance, and Habits.—The eggs are laid either in the soil or on the
surface. They are deposited in clusters containing 20 to 100 or more. Each female
deposits about 300. The eggs are covered with a somewhat sticky material and are
nearly translucent when first laid. They hatch in about 3 weeks, giving rise to young
millipedes, which differ from the adult only in size, in number of segments on the
body, and in having at first only three pairs of legs. They grow slowly, gradually
assuming the adult form. There is probably one generation each year.

Control Measures.—The poison-bran bait recommended for sowbugs is fairly
effective against these creatures. Thoroughly drenching the soil with a solution of
mercuric bichloride, at the rate of 1 ounce to 8 or 10 gallons of water, can be used with

---

[1] *Orthomorpha gracilis* Koch, *Julus* spp., and others, Class Diplopoda.

good results. For small amounts, use the mercuric bichloride at the rate of one 2-grain tablet to 1 pint of water. This poison should be mixed in glass or wooden vessels, and should not be allowed to come in contact with the bare skin.

281. GARDEN CENTIPEDE[1]

Importance and Type of Injury.—Tiny, whitish, hundred-legged-worms live in the soil of gardens and ground beds in greenhouses, eating off the fine roots and root hairs to such an extent that entire crops are sometimes ruined. Plants become stunted, grow slowly, or die, and the root system will be found to have been severely pruned.

Plants Attacked.—Lettuce, radishes, tomatoes, asparagus, and cucumbers are often severely injured, and many kinds of ornamental plants are also attacked.

Life History, Appearance, and Habits.—The worms are found in winter about the roots of plants and in cracks in the soil near the surface. The full-grown ones are only $\frac{1}{4}$ to $\frac{3}{8}$ inch long by $\frac{1}{25}$ inch thick, with a head and 14 body segments. The head bears a pair of antennae, more than one-third as long as the body, and delicate chewing mouth parts. The body has 12 pairs of short legs. The creatures spin silk, from a pair of short cerci at the tip of the abdomen, as a lining for their burrows, which follow earth worm tunnels, soil cracks, and the site of decayed roots. After the winter crops are harvested and the greenhouse soil dries out, the centipedes make their way through cracks and earth worm tunnels down to moist subsoil where they feed upon decaying vegetable matter and lay their white spherical eggs at a depth of a foot or more. Most of the eggs are laid from April to September in small clusters of 5 to 20, from which the very tiny young hatch, a week or 10 days later. At first they have only 10 body segments, six pairs of legs, and very short antennae. As they grow, an additional pair of legs is acquired each time the skin is molted, body segments are added, and the antennae elongate. As the soil is wet down and crops started in the fall, the centipedes work their way upward to feed on the plant roots. A generation may be completed in a few months. The pests are rarely seen on the surface of the soil and thrive best in deep soil that is undisturbed throughout the season. They are destructive only in ground benches, since the soil of raised benches is too shallow and too dry during the summer months for them to breed.

Control Measures.—Fumigating the infested soil with carbon bisulphide at the rate of 1 ounce to each square foot of surface is recommended. This is best done before the first fall planting, at a time when the soil temperature is above 65°F., and some days after the soil has been wet down to attract the centipedes to the surface. The carbon bisulphide may be forced into the soil under steam pressure in sterilizing pipe lines or through pieces of pipe fitted with a funnel and a center cylinder of wood. The wood is removed after the pipe is forced into the soil, and the carbon bisulfide is poured through the funnel and pipe to a depth of 18 inches. All soil and manure brought into greenhouses should be steam sterilized. Ground beds should be deeply cultivated as often as possible, taking care to disturb the retreats of the pests near foundation supports, walls, walks, and curbings.

---

[1] *Scutigerella immaculata* (Newport), Class Symphyla.

References.—Ohio Agr. Exp. Sta. Bul. 486, 1931; *Purdue Agr. Exp. Sta. Bul.* 331, 1929.

282. SLUGS[1]

Importance and Type of Injury.—The foliage of plants, particularly in the damper parts of greenhouses, is frequently fed upon and injured by grayish, or grayish-brown, slimy, legless, soft-bodied creatures, from ½ to 4 inches in length (Fig. 507). A shiny trail, composed of a sticky, viscid secretion given off from the body of the slug, will mark the course of its travels over benches and plants. Flower pots, flats, and other supplies standing on or beneath greenhouse benches harbor these pests and should be stored elsewhere.

Plants Attacked.—Many kinds of plants are attacked, particularly coleus, cineraria, geranium, marigold, and snapdragon.

Life History, Appearance, and Habits.—The eggs are laid in masses, in damp places about the greenhouse benches, underneath boards and flower pots, or in the soil. They are held together by a sticky secretion which turns yellow before the eggs hatch.

FIG. 507.—A slug, enlarged. (*From Can. Dept. Agr., Bul. 7, n.s.*)

In about 1 month the eggs give rise to very small young, which closely resemble the adult slugs except in size. They develop slowly and probably live for a year or more.

Control Measures.—Trapping and hand-picking are fairly effective under greenhouse conditions. For this purpose bits of boards or several handfuls of plant refuse may be piled about the benches in the damper parts of the greenhouses where the slugs congregate in greatest numbers. If these piles are lifted each morning, and the slugs picked up and killed, they can be kept down to a point where little damage will occur except in cases of very heavy infestations. In hot dry weather slugs often invade basements and rooms at the ground level where there is more moisture. Sprinkling the surface of the heavily infested beds with hydrated lime is somewhat effective in killing the slugs; but care will have to be used that the lime is not applied to growing plants, or injury is almost sure to result. Spraying the soil with a solution of mercuric bichloride, as recommended for the control of millipedes and sowbugs, is effective also in controlling slugs. Dusting garden plants with 1 part lead arsenate in 5 parts hydrated lime is effective. Poisoned molasses bran bait is recommended.

283. NEMATODES OR EELWORMS[2]

Importance and Type of Injury.—Plants infested by nematodes will present a weakened, sickly appearance, without visible injury to the stem or any part of the plant above ground. An examination of the roots will show numerous knots or galls; or, in cases of severe infestation, practically all roots will have a swollen, gouty appearance (Fig. 508).

Plants Attacked.—Four or five hundred different kinds of plants are known to be attacked by nematodes, including practically all plants grown in greenhouses in the United States. They are particularly a pest of tomatoes and cucumbers, and also very bad on cyclamen.

---

[1] *Limax maximus* and others, Class Gastropoda, Order Pulmonata, Family Limacidae.

[2] *Heterodera radicicola* Müll, Class Nematoda, Family Anguillulidae.

Life History, Appearance, and Habits.—The nematodes that develop in the gall are practically invisible, but the pear-shaped females may be distinguished without a microscope. Either the adults or the larvae may winter in the root galls or the soil. After mating, the females lay several hundred eggs in the galls. The minute worms that hatch from the eggs may work through the soil for a considerable time, but, when they encounter tender roots, they penetrate them and start new galls. A life cycle may be completed in about a month but is often much longer.

Control Measures.—The best method of controlling these creatures is to use soil which is known to be free from infestation. If nematodes have gained access to greenhouse benches, it will be necessary to sterilize the soil, between crops, in order to keep down injury. Probably the most effective method of soil sterilization is by steam. Gibson gives the following directions for the use of this method:

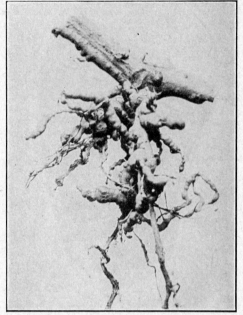

FIG. 508.—Roots of a tomato plant, showing enlargements caused by a heavy infestation of eelworms. (*From Ill. State Natural History Surv.*)

"*The Pipe Method.*—In this method the steam is applied by means of a set of 1¼ inch or 1½ inch pipes, perforated with holes ⅛ to ¼ inch in diameter, and about 1 foot apart. The number and length of the pipes should conform to the boiler capacity and length of beds. According to Selby and Humbert,[1] the perforated pipes should not be more than 40 feet in length, nor exceed seven or eight in number. With medium boiler capacity, say 50 to 60 horsepower, pipes 30 feet in length are most serviceable. The pipes should be about 16 inches apart and should be connected with a 2-inch crosshead by means of T-connections.

"In using this system, the pipes are buried at a depth of about 6 inches, care being taken to see that they lie level. The soil is then levelled and covered with canvas or sacking to check the escape of steam."

Potatoes may be used to determine the duration of the treatment. The potatoes should be placed near the surface and treatment stopped when they are well cooked.

"Steam sterilization not only eradicates nematodes, but it also rids the soil of all insect life; of pests such as sowbugs; of injurious fungi such as those which cause damping-off, lettuce drop, and Rhizoctonia; and of weed seeds. It also has the advantage of improving the soil conditions, and for this reason it is said that sterilized soil requires less fertilizer than untreated soil."

Hot-water sterilization can be used and is nearly as successful as steam; it is easier to use in most greenhouses. Where sterilization can-

[1] *Ohio Agr. Exp. Sta., Cir.* 151, 1915.

not be used, heavy dosages of carbon bisulphide will give some relief, but this has usually not been found so effective as steam sterilization.

284. Miscellaneous Greenhouse Pests

Ants (page 767), earth worms, and termites (page 773) sometimes infest the soil of greenhouses, and the latter are often destructive to the woodwork.

CHAPTER XXI

HOUSEHOLD INSECTS, AND PESTS OF STORED GRAINS, SEEDS, AND CEREAL PRODUCTS

No other insects cause so much personal annoyance and embarrassment as those that frequent our dwellings. They attempt, often successfully, to appropriate our food and clothing to their use, and sometimes even make themselves at home on our persons. Most of these insects may be cheaply and easily controlled, if the right method is employed.

A large number of insects, including many species of beetles, moths, and mites, attack grain and grain products in farmers' bins, elevators, mills, warehouses, retail stores, and the home. The damage done in this way is estimated to exceed 200 million dollars yearly. It is impossible to treat all these insects in this book. The more important insects are mentioned, and some of the points of their life history are given.

As the method of control is much the same for all the insects of stored grains and seeds, no attempt is made to give a separate control for each species, but the general measures for controlling these insects are given near the end of the chapter, pages 806 to 808.

Many of the best publications on these insects deal in a general way with the many species attacking this class of products, and, as some of these publications contain references or short bibliographies, no special references are given for some of the species. Some of the publications dealing with this class of insects are given below:

General References.—Kans. Agr. Exp. Sta. Bul. 189, 1913; *Ill. Agr. Exp. Sta. Bul.* 156, 1912; *Minn. Agr. Exp. Sta. Bul.* 198, 1921; *U.S.D.A., Dept. Buls.* 872, 1920, and 1393 and 1428, 1926, and *Farmers' Buls.* 1029, 1919, and 1260 and 1275, 1922, and 1483, 1926; Herrick, "Insects Injurious to the Household," Macmillan, 1921; *U.S.D.A., Cir.* 390, 1936.

KEY TO INSECTS INFESTING THE HOUSEHOLD AND STORED GRAIN, SEEDS, AND GRAIN PRODUCTS

A. Ant-like insects, of various sizes from $\frac{1}{16}$ to $\frac{1}{2}$ inch long, and yellow, reddish, brown, or black in color, with a slender waist (pedicel) between thorax and abdomen which always bears 1 or 2 enlargements (petiole). They generally crawl hurriedly about indoors or out and get into foods as they forage out from populous nests in the ground or in wood:

The most important American species may be determined by the following key prepared by Dr. M. R. Smith of the Bureau of Entomology and Plant Quarantine.

1a, With a 1-segmented pedicel..2.
1b—With a 2-segmented pedicel..5.
2a, Petiole poorly developed, hidden beneath base of abdomen (gaster). Tip of abdomen without a circlet of hairs. Body of freshly-crushed worker with a sweetish, nauseating, rotten cocoanut-like odor....*Odorous house ant,* page 770.

2b—Petiole well developed and erect. Body of freshly crushed worker with a different odor...3.

3a, Tip of abdomen without a circlet of hairs; practically uniform-sized ants, about $\frac{1}{10}$ inch long; color brown. Freshly crushed worker with a musty or greasy odor. An imported species occurring in southern United States and in California, where they swarm in buildings and over trees, especially citrus........ ....*Argentine ant*, page 769.

3b—Tip of abdomen with a circlet of hairs. Freshly crushed worker with a distinct acid (formic) odor..4.

4a, Large, black, variable sized ants, from $\frac{1}{4}$ to $\frac{1}{2}$ inch long. Colonies nest entirely in wood, sometimes of old buildings....*Black carpenter ant*, page 771.

4b—Small, brown, robust, almost uniform-sized ants, approximately $\frac{1}{10}$ inch long. ....*Cornfield ant*, page 770.

5a, Head and thorax with numerous spines; body rusty brown; size variable, from $\frac{1}{16}$ to $\frac{1}{2}$ inch long. Strictly a leaf-cutting species, known only from Texas and Louisiana.................................*Texas leaf-cutting ant*, page 771.

5b—Head and thorax either spineless or with only 1 pair of spines on the posterior portion of the thorax...6.

6a, Thorax and head covered with numerous, longitudinal, parallel ridges or lines. Posterior portion of thorax with a single pair of spines. Imported species most abundant along Atlantic seaboard...................*Pavement ant*, page 770.

6b—Thorax and head without parallel ridges or lines as in 6a. Posterior portion of thorax without spines..7.

7a, With a 2-segmented antennal club...............................8.

7b—With a 3-segmented antennal club................................9.

8a, Exceedingly small ants, from $\frac{1}{14}$ to $\frac{1}{10}$ inch long, approximately uniform in size, yellowish. Eyes extremely small...................*Thief ant*, page 770.

8b—Moderately large ants, highly variable in size, from $\frac{1}{16}$ to $\frac{1}{4}$ inch long. Color varying from yellowish red to black. Eyes prominent. Sting well developed. Common in Gulf Coast states and Southwestern United States.... ....*Southern fire ant*, page 771.

9a, Small shining black ants, approximately $\frac{1}{10}$ inch long.................. ....*Little black ant*, page 770.

9b—Small yellowish-red ants, $\frac{1}{12}$ to $\frac{1}{10}$ inch long. Head and thorax with dense, pitted impressions or punctulations. Nests in buildings.................... ....*Pharaoh's ant*, page 770.

B. *Insects working in the interior of wood and wood products:*

1. Large numbers of slender dark-brown winged insects, without a slender waist, emerge from cracks in floors or walls or from holes eaten through the woodwork of buildings. Small, whitish, brown-headed, soft-bodied, wingless insects, ant-like but with thick waists, feed and tunnel in the woodwork of buildings, particularly on timbers that come in contact with the ground. The tunnels are kept free of frass. Injured wood is almost completely eaten, except an outer shell. Sometimes cement-like tubes are constructed over brick or stone foundations, leading to the woodwork.................... ....*Termites* or *white ants*, page 773.

2. Wood with holes from $\frac{1}{16}$ to $\frac{1}{4}$ inch in diameter leading to the interior of timbers, the interior of which may be completely reduced to fine yellowish powder, packed in the tunnels made by small white grubs with large heads and 6 small legs. Adults which make the holes to the outside are hardshelled beetles, $\frac{1}{12}$ to $\frac{1}{5}$ inch long, brownish, elongate and cylindrical, or short and stubby, with varied sculpturing on body and wings............ ....*Powder post beetles*, page 777.

C. *Insects running very swiftly about the house in dark places or at night: rarely found in foods:*

1. Small wingless grayish-white to brownish carrot-shaped glistening insects, up to $\frac{1}{2}$ inch in length, with slender feelers projecting from the head and

from the tail, run rapidly over books, walls, and papers and feed on the
binding of books, wall paper, or foodstuffs............*Silverfish*, page 779.

2. Dark-brown, light-brown, or black, shiny, flat-bodied, long-legged insects, the
smaller ones without wings, the larger ones winged, feeding on the binding of
books, the sizing of papers, and foods of all kinds. The insects are seldom
seen in the daytime. A noticeable, sweetish, sickening odor where these
insects are present in numbers....................*Cockroaches*, page 781.

3. Slender pale-brown creatures, up to 2 inches in length, with many extremely
long legs, scurry rapidly over walls or floors when lights are suddenly turned
on. Do not feed on stored foods or fabrics......*House centipede*, page 784.

4. Rather hard-shelled cylindrical brownish creatures, from 1 to 2 inches long,
with many legs, usually 2 pairs to each principal segment of the body.
Generally found in basements or under boards or stones around foundation
walls where it is damp............................*Millipedes*, page 758.

D. *Insects feeding in tobacco, drugs, the silk upholstery of furniture, leather, etc.:*
1. Small, white, very hairy grubs and very small light-brown beetles ($\frac{1}{16}$ inch
long), with hairy but not striated wings, eat holes through the tobacco or
feed in numbers within packages of cigars, cigarettes, and package tobaccos.
Upholstered furniture with numerous holes, eaten through the covering by
these grubs or beetles........................*Cigarette beetle*, page 785.

2. Small, white, curved grubs, similar to *D*, 1, but not hairy; and small, rather
narrow-oval, reddish-brown beetles with striated wings, mine and tunnel
through packages of drugs, vegetable poisons, foods, leather, and a great
variety of other substances, which they eat.....*Drug-store beetle*, page 786.

3. Stored tobacco, nuts, chocolate, and other dried vegetable products infested
with yellowish-white to pinkish-brown very active caterpillars, up to $\frac{5}{8}$ inch
long, which eat holes through the products and contaminate them with their
silk and excrement. Moths, $\frac{1}{2}$ inch from tip to tip of wings and with base
and tip of wings darker than the middle, flutter about the storage........
....*Tobacco moth*, page 787.

E. *Insects destroying clothing, fabrics, furs, and other animal products:*
1. Rugs or carpets with holes eaten in them, or stuffed animals, birds, and
insects being fed upon by small dark-brown, very hairy larvae, about $\frac{1}{4}$ inch
in length. Small brown-and-gray, or black, beetles, sometimes flecked with
red, crawling over infested materials. Beetles oval in outline, about $\frac{1}{8}$
inch long......*Carpet beetles*, or *buffalo beetles*, and *museum pests*, page 787.

2. Wool or silk clothing, feathers, and other animal products with many silken
cases or webs on the surface, containing small whitish worms. Fabrics with
numerous holes eaten through them. Small buff-colored moths, about $\frac{1}{8}$
inch long, running rapidly over the surface of fabrics when exposed to light,
or flying aimlessly about in closets................*Clothes moths*, page 790.

F. *Insects eating in meats, cheese, and other foods:*
1. Very hairy, brown larvae, somewhat resembling *E*, 1, but with two, short,
curved hooks on the last segment, feed on meats, particularly dried or
smoked meats, other food products, feathers, skins, etc. Dark-brown
beetles, about $\frac{1}{3}$ inch long, with a pale-yellow band across the middle of the
body, crawling about on the surface of food materials where larvae are
feeding..*Larder beetle*, page 792.

2. Elongate, slender, purplish larvae, not very hairy, cylindrical but tapering
toward the head, with 6 short legs, eat through hams and bacons. Brilliant
greenish-blue beetles with reddish legs, up to $\frac{1}{4}$ inch long, run rapidly from
the food substances when disturbed........*Red-legged ham beetle*, page 793.

3. Insects similar to *F*, 1, feeding in meats, hides, tallow, and other animal
matter. The beetles are black above, whitish beneath, and from $\frac{1}{4}$ to $\frac{1}{3}$ inch
long............................*Hide beetle*, *Dermestes vulpinus* Fabricius.

4. Cheese and smoked meats with small yellowish maggots crawling over the
surface or through the product. These maggots sometimes jump or hop for

some distance, by suddenly bending and then straightening their bodies....
....*Cheese skipper*, page 794.

5. Cheese, hams, cereals, and other food products become musty and are found on close examination to be swarming with tiny, whitish, 8-legged mites, not over $\frac{1}{32}$ inch long.......*Cheese mites, ham mites,* and *flour mites*, page 795.

G. *Insects attacking stored grains and grain products, dried fruits, and some other products:*

a. *The parents are hard-shelled beetles: the larvae do not have prolegs and spin no silk.*

1. Dark-brown snout beetles, up to $\frac{1}{6}$ inch in length, crawling over and through grain products held in storage. Grain heats and becomes moist, sometimes matted together because of sprouting. Individual grains contain small fat-bodied, legless, white grubs......*Granary weevil* or *rice weevil*, page 796.

2. Small dark-red narrow-bodied beetles, about $\frac{1}{6}$ inch in length, running rapidly over the surface of grains; often found in flour, meal, breakfast foods, and other food products. Small brownish-white 6-legged larvae, about $\frac{1}{6}$ inch long, feeding in and on the grain...........................
....................*Confused flour beetle* and *red flour beetle*, page 799.

3. Insects similar to G, 2, but more slender, flatter, and somewhat darker in color, feeding on dried fruits, grain and grain products, nuts, seeds, and many other foods. Each side of the thorax shows 6 fine saw-tooth-like projections when examined under a lens. Larvae with bodies very slender, nearly bare of hairs, somewhat depressed..*Saw-toothed grain beetle*, page 801.

4. Very small dark-brown beetles with rather long, prominent antennae, and slight knobs at front angles of thorax, found in same situations as G, 3, and particularly in bins of oats and barley...................................
.............................*Foreign grain beetle, Cathartus advena* Waltl.

5. Very small, brown to black, nearly cylindrical beetles, with the head completely hidden by the thorax, only $\frac{1}{8}$ inch long. Curved-bodied, grub-like, whitish larvae, with the swollen anterior end bearing a small head and 6 small legs, are found inside grains of wheat and other seeds, which they completely hollow out. Holes often eaten in wooden or paper boxes in which seeds are stored................................*Lesser grain borer*, page 801.

6. Very similar to G, 5, but the adults a little larger, thicker, smoother, and shinier; eat holes chiefly in corn..............*Larger grain borer*, page 801.

7. Smooth shiny-bodied, uniformly brown to yellow worms, closely resembling wireworms, feeding in the lower parts of, or beneath, grain bins and where the grain is moist. Rather robust, somewhat flattened, black to brown beetles, up to nearly 1 inch in length, in the same situations..............
........................*Yellow mealworm* and *dark mealworm*, page 797.

8. Black beetles, somewhat resembling G, 7, but smaller in size, about $\frac{1}{2}$ inch in length, with the head and prothorax distinctly separated from the rest of the body by a narrow waist. Whitish to grayish-white soft-bodied larvae, with black heads and 2 short black projections from last segment of abdomen, work among grains, particularly where undisturbed for some time........
....*Cadelle*, page 798.

9. Active strong-flying beetles, $\frac{1}{8}$ inch long and half as wide, with wings much shorter than the abdomen, each bearing an amber-brown spot at the tip and a smaller one near the base; and their white shortlegged grubs, with brown head and tail end, work their way to the inside of partially dried figs, other drying fruits, fermenting watermelons, and the like, causing them to rot and sour.......................................*Dried fruit beetle*, page 802.

b. *The parents are delicate winged moths; the larvae have prolegs and often spin much silk among the grain.*

1. Wheat and corn in storage have small round holes, a little larger than a common pinhead, eaten into the grains. Whitish worms, up to nearly $\frac{1}{2}$ inch in length, living and feeding *inside* the kernels. Brownish-gray moths, a little less than $\frac{1}{2}$ inch in length, with a long fringe of hairs on the wings, fly

about the bins or cribs, or crawl over the surface of the grain.............
....*Angoumois grain moth*, page 802.

2. Flour in mills, and sometimes in storehouses, webbed and matted together in masses, by small white to pinkish worms up to ⅜ inch in length, which are usually concealed in their silk tubes. Masses of webbed flour frequently becoming so tightly packed in mill machinery as to prevent the operation of the machines. Gray, black-marked moths, resting on the surface of grain or flour, in infested mills................*Mediterranean flour moth*, page 804.

3. Flour, grain, and grain products and especially dried fruits, nuts, nut candies, and other food products, infested by small whitish to greenish-white worms, up to ½ inch in length. More or less webbing of infested material, but not so much as in *G, b, 2*. Small moths with the basal half of the front wings much lighter in color than the tip, resting on infested material, or flying about over it......................................*Indian meal moth*, page 805.

4. Grain stored while damp or in damp places, infested with whitish caterpillars, having black heads and prothorax, the body tinged with orange at each end, which rest in tough silken tubes and feed from the open end. Often cut through sacks. Adult moths 1 inch from tip to tip of wings, the forewings dark-brown on basal and distal thirds, light brown across the middle, the light- and dark-brown areas separated by transverse wavy white lines......
Meal snout moth, Pyralis farinalis Linné, Order Lepidoptera, Family Pyralididae......

H. Insects attacking stored beans and peas, also the developing seeds within pods in the field:

1. Peas in the pod contain white, nearly footless grubs or short, chunky, brownish beetles (never more than 1 to a seed), which feed exclusively inside and show no entrance hole except a small dark dot. Stored peas often show a small round hole where the beetle has emerged. Beetles ⅕ inch long with 2 black spots on the extreme tip of the abdomen that is exposed beyond the wing covers. Hind femur with a sharp tooth near apex.............
....*Pea weevil*, page 808.

2. Similar to the preceding but occurring chiefly in broad beans in addition to peas. The adult beetle smaller, without black spots on that part of abdomen exposed beyond the wing covers, and the tooth near apex of femur broader and more blunt than in the pea weevil........*Broad bean weevil*, page 810.

3. Growing beans and peas attacked in the field and in storage as by the pea weevil, by a smaller, brownish beetle, about ⅛ inch long, and by footless white grubs that feed exclusively inside the seeds. Stored beans often show many round holes and fine, powdered frass. Hind femur of beetle has 1 large and 2 small teeth near the tip................*Bean weevil*, page 810.

4. Insect and injury similar to the preceding, but prefers cowpeas. Adult with a large dark spot on apex of each wing cover and another pair of spots near the middle of costal margin of wing covers............................
.... *Southern cowpea weevil*, page 810.

5. Insect and injury similar to *H, 3*, and *H, 4*, but smaller and with two prominent ivory-white spots at the middle of hind margin of prothorax. Wing covers with a narrow dark crossband at middle. Male with comb-like antennae..................*Cowpea weevil, Callosobruchus chinensis* (Linné).

A. HOUSEHOLD INSECTS
285 (13, 23). ANTS[1,2]

Importance and Type of Injury.—There are few insects which have proved themselves more persistently exasperating to the housekeeper

[1] Many species of Order Hymenoptera, Family Formicidae.

[2] The authors make grateful acknowledgment for valuable aid in the preparation of the section on ants to Dr. M. R. Smith.

than ants. Everyone is familiar with the fact that when ants have invaded houses, the workers will be found crawling over any food that is to their liking, bits of which they cut off and carry to their nest (Fig. 161). Some species also cause injury by establishing their nests in the sills and woodwork of old houses; but as a rule, they do not attack perfectly sound wood. Other species throw up mounds of earth about the entrance to their nests, disfiguring lawns and walks. Some species obtain a part of their food from the sweet, sticky honeydew given off from the bodies of aphids, scale insects, and mealy bugs, and these species may establish and

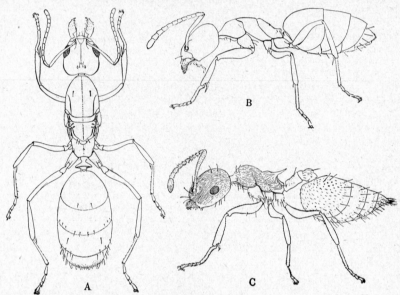

Fig. 509.—Three common and troublesome American ants. *A*, the queen Argentine ant, *Iridomyrmex humilis* Mayr, after discarding her wings. This is the worst household ant where it occurs, in the warmer parts of the United States. Actual size of the queen, about ⅕ inch long; of the workers, from 1/16 to ⅛ inch in length. *B*, Worker of the odorous house ant, *Tapinoma sessile* (Say), the commonest household ant over much of the United States. Length 1/12 to ⅛ inch; black; with a pronounced "rotten pineapple" odor when crushed. Note how the top of the abdomen overhangs the nodus of the petiole. *C*, worker of the acrobat ant, *Crematogaster lineolata* (Say), a shining black species about ⅛ inch long, that often carries its pointed abdomen turned up over the rest of the body. (*From Woodworth, Calif. Agr. Exp. Sta. Bul. 207.*)

protect colonies of these insects on plants out-of-doors or in greenhouses or residences.

Food.—The food of ants is even more varied than that of man. Various species eat particles of human foods, sweets, honeydew, seeds, fats, meats, dead insects, roots of vegetables, and fungi, which they grow in their nests. They are fond of nearly every kind of human food, and in some parts of the country they become a pest from their habit of gathering and storing quantities of grains and seeds in their nests.

Distribution.—As a group, world-wide; certain species are more abundant and annoying in some localities than in others.

Life History, Appearance, and Habits.—The wingless workers (Fig. 161,*e*), which are the form of ants generally noticed, are all sterile females and are only one of the forms that occur in the ant colony. The winged males and females (Fig. 161,*a*,*d*), which swarm out of the nests at certain seasons, are also very evident. But the wingless females or "queens" (*c*) and the helpless, maggot-like young (*f*, *b*) remain exclusively in the underground nests and are rarely seen by the layman. In some colonies even the worker caste is divided into two or three forms. Some of these forms, as in the honey ants, have abdomens which can be greatly distended and are used as a sort of jug in which to store honey for the colony. Others, as in some of the seed-gathering ants, have certain castes with enormously enlarged heads and very strong jaws which serve as seed crackers. Ants, as a group, can be distinguished from other insects by having one or two wart-like elevations on the slender pedicel that separates the thorax from the large part of the abdomen (Fig. 509).

Individual worker ants cannot exist alone. A typical colony consists of from one to several females or queens (whose duty it is to lay eggs), and thousands of workers. The workers are incompletely developed females and seldom lay eggs. The queens are cared for by the workers and usually remain in the inner chambers of the nest, seldom leaving the nest except at times when the colony may migrate to a new location. From the eggs laid by the females (Fig. 161,*g*), larvae or maggots (Fig. 161,*f*) hatch, the majority of which develop into workers after going through the pupal stage (*b*), either in a cocoon, or without such protection. A certain percentage, however, produce winged males and females, the "kings" and "queens" of the ant colonies. At certain times of the year, varying with the species, these winged males and females leave the nest or swarm. Usually this occurs over a wide area on the same day for any one species. The males mate and soon die. The female flies to a situation that is attractive to her as a nesting site, alights, tears off her wings, encloses herself in a small cave which she excavates in the soil and in which she lays a few eggs. After the eggs have hatched, the female feeds and cares for the young larvae until they have become full-grown. The larvae of ants are whitish, helpless, maggot-like grubs, without legs and with very small heads. These larvae develop into workers; and, as soon as the adult workers appear, they start taking care of the queen and her young. From this time forward, the queen confines her activities entirely to egg-laying. The queen may live from 1 or 2 to 12 to 15 years, during which time she will produce many thousands of eggs and may develop many thousands of individuals.

Ant pests of the first magnitude include the following:

1. Argentine ant, *Iridomyrmex humilis* Mayr (Fig. 509,*A*). Although first observed in the United States in 1891, and restricted in its distribution chiefly to the southeastern fourth of the United States and California, this small ant is the most annoying and destructive household insect wherever it is established. Locally in cities and towns it drives out all other kinds of ants; swarms into houses and stores; gets into all kinds of foods; is indirectly very injurious to citrus and other fruits and to shade trees, by attacking the blossoms and by distributing, establishing, and protecting aphids, mealy bugs, and scale insects; and is very troublesome in bee and poultry yards. It prevents people from sleeping and often destroys the value of real estate by rendering neighborhoods undesirable. It is believed to have been brought into New Orleans in shipments of coffee from Brazil and has subsequently been widely spread by commerce. The workers are slender, $\frac{1}{12}$ to $\frac{1}{8}$ inch long, brown with lighter appendages. There is a single segment in the pedicel, the petiole is erect, and the mandibles have about a dozen denticles on their inner margins. They have a very

slight, musty, greasy odor when crushed and always travel in definite trails. Nests are established in any dark, somewhat moist situation in orchards, lawns, or gardens. The winter is passed in large centralized colonies in particularly favorable nesting sites, such as fairly dry, warm piles of manure or other decaying vegetable matter. All stages may be present throughout the winter but activity is greatly accelerated in warm weather. Mating takes place in the nest and the nuptial flights are less evident than in most ant species. There may be several to many queens in one colony. Eggs hatch in 12 to 50 days. The larvae require about 1 month for their development, and the pupae 15 days. The time from egg to adult averages 75 to 80 days. Widespread and intensive community campaigns of eradication should be initiated wherever this pest appears.

  References.—U.S.D.A., Bur. Ento. Bul. 122, 1913, and *Cir.* 387, 1936.

  2. Pharaoh's ant, *Monomorium pharaonis* (Linné). This tiny, slender, yellowish-red ant, $\frac{1}{12}$ to $\frac{1}{10}$ inch long, generally nests in inaccessible places about the foundations and in the walls of buildings, from which it forages indoors the year round in search of food. It is very difficult to eradicate. It is an imported species, found mostly in the larger towns and cities, but usually sporadically distributed. It is practically omnivorous, showing great fondness for proteins as well as sweet foods, and is especially pestiferous about apartment houses, groceries, cafes, and hotels. The workers of this and the following species have two segments in the pedicel and three segments in the antennal club. There are no spines on the posterior part of the thorax. Most of the body is covered with minute pitted impressions.

  3. Little black ant, *Monomorium minimum* (Buckley). The workers of this species (Fig. 161) are from $\frac{1}{12}$ to $\frac{1}{10}$ inch long, slender and shining, and generally jet black in color. They typically nest out-of-doors, in the soil, where they make tiny craters, or in stumps; but sometimes indoors in the woodwork or masonry of buildings. This is a native species that is a common house pest in both town and country. It is very general in food habits, taking either sweets, honeydew, fruits, vegetables, or meats.

  4. Odorous house ant, *Tapinoma sessile* (Say). The workers of this species (Fig. 509,*B*) are from $\frac{1}{10}$ to $\frac{1}{8}$ inch long, soft-bodied, the pedicel one-segmented, the petiole somewhat horizontal and difficult to see because hidden by the over-hanging abdomen. They are darker in color than the Argentine ants, deep brown to black, have a broader abdomen, and are less agile in their movements. When crushed, they have a characteristic, nauseating, sweetish, rotten cocoanut-like odor. This is a native species, widely distributed over the United States. It generally colonizes out-of-doors in shallow nests under diverse conditions, but sometimes in foundations of buildings. It is practically omnivorous, but especially fond of sweets and is a common attendant upon honeydew-excreting insects.

  5. Tiny thief ant, *Solenopsis molesta* (Say). The workers of this species (Fig. 243) are light yellow to bronze, $\frac{1}{15}$ to $\frac{1}{10}$ inch long. They are similar to, but still smaller than, Pharaoh's ant, lighter in color, and have two segments in the antennal club, a two-segmented pedicel, and extremely small eyes. This native species usually nests out-of-doors, in soil or wood, and is troublesome only in warm weather. It is a common house pest, showing a preference for meats, butter, cheese, seeds, nuts, oils, and other protein foods, rather than sweets. It sometimes does damage to germi-nating seeds of grains (page 389).

  6. Cornfield ant, *Lasius niger* var. *americanus* Emery. This species nests mostly in open places, in the soil or in rotten wood. The workers (Fig. 230) are fond of sweets, often attending and fostering honeydew-excreting insects such as aphids (page 371). They are $\frac{1}{12}$ to $\frac{1}{10}$ inch long, robust, soft-bodied, light to dark brown. The anal opening is terminal, circular, and surrounded by a fringe of hairs. There is no antennal club, and the pedicel is one-segmented. The body when crushed has an acid (formic) odor. It is a native species, distributed over most of the United States. It sometimes makes objectionable nests in lawns.

  7. Black pavement ant, *Tetramorium caespitum* (Linné). This imported ant is a rather hairy, robust, hard-bodied species, from $\frac{1}{12}$ to $\frac{1}{7}$ inch long, with two segments in the pedicel and no antennal club. The abdomen is black, the appendages pale.

The head and thorax are grooved with fine parallel lines, and there are two small spines on the posterior part of the thorax. It is common in lawns, especially in or near towns or cities along the Atlantic seaboard. It nests under stones or the edges of pavements, from which it forages into houses and greenhouses. It is often a pest in gardens, where it gnaws at the roots of vegetables or steals planted seeds.

8. Perfumed yellow ant, *Lasius interjectus* Mayr. The workers of this species are $\frac{1}{10}$ to $\frac{1}{8}$ inch long and yellow. When crushed they have a pronounced and characteristic lemon-verbena odor. Although a common soil-inhabiting species, which fosters mealy bugs and aphids on the roots of plants, it not infrequently nests in walls or beneath the floors of basements, from which the winged, sexual individuals emerge inside or outside the house and cause concern. But it does not attack human foods.

9. Black carpenter ant, *Camponotus herculeanus pennsylvanicus* (DeGeer). This is the largest of our common ants, the workers ranging from $\frac{1}{4}$ to $\frac{1}{2}$ inch long. The body is dark brown to black. The pedicel is one-segmented, there is no antennal club, and the tip of the abdomen has a circlet of hairs. When crushed the ants have a distinct acid (formic) odor. It is a native species generally distributed over the eastern half of the United States. It nests entirely in wood, hollowing the wood out, but not eating it. It is a very common house pest, especially fond of sweet foods. It will also bite people, though it does not sting. In addition to nesting in stumps and tree trunks, it does appreciable injury to telephone and telegraph poles, and the sills and other timbers of older buildings.

10. Southern fire ant, *Solenopsis xyloni* MacCook. The workers are variable in size, from $\frac{1}{16}$ to $\frac{1}{4}$ inch long, shining, the largest ones yellowish red, the smaller ones much darker. The pedicel is two-segmented, and there is a two-segmented antennal club. The eyes are prominent, the body hard, and these ants have a sting like a bee. It is a native species common in the Gulf Coast states, less common eastward, and with subspecies ranging into the southwestern states. It nests generally in open, exposed places, often forming loose mounds or numerous scattered craters. They steal planted seeds, infest houses, and sting severely, often killing young poultry and quail just as they are hatching. While omnivorous, they show some preference for protein foods.

11. Texas leaf-cutting ant, *Atta texana* Buckley. This native species is confined to southern and eastern Texas and western Louisiana, where there is well drained soil of a loamy nature. The workers are variable in size, from $\frac{1}{16}$ to $\frac{1}{2}$ inch long, light brown, hard-bodied, with many spines on head and thorax, and a two-segmented pedicel. Nests are often enormous—10 to 20 feet deep but not very high, with numerous craters, and containing many thousands of individuals. They invade houses and steal seeds and other farinaceous food. The workers cut leaves from both wild and crop plants, macerate them and grow a fungus for food upon the macerated leaves in their nests. These ants are serious pests of garden and field crops, and also in reforested areas, where they destroy young pine seedlings.

Control Measures.—The simplest but least dependable thing to do when ants appear in houses is to scatter a dust in their paths, or to use one of the highgrade household oil sprays. The best dusts are sodium fluoride, which is also poisonous to man, or derris and pyrethrum dusts, which are not very poisonous to man. Sometimes this method works like magic. The secret of thorough ant control, however, is the destruction of the queens in the nest, following which all other forms will perish. The most effective method is to find the mounds or nests of the ants in the ground. This can usually be done by following the line of foraging workers back to the place from which they are coming. If the nest is located, it can be destroyed by applications of carbon bisulphide, calcium cyanide or, if there are no plants near, by equal parts of creosote and gasoline. Several holes should be punched with a stick into the nest to a depth of 6 to 15 inches, depending upon the volume of the nest, then, with the aid of a funnel, the poison should be poured into the holes and the holes closed with plugs of mud. Into each hole should be poured a tablespoonful or two of carbon bisulphide or of calcium cyanide dust. If a number of nests are being treated in a small area, a wet blanket or piece of canvas may be thrown over the entire area to confine the poison. In using carbon bisulphide, all fires must be kept at a distance, and in using calcium cyanide extreme care must be used not to breathe the fumes or to get the poison into the mouth or sores. The ground should be fairly dry and warm when using these treatments. When the nests are found to be in the walls or beneath the basement floor of a house, cotton wads saturated with ethylene dichloride and carbon tetrachloride (page 288), or a pyrethrum extract, should be forced into the crevices from which the ants are emerging.

If the nests cannot be located, a poison bait should be used which is attractive to the worker ants and slow acting so that they will carry it to their nests and feed it to all the developing young and the queens before they themselves are killed or suspicious of it. The bait must be carefully selected with reference to the species of ant. The best times to poison are in spring and autumn and, in the south, in winter. Among the best poisons are the following:

I. The Argentine Ant Bait.

Boil together the following materials for 30 minutes:
Granulated sugar	1¼ lb.
Water	1¼ pt.
Tartaric acid (crystallized)	1 g.
Benzoate of soda	1 g.

Dissolve sodium arsenite in hot water in the following proportions:
Sodium arsenite (C.P.)	⅛ oz.
Hot water	1 fluid ounce

When the above solutions have cooled, add the second to the first and stir well. Then add ⅔ pound of strained honey to the resulting sirup and mix thoroughly.

This bait is successful for the Argentine ant and many other species that show a preference for sweet foods, but not for Pharaoh's ant, the tiny thief ant, or the southern fire ant. For the Argentine ant, poisoning should be undertaken on a community- or city- wide basis. The poison is exposed in small aluminum or covered paperoid cups. Small holes must be provided near the top of the container for the ants to enter and the cups should have securely fastened lids or be placed high out of the reach of children. The cups should be placed about 20 to 25 feet apart. Ten to 20 such containers are needed about an isolated house or 100 to 200 for an average city block. The cost in well-organized campaigns will run about $5.00 per city block.

Thallium sulphate is a very deadly poison and must be handled with great care. It is, however, a more effective poison for certain kinds of ants than sodium arsenite. Ready prepared baits containing this poison, either in a sweet or a greasy base, may be purchased or can be made as follows.

II. Thallium Sulphate Bait for Sweet-loving Ants.

Dissolve 2 grams of thallous sulphate, Tl_2SO_4, carefully weighed, in ½ pint of lukewarm, not hot, water. In a separate container mix ½ pint of water, 1 pound of granulated sugar, 3 ounces of strained honey, and 45 cubic centimeters of glycerin. Bring this mixture to a boil and remove from the fire. Cool for about 5 minutes, add the solution of thallium sulphate, and stir thoroughly. Label "Poison" and store in a cool place.

III. Thallium Sulphate Bait for Protein-loving Ants.

Thallous sulphate, Tl_2SO_4.................................... 0.5 g.
Peanut butter... 75 g.
German sweet chocolate.............................. 25 g.
Mix together very thoroughly.

The poison baits may be put out in metal bottle caps or in short sections of soda fountain straws.

References.—WHEELER, "Ants, Their Structure, Development, and Behavior," Columbia Univ. Press, 1910; *U.S.D.A.*, *Cir.* 387, 1936, and *Leaflet* 147, 1937; *Calif. Agr. Exp. Sta. Cir.* 342, 1937.

286. TERMITES[1]

Importance and Type of Injury.—The presence of termites in dwellings or other buildings is often first indicated by swarms of winged, black, ant-like insects (Fig. 510) suddenly appearing inside the buildings, often emerging from holes in the walls, floor, or other parts of the woodwork. All termites differ from true ants in not having a slender waist. Badly infested buildings may suddenly sag, due to the weakening of the frame by the feeding of the termites on the main timbers. Occasionally their

FIG. 510.—Termite queen with wings spread; about 6 times natural size. (*From Ill. State Natural History Surv.*)

presence is indicated by the breaking through of floors or the cutting through of the rugs or floor covering. Brick or cement foundations will have small tunnels constructed over their surfaces, of a clay or cement-like substance. Growing plants may be eaten out.

Food.—The food of termites consists of wood, paper and other wood products, fungi, dried plant and animal products, and also partly digested food material passed from the mouth or anus and an oily exudate sweated through the skin, which are freely circulated from caste to caste and individual to individual throughout the colony (see page 192).

Distribution.—Termites of various species occur in nearly all the warmer parts of the world, extending in a general way around the earth between the mean annual isotherms of 50°N. and 50°S., but are much more numerous in the tropics.

Life History, Appearance, and Habits.—As to habits termites have been classified as follows:

A. *Soil-inhabiting termites,* in which the colony is always partly in the ground or the mating pairs enter the soil, or wood in the earth, following swarming. These are further subdivided into (*a*) subterranean termites, (*b*) desert termites, (*c*) mound-building termites, and (*d*) carton-nest-building termites.

[1] Several families, a number of genera, and about 1,500 species of the Order Isoptera. The commonest species in the eastern United States is *Reticulitermes flavipes* (Kollar) of the Family Rhinotermitidae.

B. *Wood-inhabiting termites*, in which the colony is confined entirely to wood, or the swarming pairs enter wood above ground. These are subdivided into (*a*) damp-wood termites, and (*b*) dry-wood termites.

The most important species in the United States belong to the soil-inhabiting, subterranean group. The kings and queens are always found in the ground, no external mounds are erected, and, although the workers may carry on a large part of their activities in wood above ground, they always maintain a connection with the ground nest, either through burrows or covered runways. The colonies of termites, or white ants, somewhat resemble those of the true ants in their organization. In the termite colony, however, the male helps to start the nest and remains with the female or queen throughout life. The latter is usually the mother of the entire colony. In some species more than one royal pair

FIG. 511.—Termites in their galleries in sill of a house. Slightly enlarged. (*From Ill. State Natural History Surv.*)

will be found in the colony. There are only three stages in the development of the termites; first, the egg; then the immature or nymphal forms; and, finally, the adults, which are always separated into several different castes, such as workers, soldiers, kings, and queens. They do not have larval and pupal stages, as the true ants do. The so-called workers and soldiers in the termites are always wingless and consist each of *both* males and females, although they never or rarely produce any young. The workers and the nymphs are the destructive forms which are active in foraging and feeding on wood (Fig. 511). The winged true males and females develop within the colony and at certain times of the year leave the nest in swarms. Several such swarms may come out from the same nest, usually within short intervals of time. Most of the termite species swarm in the spring. These males and females differ greatly from the soft-bodied, white, blind and wingless workers and soldiers. They (Fig. 510) are black, rather narrow-bodied, and possessed of rather long,

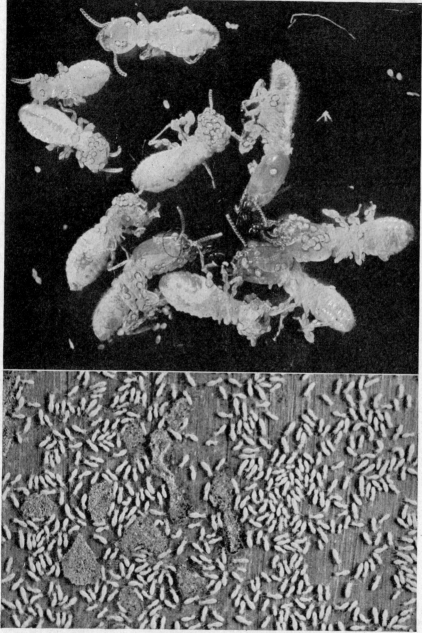

FIG. 512.—*Above*, a group of worker termites badly infested with a parasitic mite, which sucks blood from head, legs, and other parts of the body and eventually kills the termites. About 8 times natural size. *Below*, worker termites, *Reticulitermes flavipes* (Kollar), as exposed by tearing away a portion of the infested timbers, natural size. (*Original.*)

very brittle wings, which are broken off near the base when the kings and queens enter the ground after a single pairing flight. After leaving the nest, they separate in pairs and select sites for starting new colonies. They then break off their wings, and the females become much enlarged, although, in our species, still capable of moving about. They are said to lay eggs at the rate of a dozen or more a day for years. In the absence of the winged (*macropterous*) males and females, reproduction may be accomplished by short-winged (*brachypterous*) or wingless (*apterous*) males and females, which cannot, however, produce the winged caste and so cannot give rise to any new colonies, though they may start subdivisions of the parent colony in adjacent burrows. As in the case of ants, a colony may endure for many years. Except for the winged males and females, which are positive to light and air for a time and escape from the nest to start new colonies, all other castes and all developmental stages

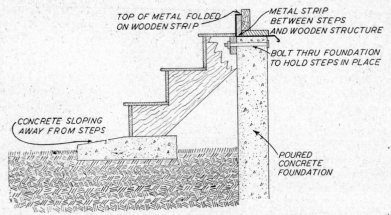

Fig. 513.—A method of installing out-of-door steps of a building to prevent termites working into the building. All wood is isolated from the soil by concrete and a metal termite shield is installed to separate the concrete foundation wall and the steps from the rest of the house. Where steps do not interrupt, the metal shield should project two inches both outside and inside the foundation wall and be bent downward at an angle of 45°. (*From Ill. State Natural History Surv.*)

live in a completely closed system of intercommunicating tunnels in soil and wood. It was long supposed that they were negatively phototropic, but it is now known that their seclusion is due to an aversion to air currents and reaction to a closely controlled humidity, which is maintained by carefully sealing themselves into their tunnels.

Control Measures.—To prevent infestation by these pests, great care should be taken to avoid having any woodwork of buildings within 18 *inches of contact with the ground,* as such points of contact are nearly always the cause of buildings becoming infested with termites. The insertion of a thin sheet of metal between the foundation and timbers of the house, running all the way around and projecting about an inch on each side, will prevent termite infestations (Fig. 513). Similar metal strips should be inserted between porch steps and adjacent wooden timbers, in all joints between concrete floors and side walls, and beneath wooden "sleepers." These shields must be carefully fitted to avoid leaving small openings through which termites may pass. They should always be placed so that it will be possible to inspect them from both the inside and outside of the building. Foundation brick- or stonework should be laid with good-grade cement mortar, well capped, and all cracks and crevices should be eliminated. All wood that must have contact with the soil should be termiteproofed by pressure impregnation with coal-tar creosote, zinc chloride, mercuric chloride, sodium silicofluoride, chlorinated naphthalene, or other tested preservative, after the wood is cut. When building, no bit of waste wood, such as grade stakes and concrete forms should be left buried in the soil near the structure. Stumps and waste wood should be removed from the premises. The treatment of the soil with kerosene or mixtures of kerosene and orthodichlorobenzene will kill termites in the soil and should be used where colonies are located. Termites may be killed and kept away from buildings for several years by removing the soil about the foundation walls from the ground level to the footings and treating it with trichlorbenzene, 1 part, and fuel oil or creosote, 2 parts, using a gallon of the mixture for each 10 cubic feet of soil as it is replaced. This mixture may also be injected into the center space of brick walls and beneath concrete floors, footings, and walks through holes drilled through them. The injection of paris green or sodium fluosilicate at intervals along infested timbers through holes bored into their tunnels will sometimes eradicate colonies, but great care must be used not to endanger human lives with these poisons. The controlling of these insects, once they have become established in a building, often means the expenditure of considerable money and labor. Passageways that lead over the foundation walls of buildings should be broken and the surface of the wall, over which these passages have been running, should be heavily treated with one of the sprays recommended above for soil treatment. All woodwork of the buildings, including timbers or parts of the framework which are badly eaten by the termites, should be replaced with metal or cement or, if this is not advisable, with wood thoroughly termiteproofed. Timbers that have not been seriously damaged, but which are exposed to infestation by termites, may be sprayed with warm creosote, using a compressed-air painting machine or other sprayer that will apply this liquid with considerable force.

References.—U.S.D.A., Farmers' Bulls. 1472, 1937, and Bur. Ento. Bul. 94, Part II, 1915, and Dept. Bul. 333, 1916; Ill. State Natural History Surv. Cir. 30, 1938; Conn. Agr. Exp. Sta. Bul. 382, 1936; Kofoid, C. A., et al., "Termites and Termite Control," Univ. Calif. Press, 1934.

287. Powder Post Beetles[1]

Importance and Type of Injury.—Second only to termites as destroyers of seasoned wood, these species do not require that the wood be in contact with the soil. The larvae eat the hard, dry wood, tunneling through and through timbers in successive generations until the interior is completely reduced to fine, packed powder and the surface shell is perforated by many small "shot holes." Some of the species (Lyctus spp.) attack only the sapwood of the hardwoods, but others work through pine and fir and the heartwood as well. They may completely destroy the timbers of build-

---

[1] Various species of the Order Coleoptera, Families Lyctidae, Ptinidae, Anobiidae, and Bostrichidae. The best-known are species of the Genus Lyctus, and the furniture beetle, Anobium punctatum DeGeer.

ings, log cabins, rustic work, ship and airplane lumber, furniture, tool handles, wheel spokes, oars, casks, and other lumber. Their attack is an insidious one, because they may live beneath the surface for months and bore out through the surface in great numbers after the lumber has been made into furniture, implements, flooring, girders, or interior finish.

Life History, Appearance, and Habits.—The winter, in unheated places, is generally passed in the larval stage, pupation occurs in spring, and the adults emerge in spring or early summer. The adults of most of the species are small, ranging from $\frac{1}{12}$ to $\frac{1}{5}$ inch in length, hard-shelled, brownish, elongate and cylindrical or short and stubby, and with varied sculpturing on body and wings. They are not often seen in the adult stage. Eggs are laid in the pores of the wood. The larvae resemble

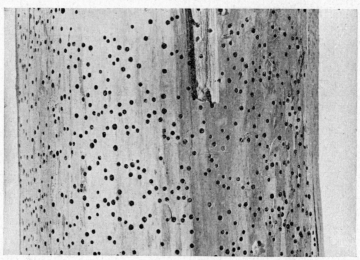

Fig. 514.—A piece of timber ruined by powder post beetles. The holes are made by the adult beetles emerging to find mates and lay eggs for another generation. The wood beneath the outer shell of such timbers will often be found reduced to powder, as fine as talcum, by the work of the larvae. (*From Ill. State Natural History Surv.*)

small white grubs ($\frac{1}{8}$ to $\frac{1}{3}$ inch long) with an unusually large head end. They eat the wood and pack their burrows with exceedingly fine flour-like frass. The holes are from $\frac{1}{16}$ to $\frac{1}{4}$ inch in diameter. Generation after generation of the insects may develop in the dry wood with little external evidence of damage until structural timbers or vehicle stock collapse, or furniture and finish are completely ruined.

Control.—The sapwood of green trees can be impregnated with solutions of copper sulphate or zinc chloride, $\frac{3}{4}$ pound in $\frac{1}{2}$ gallon of water for each cubic foot of sapwood, in the spring as leaf buds begin to swell, by sawing the tree off and setting the butt in a container holding the solution while the top is lodged against an adjacent tree. Or by various methods of banding, capping, or collaring the trunk, as described in the last reference below. Wood so impregnated resists insects and decay for many years. For the Lyctus species the complete removal of sapwood from the heartwood when lumbering, and eliminating it entirely from storage where lumber and lumber products are kept, is said to be completely effective in preventing attacks. Frequent inspection and destruction of infested stock, avoiding long storage, kiln-drying, steaming under pressure, treating with linseed oil, creosote, orthodichlorobenzene, kerosene, or turpentine, or vacuum fumigation of valuable furniture are available control measures. For the protection of furniture, finishing all surfaces with varnish or paint prevents egg-laying though it will not destroy the insects if already present.

References.—DOANE, *et al.*, "Forest Insects," pp. 216–224, McGraw, 1936; *U.S.D. A., Dept. Bul.* 1490, 1927, and *Farmers' Buls.* 778, 1917, 1477, 1926, and 1582, 1929, and *Bur. Ento. Plant Quar. Cir.* E-409, 1937.

288. SILVERFISH[1] AND FIREBRATS[2]

Importance and Type of Injury.—The bindings of books, papers, cards, boxes, and the like have the surface eaten off in irregular patches; the paper on walls is gnawed or the paste largely eaten off where the paper is attached to the walls. Small, whitish, grayish-brown, or greenish, glistening, carrot-shaped, entirely wingless insects, up to $\frac{1}{2}$ inch long (see Fig. 515), scurry rapidly about over shelves and walls on exposure to light.

Food.—These insects feed on a large variety of materials, such as starched clothes, rayon fabrics, bindings of books, book labels, the sizing of paper, or any papers on which paste or glue has been used.

Distribution.—General throughout the United States and Canada.

Life History, Appearance, and Habits.—The insects reproduce by means of eggs, the young closely resembling the adults except in size. The full-grown insect is wingless, about $\frac{1}{3}$ to $\frac{1}{2}$ inch in length, varying with the species; of a silvery, greenish-gray, or brownish color sometimes faintly spotted. The body tapers very markedly from head to tail, and is covered with thin scales which give it a silvery shiny appearance. There are two long antennae on the head and three similar appendages at the tail end, each nearly as long as the body. The preferred food of these insects is vegetable matter high in carbohydrates, such as flour and oatmeal. Most damage, however, is done by their attacks upon wall paper, card files, book bindings, rayon fabrics, starched clothing, and stocks of paper on which paste or glue has been used as a sizing. The silverfish or slicker,[1] which is uniform, silvery or greenish gray, prefers damp places next to the soil about basement rooms and porches and is less abundant in upper stories of houses. The firebrat,[2] which is mottled with patches of whitish and blackish scales, will not lay eggs at ordinary room temperatures, but at 98 to 102°F. and 70 to 80 per cent humidity it breeds rapidly. It is most abundant in very hot rooms, such as furnace rooms, leading people to suppose that it has been brought to them in deliveries of coal. The whitish oval eggs are deposited loosely in secluded places and may hatch in one to many weeks, depending upon the temperature. The young are whitish minatures of their parents. They grow slowly, with a large and indefinite number of molts, reaching the adult stage in from 3 to 24 months depending upon temperature and humidity.

Control Measures.—This insect may be controlled by liberally dusting pyrethrum powder, or sodium fluoride, about the parts of the houses

---

[1] *Lepisma saccharina* Linné, Order Thysanura, Family Lepismatidae.
[2] *Thermobia domestica* (Packard).

where they are most abundant. Fresh pyrethrum is somewhat better than the sodium fluoride for the control of this insect. Spraying with a saturated solution of paradichlorobenzene in carbon tetrachloride has

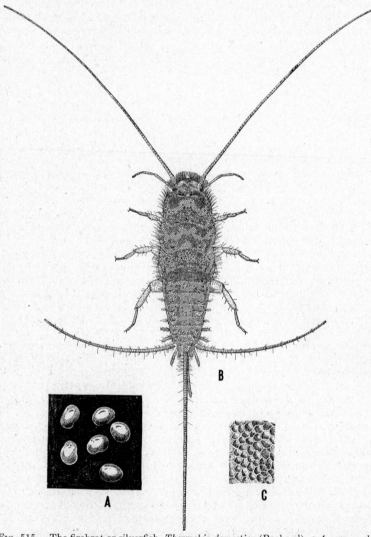

Fig. 515.—The firebrat or silverfish, *Thermobia domestica* (Packard). *A*, eggs, enlarged about 7 times; *B*, adult female, about 4 times natural size, showing the long antennae, terminal filaments, and covering of minute scales; *C*, a portion of the egg shell, showing the sculpturing which is faintly visible at high magnification. The nymphs are illustrated in Fig. 83. (*Drawings by Ruth Slabaugh and Kathryn Sommerman.*)

also been recommended. The building should be closed for 24 hours. Silverfish may also be killed by the use of poisoned baits. The following mixture makes an effective bait:

Oatmeal finely ground........... 200 parts by weight or about ⅞ pt.
Sodium fluoride or white arsenic.. 16 parts or about ¼ tsp.
Powdered confectioners' sugar.... 10 parts or about ½ tsp.
Common salt, powdered......... 5 parts or about ¼ tsp.

If white arsenic is used as the poison, enough water should be added to moisten and bind the ingredients together. When thoroughly dried, it should be broken into small particles and scattered lightly in out-of-the-way places where it need not be swept up for a long time; or scattered among loosely crumpled paper in uncovered boxes; or the moist bait may be spread upon squares of cardboard and tacked in the secluded harbors of the pests.

References.—Idaho Agr. Exp. Sta. Bul. 185, 1931; Jour. N.Y. Ento. Soc., **41**: 557-562, 1933; U.S.D.A., Leaflet 149, 1937.

289. COCKROACHES

Importance and Type of Injury.—These brown, brownish-black, or tan, shiny, flat-bodied, foul-smelling insects are well known to almost everyone. They are mainly active at night or in dark basements. Their filthy habits, repulsive appearance, bad odor, and the probability that they may spread diseases, such as tuberculosis, cholera, leprosy, dysentery, and typhoid, make them very objectionable.

Food.—They feed on many kinds of material, often becoming annoying in houses by eating the binding or leaves of books or magazines, the paper covering of boxes, various food products in pantries, kitchens, bakeries, restaurants, and like places, and by fouling with their excreta the material over which they run.

Distribution.—World-wide; especially abundant in the warmer parts of the world.

Life History, Appearance, and Habits.—There are 5 or 6 species of cockroaches which are common in houses and other buildings in the United States.

1. The small tan German cockroach[1] (Fig. 516, *lower figures*), about ½ inch long, with two dark stripes on the upper side of the prothorax, is one of the worst species. The females of this species carry their egg capsules protruding from the abdomen for about 2 weeks until they are nearly ready to hatch. There are commonly 25 to 30 eggs in each capsule, sometimes as many as 46, and a female produces 1 to 7 or more capsules during her lifetime. The nymphs pass through seven molts in 6 to 8 weeks, and the total life span is 2 to 5 months, with 2 or 3 generations a year in the average house. This species is most common in kitchens and bathrooms. It is very active, but rarely flies. The last ventral segment is entire in both sexes.

2. The large brown American cockroach[2] (Fig. 516, *upper figures*) reaches a length of 1½ inches or more. The female drops the egg capsule, or glues it in a sheltered place by secretions from her mouth, the day after it is formed, and may produce a capsule once a week until 15 to 90 capsules, averaging about 14 to 16 eggs each, have been formed. The

[1] *Blattella germanica* (Linné), Order Orthoptera, Family Blattidae (see *Jour. Kans. Ento. Soc.*, **11**: 94–96, 1938).
[2] *Periplaneta americana* (Linné) (see *Ann. Ento. Soc. Amer.*, **31**: 489–498, 1938).

nymphs hatch in 35 to 100 days and require 10 to 16 months and 13 molts before becoming adult, the total life span sometimes being as much as 2½ years.   This species often becomes abundant in city dumps and is most common in basements, restaurants, bakeries, packing houses, and groceries.   The pronotum has a yellowish posterior border and the last ventral segment in the female is notched.

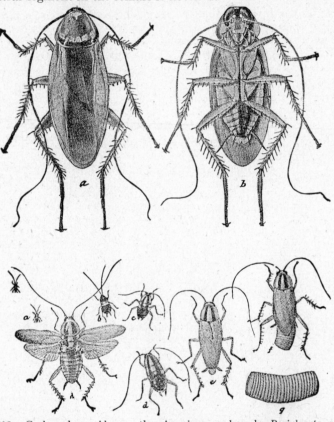

Fig. 516.—Cockroaches.   Above, the American cockroach, *Periplaneta americana* (Linné): *a*, dorsal view; *b*, ventral view.   Below, the German cockroach, *Blattella germanica* (Linné), adults and young; *f*, female with egg capsule protruding from tip of abdomen; *g*, egg capsule.   (*From U.S.D.A.*)

3. The large black Oriental cockroach[1] is about 1 inch long, uniformly black, the females nearly wingless, and the males with wings much shorter than the abdomen.   The last ventral segment in the female is long and notched.   The life cycle is about 13 months.   The female produces an average of 14 or 15 capsules averaging 12 to 16 eggs each.   This species is most prevalent in damp basements and along sewer lines and is considered the filthiest of all roaches.

4. The Australian cockroach[2] is very similar to the American roach, but only about 1¼ inches long, and has a prominent yellow stripe about

[1] *Blatta orientalis* Linné.
[2] *Periplaneta australasiae* (Fabricius).

one-third the length of the front wing, along the costal margin, and a distinct dark spot at the center of the pronotum. This is not a very important species.

5. The smoky-brown cockroach[1] is a subtropical species that lives in houses and greenhouses in the North, prefers high temperature and humidity, and completes a cycle in about a year. Its egg capsules contain up to 24 eggs each.

6. The brown-banded cockroach[2] has recently spread over many of the Gulf Coast and northern states. It is smaller than the German cockroach, being usually under $\frac{1}{2}$ inch in length. There is a crossband of light yellow at the base of the wings and another about $\frac{1}{16}$ inch farther back, and the dark stripes on the thorax are much broader than in the German roach. The female is very broad-bodied, the male more slender and lighter-colored. This species seems to prefer living rooms of dwellings and apartments where it hides in cracks in the woodwork or furniture and is particularly hard to control. The life cycle is about 200 days, and there may be two generations in a year. The egg capsules contain a maximum of 18 eggs.

7. The Surinam cockroach[3] is a large dark-brown lubberly insect, about the size of the American cockroach, but with wings only one-third the length of the body which is distinctly oval in outline. They are often common in greenhouses, where they feed upon growing plants.

8. The long-winged males of the wood cockroach,[4] $\frac{2}{3}$ to 1 inch long, may fly for long distances, and are sometimes abundant in houses; and the short-winged females also invade dwellings near woods. The life cycle is about 1 year. The egg capsules contain a maximum of 32 eggs.

The life histories of all species are much the same. The eggs are laid in pod-like or bean-like capsules called *oötheca* (Fig. 516,*g*). This pod-like case may be seen protruding from the abdomen of the female (*f*), as she moves about, especially in the case of the German cockroach. The small roaches hatching from the eggs have much the same appearance as the adults, except that they lack wings. They develop rather slowly, requiring 2 to 18 months or more, to become full-grown. Cockroaches hide in the cracks of buildings during the daytime, and their abundance is much greater than ordinarily supposed. The household species are active throughout the year in heated buildings. They prefer high temperatures and humidity and are easily killed by cold, although wood roaches live through the winter under the bark of trees and in similar situations.

Control Measures.—The first steps in control should be scrupulous cleanliness and to guard against reinfestation from reservoirs of infection, such as neighboring buildings, sewers, covers of wells and cisterns, and stores from which deliveries of groceries, laundry, and furniture are being received. All pipe lines should be tightly sealed where they enter through floors and basement walls, and cracks in which the pests hide should be filled with a crack filler. It would be well to trade only at sanitarily built stores. To eradicate cockroaches from a building the following methods will be found effective: (1) One of the most effective

---

[1] *Periplaneta fuliginosa* (Serville).
[2] *Supella supellectilium* (Serville).
[3] *Pycnoscelus surinamensis* (Linné).
[4] *Parcoblatta pennsylvanica* (DeGeer).

methods of controlling the German cockroaches, and one that will practically always give relief from the other species, is to dust thoroughly all parts of the houses with sodium fluoride[1] or sodium fluosilicate. This dust should be applied in liberal quantities in the dark corners of closets, at the base of the walls in basements, under sinks, around drain pipes, behind baseboards, upon shelves, or in any cracks in the wall, where the cockroaches are likely to hide. If this treatment is persisted in, even the most severe infestations may be entirely cleaned up. The roaches do not

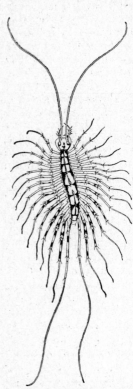

feed on the powder but get it upon their bodies, and are killed by contact; or, in cleaning their feet and legs, they take particles of the material into their mouths and so become poisoned. One should bear in mind that sodium fluoride is poisonous, and it should not be put out where children, or animal pets, will get it in their food. It retains its effectiveness indefinitely if it is kept dry. It is best applied with an electrically driven duster designed to blow the dust into cracks and hiding places rather than about the room. (2) For the American, Oriental, and brown-banded cockroaches fresh pyrethrum powder, applied in the same way, is very effective and not poisonous to man. It loses its effectiveness in a short time and should be repeated every 2 or 3 days. Sodium fluoride and pyrethrum should not be mixed and applied together, since pyrethrum is repellent and will tend to keep the roaches from running through the fluoride. Mixing lime with the sodium fluoride causes it to act much more rapidly. (3) High-grade thiocyanate or pyrethrum household sprays, applied with a good sprayer, are effective in well-constructed buildings. (4) Tested commercial phosphorus pastes are one of the best methods of controlling the American, the Oriental, and the Surinam cockroaches. The paste may be spread upon cardboard and rolled into cylinders, with the poison on the inside, and placed or tacked in out-of-the-way places; these are good to use when roaches are not numerous or to supplement other measures. Even though all roaches have been cleaned out of a house, the house will not remain free long if other houses in the immediate vicinity

FIG. 517.—House centipede, *Scutigera forceps* (Rafinesque). Adult, natural size. (*From U.S.-D.A., Bur. Ento. Cir.* 48.)

are infested, as these insects move about very freely, and actual migrations of the insects from house to house have been noted.

References.—*U.S.D.A.*, *Leaflet* 144, 1937; *Proc. Ind. Acad. Sci.*, **47**: 281–284, 1938.

290. HOUSE CENTIPEDE[2]

This creature, which is frequently encountered about houses, is really beneficial, causing no injury whatever to stored products. It becomes annoying to some people

---

[1] Sodium fluoride not to be confused with *sodium chloride*, or common salt.
[2] *Scutigera forceps* (Rafinesque), Class Chilopoda, Family Scutigeridae.

because of its appearance and habit of rushing over the walls or floors, occasionally toward a person, when it is suddenly disturbed by the turning on of lights or by other causes. There are a few cases on record where the centipede has inflicted a painful bite when handled.

In appearance these creatures are of a grayish-tan color, with long antennae and many extremely long legs extending out all around the body (Fig. 517). They move so rapidly that it is often difficult to get an accurate idea of their appearance. When full-grown, they are from 2 to 3 inches in length. Their food consists of small insects, such as roaches, clothes moths, house flies, and others which they may encounter about houses.

Unless they become extremely abundant, they should not be killed. In cases where they do become so numerous as to prove annoying, the liberal use of pyrethrum powder about water pipes, or other damper parts of the house frequented by the centipedes, will be found effective in controlling them. They are killed by fumigation, either with hydrocyanic acid gas or carbon bisulphide, but it is rarely, if ever, necessary to go to this expense for the control of these creatures.

291. Cigarette Beetle or Towbug[1]

Importance and Type of Injury.—This is the most important insect pest of tobacco in factories and cigar stores and also causes considerable

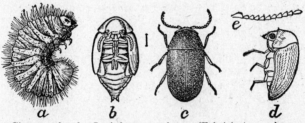

Fig. 518.—Cigarette beetle, *Lasioderma serricorne* (Fabricius). *a*, larva; *b*, pupa; *c*, adult, dorsal view; *d*, adult, side view; *e*, antenna. About 6 times natural size. Line indicates natural size. (*From U.S.D.A., Farmers' Bul.* 846).

damage to other products. Package and chewing tobaccos, cigars, and cigarettes have holes eaten through the tobacco (see Fig. 18), and contain many, small, light-colored, brown-headed grubs or brown beetles working in the tobacco. Furniture upholstered in flax, tow, or straw is often grossly infested by these beetles and sometimes has holes eaten through the covers by the larvae or beetles.

Food.—Tobacco products, especially cigarette tobaccos with high sugar content, and dried leaves; upholstered furniture, materials used in stuffing furniture, seeds and other dried plant products, especially those used as drugs, black and red pepper, and many others.

Distribution.—Throughout the United States and southern Canada.

Life History, Appearance, and Habits.—The adult beetle is rounded in outline, of a very light-brown color, and only about $\frac{1}{16}$ to $\frac{1}{10}$ inch in length, with the head and prothorax bent downward so as to give the insect a strongly humped appearance. The wing covers are not striated, and the antennae are of the same thickness from base to tip (Fig. 518,*c*,*d*,*e*). The oval whitish eggs, $\frac{1}{50}$ inch long, are laid in and about the substances on which the insects feed and hatch in 6 to 10 days at

[1] *Lasioderma serricorne* (Fabricius), Order Coleoptera, Family Anobiidae.

summer temperatures. The larvae are yellowish-white, curved-bodied, very hairy little grubs, with light-brown heads. They may become full-grown in 30 to 50 days when they are nearly ⅙ inch long. They pupate for 8 to 10 days or more in silken cocoons covered with bits of their food material. The entire life cycle may be passed in from 45 to 50 days, and there are commonly three to six generations a year.

Control Measures.—Infested tobacco factories, warehouses, or other premises should be thoroughly cleaned before a new crop is stored, removing all refuse material on which the beetles may be feeding and treating the cracks in walls and floors with a strong disinfectant. Infested tobacco in closed storage warehouses may be freed of the insects by fumigating with 10 to 16 ounces of liquid hydrocyanic acid per 1,000 cubic feet, or 2 to 2½ pounds of sodium cyanide per 1,000 cubic feet. This should be applied just after the peak of adult abundance of any generation has passed and almost reached its lowest level, as shown by the catch at light traps, so as to avoid most of the large larvae and pupae which are hardest to kill and likely to be deepest in the tobacco. Kearns suggested that the given dosage be divided and applied in four "shots" at 2-hour intervals, for best results. Ethylene oxide and carbon dioxide (see page 287) and other industrial fumigants are used for imported and manufactured, packaged tobaccos in vacuum vaults to insure freedom from this pest. Warehouses freed of the insects should have all openings covered with 24-mesh screens to prevent reinfestation by the entrance of adults from infested surroundings. Suction light traps should be operated throughout the season in warehouses, beginning in spring when the temperature reaches 65°F., one trap to each 75,000 cubic feet of space. All other lights must be kept off. Heating to 130 to 135°F. is very effective. The heat must be applied long enough to penetrate all parts of the infested material, and the temperature should be maintained for at least 6 hours. All stages of the insect can be killed in tobacco products stored for a week at 25°F. or 16 days at 36°F., but a week or two more may be required for the cold to penetrate to the center of tobacco hogsheads. In cooler areas this may be accomplished by opening doors and ventilators during the winter. The control, as a furniture pest, is much the same as for carpet beetles. Overstuffed furniture may be freed of these beetles by thoroughly wetting it with a good grade of uncolored gasoline.

References.—*U.S.D.A., Dept. Bul.* 737, 1919, and *Farmers' Bul.* 846, 1917, and *Cirs.* 356, 1925, and 462, 1938; *Furniture Manufacturer*, **34**: 3–7, 1927; *Ill. Agr. Exp. Sta. Cir.* 473, 1938.

292. Drug Store Beetle[1]

This insect closely resembles the cigarette beetle in its life history and habits. Its food is even more varied than that of the cigarette beetle and includes practically all dry plant and animal products. The adult is about ⅒ inch long, of a reddish-brown color, and densely covered with very short light hairs. The wing covers are plainly striated, and the antennae are enlarged at the end (Fig. 519,*c,d,e*). The larva (*a*) differs from the cigarette beetle in being nearly bare.

Control Measures.—The same as for the cigarette beetle.

[1] *Stegobium* (= *Sitodrepa*) *paniceum* (Linné), Order Coleoptera, Family Anobiidae.

293. Tobacco Moth[1]

Importance and Type of Injury.—Since about 1930 this relative of the Mediterranean flour moth has become an alarming pest of stored tobacco, especially the brighter grades of flue-cured tobacco. The larvae eat the tender parts of leaves, making holes much larger than the cigarette beetle does, and pollute what they do not eat, with their silk and excrement.

Food.—Cured tobacco, nuts, chocolate, and many other dried vegetable products.

Life History, Appearance, and Habits.—The elliptical, granular, grayish-white to brownish eggs are laid singly or in small clusters, usually upon the tobacco. They hatch in 4 to 7 days. The larvae are whitish, tinged with yellow, brown, or pink and peppered with small brownish, scattered spots. They measure ⅜ to ⅝ inch long, when full-grown,

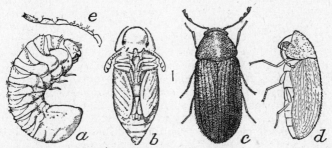

Fig. 519.—Drug store beetle, *Stegobium paniceum* (Linné): *a*, larva; *b*, pupa; *c*, adult, dorsal view; *d*, adult, side view; *e*, antenna. Enlarged about 12 times. Line indicates natural size. (*From U.S.D.A., Bur. Ento. Bul. 4.*)

and are very active. The larval or feeding period lasts 40 to 45 days. They pupate in almost any sheltering crevice, commonly for 7 to 19 days. The adults are grayish to brownish moths, ½ inch from tip to tip of the wings. The middle half of the front wings is lighter in color than the base or tip and is separated from these darker extremities by a narrow whitish crossband. They are most active at night.

Control.—The control measures recommended for the cigarette beetle are effective for this species also. Dusting empty warehouses once a week with a 200-mesh pyrethrum powder, 3 ounces to each 1,000 cubic feet is recommended.

Reference.—*U.S.D.A., Cirs.* 269, 1933, E-325, 1936, and E-422, 1937; *Tobacco,* **105**: 21, 22, 1937.

294. Carpet Beetles or Buffalo Beetles[2,3,4]

Importance and Type of Injury.—Holes are eaten in fabrics, such as clothing, rugs, the upholstery covers and interior padding of furniture, curtains, especially those containing wool, fur, feathers, or hair, and brushes made of animal bristles, by very hairy or bristly, brownish larvae

[1] *Ephestia elutella* (Hübner), Order Lepidoptera, Family Pyralididae.
[2] *Anthrenus scrophulariae* (Linné), Order Coleoptera, Family Dermestidae.
[3] *Attagenus piceus* (Olivier), Order Coleoptera, Family Dermestidae.
[4] *Anthrenus vorax* Waterhouse.

of all sizes up to more than ¼ inch long (Fig. 520). Some of the species[1] have a long tail of black hairs (Fig. 521). There is no webbing together of the fabric nor silken threads spun over the surface as is the case with clothes moths. These larvae live a large part of the time in secluded places about the rooms, out of reach of house-cleaning operations, so that they are particularly insidious and are responsible for very many holes, ravelings, and defects in clothing and other fabrics, which are not even suspected of being insect injury. Small, blackish, hard-shelled beetles, sometimes flecked with white or reddish scales, are often seen in the infested materials or about windows. Most severe injury occurs in materials that have lain undisturbed for some time.

Food.—Woolen goods of all kinds; sometimes cotton goods; leather, bristles, feathers, hair, silk; dried meat, milk or insect specimens; stuffed animals, fur; grains, flour, and many other animal and plant products.

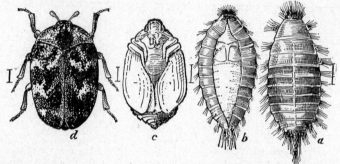

FIG. 520.—The common carpet beetle, *Anthrenus scrophulariae* (Linné); *a*, larva; *b*, pupa within larval skin; *c*, pupa removed from larval skin, ventral view; *d*, adult. All enlarged about 7 times; lines indicate natural size. (*From Riley, U.S.D.A.*)

Distribution.—Two of the most destructive species of carpet beetles have been imported from Europe, these being the common carpet beetle[2] and the black carpet beetle.[1] These and several other species[3,4] occur generally in this country.

Life History, Appearance, and Habits.—Although breeding may take place throughout the year in uniformly heated buildings, most carpet beetles winter as larvae. They pupate from April to June, for 1 to 4 weeks, as a white soft stage inside the split larval skin (Fig. 520,*b*) but without the protection of a cocoon, and the adults are most abundant during hot weather from early May to mid-June. After molting to the adult stage, the insects may further remain within the larval skin from a few days to a few weeks. The adults may live from 2 weeks to several months but never damage household goods in this stage. They are active, attracted to light, and are often found about windows and out-of-doors upon flowers, especially spiraea, eating the pollen. There is some evidence, however, that the females usually lay their eggs before they leave the building in which they developed. They lay their soft white eggs upon the food materials of the larvae or in dark secluded places, as

[1] *Attagenus piceus* (Olivier), Order Coleoptera, Family Dermestidae.
[2] *Anthrenus scrophulariae* (Linné), Order Coleoptera, Family Dermestidae.
[3] *Anthrenus vorax* Waterhouse.
[4] *Anthrenus verbasci* (Linné)

upon clothing, the angles of upholstered furniture, cracks about baseboards, in warm-air furnace pipes, and other places where dust and lint accumulate and are undisturbed. The eggs hatch in a week or two and the young larvae immediately attack such food as they can find. Unlike their parents, they avoid the light and feed voraciously in dark secluded crevices or folds of their food. Unlike the clothes moths they may crawl away from the food material to molt in cracks and crevices inaccessible to insecticides. They molt from 5 to 11 times in the course of their growth, and the cast skins closely resemble, and often are mistaken for, living larvae (*a*). It requires from less than 1 year to as much as 3 years for the larva to complete its growth, the length of time depending on climatic conditions and food.

The black carpet beetle is believed to be the most important pest of fabrics in storage in many parts of the United States. The adults are uniformly dull black, with legs and antennae brownish, elliptical in outline, $\frac{1}{8}$ to $\frac{5}{32}$ inch long by half as wide. Females produce an average of about 50 eggs which hatch in a week or 10 days. The larvae of this species, which may reach a length of nearly $\frac{1}{2}$ inch, are very different from the others, being elongate, carrot-shaped, golden to chocolate brown and with a tuft of very long brown hairs at the tail end of the body.

References.—Jour. Econ. Ento., **31**: 280–286, 1938; *U.S.D.A.*, *Leaflet* 150, 1938.

The larvae of all the other species are short, rarely over $\frac{1}{4}$ inch long, chubby, and covered with erect brown or black bristles all over the body and with three tufts of bristles at each side of the posterior third of the body which lie closely against the body like folded wings. The buffalo beetle[1] and the furniture carpet beetle[2] are covered with black bristles while the varied carpet beetle[3] has tawny or brownish bristles. The adults of the common carpet beetle are elliptical, $\frac{5}{32}$ inch long and three-fourths longer than wide, with a band of brick red scales down the center of the back and red-and-white scales forming several irregular bands across the body (Fig. 520). The furniture carpet beetle and the varied carpet beetle are scarcely a third longer than broad, are irregularly mottled with flecks of white, yellow, and black, and are white on the underside. The furniture carpet beetle is very destructive to upholstery of hairs and feathers.

Control Measures.—The control of these pests is of two distinct phases: (*a*) Measures which will tend to prevent them from becoming established in the home and (*b*) measures necessary to eradicate them after damage has begun. Good construction which does not provide hiding places for the insects in cracks of floors, baseboards, and quarter round, or the elimination of such crevices with a good crack filler, are important steps in prevention. Good housekeeping methods, which promptly eliminate

FIG. 521.—The black carpet beetle, *Attagenus piceus* (Olivier); larva, about 4 times natural size. (*From Herrick, "Insects Injurious to the Household," copyright, 1914, Macmillan. Reprinted by permission.*)

dust and lint, do not permit floor coverings, draperies, blankets, clothing and the like to remain long undisturbed, but subject them to frequent vacuum sweeping and at least semiannual sunning, beating, and brushing, give the insects little opportunity to establish themselves. The interior of furnace pipes and pianos is especially likely to be neglected. Once

[1] *Anthrenus scrophulariae* (Linné), Order Coleoptera, Family Dermestidae.
[2] *Anthrenus vorax* Waterhouse.
[3] *Anthrenus verbasci* (Linné).

the insects are found, the nature and extent of the infestation must determine the wisest procedure.   If the infestation is widespread and intense,
thorough fumigation of the entire house or of infested rooms with hydrocyanic acid or carbon bisulphide or by superheating, as described on
pages 278, 294, is the best and quickest method of stamping them out.
If this is not practicable, the next best method is the use of a high-grade
pyrethrum, rotenone, or thiocyanate household spray, applied to cracks
in floors and baseboards with a power atomizer or hand sprayer.   This
may have to be repeated for some time, for only those specimens wet by
the spray will be destroyed.   Individual pieces of badly infested furniture
should be sent to companies equipped to do vacuum fumigation or treated
with one of the moth sprays applied with nozzles in the form of perforated
needles, for reaching the interior of the upholstery.   There are a number
of good mothproofing materials which may be applied by the manufacturers.   Other processes are widely available through dry-cleaning
companies, and solutions of sodium fluosilicate, sodium aluminum silicofluoride, cinchona alkaloids, and other poisons may be applied by the
housekeeper herself as discussed on page 292.   Many companies offer the
facilities of cold storage for furs and other valuable winter clothing during
the summer.   Even though present the insects cannot injure the goods at
temperatures below 40°F.   Articles not to be used for some time may
best be stored in very tight trunks or boxes in the following manner: At
various levels among the clothing or blankets place flake naphthalene or
paradichlorobenzene between thin sheets of paper, using a pound for 20 to
100 cubic feet.   On top of the clothing place a shallow pan or dish and
pour carbon bisulphide upon it, in a warm place away from any flame,
using at least 1 pound to 100 cubic feet of space.   Quickly close the box
and seal tightly with strips of gummed paper.   The carbon bisulphide
will destroy any stages of the insects that may be present, and the other
chemical will serve as a repellent against infestation for months.

Cedar chests will not kill these insects.   Heating to 135°F. for 6
hours will kill all stages of the insects, but care must be taken in using this
method that the desired temperature is maintained in the center of the
articles treated (see page 294).

References.—*U.S.D.A., Div. Ento. Buls.* 4, 1896, and 8, 1897, and *Cir.* E-395, 1936
and *Leaflet* 150, 1938; *Ill. Agr. Exp. Sta. Cir.* 473, 1938; *Jour. Econ. Ento.*, **31**: 280–286,
1938.

295. Clothes Moths[1,2,3]

Importance and Type of Injury.—There are several species of clothes
moths that are responsible for damage in this country, the most common
being the casebearing clothes moth[1] and the webbing clothes moth.[2]
The carpet moth[3] is much less common in this country.   Fabrics
injured by clothes moths have holes eaten through them by small, white
caterpillars, and in most cases the presence of the insect is indicated by
silken cases or lines of silken threads over the surface of the materials.
Materials left undisturbed for some time or stored in dark places are most
severely injured by these insects.   Small, buff-colored moths, not over

[1] *Tinea pellionella* Linné, Order Lepidoptera, Family Tineidae.
[2] *Tineola biselliella* Hummel, Order Lepidoptera, Family Tineidae.
[3] *Trichophaga tapetzella* (Linné), Order Lepidoptera, Family Tineidae.

½ inch across the wings, will be found running over the surface of infested goods when such goods are exposed to light, or flying somewhat aimlessly about in houses or closets. The clothes moths are not attracted to lights.

Food.—The clothes moth larvae feed on wool, hair, feathers, furs, upholstered furniture (Fig. 17), occasionally on dead insects, dry dead animals, animal and fish meals, milk powders such as casein, and nearly all animal products, such as bristles, dried hair, and leather. The adults take no food.

Fig. 522.—The casebearing clothes moth, *Tinea pellionella* Linné. *a*, adult; *b*, larva in case; *c*, larva removed from its case; *d*, cocoon enclosing pupa. About 7 times natural size. (*From Ohio Agr. Exp. Sta. Bul.* 253.)

Distribution.—Clothes moths are distributed generally over the world.

Life History, Appearance, and Habits.—The adult "millers" or moths (Fig. 522,*a*) are entirely harmless and probably take no food of any sort. They lay their eggs singly on the products in which the larvae feed, each female laying from 100 to 150 eggs. Occasionally, with some species, the eggs may be laid in groups. They are small, being about one-tenth the size of a pinhead (1⁄50 inch long), of a white color, which makes them rather conspicuous when deposited on black material. They hatch in about 5 days. The larvae (Fig. 522,*b,c*) which hatch from these eggs are the only stage of the insect causing damage. They are white, and vary in size from about 1⁄16 inch long, when first hatched, up to about 1⁄3 inch when full-grown. The length of the larval period varies greatly according to the conditions and food supply. The complete development of this stage may take from 6 weeks to nearly 4 years.   Development is greatly influenced by humidity, the life cycle being shortest, in average room temperature, at about 75 per cent relative humidity.   The larvae of some species live in silken cases (b) which are dragged about with them and enlarged as they grow.   Upon completing its growth, the larva adds to the silk of the case in which it has been living, forming a rather tough cocoon.   Within this case (d) it changes to a white pupa, about ⅙ inch in length.   The pupa later turns brown and in 1 to 4 weeks the adult moth emerges.   The complete life cycle may be passed in about 2 months or may be prolonged over 4 years, including long periods of larval inactivity or dormancy.   In heated buildings, the adults may be found at any time of the year but are most abundant during the summer months.

Control Measures.—Control measures are essentially the same as for the carpet beetles.   Frequent dusting, brushing, and vacuum cleaning are very important control measures and should be extended to the most remote cracks, warm and cold air passages, and similar hiding places. Furniture upholstered in leather, silk, rayon, linen, or cotton will not be damaged by clothes moths.   Layers of cotton will not be damaged by clothes moths.   Layers of cotton batting properly installed beneath the covers of furniture or furniture springs will nearly always prevent moths from attacking mohair or woolen upholstery from the inside.   Vault fumigation and mothproofing are highly effective, when properly done, and furniture should always be so treated before it is delivered to the buyer.   When houses are to be vacant for several weeks, as during a vacation period, a simple method of fumigation consists of closing rooms very tightly, spreading papers over the floor and scattering paradichlorobenzene crystals over the papers at the rate of 8 to 10 pounds to each 1,000 cubic feet.   Clothing which is in daily use is practically never infested by the clothes moths.   It is highly important that clothing, or other fabrics placed in storage, should be free from moths.   To insure this, such clothing should be thoroughly brushed and shaken and hung out-of-doors in the bright sun for several hours before being packed away. If placed in trunks or boxes, clothing may be protected from infestation as described for carpet beetles.

Tight-fitting cedar chests will protect clothing from attacks of the clothes moths and will kill larvae hatching from the eggs within the chest. However, partly grown larvae placed in such chests may complete their development and cause considerable damage.   Heating and fumigating as recommended for the carpet beetle will kill all stages of the clothes moths.   No damage to clothing will occur if kept in cold storage at 45°F., or lower.

References.—U.S.D.A., Farmers' Buls. 1353, 1923, and 1655, 1931, and Dept. Bul. 1051, 1922; Furniture Manufacturer, vol. 35, nos. 1 (n.s.) and 5, 1928; Mon. Rev. Nat. Retail Furniture Assoc., vol. 3, nos. 6 and 8, 1929.

296. Larder Beetle[1]

Importance and Type of Injury.—Very hairy, brown larvae (Fig. 523), tapering toward both ends of the body, feed on meats and animal products of nearly all kinds.

Food.—Feathers, horn, skins, hair, beeswax, ham, bacon, dried beef, and like products.

Distribution.—World-wide.

[1] Dermestes lardarius Linné, Order Coleoptera, Family Dermestidae.

Life History, Appearance, and Habits.—The adult beetles are about ⅓ inch long, of a very dark-brown color and with a moderately wide yellowish band across the front part of the wing covers. There are six black dots in this band, three on each wing cover, usually arranged in a triangle (Fig. 523). The eggs are laid on the food or in sheltered places near by. The larvae, on hatching, increase rapidly in size. They feed chiefly near the surface of the infested materials and become full-grown (a little over ⅓ inch long) in 40 to 50 days. Pupation takes place in the larval skin. The exact number of generations and some details regarding the length of the stages are not known. The adult beetles will occasionally be found in numbers out-of-doors on flowers, where they feed on pollen.

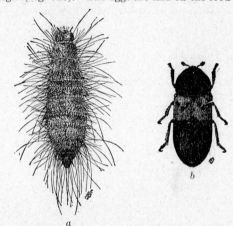

FIG. 523.—The larder beetle, *Dermestes lardarius* Linné. *a*, larva; *b*, adult. About 4 times natural size. (*From Herrick, "Insects Injurious to the Household," copyright, 1914, Macmillan. Reprinted by permission.*)

Control Measures.—Smoked meats kept in farm storehouses should be carefully sacked or wrapped in paper, muslin, or other cloth immediately after smoking, care being taken that the entire piece of meat is covered and no openings or cracks left through which the beetles may gain access to the meat. Smoked meats held in cold storage will not become infested. If meat storerooms or larders become infested, they should be heated to 130 to 135°F. for 3 hours, or fumigated with carbon bisulphide or hydrocyanic acid gas. All meat in the storeroom should be removed before treatment is applied, and the infested parts of such meat should be carefully trimmed off and burned.

References.—*U.S.D.A., Div. Ento. Bul.* 4 (n.s.), p. 107, 1902; *Can. Ento.*, **38**: 68, 1910; *N.Y. (Geneva) Agr. Exp. Sta. Bul.* 202, 1931.

297. Red-legged Ham Beetle[1]

Importance and Type of Injury.—This is the most destructive pest attacking well-cured and dried smoked meats, occasionally ruining large quantities of hams and bacons. The larvae do most of the damage by burrowing through meat, especially in fatty portions; but larvae and adults both feed, and are found, in the infested products.

Food.—Cured pork is most injured, but the larvae or adults feed also in fish, guano, bone meal, cheese, dried egg, hides, copra, dried fruits and nuts, other insects, and insect eggs including their own kind.

Distribution.—Nearly cosmopolitan, especially troublesome in the Middle Atlantic states.

Life History, Appearance, and Habits.—Winter in unheated buildings is probably passed chiefly as large larvae, the adults becoming evident in May and June. The adults are brilliant greenish-blue, convex, straight-sided, noticeably "punctured" beetles, about ⅐ to ¼ inch long, and less than half as wide, with the legs and base of antennae reddish-brown. The adults disperse mostly by rapid running, but they also fly. The elongate, cylindrical, translucent eggs are deposited in clusters in dry recesses of the food substances and hatch in 4 or 5 days in warm weather. The larvae are elongate, slender, tapering toward the head, purplish in color, with six short legs and short hairs; about ⅖ inch long when full-grown. They are strongly repelled by

[1] *Necrobia rufipes* (DeGeer), Order Coleoptera, Family Cleridae. Also called copra bug.

light.   They migrate from greasy foods to dry dark spots to pupate, lining a crevice with silk.   The entire life cycle occupies from 36 to 150 days or longer, depending upon the food.   About two-thirds of that time is spent in the growing, larval stage.

Control Measures.—Very thorough and prompt wrapping of hams, shoulders, and bacons is effective in preventing attack.   Screening with wire cloth, 30 meshes per inch, will exclude the adults from storage rooms.   Fumigation of badly infested premises with hydrocyanic acid gas is advised.

Reference.—*Jour. Agr. Res.*, **30** : 845–863, 1925.

298. Cheese Skipper[1]

Importance and Type of Injury.—Small, naked, yellowish maggots crawling in and over cheese or meat (Fig. 524), or jumping for short distances by bending their bodies nearly double and then suddenly straightening them.

Food.—Cheese and smoked, cured meats.

Distribution.—World-wide; probably imported into this country.

Fig. 524.—The cheese skipper, *Piophila casei* Linné; eggs, larvae or "skippers," pupae and adult flies.   About 1½ times natural size.   (*From E. O. Essig, Univ. of Calif.*)

Life History, Appearance, and Habits.—The adult insects are small, rather shiny two-winged flies, a little less than ⅙ inch in length (Fig. 524).   They lay their eggs singly or in clusters up to about 50.   The eggs hatch in 24 to 36 hours into small fleshy maggots which are legless and taper toward the head end.   They are yellowish-white in color, and about ⅓ inch long when full-grown.   These larvae may complete their growth in from 1 week to several months.   They then change to a very light-brown puparium and remain in this stage for about 1 week to 10 days.   The entire life cycle averages about 18 days.

Control Measures.—Infested portions of cheese or ham should be cut away and burned.   Where cheese is stored, care should be taken that the grubs are kept out of the storeroom by tightly closing all openings or by using 30-mesh screen or cloth to protect windows open for ventilation.   Frequent examinations should be made of cheese in storage to see that no infestation has started.   Bandages or coatings of paraffin used for covering cheeses should be kept as firm and smooth over the cheese as possible.   Where serious infestations have started in storerooms, they should be fumigated with hydrocyanic acid gas, at 20 ounces per 1,000 cubic feet, or thoroughly cleaned out and the walls and entire room washed with very strong soap suds, using

[1] *Piophila casei* Linné, Order Diptera, Family Piophilidae.

the soap at the rate of 3 to 4 ounces to the gallon of water. The water should be as hot as possible while the washing is done. Heating to 125°F. for one hour will kill the larvae and pupae. Meats and cheese held in storage at temperatures below 43°F. will not be infested by the cheese skipper.

References.—U.S.D.A., Dept. Bul. 1453, 1927; *Calif. Agr. Exp. Sta. Bul.* 343, 1922; *Jour. Exp. Biol.*, **12**: 384–388, 1935.

299 (365). CHEESE, HAM, AND FLOUR MITES[1]

There are a number of species of mites that infest ham, cheese, flour, stored grains, and other food products. In mushroom houses they often eat holes in the caps and stems or consume the spawn. They are all whitish in color and so small as to be barely discernible to the naked eye (Fig. 525). They vary in size, the largest being only about $\frac{1}{32}$ inch long. When they are abundant, a quantity of fine brownish powder accumulates over and through the materials, and a musty, sweetish odor is given off by the infested products, which is quite characteristic. Occasionally a rather intense irritation of the skin is caused by the mites where one has been handling food products infested by them. This has received the descriptive name of grocer's itch (see page 899).

The mites reproduce by means of eggs, which are laid promiscuously over the food materials. The young mites grow rapidly, being six-legged at first, but eight-legged on becoming adults. When conditions are unfavorable, the mites pass into a very active, nonfeeding, but very resistant stage, known as the *hypopus*. In this stage the body wall hardens and suckers are formed on the underside with which they may attach to other insects or mice and be carried to new locations. The mites may remain in

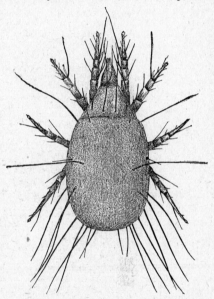

FIG. 525.—Cheese or ham mite, *Tyroglyphus* sp. Greatly magnified. *(From Minn. Agr. Exp. Sta. Bul.* 198.)

this condition for a number of months without food, but, when conditions become favorable to their growth, they molt and again become active. This peculiar adaptation enables them to survive for considerable periods, and for this reason premises once infested are often difficult to clean up.

Control Measures.—Curing cheeses slowly at temperatures of 30 to 36°F. prevents the development of both mites and skippers. Complete coating with paraffin is effective. All infested material should be carefully gone over and that which shows serious injury discarded. Infested parts of hams or cheeses should be cut off and burned. Storerooms should be thoroughly fumigated with sulphur, using the sulphur at the rate of 2 to 3 pounds per 1,000 cubic feet. In rooms where the oil is not objectionable, the walls, floor, and ceilings should be sprayed with some light oil, such as gasoline or a deodorized kerosene. As the mites thrive best where the humidity is rather high, keeping storerooms as dry as possible will tend to prevent infestation. It has been found that the mites cannot live for any length of time, except in the hypopus stage, where the humidity is less than 11 per cent. However, when mites are working on food, they give off moisture which often raises the humidity to a point

[1] *Tyroglyphus siro* Linné, *T. farinae* DeGeer, and *T. longior* Gerv., Order Acarina, Family Tyroglyphidae.

favorable to their development. Strong nicotine dusts will kill many of the mites with which they come in actual contact.

References.—*Minn. Agr. Exp. Sta. Bul.* 198, 1921; *U.S.D.A., Div. Ento. Bul.* 4 (n.s.), 1902; *Insect Life,* **4**: 170, 1892; *Calif. Agr. Exp. Sta. Bul.* 343, 1922.

B. INSECTS ATTACKING STORED GRAIN, SEEDS, AND GRAIN PRODUCTS

300. GRANARY WEEVIL[1] AND RICE WEEVIL[2]

Importance and Type of Injury.—These two true weevils (Fig. 526) are perhaps the most destructive grain insects in the world. They frequently cause almost complete destruction of grain in elevators, in farmers' bins, or on ships where conditions are favorable to their growth and the grain is undisturbed for some length of time. Infested grain will usually be found to be heating at the surface, and it may be damp, sometimes to such an extent that sprouting occurs. Many small brown beetles, with distinct snouts projecting from their heads, will be found working over and in the grain. The kernels of grain are eaten out or contain small, legless, fat-bodied, white grubs feeding on the interior. South of about 39°N. latitude, corn may be attacked by the rice weevil in the field, especially following damage by the corn earworm. Thence the pest is sure to be carried into storage.

FIG. 526.—Above, the granary weevil, *Sitophilus granarius* (Linné), adult beetle about 11 times natural size. Below, the rice weevil, *Sitophilus oryzae* (Linné), about 12 times natural size. (*From Ill. State Natural History Surv.*)

Food.—Wheat, corn, macaroni, oats, barley, sorghum, kaffir seed, buckwheat, and other grain and grain products. Adults feed upon whole seeds or flours, but the larvae develop only in seeds or pieces of seeds or cereal products large enough for them to live within them and not in flours unless they have become caked.

Distribution.—The granary weevil is probably not native to North America. The rice weevil is supposedly a native of India. Both species have been distributed all over the world in shipments of grain and are very frequently found together. The granary weevil is less prevalent, at least in tropical and semitropical climates.

Life History, Appearance, and Habits.—In unheated storages winter may be passed as larvae or adults, the latter stage surviving zero weather for several hours. The adult granary weevil is a somewhat cylindrical beetle, about 1/6 inch in length. It is dark brown or nearly black in color, with ridged wing-covers, and a prolonged snout extending downward from the front of the head for a distance of about one-fourth the length of the body. The rice weevil has much the same general appearance,

---

[1] *Sitophilus granarius* (Linné), Order Coleoptera, Family Curculionidae.

[2] *Sitophilus oryzae* (Linné), Order Coleoptera, Family Curculionidae.

although on the whole it will average somewhat smaller. There is usually a patch of somewhat lighter, yellowish color on the front and back of each wing cover. A distinguishing mark is in the shape of the small shallow pits on the prothorax of the beetles: in the rice weevil these are round (Fig. 526, *lower figure*), in the granary weevil (*upper figure*) they are oval.

The female weevil chews slight cavities in the kernels of grain or in other foods and there deposits small white eggs, one in a cavity, sealing it in with a plug of gluey secretion. The eggs hatch in a few days into soft, white, legless, fleshy grubs which feed on the interior of the grain, hollowing it out; on becoming full-grown, they are about ⅛ inch in length. They change to naked white pupae and later emerge as adult beetles. The entire life cycle may be passed under favorable conditions in from 4 to 7 weeks. Adults can withstand starvation for 2 or 3 weeks, often live 7 or 8 months, and may survive for over 2 years. In Kansas there are from four to five generations of the insect each season. The rice weevil has well developed wings, and frequently flies, especially during periods of high temperature. The granary weevil has the wing covers somewhat grown together and is unable to fly. In other respects the two insects closely resemble each other in their life histories and will very frequently be found associated and working together in the same bins.

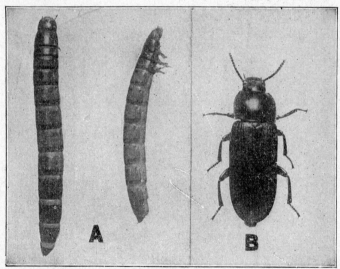

Fig. 527.—Yellow mealworm, *Tenebrio molitor* Linné. *A*, larvae about twice natural size; *B*, female adult, 2½ times natural size. (*From Kans. Agr. Exp. Sta. Bul.* 189.)

301. MEALWORMS[1]

Importance and Type of Injury.—Yellowish to brown, smooth, shiny-bodied worms, closely resembling wireworms, and large black beetles up to 1 inch long, feed in and around grain bins, particularly in dark, moist places where the grain has not been disturbed for some time.

[1] *Tenebrio molitor* Linné, the yellow mealworm, and *T. obscurus* Fabricius, the dark mealworm, Order Coleoptera, Family Tenebrionidae.

Food.—Cereals and cereal products, including meal, bran, some other food products, meat scraps, feathers, and dead insects.

Distribution.—The insects are natives of Europe. They are now distributed throughout the world. The yellow mealworm is less common in the southern states.

Life History, Appearance, and Habits.—The adults of the two mealworms rather closely resemble each other. They are black to nearly black beetles, robust, flattened, somewhat shining, from ½ to nearly 1 inch in length (Fig. 527,*B*). The female deposits her whitish oval eggs, to the number of 250 to 1,000, singly or in clusters in the food materials. The eggs hatch in from 4 to 18 days into white larvae which become yellow in color as they grow. When full-grown, they are 1 to 1½ inches in length and very closely resemble wireworms in appearance (*A*). The larval period usually requires from 6 to 9 months, and they usually winter in the larval stage, although a complete generation may range from 4 months to nearly 2 years. The pupal stage is white and is passed without any cocoon or protective covering. The two species of mealworms can be separated only by a careful examination and have practically the same life history and habits. They are always most abundant in damp grains, or grain products that have remained undisturbed for some time.

Reference.—*U.S.D.A., Tech. Bul.* 95, 1929.

302. Cadelle[1]

Importance and Type of Injury.—This insect is sometimes very important as a pest of stored grains; but it is also, to some extent, a feeder upon other insects. When abundant, however, it becomes seriously destructive in grain bins. It also causes damage in flour mills by cutting holes in flour sacks, silk bolting cloth, and other silk cloth used in the machinery, and it eats holes in cartons of packaged foods.

Food.—The insect is a general feeder on stored grains and seeds. It usually attacks the embryo, eating out only the softer parts of the grain. While the adults will often kill and feed upon other insects with which they come in contact, their general habits are not those of predaceous insects.

Life History, Appearance, and Habits.—This insect is, next to the mealworms, the largest of those attacking stored grains. Both adults and larvae, but not the eggs or pupae, may pass the winter. The adult beetle (Fig. 528,*D*) is black or nearly black, ⅓ to ½ inch in length, with the head and prothorax distinctly separated from the rest of the body, to which it is attached by a rather loose, prominent joint.

The eggs are laid on or near the food, by preference in cracks, under flaps of cartons, or in some such protected situation, commonly being deposited in groups of 10 to 60. Over 1,300 eggs have been secured from a single female. The larvae (Fig. 528,*B*) that hatch from these eggs, 1 or 2 weeks later, are rather soft-bodied, white to grayish-white in color, with prominent black heads, black spots on the three thoracic segments, just behind the head, and two short dark hooks at the rear end of the body. They are about ⅔ inch long when full-grown. Under favorable conditions the larvae may complete their development in from 70 to 90 days;

Tenebroides mauritanicus (Linné), Order Coleoptera, Family Ostomidae.

others require from 7 to 14 months, and they have been kept alive for nearly 3½ years.   Full-grown larvae and adults have the habit of boring into wood, adjoining the grain on which they have been feeding, in granaries, storerooms, and grain ships, where they may remain hiding for months, ready to attack the next consignment of foods.   The pupal stage

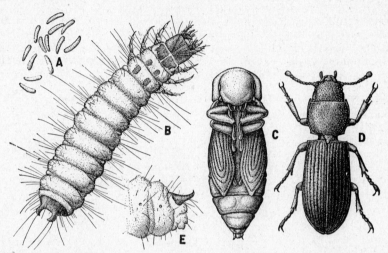

FIG. 528.—The cadelle, *Tenebroides mauritanicus* (Linné), about four times natural size. *A*, eggs; *B*, mature larva, dorsal view, showing characteristic spots on thoracic segments, and the curious rigid hooks on tip of abdomen; *C*, pupa, ventral view; *D*, adult, showing the strong constriction between prothorax and mesothorax; *E*, last three abdominal segments of larva in side view, showing terminal hooks and spiracles.   (*Original.*)

is passed in such cavities or in other secluded places.   Female adults commonly live for a year.

Reference.—U.S.D.A., Dept. Bul. 1428, 1926.

303. CONFUSED FLOUR BEETLE[1] AND RED FLOUR BEETLE[2]

Importance and Type of Injury.—This insect is one of the most common occurring in situations where grain products are stored.   It is one of the most annoying pests in retail grocery stores and warehouses and extremely serious in flour mills.   Infested material will show many elongate reddish-brown beetles, about ⅐ inch long (Fig. 529), crawling over the material when it is disturbed, and brownish-white somewhat flattened six-legged larvae feeding on the inside of the grain kernels and

[1] *Tribolium confusum* Duval, Order Coleoptera, Family Tenebrionidae.   The red flour beetle is very similar to the confused flour beetle.   The two may be distinguished in the adult stage by the following differences: as seen from the underside of the head, the eyes of the confused flour beetle are separated by about three times the width of either eye, whereas the width of each eye seen from below in the red flour beetle is about equal to the distance between them.   The confused flour beetle has the antennae gradually enlarged toward the tip, the red flour beetle suddenly enlarged at the tip; the margin of the head is notched at the eyes in the confused flour beetle, and not so notched in the other species.   The red flour beetle seems to be less common.

[2] *Tribolium castaneum* (Herbst) (= *T. ferrugineum* Fabricius).

crawling over the infested seeds. They are generally known among millers as "bran bugs."

Food.—This insect feeds on a great variety of products, including all kinds of grains, flour, starchy materials, beans, peas, baking powder, ginger, dried plant roots, dried fruits, insect collections, nuts, chocolate, drugs, snuff, cayenne pepper, and many other foods.

Distribution.—The confused flour beetle was first noted in this country in 1893. Both species occur throughout the world, but the confused flour beetle is most abundant in the northern part of the United States, while the red flour beetle is not commonly found north of the forty-first parallel.

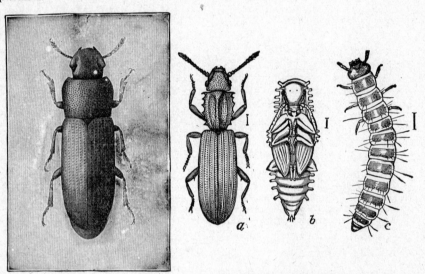

FIG. 529.—The confused flour beetle, *Tribolium confusum* Duval. Adult, about 14 times natural size. (*From Ill. State Natural History Surv.*)

FIG. 530.—The saw-toothed grain beetle, *Oryzae-philus surinamensis* (Linné): *a*, adult; *b*, pupa; *c*, larva. Enlarged about 16 times. Lines indicate natural size. (*From U.S.D.A.*)

Life History, Appearance, and Habits.—The adult beetles are very active, moving rapidly when disturbed. They may survive moderately cold winters in unheated buildings and often live 2 years or more in the adult stage, during which period the female may produce nearly 1,000 eggs. The very small clear-white sticky eggs are laid on sacks, in cracks, or directly on the food material. These hatch in 5 to 12 days into small brownish-white worms, which become full-grown in 1 to 4 months, depending upon temperature and kind of food, and are then about ⅙ inch in length. They change to white naked pupae, remaining in this stage for 1 to 2 weeks. A complete generation is passed in 3 to 4 months when the temperature is high. Under Kansas conditions four or five generations occur annually in heated storehouses or mills. All stages of the insect may be found at any time of the year in such buildings.

Reference.—*Jour. Agr. Res.*, **46**: 327–334, 1933; *Rept. Minn. State Ento.*, **17**: 73–94, 1918.

304. Saw-toothed Grain Beetle[1]

Importance and Type of Injury.—Infested material will show many very slender, much-flattened, small, dark-red beetles hurrying over the surface of the food.   Its flattened shape enables this species to work into packages of food that are apparently tightly sealed.   Feeding consists of scarring and roughening of the surface of the food.   Its attack upon stored seeds usually follows that of other insects, since it is not able to attack sound seeds.

Food.—As with the confused flour beetle, this insect feeds on a great variety of products including practically all grains and grain products, dried fruits, breakfast foods, nuts, seeds, yeast, sugar, candy, tobacco, snuff, dried meats, and, in fact, almost all plant products used for human food.

Distribution.—Throughout the world.

Life History, Appearance, and Habits.—In unheated buildings only the adult stage (Fig. 530,*a*) winters.   They are dark-brown, much flattened, slender insects, about $\frac{1}{10}$ inch long.   When examined under a lens, six saw-tooth-like projections will be seen on each side of the thorax. The adults almost never fly.   They have been kept alive for over 3 years. The eggs are laid on and near the food and hatch in from 3 to 17 days.   The larvae (*c*) are brown-headed, elongate, white, six-legged grubs with abdomens tapering to the tip.   They become full-grown in 2 to 10 weeks when they are about $\frac{1}{8}$ inch in length.   They then form a protective covering by sticking together small bits of the food material. They remain in the pupal stage (*b*) for from 6 to 21 days, emerging at the end of this time as adult beetles.   There are from four to six generations annually throughout most of the United States.   It is possible, under very favorable conditions, for the entire life cycle to be passed in from 24 to 30 days.

Reference.—*Jour. Agr. Res.*, **33**: 435–452, 1926.

305. Lesser Grain Borer[2]

This insect, which is also known as the Australian wheat weevil, is now well distributed through the south and midwest, but not in the more northern areas.   It feeds in both larval and adult stages in the interior of nearly all grains and some other substances, such as seeds, drugs, dry roots, and cork, and eats into wood and paper boxes. It is most common in wheat and one of the most destructive wheat insects.   Under favorable conditions a generation may develop in a month.   The adults are brown to black, nearly cylindrical, about $\frac{1}{8}$ inch long by $\frac{1}{4}$ as wide.   The large head is bent under the thorax, and the rear end of the body is blunt.   The antenna has a large three-segmented serrate club.   The larvae are grub-like, lie curved, and become about $\frac{1}{10}$ inch long.   The anterior end is much swollen, bears a small brown head and six short legs.

Reference.—*U.S.D.A.*, *Bur. Ento. Bul.* 96, Part III, 1911.

306. Larger Grain Borer[3]

This species is very similar in appearance and habits to the lesser grain borer. It is larger ($\frac{1}{6}$ inch long), cylindrical, nearly one-third as wide as long, and the wing

---

[1] *Oryzaephilus surinamensis* (Linné), Order Coleoptera, Family Cucujidae.

[2] *Rhizopertha dominica* Fabricius, Order Coleoptera, Family Bostrichidae.

[3] *Dinoderus truncatus* Horn, Order Coleoptera, Family Bostrichidae.

covers are smoother and more polished. It is of importance in the United States at only a few places in the south. It feeds chiefly on corn and also bores into wood.

307. DRIED FRUIT BEETLE[1]

Importance and Type of Injury.—This small beetle has been reported chiefly as a pest of the fig industry in California. The adults work their way to the inside of ripening and partially dried figs carrying upon their bodies bacteria, yeasts, and fungus spores which cause smut and souring of the fruit.

Food.—Fermenting tree fruits, raisins, and watermelons.

Distribution.—Widely distributed in California and the Gulf states and many other warm countries.

Life History, Appearance, and Habits.—Pupae winter in the soil and the adults in all kinds of cull and fermenting fruits and fruit refuse and in stored fruits, the immature stages becoming adults by March or April. The adults deposit on the average over 1,000 small white eggs on the pulp of ripe figs on the tree, piles of cull oranges, raisins, stone fruits, watermelons, and other fermenting fruits. The eggs hatch on the average in 2 days. The whitish short-legged grubs, with brown head and posterior end, grow in a week or two to ¼ inch in length, then pupate for about a week in the soil, so that there may be a generation of the insects about every 3 weeks in warm weather. The adults are very active, strong fliers, which readily find fermenting or ripe fruits and in the case of figs, enter promptly through the "eye" of the fruit. They are only ⅛ inch long by half as wide. The short elytra leave the terminal third of the abdomen exposed. Each wing cover has a prominent amber-brown spot at the tip and a smaller one near the base, on the side. The appendages are also amber or reddish, the rest of the body black.

Control.—Great numbers of beetles have been caught in traps, baited with cull dried peaches moistened with water, 1 pint to 1 pound of fruit. The fermenting bait remained attractive to the beetles for about a month. Such bait was especially attractive after it became infested by larvae of the beetle. The larvae were prevented from escaping to the soil to pupate by smearing the top 3 inches of the inside of the container with cup grease. Prompt collection of fallen fruits from beneath the trees and destruction of all culls by burning, drying thoroughly, or feeding to livestock are recommended.

References.—*U.S.D.A., Cir.* 157, 1931; *Jour. Econ. Ento.*, **28** : 396–400, 1935.

FIG. 531.—Ear of corn heavily infested by the Angoumois grain moth, *Sitotroga cerealella* (Olivier); showing emergence holes of the moths. Reduced ½. (*From Ill. State Natural History Surv.*)

308. ANGOUMOIS GRAIN MOTH[2]

Importance and Type of Injury.—Grain in bins or ear corn in storage has small buff moths flying about the bins or crawling rapidly over the surface of the grain

[1] *Carpophilus hemipterus* (Linné) and related species of Order Coleoptera, Family Nitidulidae.

[2] *Sitotroga cerealella* (Olivier), Order Lepidoptera, Family Gelechiidae.

when it is disturbed.   One or two small round holes are eaten in the kernels of infested corn (Fig. 531) or in other grain.   This insect is the most destructive grain moth occurring in this country.   It is extremely bad in the South, causing great damage to corn in cribs and also destroying ripening grain, especially wheat, in the field.

Food.—Wheat, corn, and other grains.

Distribution.—The insect received its name from having been first reported as injurious in the Province of Angoumois, France, about 1736.

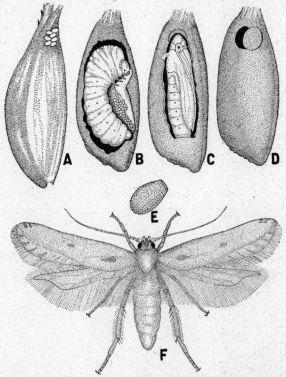

Fig. 532.—Angoumois grain moth, *Sitotroga cerealella* (Olivier).   *A*, glume or chaff covering a grain of wheat, with a cluster of eggs deposited near outer end; *B*, the larva or caterpillar inside the grain, most of the contents of which it has eaten; *C*, the pupa within a grain of wheat; *D*, hole made by moth through the thin seed coat, the round "lid" having been nearly cut through by the larva before it pupated; *E*, a single egg greatly magnified; *F*, the adult moth: note characteristic shape of tip of hind wing and the long fringe of hairs on both wings.   (*Redrawn after King, Pa. Bureau of Plant Industry.*)

It was first reported in the United States, from North Carolina, and is now distributed throughout the United States.

Life History, Appearance, and Habits.—The adult insect is a rather delicate moth of a buff color with a wing spread of ½ to ⅔ inch.   This is the only stage commonly observed, since the eggs are almost microscopic and the larvae and pupae live entirely inside of seeds.   The hind wings are uniformly light gray, with a heavy fringe of hairs which are longer than the width of the wing membrane.   The wing membrane is prolonged at the apical angle like a thumb or finger (Fig. 532,*F*).   This easily distinguishes it from the clothes moths, for which it is often mistaken. The female moths lay several hundred, very minute, white to reddish eggs, singly, or in clusters of as many as 20, on the grains where the larvae feed. Where the insects are working out-of-doors, the eggs are attached to wheat heads in the field or to the grain in the shock. They hatch in from 4 days to 4 weeks, and the worm-like larvae at once burrow into the kernels or berries of the grain, feeding upon the starchy parts of the kernel. In wheat usually only one larva is found in a kernel, although several may find sufficient food in one kernel of corn. When full-grown the larva is about $\frac{1}{5}$ inch long, white in color, with a yellowish head, six true legs, and four pairs of prolegs. It then eats an exit tunnel for the adult, leaving a very thin transparent film of the seed coat covering the emergence hole. Next it spins a thin silken cocoon within the grain and there changes to the pupal stage. The adult moth emerges through a small round hole in the seed coat. The larval stage normally lasts from 20 to 24 days, and the entire life cycle may be passed in about 5 weeks. However, the larvae may lie dormant over winter. In unheated buildings, there are probably two generations a year in the latitude of Kansas and central Illinois. In the warmer parts of the South, as many as six generations occur annually. The insect is killed by cold winter weather, but in heated buildings, such as storehouses or mills, breeding is continuous throughout the year.

References.—Pa. Bur. Plant Ind. Cir. 1, 1920; *U.S.D.A., Farmers' Bul.* 1260, 1931.

Fig. 533.—Mediterranean flour moth, *Ephestia kuehniella* Zeller. *A,* larvae, *B* and *C,* adults, dorsal view; *D,* position of moth when at rest, from side. Twice natural size. (*From Kans. Agr. Exp. Sta. Bul.* 189.)

309. Mediterranean Flour Moth[1]

Importance and Type of Injury.—The Mediterranean flour moth is, without much question, the most troublesome insect occurring in flour mills in this country. Where this insect is present, masses of flour in shoots and elevators will be found webbed together and containing small

---

[1] *Ephestia kuehniella* Zeller, Order Lepidoptera, Family Pyralididae. The closely related almond or fig moth, *Ephestia cautella* (Walker), has very similar habits. The adult is more slender and less conspicuously marked and the larva somewhat tinged with brown and faintly striped with darker dots. It is a serious pest of harvested figs but breeds in nearly all kinds of stored vegetable products.

pinkish-white caterpillars. The shoots may become entirely clogged and the machinery stopped by these webbed masses of flour. Small gray moths will also be noted flying about the infested buildings.

Food.—Flour is the favorite food of the insect, but it will attack also the whole grain of wheat, bran, breakfast foods, corn and other grains, and pollen in beehives.

Distribution.—The insect was first reported in North America in 1889, when it was found in Canada. It has now spread throughout the United States and Canada and probably throughout most of the world.

Life History, Appearance, and Habits.—The adult moth (Fig. 533, B,C,D) is of a pale-gray color and is from $\frac{1}{4}$ to $\frac{1}{2}$ inch long. The head and tail are slightly raised when the insect is resting, this being very characteristic. The wings are marked with two zigzag lines of black which are not prominent. The eggs are laid in accumulations of flour or other foods, in cracks about buildings, or on the cloth in spouts or bolters in mills. They hatch in from 3 to 6 days, depending on the temperature. The caterpillars immediately begin spinning silken threads which form into little tubes in which they live and feed. It is this web-spinning that causes the greatest amount of damage by the insect. The caterpillars when full-grown are about $\frac{3}{5}$ inch long, of a general whitish to pinkish color (Fig. 533,A) and have three lengths of crochets in a circle on the prolegs. They pupate in silken cocoons, this stage lasting from 8 to 12 days. Under conditions encountered in mills, the entire life cycle is usually passed in from 9 to 10 weeks.

Reference.—*Jour. Agr. Res.* **32**: 895–929, 1926.

310. INDIAN MEAL MOTH[1]

Importance and Type of Injury.—This is one of the most general and troublesome of the moths infesting stored products. Infested material will be more or less webbed together and often fouled with dirty silken masses containing the excreta of the larvae.

Food.—This insect attacks all kinds of grains, meal, breakfast foods, soybeans, dried fruits, nuts, seeds, dried roots, herbs, dead insects, museum specimens, powdered milk, the excrement, exuviae, and pollen in beehives, and many other substances. It is frequently a pest in candy factories where nut candies are made and has been bred from milk chocolate not containing nuts.

Distribution.—The insect is of European origin but is now generally distributed throughout the United States.

Life History, Appearance, and Habits.—In unheated buildings, the insects winter as larvae, but in warm situations feeding and breeding are continued. The adult moth is about $\frac{1}{2}$ to $\frac{3}{4}$ inch from tip to tip of wings and is active at night or in dark places. When at rest (Fig. 534,A) the wings are folded closely together along the line of the body and the antennae flat on the wings. The front or base of the forewings is a grayish-white color, and the tip half or two-thirds, as well as the head and thorax, a contrasting reddish brown; the underwings are grayish-white. The palps form a characteristic cone-like beak in front of the head. The minute whitish ovate eggs, some 40 to 350 or more, are deposited singly or in clusters of from 12 to 30 on the larval food or adjacent objects.

[1] *Plodia interpunctella* (Hübner), Order Lepidoptera, Family Pyralididae.

They hatch in from 2 days to 2 weeks into small caterpillars, of a general white color, but often with a distinct greenish or pinkish tinge, with a light-brown head, prothoracic shield, and anal plate. On becoming full-grown, in from 2 weeks to 2 years, these caterpillars are about ⅓ to ½ inch in length. They crawl up to the surface of the food material and pupate within a thin silken cocoon, from which the adult moths emerge in

Fig. 534.—Indian meal moth, *Plodia interpunctella* (Hübner). *A*, adult with wings closed; *B*, adult with wings spread; about 3 times natural size. (*From Kans. Agr. Exp. Sta. Bul.* 189.)

4 to 30 days. The entire life cycle of the insect will require from 4 to 6 weeks under conditions usually encountered in heated buildings, there being from four to six generations of the insect each year.

Reference.—U.S.D.A., Tech. Bul. 242, 1931.

METHODS OF CONTROLLING STORED-GRAIN PESTS

While the methods used for controlling insects infesting farmers' grain bins and elevators differ somewhat from those used for the control of the same insects in mills, warehouses, retail stores, and dwellings; the most effective means of preventing damage is the same. In all cases this consists of keeping the premises as clean as possible. This means the gathering up and removal of all refuse grain, seeds, grain products, flour, meal or other material in which the insects breed; thoroughly cleaning the bins, storerooms, mills, or warehouses at periods of the year when they can be emptied; the carrying over from year to year of as little material as possible; and the careful inspection of all material when brought in for storage or processing. When bins, mills, seed houses, or warehouses are built, the insect problem should be given due consideration, and the construction should be such that the stored products will be kept dry, that cracks and crevices are reduced to the minimum, and that the most effective control measures may be applied at the minimum expenditure of effort and expense. Provision should be made in mills for heat-treating, and bins and storehouses should be made sufficiently tight for effective fumigation. Any building where grains or grain products or seeds are held in storage for some time is practically sure to become infested sooner or later and to require treatment. Inaccessible dead spaces in mills and storerooms, where wastes collect, may be treated with miscible oil, 1 gallon, lye, ½ pound, and water, 9 gallons. Screenings and waste grain of all kinds should be kept away from mills and storehouses. Since insect damage is greater the higher the moisture content of the grain, storage in damp situations should be avoided.

Destruction by Heat and Cold.—The heat-treatment has given good results for treating mills, retail stores, or storerooms in dwellings, as it can usually be applied so that practically all insects are killed; and the expense of treatment is lower than that of any other method. The heat-treatment should be used during the period when the outside temperature is high. This treatment can seldom be used in elevators, farmers' bins, or other small storages. In most buildings the heating system used in the winter will be sufficient to maintain temperatures high enough to kill the insect life in the mills during the summer, when outside temperatures are from 90 to 100°F. and little wind is blowing. Repeated tests with this method have shown that, in mills of fairly tight construction under the above conditions, it is possible to maintain temperatures from 120 to 150°F. for several hours and that such temperatures are fatal to all insects exposed to them. Owing to the construction of the buildings or the lack of heating facilities, this method cannot always be applied effectively. In most mills, however, the expense of installing sufficient heating units will not be so great as that of annually fumigating with hydrocyanic acid gas or some other fumigant. Heat is not satisfactory for the treatment of nuts and dried fruits. Heat may be used for seeds intended for planting, if the exposure is not over 6 hours, the temperature not higher than 135°F., and the seeds reasonably dry (containing not more than 12 per cent moisture).

If there are large quantities of seeds to be treated, it will be found difficult to raise the temperature sufficiently in the interior of the mass, without exposing the outer layers too long or to too high temperatures. Under such conditions it is better to use a heat-treating machine. There are a number of heat-treating machines through which grain may be passed as it is moved from one bin to another, and subjected during its passage through the machine to temperatures of 130 to 140°F. for 30 minutes. Such machines are frequently very valuable, particularly for use in elevators or mills, where conditions are such that frequent infestations are sure to occur from the waste grain about the building and from grain being brought in. Where outside winter temperatures go as low as 20° below zero, opening the building to this temperature will kill most grain-infesting insects. In elevators and mill storage bins it is often possible to keep down insect damage by frequently moving the grain from one bin to another, as the insects do not thrive and multiply rapidly in grain that is frequently disturbed.

Destruction by Fumigation.—There are a number of fumigants useful for treating stored grains. Most of them are heavier than air and are applied over the top of the grain. Carbon bisulphide is by far the most generally used. It is also the cheapest. Because of the inflammable and explosive nature of this gas when mixed with air, it should not be used where there is any danger of fire. Where there is no danger from fire, it is the best fumigant to use. For the usual type of fairly tight bin, it should be used at the rate of 2 pounds to 100 cubic feet or 80 bushels of grain, for wheat, barley or oats. For corn 1 pound for 80 cubic feet should be sufficient. If the bin is very tight the amounts may be reduced slightly, but if it is poorly constructed correspondingly more will be required. Commercial mixtures of 20 per cent carbon bisulphide, 80 per cent carbon tetrachloride, and a trace of sulphur dioxide, which are said to have a greatly reduced fire hazard, can be purchased but cannot be mixed at home. The mixture is recommended to be used 2 pounds per 100 cubic feet or 4 gallons per 1,000 bushels of grain. A mixture of 3 parts ethylene dichloride and 1 part carbon tetrachloride makes a fumigant that has no fire hazard. A much higher dosage is required than with carbon bisulphide; in most farm bins, 8 pounds to each 100 bushels of grain. Products in paper-lined bags must be fumigated by injecting the proper dosage into the interior of the bag through a pointed-nozzle extension rod. The general methods of fumigating with hydrocyanic acid gas have already been described on pages 278 to 283. The same methods should be followed in the treatment of warehouses or storerooms where grain insects are causing injury. Hydrocyanic acid gas cannot be relied upon to penetrate more than a few inches into a mass of seeds, flour, or stored or sacked grain. Sulphur fumigation is the most effective means of cleaning out infestations of mites in bins or warehouses. Grain or grain products must be disposed of before the sulphur fumigation is applied, as the sulphur fumes are injurious to grains. From 2 to 3 pounds of sulphur per 1,000 cubic feet of space should be used for such fumigations.

References.—U.S.D.A., Cirs. 390, 1937, and E-419 and E-429, 1938, and *Farmers' Bul.* 1811, 1938.

C. INSECTS ATTACKING STORED PEAS AND BEANS

311. PEA WEEVIL[1]

Importance and Type of Injury.—Most housekeepers and seedsmen are familiar with "buggy" peas. The insides of the seeds in storage are

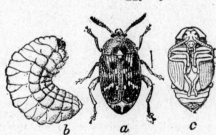

FIG. 535.—Pea weevil, *Bruchus = (Mylabris) pisorum* (Linné); *a,* adult beetle; *b,* larva or grub; *c,* pupa. Enlarged about 4 times. The line shows the natural length of the beetle. (*From Fernald's "Applied Entomology," after U.S.D.A.*)

eaten out by short chunky beetles (Fig. 535,*a*), about $\frac{1}{5}$ inch long, of a general brownish color, flecked with white, black, and grayish patches; and by the larvae (*b*) which are white all over except for the small brown head and mouth parts. In the spring and summer many of the old peas are found with neat circular holes, about $\frac{1}{10}$ inch in diameter, leading into the cavity where the insect developed. Heavily infested peas are often reduced to mere shells. Green peas are infested with the minute larvae, but there are only the dot-like entrance holes to show that they are not sound at this stage, and they are generally overlooked and eaten. If buggy peas are planted, a poor stand of weak, unproductive plants results.

Plants Attacked.—Peas.

Distribution.—Throughout the country.

Life History, Appearance, and Habits.—The winter is passed in the adult stage and, under northern conditions, chiefly in the peas either in

---

[1] *Bruchus (= Mylabris) pisorum* (Linné), Order Coleoptera, Family Mylabridae.

the field or in storage. Some, however, and in the South many, of the beetles leave the seeds in the fall and hibernate in protected places, up off the ground, out-of-doors. Some of the overwintering beetles may survive in their hibernating places until the second summer after they became adults. If infested seed is not treated, the beetles may be planted with it. They wait until the plants are in blossom, then join others that wintered out-of-doors, and start feeding upon the pollen, the petals, the leaves, or the pods of the plants. The females glue their elongate yellowish eggs to the outside of the pods, from 1 to 12 or more on a pod. The tiny yellowish larva that hatches from the egg, 5 to 18 days later, is well adapted with spines and very short legs to burrow through the pod until it reaches one of the developing seeds, which it enters. Having found a place where it no longer needs to search for food, the larva loses its spines at the first molt, and the legs become very short. The grub grows slowly, consuming a third or more of the contents of the seed and finally eating an exit passage to the surface of the pea, leaving only a thin circular lid (the outer seed coat) intact to protect its tunnel. The larva requires about 4 to 6 weeks to become full-grown. The feeding of the weevils within the peas causes them to heat. This rise in temperature aids in the development of the weevil larva. It then paints the walls of its burrow with a gluey secretion from the mouth and in this snug chamber passes the pupal stage (Fig. 535,c), which occupies about 2 weeks in late summer.

There is usually only one generation a year and only one weevil matures in a pea. Eggs are never laid on dried peas, and there is no increase in numbers in storage. The adults must get to the growing plants in the spring or perish without laying eggs.

Control Measures.—Dusts of derris, cubé, or timbo, containing at least 0.75 per cent rotenone, applied with a good duster, 20 to 30 pounds to the acre, within a few days after the peas start to bloom and at 6 to 10-day intervals if necessary, may control this insect. To destroy the weevils wintering in seeds left in the field, plowing under the pea straw and chaff to a depth of 8 inches, converting it into silage, or burning it as soon as possible after the last picking, destroys most of the insects that would winter out-of-doors. It is unwise to allow peas to ripen for seed in districts where green peas and canning peas are important crops. The best control is to tie the seed up in bags at the time of harvesting and, as soon as dry, to fumigate it with carbon bisulphide. Use 1 ounce for each bushel of seed or 3 pounds to 100 cubic feet of space in the fumigating box. This may be done in any tight barrel, metal garbage can, or box that has a close lid, or in a tight granary. The seeds should be put in first, several gunny sacks placed on top of them, the carbon bisulphide poured over the sacks, and the lid put in place. After 24 to 48 hours, remove the seeds from the fumigating box and allow them to air out. Choose a day for the work when the temperature is above 70 and below 95°F. Keep all lights and fires away from the liquid and its fumes, since the chemical is inflammable and explosive. Hydrocyanic acid may be used as a fumigant for large amounts of sacked peas held in storage; this fumigant should be used only by experienced fumigators. The weevils may be killed by suspending the seeds in a bag of cold water and heating it to 140°F., then pouring the peas out on a surface where they will dry quickly. Or the seeds may be heated dry at a temperature of 135°F. for 3 or 4 hours and thus all stages of the beetle be killed without injuring germination.   Do not plant badly infested seeds even after the weevils have been killed.

References.—U.S.D.A., Farmers' Bul. 1275, 1925, and Tech. Bul. 599, 1938, and Cir. E-435, 1938.

312. BEAN WEEVILS AND COWPEA WEEVILS

The broad bean weevil[1] is almost identical in appearance and habits with the pea weevil, but only about two-thirds as large.   It can be distinguished by the points given in the key.   It prefers the European broad bean as food but attacks also peas and vetches.   The only other important difference appears to be that several individuals occur in a single seed in contrast with the invariable one of the pea weevil.   It occurs in California.

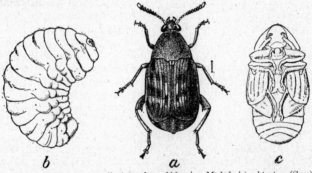

FIG. 536.—Common bean weevil Acanthoscelides (= Mylabris) obtectus (Say); a, adult; b, larva or grub; c, pupa.   About ten times natural size.   The line indicates the actual length of the beetle.   (From U.S.D.A., Farmers' Bul. 1275.)

The best-known species attacking beans is the common bean weevil,[2] which is thought to be native to the American continent.   The Southern cowpea weevil[3] is a closely related species that has very similar habits. It prefers cowpeas in which to develop, but attacks various kinds of beans and peas, at least in storage.   It can be distinguished from the bean weevil by the characters given in the key.

Importance and Type of Injury.—All kinds of beans and peas stored for seed or food, unless they are protected, are almost sure to be devoured and rendered useless by these hungry weevils.   In the field the beans may be stunted and deformed so as to be worthless, but often the infestation in green beans is not detected and the infested beans are eaten or stored by the unsuspecting.   These weevils are so destructive in some sections as practically to prevent the commercial production of beans either for seed or for food.

Plants Attacked.—Kidney beans, Lima beans, and cowpeas, in the field; and all varieties of beans, peas, lentils, and some other kinds of seeds in storage.

---

[1] Bruchus   (= Mylabris)   rufimanus Boheman, Order Coleoptera, Family Mylabridae.
[2] Acanthoscelides (= Mylabris or Bruchus) obtectus (Say).
[3] Callosobruchus maculatus (Fabricius) [= Bruchus quadrimaculatus].

Distribution.—The bean weevil occurs throughout the United States and in many other countries. The broad bean weevil is confined to the western coast, while the cowpea weevil is more abundant in the southern states.

Life History, Appearance, and Habits.—Unlike the pea weevil and the broad bean weevil, these species breed continuously in the dry seeds, if they are stored in a warm place, and all stages may be found in winter. In the spring the adults escape from storerooms, and those that hibernated out-of-doors become active. They appear upon the plants as the latter come into bloom and feed on dew and honeydew and possibly on the foliage slightly. The adults are only half as large as the pea weevil, of a general light olive-brown color, mottled with darker brown and gray (Fig. 536,*a*). They are about ⅛ inch in length, the appendages are reddish, and the body narrows evenly toward the small head. In the cowpea weevil the elytra have a large rounded spot at mid-length, and the tips are black. There is a similar pair of black areas on the part of the abdomen exposed beyond the wing covers. The bean weevil has smaller black mottlings and can also be distinguished from the former species by having one large and two small teeth at the apex of the hind femur, while the cowpea weevil has

Fig. 537.—Navy beans showing emergence holes of bean weevils, about natural size. (*From Fernald's " Applied Entomology.*")

only one such tooth. The whitish ellipsoidal eggs of the bean weevil are laid in loose groups of a dozen or more, in holes chewed by the female in the green pods along the seam where the two parts of the pod meet or in any natural crack that she finds in the pod. The cowpea weevil does not make a cavity for her eggs, but glues them securely to pods or exposed beans.

The minute, whitish, hairy grubs that hatch from the eggs, 3 to 30 days later, are equipped with short slender legs; and they scatter throughout the pod, seek out the developing seeds, and eat their way to the inside. On account of their very small size the entrance holes heal over and leave only a slight brown dot, so that growers and shippers rarely detect infestations of the weevils inside the beans. After feeding a few days the larvae molt and appear as white grubs with very small heads, no legs, and no long hairs. Feeding and growth continue until the larvae are about ⅛ inch long, nearly half as thick, wrinkled, and humpbacked (Fig. 536,*b*). The larval stage may occupy 2 weeks to 6 months or longer, depending upon the temperature and the moisture content of the beans. The pupal stage (*c*) is passed in the larval cell, which has been cemented over on the inside to exclude the larval excrement from contaminating the pupa, and which has a cylindrical extension to the outside but not penetrating the thin seed coat. Through these circular holes (Fig. 537) the adults commonly escape 2 to 8 weeks after the larvae entered the seeds. Adults emerging from seeds out-of-doors soon seek out growing beans and deposit their eggs upon them. If the weather remains warm, several generations may develop in the field. When beans are harvested, all stages may be taken into storage, and the adults as they transform continue to lay eggs on or among the beans. Breeding goes on steadily as long as there is any food left in the beans and the temperature is warm enough. Six or seven generations may be completed in a year, and as many as 28 weevils have been known to develop in one bean.

Control Measures.—When beans are harvested, they should be at once sacked up tightly and, as soon as they are dry, either fumigated or heated as described for the pea weevil.   Under no circumstances should weevily beans be planted.   A simple method of killing the weevils and preventing their destructive increase during the winter is to mix the seeds with hydrated lime, spent pyrethrum, timbo or derris powder, at the rate of 1 pound of dust to 2 pounds of beans in small quantities or 1 to 4 in larger quantities.   If stored in tight drums or barrels, an inch layer of lime or other fine dust over the top will stop nearly all increase but will not kill the weevils in the beans.   Beans so treated can be used for food after washing.   Bean vines and other refuse and left over beans should be fumigated, fed, plowed under, or burned to prevent the weevils spreading into the fields in the spring.

References.—U.S.D.A., Farmers' Bul. 1275, 1922; *Jour. Econ. Ento.,* **10**: 74–78 1917; *Ky. Agr. Exp. Sta. Bul.* 213, 1917; *Jour. Agr. Res.,* **24**: 606–616, 1923 and **28**: 347–356, 1924; *Tex. Agr. Exp. Sta. Bul.* 256, 1919; *U.S.D.A., Tech. Bul.* 593, 1938.

CHAPTER XXII

INSECTS INJURIOUS TO DOMESTIC ANIMALS

For insects to attack living animals is a hazardous method of securing food. Unlike plants and stored foods, animals actively retaliate when insects attack them. Yet thousands of species of insects are specialized in structure or habit to secure all their food from the bodies of animals, and there is not a kind of wild or domesticated animal living that is not attacked by from one to many kinds of external or internal parasites.

So far as we know, no one has given us an accurate estimate of the loss that this attack upon our useful animals causes to man. We know that valuable animals, from the smallest chicks to the largest beef animals, are sometimes killed outright; that the flow of milk from dairy cattle decreases greatly during "fly time;" that beef and show cattle lose flesh and condition when flies or ticks are bad; that work horses are less efficient, often unmanageable, when annoyed by these pests; that lousy poultry, cattle, or sheep cannot be healthy; and that a number of deadly diseases are carried from animal to animal solely by the bites of insects and ticks. So it seems safe to say that this group of insects is deserving of much more careful study and active opposition than has been accorded to it.

General References.—Can. Dept. Agr., Pub. 604, 1938; U.S.D.A., Div. Ento. Bul. 5(n.s.), 1896.

A. INSECTS ATTACKING HORSES, MULES, AND DONKEYS

FIELD KEY FOR THE IDENTIFICATION OF INSECTS INJURING HORSES, MULES, AND DONKEYS

A. *Free-flying insects that alight on the animals to suck blood, coming and going repeatedly:*
  1. Large, heavy-bodied, swift-flying, black, gray, or brownish, two-winged, often green-eyed flies, ½ to 1 inch long, with wings clear or banded, usually alight on head or shoulders and suck blood on warm sunny days. A drop or two of blood exudes from the puncture after they leave..........................
  ....*Horse flies* or *deer flies*, page 814.
  2. Flies about the size and general appearance of the house fly, suck blood, especially about the legs, causing animals to stamp their feet. Distinguished from the house fly by stiff, pointed beak or proboscis that projects forward from lower side of head, by broader black-spotted abdomen and grayer appearance, and by arista having hairs on upper side only. Palps less than ½ as long as proboscis. Sits generally with head up.................*Stable fly*, page 816.
  3. Small gnat-like humpbacked flies, not more than ¼ the size of the house fly, attack the animals, especially about the eyes, ears, and nostrils, and crawl into the hair to suck blood. Most abundant in spring...............................
  ....*Black flies* or *buffalo gnats*, page 818.
  4. Slender-bodied scaly-winged long-legged two-winged insects, up to ½ inch long, especially abundant about lowlands and swamps at dusk or at night....
  ....*Mosquitoes*, page 875.
B. *Insects that stay on the animals all of the time:*
  1. Wingless, flattened lice, with chestnut-colored head and thorax and yellowish black-banded abdomen, about ¹⁄₁₂ inch long, that feed on the dry skin and

hairs, but do not pierce the skin or draw blood. Head large, short, and broad, with chewing mouth parts; legs slender, abdomen not very broad............
....*Horse chewing louse*, page 819.

2. Wingless, flattened, grayish bloodsucking lice, about ⅛ inch long, with long, slender head, broad abdomen and legs much thickened toward the end.......
....*Horse bloodsucking louse*, page 820.

3. Leathery-skinned flattened long-legged two-winged flies, about ⅓ inch long, cluster under hairs and suck blood..
Horse "tick" or forest fly, (see Hindle, "Flies and Disease: Blood-sucking Flies").

4. Animals rub their bodies vigorously, hairs stand erect, areas of skin with only scattered hairs remaining; skin with a fine eruption, scurfy, or covered with small dry yellow scabs. Minute, very long-legged, rounded mites, hardly visible without a lens, burrow into the skin, causing intolerable itching.......
....*Mange mite*, page 821.

5. Hair comes out in patches; skin with large moist scabs under which live myriads of minute, longer-legged oval-bodied mites, scarcely visible without a lens, that pierce the skin with their mouth parts but do not tunnel beneath it..........
....*Scab mite*, page 822.

C. *Short, chunky, spiny maggots that live in the alimentary canal and, during the spring, are passed in the droppings. Animals become run-down, unkempt, colicky:*

1. Elongate, whitish or yellowish eggs or "nits," glued by half their length to base of hairs, chiefly on the front legs. Light-brown bee-like flies, with faintly spotted wings, hover about forequarters of animals, laying these eggs in summer. Wing venation like that of the throat botfly, and fly has a prominent spur on the trochanter of the hind legs. During winter months thick, spiny, yellowish or pinkish maggots, up to ⅔ inch long, attach to walls of stomach or duodenum; circlets of spines double, the circlet on third from last segment (the tenth) lacking 1 or 2 pairs of spines at the middorsal line; next to last segment (the eleventh) with 1 to 5 spines visible from above at each side.
....*Common horse botfly*, page 823.

2. Elongate whitish eggs, glued by nearly their full length to hairs, chiefly under the jaws, by a bee-like fly with rust-colored thorax and unspotted wings in which the discal cell (1st M_2) is only a little longer than the anterior basal cell (M) and the medial cross-vein is nearly in line with the radiomedial. During winter, short, thick, spiny, pale-yellowish maggots attach to walls of pharynx when small, sometimes cutting off breathing, and to stomach or duodenum; circlets of spines on larval segments in a single row.... *Chin fly or throat bot*, page 823.

3. Elongate black eggs, glued to base of the minute dark hairs of horses' lips by a brightly marked bee-like fly with its abdomen white at base, black across the middle, and bright orange-red at the end; the thorax with a black, shining area between the wings; and unspotted wings in which the anterior basal cell (M) is not more than ½ as long as the discal cell (1st M_2), and the medial cross-vein is distad of the radiomedial. During the winter, moderately spiny, pinkish maggots attach to stomach, duodenum, or rectum (often causing an obstruction) or cling exposed about margin of anus; circlets of spines on larval segments in double rows, but the circlet on third from last segment (the tenth) interrupted by at least half the width of the segment; next to last segment (the eleventh) without any spines visible from above............................
....*Lip fly or red-tailed botfly*, page 823.

General References.—U.S.D.A., Cir. 148, 1930; *Ill. Agr. Exp. Sta. Cir.* 378, 1931.

313 (322, 333, 354). HORSE FLIES[1]

Importance and Type of Injury.—Horses and mules strike with their heads, twitch the skin, shake their bodies, or otherwise evince sharp pain as large heavy-bodied black or brown flies (Figs. 539 and 540) alight

[1] *Tabanus* and *Chrysops* spp., Order Diptera, Family Tabanidae.

on the head, neck, shoulders, or back. These insects fly alongside the animals, even when they are running swiftly, and bite repeatedly, but only in the daytime. If they are not dislodged by the animal, they cut through the skin with their knife-like mouth parts and suck the blood for several minutes. When they finally fly away, a drop or two of blood usually exudes from the hole they made. In swampy, wooded sections animals in harness often become unmanageable and run away. Webb and Wells estimated that, in areas where horse flies are prevalent, animals commonly lose an average of more than 3 ounces of blood a day. They are forced to cease pasturing during most of the day, and as they congregate together, often wound each other by kicking or goring. Horse flies are carriers of tularaemia, of a kind of filariasis (Calabar swellings) among human beings in West Africa, and of *el debab*, an Algerian disease of horses and camels. They are also suspected carriers of anthrax, of equine infectious anemia or "swamp fever" of horses, and of surra.

Animals Attacked.—Horses, mules, cattle, hogs, man, dogs, deer, and other wild and domesticated animals.

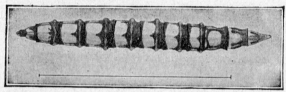

Fig. 538.—Larva of the horse fly, *Tabanus atratus* Fabricius. Line indicates natural size. (*From Ill. State Natural History Surv.*)

Distribution.—Horse flies of various species occur throughout the world, being especially abundant in moist wooded areas and up to at least several thousand feet elevation in the mountains.

Life History, Appearance, and Habits.—The life histories of only a few of the many species have been studied. Apparently the winter is generally passed as nearly full-grown larvae in the mud about lakes, streams, or wet areas of land. These maggots (Fig. 538) are 2 inches or less in length. They are pointed at each end, whitish or banded with black or brown, and with a fleshy elevated ring on each body segment. They are very tough-skinned and in some sections are much prized as bait for fish. They become full-grown in late spring, when they pass through a pupal stage of several weeks in drier mud and then the flies begin to appear in early summer. The eggs are laid on the leaves or stems of aquatic plants or trees, on stones and other objects that overhang water, or on grasses in moist swampy places; in dark-colored, wedge-shaped masses of several hundred. Nearly a week later the very small maggots hatch, drop into the water, sink to the bottom, and bury themselves in the mud or sand. Their food is small animals such as other insects, earth worms, small crustacea, snails, and other horse fly larvae. Some have a single generation and others apparently two each year.

Schwardt found that the big black horse fly, *Tabanus atratus* Fabricius, usually spends one winter and sometimes two in the larval stage, this larval period averaging 9 months but varying from 49 to 410 days, before producing the adults. The lined horse fly, *Tabanus lineola* Fabricius, on the other hand normally has at least two generations a year in Arkansas. There is thus great variation in the life cycles, habits, and occurrence of different species in correlation with vegetation, drainage, and topography.

There are several hundred species of horse flies in the United States, ranging in size from about ⅓ inch long to nearly 1 inch long. They are mostly black or brown, sometimes striped or spotted on the body, and many of the smaller species have the wings banded with brown. The eyes are often brilliantly colored and banded, while alive. Horse flies can usually be told from other flies by their antennae, which are divisible into three parts, the third being long and composed of five to eight rings and often with a short thumb-like projection at one side near its base. Only the females bite, the males feeding on nectar, honeydew, and the like.

Control Measures.—No control has been discovered that is very effective. Draining marshes and wet meadows where the flies develop is of the greatest value but should be done in such way as to preserve the desirable wild life of such areas, if possible. Care should be taken to avoid overirrigation. Oiling stagnant pools as for mosquitoes (page 879) is of value in certain cases. It may kill the adults which come to quiet pools and dip into them and also kills the young maggots as they hatch from the eggs and drop into the water. Stabling animals during the day or covering them with light blankets, fly nets or ear nets, helps to ward off the

Fig. 539.—Adult of the horse fly, *Tabanus atratus* Fabricius; male, natural size. (*From Ill. State Natural History Surv.*)

Fig. 540.—Adult of the horse fly, *Tabanus sulcifrons* Macquart, female. Enlarged ½. (*From Ill. State Natural History Surv.*)

attacks. The following repellent mixture has been widely recommended: pine tar, 1 gallon, fish oil or crude carbolic acid, 1 quart, flowers of sulphur, 2 pounds. It may be smeared over the ears and head or applied as a spray.

References.—*U.S.D.A., Bur. Ento., Tech. Bul.* 12, Part, II, 1906 and *Dept. Bul.* 1218, 1924; *Nev. Agr. Exp. Sta. Bul.* 102, 1921; *Ann. Ento. Soc. Amer.*, **24**: 409–416, 1931, and **25**: 631–637, 1932; *Ark. Agr. Exp. Sta. Bul.* 256, 1930; *La. Agr. Exp. Sta. Bul.* 93, 1907; *Ohio State Acad. Sci. Spcl. Paper* 5, 1903; *Jour. Econ. Ento.*, **30**: 214, 1937.

314 (322, 333, 355). Stable Fly[1]

Importance and Type of Injury.—The most injurious insect attacking horses and mules is a small fly, very similar in appearance to the common

---

[1] *Stomoxys calcitrans* (Linné), Order Diptera, Family Muscidae.

house fly, which bites the animals especially on the legs. It sometimes comes into houses, especially in stormy weather, and bites people about the ankles. Animals stamp their feet continually to dislodge these tormenting pests. The stable fly takes one or two drops of blood at a meal, several such meals in a day, so that each animal probably supplies hundreds, if not thousands, of fly meals a day when stable flies are abundant. As a result of the pain from the bites, the constant worry, and the loss of blood, animals lose weight, milk yield of dairy cattle is reduced, work animals become unmanageable, and sometimes animals are killed, either as the direct result of the flies or from disease induced by the flies. While the stable fly has been suspected of transmitting infantile paralysis, anthrax, leprosy, surra, and swamp fever (or infectious anemia of horses), it has not been proved to be the usual carrier of any animal disease in America.

The stable fly, mosquitoes, or other bloodsucking insects may be carriers of a disease of horses and mules known as equine encephalomyelitis, which causes the animals to be weak, drowsy, and unable to stay on their feet. This disease caused a loss of $11,000,000 in western Kansas, alone, in 1912, and caused the death of 33,000 horses and mules in 1937. Keeping horses in screened stables, at night, the use of muslin covers for the animals, and fly-repellent sprays have been suggested for control. There is some evidence that the virus of equine encephalomyelitis is the same as that of human encephalomyelitis. The mosquitoes may be the agency that spreads the disease from horse to man as well as among horses.[1]

Animals Attacked.—Horses, mules, cattle, hogs, dogs, cats, sheep, goats, guinea pigs, rabbits, rats and man.

Distribution.—The stable fly occurs in all parts of the United States and throughout most of the world. In the United States it appears to be most abundant "in the Central States from Texas to Canada, where grain is grown extensively" (Bishopp).

Life History, Appearance, and Habits.—In the northern states the stable fly is believed to winter as larvae and pupae in wet straw piles or strawy manure. Farther south development continues throughout the year, and all stages may be found in winter. During the warm months of the year, breeding is continuous. The length of the several life stages has been given as follows: egg, commonly 2 or 3 days; larva, commonly 2 to 4 weeks; pupa, 1 to 3 weeks; and adult probably 3 weeks. Since the female is commonly 2 or 3 weeks old and must take several meals of blood before she begins laying eggs, the total average life cycle may be from 20 to 60 days, the longer period in cool weather. Five or six hundred elongate whitish eggs are deposited by the female in 4 or 5 batches. The stable fly (Fig. 541) is about $\frac{1}{4}$ inch long, of a general grayish color, like the house fly. It can be distinguished from the house fly by its habit of biting, and by its mouth parts, which stick forward from under the head as a stiff, somewhat pointed, slender beak, about twice as long as the head. The abdomen has seven rounded dark spots on the upper side, arranged in a figure 8 (Fig. 541). The cell nearest the tip of the wing (R_5) is open more than half its width. Both males and females suck blood as their chief food. They are active only during the day, either in the stable or in the field. The yellowish-white maggots of the stable fly develop in masses

---

[1] *References.—Nev. Agr. Exp. Sta. Bul.* 132, 1933; *Jour. Amer. Vet. Med. Assoc.,* **39,** (n.s.): 662–666, 1935, and **42,** (n.s.): 187–196, 1936.

of straw, grain, hay, piles of grass, weeds, and other materials that have become water-soaked or contaminated with manure, and in the excrement of animals only if it contains much hay or straw.   They have been found developing in great numbers on the trickling filters of sewage-disposal plants.   In such cases flooding every 12 hours has been effective in cleaning out the larvae.   Most serious outbreaks follow periods of excessive rainfall.   The full-grown maggot is about ¾ inch long, tapering almost to a point at the head, and the posterior end is cut off squarely.   It can be told from the house fly larva by looking at the spiracles on the last segment of the body.   In the stable fly these are small, somewhat triangular, and the two separated by twice their own width; in the house fly they are almost touching, larger, and D-shaped (Fig. 584,*f*), with the three slits more sinuous.   The insect passes a pupal stage of a week or two, in a brown puparium among the straw, and then emerges as the adult.

FIG. 541.—The stable fly, *Stomoxys calcitrans* (Linné).   Adult as seen from above.   About 5 times natural size. (*From U.S.D.A., Farmers' Bul. 1097.*)

Control Measures.—The destruction and avoidance of conditions in which the maggots thrive offer most promise, and, while this method cannot be expected to eliminate the fly, it should be possible to reduce its numbers. Straw from threshing should be scattered over the ground and plowed under, or burned, if it is not needed for bedding or feed. If it is to be preserved for use, it should be baled and stored in a dry place or else stacked carefully so that it does not become water-soaked and rotten. Masses of water-soaked feed should not be allowed to accumulate around stalls or feed troughs.   The prompt disposal of manure and all other accumulations of fermenting organic matter, as explained for the house fly (page 909), will also keep this species in check.

Animals may be covered with blankets or old trousers pulled over the legs; sprayed with a tested commercial or home-mixed livestock spray; or allowed to run into darkened stables with nets, brush, or sacking so arranged over the doorway as to brush off the flies as the animals enter. About dairy barns and other stables, electrical or mechanical window traps may be used to catch myriads of the stable fly as well as house flies. Bait may be placed in the traps to attract the house flies but is of no value for stable flies.   While repellent sprays and traps help in lessening the annoyance to animals, cleaning up the breeding places is of far greater importance. .

Reference.—U.S.D.A., Farmers' Bull. 1097, 1920; Jour. Econ. Ento., **27**: 1197–1198, 1934.

315 (322, 352).   BLACK FLIES OR BUFFALO GNATS

These very small gnats (Fig. 573) hover about ears, eyes, and nostrils, alighting frequently and puncturing the skin with a very irritating bite.

[1] *Simulium* spp., Order Diptera, Family Simuliidae.

The young develop in the water, especially of rocky, swift-flowing streams. They are very difficult to destroy without destroying the fish in the stream and about the only means of control is to provide smudges, in the smoke of which animals can get relief from attack, or to spray animals with repellent oils. These flies are more fully discussed under Insects that Attack Man (page 881).

316. Horse Chewing Louse[1]

Importance and Type of Injury.—At least two very different species of lice attack the horse, and it is important to recognize which is present in any case of infestation, since the control measures will be somewhat different. This species does not pierce the skin or suck blood but runs freely about over the animal, nibbles at the dry skin and hairs and causes great irritation. In the spring of the year, horses that have not wintered well rub against fences, stalls, and other objects. The coat, especially on the head, withers, and about the base of the tail becomes unkempt, hairs stand erect, and the skin is dry and full of scurf. When infested horses are brought from stables into the warm sunlight, the lice often are seen by hundreds clinging to the tips of the hairs.

Animals Attacked.—Horses, mules, and donkeys.

Life History, Appearance, and Habits.—These lice generally become noticeable in late winter or early spring, when all stages can usually be found on the animal. The full-grown ones (Fig. 542), are about $\frac{1}{10}$ inch long, of a chestnut-brown color except on the abdomen, which is yellowish with dark crossbands. The head is much broader and shorter than that of the following species and is rounded in front, forming a full semicircle in front of the antennae. The legs are slenderer than those of the blood-sucking horse louse. The eggs are glued to the hairs close to the skin, especially around the angle of the jaw and on the flanks. The eggs hatch in 5 to 10 days, into very small pale-colored lice, of the same general shape as the full-grown ones, and they become full-grown in 3 or 4 weeks. Breeding is continuous throughout the year, but the numbers become fewer in summer.

Fig. 542.—The horse chewing louse, *Trichodectes equi* (Linné); female, about 25 times natural size. (*From U.S.D.A., Farmers' Bul.* 1493.)

Control Measures.—Horses are not usually troubled with lice unless they have been neglected in feeding, stabling, and grooming. Sodium fluoride dusted into the coat of the horse using about 2 ounces per animal has given complete control. It should be dusted on the coat but not rubbed in and not applied about wounds nor too freely about body openings. It remains in the coat, if the animal is not groomed, and kills the

[1] *Trichodectes equi* (Linné) (= *T. parumpilosus* Piaget); and *T. pilosus* Giebel, the European horse chewing louse, Order Mallophaga, Family Trichodectidae.

young lice after they hatch from eggs.   Therefore, one good application should be sufficient.   Instead of sodium fluoride, wettable sulphur of 300- to 400-mesh fineness, 1 or 2 pounds per animal, may be rubbed into the coat.   Sulphur has the advantage over sodium fluoride that it is not poisonous; but a second treatment will need to be given 2 weeks after the first.   Dusts containing not less than 0.125 per cent rotenone, used 3 or 4 ounces per animal, are effective, and will also control any sucking lice present; they must, however, be repeated in 2 weeks to kill lice hatching from eggs.   The animals may be washed or dipped in a 2 or 3 per cent solution of creolin in water, or in one of the well-known coal-tar dips, or they may be rubbed with raw linseed oil.   If linseed oil is used, do not work the animal or expose to hot sun for a day after treatment.

Reference.—U.S.D.A., Farmers' Bul. 1493, 1926.

317. HORSE BLOODSUCKING LOUSE[1]

Importance and Type of Injury.—This seems to be the most important louse infesting the horse, being commoner and also more irritating, because it feeds by piercing the skin and sucking the blood.   The bites are painful and, when the lice become abundant, the loss of blood is a severe drain on the vitality of the host.   The horse shows the same symptoms of scurfy skin, unkempt coat, and scratching or rubbing its body as in the case of the chewing louse.   Often parts of the body will be rubbed raw, and the lice may feed about these wounds. Only an examination of the lice themselves will determine which species is present.

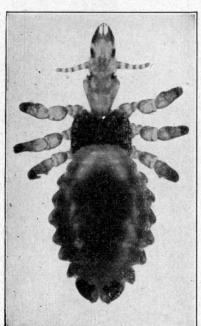

FIG. 543.—The horse bloodsucking louse, *Haematopinus macrocephalus* Burmeister; female, about 25 times natural size. (*From U.S.D.A., Farmers' Bul.* 1493.)

Animals Attacked.—Horses, mules, and donkeys.

Life History, Appearance, and Habits.—All sizes of lice and eggs will usually be found during the winter, when these insects are troublesome. The full-grown lice (Fig. 543) are of a dirty grayish or yellowish-brown color, about ⅛ inch long, by half as broad at the middle of the abdomen.   The thorax is only half as wide, and the head less than a third as wide, as the abdomen, distinctly narrowed toward the front.   The legs are short and very clumsy, fitted for grasping about hairs.   The lice are commonest about head and neck and at the base of the tail.   The egg stage is normally from 11 to 20 days, but some may hatch as long as a

[1] *Haematopinus macrocephalus* Burmeister (= *H. asini* Linné), Order Anoplura, Family Haematopinidae.

month after they were laid. The young lice are similar to the large ones except paler in color, and gradually grow to the size and color of the adults in 2 to 4 weeks. There are several generations a year.

Control Measures.—Sodium fluoride is not effective for this species, but rotenone dusts, as recommended for the horse chewing louse, may be used. Raw linseed oil, 2 or 3 per cent creolin in water, one of the commercial coal-tar or nicotine dips, or standard arsenical dip should be applied. Dips should not be applied in cold weather. If linseed oil is used, the animal should be kept out of the sun and overheating avoided for a day or two. Since the eggs hatch over so long a period, two to four treatments at intervals of 10 to 20 days, and one of these treatments at least 5 weeks after the first, are recommended. Keeping the animals in good thrifty condition helps greatly to lessen the injury from lice.

Reference.—*U.S.D.A., Farmers' Bul.* 1493, 1926.

318 (335, 366). MANGE MITE[1] or SARCOPTIC MANGE

Importance and Type of Injury.—Animals rub and scratch their bodies vigorously. Areas on the head, neck, back, or at the base of the tail become inflamed, pimply and scurfy, with the hairs bristling and only scattered hairs remaining. Later the infestation may spread over the entire body, and large, dry, cracked scabs form on the thickened skin. To distinguish it from lousiness, some scrapings from the affected skin should be examined under a microscope for the mites which cause the trouble. The mange of the horse may spread to man, causing "cavalryman's itch," but it does not persist on man, dying out in a few weeks (see also pages 848 and 900, and Fig. 9).

Animals Attacked.—Horses, hogs, mules, men, dogs, cats, foxes, rabbits, squirrels, sheep, and cattle are attacked by different varieties of the same species.

Life History, Appearance, and Habits. Mange is caused by a very small ovoid mite (Fig. 544), scarcely as big around as the cross-section of an ordinary pin ($\frac{1}{60}$ inch in diameter), and with very short legs that barely extend beyond the margins of the body. It is not a true insect, but an eight-legged form related to the ticks and spiders. The mites themselves will seldom be seen except by the specialist. They burrow beneath the skin on less-hairy parts of the body, making very slender winding tunnels, from $\frac{1}{10}$ inch to nearly 1 inch long. The serum discharged from the mouth of the burrows dries to form small dry papules. Within such a tunnel the female lays about 24 eggs and dies at the end of the tunnel. Within 3 to 10 days the eggs hatch to minute nymphs, which

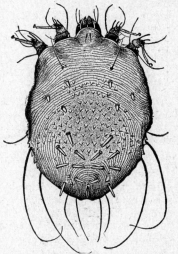

FIG. 544.—The mange or itch mite, *Sarcoptes scabiei* DeGeer; female about 100 times natural size. (*From U.S.D.A., Farmers' Bul.* 1493.)

[1] *Sarcoptes scabiei equi* Gerlach, Order Acarina, Family Sarcoptidae.

are at first six-legged. At the first molting they acquire an additional pair of legs, and after two more molts they are full-sized and ready to mate. After mating, the males die; the females again shed their skin and begin new tunnels 10 days to a month after hatching from the eggs. The mites secrete an extremely irritating poison, which, together with their tunneling, causes the excruciating itching. The trouble is most evident during the winter months, but some of the mites live on the animals the year round, unless treated. A generation may be completed in 2 weeks.

Control Measures.—Mange is contagious, and the most important control is the isolation and quarantine of infected animals to prevent spread. Treatment for horses consists in dipping; or clipping, prolonged bathing or massaging in soap and water to soften the scabs and open up the tunnels, and applying one of the many dressings recommended for this pest. Hadwen recommends a hot (not boiled) mixture of raw linseed oil, 1 gallon, oil of pine tar, 8 ounces, and sulphur, 2 pounds. Leave on the skin for 10 days; then repeat. Another preparation especially recommended not to cause loss of hair is equal parts of sodium carbonate, water, soap, sulphur, and creolin. Dissolve the soda in the water and add the others in order. Stir well and dilute with 5 more volumes of water. Rub into the affected spots for ½ hour, then wash with clean water. Apply four times, 3 days apart. Repeated dippings in lime-sulphur solution, for 2 or 3 minutes, at 105°F., is the best and surest way to stop infestation. The dip can be made as follows:

Slake 10 pounds of fresh stone lime, thoroughly mix into the lime paste 24 pounds of sulphur, put through a screen to break up all lumps. Add 30 gailons of boiling water and boil for 2 hours with frequent stirring. Allow to stand overnight, then siphon off the clear liquid, without disturbing the sediment, and add water to make 100 gallons ready to use. Care should be taken not to get any of the sludge or sediment into the dip.

Repeated applications of any treatment are necessary. The walls and floors of stalls, curry combs, harness, and other objects about the horse should be treated with a good coal-tar-creosote disinfectant, since the mites may live on these objects for several weeks off the animal. Treated animals should not be returned to their former pasture for a month or two. Dogs and cats infested with mange mites should be clipped, subjected to a prolonged washing in warm water and green soap until all scabs are softened and loosened, and then an ointment consisting of lard, 8 ounces, sulphur, 1 ounce, and balsam of Peru, 1 dram, should be applied to the skin. It is sometimes recommended that one-fourth of the animal's body be treated on each of four succeeding days with this ointment. The bathing and ointment must be repeated at weekly intervals until all mites have been killed and all scabs have healed.

Reference.—*U.S.D.A., Farmers' Bul.* 1493, 1926.

319 (326, 339). Scab Mite[1]

The scab mite is less serious on horses than on cattle and sheep. It is sometimes spoken of as the "wet mange," in contrast with the drier scabs that result from the attacks of the mange mite. It usually starts among

[1] *Psoroptes communis bovis* Hering, Order Acarina, Family Sarcoptidae.

the longer hair of the neck, withers, or base of tail. Treatment is the same as for mange mite (see also page 855).

320. HORSE BOTS

There are three kinds of botflies that commonly molest horses in this country. They are known as the common horse botfly,[1] the chin fly or throat bot,[2] and the lip or nose botfly.[3] A fourth species *Gastrophilus inermis* Brauer, was reported by Knipling in 1935 (see *Entomological News*, **46** : 105–107, 1935). The adult is very similar to the common horse bot, but lacks the spur on the hind trochanters. The larva is most like the lip bot, but, unlike that species, the circlet of spines on segment 3 is not interrupted on the ventral side and has as many spines as segment 4. The egg of this species is much shorter and thicker than any of the other botflies and does not flare away from the hair to which it is attached any more at the head end than it does at the end toward the base of the hair.

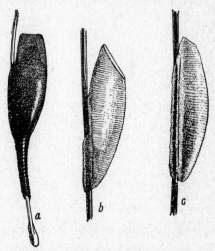

Importance and Type of Injury. All three kinds of botflies, but especially the lip botfly, are dreaded by horses, which fight them viciously, although we know from the structure of the flies that they can neither bite nor sting the animals. Their sole object in flying about horses is to glue their eggs (Fig. 545) fast to the hairs. This act results in a slight pull on the hair, but it is believed that the alarm caused by the flies is mostly due to fear or nervous excitation caused

FIG. 545.—Eggs of horse bots; *a*, the lip botfly; *b*, the common horse botfly; and *c*, the throat botfly. About 35 times natural size. (*From U.S.D.A., Farmers' Bul.* 1503.)

by buzzing and striking. The most injurious stage of the flies is the maggot or larval stage, which lives in the digestive tract of the horse, causing mechanical injuries to the tongue, the lips, and the lining of the stomach and intestine, interfering with glandular activity, causing an inflamed ulcerous condition, absorbing food, which progressively starves the host, probably secreting toxic substances, and frequently causing complete obstruction to the passage of food substances from stomach to intestine (Fig. 546). A horse badly infested with bots presents a run-down condition, caused by digestive disturbances, and has a rough coat. The presence of bots may also be detected by finding the full-grown larvae in the feces in the spring months, although this occurs too late to apply effective control measures.

Animals Attacked.—Horses, mules, donkeys, and rarely dogs, rabbits, hogs, and man.

[1] *Gastrophilus* (= *equi*) *intestinalis* DeGeer, Order Diptera, Family Oestridae.
[2] *Gastrophilus* (= *veterinus*) *nasalis* Linné.
[3] *Gastrophilus haemorrhoidalis* Linné.

Distribution.—The common horse bot and the throat bot occur throughout the United States. The throat bot is said to be especially abundant in the Rocky Mountain region. The nose or lip botfly is recorded from northern Illinois, westward to Idaho, and from northern Kansas and Colorado to Manitoba, Saskatchewan, and Alberta. It has been spreading in all directions.

Life History, Appearance, and Habits.—The bots winter as larvae (Fig. 546) in the alimentary canal of the host, usually becoming full-grown by late winter or spring. They may be found in the digestive tract every month of the year; but after the first of October only small larvae will be found, the mature larvae having all been passed by that date. It is believed that the larvae feed upon the inflammatory products induced in the mucosa of the stomach by their presence. The full-grown horse bots are thick tough-skinned maggots, blunt at the posterior end, tapering in front to the two strong mouth hooks, and with a circlet of prominent spines around each body segment, giving them a screw-like appearance. They are about ½ to ⅔ inch long when full-grown, yellowish white or pinkish in color. The three kinds may be told by the characters given in the key. When growth is complete, they release their hold on the walls of the stomach and pass to the ground, usually with the excrement, during the first half of May, in the case of the lip and throat bots, and somewhat later for the common horse bot. This fact has been taken advantage of by quacks who recommend various treatments to be given in the late winter and cite the normal appearance of bots in the excrement as a result. The bot larvae burrow into the soil a short distance and inside their hardened skins form the pupae, from which adults appear 3 to 10 weeks later. The adults are abroad commonly, in the case of the lip bot from early June until late summer in the northern states, being rare in the fall; while the throat bot appears early in June and the common horse bot about the last of June. These two species are abundant until heavy frosts, commonly about November 1. The adults look somewhat like bumblebees, being very hairy, two-winged, of a general brownish color, and about ⅔ inch long. They take no food and live only from 3 days to 3 weeks. The common horse bot may easily be told by the faint smoky spots on the wings. The throat bot has a rust-red thorax, clear wings, and a prominent band of black hairs at mid-length of the abdomen. The lip botfly is somewhat smaller than the others, and the tip of the abdomen has reddish hairs.

The common horse bot,[1] hovers about the animal without causing much excitement, bends her abdomen forward under the body, and, darting in, quickly glues an egg to a hair, usually on the front leg but also to a lesser extent on shoulders, belly, and hind legs. Egg after egg (Fig. 545,*b*) is attached in this way, so that the legs of horses on pasture or in corrals often assume a yellowish-gray cast from the large number of "nits" cemented to the hairs. The females average over 800 eggs. A very curious fact about the eggs of this fly is that they cannot hatch without some artificial aid. The necessary stimulus to hatching has long been supposed to be moisture and friction. It has been discovered, however, that it is a sudden rise in temperature, the optimum temperature for hatching being between 114 and 118°F. The fully formed, tiny larva

---

[1] *Gastrophilus* (= *equi*) *intestinalis* DeGeer, Order Diptera, Family Oestridae.

is ready to hatch 10 to 14 days after the egg was laid, but may remain alive in the eggshell for 100 to 140 days. At any time during this long period when the warm, moist lips or tongue of the horse pass over them in scratching or nibbling at the legs, they hatch very quickly and the larva clings to the lips or tongue. Upon entering the mouth, the first-instar larva burrows into the mucous membrane of the tongue and works its way backward toward the base of the tongue for 24 to 28 days, causing linear lesions. The molt to the second instar occurs when they leave the tongue. The small larvae then usually pass directly to the stomach, where they molt to the third instar after about 5 weeks. Unless the eggs on the hair of the horse are destroyed, larvae may continue to be taken into the mouth until about the middle of December. Having reached

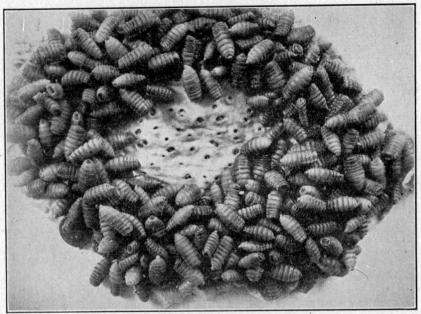

Fig. 546.—Horse bots. Larvae attached to lining of stomach of horse; at center, lesions where bots have been removed. About two-thirds natural size. (*From U.S.D.A., Dept. Bul.* 597.)

the stomach, 9 to 10 months are commonly spent there, during which growth is completed, there being but one generation a year.

The throat bot[1] hovers in front of the horse and, at intervals, darts upward and glues its eggs (Fig. 545,c), to the bases of hairs under the head or jaw of the horse or down on the throat. This causes the animals to nod their heads, drawing the nose in toward the breast and, when loose in pastures, to stand with their heads across the shoulders of another horse. These eggs are also whitish but can be distinguished from those of the common bot by the characters given in the key. The throat botfly averages 400 to 500 eggs per female. These eggs do not require heat, moisture, or rubbing to cause hatching. After 10 to 12 days, the larvae emerge and are said to crawl down the hairs and wriggle through the hairs

[1] *Gastrophilus* (= *veterinus*) *nasalis* Linné.

over the skin to enter between the lips.   They are probably positively geotropic and hygrotropic and are said to reach the horse's mouth from eggs laid at least 8 inches away.   It is said that they burrow in the mucosa of the cheeks during the first instar, reaching the duodenum 17 to 19 days after entering the mouth.   Other investigators think they penetrate the skin near where the egg was laid and tunnel through the flesh to the alimentary canal.   They tend to attach to the pharynx, paralyzing the muscles that control swallowing, or in the duodenum, and are less common in the stomach.   The rest of the life cycle is similar to that of the common horse bot.   The mature larvae pass from the host by the middle of May, the freshly dropped ones being whitish, whereas those of all the other species are reddish.   Adults may be ovipositing by the first of June.

The lip or nose botfly[1] is generally considered the worst species of the three.   It lays its eggs, not about the nostrils, as the name nose botfly wrongly implies, but attached very close to the skin on the small hairs of the lips mostly on the upper lip.   Striking the horse on this sensitive spot causes it to jerk and toss its head, and repeated attacks often cause the animals to become frantic or to rub their lips over fences, or to stand with the lips appressed to another animal's back.   The eggs (Fig. 545,a) are very different from the other species, being blackish and with a slender screw-like "tail" nearly as long as the egg.   Only about 160 eggs per female are laid by this species.   Moisture is apparently the stimulus to hatching of the eggs of the lip botfly and temperature of little significance. Eggs attached to hairs of lips where they are moistened by the saliva, hatch in 2 to 6 days; and it seems probable that those laid far from the lips never hatch.   Immediately upon hatching the larvae are said to tunnel beneath the epidermis of the lips and migrate in this way into the mouth, thus causing the great irritation always noted when this fly is prevalent. Eventually they are swallowed with food or drink.   From this point on, the life cycle is similar to that of the other species, except that these larvae, instead of passing with the excreta, have the habit of attaching to the walls of the rectum or to the skin about the anus for a time, after midwinter, before dropping to the soil, causing rubbing and switching of the tail.   The adults begin to appear in early June, and some are present until frost, although any individual lives only a few days.

Control Measures.—Botflies are active only in the open and during the daytime, so that animals kept stabled during the day and pastured only at night, if at all, do not become infested.   A piece of canvas stretched across the underside of the head and front of the neck, from the throatlatch or collar to the bit rings, effectively checks the egg-laying of throat botflies.   The lip fly may be similarly prevented from laying eggs by a belting lip protector, constructed as described by Bishopp and Dove in *U. S. Department of Agriculture Farmers' Bulletin* 1503.   The application of equal parts of pine tar and lard will prevent egg-laying on the treated parts for about 4 days.   By November first, in the latitude of central Illinois and Iowa, killing frosts will usually have destroyed all adult flies, so that no more eggs will be laid.   The legs and even the under jaw and breast of all horses that bear eggs of botflies should be promptly sponged, on a day when the temperature is below 60°F., with warm water (104 to 118°F.) containing 2 per cent phenol or coal-tar creosote.   This causes the eggs to hatch and the larvae to die.   No more larvae will then enter

[1] *Gastrophilus haemorrhoidalis* Linné.

the body of the horse until the following summer. It is recommended that after delaying about 28 days more to allow any larvae burrowing in lips or tongue to reach the stomach, the digestive tract be fumigated in the following manner. After withholding all food for 24 hours and water for 5 hours, the horse is given a dose of carbon bisulphide, by means of a stomach tube or in such a way that it reaches the stomach unbroken. The proper dose is 2 to 2.5 cubic centimeters of carbon bisulphide for each 100 pounds of body weight of the horse or 6 drams per 1,000 pounds. It should be administered only by an experienced veterinarian and should not be attempted by inexperienced persons. Feeding and watering may be resumed 3 hours after the treatment. The carbon bisulphide forms a gas that kills the larvae, and they are passed in the feces beginning within 6 hours after treatment and continuing for 2 weeks or more. If the larvae are destroyed early in fall in this manner, the animals are spared most of the damage caused by the growth of the larvae in their stomachs, and, if communities cooperate in the treatment of all horses, the number of flies developing in successive years should be greatly reduced. In Illinois, 130,000 horses have thus been treated in a single year. This treatment also destroys the roundworm of the horse,[1] another serious intestinal parasite.

References.—U.S.D.A., Dept. Bul. 597, 1918, and Farmers' Bul. 1503, 1926; Bul. Ento. Res. vol. 9, Part 2, pp. 91–106, September, 1918; Iowa State Coll. Jour. Sci., **12**: 181–203, 1938.

B. INSECTS ATTACKING CATTLE

FIELD KEY FOR THE IDENTIFICATION OF INSECTS INJURING CATTLE

A. *Free-flying insects that alight on the animals to suck blood, coming and going repeatedly:*
1. Heavy-bodied, swift, brownish to black flies, ½ to 1 inch long, with 2 clear or banded wings, alight on animals and suck their blood, usually leaving a drop or two of blood exuding from the puncture. Especially active about marshes, swampy woods, or meadows and on warm sunny days......................
   ....*Horse flies,* pages 814 and 830.
2. Flies of the size and general appearance of the house fly, suck blood especially from the legs, causing the animals to stamp their feet. Distinguished from the house fly by its stiff, slender beak, sticking straight forward from underside of head, by its broader black-spotted abdomen, grayer appearance, and by the bristle on the antenna having hairs on its upper side only. Palps less than half as long as proboscis. Usually rests with its head upward..................
   ....*Stable fly,* pages 816 and 831.
3. Flies similar to the stable fly, but only half as large, stand on their heads among the hairs especially of the back and sides of cattle and suck blood. Palps nearly as long as proboscis. They sometimes cluster in a mass about the base of the horns to rest...*Horn fly,* page 829.
4. Very small, stout, black flies, less than ⅙ inch long, with a humped thorax, thick 11-segmented antennae as long as the head, 2 broad, delicately veined wings, and stout legs, pierce the skin and suck the blood, especially about the eyes, ears, and nostrils, causing extreme pain...........................
   ..../....................*Black flies, turkey gnats,* or *buffalo gnats,* page 881.
5. Very slender-bodied scaly-winged bloodsucking flies swarming over the animals, especially at dusk and at night near swampy wooded sections...............
   ....*Mosquitoes,* pages 830 and 875.

[1] *Habronema* spp., Class Nematoda.

B. *Insects that stay on the animals all the time:*
  1. Very small lice, about $\frac{1}{13}$ inch long, reddish in color, with distinct dark cross-bands on the abdomen; head broad, bluntly rounded in front. Chew hairs, epidermal scales, etc.; do not suck blood. Run about actively among the hairs. Eggs whitish, glued to hairs..................*Cattle chewing louse*, page 831.
  2. Slate-gray, wingless, broad, flat, bloodsucking lice, up to $\frac{1}{8}$ inch long and half as broad; head pointed in front. The white eggs are glued to the hairs, especially about the fore part of the body...........*Short-nosed ox louse*, page 832.
  3. Similar to the short-nosed ox louse but only about half as large when mature, often attaching in dense groups about head and neck. Eggs yellowish.......
  ....*A bloodsucking louse, Solenopotes capillatus*, page 832.
  4. Bluish, shiny, wingless lice, a little smaller, darker, much more slender and with a longer head, than the preceding. Move about but little, usually stand on their heads among the hairs. Eggs nearly black, glued to the hairs.......
  ....*Long-nosed ox louse*, page 832.
  5. Hair comes out in patches; skin forms large scabs, under which may be found myriads of minute 8-legged oval-bodied pale mites, just visible to the naked eye, which pierce the skin with their mouth parts but do not tunnel beneath it..*Scab mite*, page 834.
  6. Hard lumps, up to the size of a pea, form on skin of muzzle, head, and shoulders and burst open, disclosing a yellowish pus in which are many microscopic worm-like mites, with 8 short stubby legs toward one end of their short slender bodies...*Follicle mite*, page 835.
  7. Eight-legged, leathery, glossy, ovate seed-like, brown to bluish-gray ticks, attached to the skin by their mouth parts and seldom moving about; up to $\frac{1}{2}$ inch long. Drop off host to lay eggs in a cluster on the ground. Minute, first-instar nymphs or "seed ticks" have only 6 legs......................
  Cattle tick (carries the disease known as Texas fever), *dog tick*, and others, pages 835 and 838.
C. *Spiny worms or maggots that live in the flesh or in wounds:*
  1. During the winter, tumorous swellings or "warbles" appear along the back, in which fat, wrinkled, white or blackish maggots, from $\frac{1}{3}$ to 1 inch long, live; and from which they can be squeezed out. The full-grown larvae have minute spines on the posterior border of the ventral surface of segments 2 to 10, inclusive, *i.e.*, on all but the last larval segment. The surface of the stigmal plates is flat and the inner edges of the C-shaped stigmal plates are farther apart than in the bomb fly. These are the young of hairy, swift flies (about the size and appearance of a honeybee, $\frac{1}{2}$ inch long, with hairy legs, uniformly white tegulae and orange hairs at end of abdomen only) that glue small, slender, yellowish eggs in rows to hairs of legs, belly, etc., in early spring.....
  ....*Heel fly* or *striped ox-warble fly*, page 838.
  2. Like the above, but the last 2 segments of the larva without spines, *i.e.*, the ventral surface of segments 2 to 9 (but not 10) with small spines on the posterior border. The stigmal plates cup-shaped or funnel-shaped, and the inner edges of the C-shaped plates closer together than in the heel fly. Adults a little larger ($\frac{3}{5}$ inch long, with legs smoother, tegulae with a brown border, and yellowish hairs on front half of thorax as well as on base and tip of abdomen), appear later in spring and annoy cattle greatly, causing them to stampede or run madly, though the flies neither bite nor sting............................
  ....*Bomb fly* or *northern ox-warble fly*, page 838.
  3. Slender, whitish maggots, up to $\frac{3}{4}$ inch long, blunt behind, tapering in front, and with an elevated ring at each segment, bearing short spines, burrow in wounds and *feed only in the living tissues of the animals*. The eggs are laid about wounds and fly or tick punctures, by a metallic, bluish-green fly, larger than the house fly, wi h three black stripes on thorax and a reddish-yellow face.......
  ....*Screwworm fly*, page 843.
  4. Slender whitish maggots, blunt behind, tapering in front, and $\frac{3}{4}$ inch or less in length, work in the flesh about sores, especially following dehorning........
  ....*Greenbottle fly* and *black blowfly*, page 860.

321. Horn Fly[1]

Importance and Type of Injury.—The horn fly is a close relative of the stable fly, and its harmful effects on cattle are very similar. It pierces the skin to suck blood, causing pain and annoyance and interfering with the feeding and resting of the animals, so that they lose weight, yield less milk, develop indigestion, and suffer other disorders. These will be recognized as the small flies (Fig. 547), about half as big as the house fly

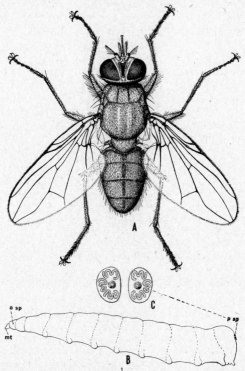

Fig. 547.—The horn fly, *Haematobia irritans* (Linné). *A*, adult female, 9 times natural size. *B*, larva in side view; *a sp*, anterior spiracle; *mt*, mouth; *p sp*, posterior spiracles or stigmal plates. *C*, the stigmal plates greatly magnified, showing median "button" and three sinuous spiracular slits. (*Original.*)

or stable fly, that hover over the backs of cattle all summer long, especially out-of-doors, crawling down among the hairs on the withers, back, or belly and, with wings partly spread, suck blood until a swish of the head or tail scares them up, temporarily. These flies have been suspected of transmitting anthrax.

Animals Attacked.—Cattle chiefly; also goats and sometimes horses, dogs, and sheep. Annoying to people working about cattle.

Distribution.—This insect was first found in this country near Philadelphia in 1887 and is believed to have been brought to this country with shipments of cattle a year or two before. Within 10 years it had spread over all of the United States east of the Rocky Mountains, and to California and Hawaii. It is now generally distributed throughout America.

[1] *Haematobia irritans* (Linné), Order Diptera, Family Muscidae.

Life History, Appearance, and Habits.—Horn flies winter as larvae or pupae within puparia, in or beneath the droppings of cattle. Toward the end of April in the latitude of central Illinois they begin to appear about cattle, and their numbers rapidly increase. Apparently they develop only in fresh droppings of cattle, the flies darting from an animal to the fresh dung and depositing a few eggs on the surface of the mass, nearly always within a minute or two of the time it is dropped. Within about 2 minutes the dung has lost its attraction for the egg-laying flies. In masses of cattle dung dropped from 9 A.M. to 4 P.M., Mohr found an average of 150 larvae per mass, whereas dung dropped at night was uninfested. The eggs are brown in color and not easy to see. The maggots hatch from these eggs in a day or so, feed in the dung and become full-grown (about ⅜ inch long), in 3 to 5 days. They then form pupae inside brown seed-like puparia, either in the soil or in the dung, and emerge as flies about a week later. The entire life-cycle may be completed in 10 days to 2 weeks. The adults look like half-sized stable flies but do not have the spotted abdomen, and the palps at the sides of the beak are about two-thirds as long as the beak. Although the flies are said to feed only once a day, they roost on the bodies of cattle both day and night. When not feeding, they often rest about the head, especially on the base of the horn, if the animal has horns, sometimes so many of them that they make a black ring around the horn. It should be clearly understood that no injury is done to the horn. The flies remain abundant until frosts kill them, commonly toward the middle of October in central Illinois, when the immature stages in the dung go into hibernation.

Control Measures.—Recently it has been shown that, when cattle are fed 4 to 5 grams of phenothiazine per hundred pounds of body weight, daily, the horn fly will not breed in their dung. This offers a cheap and sure method of reducing the numbers of horn flies about dairy herds or beef cattle, under conditions where the drug may be fed. Darkened stables, with curtains or brush arranged over the entrance to brush the flies off as the cattle go in, give a measure of relief. If the cattle are being fed grain, hogs and chickens running with them help to control the flies by scattering the dung and destroying the maggots. The dung may be scattered with a rake, hoe, or brush. Some countries have introduced dung beetles, which feed in the manure and help to keep down the flies. Repellent livestock sprays (see page 291) give temporary relief from these flies and are useful at milking time. Goats are annoyed much less if they are pastured apart from cattle.

References.—U.S.D.A., *Bur. Ento. Cir.* 115, 1910; *Jour. Econ. Ento.*, **21**: 494–501, 1928, and **31**: 315, 316, 1938; *Proc. La. Acad. Sci.*, **4**: 129–135, 1938.

322. Other Bloodsucking Flies

The large bloodsucking horse flies or greenheads (Figs. 539 and 540), often punish cattle severely, especially if the pastures border woods or wet areas of land. Mosquitoes may dry up dairy cattle and cause loss of flesh in other animals, especially in wet seasons, when they come out of the water in swarms and bite the animals day and night. Draining wet land, screening stables, and providing smudges will give relief from horse flies and mosquitoes. Black flies (Fig. 573) attack about the eyes, nostrils, ears, and under the belly and may be sucked into the mouth or nostrils by snorting animals, which sometimes die as though suffocated by the innumerable gnats. Smudges appear to be the only practical help for this scourge. Further discussion of black flies is given on page 881. The stable fly (Fig. 541), while generally less serious on cattle than the horn fly, in times of great epidemics may exceed the latter as a cattle pest. It bites chiefly about the legs. Control measures are the same as given under Horses.

323. OTHER FLIES ABOUT CATTLE

A number of flies, such as the house fly, that do not pierce the skin, visit the animals to suck up blood exuding from wounds made by horn flies, stable flies, horse flies, and others. They annoy the animals and may easily have a connection with the transmission of blood parasites from one animal to another. Other flesh flies besides the screwworm fly sometimes deposit eggs or maggots in wounds where the maggots develop.

CATTLE LICE

As in the case of the horse, two very different kinds of lice are common on cattle, some of them sucking the blood, and others, which cannot pierce the skin, living off the dry skin scales, hairs, and scabs. They are especially injurious to calves and to poorly fed, unhoused, old animals, during the winter months. The lack of oiliness of the skin of such animals makes conditions ideal for lice. Holsteins are said to be the worst infested and Jerseys the least so.

324. CATTLE CHEWING LOUSE OR LITTLE RED LOUSE[1]

Importance and Type of Injury.—When cattle rub and scratch against stanchions, fences, and other objects, and show areas of the skin which are full of scurf and partly denuded of hairs, or raw and bruised from rubbing, they are almost certainly infested either with lice or with scab mites. If lice are the cause of the irritation, parting the hairs and folds of the skin on head, neck, and shoulders will reveal the lice, while if the trouble is due to scab, no insects will be visible to the naked eye. The lice which suck blood (Fig. 549) are all a bluish slate color, while the chewing louse (Fig. 548) is yellowish white with a reddish head and eight dark crossbands on the abdomen, giving a somewhat ladder-like appearance to this part of the body. The chewing lice crawl about freely over the skin between the hairs, irritating the skin both with their sharp claws and with their sharp chewing mandibles. When very abundant, they form colonies about the base of the tail or on the withers, which may become covered over with a light scurf, in patches as big as the hand. Under this scurf the lice are feeding on the raw skin. Such gross attacks weaken the animals, check growth, and predispose the animals to other diseases.

Animals Attacked.—It is the general rule that each kind of animal has its own kinds of lice that do not feed on any other animal. Cattle lice have also been recorded from deer but do not attack other domestic animals.

Life History, Appearance, and Habits.—Lice are most abundant during the winter, when the coat becomes thick and long and the skin is relatively dry of oil. At this season all sizes of lice and the eggs are to be found on an infested animal. The eggs are delicate, white, barrel-shaped objects, glued by one end to a hair while the other end has a slight rim within which fits a lid that is pushed off when the egg hatches. The young

[1] *Bovicola bovis* (Linné) (= *Trichodectes scalaris* Nitzsch), Order Mallophaga, Family Trichodectidae.

louse is paler in color but of the same form and structure as the old ones. It appears from the egg within a week after laying, under favorable conditions, and within 2 weeks the young louse may have become full-grown (about $\frac{1}{12}$ inch long) and be laying eggs for another generation. Consequently they usually increase to great numbers during the winter and early spring, if not controlled. When the coat is shed in spring and during the heat of summer, the lice seem to disappear, but enough remain to carry the species over until favorable conditions permit them to increase again. In one series of 692 lice collected from calves in Illinois, only 6 were males.

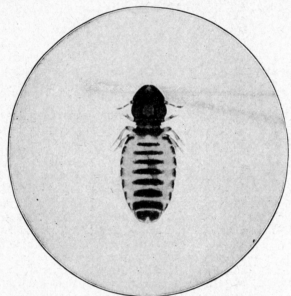

Fig. 548.—Cattle chewing louse or little red louse, *Bovicola bovis* (Linné); female about 20 times natural size. (*From U.S.D.A., Farmers' Bul.* 909.)

Control Measures.—Lice cannot live off the animal for more than 7 or 8 days, but the eggs that become scattered about on shed hairs may hatch at any time up to 3 weeks and the nymphs crawl back upon the animal. Thoroughly dusting the animal's coat with sodium fluoride, wettable sulphur, or powdered derris, timbo, or cubé, as recommended for the horse chewing louse (page 819), will greatly check if not eradicate the chewing lice. Treatment with raw linseed oil, as described under blood-sucking lice, is probably the best control for a limited number of animals, and dipping, with one of the arsenical or nicotine dips, or, if good soft water is available, a coal-tar creosote preparation, in specially constructed vats, is practical for large herds.

References.—U.S.D.A., *Farmers' Bul.* 909, 1918; *Conn. (Storrs) Agr. Exp. Sta. Bul.* 97, 1918.

325. Cattle Bloodsucking Lice

Importance and Type of Injury.—The symptoms of bloodsucking lice are similar to those shown when the little red louse is present, but the irritation is greater because of their habit of piercing the skin or "biting" to get the blood. The loss of blood keeps young animals runty and prevents normal production of milk or meat in older ones. The animals scratch and rub persistently and patches of the skin become bare of hairs and are sore.

Life History, Appearance, and Habits.—If an animal infested with bloodsucking lice is examined during the winter, dark-blue patches on the skin, often as big as a half dollar, which at first look like dirt, may be found on folds of the skin, on head, neck, withers, or along the inner surfaces of the legs. Examining more closely will show that these spots are composed of clusters of little lice of all sizes (Fig. 549), standing on

Fig. 549.—Short-nosed cattle louse, *Haematopinus eurysternus* (Nitzsch); female, about 20 times natural size. (*From U.S.D.A., Farmers' Bul.* 909.)

their heads, clinging by their claw-like legs to the hairs, and with noses appressed to the skin, from which they draw blood. They move about very little except when laying their eggs. The largest of them are only ⅛ inch long, and successively smaller ones are generally present, down to the size of the egg. All stages are passed on the skin of the host, the eggs usually hatching in 10 days to 2 weeks. The nymphs feed frequently, and the females begin to lay eggs when they are about 2 weeks old. The eggs are glued fast to the base of the hairs by one end, and are somewhat keg-shaped.

The long-nosed cattle louse,[1] which is especially prevalent on young cattle, is a slender species about ⅒ inch long, the body one-third as wide as long, and the head nearly twice as long as broad and pointed in front.

---

[1] *Linognathus* (= *Haematopinus*) *vituli* (Linné), Order Anoplura, Family Haematopinidae.

Its eggs are nearly black. The short-nosed cattle louse[1] (Fig. 549) is the largest louse found on cattle, being ⅛ inch long and much broader than the long-nosed louse, and the head is only a half longer than broad and bluntly pointed in front. It is said to be more common on mature cattle. The eggs are white. The third species[2] is similar to the short-nosed cattle louse, with the head still shorter and rounded in front. It is only about half as large as the short-nosed species,[3] and its eggs, which are yellowish, are said not to hatch if removed from the host.

Control Measures.—Lice are most prevalent on poorly fed animals. Cattle should be well fed, kept in clean light well-ventilated stables and not overcrowded. As soon as they are brought off pasture in the fall, the neck and withers should be examined for the presence of lice. These will be especially easy to see on white or light-coated animals. If even a few lice are found, all animals in the herd should be treated with raw linseed oil or commercial summer-oil spray emulsions. Many owners make it a practice to give the treatment even if no lice are seen at this time. The oil may be applied with a hand brush that has some bristles long and some short. A pint of oil is enough to groom 4 or 5 cows, and 10 or 12 animals can be treated in an hour. At least one more application should be given, 12 to 15 days later; and three or four additional ones at similar intervals throughout the winter will assure a herd free from injury by lice. The oil should not be rubbed in too vigorously, but just enough applied so that it reaches the skin. The animals should be kept out of strong sunlight for at least 12 hours after treatment and not overheated in any way or exposed to extreme cold. Boiled or refined linseed oil should not be used. Failure to observe these precautions may result in burning the skin. Where the number of animals is too large to permit hand applications, they may be sprayed lightly, with raw linseed oil or dipped in nicotine or creosote dips. This should not be done in cold weather, but two or three dippings in the fall, 2 weeks apart, will practically insure clean animals the following winter. Complete directions for dipping are given in *U. S. Department of Agriculture, Farmers' Bulletin* 909, which should be studied before dipping is undertaken. Before treated animals are returned to their quarters, the stalls and pens should be thoroughly cleaned and disinfected by spraying with a coal-tar preparation or with kerosene emulsion.

Reference.—*U.S.D.A.*, *Farmers' Bul.* 909, 1918, and *Cir.* E-447, 1938.

326 (319, 339). SCAB MITE OR PSOROPTIC SCAB[4]

If animals that are rubbing and scratching, twitching the skin, and shaking their heads, do not show lice upon examination, cattle scab should be suspected. An examination of skin scrapings under a microscope will usually reveal the minute whitish eight-legged mites (Fig. 563) that cause this trouble by puncturing the skin with their sharp mouth stylets. They are similar to the mange mite but can be distinguished by their longer legs and minor differences in the appendages. They do not burrow into the skin but rest and feed upon the raw skin, completely

---

[1] *Haematopinus eurysternus* (Nitzsch), Order Anoplura, Family Haematopinidae.
[2] *Solenopotes capillatus* Enderlein.
[3] BISHOPP, F. C., *Jour. Agr. Res.*, **21**: 797–801, 1921.
[4] *Psoroptes communis bovis* Hering, Order Acarina, Family Sarcoptidae.

covered over by scabs. Infestations usually begin on thickly haired parts of the body. The earliest symptoms are small reddened pimples that ooze pus. As the mites increase, larger areas become covered with yellowish crusts filled with serum. Large scabs form on the skin over the mites and the hair comes out in great patches. This mite and its control are further discussed under Sheep, page 855. Valuable animals should be dipped a number of times.

327. Follicle Mite[1]

A minute worm-like mite sometimes attacks cattle, burrowing deep into the natural pores of the skin and causing lumps as large as peas to form about the head and over the shoulders. It injures the hides. The mites cannot easily be killed, and it is best to kill or market badly infected animals. Very valuable animals may be cured by persistent dipping, as for mange.

328. Cattle Tick[2]

Importance and Type of Injury.—The most destructive parasite of domestic animals throughout the southern states is the cattle tick (Fig. 550). For more than a century this pest has held back the cattle and dairy interests of the southern states and, indirectly, the entire agricultural development of these states. The cattle tick is a bloodsucking parasite and is injurious to cattle in the several ways described above for lice and flies. In addition to this it is the only means of spread of the disease known as cattle fever, tick fever, splenetic fever, or Texas fever, from sick animals to healthy ones. The principal symptoms of tick fever are a high fever, reddish discoloration of the urine, enlarged spleen, congested liver, loss of flesh and condition, dry muzzle, arched back, and drooping ears. Death results in from 10 per cent of the cases occurring during the summer months, in southern cattle, to 90 per cent of those during the autumn and early winter, especially among northern or imported stock. The cause of the disease is a minute form of animal,[3] belonging to the phylum Protozoa, which lives inside the red corpuscles of the blood and destroys them, thus causing the disease. It should be clearly understood that while the *direct* cause of the disease is a protozoon, the only method of spread of the disease is by the bite of the tick. If there are no ticks in a given section, there can be no Texas fever. Ticks do not pass from one animal to another, but spend their entire feeding period on a single animal. The germs of Texas fever pass through the eggs of diseased ticks to their young, and the young later get on new animals and so infect them. The presence of this disease has meant incalculable loss to the southland from the failure of diseased cattle to make beef production or dairying profitable, and especially from the "one-crop system" which has prevailed largely because of the difficulty and discouragements of cattle-raising in tick-infested territory. According to the Department of Agriculture,[4] the average price of cattle in the

---

[1] *Demodex folliculorum bovis* Stiles, Order Acarina, Family Demodecidae.

[2] *Margaropus* (= *Boophilus*) *annulatus* Say, Order Acarina, Family Ixodidae.

[3] *Piroplasma bigeminum* (Smith and Kilbourne).

[4] *U.S.D.A., Weekly News Letter*, Sept. 15, 1915.

tick territory is only about half that for similar grades in tick-free states. The hides are damaged from 50 cents to $1.25 each by the punctures of the ticks, and milk yield is reduced from 18 to 42 per cent, depending on the severity of the infestation.   All in all, it is generally estimated that the tick costs southern agriculture about $2,500,000 a year.   The punctures made by ticks attract the screwworm fly (page 843) to lay her eggs upon the cattle.

Animals Attacked.—The cattle tick is chiefly a pest of cattle but is found also on horses, mules, sheep, and deer.   It does not transmit Texas fever to the other animals.

Distribution.—The cattle tick occurs only in the southern half of North America, in the states south of West Virginia, Kentucky, Missouri, Kansas, Arizona, New Mexico, and Nevada.   The disease that it transmits is prevalent also in Central and South America, Africa, Europe, and the Philippines, where it is carried by other kinds of ticks.

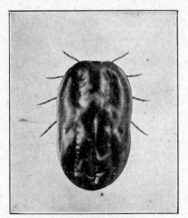 

Fig. 550.—Cattle tick or Texas-fever tick, *Margaropus annulatus* Say.   At left, full-grown engorged, female tick, ready to drop from animal and deposit eggs; at right, female laying eggs on the ground.   Each female may lay as many as 5,000 eggs.   Both figures 3 times natural size.   (*From U.S.D.A., Farmers' Bul.* 498.)

Life History, Appearance, and Habits.—The cattle tick winters in two distinct ways in the south.   Some individuals winter as eggs or minute six-legged "seed ticks," about half the diameter of a pinhead ($\frac{1}{32}$ inch). They may be found on the ground or on grass, weeds, and other objects in fields where infested cattle have pastured.   Other individuals winter as nymphs or adults on the skin of the cattle.   When the ticks reach maturity they mate on the animal, and then the female, having fed to repletion on blood, loosens her mouth parts from the skin and drops to the ground. At this time she is nearly $\frac{1}{2}$ inch long, the body olive-green to bluish-gray, bean-shaped, with hard, glossy, and somewhat wrinkled skin (Fig. 550, *left*).   Within 2 to 6 weeks after dropping from the animal, each female ordinarily lays from a few hundred to 5,000 eggs, pushing them out in front of her head in a large brownish mass on the soil.   The eggs usually hatch in from 13 days to 6 weeks, but eggs may remain viable for 7 or 8 months.   The young that hatch from the eggs are known as larvae or "seed ticks," and have only 6 legs (Fig. 551).   They crawl about actively on the ground and climb up on vegetation, often in large bunches, and remain there waiting for an animal to come along.   If an animal brushes by them they cling fast and at once insert their mouth parts and feed in that one spot for a week or two, then shed their skins and appear as nymphs with an extra pair of legs, now having eight for the rest of their lives.   Another period of feeding for about a week enables them to molt and transform to adult males and females.   After mating, the female, still feeding, increases rapidly in size for a few days to a few weeks before she drops to the ground to lay eggs for another generation.   If the seed tick is not brushed off the vegetation, it must starve to death but may live 200 to 250 days before it perishes.   There are thus four stages in the life of a tick: the egg, the seed tick, the nymph, and the adult.   The life cycle is in two periods, a *parasitic* and a *nonparasitic* period.   The parasitic period begins when the tiny seed tick is picked up by an animal, and ends when the fully-engorged female drops from the host.   Part of the adult female lifetime, all the egg stage, and most of the seed tick's life are spent on or near the ground.   A life cycle requires about 60 days, at the least, and there are three generations throughout most of the South.

FIG. 551.—Larvae or seed ticks of the cattle tick, just after hatching. Eight times natural size. (*From U.S.D.A., Farmers' Bul.* 498.)

Control Measures.—A quarantine maintained by state and federal governments prevents the movement of southern tick-infested cattle out of the quarantined area. The cattle tick cannot survive the winter in the northern states.   Without the tick there can be no fever, so that cattle in the north, which are very susceptible, are protected from the disease by the quarantine.   In the South two measures of control have gone hand in hand: (1) dipping or spraying animals with arsenical dips or oil emulsions. It has been found that a farm can be completely freed of ticks by dipping all cattle, horses, and mules on the farm every 2 weeks, all spring and summer, and allowing them to range over the pastures and pick up the seed ticks, between dippings; (2) rotating pasture lands, *i.e.*, keeping all animals off certain fields until the seed ticks will all certainly have starved, and then putting on such land only cattle freed of ticks by dipping.   Both of these measures have been worked out with the greatest of care and are fully described in *U. S. Department of Agriculture, Farmers' Bulletin* 1057.   By a combination of quarantines, dipping, and pasture rotations, in operation since 1906, over 700,000 square miles of southern land (98 per cent of the area originally quarantined) have been freed of the curse of Texas fever.   With the exception of about 9 per cent of the area of Florida and about 6 per cent of Texas, all the originally quarantined territory had been released from this embargo by December, 1938. It is expected that this work will continue until the entire country is free of the disease.

References.—*U.S.D.A., Farmers' Buls.* 1057, 1926, and 1625, 1930, and *Bur. Ento. Bul.* 72, 1907, and *Misc. Publ.* 2, 1927; *Tenn. Agr. Exp. Sta. Bul.* 113, 1915.

329 (340, 361, 362). Other Ticks

Several other species of ticks attach to cattle and suck their blood, but none of them carry Texas fever or other cattle diseases so far as known. The most serious species is the spinose ear tick[1] which is prevalent in the semiarid sections of southwestern United States. These ticks have the curious habit of entering the outer ears of cattle, horses, mules, dogs, sheep, and other animals, where they live and feed for from 1 to 7 months. The ears may be literally packed full of these vermin, which range up to ⅓ inch or more in length. After this period of feeding in the ear of the animal, the full-fed nymphs (Fig. 552), which have the skin covered with short heavy peg-like hairs, leave the host, crawl into cracks about stables, fences, or trees, and there shed their skins and become adults. The adult stage is not found on animals but, after mating, the females lay eggs in dry places usually up off the ground. The newly hatched "seed ticks" subsequently attach to the skin of the animals which they infest. The Gulf Coast tick[2] has very similar habits. These ticks are so irritating to animals that they run about shaking their heads or roll on the ground until they become exhausted, ill, deaf, or die from the attack. The wounds made by the tick in feeding often become infested by the screwworm fly. Nymphal Gulf Coast ticks are frequently found on quail, meadow larks, and other ground birds and may cause their death. The ticks may be killed by injecting ½ ounce of a warm mixture of 3 parts pine-tar oil and 1 part linseed oil, by means of a syringe, into each ear. This should be repeated every 30 days to kill the ticks present and repel others. The dog tick,[3] which is often found on dogs, commonly attaches to cattle pastured in woodlands. The Rocky Mountain spotted-fever tick (page 892) and the lone-star tick[4] also attack cattle, in regions where these ticks occur.

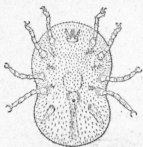

Fig. 552.—Spinose ear tick, *Otobius megnini* (Duges), nymph, ventral view, greatly enlarged. Note the mouth parts or capitulum, consisting of hypostome, chelicerae, and pedipalps (*cf.* Fig. 72) on the under side of body between front legs; and the covering of short peg-like setae which give the name to the species. (*In part after Nuttall, Monograph of the Ixodoidea.*)

References.—*U.S.D.A., Farmers' Buls.* 980, 1918, and 1150, 1920; *Jour. Econ. Ento.*, **29**: 1068–1076, 1936.

330. Ox Warbles or Cattle Maggots

Two different species of botflies attack cattle in America: the smaller one is known as the heel fly,[5] or common cattle maggot,[5] or striped ox-warble fly;[5] the other is known as the bomb fly[6] or northern cattle maggot.[6] There are very important differences in the habits of the two species, but their life histories and control are sufficiently similar to make their consideration together advantageous.

Importance and Type of Injury.—Tumors under the skin of the back, as big as the end of one's thumb, contain each a large fat maggot (Fig. 553), which may be squeezed out, but which emerges when "ripe" and falls to the ground. Hairy flies about as big as honeybees chase the animals in pastures, while laying their eggs on them. The cattle usually run

---

[1] *Otobius* (= *Ornithodorus*) *megnini* (Duges), Order Acarina, Family Argasidae.
[2] *Amblyomma maculatum* Koch.
[3] *Dermacentor variabilis* (Say), Order Acarina, Family Ixodidae.
[4] *Amblyomma americanum* (Linné), Order Acarina, Family Ixodidae.
[5] *Hypoderma lineatum* (De Villiers), Order Diptera, Family Oestridae.
[6] *Hypoderma bovis* DeGeer.

wildly (Fig. 6), with tails in the air, and often injure themselves in their attempts to get away. The small maggots from these eggs tunnel into the skin and migrate through the body for about 6 months before they find lodgment in the back, causing inflammation and suffering; milk

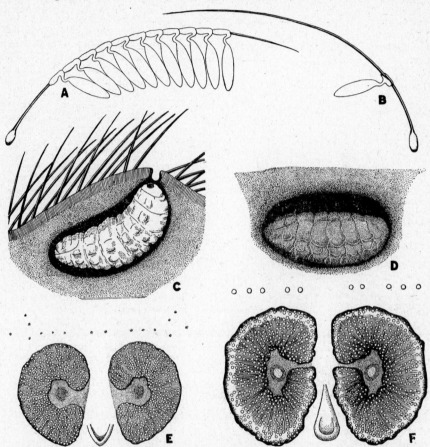

FIG. 553.—Ox warble, life stages. *A*, Eggs of the heel fly as often laid in a row on the hairs; note curved petiole arising near distal end of clamp. *B*, eggs of the bomb fly, as usually laid one to a hair; note straighter petiole arising near middle of clamp. *C*, larva in last instar in the tumor beneath skin of back; note hole through skin for respiration, with posterior end of larva bearing stigmal plates next to the hole. *D*, puparium in the soil in which the transformation from larva to fly takes place. *E*, posterior stigmal plates of the heel fly, *H. lineatum*, from fifth-instar larva; note that the plates are flat, with a rather wide gap on their inner margins. *F*, posterior stigmal plates of fifth-instar larva of the bomb fly, *H. bovis;* note that the plates are funnel-shaped, and have a much narrower gap on their mesal margins. *A* and *B*, about 20 times natural size; *C* and *D*, enlarged about one-third; *E* and *F*, greatly magnified. (*Drawn by Kathryn Sommerman, in part after U.S.D.A. publications.*)

production is reduced; growth and fattening checked; the quality of the meat is lowered, where the maggots tunnel through the flesh of the back; and the value of the hides for leather is greatly depreciated. The loss from this insect has been estimated to run from $50,000,000 to $120,000,-

000 a year in the United States alone. Bishopp estimates that 75 per cent of the cattle in the United States are infested with from 1 to about 100 warbles each. In certain experiments a gain in milk flow of 25 per cent resulted when the grubs were removed from the back, and 5 per cent gain in weight was made by animals from which grubs were extracted, as compared with similar animals in which they were allowed to develop to maturity.

Animals Attacked.—Cattle and bison and possibly some other wild ruminants and very rarely horses, goats, and man.

Distribution.—The heel fly is the more prevalent and widely distributed species, occurring in every state in the union. The bomb fly does not occur in the southern states but is recorded as the more common species in the northeastern states. It ranges from coast to coast in southern Canada and the northern United States.

FIG. 554.—Heel fly or striped ox-warble fly, *Hypoderma lineatum* (De Villiers); male, about 4 times natural size. (*From U.S.D.A., Dept. Bul.* 1369.)

Life History, Appearance, and Habits.—Many erroneous theories and statements about the life cycle of ox warbles are current in print and among herdsmen. Consequently the true life cycle, which has now been established after more than 200 years of study, should be carefully noted. All ox warbles winter as maggots (Fig. 553,*C*), in the backs of animals. In the latitude of central Illinois the first ones begin to appear under the skin of the back in late December. From that time on until April the swellings can easily be felt by passing the hand over the backs of infested cattle. The presence of the maggots causes the formation of a cyst about them, and constant irritation by the spiny maggot keeps the walls of the cyst inflamed, and the warble lives on the secretions or suppuration inside the cyst wall. Soon after reaching the back, the larva cuts a small hole through the skin through which it takes air. Since maggots have spiracles at the posterior end of the body, this "tail" end lies against the hole in the hide. Any one larva spends about 2 months under the skin of the back, growing rapidly until it is about 1 inch long and very thick. It is significant that the larvae at this time live in that part of the body from which the most valuable steaks are secured and the holes they make are through that part of the skin which makes the best leather. When mature, the larva changes from white to dark brown or black and finally squeezes its body, tail foremost, through the hole in the skin and drops to the ground, usually in the morning. The larvae become mature and leave the back over a period of 5 or 6 months, from January to June in the

North, and from November to March in the South. Having freed itself from the animal the maggot seeks protection on or in the soil and changes, through a hard-skinned pupal period, to the adult fly, which splits open the end of its case and emerges about 5 weeks later. The adults of the heel fly are found in Illinois from early April to the end of June and the bomb fly chiefly from early June to mid-August. Some of the flies (Fig. 554), therefore, may be found about cattle from April to September although any one fly probably lives only a few days and takes no food. In southwest Texas Bishopp reports the heel fly as active throughout the warm periods of the winter.

The eggs of these flies are never laid on the back near the spot where the tumors appear, but mostly about the legs and lower part of the belly. The bomb fly is active chiefly on hot sunny days and is very tactless in laying its eggs. It darts at the animal with much buzzing and clings to the skin for a second while it glues an egg fast to the base of a hair. Then it retreats and in a few minutes strikes again. This quickly excites the animals, though it does not seem that it could hurt them, and they throw their tails in the air and start to run wildly. The fly follows until it loses its victim in the underbrush or in water. As many as 800 eggs may be laid by one fly, one to a hair here and there over the hind legs, especially near the hock and the knee, and also on the belly. This sudden stampeding of the herd is one of the serious phases of injury by this fly and the thing which suggested the name "bomb fly." The heel fly is sneaking in its egg-laying. It generally alights in the shadow near an animal, and if the animal is standing, quietly backs up and tucks its eggs among the hairs of the "heels" or, if the animal is lying down, attaches them to such hairs as it can reach along the flanks, while standing on the ground. Egg-laying causes no pain and the fly is usually unnoticed, so that it may lay a dozen or more eggs on one hair in a row like the teeth of a comb (Fig. 553,*A*). In from 3 days to 1 week the eggs of either species hatch, and the minute larva crawls down the hair and bores into the skin near its base. Penetration of the skin takes several hours and causes much irritation to the animal. The tiny larvae then disappear from view, at a point usually on the legs, and the heel fly larvae are commonly next seen in butchered animals, lying in the walls of the pharynx, gullet and esophagus, especially from August to November. By this time they are second- and third-instar larvae, one-fourth to one-third grown. The maggots are constantly on the move, thus avoiding the attacks of white blood corpuscles, and there is every reason to believe their visit to the esophagus is only incidental to their general wandering about in the connective tissue of the body (Fig. 555). The larvae of the bomb fly are rarely found in the gullet, probably taking a more direct route to the back where they are frequently found inside the neural canal. At the approach of winter the heel fly maggots begin to disappear from the gullet and by the end of December, the warbles begin to appear under the skin of the back as third-instar larvae about ½ inch long. In this stage the hole is cut through the skin, an operation requiring about 4 days. The larva then molts to the fourth instar and, 2 to 8 weeks later, to the fifth and final instar during which most of the growth takes place. According to Bishopp, the heel fly larva spends an average of 58 days in the larval stage after reaching the back, and the bomb fly larva an average of 73 days in this situation. There is but one generation a year.

Control Measures.—It will be clear from the life cycle of ox warbles that they can most easily be attacked when the maggots are in the backs of cattle, from December to June. It should be clearly understood, however, that destroying the maggots at this time is destroying them after they have done most of the damage that those particular individuals will do. Unless, therefore, the job is done thoroughly enough and over wide enough territory to cut down the fly population materially during the following summer little good will result. All the stockmen over a considerable area should cooperate to make the cleanup most effective.

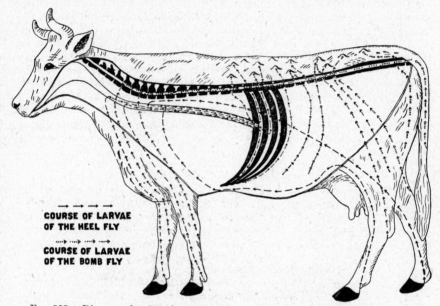

COURSE OF LARVAE
OF THE HEEL FLY

COURSE OF LARVAE
OF THE BOMB FLY

Fig. 555.—Diagram showing the course of ox-warble larvae through the body of the animal, from the position where the eggs are laid on the legs or flanks of the animal to their final feeding place beneath the skin on the back. The solid arrows suggest the course of the larvae of the heel fly, *H. lineatum*, to the gullet, thence backward to the diaphragm and up to the back, or to the ventral ends of the ribs and thence beneath the pleura up to the back. The dotted arrows suggest the course of the larvae of the bomb fly, *H. bovis*, from the point where the eggs were laid, directly to the back, or through the neural canal inside the backbone and thence to the back. This species is not found in the gullet. (*Drawn by Kathryn Sommerman.*)

The backs of all cattle in the community should be treated with a chemical that will kill the larvae and, in some cases, cause them to desert the host before they die. Many materials have been recommended for this purpose but at present the one in highest favor the world over is derris or other rotenone-bearing material. These treatments are applied either in the form of ointments, powders, or washes. An effective wash may be made of derris or other 4.5 per cent rotenone powder, 1 pound, mixed in a gallon of water in which has been dissolved ¼ pound of soap. The wash or powder may be applied liberally, all over the backs of the cattle, with a brush or sponge, as soon in midwinter as the first lumps appear under the skin, which will vary from early December in the South to the middle of February in the most northern areas. Four more treatments should be given at intervals of 4, 4, 5, and 5 weeks. The ointment containing derris root is applied by rubbing a small quantity into the top of each lump. Stabling of cattle during the daytime and running them on pasture only at night is a complete preventive measure, since the eggs are laid only in daytime, and the flies do not enter stables. Darkened sheds or brush shelters, into which the animals can retreat when flies are pestering them, are highly recommended. It is a common practice to squeeze the larvae out of the backs of animals by pressure from the hands, or remove them with sterilized forceps, but this is likely to crush the maggots and it is considered dangerous to burst them under the skin because of the poisonous nature of the body fluids of the maggots, which may cause anaphylactic shock. Suction pumps have been used to draw out both the maggots and the pus that surrounds them. If the warbles are removed mechanically they must be killed before being thrown down. No practical method of control is known for cattle on the range but if herds can be kept on the move, as from lower to higher ranges as the season progresses, they will leave the larvae which drop from their backs far behind and escape attack by the egg-laying flies that develop from them.

References.—U.S.D.A., Dept. Bul. 1369, 1926; *Can. Dept. Agr., Health of Animals Branch Bul.* 16, 1912, and *Sci. Ser. Buls.* 22, 1916, and 27, 1919; *Jour. Agr. Res.,* **21:** 439–457, 1921.

331 (333, 369). SCREWWORM FLY[1]

Importance and Type of Injury.—The presence of the maggots of flies in the living bodies of man or other animals is called *myiasis.* While a number of kinds of flies such as the horse bots and ox warbles attack

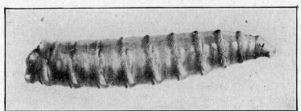

FIG. 556.—Larva of screwworm fly, side view, about 4 times natural size.   (*From U.S.D.A., Farmers' Bul.* 857.)

perfectly healthy animals in this way, certain others parasitize animals only when there is a wound or diseased body opening to attract the egg-laying female flies. Among the most serious of these pests is a dark, shiny blue-green blow fly, about twice as large as the house fly, with three black stripes on the back between the wings, and a reddish-yellow face (Fig. 557), which lays eggs only about the edges of wounds on animals, such as barbed-wire cuts, scratches from fighting, blood spots where ticks or other flies have bitten the animal, brand marks, sore eyes and wounds from dehorning or castrating. The maggots of these flies (Fig. 556) start to feed in the wounds but soon invade the sound tissue, tearing it with their mouth hooks. Wounds are prevented from healing, and the sickened animal hides away in the woods or brush, refusing to eat and usually dying if not found and treated. The odor of an infested wound attracts

[1] *Cochliomyia americana* Cushing and Patton (= *C. hominivorax* Coquerel), Order Diptera, Family Muscidae.

additional flies to lay eggs about it and hundreds of maggots may produce a terrible sore. Other species, such as *Cochliomyia macellaria* (Fabricius) and blow flies (see Fig. 11), which normally lay their eggs upon the carcasses of dead animals, may also attack wounds already infested by the screwworm, but this species is said to cause 90 per cent of the primary invasions

in the southern states. How long the animal lives depends upon the number of larvae present and the location of the wound. If the infestation is in the eyes or nasal passages or follows dehorning, meningitis frequently follows and kills the animal. Infestations about the navel of a newborn animal frequently result in peritonitis and death. A very serious epidemic, beginning in 1932, resulted by 1934 in over 1,350,000 cases of infestation and the death of over 200,000 animals in the Gulf states alone. Losses are estimated to reach $10,000,000 in certain years.

Animals Attacked.—Cattle, hogs, horses, mules, sheep, goats, man, dogs, and other domestic and wild animals.

Distribution.—Resident only in the most southern parts of the United States and southward to Argentina, this insect may range

Fig. 557.—Adult screwworm fly, *Cochliomyia americana* C. and P.; about 5 times natural size. (*From U.S.D.A., Farmers' Bul.* 857.)

northward in summer to the Carolinas, Illinois, Kansas and California and even southern Canada. It is probably always brought into these northern areas in infested animals, but may then breed there for several months, until killed by cold weather.

The screwworm situation in America illustrates forcefully the absolute necessity of taxonomic work, by which entomologists learn to distinguish closely related kinds of insects. It was not until 1933 that Cushing and Patton (*Annals of Tropical Medicine and Parasitology,* **27**: 539–551, 1933), discovered that two species, almost indistinguishable in appearance but very different in habits, have been confused under the name of the screwworm fly. It used to be believed that screwworms become epidemic and attack living animals only after they have increased to prodigious numbers by breeding in unburied carcasses. It is now believed that one of these species, *Cochliomyia macellaria* (Fabricius), never starts an infestation of wounds and is of little importance, though it may lay eggs in wounds already infested by other maggots. This species normally breeds in unburied carcasses and transfers its attack to living animals only when it has become excessively abundant. The important species is the newly discovered American screwworm fly, *Cochliomyia americana*, which is responsible for most of the maggot-infested wounds of warm-blooded animals but which apparently never lays its eggs except about wounds of living animals.

Life History, Appearance, and Habits.—The adults are active throughout the winter in extreme south Texas and Florida and begin to appear on the wing from early April to mid-June farther north. The eggs of the

American screwworm fly are laid only upon the dry skin near a wound or infected body opening, in regular shingle-like masses, firmly cemented to the skin and to each other. They hatch within 10 to 20 hours, and the small maggots tear out pockets in the healthy flesh adjacent to the wound with their sharp mouth hooks, severing minute blood vessels, and continually secreting a toxin that completely prevents healing and promotes contamination resulting in very foul-smelling pus-discharging wounds. The larvae are of typical maggot shape (Fig. 556) about ⅔ inch long when full-grown, with elevated spinose circlets at each segment, somewhat suggesting the ridges of a screw. Since they breathe through spiracles located at the large, blunt, posterior end of the body, they must frequently back out of the deeper recesses of a wound until the spiracles contact air. The larvae commonly become full-grown in from 4 to 10 days, when they drop to the ground and spend 3 to 14 days or longer in a brown seed-like puparium in the soil, transforming to adults. Under favorable conditions a generation averages about 3 weeks, and it is believed that there are commonly 8 to 10 generations during a summer.

Control Measures.—The most important control measure is to prevent the breaking of the skin of animals or the flowing of blood during the warm periods of the year when the egg-laying flies are active. Dehorning, castrating, branding, earmarking, docking of lamb tails, and similar operations should not be performed in spring or summer. Dogs should not be allowed to bite livestock. Barbed wire, projecting nails, and similar snags that may tear the skin should be avoided. In areas where the flies hibernate in winter, animals should be bred so that the young will be born between mid-November and April first. In the most southern areas and for all summer births, the navels of the young and the vulva of the dams should be treated with pine tar oil, the latter both before and after the birth. Every effort should be made to prevent insect bites which are often followed by screwworm attack. Unavoidable wounds should be treated immediately by covering with pine tar oil repeated daily until healed or covered with finely ground crystals of diphenylamine, every third day. The Bureau of Entomology and Plant Quarantine recommends (1) that infested wounds be very carefully cleaned and made as dry as possible by swabbing with cotton and not probed with sharp instruments to get out living or dead maggots. (2) That the wound be next syringed with commercial 90 per cent benzol to retard the flow of blood and pus. (3) That a plug of cotton moistened with the benzol be carefully inserted in the wound. The fumes of benzol kill the larvae. (4) That "acid free dehydrated pine tar oil of specific gravity 1.065 to 1.085" be applied over the cotton plug and around the wound and repeated every day or two, to aid in healing and keep the flies from laying more eggs. Recently the Bureau of Entomology and Plant Quarantine has recommended the use of finely ground, crystaline diphenylamine applied liberally in and over wounds, instead of pine tar oil. Although not repellent to the flies, this material kills the larvae unless they are more than three days old. (5) If a bloody discharge appears after the first treatment it indicates that all larvae have not been killed and steps (1) to (4) or an application of diphenylamine, should be repeated. Persons buying cattle from southern states should have the animals inspected by a competent agent to make sure they are not infested or that they are treated before they are shipped.

References.—*U.S.D.A.*, *Farmers' Bul.* 857, 1926, and *Dept. Bul.* 1472, 1927; *Jour. Agr. Res.*, **31**: 885–888, 1925; *Jour. Econ. Ento.*, **19**: 536–539, 1926, and **30**: 735–743, 1937; *U.S.D.A.*, *Tech. Bul.* 500, 1936, *Leaflet* 162, 1938, and *Cir.* E-481, 1939.

C. INSECTS ATTACKING HOGS

So far as its insect parasites are concerned, the hog sustains its reputation of being a hardy and healthy animal. It has only two serious insect parasites, the hog louse and the mange mite, besides which it is attacked to some extent by several kinds of flies.

FIELD KEY FOR THE IDENTIFICATION OF INSECTS INJURING HOGS

A. *Free-flying insects that alight on the animals to suck blood, coming and going repeatedly:*
1. Flies of the size and appearance of the house fly often cluster about the ears and in other places on the body and suck blood from the animals, causing much pain. They can be recognized by the slender, stiff beak that projects forward from the lower side of the head; by the broad abdomen, that is gray with 4 to 6 black spots; and by the palps, that are less than half as long as the beak....
....*Stable fly*, page 816.
2. Heavy-bodied, swift, brownish to black flies, ½ to 1 inch long, with two clear or banded wings, alight on animals and suck their blood, usually leaving a drop or two of blood exuding from the puncture. Especially troublesome about marshes, swampy woods, or meadows, and on warm sunny days.............
....*Horse flies*, page 814.
3. Very small, stout, black flies, less than ⅙ inch long, with a humped thorax, thick 11-segmented antennae as long as the head, 2 broad delicately veined wings, and stout legs, pierce the skin and suck the blood especially about the eyes, ears, and nostrils, causing extreme pain.............................
.......................*Black flies*, turkey gnats or buffalo gnats, page 881.
B. *Wingless insects that crawl or jump upon the body to bite and suck blood, but do not spend their entire lives on the animal:*
1. Small, brown, wingless insects, about 1⁄16 inch long, very flat from side to side and with long hind legs, jump vigorously when disturbed...................
..........................*Cat flea, dog flea, human flea*, and others, page 889.
C. *Wingless insects that stay on the animals all the time:*
1. Hogs rub and scratch. Large, flattened, grayish, wingless lice, up to ¼ inch long, with head, thorax, abdomen, and legs bordered with black; head and legs long, the latter with a peculiar hook at the end to clasp about hairs; mouth parts withdrawn into the head when not sucking blood; especially abundant in folds of the skin. The large yellowish-white eggs are glued to the hairs on lower half of body. The only louse of the hog..............*Hog louse*, 847.
2. Hogs rub vigorously. The hair stands erect, the skin about the ears, on top of the neck, on the withers, and down the back to the base of the tail becomes cracked and scabby. Caused by minute short-legged rounded mites, just big enough to be seen, that burrow into the skin..........*Mange mite*, page 848.
3. Tender skin about muzzle, eyes, or inner side of legs forms small hard pimples from the size of a pinhead to that of a small marble, which are filled with a yellowish pus....................................*Follicle mites*, page 850.
D. *Spiny worms or maggots that live in the flesh of wounds:*
1. Slender, whitish maggots, up to ¾ inch long, blunt behind and tapering to a point in front, with elevated spiny rings about the segments giving them a screw-like appearance, and the dark, posterior tracheal trunks visible through the skin, burrow into the flesh of wounds and abrasions or into the body through its natural openings. The eggs are laid about scratches, wounds, or natural body openings by a dark bluish-green fly, about ⅜ inch long, with three black stripes on thorax and with a reddish face...*Screwworm fly*, pages 843 and 848.

332. Hog Louse[1]

Importance and Type of Injury.—The only louse found on the body of hogs is a bloodsucking louse (Fig. 558), very similar in appearance to the short-nosed cattle louse, but about twice as large when mature. It reaches a length of nearly ¼ inch and is the largest bloodsucking louse found on any farm animal. On account of its size, it is easily seen, although its color is a dirty gray-brown, almost matching the skin of the hog. The margins of the body and appendages are bordered with black. The lice torment the hogs by piercing the skin to suck the blood. This causes the animals to rub vigorously against feed troughs and fences and to scratch with their feet. The skin becomes thick, cracked, tender, and sore, the animals restless and unprofitable.

Animals Attacked.—Hog lice do not infest other kinds of livestock.

Life History, Appearance, and Habits.—The lice (Fig. 558) are most noticeable on hogs in cold weather. In winter they usually cluster in small clumps on the inside of the ears or in folds of skin about the neck or on the inside of the upper part of the legs. Big and little together, they cling to the hairs by their legs, which are adapted to clamp about the hair very securely. They feed frequently, puncturing the skin each time with very slender stylets, which are completely withdrawn into the head when not feeding. Egg-laying goes on all winter long, a female laying from three to six eggs a day, gluing each fast to a hair close to the skin. The eggs (Fig. 558) are big enough to be seen, elongate, the smaller end glued by one side to the hair, the other end with a rounded cap. They are whitish in color when fresh, but after a few days become stained yellow or brownish. Most of the eggs are found on the lower half of the body. In 2 or 3

Fig. 558.—Female hog louse, *Haematopinus suis* (Linné), and egg attached to hair. About 6 times natural size. (*From U.S.D.A., Farmers' Bul.* 1085.)

weeks the small louse pushes off the cap of the egg, seeks a tender place on the skin, and sucks blood until satisfied. It then withdraws its mouth parts and soon bites in another place. The young are of the same shape as the adults, but pale-colored. In 2 weeks, during which time they molt three times, they are full-grown, and mating and egg-laying begin again. The females live about 5 weeks, during the last 3 of which they lay eggs almost every day. There are probably six to a dozen or more generations a year. All stages are passed on the host. The lice never voluntarily leave a hog except when they can crawl directly upon the body of another hog. If dislodged from the animal they rarely live more than 3 days.

Control Measures.—Where the number of hogs in the herd is not too large, the best method of treatment is to apply a thin even coat of oil all

---

[1] *Haematopinus suis* (Linné), Order Anoplura, Family Haematopinidae.

over their bodies with a fine-bristled brush. Especial attention should be given to the inside of the ears, the folds of the skin about the neck, and the inner surface of the thighs, to be sure that all lice and eggs are wet with the oil. Any of the following mixtures is satisfactory: (*a*) crude petroleum, (*b*) raw linseed oil, (*c*) half and half kerosene and lard, or (*d*) equal parts kerosene and cottonseed oil. Every animal in the herd should be treated and again in 2 weeks. Hogs must be kept out of the sun and not driven or excited, for a day after oiling. A simple and effective method of treatment is to apply crude oil or used crankcase oil directly to the hogs by means of a sprinkling can. Two or three applications should be made at intervals of 2 or 3 weeks. If animals cannot be treated by hand, they should be dipped in natural crude petroleum or the processed fuel oil or in one of the commercial coal-tar-creosote dips. Suggestions for dipping and detailed plans for a dipping vat are given in *U. S. Department of Agriculture, Farmers' Bulletin* 1085. The Department of Agriculture also recommends the use of medicated hog wallows. These should be made of concrete and contain 3 or 4 inches of water, to which is to be added, once a week, a quart of crude petroleum for each hog. A day or two after each application of oil, the wallow is drained and the hogs are given water without oil the rest of the week. A 1 per cent solution of pine tar in water may be used instead of the petroleum on water. Such a wallow should be built in a shady place, all other wallows done away with, and the oil added in the evening, but not until the hogs are accustomed to using the wallow with water alone. The so-called hog oilers, consisting of an oil-saturated fabric of some kind wrapped about a post, or commercial metal hog oilers, against which the hog may rub and so wet its skin with oil, are not considered very effective, because they cannot treat the parts of the body worst infested. The lice cannot live off the host more than a few days, and do not breed in the bedding.

References.—U.S.D.A., Farmers' Bul. 1085, 1920; *Tenn. Agr. Exp. Sta. Bul.* 120, 1918.

333. FLIES

The stable fly (Fig. 541) bites hogs severely about the head and ears or at cracks or scratches in the skin. Besides the controls given on page 818, a deep, dusty furrow in which hogs may lie should be provided. Horse flies, (Fig. 540), black flies, (Fig. 573), and mosquitoes also bite hogs when abundant. These pests are discussed under Horses and Cattle. The screwworm fly (Fig. 557) attacks all kinds of wounds, scratches, and sores on hogs, laying its eggs around them, and the maggots invade the sound flesh and prevent the wound from healing. Treatment is the same as given under Cattle.

334 (359). FLEAS

These well-known little pests (Fig. 578) often increase to great numbers about hog lots, doubtless making life very uncomfortable for the hogs, as they do also for any persons who come near. Mules and horses kept near infested hog lots are often annoyed. A thorough cleaning of the hog lots and spraying the soil with a dormant tree-spray oil, 1 part to 9 parts of water, or a good stock dip, will be effective in destroying the fleas on the ground. If preferred, a coating of salt or crude flake naphthalene may be used. Fleas are further discussed as pests of man, page 889.

335 (318, 366). MANGE MITES[1]

Importance and Type of Injury.—When hogs are scratching and rubbing vigorously and their hair is standing erect, if an examination does

[1] *Sarcoptes scabiei suis* Gerlach, Order Acarina, Family Sarcoptidae.

not reveal the large gray hog lice, it is probable that the animals are infested with mange mites (Fig. 559). If the skin about the eyes, ears, and along the top of the neck and back is inflamed, scurfy, scabby, with small pimples, or cracked and raw, scrapings from the skin should be made. It is necessary to scrape with a dull knife until the blood starts, to get the mites out of their burrows. Spread the scrapings out over some

Fig. 559.—A severe case of hog mange, showing pimple-like scabs caused by the mites. (*From U.S.D.A., Farmers' Bul.* 1085.)

dark surface and examine under a hand lens or magnifying glass, in a warm place. If minute, pale-colored, nearly round, eight-legged mites (Fig. 544) are found crawling among the scrapings, the hog is infested with common mange, which is further discussed under the insects attacking horses. This mite may also attack man (see page 900).

Control Measures.—The control of mange is the same as for hog lice, but very thorough applications are necessary. One dipping in crude petroleum is said to eradicate both lice and mange. Lime-sulphur dips are also effective for mange but are not very effective against hog lice.

Mange is very contagious, and all animals in the herd should be treated, whether they seem to be infested or not. Healthy animals must be kept so that they cannot touch infested ones until the latter are treated, and pens harboring infested animals must be disinfected, since the mites or their eggs may live off the animal for several weeks.

Reference.—U.S.D.A., Farmers' Bul. 1085, 1920.

336. OTHER MITES AFFECTING HOGS, DOGS, CATS, RABBITS, AND FOXES

Sometimes hogs, dogs, cattle, and man become infested with another kind of mite[1] which lives in the hair follicles about the muzzle, eyes, base of tail, or on the tender skin on the inner sides of the legs. The skin becomes red and inflamed, and small hard pimples ranging in size from that of a pinhead to lumps as big as a small marble form and break, discharging a yellowish cheesy pus. The trouble is caused by a very minute, slender, worm-like mite, with four pairs of very short legs. Secondary infection by Staphylococcus bacteria frequently adds to the seriousness of the infection by these mites. A related species[2] causes the serious red mange of dogs, which is characterized by bare inflamed spots about the eyes, ears, and joints of the legs. The greatly elongate worm-like mites live deep in the hair follicles and cure is almost impossible. Ear mange mites,[3] affecting cats, dogs, rabbits, and foxes attack the skin inside the ears, causing great irritation, deafness, and inability to coordinate movements. Sometimes the ears become packed with numerous laminae of dry scabs.

Control Measures.—On account of the fact that the mites burrow so deeply in the skin, there is no practical way to cure an infested herd. Badly infested animals should be killed, the others fattened for butchering and disposed of as soon as practicable. Since the mites may live off the host for several days, the premises should be disinfected with a 1-to-15 lime-sulphur spray, kerosene emulsion, or one of the commercial dips, before restocking with healthy hogs. For the ear mites, careful swabbing of the ears with glycerin containing 1 per cent phenol or sulphur in sweet oil is recommended to destroy the mites.

D. INSECTS AFFECTING SHEEP AND GOATS

FIELD KEY FOR THE IDENTIFICATION OF INSECTS INJURING SHEEP AND GOATS

A. *Free-flying insects that alight on the animals to suck blood, coming and going repeatedly:*
  1. Flies of the size and general appearance of the house fly suck blood especially from the legs, causing the animals to stamp their feet. Distinguished from the house fly by its stiff, slender beak, sticking straight forward from the underside of the head, by its broader black-spotted abdomen and grayer appearance, and by the bristle on antenna having hairs on its upper side only. Palps less than half as long as proboscis. Usually rests with its head upward.............
  ....*Stable fly*, page 816.
  2. Very small, stout, black flies, less than ⅙ inch long, with a humped thorax, thick 11-segmented antennae, as long as the head, 2 broad, delicately veined wings, and stout legs, pierce the skin and suck the blood, especially about the eyes, ears, and nostrils, causing extreme pain...............................
  ....*Black flies* or *buffalo gnats*, page 881.
B. *Insects that stay on the animal all the time:*
  1. A very small brownish louse with a broad red head, only about 1/20 inch long, is found among the wool and on the skin of sheep, causing them to rub and scratch.................................*Red-headed sheep louse*, page 851

---

[1] *Demodex phylloides* Csokor or *Demodex folliculorum suis* Stiles, Order Acarina, Family Demodicidae.

[2] *Demodex canis* Leydig.

[3] *Otodectes cynotis* Hering, *Chorioptes cuniculi* Zurn, and others.

2. A louse, up to $\frac{1}{12}$ inch long, the body tapering toward the front, the head as broad as long, and 2 transverse rows of hairs on each abdominal segment, is found among the short coarse hairs of the legs below the true wool..........
....*Bloodsucking foot louse*, page 851.
3. About the same size and appearance as *B*, 2, but occurs over entire body and face under the wool, sometimes in clusters. Slightly more slender than the foot louse, and the head twice as long as broad. Wool about the clusters of lice generally discolored by the small pellets of excrement......................
....*Bloodsucking body louse*, page 851.
4. Chestnut-brown wingless flattened tick-like insects, about $\frac{1}{4}$ inch long, with a leathery spiny skin, a sack-like unsegmented abdomen, short head, and 6 tapering, widespread legs, cling to the skin beneath the wool and suck blood. Often very severe on lambs. No exposed egg or larval stage; the rounded, brownish, ovate puparia, about $\frac{1}{8}$ inch across, are often wrongly called "nits." The maggots are born full-grown and glued to the wool, especially about the neck...............................*Sheep tick, louse fly* or *ked*, page 853.
5. The wool comes out in patches, "tagging" to weeds, fences, etc., on which the animals rub, leaving bare, scabby places on the skin. Under these scabs may be found thousands of minute 8-legged oval-bodied mites, just visible to the naked eye, which pierce the skin with their mouth parts, causing an exudation that hardens over them as a scab. They do not burrow into the skin........
....*Scab mite*, page 855.
C. *Maggots that live in the bony cavities of the head or under the wool or in sores:*
1. Animals are nervous, push their noses between other sheep or into the dust, paw with their feet, shake their heads, run with heads held low, and sneeze. There is a copious, foul discharge from the nostrils. This trouble is caused by a grayish-brown fly, about $\frac{1}{2}$ inch long, which deposits small active maggots in the nostrils. The maggots work upward through the nostrils and then tunnel through the nasal and frontal sinuses, frequently causing death. These larvae are fleshy, creamy-white, and brownish banded, finally reaching a length of about 1 inch...............
...........*"Grub-in-the-head," "staggers," sheep bot* or *sheep nose fly*, page 858.
2. Slender, whitish maggots, blunt behind, tapering in front, up to $\frac{3}{4}$ inch long, burrow in matted, soiled wool, or into the flesh, especially at points injured by "ticks" or other piercing insects or by dogs. Eggs are laid by a brilliant-green or a blue, metallic fly, larger than the house fly, about $\frac{3}{8}$ inch long..........
...................*Greenbottle fly, screwworm fly* or *black blow fly*, page 860.
3. Slender, whitish maggots, up to $\frac{3}{4}$ inch long, pointed at the head end and truncate behind, with elevated, spiny rings about the segments giving them a screw-like appearance and the posterior tracheal trunks dark and showing through the skin, feed and eat into the living tissue next to wounds..........
....*Screwworm fly*, page 843.

General References.—"Parasites and Parasitic Diseases of Sheep," *U.S.D.A., Farmers' Bul.* 1150, 1920; "Insect Parasites of Goats in the United States," *National Angora Record Jour.*, vol. 1, no. 1, September, 1922.

337. Sheep and Goat Lice

Importance and Type of Injury.—Three species of lice are known to live on sheep in this country: the bloodsucking body louse,[1] the bloodsucking foot louse,[2] and the red-headed or chewing sheep louse.[3] They are by no means so prevalent as the sheep tick. The usual symptoms of scratching and rubbing are caused by these lice biting and running over

[1] *Haematopinus ovillus* (Neumann), Order Anoplura, Family Haematopinidae.
[2] *Linognathus pedalis* (Osborn), Order Anoplura, Family Haematopinidae.
[3] *Trichodectes ovis* (Linné), Order Mallophaga, Family Trichodectidae.

the skin. The chewing lice eat off the wool fibers and tangle and soil the hair; the bloodsucking kinds rob the host of nutrition, and the bloodsucking body louse stains the wool with its small brown fecal spots. The red-headed sheep louse is said to be the most irritating, and the foot louse to be comparatively innocuous. Angora goats are injured and the clip of mohair and kid hair sometimes reduced as much as 10 to 25 per cent in value by at least five different kinds of lice. There are two kinds of bloodsucking lice,[1] which are blue-gray in color; and three species of chewing lice: a large yellowish one[2] and two reddish or orange species.[3]

Life History, Appearance, and Habits.—The red-headed sheep louse is the smallest and apparently the worst of the three species. It is only $\frac{1}{20}$ inch long, pale-brownish in color, with a broad, reddish head, broadly rounded in front. Each segment of the abdomen has a single transverse row of hairs. These lice crawl about among the wool and on the skin, eating wool fibers and skin scales. They do not suck blood but, when they cluster on the skin in great numbers, may cause raw sores. The bloodsucking foot louse is similar in form to the short-nosed cattle louse but only about $\frac{1}{12}$ inch long and somewhat slenderer and paler colored. The head is about as wide as long, and each abdominal segment has two transverse rows of hairs. The bloodsucking body louse is similar to the foot louse, but the head is twice as long as broad and bluntly pointed in front. All these lice spend all stages, eggs, nymphs of various sizes, and adults, on the body of the animal though they may live 3 days to a week if dislodged, and their eggs may hatch several weeks after being separated from the host. The eggs are glued fast to the hairs, and hatch in 5 to 10 days in the case of the chewing louse, or 10 to 18 days in the other species. The young are said to complete growth and begin laying eggs in about 2 weeks after hatching.

Control Measures.—All kinds of goat and sheep lice are perhaps best controlled by dipping every animal in the flock twice, with an interval of 11 days, and turning them into pens or pastures that have not been occupied for at least 4 weeks. One of the most satisfactory dips for lice may be made of 300-mesh wettable sulphur, 10 pounds, in water 100 gallons. If hard water is used, add trisodium phosphate, 10 to 25 ounces, depending upon the hardness of the water. Coal-tar-creosote dips are difficult to make with hard water and may stain the wool or mohair; the standard arsenical dips are poisonous, should the animals swallow any of the dip. In cold weather, dusts containing rotenone from powdered derris or timbo root may be used to reduce the number of both chewing and sucking lice. The chewing lice may be controlled by dusting sodium fluoride into the wool all over the animal. About 1 ounce is sufficient for each sheep. This stomach poison is not effective for bloodsucking lice. All new animals should be examined before they are turned into a healthy flock, and especial attention should be given to males, which often infect a flock of ewes at breeding time. There is also the danger of infestation from neighboring flocks along line fences.

References.—*U.S.D.A.*, *Leaflet* 13, 1928, and *Bur. Ento. Cir.* E-394, 1936, and *Farmers' Bul.* 1330, 1932.

[1] *Linognathus stenopsis* (Burmeister) and *L. africanus* Kellogg and Paine, Order Anoplura, Family Haematopinidae.

[2] *Bovicola penicellata* Piaget, Order Mallophaga, Family Trichodectidae.

[3] *Bovicola caprae* Gurlt and *B. limbatus* Gervais.

338. Sheep Tick, Louse Fly, or Ked[1]

Importance and Type of Injury.—The sheep "tick" (Fig. 560) is one of the most remarkable insects known. It is not a tick, like the Texas-fever tick or spotted-fever tick, but is a degenerate louse-like fly that has completely lost its wings. It crawls about over the skin among the wool and feeds by thrusting its sharp mouth parts into the flesh and sucking blood. It causes sheep to rub, bite, and scratch at the wool, thus spoiling the fleece; when this tick is abundant, the animals are unthrifty and

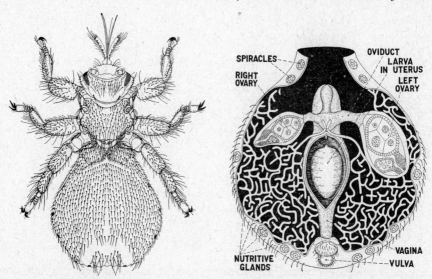

Fig. 560.—The sheep tick or ked. *Left,* dorsal view of adult, about 8 times natural size; note mouth parts, shielded by the maxillary palps, the compound eyes, and, in front of them, the concealed antennae; and, along the sides of the thorax and abdomen, the spiracles. (*Drawn by Kathryn Sommerman.*)

Right, dissection of the abdomen of the female, ventral view, to illustrate the viviparous reproduction. At center is a larva lying in the uterus, with its mouth toward the small nipple through which a nutritive fluid is passed from the elaborately branched, cylindrical glands that furnish the only food of the insect until the adult stage is reached. This larva originated from an egg produced in the right ovary (on the reader's left); while another egg is being produced by the left ovary, the two functioning in turn. When full grown the larva will pass from the vulva in the condition shown in Fig. 562,*A*. (*Drawn by Kathryn Sommerman, in part after Pratt.*)

unprofitable. Estimates by a large number of sheep owners some years ago indicate that the presence of this insect causes a loss amounting to 20 to 25 cents a head a year, on the average, in weight and wool production. If this estimate is correct, the sheep tick taxes the sheep industry of this country several million dollars a year on the approximately 50 million head of sheep kept by American farmers. They are especially severe on lambs, to which they migrate readily from the ewes especially at shearing time.

Animals Attacked.—Sheep and sometimes goats.

Distribution.—Where sheep and goats are kept throughout the world.

[1] *Melophagus ovinus* Linné, Order Diptera, Family Hippoboscidae.

Life History, Appearance, and Habits.—The second remarkable thing about the sheep tick is its method of reproduction. Unlike true ticks, which always drop to the ground to lay eggs, the sheep tick spends its whole life on the animals. Two life stages are commonly found on sheep at any season of the year. The adults (Fig. 560), which are grayish-brown, wingless, and six-legged, have a broad, leathery, somewhat flattened, unsegmented, sac-like abdomen covered with short spiny hairs. The thorax and head are much narrower; the legs are widespread, the first pair appearing to come out at the sides of the head. The body is about ¼

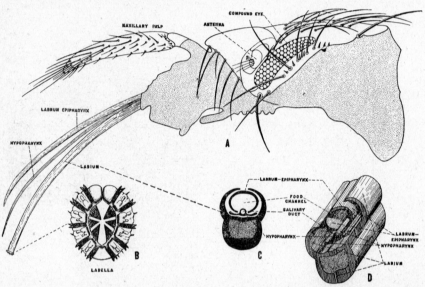

Fig. 561.—Mouth parts of the sheep tick—special biting fly subtype, essentially like those of the stable fly (*cf.* Fig. 67). *A*, side view of the head with the mouth parts protruded and separated to show the two stylets. Note the curious, concealed antenna in front of the compound eye. *B*, the labella greatly enlarged to show the prestomal, cutting teeth. *C*, diagrammatic cross-section and *D*, isometric projection of the proboscis with parts in normal position, showing how the labrum-epipharynx and the hypopharynx form the food channel, the salivary duct within the hypopharynx, and both stylets enclosed by the labium. (*Drawn by Kathryn Sommerman, in part after Jobling.*)

inch long and covered with short, spiny hairs. The other exposed stage of the sheep tick is the so-called "nit." It (Fig. 562,*E*) is a nearly round, chestnut brown, seed-like object that is glued fast to the hair, especially about the neck, inside of thighs, and along the belly. It is not an egg, but a pupal case or puparium enclosing the pupal stage of the fly. The sheep tick does not lay its eggs. The maggots are nourished within the body of the female until they are full-grown, never feeding externally. When born, they are whitish in color, oval, about ⅛ inch long, and without appendages. The female secretes a gelatinous glue which sticks the larva to the hair near the skin, especially about the neck and belly. Within 12 hours the skin turns brown and forms a hard puparium about the larva. Within this case, pupation takes place, and in summer the adult tick breaks out of its puparium from 19 to 23 days after it was born. In cold weather 3 to 5 weeks, or even longer, may be spent in the puparium. After emerging, the adults may mate in 3 or 4 days or not for several months. After mating, the females begin depositing full-grown larvae in from 8 to 30 days. Only one young is developed at a time, and each female produces from 10 to 20 maggots, which are born at the rate of about 1 a week. The female may live as long as 5 or 6 months. Sheep ticks never normally leave their hosts. If separated from the animal, they may live for a week or slightly longer, but most of them die in 3 or 4 days. The pupae, however, may live 1 or 2 months apart from the host, in warm weather. Freezing kills this stage. Breeding is continuous, though slower in winter, and there are probably several generations a year.

Control Measures.—The best control for sheep ticks is to dip the sheep in early fall in one of the coal-tar-creosote or nicotine dips used according to directions furnished with the materials. No dip which can be used will

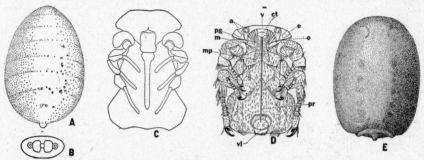

Fig. 562.—The sheep tick. *A*, new-born larva, as first glued to the wool of the sheep. *B*, the posterior stigmal plates of the larva. *C*, an early stage pupa, dissected from the puparium. *D*, an adult dissected from its puparium shortly before emergence: *a*, antenna; *e*, compound eye; *mp*, maxillary palp; *o*, occiput; *pg*, postgena; *pr*, proboscis; *vl*, vulva. *E*, puparium, formed by hardening of the larval skin, within which the larva transforms from larval to adult stage. All enlarged about 9 times. (*A*, *B*, *and E drawn by Kathryn Sommerman; C and D by Carl Weinman.*)

kill all the pupae. Still, it is possible by carefully timing the dips, completely to eradicate the ticks in two dippings, 24 days apart, in warm weather. The cost of dipping should not exceed a few cents a head for each treatment. Dipping cannot be done in cold weather and should not be done following shearing until all cuts have healed. After dipping, sheep should be kept from fields and pens previously used for from 1 month to 6 weeks. If sheep have not been dipped in the fall, and ticks become abundant in winter, they may be checked by several dustings of pyrethrum or derris powder sifted into the wool. Since the ticks may crawl a considerable distance in search of animals, it is not wise to leave wool that has been sheared near the flock. It should be stored at least 50 feet away.

References.—*U.S.D.A.*, *Farmers' Bul.* 798, 1932; *Wyo. Agr. Exp. Sta. Buls.* 99, 1913, and 105, 1915.

339 (319, 326). Scab Mite[1]

Importance and Type of Injury.—One of the most injurious and contagious of sheep diseases is the trouble known as scab or scabies. It is

[1] *Psoroptes communis ovis* Hering, Order Acarina, Family Sarcoptidae.

usually first indicated by "tagging," *i.e.*, the loss of bits of wool on weeds, fences, and other objects against which the animal has rubbed its body (Fig. 564).   The cause of this tagging or loss of hair is a very small eight-legged mite that punctures the skin with its sharp mouth parts until the lymph exudes and flows over it.   As this serum hardens to form a scab, the mites remain underneath on the raw skin where their continued feeding results in successive layers of scabs that eventually lift the hair out by the roots.   The irritation and loss of blood due to the mites and an extremely irritating poison introduced as they feed cause the sheep to

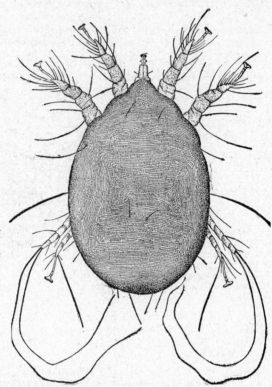

FIG. 563.—The scab mite, *Psoroptes communis ovis* Hering, female, greatly magnified. (*From U.S.D.A., Farmers' Bul.* 713.)

lose condition, become sickly, and, if not treated, to die.   In less severe cases the loss of wool is a considerable item.   The skin becomes first reddened, then white and glistening, then hardened and uniformly thickened over the infested part, and eventually bare of wool, cracked, and bleeding or oozing serum.   This trouble is sometimes called "wet mange," in contrast with the dry mange caused by Sarcoptic mites (page 821). Sheep which are rubbing or biting the skin or show a tangled condition of the fleece, should be examined for scab by a competent authority.

Animals Attacked.—Sheep and cattle suffer most, while horses and goats are attacked rarely.   Fine-wooled varieties of sheep are most seriously affected.

Distribution.—Nearly cosmopolitan; but quarantines and dipping campaigns have eradicated it from many sections.

Life History, Appearance, and Habits.—This mite feeds on the surface of the skin and does not burrow through the outer layers of the flesh, as does the mange mite. All stages of the scab mite will be found upon lifting a scab from an infested animal, the eggs, young, and adults swarming together under this protection. The largest mites (Fig. 563) are only as big as the cross-section of an ordinary pin or needle ($\frac{1}{50}$ inch in diam-

Fig. 564.—A severe case of sheep scab due to the attacks of the scab mite.   (*From U.S.D.A., Bur. Animal Industry.*)

eter), and may be seen as minute gray crawling specks against a black surface. A dozen or two of eggs are laid by the female in small clusters at the base of the hairs. They hatch in 2 to 10 days to six-legged nymphs. After the skin is shed the first time, the nymphs have an additional pair of legs. They grow rapidly and are mature in from 1 to 2 weeks. A female may live 1 month or more (may even survive away from a host for 2 or 3 weeks) and deposit up to 100 eggs. In wet cold weather generations succeed each other rapidly; the mites spread from sheep to sheep and whole flocks are soon affected.

Control Measures.—For a number of years quarantine measures have been in effect against the shipment of scabby sheep, and the disease has been greatly reduced in prevalence. Dipping with lime-sulphur, creosote, or nicotine dips is the most effective way to stop injury by this mite. The homemade lime-sulphur dip, described on page 822, may be used. Many very dependable commercial dips are available. These should always be used strictly as directed by the manufacturer. At least two dippings are necessary in cases of heavy infestation. They should be made at 10- to 14-day intervals. The sheep should be held in the dip for at least 2 minutes and the head ducked several times. Dipping is most effective shortly after shearing. After dipping, the animals should be kept away from pens or pastures, recently occupied, for at least 1 month. The pens should be sprayed with a strong phenol or creosote wash, such as 2 per cent cresol or 5 per cent carbolic acid, after having been thoroughly cleaned of all droppings, straw, and other refuse.

References.—U.S.D.A., Farmers' Bul. 713, 1916; *U.S.D.A., Farmers' Bul.* 1330, 1932, and *Bur. Ento. Plant Quar. Cir.* E-406, 1937.

340 (329, 361). TRUE TICKS

Although several species of ticks may attach to sheep to feed, they are not so injurious to sheep as to the less hairy animals. In fact, pasturing infested ground with sheep has been suggested as a control for ticks, since the latter may become entangled in the wool and killed. However, at least two species of ticks are injurious to sheep. The spotted-fever tick, which is discussed as a pest of man, page 892, causes a kind of paralysis in sheep by biting and engorging along the backbone. The removal of the large female ticks from the head, neck, and back, by hand, or the use of an arsenical dip results in rapid recovery from the disease. The spinose ear tick (page 838) also attacks sheep, feeding especially inside the outer ear. The lone-star tick[1] is very troublesome to goats, especially in the southwest.

341. SHEEP BOT OR NOSE FLY[2]

Importance and Type of Injury.—Sheep shake their heads, stamp their feet, and crowd together, holding their noses to the ground, especially in bare dusty places; or run away, with noses held low, in efforts to keep the fly from striking at their nostrils. The presence of the maggots in the nostrils and head sinuses causes inflammation, and a copious "catarrhal" discharge. The excess of mucus, together with irritating dust drawn into the air passages, causes sneezing and labored breathing. The presence of the maggots may cause giddiness or "blind staggers." Sometimes the flies may deposit their larvae in the eyes, nostrils, or mouth of man, sometimes causing blindness.

Animals Attacked.—Sheep, goats, wild deer, and, very rarely, man.

Distribution.—General throughout North America; especially troublesome from Idaho and Montana to New Mexico and Texas.

Life History, Appearance, and Habits.—The life history of this fly appears not to be well known. In Canada both small and full-grown maggots have been found in the heads of sheep, both in winter and in summer, indicating that some larvae may remain in the head for more than a year. In Russia the larvae are said to pass the winter in the second instar in a state of rest with little change, forming the third- and last-stage larvae in spring. The larvae may complete their development

---

[1] *Amblyomma americanum* (Linné), Order Acarina, Family Ixodidae.

[2] *Oestris ovis* Linné, Order Diptera, Family Oestridae.

in 2½ to 3½ months in lambs, but require much longer—at least 10 months—in older animals. The full-grown larva is nearly 1 inch long by a third as thick, without definite head or legs, and the segments prominently marked with blackish crossbands. Probably the typical cycle is for different larvae to become mature from early spring to late summer, when they retreat from the deeper tunnels they have made and drop or are sneezed out of the nostrils to the ground. Here a transformation period of 20 to 60 days is passed in the hard, blackish puparia before the flies emerge. In Texas adults are active every month of the year; farther north they are often present during 4 to 6 months in summer. They are about ½ inch long, of a general grayish color, with minute dark spots peppered over the back and with eyes rather small and a broad space

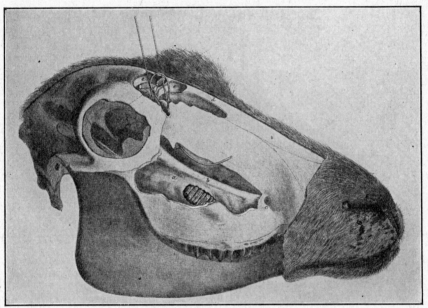

Fig. 565.—Sheep botfly, larvae (at *a* and *e e*) tunneling through sinuses of head of sheep. (*From U.S.D.A., Bur. Animal Industry.*)

between them. They take no food, but follow sheep on hot, still, sunny days and hide away in crevices of fences and walls mornings, evenings, and in inclement weather. The eggs hatch in the body of the female before they are laid, and a number of very active, small maggots, ⅒ inch long, enclosed in a little drop of sticky liquid, are deposited up the nostrils, as the female flies rapidly past the sheep's head without alighting. The maggots work their way upward through the nostrils until they reach the sinus cavities between the bony plates of the skull in the region of the forehead (Fig. 565). In other cases they find their way into the bronchi or into cavities in the horns or bones of the nose or jaw. They lacerate the tissues, growing slowly and sometimes migrating far about in the head but probably do not penetrate through the skull into the brain, as often supposed. As many as 80 larvae have been reported from one head, but the usual number is from 5 to 12.

Control Measures.—To prevent infestation by the sheep bot, the nostrils may be kept smeared with pine tar, applied weekly during the period when adults are active.   In small flocks the pine tar can be applied by hand with a brush.   The first application should be made about mid-April in the latitude of Illinois.   The sheep may be made to smear their own noses by an ingenious salt log, in which salt is placed in the bottom of 2- or 3-inch holes bored in the log and the edges of the holes smeared with pine tar.   Hadwen recommends the provision of a dark shed with a curtain over the door, into which the sheep may retreat when the flies begin to attack them.   Since the flies are short-lived and not capable of long flights, mobile herding or frequent change of pastures, which avoids concentrating the animals for grazing or watering in any place for more than a short time, is of some value in reducing infestations. The application of chemicals in the nostrils of sheep is of doubtful value and may be dangerous to the sheep.   A skilled veterinarian may remove larvae from the heads of valuable animals by opening the sinuses above the nasal fossae with a trephine.

342. Sheep Maggots or Wool Maggots

Importance and Type of Injury.—When the wool of sheep becomes soggy from warm rains, or soiled with urine, feces, or blood from wounds or from lambing, certain blow flies are attracted to the animal and deposit their eggs in the dirty wool, most commonly about the rump, but also about the horns where wounds have resulted from fighting.   The maggots feed in the wet wool and the adjacent skin, causing the latter to fester and the wool to loosen and become putrid; the inflamed, raw flesh, with the whitish maggots tunneling in it, is exposed.   The dirty wounds readily become infected and the sheep may die of blood poisoning.

Animals Attacked.—Sheep, goats, cattle, and other animals if they have putrid sores.   Greasy fine-wooled sheep, such as Merinos, are more likely to be infested.

Life History, Appearance, and Habits.—There are several kinds of flies that attack soiled wool, but two of the most important are the greenbottle fly[1] and the black blow fly[2].   The first species is about twice the bulk of the common house fly and of a brilliant metallic, bluish-green color with bronze reflections, without stripes, and with a fine whitish bloom on the front of the thorax just behind the head.   The black blow fly is a little larger, very dark greenish black in color all over, without stripes or grayish markings and not very bristly.   Both of these flies probably winter as larvae or pupae in soil beneath carcasses, or in manure.   At any rate the flies appear very early in spring, and from that time on, breeding is continuous except as checked by dry weather.   The larvae of the black blow fly are said to live chiefly in carcasses or carrion; the greenbottle fly develops also in garbage.

Either fly may complete a generation from egg to egg in about 3 weeks.   Their numbers increase as the season progresses and, during warm, rainy, or foggy weather, they are especially likely to lay eggs in the wool.   The pupal stage is passed in the ground.

Control Measures.—Since the flies attack animals chiefly after having become abundant by breeding in carcasses, the control measures are to destroy all carrion by promptly burning or burying it to a depth of at least 4 feet.   If possible, the carcass should be sprinkled with lime before covering it with earth.   Burning is better than burying.   Sweetened arsenical baits and trapping are of some value in reducing the numbers of flies.   Breeding hornless sheep; having the lambs come as early in spring as practicable; shearing before lambing occurs; docking of lambs; "tagging" dirty sheep; and applying pine tar to wounds are recommended as preventive measures. When sheep are infested, the wool should be clipped close around the infested area

---

[1] *Lucilia sericata* Meigen, Order Diptera, Family Muscidae.
[2] *Phormia regina* Meigen, Order Diptera, Family Muscidae.

and benzol applied to the infested areas to kill the maggots. The wound should then be dressed with pine tar oil as recommended for the screwworm (page 845), or treated with copper sulphate or air-slaked lime to dry up suppuration and deodorize it (see also page 905).

References.—U.S.D.A., Farmers' Bul. 857, 1922; *U.S.D.A., Tech. Bul.* 270, 1931.

E. INSECTS THAT ATTACK POULTRY

FIELD KEY FOR THE IDENTIFICATION OF INSECTS INJURING POULTRY

A. *Pests that visit the fowls only to secure food, coming and going repeatedly; or live on the fowls at night, hiding away in the daytime; or spend only part of the life-cycle on the fowls, being free-living or intermittent parasites during other life stages:*
  1. Mahogany-brown, broad, very flat or thin, oval, wingless bugs, of all sizes up to ⅕ inch long, live in nests, behind boards, and in cracks of houses during the daytime, and crawl out upon fowls at night and suck blood. Bugs have a bad odor. Small black spots of excreta from the bugs often seen on the eggs and about cracks...*Bedbug,* page 866.
  2. Small gnat-like humpbacked flies, less than ⅙ inch long, with thick antennae, as long as the head; 2 broad delicate wings and stout legs, hover about the heads of fowls on the roost, suck blood, and appear to smother the birds, when very abundant in spring.......*Black flies, turkey gnats,* or *buffalo gnats,* page 881.
  3. A tiny, hard, long-legged, jumping insect, about ½₀ inch long and flattened from side to side; the females attach to, or burrow into, the skin about the eyes, comb, wattles, or vent in clusters, often forming dark areas visible from some distance; ulcers, blindness, and death, especially of young chicks, often result. Immature stages are passed in cracks of the henhouse or in the soil. A southern species......................*Sticktight* or *southern chicken flea,* page 866.
  4. Small grayish to dark-red pear-shaped or ovate mites, from ¼₀ to ½₀ inch long, with 8 slender legs, remain in cracks under the roosts or in nest boxes during the day, except on sitting or laying hens in dark places. At night they swarm over the birds and suck the blood. .*Poultry mite* or *roost mite,* page 868.
  5. Similar to the poultry mite but females with tip of body slightly notched. The mites live on the fowls day and night, laying their eggs among the fluff feathers. Nymphs and adults suck blood............................... ....*Feather mite* or *tropical fowl mite,* page 869.
  6. In the South, poultry are attacked by a larger 8-legged oval-bodied brown tick, up to ⅓ inch long, which in the adult stage attacks the host only at night, when it sucks blood in quantities, and hides in cracks during the day like the poultry mite. In its younger stages, however, it is a permanent parasite, remaining on the fowl day and night until ready to molt.............................. ....*Fowl tick, adobe tick,* or *bluebug,* page 867.
B. *Small, wingless, flattened chewing lice that stay on the skin or feathers of the fowls all the time:*
  1. Ovate yellow lice, less than 1/16 inch long, with a single transverse row of hairs on each abdominal segment, above. Found along the shafts of the feathers of chickens rather than on the skin; when the feathers are parted, they run toward the body along the shaft. Chew at the feathers. Eggs glued to base of feathers. Not on young chicks...*Small body louse* or *shaft louse,* page 862.
  2. Similar lice, from ¹⁄₁₀ to ⅛ inch long, darker yellow and more hairy; the hairs on upper side of abdomen in 2 transverse rows on each segment. Found running rapidly over the skin of chickens, turkeys, and pheasants in less feathered parts; not on the feathers. The most injurious species on grown chickens. Eggs attached especially to small feathers below the vent........ ....*Large body louse,* page 862.
  3. Lice about the size of the shaft louse, but dark grayish in color and with a longer head, found standing on their heads among the feathers of the head on chickens; move but little. Hairs on upper side of abdomen mostly confined to

a wide median stripe.   Antennae of male unusually large.   The most injurious louse to young chicks.   Eggs laid singly on the down of the head.......
....*Head louse*, page 862.
4. Similar to the head louse but more slender and darker in color.   The only
   species found commonly on the large wing feathers of chickens, where it
   often lies between the barbules on the underside of the shaft, showing no
   signs of life.   Eggs between barbules of large feathers....................
   ....*Wing louse* or *variable louse, Lipeurus variabilis* Nitzsch.
5. A very large species, ⅛ inch long and more than half as wide.   Smoky gray
   to black in color; found on feathers of chickens and very active.   Not
   common.................*Large chicken louse, Goniocotes abdominalis* Piaget.
6. A large chewing louse, ⅛ inch long, with each hind angle of the head prolonged
   into a sharp process, at the end of which is a very long bristle; occurs on the
   feathers of the turkey, especially on the neck and breast...................
   ....*Large turkey louse, Goniodes stylifer* Nitzsch.
7. A louse of equal length with the preceding, but only one-sixth as wide as long;
   pale yellowish, with a black margin around the body; especially common on the
   primary wing feathers of turkeys..
   ....*Slender turkey louse, Lipeurus polytrapezius* Nitzsch.
8. A small species, only 1/25 inch long, with head curiously expanded and rounded
   in front; dark red in color, with a white region in middle of abdomen; is common at the base of the large wing feathers of ducks and geese.............
   ....*Chewing louse of ducks and geese, Docophorus icterodes* Nitzsch
9. A larger species, ⅙ inch long, slender, light yellow in color with a dark margin
   to the body and squarish dark spots on the abdomen.   Infests ducks and
   geese, especially at the base of large wing feathers.....................
   ....*Squalid duck louse, Lipeurus squalidus* Nitzsch.
10. A short broad species, 1/25 inch long, with abdomen squarish behind; whitish
    with a brown margin.   Infests pigeons........*Broad pigeon louse*, page 865.
11. An exceedingly slender louse about 1/12 inch long, with dark abdomen and
    reddish-brown head and thorax; very abundant on old pigeons and partially
    feathered squabs..........................*Slender pigeon louse*, page 865.
C. *Minute, almost invisible, 8-legged rounded mites that burrow into the skin beneath
   scales of legs or at base of feathers and feed and reproduce in the tunnels:*
1. Chickens, turkeys, pheasants, and other birds walk painfully or refuse to walk.
   Legs are encrusted with elevated scales from which a fine white powder and
   serum exude from the irritated and inflaméd skin, and the legs become much
   swollen.   Numerous minute, circular, very short-legged mites less than 1/50 inch
   long burrow under the scales............................ ....*Scaly leg mite*, page 870.
2. A similar but still smaller mite, burrows into the skin at the base of the feathers
   of the rump, back, abdomen, head, and neck of chickens and pigeons, causing
   the feathers to fall or to be pulled out by the bird.   If the stumps of such
   feathers are examined, an abundance of dry scales, crusts, and mites will be
   found.......................................*Depluming mite*, page 871.

General References.—N.Y. (Cornell) Agr. Exp. Sta. Bul. 359, 1915; *Ohio Agr. Exp.
Sta. Bul.* 320, 1917; *Conn. (Storrs) Agr. Exp. Sta. Bul.* 86, 1916; *U.S.D.A., Farmers'
Buls.* 801, 1931, and 1110, 1920.

343. POULTRY LICE[1]

Importance and Type of Injury.—Contrary to the belief of most
poultrymen, the lice that live on fowls do not suck blood.   They feed by
nibbling or chewing the dry skin scales, feathers, or scabs on the skin.
The irritation from the mouth parts, together with that of the sharp
claws on their feet in running about over the skin, results in a nervous

[1] Order Mallophaga.

condition of the infested birds that prevents sleep, causes loss of appetite and diarrhea, and renders the weakened fowls easy prey for various poultry diseases. Young chickens and turkeys that are brooded by lousy hens are often killed in great numbers by the swarming of lice from the hen to them almost as soon as they hatch from the eggs. The most serious effect upon older fowls is a reduction in the number of eggs laid. Infested fowls are in a mopey, drowsy condition with droopy wings and ruffled feathers, refuse to eat, and gradually become emaciated. If the feathers of such a fowl are parted, the lice will often be found running about on the skin in great numbers, particularly below the vent, on the head, or under the wings.

Animals Attacked.—Every kind of domestic fowl (and probably every kind of wild bird as well) has from one to several kinds of lice. In general, each species of bird has lice peculiar to it. The exceptions to this will be noted in discussing the different lice. At least a dozen kinds attack chickens and three to five different kinds are found on ducks, pigeons, and turkeys.

Distribution.—Wherever fowls are kept.

Life History, Appearance, and Habits.—Poultry lice generally breed faster and become more abundant in summer than in cold weather, but all stages can usually be found on the host in winter. All these chewing lice are permanent parasites, spending all life stages, generation after generation, on the same bird, and never normally leaving its body, except as they pass from one fowl to another, particularly from old to younger birds. The eggs are cemented fast to some part of the feathers. They are oval in shape, generally white in color, and often beautifully ornamented with spines and hairs (Fig. 78,*J*). While laid singly they may be abundant enough to form dense clusters on the fluffy feathers of badly infested chickens. In a few days or weeks the young nymph hatches from the egg in a form much like the parent lice only much smaller and paler in color. It at once begins running about and feeding, and in the course of the next few weeks passes through several molts, gradually assuming the size, form, and coloration of the adult.

Poultry lice are entirely wingless, six-legged insects with a much flattened body and broad head rounded in front. The mouth parts are near the middle of the underside of the head, the most prominent parts being two sharp-pointed teeth or mandibles. The legs are good-sized, and, in all of the species that live on birds, they have two claws at the end of the tarsus. Their relatives that live on hair-bearing animals, such as the little red louse of cattle or the chewing lice of horses and sheep, have only one tarsal claw, fitted for grasping about the hairs.

The head louse[1] (Fig. 566) is especially noticeable and injurious on young chicks and turkeys. The dark-gray large-headed adults, about $\frac{1}{10}$ inch long, and the paler young ones are found standing head down along the base of the feathers on top of the head with their mouth parts against the skin. They constantly nibble at the skin scales but apparently never eat through the skin or into the flesh. Although they move about only a little, they pass very early from brooding hens to little chicks, which are often killed by them. The eggs are cemented to the barbs of the down or small feathers of the head or neck. They hatch in about 5 days and the young are full-grown in about 10 days more. In an incuba-

---

[1] *Lipeurus heterographus* Nitzsch, Order Mallophaga, Family Liotheidae.

tor at a constant temperature of 33–34°C. the egg stage required 5 to 7 days, and there were 3 nymphal instars requiring a minimum of 6, 8, and 11 days, respectively, or a minimum of 30 days for a generation. It is generally believed that the young become full-grown in about 10 days upon the host (see *Journal Parasitology*, **6**: 350–415, 1934).

The chicken body louse[1] (Fig. 567) lives most of the time on the skin of either chickens or turkeys, being especially abundant about the vent and under the wings, and is common on both young and old fowls. When the feathers are parted, all sizes of the lice run rapidly to cover. The smaller ones are pale yellowish white but the larger ones, which reach a length of nearly ⅛ inch, appear brownish. The body is covered with fine long

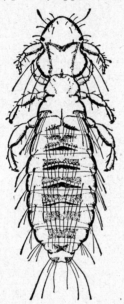

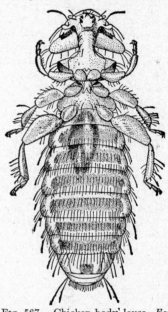

FIG. 566.—Chicken head louse, *Lipeurus heterographus* Nitzsch; male, about 25 times natural size. (*From U.S.D.A., Farmers' Bul.* 801.)

FIG. 567.—Chicken body louse, *Eomenacanthus stramineus* (Nitzsch); female, underside, about 25 times natural size. (*From U.S.D.A., Farmers' Bul.* 801.)

hairs. Bishopp and Wood consider this the most injurious louse of grown chickens, because it is constantly on the skin. The eggs are fastened to the basal barbs from the shaft of the feathers, especially below the vent. They are said not to hatch on feathers dislodged from the host. On the body they hatch in about a week, and 10 or 12 days of growth brings the nymphs to the adult stage. They increase very rapidly: Lawson and Manter record having counted over 35,000 lice on one chicken, which they think was not half of those actually present.

The shaft louse, small body louse, or common body louse[2] is similar to the large body louse in appearance, but it is distinctly smaller (1/16 inch

[1] *Eomenacanthus stramineus* (Nitzsch) (= *Menopon biseriatum* Piaget), Order Mallophaga, Family Liotheidae.

[2] *Menopon gallinae* Linné (= *M. pallidum* Nitzsch), Order Mallophaga, Family Liotheidae.

long), paler colored, and less hairy. It has commonly been considered the most injurious louse of chickens, but Bishopp and Wood contend that it lives mostly on the feathers, lying along the shaft and running down the feathers to the skin when the feathers are parted. It is very common about the vent, also on the back and breast. It does not infest young chicks, presumably because of the lack of well-developed feathers. The eggs are fastened to the base of feathers and hatch in 2 or 3 weeks. These lice are very hardy, having been kept alive for 9 months. This species occurs on ducks, turkeys, and guineas, at least when they are housed with chickens, and is sometimes troublesome to horses stabled near badly infested poultry.

The other lice on chickens are less abundant and less important, and live chiefly among the feathers, where their nibbling and crawling do not cause great annoyance. The lice that attack turkeys, geese, and ducks are said to be less abundant and generally not sufficiently injurious to require special treatment. Pigeons, however, often become grossly overrun with the broad pigeon louse[1] or the slender pigeon louse,[2] which live among the feathers of both old and young birds and doubtless interfere with the profitable raising of these fowls.

Control Measures.—It should be evident from the different life habits of lice and mites that to control *lice*, with the measures discovered up to the present time, as described below, it is necessary to apply some substance directly to the *body* of the fowl; whereas, with poultry *mites*, treatment of the *house* is essential. In both cases cleanliness, an abundance of fresh air and light, a good supply of drinking water, and the provision of good clean dust baths are important factors in keeping these external parasites and the injurious effects from them in subjection.

In the past 10 years commercial sodium fluoride[3] has been found to be so effective in the control of poultry lice that it is now generally accepted as the most important remedy. Fowls should be examined every few months, and, if found infested, every fowl on the premises should be treated, since a single one that escapes the cleanup may soon reinfest the entire flock. The sodium fluoride may be used either as a dust or as a dip (the latter should always be done early in the day), but in cold weather dipping is not advisable. The dip should be prepared in a wooden tub or large earthenware jar, dissolving 1 ounce of sodium fluoride and 1 ounce of fish-oil or laundry soap to each gallon of water. Hold the fowl by both wings with one hand, lower it in the water and ruffle the feathers for about ½ minute until they are soaked to the skin; then duck the head under twice, and the fowl is ready to release. In dusting with sodium fluoride, it is best to lay the fowl over a shallow pan or paper on a table. A "pinch" of the powder is sifted among the feathers next to the skin on the head, another on the neck, two on the back, one on the breast, one below the vent, one on the tail, one on each thigh, and one on the underside of each wing. The cost for materials in such treatment is not over a half cent per fowl. Instead of using the pinch method, the powder may be sifted among the feathers, next to the skin, on all parts of the body from a talcum box or other can with small holes in the top, giving especial

---

[1] *Goniocotes compar* Nitzsch, Order Mallophaga, Family Philopteridae.

[2] *Lipeurus baculus* Nitzsch, Order Mallophaga, Family Philopteridae.

[3] Care must be taken not to confuse this with sodium chloride, which is ordinary table salt, since the two names sound so much alike.

attention to the region around the vent. An ounce of the material is sufficient to treat from 50 to 100 hens. One treatment carefully given should completely eradicate the lice. In dusting, the lice are not all killed for several days, but when the fowls are dipped, the death of the lice occurs very promptly. An alternative treatment consists of applying a line of nicotine sulphate along the top of all roosts or perch poles shortly before the fowls go to roost, using 1 ounce to 15 or 20 linear feet. It may be applied with an oil can or paint brush to clean perches. This requires much less labor than treatment with sodium fluoride, and avoids handling of the fowls, but it costs considerably more for materials, does not kill the head louse, and *must be repeated* 8 to 14 days later to destroy lice hatching from eggs present at the first treatment. This measure cannot be used in freezing weather. The directions of the manufacturer should be carefully followed in using this material. Certain rotenone dusts and dips are effective if properly used. Chemicals, put in the drinking water to rid poultry of lice, have not been found of value. Mercurial ointment (50 per cent metallic mercury) thoroughly mixed with 1 to 2 parts of vaseline makes an effective ointment for head lice. A lump not larger than a pea should be thoroughly rubbed over the feathers on top of the head. Care should be used not to apply too much. Carbolated vaseline or even lard is recommended to be used in the same way.

344 (357). BEDBUG[1]

The common bedbug (see Fig. 576) and several of its close relatives[2] are frequently pests in poultry houses. They hide, breed, and lay their eggs in nests, behind nest boxes, under loose boards, and in other cracks about the walls, roosts, and roof of the building. At night the nymphs and adults find their way upon the sleeping hens and suck their blood. They are almost never found on the fowls in daytime. Sitting hens suffer especially from these pests and may be driven to leave the nests. The small black spots of excreta from the bedbugs may often be seen on the eggs and about cracks.

Control Measures.—Because of the likelihood of carrying these bugs into the house, quite as much as on account of their injury to the fowls, vigorous control measures should be applied. Effective measures are given in the chapter on insects affecting man (page 888). In most cases chicken houses can be rid of these pests by spraying all cracks thoroughly with creosote oil or crude petroleum. Dormant miscible oil emulsions such as are used on fruit trees, used 1 part to 9 parts of water, containing ½ pound of lye to each 10 gallons, make effective sprays.

345 (359). STICKTIGHT OR SOUTHERN CHICKEN FLEA[3]

Importance and Type of Injury.—In the South and Southwest poultry sometimes show clusters of dark-brown objects about the face, eyes, ear lobes, comb, and wattles made by hundreds of small flattened fleas that have their heads embedded in the skin so that they cannot be brushed off. Young fowls are often killed, and egg-laying and growth are greatly checked by the loss of blood and the great irritation caused by the bites. The ears of dogs and cats often become lined with them.

Animals Attacked.—Chickens, turkeys, and other poultry, and also cats, dogs, horses, and man.

Distribution.—Southern and southwestern United States from South Carolina to California.

[1] *Cimex lectularius* Linné, Order Hemiptera, Family Cimicidae.

[2] The Mexican chicken bug, *Haematosiphon inodorus* (Duges), the European pigeon bug, *Cimex columbarius* and the barn swallow bug, *Oeciacus vicarius* Horvath.

[3] *Echidnophaga gallinacea* (Westwood), Order Siphonaptera, Family Sarcopsyllidae.

Life History, Appearance, and Habits.—Adult males and females are found, often *in copula*, on the heads of fowls. The females, at least, remain attached by their mouth parts in the same spot sometimes as long as 2 or 3 weeks. During this time the eggs are laid, being thrown with considerable force from the vagina of the female. The eggs hatch on the ground in from 2 days to 2 weeks, and the slender white larvae feed on the excreta of the adult fleas and possibly other filth in the cracks and litter about the floor of henhouses or on the ground in dry protected places. After a growing period of 2 weeks to 1 month, they spin silken cocoons covered with dust and dirt, in which the pupal transformation occurs. The adults do not attach to the host for several days or 1 week and then a second period of about 1 week elapses before the females begin laying eggs. Only a few eggs, one to five, are laid at a time. The life cycle may be completed in from 1 to 2 months. The pest thrives best in dry cool weather, and the adults may live for several months under such conditions.

Control Measures.—Thoroughly clean the house and yards, burning the dirt and litter on the spot, or soaking it with kerosene or creosote oil; or cover the ground with a uniform thin layer of lime; or treat with kerosene or boiling water. Spray floors, dropping boards, and nest boxes with creosote oil. Derris powder may be applied directly to the fleas on the animal. Keep dogs, cats, and rats away from poultry houses, as they may spread this flea. Exclude fowls from beneath buildings.

Reference.—*Okla. Agr. Exp. Sta. Bul.* 123, 1919; *Jour. Agr. Res.*, **24**: 1007–1009, 1923; *Jour. Econ. Ento.*, **25**: 164–167, 1932; *U.S.D.A., Cir.* 388, 1934; *Jour. Agr. Res.*, **24**: 1, 1923.

346. Fowl Tick or Bluebug[1]

Importance and Type of Injury.—The injury and symptoms of attack by fowl ticks are much as in the case of poultry mites; weakness of the legs, droopy wings, pale comb and wattles, cessation of egg-laying and death from loss of blood. Small, rounded, reddish or dark-colored objects attach in clusters to the skin on neck, breast, thighs, or under wings, sucking blood and causing great irritation. Large, reddish or blue ticks up to ½ inch long are found in daytime under bark of trees where the fowls roost or under loose boards about the henhouse or where roost poles meet the walls; black spots of excrement stain the woodwork near such cracks. These large ticks suck their fill of blood in about ½ hour, and, when they are abundant, profitable poultry husbandry is out of the question unless control measures are applied. In addition to bleeding fowls, the fowl tick is the proved carrier of a highly fatal poultry disease known as fowl spirochaetosis in many parts of the world. There is danger that this disease may be introduced into North America.

Animals Attacked.—Chickens are the preferred host, but all other domestic fowls and some wild fowls are attacked, and rarely domestic mammals and man.

Distribution.—In the southwestern states from Texas westward to California, and in Florida. Apparently limited to warm, semiarid regions, around the world.

Life History, Appearance, and Habits.—In cooler regions the ticks appear to winter chiefly as adults and half-grown nymphs. Throughout most of its range breeding may continue slowly even in winter. They thrive best, however, in hot dry weather. The adults are flattened, leathery, eight-legged, with thin edges to the body, egg-shaped in outline, and red to blue-black in color. They range in size from about ¼ to ½ inch long. The brownish eggs are laid in cracks about the house. They hatch in from 10 days to 3 months, and the grayish six-legged nymphs seek a fowl, particularly at night, and attach in bunches to the skin, sucking blood from 3 to 10 days in one spot. They then release their hold and hide away in a protected place for about a week, molting and acquiring the fourth pair of legs. From this time on they suck blood only at night, hiding in cracks during daylight and alternately engorging and molting at intervals of a week or two until they become adult. The females deposit several lots of eggs, a total of 500 to 900. There is normally one generation a year, but if deprived of a host the adults may survive for 2 or 3 years without food or water.

[1] *Argas miniatus* Koch (= *A. persicus* Oken), Order Acarina, Family Argasidae.

Control Measures.—Control is much the same as for the poultry mite but must be given with extreme thoroughness because these flattened ticks can hide deeply in cracks and are most hardy and resistant creatures. Roosts should be suspended by wires from the ceiling or supported entirely from the floor so that they do not touch the walls. Nests should be so constructed that they can be easily removed and treated. Poultry must not be allowed to roost in trees, barns, and other shelters which cannot be treated effectively. Any new fowls brought into the flock should first be isolated for 10 days to allow the seed ticks to drop off, and the temporary quarters then treated or burned.

Reference.—*U.S.D.A., Farmers' Bul.* 1653, 1933.

347. Poultry Mite or Chicken Mite[1]

Importance and Type of Injury.—These mites (Fig. 568) live in cracks about the roosts, floors, walls, or ceiling of the houses in the daytime and

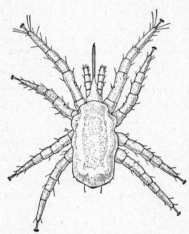

crawl upon the fowls at night or when they are on the nests. Only a very few mites are found on fowls during the daylight hours. Their only normal food is the blood of fowls, which they draw through their sharp piercing mouth parts at night. Since they rarely stay on poultry during the daytime, a flock may be badly run down by them without the owner's being aware of the cause of the trouble. Small areas about the roosts or elsewhere in the house show patches of gray, brown, or red mites or fine, black-and-white speckling as though dusted with pepper and salt—the excrement of the mites. When they become extremely abundant, the litter and manure may appear to be literally crawling with tiny gray-and-brown specks. The mites are especially abundant where the roost timbers rest on their supports. The fowls become droopy, pale about the head, and listless, and they stop laying. Sitting hens and chicks often die.

Fig. 568.—The poultry mite, *Dermanyssus gallinae* (DeGeer); nymph before engorging on blood. Greatly magnified. (*From U.S.D.A., Farmers' Bul.* 801.)

Animals Attacked.—Chickens are preferred, and other poultry is not likely to be badly attacked if chickens can be reached. The mites often greatly irritate persons or animals about infested houses.

Distribution.—Practically cosmopolitan; more troublesome in warmer and drier regions.

Life History, Appearance, and Habits.—The mites become inactive and greatly reduced in numbers during cold weather, except in heated houses, although feeding and breeding to some extent may take place during warm spells, even in winter. With the advent of spring, activities are greatly accelerated. The females deposit their pearly-white elliptical eggs in dark protected places, such as cracks between timbers or in dry manure under the roosts. Two or three dozen eggs are laid over a period

---

[1] *Dermanyssus gallinae* (DeGeer), Order Acarina, Family Gamasidae.

of several weeks. The eggs hatch in 2 to 4 days on the average. Several days more elapse, and the young six-legged nymph sheds its skin once and becomes eight-legged before it seeks a fowl and fills up on blood (Fig. 568). After engorging, it hides away in some dark place for a day or two or until it sheds its skin again. It then finds a host, engorges on blood a second time and again hides away for a day or two before it makes the third and final molt and becomes an adult. It requires only 1 week to 10 days to pass through the stages from egg to adult. The adults are $\frac{1}{40}$ to $\frac{1}{30}$ inch long, grayish in color but, when filled with blood, appearing bright red to nearly black. It is believed that the same adult does not feed every night, but after filling up on blood, leaves the fowl and during the next few days lays three to seven eggs, then feeds and repeats the process until all the eggs have been laid.

Control Measures.—For the control of these mites attention must be given to the house rather than to the birds themselves. A very thorough cleanup of the poultry house is essential, since the mites abound in filth and infest any crack about the house big enough for them to crawl into. This should be followed with a thorough spraying of crude petroleum, carbolineum, anthracene oil, crankcase oil mixed with half its volume of kerosene, or creosote. Certain oils used for dormant spraying of fruit trees are very effective, diluted 1 part to 9 of water. Use a pump that gives a strong penetrating spray, and treat the floor, the walls, nest boxes, and, in badly infested houses, the roof and outside of the walls. If the above oils are not available, kerosene alone may be used, but several treatments a week or 10 days apart will be necessary. Arsenical or coaltar stock dips are also recommended. The spray should be applied in the morning so the oil will have dried before the fowls go to roost. After the cleanup, the cracks of the roosts and the supports on which they rest should be brushed over once a month with crude petroleum or anthracene oil to catch any mites that escaped the first treatment. Mites may live 4 or 5 months in a house after all poultry has been removed, especially if surroundings are moderately moist. These pests may also stay on the bodies of fowls in small numbers for at most about 3 days. When moving fowls to a new building or introducing new birds to a flock, they should be isolated for 3 days and nights in pens so that the mites will all crawl away from them. Poultry houses should be so constructed as to receive plenty of light and air, to be dry, and to be easily cleaned. It is important that roosts, nest boxes, and all other boards and fixtures be arranged so they can easily be removed for cleaning and spraying outside the house. Roost poles are sometimes suspended by wires from the roof to prevent the mites crawling upon the roosting birds.

References.—*U.S.D.A., Dept. Buls.* 1652, 1933, 801, 1931, and 1228, 1924.

348. FEATHER MITE OR TROPICAL FOWL MITE[1]

Importance and Type of Injury.—Small, reddish or brown, eight-legged mites swarm over the skin, congregating about the vent, where they may cause bloody scabs to form. The feathers in this region become dirty from the eggs and excreta of the mites. Heavy infestations kill the fowls from loss of blood and irritation. This

---

[1] *Liponyssus bursa* (Berlese), Order Acarina, Family Liponyssidae. The very similar northern fowl mite, *L. sylviarum* C. and F., and *L. canadensis* Banks occur in parts of the northern states and Canada.

mite has been suspected of transmitting the dangerous tropical disease, fowl spirochaetosis.

Animals Attacked.—Chickens and other domestic fowls; English sparrow, starling, and other wild birds.

Distribution.—This pest has been known in this country only since 1916, but it is already widely distributed throughout the northern, eastern, and central states. The infestations have been local and many of them have been cleaned up.

Life History, Appearance, and Habits.—These mites are very similar, in a general way, to the poultry mite. They are slightly smaller, more hairy, and with smaller legs; they move more rapidly, and the tip of the abdomen of the female is slightly notched. In habits, however, they are very different, since they live day and night on the body of the fowl, instead of harboring in cracks about the house, and also lay their eggs on the fluff feathers. The eggs adhere to the barbules of the feathers and are also found in the nests.

Control Measures.—Fowls should be dipped during warm weather. Wood and Cleveland recommend a dip made by dissolving 1 ounce of soap in a gallon of water and stirring into it 2 ounces of fine flowers of sulphur. If the fowls are also infested with chewing lice, add 1 ounce of sodium fluoride. The fowls must be entirely submerged in the dip and the feathers ruffled to permit the liquid to reach the skin. Applications of nicotine sulphate to the perches as described for poultry lice (page 866) will control the mites if three applications are given 3 days apart. If an infestation is discovered in cold weather, the fowls should be dusted very liberally and thoroughly with very fine flowers of sulphur, to hold the mites in check until the weather is warm enough to permit dipping. Since the mite is known to live in the nests and on the bodies of English sparrows, these birds should be fought and their nests burned to prevent spread. New birds or show animals should be isolated before introducing them to a clean flock, and carefully examined to make sure they are not infested. If any mites are found, treat as suggested above. The nests used by infested fowls, and in cases of bad infestations the entire house, should be treated as described for the poultry mite.

Reference.—*Poultry Sci.*, **2**: 129–135, 1923; *U.S.D.A., Dept. Cir.* 79, 1920.

349. SCALY-LEG MITE[1]

Importance and Type of Injury.—This species and the following one are close relatives of the mites and ticks discussed earlier in this chapter, but their habits are quite different. They are more like mange or itch mites, since they remain on the body all the time and make tunnels in the skin, in which the eggs are laid. The scaly-leg mite attacks especially the feet and lower part of the legs but is also found about the comb and neck. The scales of the legs (Fig. 569) become elevated, and a fine white dust sifts from beneath them. Lymph and blood exude and red blotches form on the legs. The birds become crippled or even unable to walk at all. Great irritation must result from the burrowing of the mites.

Animals Attacked.—Poultry and wild fowls; and rabbits, guinea pigs, and other animals housed near infested birds.

Life History, Appearance, and Habits.—This trouble is caused by a tiny, eight-legged mite, measuring from $\frac{1}{100}$ to $\frac{1}{50}$ inch across, pale gray in color and nearly circular in outline. They are not seen by the naked eye. If the fowl's legs are soaked in warm soapy water, the scales may be lifted and the mites found by use of a microscope among the powder and lymph from beneath the scale. They have very short legs and the skin is traversed with fine lines as is the palm of one's hand. It is believed that the eggs are laid in the tunnels made by females beneath the skin

---

[1] *Cnemidocoptes mutans* Robin and Lanquentin, Order Acarina, Family Sarcoptidae.

scales.   Like other mites, the young are at first six-legged and the development from that point on is a simple metamorphosis.

Control Measures.—Poultrymen should be on the lookout for this trouble, especially on newly purchased birds, and should treat fowls as soon as the symptoms show up to prevent spread to others.   The legs may be brushed with, or dipped in, crude petroleum or a mixture of raw linseed oil, 2 parts, and kerosene, 1 part.   If the swollen scales are not

Fig. 569.—Rooster severely attacked by scaly-leg mite, *Cnemidocoptes mutans* Robin and Lanquentin.   (*From U.S.D.A., Farmers' Bul.* 1337.)

largely shed within a month, the treatment should be repeated; but one treatment is usually sufficient.

350. Depluming Mite[1]

This mite is similar to the scaly-leg mite but is still smaller, and it burrows into the skin at the base of the feathers on the rump, back, head, abdomen, and legs.   The oval mites may be found on the fallen feathers or among the dry powdery material in the skin.   The irritation from the mites causes the fowls to pull out their feathers.

Control Measures.—Sulphur ointment carefully applied to the affected parts, or a dip made by mixing 2 ounces of wettable sulphur and 1 ounce laundry soap to the gallon of water.   Repeated applications will be necessary.   The affected birds should be isolated.

[1] *Cnemidocoptes gallinae* Railliet, Order Acarina, Family Sarcoptidae.

CHAPTER XXIII

INSECTS THAT ATTACK AND ANNOY MAN AND AFFECT HIS HEALTH

The various ills man suffers from insects reach their climax in their attacks upon his person. "Bugs" are no respecters of persons. While for the most part clean, sanitary living and reasonable precautions about associating with less sanitary persons and surroundings will prevent these insects from attacking a given individual, nevertheless some of these unwelcome visitors are likely to come into any household at any time. In such a case, to be forewarned is to be forearmed. The housekeeper who knows when to ignore a newfound insect in her house as a creature of no real significance and when to fight one that may start an infestation which could cause trouble for months has the battle half won. Brues[1] says:

"The importance of insects as detrimental to public health is well known to professional zoologists, medical men, and laymen alike, but is usually emphasized only under the stress of particular circumstances, such as the safety of soldiers in war or of unusual outbreaks of diseases for which insects are directly responsible. Insect borne diseases present a *constant* menace to the world, and aside from the actual toll of lives which they exact, they impair its efficiency by enfeebling the health of its human population."

W. D. Hunter, in his address as President of the American Association of Economic Entomologists,[2] draws the conclusion that the losses caused by diseases transmitted by insects is approximately one-half as great as the losses caused to all farm products. He concludes: "Surely this is a sufficient argument for greater attention to medical entomology."

FIELD KEY FOR THE IDENTIFICATION OF INSECTS THAT ATTACK AND ANNOY MAN AND AFFECT HIS HEALTH

A. *Free-flying insects that alight on face, arms, and other exposed parts of the body to bite and to suck blood:*
  1. Slender-bodied long-legged insects, up to $\frac{1}{2}$ inch long, with delicate wings fringed with scales, long slender mouth parts, and bushy antennae, make a high-pitched humming noise as they alight to suck blood. Especially abundant at dusk or at night and about swamps and woodlands...*Mosquitoes*, page 875.
  2. Small, chunky, humpbacked gnats, not over $\frac{1}{5}$ inch long, with broad clear wings and short heavy mouth parts and antennae, alight on the body, crawl into eyes, ears, hair, or under the clothing and suck blood. They make comparatively little noise, and the bite at first is not very painful. Especially troublesome during the daytime.......................*Black flies*, page 881.
  3. Very small midges, not over $\frac{1}{10}$ inch long, with hairy but not scaly, sometimes mottled, wings, bite from early evening to early morning. They fly quietly and are seldom seen or heard until the hot, very painful bite is inflicted.......... ....*No-see-ums* or *Punkies*, page 883.

[1] BRUES, C. T., "Insects and Human Welfare," Harvard Univ. Press, 1920.
[2] HUNTER, W. D., *Jour. Econ. Ento.*, **6**: 27–39, February, 1913.

4. Large, heavy-bodied, brown, black, and orange, often green-eyed flies, about ½ inch long, with wings clear or banded, fly wildly about the head with much noise, and cause a very painful, bloody bite on arms, head, or neck. Encountered chiefly in woods or marshes on warm sunny days....................
....*Deer-flies* or *horse flies*, page 884.

5. A fly about the size and general appearance of the house fly often comes indoors in lowering weather and bites especially about the ankles. It differs from the house fly in having a stiff, pointed beak that projects forward from lower side of the head, a broader black-spotted abdomen, grayer appearance, the apical cell (R_5) of wings more widely open, and hairs on upper side of arista only...
....*Stable fly*, pages 816 and 884.

B. *Insects that crawl upon the body to bite and suck blood, not spending their entire life on the human host:*

1. Mahogany-brown, broad, very flat or thin, oval, wingless bugs, of all sizes up to ⅕ inch long, live in beds and cracks about the room and crawl over the body and bite at night. Bugs have a distinct disagreeable odor..*Bedbug*, page 886.

2. Large, somewhat flattened, oval bugs, brownish or black, sometimes marked with pink or red; between ½ and 1 inch long, tapering to a slender head in front, with well-developed wings crossing over on the back, 2 or 3 large cells in the membrane of the wing, 4-segmented antennae, and 3-segmented labium. They bite when picked up and, in some localities, regularly visit the bodies of sleeping persons to suck their blood...
..............*Mexican bedbug, kissing bug*, and other *assassin bugs*, page 888.

3. Small, brown, wingless insects, about 1/16 inch long, very flat from side to side and with long hind legs, slip into the clothing or jump vigorously when disturbed. They bite especially about the legs and waist. Troublesome chiefly in basements or houses where dogs or cats are allowed......................
......................*Cat flea, dog flea, human flea*, and others, page 889.

4. In southern states or tropical countries, painful, inflamed sores develop between the toes or under the toenails, each containing an ulcerated body that may become as big as a small pea. This is due to a flea that buries its body in the skin...*Chigoe flea*, page 892.

5. Tough-skinned, 8-legged, reddish-brown to bluish-gray, seed-like bodies up to ½ inch long, often with silvery-white markings on the hard plate near the head, sink their mouth parts deeply and firmly in the skin and suck blood. No particular pain is felt at the time, but motor paralysis, infected sores, or spotted fever may follow the attack.................................
....*Spotted-fever tick, castor bean tick, dog tick*, and others, pages 892 and 896

6. An eruption on the skin like hives or chicken pox, accompanied by severe itching, and sometimes nausea, headaches, chills, or fever, results from handling straw or flour, dried fruits, meat, and other groceries, or follows a visit to grassy, brambly spots out-of-doors. Caused by nearly microscopic mites that crawl upon the skin and bite...
Chiggers, harvest mites, louse-like mites or *flour* and *meal mites*, pages 897 to 899.

C. *Insects that live upon the body all their lives, generation after generation:*

1. Painful burning and itching bites, which become whitish scars ringed with brown pigmented skin, occur anywhere on parts of body covered by clothing. No "cause" found on the skin but examination of clothing about the neckband, armpits, waist, or crotch of trousers, reveals elongate, oval, flattened, wingless gray lice, up to ⅛ inch long, with 6 legs, each bearing a single curved claw at the end. Whitish keg-shaped eggs are laid in the seams of the clothing......
...................................*Body louse, grayback*, or *cootie*, page 901.

2. Hair of the head, especially back of the ears and at the nape of the neck, infested with crawling, grayish, 6-legged lice, similar to the above; the whitish eggs fastened to the hairs.........................*Human head louse*, page 901.

3. Painful burning and itching bites with small inflamed spots among the hairs between the legs. Small, broad, grayish, 6-legged lice up to 1/16 inch long, that

look something like miniature crabs and tend to remain fixed in one spot.....
....*Pubic louse* or *crab louse* page 904.

4. Extreme itching of tender places on the skin, such as between the fingers, behind the knee, and inside the elbow, without visible cause. Careful examination may reveal delicate, tortuous, gray thread lines just beneath the skin. Hard pimples as big as pinheads, containing a yellow matter, form on the skin. Scraping the affected spots to the "quick" and examining the scrapings under a lens reveals very small, whitish 8-legged, nearly round mites..............
....*Itch mite*, page 900.

D. *Short, whitish, legless, segmented maggots or "worms"*[1] *live under the skin in sores, or in the natural body cavities, or in the alimentary canal, and are not infrequently passed in the excreta:*

    1. The maggots taper gradually from a blunt posterior end, which bears 2 rounded plates or short tubes for breathing, to a pointed head end which has short mouth hooks.....................
.......Various *flesh flies, house flies, fruit flies*, and *root-maggot flies*, page 905.

    2. The maggots are thickest near mid-length and narrow strongly toward either end; the skin is very tough, and usually beset with many, minute, short, sharp spines...*Human botflies*, page 905.

    3. The maggots are flattened, narrowing toward either end, each segment with prominent, fleshy, pointed processes, some of which have minute side spines...
....*Latrine fly* and *little house fly*, page 905.

    4. The maggots are nearly cylindrical, with head end rounded off and the opposite end prolonged into a long slender "tail" that is extensible like a telescope.....
....*Rattailed maggots*, page 905.

E. *Insects that frequent both filthy materials and human habitations and carry diseases to man without inflicting pain on the body:*

    1. Flattened, oval, running, brown or black, nocturnal insects from ½ to 1½ inches long, with 6 long, very spiny legs, antennae longer than the body, and usually 4 finely veined wings that are seldom used; occur in kitchens, bakeshops, restaurants, public buildings, ships and other moist warm places............
....*Cockroaches*, page 781.

    2. Two-winged flies, about ¼ to ⅓ inch long, of a general grayish color, with 4 equal black stripes on the back between the wings; the arista or antennal bristle feathered, *i.e.*, hairs coming off on two sides of it; the mouth parts soft, spongy, and retractile; the vein that ends nearest the wing bent forward so as nearly to meet the vein in front of it at the wing margin; and no large, bristly hairs on front segments of the abdomen.......................*House fly*, page 906.

F. *Insects and other arthropods that accidentally or occasionally hurt man, in defense of themselves or their nests, by stinging, biting, nettling, or blistering the skin:*

    1. Bugs of various sizes and shapes up to 2 inches long, but all with a slender tubular beak projecting from the lower side of the head, 6 legs, and the front pair of wings thicker at base and thinner and overlapping at the tip; bite painfully when handled or when they fly against the face, especially at night.....
Assassin bugs, water-scorpions, electric-light bugs, and many other Hemiptera, pages 200 and 888.

    2. Elongate, rather soft-shelled, 6-legged beetles, commonly ½ to ¾ inch long, blister or corrode the skin if crushed upon it...........................
.....................*Spanish fly*, and other *blister beetles*, pages 49 and 509.

    3. Caterpillars ("worms") of a variety of sizes and colors, but always with 3 pairs of jointed legs and 5 pairs of fleshy prolegs, generally more or less spiny or woolly, nettle the skin when they brush against it, causing pain, itching, and inflamed spots..................*Various caterpillars* or *moth larvae*, page 21.

    4. Four-winged, swift-flying, smooth or very hairy insects, usually conspicuously marked with yellow, insert the ovipositor or "stinger" from the posterior end

[1] Not to be confused with the parasitic *roundworms* and *flatworms*, which are usually much longer. These, if cylindrical, are not segmented, and if segmented are much flattened. Insect larvae never show more than 13 segments.

of the body, and inflict a painful sting. . . . .
. . . .*Bees, wasps,* and *hornets,* pages 20 and 220.
5. Wingless, crawling, 6-legged, slender-waisted "ants," the hairy kinds often brilliantly colored, sting as in *F,* 4. . . . .
. . . .*Stinging ants* and *velvet ants,* pages 21 and 228.
6. Wingless, many-colored, 8-legged creatures with the body in 2 regions, the part bearing the long legs separated from the abdomen by a slender pedicel or stalk, rarely puncture the skin and introduce a venom with the mouth parts. . . . .
. . . .*Tarantulas* and other *spiders,* page 166: *black widow spider,* page 884.
7. Elongate creatures, commonly 2 or 3 inches long, with a pair of "pincher legs" in front, eight long walking legs at the middle and a very long abdomen in two parts, the posterior slender part with a short swollen sting on the end; cause great pain by thrusting this sting into the flesh. . . . .*Scorpions,* page 168.
8. Elongate, worm-like "hundred-legs," from 1 to 8 inches long, with distinct head and antennae, somewhat flattened body, and 15 or more pairs of legs, "bite" with a pair of poison claws just back of the head. .*Centipedes,* page 165.

General References.—Herms, "Medical and Veterinary Entomology," 3d ed., Macmillan, 1939; Riley and Johannsen, "Medical Entomology," McGraw, 1938; Herrick "Insects Injurious to the Household and Annoying to Man," Macmillan, 1926; Matheson, "Medical Entomology," Comstock Publishing Co., 1936; Ewing, "Manual of External Parasites," Thomas, 1929; Patton and Evans, "Insects, Ticks Mites, and Venomous Animals," Vols. I and II, Liverpool School of Tropical Medicine, 1929; Hindle, "Flies in Relation to Disease: Blood-sucking Flies," Cambridge Univ. Press, 1914; Graham-Smith, "Flies in Relation to Disease: Non-blood-sucking Flies," Cambridge Univ. Press, 1914.

351 (322). Mosquitoes

Importance and Type of Injury.—Besides the well-known painful bites inflicted by the females of mosquitoes, these insects are the proved carriers of four distinct human diseases. There is no other known method of acquiring malaria, yellow fever, dengue, and certain forms of filariasis except by the bites of mosquitoes which have previously bitten persons that had these diseases. Considering the entire world, malaria ("ague" or "chills and fever") has been said to be the most important disease. It causes a large percentage of the deaths among mankind. In the United States (which is only mildly malarious) there are five or six million cases and 10,000 or 12,000 deaths a year, in addition to which several millions suffer illness, loss of time, and often prolonged inefficiency from this disease. The proper agricultural development of certain sections of our country has been greatly retarded. In the tropics entire countries have been practically barred from civilization, by this disease. As pointed out by Ross, these are unfortunately "more especially the fertile, well-watered and luxuriant tracts; precisely those which are of the greatest value to man." The several kinds of malaria are caused by microscopical animals[1] that live in the blood, destroying the red corpuscles and causing anemia, accompanied by the characteristic alternating chills, fever, and sweating. Once introduced to a human body by a mosquito, the parasite may increase rapidly until as many as three billion are present in the blood of one patient; but it cannot get from that person to another without the help of certain kinds of mosquitoes[2] (Figs. 570 and 572), in the bodies of which a necessary part of its life cycle is completed. Every

[1] Three species of Plasmodium, Phylum Protozoa, Class Sporozoa.
[2] Fifty or more species of the genus Anopheles, especially *A. quadrimaculatus* Say and *A. maculipennis* Meigen, Order Diptera, Family Culicidae.

step of the life cycle of this little animal has been followed in both the human and insect hosts, and we know that both man and mosquito are necessary to its continued existence.

Until the end of the nineteenth century yellow fever was one of the most dreaded diseases in the world. Its cause was not known, and terrible epidemics swept tropical countries and our seaport towns such as New Orleans, Philadelphia, and Havana. This disease was the most potent factor in the failure of the French to build the Panama Canal, in the latter part of the last century. In 1900 American army surgeons working in Cuba discovered that the disease is spread by a particular kind of mosquito,[1] since known as the yellow-fever mosquito. Since this discovery the disease has been rapidly stamped out of country after country until it has now been nearly eradicated from many parts of the earth. Until 1929 it seemed certain that yellow fever could be annihilated from the entire world. Recently, however, a type of the disease known as "jungle fever" has been found in wild areas in South America and Africa, where it is believed that monkeys constitute a permanent reservoir or alternate hosts, and where it is spread by the bites of several other mosquitoes. From this ineradicable source it may continually spread to towns and cities and start epidemics. There is continual grave danger of transporting infective mosquitoes or an incipient human case of yellow fever by airplane to many parts of the world where the mosquito carriers are present. The cause of the disease is an extremely minute organism, or filterable virus, which has so far eluded all efforts to discover and isolate it. While its transformations in man and the mosquito are not known, there is evidence that it passes an essential part of its life cycle in each host. A method of vaccination against yellow fever has recently been perfected. The conquering of this disease constitutes one of the greatest triumphs of modern science.

Filariasis is caused by minute worms[2] that live in the blood and lymph. This disease is seldom fatal but is often followed by terrible enlargements or deformities of the legs, arms, genital organs, or other parts of the body, known as "elephantiasis." The embryo or larval worms are found in the lungs and deeper blood vessels during the daytime, but when the patient is resting, usually during the night, these worms swarm into the superficial blood vessels, and mosquitoes feeding at this time may draw in some of them with the diseased blood. The worms undergo a transformation in the muscles of the mosquito, and 2 or 3 weeks later, when about $\frac{1}{16}$ inch long, work out through the labium, as it is feeding on a new victim. They bore through the labella and into the skin of man, migrate to the lymphatics where they grow to a length of about 3 inches and a diameter of $\frac{1}{100}$ inch. They mate, and the females produce the young ovoviviparously. Several kinds of mosquitoes[3] carry this disease. It occurs in many parts of the tropics.

The fourth disease of which mosquitoes are the known carriers is dengue or breakbone fever. It occurs in our southern states, but is more prevalent in the tropics. It is seldom fatal but may attack practi-

---

[1] *Aedes aegypti* (Linné), Order Diptera, Family Culicidae.

[2] *Filaria bancrofti* Cobbold, Phylum Nemathelminthes, Class Nematoda, Family Filariidae.

[3] The most important one is *Culex quinquefasciatus* Say, the common house mosquito of the tropics.

cally the whole population of a village or community, so that temporarily great inconvenience and suffering are experienced. The symptoms are a very high, intermittent fever accompanied by terrible aches in the bones and joints and a skin rash. The cause of the disease has never been found and is believed to be an ultramicroscopic organism. It is spread chiefly by the bites of the yellow-fever mosquito. There were half a million cases of this disease in Texas in 1922.

Animals Attacked.—Mosquitoes bite all kinds of warm-blooded animals, especially domestic and wild mammals and even such creatures as snakes and turtles. While the diseases just discussed are known to be troublesome only in man, mosquitoes transmit diseases of other animals, such as equine encephalomyelitis (page 817) and bird malaria.

Distribution—While yellow fever has been eradicated from many parts of the world, the mosquito which carried this disease is still common in the southern part of the United States and the tropics. Other mosquitoes range from the equator nearly to the poles and from sea level to at least 7,000 feet altitude.

Life History, Appearance, and Habits.—Since there are more than 350 kinds of mosquitoes in North America, we shall attempt to give only certain general features of their life cycle and habits. Many species of mosquitoes go through the winter in the egg stage. Some winter as adult, fertilized females in washrooms, cellars, outbuildings, hollow trees, and other shelters, where they are often seen hiding in fall and spring. Others survive the winter in the larval stage, either freezing up with the water or remaining dormant at the bottom of ponds and puddles. Mosquitoes always develop in water, which contains microscopic plants and animals that serve as food for the larvae, and their eggs are laid on the water or

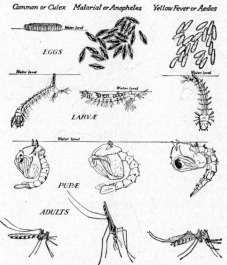

FIG. 570.—Three important kinds of mosquitoes, showing egg, larval, pupal and adult stages of each: in the left-hand column the life stages of a common house mosquito, in the center column of a malarial mosquito, and in the right-hand column of the yellow-fever mosquito. Two or three times natural size. (*From "Everyday Problems in Science," by Pieper and Beauchamp, copyrighted, 1925, by Scott, Foresman & Company*).

in places where water is likely to accumulate, as on ice or snow or in dry depressions. The eggs of the common Culex mosquitoes are built in minute rafts that look like a bit of soot floating on the water but are really composed of several hundred eggs standing on end. The eggs of malarial mosquitoes and yellow-fever mosquitoes are laid singly, the former having curious hollow expansions at the middle like a life belt, which keep them floating (Fig. 570).

The larvae of mosquitoes (Figs. 570 and 571) are the common "wrigglers" of rain barrels and quiet pools. The head is large, and has complex mouth brushes that, constantly in motion, waft food into the mouth. The mouth parts are of the chewing type, and they feed on algae and other small plant or animal life, either living or dead. The thorax is swollen and appears as one segment, but has no trace of legs in this stage. The abdomen is slenderer and bears on the eighth or next-to-last segment a short tube, known as a siphon, which the larva must thrust up into the air at intervals to breathe. This supply of air is supplemented by four finger-like, tracheal gills attached to the last segment on the body, by which oxygen is taken from that dissolved in water. The gills alone will

not keep the larvae alive, and they must come often to the surface to breathe. Indeed, they usually lie with the siphon projecting up through the surface film and the rest of the body hanging down at an angle in the water. Most species descend to the bottom of shallow water to feed upon organic

FIG. 571.—Larva of a malarial mosquito, *Anopheles quadrimaculatus* Say, in resting position at the surface of the water. About 5 times natural size. (*From U.S.D.A., Farmers' Bul.* 450.)

matter there. In the larvae of malarial mosquitoes, the siphon is very short, and the larvae generally lie parallel to the surface just below the surface film feeding upon organisms, such as algae, floating on the surface of the water. When disturbed they swim down into the water by lashing the abdomen from side to side.

In as short a time as 2 days to 2 weeks the larvae may be full-grown, about ⅜ inch long in common species. The change to the pupal stage takes place quickly at the fourth molt. This is a very unusual kind of a pupa (Fig. 570). It swims about actively in the water, avoids enemies, and does nearly everything the larva does except to feed. It breathes through two trumpet-like tubes on the thorax. The eyes, legs, and wings can be seen developing through the body wall on the large combined head and thorax. The pupal stage is often called a "tumbler." After a few hours to a few weeks in this condition, the insect splits its skin down the back and the adult (Fig. 572) crawls out, balances for a few moments on the empty pupal shell until its wings spread and dry, and then flies away. Most species pass through a number of generations each year.

The larvae and pupae of many common species breed in stagnant water of large or small quantity—anything from a bit of rainwater, caught in a discarded tin can, to the acres and acres of marsh water along our coasts and streams. The larvae of malarial mosquitoes breed in a great variety of situations, including especially slowly moving or standing water in which green algae abound. The yellow-fever mosquito breeds especially in and about dwellings and in cities wherever a bit of water, clean or foul, is left exposed long enough for it to complete its aquatic cycle. The adult of this species bites chiefly in the late afternoon and early morning. It is stealthy, does not make a very loud hum, is fond of crawling up under the clothing to bite, and prefers to stay in dwellings and other buildings. It has a peculiar, white, lyre-like pattern on the thorax. The malarial mosquitoes bite chiefly in the evening and early morning. Many of them can be distinguished from nonmalarial kinds by the white-and-black or rusty-red spotting of the wings. The females also differ in having the palps about as long as the labium, whereas other mosquitoes have the palps, in the female, not over a third or fourth as long as the other mouth parts. When biting or resting, the long axis of the body in the malarial mosquitoes is at an angle to the surface on which they are standing, while the common Culex mosquitoes hold the body parallel to the surface. Other kinds of mosquitoes bite mostly at night, but there are enough that venture forth in daytime in most sections to make life uncomfortable at any hour of the day or night, especially in the woods or about lowlands.

Control Measures.—Generally speaking, mosquito abatement should be carried out on a community-wide basis, covering several square miles, under the direction of experts who will determine the exact species

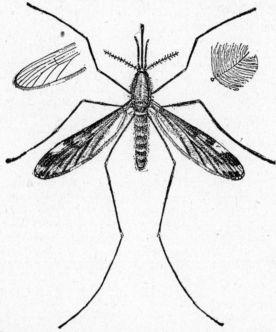

Fig. 572.—Female malarial mosquito, *Anopheles punctipennis* (Say), about 4 times natural size. Antenna of male at right, wing with scales removed to show venation, at left. (*From U.S.D.A.*)

involved, the places where they are breeding, and the most effective and economical methods of suppression. Since no mosquito can come into its winged existence without water in which to grow, the best control is the removal of excess standing water by draining swamps, pools, and open ditches, by doing away with trash and all useless receptacles which may hold water, and by seeing that water does not stand in roof or street gutters, catch basins, and drains. Water which cannot be drained may be rendered mosquito-free by applying a thin film of oil to the surface, using about 1 pint (preferably an oil specially designed for the purpose) to each 250 square feet of surface every 12 days. Used crankcase oil, to each 100 gallons of which is added 1 gallon of cresylic acid, makes a cheap and highly toxic oil, but it is very injurious to plants and fish and leaves an unsightly residue. Where oils cannot be used, because of injury to plants and waterfowl, certain pyrethrum larvicides may be employed. In water to be used for laundry purposes but not for drinking, 2½ ounces of borax per gallon, will render it permanently mosquito-free. In cases of great necessity, several hours protection from mosquitoes for outdoor gatherings, from a few square feet to a number of acres in area, can be provided by thoroughly fogging the foliage and atmosphere with a fine spray of pyrethrum extract, a half hour in advance of the meeting. The malarial mosquitoes (only) may be killed by dusting paris green (or crude sodium arsenite) over the water, using about 2 level teaspoonfuls of paris green in 100 times as much fine dry dust for each 1,000 square feet of surface. It may be distributed by hand or, for large areas and impenetrable swamps, from airplanes.

The introduction of top-feeding minnows, goldfish, and other kinds of fish which feed on the larvae is a practicable control in ornamental pools, ponds or slow streams where the aquatic plants or shallowness of the water does not keep them away from the margins. A modification of this method may be used where it is desirable to retain the marsh as a wild-life area. Excavations 8 to 10 feet deep and of about the same diameter are made in the lower parts of the marsh. Radiating open ditches are dug from these to the higher points. These are stocked with top minnows. When the water is high, the minnows work out through the ditches and eat the mosquitoes. When it is low they retreat into the excavated pools. The complete and careful screening of houses or beds is essential to comfort and health in most parts of the world. Mosquitoes may be cleaned out of a house where they are hibernating or attacking the occupants by the use of certain commercial oils, by pyrethrum sprays, by generating hydrocyanic acid gas, or by burning sulphur as a fumigant (see pages 278 and 285). When one must be exposed to mosquitoes, gloves, veils, and leggings give a partial protection, as does the application of repellent dopes. Among those found most effective by the writers are the following:

I

Gum camphor	15 g.
Oil of pennyroyal	15 cc.
Oil of citronella	15 cc.
Pine oil	15 cc.
Stearic acid, sufficient to solidify the above, about	30 g.

II

Olive oil	30 cc.
5% phenol	5 cc.
Oil of pennyroyal	4 g.
Oil of cedar	4 g.
Oil of camphor	2 g.

Household sprays such as Lethane, derris, or pyrethrum may be applied to the clothing or about porches and chairs where one wishes to sit out-of-doors. For the relief of the bites of mosquitoes and other insects, the first rule is not to break the skin by scratching, on account of the danger of secondary infection. Heat from an electric light or the application of a paste of baking soda or salicylic acid, or ammonia, soap and vinegar, or iodine in glycerin are suggested.

References.—"The Mosquitoes of North and Central America and the West Indies," 4 Vols., Carnegie Inst., Washington, 1912; *U.S.D.A., Bur. Ento. Bul.* 88, 1910, and *Farmers' Bul.* 1354, 1923; *N.J. Agr. Exp. Sta. Buls.* 276, 1915, and 348, 1921; *Jour. Econ. Ento.,* **30**: 10–28, 1937; MATHESON, "Mosquitoes," Thomas, 1929.

352 (315, 322). BLACK FLIES, BUFFALO GNATS, OR TURKEY GNATS[1]

Importance and Type of Injury.—During the daytime, out-of-doors, small, clear-winged, humpbacked, chunky, blackish gnats (Fig. 573) hover about the eyes, ears, nostrils, and other parts of the body. They make little noise but promptly alight, run greedily over the skin and suck blood through their short, sharp mouth parts. They often appear in great numbers and may be drawn into the air passages of animals, as poultry and large animals are apparently sometimes smothered by the swarms of flies. The bites are very irritating, and in many sections of our northern states and Canada these little bloodsucking gnats make life unendurable for a definite season, mostly in spring. They bite especially about the face and neck but also on arms and any other part of the body that is exposed; they do not hesitate to squeeze under the clothing or into the hair to bite. The bites are not especially painful when made but become increasingly itching, swollen, and irritating for some days. The venom has a specific effect upon the glands about the ears and neck causing symptoms similar to mastoiditis. Black flies were at one time suspected of being carriers of the human disease, pellagra, but this has been disproved. They are carriers of a filarial parasite causing "Mexican blindness."

FIG. 573.—Adult buffalo gnat, *Prosimulium pecuarum* (Riley), about 5 times natural size. (*From H. Garman, Ky. Agr. Exp. Sta.*)

Animals Attacked.—All warm-blooded animals.

Distribution.—Some species occur in nearly all parts of the United States and Canada. Especially troublesome in the northern woods and mountains. They have a definite season after which little trouble is experienced for the rest of the summer.

Life History, Appearance, and Habits.—There are many species of black flies which differ considerably in life cycle. They generally winter as maggots below the water surface in swift, rocky streams, and the adult flies in many sections are most abundant in spring. The adults are from $\frac{1}{25}$ to $\frac{1}{5}$ inch long, with broad clear wings, short, stiff horn-like antennae, and sooty black, chunky body, from the front end of which the head hangs downward, giving them a curious humpbacked appearance. The yellowish to black eggs are deposited on the surface of rocks, sticks, or vegetation at, or actually below, water in swiftly flowing streams. The eggs are often subtriangular in outline and laid, a single layer deep, in large patches covering many square inches. The eggs may hatch in 4 to 12 days and the grayish-black, legless maggots attach themselves to rocks, sticks, or other obstructions in streams where the water churns or boils over them, or to the slightly submerged leaves of sedges or trees. Here they may

[1] Various species of *Simulium* and related genera, Order Diptera, Family Simuliidae.

be seen attached by their tail ends, standing erect or stretched downstream like a banner in the wind, or writhing over their submerged support by bending the body from side to side.   From a single stone, 10 inches in diameter, 2,880 larvae have been taken.   The individual larva is less than ½ inch long, shaped like an Indian club or bowling pin, with a brownish head at the smaller end that bears a pair of marvelous mouth rakes each consisting of 30 to 60 long curved bristles, palmately arranged and constantly in motion, straining microscopic plants and animals from the current and scooping or fanning them toward the chewing mouth parts. The larvae hang on to the rocks or loop about by means of a complex disk, bearing hundreds of minute hooked hairs, near each end of the body, and also by clinging to silken threads that they spin over the surface of the rocks or sticks.   They respire through three retractile lobes or clusters of soft, white, blood gills near the tail.   The food is minute plant and animal life such as algae, protozoa, and diatoms.   The larvae are therefore entirely harmless creatures that develop in streams often far from human habitations.   After 2 to 6 weeks in this stage, or in some species not until the following spring, growth is completed, and the larva makes a slipper- or vase-shaped silken cocoon, wide open at the downstream end, fastened to the rock where the larva lived.   In this pocket the pupa is hooked, with its 4 to 60 long thread-like tracheal gills which attach to the prothorax.   From about 2 days to 3 weeks later the adults emerge, float to the surface of the water, and quickly take wing before being drowned in the current.   Simultaneously with their emergence from the pupae they change from harmless aquatic curiosities to bloodthirsty plagues (Fig. 573) that seek warm-blooded animals of all kinds, pierce the skin and draw the blood, leaving at the same time an irritating venom that causes extreme pain.   It is only the females that suck blood.   Sometimes black flies appear in sections remote from swift rocky streams.   This may be due to migration (aided by winds) from a considerable distance, but it must be remembered that there are many species of these flies, and the life habits of most of them are imperfectly known.   Their favorite breeding grounds are the ripples of cold swift streams.   Some kinds develop in large streams, ditches, and other slowly moving water, but they cannot live in standing water.

Control Measures.—When one must venture into territory preoccupied by these gnats, the clothing should be securely closed at boot tops, neck, and wrists, and gloves and head veils worn if possible.   A double layer of clothing, however light, is desirable; and long-sleeved full-length underwear is recommended.   Repellents such as recommended for mosquitoes (page 880) are also of some importance but must be applied frequently. The locating of tents or camps on high dry spots, away from trees and underbrush, and where the wind strikes, will avoid the worst of their attacks.   One of the most important measures for the protection of livestock is to provide smudges in the fields or before the doors of barns or poultry houses by burning bark, moist punky wood, old leather, or green grass.   The flies will not attack in the smoke.   It is extremely difficult to destroy them in their breeding grounds.   The application of a few gallons of miscible oil of the right grade to a small stream will kill the larvae and pupae down stream for many rods but there is danger that it may also kill the fish.   Cleaning and deepening channels to remove logs, roots, stones, and other obstructions that cause ripples and waterfalls helps to reduce the numbers of these flies.   The outlets of lakes, dams, and spillways should have a clear unobstructed drop into a deep pool.

References.—U.S.D.A., Bur. Ento. Bul. 5 (n.s.), 1896, and *Dept. Bul.* 329, 1916, and *Bur. Ento., Tech. Bul.* 26, 1914; *N.Y. State Museum Bul.* 289, 1932.

353. Punkies, No-see-ums, and Sand Flies[1]

Importance and Type of Injury.—These blood-sucking midges (Fig. 574) are not uniformly distributed but occur locally in numbers sufficient to make them almost intolerable.   They bite chiefly in the evening and very early in the morning.   Most species are attracted to artificial lights.   They are most prevalent from mid- to late summer.   In some species the bite is very burning and painful, and the victim is likely to be astonished when he notes the extremely small size of the creature that inflicted it.

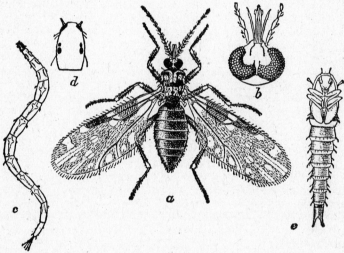

Fig. 574.—A punkie or no-see-um, *Culicoides guttipennis* (Coquillett).   *a*, adult, 15 times natural size; *b*, head of adult, more enlarged; *c*, larva; *d*, head of larva; *e*, pupa. (*From Riley and Johannsen, "Handbook of Medical Entomology," after Pratt, Comstock Publishing Co.*)

Life History, Appearance, and Habits.—Punkies are very small two-winged midges, commonly about $\frac{1}{15}$ to $\frac{1}{25}$ inch long, the wings hairy and sometimes pictured, but never with flattened scales.   Two long veins near the front of the wings are distinct, the others very faint.   The mouth parts are short, the body moderately heavy, and the legs rather stout.   The larvae develop in decaying leaves and silt, where salty tidal water backs up into fresh water streams, along the margins of ponds, streams and pools of various kinds, and in rot holes of trees.   They are extremely slender and have a small brown head and a tuft of hairs at the opposite end.   They breathe by means of blood gills after the manner of fish.   The pupae of some species are said to float in the water breathing from above the surface film.

Control Measures.—Tree holes near residences should be drained, repaired by tree surgery, or saturated with creosote.   In restricted areas about drainage ditches, creosote or carbolic acid larvicides are recommended.   Near cities where the problem is acute, the building of dikes or bulkheads, with pumps and tide gates to prevent flooding by tides or surface drainage, may be the only method of permanent relief. One part pyrethrum extract concentrate (20 to 1) in 20 parts lubricating oil (S.A.E. 5)

[1] *Culicoides* spp. Order Diptera, Family Chironomidae.

is recommended to apply to exposed parts of the body every 2 hours, or to window and door screens every 24 hours. Sixty-mesh silk bolting cloth over doors and windows is required to exclude the smallest of these pests.

References.—U.S.D.A., Cir. E-441, 1938; *Jour. Parasitol.,* **20**: 162–172; 1934; *N.Y. State Museum Bul.* 289, 1932.

354 (313). Horse Flies or Deer Flies[1]

Many of the horse flies do not attack man, but a number of species of banded-winged smaller deer flies descend upon him when he ventures into their haunts in woods or marshes. They fly threateningly about the head in wild circles and, if not constantly warded off, alight and sink their mouth blades into the exposed skin. The bite is instantly very painful and considerable blood is drawn. Horse flies are the known carriers of *Loa loa,* tularaemia, and anthrax, which they may spread by biting a person or animal soon after feeding on a diseased animal or carcass.

Control Measures.—General measures are discussed under Horses (see page 816). The same repellents recommended for mosquitoes are useful in preventing the bites of horse flies.

355 (314, 322, 333). Stable Fly[2]

This insect, which has been fully discussed under Horses, commonly comes indoors or on porches, especially before a storm, and bites people but is also troublesome out-of-doors. It selects especially the legs and ankles on which to feed and rarely bites elsewhere. It is noiseless and stealthy. Its bite is instantly very painful but has no prolonged after-effects. It is often called "the biting house fly" but should not be confused with the true house fly. The two species can be distinguished by the characters given in the key.

356. Black Widow Spider[3]

Importance and Type of Injury.—This is the most seriously venomous spider native to North America. Its bites commonly give rise to very severe symptoms, but death results in only about 5 per cent of the known cases, and recovery is usually complete in from 2 to 5 days. The bite itself is usually a mere pinprick, but excruciating pains usually begin in a few minutes and spread from the point of the bite to arms, legs, chest, back, and abdomen; and within a few hours symptoms, such as chills, vomiting, difficult respiration, profuse perspiration, delirium, partial paralysis, violent abdominal cramps, pains, and spasms, frequently result. The pain is so severe as to lead frequently to a diagnosis as appendicitis, colic, or food poisoning. The venom is described as about fifteen times as potent as rattlesnake venom.

Food.—A great variety of insects and other very small animals.

Distribution.—General throughout most of the Western Hemisphere; extremely common in many of the southern states.

Life History, Appearance, and Habits.—Young adults, which have not yet laid eggs, winter in buildings, rodent burrows, and other sheltered places, becoming mature in late spring. The female is a shiny, coal-black, eight-legged, slender-waisted creature, nearly ½ inch long (Fig. 575), the slim glossy black legs having a reach of about 1½ inches. The best recognition mark is an hourglass-shaped bright-reddish spot on the *under*side of the globular abdomen. There may be red or yellow marks

---

[1] *Chrysops* spp., Order Diptera, Family Tabanidae.

[2] *Stomoxys calcitrans* (Linné), Order Diptera, Family Muscidae.

[3] *Latrodectus mactans* (Fabricius), Order Araneida, Family Theridiidae. The closely related *L. geometricus* Koch occurs in the subtropical parts of the United States.

on the upper side of the abdomen also, especially in the male. There is a "comb" or row of toothed setae on the tarsus of the fourth pair of legs. The male is very much smaller than his mate and has greatly swollen pedipalps. During the warm months of the year the spiders are found in sheltered, dimly lighted places such as barns, garages, basements, outdoor toilets, hollow stumps, rodent holes, and among trash, brush and dense vegetation. Cold and drought drive them into buildings. The female constructs an irregular, tangled web of tough silk with a funnel-shaped retreat extending to or toward the ground, from which she forages out to entangle, paralyze, and devour such prey as blunder into it. After a prolonged courtship the male transfers sperms from his gonopore to that

Fig. 575.—The black widow spider, *Latrodectus mactans* (Fabricius). At left, the female from the underside, showing the characteristic hourglass-shaped spot, which is red and on the *under*side of the abdomen. At right, the male, showing the markings that often occur upon the upper side of the male but not the female abdomen; note his enlarged pedipalps. Female, enlarged about ½; male, several times natural size. (*After Baerg, Ark. Agr. Exp. Sta.*)

of the female by means of his pedipalps. It has been said that the female often kills and eats her mate after mating, hence the name black "widow." During spring and summer, the female lays many eggs, enclosing them in grayish silken egg balls attached to her web. Each ball contains from 200 to 900 eggs, and a female may construct from 5 to 15 such egg cases. The rate of increase is greatly curtailed by the cannibalistic habits of the young, so that from 1 to 12 young may be all that survive from each egg case. From 10 to 30 days after the egg ball was made, the young spiders emerge from it. Growth requires 2 or 3 months, during which the male molts three to six times and the female six to eight times. The old ones apparently die in summer or autumn after having laid their eggs. In late summer young males and females live together in the same web.

Control Measures.—The spiders can best be killed with sprays of kerosene, crude oil, or creosote. These materials sprayed liberally about basements, outdoor toilets, and outbuildings, will kill all spiders which are wet by them, and the creosote will keep them away for a time. The only way in which these materials can be used on plants, without injury, is to make them into an emulsion with soap and hot water with much agitation. In case one has the misfortune to be bitten by one of the spiders, the best available physician should be called at once and advised to administer intravenous injections of a 10 per cent solution of calcium gluconate in 10-cubic centimeter doses. Sodium amytal has also been recommended to be used in the same way and, if neither of the above is available, magnesium sulphate solution has been used with varying success. Perhaps the best remedy to use while awaiting the arrival of a physician is frequent baths as hot as can be endured. The poisons spread so rapidly that the application of a tourniquet and making incisions through the site of the bite and sucking out the venom are of doubtful value. The sterilizing of the skin about the bite with tincture of iodine should, however, be performed. The physician might be advised that strychnine may be useful as a heart stimulant and sedatives may be employed, although hypodermics of morphine have frequently been found to be practically useless in relieving the pain. It should be clearly understood that this spider is not an aggressive one but is really very shy and retiring. There appear to be two circumstances under which they bite people. (1) When they are squeezed, as by picking them up or while donning shoes or clothing in which they are hiding, or when they are encountered among sheets or blankets in the bed. (2) Especially when the female is guarding her egg case, if she is disturbed in her retreat, she may rush out and bite the intruder. This seems to be a reaction largely to a vibration of her web and is the response which she would normally give when some insect blunders against her web. Such prey is usually entangled with the silk, paralyzed by the bite and subsequently devoured. In a large number of cases which have been studied, it has been found that by far the majority of the bites have been inflicted while the person was using an outdoor toilet. In such cases apparently the web spun across the toilet seat had been vibrated and the spider has rushed out and inflicted the bite in the typical manner. In some areas and in some seasons the black widow spider has become so abundant among crops such as tomatoes and grapes that there is some hazard in picking the fruit. Under such circumstances workmen should wear gloves while working among the plants where they are hiding.

References.—*Calif. Agr. Exp. Sta. Bul.* 591, 1935; *Ark. Agr. Exp. Sta. Bul.* 325, 1936; *Ore. Agr. Exp. Sta. Cir.* 112, 1935; *Quart. Rev. Biol.,* **11**: 123–160, 1936.

357 (344). BEDBUG[1]

Importance and Type of Injury.—The bite of a bedbug is not generally felt immediately, but the venom introduced soon causes itching, burning, and swelling to a variable degree, depending on individual susceptibility. On many people the bites become increasingly painful for a week or more. To many who live in well-cared-for homes and travel but little, the bedbug may seem a joke, but there are thousands who live in con-

---

[1] *Cimex lectularius* Linné, Order Hemiptera, Family Cimicidae.

gested city districts, who never get a night's rest free from the attack of this loathsome pest. On account of the ease with which the insect is carried in baggage or on the clothing, hotelkeepers and theater managers find it one of their greatest plagues, and anyone who travels extensively must expect to meet this bug frequently, not only in the cheap but also in the best of hotels and in sleeping cars. Under such conditions the bugs probably feed on different persons nearly every time they draw blood, and the conditions are very favorable for the spread of any blood-infesting disease organism. These insects have been suspected as carriers of leprosy, bubonic plague, oriental sore, relapsing fevers, and of a very fatal disease in India known as *kala azar;* but so far the proof that they are important in the transmission of any disease is not very conclusive.

Animals Attacked.—Besides feeding on man, the bedbug will attack mice, rabbits, guinea pigs, horses, cattle, and poultry. They are often abundant in neglected poultry houses.

Distribution.—The true bedbug is found all over North America and is nearly cosmopolitan.

Life History, Appearance, and Habits.—In warmer regions and in rooms that are kept uniformly heated, the bedbug probably breeds throughout the year; in rooms where the temperature lowers perceptibly in winter it apparently winters chiefly as adults and nymphs, and egg laying is suspended until spring. The eggs may be found at any time during the warm months of the year. They are laid in cracks of furniture, behind base-

Fig. 576.—A bedbug, *Cimex lectularius* Linné, 8 times natural size. (*From Herrick, "Insects Injurious to the Household," copyright,* 1914, *Macmillan. Reprinted by permission.*)

boards, under loose edges of wall paper and all such crevices, where the adults hide during the day. They are elongate, whitish, big enough to be easily seen, and have a distinct cap at one end. Each female may lay from 75 to over 500 eggs, at the rate of 3 or 4 a day, gluing them to wood or fabrics in their hiding places. They hatch in 6 to 17 days. The young bugs are similar to the adult when they hatch, but paler, yellow in color. They molt five times, and at the last molt the abbreviated, useless, rudimentary wings appear and the insect is adult. Any one bug probably does not feed every night but, at intervals of several days to a week, usually once before each molt. They may become adult within a month or two after hatching, but ordinarily growth is much slower, and there are from one to four generations a year. They may live long periods (4 to 12 months) without food. They may also feed on mice and other animals than man, so that empty houses may remain infested for long periods. Both males and females live on blood alone.

Any very flat brown bug is likely to be mistaken for the bedbug. There are many species living under bark of trees[1] that look something like bedbugs, but do not bite, and there are several other species that live in the nests of birds and bats, sucking their blood. Many false notions about bedbugs are current because people do not distinguish these similar-

[1] Flat bugs, Order Hemiptera, Family Aradidae.

looking insects.   The true bedbug (Fig. 576) may be recognized by the very slender third and fourth segments of its antennae (the second segment being shorter than the third), by its almost lunate-shaped prothorax into which the head is sunken and by being covered with very short, curved, serrate hairs.   The adults are about 1/5 inch long.   They emit an oily substance which has a very offensive odor.

Control Measures.—A house infested with bedbugs should be fumigated with hydrocyanic acid gas, using 10 ounces of sodium cyanide, or the equivalent of calcium cyanide, to each 1,000 cubic feet of space (see page 278).   This should never be done by anyone but an experienced operator.   Superheating to 130°F., holding that temperature for 3 to 6 hours; or, in empty houses, burning sulphur at the rate of 4 pounds to each 1,000 cubic feet are also effective.   In congested places individual rooms or apartments can more safely be fumigated with commercial gases containing methyl bromide, methyl formate, or ethylene oxide mixed with carbon dioxide (see pages 287, 289); but community cooperation should be enlisted because the bugs will migrate from house to house.   In barns, poultry houses, basements, and other unfinished rooms, a thorough spraying with crude oil, kerosene, gasoline, or creosote is very effective.   If the bugs are believed to be localized in some one room or piece of furniture, they may be killed by spraying thoroughly all cracks and other hiding places with a good Lethane or rotenone spray, applied in sufficient amounts to wet the articles thoroughly.   Electric-driven power sprayers will force the vapor of these sprays into the remote hiding places of the bugs more effectively.   Mattresses, rugs, and upholstery can be effectively treated by steam cleaning or by having them vacuum fumigated by competent persons.   New, and especially secondhand furniture, laundry, traveling bags, and similar articles delivered to the home should be watched that the bedbug be not introduced with them.   If one is traveling extensively, it is advisable to carry a small bottle of pyrethrum powder to be sprinkled over the bed under the sheets if these pests are found or suspected in the room.   This will usually keep them from attacking one for the single night.

References.—U.S.D.A., *Farmers' Bul.* 754, 1916, and *Leaflet* 146, 1938.

358. Assassin Bugs or Kissing Bugs

Importance and Type of Injury.—The bites inflicted by the Mexican bedbugs or big bedbugs,[1] China bedbugs,[2] and the so-called "kissing bugs" or bloodsucking cone-noses[3] are scarcely exceeded in severity by any other insect.   They have been likened in effects to snake bites.   The pain is intense and usually affects a considerable part of the body; swelling generally follows; and in the worst cases faintness, vomiting, and other ill effects are experienced that may last weeks or even months.   These bites may be experienced when one picks up the bugs or when they fly against the face. In the South and Southwest, the "big bedbugs," mentioned above, are aggressive and come into houses and bite at night to secure a meal of blood.   It is probable that many of the painful bites of which spiders are accused are caused by these bugs. In South and Central America *Triatoma megista* and several other species are carriers of a highly fatal human disease known as Chagas disease (see Table II, page 26).   In some of these species which are habitual bloodsuckers, the bites are entirely painless.

[1] *Triatoma sanguisuga* Leconte, Order Hemiptera, Family Reduviidae.
[2] *Triatoma protracta* (Uhler), Order Hemiptera, Family Reduviidae.
[3] *Reduvius personatus* (Linné) and *Melanolestes picipes* Herrick-Schaeffer.

Animals Attacked.—Man, domestic mammals, and poultry.

Life History, Appearance, and Habits.—The eggs of these bugs are mostly laid out-of-doors under stones, logs, or other shelter or, in some species, on plants. The young bugs probably feed chiefly on other insects but may attack warm-blooded animals. One species is known as the masked bedbug hunter because the nymph has a sticky secretion all over the body to which dust and lint adhere so that, as it crawls along the floor, it looks like a bit of lint being blown along. It is believed to catch bedbugs and suck the blood from them. So far as these species have been studied, it appears that most species have but one generation a year.

These bugs (Fig. 577), which are from $2/3$ to 1 inch long, have the characteristics of the order Hemiptera. The head is long, somewhat conical, the prothorax narrows in front, the wings cross over flat on the back and have 2 or 3 large cells in the membrane; and the edges of the abdomen are produced as thin flat plates at the sides of the wings. These plates are in some species marked with pink or red bars, though the general color is dark-brown or black.

Control Measures.—No control is known except to screen the bugs out of houses and to be very careful about picking them up. If they alight on the face flip them off quickly and do not take hold of them.

359 (334, 360). FLEAS

Importance and Type of Injury. Everyone knows that fleas commonly infest dogs and cats, and many have experienced the painful irritating bites that result when they suck the blood

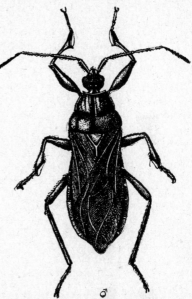

FIG. 577.—An assassin bug or "kissing bug," *Melanolestes picipes* Herrick-Schaeffer, male, about 4 times natural size. (*From Ill. State Natural History Surv.*)

of man. The bites are likely not to be felt immediately but become increasingly irritating and sore for several days to a week or more afterward. It is rather characteristic of flea bites that there are frequently 2 or 3 in a row. They bite mostly about the legs. Unlike lice and ticks which, having found a host, cling to it for dear life, fleas shift from host to host and feed indifferently on several kinds of animals. The cat flea[1] is nearly as likely to be found on a dog or a man as on a cat, and the so-called human flea[2] has been taken from dogs, skunks, rats, mice, and deer. The rat flea and those of ground squirrels also bite man. This promiscuous-feeding habit makes possible the most serious injury that fleas inflict, *viz.*, the transmission of bubonic plague from man to man. This is not the only way that plague may be contracted, but, in the great epidemics, fleas have been the most important factor in the spread of the disease. Plague is a bacterial disease caused by *Pasteurella* (= *Bacillus*) *pestis*. This organism causes a fatal disease in rats as well as in man. In man these bacilli rapidly increase in numbers in the blood, lungs, or lymph glands, and, in the most prevalent form of the disease, swellings from the size of a golf

[1] *Ctenocephalus felis* (Bouché), Order Siphonaptera, Family Pulicidae.
[2] *Pulex irritans* Linné, Order Siphonaptera, Family Pulicidae.

ball to that of an orange appear on the body, especially about the groin and armpits. From 20 to 95 per cent of the cases terminate fatally. Unlike the mosquito-borne diseases, no essential part of the life cycle of the plague bacillus takes place in the body of the flea. At least nine different species of fleas may carry the disease. Fleas are also carriers of typhus, among rodents, kala azar in children, and are intermediate hosts of tapeworms that may be transferred to man.

Animals Attacked.—Man, hogs, dogs, cats, rabbits, foxes, and other fur-bearing animals and rats, mice, and all other kinds of rodents as well as many other animals are bitten by fleas. Bubonic plague occurs in man, rats, mice, certain ground squirrels, and some other rodents.

Distribution.—Fleas, as a group, are cosmopolitan. In the eastern United States the cat flea[1] and the dog flea[2] are most often found in dwellings. In the western states the human flea[3] is most often encountered in houses. Since 1936 the Indian rat flea,[4] worst carrier of the plague in

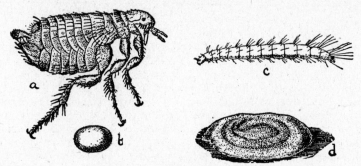

Fig. 578.—A flea. *a*, adult; *b*, egg; *c*, larva; *d*, cocoon. All much enlarged. (*From U.S.D.A.*)

the world, has been found in widely scattered places in the central United States. Bubonic plague has been most serious in Asia, Europe, and Africa. In the fourteenth century a great epidemic swept the Old World during which it is estimated that about 25 million people died of this disease. Bubonic plague has been present in the western part of the United States since 1900, where it has become established in ground squirrels and has spread from San Francisco to several other Pacific and mountain states. In 1914 a mild outbreak of this disease occurred in New Orleans.

Life History, Appearance, and Habits.—Fleas, like most of the parasites of the large animals, continue their activity and reproduction throughout the winter, but breeding and all life processes are somewhat slowed down by cold weather. The adults are hard-skinned, very spiny insects, extraordinarily thin from side to side, with piercing-sucking mouth parts retaining both pairs of palps, concealed antennae, no trace of wings, and very large legs fitted for jumping. Their only food is blood. Although the adults are the only stage most of us see, fleas pass through all the stages of a complete metamorphosis. The eggs (Fig.

[1] *Ctenocephalus felis* (Bouché), Order Siphonaptera, Family Pulicidae.
[2] *Ctenocephalus canis* (Curtis).
[3] *Pulex irritans* Linné, Order Siphonaptera, Family Pulicidae.
[4] *Xenopsylla cheopis* Rothschild.

578,*b*) are deposited in either the dust, dirt, or bedding of the host, or laid while the female is on an animal. They are never glued fast to the hairs, and those laid on the host usually sift readily through the hairs to the ground. The eggs are white, relatively large, short ovoid, $\frac{1}{50}$ inch long. Only a few are laid at a time, though the total number may be several hundred. The length of the egg stage is given as 2 to 14 days. The young flea (Fig. 578,*c*) is a very slender whitish larva, about $\frac{1}{6}$ to $\frac{1}{4}$ inch long, with a small pale-brown head but without legs or eyes. The body is plainly segmented and is covered with scattered, very long hairs. The larva has chewing mouth parts and feeds on a variety of dry organic matter such as the excreta of the adult fleas or that of mice, rats, and other rodents. Larvae can be reared successfully on the dirt scraped from the cracks of a wooden floor. The larval stage occupies from 1 to 5 weeks or more, and two molts have been recorded. However, since the larvae generally eat their molted skins, there may be additional instars that have not been observed. When full-grown, the larva forms a small oval cocoon of white silk to which adhere particles of dust and trash so as to give it a dirty, obscure appearance. From this cocoon (*d*) the flea emerges in 5 days to 5 weeks although it may pass the winter in this way. Thus 2 or 3 weeks to 2 or 3 months and rarely as long as 2 years are required for a generation of fleas of different species and under varied conditions. It thus happens that a house or basement in which cats or dogs have been permitted may be closed during an absence of the occupants and found upon their return to be overrun with these pests, although all animals have been excluded for a considerable time. Even a stray cat or dog, sleeping on one's doorstep or under a porch, may peddle enough fleas or flea eggs to start an epidemic. Adult fleas, after having fed, have been kept alive for 50 to 100 days without further food, while unfed adults may remain alive from 1 to 2 years.

Control Measures.—The habits of fleas make it plain that either to rid an animal of the pests or to clean up infested premises, the control measures must be of two kinds: (1) The destruction of rats and other rodents and the treatment of any cats, dogs, hogs, or other infested animals, to kill the adults on them. (2) Thorough and vigorous treatment of the kennels, pens, or sleeping quarters and even of the dry soil, dry manure, and other litter in hog lots and barnyards, under porches, and in or under buildings to which the infested animals have had access. The best treatment of pets is frequent dusting with derris powder containing at least $\frac{1}{2}$ per cent rotenone, or washing with derris soaps. Although derris kills slowly, it is very deadly to fleas and harmless to the pets, should they lick it off their coats. On hogs, which often become grossly infested, a cheaper treatment by sprinkling their bodies lightly with crude petroleum, fuel oil, miscible spray oil, or crankcase oil may be used. A thorough spraying of rooms with a high-grade rotenone, pyrethrum, or Lethane spray will often check an infestation in its incipiency. These sprays, or dusts of derris or pyrethrum, will also help to keep fleas off of one's legs or bed, temporarily. Houses or basements in which fleas are established should be fumigated with hydrocyanic acid gas (see page 278); or by burning sulphur (see page 285); or by closing the rooms tightly, one after another, and sprinkling over the floor 1 pound of flake naphthalene, or better, crude naphthalene, to each 100 square feet. The room should be tightly closed for 24 to 48 hours. Any remaining naphthalene may be swept up and enough more added to treat the next room, and so on until the entire house has been treated.   After such treatment the floors should be thoroughly scrubbed with hot soapsuds to kill the eggs, or with an oil mop wet in kerosene.   Mats or rugs on which pets sleep should be laundered or shaken frequently or dusted at intervals with derris or pyrethrum powder. It should be realized that cats and dogs cannot be allowed regular access to the house without starting an infestation of fleas sooner or later, unless they are treated at regular intervals with a good flea powder or bath made with flea soap.   Where fleas have become established about barns, hog lots, and outbuildings, the infestation may be cleaned up by removing the manure and litter and spreading it in the fields at some distance from buildings and covering the ground in and around the infested buildings with a layer of salt, or spraying thoroughly with creosote oil, stock dip, or dormant-tree spray oil.

References.—U.S.D.A., Dept. Bul. 248, 1915, and Farmers' Bul. 897, 1917.

360 (359). Chigoe Flea[1]

Importance and Type of Injury.—In the southern part of the United States, Mexico, the West Indies, and other tropical regions, a small reddish-brown flea, about $\frac{1}{25}$ inch long, has the despicable habit of burrowing into the skin, especially between the toes and under the toenails.   It causes much pain and itching and, as the female enlarges beneath the skin, a pus-filled ulcer is formed.   The sore is very likely to become infected with bacteria, and the entire toe, foot, or limb may be lost by blood poisoning. This insect must not be confused with the chigger mites, since both are called chiggers.

Animals Attacked.—Man; hogs, and other domestic animals.

Life Cycle, Appearance, and Habits.—Chigoes, like other fleas, develop in the soil or in filth, through a slender larval stage and a pupal period spent in a cocoon.   When they become adult, they attach to warm-blooded animals and suck blood.   After mating, the female, aided by her long mouth parts, works her way into the skin, and her body becomes enormously enlarged, as her eggs develop, until she may be as large as a small pea.   The eggs are generally extruded through the entrance hole and drop to the ground, where they develop to adults in about 1 month.

Control Measures.—Infested sores should be opened with sterile instruments, the fleas removed, and the wound given an antiseptic dressing.   Shoes or boots should always be worn in the infested territory.   Where the pest abounds, pigs and all other animals should be kept away from dwellings.

361 (340). Spotted-fever Tick[2]

Importance and Type of Injury.—The spotted-fever tick is the most important tick in the United States.   It is very annoying to domestic and wild animals and is commonly found attached securely to the skin of children, hunters, and other persons who are much out-of-doors, especially in brush- and scrub-covered areas.   The bites are not felt at the time the tick attaches but cause more or less inflammation later.   If the tick is forcibly removed, its mouth parts usually remain in the skin and cause an ulcer that is in danger of bacterial infection and serious complications. Another somewhat mysterious injury caused by ticks is a paralysis of the motor nerves affecting first the legs, and a few days later the arms, and gradually spreading, if not checked, until death may result.   This disease,

[1] Tunga (= Dermatophilus) penetrans (Linné), Order Siphonaptera, Family Sarcopsyllidae.

[2] Dermacentor andersoni Stiles (= D. venustus Banks), Order Acarina, Family Ixodidae.

which is known as tick paralysis, results apparently only when the tick attaches at the back of the neck or base of the skull and feeds very greedily, a week or so after it first attaches. The careful removal of the tick usually results in a speedy recovery. It is believed to be due to a poisonous substance introduced by the mouth parts of the tick.

The most serious injury to man is the transmission of Rocky Mountain spotted fever. This is a highly fatal, continuous fever that begins a few days to a week after the attachment of an infected tick to the skin and often results in death within 2 weeks. A peculiar skin rash of grayish or brownish spots usually appears on the arms, legs, and other parts of the body a few days after the fever begins. There have been about 1,000 cases a year in the United States. The fatality in the Bitter Root Valley of Montana runs from 70 to 90 per cent. In the other western states and in the states east of the Rockies (where the American dog tick is the principal carrier) the mortality is less than 25 per cent. The cause of the disease is a minute, intracellular protozoon, discovered by Wolbach in 1916, and called *Dermacentroxenus* (= *Rickettsia*) *rickettsi* Wolbach, which attacks especially the peripheral blood vessels. Another offence by the spotted-fever tick is the dissemination of tularaemia, which, like spotted fever, is a rodent disease that occasionally attacks man and domestic animals. This disease is said to have killed 5,000 sheep in a single year in Idaho. It also affects cattle, goats, rabbits, many other rodents, and certain game birds. Many kinds of insects and other ticks transmit the organism, *Bacterium tularense*, among wild animals and from the wild animal hosts to man and domestic animals (see Table II, page 26).

Animals Attacked.—Spotted fever affects man and many kinds of rodents. Tularaemia is also prevalent among men, domestic and wild animals, and certain game birds. Tick paralysis has been noted only in man, sheep, and cattle, although it has been produced experimentally also in horses, dogs, rabbits, and guinea pigs. The spotted-fever tick feeds in its nymphal stages upon small wild animals, mostly rodents, such as ground squirrels, woodchucks, rabbits, rats, and mice and, when adult, upon large animals such as men, dogs, horses, cows, mules, sheep, deer, mountain goats, and many others.

Distribution.—The spotted-fever tick occurs in the eight or nine states centering around southern Idaho, and rarely in the states bordering these. Spotted fever and tularaemia, however, are now known to occur throughout the United States (see page 896). Tick paralysis is reported from Montana, Oregon, British Columbia and Saskatchewan.

Life History, Appearance, and Habits.—The spotted-fever tick has four distinct life stages: the egg, the "seed tick" or "larva," the nymph, and the adult. The adults of the two sexes are also very different in appearance. The females (Fig. 579, *upper*) are dark reddish brown, the anterior third of the body covered with a hard shield which is white, splotched with several small reddish dashes and dots each side of the center. In the male (Fig. 579, *lower*) the shield or *scutum* covers the entire back, and the males are grayish white, splotched with bluish-gray markings. Before feeding, the males and females are about $\frac{1}{16}$ inch long, but, when fully engorged, the female is much larger, $\frac{1}{2}$ inch long by $\frac{1}{3}$ inch wide. The larvae or seed ticks, upon hatching, are only about $\frac{1}{40}$ inch long, have only three pairs of legs, and are pale yellow in color, but they increase greatly in size and become slate gray after feeding. The nymphs are easily distinguished from the larvae by their larger size (¹⁄₁₆ to ⅙ inch long), darker color, and four pairs of legs. Adults can always be told from nymphs by the presence of the genital opening at the anterior third of the body on the underside.

Each tick during its lifetime requires three hosts, which are probably always three different individuals and commonly three different species of animals. Only the adults feed on man and the large animals. The larvae and nymphs feed on small wild animals, especially rodents. Each stage remains anchored by its mouth parts to one spot on its host for a week or two, then drops to the ground and spends a week or two, or some-

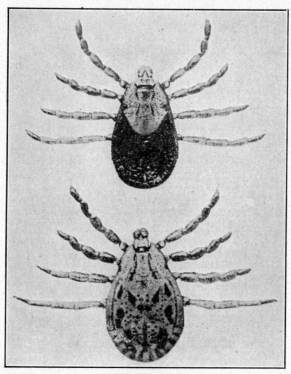

Fig. 579.—The spotted-fever tick, *Dermacentor andersoni* Stiles. Adult, unengorged female, above; adult male, below. About 5 times natural size. (*From U.S.D.A., Bur. Ento. Bul.* 105.)

times the entire winter, resting, digesting the blood it has swallowed, and molting. Two, three, or four years may be occupied completing a generation, 2 years being the most common. The winter is passed as unfed males and females or as nymphs, among grass and leaves on the ground. The adults and nymphs feed only during the spring months, from about the middle of March to the middle of July; and this is the only period when man contracts spotted fever. During the first warm days of spring they climb up on the brush and may attach to any passing object. If this is a man, horse, cow, or other large animal, they insert their mouth parts and draw blood for the next week or two. During this time the males visit the females, mating occurs on the host animal, and, when fully engorged, the female drops off the host. During the next month she normally deposits 4,000 to 7,000 small brown eggs, all in one mass, under stones and other trash on the ground, laying several hundred each day. From 2 to 7 weeks later the eggs hatch, especially during June and July. These "larvae" have only six legs and are much smaller than pinheads. They climb up on grass, brush, and other objects, and the fortunate few that are brushed off by some small wild rodent may suck its blood for a week, then drop off the host and spend 1 to 4 weeks on the ground before they molt to the eight-legged nymphal stage. By this time hot weather usually causes the nymphs to go into a period of rest under the shelter of grass or leaves where they remain until the following spring. The first warm spring days bring them up on the vegetation again, where, if fortunate, they attach to a rabbit, mouse, ground squirrel, or woodchuck, suck its blood for a week, drop to the ground, and devote the next 6 or 8 weeks to the change to the adult stage. The hot weather of summer now prevents these adults from seeking a host and they remain in estivation and hibernation until the following spring. Such is the normal life cycle of the lucky ones—nymphs and adults being present the year round, but seeking hosts and feeding only in the first half of the growing season, and the larvae feeding only during midsummer. The larvae from late-laid eggs or ones that do not find a host perish during the winter. Nymphs and adults, however, which fail to find a host before midsummer, may survive an additional year before feeding, and the adults even a third and fourth year before they find a host or finally perish.

Control Measures.—Every effort should be made to discover the presence of ticks on the body and the clothing by careful examination, especially of the head, promptly after exposure to tick-infested brush. A change of clothing after walking in ticky areas, and keeping the outdoor clothing away from sleeping quarters are important. Rotenone dusts or sprays applied to clothing will help to prevent ticks from attaching. If ticks are found attached to the skin, they should be touched with a few drops of chloroform, gasoline, or turpentine or a hot needle to cause them to relax their mouth parts, or grasped firmly with tweezers or fingers and removed by a steady pull so as not to break off the mouth parts in the skin. The point of attachment should then be sterilized by dipping the point of a round toothpick in carbolic acid, silver nitrate, or tincture of iodine and drilling it into the hole made by the mouth parts. Since the tick does not transmit the disease unless it feeds for 4 to 8 hours, thorough examination of the body, morning, noon, and evening when ticks are prevalent, is very important. Vaccines that will protect from the disease for a single season have been discovered but are not generally available. The very complicated life cycle of the tick, its many diverse hosts, and the wild nature of much of the infested territory make destruction extremely difficult. Control measures suggested include the dipping or spraying of horses, cattle, and sheep, three times at 10-day intervals, during the spring months, as directed for the cattle tick; restricted grazing of domestic animals in tick-free pastures; the use of repellent oils on grazing animals; the destruction of the rodents which serve as hosts for the larvae and nymphs, with poisoned grain scattered around their burrows or with calcium cyanide introduced into the burrows.

References.—*U.S.D.A.*, *Bur. Ento. Buls.* 105, 1911, and 106, 1912; *Mont. State Bd. Ento. Seventh Bien. Rept.*, 1929; *Arch. Path.*, **15**: 389–429, 1932; *Jour. Med. Res.*, **41**: 1–197, 1919; *Jour. Econ. Ento.*, **30**: 51–69, 1937.

362. American Dog Tick or Wood Tick[1]

Importance and Type of Injury.—Previous to 1931 this tick was considered merely an annoying parasite of dogs and human beings who roamed woody places; but with the discovery that this common wood tick is the carrier in the central and eastern states of Rocky Mountain spotted fever, it has taken on real significance. Although there have been fewer than 150 cases a year of this disease east of the Rocky Mountains, and the death rate is only about 25 per cent, it is a dreaded disease and the tick itself a repulsive bugbear of the out-of-doors. This tick is also a carrier of tularaemia and of bovine anaplasmosis, as well as a troublesome parasite of dogs and horses.

Animals Attacked.—Dogs, man, cattle, horses, hogs, mice, and many other domestic and wild mammals.

Distribution.—Throughout most of the United States except the Rocky Mountain area; especially abundant in areas of considerable humidity which are covered with grass or underbrush, where its mice hosts are abundant.

Life History, Appearance, and Habits.—In the most southern states all stages of the tick may be found the year round, although reproduction is slowed down by both the cool weather of winter and the hot dry weather of midsummer. In the North all stages, except the eggs, apparently winter successfully; the larvae and nymphs on mice and the unengorged adults in clumps of grass and similar shelter. The adult, which is the stage that attacks man, is most abundant in spring and early summer and is rare after the first of August. This is the only stage that attacks man, the dog, and the domestic animals. They are flattened, chestnut brown to blue gray, very tough-skinned creatures with eight large legs. When unengorged, they measure about $3/16$ inch in length. The male has a hard white-marked shield all over his back, whereas in the female this scutum reaches only half the length of the body and scarcely $1/10$ its length after engorgement. The fully fed female reaches a length of $1/2$ inch and becomes bluish gray in color. If they do not feed, adults may live in moist situations nearly 3 years but normally die in from 3 weeks to 3 months. Females may remain attached in one spot to the host for 5 to 13 days. On the fourth to sixth day the female may be visited by a male, mating occurs, and the female then rapidly fills with blood to repletion, releases her mouth parts, and drops to the ground. She may spend 2 weeks to a month laying from 4,000 to 6,500 round brownish eggs, in some protected place on the ground, in an elongate mass in front of her. Then she dies. The eggs hatch in about a month. The tiny six-legged, yellow to grayish larvae, $1/40$ inch long, climb up to the ends of grass blades and other leaves, ready to attach to some passing animal. They may survive in moist places for nearly a year if they do not find a host sooner; but, if fortunate, they may attach to a host a few days after hatching and be full fed in 3 to 12 days. They then drop from the host, and after 1 to 12 weeks on the ground (the longer interval in cold weather) they shed their skins and become eight-legged nymphs. They are $1/16$ inch long before feeding, may attach, if they find a host, in from 4 days to 9 months, feed for 3 to 10 days, and again drop to the ground where they may undergo the final molt to the adult stage in from 16 to over 100 days.

---

[1] *Dermacentor variabilis* (Say), Order Acarina, Family Ixodidae.

The nymphs may live more than 2 years, if they do not find a host sooner. Many individuals probably spend 2 or 3 years completing their life cycle.

Control Measures.—Complete destruction of the ticks is almost impossible. In most cases avoidance of moist areas where grass and underbrush abound during spring and early summer or the careful examination of the body and especially the hair of the head at least twice a day and removal of any ticks as promptly as possible after exposure will prevent serious consequences. Tincture of iodine should be forced into the minute hole made by the tick's mouth parts. Care should be taken not to get the blood of crushed ticks into the eyes or into scratches on the skin. Dogs should be kept from roaming in ticky places or treated with derris, cubé, or timbo powder having a rotenone content of at least 4 per cent, or about 14 per cent total extractives, to kill the ticks they collect. A wash or dip should be made by dissolving 1 ounce of neutral soap in 1 gallon of water and adding 4 ounces of the rotenone-bearing powder; this wash should be applied every 5 days to dogs that run out-of-doors. Or a dust may be made by mixing the powder with equal parts of fine talc or flour and rubbed all through the coat of the dog every 2 or 3 days. Clearing away underbrush, keeping grass closely mowed, poisoning or trapping meadow mice, and wearing boots laced up over trouser legs, and clothing securely fastened about neck and wrists are aids in control.

References.—*U.S.D.A., Cirs.* 478, 1938, and E-454, 1938, and *Bur. Ento. Bul.* 106, 1912; *Jour. Econ. Ento.*, **30**: 51–69, 1937; *Arch. Path.*, **15**: 389–429, 1933.

MITES

A number of species of mites are troublesome to man, among them being the chigger or jigger (not to be confused with the chigoe flea, see page 892), harvest mites, louse-like mites, and flour and meal mites. Mites are close relatives of ticks and spiders and are not true insects. They represent distinct families of the class Arachnida and order Acarina, the larger ones of which are called ticks and the smaller ones mites.

363. CHIGGERS, JIGGERS, OR RED BUGS[1]

Importance and Type of Injury.—From 12 to 24 hours after one has been on an outing in summer or autumn, particularly to spots where tall grass, weeds, and brambles abound, the skin may become inflamed in spots, especially where the clothing closely pressed the skin. Scattered red blotches of varied size appear, accompanied by a most intense itching that may not subside for a week or more. Some persons become feverish, extremely nervous, and seriously disturbed by such infestations.

Animals Attacked.—Man, domestic animals, poultry, certain ground-nesting birds, various snakes, land turtles, frogs, toads, squirrels, rabbits. and possibly other rodents.

Distribution.—The distribution of chiggers is not uniform and is somewhat peculiar. Some local areas are badly infested, while others apparently similar and not far distant are practically free.

---

[1] *Trombicula irritans* (Riley) [= *Eutrombicula alfreddugesi* (Oudemans)] and others, Order Acarina, Family Trombidiidae.

Life History, Appearance, and Habits.—The causes of this trouble are small "larval" or six-legged mites, less than $\frac{1}{150}$ inch in diameter and hence almost invisible to the naked eye. According to Miller[1] at least one species of chigger,[2] spends the winter in earthen cells $\frac{1}{2}$ to 1 inch below ground in the adult stage. The following spring they come out of the ground and lay their eggs. The young six-legged mites that hatch from the eggs normally live on land turtles or on snakes, under the overlapping scales of the back, according to Miller,[1] or on rabbits, according to Ewing,[3] and it is only these first-stage "larvae" that attack man. These are rounded, oval, bright orange-yellow in color, blind, and run very rapidly. When one walks through grass and underbrush, these young chiggers may swarm over the body for several hours but are not felt until they settle down and begin to feed (Fig. 580). They undoubtedly introduce a definite poison that causes the irritation. It has often been stated that chiggers burrow into the skin and suffer a speedy death as a penalty for trespassing upon man; but we are apparently to be deprived of this consolation, for Ewing[3] contends that they do not burrow beneath the skin but only insert the mouth parts, sometimes in a skin pore or hair follicle. When full-fed they drop off man. According to Miller, the full-fed, first-stage larvae fall from their hosts in late September, in southern Ohio, and go into the loose soil to pass the winter. After molting and spending 2 or 3 weeks as quiescent nymphs, they molt to the adult stage but remain in their earthen cells until spring. There is only one generation a year. Ewing states that about 10 months are spent in the adult stage. The adults do not attack man but are scavengers on the excrement of other arthropods or on decaying wood.

Fig. 580.—Chigger engorging, or feeding, at the base of a hair. (*From U.S.D.A., Dept. Bul.* 896.)

Control Measures.—If one anticipates a visit to the domains of these little tormentors, trouble may be prevented by dusting finely powdered sulphur or naphthalene liberally into the clothing. Special, scented and colored sulphurs may be purchased for the purpose. Dipping the hiking clothes in a solution of 4 ounces of soap and 2 ounces of wettable sulphur per gallon of water, or spraying them with a pyrethrum or Lethane household spray is equally effective. If this has not been done, or even if it has, a soapy bath within a few hours after exposure, allowing the soap to dry on the skin, will usually prevent infection. After itching begins, little can be done except to avoid infecting the bites by scratching. Some relief may be had by applying cooling ointments, such as ammonia or salicylic acid in alcohol with a little olive oil. Premises infested with these mites may be freed by cutting out the underbrush, especially berry brambles, by keeping grass closely trimmed, by pasturing with sheep, and by dusting with sulphur at the rate of 50 pounds to the acre. Infested lawns have been cleaned of chiggers by dragging over them a piece of canvas or sacking wrung out of kerosene. The oil-soaked canvas should

[1] Miller, A. E., *Science*, **61** (no. 1578): 345–346, 1925.

[2] *Trombicula thalzahuatl* (Murray).

[3] Ewing, H. E., *Science, Supplement* 2, **59**: xiv, 1924.

not be allowed to drip or remain long in one place, or the grass may be killed.

Reference.—U.S.D.A., Dept. Bul. 986, 1921.

364. Louse-like Mites or Harvest Mites[1]

Importance and Type of Injury.—Workmen while threshing wheat and other small grains, or handling beans or cotton, and occasionally people that sleep on straw-filled mattresses, may be overrun with microscopic mites that produce symptoms much like chiggers and are often called by that name. Within a day after exposure a hive-like eruption appears over much of the body (Fig. 581). These spots itch intolerably for several days to a week, and vomiting, headache, and fever may occur.

Animals Attacked.—Man and insects only, so far as known.

Distribution.—This mite has been most troublesome in the central grain-growing states.

Life History, Appearance, and Habits.—This mite is usually predaceous on other insects, among which are the wheat jointworm, granary and rice weevils, and the Angoumois grain moth. Like the chiggers, they are practically microscopic, except that the female after mating becomes greatly enlarged with developing young so that her body, originally about $\frac{1}{125}$ inch long, then measures nearly $\frac{1}{16}$ inch. The young mites hatch from the eggs, and the nymphs develop to the adult condition within the abdomen of the female. The adults are then born viviparously; they mate soon after birth, and within a week may have matured a second generation in the same manner. Each female may produce as many as 200 or 300 adult mites. The predaceous

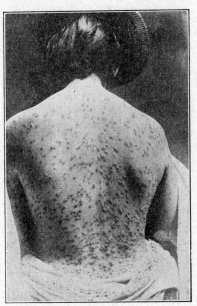

Fig. 581.—Eruptions caused by bites of the harvest mite, *Pediculoides ventricosus* (Newport). (*From U.S.D.A., Bur. Ento., Cir.* 118.)

activities of the mites and their attacks on man, unlike that of chiggers, are therefore confined to the adult stage.

Control Measures.—The free use of sulphur dust among the clothing or the application of a greasy ointment to the body before working among infested straw, stubble, or seeds, followed by a soapy bath and change of clothes promptly afterward, should prevent any unpleasant results. Infested mattresses and other material may be cleaned by fumigation, by heating to at least 130°F. for about 6 hours, or by steaming. After the irritation has begun, slight relief may be secured from the application of cooling ointments.

Reference.—U.S.D.A., Bur. Ento., Cir. 118, 1910.

365 (299). Flour, Meal and Cheese Mites[2]

Trouble similar to that from chiggers and louse mites is sometimes experienced by persons who work with flour, meal, sugar, dried fruits, copra, cheese, hams, and the like, and is called "grocer's itch." This is caused by small mites (Fig. 525), which normally feed on these stored products (see page 795), crawling upon the body and inserting their mouth parts. The cheese mites are often eaten and sometimes cause

[1] *Pediculoides ventricosus* (Newport), Order Acarina, Family Pediculoididae.

[2] Order Acarina, Family Tyroglyphidae.

digestive troubles. The dried powdered bodies or excrement of these mites may cause skin irritations or allergic disturbances.

They develop very rapidly under favorable conditions and have a remarkable non-feeding stage known as a *hypopus* that sometimes intervenes between nymphs and adults. In this stage the mouth parts are wanting, and there is a minute sucker on the underside of the body with which they attach to the bodies of such insects as occur in their feeding places, and probably to mice, and may thus be transported considerable distances to start a new infestation.

Control Measures.—When workmen are suffering from grocer's itch, the source of the mites should be found and the infested material or storeroom freed of them by fumigation (see pages 275 and 795) or by superheating (see page 294) and thorough cleaning.

366 (318, 335). Itch Mite[1]

Importance and Type of Injury.—The human itch mite (Fig. 544) is the same creature which, in horses and dogs, causes mange. There are several strains or varieties, each somewhat adapted to its particular host. The itch mite differs from the mites just discussed in spending its entire life cycle on the host. The eggs are laid in tunnels made by the burrowing females beneath the skin (Fig. 9). Because of their internal position and very small size, the cause of itch was for centuries unknown, the trouble was attributed to improper living, immorality, or "bad blood." Since no proper treatment was known, the mites multiplied for generations and years, on the body of their victim, thus giving rise to the expression "seven-year itch." The female mites (Fig. 544) are only about $\frac{1}{60}$ inch in diameter, the males half as large, with four pairs of short stubby legs, the third pair in the male and the third and fourth pairs in the female ending in a bristle nearly as long as the body, the other pairs terminating in small stalked suckers. The skin is pale colored and striated like the palm of one's hand and with scattered tack-like spines on the upper side. The mites have neither eyes nor tracheae. The newly mated females dig into the skin, making tunnels about $\frac{1}{50}$ inch in diameter and up to 1 inch in length, in which about 24 eggs are laid. The egg stage, the six-legged "larval" stage, and two eight-legged nymphal stages each lasts only 2 or 3 days. The young females start to dig new burrows in which mating takes place, and egg-laying may begin in 10 days to 2 weeks after the female hatched from the egg. The adults are said to live a month or more. This burrowing and more especially the feeding of the mites cause the extreme itching that is the chief symptom of the disease. The tunnels are parallel with the surface of the skin and not very deep. They can often be seen as delicate gray thread lines beneath the skin, between the fingers and toes, behind the knee, on the external genitalia and, in prolonged cases, over most of the body except the head. Hard pimples about the size of pinheads containing a yellow fluid form over the affected skin, and, as these are scratched, they usually become infected and cause large ugly sores and scabs. The nervous strain and the venom introduced by the feeding of the mites may greatly depress the individual and disturb the health.

Animals Attacked.—Man, dogs, rabbits and ferrets. Varieties of the same species cause mange of many other animals (see pages 821, 848).

Control Measures.—Since the mites crawl about over the body chiefly at night, infection usually occurs by occupying the same bed with one who

[1] *Sarcoptes scabiei hominis* (Hering), Order Acarina, Family Sarcoptidae.

has the itch. The use of the same towels or clothing with a victim may also spread the disease, since the mites can live for 10 days or more off the body in moist conditions. Infected individuals should be isolated from the rest of the household as much as possible. Treatment consists in (1) thoroughly massaging the skin over the entire body except the head, with green soap and hot water, keeping this up for about 1 hour in order to soften the scabs and open the burrows of the mites; (2) the application, with prolonged rubbing in, of balsam of Peru, sulphur ointment, pyrethrum ointment, or, as recommended by Chandler,[1] a mixture of "betanaphthol 75 grains, olive oil 2½ fluid grams, sulphur 1 ounce, lanolin 1 ounce, and green soap 1 ounce;" (3) the sterilizing, by superheating or otherwise, of the underwear, other clothing, bed clothing, and towels used by a patient; (4) the repetition of the treatment in from 3 to 10 days, to destroy mites hatching from unkilled eggs.

367. HUMAN BODY LOUSE[2] AND HEAD LOUSE[3]

Importance and Type of Injury.—Human lice, like the bedbug, are fortunately nothing but a loathsome thought to the vast majority of cleanly people. But to thousands of the world's less happily situated souls they are inseparable companions. In time of war troops have always suffered unspeakably from these pests. During the Civil War they were known as "greybacks" but during the World War they became notorious under the name of "cooties." In times of peace lice are largely confined to jails, prisons, laborers' camps, slums of cities, or other places where unsanitary living prevails. In the great melting pot of our public schools, lice may easily gain access to any child and so be introduced to any household. It is as well, therefore, to swallow our loathing, and learn enough about these parasites to enable us to help some unfortunate, if occasion arises, and to protect ourselves or our homes if chance or the stress of war should subject us to so unpleasant an experience. All the lice found on the body of man are of the bloodsucking kind,[4] no chewing lice[5] of man ever having been discovered. The bites of these lice, although scarcely felt at the time, are as irritating as those of fleas and bedbugs. Their crawling on the skin is also very annoying. Scratching is almost inevitable, and there is great danger of streptococcic infection. The skin becomes scarred, thickened, and bronze colored, with brownish spots. A generally tired, "grippy" feeling, fever, and an irritable and pessimistic state of mind are attributed to the feeding of lice. The most serious injury, however, is the proved connection of the body louse with several human diseases, such as typhus fever, trench fever, and relapsing fevers. The most serious of these, and the only one so far occurring in America, is typhus fever. Typhus fever (not to be confused with typhoid) is a continuous fever accompanied by a spotted skin eruption like spotted fever to which it is very closely related. The mortality is ordinarily from 15 to 30 per cent but, under war conditions, more often 50 to 70 per cent. There have been outbreaks in New York and Philadelphia, but the United States has never experienced a devastating epi-

---

[1] "Animal Parasites and Human Disease," Wiley, 1918.
[2] *Pediculus humanus corporis* DeGeer, Order Anoplura, Family Pediculidae.
[3] *Pediculus humanus capitis* DeGeer, Order Anoplura, Family Pediculidae.
[4] Order Anoplura.
[5] Order Mallophaga.

demic of this disease such as occurred in Russia, 1905 to 1918, with over 2,000,000 deaths; or in Serbia during 1915, when there were as high as 9,000 deaths a day and a total of 150,000 in that small country. The disease occurs every year in the higher and cooler parts of Mexico and is in general a disease of cool climates and of winter weather. Typhus rarely occurs in the absence of human lice and may be acquired either by the bites of diseased lice or by crushing them upon the skin near the punc-

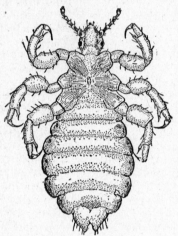

tures they make or by scratching with fingernails infected with the body fluids or feces of lice. A variety of typhus known as Brill's disease may be spread also by fleas (page 889) and the tropical rat mite.[1]

Animals Attacked.—This species lives only on the various races of men, except that it occasionally occurs on monkeys or apes in confinement.

Distribution.—Lice occur in all countries and upon all races of people. The body louse is less abundant in the tropics; probably because, as Nuttall suggests, of the lesser amount of clothing worn as well as of temperatures too high for them.

Life History, Appearance, and Habits. Head lice and body lice are almost indistinguishable in appearance, but their habits are very distinct. They have usually been considered two different species; but, since Bacot[2] has shown that they produce fertile offspring when crossed, and Nuttall[3] has shown that there are no constant morphological differences, and Keilin

Fig. 582.—Human body louse, *Pediculus humanus* Linné. Underside, 20 times natural size. (*From Herrick's "Insects Injurious to the Household," copyright, 1914, Macmillan. Reprinted by permission.*)

and Nuttall[4] have shown that the head lice lose all their usual differences when reared for a few generations under conditions that the body louse normally experiences, we must consider that they are only races or varieties of one species.

Human lice (Fig. 582) are grayish, flattened, wingless six-legged insects, $\frac{1}{16}$ to $\frac{1}{6}$ inch long by nearly one-half as wide, with short five-segmented antennae, simple eyes, rather heavy legs that terminate in a single, sharp, curved claw for grasping about hairs, and piercing mouth parts that, when not in use, disappear completely into the front of the head (see Fig. 69). The body louse is typically larger, lighter in color, with slightly longer antennae, longer, more slender legs, and the constrictions between the abdominal segments less conspicuous than in the head louse. In spite of these differences it is usually difficult to tell from an isolated specimen which form it is. Normally, however, the habits are very distinct. The head louse lives *on the skin and among the hairs of the head*, and glues its eggs to hairs in the typical louse manner. The body louse lives *in the clothing*, only going upon the skin to feed, and even then

[1] *Liponyssus bacoti* (Hirst), Order Acarina, Family Dermanyssidae.

[2] Bacot, A., *Parasitology,* **9**: 228–258, 1917.

[3] Nuttall, G. H. F., *Parasitology,* **11** (nos. 3, 4): 329–346, 1919.

[4] Keilin, D., and G. H. F. Nuttall, *Parasitology,* **11** (nos. 3, 4): 279–328, 1919.

retaining its hold on the adjoining part of the clothing; *it lays its eggs in seams of the clothing.* Both of these species occur associated with man the year round but become most prevalent in winter or at any time when their victims crowd together in warm, poorly ventilated houses, which furnish conditions best suited for the increase of lice. All stages may be found at any season of the year. The head louse lays its eggs chiefly glued to the hairs of the head, back of the ears, and at the back of the neck. The body louse is remarkable in being the only louse that lays its eggs among the seams of clothing, to the fibers of which they adhere. The eggs are elongate oval in outline, about $\frac{1}{25}$ inch long, with a distinct pebbled lid at one end and whitish in color. They hatch in about a week under favorable conditions, and the very tiny, slender lice at once begin feeding. They continue to feed from two to six or more times a day for the next 1 to 4 weeks, growing meantime and molting three times before they become adult. The adults live for about 1 month more, the females laying an average of 8 to 10 eggs a day until a total of 50 to 100 are deposited by the head louse and 200 to 300 by the body louse. Under conditions where the clothing is not, or cannot be, changed for long periods, the number of lice on one person may sometimes range from 1,000 to 10,000. The body louse may survive as adults, eggs, and nymphs a total of a month or more in moist clothing off the host.

Control Measures.—On account of the abundance of lice in Europe during the World War and their announced connection with the spread of typhus and trench fevers, the most extensive investigations were undertaken and hundreds of remedies and very elaborate plans for delousing troops were devised. For an excellent discussion of this subject, the reader is referred to the article by Nuttall.[1] In peace times most persons will be interested chiefly in methods of avoiding infestation by these revolting creatures and possibly in helping some unfortunate child or wretched tramp to get rid of an infestation. The indiscriminate use of public combs, brushes, and towels, trying on of headwear, sleeping in infested rooms or sleeping cars, and all contact with unclean persons should be avoided. An infested head should be treated repeatedly with full-strength grain alcohol; with Lethane; with derris extracts; with $2\frac{1}{2}$ per cent phenol in water; with equal parts of kerosene and olive oil overnight and thoroughly shampooed in the morning; or with xylene, 95 per cent, and vaseline, 5 per cent, for $\frac{1}{2}$ hour followed by a shampoo. Cider vinegar or 25 per cent acetic acid should be applied liberally to dissolve the glue fastening the eggs to hairs. Clipping the head and burning the hair will make control easier. A second treatment will not usually be necessary.

When the infestation is of body lice, especial attention must be given to killing the lice and their eggs in the clothing and baggage. The surest and safest method is by vacuum fumigation with hydrocyanic acid gas, which is the only method effective without spreading out articles of baggage. Destruction may be accomplished by steam sterilization or by a dry heat of 140 to 160°F. for 6 to 12 hours, or by fumigation with carbon bisulphide or carbon tetrachloride in a tight box. During this time the victim should take a thorough bath in hot water and soap; or the kerosene and olive oil mixture or a gasoline and soap mixture may be applied all over the body. Sleeping quarters of infested persons may be disinfected

---

[1] NUTTALL, G. H. F., *Parasitology,* **10** (no. 4): 411–588, 1918.

by live steam or by fumigating as for bedbugs.   Zinc oxide ointment is soothing to the skin.

References.—Parasitology, **9**: 228–258, 1917, **10**: 411–588, 1918, and **11**: 279–346, 1919; *Res. Publ., Univ. Minn.,* vol. 8, no. 4, 1919.

368. Crab Louse or Pubic Louse[1]

Importance and Type of Injury.—Although this species is very closely related to the head louse and body louse, it presents a very different

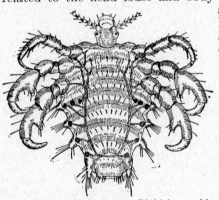

appearance (Fig. 583), and its habits are quite distinct. It is almost entirely limited to the part of the body at the crotch of the legs and to the armpits, but it rarely occurs among the eyelashes, eyebrows, and beard. Nuttall[2] shows that the reach of the two extended legs of the louse corresponds closely to the distance apart of the coarse hairs in these places and concludes that this louse so seldom infests the head because the hairs there are finer and closely crowded together. Severe itching, especially in the hairy parts of the pubic region, accompanied by inflamed spots, is caused by the feeding of these small lice.   When these spots are

Fig. 583.—Crab louse, *Phthirius pubis* (Linné); underside, about 20 times natural size. (*From Herrick's "Insects Injurious to the Household," copyright,* 1914, *Macmillan. Reprinted by permission.*)

scratched, a more or less severe eczema may develop.   According to Nuttall[3] faint bluish-gray spots of varying size characteristically appear beneath the skin where these lice have fed.   The same author finds that fever, headaches, and other disturbances may occasionally result from the poison introduced by the feeding of the lice.   This louse is not known to transmit any disease.

Animals Attacked and Distribution.—Known to attack only the white and negro races of men and probably occurs wherever these races live. It has been taken from dogs.

Life History, Appearance, and Habits.—The eggs are attached to the coarse hairs of the body among which the lice feed.   Nuttall[2] thinks that as many as 50 eggs are laid by one female, although other authors record only about a dozen.   The same author finds that the egg stage lasts about a week and the nymphs molt three times and are mature in 2 or 3 weeks after hatching.   The full-grown female louse is about $\frac{1}{12}$ to $\frac{1}{16}$ inch long by about two-thirds as broad, the male about half as large, but the legs sticking out at the sides give them a still broader appearance. They are grayish white in color with darker legs and shoulders.   The general appearance of the louse is that of a minute gray speck which, when magnified, suggests a miniature crab.   The legs are similar to those of the body louse except that those of the first pair lack the "thumb" against which the curved claw closes in the other legs.   The thorax is very broad

[1] *Phthirius pubis* (Linné), Order Anoplura, Family Pediculidae.

[2] Nuttall, G. H. F., *Parasitology,* **10** (no. 3): 383–405, 1918.

[3] *Ibid.,* 375–382, 1918.

and the margins of the abdomen have short, finger-like projections, the ends of which are provided with bristles.

Control Measures.—One should use care to avoid contamination with this pest in public toilet rooms, baths, and unclean rooming houses, and especially by close association with infested persons. Frequent and repeated spraying of the infested part of the body with full-strength commercial grain alcohol or a livestock spray containing Lethane is probably the simplest control measure. Derris ointments are effective and nonpoisonous. The mercurial ointments formerly recommended are very poisonous and messy to use.

Reference.—*Parasitology*, **10**: 375–405, 1918.

369. INTERNAL INSECT PARASITES

Importance and Type of Injury.—While there are in the tropics several species of maggots that live in the flesh of the human body, such as the Congo floor maggot,[1] the tumbu fly,[2] and the human botfly,[3] myiasis (as the presence of fly maggots in the body is called) is really exceptional and accidental under normal conditions in temperate America.

About 50 different species of flies have been recorded as occasionally infesting the human body in their larval stages, 19 of them in North America. They penetrate the unbroken skin, occur in wounds, in the cavities leading from the mouth and nose, in the eyes, in the alimentary tract, and very rarely in the urinogenital passages. They reach these positions by being swallowed with water or food containing the eggs or young maggots, or the latter are deposited in wounds or body openings by the adult flies. In either case the larvae of the species concerned find conditions such within the body that they can live for a long time and often until full-grown. The results of the infection are apt to be very pronounced, and the condition is so offensive that it is likely to assume undue importance from the vivid accounts of the sporadic cases in medical literature.

The screwworm fly, which has already been discussed as a pest of cattle (page 843), sometimes attacks men, especially through the nostrils or in wounds, with dreadful results. A flesh fly known as *Wohlfahrtia vigil* has been taken working under the skin. Horse bots (page 823) and ox warbles (page 838) have been found living beneath the skin of man, and the sheep bot (page 858) may cause blindness by penetrating the eye. The little house fly[4] has been recorded a number of times from the urinary tract especially of women. The largest number of cases is recorded from the intestines. These are caused by the little house fly, the latrine fly,[5] the pomace fly,[6] the house fly, several kinds of flesh flies, and species of rattailed maggots.[7] The usual symptoms are nausea, vomiting, severe abdominal pains, a bloody diarrhea, and sooner or later the passage of maggots with the excrement.

Control Measures.—Reasonable care about eating uncooked meats, vegetables, and fruits, or using poorly cleaned milk bottles, and the protection of the mouth, nostrils, and other body openings, especially in the case of children and others when they are afflicted with catarrh, bad breath, or otherwise unclean, is important in avoiding the attacks of these flies. Control of the screwworm in animals is the best safeguard from attack upon man. The presence of fly larvae in any part of the body should receive the immediate attention of a skilled physician to avoid serious injury and possibly a horrible death.

Reference.—*Jour. Econ. Ento.*, **30**: 29–39, 1937.

[1] *Auchmeromyia luteola* Fabricius, Order Diptera, Family Muscidae.
[2] *Cordylobia anthropophaga* Blanch., Order Diptera, Family Muscidae.
[3] *Dermatobia cyaniventris* Mac Q., Order Diptera, Family Oestridae.
[4] *Fannia canicularis* Linné, Order Diptera, Family Anthomyiidae.
[5] *Fannia scalaris* Fabricius, Order Diptera, Family Anthomyiidae.
[6] *Drosophila* spp., Order Diptera, Family Drosophilidae.
[7] Several genera of Order Diptera, Family Syrphidae.

370. House Fly[1]

Since the discovery of the importance of the common house fly (Fig. 584) as a carrier of various diseases, a few decades ago, several entire books and many special articles have been written about it. Only the more important points about the fly, that every citizen should understand, are explained in the following account. For a more complete exposition of its revolting and dangerous habits and the details of its morphology, life cycle, and control, the reader should refer to some of the special references.

Importance and Type of Injury.—The common house fly is the most dangerous animal living within the boundaries of many of our states. This in spite of the fact that it cannot bite or sting or otherwise *cause* disease *of itself* in any life stage. Its importance lies in the fact that it is the *carrier* of millions of bacteria and protozoa among which have been found the germs or pathogens of the following human diseases:

Typhoid fever	Anthrax	Whipworm	Erysipelas
Diarrhea	Leprosy	Bubonic plague	Gonorrhea
Asiatic cholera	Tapeworms	Yaws	Septicemia
Amoebic dysentery	Hookworms	Ophthalmia	Abscesses
Tuberculosis	Roundworms	Trachoma	Gangrene

House flies also serve as intermediate hosts of roundworms[2] and transfer them to sores or to lips and eyes of horses; and of tapeworms which infest poultry when the poultry eat the flies or their maggots. It must be understood that the house fly is not the sole transmitter of any of these diseases in the way that mosquitoes are related to malaria or lice to typhus. Practically all authorities are agreed, however, that the house fly is an important factor in the spread of diseases such as typhoid fever, epidemic or summer diarrhea, dysenteries, cholera, and other bacterial infections of the stomach and intestines. There is also strong evidence that the fly may be a dangerous carrier of tuberculosis, anthrax, parasitic worms, and many other diseases, especially of poultry. In all these cases the germs of the diseases are picked up by the fly from human excrement, sputum, the carcasses of diseased animals, manure, and other filth, which the fly visits; or are taken up by the larvae and remain in a living condition in its body during the pupal stage and may then be scattered about by the deposits from the body of the adult fly. That the hairy body of the house fly is an admirable carrier of bacteria is attested by Esten and Mason who found an average of 1,250,000 bacteria on each of 414 flies examined, the maximum on a single fly being 6,600,000. The house fly carries disease germs either on the outside of the body clinging to its mouth parts, on its sticky foot pads (pulvilli), wings, and other body surfaces, or in its alimentary canal where they resist digestion and may remain in a living condition for many days. If carried externally, they may be dropped or rubbed off or washed off by the fly, on foods, drinks, wounds, or the eyes or lips of man. If carried internally, they are deposited either by regurgitation or in the feces ("fly specks") in such

---

[1] *Musca domestica* Linné, Order Diptera, Family Muscidae.
[2] *Habronema* spp., Phylum Nemathelminthes, Class Nematod.

situations that they may be taken into the mouth or upon some delicate body surface, like the eye or a sore, and start disease. A minor but not unimportant injury from the house fly is as a nuisance about men and other animals where rest and sleep are disturbed by its ubiquitous buzzing and alighting on the skin.

Animals Attacked.—Man is not the only animal injured by the house fly. Although it does not bite, it annoys domestic animals by alighting on their bodies and serves to transmit certain diseases by sipping blood from wounds made by other insects.

Distribution.—The house fly is practically cosmopolitan and is the commonest fly found in houses throughout most of the world. Explorers in parts of the earth rarely frequented by man have been greatly annoyed by the house fly.

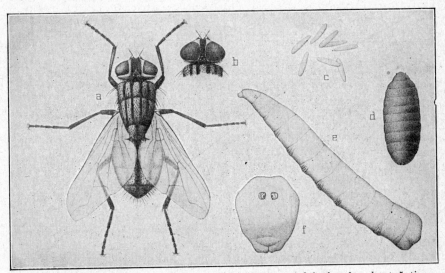

Fig. 584.—The house fly, *Musca domestica* Linné; *a*, adult female, about 5 times natural size; *b*, head of male; *c*, eggs about 6 times natural size; *d*, puparium, about 4 times natural size; *e*, larva or maggot, about 7 times natural size; *f*, last segment of larva to show spiracles. (*From Ill. State Natural History Surv.*)

Life History, Appearance, and Habits.—In the more northern parts of the United States, house flies winter chiefly as larvae or pupae in their puparia (Fig. 584,*d*), beneath manure piles and in other breeding grounds. In heated buildings some adults probably survive, and in warmer parts of the earth it continues to breed slowly throughout the winter. The house fly combines a large family with one of the shortest life cycles known among insects. From 100 to 150 eggs (Fig. 584,*c*) are laid at a time, and from two to seven such batches may be deposited by one female, so that a total of 500 eggs from one fly is probably a normal production. The record so far as known is 21 batches, or a total of 2,387 eggs, from one fly. The whole life cycle from egg to adult may be completed in from 6 to 20 days and a new generation started in from 2 to 20 days more. The length of the various stages under favorable conditions is somewhat as follows:

Egg	Larva	Pupa	Adult to Eggs (Preoviposition Period)
8 to 30 hours	5 to 14 days	3 to 10 days	2¼ to 23 days

Most house flies come from horse manure, as it is in this substance that they prefer to lay their whitish elongate eggs, 25 of which end to end reach 1 inch. A young house fly (Fig. 584,*e*) is not a miniature of its parents but is the legless, footless, and almost headless, white maggot familiar to every boy who has cleaned the manure from stables. The larvae of house flies are believed to feed upon the microorganisms which cause fermentation and decay and hence live only in moist masses of organic matter which are warm enough to promote the growth of these organisms. Feeding on this material, the maggot reaches a length of ⅓ to ½ inch in a few days, and then forms a seed-like, chestnut-colored puparium (*d*) of its third larval skin, within which the change to adult takes place. Before forming the puparium, the larva migrates to some drier part of the substance on which it fed. Every house fly thus passes a wingless egg, larval, and pupal existence in some fermenting material before it can appear in the familiar winged condition. While horse manure is preferred, they can develop in any moist, warm, fermenting, organic matter, ranging from piles of grass, decaying fruits or vegetables, garbage, and waste about feed troughs, to human and animal excrement of all kinds. While it was estimated a few years ago that 90 per cent of our flies develop in horse manure under average conditions, and the replacement of horses by motor cars has greatly reduced their numbers about cities and villages, we may be sure that the complete extermination of the horse would not remove the house fly pest from our midst.

Professor C. F. Hodge[1] has made the following remarkable statement about the possible increase of house flies if all were to continue to reproduce at the normal rate: "A pair of flies beginning operations in April may be progenitors if all were to live, of 191,010,000,000,000,000,000 flies by August. Allowing ⅛ cubic inch to a fly, this number would cover the earth 47 feet deep." Of course such increase never occurs, but this statement serves to show why flies increase so rapidly when summer comes. In one series of experiments, the average adult life of 3,000 flies kept in cages was 19 days. Another investigator found 30 days to be a normal lifetime. One fly was kept alive 70 days.

The thing above all else that makes the house fly such a dangerous messmate is its inordinate curiosity, which leads it to go everywhere and to feed on almost anything. From manure piles, sewage, garbage, sputum, and carcasses of animals to food of all kinds in dining room, kitchen, restaurant, and grocery and to the lips, eyes, and nursing bottles of sleeping children is "all in the day's work" for a busy house fly. If they stayed in filthy places or remained all the time in our houses, the objection to them would not be so great. But their promiscuous habits result in polluting their bodies with filth and shedding it in their paths, wherever they go. House flies are characteristically day active, showing preference for direct sunshine. The adults are almost omnivorous but, because of their sponging mouth parts (Fig. 74), can take food only in the liquid state. Solids must be dissolved in their saliva or the regurgitated

[1] HODGE, C. F., in *Nature and Culture*, July, 1911.

contents of their stomachs. They have a keen sense of smell and thirst for liquids of nearly every kind.

The point as to how far a house fly can fly becomes an important one when the question of control in restricted areas arises. In a number of experiments marked flies have been recovered at distances ranging from 50 yards to 13 miles from the point where they were released. In general it seems that few flies will range farther than is necessary to secure food and a place to lay their eggs. When abundant in a city, it is fairly certain that they are breeding in the same or an adjoining block. Their roving habits, however, point to the necessity in general of community action to control them. It must not be assumed that any fly caught in a house is the house fly. Only a checking by *all* the characteristics given in the key (page 874) will determine it certainly.

Control Measures.—The presence of house flies means that our sanitation is defective. Failure properly to dispose of manure, garbage, sewage, food wastes, human excrement, dead animals, and other organic waste materials is the only thing that will account for an abundance of these insects. Stables built with concrete floors, properly constructed to prevent liquid manure from leaching into them and to make it possible to clean them thoroughly by washing down everyday, and the sanitary disposal of sewage and garbage are the first steps in control. Manure removed from stables should be carted away every 2 or 3 days, and scattered thinly over fields. Manure is never so valuable as when perfectly fresh. The common practice of storing manure in the barnyard, results in a loss of from 25 to 65 per cent of its fertilizing value in 6 months' time, owing to leaching and hot fermentation. But more important is the fact that flies cannot breed in the thinly scattered material, because it dries out and there is no fermentation. A half-ton manure pile, after being exposed for only 4 days, was found to contain (estimated) over 400,000 fly larvae, or 400 larvae to the pound of fresh manure.[1] After manure is well rotted, flies are not attracted to it to lay their eggs. If daily scattering of manure cannot be practiced, the maggot trap may be used to destroy most of the maggots, without depreciation of the fertilizing value of the manure. The maggot trap is a slatted platform about 10 by 20 feet in size, supported within and over a shallow cement vat. The manure is removed from the stable each day and dumped upon this platform. It is kept wet by pumping water over it from a cistern adjoining the vat, into which the water from the vat can be drained at intervals. A few inches of water, with a thin film of oil on top, are kept in the vat all the time. Under such circumstances, fly larvae grow in the manure to full size. Then, driven by a negative hydrotropism to seek a dry place before pupating, they desert the manure if it is kept wet, fall through the slats, and are drowned in the water. If manure or garbage and similar wastes must be allowed to accumulate in the stable or barnyard or about dwellings, they may be treated chemically to destroy the eggs and larvae. Considering the factors of safety about the premises, noninjury to the legs of animals, high toxicity to fly larvae, and safety to plants on soil where it is applied, *borax* appears to be best for this purpose. A pound of borax should be dissolved in 25 gallons of water and about 3 gallons of the solution sprinkled over the manure from each horse each day so as to wet all parts of it. Sodium fluosilicate or hellebore, 1 pound in 20 gallons of

[1] Herms, "Medical and Veterinary Entomology," p. 209, 1923.

water, also makes an effective larvicide, but great care must be taken that animals do not drink the sodium fluosilicate.   Open outdoor toilets are the greatest menace from the standpoint of fly-borne human diseases.   Where they cannot be avoided, they should be built as flyproof as possible (see *U. S. Department of Agriculture, Farmers' Bulletin* 463) and the refuse kept covered by daily applications of waste crankcase oil (if not to be used as fertilizer) or with borax, or liberal amounts of lime.   If manure must be stored in piles, the breeding of house flies can be largely prevented by making piles with straight sides, clean margins and covering the entire pile with heavy building or roofing paper.   The heat thus developed in the sunshine kills the larvae.

In addition to these control measures, which go to the heart of the problem and get the fly before it reaches the dangerous winged condition, some help may be secured by the use of traps and fly poisons.   The electrified screens for doors and windows and electrified fly traps are valuable about dwellings and barns.   Various homemade traps are described in *U. S. Department of Agriculture, Farmers' Bulletin* 734. Bananas and milk or brown sugar and cheese, equal parts, kept wet and fermenting, make attractive baits for traps.   As repellents, the electric fan over entrances; creosote oil about stables; and a mixture of coal tar, 3 parts, and carbon bisulfide, 1 part, kept stoppered and applied with a brush to wounds of animals, may be found useful.   Various commercial household sprays containing synthetic or organic sprays, such as Lethane, pyrethrum, derris, and the like, are helpful in destroying flies that get into houses, but wherever possible control efforts should be directed against the developing immature stages.   The control of the house fly presents no greater difficulties than that of the southern cattle tick, the codling moth, or other widespread and abundant pests.   It can and will be controlled when the general public becomes educated to view it as the real menace to health that scientists know it to be.

References.—Among the most important are HOWARD, L. O., "The House Fly, Disease Carrier," Stokes, 1911; HEWITT, C. GORDON, "The House Fly; A Study of its Structure, Development, Bionomics and Economy," Manchester Univ. Press, 1910; GRAHAM-SMITH, "Flies and Disease: Non-blood-sucking Flies," Cambridge Univ. Press, 1914; *U.S.D.A., Farmers' Buls.* 734, 1936, and 1408, 1926, and *Dept. Buls.,* 118, 1914, 245, 1915, and 408, 1916; *Iowa Agr. Exp. Sta. Bul.* 345, 1936.

INDEX

The most important page references are indicated by **bold-face type**.

A

Abdomen, 77, 78
Abscesses, 906
Absorption, 88–91
Abundance of insects, 5, 60, 70, **71, 141,** 171, 341
Abutilon, 743
Acacia, 677
Acalyptratae, 235
Acanthocephala femorata, 506
Acanthoscelides obtectus, 810
Acarina, 163, 164, **167, 482,** 616, 664, 682, 683
Accessory glands, 93, 105
  veins, 221
Acetate, amyl, 254, 339
Aceto-arsenite of copper, 248
Acid, acetic, 903
  carbolic, 291, 895
  hydrocyanic, **278–283** (see *Hydrocyanic acid*)
  salicylic, 898
  sulphuric, 280
  tannic, 39, **49**
  tartaric, 772
Acid gas, hydrocyanic, **278–283**
Acids, arsenic, 249
  arsenious, 249
Acrididae, 186 (see *Locustidae*)
Acrobatic ant, 768
Acrosternum hilaris, 417, 443, 486
Activators for nicotine, 258
Active ingredients of insecticides, 309
Adalia bipunctata, 59
Adaptability of insects, 61
Adephaga, 209
Adfrontal areas, 215, 217
  sclerite, 215, 217
  suture, 215, 217
Adhesion of stomach poisons, 242, 244, 246, 248, 250
Adipose tissue, 94
Adobe tick, 861
Adoretus caliginosus, larva, 150
Adoxus obscurus, 653
Adult, 145, 152, **155,** 173, 601, 908
Adulteration of insecticides, 240, 306, **308**
Adventitious mouth, **124**
Aecia and insects, 17
Aedes aegypti, 26, **876**
  albopictus, 26
Aedes mosquito, 26, **876**
Aegeria (see *Conopia; Synanthedon*)
Aegeriidae, 219, 502, 629, 632, 659, 725
Aegerita webberi, 304
Aesthetic value of insects, 39, **66**

Afferent neuron, 99
African sleeping sickness, 26–28
Agallia sanguinolenta, 431
  stictocollis, 15
Ageratum, 462, 733, 747
Aggressive mimicry, 107
  resemblance, 107
Agitation of sprays, 315
Agitator, 319, 320
Agonoderus pallipes, 335, 388
Agricultural development, effect of cattle tick on, 835, 836
Agrilus anxius, 712
  ruficollis, 661
  sinuatus, 618
Agriotes mancus, **411**
Agromyza pusilla, 539
  simplex, 552
Agromyzidae, 731
Agrotis c-nigrum, 347, 348, 460
  ypsilon, 347, 348, 460
Ague (see *Malaria*)
Aim, of dusting, 315, 326
  of spraying, 315
Air brake, Westinghouse, 30
  chamber, 318, 319
Airplane dusting, **330, 331,** 451, 636, 671
  spraying, **330, 331**
Airplanes, use of shellac in, 46
Air-slaked lime, 219, 861
Alabama argillacea, **444**
Alcohol, allyl, 256
  grain, 903, 905
  wood, 374
Aleppo gall, 49
Aleurocanthus woglumi, 681
Aleurothrixus howardi, 681
Aleyrodidae, 200, 304, 681, **747**
Alfalfa, 34, 52, 299, 344, 364, 374, 376, 378, 388, 406, 412, 416, 418, 431, 474, 475, 480, 488, 491, 497, 511, 512
  aphid, 223
  caterpillar, 151, 416, **425**
  hay, 488
  looper, larva, 150
  meal, 489
  webworm, 417, **427,** 465, **474–475,** 542
  weevil, 303, 307, 416, **418, 420–422**
Alimentary canal, 91, 94
  development of, 90, 91
Alkaloids, cinchona, 790
Allantoin, 49
Alligators, 158
Allocoris pulicaria, 546

911